ENCYCLOPEDIA OF STATISTICAL SCIENCES

VOLUME 2

**Classification
to Eye Estimate**

ENCYCLOPEDIA OF STATISTICAL SCIENCES

VOLUME 2

**CLASSIFICATION
to EYE ESTIMATE**

1807 1982

A WILEY-INTERSCIENCE PUBLICATION

John Wiley & Sons

NEW YORK · CHICHESTER · BRISBANE · TORONTO · SINGAPORE

Library of Congress Cataloging in Publication Data:

Main entry under title:

Encyclopedia of statistical sciences.

   "A Wiley-Interscience publication."
   Contents: v. 1. A–Circular probable error—
v. 2. Classification to eye estimate.
   1. Mathematical statistics—Dictionaries.
2. Statistics—Dictionaries. I. Kotz, Samuel.
II. Johnson, Norman Lloyd. III. Read, Campbell B.
QA276.14.E5   519.5'03'21   81-10353
ISBN 0-471-05546-8 (v. 1)   AACR2
ISBN 0-471-05547-6 (v. 2)

Printed in the United States of America

10 9 8 7 6 5 4 3 2 1

CONTRIBUTORS

R. A. Bailey, *Rothamsted Experimental Station, Harpenden, England*. Confounding

C. R. Baker, *University of North Carolina, Chapel Hill, North Carolina*. Communication Theory, Statistical; Dirac Delta Function

O. Barndorff-Nielsen, *Aarhus Universitet, Aarhus, Denmark*. Exponential Families

W. A. Barnett, *Federal Reserve Board, Washington, D.C.* Divisia Indices

E. M. L. Beale, *Scicon Computer Services, Ltd., Milton Keynes, England*. Elimination of Variables

J. R. Blum, *University of California, Davis, California*. Ergodic Theorems

W. R. Boggess, *University of Arizona, Tucson, Arizona*. Dendrochronology

L. Bondesson, *University of Umeå, Sweden*. Equivariant Estimators

G. E. P. Box, *University of Wisconsin, Madison, Wisconsin*. Evolutionary Operation (EVOP)

N. E. Breslow, *University of Washington, Seattle, Washington*. Clinical Trials

I. W. Burr, *Ocean Park, Washington*. Engineering Statistics

P. Burridge, *University of Warwick, Coventry, England*. Cliff-Ord Test

T. Caliński, *Academy of Agriculture, Poznán, Poland*. Dendrograms; Dendrites

S. Cambanis, *University of North Carolina, Chapel Hill, North Carolina*. Conditional Probability and Expectation

J. M. Cameron, *Wheaton, Maryland*. Error Analysis

I. M. Chakravarti, *University of North Carolina, Chapel Hill, North Carolina*. Estimability

D. R. Cox, *Imperial College, London, England*. Combination of Data

P. R. Cox, *Sussex, England*. Demography

E. M. Cramer, *University of North Carolina, Chapel Hill, North Carolina*. Collinearity

R. B. D'Agostino, *Boston University, Boston, Massachusetts*. Departures from Normality, Tests for

H. A. David, *Iowa State University, Ames, Iowa*. Cyclic Designs

P. J. Davy, *Macquarie University, North Ryde, New South Wales, Australia*. Coverage

M. H. DeGroot, *Carnegie-Mellon University, Pittsburgh, Pennsylvania*. Decision Theory

J. M. Dickey, *State University of New York, Albany, New York*. Conjugate Families of Distributions

J. B. Douglas, *University of New South Wales, Kensington, Australia*. Contagious Distributions

N. R. Draper, *University of Wisconsin, Madison, Wisconsin*. Evolutionary Operation (EVOP)

J. F. Early, *General Research Corporation, McLean, Virginia*. Consumer Price Index

R. C. Elandt-Johnson, *University of North Carolina, Chapel Hill, North Carolina*. Concomitant Variables

v

O. M. Essenwanger, *University of Alabama, Huntsville, Alabama.* Curve Fitting

W. T. Federer, *Cornell University, Ithaca, New York.* Data Collection

S. E. Fienberg, *Carnegie-Mellon University, Pittsburgh, Pennsylvania.* Contingency Tables

W. A. Fuller, *Iowa State University, Ames, Iowa.* Cluster Sampling

M. H. Gail, *National Cancer Institute, Bethesda, Maryland.* Competing Risks

J. Galambos, *Temple University, Philadelphia, Pennsylvania.* Exchangeability; Exponential Distribution

J. Gani, *University of Kentucky, Lexington, Kentucky.* Dam Theory

J. D. Gibbons, *University of Alabama, University, Alabama.* Distribution-Free Methods

F. G. Giesbrecht, *North Carolina State University, Raleigh, North Carolina.* Effective Degrees of Freedom

A. L. Goel, *Syracuse University, Syracuse, New York.* Cumulative Sum Control Charts

I. J. Good, *Virginia Polytechnic Institute & State University, Blacksburg, Virginia.* Degrees of Belief

H. L. Gray, *Southern Methodist University, Dallas, Texas.* Cornish-Fisher and Edgeworth Expansions

R. B. Griffiths, *Carnegie-Mellon University, Pittsburgh, Pennsylvania.* Critical Phenomena

G. Hahn, *General Electric Company, Schenectady, New York.* Design of Experiments, Industrial and Scientific Applications; Design of Experiments, An Annotated Bibliography

B. Harris, *University of Wisconsin, Madison, Wisconsin.* Entropy

J. A. Hartigan, *Yale University, New Haven, Connecticut.* Classification

H. O. Hartley*, *Duke University, Durham, North Carolina.* Computers and Statistics

I. D. Hill, *Clinical Research Centre, Harrow, England.* Double Sampling

M. O. Hill, *Institute of Terrestrial Ecology, Bangor, Wales.* Correspondence Analysis

D. C. Hoaglin, *Harvard University, Cambridge, Massachusetts.* Exploratory Data Analysis

M. Hollander, *Florida State University, Tallahassee, Florida.* Dependence, Tests for

J. S. Hunter, *Princeton University, Princeton, New Jersey.* Composite Design

W. G. Hunter, *University of Wisconsin, Madison, Wisconsin.* Environmental Statistics

K. Itô, *Gakushuin University, Tokyo, Japan.* Diffusion Processes

J. N. R. Jeffers, *Institute of Terrestrial Ecology, Grange-over-Sands, England.* Component Analysis

K. Jogdeo, *University of Illinois, Urbana, Illinois.* Dependence, Concepts of

B. L. Joiner, *University of Wisconsin, Madison, Wisconsin.* Consulting, Statistical

P. M. Kahn, *San Francisco, California.* Credibility

D. Kannan, *University of Georgia, Athens, Georgia.* Embedded Processes

R. Kay, *University of Sheffield, Sheffield, England.* Cox's Regression Model

J. Kiefer*, *University of California, Berkeley, California.* Conditional Inference

L. Kish, *University of Michigan, Ann Arbor, Michigan.* Design Effect

P. A. Lachenbruch, *University of Iowa, Iowa City, Iowa.* Discriminant Analysis

H. O. Lancaster, *University of Sydney, Sydney, New South Wales, Australia.* Dependence, Measures and Indices of

E. L. Lehmann, *University of California, Berkeley, California.* Estimation, Point

D. V. Lindley, *University College, London, England.* Coherence

R. F. Link, *Artronic Information Systems, New York, NY.* Election Projections

E. Lukacs, *American University, Washington, D.C.* Convergence of Sequences of Random Variables

E. D. McCune, *Stephen F. Austin State University, Nacogdoches, Texas.* Cornish-Fisher and Edgeworth Expansions

N. R. Mann, *University of California, Los Angeles, California*. Extreme-Value Distributions

K. G. Manton, *Duke University, Durham, North Carolina*. Compartment Models, Stochastic

K. V. Mardia, *The University of Leeds, Leeds, England*. Directional Distributions

E. A. Maxwell, *Trent University, Peterborough, Ontario, Canada*. Continuity Corrections

P. W. Mikulski, *University of Maryland, College Park, Maryland*. Efficiency, Second Order

S. Mustonen, *University of Helsinki, Helsinki, Finland*. Digression Analysis

N. K. Namboodiri, *University of North Carolina, Chapel Hill, North Carolina*. Cohort Analysis

J. I. Naus, *Rutgers University, New Brunswick, New Jersey*. Editing Statistical Data

L. S. Nelson, *Nashua Corporation, Nashua, New Hampshire*. Control Charts

D. B. Owen, *Southern Methodist University, Dallas, Texas*. Communications in Statistics

E. Parzen, *Texas A & M University, College Station, Texas*. Cycles

C. Payne, *University of Oxford, Oxford, England*. Election Forecasting in the United Kingdom

A. N. Pettitt, *University of Queensland, Brisbane, Queensland, Australia*. Cramer-von Mises Statistic; Durbin-Watson Test

E. C. Pielou, *Lethbridge University, Alberta, Canada*. Diversity Indices

F. Proschan, *Florida State University, Tallahassee, Florida*. Coherent Structure Theory

M. B. Rao, *University of Sheffield, Sheffield, England*. Damage Models

S. I. Resnick, *Colorado State University, Fort Collins, Colorado*. Extremal Processes

G. K. Robinson, *Department of Agriculture—Victoria, Melbourne, Victoria, Australia*. Confidence Intervals and Regions

D. S. Robson, *Cornell University, Ithaca, New York*. Ecological Statistics

R. N. Rodriguez, *General Motors Research Lab, Warren, Michigan*. Correlation

S. C. Saunders, *Washington State University, Pullman, Washington*. Cumulative Damage Models

P. Schmidt, *Michigan State University, East Lansing, Michigan*. Econometrics

E. Seneta, *University of Sydney, Sydney, Australia*. Criticality Theorem; de Moivre, Abraham; Dispersion Theory; English Biometric School

D. N. Shanbhag, *University of Sheffield, Sheffield, England*. Damage Models

G. A. Shea, *Exxon Research, Linden, New Jersey*. Efficient Score

V. J. Sheifer, *U.S. Department of Labor, Washington, D.C.* Employee Cost Index

G. Simons, *University of North Carolina, Chapel Hill, North Carolina*. Contiguity

N. D. Singpurwalla, *George Washington University, Washington, D.C.* Extreme-Value Distributions

F. J. Smith, *Queen's University, Belfast, Northern Ireland*. Election Forecasting, Transferable Vote System

G. W. Somes, *East Carolina University, Greenville, North Carolina*. Cochran's Q-Statistic

H. Sonnenschein, *Princeton University, Princeton, New Jersey*. Econometrica

M. A. Stephens, *Simon Fraser University, Burnabay, British Columbia, Canada*. EDF Statistics

W. Stout, *University of Illinois, Urbana, Illinois*. Domain of Attraction

W. D. Sudderth, *University of Minnesota, Minneapolis, Minnesota*. Dynamic Programming

V. Susarla, *Michigan State University, East Lansing, Michigan*. Empirical Bayes Theory

L. Takacs, *Case Western Reserve University, Cleveland, Ohio*. Combinatorics

M. M. Tatsuoka, *University of Illinois, Urbana, Illinois*. Educational Statistics

G. S. Watson, *Princeton University, Princeton, New Jersey*. Directional Data Analysis

E. J. Wegman, *Office of Naval Research, Arlington, Virginia*. Density Estimation; Exponential Smoothing

M. E. Wise, *Rijksuniversiteit, Leiden, The Netherlands*. Epidemiological Statistics

M. A. Woodbury, *Duke University Medical Center, Durham, North Carolina*. Compartment Models

S. L. Zabell, *Northwestern University, Evanston, Illinois*. Cournot, Antoine Augustin

ENCYCLOPEDIA OF STATISTICAL SCIENCES

VOLUME 2

**Classification
to Eye Estimate**

continued

CLASSIFICATION

Classification is the grouping together of similar things. The words of a language are an example of a classification. The word "dog" is the name of a class of similar objects that we need to refer to simply. There are numerous examples of technical special-purpose classifications, such as those for animals and plants, for races, for stars and comets, for books, for diseases of body and mind, for new substances, for jobs, for geography; these are subsets of the words of the language, but they are worthy of special study because their words have been constructed rationally.

Why classify things? There are two types of use of a classification, descriptive and predictive. The descriptive use includes naming and summarizing. For example, if a job classification is prepared for a factory, names such as supervisor, foreman, management trainee, welder, and puddler will be assigned to the various job classes; and summary counts, average wages and fringe benefits, and probable health problems will be developed for each class. The predictive use includes discovery, prediction, and action; if a number of students at a university clinic present similar combinations of respiratory symptoms, it will be concluded that a new

flu virus is about the campus; the future course of symptoms will be predictable from the course of symptoms of the first few students; and those actions that prove effective for the first few patients will be carried out on later patients.

It will be seen that classification intertwines intimately with statistics and probability. Some classification procedure, formal or informal, is necessary prior to data collection; we must decide which things are similar enough to be counted as the same. Further classification may be done during data processing if populations appear heterogeneous; for example, a questionnaire might be summarized separately on sex and age classes if these appear to matter. We must also classify in making inferences; in tossing a coin many times and predicting a head if the frequency of heads is higher, we must classify the new toss of the coin as being similar to past tosses, and the predicted result, heads, as similar to past results. We expect the future to be like the past, so future events must be classified as being like some past events.

Aristotle developed new classifications in logic, ethics, politics, and of animals and parts of animals. He classified some 500 animals, from marine organisms through man, into species and genera not far different from ours. Our present classification of animals and plants was well established by the time of Linnaeus in 1753 [45], who thought of it as a descriptive device, a method of naming and storing knowledge [55]. It was Darwin [10] who interpreted the classificatory tree as corresponding to an actual physical process of evolution; ancestral animals exist at each node of the tree, branching out to form new species by a process of mutation and natural selection. There is a physical reality behind the classification of animals and plants which gives it special interest.

The beginnings of a formal classification procedure may be seen in Kulczynski's work on plant sociology [41], Zubin [61] in psychiatric classification, and Tryon [57] in factor analysis*. Many of the now standard algorithms were developed in the 1950s as computing resources became more able to handle the immense calculations required.

The book *Principles of Numerical Taxonomy* by Sokal and Sneath [53] must be regarded as the beginning of a new era in classification. In this book, a new system of biological classification is proposed, covering the collection of data, the selection and coding of characters, the computation of similarity, the construction of hierarchical clusters, and their evaluation through statistical techniques. The same general program may of course be applied to other disciplines. The program has not been universally welcomed by traditional taxonomists; several key questions remain unsettled—the weighting and selection of variables, the choice of similarity coefficients, the choice of algorithm, the validation of the clusters. But many of the old arguments between proponents of opposing classifications have now become arguments between proponents of opposing algorithms. The outline of the Sokal and Sneath program has been accepted, although the details remain controversial.

DATA AND DISTANCES

The standard form of data used in classification studies is the data matrix,—whereby a number of variables are measured on a number of cases or objects which are to be classified. Clearly, the selection of variables to measure will determine the final classification. Some informal classification is necessary before data collection: deciding what to call an object, deciding how to classify measurements from different objects as being of the same variable, deciding that a variable on different objects has the same value. For example, in classifying political units in the United States, we might choose the objects to be counties or to be census tracts or to be states; we might measure Democratic party strength by the number of registered Democrats, the proportion of registered Democrats in the population, or the proportion of Democrats to major party registrants. Thus important decisions affecting the final classification must be made informally and subjectively.

We cannot be expected to describe an ant and an elephant in the same terms, or to evaluate the job duties of an automobile plant executive by the same tasks as those of a construction worker. Yet we seek classification schemes that encompass very wide ranges of objects. If we use a cases × variables data structure, we must expect many "inapplicable" or "missing" responses. We must seek "global" variables that will provide comparisons across a wide range of objects, as well as local variables that will discriminate well within a narrow range of objects; to range from an ant to an elephant, we might use protein sequences [12] as global variables, and, say, size of ears or length of antennas as local variables.

It is important to allow structure on the variables; several variables may measure more or less the same thing (a variety of dimensions of an insect's head) or some variables may apply only when other variables have a given value (the venation of an insect's wing does not apply if it has no wings; the frequency of coughing is irrelevant if no respiratory symptoms are present).

Since there are many classification problems in selecting and treating variables to be used in classifying objects, it is tempting to consider classifying variables and objects at the same time, as is done in the section on block clustering. Another approach is the linguistic approach, in which objects, variables, and clusters are all words in the system; after all, a dictionary defines words in terms of other words. The dictionary procedure is circular, but we have seen that the objects-by-variables procedure for classification is also circular in requiring prior classification. The data in a linguistic approach consist of sentences in an artificial language of known structure; the clustering consists of constructing new words corresponding to concepts expressed in the sentences. The new words require new sentences to define them, but may simplify the original sentences. For example, in the United Nations in 1970:

1. The Soviet Union voted yes on admitting China, no on China's admission being an important issue (requiring a two-thirds majority for passing).
2. Bulgaria voted yes on admitting China, no on China's admission being an important issue.
3. The United States voted no on admitting China, yes on China's admission being an important issue.

New words are defined as follows:

1. The *Eastern bloc* consists of the Soviet Union and Bulgaria.
2. The *China votes* are the admission of China, and that China's admission is not an important issue.

The simplified sentences are the following:

1. The Eastern block votes yes on China votes.
2. The United States votes no on China votes.

The clustering here lies in construction of new words, *Eastern bloc* and *China votes*, which permit expression of complex sentences as simple ones. Linguistic approaches have been followed in the pattern-recognition* literature (e.g., Cohen [9]), but formal extraction of concepts from given sentences awaits us.

Whatever the data, many standard clustering techniques require computation of distances between all pairs of objects before passing to a clustering algorithm. Numerous measures of distance for various types of variables have been proposed; see, for example, Sneath and Sokal [51]. Suppose that matching measurements for two objects on p variables are $x_1, x_2, \ldots, x_p$ and $y_1, y_2, \ldots, y_p$. For continuous variables, the simplest measure of distance is the *Euclidean distance*:

$$d(x, y) = \left[\sum (x_i - y_i)^2 \right]^{1/2}.$$

For categorical variables, the simplest measure is *matching distance:*

$$d(x, y) = \text{number of times } x_i \neq y_i.$$

If variables are measured on different scales, they must be standarized somehow; for example, continuous variables might be linearly transformed to have constant variance. It is sometimes suggested that the Mahalanobis distance*

$$d(x, y) = (\mathbf{x} - \mathbf{y})' \mathbf{\Sigma}^{-1} (\mathbf{x} - \mathbf{y})$$

be used, where $\mathbf{\Sigma}$ is the covariance matrix of the variables. Both the constant variance and Mahalanobis rescalings may act to downweight variables which effectively discriminate between distinct clusters; the variance and covariance calculations should be based on the behavior of variables within clusters, so this type of scaling should be done iteratively: scaling, clustering, rescaling, and reclustering.

For discrete variables, if the measure of distance is to give equal weight to all variables, use the scaled matching distance

$$d(x, y) = \sum_{x_i \neq y_i} \alpha_i,$$

where $1/\alpha_i$ is the probability that two randomly selected objects are different on the ith variable. Discrete and continuous variables may be used together by adding the scaled matching distance for the discrete variances to the squared Euclidean distance for the continuous variables scaled to have variance 1.

Classification is the grouping of similar objects; we cannot classify without making similarity judgments and similarity calculations. In a cases × variables data structure, the similarity information is carried by the values of the variables for the various cases. We should not expect to be able to combine this similarity information across variables with any simple distance calculation between pairs of objects. We may be forced to use such distances along the way to a final clustering, but weighting, scaling, transformation, and clustering of variables will interact with the clustering. We shall want to refer back to the original data matrix to evaluate the clusters. For this reason, we should avoid techniques that assume the distances between pairs of objects as given; we can only convert data to that form by important premature decisions about the variables; we should seek techniques that recognize the form of the data as collected.

PARTITIONS

Suppose that we desire to partition the data into k disjoint clusters. If a criterion function measuring the fit of a partition to the data is specified, we might in principle find that partition which best fits the data. In practical cases, the search is too expensive and some sort of "local optimum" is sought.

Let the objects be numbered $1, 2, \ldots, N$, and let $\rho(i)$ be an integer between 1 and k that specifies which cluster the ith object belongs to. The partition is determined by the function ρ. Suppose that p variables are measured on each object, giving a vector $\mathbf{x}_i$ of p values for the ith object. Let

$$D(\mathbf{x}, \mathbf{y}) = \sum_{i=1}^{p} (x_i - y_i)^2.$$

The k-means criterion is the within-cluster sum of squares

$$W(x, \rho) = \inf_y W(y, x, \rho)$$
$$= \sum_{i=1}^{N} D(\mathbf{x}_i, \mathbf{y}_{\rho(i)}),$$

where $\mathbf{y}_j$ denotes a vector of values for the jth cluster.

The k-means algorithm developed by Sebestyn [49], MacQueen [46], and Ball and Hall [4] uses the criterion function $W(y, x, \rho)$. It moves in two steps:

1. Given $\mathbf{y}_1, \ldots, \mathbf{y}_k$, choose ρ to minimize $W(y, x, \rho)$.
2. Given ρ, choose $\mathbf{y}_1, \ldots, \mathbf{y}_k$ to minimize $W(y, x, \rho)$.

In the first step, each case is assigned to that cluster whose cluster center it is closest to; in the second step the cluster centers are changed to be the mean of cases assigned to the clusters. The steps are repeated until no further decrease of $W(y, x, \rho)$ occurs; the stationary point is reached after a finite number of steps. There is no guarantee that the local optimum obtained is the global optimum of $W(y, x, \rho)$ for all choices of y and ρ; for $p = 1$, Fisher [19] presents a dynamic programming* technique that obtains the global optimum in $O(N^2k)$ operations.

To evaluate k-means clusters, MacQueen [46] shows that a variation of the k-means algorithm described above converges, when the N cases are sampled from some population, to a partition of the population that minimizes within-cluster sum of squares on the population. The asymptotic distribution of $W(y, x, \rho)$ has been shown by Hartigan [30] to be normal under general conditions; the asymptotic distribution theory allows a test for bimodality based on the projection of the case vectors onto the line between the two cluster means, when $k = 2$.

Day [11] describes a probabilistic version of k-means in which clusters are components of a multivariate normal* mixture distribution. Given the parameters of the model, each case $\mathbf{x}_i$ has probability $p(j|\mathbf{x}_i)$ of belonging to the jth cluster; this "belonging" probability array corresponds to the partition function ρ in k-means. The cluster means and covariances, given the belonging probabilities, are computed by weighted averaging for each cluster, weighting by the belonging probabilities. This corresponds to the second step of the k-means algorithm. Indeed, the two algorithms give the same results for widely separated clusters.

TREES

A *tree* is a set of clusters (subsets of a set of objects) such that any two clusters are disjoint, or one includes the other. A tree is more general than a partition; it may encompass many partitions of varying degrees of fineness.

Algorithms for constructing trees (or hierarchical clustering) preceded algorithms for partitions. One type of algorithm is the joining method. Distances between all pairs of objects are assumed to be given. (*Warning*: That can be quite an assumption!) The algorithm proceeds in the following steps:

1. Find the closest pair of objects and join them to form a new cluster.
2. Define in some way the distance between the new cluster and all other objects.
3. Ignore the two joined objects, and consider the new cluster as an object.
4. Repeat steps 1 through 3 until a single object remains. All clusters obtained along the way are members of the tree.

Important variations in the algorithm occur according to the choice of the distance between clusters in step 2. For example, Sørenson [54] suggests using the distance between clusters as the maximum distance of all pairs of points, one in each cluster, a technique now known as complete linkage clustering, which tends to give compact clusters. Florek et al. [22] use a minimum spanning tree (see DENDRITE) for classification, which is equivalent to defining distance to be the minimum distance over all pairs of points, one in each cluster, a technique now know as *single linkage* clustering, which may easily give long straggly clusters. Sokal and Michener [52] suggest *average linkage*, in which the distance between clusters is the average distance over all pairs of points, one in each cluster. Lance and Williams [43] give a parametric family of distances between clusters which include the three measures noted above.

There are a number of ways to evaluate the algorithms. One method is to assume that perfect clustering corresponds to distances satisfying the ultrametric inequality (also called *triangular inequality*) for any cases i, j, and k:

$$d(i, j) \leqslant \max\left[d(i, k), d(j, k) \right]$$

and then to ask how well the various techniques recover the tree corresponding to the ultrametric when the distances are perturbed a little. See Jardine and Sibson [36], Baker and Hubert [3], and Fisher and Van Ness [18], for leads to the literature. Performances of various joining techniques for samples from multivariate normal mixtures have been examined by Everitt [16], Kuiper and Fisher [40], and others.

An alternative model suggested by Hartigan [27] assumes a population to have a probability density* in p dimensions with respect to some underlying measure, and takes population clusters to be *high-density* clusters—maximal connected sets with density $\geqslant f_0$, with different levels of clustering obtained by varying f_0. Clustering techniques may be evaluated by how well clusters based on a sample of N points from the population approximate high-density clusters for the population. Complete linkage is asymptotically worthless, in that the large complete linkage clusters are not influenced by the population density; average linkage clusters are quite poor, in that the large clusters may or may not approximate population clusters; and single linkage is consistent in the following sense. For each population cluster C, as the sample size N increases, there is a sequence of single linkage clusters C_N such that every point in C_N is within ϵ_N of a point in C, and every point in C is within ϵ_N of a point in C_N, where $\epsilon_N \to 0$ in probability (see Hartigan [31]).

We may estimate a density from the sample, and then construct high-density clusters from the density estimate. Single linkage corresponds to a rather poor density estimate, the nearest-neighbor density estimate, in which the density at any point is inversely proportional to the distance to the nearest sample point. An improved density estimate takes the density to be inversely proportional to the distance to the k_N nearest neighbor, where $k_N \to \infty$ as $N \to \infty$. Hierarchical algorithms corresponding to kth-nearest-neighbor density estimates are proposed by Wishart [60]. The corresponding joining algorithm would define the distance between two clusters as the kth-closest distance over pairs of points, one in each cluster.

BLOCK CLUSTERING

A cluster is a set of cases that take similar values for some set of variables. If the data come as a data matrix, a cluster will correspond to a submatrix of the data; associated with the cluster will be both a cluster of cases, and a cluster of variables on which the cases take similar values. Such a submatrix will be called a *block*. Block clustering requires simultaneous clustering of rows and variables. Techniques for such simultaneous clustering are suggested by Lambert and Williams [42] and by Tharu and Williams [56], who split a binary data matrix simultaneously on rows and columns. Good [24] suggests a simultaneous split of rows and columns analogous to the first term in a singular-value decomposition of a matrix. (If A_{ij} is the data matrix, a row vector r_i and column vector c_j consisting of ± 1's is chosen to minimize $\sum_{i,j} r_i c_j A_{ij}$.)

Somewhat the same techniques may be used in clustering cases and variables simultaneously as in clustering cases alone, but the technical difficulties are multiplied. For example, there are now three clustering structures to be considered, on sets of cases, sets of variables, and subsets of the data matrix. In principle there is no reason why the blocks should not be overlapping, but overlapping blocks certainly make it difficult to represent and interpret the clusters. See Hartigan [27] for a number of block clustering techniques.

The modal block algorithm [28] is presented as an example. A sequence of blocks $B_1, B_2, \ldots, B_k$ is constructed, together with block codes that specify a modal value for each variable in the block. The goodness-of-fit criterion is the number of data values that agree with the block code for the latest block that includes them.

After i blocks have been selected, all values that agree with the appropriate block code are said to be "coded." The $(i + 1)$th block is selected as follows:

1. The initial block code is a modal value for each variable, over all uncoded values for the variable. The initial block case is that case which has most values in agreement with the block code. The initial variables in the block are those variables for which the initial case agrees with the block code.

2. For each case in turn, place the case in the block if it has more values agreeing with the block code than coded values, over variables in the block.

3. Redefine the block code to be a modal value for each variable, over uncoded values in the cases in the block.

4. For each variable in turn, place the variables in the block if it has more values agreeing with the block code than coded values, over cases in the block.

5. Repeat steps 2 through 4 until no further change occurs.

GRAPHICAL TECHNIQUES

Formal statistical evaluations, such as significance tests* or exact distributions, are rare in clustering; see Hartigan [29] for a brief survey. Too frequently we rely on hearsay evidence that "this technique seems to have worked well" or on customer testimonials that "the clusters make a lot of sense." Perhaps the inherent subjectivity of similarity judgments will prevent convincing evaluative theory; we cannot expect to say "these are the best clusters" or "these are the only clusters."

There is no generally accepted concept for a cluster, yet people frequently have strong opinions about what a cluster should be. Accordingly, there are available some graphical techniques that present the data in various ways so that the user may make his or her own judgments about the presence and construction of clusters.

Continuous variables, if they are well chosen by a sensible investigator, will frequently show modes* in their histogram* corresponding to clusters in the cases. (Here again is the characteristic circularity of classification problems, where measurements are selected with an eye to assumed clusters, and the same clusters then computed "objectively" from the data.) Histograms and two-dimensional plots for all pairs of variables may therefore reveal clusters. An eigenvector* analysis, with the first few pairs of eigenvectors plotted against each other, should give some picture of the scatter of points in many dimensions, and of the realtionships between variables. If all the variables are discrete, say binary, there is no value in doing histograms or two-variable plots, but eigenvector analysis of binary data* (the discrete data may be converted to binary) may be suggestive.

An alternative approach is the method of constructing objects corresponding to the cases; the objects appeal to our intuition better than the dimensional representations; there are stars (e.g., Goldwyn et al. [23]), glyphs [2], faces* [7], boxes [27], trees and castles [38], and no doubt a menagerie of other objects waiting to appear.

APPLICATIONS

Classification is applied, formally or informally, in every field of thought. Often, the informal classification stands because formal techniques are not appropriate or sufficiently well developed.

In psychiatry, formal techniques have been used to discover both clusters of patients [17, 34], and clusters of symptoms [32]. In archaeology*, large numbers of objects, such as stone tools, pottery, skulls, or statues, found in diggings must be clustered [35]. In phytosociology, the spatial distribution of plant and animal species is studied, requiring clustering of both species and habitats [44].

Fisher [20] uses input–output matrices for clustering industries. In market research, King [37] clusters firms by stock price behavior. Dyen et al. [15] compare the words for 196 standard concepts in different languages to construct an envolutionary tree of languages (see LINGUISTICS, STATISTICS IN). Weaver and Hess [58] use clustering techniques to establish legislative districts. Abell [1] clusters galaxies.

A problem area that deserves much more study is the classification of disease. To maintain statistics, an international statistical classification of diseases exists, based on no particular principle. Small-scale studies within well-defined groups have been carried out: Knusman and Toeller [39] on diabetes, Winkel and Tygstrup [59] on cirrhosis of the liver, Hayhoe et al. [33] on leukemia, Manning and Watson [47] on heart disease, and others. For diseases that are known to be caused by viruses or bacteria, the disease classification follows the classification of the agent [25].

The workhouse of formal clustering techniques remains taxonomy*, the classification of animals and plants. Much of the development of modern classification techniques is due to "numerical taxonomists"; the journal *Systematic Zoology* is perhaps the principal journal for classification research and for debates on the philosophy and principles of a classification. Some examples of application are to yeasts [13], pollen [50], fish [5], and horses [6]. To me, the most interesting area of application in all of classification is the development of evolutionary trees using DNA, RNA, and protein sequences, as followed by Dayhoff [12], Fitch [21], and others. The genetic material used as data is comparable across very wide ranges of species; the data do not come in cases-by-variables form, but as sequences of amino acids or nucleic acids that must be matched between species [48]. A crucial part of the problem is development of a model for mutation of the sequences. The standard distance techniques have been used, with some success, but great gains remain to be made by development of appropriate probability models and corresponding algorithms.

OUTSTANDING PROBLEMS

There are many clustering techniques, but not many ways of deciding among them. We need to develop theories of clustering and of methods of evaluating and comparing clusters. In a statistical approach, we assume that the objects to be studied are a random sample from some population, and ask how well the clusters in the sample conform to the population clusters; see for example, MacQueen [46], Hartigan [30], and Everitt [16]. Within the statistical approach there are many difficult distribution theory problems: the appropriate definition of consistency, the detection of modes (or deciding the number of clusters), the reliability of cluster membership, the choice of probability models, and models for the simultaneous clustering of cases and variables.

The assumption that the objects form a random sample is itself questionable; for example, the objects might be the counties in the United States, the jobs in the factory, the species of birds in the world, or other complete sets. Or the objects might obviously not be selected randomly—we collect many specimens of ants, many groups of which are identical, so we study only one specimen from each group. What alternative evaluations are there to the sampling approach?

Present clustering theory and techniques are inadequate for the scale of many clustering problems. Our data structures and algorithms work for a few hundred objects, but we need classification schemes for thousands of birds, millions of animals, and billions of people. We need to develop types of data structures and computing techniques and evaluative techniques appropriate for very large numbers of diverse objects.

We expect the future to be like the past. The foundations of probability and statistics depend on a proper formulation of the similarity between the known and unknown, of the grouping together of partly known objects so that known properties for some objects in a class may be extended to other objects in a class. I expect that theories of classification will do as much to clear up the murky foundations of probability and statistics as theories of probability and statistics will do to clear up the murky foundations of classification.

Further Reading

I would suggest Everitt [16], Clifford and Stephenson [8], or Duran and Odell [14] for introductory reading. Sneath and Sokal [51] is oriented to biology, but contains many valuable ideas. Hartigan [27] contains the standard algorithms and many exotic ones besides. All books lack a strong theoretical backbone. For applications, try Sankoff and Cedergren [48] for a fascinating exercise in inference of evolutionary sequences of DNA, and Henderson et al. [34] for a classification of attempted suicides into three types of motivation for the attempt.

Bibliography

Anderburg, M. R. (1973). *Cluster Analysis for Applications*. Academic Press New York.

Ball, G. H. (1970). *Classification Analysis*. Stanford Research Institute, Menlo Park, Calif.

Benzecri, J. P., et al. (1976). *L'Analyse des données*, 2 vols. Dunod, Paris.

Bijnen, E. J. (1973). *Cluster Analysis*. Tilburg University Press, Tilburg, Netherlands.

Blackith, R. E. and Reyment, R. A. (1971). *Multivariate Morphometrics*. Academic Press, New York.

Clifford, H. T. and Stephenson, W. (1975). *An Introduction to Numerical Classification*. Academic Press, New York.

Cole, A. J. (1969). *Numerical Taxonomy*. Academic Press, London.

Duran, B. S. and Odell, P. L. (1974). *Cluster Analysis: A Survey*. Springer-Verlag, Berlin.

Everitt, B. S. (1974). *Cluster Analysis*. Halstead Press, London.

Fisher, W. D. (1969). *Clustering and Aggregation in Economics*. Johns Hopkins University Press, Baltimore, Md.

Hartigan, J. A. (1975). *Clustering Algorithms*. Wiley, New York.

Jardine, N. and Sibson, R. (1971). *Mathematical Taxonomy*. Wiley, New York.

Lerman, I. C. (1970). *Les Bases de la classification automatique*. Gauthier Villars, Paris.

Lorr, M. (1966). *Explorations in Typing Psychotics*. Pergamon, New York.

MacNaughton-Smith, P. (1965). *Some Statistical and Other Numerical Techniques for Classifying Individuals*. Her Majesty's Stationery Office, London.

Ryzin, J. Van (1977). *Classification and Clustering*. Academic Press, New York.

Sneath, P. H. A. and Sokal, R. R. (1973). *Numerical Taxonomy*. W. H. Freeman, San Francisco.

Sokal, R.R. and Sneath, P. H. A. (1963). *Principles of Numerical Taxonomy*. W. H. Freeman, San Francisco.

Tryon, R. C. (1939). *Cluster Analysis*. Edwards Brothers, Ann Arbor, Mich.

Tryon, R. C. and Bailey, D. E. (1970). *Cluster Analysis*. McGraw-Hill, New York.

Watanabe, S. (1969). *Methodologies of Pattern Recognition*. Academic Press, New York.

REFERENCES

[1]   Abell, G. O. (1960). *Astrophysical J. Supp. Series*, **32**, 211–288.

[2]   Anderson, E. (1960). *Technometrics*, **2**, 387–392.

[3]   Baker, F. B. and Hubert, L. J. (1975). *J. Amer. Statist. Ass.*, **70**, 31–38.

[4]   Ball, G. H. and Hall, D. J. (1967). *Behav. Sci.*, **12**, 153–155.

[5]   Cairns, J. and Kaesler, R. L. (1971). *Trans. Amer. Fish. Soc.*, **100**, 750–753.

[6]   Camin, J. H. and Sokal, R. R. (1965). *Evolution*, **19**, 311–326.

[7]   Chernoff, H. (1973). *J. Amer. Statist. Ass.*, **68**, 361–368.

[8]   Clifford, H. T. and Stephenson, W. (1975). *An Introduction to Numerical Classification*. Academic Press, New York.

[9]   Cohen, B. L. (1978). *Artif. Intell.*, **9**, 223–255.

[10] Darwin, C. (1859). *The Origin of Species*.

[11] Day, N.E. (1969). *Biometrika*, **56**, 463–474.

[12] Dayhoff, M. O. (1976). *Atlas of Protein Structure and Sequence*, Vol. 5. National Biomedical Research Foundation, Washington, D.C.

[13] Dupont, P. F. and Hedrick, L. R. (1971). *J. Gen. Microbiol..*, **66**, 349–359.

[14] Duran, B. S. and Odell, P. L. (1974). *Cluster Analysis: A Survey*. Springer-Verlag, Berlin.

[15] Dyen, I., James, A. T., and Cole, J. W. L. (1967). *Language*, **43**, 150–171.

[16] Everitt, B. S. (1974). *Cluster Analysis*. Halstead Press, London.

[17] Everitt, B. S., Gourlay, A. J., and Kendell, R. E. (1971). *Brit. J. Psychiatry*, **119**, 399–412.

[18] Fisher, L. and Van Ness, J. W. (1971). *Biometrika*, **58**, 91–104.

[19] Fisher, W. D. (1958). *J. Amer. Statist. Ass.*, **53**, 789–798.

[20] Fisher, W. D. (1969). *Clustering and Aggregation in Economics*. Johns Hopkins University Press, Baltimore, Md.

[21] Fitch, W. M. (1971). *Syst. Zool.*, **20**, 406–416.

[22] Florek, J., Lukaszewicz, J., Perkal, J., Steinhaus, H., and Zubrzycki, S. (1951). *Colloq. Math.*, **2**, 282–285, 319.

[23] Goldwyn, R. M., Friedman, H. P., and Siegel, J. H. (1971). *Computer Biomed. Res.*, **4**, 607–622.

[24] Good, I. J. (1965). *Mathematics and Computer Science in Biology and Medicine*. Her Majesty's Stationery Office, London.

[25] Goodfellow, N. (1971). *J. Gen. Microbiology*, **69**, 33–80.

[26] Hartigan, J. A. (1975). *J. Statist. Comp. Simul.*, **4**, 187–213.

[27] Hartigan, J. A. (1975). *Clustering Algorithms*. Wiley, New York.

[28] Hartigan, J. A. (1976). *Syst. Zool.*, **25**, 149–160.

[29] Hartigan, J. A. (1977). In *Classification and Clustering*, J. Van Ryzin, ed. Academic Press, New York.

[30] Hartigan, J. A. (1978). *Ann. Statist.*, **6**, 117–131.

[31] Hartigan, J. A. (1979). *J. Amer. Statist. Ass.*, **76**, 388–394.

[32] Hautaloma, J. (1971). *J. Consult. Clin. Psychol.*, **37**, 332–344.

[33] Hayhoe, F. G. J., Quaglino, D., and Doll, R. (1964). *Med. Res. Council Special Report Series*, **304**, H.M.S.O., London.

[34] Henderson, A. S., et al. (1977). *Brit. J. Psychiatry*, **131**, 631–641.

[35] Hodson, F. R. (1969). *World Archaeol.*, **1**, 90–105.

[36] Jardine, N. and Sibson, R. (1971). *Mathematical Taxonomy*, Wiley, New York.

[37] King, B. F. (1966). *J. Bus.*, **39**, 139–190.

[38] Kleiner, B. and Hartigan, J. A. (1979). Representing points in many dimensions by trees and castles. Unpublished.

[39] Knusman, R. and Toeller, M. (1972). *Diabetologia*, **8**, 53.

[40] Kuiper, F. K. and Fisher, L. (1975). *Biometrics*, **31**, 777–784.

[41] Kulczynski, S. (1928). *Bull. Int. Acad. Pol. Sci. B*, **2**, 57.

[42] Lambert, J. M. and Williams, W. T. (1962). *J. Ecol.*, **50**, 775–802.

[43] Lance, G. N. and Williams, W. T. (1966). *Computer J.*, **9**, 373–380.

[44] Lieth, H. and Moore, G. W. (1970). In *Spatial Patterns and Statistical Distributions*, G. P. Patil, E. C. Pielou, and W. E. Waters, eds. Pennsylvania State Statistics Series. Pennsylvania State University Press, University Park, Pa.

[45] Linnaeus, Carolus (1753). *Species Plantarum.*

[46] MacQueen, J. (1967). *Proc. 5th Berkeley Symp. on Math. Statist. Pro.*, Vol. 1. University of California Press, Berkeley, Calif. pp. 281–297.

[47] Manning, R. T. and Watson, L. (1966). *J. Amer. Med. Ass.*, **198**, 1180–1188.

[48] Sankoff, D. and Cedergren, R. J. (1973). *J. Mol. Biol.*, **77**, 159–164.

[49] Sebestyen, G. S. (1962). *Decision Making Process in Pattern Recognition.* Macmillan, New York.

[50] Small, E., Bassett, I. J., and Crompton, C. W. (1971). *Taxon*, **20**, 739–749.

[51] Sneath, P. H. A. and Sokal, R. R. (1973). *Numerical Taxonomy.* W. H. Freeman, San Francisco.

[52] Sokal, R. R. and Michener, C. D. (1958). *Univ. Kans. Sci. Bull.*, **38**, 1409–1438.

[53] Sokal, R. R. and Sneath, P. H. A. (1963). *Principles of Numerical Taxonomy.* W. H. Freeman, San Francisco.

[54] Sørenson, T. (1948). *Biol. Skr.*, **5**, 1–34.

[55] Stearn, W. T. (1959). *Syst. Zool.*, **8**, 4–22.

[56] Tharu, J. and Williams, W. T. (1966). *Nature, (Lond.)*, **210**, 549.

[57] Tryon, R. C. (1939). *Cluster Analysis.* Edwards Brothers, Ann Arbor, Mich.

[58] Weaver, J. B. and Hess, S. W. (1963). *Yale Law J.*, **73**, 288–308.

[59] Winkel, P. and Tygstrup, N. (1971). *Computer Biomed. Res.*, **4**, 417–426.

[60] Wishart, D. (1969). In *Numerical Taxonomy*, A. J. Cole, ed. Academic Press, London.

[61] Zubin, J. (1938). *Psychiatry*, **1**, 237–247.

(DENDRITES
DENDROGRAMS
GENETICS, STATISTICS IN
HIERARCHAL CLASSIFICATION
MIXTURES
MULTIDIMENSIONAL SCALING)

J. A. HARTIGAN

CLASS L LAWS   *See* L CLASS LAWS

CLIFF–ORD TEST

Cliff and Ord [7] have developed tests for independence of variates observed on the vertices of a lattice or in regions of the plane, against autocorrelated alternatives. The tests fall into two classes, depending on whether the variates in question are measured on ordinal/interval or nominal scales (although binary variates can be treated in either class; *see* CATEGORICAL DATA). In both cases the pattern of autocorrelation* under the alternative hypothesis must be specified a priori by a matrix of "weights" $\{w_{rs}\}$ representing the strength of influence of site s on site r. In the remainder of this article the "join-count" (nominal) and "I" (ordinal-/interval) statistics are discussed separately, followed by reference to problems, related work, and one or two applications.

I STATISTIC

Let the set of random variables $\{X_{ij}\}$ be associated with the vertices $\{i = 1, \ldots, M, j = 1, \ldots, N\}$ of a regular rectangular lattice. Moran [18] proposed that the statistic

$$
\begin{aligned}
I = \Big\{ &\sum_{i=1}^{M} \sum_{j=1}^{N-1} (x_{ij} - \bar{x})(x_{i,j+1} - \bar{x}) \\
&+ \sum_{i=1}^{M-1} \sum_{j=1}^{N} (x_{ij} - \bar{x})(x_{i+1,j} - \bar{x}) \Big\} \\
&\div \Big\{ \sum_{i=1}^{M} \sum_{j=1}^{N} (x_{ij} - \bar{x})^2 \Big\}
\end{aligned}
$$

be employed to test for dependence between adjacent sites, commenting that when scaled by $MN/(2MN - N - M)$, this measure was naturally interpreted as a correlation coefficient* between nearest neighbors. Moran suggested that a test be based on the asymptotic normality* of I under the null hypothesis of independence. Geary [12] proposed a generalization of the von Neumann ratio* applicable to irregular lattices:

$$
C = \frac{n-1}{2K} \left[\sum_{i=1}^{n} \sum_{j=1}^{n} \delta_{ij}(x_i - x_j)^2 \right] \times \left[\sum_{i=1}^{n} (x_i - \bar{x})^2 \right],
$$

where $\delta_{ij} = 1$ if i and j are neighbors, zero otherwise, and $\sum_{i=1}^{n} \sum_{j=1}^{n} \delta_{ij} = 2K$. To test for dependence when the $\{X_i\}$ were regression residuals*, and thus (generally) correlated under the null, a randomization* procedure was employed in which the arrangement of the residuals was permuted over the lattice, and the observed value of C located in the resulting pseudo-sampling distribution. The I statistic can, of course, be generalized in the same way.

These statistics suffer from the defect that the range of alternatives against which they may be expected to have high power is very restricted, and this led Cliff and Ord [7] to further generalization. Consider the regression model* $Y = X\beta + \epsilon$, where the disturbance process, ϵ, follows a simultaneous autogressive scheme: $\epsilon = \rho W\epsilon + \eta$, with $E(\eta) = 0$, $E(\eta\eta') = \sigma^2 I_n$. To test for independence of ϵ (i.e., $\rho = 0$) Cliff and Ord propose the generalized Moran statistic $T = Y'MWMY/Y'MY$, where M is the matrix $|I_n - X(X'X)^{-1}X'|$. That is, $T = e'We/e'e$, where e is the vector of Ordinary Least Squares (O.L.S.) residuals from the regression of Y on X. This quantity can be shown to be asymptotically normally distributed under fairly mild conditions on the eigenvalues of the $n \times n$ matrix of weights, W. In small samples, however, the distribution of T is very lumpy, so that no standard approximation is likely to perform well (Imhof's method [14] can, of course, be used to obtain the exact distribution, although at the cost of some labor); furthermore, for models in which the elements of few columns of W dominate the rest in magnitude, a two-parameter beta distribution* is superior to the normal. Cliff and Ord [6] report a simulation study of the small-sample properties of I and make a number of suggestions for obtaining appropriate critical regions.

If η is normally distributed, it can be shown that W is the best choice of matrix for the numerator of I (for both autoregressive* and moving-average* alternatives), either by examining the asymptotic relative efficiency* of the test [7], or, more simply, by noting that I is in fact the likelihood derivative* test [4]. A Monte Carlo study of the power of tests based on O.L.S., recursive, or Theil's Best Linear Unbiased Scalar (B.L. U.S.) residuals can be found in Bartels and Hordijk [1]. Haining [13] compares the power of the I statistic with that of the likelihood ratio for moving-average models.

The Asymptotic Relative Efficiency (A.R.E.) of the generalized Geary statistic is shown by Cliff and Ord to be lower than that of T, except for regular lattices, so its use is not recommended.

JOIN COUNT STATISTICS

Consider a map consisting of a set of n zones, which are colored black (B) or white (W) depending on the presence or absence of some quality. Moran [16, 17] proposed a test for randomness in the resulting pattern based on counting the numbers of times black zones had boundaries in common with black and with white zones. The resulting statistics, BB, and BW, were shown to be asymptotically normally distributed, both under the assumption that the numbers of B and W zones were fixed, and alternatively that the map represented the outcome of n independent drawings from a population in which B occurred with probability p. For small samples, Moran recommended the fitting of a Pearson curve* using the third and fourth moments. Evidently, a choice has to be made between BB, BW, or a combination of both in constructing a test, but this problem was not addressed until the work of Cliff and Ord [7]. For mapped variables with more than two categories, there are several basic statistics: the number of joins between zones of the same (given) color, of two (given) different colors, and of any different colors.

The generalization of the join counts proposed by Cliff and Ord is exactly analogous to that for the Moran and Geary statistics, and is as follows for binary data. Let the binary variable X take the values 1 (B) and 0 (W); then if $\{w_{ij}\}$ are as before,

$$BB^* = \frac{1}{2} \sum_{i \neq j} \sum w_{ij} x_i x_j$$

and

$$BW^* = \frac{1}{2} \sum_{i \neq j} \sum w_{ij} (x_i - x_j)^2.$$

Cliff and Ord prove the asymptotic normality of these quantities, give their small-sample moments, and carry out extensive simulations. Their more important findings are as follows: (1) where the number of black or white zones (or sites) is at choice, the tests have the most power for a ratio of 1 : 1; (2) except on regular lattices, BW^* has more power than BB^*, or any combination therewith (this is not surprising since BW^* contains BB^*); (3) the power of the tests is greater for relatively sparse **W** matrices (this is also true of the Moran and Geary tests for interval data); (4) in small samples chi-squared* is a slightly better approximation than the normal distribution for most cases.

PROBLEMS, APPLICATIONS, AND RELATED WORK

An underlying theoretical difficulty concerns the passage to the limit in obtaining asymptotic results. It is clear what this entails on a regular lattice, but not on an irregular one. In time series, the natural ordering of the observations restricts the alternatives that need to be considered, so that a portmanteau test with tabulated critical regions, such as the Durbin–Watson test* [10] for serial correlation*, may be widely employed; unfortunately, the null distribution of the Cliff–Ord test depends on the matrix **W**, so that this is no longer possible. In a regression context, Brandsma and Ketellapper [3] have attempted to introduce greater flexibility by specifying an alterntive of the form $\epsilon = (\rho_1 \mathbf{W}_1 + \rho_2 \mathbf{W}_2)\epsilon + \boldsymbol{\eta}$, but they use a likelihood ratio test* rather than employing an explicit function of the data, and nothing is known of the comparative performance of their method and the Cliff–Ord test. Kooyman [15] has sought a more powerful test by maximizing T as a function of **W** subject to constraints. A number of authors have used the autocorrelation statistics to identify the most significant interaction pattern, but, as Cliff and Haggett [5] note, this is made hazardous by the dependence of the power of the tests on the form of W. Cliff et al. [9] have proposed a test for binary variables on a regular lattice based on the distri-

bution of the number of B zones with $0', 1, 2, \ldots, k$ BW or BB joins. Being based on a modified χ^2 text of goodness of fit, this statistic has the dual merits of possessing good small-sample properties, and power against a wider range of alternatives than the join count tests. Sen and Sööt [19] define a class of locally most powerful tests based on ranked data, but their specification of the second-order properties under the alternative is restrictive, except for normally distributed variables. Besag and Moran [2] discuss the properties of a test for dependence in a conditional Gaussian model, for a rectangular lattice, of the form

$$E\left(X_{ij} \mid X_{rs} = x_{rs}, r, s \neq i, j\right)$$
$$= \mu + \rho\left(x_{i-1, j} + x_{i+1, j} + x_{1, j-1}\right.$$
$$\left. + x_{i, j+1} - 4\mu\right).$$

The relationship between autocorrelation and information content of binary maps has been investigated empirically by Gatrell [11], but there appears to be a dearth of theoretical results in this area. Finally, estimation methods for autocorrelated models have received much attention recently, and these developments have been reviewed at length by Cliff and Ord [8].

References

[1] Bartels, C. P. A. and Hordijk, L. (1977). *Reg. Sci. Urban Econ.*, **7**, 83–101.

[2] Besag, J. E. and Moran, P. A. P. (1975). *Biometrika*, **62**, 555–562.

[3] Brandsma, A. A. and Ketellapper, R. H. (1979). *Environ. Plann. A*, **11**, 51–58.

[4] Burridge, P. (1980). *J. R. Statist. Soc. B*, **42**, 107–108.

[5] Cliff, A. D. and Haggett, P. (1979). In *Statistical Applications in the Spatial Sciences*, Wrigley, N., ed. Pion, London.

[6] Cliff, A. D. and Ord, J. K. (1971). *Geogr. Anal.*, **3**, 51–62.

[7] Cliff, A. D. and Ord, J. K. (1973). *Spatial Autocorrelation*, Pion, London. (The material in this book has been revised and considerably expanded in ref 8.)

[8] Cliff, A. D. and Ord, J. K. (1980). *Spatial Processes: Models, Inference and Applications*, Pion, London.

[9] Cliff, A. D., Martin, R. L., and Ord, J. K. (1975). *Inst. Brit. Geogr. Trans.*, **65**, 109–129.

[10] Durbin, J. and Watson, G. S. (1950). *Biometrika*, **37**, 409–428.

[11] Gatrell, A. C. (1977). *Geogr. Anal.*, **9**, 29–41.

[12] Geary, R. C. (1954). *Statistician*, **5**, 115–145.

[13] Haining, R. P. (1977). *Inst. Brit. Geogr. Trans.*, New Ser., **3**, 202–225.

[14] Imhof, P. J. (1961). *Biometrika*, **48**, 419–426.

[15] Kooyman, S. A. L. M. (1976). *Ann. Syst. Res.*, **5**, 113–132.

[16] Moran, P. A. P. (1947). *Proc. Camb. Philos. Soc.*, **43**, 321–328.

[17] Moran, P. A. P. (1948). *R. Statist. Soc. B*, **10**, 243–251.

[18] Moran P. A. P. (1950). *Biometrika*, **37**, 17–23.

[19] Sen, A. K. and Sööt, S. (1977). *Environ. Plann. A*, **9**, 897–903.

P. BURRIDGE

CLINICAL TRIALS

Ancient medical practice was based on a reverence for authority, often of a religious nature. Even as late as the eighteenth century, the value of such therapies as bleeding, purging, and starvation was regarded as self-evident. With the development of the scientific method, however, increasing emphasis began to be placed on experience rather than on a priori principles as a basis for selecting medical treatments. The need for a careful statistical approach was demonstrated by P. C. A. Louis in 1835. Influenced by Laplace's writings on probability theory, he demonstrated by numerical methods that bleeding had no effect on the morbidity* or mortality of patients with pneumonia [2]. Modern surgical procedures that have similarly proved to do more harm than good include portacaval shunts for esophageal varices and gastric freezing for duodenal ulcer.

These and other examples illustrate the importance to contemporary medicine of therapeutic evaluation based on sound scientific methods. Thanks largely to the pioneering efforts of Bradford Hill [5], who used the new methodology to help demonstrate the value of streptomycin therapy for tuberculosis in the late 1940s, clinical trials incorporating the classical Fisherian principles of randomization* and replication* now play a key role in such evaluation. Nevertheless, most of the statistical ideas that underlie such trials are elementary, certainly in comparison with those employed in industry and argiculture. Thus the discussion here deals primarily with practical issues which arise in their application to the clinical setting.

SELECTION OF PATIENTS AND TREATMENTS

If the goal of a clinical trial were to answer a purely biological question, the study population would ideally be chosen to be homogeneous, so as to reduce between-subject variability. However, for most trials, particularly those cooperative endeavors organized at a regional or national level to compare major therapies, it is wise to adopt a more *pragmatic* attitude toward their design [10]. Although it is necessary to restrict entry to patients who have the proper diagnosis, who are eligible for any of the treatments being compared, and who have a reasonable expectation of continued follow-up* for the end points under investigation, the trial will ultimately have a greater impact on medical practice if the patient sample is large and relatively unselected. Similarly, when one recognizes that a certain amount of tailoring of the treatment according to individual needs is an integral part of patient care, it is clear that the comparison is between two or more treatment *policies* in the hands of representative practitioners.

Limited use has been made in clinical trials of factorial designs*. This is due partly to a desire to keep the treatment protocol as simple as possible, and to the difficulty of getting a number of participants to agree to randomize patients to more than two or three treatment groups. Nevertheless, in view of the large costs of conducting collaborative clinical trials, more attention should be given to the possibility of using simple 2×2

or perhaps even $2 \times 2 \times 2$ designs. Only then can one obtain systematic information about the interaction between two treatments. In their absence, such designs provide almost as much information about treatment effects as would be obtained from trials of like size conducted separately for each treatment. Factorial designs could be especially helpful in answering secondary questions which could be added on to an existing trial at minimum cost.

New experimental therapies are usually compared with the best standard treatment for the particular condition. However, situations sometimes arise when it is desirable to decide whether the new treatment has any value at all. In this case, in order that the comparisons not be biased by the patient's knowledge of his or her treatement, the controls may be given a *placebo* or dummy treatment, for example an inert, sugar-coated pill. The well-known "placebo effect" consists in the fact that patients receiving such dummy treatments often respond better than those receiving nothing at all.

EVALUATION OF RESPONSE

Even the most innocuous medical treatments carry some degree of risk as well as the hope of benefit. Consequently, it is usually not possible to evaluate them in terms of a single criterion of response. Rather, there will be multiple end points representing various therapeutic and toxic effects. A full evaluation often involves a rather complicated weighing of costs, risks, decision theory, and benefits.

More accurate comparisons of treatment effects are possible if one records the *time to occurrence** of the various end points in addition to the fact of their occurrence. Since the detection of many events depends on how hard one looks for them, it is important that the trial protocol specify a regular sequence of follow-up examinations and that these be administered uniformly to all treatment groups.

Many of the responses recorded in clinical trials involve subjective judgments. The patient is asked to rate the degree of pain on a scale from 1 to 5, or the physician to scan a radiograph for evidence of recurrent tuberculosis. So that knowledge of the patient's treatment not influence these responses, it may be necessary to conduct the trial in such a way that the treatment assignment is not known to the person responsible for the evaluation. It is sometimes possible to arrange that neither the patient nor the physician is aware of the actual treatment, in which case one talks of a *double-blind* trial. For example, a central office may supply coded boxes of identical-looking pills. Of course, provision must be made for breaking the code, and identifying the treatment received, if in the physician's judgment this is required.

RANDOMIZATION*

The foundation of the modern clinical trial, and the feature that generates the greatest confidence in the validity of its conclusions, is its insistence on randomization as the basis for treatment assignment. This ensures (1) that neither the patient nor the physician knows the treatment at the time of entry into the trial, and thus cannot bias the result through the decision to participate; (2) that factors which influence prognosis, whether or not they are known, are evenly distributed among the various treatment groups; and (3) that the assumption of a random-error term, which underlies such quantitative statistical measures as p-values and confidence intervals, is well founded.

Unrestricted randomization may by chance result in a serious imbalance in the number of patients assigned to a particular treatment. The *randomized blocks** design avoids this possibility. For example, with two treatments A and B randomized in blocks of six, the first two blocks might be $BAABBA$ and $AABABB$. Unfortunately, a physician who is acquainted with this scheme may be able to predict the assignment for patients arriving at or toward the end of each block, which suggests that blocks of size 6 or 8 be used in preference to those of size 2 or 4.

A compromise between unrestricted randomization and randomized blocks is adaptive randomization. For 1 : 1 randomization of two treatments A and B, the adaptive procedure assigns the next patient to A with probability $\frac{1}{2}$ provided that the number already assigned to A equals the number assigned to B. If there are already more patients on B than A, the next assignment is to A with probability $p > \frac{1}{2}$; whereas if there are more on A, the assignment to B is with probability p. Using $p = \frac{2}{3}$, this procedure makes severe imbalance extremely unlikely while avoiding deterministic assignments.

Randomization should generally be carried out just before the treatments are administered. In cases where the treatments are identical up to a certain point, e.g., 6 months of drug therapy vs. 12 months, the randomization should be performed just prior to divergence. The reason is that all events that occur after the time of randomization must be counted in the analysis to avoid bias. Events that occur before the two treatments actually differ tend to dilute the treatment effect.

HISTORICAL CONTROLS*

Several alternatives to randomization have been proposed as a basis for treatment assignment and comparison. Indeed, most of these were in widespread use before the advent of the clinical trial and results based on such comparisons continue to be reported in the medical literature. They include (1) comparison of the treatment used by one hospital on its patients with the treatment used by another hospital; (2) allowing the patient to choose his or her own treatment, as with volunteers; (3) comparison with the results of case reports or series assembled from the medical literature; and (4) comparison of current patients receiving the new treatment

with unselected controls assembled from the immediate past experience of the same investigator(s). Proponents of such *historical controls* suggest that any imbalance between the comparison groups can be corrected by using an appropriate analysis of covariance*.

Although the hazards of nonrandomized comparisons should be obvious to the trained statistician, they seem to need frequent and repeated emphasis. Even patients treated at the same institution can show notable changes in response over time due to subtle variations in referral patterns or improvements in supportive care. Adequate adjustment for imbalance by covariance analysis presumes first that one knows all the relevant prognostic factors and how they influence outcome, so that the residual variation is reduced to a minimum, and second that sufficient details are available about such factors in the medical record. Both presumptions are often false, owing to the discovery of new prognostic factors, advances in diagnostic techniques, or the use of a different follow-up schedule in the historical series.

The objections to historical controls are by no means purely theoretical or speculative. Most statisticians who have worked in clinical trials for some years will be able to recall *from their own experience* situations in which the outcome changed over time. Pocock [9] identified 19 instances in which the same treatement was used in two consecutive trials by a cooperative group of investigators. Changes in death rates ranged from -46% to $+24\%$, the differences being significant at the 0.02 level or less in four instances. Other investigators have collected from the medical literature samples of reports concerning the same treatement, and have compared the results obtained according to the methodology used. The general tendency seems to be for the degree of enthusiasm for a procedure to vary inversely with the degree of rigor used in its evaluation.

Of course, historical data provide valuable background for interpreting the results of any particular trial. If there is no variation in outcome in the historical series, as for example with a 100% fatality rate, all the new treatment need do is result in cure for the few cases with proven diagnosis. Rare medical "breakthroughs" will sometimes result in a treatment effect so large that it cannot reasonably be ascribed to changes in patient characteristics. However, most therapeutic progress is in the form of small steps which demand a rigorous methodology to ensure that one is headed in the right direction.

A recent compromise proposal is to alter the randomization ratio so that more patients receive the new treatment than the standard one. With a 2 : 1 ratio, for example, there is little loss in the power of the test of the null hypothesis as compared with 1 : 1 randomization [8]. Those who have confidence in the historical controls can use the additional information they provide to improve the estimate of the treatment effect. A Bayes* solution to the problem of combining information from the concurrent randomized and historical control series includes a formula for the optimum randomization ratio. However, its practicality is limited by the requirement that the prior distribution on the response bias have mean 0 and known variance [9].

STRATIFICATION*

Unrestricted randomization may occasionally result in a major imbalance in the distribution of prognostic factors among the treatment groups. Even if accounted for in the analysis, this affects the efficiency of the treatment comparison. It also leads to suspicions that something has gone wrong with the trial and a tendency for its results to be ignored. Various methods of restricted randomization have been put forth to cope with this dilemma.

The classical solution is to construct relatively homogeneous strata on the basis of one or more prognostic factors. Patient assignment is then by randomized blocks within strata. This works fine as long as the number of strata is quite small in relation to the number of patients. For then, approximate if not perfect balance will have been achieved even if the last block in each stratum is left incomplete at the time patient entry is stopped. However, because patients arrive sequentially and must be randomized before the characteristics of the entire study sample are known, it is not possible to achieve perfect balance* using this technique. As the number of strata increase, it tends to break down altogether, yielding scarcely more balance than with unrestricted randomization. Hence it is wise to limit the variables used for stratification to one or at most two of the major prognostic factors. If good information about the effect of prognostic variables is available from previous trials, one could construct a multivariate *prognostic score* for each patient and use this to define a small number of strata.

Rather than attempt to achieve balance with respect to the joint distribution of several prognostic factors, an alternative is to balance simultaneously all the marginal distributions. Such a design is reassuring to clinicians, as they can readily see its effect in maintaining comparability among the treatment groups. It is also defendable on statistical grounds unless there are strong interactions among the prognostic variables in their effect on response. Increasing use is therefore being made of a type of adaptive randomization known as *minimization*. When a new patient is entered in the trial, possible treatment assignments are considered in terms of their effect on the degree of balance achieved for each factor. Suppose that x_{ij} assignments to treatment j have already been made at the new patient's level of factor i. One suggestion is to select (with high probability) the next treatment so as to minimize the sum of the variances in the treatment totals at each factor level. This is equivalent to putting the highest probability on the treatment that minimizes the quantity $\sum_i x_{ij}$.

Unfortunately, stratification or minimization complicates the administration of a trial and thus increases the chances for something to go wrong. Also, to achieve the anticipated gains in efficiency, it is necessary to account for the design in the analysis of the data. This may constrain the amount of adjustment that is possible for factors not considered in the design. Methods of analysis of data collected using the complicated adaptive randomization schemes have yet to be fully developed.

For these reasons clinical trials statisticians continue to use the simple randomized blocks design, relying on post hoc stratification and adjustment to account for any imbalances. On average, the loss in efficiency* amounts to no more than one observation per stratum. However, some form of restricted randomization may be desirable in small trials to guard against the occasional badly skewed design.

CROSSOVER DESIGNS

In view of the large patient-to-patient variability in response to treatment, it is natural that an investigator would hope wherever possible to use the patient "as his or her own control." With two treatments A and B, the *crossover* design randomizes half the patients to receive A during the first treatment period and B during the second, while for the other half the order is reversed. Clearly, such designs are appropriate only if the treatment outcome is available quickly, as with trials of analgesics. They are of no use in follow-up studies, where the major end points may occur only after long and variable periods (*see* CHANGEOVER DESIGNS).

With the crossover design the treatment effect is estimated by comparing the differences in response between the two periods for the A, then B, group to the between-period differences for the B, then A, group. However, this is valid only if the treatment effect in the two groups is the same. Interactions* between treatment and period on response might arise if there was a *carryover* effect of the first treatment into the second period, or if the patient's response in the

second period was somehow conditioned by his or her response during the first. This would mean that the patients were in a qualitatively different state at the beginning of the second period, depending on the treatment received in the first, in which case the two groups would be comparable no longer. Many crossover trials incorporate a "washout" period in an attempt to alleviate this problem and return the patient to his or her original state.

The paradox of the crossover design is that the test for interaction involves sums rather than differences in the responses for the two periods. The assumption necessary for use of the sensitive within-patient test of treatment effect is itself tested by an insensitive between patient comparison. Unless the assumption can be verified a priori, therefore, one might just as well use a randomized blocks design to start with [6].

One means of recovering some of the sensitivity of the crossover design is to record responses for each patient during a pretreatment period prior to administration of either A or B. One can then use the difference in responses between treatment and pretreatment periods as a measure of effect, or otherwise use the pretreatment data in the analysis as a covariate.

SAMPLE SIZE*

The most critical statement that can be made about sample sizes for clinical trials is that most of them are too small. Clinical investigators as a rule are overly optimistic about the gains to be expected from new treatments and are consequently prone to launch into investigations that have little chance of detecting the small but nevertheless meaningful and important differences that may be present. The result is a large number of indecisive trials, many of which are never reported in the medical literature. Those that do get reported tend to be those for which a positive result was found, a type of selection that results in distortion of the published P-values. Since the number of new medical treatments that actually do result in substantial improvement is likely to be limited, the result is that the large number of the reported positive results are misleading. Reported results for large trials are less likely to be misleading because of the greater power of the large trial to detect real differences, and also because such trials tend to get published whether or not they yield positive results.

Sample-size calculations for clinical trials need not be terribly precise. Mostly they are carried out to give broad indications as to the feasibility of answering a particular therapeutic question with the resources available. If the question appears unlikely to be resolved, one can consider increasing the number of investigators, prolonging the period of entry, or (in some cases) prolonging the period of follow-up. The first option, finding additional participants, is usually the best. Trials that are long and drawn out lead to declining morale among both the statistical staff and clinical participants, and run the risk of having the original question become outmoded.

Since the usual goal of sample-size calculations is to provide "ballpark" estimates, it makes no sense to spend a lot of effort on their refinement. The slight theoretical gains expected by accounting for the stratified sampling scheme or planned covariance analysis are likely to be more than offset by unexpected losses due to patient withdrawals or competition from other trials. However, it is important to decide at the outset whether one wants a separate evaluation of treatment effects within each of several subgroups, or whether to settle instead for a general statement that the new therapy is or is not beneficial for the patient population as a whole. A much larger trial may be required to identify accurately those types of patients who respond particularly well to the new treatment.

One way to increase efficiency is to select end points that are as sensitive as possible to differences between treatments. Although death is the "ultimate" end point, it may take a long time to occur. Moreover, treatment effects on mortality are often diluted by the fact that the treatment is modified after the first indication of failure. Hence the time to first relapse is often a better indicator.

The mechanics of sample-size calculation are straightforward. If the new treatment has a reasonable chance of resulting in complete recovery after a definite time period, it is well to use the proportion "cured" as the major end point and to base sample-size calculations on the comparision between two or more proportions. For chronic diseases, however, relapses are likely to continue more or less indefinitely into the future. Then the design and analysis are better based on comparison of the instantaneous *rates* * *of occurrence* of the major end point(s). Roughly speaking, the information available for comparing such rates is determined by the total number n of events that have occurred [8]. Suppose that one adopts the *proportional hazards* * assumption, such that the time t failure rates in treatment groups A and B are related by $\lambda_B(t) = \theta\lambda_A(t)$. Provided that equal numbers of patients are entered on both treatments, and that approximately equal numbers continue to remain *at risk* during the course of the follow-up period (which means that the failure rates cannot be too disparate), the usual "log rank" * statistic used to test $\theta = 1$ will have an approximate normal distribution with mean $n \log \theta$ and variance $n/4$. It follows that the two-sided test of size α will have approximate power $1 - \beta$ against a nearby alternative θ provided that

$$n = \frac{4(Z_{\alpha/2} + Z_{\beta})^2}{\log^2(\theta)},$$

where Z_p denotes the upper $100p$ percentile of the standard normal distribution. The number of years one must wait to accumulate the required n events will depend on rates of patient accrual as well as of failure.

THE ETHICAL DILEMMA

Good general statements concerning the investigator's responsibility in clinical research are those of the Medical Research Council (1962–1963) and the World Health Assembly (1964), both of which are reproduced in Hill [5]. One ethical issue that particularly concerns the statistician is the problem of *accumulating evidence*. Having designed the trial at the outset to accrue patients for a certain number of years, unexpected differences between regimens may start to appear early. This is less likely to occur with follow-up studies, where the majority of the events needed to distinguish treatments are often delayed until after the study is closed to new entries. However, if patients already in the trial continue to receive treatment, the question will arise as to whether they should not all be switched over to the apparently superior regimen. This presents the trial participants with a real conflict of interest: of choosing to treat their patients with what appears at the moment to be the best therapy versus continuing to assign some patients to an apparently inferior regimen in the hopes that the information gained will be of benefit to future patients with the same disease. One technique used to relieve pressure for early termination is simply to keep secret the interim results of the trial. An external monitoring committee is given sole access to the data as they accrue and is charged with deciding when sufficient information has been collected to justify closing the trial.

Considerable effort has been expended by statisticians during the past two or three decades in attempting to design rational *stopping rules** that resolve the dilemma. Armitage [1] is concerned especially with the fact that repeated examination of accumulating data, using a conventional test of significance, will lead under the null hypothesis to the finding of a "positive" result on at least one occasion with probability higher than the nominal size of the test. He proposes a truncated version of the two-sided sequential probability ratio test* (SPRT) which has a wedge-shaped stopping region. This maintains control of the α and β error probabilities, yet avoids the unacceptable uncertainty regarding the ultimate size of the investigation that accompanies the SPRT.

Alternatively, one may reduce the size of the conventional test used to take periodic "peeks" at the data in order that a specified overall significance level be maintained.

Such stopping rules have not been widely adopted, partly because of their lack of flexibility. Rather arbitrary choices of α and β, formerly used only to provide rough guidelines as to feasibility, are translated into a rigid prescription for when to stop the sequential trial. In actual practice, the decision to terminate tends instead to be made jointly by statisticians and clinicians in a more informal framework. Important considerations, in addition to accumulating evidence regarding the multiple therapeutic and toxic end points, are the enthusiasm for the trial as reflected in patient entry rates, the results of similar trials, and the degree to which the original therapeutic question continues to be relevant.

Other objections to the use of classical sequential designs* have come from the statistical community. Cornfield [3] argues that *p*-values are inappropriate expressions of the uncertainty concerning the difference in treatment effects, since they take no account of alternative hypotheses. He proposes that one calculate instead a type of weighted likelihood ratio* called the relative betting odds (RBO) for use in informal deliberations. In the simplest case, with θ representing the true difference in outcome and x the accumulated data, the RBO is defined as

$$\mathrm{RBO} = \frac{\int f(x;\theta)g(\theta)\,d\theta}{f(x;\theta)},$$

where $f(x;\theta)$ is the density function of x given θ and $g(\theta)$ is a prior density on θ.

Further formulations of the ethical dilemma as a *statistical decision** problem explicitly acknowledge that the interests of patients in the trial are being traded off against those of future patients with the same disease. Suppose that there is a certain *horizon* of N patients who must ultimately be treated. In one design the first $2n$ patients are randomized equally to treatments A and B, after which the remaining $N - 2n$ are assigned to the "better" treatment. Using a prior distribution on the treatment difference, the trial size $2n$ is chosen (depending on the data) in such a way as to maximize the total expected number of treatment "successes." A more complicated version, known as the *two-armed-bandit** problem, allows the treatment assignment for each of the N patients to depend on all the preceding data. Other procedures that attempt to ensure in a rather ad hoc manner that more patients are assigned the better treatment are known as *play-the-winner** rules [7].

So far, none of the decision formulations have come to grips sufficiently with the actual concerns of those responsible for clinical trials to be put into operation. One reason is uncertainty regarding specification of the prior and of the patient horizon, which should undoubtedly be chosen to *discount the future*. Another is that, as yet, they make no provision for multiple end points, some of which may be delayed, or for concomitant information. Many of them use deterministic treatment assignments which are dangerously sensitive to selection bias or to secular changes in the study population. Finally, they ignore the fact that the clinical trial in truth is a scientific investigation, carried out not only to enable the participants to determine the best treatment for their own patients, but also to provide well-documented information about treatment risks and benefits to an international community of medical scientists.

ANALYSIS AND INTERPRETATION

One source of misunderstanding between statistician and clinician concerns the necessity of allowing *no exclusions* of randomized patients from the analysis. Three categories of patients often considered for such exclusion are: (1) "inadequate trials," who fail before treatment can be completed; (2) those not treated according to the randomly assigned regimen because of last-minute refusals or contraindications; and (3) patients withdrawn from the trial or who otherwise had severe "protocol violations." The reasons for allowing no exclusions are clear.

Differences in patients excluded from one treatment group as opposed to another may bias the comparison. The pragmatic view of the trial recognizes that similar protocol deviations and early failures will take place in actual practice. Exclusions also make it extremely difficult to compare the published results of one trial with those of another.

Multivariate analyses involving treatment and prognostic factors are important not only as a means of adjusting the treatment comparisons but also to identify those patients who may be especially helped or harmed by a particular therapy. By exploiting the interactions* between treatment and prognostic variables, it may eventually be possible to determine the *optimum* treatment for each patient based on his or her own individual characteristics. However, in view of the multiplicity of comparisons involved, it is probably wise to adopt rather stringent criteria for deciding that individual interaction terms are real.

Many clinical trials record times to occurrence of various events, not all of which will be observed for all patients. Considerable progress has been made in recent years in developing appropriate statistical methodology for such censored survival data*, notably the proportional hazards regression model [4] and related life-table* techniques. Although these methods are extremely useful, it is important that their properties be clearly understood, lest mistaken inferences be drawn. Caution must be exercised when analyzing interim data before follow-up on all patients is complete. For example, the new treatment may have lower relapse rates and thus fewer failures than the standard during the first year of follow-up, yet at the end of two years the proportions who have failed are found to be identical. One must then decide whether the apparent benefit, which consists more in delaying relapse than in preventing it, is worth the added toxicity and complications that may be involved. Similarly, it is important to realize that both treatment and prognostic factors may have greater effects on event rates during the initial follow-up period than they do in later ones, and appropriate modifications in the proportional hazards analysis should then be made.

References

[1] Armitage, P. (1975). *Sequential Medical Trials*, 2nd ed. Blackwell, Oxford. (Written for the practioner, this introductory work provides details about the author's own sequential plans, and discusses briefly more recent proposals based on statistical decision theory.)

[2] Bull, J. P. (1959). *J. Chronic Dis.*, **10**, 218–248.

[3] Cornfield, J. (1976). *Amer. J. Epidemiol.*, **104**, 408–421.

[4] Cox, D. R. (1972). *J. R. Statist. Soc. B*, **34**, 187–202.

[5] Hill, A. B. (1971). *Principles of Medical Statistics*, 9th ed. The Lancet, London. (Generations of medical students have learned their statistics from this classic text. The chapter on clinical trials discusses general principles of randomization, double blinding, the use of historical controls, and medical ethics.)

[6] Hills, M. and Armitage, P. (1979). *Brit. J. Clin. Pharmacol.*, **8**, 7–20.

[7] Hoel, D. G., Sobel, M., and Weiss, G. H. (1975). *Perspectives in Biometrics*, R. Elashoff, ed. Academic Press, New York, pp. 29–61.

[8] Peto, R., Pike, M. C., Armitage, P., Breslow, N. E., Cox, D. R., Howard, S. V., Mantel, N., McPherson, K., Peto, J., and Smith, P. G. (1976). *Brit. J. Cancer*, **34**, 585–612; *ibid.*, **35**, 1–39 (1977). (This extremely popular two-part article discusses basic principles of the design and analysis of follow-up trials used to evaluate cancer therapies. Topics include the determination of sample size, unequal randomization ratios, historical controls, prognostic factors, and a recipe for the "log rank" test.)

[9] Pocock, S. J. (1979). *Biometrics*, **35**, 183–197.

[10] Schwartz, D., Flamant, R., and Lellouch, J. (1970). *L'Essai thérapeutique chez l'homme.* Flammarion, Paris. (A full-length text devoted to practical issues in the design and analysis of clinical trials, this emphasizes the authors' "pragmatic" viewpoint that one should compare broad treatment policies rather than narrowly prescribed protocols.)

(BIOSTATISTICS
CONCOMITANT VARIABLES
DESIGN OF EXPERIMENTS
EPIDEMIOLOGICAL STATISTICS
FOLLOW-UP
HISTORICAL CONTROLS
LIFE TABLES
RANDOMIZATION
SURVEY SAMPLING
SURVIVAL ANALYSIS)

N. E. BRESLOW

CLISY

The skewness measure $\sqrt{\beta_1}$ * of the conditional distribution of a variate, given the values of other variables.

The *clisy curve* (or *clisy surface*) is the graph of the clisy against these latter values. The terms are not common in current usage. Examples of clisy curves can be seen in Pretorius [1].

Reference

[1] Pretorius, S. J. (1931). *Biometrika*, **22**, 109.

CLOPPER–PEARSON CONFIDENCE INTERVALS

Confidence intervals for the parameter p of a *binomial distribution** were proposed by Clopper and Pearson [1]. From an observed value of X successes in n trials, the $100 \cdot (1 - \alpha)\%$ confidence interval is obtained by including all values of p such that

$$\sum_{x=0}^{X-1} \binom{n}{x} p^x (1-p)^{n-x} \geqslant 1 - \tfrac{1}{2}\alpha \quad (1)$$

and

$$\sum_{x=X+1}^{n} \binom{n}{x} p^x (1-p)^{n-x} \geqslant 1 - \tfrac{1}{2}\alpha, \quad (2)$$

that is, all values of p such that neither "tail" of the distribution (either below X or above X) is less then $\tfrac{1}{2}\alpha$. This leads to an interval with upper and lower limits given by solution of the equations (in p) obtained by replacing (1) and (2) by corresponding equalities.

The actual *confidence coefficient** varies with p; it cannot be less than $100(1 - \alpha)\%$ for any p. The same would be true if the right-hand sides of (1) and (2) were replaced by $1 - \alpha'$ and $1 - \alpha''$ with $\alpha' + \alpha'' = \alpha$. The symmetrical choice ($\alpha' = \alpha'' = \frac{1}{2}\alpha$) used by Clopper and Pearson is sometimes termed a *central* confidence interval.

Reference

[1] Clopper, C. J. and Pearson, E. S. (1934). *Biometrika*, **26**, 404 – 413.

(BINOMIAL DISTRIBUTION CONFIDENCE INTERVALS)

CLOSENESS OF ESTIMATORS

A criterion (suggested by Pitman [1]) for comparing two competing estimators $\hat{\theta}$ and $\hat{\theta}'$, say, of a parameter θ. The estimator $\hat{\theta}$ is said to be a *closer* estimator of θ than $\hat{\theta}'$ if

$$\Pr\left[\,|\hat{\theta} - \theta| < |\hat{\theta}' - \theta|\,\right] > \tfrac{1}{2},$$

and conversely.

This method of comparison has the advantage that it does not require knowledge of (or even the existence of) the expected value or variance of the estimators, although it does call for knowledge of their joint distribution. On the other hand, "closeness" does not define a unique order of preference —it is possible for $\hat{\theta}$ to be closer than $\hat{\theta}'$, and $\hat{\theta}'$ closer than $\hat{\theta}''$, but for $\hat{\theta}''$ to be closer than $\hat{\theta}$.

It is interesting to note that this criterion does not correspond to any specific loss function*.

Reference

[1] Pitman, E. J. G. (1937). *Proc. Camb. Philos. Soc.*, **33**, 212–222.

(ESTIMATION, POINT)

CLUSTER ANALYSIS, GRAPH THEORETICAL *See* GRAPH THEORETICAL CLUSTER ANALYSIS

CLUSTER SAMPLING

Cluster samples are a particular kind of probability sample. As the name suggests, they are characterized by units selected in groups or "clusters." If any primary sampling unit in the frame contains more than one or less than one observation unit (element), the primary sampling units are called *clusters* and a sample of such primary sampling units is called a *cluster sample. See* SAMPLING for a discussion of terms such as "primary sampling unit."

Cluster sampling is used for two reasons.

1. It may be impossible or prohibitively expensive to construct a list of observation units. For example, lists of residents for even the smaller political subdivisions of the United States, such as Ames, Iowa, do not exist.

2. For a fixed expenditure, it is often possible to obtain a smaller mean square error for an estimator by observing groups of observation units. For example, it is clear that, on the average, it would cost more to travel to a sample of 40 households located at random in the state of Iowa than it would to travel to a sample composed of 10 clusters of four households.

If only a subset of the observation units in each cluster is observed, the sample is called a multistage cluster sample, or a sample with subsamples. In many applications samples with more than two stages are used. Estimators of totals are constructed by sequentially using each subsample to estimate the total of the unit from which it was selected. For a three-stage sample with simple random sampling at each stage, the unbiased estimator of the population total of y is

$$
\begin{aligned}
\hat{Y} &= Nn^{-1} \sum_{i=1}^{n} \hat{Y}_i \\
&= Nn^{-1} \sum_{i=1}^{n} M_i m_i^{-1} \sum_{j=1}^{m_i} \hat{y}_{ij} \\
&= Nn^{-1} \sum_{i=1}^{n} M_i m_i^{-1} \sum_{j=1}^{m_i} B_{ij} b_{ij}^{-1} \sum_{k=1}^{b_{ij}} y_{ijk},
\end{aligned}
$$

(1)

where y_{ijk} is the kth observation within the ijth second-stage unit within the ith primary sampling unit, b_{ij} third-stage units are selected from the B_{ij} third-stage units in the

ijth second-stage unit, m_i second-stage units are selected from the M_i second-stage units in the ith primary sampling unit, and n primary sampling units are selected from the N primary sampling units in the population.

Note that $\hat{y}_{ij}$ is the estimated total for the ijth second-stage unit and that $\hat{Y}_i$ is the estimated total for the ith primary unit. The variance of $\hat{Y}$ is

$$V\{\hat{Y}\} = N(N-n)n^{-1}S_1^2 + n^{-1}N^{-1}$$
$$\times \sum_{i=1}^{N} w_i^2 M_i^{-1}(M_i - m_i)m_i^{-1}S_{2i}^2$$
$$+ n^{-1}N^{-1}\sum_{i=1}^{N} m_i^{-1}M_i^{-1}w_i^2$$
$$\times \sum_{j=1}^{M_i} v_{ij}^2 B_{ij}^{-1}(B_{ij} - b_{ij})b_{ij}^{-1}S_{3ij}^2,$$
$$(2)$$

where

$$w_i = \left[N^{-1}\sum_{i=1}^{N}\sum_{j=1}^{M_i}B_{ij}\right]^{-1}\sum_{j=1}^{M_i}B_{ij},$$

$$v_{ij} = \left[M_i^{-1}\sum_{j=1}^{M_i}B_{ij}\right]^{-1}B_{ij},$$

$$S_1^2 = (N-1)^{-1}\sum_{i=1}^{N}\left(Y_i - \overline{Y}\right)^2,$$

$$S_{2i}^2 = (M_i - 1)^{-1}\sum_{j=1}^{M_i}\left(y_{ij} - M_i^{-1}Y_i\right)^2,$$

$$S_{3ij}^2 = (B_{ij} - 1)^{-1}\sum_{k=1}^{B_{ij}}\left(y_{ijk} - B_{ij}^{-1}y_{ij}\right)^2,$$

$$y_{ij} = \sum_{k=1}^{B_{ij}}y_{ijk}, \qquad Y_i = \sum_{j=1}^{M_i}y_{ij},$$

$$\overline{Y} = N^{-1}\sum_{i=1}^{N}Y_i.$$

The three terms represent the contribution to the sampling variance from each of the three stages. For example, the third term will be zero if all third-stage units are observed in each of the selected second-stage units, for then the sample is a two-stage sample. In a similar way the first term will be zero if every first-stage unit is included in the sample, for then the sample is a stratified sample (*see* STRATIFICATION). Estimators of the variance of cluster samples are given in texts such as those of Cochran [1, p. 278] and Sukhatme and Sukhatme [17, p. 303]. Formula (2) was taken from Sukhatme and Sukhatme [17].

The population mean *per observation unit* is

$$\overline{\overline{Y}} = \left(\sum_{i=1}^{N}M_i\right)^{-1}\sum_{i=1}^{N}Y_i = M^{-1}Y,$$

where M is the total number of observation units in the population. An estimator of $\overline{\overline{Y}}$ for a three-stage sample is

$$\overline{\overline{y}}_n = \left(\sum_{i=1}^{n}M_i\right)^{-1}\left(\sum_{i=1}^{n}\hat{Y}_i\right),$$

If the M_i are exactly equal to a constant, $\overline{M}$, the estimator is unbiased. If the M_i are not all equal, $\overline{\overline{y}}_n$ is a ratio estimator (*see* RATIO ESTIMATOR). As such, it is biased in small samples and it is only possible to obtain the approximate variance of the estimator.

In some situations the total number of observation units, M, in the population is known. This information may be used to construct alternative estimators. The ratio estimator of the total of Y is $\hat{Y}_r = M\overline{\overline{y}}_n$. If M is known, $M^{-1}\hat{Y}$ is an unbiased estimator of the mean per observation unit. The estimator $M^{-1}\hat{Y}$ is seldom used in practice because the variance is usually larger than the variance of the ratio estimator $\overline{\overline{y}}_n$.

If a variable x_i is available for each cluster and is correlated with the M_i, the information of x_i can be used in the sample design to increase the efficiency of the sample design. Hansen and Hurwitz [4] suggested the selection of samples of clusters with probabilities proportional to x_i as a method of increasing efficiency (*see* UNEQUAL PROBABILITY SAMPLING). This approach was developed further by Sampford [14].

Clusters may be formed at the design stage either for convenience or to increase efficiency. Systematic sampling is one method of forming cluster samples (*see* SYSTEMATIC SAMPLING). If only one start value is used for a systematic sample, the systematic sample is formally equivalent to a sample of one cluster. By using auxiliary information to arrange the population, it is sometimes possible to select systematic (cluster) samples that are more efficient than stratified samples of the same size.

The design of cluster samples can involve a number of decisions.

1. Definition of the primary sampling units. The designer often has some choice with respect to the size (number of secondary units), composition, and the shape of and boundaries for area clusters of primary units.
2. Determination of the number of stages.
3. Allocation of sample between primary sampling units and secondary units within primary units, and so on.

Texts, such as Jessen [7], Cochran [1], Sukhatme and Sukhatme [17], and Hansen et al. [5], discuss these design problems.

The comparative analysis of survey data using techniques such as regression* equations and contingency tables* is complicated by the use of cluster sampling. Konijn [11], Kish and Frankel [9], and Fuller [3] have studied the estimation of regression equations using cluster samples. Hidiroglou et al. [6] provide computer software for the computation of regression equations from cluster samples. Koch et al. [10], Cohen [2], and Rao and Scott [13] have studied the behavior of chi-square tests* for contingency tables under cluster sampling.

Articles that treat other specific problems associated with cluster sampling include Sedransk [15], Joshi [8], and Levy [12].

References

[1] Cochran, W. G. (1977). *Sampling Techniques*. Wiley, New York.

[2] Cohen, J. E. (1976). *J. Amer. Statist. Ass.*, **71**, 665–670.

[3] Fuller, W. A. (1975). *Sankhyā C*, **37**, 117–132.

[4] Hansen, M. H. and Hurwitz, W. N. (1943). *Ann. Math. Statist.*, **40**, 1439–1448.

[5] Hansen, M. H., Hurwitz, W. N., and Madow, W. G. (1953). *Sample Survey Methods and Theory*, Vols. 1 and 2. Wiley, New York.

[6] Hidiroglou, M. A., Fuller, W. A., and Hiekman, R. D. (1979). *SUPER CARP*, Statistical Laboratory, Iowa State University, Ames, Iowa.

[7] Jessen, R. J. (1978). *Statistical Survey Techniques*. Wiley, New York.

[8] Joshi, V. M. (1968). *Ann. Math. Statist.*, **39**, 278–281.

[9] Kish, L. and Frankel, M. R. (1974). *J. R. Statist. Soc. B.*, **36**, 1–37.

[10] Koch, G. C., Freeman, D. H., Jr., and Freeman, J. L. (1975). *Int. Statist. Rev.*, **43**, 59–78.

[11] Konijn, H. S. (1962). *J. Amer. Statist. Ass.*, **57**, 590–606.

[12] Levy, P. S. (1977). *Proc. Social Statist. Sect. Amer. Statist. Ass.*, 1977, pp. 963–966.

[13] Rao, J. N. K. and Scott, A. J. (1979). *Proc. Surv. Res. Methods Sect. Am. Statist. Ass.*, 1979.

[14] Sampford, M. R. (1962). *Biometrika*, **49**, 27–40.

[15] Sedransk, J. (1965). *J. R. Statist. Soc. B*, **27**, 264–278.

[16] Sirken, M. G. (1970). *J. Amer. Statist. Ass.*, **65**, 257–266.

[17] Sukhatme, P. V. and Sukhatme, B. V. (1970). *Sampling Theory of Surveys with Applications*. Iowa State University Press, Ames, Iowa.

(SAMPLING)

Wayne A. Fuller

C_N **TEST**   *See* Capon test

COCHRAN'S *Q*-STATISTIC

When comparing proportions in independent samples, a large-sample chi-square* statistic is quite well known. Often, however, each member of a sample is matched with a corresponding member of every other sample in order to increase the precision of comparison. When matching occurs, the samples are correlated and the ordinary Pearson chi-square statistic is no longer valid.

One way of creating matched samples is to divide *nc* individuals into *n* groups of *c* individuals matched according to one or more characteristics or variables and then to randomize the *c* treatment assignments independently within each matched group. The individuals should be matched on characteristics that are associated with the response being studied. Matched samples also occur when one observes n individuals under c different treatments (e.g., c different questions or one question at c different times). No matter how the matching occurs, matched c-tuples are used for comparing c samples to gain efficiency.

McNemar [10] developed a test for comparing two matched samples when the response variable is a dichotomy. Cochran [6] extended McNemar's results to the case of several matched samples. To derive Cochran's Q-statistic, consider comparing c matched treatments (i.e., samples) with n observations per treatment. Let $X_{ij} = 1$ if the outcome for the jth treatment in the ith matched observation is a "success" and $X_{ij} = 0$ otherwise, for $j = 1, 2, \ldots, c$ and $i = 1, 2, \ldots, n$. The total number of successes for the jth treatment is $T_j = \sum_{i=1}^{n} X_{ij}$, and the total number of successes for the ith matched observation is $u_i = \sum_{j=1}^{c} X_{ij}$.

In deriving the test statistic, Cochran regarded the u_i as fixed. He then argued that under the "null hypothesis" each of the ways of distributing the u_i successes among the c treatments is equally likely. Therefore, for the ith matched observation, each treatment has an equal probability of success, and this probability is dependent on the fixed u_i and is allowed to vary from observation to observation. Note that Cochran's Q-statistic reflects the idea that the probability of the response is related to the values of the matching variables.

The statistic is given as

$$Q = \frac{c(c-1)\sum_{j=1}^{c}\left(T_j - \bar{T}\right)^2}{c\sum_{i=1}^{n} u_i - \sum_{i=1}^{n} u_i^2};$$

it provides an exact conditional test and, also, a large-sample chi-square test of the hypothesis of permutational symmetry. Bhpkar [3] formally states this hypothesis as

H(c): $(X_{ij_1}, X_{ij_2}, \ldots, X_{ij_c})$ has the same conditional distribution, given u_k, as $(X_{i1}, X_{i2}, \ldots, X_{ic})$ for every permutation $(j_1, j_2, \ldots, j_c)$ of $(1, 2, \ldots, c), i = 1, 2,, \ldots, n.$

Although Cochran [6] never formally states $H(c)$ as the null hypothesis of interest, he does show that Q has a limiting $\chi^2(c-1)$ distribution under the "null hypothesis."

An application of Cochran's Q-statistic is seen in the following data originally given by Somes [13]. Shown in Table 1 are the results of observing 37 horses over a 4-day period and noting their responses as either correct or incorrect, represented by 0 or 1. The plausible alternative to rejection of $H(c)$ would be that the horses have learned.

To simplify the calculations, Q may be expressed as

$$Q = \frac{(c-1)\left(c\sum_{j=1}^{c} T_j^2 - S^2\right)}{cS - \sum_{i=1}^{n} u_i^2},$$

where

$$S = \sum_{j=1}^{c} T_j = \sum_{i=1}^{n} u_i = \sum_{i=1}^{n} \sum_{j=1}^{c} X_{ij}.$$

For these data, Q has a value of 15.77, which is significant at the 0.005 level in a $\chi^2(3)$ table.

Madansky [11] mentions Cochran's Q-statistic when he deals with a test of an hypothesis of interchangeability. He considers a more general situation of n individuals holding one of s opinions at t successive

Table 1 Responses of Horses as Correct (0) or Incorrect (1)

	Day				Number
	1	2	3	4	of Cases
	1	1	1	1	3
	1	1	0	1	4
	1	0	1	1	1
	0	1	1	1	1
	1	1	0	0	6
	1	0	1	0	1
	0	1	1	0	1
	1	0	0	0	8
	0	0	1	0	3
	0	0	0	1	1
	0	0	0	0	8
T_j	23	15	10	10	

time intervals. He derives a conditional test statistic which for $s = 2$ reduces to that given by Cochran by letting $c = t$. Fleiss [7] demonstrates the applicability of Cochran's Q-statistic for situations where the ordering of the c treatments is not necessarily random for any of the n c-tuples.

Cochran's Q-statistic has been used to test hypotheses other than the one Cochran had in mind. If the probability of success is assumed to be constant from observation to observation within a sample, then the hypothesis concerning equality of proportions in the samples may be formally stated as

$$H_0 : \pi_1 = \pi_2 = \cdots = \pi_c,$$

where π_j is the probability of success for the jth treatment. It is important to recognize that in H_0 the probability of success is assumed to be constant from matched observation to observation, whereas in $H(c)$ it is allowed to vary, conditional on u_i. Cochran's Q-statistic has been used to test H_0 and it was erroneously assumed that Q had an asymptotic $\chi^2(c - 1)$ distribution when H_0 was true. The conditions necessary for Q to have an asymptotic $\chi^2(c - 1)$ distribution when H_0 is true were derived by Bhapkar [2]. This necessary and sufficient side condition may be expressed as

$$H_1 : \pi_{1,2} = \pi_{1,3} = \pi_{1,c} = \pi_{2,3} = \cdots$$
$$= \pi_{c-1,c},$$

where $\pi_{j,k}$ is the probability of simultaneous success for the jth and kth treatments. Berger and Gold [1] obtained the asymptotic distribution for Q under H_0 for $c = 3$, and more recently Bhapkar and Somes [4] derived the asymptotic distributions of Q under H_0 for any c and demonstrated that the limiting rejection value of Q under H_0 is always at least as large as the rejection value obtained from the $\chi^2(c - 1)$ distribution and often much larger, depending on the violation of H_1.

If the hypothesis of interest is H_0, then Bhapkar [2] offers a Wald statistic [14] which is asymptotically $\chi^2(c - 1)$ under H_0. Even though this test statistic is more cumbersome than Cochran's Q-statistic, it may be preferable, since the added calculations

necessary to arrive at an approximation to the asymptotic distribution of Q under H_0 are also considerable. Naturally, in testing $H(c)$ or if H_1 is assumed when testing H_0, then Q is the correct test statistic.

When Cochran [6] introduced his Q-statistic he also indicated how it may be broken down into components for more detailed tests comparing mutually exclusive subgroups of the c treatments. It can be shown that the statistic given by Miettinen [12] for comparing a case to multiple matched controls is a special case of breaking Q into its components. That is, Miettinen's statistic* [12] is identical to that one given by Cochran [6] for comparing the first treatment (say) with the $(c - 1)$ other treatments. Along these same lines it should be noted (see, e.g., Brownlee [5]) that Cochran's Q-statistic is a special case of Friedman's rank test statistic* [9] when there are ties in the ranks.

For further readings on Cochran's Q-statistic, one should refer to Cochran's original article [6] or possibly to the appropriate section of books by Brownlee [5] and Fleiss [8]. The latter is a good reference for many hypotheses dealing with categorical data*. For more information regarding the applicability of Cochran's Q-statistic for hypotheses other than $H(c)$, refer to Bhapkar [2, 3] and Bhapkar and Somes [4].

References

[1] Berger, A. and Gold, R. F. (1973). *J. Amer. Statist. Ass.*, **68**, 984–993.

[2] Bhapkar, V. P. (1970). In *Random Counts in Scientific Work*, Vol. 2, Patil, G. P. ed. Pennsylvania State University Press, University Park, Pa., pp. 255–267.

[3] Bhapkar, V. P. (1973). *Sankhyā A*, **35**, 341–356.

[4] Bhapkar, V. P. and Somes, G. W. (1977). *J. Amer. Statist. Ass.*, **72**, 658–661.

[5] Brownlee, K. A. (1965). *Statistical Theory and Methodology in Science and Engineering.* Wiley, New York, pp. 262–265.

[6] Cochran, W. G. (1950). *Biometrika*, **37**, 256–266.

[7] Fleiss, J. L. (1965). *Biometrics*, **21**, 1008–1010.

[8] Fleiss, J. L. (1973). *Statistical Methods for Rates and Proportions.* Wiley, New York, pp. 83–87.

[9]   Friedman, M. (1937). *J. Amer. Statist. Ass.* **32**, 675–701.

[10] McNemar, Q. (1947). *Psychometrika*, **12**, 153–157.

[11] Madansky, A. (1963). *J. Amer. Statist. Ass.*, **58**, 97–119.

[12] Miettinen, O. S. (1969). *Biometrics*, **25**, 339–355.

[13] Somes, G. W. (1975). Some Contributions to Analysis of Data from Matched Samples. Ph.D. dissertation, University of Kentucky.

[14] Wald, A. (1943). *Trans. Amer. Math. Soc.*, **54**, 426–482.

(CATEGORICAL DATA
CHI-SQUARE TESTS
FRIEDMAN'S RANK TEST
MATCHED PAIRS
MATCHED SAMPLES
McNEMAR'S TEST
MIETTINEN'S STATISTIC)

GRANT W. SOMES

COCHRAN'S (TEST) STATISTIC

Used for testing equality of variances. For an array $\{ Y_{ij}, i = 1, \ldots, r, j = 1, \ldots, n \}$ of independent normal random variables representing random samples of size n from r populations, let

$$E\{ Y_{ij} \} = \mu_i, \quad \text{var}(Y_{ij}) = \sigma_i^2,$$

$$i = 1, \ldots, r, j = 1, \ldots, n.$$

let the sample variance within population i be s_i^2, $i = 1, \ldots, r$ [i.e., $s_i^2 = 1/(n - 1) \sum_j (Y_{ij} - \overline{Y}_{i\cdot})^2$]. Let $s_{\max}^2 = \max\{ s_1^2, \ldots, s_r^2 \}$. Cochran's statistic for testing $H_0 : \sigma_1^2 = \sigma_2^2 = \cdots \sigma_r^2$ is

$$C = S_{\max}^2 / \sum_{i=1}^{r} S_i^2. \tag{1}$$

Upper 5% points of (i) for $r = 3(1)10$ and $\nu = n - 1 = 1(1)6(2)10$ are given in Cochran [3]. As is the case with Bartlett's M-test* [1], Cochran's test is very sensitive to nonnormality due to the fact that both these tests do not utilize the variance variability within the samples (see Box [2]).

For additional information on this topic, see Miller [4].

References

[1]   Bartlett, M. S. (1937). Properties of Sufficiency and Statistical Tests, *Proc. R. Soc. A*, **160**, 268–282.

[2]   Box, G. E. P. (1953). *Biometrika*, **40**, 318–335.

[3]   Cochran, W. G. (1941). *Ann. Eugen. (Lond.)*, **11**, 47–52.

[4]   Miller, R. G. (1966). *Simultaneous Statistical Inference*. McGraw-Hill, New York, pp. 222–223.

(HETEROSCEDASTICITY)

CODE CONTROL METHOD   *See* EDITING STATISTICAL DATA

CODED DATA

Generally, this means data recorded not as originally observed, but after applying a (usually monotonic) transformation. Although coded data are sometimes used to preserve confidentiality (for personal, industrial, or national security reasons), the term is not commonly applied to values used for convenience in computation.

As an example, suppose that one wishes to compute the mean* and variance* of the 10 values $x = 9750$, 9900, 10,500, 10,350, 10,250, 9950, 10,300, 10,150, 10,100, and 10,400. By using the transformed variable $y - (x - 10,000)/50$, we obtain the values $y = -5$, -2, 10, 7, 5, -1, 6, 3, 2, and 8, which are easier to use than the original y-values.

The value of x giving $y = 0$ is called an *arbitrary origin*.

(CODING THEOREM)

CODING THEOREM

If $y_i = a + bx_i$, then

(arithmetic mean* of $y_1, y_2, \ldots, y_n$)

$= a + b$ (arithmetic mean

of $x_1, x_2, \ldots, x_n$)

and

(standard deviation* of $y_1, y_2, \ldots, y_n$)

$= |b|$(standard deviation

of $x_1, x_2, \ldots, x_n$).

These relationships are used to derive values of mean and standard deviation of original data, from corresponding values for coded data*.

The name "coding theorem" is a convenient way of referring to these results.

(CODED DATA)

COEFFICIENT OF ALIENATION (*A*)

The ratio of the unexplained variance [the part that cannot be predicted from the knowledge of the independent variable(s) in linear regression] to the total variance. It is the percentage of variation in "*y*" that is independent of the variation in "*x*." It is equal to 1 minus the coefficient of determination*.

(COEFFICIENT OF DETERMINATION CORRELATION)

COEFFICIENT OF CONCORDANCE (*W*)

To measure the degree of agreement among m observations ranking n individuals according to some specific characteristic, Kendall and Smith [3] proposed the *coefficient of concordance*

$$W = 12S / \{ m^2(n^3 - n) \},$$

where S is the sum of squares of the deviations of the total of the ranks* assigned to each individual from $m(n + 1)/2$. Since the total of all ranks assigned is $m(1 + 2 + \ldots + n) = mn(n + 1)/2$, this is the average value of the totals of the ranks, and hence S is the sum of squares of deviations

from the mean. W can vary from 0 to 1; 0 represents no "community of preference" and 1 represents perfect agreement. To test the hypothesis that the observers have no community of preference, one uses tables given in Kendall [1, 2], where the values of S are tabulated. If $n > 7$, the distribution of $m(n - 1)W = 12S/[mn(n + 1)]$ is approximately χ^2_{n-1} if there is indeed no community of preference.

A somewhat more accurate approximation is to take

$$(m - 1)W/(1 - W) \tag{1}$$

to have an F distribution* with $v_1 = n - 1 - 2/m$, $(m - 1)v_1$ degrees of freedom.

If there are ties in some ranking(s), the formulas are somewhat modified. If W is not significant, this indicates that it is unjustifiable to attempt to find an average or "pooled" estimate of true ranking, since there is insufficient evidence that this exists. If W is significant, it is reasonable to estimate a supposed "true" ranking of the n individuals. This is done by ranking them according to the sum of ranks assigned to each, the one with the smallest sum being ranked first, the one with the next smallest sum being ranked second, etc. Kendall [2] discusses in detail the modification when ties occur.

The distribution of W in the *nonnull* case (under the assumption that the rankings have been generated by taking n observations from a m-variate normal distribution* with all the correlations equal) has been investigated by Wood [4]. (*See also* SPEARMAN RANK CORRELATION COEFFICIENT.)

References

[1]   Kendall, M. G. (1948). *Rank Correlation Methods*, 1st ed. Charles Griffin, London.

[2]   Kendall, M. G. (1962). *Rank Correlation Methods*, 3rd ed. Charles Griffin, London.

[3]   Kendall, M. G. and Smith, B. B. (1939). *Ann. Math. Statist.*, **10**, 275–287.

[4]   Wood, J. T. (1970). *Biometrika*, **57**, 619–627.

(ASSOCIATION, MEASURES OF)

COEFFICIENT OF CORRELATION

This term is usually understood to mean either the sample coefficient

$$\frac{n^{-1}\sum_{i=1}^{n}(X_i - \overline{X})(Y_i - \overline{Y})}{\sqrt{\left[n^{-1}\sum_{i=1}^{n}(X_i - \overline{X})^2\right]\left[n^{-1}\sum_{i=1}^{n}(Y_i - \overline{Y})^2\right]}},$$

conventionally denoted r_{XY} or r, calculated from a set of n pairs of observed values (X_i, Y_i), or the population coefficient

$$\text{cov}(X, Y)/\sqrt{\text{var}(X)\,\text{var}(Y)},$$

conventionally denoted ρ_{XY} or ρ, where $\text{cov}(X, Y)$ is the covariance* of X and Y and $\text{var}(X)$ and $\text{var}(Y)$ are the variances* of X and Y respectively.

These measures (also called *product moment* correlation) are rather specially related to linear relations between the variables and would be more appropriately defined as coefficients of *linear* correlation.

(ASSOCIATION, MEASURES OF
CORRELATION
REGRESSION)

COEFFICIENT OF DETERMINATION (*D*)

D, the square of the correlation coefficient*, is defined as the coefficient of determination. In a linear regression model $Y = a + bX$, it is the proportion of the total variation (or variance) that can be explained by the linear relationship existing between X and Y. When multiplied by 100, the proportion is converted to percentage. For example, if the correlation coefficient is 0.952, this shows that $(0.952)^2 \times 100 = 90.6\%$ of the variation in Y is due to the linear relation existing between Y and X; the rest of the variation is due to unexplained factors and is called *experimental error*. We emphasize that the interpretation in terms of percentages applies only to the *variance* of Y, not to the standard deviation of Y.

Bibliography

Johnson, N. L. and Leone, F. C. (1977). *Statistics and Experimental Design*, 2nd ed. Vol. 1. Wiley, New York.

(COEFFICIENT OF ALIENATION
CORRELATION
LINEAR REGRESSION)

COEFFICIENT OF PROPORTIONAL SIMILARITY   *See* COS THETA (θ)

COEFFICIENT OF VARIATION

A measure of relative dispersion* equal to the ratio of standard deviation to mean. It is often expressed as a percentage $\theta = 100 \cdot \sigma/\mu\%$, where σ and μ are, respectively, the standard deviation and the mean of the distribution under consideration. For exponential distributions the coefficient of variation is 1 (or 100%) since for these distributions $\sigma = \mu$. The sample coefficient of variation V is $V = 100(S/\overline{X})$, where S is the sample standard deviation and $\overline{X}$ is the sample arithmetic mean.

The standard error of the sample coefficient of variation for a normal distribution is *approximately* $\sigma_V = \theta/\sqrt{2n}$, where n is the sample size (obtainable by the method of statistical differentials*).

Bibliography

Johnson, N. L. and Leone, F. C. (1977). *Statistics and Experimental Design*, 2nd ed., Vol. 1. Wiley, New York.
Mack, C. (1970). *New J. Statist. Operat. Res.*, **6**(3), 13–18.

COGRADIENT ESTIMATOR   *See* EQUIVARIANT ESTIMATOR

COHERENCE

The historical development of statistics has consisted largely of the invention of techniques followed by a study of their properties: for example, the ingenious concept of a confidence interval* is studied for optimality features. As a result, modern statistics consists of a series of loosely related methods and the practitioner has to choose which to use: point estimate or hypothesis test*. There is another way of proceeding which reverses the order and begins by asking what properties are required of statistical procedures, only then going on to develop techniques that possess them. This method is the one used in mathematics, where the basic properties are taken as axioms for the subject, and it is surprising that it has only recently been tried in statistics.

A basic property surely required of statistical procedures is that they not only have reasonable properties of their own but that they fit together sensibly, or, as we say, cohere. The sort of situation one wants to avoid is illustrated by the behavior of the usual significance tests* for the mean (μ) of a normal distribution* in their univariate and bivariate forms, where the separate univariate tests of $\mu_1 = 0$ and of $\mu_2 = 0$ can lead to rejection but the bivariate test of $\mu_1 = \mu_2 = 0$ to acceptance. Significance tests are incoherent ways of processing data. Interestingly it turns out that coherence on its own is enough to develop a whole statistical system. We now proceed to make this idea precise. The best exposition is contained in Part 2 of DeGroot [1], and the reader is referred there for details.

Statistical inference* is fundamentally concerned with situations of uncertainty; as with the normal mean, is it zero? The elements of the axiomatic system are therefore events about which we are unsure whether they are true or false. The first axiom says that any two events can be compared in respect of how likely they are to be true, and the second that this comparison is transitive: that is, if A is more likely to be true than B, and B more likely than C, then A is more likely than C. Within the narrow context of a specific statistical situation, these requirements are compelling, although the first is perhaps less so on a wider scale. The next two axioms are more in the nature of "housekeeping" requirements. The first rules out the trivial possibility that all events are

equally likely, and gets "more likely" separated from "less likely" by having the impossible event as the least likely. The second enables an infinity of events to be discussed. As scientists we are concerned with measuring this uncertainty, and to do this a standard for uncertainty is introduced, just as we use a standard for length: the simplest is an event judged as likely to be true or false, although DeGroot uses a more sophisticated device. It is then presumed that any event can be compared with the standard with respect to how likely it is to be true. From these assumptions it is possible to prove that there exists a unique probability distribution* over the events. The axioms are usually called the axioms of coherence because each of them says something about how one judgment coheres with another—transitivity is the most obvious one—and the conclusion can be expressed roughly by saying that coherence requires probability. Thus there is, for a coherent person, a probability that the normal mean is zero: not a significance level. A further axiom is needed for conditional probabilities* and says that comparison of how likely A and B are, were C to be true, can be effected by comparing $A \cap C$ and $B \cap C$.

With the basic properties established in the form of axioms, we can now see how to achieve them: by using the probability calculus, a calculus that is extremely rich in results. This leads to Bayesian inference*, in which all uncertain quantities are assigned probability distributions. It is a historically curious fact that the first modern development of coherence [2], was an attempt to use the concept as a foundation for the apparatus of modern statistics, whereas in fact coherence conflicts at almost all points with that apparatus, in that the techniques of modern statistics are largely incoherent, as our significance-test example illustrates.

The axioms can usefully be extended to go beyond inference and embrace decision making (*see* DECISION THEORY). For this we introduce a space of consequences that could result from the action of taking a decision. The event of obtaining a consequence has, by the earlier argument, a probability, and the axioms concern probability distributions over the consequences. The first two axioms parallel those for events and postulate the transitive comparison of probability distributions. The next says that if three distributions are such that P is intermediate between P_1 and P_2 in this comparison, there exists a unique α, $0 \leqslant \alpha \leqslant 1$, such that P is equivalent to the new distribution formed by taking P_1 with probability α and P_2 with probability $(1 - \alpha)$. This axiom enables a number, α, to be associated with any distribution between P_1 and P_2, and then with any of the wide class of distributions. Next, it is assumed that these α-values are bounded, to avoid the concept of "Heaven" —whose perfection is so great that its presence dominates all else; and similarly for "Hell." A final axiom says that we may, in all cases, substitute the α-mixture of P_1 and P_2 for P. The major result that follows says that there exists a utility* function, u, over the consequences (essentially the α-values) and that P_1 is preferred to P_2 if the expectation of u for P_1 exceeds that for P_2: or the best act is that of maximum expected utility* (MEU).

The axioms of coherence therefore lead to the following conclusions:

1. Uncertainties are described by probabilities.
2. Consequences are described by utilities.
3. The best decision is that which maximizes expected utility.

We now have a complete system for inferences and decision making involving only the probability calculus. (Notice that even utility is obtained in terms of probabilities, and so obeys the rules of that calculus, as is clear from its essential derivation as an α-value above.) Inference is accomplished by calculating the probability of the uncertain event of interest conditional on the data: a decision is made by calculating the expectation, according to this probability, of the utilities for the consequences. Simple and

compelling as these results are, they are in conflict with almost all statistical theory.

An important limitation of coherence is that it only applies to a single decision maker who makes the judgmental comparisons. This decision maker need not be an individual; it could be a government or even the scientific community. The ideas do not necessarily apply to the situation with several decision makers, where conflict can arise. There appears to be no corresponding theory for this except in the case of zero-sum, two-person games*. For a bibliography, *see* BAYESIAN INFERENCE.

References

[1] DeGroot, (1970). *Optimal Statistical Decisions.* McGraw-Hill, New York.

[2] Savage, (1954). *The Foundations of Statistics.* Wiley, New York.

(AXIOMS OF PROBABILITY
BAYESIAN INFERENCE
CONFIDENCE INTERVAL
DECISION THEORY
ESTIMATION, POINT
FIDUCIAL INFERENCE
FOUNDATIONS OF PROBABILITY
STATISTICAL INFERENCE)

D. V. LINDLEY

COHERENT STRUCTURE THEORY

Modern system reliability theory is based on coherent structure theory. In 1961, Birnbaum et al. inspired by the brilliant two-part paper of Moore and Shannon [8] on relay networks, published a paper laying the foundations of coherent structure theory [3]. The main idea of their paper was to show that practically all engineering systems could be treated in a simple, unified fashion in determining the probability of system functioning in terms of the probabilities of functioning of the components.

Since the publication of this basic paper,

some of the definitions have been changed slightly and some of their results have been proven in a different fashion. We summarize the theory, using the most recent definitions. A comprehensive discussion of the theory, together with a discussion of the key references, is presented in Barlow and Proschan [1]; moreover, the intimate connection between coherent structure theory and fault-tree analysis* is brought out in the appendix of the book.

To define a coherent structure having n components, we first indicate the state x_i of component i, setting $x_i = 1$ if component i is functioning and $x_i = 0$ if component i is failed, $i = 1, \ldots, n$. Similarly, the corresponding state ϕ of the system is 1 if the system is functioning and 0 if the system is failed. Since the state of the system is determined completely by the states of the components, we write $\phi = \phi(\mathbf{x})$, where $\mathbf{x} = (x_1, \ldots, x_n)$; $\phi(\mathbf{x})$ is known as the structure function of the system.

Very few systems are designed with irrelevant components. Component i is *irrelevant* to the structure ϕ if ϕ does not really depend on x_i; i.e., $\phi(\mathbf{x})$ is constant in x_i for each of the 2^{n-1} possible combinations of outcomes of the remaining components of the system. Otherwise, component i is *relevant* to the structure.

DEFINITION

A system of components is *coherent* if (a) its structure function is nondecreasing in each argument, and (b) each component is relevant.

Requirement (a) states essentially that replacing a failed component by a functioning component will not cause a functioning system to fail—a reasonable requirement. Requirement (b) rules out trivial systems not encountered in engineering practice.

From this deceptively simple definition, a wealth of theoretical results may be derived, many of which yield fruitful applications in reliability* practice.

TYPES OF SYSTEMS

Some basic coherent systems, shown in Fig. 1, are: (a) A *series system** of n components—the structure function is $\phi(\mathbf{x}) = \prod_1^n x_i \equiv \min(x_1, \ldots, x_n)$. (b) A *parallel system** of n components—the structure function is $\phi(\mathbf{x}) = \coprod_{i=1}^n x_i \overset{\text{def}}{=} 1 - \prod_{i=1}^n (i - x_i) \equiv \max(x_1, \ldots, x_n)$. (c) A *k-out-of-n-system**—the structure is $\phi(\mathbf{x}) = 1$ if $\sum_1^n x_i \geqslant k$, and 0 if $\sum_1^n x_i < k$. Note that the series (parallel) system is a special case of the k-out-of-n system

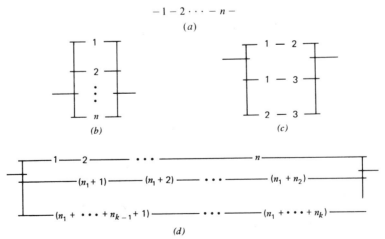

Figure 1. Diagrammatic representation of basic systems. (*a*) Series system. (*b*) Parallel system. (*c*) 2-out-of-3 system (a special case of a *k*-out-of-*n* system. (Note the replication of identical conponents.) (*d*) A parallel series system.

with $k = n(1)$. Figure 1c shows a two-out-of-three system, another special case of a k-out-of-n system. Note the replication of identical components. (d) A parallel–series (series–parallel) system; the system consists of a parallel (series) arrangement of series (parallel) subsystems.

The structure function of every coherent system is bounded below by the structure function of the series system formed from its components and bounded above by the structure function of the parallel system formed from its components. Stated formally, we have:

Theorem 1.   Let ϕ be a coherent system of n components. Then

$$\prod_1^n x_i \leq \phi(\mathbf{x}) \leq \coprod_{i=1}^n x_i. \qquad (1)$$

Design engineers have long followed the rule: *Redundancy at the component level is superior to redundancy at the system level.* Coherent system theory proves the corresponding

Theorem 2.   Let ϕ be a coherent system. Then $\phi(x_1 \amalg y_1, \ldots, x_n \amalg y_n) \geq \phi(\mathbf{x}) \amalg \phi(y)$; [$x \amalg y$ denotes $1 - (1 - x)(1 - y)$]. Equality holds for all $\mathbf{x}$ and $\mathbf{y}$ if and only if the structure is parallel.

The variety of types of coherent systems is very large, especially for large n. Thus it is reasuring to know *that every coherent system may be represented as a parallel–series system and as a series–parallel system if replication* of components is permitted.* (A small-scale example of this general result is shown in Fig. 1C.) These representation results not only conceptually simplify the theory of coherent systems; they also yield simple upper and lower bounds on coherent system reliability, as we shall see shortly.

To describe these representations, we need some terminology and notation.

A *minimal (min) path set* of a coherent structure is a set of components satisfying: (a) if each component in the set functions, the system functions; (b) if all remaining components fail and any one or more of the

components of the min path set fails, the structure fails. The corresponding *min path series structure* is the series structure formed from the components of the min path set. Given a coherent structure ϕ with p min paths, denote the ith *min path series structure function* by $\rho_i(\mathbf{x})$, $i = 1, \ldots, p$. The *min path representation* is given by

$$\phi(\mathbf{x}) = \coprod_{i=1}^p \rho_i(\mathbf{x}), \qquad (2)$$

corresponding to a *parallel arrangement of the p min path series structures.*

For example, the two-out-of-three system has $p = 3$ min path series structures:

$$\rho_1(\mathbf{x}) = x_1 x_2, \quad \rho_2(\mathbf{x}) = x_1 x_3, \quad \rho_3(\mathbf{x}) = x_2 x_3.$$

The min path representation is

$$\phi_{2|3}(\mathbf{x}) = \coprod_{i=1}^3 \rho_i(x),$$

diagrammatically displayed in Fig. 1c. Note that each of the components appears twice.

Next we develop the dual min cut representation of a coherent structure. A min cut set is a set of components satisfying: (a) if each component in the min cut set fails, the system fails; (b) if all remaining components function and one or more of the components in the min cut set function, the structure functions. The corresponding *min cut parallel structure* is the parallel structure formed from the components in the min cut set. Given a coherent structure ϕ with k min cuts, denote the ith *min cut parallel structure function* by $\kappa_i(\mathbf{x})$, $i = 1, \ldots, k$. The *min cut representation* is given by

$$\phi(\mathbf{x}) = \prod_{i=1}^k \kappa_i(\mathbf{x}), \qquad (3)$$

corresponding to a *series arrangement of the k min cut parallel structures.*

In the two-out-of-three system,

$$\kappa_1(\mathbf{x}) = x_1 \amalg x_2, \quad \kappa_2(\mathbf{x}) = x_1 \amalg x_3, \quad \kappa_3(\mathbf{x}) = x_2 \amalg x_3.$$

The min cut representation is

$$\phi_{2|3}(\mathbf{x}) = \prod_{i=1}^3 \kappa_i(\mathbf{x}).$$

The diagram of the min cut representation is

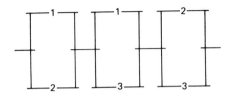

Again, each component appears twice. Note further that the diagram just above and the diagram in Fig. 1(c) are alternative representations of the same system.

Thus far we have confined our discussion to the *deterministic* aspects of coherent systems. Next we summarize the *probabilistic* properties of coherent systems. These properties are directly relevant to the prediction* (probabilistically) and the estimation* (statistically, i.e., from data) of system reliability.

SYSTEM RELIABILITY

Assume first that component states, $X_1, \ldots, X_n$, are random but statistically independent*. Thus let X_i be a Bernoulli random variable* indicating the state of component i:

$$X_i = \begin{cases} 1 & \text{(component } i \text{ is functioning)} \\ & \text{with probability } p_i \\ 0 & \text{(component } i \text{ is failed)} \\ & \text{with probability } q_i \equiv 1 - p_i, \end{cases}$$

where $p_i (0 \leqslant p_i \leqslant 1)$ is called the *reliability** of component i, $i = 1, \ldots, n$. The corresponding *system reliability* h is given by

$$h = \Pr[\phi(\mathbf{X}) = 1] \equiv E\phi(\mathbf{X}).$$

Since component states $X_1, \ldots, X_n$ are mutually independent, system reliability h is completely determined by component reliabilities $p_1, \ldots, p_n$; thus we write $h = h(\mathbf{p})$, where $\mathbf{p} \stackrel{\text{def}}{=} (p_1, \ldots, p_n)$. In the special case of interest, $p_1 = p_2 = \cdots = p_n = p$, we write $h(p)$. We call $h(\mathbf{p})(h(p))$ the reliability function; it expresses system reliability as a func-

tion of component reliabilities (common component reliability).

As examples, for the series system, $h(\mathbf{p}) = \prod_{i=1}^{n}(1 - p_i)$, and for the k-out-of-n system with common component reliability p, $h(p) = \sum_{i=k}^{n} \binom{n}{i} p^i (1 - p)^{n-i}$, the binomial right-hand tail.

Basic properties of the reliability functions $h(\mathbf{p})$ and $h(p)$ are:

1. $h(\mathbf{p})$ is multilinear in $p_1, \ldots, p_n$.
2. $h(p)$ is a polynomial in p, with all coefficients nonnegative.
3. $h(\mathbf{p})$ is strictly increasing in each p_i on the domain $0 < p_i < 1$, $i = 1, \ldots, n$.
4. $h(p_1 \amalg p_1', \ldots, p_n \amalg p_n') \geqslant h(\mathbf{p}) \amalg h(\mathbf{p}')$.

Equality holds for all $\mathbf{p}$ and $\mathbf{p}'$ if and only if the system is parallel.

Property 4 states that *redundancy at the component level yields higher system reliability than redundancy at the system level*. This is the probabilistic version of Theorem 2, which gives the deterministic version of this familiar design engineer's rule.

Computation of System Reliabilities

A basic problem is to compute system reliability in terms of component reliabilities. Alternative *exact* methods are:

1. By means of *min cut and min path representations*, based on (2) and (3):

$$h(\mathbf{p}) = E \coprod_{i=1}^{p} \prod_{i \in P_j} X_i$$

$$\equiv E \prod_{j=1}^{k} \coprod_{i \in K_j} X_i, \qquad (4)$$

where P_j denotes the jth min path set and K_j the jth min cut set.

2. By *examining all 2^n possible outcomes of the n components* and using the obvious formula

$$h(\mathbf{p}) = \sum_{x} \phi(\mathbf{x}) \prod_{i} p_i^{x_i} q_i^{1 - x_i}, \qquad (5)$$

the summation being taken over all 2^n vectors x with 0 or 1 coordinates.

3. By using the special features of *certain types of systems*:
    (a) Many systems are designed to consist of distinct subsystems, which in turn consist of distinct subsystems, etc. By computing the reliability of each of the lowest-level groupings, by then computing the reliability of each of the next level groupings from the lowest level groupings, etc., it becomes possible to compute system reliability.
    (b) If the components have common reliability p, we may use the formula

$$h(p) = \sum_{i=1}^{n} A_i p^i q^{n-i}, \qquad (6)$$

   where A_i denotes the number of vectors $\mathbf{x}$ for which $\phi(\mathbf{x}) = 1$.
    (c) By eye, for simple or small systems, using basic probability rules.

Approximations and Bounds

Clearly, there is a need for approximations and bounds for system reliability, since the exact computations for large systems can become formidable or even intractible. Next, we list bounds and methods for obtaining them, which apply even when *component states may be mutually statistically dependent*.

INCLUSION–EXCLUSION* METHOD. Let E_r be the event that all components in min path set P_r work. Then $\Pr[E_r] = \prod_{i \in P_r} p_i$. System success corresponds to the event $\cup_{r=1}^{P} E_r$. Thus $h = \Pr[\cup_{r=1}^{P} E_r]$. Let

$$S_k = \sum_{1 \le i_1 < i_2 < \cdots < i_k \le p} \Pr[E_{i_1} \cap E_{i_2} \cap \cdots \cap E_{i_k}].$$

By the inclusion–exclusion principle,

$$h = \sum_{i=1}^{p} (-1)^{k-1} S_k,$$

and

$$h \le S_1 = \sum_{r=1}^{p} \prod_{i \in P_r} p_i,$$
$$h \ge S_1 - S_2,$$
$$h \le S_1 - S_2 + S_3,$$
$$h \ge S_1 - S_2 + S_3 - S_4,$$

etc. Although it is not true that the successive upper bounds decrease necessarily and the successive lower bounds increase necessarily, in practice it may be necessary to calculate only a few S_k's to obtain a close approximation.

BOUNDS FOR SERIES AND PARALLEL SYSTEMS. If $X_1, \ldots, X_n$ are *associated* random indicators of the respective component states, then

$$\Pr\left[\prod_{i=1}^{n} X_i = 1\right] \ge \prod_{i=1}^{n} \Pr[X_i = 1] \qquad (7)$$

$$\Pr\left[\coprod_{i=1}^{n} X_i = 1\right] \le \coprod_{i=1}^{n} \Pr[X_i = 1]. \qquad (8)$$

Note that (7) [(8)] states that the reliability of a series (parallel) system of positively dependent components is bounded below (above) by system reliability computed under the assumption of independent components.

CRUDE BOUNDS FOR COHERENT SYSTEMS. Let ϕ be a coherent structure of associated components with respective reliabilities $p_1, \ldots, p_n$. Then

$$\prod_{1}^{n} p_i \le \Pr[\phi(\mathbf{X}) = 1] \le \coprod p_i. \qquad (9)$$

These bounds are generally rather crude since each arises as the result of two successive bounding operations.

Next we present improved bounds on system reliability using additional information: the minimal path and minimal cut representation of the structure given in (2) and (3), respectively.

BOUNDS FOR COHERENT STRUCTURES BASED ON MIN PATH AND MIN CUT SETS

1. Let ϕ be a coherent structure of *associated* components. Let $\rho_1(\mathbf{x}), \ldots, \rho_k(\mathbf{x})$ be the minimal path series structures, and $\kappa_1(\mathbf{x}), \ldots, \kappa_k(\mathbf{x})$ be the minimal cut parallel structures of ϕ. Then

$$\prod_{j=1}^{k} \Pr\left[\kappa_j(\mathbf{X}) = 1\right] \leqslant \Pr\left[\phi(\mathbf{X}) = 1\right]$$

$$\leqslant \prod_{j=1}^{p} \Pr\left[\rho_j(\mathbf{X}) = 1\right]. \tag{10}$$

2. If components are *independent*, the bounds become more explicit:

$$\prod_{j=1}^{k} \coprod_{i \in K_j} p_i \leqslant \Pr\left[\phi(\mathbf{X}) = 1\right]$$

$$\leqslant \coprod_{j=1}^{p} \prod_{i \in P_j} p_i. \tag{11}$$

A numerical example of such bounds is presented in Barlow and Proschan [1, pp. 35–36].

MIN–MAX BOUNDS FOR COHERENT STRUCTURES

1. Regardless of the joint distribution of component states, the following bounds hold:

$$\max_{1 \leqslant r \leqslant p} \Pr\left[\min_{i \in P_r} X_i = 1\right]$$

$$\leqslant \Pr\left[\phi(\mathbf{X}) = 1\right]$$

$$\leqslant \min_{1 \leqslant s \leqslant k} \Pr\left[\max_{i \in \kappa_s} X_i = 1\right]. \tag{12}$$

2. If components are *associated*, the more explicit bounds hold:

$$\max_{1 \leqslant r \leqslant p} \prod_{i \in P_r} p_i \leqslant \Pr\left[\phi(\mathbf{X}) = 1\right]$$

$$\leqslant \min_{1 \leqslant s \leqslant k} \coprod_{i \in \kappa_s} p_i. \tag{13}$$

Coherent structure theory is currently being generalized so that instead of only the functioning and failed states being possible for both components and system, a finite or even infinite number of states are now possible. These states correspond to levels of performance of component and system. Different axiomatic treatments are presented by El-Neweihi [4], Barlow and Wu [2], and Ross [9]. This current research on *multistate coherent systems* represents just the initial phase of a flood of research to come.

References

[1] Barlow, R. E. and Proschan, F. (1975). *Statistical Theory of Reliability and Life Testing: Probability Models*. Holt, Rinehart and Winston, New York. (A clear, comprehensive, and detailed treatment of coherent system theory is presented in Chaps. 1 and 2. The appendix presents a brief treatment of fault-free analysis, showing the connection with coherent systems theory.)

[2] Barlow, R. E. and Wu, A. S. (1978). *Math Operat. Res.*, **3**, 275–281.

[3] Birnbaum, Z. W., Esary, J. D., and Saunders, S. C. (1961). *Technometrics*, **3**, 55–77. (This paper constitutes the first systematic treatment of coherent system theory.)

[4] El-Neweihi, E., Proschan, F., and Sethuraman, J. (1978). *J. Appl. Prob.*, **15**, 675–688.

[5] Esary, J. D. and Proschan, F. (1962). *Redundancy Techniques for Computing Systems*. Spartan Books, Washington, D. C., pp. 47–61.

[6] Esary, J. D. and Proschan, F. (1963). *Technometrics*, **5**, 191–209.

[7] Fussell, J. B. (1973). *Nucl. Sci. Eng.*, **52**, 421–432.

[8] Moore, E. F. and Shannon, C. E. (1956). *J. Franklin Inst.*, **262**, Part I, 191–208; Part II, 281–297. (This brilliant paper shows that starting with arbitrarily "crummy" relays, a relay network may be designed to achieve arbitrarily high reliability. It inspired the Birnbaum et al. paper.)

[9] Ross, S. M. (1979). *Ann. Prob.*, **7**, 379–383.

F. PROSCHAN

COHORT ANALYSIS

DEFINITION

Ancient Romans seem to have understood by the term "cohort" a military unit consisting of several hundred soldiers. Now, however, the dictionary definitions of the term include one's accomplice, associate, or supporter, or in the collective sense a band or a group. Demographers adapted the term to denote a group of persons in a geographically or otherwise defined population who experienced a given significant life event (e.g., birth, marriage, or graduation) during a given time interval (e.g., a calendar year or a decade). Thus persons born in the United States during the calendar year 1935 form the U.S. birth cohort 1935. Marriage cohorts, high school graduation cohorts, and others are similarly defined. It may be noted in passing that the concept can be extended to include nonhuman populations. For example, the housing starts in a country in a given year, the passenger cars manufactured by a company in a given period, and similar aggregates are also cohorts.

Cohort analysis is a quantitative research orientation that emphasizes intracohort and intercohort comparisons of aggregate measures of life experiences, classified by age (duration from the start of the life history). Cohort analysis has been used by demographers primarily in fertility studies, but it has also been applied to the analysis of mortality, nuptiality, migration, and other demographic phenomena; *see* DEMOGRAPHY. In recent years its field of application has been extended to include nondemographic topics, such as the effects of human aging and the dynamics of political behavior.

METHODOLOGY

Cohort analysts often use the presentation known as a Lexis* diagram to locate in a two-dimensional space the life experiences of cohorts. Figure 1 is a Lexis diagram in the age–time plane for birth cohorts, with death as the relevant experience. Time (e.g., calendar year) is laid out on the horizontal axis and age on the vertical. Both axes use the same time unit (e.g., year). Each individual member of each cohort is represented by a life line inclined at 45° to either axis, starting on the horizontal axis at the moment of birth, and terminating at a point corresponding to the moment of death and the age at that moment. Events such as marriage, high school graduation, or entry into the labor force can be marked on the life lines by points corresponding to the time and age when each such event occurs.

Consider a particular cohort in Fig. 1, say the one with its origin in the time interval t to $t + 1$. If E_0 is the number of births that occur during the period t to $t + 1$, the cohort in question starts with E_0 life lines between t

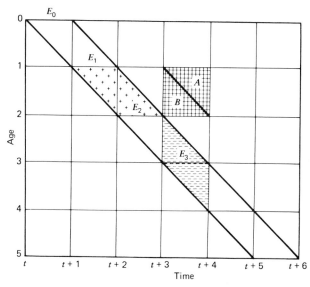

Figure 1 Lexis diagram showing birth cohorts in a time–age plane.

and $t + 1$. Suppose that E_x of those born during the period survive to their xth birth day, $E_x - E_{x+1}$ dying between their xth and $(x + 1)$st birth days. Then the diagram will show E_x of the original E_0 life lines reaching the horizontal at age x, and $E_x - E_{x+1}$ life lines terminating between the horizontals at x and $x + 1$. The sequence of numbers E_0, E_1, . . . provide the full information regarding the experience of the cohort, as far as death or survival is concerned.

There are several ways in which the information in a Lexis diagram can be aggregated. One mode of aggregation* is by age and cohort, i.e., by parallelograms of the type marked by $+$'s in Fig. 1. An example is the sequence $(E_0 - E_1)$, $(E_1 - E_2)$, . . . , referred to above. This mode of aggregation mixes adjacent periods, e.g., t to $t + 1$ with $t + 1$ to $t + 2$.

Another mode of aggregation is by period and cohort, i.e., by parallelograms of the type marked by $-$'s in Fig. 1. An example of this type of aggregation is a table that presents for each marriage cohort, i.e., persons who got married in a given year, the number of marriages that end in divorce, separation, or widowhood during a given year after marriage. This mode of aggregation mixes adjacent age groups.

A third form of aggregation is by age and period, i.e., by squares of the type crosshatched in Fig. 1. A common example of this is the tabulation of deaths in a population by year of occurrence and age at last birth day of the deceased. This mode of aggregation mixes the experiences of adjacent cohorts, as can be seen by noting that the triangle marked A in the crosshatched square (Fig. 1) belongs to one cohort and the one marked B to another.

Instead of squares or parallelograms, one may use lines to define aggregation units. The sequence E_0, E_1, . . . mentioned above is an example. This sequence represents, for a given birth cohort of initial size E_0, the numbers of persons who survive to the first, second, . . . birthdays or anniversaries. The unit of aggregation in this case is a period

segment on the age (horizontal) line. Similarly, it is possible to use age segments on time (vertical) lines as units of aggregation. To give an example, suppose that a particular interview question was repeated in opinion surveys conducted on November 1 of each presidential election year in the United States. The date of each survey is thus 4 years after that of the preceding one. If the responses to the interview questions in the successive surveys are tabulated using 4-year age groups, the resulting aggregation is by age segments on time lines.

It should be noted that often a particular aggregation used may only be an approximation to one or another of the forms just described. For example, the field work of an opinion survey may have been completed over a short interval of time such as a week, rather than in a day or in one instant. In such situations, the aggregation involved will not be exactly by the age segments on a single time (vertical) line, but by those on a small strip of time interval. For practical purposes one often regards such aggregations as if age segments on a single time line had been used.

Data useful for cohort analysis may be obtained from any of a number of sources, including registration systems, censuses*, and sample surveys*. In nondemographic studies, the required data are drawn more often than not from sample surveys or censuses. Survey (census) data may be classified into three different types: data from reinterviewing a panel, retrospective data from a single survey (census), and comparable data from two or more independent surveys (censuses).

Whatever the source of data, a cohort analyst will have to first assemble the available information in a form that permits tracing the experience of each cohort as it ages. Often this is done by constructing a "standard" cohort table in which counts, rates, proportions, averages, or probabilities pertaining to the phenomenon taken for study are arranged by time and age, such that the width of the age intervals bears a

1:1, 1:2, 2:1 or similar relation to the intervals between the time points for which there are data. It is not infrequent, however, that the available data may be spaced irregularly in time or that the width of the age intervals used for aggregation for the data may not be the same as those used for the data collected at another point in time. In such situations it becomes difficult, if not impossible, to trace the experiences of cohorts as they age. A common strategy used by cohort analysts in such situations is to derive a table in the standard form from the available data, using techniques such as interpolation. This strategy has its own drawbacks, however. Consider, for example, using interpolation to estimate rates or probabilities for regular 5-year age groups from the data available for irregular age groups. The technique is based on certain assumptions, of unknown validity, concerning the underlying age pattern of the rates or probabilities. Inferences based on interpolated data are more likely to be a function of the assumptions underlying the interpolation* technique than a reflection of any basic features of the phenomenon studied.

In analyzing cohort data, attention is very often focused on cohort, age, and period effects. It is logical to assume that each environmental circumstance or stimulus, such as a change in the economic conditions, evokes from each cohort an immediate response as well as delayed response. (These correspond to the "direct" and "carryover" effects referred to in the literature of experimental designs.) They vary according to the age or stage of development of the cohort at the moment when the stimulus is received, the reason for the age dependence being that the response to any stimulus is likely to vary according to the history of past experiences. To illustrate the point, consider a sequence of three decades: the first and the last being prosperous ones, whereas the middle decade represents a depression period. Imagine birth cohort A entering young adult ages (say, 20 to 29 years) during decade 1, cohort B entering the same age group in decade 2,

and cohort C doing so in decade 3. These cohorts are likely to show different responses to the prosperity of decade 3: cohort A is unlikely to show any significant fertility response to the prosperity of the decade, having almost reached the end of its childbearing period by the beginning of the decade. The response of cohort B might very well be to show a tendency toward later marriage and later childbearing, because of a transfer of marriages and births from the depression decade 2 to the prosperous decade 3. Cohort C, on the other hand, since it enters young adult ages during a prosperous decade, might very well show a tendency toward early marriage and early age of childbearing. Thus the same environmental stimulus (prosperity in decade 3 in the illustration above) may evoke different "immediate" responses from different cohorts, depending upon the stage of the life history at which they receive the stimulus and on their life experiences until then. Similar comments apply also to "delayed" responses. Such intercohort differences in responses constitute a cohort effect. The period effect and age effect are defined in the same way as column and row effects are defined in the analysis of variance.

There are certain problems in identifying the cohort, period, and age effects from cohort tables. One source of the difficulty lies in the fact that as a (birth) cohort ages, its composition may change because of death or outmigration. Also, the survivors of a cohort may get mixed up with inmigrants. Those who die or migrate may differ from the rest of the cohort in regard to the phenomenon being studied, and if so, some of the observed intracohort and intercohort variations may be due to the compositional changes within cohorts, resulting from death or migration. Standardization techniques can be used to separate some of the compositional effects, but this could only be tentative, because it is impossible to know how the deceased or the outmigrants would have behaved in regard to the phenomenon being studied if they had survived or not migrated.

Table 1   Cohort Table Involving Three Periods and Three Age Groups

Age	Period		
Group	1	2	3
1	Y_{113}	Y_{122}	Y_{131}
2	Y_{214}	Y_{223}	Y_{232}
3	Y_{315}	Y_{324}	Y_{333}

Another source of the difficulty in dealing with cohort, period, and age effects is that in a cohort table, each effect is to a certain degree confounded with either of the other two. To see this, consider Table 1, in which (Y_{ijk}) measures on a dependent variable are displayed by age for three successive, evenly spaced cross-sectional surveys. The dates of data collection, which are taken to represent instants rather than time intervals, define the periods. The interval between each two consecutive periods match the width of the age groups used for aggregation. The cohort membership is defined in terms of the age group reached in a given period. The youngest cohort in Table 1 is the one that reaches the first age group in period 3; the oldest is the one that reaches age group 3 in period 1; and so on. The third subscript of Y_{ijk} denotes cohort, the second the period, and the first the age group. It is easily seen that cohort membership $(k) =$ period $(j) -$ age $(i) + 3$, which is a perfect linear relationship. This situation resembles the one in the regression analysis when one regressor is a perfect linear function of two or more other regressors. In such situations, some of the effects (regression coefficients) are not estimable. This is known as the identification problem*. Situations of this type arise in the analysis of social mobility, status inconsistency, and similar topics.

To see the nature of the identification problem just referred to, suppose for the data in Table 1 that information is available on the exact age of each respondent. One may then treat period, age, and cohort as continuous variables of time and write a cohort–age–period model, considering only

main effects, thus:

$$E(Y) = \alpha + \sum_1^m \beta_r A^r + \sum_1^k \gamma_s C^s$$
$$+ \sum_1^2 \delta_t P^t, \tag{1}$$

where A stands for age, C for cohort, and P for period. (The maximum degree of the polynomial in P is constrained to 2 because there are only three periods represented in the data. The degrees of the polynomials in A and C are similarly constrained by the available number of data points.)

Now, since by definition $C = P - A$, one may substitute $(A + C)$ for P in the model, thus getting

$$E(Y) = \alpha + (\beta_1 + \delta_1)A$$
$$+ (\beta_2 + \delta_2)A^2 \sum_3^m \beta_r A^r$$
$$+ (\gamma_1 + \delta_1)C + (\gamma_2 + \delta_2)C^2$$
$$+ \sum_3^k \gamma_s C^s + 2\delta_2 AC. \tag{2}$$

Clearly, except for β_1 and γ_1, which are both indistinguishable from δ_1, all parameters in the foregoing model are estimable. It follows that it is the linear components of the cohort, age, and period main effects that are indistinguishable; all higher-order components are estimable [2, 10] (*see* ESTIMABILITY).

By restricting δ_1/δ_2, δ_1/β_1, or δ_1/γ_1 to equal a known constant, one can make the model (2) exactly identified. Obviously, except in very limited contexts, such restrictions may not be justifiable. For this reason it is not advisable to base one's interpretation of cohort data exclusively on the results given by formal models fitted to the data, unless the models, including the restrictions imposed on the parameters, have been carefully selected on the basis of substantive considerations. In this connection, a debate that has been going on for some time now between two camps of cohort analysts may be worth mentioning. The issue of the debate is whether it is appropriate to use additive age–period–cohort models for interpreting data of the type shown in Table 1. One camp (see, e.g., Glenn [4]) holds that there is likely to be age–period, age–cohort, or period–cohort interaction in cohort tables, and hence additive models, if applied to such tables, may suffer from specification errors, because of ignoring interactions*. The opposite camp (see, e.g., Fienberg and Mason [2]) holds that all models are simplifications (abstractions) of reality; that whether a given simplification is acceptable in a given context is to be decided on substantive grounds; and that this should be decided on a case-by-case basis rather than by invoking a general principle such as "interactions should always be included in the model".

APPLICATIONS

Fertility*

As already mentioned, the cohort method has been used more extensively in fertility analysis than in any other demographic studies. The cohort analysis of fertility may employ birth cohorts or marriage cohorts or both in combination. Confining attention to birth cohorts, imagine that schedules of age-specific fertility rates are available for a number of consecutive years. A table of these rates arranged in columns for years (and ages in rows) can be summarized in two obvious ways: by column or diagonally. For each column the sum of the age-specific fertility rates gives the period total fertility rate, and the arithmetic mean of the age distribution of the birth rates gives the period mean age of fertility. The cohort schedule (the figures in the diagonal) can be summarized similarly to give the cohort total fertility rate and the cohort mean age of fertility. It can be shown that the time series of period total fertility rate diverges from that of the cohort total fertility rate to the extent that there is temporal change in the

period or cohort mean age of fertility, and that the time series of the period mean age of fertility will diverge from the cohort mean age of fertility to the extent that there is a temporal shift in the period or cohort total fertility rate. To take a specific example, which is only slightly exaggerated, suppose that in a large population, for a long period of time, a constant age pattern of childbearing has been prevailing, with each cohort having on an average 3 children per woman by the end of the childbearing period. Suppose that the younger cohorts at time t suddenly change the timing of births in such a way that they bear their children earlier than usual, still having the same cohort total fertility rate as the older cohorts. Then an examination of the period total fertility rate for several successive years will show a rise in the birth rate. But such a rise does not really constitute increased (cohort) fertility. Here the divergence between the period and cohort total fertility rates is attributable to the temporal shift in the cohort mean age of fertility.

Suppose again that women vary the time at which they have children, delaying pregnancies during recessions and depressions and advancing them when times are prosperous, but with no change in the average number of children borne by the end of childbearing period. The period data would show considerable fluctuations, but they do not reflect variations in total cohort fertility.

Referring to Fig. 1, it may be noted that the childbearing of women is represented by density along the life lines. If the net maternity function, which is the product of survivorship to a given age and reproduction rate at that age, is represented by $\varphi(a, t)$, a function of age and time, then the integral (or total) along the vertical strip will give the period net reproduction rate and that along the diagonal strip the cohort net reproduction rate. It is possible to relate the integral (or total) along the diagonal to that along the vertical (see Ryder [7]), thereby enabling one to translate cohort information into period information, and vice versa.

Mortality

The sequence of numbers $E_0, E_1, \ldots$ in Fig. 1 shows that of E_0 births that occur in the time interval $(t, t + 1)$, E_x survive to the xth birthday and $(E_x - E_{x+1})$ die between the xth and $(x + 1)$st birthdays. From the sequence $(E_0, E_1, \ldots)$, mortality rates* according to age in successive years can be computed using the formula $q_x = (E_x - E_{x+1})/E_x$. The life table* completed on the basis of the q_x-values thus computed is known as the cohort or generation life table, since it reflects the actual mortality experience of a cohort as it ages in successive years. The cohort life table shows among other things e_x^0, the average length of life remaining after the xth birthday for those who survive to that birthday. The cohort life table is to be distinguished from the current life table, which is derived from the mortality rates for a single calendar year or period. In Fig. 2, the life table based on the q_x-values in any vertical line is the current life table, whereas the one based on the q_x sequence on any diagonal is a cohort life table.

The current life table obviously is derived from a mixture of the mortality rates that different cohorts experience at different stages in their life history, one cohort during its first year of life, another during its second year, and so on. Consequently, given the usual downward trend in mortality over time, the current life table understates the average length of life of the newborn, and the average remaining life of those who survive to a given age, during the period of the life table.

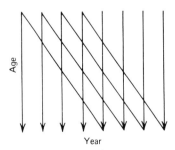

Age

Year

Figure 2

That cohort analysis helps to avoid erroneous inferences from data is more dramatically illustrated by Frost's analysis of the age pattern of tuberculosis mortality in Massachusetts during the years 1880, 1910, and 1930 [3]. An examination of the age curve of tuberculosis mortality for the three years showed:

1. The curve for 1930 was consistently below that for 1910, which in turn was consistently below that for 1880.
2. For each year the curve started at a high level in infancy, dropped to a low level in childhood, then rose again to reach higher levels in adult ages.
3. The curve for 1930 peaked between ages 50 and 60, whereas that for 1910 peaked between ages 30 and 40, and that for 1880 between ages 20 and 30.

Earlier explanations for the apparent shift in the peak of the age curve of tuberculosis mortality were based on the notion that, over time, tuberculosis lost its killing power, shifting fatality to older ages at which the vital resistance of the body characteristically is low. When Frost examined the data in cohort form, however, he noticed that all the cohorts exhibited similarly shaped age curves of mortality, peaking at about the same age (20 to 29 years), each cohort having its age curve consistently above those of all younger cohorts. Frost concluded that the apparent shift in the peak age of tuberculosis mortality (from between 20 and 30 in 1880 to between 50 and 60 in 1930) was an artifact of the period method of organizing data. The period method, it may be noted, juxtaposes the experience of an older cohort at an older age with that of younger cohorts at younger ages. Consequently, given the consistently higher mortality of older cohorts at all ages, this method produces the artifact of the age curve peaking at higher ages for more recent periods.

Frost's work on tuberculosis mortality illustrates why the experiences of real cohorts rather than the age pattern observed in period data should be the basis for inferences regarding the age pattern of mortality or survival functions.

The difficulty with the cohort method in mortality studies, of course, is that 100 years or more must elapse before the death of the last survivor of a birth cohort. This difficulty can be partially overcome by confining attention to a part of the cohort experience, for example, mortality up to age 50, or by studying cohorts between two age points, say, between ages 20 and 50 [8].

Nuptiality

In the area of nuptiality, cohort analysis has frequently been applied to the study of the age patterns of first marriage. From cohort data, one may construct the nuptiality table along the lines of the usual life table or in the fashion of the double-decrement table, the latter if mortality is recognized as an additional factor responsible for the decrement of never-married persons in the cohort as it advances in age.

Mathematical functions, parallel to the survival functions in mortality studies, have been developed for the age pattern of cumulative proportions ever married [1, 5].

The fluctuations in the number of marriages from period to period, with changing economic conditions despite stability in the likelihood of getting married eventually, have been interpreted in terms of temporary shifts in the time pattern of cohort nuptiality [6].

Migration*

The cohort approach to migration studies has taken two forms: one in which the focus is on birth cohort and the other on migration cohort. In the former, migration is viewed as an event in the life history of the individual or cohort, similar to marriage or childbearing. When the focus is on migration cohorts, the emphasis is on duration of residence in a given area.

Effects of Aging

In the next few years, given the trend toward older age structures in almost all the developed populations, interest in the study of the effects of aging on responses to forces of change is likely to be on the increase. A basic issue for examination would be whether aging individuals tend to become so inflexible in their attitudes and behavior that they will be a formidable obstacle to social and economic change. The method appropriate for investigation of this issue involves separating cohort effect from age effect in intercohort differences in susceptibility to change.

Other Fields

The cohort approach has also been applied to the study of educational careers from the time of entry into school, of occupational careers from the time of entry into the labor force, and of morbidity* history beginning with the first exposure to a condition; it has also been used in many other fields (see, e.g., United Nations [9]).

Underlying these and other applications of the cohort method is the hypothesis that in the macro-biography of any group, what happens today is contingent on what happened yesterday, and what happens tomorrow depends upon what happens today. It is the plausibility of this hypothesis on which rests the utility of the cohort method.

References

[1] Coale, A. J. (1971). *Popul. Stud.*, **25**, 193–196.
[2] Fienberg, S. E. and Mason, W. M. (1979). In *Sociological Methodology*, Karl F. Schuessler, ed., Jossey-Bass, San Francisco, Calif., 1–67.
[3] Frost, W. H. (1939). *Amer. J. Hyg. A*, **30**, 91–96.
[4] Glenn, N. D. (1977). *Cohort Analysis*. Sage Univ. Paper Ser. Quant. Appl. Social Sci.: Ser. No. 07-005. Sage Publications, Beverly Hills, Calif. (Contains a useful set of references.)
[5] Hernes, G. (1972). *Amer. Sociol. Rev.*, **37**, 173–182.
[6] Ryder, N. B. (1956). *Population (Paris)*, **11**, 29–46.

[7] Ryder, N. B. (1964). *Demography*, **1**, 74–82.

[8] Spiegelman, M. (1969). *Demography*, **6**, 117–127.

[9] United Nations (1968). *Methods of Analysing Census Data on Economic Activities of the Population.* Ser. A. Popul. Stud. No. 43. United Nations, New York.

[10] Winsborough, H. H. (1976). Discussion. *Proc. Social Statist. Sect. Part 1*, American Statistical Association, Washington, D.C.

N. KRISHNAN NAMBOODIRI

(CENSUSES
DEMOGRAPHY
EPIDEMIOLOGICAL STATISTICS
FERTILITY
LIFE TABLES
MARRIAGE
MIGRATION
MORBIDITY
MORTALITY
POPULATION PROJECTION
SOCIOLOGY, STATISTICS IN
SURVEY SAMPLING
SURVIVAL ANALYSIS)

COLD DECK METHOD   *See* EDITING STATISTICAL DATA

COLLINEARITY

From a mathematical point of view, a set of points are said to be collinear if they lie on the same straight line. This definition has been adapted to multiple discriminant analysis* where one has a number of multivariate observations in each of a number of groups. Each observation consists of measures on p variables. Under standard assumptions, the observations from each group are sampled from the same multivariate normal* population, where the populations differ only with respect to the means on the p variables. We may regard the multivariate group mean (or centroid) as a point in a p-dimensional space, and the only basis for discriminating among these populations is the separation among the centroids.

Given k populations, the k centroids may lie in a subspace whose dimensionality is less than p. If k is equal to 2, for example, the points must lie on a straight line; that is, they must be collinear. In general, the dimensionality of the subspace is at most equal to the smaller of $k - 1$ and p. It is always possible that for any value of k, the centroids may be collinear. This obviously provides a simple basis for discriminating among the populations. If the dimensionality of the subspace is 2, the centroids are said to be coplanar.

A statistical test of significance developed by Bartlett has been applied to test the dimensionality of the subspace, in particular to test for collinearity [1, p. 406].

Reference

[1] Bock, R. D. (1975). *Multivariate Statistical Methods in Behavioral Research.* McGraw-Hill, New York.

(MULTICOLLINEARITY)

ELLIOT M. CRAMER

COLTON'S MODEL

This is a model for the loss function* when:

1. A choice is to be made between two populations π_1 and π_2 ("treatments") on the basis of estimates of the values θ_1 and θ_2 of a parameter θ in the two populations.

2. There are N individuals, of which $2n$ will be used in test samples, n for each treatment.

3. The "better" treatment will be used for the remaining $(N - 2n)$ individuals.

Maurice [4] pointed out that, in addition to the loss incurred as a consequence of the wrong decision in item 3, which will be proportional to

$$(N - 2n)\Pr[\text{choose wrong population} \mid \text{sample size } n; \theta_1\theta_2]$$
$$= (N - 2n)p(n, \boldsymbol{\theta}),$$

one should include the loss arising from the fact that n individuals will be given the worse treatment, so that the total loss will be equal to

$$n + (N - 2n)p(n, \boldsymbol{\theta})$$

multiplied by some function $g(\boldsymbol{\theta})$ of θ_1 and θ_2. Sampling costs (e.g., $2nc$) would be added to this. Maurice formulated this model in terms of industrial and agricultural settings where "treatment" could mean production method, fertilizer, etc.

Colton [2] adapted the model to clinical trials for the selection of one out of two medical treatments.

This model is used to assist in assessing a desirable value for n—the size of the trial with each treatment. There is generally no minimax* solution if the function $g(\boldsymbol{\theta})$ is unbounded. Maurice [4] shows how to determine a minimax solution when $g(\cdot)$ is bounded, in particular when (1) the observed variables are normally distributed with standard deviation σ and expected values θ_1 and θ_2 in π_1 and π_2, respectively, and (2) $g(\boldsymbol{\theta})$ is proportional to $|\theta_1 - \theta_2|$. She also obtains an (approximately) optimal sequential procedure. Colton [2] also uses a different criterion to obtain an optimal value for n —"maximin expected net gain". "Gain" is defined as $+(-)g(\boldsymbol{\theta})$ if the better (worse) treatment is received, so that expected net gain is proportional to

$$(N - 2n)\left[(1 - p_n(\boldsymbol{\theta})) - p_n(\boldsymbol{\theta})\right]g(\boldsymbol{\theta}).$$

Colton [2] also obtains an (approximately) optimal sequential procedure (see also Anscombe [1]), and later [3] describes optimal two-stage sampling procedures based on the same model.

In more recent investigations, methods of utilizing early results in a sequential trial to assist in determining treatment assignments so as to reduce the expected number of individuals given the worse treatment during the trial has been given considerable attention (*see* PLAY-THE-WINNER).

Oudin and Lellouch [5] consider another aspect, in which it is desired to combine the results of parallel trials on groups of individuals in different categories according to values of some concomitant variables* (age, sex, etc.).

References

[1] Anscombe, F. J. (1963). *J. Amer. Statist. Ass.*, **58**, 365–383.

[2] Colton, T. (1963). *J. Amer. Statist. Ass.*, **58**, 388–400.

[3] Colton, T. (1965). *Biometrics*, **21**, 169–180.

[4] Maurice, R. J. (1959). *J. R. Statist. Soc. B*, **21**, 203–213.

[5] Oudin, C. and Lellouch, J. (1975). *Rev. Statist. Appl.*, **23**, 35–41.

(CLINICAL TRIALS
DECISION THEORY
PLAY-THE-WINNER
SEQUENTIAL ANALYSIS)

COMBINATION OF DATA

"Combination of Observations" is an old term for the numerical analysis of data. In a sense, the words "combination of data" encompass the whole of the statistical analysis of data, with an implied emphasis on condensation and summarization, two important general themes. The present article, however, concentrates on the combination of information from separate sets of data, each of which has a viable analysis in its own right. What issues arise in putting together conclusions from separate analyses?

Such combination can arise in two rather different ways. First, the different sets of data may be obtained from quite distinct investigations, possibly even using dissimilar designs and different experimental techniques. Secondly, a sensible strategy for handling data of relatively complex structure is often to divide the data into simpler subsections, to analyze these separately, and then, in a second stage of analysis, to merge the conclusions from the component analyses. While the technical statistical problems in these two kinds of application may be identical, assumptions of homogeneity, for example of error variance, may be made with greater confidence in the second situation and the need to produce a single final conclusion is greater.

In general, the drawing together and comparison of interrelated information from different sources is very important. Often, however, it is enough to proceed qualitatively.

WEIGHTED MEANS

Suppose first that a parameter θ of clear interest can be estimated from m independent sets of data. Thus θ might be a mean, a difference of means, a factorial contrast*, a log odds ratio* in a comparison of binomial probabilities, etc., and the m sets might correspond to m independent studies by different investigators. Let $t_1, \ldots, t_m$ be the estimates, assumed for the moment normally distributed around θ with known variances $v_1, \ldots, v_m$, calculated from the internal variability within the separate sets of data.

If these assumptions are a reasonable basis for the analysis, the "best" combined estimate of θ is

$$\tilde{t} = (\sum t_j/v_j)/(\sum 1/v_j); \qquad (1)$$

this is called a weighted mean of the t_j's and $1/v_j$ is called the weight attached to t_j. Then $\tilde{t}$ is normally distributed around θ with standard error

$$(\sum 1/v_j)^{-1/2}, \qquad (2)$$

from which confidence limits for θ can be calculated. Equation (1) is summarized by the slogan "Weight a value inversely as its variance."

Contrast (1) with the unweighted mean,

$$\bar{t} = (\sum t_j)/m, \qquad (3)$$

the two formulae being the same if and only if all the variances v_j are equal. It can be shown that the general weighted mean, with weights w_j,

$$\tilde{t}_w = \sum w_j t_j / \sum w_j, \qquad (4)$$

has variance $\sum w_j^2 v_j/(\sum w_j)^2$, close to that of its minimum $(\sum 1/v_j)^{-1}$, achieved for $\tilde{t}$, for a fairly wide range of weights. Thus the choice of weights is usually not critical provided that extreme weights are avoided. This is some comfort when there is doubt about the v_j.

A number of assumptions are made in (1) and (2). The principal one is that the parameter θ is indeed the same for all sets of data. It is an important general principle that, before merging information from different sources, mutual consistency should be checked. Sometimes informal inspection is enough, but if a formal test is required,

$$\tilde{d} = \sum v_j^{-1}(t_j - \tilde{t})^2$$
$$= \sum t_j^2/v_j - (\sum t_j/v_j)^2(\sum 1/v_j)^{-1} \qquad (5)$$

can be treated as chi-squared* with $m - 1$ degrees of freedom.

Incidentally, the quickest derivation of (1), (2), and (5), and their numerous generalizations, is via a standard least-squares analysis of the linear model

$$E(t_j/\sqrt{v_j}) = (1/\sqrt{v_j})\theta, \qquad \text{var}(t_j/\sqrt{v_j}) = 1. \qquad (6)$$

Then (1) is the "ordinary" least-squares* estimate of θ and (5) is the "ordinary" residual sum of squares corresponding under the model to a true variance of 1.

Too extreme a value of (5) suggests that something is wrong; usually, it will be that large values of (5) are obtained, indicating either that the v_j underestimate the real error in the t_j or that $E(t_j)$ is not constant.

POOLING* IN THE PRESENCE OF OVERDISPERSION

Suppose that (5) shows that the separate estimates t_j differ by more than they should under the assumptions of homogeneity outlined above (1). Here are some possible next steps:

1. It may happen that the v_j are based on unrealistic assumptions about the individual sets of data. Revision of the val-

ues of the v_j will in general give a new estimate (1), although as noted above, minor adjustments to weights are usually unimportant. If, however, all the v_j are assumed in error by an unknown common factor, the estimate (1) is still optimal, although its standard error (2) has to be multiplied by $\sqrt{\{\tilde{d}/(m-1)\}}$, a reasonable thing to do only if m is not very small. This assumption about variances may sometimes be reasonable in connection both with superficially Poisson or binomial variation and also with cluster sampling*.

2. Interest may switch from providing a single estimate to a study of the differences between the individual estimates. Such a move led to the discovery of the inert gases.

3. It may be reasonable to suppose that the property measured by θ is the same for all sets of data, but that some or all of the determinations are subject to systematic errors. For example, θ may be a fundamental constant, such as the velocity of light, or the rate constant of a well-defined chemical reaction. Then

$$E(t_j) = \theta + \Delta_j, \qquad (7)$$

where Δ_j is an unknown constant. Sometimes bounds for the Δ_j, can be calculated from detailed examination of the measurement techniques. We can, by adding suitable constants to the t_j's, take such bounds in the form $|\Delta_j| \leqslant a_j$. Then for the weighted mean $\tilde{t}_w = \sum w_j t_j / \sum w_j$, cautious $1 - 2\epsilon$ confidence limits are

$$-\frac{\sum w_j a_j}{\sum w_j} - k_\epsilon^* \frac{\left(\sum w_j^2 v_j\right)^{1/2}}{\sum w_j} + \tilde{t}_w,$$

$$\frac{\sum w_j a_j}{\sum w_j} + k_\epsilon^* \frac{\left(\sum w_j^2 v_j\right)^{1/2}}{\sum w_j} + \tilde{t}_w, \qquad (8)$$

where k_ϵ^* is the upper ϵ-point of the standardized normal distribution. The adjustment for systematic error covers the possibility that all systematic errors are simultaneously extreme. For given a_j, v_j, and ϵ, the w_j's can be chosen to minimize the width of the interval (8); in the extreme case where systematic errors predominate, concentrate on the estimates with smallest a_j.

Some of the sets of data, for example "high-quality" determinations, have, or can be assumed to have, or can even be defined to have, zero Δ_j. For example, some sets of data might compare a new treatment with a concurrent control, with due randomization*, whereas other sets might use historical controls*. One reasonable procedure is then as follows. Examine the "bias-free" estimates for mutual consistency. Subject to this, pool them by (1) to give $\tilde{t}_{BF}$. Then examine the remaining t_j for consistency with $\tilde{t}_{BF}$, pooling into $\tilde{t}_{BF}$ those clearly consistent with it. The remaining values are studied and if possible the biases in them explained, for example by adjustment for discrepant concomitant variables. Care is needed in case, for example, all the possibly biased values in fact err in the same direction. A final statement of conclusions should include both $\tilde{t}_{BF}$ and the final pooled estimate.

If the Δ_j can be modeled as random variables, analyses similar to those discussed in step 6 below become available, but such assumptions about systematic errors should be treated with due scepticism. An assumption that systematic errors in different sections of the data are independent random variables of zero mean deserves extreme scepticism.

4. Inspection, or more detailed analysis, may show that the parameter θ is inappropriately defined. Thus θ might be taken initially as the difference in mean response between two treatments. Yet the data may show that while the difference between means varies between the sets of data, the ratio of means is stable. This suggests redefining θ before combination.

5. It may be necessary to abandon the notion of a single true θ, supposing then

that

$$E(t_j) = \theta_j, \qquad (9)$$

where θ_j is the parameter value of interest in the jth set of data. Note that although (9) is superficially similar to (7), the interpretation is different, in that in (7) only θ is really of concern. If m is very small (e.g., 2) and the discrepancy cannot be explained, it may be best to abandon the notion of combining the sets of data. With larger values of m one would normally first attempt an empirical explanation of the variation of θ_j. For example, if z_j is a scalar or vector explanatory variable attached to the whole of the jth set of data (e.g., a characteristic of the center or laboratory in which the jth set of data is obtained), one might hope that

$$\theta_j = \theta + \boldsymbol{\beta}^T(\mathbf{z}_j - \mathbf{z}_0), \qquad (10)$$

where θ is the value at some reference level $\mathbf{z}_0$, possibly $\bar{\mathbf{z}} = \sum \mathbf{z}_j/m$, and $\boldsymbol{\beta}$ is a vector of unknown parameters to be estimated, for example via the generalization of (6).

A less ambitious representation is

$$\theta_j = \theta + \boldsymbol{\beta}^T(\mathbf{z}_j - \mathbf{z}_0) + \eta_j, \qquad (11)$$

where $\eta_1, \ldots, \eta_m$ are independent random variables of zero mean and variance σ_η^2, in which only a portion of the variability of θ_j can be explained. Equations (10) and (11) lead in effect to empirically estimated "corrections" to apply to the t_j to adjust to the reference level, $\mathbf{z} = \mathbf{z}_0$.

Another possibility is that variation among the θ_j can be explained by one, or a small number, of anomalous data sets. Then the inconsistent set or sets can be studied separately and the remainder combined by (1). Occasionally, it may be wise to replace the mean (1) by some robust* estimate of location in which isolated extreme values are automatically discounted.

6. Finally, in the absence of a specific explanation of the variation in θ, it may

sometimes be taken as random; i.e., we may take a representation

$$\theta_j = \theta + \eta_j, \qquad (12)$$

where $\eta_1, \ldots, \eta_m$ are, as in (11), independent random variables normally distributed with zero mean and variance σ_η^2. This is equivalent to supposing $t_1, \ldots, t_m$ to be independently normally distributed around θ with variances $v_1 + \sigma_\eta^2, \ldots, v_m + \sigma_\eta^2$. For the moment assume, usually quite unrealistically, that σ_η^2 is known. Then the discussion of (1) and (2) holds, with changed variances, so that the weighted mean (4) is indicated with $w_j = (v_j + \sigma_\eta^2)^{-1}$. Note that the new mean is intermediate between the original weighted mean $\tilde{t}$ of (1) and the unweighted mean $\bar{t}$ of (3). Of course, if the v_j are all equal, the estimate is unchanged, although its standard error now includes a contribution from the variation between sets.

Representation (12) is a version of a components of variance* or random-effects model, unbalanced if the v_j are unequal, balanced if they are equal. The difference from the standard form of such models is that the "within-group" variances v_j are possibly different but assumed known, i.e., estimated with relatively high precision. Maximum likelihood estimation* of θ and σ^2 from $t_1, \ldots, t_m$ is straightforward. The maximum likelihood estimate $\hat{\theta}$ is a weighted mean with weights obtained by replacing σ_η^2 by its maximum likelihood estimate $\hat{\sigma}_\eta^2$. Asymptotically, $\hat{\theta}$ and $\hat{\sigma}_\eta^2$ are independent.

Alternatively, σ_η^2 can be estimated by equating d of (5) to its expectation, or indeed by equating other quadratic forms*, such as $\bar{d} = \sum (t_j - \bar{t})^2$, to their expectations. Of course, m must not be too small or this will be ineffective. Having estimated σ_η^2 in some such way, we may use a weighted mean with estimated weights.

The use of a random-effects representation (12) needs careful consideration. If the sets of data arise say from different centers or laboratories or batches of material randomly sampled from a well-defined universe of centers or laboratories or batches, a model of this broad type is appropriate; θ can be regarded as an average applying to the universe in question. Unfortunately, such clear-cut justification and interpretation is relatively rare. More commonly, we have a number of estimates that are inexplicably different and if, as is frequently the case, a single estimate is required, we have little choice but to assume that the additional variation is in some loose sense random and to adopt (12): note that (12) is in effect a definition of the target parameter θ as the value achieved when this additional source of variability averages out to zero, a somewhat hypothetical construct.

SUMMARY DISCUSSION

If the separate estimates are mutually consistent, to merge results, calculate a weighted mean. If the separate estimates are not consistent in the light of the internal estimates of error, explain the discrepancies, if possible. If this is not possible and merging is desirable, with due caution treat the superfluous variation as random. If m is reasonably small, it will usually be desirable to report the individual estimates as well as the pooled one.

RELATION WITH SIGNIFICANCE TESTS

The discussion above has concentrated on estimation rather than significance testing, in the belief that this is normally more fruitful. If a particular value of θ, say θ_0, is of special interest this can be examined via the final estimate and its standard error.

The practice of testing the hypothesis $\theta = \theta_0$ separately for each set of data, dividing the sets into two sections, those consistent with θ_0 and those inconsistent at some selected level, is in general bad. If, in fact, all sets of data correspond to the same value of θ deviating modestly from θ_0, quite mis-

leading conclusions result from such a split of the data.

There is sometimes the following related difficulty of interpretation. Suppose that essentially the same topic is investigated independently in a number of centers and the hypothesis $\theta = \theta_0$ tested in each: the hypothesis might, for example, correspond to a new treatment being equivalent to a control. Suppose that one of the centers finds significant evidence against $\theta = \theta_0$ and the others do not. It is clear that all centers should report their results. Next suppose that the results from the different centers are mutually consistent and that the pooled estimate leads to confidence limits for θ containing and sharply concentrated around θ_0. What should be concluded?

The significance level attaching to the results from the center reporting a significant effect is concerned with what can reasonably be concluded from that center's results in isolation. Because interpretation should be based, as far as reasonably feasible, on all relevant available information, this one significance level is unimportant when all the data become available, unless strong rational arguments can be produced for singling out the one center as different in some crucial respect from the others. Of course, it may be hard to get agreement on this last point! The morals are partly the need to report all results significant or not, partly the desirability of avoiding undue emphasis on significance tests and partly the importance of assessing all available relevant information.

MORE TECHNICAL COMMENTS

The discussion above has focused on general ideas, and by concentrating on the simplest situations in which the t_j are normally distributed with known variances, detailed formulae have been largely avoided. Broadly similar problems arise when the t_j have specific nonnormal distributions and when there are nuisance parameters present, in particular in normal theory when the variances of the t_j are unknown, but can be estimated.

Theoretical discussion of such problems can be hard. For example, suppose that $t_1, \ldots, t_m$ are independently normally distributed with mean θ and unknown variances; let these variances be estimated by $v_1, \ldots, v_m$, which are now normal theory estimates of variance with $f_1, \ldots, f_m$ degrees of freedom, independently of one another and of the t_j. Thus if t_j is the mean of a random sample of n_j observations, $f_j = n_j - 1$ and v_j is the usual sample estimate of population variance divided by n_j.

The simplest solution is to use (1), which we now call a weighted mean with empirically estimated weights. This is satisfactory if all the degrees of freedom are reasonably large (e.g., provided that none is smaller than 10). If m is large and some or all of the f_j are small, an asymptotically efficient estimate is constructed by solving for $\tilde{\theta}$ the estimating equation

$$\sum \frac{(f_j - 1)(t_j - \tilde{\theta})}{f_j v_j + (t_j - \tilde{\theta})^2} = 0 \qquad (13)$$

In principle, exact confidence limits can be derived. Note that (13) has the odd feature that samples with $f_j = 1$, and indeed those with $f_j = 0$, do not contribute. If $f_j \gg 1$ and the second term in the denominator of (13) is ignored, (13) defines the weighted mean (1).

The formal optimal properties of $\tilde{\theta}$ arise from a limiting operation as $m \to \infty$, involving a model with $m + 1$ parameters. Standard maximum likelihood theory is inapplicable and indeed (13) is not quite the maximum likelihood estimating equation, which, when the t_j are means of random samples and the v_j corresponding estimated variances, has $(f_j + 1)$ instead of $(f_j - 1)$ in the numerator. Models with large numbers of incidental parameters are in any case best avoided, and if the variation in the true variances cannot be rationally explained, a reasonable procedure for large m would often be to take the true variances as random

variables (i.e., to have a random-effects model for the variances). Under suitable special assumptions this leads to a modification of (13) in which the weights are shrunk toward their mean.

POOLING OF DATA AND EMPIRICAL BAYES* ESTIMATION

A different application may be made of the random-effects model (12). Instead of combining data to reach conclusions about θ, one may wish to study θ_m, the parameter value in one of the sets, taken without loss of generality to be the last. In particular, if the sets are ordered in time, we may wish to study the current set, making some use of historical data.

Because θ_m is a random variable, Bayesian arguments are applicable, provided that the prior distribution is known or can be estimated. With the new focus of interest, (12) is called an empirical Bayes model. If θ, v_m, and σ_η^2 are all known, and normality assumptions hold, the posterior distribution is normal with mean

$$\frac{t_m/v_m + \theta/\sigma_\eta^2}{1/v_m + 1/\sigma_\eta^2} \qquad (14)$$

and variance

$$\left(1/v_m + 1/\sigma_\eta^2\right)^{-1}. \qquad (15)$$

This involves the data only through t_m, and (14) is formally a weighted mean of two "observations," t_m and θ with variances v_m and σ_η^2, respectively. In parametric empirical Bayes estimation θ, σ_η^2, and, if necessary, v_m are replaced by estimates, so that all the data become involved. The proportional "shrinkage"* of even extreme values in (14) has to be viewed critically because it depends strongly on the normality of η_m in (12).

More elaborate representations of the η_j's, for example involving serial dependence, may be useful.

COMBINATION OF DATA ABOUT NUISANCE PARAMETERS*

Sometimes, for example in repetitive routine tests, sets of data are obtained each with its own value of the parameter, θ, of interest (e.g., a mean). The possibility of an empirical Bayes analysis of the θ's then arises. Suppose that this is undesirable, for example because of doubts about an appropriate form of prior distribution. It may nevertheless be useful to use previous information about a nuisance parameter, for example a variance, especially if this is estimated poorly from each individual set of data. Such an approach is sometimes called partially empirical Bayes. Detailed formulae will not be given here. An extreme case is when a fixed "historical" value is used for, say, the variance, the "fixed" value preferably being updated from time to time.

Even in contexts, such as agricultural field trials, with satisfactory internal estimates of error, it will be wise to compare qualitatively with historical values. Often the coefficient of variation of yield for a particular crop and plot size will be known roughly, and comparison with this gives some check on reliability.

MORE COMPLEX CONTEXTS

The discussion so far has assumed one or more clearly defined parameters of interest. A conceptually more difficult problem arises if we have a number of sets of similar data and wish to describe as concisely as possible the common features of and differences between the sets.

First, if separate fits are made to the individual sets of data, it is normally desirable, unless there are specific arguments to the contrary, to fit all sets in compatible fashion. To take a simple example, suppose that a regression of y on x is fitted to each set and that judged in isolation some sets are satisfactorily fitted by a straight line and that others need a quadratic. Then for comparative purposes a quadratic should be fitted to

all sets. As pointed out above in a slightly different context, it is possible that all sets are consistent with the same modest curvature.

Having fitted the separate sets in some compatible fashion, we look in principle for a parameterization in which as many component parameters as possible are the same for all sets of data. Thus, if the data are not totally homogeneous, one might hope that all except one of the component parameters are constant and that the variation in the anomalous component can be explained or modeled as in the section "Pooling in the Presence of Overdispersion". For example, if separate straight lines are fitted to the separate sets of data, are the lines either parallel or concurrent? Canonical regression analysis might be applied to the estimates in the original parameterization to discover combinations that are nearly constant, although often a more informal approach will do.

SERIES OF EXPERIMENTS OF SIMILAR DESIGN

A special problem that has been studied in considerable detail is the combination of data from a series of experiments of similar or even identical design, methods based on analysis of variance being appropriate. For simplicity, suppose that the same design, say in b randomized blocks, is used to compare t treatments in m "places." Very often the treatments will have factorial structure.

For one set of data, the analysis of variance has the form shown in Table 1a, where a decomposition of treatment comparisons into component contrasts is hinted at. Subject to homogeneity of error, which can be examined, the composite data have the conventional analysis of variance shown in Table 1b. Note that the treatments $\times$ places interaction can be partitioned as in Table 1c.

For a particular single-degree-of-freedom contrast, the analysis suggested in the first two sections corresponds in effect to Table 1c. That is, the constancy of a particular

Table 1   Some Analysis-of-Variance Tables for a Series of Similar Experiments: (a) Single Place; (b) Several Places; (c) Several Places, with Decomposition of Interaction[a]

(a)		(b)		(c)
Mean	1	Mean	1	
		Places	$m - 1$	
Blocks	$b - 1$	Blocks within places	$m(b - 1)$	As for
Treatments	$t - 1 : C_1$	Treatments	$t - 1 : C_1$	(b)
	$\vdots$		$\vdots$	with
	C_k		C_k	Treatments $\times$ places:
		Treatments $\times$ places		$C_1 \times$ places
			$(t - 1)(m - 1)$	$\vdots$
Residual	$(b - 1)(t - 1)$	Residual	$m(b - 1)(t - 1)$	$C_k \times$ places

[a] $C_1, \ldots, C_k$ treatment contrasts.

contrast C is examined by comparing the component of interaction $C \times$ places with error. If interaction is present, cannot be explained, and a random-effects representation is adopted, the standard error of the overall mean contrast is estimated via the interaction mean square, $C \times$ places.

An alternative is to use the analysis of Table 1b, in which homogeneity of the interaction mean squares is assumed. If m is small, one must use either this or something intermediate between Tables 1b and 1c. In a general way, it is likely that larger effects will show larger interactions and hence, for example, treatment main effects $\times$ places are likely to be bigger than high-order-treatment interactions $\times$ places, so that the analysis of Table 1c is preferable when enough degrees of freedom are available.

There are numerous generalizations of the above, for example to experiments with different error variances and different designs. If a second factor (e.g., times) is included with levels regarded as random, the error variance for treatment contrasts has to be estimated by synthesis of variance.

DIVISION OF DATA FOR ANALYSIS

Most of the previous discussion assumes a number of independent sets of data for pos-

sible comparison. As indicated in the introduction, the sets may arise, not from separate investigations, but as part of a plan of statistical analysis in which the data are divided into rational sections for separate analysis and ultimately combination.

It is impossible to say at all precisely when such split analysis is advisable. Computational considerations may demand it. The investigators' lack of experience with complex techniques of analysis may make splitting an aid to comprehension. Lack of homogeneity, especially with time series data, may be best detected by splitting. Finally, a direct empirical demonstration of reproducibility of conclusions can often be nicely displayed by splitting the data.

Bibliography

Cochran, W. G. (1954). *Biometrics*, **10**, 101–129.

Cochran, W. G. and Cox, G. M. (1957). *Experimental Design*, 2nd ed. Wiley, New York. (Chapter 14 concerns series of experiments.)

Cox, D. R. (1975). *Biometrika*, **62**, 651–654.

James, G. S. (1956). *Biometrika*, **43**, 304–321. (Theoretical discussion.)

Patterson, H. D. and Silvey, V. (1980). *J. R. Statist. Soc. A*, **143**. (Discussion of specific application and numerous practical complications.)

Yates, F. and Cochran, W. G. (1938). *J. Agric. Sci.*, **28**, 556–580. (General discussion, methods, and practical examples.)

(ANALYSIS OF VARIANCE INTERACTION)

D. R. Cox

COMBINATORICS

The oldest combinatorial problems of any importance are connected with the notion of binomial coefficients. For any a the kth $(k = 1, 2, \ldots)$ binomial coefficient is defined as

$$\binom{a}{k} = \frac{a(a-1) \cdots (a-k+1)}{k!} = \frac{a^{(k)}}{k!},$$

and $\binom{a}{0} = 1$.

The notion of binomial coefficients originates from that of figurate numbers F_n^k $(n \geqslant 0, k \geqslant 1)$. The numbers $F_0^k, F_1^k, \ldots$ can be obtained from the sequence $F_0^1 = 1, F_1^1 = 1, \ldots$ by repeated summations, as illustrated below:

$$F_n^1: \quad 1, 1, 1, 1, 1, 1, \ldots$$

$$F_n^2: \quad 1, 2, 3, 4, 5, 6, \ldots$$

$$F_n^3: \quad 1, 3, 6, 10, 15, 21, \ldots.$$

Here $F_n^1 = 1$ for $n \geqslant 0$ and $F_n^{k+1} = F_0^k + F_1^k + \cdots + F_n^k$ for $k \geqslant 1$ and $n \geqslant 0$. The formation of triangular numbers, $F_n^3 (n \geqslant 0)$, goes back to Pythagoras (ca. 580–500 B.C.). Pyramidal numbers, $F_n^4 (n \geqslant 0)$, and others were studied by Nicomachus of Gerasa, who lived in the first century. (see Dickson [16]). Obviously, $F_n^k = \binom{n+k-1}{n}$ and F_n^k can be interpreted as the number of different ways in which n pearls (or other indistinguishable objects) can be distributed in k boxes.

It has been known for a long time that n coins can be arranged in 2^n ways in a row if each coin may be either head up or tail up. If k coins are head up and $n - k$ coins are tail up, the number of possible arrangements is $\binom{n}{k}$. In a different context and for $n \leqslant 6$, the foregoing results are in a 3000-year-old Chinese book, "I Ching" (Book of Changes). A tenth-century commentator, Halāyudha, derived the numbers $\binom{n}{k}$ $(0 \leqslant k \leqslant n)$ from a passage in the manuscript "Chandah-sūtra"

by Piṅgala (ca. 200 B.C.), a Hindu writer on metrical forms.

The numbers $\binom{n}{k}$, $0 \leqslant k \leqslant n$, gained importance when it was recognized that they appear as coefficients in the expansion of $(1 + x)^n$. Apparently, this was known to Omar Khayyam, a Persian poet and mathematician who lived in the eleventh century, (see Woepcke [49]). He did not give a law, but in his "Algebra" (ca. 1100) he mentioned that the law can be found in another work by him, which work, if it ever existed, has been lost. In 1303, Shih-chieh Chu [39] refers to the numbers $\binom{n}{k}$, $0 \leqslant k \leqslant n$, as an old invention and he mentions several surprising identities for these numbers (see Mikami [27], Needham [29], and Takács [43]). The numbers $\binom{n}{k}$, $0 \leqslant k \leqslant n$, arranged in the form of a triangular array, appeared in 1303 at the front of the book of Shih-chieh Chu. Unfortunately, the original book was lost, but in 1839 Shin-lin Lo discovered a copy made in Korea in 1660. The triangular array first appeared in print in 1527 on the title page of the book of P. Apianus (see Smith [40, p. 509]). In 1544, M. Stifel showed that in the binomial expansion

$$(1 + x)^n = \sum_{k=0}^{n} C_n^k x^k,$$

the coefficients $C_n^k (0 \leqslant k \leqslant n)$ can be obtained by the recurrence equation $C_{n+1}^{k+1} = C_n^{k+1} + C_n^k$, where $C_n^0 = C_n^n = 1$ for $n = 0, 1, 2, \ldots$. He arranged the coefficients C_n^k $(0 \leqslant k \leqslant n)$ in the following triangular array, which is known as the arithmetic triangle:

$$
\begin{array}{ccccccccc}
 & & & & 1 & & & & \\
 & & & 1 & & 1 & & & \\
 & & 1 & & 2 & & 1 & & \\
 & 1 & & 3 & & 3 & & 1 & \\
1 & & 4 & & 6 & & 4 & & 1 \\
\end{array}
$$

In 1556, N. Tartaglia claimed this triangular array as his own invention.

The numbers $C_n^k (0 \leqslant k \leqslant n)$ appeared in the seventeenth century in connection with combinations. The number of combinations without repetition of n objects taken k at a time can be expressed as C_n^k. In 1634, P.

Hérigone knew that $C_n^k = n(n - 1) \cdots (n - k + 1)/k!$. The same formula appeared also in 1665 in a treatise by B. Pascal* [33].

In those early days no mathematical notation was used for the binomial coefficients. It seems that L. Euler (1707–1783) was the first (1781) who used $(\frac{a}{k})$ and later $[\frac{a}{k}]$ for $(\frac{a}{k})$. The notation $(\frac{a}{k})$, which is a slight modification of Euler's first notation, was introduced in 1827 by A. V. Ettingshausen.

The combinatorial formula $C_n^k = \binom{n}{k}$ gained significance in the middle of the seventeenth century, when P. Fermat* (1601–1665) and B. Pascal (1623–1662) solved the classical problem of points* (division problem). In this problem two players A and B play a series of games. In each game, independently of the others, either A wins a point with probability $\frac{1}{2}$, or B wins a point with probability $\frac{1}{2}$. The players agree to continue the games until one has won a predetermined number of games. However, the match has to stop when A still needs a points and B still needs b points to win the series. In what proportion should the stakes be divided?

According to Ore [31] it seems likely that the problem is of Arabic origin. He found some particular versions of this problem in Italian manuscripts dating from as early as 1380. The problem appears for the first time in printed form in 1494 in the book of Lucas de Burgo Pacioli [32, p. 197]. In Pacioli's version $a = 1$ and $b = 3$. L. Pacioli (1494), N. Tartaglia (1556), F. Peverone (1558), and others tried to solve this problem without success. In 1654, Antoine Gombauld Chevalier de Méré, a distinguished philosopher and a prominent figure at the court of Louis XIV, called the attention of Blaise Pascal to the problem of points. It seems that Pascal started working on a solution and communicated the problem to Pierre de Fermat. In reply, Fermat found a remarkably elegant solution. He demonstrated that

$$P(a, b) = \frac{1}{2^{a+b-1}} \sum_{k=a}^{a+b-1} \binom{a+b-1}{k}$$

is the probability that A wins the series. As a second solution, Fermat also proved that

$$P(a, b) = \sum_{n=a-1}^{a+b-2} \binom{n}{a-1} \frac{1}{2^{n+1}}.$$

In the meantime Pascal discovered the recurrence formula

$$P(a, b) = \tfrac{1}{2} P(a - 1, b) + \tfrac{1}{2} P(a, b - 1)$$

and found the same probabilities as Fermat.

The term *combinatorics* was introduced by G. W. Leibniz (1646–1716) in 1666 [23] and he gave a systematic study of the subject. The basic combinatorial problems are concerned with the enumeration of the possible arrangements of several objects under various conditions. The number of ordered arrangements of n distinct objects, numbered $1, 2, \ldots, n$, without repetition is $n! = 1 \cdot 2 \cdots n$. Each such arrangement is called a permutation without repetition. If only k objects are chosen among n distinct objects, numbered $1, 2, \ldots, n$, the number of ordered arrangements without repetition is $n(n - 1) \cdots (n - k + 1)$, and each arrangement is called a variation without repetition. The number of unordered arrangements of k objects chosen among n distinct objects, numbered $1, 2, \ldots, n$, is $\binom{n}{k}$, and each arrangement is called a combination of size k of n elements without repetition. Permutations, variations, and combinations with repetition are defined in a similar way.

In about 1680 [4] Jacob Bernoulli* (1654–1705) found that the probability that in n independent and identical trials, a given event, which has probability p, occurs exactly k times is given by

$$P(n, k) = \binom{n}{k} p^k (1 - p)^{n-k}$$

for $k = 0, 1, \ldots, n$.

In 1693, John Wallis (1616–1703) published a textbook of algebra [48] and devoted a considerable part of the book to combinatorial problems.

At the beginning of the eighteenth century the applications of combinatorial methods expanded with the rapid growth of probability theory. The works of P. R. Montmort (1678–1719), Nicolas Bernoulli (1687–1759), and A. De Moivre* (1667–1754) contain many significant contributions to combinatorics. For a detailed account of their works, see Todhunter [46]. Here are a few samples.

In 1708, A. De Moivre studied the problem of duration of plays in the theory of games of chance and obtained the following significant result in combinatorics: One can arrange $k + j$ letters A and j letters B in

$$L(j, k) = \frac{k}{k + 2j} \binom{k + 2j}{j}$$

ways such that for every $r = 1, 2, \ldots$, $k + 2j$ among the first r letters, there are more A's than B's.

In 1713, P. R. Montmort [28] studied the "problème de rencontre" and obtained an important result in combinatorics. Denote by $Q(n, k)$ the number of permutations of the elements $1, 2, \ldots, n$ in which exactly k matches occur. If the ith element of a permutation of $1, 2, \ldots, n$ is i, then it is said that there is a match at the ith place. Montmort proved that $Q(n, k) = \binom{n}{k} Q(n - k, 0)$ and if $Q(n) = Q(n, 0)$, then $Q(n + 2) = [Q(n + 1) + Q(n)](n + 1)$, where $Q(0) = 1$ and $Q(1) = 0$. Hence it follows that

$$Q(n) = n! \sum_{i=0}^{n} \frac{(-1)^i}{i!}.$$

In 1708 A. De Moivre, and in 1710 P. R. Montmort, discovered that the number of ways in which one can obtain a total number of k points by throwing n dice is

$$C(n, k) = \sum_{j=0}^{[(k-n)/6]} (-1)^j \binom{n}{j} \binom{k - 1 - 6j}{n - 1}$$

if $n \leqslant k \leqslant 6n$. This problem had already been studied by G. Cardano (1501–1576), and Galileo (1564–1642) for $n = 2$ and $n = 3$. In 1730, A. De Moivre [14] proved the foregoing formula by observing that $C(n, k)$ is the coefficient of x^k in the expansion of $(x + x^2 + \cdots x^6)^n$. Today De Moivre's method is called the method of generating functions*. In the hands of L. Euler (1707–1783), J. L. Lagrange (1736–1813), and P. S. Laplace* (1749–1827), the method of generating functions became a powerful tool of combinatorics.

In 1730, J. Stirling (1692–1770) introduced the numbers $S(n, k)$ $(0 \leqslant k \leqslant n)$ and $\mathfrak{S}(n, k)$ $(0 \leqslant k \leqslant n)$ defined by the equations

$$x^{[n]} = \sum_{j=0}^{n} S(n, j) x^j$$

and

$$x^n = \sum_{j=0}^{n} \mathfrak{S}(n, j) x^{(j)}.$$

(See DIFFERENCE OF ZERO.) The numbers $S(n, k)$ $(0 \leqslant k \leqslant n)$ and $\mathfrak{S}(n, k)$ $(0 \leqslant k \leqslant n)$, which are called Stirling numbers* of the first kind and second kind, respectively, have significant importance in combinatorics (see Ch. Jordan [21]).

In 1736, L. Euler introduced some numbers $A(n, k)$ $(0 \leqslant k < n)$ which are now called Eulerian numbers. In 1883, J. Worpitzky demonstrated that

$$x^n = \sum_{k=0}^{n-1} A(n, k) \binom{x + k}{n}$$

for $n \geqslant 1$ and for every x. In an $n \times n$ triangular grid in which the ith row contains i squares $(i = 1, 2, \ldots, n)$, one can put a mark in a square in each row of the grid in $A(n, k)$ ways so that there are exactly k empty columns (see Takács [44]).

In 1908, P. A. MacMahon, by solving a problem of Simon Newcomb concerning a card game, proved that a deck of n cards numbered $1, 2, \ldots, n$ may be dealt into $k + 1$ piles in $A(n, k)$ ways if cards are placed in one pack as long as they are in decreasing order of magnitude.

In 1879, D. André proved that if $A(n)$ is the number of such permutations i_1, $i_2, \ldots, i_n$ of $1, 2, \ldots, n$ for which $i_1 < i_2$, $i_2 > i_3$, $i_3 < i_4, \ldots$, then

$$\sum_{n=0}^{\infty} A(n) \frac{x^n}{n!} = \sec x + \tan x.$$

Extending the work of A. M. Legendre (1798) on the method of inclusion and exclusion*, in 1867 C. Jordan [20] proved the following general theorem: Let Ω be a finite set and $A_1, A_2, \ldots, A_n$ be subsets of Ω. Denote by H_k the set that contains all those elements of Ω that belong to exactly k sets among $A_1, A_2, \ldots, A_n$. If $N(A)$ denotes the number of elements in any set A, then

$$N(H_k) = \sum_{j=k}^{n} (-1)^{j-k} \binom{j}{k} S_j$$

for $k = 0, 1, \ldots, n$, where $S_0 = N(\Omega)$ and

$$S_j = \sum_{1 \leqslant i_1 < i_2 < \cdots < i_j \leqslant n} N(A_{i_1} \cap A_{i_2} \cap \cdots \cap A_{i_j})$$

for $1 \leqslant j \leqslant n$. In the particular case where $k = 0$, the method of inclusion and exclusion was used by A. M. Legendre (1798), D. A. da Silva (1854), and J. J. Sylvester (1883) in number theory. In combinatorics and in probability theory the method of inclusion and exclusion has been used extensively (see Takács [41]). Such problems as the "problème des ménages," which was proposed by E. Lucas in 1891, have an easy solution by using the method of inclusion and exclusion. The "problème des ménages" is as follows. At a dinner party n married couples are seated in $2n$ seats at a round table according to the following pattern: the women take alternate seats and the men choose the remaining seats. The problem is to determine the number of sitting arrangements in which exactly k men are sitting next to their wives.

Here are a few famous problems that originated in the eighteenth and nineteenth centuries.

In 1751, L. Euler discovered that the number of different ways of dissecting a convex polygon of n sides into $n - 2$ triangles by $n - 3$ nonintersecting diagonals is

$$D_n = \binom{2n - 3}{n - 2} \frac{1}{2n - 3}$$

for $n \geqslant 3$. This formula was proved by J. A. de Segner in 1758. Other proofs and generalizations were given by N. v. Fuss (1793), G. Lamé (1838), E. Catalan (1838), O. Rodrigues (1838), J. Binet (1839), J. A. Grunert (1841), J. Liouville (1843), E. Schröder (1870), A. Cayley (1890), and others. In sev-

eral papers, T. P. Kirkman (1857, 1860) gave far-reaching generalizations of Euler's problem. Some recent developments of the subject are discussed by T. Motzkin (1948), W. T. Tutte (1962, ref. [47]), W. G. Brown (1964; 1965, ref. [10]), and others.

In 1838, E. Catalan proved that the number of ways a product of n factors can be calculated by pairs is D_{n+1}. This result was extended by J. H. M. Wedderburn (1923), I. M. H. Etherington (1937), G. N. Raney (1960), and others.

In the years 1859–1889, A. Cayley solved several combinatorial problems concerning some special graphs* which he appropriately called trees. His results made it possible to determine the number of possible isomers of certain hydrocarbons. Cayley's work was an inspiration to Pólya [34], who in 1937 worked out a general method for the enumeration of trees, graphs, groups and chemical structures (see also de Bruijn [13]).

In 1748, L. Euler proved that if $p(n)$ is the number of ways in which n can be represented as a sum of any number of positive integers, then

$$1 + \sum_{n=1}^{\infty} p(n) x^n = \prod_{n=1}^{\infty} (1 - x^n)^{-1}$$

for $|x| < 1$. By using this equation G. H. Hardy and S. Ramanujan (1918) expressed $p(n)$ in the form of an asymptotic series which made it possible to calculate $p(n)$ for every n. In 1937, H. Rademacher expressed $p(n)$ in the form of a convergent series.

In 1779, L. Euler proposed the following problem. n officers of n different ranks are chosen from each of n different regiments. It is required to arrange them in a square array so that no officer of the same rank or of the same regiment shall be in the same row or in the same column. In the case of $n = 6$, this problem is known as "the problem of the 36 officers." Euler conjectured that the problem of the n^2 officers has no solution if $n \equiv 2$ (mod 4). If $n = 2$, then obviously there is no solution. In 1901, G. Tarry proved that there is no solution for $n = 6$. In 1959, Bose and Shrikhande [6] demonstrated that Euler's conjecture* is false for $n = 22$, and in 1960,

Bose et al. [9] proved that Euler's conjecture is false for $n > 6$. Euler's problem led to the problems of orthogonal Latin squares* and combinatorial designs which have great importance in the theory of design of experiments*.

A particularly important topic is that of balanced incomplete block designs. A balanced incomplete block* design is an arrangement of v distinct objects (varieties) into b blocks (sets) in such a way that each block contains exactly k objects, each object occurs in exactly r different blocks, and every pair of distinct objects occurs together in exactly λ blocks. Such a design is a (b, v, r, k, λ)-configuration. The integers b, v, r, k, λ cannot be chosen arbitarily. They should satisfy the requirements $\lambda > 0$, $v - 1 > k$, $r(k - 1) = \lambda(v - 1)$ and $bk = vr$. The main problem is to find conditions for the existence of various types of configurations and to give methods for the construction of an actual design. The aforementioned problem is of recreational origin and is associated with the triple systems of T. P. Kirkman (1847, 1850) and J. Steiner (1853). Kirkman's 15 schoolgirls problem is famous and it runs as follows. A schoolteacher takes her class of 15 girls on a daily walk. The girls are arranged in 5 rows of 3 each, so that each girl has 2 companions. The problem is to arrange the girls so that for 7 consecutive days no girl will walk with one of her companions in a triplet more than once.

The basic paper on block designs is Bose [5]. Further references are Bose and Shrikhande [7], Fisher [18], Mann [26], Ryser [38], Hall [19], and several others. The book by Dénes and Keedwell [15] is a comprehensive study of Latin squares.

Recreational combinatorial problems are discussed in many books. The two-volume work of Ahrens [1] and the book of Rouse Ball and Coxeter [37] are inspiring. Here is an example for recreational problems. In 1850, F. Nauck proposed the problem of finding the number of ways in which eight queens can be placed on a chessboard so that no queen can take any other. There are 92 solutions which can be obtained from 12 fundamental solutions by rotation and reflection.

In 1901 the first textbook on combinatorics by Netto [30] appeared. The book is an excellent collection of classical problems and methods.

One of the most active researchers in combinatorics at the turn of the century was MacMahon (1854–1929). His two-volume book *Combinatory Analysis* [24] appeared in 1915 and 1916. His 1500-page *Collected Papers* [25] on combinatorics was published in 1978. Symmetric functions*, lattice permutations, and partition problems form the bulk of his work in combinatorics.

In the twentieth century, combinatorics penetrated many fields of mathematics. In probability theory, random walk*, occupancy problems* and fluctuation problems require the use of combinatorial methods. In mathematical statistics, order statistics* and theory of sampling are based on combinatorial methods. In graph theory* and in number theory, combinatorial methods are abundant. Combinatorial geometry is one of the recent fields of mathematics. The outstanding work of Gian-Carlo Rota and his collaborators on the foundations of combinatorial theory amply demonstrates the importance of combinatorics in all fields of mathematics.

Most of the articles in two recent journals, the *Journal of Combinatorial Theory* (started in 1966) and the *Journal of Graph Theory* (started in 1977), are devoted to new developments in combinatorics.

References

[1] Ahrens, W. (1910, 1918). *Mathematische Unterhaltungen und Spiele*, 2nd ed., Vols. 1 and 2. B. G. Teubner, Leipzig.

[2] Andrews, G. E. (1976). "The Theory of Partitions." *Encyclopedia of Mathematics and Its Applications*, Vol. 2. Addison-Wesley, Reading Mass.

[3] Beckenbach, E. F., ed. (1964). *Applied Combinatorial Mathematics*. Wiley, New York.

[4] Bernoulli, J. (1713). *Ars Conjectandi* (Opus posth.). Basel. (Reprint: *Culture et Civilisation*, Bruxelles, 1968.)

[5] Bose, R. C. (1939). *Ann. Eugen. (Lond.)*, **9**, 353–399.

[6] Bose, R. C. and Shrikhande, S. S. (1959). *Proc. Natl. Acad. Sci. USA*, **45**, 734–737.

[7] Bose, R. C. and Shrikhande, S. S. (1960). *Trans. Amer. Math. Soc.*, **95**, 191–209.

[8] Bose, R. C. and Shrikhande, S. S. (1960). *Canad. J. Math.*, **12**, 177–188.

[9] Bose, R. C., Shrikhande, S. S., and Parker, E. T. (1960). *Canad. J. Math.*, **12**, 189–203.

[10] Brown, W. G. (1965). *Canad. J. Math.*, **17**, 302–317.

[11] Cayley, A. (1875). *Rep. Brit. Ass. Adv. Sci.*, pp. 257–305. (Reprinted in *The Collected Mathematical Papers of Arthur Cayley*, Vol. 9. Cambridge University Press, Cambridge, 1896, pp. 427–460.)

[12] Crapo, H. H. and Rota, G.-C. (1970). *On the Foundations of Combinatorial Theory: Combinatorial Geometries*. MIT Press, Cambridge, Mass.

[13] de Bruijn, N. G. (1959). *Indag. Math.*, **21**, 59–69.

[14] De Moivre, A. (1730). *Miscellanea Analytica de Seriebus et Quadraturis*. London.

[15] Dénes, J. and Keedwell, A. D. (1974). *Latin Squares and Their Applications*. Akadémiai Kiadó, Budapest.

[16] Dickson, L. E. (1919). *History of the Theory of Numbers*, Vol. 2. Carnegie Institute, Washington, D.C. (Reprint: Chelsea, New York, 1952.)

[17] Erdös, P. and Spencer, J. (1974). *Probabilistic Methods in Combinatorics*. Akadémiai Kiadó, Budapest.

[18] Fisher, R. A. (1966). *The Design of Experiments*, 8th ed. Oliver & Boyd, Edinburgh.

[19] Hall, M., Jr. (1967). *Combinatorial Theory*. Blaisdell, Waltham, Mass.

[20] Jordan, C. (1867). *C. Acad. Sci. Paris*, **65**, 993–994.

[21] Jordan, Ch. (1939). *Calculus of Finite Differences*. Budapest. (Reprint: Chelsea, New York, 1947.)

[22] König, D. (1936). *Theorie der endlichen und unendlichen Graphen*. Teubner, Leipzig. (Reprint: Chelsea, New York, 1950.)

[23] Leibniz, G. W. (1666). *Dissertatio de Arte Combinatoria*.

[24] MacMahon, P. A. (1915–1916). *Combinatory Analysis*, Vols. 1 and 2. Cambridge University Press, Cambridge. (Reprint: Chelsea, New York, 1960.)

[25] MacMahon, P. A. (1978). *Collected Papers*, Vol 1: *Combinatorics*. MIT Press, Cambridge, Mass.

[26] Mann, H. B. (1949). *Analysis and Design of Experiments*. Dover, New York.

[27] Mikami, Y. (1913). *The Development of Mathematics in China and Japan*. Teubner, Leipzig. (Reprint: Chelsea, New York, 1961.)

[28] Montmort, P. R. (1713). *Essai d'analyse sur les jeux de hazard*, 2nd ed. Paris. (Reprint: Chelsea, New York, 1981.)

[29] Needham, J. (1959). *Science and Civilisation in China*, Vol 3: *Mathematics and the Sciences of the Heavens and the Earth*. Cambridge University Press, Cambridge.

[30] Netto, E. (1901). *Lehrbuch der Combinatorik*. Teubner, Leipzig, (2nd ed.: 1927; Reprint: Chelsea, New York, 1964.)

[31] Ore, O. (1960). *Amer. Math. Monthly*, **67**, 409–419.

[32] Pacioli, L. de B. (1494). *Summa de Arithmetica, Geometria, Proportioni e Proportionalita*. Venice.

[33] Pascal, B. (1665). *Traité du triangle arithmétique*. Paris.

[34] Pólya, G. (1937). *Acta Math.*, **68**, 145–254.

[35] Riordan, J. (1958). *An Introduction to Combinatorial Analysis*. Wiley, New York.

[36] Riordan, J. (1968). *Combinatorial Identities*. Wiley, New York.

[37] Rouse Ball, W. W. and Coxeter, H. S. M. (1974). *Mathematical Recreations and Essays*, 12th ed. University of Toronto Press, Toronto.

[38] Ryser, H. J. (1963). *Combinatorial Mathematics*. Carus Math. Monogr. No. 14. Mathematical Association of America, Washington, D.C.

[39] Shi-chieh Chu (1303). *Ssu Yuan Yü Chien*. (*Precious Mirror of the Four Elements*.)

[40] Smith. D. E. (1925). *History of Mathematics*, Vol. 2. (Reprint: Dover, New York, 1958.)

[41] Takács, L. (1967). *J. Amer. Statist. Ass.*, **62**, 102–113.

[42] Takács, L. (1969). *J. Amer. Statist. Ass.*, **64**, 889–906.

[43] Takács, L. (1973). *Acta Sci. Math. (Szeged)*, **34**, 383–391.

[44] Takács, L. (1979). *Publ. Math. (Debrecen)*, **26**, 173–181.

[45] Takács, L. (1980) *Arch. Hist. Exact Sci.*, **21**, 229–244.

[46] Todhunter, I. (1865). *A History of the Mathematical Theory of Probability from the Time of Pascal to that of Laplace*. Cambridge University Press, Cambridge. (Reprint: Chelsea, New York, 1949.)

[47] Tutte, W. T. (1962). *Canad. J. Math.*, **14**, 708–722.

[48] Wallis, J. (1693). *De Algebra Tractatus*. Oxford. (Reprinted in John Wallis, *Opera Mathematica*, Vol. 2. Georg Olms Verlag, Hildesheim, 1972.)

[49] Woepcke, F. (1851). *L'Algèbre d'Omar Alkâyyami* (French translation). Duprat, Paris.

(FINITE GEOMETRY IN STATISTICS
HYPERGEOMETRIC DISTRIBUTIONS

OCCUPANCY PROBLEMS
URN MODELS)

LAJOS TAKÁCS

COMMITTEE ON NATIONAL STATIS-TICS *See* NATIONAL STATISTICS, COMMITTEE ON

COMMONALITY ANALYSIS

A method of splitting up the variation in a predicted variable X_0, according to contributions from each possible subset of a number of predicting variables $X_1, \ldots, X_m$. It is implicitly assumed that a multiple linear regression model*, with homoscedastic* variation within arrays* is applicable, so that the variance of X_0 given a set $X_{a_1}, \ldots, X_{a_s}$ of predictor variables is proportional to $(1 - P^2_{0 \cdot a_1, \ldots, a_s})$, where $P^2_{0 \cdot a_1 \ldots a_s}$ is the (squared) multiple correlation* of X_0 on $X_{a_1}, \ldots, X_{a_s}$. The proportion of variance of X_0 accounted for by $X_{a_1}, \ldots, X_{a_s}$ is thus $P^2_{0 \cdot a_1 \ldots a_s}$, of which C_{a_i} is the "unique contribution" of X_{a_i} $(i = 1, \ldots, s)$, $C_{a_i a_j}$ is the "common contribution" (*commonality*) of X_{a_i} and X_{a_j}, and $C_{a_i a_j a_k}$ is the commonality of X_{a_i}, X_{a_j}, and X_{a_k}.

The equations

$$P^2_{0 \cdot a_1 \ldots a_s} = \sum_{i=1}^{s} C_{a_i} + \sum \sum_{i<j} C_{a_i a_j} + \ldots$$
$$+ C_{a_1 a_2 \ldots a_s}$$

(for all possible $2^m - 1$ subsets of 1, 2, $\ldots$, m) can be inverted, leading to

$$C_{a_1 \ldots a_s} = \sum_{i=1}^{s} P^2_{0 \cdot a_i} - \sum \sum_{i<j} P^2_{0 \cdot a_i a_j} + \ldots$$
$$+ (-1)^{s+1} P^2_{0 \cdot a_1 \ldots a_s}. \quad (1)$$

(*See* INCLUSION–EXCLUSION PRINCIPLE.)

Newton and Spurrell [5, 6] introduced this technique as "elements analysis." Olkin and Hedges [7] gave approximate formulas for variances of the estimators of the C's obtained by replacing the P's in (1) by sample correlations (r), and use these in forming approximate confidence intervals* for the C's.

The technique has been used especially in educational research (e.g., Coleman et al. [1], Mayeske et al. [3], Mood [4], and Kerlinger and Pedhazur [2]. It differs from principal component analysis* and factor analysis* in that (a) there is no predicted variable in the latter, and (b) the analysis of variation is in terms of combinations of the measured and not linear functions thereof.

References

[1] Coleman, J. S., Campbell, E. Q., Hobson, C. J., McPartland, J., Mood, A. M., Weinfeld, F. D., and York, R. L. (1966). *Equality of Educational Opportunity*. U.S. Government Printing Office, Washington, D.C.

[2] Kerlinger, F. N. and Pedhazur, E. J. (1973). *Multiple Regression in Behavioral Research*. Holt, Rinehart and Winston, New York.

[3] Mayeske, G. W., Wisler, C. E., Beaton, A. E., Weinfeld, F. D., Cohen, W. M., Okada, T., Proshek, J. M., and Tabler, K. A. (1972). *A Study of Our Nation's Schools*. U.S. Government Printing Office, Washington, D.C.

[4] Mood, A. M. (1971). *Amer. Educ. Res. J.*, **8**, 191–202.

[5] Newton, R. G. and Spurrell, D. J. (1967). *Appl. Statist.*, **16**, 51–64.

[6] Newton, R. G. and Spurrell, D. J. (1967). *Appl. Statist.* **16**, 165–172.

[7] Olkin, I. and Hedges, L. V. (1979). The Asymptotic Distribution of Commonality Components. *Tech. Rep. No. 140*, Dept. of Statistics, Stanford University, Stanford, Calif.

(COMPONENT ANALYSIS
CORRELATION
GENERAL LINEAR MODEL
MULTIPLE CORRELATION
MULTIPLE LINEAR REGRESSION
PARTIAL CORRELATION
PRINCIPAL COMPONENT ANALYSIS)

COMMUNALITY

For a variable in a factor analysis* model, the communality is defined as the sum of squares of its factor loadings* over all the factors. These values give an indication of the extent to which the variables overlap with the factors or, equivalently, they give the proportion of the *variance* in the variables that can be accounted by the *scores* in the factors. In many methods of factor extraction the estimates of communality values are inserted in the diagonal cells of the original correlation matrix* before the factor analysis is carried out.

Bibliography

Comrey, A. L. (1973). *A First Course in Factor Analysis.* Academic Press, New York.

(FACTOR ANALYSIS)

COMMUNICATIONS IN STATISTICS

Communications in Statistics is the name of a journal in which articles are published on statistical research and application of statistical methods. The journal is divided into two parts. Part A includes articles on theory and methods; and Part B deals with articles on simulation and computation, including a section for algorithms. Each issue is approximately 100 pages in length, and the number of issues for Part A, which was 12 per year when the journal first started in 1972, grew to 18 per year in 1980, while Part B started with four issues per year in 1976 and grew to six per year in 1980, i.e., to approximately 2400 pages per year for the combined journals.

Articles that have immediate real-world applications are sought for both Part A and Part B. Routine and good theoretical papers that have no apparent application are not acceptable. Outstanding theoretical papers are welcome. The emphasis is on methodology and application. Also, if there is a controversy over the effectiveness or usefulness of a proposed technique, the paper will be published in *Communications in Statistics*, provided that one of the members of the Editorial Board will recommend it, together with a referee. That is, *Communications in Statistics* is available for all legitimate points of view, even if there is a segment of the statistical community that does not approve of the concept being studied.

This journal was started in response to the critical need that was felt in the late 1960s and early 1970s for more rapid publication of articles on statistics. To accomplish this, a different editorial approach was adopted from that of most other statistical journals being published in the United States in the early 1970s. However, the approach is consciously and truly international. A sincere attempt is made to bring the international statistical community together.

In order to understand the difference in approach, a description of the editorial procedures for most other journals will be described. For most journals, the author of a paper will send the paper to the editor of the journal he selects. The editor then sends the paper to an associate editor who has indicated an interest in that particular subject matter. The associate editor picks two or more referees (usually two) and asks them to read the paper. The associate editor waits until both referees report on the paper and then evaluates their reports and makes a recommendation to the editor, usually after reading the paper in conjunction with the referees' reports. Then the editor decides what is to be done with the paper and writes the author accordingly. If the paper is judged poor or not appropriate, that will end its consideration by that journal. All of this typically was taking a year or more in the early 1970s. If the paper is judged acceptable, almost invariably some revision is required, which usually means that the paper goes back through the system to the referees, who read the revision. This can take another six months for a good paper (and longer for marginal papers) before the author knows for sure that the paper will be published. Then there follows a lapse of six or more months while the article is set in type. Hence, on the average two years or more was being consumed from submission to publication and still is being consumed in

1979, although there appears to be some shortening of this time for the journals with which *Communications in Statistics* competes.

One of the goals of *Communications in Statistics* is to shorten this interval. Two steps toward this end were instituted from the first issue. The first step eliminated typesetting and the six-month lag this typically created. Papers are required to be typed by the authors in a strict format which allows direct offset copy to be made from the author's submission. The second step taken is designed to shorten appreciably the refereeing time. Articles may be sent directly to anyone on a large editorial board (over 50 for Part A and over 20 for Part B) who will handle the paper. The editorial board member typically contacts a referee and both eventually agree to have their names added to acceptable papers as recommending the paper for publication. This strengthens the process because eventually the author usually learns who handled his published paper. Hence both referee and editorial board member are more careful about the quality of the paper and about unnecessary delays. The referee and editorial board member may remain anonymous if they request that status, but this is not usually requested after the revision process is complete because almost everyone involved understands the advantage of the new system. An analysis of the effectiveness of the system was made for 1972–1976 by Gunst and Hua [1]. That analysis showed that (from first submission to journal date) for the years 1973–1976 the median publication time for articles submitted to *Communications in Statistics* was nine months after submission, whereas the median publication times for three other leading statistical journals were between 17 and 21 months. Over 75% of the articles accepted by *Communications in Statistics* were published within one year of their first submission. The number of articles published in *Communications in Statistics* in successive years from 1973 to 1976 is, respectively, 86, 109, 93, and 96, exclusive of special issues.

Editorial Board members are invited to edit special issues of *Communications in Statistics* in areas of great interest to themselves. They propose the area to the editor and then either invite people who are working in that particular area to write special survey papers or they have a notice published that a special issue is underway and authors are invited to submit papers especially for the special issue. This had led to some very interesting summaries of the current status of several fields. Special issues and the editors for these issues include the following:

1. "Statistical Education," edited by W. T. Federer, Vol. A5, No. 10, 1976.

2. "Remote Sensing," edited by P. L. Odell, Vol. A5, No. 12, 1976.

3. "Nonparametric Statistics," edited by T. L. Bouillion and T. O. Lewis, Vol. A5, No. 14, 1976.

4. "Robustness," edited by R. V. Hogg, Vol. A6, No. 9, 1977.

5. "Selection and Ranking," edited by S. S. Gupta, Vol. A6, No. 11, 1977.

6. "Computations for Least Absolute Values Estimation," edited by J. E. Gentle, Vol. B6, No. 4, 1977.

7. "Time-Sequential Statistical Procedures," edited by P. K. Sen, Vol. A7, No. 4, 1978.

8. "Theory and Practice for Categorical Data Analysis," by N. Sugiura, Vol. A7, No. 10, 1978.

9. "Optimal Design Theory," edited by A. Hedayat, Vol. A7, No. 14, 1978.

10. "Optimization in Statistics," edited by J. S. Rustagi, Vol. B7, No. 4, 1978.

11. "Covariance," edited by W. T. Federer, Vol. A8, No. 8, 1979.

12. "Weather Modification Experiments," edited by J. J. Wiorkowski and P. L. Odell, Vol. A8, Nos. 10 and 11, 1979.

Reference

[1] Gunst, R. F. and Hua, Tsushung, A. (1978). *Communications in Statistics—Theory and Methods* **A7**(3), 197–209.

D. B. OWEN

COMMUNICATION THEORY, STATISTICAL

Statistical communication theory (SCT) can be characterized as one of the most active and significant areas of contemporary applied mathematics. The achieved and potential practical benefits of the theory, the variety of applications, and the depth of the mathematical problems have combined to draw the attention of numerous engineers, statisticians, and mathematicians. SCT is generally considered to consist of three principal theoretical areas: information theory*, signal detection, and stochastic filtering. Each of these draws heavily on various areas of statistics and probability. In addition, there are major areas of research and applications, such as data communications, which are a mixture of the three SCT disciplines and other disciplines (e.g., optimization*, queueing theory*, and linear systems theory).

It would be impractical to attempt a discussion of all the major problem areas in SCT. For the purposes of this article, attention will be restricted to the first two of the three fundamental areas mentioned above. Stochastic filtering will be omitted, since this topic is discussed elsewhere in this encyclopedia (*see*, e.g., KALMAN–BUCY FILTERING; LINEAR FILTERING). Although information theory is also a separate entry, a review will be given here of some recent research and open problems which have interesting connections with signal detection* and stochastic filtering. In addition, a discussion of probability models as they frequently appear in SCT problems is included.

The bibliography contains a number of references to work in more specialized areas of SCT, such as data communications and various applications of stochastic processes. For example, the reader interested in modulation–demodulation techniques can refer to Lucky [48] for a recent survey.

PROBABILITY MODELS IN COMMUNICATION THEORY

Some SCT problems can be viewed as simply special cases of classical problems in mathematical statistics. However, to do so is frequently incorrect, and (when correct) often misses the point. That is, SCT problems are typically generated by specific practical applications, in which the physical model and constraints play an important role in defining the problem. Moreover, SCT problems are frequently more complicated and require more sophisticated mathematical methods than those typically encountered in classical mathematical statistics. For example, classical multivariate analysis* includes the problem of testing whether an observed sample vector comes from a Gaussian vector process having one of two specified distributions (covariance matrices and mean vectors), and specification of the likelihood ratio. The corresponding signal detection problem frequently considers sample functions on an interval, generated by one of two specified Gaussian processes*. The solution then involves mathematically advanced considerations of absolute continuity* and the structure of the Radon–Nikodym derivative. In order to fully describe such problems, a summary of probability models that frequently arise in SCT applications will now be given.

Typical applications of SCT involve stochastic processes with sample paths belonging to a real separable Banach space; see Hewitt and Stromberg [21] for basic definitions and properties of these spaces. The spaces $L_2[0,1]$, l_2, and $C[0,1]$ are as usually defined [21]. The space $C^{(n)}[0,1]$ is the space of all real-valued functions on $[0,1]$ which have n continuous derivatives; $C^{(n)}[0,1]$ is a real separable Banach space under the norm $\|f\| = \sum_{i=0}^{n} \sup_{t \in [0,1]}$

$\cdot |f^{(i)}(t)|$, where $f^{(i)}(t)$ is the ith derivative of f at the point t.

Let (Ω, β, P) be a probability space and $\{X_t, t \in T\}$ a family of real random variables on (Ω, β); we denote this stochastic process as (X_t). The index set usually represents time or spatial position, although in some applications (such as array processing) T represents both position and time. For ease of exposition, suppose now that $T = [0, 1]$. If (X_t) is a measurable stochastic process with sample paths almost surely (a.s.) in $L_2[0, 1]$, an application of the Fubini theorem shows that (X_t) induces a probability measure μ on the Borel σ-field of $L_2[0, 1]$. $\mu(A) = P\{\omega : X(\omega) \in A\}$, where $X(\omega)$ is the sample path of (X_t) at the point $\omega : X(\omega) = \{X_t(\omega) : 0 \leqslant t \leqslant 1\}$. Similarly, if (X_t) has paths a.s. in $C[0, 1]$, then (X_t) induces a probability measure on the Borel σ-field of $C[0, 1]$.

The assumption that the paths of (X_t) belong a.s. to $L_2[0, 1]$ is typically satisfied in SCT applications. $X(\omega)$ is usually an observed waveform in units of voltage or current, and $\int_0^1 X_t^2(\omega) \, dt < \infty$ is then equivalent to a requirement for finite energy of the sample function $X(\omega)$. A similar consideration holds for l_2 when the index set is the integers. Moreover, when the index set is $T = [0, 1]$, the sample paths of the processes of interest are frequently very smooth; they may be elements of $C^{(n)}[0, 1]$ for some $n \geqslant 1$, due to the presence of various inertia effects in the sensing equipment. If one wishes to analyze the behavior of such a smooth process at specified instants of time, the induced measure on $L_2[0, 1]$ may not be adequate, since the sampling functionals $\{\pi_t, t \in [0, 1]\}$, $\pi_t(X) = X_t$, are not well defined. Thus one may wish to consider the measure induced on the Banach space $C[0, 1]$. Or one may prefer to consider measures induced on $\mathbb{R}^{[0, 1]}$, the space of all real-valued functions on $[0, 1]$.

A measure μ on $L_2[0, 1]$ is said to be Gaussian if for every element y in $L_2[0, 1]$ the distribution F_y is Gaussian, $F_y(\alpha) \equiv \mu\{x : \int_0^1 x_t y_t \, dt \leqslant \alpha\}$. In the case where (X_t) is a measurable Gaussian process with

paths a.s. in $L_2[0, 1]$, the induced measure on $L_2[0, 1]$ can easily be seen to be Gaussian. Similar statements hold for Gaussian processes having sample paths a.s. in l_2, or a.s. in $C^{(n)}[0, 1]$ for some $n \geqslant 0$. Gaussian measures are completely described by their covariance operators and mean elements; when the measure is induced by a stochastic process* (X_t), these parameters are defined by the covariance and mean functions of the process (X_t).

Many SCT problems involve stochastic processes that can be reasonably modeled as Gaussian. This includes processes due to shot noise or thermal noise in electronic equipment (see Davenport and Root [6]), reverberation noise in underwater acoustics or radar [11], and various other applications. Normality for such problems is usually justified on the basis of the central limit theorem*.

"White noise" models are often used in engineering analyses. This usually refers to a stationary stochastic process having spectral density* that is constant at all frequencies. Of course, this usually implies infinite energy for the sample waveforms, and the model is then not valid even from the applications viewpoint (the mathematical inconsistencies are sometimes overcome through the use of distributions in the Sobolev–Schartz sense; see Gel'fand and Shilov [14] and DIRAC DELTA FUNCTION). The justification for analyses based on a white noise model is (in addition to ease of obtaining solutions) that many problems involve a noise stochastic process with an energy spectrum that is approximately constant across the frequency range of interest. For example, a passive sonar detection problem may involve a signal process with sample functions having 90% of their energy in a frequency interval of $f_0 \pm \Delta$ hertz, embedded in an additive Gaussian noise due to electronic effects, with the noise having approximately constant energy in a frequency range much larger than $f_0 \pm \Delta$ hertz. The use of white noise* models enables one to use the Dirac delta function* as the covariance, in formal manipulations. Such methods of analysis sometimes give quite satisfactory solutions; they are very inappropriate in other problems. Many problems involve a Gaussian noise which has a component that can be reasonably modeled as white noise; however, the energy in this component is frequently small compared to the total noise energy. An example is in active sonar, when there is usually a Gaussian "white noise" component due to electronic effects, but the principal source of noise is frequently due to reverberation, which is highly nonstationary and has an energy vs. frequency distribution that is strongly dependent on the shape of the transmitted signal [32, 33].

The prevalence of the white noise model is also frequently cited to justify the use of the Wiener process* in SCT applications. Formal differentiation of the (nondifferentiable a.e. dt, a.s.) sample paths of the Wiener process yields a Gaussian white noise process. Use of the Wiener process has led to a number of results in each of the three main areas of SCT, particularly in stochastic filtering (see Wong [42], Kailath [29], Davis [7], and Kallianpur [30]).

Some problems of interest in communication theory involve counting and point processes*. One example is in the analysis of electroencephalographic (EEG) recordings of brain activity [27], in which the brain activity appears as a stationary Gaussian process. The EEG recording also includes additive Poisson-distributed impulses due to nerve action. To analyze the brain activity, one wishes to filter out the impulses. Other examples include data communication networks (e.g., those used to analyze computer communications). For such problems, stochastic calculus methods using martingale theory are being extensively employed (see Segall [37]). In fact, martingale methods have recently become increasingly prevalent in communication theory research, especially in stochastic filtering.

In addition to problems in which martingales appear explicitly in the description of the observed waveform, there is a more general framework in which martingale methods are useful. This involves the use of the *Cra-mér–Hida representation* of stochastic processes [5, 22], in which a stochastic process can be represented (under mild restrictions) as the sum of a number of mutually orthogonal stochastic processes, with each of the latter being represented in terms of a linear operation on a process with orthogonal increments. In the case of Gaussian processes, these orthogonal increment processes become Gaussian martingales; in the non-Gaussian case, they still retain some of the features of martingales* that are useful in stochastic filtering and signal detection (see Segall [38]).

INFORMATION THEORY

Research in information theory can be divided into two broad areas: algebraic coding* and the Shannon theory. The "Shannon theory" refers to a class of problems dealing primarily with channel capacity and source coding (rate distortion theory). A detailed exposition of these topics is given in the article on information theory*. The present discussion will be limited to an area of the Shannon theory that has attracted substantial attention from mathematicians since being introduced by Shannon in 1948 [39], with significant results obtained only recently and with especially interesting connections to other areas of SCT. The specific problem to be discussed is that of channel capacity for waveform channels, especially those involving Gaussian noise. We begin with some basic definitions.

Let (Ω_1, β_1) and (Ω_2, β_2) be two measurable spaces, μ_{12} a probability measure on $\beta_1 \times \beta_2$ with marginal measures μ_1 and μ_2 (on β_1 and β_2); $\mu_1 \otimes \mu_2$ the product measure (see Hewitt and Stromberg [21]). Then the average mutual information (AMI) of the measure μ_{12} can be defined as (*see* THEORY OF INFORMATION) $I(\mu_{12}) = +\infty$ if μ_{12} is not absolutely continuous with respect to $\mu_1 \otimes \mu_2$; otherwise,

$$I(\mu_{12}) = \int_{\Omega_1 \times \Omega_2} \left[\log \frac{d\mu_{12}}{d\mu_1 \otimes \mu_2} (x, y) \right] d\mu_{12}(x, y).$$

In SCT applications, the mutual information is frequently defined in terms of a communication channel. A typical channel model is as follows. The message is a sample function from a stochastic process (X_t), t in T_1, and is encoded into a (transmitted) signal waveform by a coding operation A. The parameter sets can be regarded as representing time. The channel adds to the signal a sample function from a noise stochastic process (N_t), t in T_2, which is independent of the message process (X_t). The received waveform at the channel output is then a sample function from the stochastic process (Y_t), $Y_t = [AX]_t + N_t$, $t \in T_2$. The channel is said to be Gaussian if the noise (N_t) is a Gaussian process. The channel is said to have feedback if the coding operation A is a function of the received waveform from the process (Y_t). Most feedback channels of interest involve casual feedback, wherein A is a function only of past values of the received waveform; thus, if $Y(\omega)$ is the received waveform, then the sample path $[AX]_t(\omega)$ is a function of $\{Y_s(\omega), 0 \leqslant s < t\}$. As discussed in the preceding section, the message process induces a probability measure μ_x on a σ-field of a function space Ω_X containing the (X_t) sample functions, the channel output process induces a probability measure on a σ-field of a function space Ω_Y containing the (Y_t) sample functions, and the pair of processes induces a probability measure μ_{XY} on the product σ-field in $\Omega_X \times \Omega_Y$. μ_{XY} has μ_X and μ_Y as marginal measures. The quantity of interest is then $I(\mu_{XY})$, the average mutual information between the message and channel output processes.

The channel-capacity problem is that of determining $\sup I(\mu_{XY})$, when the noise process (N_t) is fixed and the supremum is taken over all message processes (X_t) and coding operations A that satisfy a specified set of constraints. The constraints are usually determined by physical considerations. $\operatorname{Sup} I(\mu_{XY})$ is defined to be the capacity (for the specified constraints). Channel capacity is a very important quantity in applications. Its importance arises from the fact that the value of the capacity (for a large class of

communication channels and constraints) is the critical factor in the ability to communicate over a channel. In general, one wishes to have a high capacity; *see* THEORY OF INFORMATION for further development of these concepts, including a more useful definition of capacity. Theoretical studies of channel capacity involve the following problems: (1) determining the value of the capacity; (2) determining whether the capacity can be attained; (3) if the capacity can be attained, determining a message process (X_t) and coding A, which together achieve capacity; and (4) determining the minimum decoding error.

Since many communication channels involve a Gaussian noise process (N_t), capacity of the Gaussian channel has long been a subject of much interest among engineers and mathematicians. We consider only the capacity for a finite observation time and a generalized energy constraint; *see* THEORY OF INFORMATION for a more general discussion.

Capacity of the Gaussian channel without feedback was considered by Shannon [39] under the following assumptions:

1. The noise is white, with spectral density equal to $N_0/2$.
2. The coding (A) is the identity.
3. The signal is limited to a bandwidth of W hertz and a finite time interval $[0, T]$.
4. $E \int_0^T X_t^2 \, dt \leqslant P_0$.

Under these assumptions, Shannon showed that the capacity is equal to $WT \log(1 + P_0/WTN_0)$. This formula is one of the most famous results of early information theory. However, the treatment in Shannon [39] was necessarily heuristic; in addition to the use of white noise, a signal waveform cannot be simultaneously time and frequency limited. However, Shannon actually used these assumptions to represent the signal waveform as a point in a space of dimension $2WT$. Gallager [13] obtained the Shannon result with the latter assumption, through representing the channel by $2WT$ parallel discrete-time channels, and also showed that as the dimension of the signal space becomes infinite ($W \rightarrow \infty$ in Shannon's setup), the capacity is (in the limit) equal to $P_0/2$.

Results on absolute continuity* for Wiener measure led to the next advance in determining capacity of the Gaussian channel. For a finite observation interval $[0, T]$, for both no feedback and with causal feedback, and the Wiener process as the noise, Kadota et al. [28] determined the capacity under the constraint $E \|A(X)\|_N^2 \leqslant P_0$, where $\| \cdot \|_N$ denotes the reproducing kernel Hilbert space (RKHS) norm for the noise. This can be regarded as a generalized signal-to-noise ratio constraint ($\|X\|_N^2 = (1/N_0) \int_0^T X_T^2 \, dt$ if the noise is white with spectral density N_0). The capacity was found to be $P_0/2$, with and without feedback; the question of attaining capacity was not considered. Hitsuda and Ihara [23] extended the result of Kadota et al. [28] to a large class of Gaussian channels with causal feedback by using the Cramér–Hida representation of the noise, together with a result on transformation of measures [15]. With the constraint

$$E \|A(X)\|_N^2 \leqslant P_0,$$

they found the capacity again to be $P_0/2$; moreover, for this class of channels, they showed that capacity could be attained. Further results on specifying coding operations to attain capacity were given by Ihara [25]; he gave codings that yield capacity and also minimize the error in the least-squares estimate of the message.

The capacity of the general Gaussian channel without feedback was obtained by Baker [2], again for the generalized signal-to-noise ratio constraint $E \|AX\|_N^2 \leqslant P_0$. He showed that although the capacity is the same ($P_0/2$) as that obtained in special cases for the channel with causal feedback (provided that the noise measure has infinite-dimensional support), when the channel does not have feedback then capacity cannot be attained. If the message process is also constrained to have sample paths in a linear space of dimension $M < \infty$, and the noise sample paths lie in a space with

dimension $\geqslant M$, then the capacity of the Gaussian channel without feedback (still with $E \|AX\|_N^2 \leqslant P_0$) was shown to be $\frac{1}{2} M \log(1 + P_0/M)$. As can easily be seen, this gives the Shannon result using $M = 2WT$, if the noise is formally assumed to be white.

The capacity problem for the general Gaussian channel with causal feedback remains unsolved as of the writing of this article. There has been considerable speculation among members of the information theory community that causal feedback will actually increase capacity for some Gaussian channels (with infinite bandwidth, finite observation time), and that the capacity with feedback can be no greater than twice the capacity without feedback (for any Gaussian channel). A complete and rigorous proof of this general hypothesis has not been obtained. This is one of the more interesting open mathematical problems in the general Shannon theory.

For the case of channels with non-Gaussian noise, relatively few results are available. Ihara has shown [26] that the capacity of the nonfeedback channel with non-Gaussian noise of given covariance cannot be less than the capacity of the nonfeedback channel with Gaussian noise of the same covariance, and also obtained an upper bound. A recent result on asymptotic capacity for large values of the constraint on generalized signal-to-noise ratio $E \|AX\|_N^2$ has been obtained by Binia [3].

One of the more interesting aspects of recent work on capacity of Gaussian channels with casual feedback, as exemplified in Kadota et al. [28] and Hitsuda and Ihara [23], has been the connection exhibited between mutual information and stochastic nonlinear filtering. Further progress in either area is thus likely to be beneficial to both. In addition, relations between analytical properties of the signal process and the mutual information, for which some results have already been obtained (see Ibragimov and Khaz'minski [24]) form a promising avenue of further research. Finally, we mention that absolute continuity* and properties of the Radon–Nikodym derivative are important components in the mathematical treatment of channel-capacity problems, so that interesting connections exist with the signal detection problem (see Davis [8]).

Journals frequently containing research articles on information theory include the *IEEE Transactions on Information Theory*, *Information and Control*, and *Problemy Peredachi Informatskii* (translated as *Problems of Information Transmission*).

SIGNAL DETECTION

Signal detection can be viewed as inference on stochastic processes. Sonar and radar are areas in which signal detection theory as an essential component has been applied for many years. Other areas of substantial importance include such diverse and significant problems as seismic exploration and the detection of brain tumors. The specific nature of the applications leads to consideration of specific mathematical models.

The basic signal detection problem is as follows. A waveform $\{X_t, t \in T\}$ is observed. The index set most frequently represents time. Under hypothesis H_0, the waveform is a sample function from a noise stochastic process (N_t). Under hypothesis H_1, the waveform is a sample function from a signal-plus-noise process $(S_t + N_t)$. The observer is to decide in favor of one of the hypotheses. Using an appropriate optimality criterion, a test statistic Λ is determined, with acceptance region A; if $\Lambda(X) \in A$, H_1 is accepted; otherwise, H_0 is accepted.

A number of variations on the foregoing model occur in engineering applications. For example, the nonacceptance region A^c may be divided into the union of two disjoint sets B_1 and B_2; $\Lambda(X) \in B_1$ gives a decision in favor of H_0; $\Lambda(X) \in B_2$ results in observing another sample path (sequential detection). The sample function may be from a vector process: $\mathbf{X}_t = (X_{1;t}, X_{2;t}, \ldots, X_{N;t})$. This occurs, for example, when the sample function

represents the output of an array consisting of N sensors. Finally, the signal-plus-noise process is not always additive in signal and noise.

If the probability distribution of the processes (N_t) and $(S_t + N_t)$ are completely known, and if the two induced measures μ_N and μ_{S+N} are such that μ_{S+N} is absolutely continuous* with respect to $\mu_N (\mu_{S+N} \ll \mu_N)$, then the solution to the basic problem is conceptually simple. Using either minimum Bayes risk* or the Neyman–Pearson criterion* as an optimality criterion, an optimum procedure is to form the likelihood ratio test* statistic $\Lambda(X) = (d\mu_{S+N}/d\mu_N)(X)$. A constant c_0 is selected, whose value depends on the criterion used and the properties of (N_t) and $(S_t + N_t)$. If $(d\mu_{S+N}/d\mu_N)(X) \geqslant c_0$, hypothesis H_1 is accepted; otherwise, H_0 is accepted. The Neyman–Pearson criterion is most frequently used; in this case, Pr[error | H_0 true] is called (by detection theorists!) the probability of false alarm (P_{FA}), and c_0 is selected so that $P_{FA} \leqslant \alpha$ for some acceptable value α. The decision criterion then guarantees that the probability of detection Pr[accept H_1 | H_1 true] is maximized over all test procedures satisfying $P_{FA} \leqslant \alpha$. The use of such fully parametric tests has been most successful when both (N_t) and $(S_t + N_t)$ can be assumed Gaussian. A treatment of the Gaussian vs. Gaussian problem from the viewpoint of absolute continuity is contained in Root [36]. A detailed analysis of this problem, with emphasis on several practical models for the signal and noise processes and on the practical implementation of the test statistic, is given in Van Trees [40], which also has an extensive bibliography. For a discussion of absolute continuity and likelihood ratio, especially for Gaussian processes, *see* ABSOLUTE CONTINUITY.

Unfortunately, in many applications the fully parametric procedure described above cannot be used (or approximated), frequently because the probability distributions of $(S_t + N_t)$ are not known; sometimes because the distributions of (N_t) are not known. For example, in active sonar an acoustic waveform is transmitted; each inhomogeneity in the ocean upon which this waveform is incident scatters energy; a portion of the scattered energy is reflected back to a receiver, which converts acoustic energy to electrical energy. The output of this transducer is the observed waveform. If the inhomogeneities are sufficiently dense, and their spatial distribution is Poisson, the received waveform can be reasonably assumed to be Gaussian (see Faure [11] and Middleton [32, 33]). This type of noise is called *reverberation*. The total noise process at the receiver then consists of the reverberation noise and an additive component of independent stationary Gaussian noise generated in the receiving equipment. The total noise process is then Gaussian, its mean is typically zero, and its covariance can be estimated. However, a target, when present, will usually be represented by only a few large scatterers, and the receiver output due to reflections from the target (the signal process) cannot reasonably be assumed to be Gaussian. Thus the active sonar detection problem is frequently that of detecting a non-Gaussian signal in additive Gaussian noise. In some applications there will be specific characteristics of the signal that can be very helpful in the detection problem; for example, if the target is moving, a doppler shift in the signal energy distribution (relative to the reverberation noise) will be present. Using this information, detection of high-speed targets can be reasonably reliable.

Such considerations require efficient signal detection methods for many practical problems to make use of criteria other than Neyman–Pearson, since the Radon–Nikodym derivative cannot be determined. One of the criteria most frequently used is that of maximizing signal-to-noise ratio over a specified class τ of test statistics. The signal-to-noise ratio, or deflection, for Λ in τ is frequently defined as $D_{10}(\Lambda) = [E_1 \Lambda(X) - E_0(\Lambda(X))]^2 / [E_0(\Lambda(X) - E_0 \Lambda(X))^2]$, where $E_i(\cdot)$ denotes expectation under hypothesis H_i. For example, if the observations are

assumed to belong to a Hilbert space such as $L_2[a,b]$ or l_2, then τ could be taken as the class of all quadratic-linear functionals: $\Lambda(X) = \langle X, WX \rangle + \langle X, h \rangle$, where $\langle \cdot, \cdot \rangle$ is the inner product, W a bounded linear operator, and h an element of the Hilbert space. Computation of D_{10} then requires only knowledge of the first four moments and cross-moments for the family of random variables $\{\langle X, e_n \rangle, n \geq 1\}$, where $\{e_n, n \geq 1\}$ is a complete orthonormal set in the Hilbert space. This information can frequently be obtained (approximately) in one of the most prevalent practical situations: (N_t) is a Gaussian process of zero mean and known covariance, $(S_t + N_t)$ has known covariance and mean. A detailed discussion of the deflection problem for quadratic-linear test statistics under these hypotheses is contained in Baker [1].

Another signal detection problem that arises frequently in practice involves signal and noise processes which can be assumed stationary with spectral densities. The noise can be taken to be Gaussian, with its spectrum "broadband," i.e., distributed over the relevant frequency range with significant energy in all (or nearly all) increments of some small width Δ. The signal, however, need not be Gaussian; moreover, it has largely "line" spectra, in the sense that the energy is concentrated in a few intervals of width Δ. The observer may, or may not, know the location of the frequency increments containing the signal spectra. In either case, methods of estimating spectra become important tools in devising detection procedures (see Haykin [18]).

Detection of signals in non-Gaussian noise is an area of considerable recent interest. Spherically invariant noise, which is a process of the form (aN_t), where (N_t) is Gaussian and a is a random variable independent of (N_t), can be used to model some practical detection problems. Likelihood ratios have been obtained for these types of detection problems; see Gualtierotti [17].

Nonparametric methods are sometimes advocated for signal detection. However, it is seldom that all of the necessary assumptions for application of the usual nonparametric methods are satisfied; in discrete-time problems, neither noise nor signal-plus-noise samples can usually be assumed to be independent, identically distributed (i.i.d.) random variables. Nevertheless, nonparametric or partially parametric methods can be considered as one approach to doing away with the necessity of knowing the likelihood ratio, and there have been many investigations by detection theorists using these methods. A collection of papers on this approach is contained in Papantoni–Kazakos and Kazakos [35]; those papers also contain many references to the literature.

Impulsive noise (see Middleton [34] and El Sawy and Van de Linde [9]) frequently arises in detection of low-frequency signals; e.g., atmospheric noise is impulsive in nature and becomes important in radio communications. Impulsive noise is a generalized point process composed of a sequence of "impulses," which are very short (in time) pulses with random arrival time and random amplitude. It is frequently represented in the form $U(t) = \sum_{i=1}^{N(t)} a_i g(t - \tau_i)$, when a_i is a family of i.i.d. random variables, τ_i the family of random arrival times (independent of a_i), $N(t)$ the number of arrivals in the interval $[0, t]$, and $g(t)$ the impulse function: $g(t) = 0$, $t < 0$; $g(t) = c(t)$, $0 \leqslant t \leqslant T$, $g(t) = 0$, $t > T$. The pulse function $c(t)$ is often idealized as the Dirac delta function* and a considerable amount of analysis has been done under this assumption. Of course, there will usually be an additive component of Gaussian noise, $G(t)$, so that the total noise waveform is given by $N(t) = U(t) + G(t)$. There has been substantial analysis of the detection of signals in impulsive (or impulsive-plus-Gaussian) noise; e.g., a robust detection procedure is given in El Sawy and Van de Linde [10]. However, such analyses have typically been for the case of a known signal; when the noise also has a Gaussian component, it is usually assumed that the index set is discrete and that the Gaussian random variables $G(t_n)$ are i.i.d.

Adaptive detection refers to signal detection methods that adjust the operations used to obtain the test statistic as a function of the observed data. Such procedures are useful when data parameters are incompletely known. A number of adaptive detection schemes are actually procedures for minimizing a mean square error, using approaches such as Kalman–Bucy filtering* or steepest descent*.

Adaptive array processing is described in Gabriel [12], which also contains a large bibliography of work in this area. An adaptive approach using nonparametric* techniques is given in Groginsky et al. [16].

Sequential detection, as usually understood, does not involve changes in the form of the operation on the data used to obtain the test statistic (although the acceptance region may change). Since the number of observations depends on the data (and is thus a random variable), a new parameter is introduced: expected sample size. One of the oldest sequential detection procedures involves the Wald sequential probability ratio test* (SPRT). The SPRT usually involves the assumption of i.i.d. observations; even so, calculation of the acceptance/rejection regions is not easy. Approximations for the thresholds needed to define these regions in order to achieve specified error probabilities were obtained by Wald [41]. For a general discussion of sequential testing, *see* SEQUENTIAL METHODS; applications to signal detection have been discussed by Helstrom [19] and Bussgang [4].

Finally, one should mention the detection of optical signals, for which the techniques of quantum mechanics have been used in analyses (see Helstrom [20]).

As a source of papers on current and past research on signal detection, the *IEEE Transactions on Information Theory*, the *IEEE Transactions on Communications*, the Soviet journal *Problemy Peredachi Informatskii* (available translated as *Problems of Information Transmission*), and the *Bell System Technical Journal* are especially recommended. Special mention is due the *IEEE*

Proceedings of May 1970 (vol. 58, no. 5, pp. 610–785), a special issue on detection theory.

Bibliography

In addition to the cited references, see the following:

Ash, R. B. (1965). *Information Theory*. Wiley-Interscience, New York.

Balakrishnan, A. V., ed. (1968). *Communication Theory*. McGraw-Hill, New York.

Berger, T. (1971). *Rate-Distortion Theory*. Prentice-Hall, Englewood Cliffs, N.J.

Capon, J. (1970). *Proc. IEEE*, **58**, 760–770.

Carlyle, J. W. (1968). In *Communication Theory*. A. V. Balakrishnan, ed. McGraw-Hill, New York.

Dobrushin, R. L. (1959). *Uspekhi Mat. Nauk*, **14**, 3–104; translation: *Amer. Math. Soc. Transl.* **33**(2), 323–438 (1963).

Dobrushin, R. L. and Pirogov, S. A. (1975). *Proc. 1975 IEEE–USSR Joint Workshop Inf. Theory*, pp. 39–49.

Ephremides, A. and Thomas, J. B., eds. (1973). *Random Processes: Multiplicity Theory and Canonical Decompositions*. Dowden, Hutchinson, & Ross, Stroudsburg, Pa.

Fortét, R. (1961). *Proc. 4th Berkeley Symp. Math. Statist. Prob.*, Vol. 1. University of California Press, Berkeley, Calif. pp. 289–305.

Franks, L. E. (1969). *Signal Theory*. Prentice-Hall, Englewood Cliffs, N.J.

Helstrom, C. W. (1972). *Statistical Theory of Signal Detection*, 2nd ed. Wiley, New York.

Kadota, T. T. and Romain, D. M. (1977). *IEEE Trans. Inf. Theory*, **23**, 167–178.

Kailath, T. (1980). *Linear Systems Theory*. Prentice-Hall, Englewood Cliffs, N.J.

Larimore, W. F. (1977). *Proc. IEEE*, **65**, 961–970.

Middleton, D. (1960). *An Introduction to Statistical Communication Theory*. McGraw-Hill, New York.

Osteyee, D. B. and Good, I. J. (1974). *Information, Weight of Evidence, the Singularity between Probability Measures and Signal Detection. Lect. Notes in Math.*, **376**.

Schwartz, M. (1980). *Information Transmission, Modulation, and Noise*, 3rd ed. McGraw-Hill, New York.

Spaulding, A. D. and Middleton, D. (1977). *IEEE Trans. Commun.* **26**, 910–930.

Van Trees, H. L. (1968). *Detection Estimation and Modulation Theory, Part I*. Wiley, New York.

Viterbi, A. J. and Omura, J. K. (1979). *Principles of Digital Communications and Coding*. McGraw-Hill, New York.

Weber, C. L. (1968). *Elements of Detection and Signal Design*. McGraw-Hill, New York.

Wyner, A. (1974). *IEEE Trans. Inf. Theory*, **21**, 2–10.

References

[1]   Baker, C. R. (1969). *IEEE Trans. Inf. Theory*, **15**, 16–21.

[2]   Baker, C. R. (1978). *Inf. Control*, **37**, 70–89.

[3]   Binia, J. (1979). *IEEE Trans. Inf. Theory*, **25**, 448–452.

[4]   Bussgang, J. (1970). *Proc. IEEE*, **58**, 731–743.

[5]   Cramér, H. (1961). *Ark. Mat.*, **4**, 249–266.

[6]   Davenport, W. B., Jr., and Root, W. L. (1958). *An Introduction to the Theory of Random Signals and Noise*. McGraw-Hill, New York.

[7]   Davis, M. H. A. (1977). *Linear Estimation and Stochastic Control*. Chapman & Hall, London.

[8]   Davis, M. H. A. (1977). In *Proceedings of the NATO Advanced Study Institute on Communication Systems and Random Process Theory*, 2nd ed., J. Skwirzynski, ed. Sijthoff & Noordhoff, Alphen aan den Rijn, The Netherlands.

[9]   El-Sawy, A. H. and Van de Linde, V. D. (1978). *Proc. 1978 Conf. Inf. Sci. Syst.*, Johns Hopkins University, Baltimore, Md., pp. 7–12.

[10] El-Sawy, A. H. and Van de Linde, V. D. (1979). *IEEE Trans. Inf. Theory*, **25**, 346–353.

[11] Faure, P. (1964). *J. Acoust. Soc. Amer.*, **36**, 259–268.

[12] Gabriel, W. F. (1976). *Proc. IEEE*, **64**, 239–272.

[13] Gallager, R. G. (1968). *Information Theory and Reliable Communication*. Wiley, New York.

[14] Gel'fand, I. M. and Shilov, G. E. (1964). *Generalized Functions*, Vol. 1: *Properties and Operations*. Academic Press, New York.

[15] Girsanov, I. V. (1960). *Theory Prob. Appl.*, **5**, 285–301.

[16] Groginsky, H. L., Wilson, L. R., and Middleton, D. (1966). *IEEE Trans. Inf. Theory*, **12**, 337–348.

[17] Gualtierotti, A. F. (1976). *IEEE Trans. Inf. Theory*, **22**, 610.

[18] Haykin, S., ed. (1979). *Nonlinear Methods of Spectral Analysis*. Springer-Verlag, Berlin.

[19] Helstrom, C. W. (1968). In *Communication Theory*, A. V. Balakrishnan, ed. McGraw-Hill, New York.

[20] Helstrom, C. W. (1976). *Quantum Detection and Estimation Theory*. Academic Press, New York.

[21] Hewitt, E. and Stromberg, K. (1965). *Real and Abstract Analysis*. Springer-Verlag, New York.

[22] Hida, T. (1960). *Mem. Coll. Sci., Univ. Kyoto*, **A33**(1), 109–155.

[23] Hitsuda, M. and Ihara, S. (1975). *J. Multivariate Anal.*, **5**, 106–118.

[24] Ibragimov, I. and Khaz'minski, R. Z. (1979). *Problems Inf. Transmission*, **15**, 18–26.

[25] Ihara, S. (1974). *Trans. 7th Prague Conf. Inf. Theory, Statist. Dec. Funct. Random Processes*, pp. 201–207.

[26] Ihara, S. (1978). *Inf. Control*, **37**, 34–39.

[27] Johnson, T. L., Feldman, R. G., and Sax, D. S. (1973). *Amer. J. EEG Technol.*, **13**, 13–35.

[28] Kadota, T. T., Zakai, M., and Ziv, J. (1971). *IEEE Trans. Inf. Theory*, **17**, 368–371.

[29] Kailath, T. (1968). *IEEE Trans. Inf. Theory*, **15**, 350–361.

[30] Kallianpur, G. (1980). *Stochastic Filtering Theory*. Springer-Verlag, New York.

[31] Lucky, R. W. (1973). *IEEE Trans. Inf. Theory* **19**, 725–739.

[32] Middleton, D. (1967). *IEEE Trans. Inf. Theory* **13**, 372–392.

[33] Middleton, D. (1972). *IEEE Trans. Inf. Theory*, **18**, 35–67.

[34] Middleton, D. (1979). *IEEE Trans. Electromagnetic Compatibility*, **21**, 209–220.

[35] Papantoni-Kazakos, P. and Kazakos, D., eds. (1977). *Nonparametric Methods in Communications*. Marcel Dekker, New York.

[36] Root, W. L. (1963). *Proc. Symp. on Time Series Anal.*, M. Rosenblatt, ed. Wiley. New York, pp. 292–346.

[37] Segall, A. (1975). *Proc. 1975 IEEE-USSR Joint Workshop on Inf. Theory*, pp. 39–49.

[38] Segall, A. (1976). *IEEE Trans. Inf. Theory*, **22**, 275–286.

[39] Shannon, C. E. (1948). *Bell System Tech. J.*, **27**, 379–423, 623–656.

[40] Van Trees, H. L. (1971). *Detection Estimation and Modulation Theory, Part III*. Wiley, New York.

[41] Wald, A. (1947). *Sequential Analysis*. Wiley, New York.

[42] Wong, E. (1971). *Stochastic Processes in Information and Dynamical Systems*. McGraw-Hill, New York.

(ABSOLUTE CONTINUITY
HYPOTHESIS TESTING
INFORMATION AND CODING
  THEORY
LINEAR FILTERING
STOCHASTIC PROCESSES
THEORY OF INFORMATION)

CHARLES R. BAKER

COMPACT DERIVATIVE *See* STATISTICAL FUNCTIONALS

COMPARTMENT MODELS, STOCHASTIC

A discussion of stochastic compartment models is fraught with the difficulty of providing a precise definition. This is a difficult task because of the wide variety of problems to which compartment models have been applied and their even broader range of potential applications. Indeed, in conceptual terms, the discussion of compartment models is nearly as general as the discussion of systems. As a consequence we shall view compartment models as a general analytic strategy rather than as a specific set of analytic procedures. The crux of the analytic strategy of compartment modeling is the explicit consideration of the mathematical model of the time-directed behavior of a given system. Once the mathematical structure of the system behavior is described, statistical issues may be addressed. In order to evaluate the compartment model strategy in greater depth, in the remainder of this paper, we discuss (a) the origin of compartment analysis in the biological sciences, (b) the formal aspects of the general linear compartment system, and (c) issues of parameter identifiability and estimability. As will be seen, identifiability* and estimability are two crucial concerns in the development of specific compartment models.

ORIGIN OF COMPARTMENT MODELS IN THE BIOLOGICAL SCIENCES

Compartment models were originally developed for the analysis of complex biological systems—a problem that was not amenable to analysis by classical statistical procedures. The crucial feature in the analysis of complex biological systems is that certain parameters in the system could not be experimentally manipulated without distorting the behavior of the system. Indeed, much of the system behavior was unobservable as well as unmanipulatable. As a result, analysis of the behavior of the system was confined largely to the study of the inputs and outputs of the system. The only practical approach to analysis under such constraints was to posit a mathematical model, based on ancillary information, to link inputs and outputs. Inference in this case is restricted to determining if the mathematical structure consistently linked the temporal schedule of inputs and outputs.

Biologists, in developing such compartment models, found that certain classes of system structure facilitated computation and were consistent with a number of theoretical precepts about biological systems. In particular, biologists found it convenient to model complex systems as a series of discrete states or compartments linked by flows of particles governed by partially observed transition rates that were functions of continuous time. This basic type of compartment system could be described alternatively as a discrete-state, continuous-time stochastic process with partially observed transition parameters. Because the intercompartment transfers are observed only partially, a critical feature of such analyses was the generation of sufficient restrictions on parameters to achieve structural identifiability of the equations relating compartment interchanges to time. Another important feature of this particular formulation of the compartment model is that it had a dual nature, being describable either probabilistically, in terms of the waiting-time distributions of particles within compartments, or in terms of the transition-rate functions themselves. For biologists the most familiar description was in terms of the transition-rate functions, because they led naturally to solutions for linear ordinary differential equation systems [6].

LINEAR COMPARTMENT SYSTEMS

It is a relatively recent development that statisticians have realized that there are broad classes of problems that could only be approached by some "modeling" strategy such as compartment analysis. From this realization there have emerged a series of attempts to develop general analytic strategies for the analysis of compartment systems. One such attempt involves the development of strategies appropriate to the analysis of the linear compartment systems, or

$$\dot{x}(t) = A(t)x(t) + B(t) \qquad (1)$$

where $\dot{x}(t)$ represents the change in the number of particles in each of n compartments, $A(t)$ is an $n \times n$ matrix of time-dependent transition rates, $x(t)$ the vector of the number of particles in each of n compartments, and $B(t)$ the vector of inputs to each compartment. Often, these equations are simplified by assuming $A(t)$ and $B(t)$ to be time invariant. Bellman [1] has derived a solution using a matrix exponential form for the time-invariant form of (1). When the eigenvalues of A are real and distinct, a solution is available using the "sums of exponential" model [5].

Matis and Wehrly [9] have pointed out that the linear compartment system represented in (1) is often applied deterministically. They propose that this model may be usefully generalized by introducing several different types of stochasticity. They identify two types of stochasticity which they feel are frequently present in applying the linear compartment model. The first type of stochasticity is associated with individual units or particles. Particle "stochasticity" is further divided into that due to sampling from a random process and that due to particles having random rate coefficients. The second type of stochasticity they identify is associated with a replication of the entire experiment. "Replicate" stochasticity can also be divided into two types. The first is replicate stochasticity due to the initial number of particles being random. The second type of replicate stochasticity is due to the variability of rate coefficients across replicates. Naturally, in any given situation one, or a combination, of these four types of stochasticity may be present.

The basic rationale for dealing with the complications produced by including consideration of the four types of stochasticity in the linear compartment system is that a deterministic system may not adequately represent the behavior of individuals within the system. For example, the age-specific mortality* probabilities from a cohort life table* are often used to describe the trajectory of individuals through different "age" compartments. However, if individuals within a cohort have different "susceptibility" to death, individuals will be systematically selected by mortality and the cohort life table will no longer be a valid model for the age trajectory of mortality risks for individuals. Instead, it will only be descriptive of the age-specific mean risk of death among survivors to a given age [7, 10]. *See* COHORT ANALYSIS.

ESTIMATION AND IDENTIFICATION OF STOCHASTIC COMPARTMENT MODEL PARAMETERS

Although it is clear that appropriately representing various types of stochasticity in compartment models will greatly improve their theoretical and analytic worth, it is also clear that estimation of the parameters in such models will be far more difficult. For example, estimation of the parameters of a stochastic compartment system often runs into identifiability problems because a consideration of only the means is insufficient information to identify the parameters for individual transitions. Two basic approaches are employed to deal with these issues in the analysis of stochastic compartment systems. The first approach involves the development of basic compartment models with standard computational procedures that will be applicable to broad classes of analytic problems. In this approach statistical information, say as contained in the error covariance structure, is used to achieve identifiability. Representative of such approaches is the nonlinear least-squares* strategy proposed by Matis and Hartley [8]. The second basic approach is to develop compartment models specific

to individual problems. In this approach it is explicitly recognized that each compartment model is composed of a substantively determined mathematical structure and a statistical model and that each needs to be developed for the specific problem at hand. Thus identification is achieved by imposing restrictions derived from substantive theory or ancillary data on the parameter space.

In either approach one must deal with the central analytic issue of parameter identification i.e., that each parameter be observationally independent of the set of other parameters. Estimability is a related concept suggesting that the available data contain sufficient statistical information so that precise estimates of parameters can be made. In general, identifiability must be achieved with information on certain measurable outputs (i.e., the rate of exit of particles into one of a set of external, observable compartments) which we can identify as the vector y, certain (possibly manipulatable) inputs (rate of entry of particles to the system from controllable external compartment) which we can identify as the vector u, and the temporal relations of u and y. From these observed quantities one hopes to be able to determine the number of particles in each theoretically specified internal compartment, which we will represent as a vector x, and the matrix F, of transfer coefficients governing the flow of particles between compartments. As can be seen, we are restricting our discussion to linear, time-invariant systems. Let us assume further that the transfer of inputs to internal compartments is governed by an observed matrix B, and that another transfer matrix, C, determines the measured output vector y, or

$$\dot{x} = Bu + Fx \qquad (2)$$

$$y = Cx \qquad (3)$$

as functions of time.

By taking the Laplace transform on both sides of (2) and (3), and under the assumption that the system is empty at time $t = 0$, we obtain

$$sX = BU + FX \qquad (4)$$

$$Y = CX, \qquad (5)$$

where U, X, and Y are the Laplace transforms* of u, x, and y and s is the transformation-domain variable. Equations (4) and (5) can be solved to determine the relation between U and Y, or

$$Y = C(sI - F)^{-1}BU. \qquad (6)$$

In (6), U and Y are vectors, so that, to consider a single transfer coefficient, T_{ij}, we have to consider the ith element of Y and the jth element of U, as in

$$T_{ij} = e_i^T C(sI - F)^{-1}Be_j. \qquad (7)$$

Here T_{ij} represents the transfer coefficient between input state j and output state i and where e_i and e_j are unit vectors with a 1 in the appropriate ith and jth position and zeros elsewhere. In (7), U is, in effect, replaced by the identity matrix and the vector Y is replaced by the elements of the matrix of transfer coefficients, T. This situation corresponds to the analysis of cases where the time-domain input of particles to the system is in the form of a single "pulse." It can be seen that the relation of F and T is nonlinear. In this case identifiability is achieved only if it is possible to generate empirical estimates of the T_{ij} from temporal measurements of output i of the system resulting from a "pulse" of particles directed through input j by computing its Laplace transform (*see* INTEGRAL TRANSFORMS). To consider this in greater detail, let us write the expression that relates a change in a T_{ij} to a "small" change, df_{kl}, of a given transfer coefficient, f_{kl} (for k, l such that $f_{kl} \neq 0$), or

$$\frac{\partial T_{ij}}{\partial f_{kl}} = e_i^T C(sI - F)^{-1}e_k e_l^T (sI - F)^{-1}Be_j. \qquad (8)$$

If f_{kl} is not identifiable, then, for all i and j, $(\partial T_{ij}/\partial f_{kl})$ is identically zero in s. This implies that f_{kl} is not identifiable if there does not exist an i such that $e_i^T C(sI - F)^{-1}e_k \neq 0$ and a j such that $e_l^T(sI - F)^{-1}Be_j \neq 0$. To understand what this implies for a structure of a compartment system we should note

that, for sufficiently large s,

$$(sI - F)^{-1} = \frac{I}{s} + \frac{F}{s^2} + \frac{F^2}{s^3} + \frac{F^3}{s^4} + \cdots, \qquad (9)$$

so that a given entry $(sI - F)_{kl}^{-1}$ is not identically zero in s if there is some power F^P of F that has $F_{kl}^P \neq 0$. It can be seen that this is so if there is a "path" from k to l in F^P, i.e. if there exist $k = k_1, k_2, k_3, \ldots, k_p = l$ such that $f_{k_1,k_2} \neq 0, f_{k_2,k_3} \neq 0, \ldots, f_{k_{p-1},k_p} \neq 0$. This can be determined readily from the "flow chart" of the compartment system [2–4].

We now introduce the notion of input-connectable and output-connectable systems. A compartment l is input-connectable if there is a path from an input to the compartment l. This is equivalent to the "nonvanishing" of $e_l^T(sI - F)^{-1}B$. Similarly, the nonvanishing of $C(sI - F)^{-1}e_k$ is equivalent to the output connectability of compartment k. Hence a necessary condition for the identifiability of a compartmental system is that for every $f_{kl} \neq 0$, k is output-connectable and l is input-connectable [3].

Structural identifiability for a given compartment system can be achieved either by introducing statistically imposed constraints (e.g., by developing constraints on the error covariance structure) or by deriving parameter restrictions from ancillary data and theory. Estimability will, practically, be the more difficult condition to achieve because of the near collinearity of parameters in the nonlinear functional forms often employed.

MODEL BUILDING* IN STATISTICAL ANALYSIS

The foregoing discussion is meant to sensitize the statistician to the fact that compartment analysis represents an effort to formalize a series of mathematical model building strategies for analytic problems where direct information is not available to estimate all parameters of a model. As a consequence of such problems it is necessary to assess explicitly the detailed mathematical form of the response model assumed in the analysis. As such, this type of analysis represents an interesting synthesis of mathematical modeling, statistical inference, and substantative theory in an attempt to model behavior that is only partially observable. It also requires careful attention to aspects of computational theory and numerical analysis. It represents a very flexible tool for the statistician faced with a problem whose complexity makes a more traditional approach impossible. As the statistician becomes involved with a variety of important policy questions involving the behavior of complex human systems, the need to employ a modeling strategy of this type increases.

References

[1] Bellman, R. (1960). *Introduction to Matrix Analysis*, McGraw-Hill, New York. (A general reference work on the mathematical aspects of matrix applications.)

[2] Cobelli, C. and Jacur, G. R. (1976). *Math. Biosci.* **30**, 139–151. (An analysis of the conditions for establishing the identifiability of the parameters of a compartment model.)

[3] Cobelli, C., Lepschy, A., and Jacur, G. R. (1979). *Math Biosci.*, **44**, 1–18. (Extends their earlier 1976 results and responds to questions raised by Delforge.)

[4] Delforge, J. (1977). *Math. Biosci.*, **36**, 119–125. (Illustrates that Cobelli and Romanin Jacur had provided only necessary and not sufficient conditions for identifiability.)

[5] Hearon, J. Z. (1963). *Ann. N.Y. Acad. Sci.*, **108**, 36–68. (Provides some useful results for the analysis of compartment systems.)

[6] Jacquez, J. A. (1972). *Compartmental Analysis in Biology and Medicine*. Elsevier, New York. (A broad overview of the practice and theory of the applications of compartment models in the biological sciences.)

[7] Manton, K. G. and Stallard, E. (1980). *Theoretical Population Biology*, **18**, 57–75. (An analytic model of disease dependence represented via a compartment system with a heterogeneous population.)

[8] Matis, J. H. and Hartley, H. O. (1971). *Biometrics*, **27**, 77–102. (A basic methodological exposition of the application of nonlinear least squares to the estimation of the parameters of a compartment system.)

[9] Matis, J. H. and Wehrly, T. E. (1979). *Biometrics*, **35**, 199–220. (A useful and comprehensive review article on the application of linear compartment systems and of various types of stochastic formulations of such systems.)

[10] Vaupel, J. W., Manton, K. G., and Stallard, E. (1979). *Demography*, **16**, 439–454. (A presentation of a model based on explicit mathematical assumptions to represent the effects of heterogeneity on the parameters of a basic human survival model.)

(BIOLOGY, STATISTICS IN STOCHASTIC PROCESSES)

M. A. Woodbury
K. G. Manton

COMPETING RISKS

The term "competing risks" applies to problems in which an object is exposed to two or more causes of failure. Such problems arise in public health, demography, actuarial science, industrial reliability applications, and experiments in medical therapeutics. More than three decades before Edward Jenner published his observations on inoculation against smallpox, Bernoulli* [6] and d'Alembert [13] estimated the effect eliminating smallpox would have on population survival rates. This is a classic competing-risk problem, in which individuals are subject to multiple causes of death, including smallpox. Other examples follow. The actuary charged with designing a disability insurance plan must take into account the competing risks of death and disability. In offering joint life annuities, which cease payment as soon as either party dies, an insurance company must consider the competing risks of death of the first and second parties. An epidemiologist trying to assess the benefit of reducing exposure to an environmental carcinogen must analyze not only the reduced incidence rate of cancer but also effects on other competing causes of death. These examples illustrate the pervasive importance of competing-risk problems.

The actuarial literature on competing risks has been reviewed by Seal [29]. Actuarial methods for competing risks are described in MULTIPLE-DECREMENT LIFE TABLES. Statisticians interested in reliability theory or survival analysis have rediscovered many actuarial methods, emphasized certain theoretical dilemmas in competing-risk theory, and introduced new estimation procedures. The statistical literature has been reviewed by Chiang [10], Gail [20], David and Moeschberger [14], and Prentice et al. [28]. *See also* ACTUARIAL STATISTICS, LIFE *and* EPIDEMIOLOGICAL STATISTICS.

OBSERVABLES

We suppose that an individual is subject to m causes of death, and that for each individual we observe the time of death T and the cause of death J. This process is described in terms of the "cause-specific hazard functions"

$$g_j(t) = \lim_{\Delta t \downarrow 0} \Pr[t \leqslant T < t + \Delta t,$$

$$J = j \mid T \geqslant t]/\Delta t. \quad (1)$$

The term "cause-specific hazard function" is used by Prentice et al. [28]. These functions are identical to the "force of transition" functions in Aalen's Markov formulation (*see* MARKOV PROCESSES) of the competing-risk problem [1] and to the functions $g_j(t)$ in Gail [20]. Indeed, the functions $g_j(t)$ are the "forces of mortality*" which an actuary would estimate from a multiple-decrement table and were termed "decremental forces" by the English actuary Makeham [25]. Assuming the existence of the quantities $g_j(\cdot)$, which we assume throughout, the hazard

$$\lambda(t) = \lim_{\Delta t \downarrow 0} \Pr[t \leqslant T < t + \Delta t \mid T \geqslant t]/\Delta t \quad (2)$$

satisfies

$$\lambda(t) = \sum_{j=1}^{m} g_j(t). \quad (3)$$

Prentice et al. [28] emphasize that only probabilities expressible as functions of $\{g_j(\cdot)\}$ may be estimated from the observable data (T, J). We define a probability to be *observable* if it can be expressed as a function of the cause-specific hazards corresponding to the original observations with all m risks acting. For example,

$$S_T(t) \equiv \Pr[T > t] = \exp\left[-\int_0^t \lambda(u)\,du\right],$$

the probability of surviving beyond time t, is observable. The conditional probability of dying of cause j in the interval $(\tau_{\Delta-1}, \tau_i]$, given $T > \tau_{i-1}$, is computed as

$$Q(j;i) = \left[S_T(\tau_{i-1})\right]^{-1} \int_{\tau-1}^{\tau} g_j(u) S_T(u)\,du. \quad (4)$$

This probability is a function of $\{g_j(\cdot)\}$ and is, by definition, observable. The probabilities $Q(j;i)$ are termed "crude" probabilities by Chiang [10] because the events represented occur with all the m original risks acting.

The conditional probabilities $Q(j;i)$ are the basic parameters of a multiple-decrement life table. If the positive time axis is partitioned into intervals $(\tau_{i-1}, \tau_i]$, for $i = 1, 2, \ldots, I$, with $\tau_0 = 0$, then $Q(j;i)$ gives the conditional probability of dying of cause j in interval i given $T > \tau_{i-1}$. Such a partitioning arises naturally in actuarial problems in which only the time interval and cause of death are recorded, and multiple-decrement life tables also arise when exact times to death T are grouped. The conditional probability of surviving interval i, $\rho_i = 1 - \sum_{j=1}^{m} Q(j;i)$, may be used to calculate such quantities as $S_T(\tau_i) = \prod_{l=1}^{i} \rho_l$, which is the probability of surviving beyond τ_i, and $\sum_{i=1}^{k} S_T(\tau_{i-1}) Q(j;i)$, which is probability of dying of risk j in the interval $(0, \tau_k]$. Such quantities are observable because they are functions of the observable probabilities $Q(j;i)$. The maximum likelihood estimate of $Q(j;i)$ is $\hat{Q}(j,i) = d_{ji}/n_i$, where d_{ji} is the number who die of risk j in time interval i, and n_i is the number alive at τ_{i-1}. The corresponding estimate of ρ_i is $\hat{\rho}_i = s_i/n_i$, where $s_i = n_i - \sum_j d_{ji}$ survive interval i. Maximum likelihood* estimates of other observable probabilities related to the multiple-decrement life table are obtained from $\hat{Q}(j;i)$ and $\hat{\rho}_i$.

BEYOND OBSERVABLES

Observable probabilities can be thought of as arising in the original observational setting with all risks acting. Many interesting probabilities arise when one tries to predict what would happen if one or more risks were eliminated. For example, Chiang [10] defines the "net" probability q_{ji} to be the conditional probability of dying in interval $(\tau_{i-1}, \tau_i]$ given $T > \tau_{i-1}$ when only risk j is acting, and he defines $Q_{ji \cdot \delta}$ to be the corresponding "partial crude" probability of dying of risk j when risk δ has been eliminated. Such probabilities are of great practical interest, yet they are not observable because they are not functions of $\{g_j(\cdot)\}$, as estimated from the original experiment. Such calculations depend on assumed models of the competing-risk problem.

Suppose that each individual has m failure times $T_1, T_2, \ldots, T_m$ corresponding to risks $1, 2, \ldots, m$. Only $T = \min_{1 \leqslant l \leqslant m} T_l$ and the risk j such that $T_j = T$ are observed. This is termed the latent-failure time model. To use this model to compute nonobservable probabilities, a joint survival distribution $S(t_1, \ldots, t_m) = P[T_1 > t_1, \ldots, T_m > t_m]$ must be assumed. The distribution of T is

$$S_T(t) = S(t, t, \ldots, t). \tag{5}$$

Whenever the following partial derivatives exist along the equiangular line $t_1 = t_2 = \ldots t_m$, the cause-specific hazards are given by

$$g_j(t) = -(\partial \ln S(t_1, \ldots, t_m)/\partial t_j)_{t_1 = t_2 = \ldots t_m = t} \tag{6}$$

as in Gail [20]. From its definition in (2), the hazard $\lambda(t)$ satisfies

$$\lambda(t) = -d \ln S_T(t)/dt, \tag{7}$$

and (3) can be derived by applying the chain rule to $-\ln S(t_1, t_2, \ldots, t_m)$. Gail [20] points

out that (3) may be invalid if $S(t_1, \ldots, t_m)$ is singular, so that $\Pr[T_i = T_j] > 0$ for some $i \neq j$.

Suppose that $T_1, T_2, \ldots, T_m$ are independent with marginal survival distributions $S_j(t) = \Pr[T_j > t]$ and corresponding marginal hazards $\lambda_j(t) = -d \ln S_j(t)/dt$. Under independence, $S(t_1, \ldots, t_m) = \prod S_j(t_j)$, from which it follows that

$$\lambda_j(t) = g_j(t) \qquad \text{for } j = 1, 2, \ldots, J. \tag{8}$$

We note parenthetically that the converse is false. For suppose that the density of T_1 and T_2 is constant on the triangular region with vertices $(0,0)$, $(1,0)$, and $(0,1)$. Then from an example in Gail [20], $g_1 = g_2 = \lambda_1 = \lambda_2 = 1/(1-t)$, yet T_1 and T_2 are dependent. The relationships (8) allow one to compute nonobservable probabilities as follows. First estimate $g_j(\cdot)$ and hence $\lambda_j(\cdot)$ from the original data. Second, invoke Makeham's assumption [25] that the effect of eliminating cause J is to nullify $\lambda_j(\cdot)$ without altering other marginal hazards. Taking $m = 3$ risks for ease of exposition, the net probability of surviving exposure to risk 1 alone until time t is $S_1(t) = S(t, 0, 0) = \exp[-\int_0^t \lambda_1(u) \, du]$, and the partial crude probability of dying of risk 3 in $[0, t]$ with risk 1 eliminated is $\int_0^t \lambda_3(v) S_2(v) S_3(v) \, dv = \int_0^t \lambda_3(v) S(0, v, v) \, dv$.

Three important assumptions are made in the previous calculations:

1. A structure for $S(t_1, \ldots, t_m)$ is assumed.

2. It is assumed that the effect of eliminating a set of risks is known and expressible in terms of $S(t_1, \ldots, t_m)$.

3. It is assumed that the experimental procedures used to eliminate a risk will only produce the changes specified in assumption 2 without otherwise altering $S(t_1, \ldots, t_m)$.

Makeham [25] questioned assumption 3 in his discussion of smallpox. It seems self-evident, for example, that a campaign to reduce lung cancer by banning cigarettes will have wide-ranging effects on other health hazards. We turn now to an examination of assumptions 1 and 2.

In the previous case of independence, the net survival distribution for risk 1 was taken to be the marginal distribution $S_1(t) = S(t, 0, 0)$. One might generalize this result to assume that the effect of eliminating risk j is to nullify the corresponding argument in $S(t_1, t_2, \ldots, t_m)$. This is an example of assumption 2 for modeling elimination of a risk. The implications of this viewpoint have been discussed by Gail [20], who shows how various observable and nonobservable probabilities may be computed given $S(t_1, \ldots, t_m)$ and given this version of assumption 2. These methods may be used whether $T_1, T_2, \ldots, T_m$ are independent or dependent. Elandt-Johnson [16] gives an alternative preferable model for eliminating risks. For example, she asserts that the appropriate net survival distribution in the previous case is the limiting conditional distribution

$$\lim_{t_2, t_3 \to \infty} \Pr[\, T_1 > t \mid T_2 = t_2, T_3 = t_3 \,]. \quad (9)$$

Another similar version of assumption 2 might be to take

$$\lim_{t_2, t_3 \to \infty} \Pr[\, T_1 > t \mid T_2 > t_2, T_3 > t_3 \,] \quad (10)$$

as the net survival distribution from risk 1. If T_1, T_2, and T_3 are independent, the net survival distributions (9), (10), and $S_1(t) = S(t, 0, 0)$ are equivalent. Thus under independence it is reasonable to take the net survival distributions as corresponding marginal distributions of $S(t_1, \ldots, t_m)$ and, more generally, to model elimination of risk j by nullifying the jth argument in $S(t_1, t_2, \ldots, t_m)$. With dependent models, however, a good argument can be made for modeling elimination of risks as in (9) or (10). Some assumptions 2 defining the effect of elimination are inevitably required.

Although $S(t_1, \ldots, t_m)$ defines $\{g_j(\cdot)\}$ and hence all observable probabilities, the observables (T, J) do not uniquely define $S(t_1, t_2, \ldots, t_m)$. For suppose that the data are so numerous as to permit perfect estimation of $\{g_j(\cdot)\}$. Whatever the distribution of $T_1, T_2, \ldots, T_m$, define a new set of independent random variables $T_1^*, T_2^*, \ldots, T_m^*$ with marginal hazards $\lambda_j^*(t) \equiv g_j(t)$ and marginal distributions

$$S_j^*(t) = \Pr[\, T_j^* > t \,] = \exp\left[-\int_0^t \lambda_j^*(u)\, du \right].$$

It is clear that the distribution $S^*(t_1, t_2, \ldots, t_m) = \prod_{j=1}^m S_j^*(t_j)$ has the same cause-specific hazards $g_j^*(t) = \lambda_j^*(t) \equiv g_j(t)$ as the original distribution $S(t_1, t_2, \ldots, t_m)$. Thus even if the data are so complete as to specify $\{g_j(\cdot)\}$ exactly, they do not define $S(t_1, t_2, \ldots, t_m)$ uniquely. This conundrum was noted by Cox [11] and termed nonidentifiability by Tsiatis [30], who gave a more formal construction.

Nonidentifiability has two important practical implications. First, assumption 1 specifying the structure $S(t_1, t_2, \ldots, t_m)$ cannot be verified empirically. Second, one can estimate $\{g_j(\cdot)\}$ from any distribution $S(t_1, \ldots, t_m)$ by pretending that the independent process $S^*(t_1, \ldots, t_m)$ is operative and using methods for estimating the marginal hazards $\{\lambda_j^*(\cdot)\}$ corresponding to independent random variables $T_1^*, \ldots, T_m^*$. These estimation methods for independent processes are discussed in the next section.

It is apparent that estimates of nonobservable probabilities, which must depend on assumptions 1, 2, and 3, are suspect.

ESTIMATION ASSUMING INDEPENDENT $T_1, T_2, \ldots, T_m$

First we consider parametric estimation in which each marginal distribution $S_j(t; \theta_j) = \Pr[T_j > t]$ depends on parameters θ_j. Let $\{t_{ij}\}$ denote the times of death of the d_j individuals dying of cause j for $J = 1, 2, \ldots, m$ and $i = 1, 2, \ldots, d_j$, and let $\{t_l^*\}$ denote the s follow-up times of those who survive until follow-up ceases. Then the likelihood is given by

$$\left[\prod_{j=1}^m \prod_{i=1}^{d_j} \lambda_j(t_{ij}) S_T(t_{ij}) \right] \prod_{l=1}^s S_T(t_l^*), \quad (11)$$

where $S_T(t)$ is obtained from (5) and $\lambda_j(t) = -d \ln S_j(t; \theta_j)/dt$. David and Moeschberger [14] give a detailed account for exponential*, Weibull*, normal*, and Gompertz* marginal survival distributions, and they give a similar likelihood for grouped data*. Once $\{\theta_j\}$ have been estimated, estimates of observable and nonobservable probabilities may be obtained from $\hat{S}(t_1, \ldots, t_m) = \prod S_j(t_j; \hat{\theta}_j)$.

Estimates based on the multiple-decrement life table are essentially nonparametric. The quantities $\hat{Q}(j; 1) = d_{ji}/n_i$ and $\hat{\rho}_i = 1 - \sum_{j=1}^{m} \hat{Q}(j; i)$ defined in the preceding section may be used to estimate observable probabilities. To estimate the net conditional probability of surviving interval i with risk j alone present, $p_{ji} = S_j(\tau_i)/S_j(\tau_{i-1})$, we invoke the piecewise proportional hazards model*

$$\lambda_j(t) = \omega_{ji}\lambda(t) \qquad \text{for } t \in (\tau_{i-1}, \tau_i), \quad (12)$$

where $\omega_{ji} \geqslant 0$ and $\sum_j \omega_{ji} = 1$. This assumption was used by Greville [21] and Chiang [10] and is less restrictive than the assumption that the marginal hazards are constant on time intervals introduced in the actuarial method. The relation (12) will hold for fine-enough intervals, provided that $\{\lambda_j(\cdot)\}$ are continuous. It follows from (12) that $\omega_{ji} = Q(j; i)/(1 - \rho_i)$, which yields the estimate $\hat{\omega}_{ji} = d_{ji}/\sum_{j=1}^{m} d_{ji}$. Also, (12) implies that

$$p_{ji} = \rho_i^{\omega_{ji}}. \qquad (13)$$

Hence the net conditional probability may be estimated from $\hat{p}_{ji} = (s_i/n_i)^{\hat{\omega}_{ji}}$. The corresponding estimate of $S_j(\tau_i)$ is $\prod_{i=1}^{i} \hat{p}_{ji}$, from which other competing risk calculations follow. Gail [20] shows that the actuarial estimate $p_{ji}^* = d_{ji}/(n_i - \sum_{l \neq j} d_{li}/2)$ is an excellent approximation to $\hat{p}_{ji}$.

A fully nonparametric approach to estimation of $S_j(t)$ under independence is outlined by Kaplan and Meier [23], who refine the actuarial method to produce a separate interval at the time of each death. They credit this approach to Böhmer [7]. The resulting product limit estimate of $S_j(t)$ is

$$S_j^{PL}(t) = \prod_r \left[1 - d_j(t_r)/n(t_r) \right], \quad (14)$$

where r indexes the distinct times $t_r \leqslant t$ at which deaths from cause j occur, $d_j(t_r)$ is the number of such deaths at t_r, and $n(t_r)$ is the number known to be at risk at t_r. The asymptotic theory has been studied by Breslow and Crowley [9], who proved that $S_j^{PL}(t)$ converges to a Gaussian process* after normalization. (*See* KAPLAN–MEIER ESTIMATE for additional details.) Aalen [1, 2] generalized these ideas to estimate the partial crude probabilities $p_j(t; A)$ of death from cause j in $[0, t]$ when an index set of risks $A \subset \{1, 2, \ldots, m\}$ defines the only risks present. For the special case $A = \{j\}$, $P_j(t; A) = 1 - S_j(t)$. For the case $A = \{1, 2, \ldots, m\}$, $P_i\{t, A\}$ is the crude probability given by (4) with $\tau_{i-1} = 0$ and $\tau_i = t$. Aalen's estimates $\hat{P}_j(t, A)$ are uniformly consistent, have bias tending to zero at an exponential rate, and tend jointly to a Gaussian process after normalization.

In this section we have emphasized estimation. Parametric and nonparametric methods for testing the equality of two or more net survival curves are found in the statistical literature on survival analysis*. Suppose that each individual in a clinical trial* has a death time T_1 and a censorship time T_2, indicating when his or her follow-up* ends. If patients are assigned to treatments 1 or 2 with net survival curves $S_1^{(1)}(t)$ and $S_1^{(2)}(t)$, respectively, treatments are compared by testing the equality $S_1^{(1)}(t) = S_1^{(2)}(t)$. Such tests are made under the "random censorship" assumption that T_1 and T_2 are independent. Pertinent references are in Breslow [8] and Crowley [12].

To summarize, standard parametric methods and nonparametric extensions of the life-table method are available under the independence assumption. One can estimate both observable and nonobservable probabilities using these methods, and one can test equality of net survival curves from different treatment groups, but the unverifiable independence assumption is crucial.

ESTIMATION ASSUMING DEPENDENT $T_1, T_2, \ldots, T_m$

The nonparametric methods just described may be used to estimate observable probabilities even when $T_1, T_2, \ldots, T_m$ are dependent because the functions $g_j(t)$ may be regarded as marginal hazards $\lambda_j^*(t)$ from an independent process $T_1^*, \ldots, T_m^*$ as mentioned in the section "Beyond Observables." Thus crude probabilities such as $P_j(t, A)$ with $A = \{1, 2, \ldots, m\}$ may be estimated by pretending that the death times are independent and using the methods of Aalen [2] or related actuarial methods.

In contrast, nonobservable probabilities, such as net or partial crude probabilities, depend on a hypothesized structure $S(t_1, \ldots, t_m)$ and cannot be estimated as in the preceding section. Once this joint distribution has been estimated, the computation of nonobservable probabilities proceeds as in the second section. If a specific parametric form $S(t_1, \ldots, t_m; \theta)$ is posited, θ may be estimated from the likelihood (11) with $g_j(t_{ij})$ replacing $\lambda_j(t_{ij})$, provided that $S(t_1, \ldots, t_m)$ is absolutely continuous. David and Moeschberger [14] discuss the bivariate normal model and a bivariate exponential model proposed by Downton [15], as well as the bivariate exponential model of Marshall and Olkin [26], which is singular.

Peterson [27] obtained bounds on $S(t_1, t_2)$, $S(t, 0)$, and $S(0, t)$ when the observable probabilities $\Pr[T_1 > t, T_1 < T_2]$ and $\Pr[T_2 > t, T_2 < T_1]$ are known. These bounds were derived without special assumptions on the structure $S(t_1, t_2)$, but the bounds are too wide to be useful.

MARKOV MODEL AND EXTENSIONS

Aalen [1] modeled the classical competing risk problem as a continuous-time Markov process* with one alive state $j = 0$ and m absorbing death states $\{1, 2, \ldots, m\}$. The only permissible transitions in this model are $0 \rightarrow j$ for $j = 1, 2, \ldots, m$, and the corresponding "forces of transition" are the functions $g_j(t)$ in (1). It is further supposed that elimination of risk j merely nullifies $g_j(t)$. This model of risk elimination, which was adopted by Makeham [25], is equivalent to the latent-failure-time model with independent latent failure times (we call this the ILFT model) and marginal hazards $\lambda_j(t) = g_j(t)$. To see this, let $p(t, j)$ be the probability of being in state j at time t. The governing differential equations for the Markov process are $dp(t, 0)/dt = -\sum_{l=1}^{m} g_j(t)$ and $dp(t, j)/dt = p(t, 0) g_j(t)$ for $j = 1, 2, \ldots, m$. Hence

$$p(t, 0) = \exp\left[-\int_0^t \sum_{l=1}^{m} g_l(u) \, du \right]$$

and $p(t, j) = \int_0^t p(u, 0) g_j(u) \, du$. In terms of the ILFT model with $\lambda_j(t) = g_j(t)$, these probabilities are $S_T(t)$ and $\int_0^t S_T(u) \lambda_j(u) \, du$, respectively. In the Markov formulation with $m = 3$ risks, the probability of being in state 0 at time t when risks 2 and 3 have been eliminated is seen to be

$$\exp\left[-\int_0^t g_1(u) \, du \right]$$

by solution of the first differential equation above with $g_2(t) = g_3(t) = 0$. This is, however, precisely equal to $S_1(t) = S(t, 0, 0) = \exp[-\int_0^t \lambda_1(u) \, du]$ in the ILFT model. Extensions of these arguments show that the Markov method for competing risks is entirely equivalent to the ILFT model for the classical multiple-decrement problem.

However, if one considers more general transition structures, one is led to a new class of problems which admit new analytical approaches because more data are available. For example, suppose that following cancer surgery a patient is in a cancer-free state (state 0), a state with recurrent cancer (state 1), or the death state (2), and suppose only the transitions $0 \rightarrow 1$, $1 \rightarrow 2$, and $0 \rightarrow 2$ are possible. In a patient who follows the path $0 \rightarrow 1 \rightarrow 2$, one can observe transition times t_{01}, and t_{12} but not t_{02}. In one who dies directly from state 0, only t_{02} is observable. Such data allow us to answer such questions as: "Is the risk of death at time t higher in a patient with recurrent disease than in a patient in state 0?" The essential feature of these models is the inclusion of intermediate nonabsorbing states, such as state 1 above. The work of Fix and Neyman [17], Chiang [10], and Hoem [22] assumes that a stationary Markov process governs transitions among states. Nonparametric methods of Fleming [18, 19], Aalen [3], and Aalen and Johansen [4] allow one to estimate transition probabilities even when the Markov process is not homogeneous. Berlin et al. [5] discuss a Markov model for analyzing animal carcinogenesis experiments. The nonabsorbing intermediate states in this model are defined by the presence or absence of certain diseases which can only be diagnosed at autopsy. Thus one does not know the state of an animal except at the time of its death. To surmount this difficulty, which is not present in the applications treated by Fix and Neyman [17] or Chiang [10], the experiment is designed to obtain additional data by sacrificing animals serially. These extensions of the simple Markov model for multiple-decrement life tables, and the results of Prentice et al. [28] and Lagakos et al. [24], indicate the variety of methods that may be of use when the competing-risk structure is relaxed to include additional data.

References

[1] Aalen, O. (1976). *Scand. J. Statist.*, **3**, 15–27.

[2] Aalen, O. (1978). *Ann. Statist.*, **6**, 534–545.

[3] Aalen, O. (1978). *Ann. Statist.*, **6**, 701–726.

[4] Aalen, O. and Johansen, S. (1977). *An Empirical Transition Matrix for Non-homogeneous Markov Chains Based on Censored Observations.* Preprint No. 6, Institute of Mathematical Statistics, University of Copenhagen.

[5] Berlin, B., Brodsky, J., and Clifford, P. (1979). *J. Amer. Statist. Ass.*, **74**, 5–14.

[6] Bernoulli, D. (1760). Essai d'une nouvelle analyse de la mortalité causeé par la petite vérole, et des avantages de l'inoculation pour le prévenir. *Historie avec les Mémoirs, Académie Royale des Sciences*, Paris, pp. 1–45.

[7] Böhmer, P. E. (1912). Theorie der unabhängigen Wahrscheinlichkeiten. Rapport. *Mémoires et Procès-verbaux de Septième Congrès International d'Actuaires*, Amsterdam, Vol. 2, pp. 327–346.

[8] Breslow, N. (1970). *Biometrika*, **57**, 579–594.

[9] Breslow, N. and Crowley, J. (1974). *Ann. Statist.*, **2**, 437–453.

[10] Chiang, C. L. (1968). *Introduction to Stochastic Processes in Biostatistics.* Wiley, New York, Chap. 11.

[11] Cox, D. R. (1959). *J. R. Statist. Soc. B*, **21**, 411–421.

[12] Crowley, J. (1973). Nonparametric Analysis of Censored Survival Data with Distribution Theory for the *k*-Sample Generalized Savage Statistic. Ph.D. dissertation, University of Washington.

[13] D'Alembert, J. L. R. (1761). Onzième mémoire: sur l'application du calcul des probabilités à l'inoculation de la petite vérole. *Opusc. Math.*, **2**, 26–95.

[14] David, H. A. and Moeschberger, M. L. (1978). *The Theory of Competing Risks.* Griffin's Statist. Monogr. No. 39, Macmillan, New York.

[15] Downton, F. (1970). *J. R. Statist. Soc. B*, **32**, 408–417.

[16] Elandt-Johnson, R. C. (1976). *Scand. Actuarial J.*, **59**, 37–51.

[17] Fix, E. and Neyman, J. (1951). *Hum. Biol.*, **23**, 205–241.

[18] Fleming, T. R. (1978). *Ann. Statist.*, **6**, 1057–1070.

[19] Fleming, T. R. (1978). *Ann. Statist.*, **6**, 1071–1079.

[20] Gail, M. (1975). *Biometrics*, **31**, 209–222.

[21] Greville, T. N. E. (1948). *Rec. Amer. Inst. Actuaries*, **37**, 283–294.

[22] Hoem, J. M. (1971). *J. R. Statist. Soc. B*, **33**, 275–289.

[23] Kaplan, E. L. and Meier, P. (1958). *J. Amer. Statist. Ass.*, **53**, 457–481.

[24] Lagakos, S. W., Sommer, C. J., and Zelen, M. (1978). *Biometrika*, **65**, 311–317.

[25] Makeham, W. M. (1874). *J. Inst. Actuaries*, **18**, 317–322.

[26] Marshall, A. W. and Olkin, I. (1967). *J. Amer. Statist. Ass.*, **62**, 3–44.

[27] Peterson, A. V. (1976). *Proc. Natl. Acad. Sci. USA*, **73**, 11–13.

[28] Prentice, R. L., Kalbfleisch, J. D., Peterson, A. V., Jr., Flournoy, N., Farewell, V. T., and Breslow, N. E. (1978). *Biometrics*, **34**, 541–554.

[29] Seal, H. L. (1977). *Biometrika*, **64**, 429–439.

[30] Tsiatis, A. (1975). *Proc. Natl. Acad. Sci. USA*, **72**, 20–22.

(ACTUARIAL STATISTICS, LIFE BIOSTATISTICS
CLINICAL TRIALS
COX MODEL
EPIDEMIOLOGICAL STATISTICS
HAZARD RATE
LIFE TABLES
MULTIPLE-DECREMENT
   LIFE TABLES
SURVIVAL ANALYSIS)

M. GAIL

COMPLEMENTARY EVENTS

Two events are complementary if each is equivalent to absence of the other. In ordinary language they are "opposites." If A and B are complementary,

$$\Pr[A] + \Pr[B] = 1.$$

Usually, the complement of an event E is denoted $\bar{E}$. The notations E', $\tilde{E}$, and $\sim E$ are also used.

COMPLETENESS

Let $\mathscr{P} = \{P_\theta : \theta \in \Omega\}$ be a family of probability distributions of a random variable (or a statistic) X, indexed by a parameter set Ω. The family $\mathscr{P}$ is complete if for any function ϕ satisfying

(a) $E_\theta[\phi(X)] = 0$ for all $\theta \in \Omega$

$\phi(x) = 0$ for all x (except possibly on a set of probability zero for all $\theta \in \Omega$). Here E_θ denotes the expectation with respect to the distribution P_θ.

Informally, one might say that a family $\mathscr{P}$ of distributions indexed by a parameter θ is complete if there is no *unbiased* "estimator of zero" except the trivial one $\phi(x) \equiv 0$. The term "complete" is borrowed from functional analysis and is related to a complete set of elements of a Hilbert space (see, e.g., Gordesch [1]).

Intuitively, condition (a) represents restriction on the function ϕ; the larger the set Ω, or equivalently the larger the family $\mathscr{P}$, the greater the restriction on ϕ. When the family $\mathscr{P}$ is so large that condition (a) eliminates all ϕ except the trivial $\phi(x) \equiv 0$, the family $\mathscr{P}$ of distributions becomes a complete family. Completeness is applied via the Rao–Blackwell* and Lehmann–Scheffé theorems* to construction of the "best" unbiased estimators in numerous distributions of wide applicability, such as binomial*, Poisson*, exponential*, and gamma*. It should be noted that completeness is a property of a family of distributions rather than of a random variable and that removing even one point from the parameter set may alter the completeness status of the family [5]. Similarly, a complete family may lose its property if a distribution with a different support is added to the family, in spite of claims to the contrary appearing in some standard textbooks (see, e.g., Johnson and Kubicek [3] for more details).

For additional information on this topic, see refs. 2, 4, 6, and 7.

References

[1] Gordesch, J. (1972). *Amer. Statist.*, **26**(5), 45–46.

[2] Hogg, R. V. and Craig, A. T. (1970). *Introduction to Mathematical Statistics*, 3rd. ed. Macmillan, New York.

[3] Johnson, D. E. and Kubicek, J. D. (1973). *Amer. Statist.*, **27**(5), 240–241.

[4] Lehmann, E. L. (1959). *Testing Statistical Hypotheses*. Wiley, New York.

[5] Stigler, S. M. (1972). *Amer. Statist.* **26**(2), 28–29.

[6] van der Waerden, B. L. (1965). *Mathematical Statistics*, 2nd ed. Springer-Verlag, Heidelberg.

[7] Zacks, S. (1971). *The Theory of Statistical Inference*. Wiley, New York.

(LEHMANN–SCHEFFÉ THEOREM
MINIMUM VARIANCE UNBIASED
  ESTIMATOR
RAO–BLACKWELL THEOREM
SUFFICIENCY
UNBIASEDNESS)

COMPLEXITY   *See* ALGORITHMIC INFORMATION THEORY

COMPONENT ANALYSIS

Component analysis, or principal component analysis, is a method of transforming a set of variables $x_1, x_2, \ldots, x_p$ to a new set $y_1, y_2, \ldots, y_p$ with the following properties:

1. Each y is a linear combination of the x's, i.e.,

$$y_i = a_{i1}x_1 + a_{i2}x_2 + \cdots + a_{ip}x_p.$$

2. The sum of the squares of the coefficients a_{ij}, where $j = 1, 2, \ldots, p$, is unity.

3. Of all possible linear combinations un-correlated with y_1, y_2 has the greatest variance. Similarly, y_3 has the greatest variance of all linear combinations of the x_i uncorrelated with y_1 and y_2, etc.

The new set of p variables represents a transformation of the original variables such that the new variables are uncorrelated and are arranged in order of decreasing variance. The method is perfectly general, and, apart from the assumption that the variables are somehow relevant to the analysis, invokes no underlying model and hence no hypothesis that can be tested. It is simply a different, and possibly more convenient, way of reexpressing a set of variables. The method has been known for some years, but has been applied widely only since electronic computers have been available for general use.

MATHEMATICAL DERIVATION

The essential nature of multivariate data is illustrated in Table 1, which supposes that p variables are observed on each individual in a sample of n such individuals. The observed variables may be variates; i.e., they may be values of a defined set with a specified relative frequency or probability. In matrix terms, the values given to the variables for the ith individual are the scalars x_{ij}, where j can take all the values from 1 to p. The whole series of values for the ith indi-

vidual is given by the vector

$$\mathbf{x}_i' = (x_{i1}, x_{i2}, x_{i3}, \ldots, x_{ip}).$$

The complete set of vectors represents the data matrix $\mathbf{X}$ with n rows and p columns:

$$\mathbf{X} = (\mathbf{x}_1, \mathbf{x}_2, \mathbf{x}_3, \ldots, \mathbf{x}_n)'$$

and is a compact notation for the whole sample. The ith row of the matrix gives the values of the p variables for the ith sample: the jth column of the matrix gives the values of the jth variable for each of the n individuals in the sample. No assumptions are made about the extent to which the individuals can be regarded as representative of some defined population. In some instances, the individuals may be the whole population.

This compact notation for the data matrix has some convenient computational properties. Without loss of generality, we may assume that the variables are measured about the means for the set, so that all the column means are zero. Then the variance–covariance matrix may be calculated as

$$\mathbf{X}'\mathbf{X}/(n-1),$$

where $\mathbf{X}'$ is the transpose of the original data matrix. The matrix of sums of squares and products, $\mathbf{X}'\mathbf{X}$, and hence the variance–covariance matrix, has the mathematical property of being real, symmetric, and positive semidefinite. Geometrically, the data of Table 1 can be represented as n points in p dimensions, with the values of the jth variable ($j = 1, 2, \ldots, p$) for each unit referred to the jth of p rectangular coordinate axes. When the number of variables is large, the resulting geometric representation is in many dimensions and cannot easily be visualized, especially as these dimensions are not orthogonal, or at right angles to each other.

Where no a priori structure is imposed on the data matrix, component analysis seeks a rotation of the axes of the multivariate space such that the total variance of the projections of the points on the first axis is a maximum. It seeks a second axis orthogonal to the first, and which accounts for as much as possible of the remaining variance, and so on.

Table 1 p Variables Observed on a Sample of n Individuals

Individuals (Samples)	Variables (Variates)			
	V_1	V_2	V_3	V_p
I_1	x_{11}	x_{12}	x_{13}	x_{1p}
I_2	x_{21}	x_{22}	x_{23}	x_{2p}
I_3	x_{31}	x_{32}	x_{33}	x_{3p}
$\vdots$	$\vdots$	$\vdots$	$\vdots$	$\vdots$
I_n	x_{n1}	x_{n2}	x_{n3}	x_{np}

If $\mathbf{X}' = (x_1, x_2, \ldots, x_p)$ represents a point in the p-dimensional space, the linear combination of $l'\mathbf{X}$ of its coordinates represents the length of an orthogonal projection onto a line with a direction cosine l, where $l = (l_1, l_2, \ldots, l_p)$ and $l'l = 1$. The sample variance of all n elements is given by

$$\mathbf{V} = l'\mathbf{X}'\mathbf{X}l$$

and to maximize $\mathbf{V}$ subject to the constraint of orthogonality, we maximize the criterion

$$\mathbf{V}' = l\mathbf{X}'\mathbf{X}l - \lambda(l'l - 1)$$
$$= l'\mathbf{W}l - \lambda(l'l - 1),$$

where $\mathbf{W} = \mathbf{X}'\mathbf{X}$.

It can be shown that the p equations in p unknowns $l_1, l_2, \ldots, l_p$ have consistent solutions if and only if $|\mathbf{W} - \lambda\mathbf{I}| = 0$. This condition, in turn, leads to an equation of degree p in λ with p solutions $\lambda_1, \lambda_2, \ldots, \lambda_p$. These solutions are variously designated as the *latent roots*, *eigenvalues*, or *characteristic roots* of $\mathbf{W}$.

Substitution of each solution $\lambda_1, \lambda_1, \ldots, \lambda_p$ in

$$(\mathbf{W} - \lambda\mathbf{I})l = 0$$

gives corresponding solutions of l which are uniquely defined if the λ's are all distinct, and these solutions are designated as the *latent vectors*, *eigenvectors*, *or characteristic vectors* of $\mathbf{W}$.

The extraction of the eigenvalues and eigenvectors of the variance–covariance matrix of the original data matrix representing n points in p dimensions neatly defines the linear combination of the original variables which account for the maximum variance while remaining mutually orthogonal. The elements of the eigenvectors provide the appropriate linear weightings for the components, and the eigenvector, expressed as a proportion of the number of dimensions (p), gives the proportion of the total variance accounted for by the component.

The sum of the eigenvalues is equal to the sum of the elements of the principal diagonal of the variance–covariance matrix, so that the sum of the variances of the compo-nents is the same as that of the original variables.

Component analysis also has a simple geometrical interpretation. The equation $\mathbf{x}'\mathbf{W}^{-1}\mathbf{x} = K$ represents an ellipsoid in p dimensions. If the x's are variates with a multivariate normal distribution, these ellipsoids are the contours of equal probability density, centered on a common mean. The calculations involved in finding the components then correspond to the search for the principal axes of the ellipsoid, in order of length.

Because component analysis involves no assumptions about the relationships between the variables, there are no important tests of significance that can be applied. Bartlett [1], however, provides some approximate tests of interest when it can be assumed that the variables are variates which are normally and independently distributed.

1. If $\mathbf{W}$ is the sample dispersion matrix of p variates which are normally and independently distributed with the same variances,

$$-n \ln\left[|\mathbf{W}| \{ \operatorname{tr}(\mathbf{W})/p \}^{-p} \right]$$

is distributed approximately as χ^2 with $\frac{1}{2}(p + 1) - 1$ degrees of freedom.

2. If this test is applied to the correlation matrix $\mathbf{R}$, $-n \ln|\mathbf{R}|$ is approximately distributed as χ^2 with $\frac{1}{2}p(p + 1)$ degrees of freedom.

If the value of χ^2 derived from either test is not significant, the assumption that the variables are all normally and independently distributed cannot be rejected, and there is no point in calculating components.

3. The test may be extended to test whether the first k eigenvalues account for all of the interdependence between the variates. If K is the ratio of the arithmetic mean of the remaining eigenvalues $l_{k+1} \cdots l_p$ to their geometric mean, then $n(p - k)\ln K$ is approximately distributed as χ^2 with $\frac{1}{2}(p - k - 1)(p - k + 2)$ degrees of freedom.

TRANSFORMATIONS

Component analysis is not independent of scale, so that, for example, the multiplication of one or more of the variables by a constant will result in a different set of components. In practical applications, therefore, it is always important to consider the possibility, or desirability, of transforming the data before analysis. One important and frequently used transformation is that of natural or common logarithms. The effect of the logarithmic transformation is to give measures with the same proportional variability the same variance.

The linear combinations of the transformed x-values are then equivalent to complex ratios of the original x's. Measures that are relatively more variable have a higher variance and are given more weight in the analysis.

When the original data matrix consists of variables of quite different kinds, with different units or measured on widely different scales, the common practice is to divide each variable by its standard deviation, so reducing the variances to unity. This procedure is equivalent to finding the eigenvalues and eigenvectors of the correlation matrix instead of the covariance matrix.

PRACTICAL APPLICATIONS

Component analysis is widely used as a practical tool of statistical analysis. The practical reasons for using an orthogonal transformation of this kind include the following:

1. Examination of the correlations between the variables in a large set of data is often helped by an indication of the weight given to each variable in the first few components. It is then possible to identify groups of variables which tend to vary in much the same way across the set of individuals.

2. Reduction of dimensions of the variability in a measured set to the smallest number of meaningful and independent dimensions. It is frequently useful to know just how many orthogonal axes are needed to represent the major part of the variation of the data set.

3. Elimination of redundant variables. Various methods have been developed from component analysis to aid the elimination of redundant variables from large sets of variables before subjecting them to more critical analysis: some of these methods are described in papers listed in the bibliography.

4. Examination of the clustering of individuals in n-dimensional space. Although component analysis is not an essential prerequisite of *cluster analysis**, some analysts prefer to remove variability which may be regarded as extraneous or as "noise" by transforming the data to orthogonal axes and then omitting the eigenvectors corresponding to the lowest-valued eigenvalues.

5. Determination of an objective weighting of measured variables in the construction of meaningful indices as an alternative to indices or composite measures based on a priori judgments, component analysis may provide interpretable linear functions of the original variables which may serve as valuable indices of variation.

6. Allocation of individuals to previously demarcated groups is, strictly speaking, the purpose of *discriminant (function) analysis** or *canonical (variate) analysis**. Where the basic assumptions of these analyses cannot be satisfied, however, the plotting of component values on their orthogonal axes often helps in the allocation of individuals to groups, or the recognition of the existence of a new group.

7. Similarly, plotting of component values often helps in the recognition of misidentified individuals, or in the recognition of "outliers," individuals who differ markedly in one or more dimensions from the rest of the set.

8. Orthogonalization of regressor variables as a preliminary to *multiple regression analysis**. It may sometimes be helpful to use component analysis to identify the number of orthogonal dimensions measured by a set of regressor variables, and to use the resulting components as new regressor variables in subsequent regression analyses. Where component analysis is applied to sets of independent and regressor variables, the correlations between the two sets of components are sometimes more readily interpretable than the results of *canonical correlation analysis*.

RELATED TECHNIQUES

In the simplest cases of multivariate analysis, the basic data matrix of Table 1 has no *a priori* structure imposed upon it, and it is in such cases that *principal component analysis**, as defined below, is appropriate. An alternative mode of analysis, *factor analysis**, depends upon some additional assumptions, and under certain conditions gives similar results to principal component analysis. Again, when the basic data matrix is used to define the interpoint distances between the *n* points in Euclidean space, *principal coordinate analysis* finds the *n* points relative to principal axes that will give rise to these distances. Various forms of *cluster analysis* may also be derived from the unstructured matrix, and the most important of these methods depend upon the concept of the minimum spanning tree, defined as the set of lines in *p*-dimensional space joining the *n* points in such a way that the sum of the lengths of the lines is a minimum.

The imposition of certain kinds of a priori structure on the basic data matrix defines some alternative methods of analysis. Thus when the individuals or samples of the matrix can be assigned to separate groups or populations, the problem becomes one of discrimination, requiring the use of a *discriminant* function, in the case of two groups, or either *generalized distance* or *canonical variate analysis* when there are more than two groups. A priori allocation of the variables of the matrix to groups may similarly lead to *multiple regression analysis*, *orthogonalized regression*, or *canonical correlation analysis*.

Bibliography

In addition to the Bartlett reference, see the following:

Beale, E. M. L., Kendall, M. G., and Mann, D. W. (1967). *Biometrika*, **54**, 357–365.

Gabriel, K. R. (1971). *Biometrika*, **58**, 453–467.

Harris, R. J. (1975). *A Primer of Multivariate Statistics*. Academic Press, New York.

Hawkins, D. M. (1973). *Appl. Statist.*, **22**, 275–286.

Jolliffe, I. T. (1972). *Appl. Statist.*, **21**, 160–173.

Jolliffe, I. T. (1973). *Appl. Statist.*, **22**, 21–31.

Kendall, M. G. (1957). *Multivariate Analysis*. Charles Griffin, London. (Revised 1975.)

Krzanowski, W. J. (1971). *Statistician*, **20**, 51–61.

Mansfield, E. R., Webster, J. T., and Gunst, R. F. (1977). *Appl. Statist.*, **26**, 34–40.

Marriott, F. H. C. (1974). *The Interpretation of Multiple Observations*. Academic Press, New York.

Reference

[1] Bartlett, M. S. (1954). *J. R. Statist. Soc. B*, **16**, 296–298.

(CLASSIFICATION
CORRESPONDENCE ANALYSIS
DENDRITE
DISCRIMINANT ANALYSIS
FACTOR ANALYSIS
MULTIVARIATE ANALYSIS)

J. N. R. JEFFERS

COMPONENTS OF VARIANCE   See VARIANCE COMPONENTS

COMPOSITE DESIGN

To construct an experimental design one must have, however humble, a mathematical model descriptive of the response under study. Thus to discuss composite designs we first assume a measurable response variable $\eta = g(\xi, \theta)$ where $g(\cdot)$ is a continuous unknown function of k controlled variables

$\boldsymbol{\xi}' = (\xi_1, \xi_2, \ldots, \xi_k)$, containing p parameters $\boldsymbol{\theta}' = (\theta_1, \theta_2, \ldots, \theta_p)$. The objective is to approximate the unknown function g using a low-order polynomial model $f(\mathbf{x}, \boldsymbol{\beta})$ derived from a Taylor series* expansion of η about some fixed point ξ_0. The first-order approximation $f(\mathbf{x}, \boldsymbol{\beta})$ of $g(\boldsymbol{\xi}, \boldsymbol{\theta})$ is given by

$$\eta = \beta_0 + \sum_{i=1}^{k} \beta_i x_i,$$

where $x_i = (\xi_i - \xi_{i0})/C_i$ is a convenient coded (standardized) value of the controlled variable ξ_i with the ξ_{i0} usually chosen so that $\sum x_i = 0$. The initial parameter in the approximating polynomial model, β_0, is a constant and the k first-order polynomial coefficients β_i are the first-order derivatives in the Taylor series* expansion. The second-order approximation is given by

$$\eta = \beta_0 + \sum_i \beta_i x_i + \sum_i \beta_{ii} x_i^2$$
$$+ \sum_{i \neq j} \sum \beta_{ij} x_i x_j,$$

where k terms β_{ii} (the quadratic coefficients) and the $k(k-1)/2$ terms β_{ij} (the cross-product or two-factor interaction* coefficients) are the second-order derivative terms in the series expansion. Expansions to third- and higher-order terms are possible but little used. Our initial discussion of composite designs employs the approximating polynomial model $\eta = f(\mathbf{x}, \boldsymbol{\beta})$, where $\mathbf{x}' = (x_1, x_2, \ldots, x_k)$ is equivalent to $\boldsymbol{\xi}'$ and $\boldsymbol{\beta}$ is a vector of unknown coefficients.

The application of first- and second-order approximations to an unknown response function is called response surface methodology*, and has its origins in a 1951 paper by Box and Wilson [8]. The first discussion of composite designs, both central and noncentral, is found in this paper. The need for composite designs originates in the initial desire of the experimenter to use the simplest empirical model, usually a first-order model. This desire is balanced against the possibility that a higher-order model may be needed. If the need for additional terms in the initial model unfolds, the design is augmented with additional points to form a composite design. Excellent examples are described in a second early paper by Box [3].

The first-order model may be written $\eta = \mathbf{X}_1 \boldsymbol{\beta}_1$, where η is the $n \times 1$ vector of responses, $\mathbf{X}_1$ the $n \times (k+1)$ matrix of "independent variables," and $\boldsymbol{\beta}_1$ a vector containing $(k+1)$ coefficients. For the second-order model we write $\eta = \mathbf{X}_1 \boldsymbol{\beta}_1 + \mathbf{X}_2 \boldsymbol{\beta}_2$, where $\mathbf{X}_2$ is the $n \times [k^2 + k(k+1)/2]$ matrix of second-order "independent variables" and $\boldsymbol{\beta}_2$ is the vector of second-order coefficients containing $k + k(k+1)/2$ elements. The full second-order model will contain $(k+2)(k+1)/2$ coefficients.

A collection of n experimental points provides the $n \times k$ design matrix* $\mathbf{D}$. Let the row vector $\mathbf{d}_u' = (x_{1u}, x_{2u}, \ldots, x_{nu})$ indicate the settings of the k standardized variables used in the uth experimental trial, and let the single observed value be y_u. For the linear model $\eta = \mathbf{X}\boldsymbol{\beta}$ and vector of observations $\mathbf{Y}$, we have $E(\mathbf{Y}) = \eta = \mathbf{X}\boldsymbol{\beta}$ and $E[(\mathbf{Y} - \eta)(\mathbf{Y} - \eta)'] = \mathbf{I}_n \sigma^2$. The least-squares estimates $\mathbf{B}$ of $\boldsymbol{\beta}$ are given by

$$\mathbf{B} = (\mathbf{X}'\mathbf{X})^{-1}\mathbf{X}'\mathbf{Y}.$$

The variance–covariance matrix* of these estimates is $E[(\mathbf{B} - \boldsymbol{\beta})(\mathbf{B} - \boldsymbol{\beta})'] = (\mathbf{X}'\mathbf{X})^{-1}\sigma^2$, and an unbiased estimate of σ^2 by $s^2 = [\mathbf{Y}'\mathbf{Y} - \mathbf{B}'\mathbf{X}'\mathbf{Y}]/(n-p)$, where p is the total number of parameters in $\boldsymbol{\beta}$.

If a first-order model is appropriate, the design matrix $\mathbf{D}$ is chosen so as to "minimize" (in some way) the variances of the coefficients $\boldsymbol{\beta}_1$. As shown in Box [2], designs of best precision for $k = N - 1$ factors are provided by the vertices of a $N - 1$ dimensional regular simplex*, e.g., the equilateral triangle for $k = 2$, the tetrahedron for $k = 3$. In general, the k coordinates of the $N = k + 1$ points of a simplex* are given by the row elements in the following $(N \times k)$ design matrix $\mathbf{D}$:

$$\mathbf{D} = \begin{bmatrix} a & b & c & \ldots & k \\ -a & b & c & \ldots & k \\ 0 & -2b & c & \ldots & k \\ 0 & 0 & -3c & \ldots & k \\ \vdots & \vdots & \vdots & \ldots & \vdots \\ 0 & 0 & 0 & \ldots & -k^2 \end{bmatrix},$$

where for a regular simplex, the sum of squares of the elements in any column should equal N. For the special case of $k = 2^p - 1$, where p is any integer, the coordinates of the simplex can also provide the coordinates of the saturated two-level fractional factorial* designs (the Plackett–Burman designs [14]). Examples of the coordinates of first-order designs for $k = 2, 3$ are given in the following design matrices:

$$
\begin{array}{cc}
x_1 & x_2 \\
\end{array}
$$

$$
\begin{bmatrix}
\sqrt{3/2} & \sqrt{1/2} \\
-\sqrt{3/2} & \sqrt{1/2} \\
0 & -2\sqrt{1/2}
\end{bmatrix}
$$

$$
\begin{array}{ccc}
x_1 & x_2 & x_3 \\
\end{array}
$$

$$
\begin{bmatrix}
-1 & -1 & -1 \\
1 & -1 & -1 \\
-1 & 1 & -1 \\
1 & 1 & 1
\end{bmatrix}
$$

$k = 2$, equilateral $k = 3$, regular tetra-
triangle design   hedron as a frac-
                              tional factorial

$$
\begin{array}{ccc}
x_1 & x_2 & x_3 \\
\end{array}
$$

$$
\begin{bmatrix}
\sqrt{2} & \sqrt{2/3} & \sqrt{1/3} \\
-\sqrt{2} & \sqrt{2/3} & \sqrt{1/3} \\
0 & -2\sqrt{2/3} & \sqrt{1/3} \\
0 & 0 & -3\sqrt{1/3}
\end{bmatrix}
$$

$k = 3$, regular
tetrahedron

Suppose that an experimenter plans to fit the first-order model $\eta = \mathbf{X}_1 \boldsymbol{\beta}_1$, when in fact the second-order model $\eta = \mathbf{X}_1 \boldsymbol{\beta}_1 + \mathbf{X}_2 \boldsymbol{\beta}_2$ is appropriate. For $k = 2$, the equilateral triangle design contains three points and the model three parameters. Under these circumstances no indication of the inadequacy of the first-order model is possible. Further, when the second-order model is appropriate, the least-squares estimates of the coefficients in the first-order model will be biased since $E(\mathbf{B}_1) = \boldsymbol{\beta}_1 + \mathbf{A}\boldsymbol{\beta}_2$, where $\mathbf{A} = (\mathbf{X}_1'\mathbf{X}_1)^{-1}\mathbf{X}_1'\mathbf{X}_2$. Further, even assuming that the initial design is replicated, the estimate of variance obtained from the residual sum of squares will also be biased, because

$$
E[\mathbf{Y}'\mathbf{Y} - \mathbf{B}_1'\mathbf{X}_1'\mathbf{Y}]
$$
$$
= (n - p)\sigma^2 + \boldsymbol{\beta}_2'\mathbf{X}_2'\big[\mathbf{I} - \mathbf{X}_1(\mathbf{X}_1'\mathbf{X}_1)^{-1}\mathbf{X}_1'\big]\mathbf{X}_2\boldsymbol{\beta}_2.
$$

To provide some measure of the lack of fit of the postulated first-order model, one or more center points $[0, 0]$ can be added to the equilateral-triangle design to form the simplest composite design. The 1-degree-of-freedom contrast* between the average of the observed responses at the peripheral points $\bar{y}_p$ minus the average at the center of the design $\bar{y}_0$ provides a separate measure of the $(\beta_{11} + \beta_{22})$ terms in the second-order model. Should this 1-degree-of-freedom contrast prove statistically significantly different from zero, the first-order model can be declared inadequate to represent the response functions. (*See* RESPONSE SURFACE METHODOLOGY.)

An alternative four-point design useful for fitting the first-order model, $k = 2$, is the 2^2 factorial, whose design matrix is

$$
\begin{array}{cc}
x_1 & x_2 \\
\end{array}
$$

$$
\begin{bmatrix}
-1 & -1 \\
1 & -1 \\
-1 & 1 \\
1 & 1
\end{bmatrix}.
$$

This design also provides orthogonal* estimates of all the coefficients in the first-order model. The alias* structure with respect to unestimated second-order coefficients leaves only the coefficient b_0 biased, i.e., $E(b_0) = \beta_0 + \beta_{12} + \beta_{22}$, $E(b_1) = \beta_1$, $E(b_2) = \beta_2$. The extra degree of freedom for sensing the lack of fit of the first-order model can be used to provide the estimated second-order coefficient $E(b_{12}) = \beta_{12}$. When one or more center points $[0, 0]$ are added to the 2^2 design, the contrast $\bar{y}_p - \bar{y}_0$ again provides a measure of the combined second-order terms $(\beta_{11} + \beta_{22})$. The 2^2 factorial with center points thus provides 2 degrees of freedom for measuring the inadequacy of the first-order model, one degree of freedom sensitive to β_{12} and the second to $(\beta_{11} + \beta_{22})$.

The equilateral triangle with n_0 center points and the $(2^2 + n_0)$ designs are basic building blocks when $k = 2$. For example, these first-order designs can each be symmetrically augmented to provide the hexagon, and octagon designs as illustrated in Fig. 1. The $(2^2 + n_0)$ may also be augmented to give the 3^3 factorial. These composite designs are highly recommended whenever the second order is postulated.

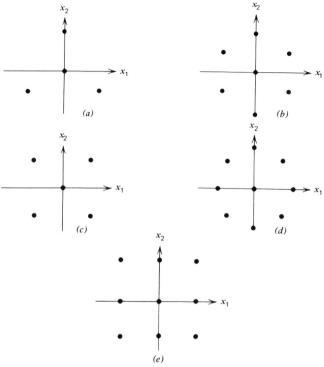

Figure 1 First- and second-order experimental designs: (*a*) first-order design (equilateral with center point), (*b*) second-order design (hexagonal), (*c*) first-order design (2^2 with center point), (*d*) second-order design (octagonal), (*e*) second-order design (3^2 factorial).

With the appropriate number of center points, both the hexagon and octagon designs are rotatable*; i.e., given that the second-order model is appropriate, the variance of the estimated response at any distance ρ from the center of the design is given by

$$V(\hat{y}) = V(b_0) + \text{cov}(b_0, b_{ii})\rho^2 + V(b_i)\rho^2$$
$$+ V(b_{ii})\rho^4.$$

CENTRAL COMPOSITE DESIGNS $k \geqslant 3$

Composite designs are commonly built up sequentially, usually beginning with a first-order design and associated model. To obtain estimates of the $(k + 1)$ parameters in the first-order model $\eta = \beta_0 + \sum_{i=1}^{k} \beta_i x_i$, the 2_{III}^{k-p} or 2_{IV}^{k-p} resolution III or IV fractional factorial designs* with center points n_0 are employed. If the fitted first-order

model proves inadequate, the initial fractional factorial is augmented by additional fractionals until either a full 2^k factorial or 2_{V}^{k-p}, (resolution V) fractional factorial is in hand. A full 2^k or 2_{V}^{k-p} allows the orthogonal estimation of all β_i and β_{ij} terms in the full second-order model. To obtain estimates of all β_{ii} terms, a "star" design consisting of axial points is symmetrically added, i.e., two additional points along each coordinate axis of the design in addition to center points. The set of points at the vertices of the cube (the 2^k or 2^{k-p} design) and the axial points of the star design are each first-order rotatable designs*. Combining these first-order designs symmetrically provides the most frequently employed central composite design. By adding varied numbers of center points, each of these designs may be partitioned in orthogonal blocks that may be used sequentially.

The design coordinates, partitioned into three blocks, for the central composite design for $k = 3$ are

Block I	Block II
x_1 x_2 x_3	x_1 x_2 x_3

$$\begin{bmatrix} - & - & + \\ + & - & - \\ - & + & - \\ + & + & + \\ 0 & 0 & 0 \\ 0 & 0 & 0 \end{bmatrix} \qquad \begin{bmatrix} - & - & - \\ + & - & + \\ - & + & + \\ + & + & - \\ 0 & 0 & 0 \\ 0 & 0 & 0 \end{bmatrix}$$

Block III

x_1 x_2 x_3

$$\begin{bmatrix} \alpha & 0 & 0 \\ \alpha & 0 & 0 \\ 0 & \alpha & 0 \\ 0 & \alpha & 0 \\ 0 & 0 & \alpha \\ 0 & 0 & \alpha \\ 0 & 0 & 0 \\ 0 & 0 & 0 \end{bmatrix}$$

For $k = 3$ and rotatability, we set $\alpha = 1.68$, and for orthogonal blocking, $\alpha = 1.63$. Table 1 gives a listing of the blocking and near-rotatable central composite designs for $k \leqslant 7$ [6].

For $k \geqslant 2$ a second-order rotatable design can always be derived from a regular simplex design. This is accomplished by first imagining the $k + 1$ vertices of a simplex symmetrically located about the origin of a k-dimensional coordinate system. The $k + 1$ vectors joining the vertices to the origin may now be added in pairs, triplets, and so on, to form additional vectors. The totality of vectors provides a basis for developing second-order rotatable designs, called the "simplex-sum" or Box–Behnken designs* [4]. Although these designs usually contain more points than those described in Table 1, they may be run in orthogonal blocks. An important by-product of the simplex-sum designs are the three-level second-order rotatable designs. The designs are formed by combining two-level factorial designs in balanced incomplete block design* arrangements. A complete listing of the designs for $k \leqslant 10$ is given in ref. 6. The design matrices for

$k = 3$, 4, and 5 are

$k = 3$

x_1	x_2	x_3	n
± 1	± 1	0	4
± 1	0	± 1	4
0	± 1	± 1	4
0	0	0	3
			$N = 15$

$k = 4$

x_1	x_2	x_3	x_4	n
± 1	± 1	0	0	4
0	0	± 1	± 1	4
0	0	0	0	1
± 1	0	± 1	0	4
0	± 1	0	± 1	4
0	0	0	0	1
± 1	0	0	± 1	4
0	± 1	± 1	0	4
0	0	0	0	1
				$N = 27$

$k = 5$

x_1	x_2	x_3	x_4	x_5	n
± 1	± 1	0	0	0	4
0	0	± 1	± 1	0	4
0	± 1	0	0	± 1	4
0	0	0	± 1	± 1	4
0	0	0	0	0	3
0	± 1	± 1	0	0	4
± 1	0	0	± 1	0	4
0	0	± 1	0	± 1	4
± 1	0	0	0	± 1	4
0	± 1	0	± 1	0	4
0	0	0	0	0	3
					$N = 56$

Note: $[\pm 1 \quad \pm 1]$ equals

$$\begin{matrix} -1 & -1 \\ +1 & -1 \\ -1 & +1 \\ +1 & +1 \end{matrix}$$

The design for $k = 4$ can be performed in three orthogonal blocks of nine runs each. The $k = 5$ design partitions into two orthogonal blocks of 23 runs each.

It is not necessary to augment a design symmetrically. Configurations of points comprising a composite design are constrained only by the imagination of the ex-

Table 1   Blocking Arrangements for Rotatable and Near-Rotatable Central Composite Design

Block	(2^{k-p})	Number of Controlled Variables					
		$k = 3$ (2_{III}^{3-1})	$k = 4$ (2_{IV}^{4-1})	$k = 5$ (2_{V}^{5-1})	$k = 6$ (2_{III}^{6-2})	$k = 7$ (2_{VI}^{7-1})	$k = 8$ (2_{V}^{8-2})
I, 2^{k-p} block	n_c	4	8	16	16	64	64
	n_0	2	2	6	4	8	8
II, 2^{k-p} block	n_c	4	8		16	a	b
	n_0	2	2		4		
III, axial block	n_a	6	8	10	12	14	16
	n_0	2	2	1	2	4	2
Total number points		20	30	33	54	80	80
α for rotatability		1.414	2.000	2.000	2.378	2.828	2.828
α for orthogonal		1.414	2.000	2.000	2.367	2.828	2.828

[a] The 2_{VI}^{7-1} portion may be partitioned into eight blocks, each a 2_{III}^{7-4} with a single center point. Generators: 124, 135, 236, 1237.

[b] The 2_{V}^{8-2} portion may be further partitioned into four blocks, each a 2_{III}^{8-4} with a single center point. Generators: 125, 236, 347, 1248.

perimenter. Examples of noncentral composite designs can be found in refs. 3 and 8. For the general polynomial model $\eta = X\beta$, the matrix $X'X$ is sometimes called the "moment matrix"* of the design. For rotatability the elements of the moment matrix $[X'X]$ are identical to those of a symmetric multivariate normal distribution* up to order $2d$, where d is the order of the fitted approximating polynomial. When over a region R, a model of order d_1 is used to approximate a response when the true function is a polynomial of degree $d_2 > d_1$, the bias caused by the neglected terms can be minimized by making the design moments equal to those of the uniform distribution up to order $d_1 + d_2$ [5]. If the region R is (hyper) spherical*, the design remains rotatable.

Many alternative "optimal" qualities have been proposed for $X'X$ [13]. In practical terms, the experimenter first chooses a rotatable or near-rotatable design, spreads the design points out to include 75 to 85% of a reasonably symmetric region of interest so as to minimize bias, and then adds, perhaps later, additional experimental points, often in asymmetric fashion. The final design thus becomes a composite of arrays of points

chosen for both their statistical merits and the practical needs of the experimenter. Parameter estimation is accomplished through least squares*.

Occasionally, controlled variables are constrained, as in a mixture experiment* wherein the mixture components (x_i) must sum to 100%. This problem was first discussed by Scheffé in 1958 [15]. The associated design D thus consists of ρ row vectors d_u, for each of which the elements sum to unity; i.e., $\sum_i x_{ui} = 1$. Thus, when $k = 3$, the design points will fall within the two-dimensional simplex bounded by the points $(1,0,0)$, $(0,1,0)$, $(0,0,1)$. In general, for mixture experiments, the arrays of design points fall on a $(k - 1)$-dimensional (hyper)plane. The objective is to arrange arrays of experimental points within the admissible region useful for fitting first- or second-order models on the subspace. In fact, all the usual strategies for employing composite designs and models can be employed in the subspace. The fitted models are also used to plot response contours in the $(k - 1)$-dimensional constrained space. Of course, many alternative interpretations of these fitted models are possible in terms of the original k-space. This is an excellent review and bibliography on the subject of composite constrained designs by Cornell [10]. Constrained composite designs for varied alternative models appears in the paper by Draper and St. John [11].

Another form of composite design occurs when experiments may be added one at a time to some original pattern, analysis following each experiment. For the case of the two-level factorials and fractional factorials, a particularly simple algorithm exists for updating the coefficients in the factorial model; see Hunter [12]. The construction of composite designs through the sequential addition of points for the case of nonlinear models has been described in a paper by Box and Hunter [7]. An excellent text describing the entire problem of linear and nonlinear models and design is by Beck and Arnold [1]. A recent exposition of the philosophy and practice of experimental design and models appears in ref. 9.

References

[1]   Beck, J. V. and Arnold, K. J. (1977). *Parameter Estimation in Engineering and Science*, Wiley, New York.
[2]   Box, G. E. P. (1952). *Biometrika*, **39**, 49–67.
[3]   Box, G. E. P. (1954). *Biometrics*, **10**, 16–60.
[4]   Box, G. E. P. and Behnken, D. W. (1960). *Ann. Math Statist.*, **31**, 838–864.
[5]   Box, G. E. P. and Draper, N. R. (1959). *J. Amer. Statist. Ass.*, **54**, 622–654.
[6]   Box, G. E. P. and Hunter, J. S. (1957). *Ann. Math. Statist.*, **28**, 195–241.
[7]   Box, G. E. P. and Hunter, W. G. (1965). *IBM Sci. Comput. Symp. Statist.*, p. 113.
[8]   Box, G. E. P. and Wilson, K. B. (1951). *J. R. Statist. Soc. B*, **13**, 1–45. (The original landmark paper, including discussion.)
[9]   Box, G. E. P., Hunter, W. G., and Hunter, J. S. (1978). *Statistics for Experimenters*. Wiley, New York.
[10] Cornell, J. A. (1973). *Technometrics*, **15**, 437–455.
[11] Draper, N. R. and St. John, R. C. (1977). *Technometrics*, **19**, 117–130.
[12] Hunter, J. S. (1964). *Technometrics*, **6**, 41–55.
[13] Kiefer, J. (1975). *Biometrika*, **62**, 277–288.
[14] Plackett, R. L. and Burman, J. P. (1946). *Biometrika*, **33**, 305–325.
[15] Scheffé, H. (1958). *J. R. Statist. Soc. B*, **20**, 344–360.

(DESIGN AND ANALYSIS OF
     EXPERIMENTS
GENERAL LINEAR MODEL
RESPONSE SURFACES)

J. S. HUNTER

COMPOSITE HYPOTHESIS

A statistical hypothesis* that does not completely determine the joint distribution of the random variables in a model*. It is usually possible, and often enlightening, to regard a composite hypothesis as being composed of a number of simple hypotheses*, each of which does completely determine the joint distribution. For example, the hypothesis that two binomial distributions* have the same value of p is composite—it can be regarded as being composed of simple hypotheses that the common value exists *and* is equal to a specified value p_0, say, for some p_0 in the interval [0, 1].

(HYPOTHESIS TESTING
SIMPLE HYPOTHESIS)

COMPOUND DISTRIBUTION

A special kind of mixture distribution* in which distributions of a particular family* are mixed by assigning a distribution to one or more parameters of the family. For example, one might assign a normal distribution* to the expected value of a normal distribution, or a gamma distribution* to the reciprocal of the variance of a normal distribution.

There is a convenient notation for compound distributions. The symbol $\wedge$ is used to denote compounding, and symbols representing the *compounding distribution* and the *compounded distribution* are placed to the right and left of $\wedge$, respectively. The compounded parameter(s) is(are) denoted by symbol(s) placed under the $\wedge$.

In this notation the symbols for the two examples in the first paragraph would be

$$N(\mu, \omega^2) \bigwedge_{\xi} N(\xi, \sigma^2)$$

and

$$\Gamma(\alpha, \beta) \bigwedge_{\sigma^{-2}} N(\xi, \sigma^2).$$

The term "compound distribution" is sometimes used rather loosely, for distributions obtained by the "generalizing" (random sum*) process. Some, but not all of the latter do also happen to be compound distributions.

(CONTAGIOUS DISTRIBUTIONS
MIXTURE DISTRIBUTIONS
NEGATIVE BINOMIAL
NEYMAN TYPE A
NONCENTRAL CHI-SQUARE
POISSON–BINOMIAL DISTRIBUTION
PÓLYA–EGGENBERGER
    DISTRIBUTION)

COMPUTER ASSISTED TELEPHONE SURVEYS *See* TELEPHONE SURVEYS, COMPUTER ASSISTED

COMPUTERS AND STATISTICS

Modern statistics is heavily dependent on high-speed computers. Their impact on statistical methodology goes far beyond that of mechanical aids. Indeed, they affect the heart and soul of statistical science and technology, influencing the outlook of statisticians.

This brief essay will select certain examples of statistical methodology, and these are discussed in order of frequency of occurrence rather than the sophistication of computer scientific technology employed.

QUALITY CONTROL* AND EDITING OF DATA*

The ever-growing number of studies and surveys in the social, engineering, and life sciences is producing an overwhelming amount of data. A statistical analysis of such information usually consists of the computation of summaries or estimates from such data. For example, we may wish to check on the accuracy of the inventory records of an arsenal by a direct sample survey* in the warehouses. This will lead to statistical estimates of the *actual* inventory as opposed to that on record. Again, we may wish to estimate incidence rates of certain diseases for certain communities from data collected in the National Health Interview Survey. If the task of a computer were confined to the mere doing of sums and percentages, this would be an easy matter and one on which a high-powered computer would probably be wasted. However, it would be reckless to assume that there are no errors in the data as recorded. It is, therefore, of paramount importance that at all stages of data collection and processing the quality of the data be controlled. One such control is the scrutiny of data for internal consistency. As a trivial example we may quote the famous "teenage grandmothers" who turn up occasionally on census questionnaires and are duly eliminated. Here the inconsistency of "relation to head of household = mother" and "age = 16" is clearly apparent. There are other inconsistencies that are not as obvious; others that are conceivably correct, such as a "son" whose age exceeds that of the wife of his father. She may be his stepmother. Until quite recently such *common-sense data* scrutiny and consequential editing was performed by hosts of clerks. For studies of a more *specialized* nature, however, inconsistencies can, of course, be discovered only by personnel with the required expert knowledge. For example, in a study involving clinical examinations of cancer of the breast the classification "stage" may have to be checked against the recorded anatomical or histological division codes. Only personnel completely familiar with clinical concepts will be able to scrutinize such records.

With the advent of high-speed computers more and more of these functions of data scrutiny are being taken over by these giant machines. For this to be feasible we must convey to the computer in minute detail the complete logical sequence of the involved check procedure, including all the know-how as to what the expert would be looking for. Moreover, for it to deal with cases that may be errors or may be correct data of an unusual nature, the computer must be able to refer to statistical information so that it can gauge a suspect discrepancy against a statistical tolerance.

After all such information has been stored in the computer's memory, the data from the particular study, survey, or census are passed through the computer for automatic scrutiny. As soon as the computer encounters an inconsistency it is instructed to either (a) record (on tape or punched card) the details of the suspected inconsistency in the data and list it for human inspection and reconciliation, or (b) immediately correct any inconsistent item (or compute a missing item) with the help of statistical estimation procedures and using the data that it has already accepted. This procedure is generally known as "computer imputation" of rejected or missing content items. The perhaps best known method is based on the "hot deck" widely used by federal agencies. This method assumes that the data tape is passed through the computer in a logical order such that the units (say households) just inspected (hot units) are closely correlated with the currently inspected unit. The hot deck consists of a multivariate classification table using important content items such as age of head of household, race, number of children in household, etc., as classifiers. In the body of the hot deck are stored all the content items for the last unit, with a complete record falling into the hot deck cell, and these are used as estimators of rejected and missing items for any unit falling into this cell.

More recently, considerable improvements in the hot deck method have been evolved, giving rise to an extensive literature on "incomplete data analysis."

Most organizations, such as the Bureau of the Census, using automatic data scrutiny and editing, employ a judicious combination of (a) and (b). Procedure (a) is usually preferred in situations when human eliminations of reconciled inconsistencies are administratively feasible, as is true with smaller and/or rather specialized studies. Method (b), imputation of suspect data, is adopted when the merging of a correction tape with the original data tape becomes practically infeasible, as is the case with certain census operations. With adequate control of the frequency with which imputations are made, such a method has in fact been in successful use during the 1960 and 1970 population censuses. Today, the Bureau of the Census* uses this and similar methods of data editing as an integral part of tight quality control of data from which its releases are tabulated. It can be said that these activities constitute one of its main uses of high-speed computers. In other situations it is always regarded as necessary to follow up a suspected error in the data by human inspection. For example, if a statistical scrutiny of inventory records encounters "number of parts in warehouse = 1360," which on comparison with the previous inventory is about 10 times too large and has a unit figure 0, one would be hesitant to instruct the machine automatically to divide that number by 10, assuming that a column shift has occurred in punching. Here one would prefer to record the suspect number and instruct personnel to chase the trouble or to satisfy themselves about the correctness of the unusually large record.

It may be argued that such uses of computers are not a breakthrough in research. After all, the computer is used for functions that could (with great effort) be performed by other means. Moreover, it has to borrow its intelligence from its human programmers. However, we must recall the tremendous speed and accuracy with which the computer scrutinizes data. By freeing trained personnel for more challenging tasks it enormously enhances the potentialities of a research team engaged in studies that involve extensive data analysis. Moreover, it permits the analysis of data which in an uncontrolled and unedited state would have been too unreliable for the drawing of inferences.

With the advent of bigger and faster computers our systems of automatic quality control of data will become more and more ambitious. Although this will result in more searching error scrutiny, it is clearly impossible to provide a control system that will detect *any* error, however unusual. Much ingenuity is therefore needed by using the knowledge of the experts to guide the computer logic to search for the errors and error patterns most likely to be found in any given body of data.

ANALYSIS OF EXPERIMENTAL DATA

The techniques most frequently used in this activity are analysis of variance and regression analysis. Many excellent computer systems are now available for performing these computations. Undoubtedly, the availability of computers has increased the capabilities of research teams of having their data analyzed where previously desk computers and punched card equipment could only cover the analysis of a fraction of their data. Moreover, computers have more or less eliminated "shortcut" analysis (such as an analysis of variance* based on range* in place of mean squares). Such shortcut methods had justified their lower statistical efficiency by their rapid execution on desk computers or by pencil-and-paper methods. Unfortunately, together with the advantages of computers, there are associated serious pitfalls, of which two are mentioned very briefly.

Use of Inappropriate "Canned Programs"

As statisticians we find all too frequently that an experimenter takes data directly to a computer center programmer (usually called an analyst) for "statistical analysis." The programmer pulls a canned statistical program out of the file and there may result extensive machine outputs, all of which are irrelevant to the purpose of the experiment. This deplorable situation can be avoided only through having competent statistical advice, preferably in the design stage and certainly in the analysis stage. Often, the

statistical analysis appropriate to the purpose of the experiment is not covered by a canned program and it is appreciated that with time schedule pressures it may be necessary to use a canned program that gives at least a relevant basic analysis. For example, it may be decided to use a basic factorial analysis of variance and subsequently, pool certain components on a desk computer to produce the "appropriate ANOVA." Or again it may be decided to use a general regression program to analyze unbalanced factorial data, although certain factors are known to be random and not fixed.

This brings up the question of how many and what kind of programs should be "canned." Such questions are so intricately linked with the nature of the research arising at the respective institutions that general guidelines are difficult. However, there is one general question that may well be raised: should there be a general analysis-of-variance system (such as AARDVAK) making provision for a great variety of designs that may be encountered, or should there be a large number of special-purpose programs "custom made" for a particular design? Undoubtedly, the best answer is that *both* should be available: the custom-made programs should be used when "they fit the bill," the general-purpose program, which must obviously take more computer time, when custom-made programs *do not* fit the bill. In a sense, therefore, the general-purpose program is an answer to the question: How should we analyze the unusual experiment? However, we must remember that even general-purpose analysis-of-variance systems are restricted in their scope; many unbalanced data situations are not covered in such general programs.

Loss of Contact between the Experimenter and the Data Analysis

Here it is argued that "in the good old days, when experimenters did their sums of squares, they were learning a lot about the data, and the computer destroys this intimate contact."

Now we must clearly distinguish between

performing sums of squares on desk computers as opposed to an intelligent scrutiny of the data, preferably an inspection of error residuals. The former is clearly pointless, the latter highly desirable. Indeed, all analysis-of-variance and regression programs should provide options for both tabulation of all individual error residuals (for inspection) as well as statistical outlier tests on all residuals that flag unusually large residuals, in case the experimenter overlooks them. It is very strange that the possibility of faulty records is clearly recognized in the area of censuses and surveys and all too often overlooked in the analysis of experimental data. But the intelligent inspection of residuals should not only provide a monitor for faulty records; it should also be used by the experimenter to learn something about the data. Systematic patterns of large errors residuals often provide useful pointers for the modification of models. For example, the form of the residuals in linear regression may indicate neglected quadratic or higher-order terms. A factorial analysis of variance of a response y, in which the main effects of two quantitative factor inputs x_1 (e.g., the temperature of exposure) and x_2 (e.g., the time of exposure) are insignificant or not very significant, but their interaction is significant, often suggests that the relevant input is a function of the two inputs x_1 and x_2. In the example above, the product $x_1 \times x_2$, representing the amount of heat administered, may well be the relevant input and an inspection of a table of residuals will often reveal such features.

We now turn to the impact of computers on statistical research computations.

SOLUTION OF STATISTICAL DISTRIBUTION PROBLEMS BY MONTE CARLO METHODS

Monte Carlo methods may be briefly described as follows. Given a mathematical formula that cannot be easily evaluated by analytic reduction and the standard procedures of numerical analysis, it is often possible to find a stochastic process generating statistical variables whose frequency distributions can be shown to be simply related to the mathematical formula. The Monte Carlo method then actually generates a large number of variables, determines their empirical frequency distributions, and employs them in a numerical evaluation of the formula.

An excellent and comprehensive account of these methods is given in a book edited by Meyer [2] as well as in numerous articles referenced by him. The more recent literature is too abundant to be comprehensively referenced.

In view of the fast-growing literature on these techniques this section is confined primarily to a very special area of their application: the numerical solution of statistical distribution problems. Moreover, our definitions of statistical distributions do not aim at any generality in terms of measure theory but are, for purposes of simplicity, confined to distribution density functions which are all integrable in the classical Riemann sense. The concepts are explained in terms of statistics depending on independent univariate samples.

Role of Monte Carlo Methods in Solving Statistical Distribution Problems

In the special case when Monte Carlo methods are used for the solution of statistical distribution problems, the mathematical formula to be evaluated is the frequency distribution of what is known as "a statistic":

$$h = h(x_1 x_2 \cdots x_n), \qquad (1)$$

i.e., a mathematical function (say a piecewise continuous function) of a random sample of n independent variate values x_i, drawn from a "parental" distribution with ordinate frequency $f(x)$ and cumulative distribution

$$F(x) = \int_{-\infty}^{x} f(v)\, dv. \qquad (2)$$

In this particular case the mathematical formula to be evaluated is the n-dimensional integral

$$G(H) = \Pr[h \leqslant H]$$

$$= \int \cdots \int \prod_{i=1}^{n} f(x_i) \, dx_1 \cdots dx_n,$$

(3)

where the range of the n-dimensional integration in (3) is defined by

$$h(x_1 x_2 \cdots x_n) \leqslant H.$$

(4)

An analytic solution of the distribution problem (3) would consist in a simplification of (3) to make it amenable to numerical evaluation, a concept not clearly defined since it depends on the tabular and mechanical aids available for evaluation. A solution of (3) by Monte Carlo methods would consist of generating a large number of samples $x_1, x_2, \ldots, x_n$, of computing (1) for each sample, and using the proportion of statistics $h \leqslant H$ as an approximation to (3). With statistical distribution problems the stochastic process mentioned above is therefore trivially available by the definition of the problem. In fact, it is the process of generating variables x_i from the parental distribution. To illustrate the foregoing concepts by a simple example for which an analytic solution for (3) is well known, consider a random sample of independent values from the Gaussian $N(0, 1)$, so that

$$f(x) = (2\pi)^{-1/2} \exp\left(-\tfrac{1}{2}x^2\right)$$

(5)

and consider the X^2-statistic

$$h(x_1 \cdots x_n) = X^2 = \sum_{i=1}^{n} x_i^2.$$

(6)

Then

$$\Pr[h \leqslant H] = \left[\Gamma(\tfrac{1}{2}n)\right]^{-1}$$

$$\times \int_0^H \exp\left(-\tfrac{1}{2}h\right)\left(\tfrac{1}{2}h\right)^{(1/2)n-1} d\left(\tfrac{1}{2}h\right),$$

(7)

which will be recognized as the incomplete gamma function extensively tabulated for statisticians under the name of the probability integral of χ^2 (see Pearson and Hartley

[3, E6]). Whereas in the example above an analytic reduction of (3) to a simple form (7) (which can be expanded in a Poisson series for even n) enabled its numerical evaluation, there are numerous instances when no exact analytic reduction is possible, but the approximations of numerical analysis such as the Euler–MacLaurin formula of numerical integration can be used effectively.

Monte Carlo Procedures for Evaluating Statistical Distributions

It is clear from the description of Monte Carlo procedures just given that the principal steps of computing estimates of frequency distributions for statistics $h(x_1 \cdots x_n)$ are as follows:

1. The generation of random samples $x_1 \cdots x_n$ drawn from the parent population with ordinate frequency $f(x)$.
2. The computation of the statistic h for each sample and computation of a frequency distribution (3) for varying H by counting the proportion of h values with $h \leqslant H$.

The standard procedure in step 1 is to first generate sets of random numbers or digits and interpret these as the decimal digits of a uniform variate u_i. The most frequently used method of generating the u_i is a method well known under the name "power residue method."

To compute from the uniform variates u_i random variates x_i following a given distribution $f(x)$, it is customary to employ the inverse $F^{(-1)}$ to the probability integral $F(x)$ given by (2), and compute the random variates x_i from

$$x_i = F^{(-1)}(u_i)$$

(8)

using either a table of $F^{(-1)}(u)$ or a computer routine. No *general* guidelines can be given for the computation of $h(x_1 \cdots x_n)$, but effective methods of reducing the computational labor are available in special cases.

Methods of Reducing "Sample Sizes" (Number of Simulations)

As is well known, a very large number N of random values of the statistic $h(x_1 \cdots x_n)$ are required in order that the empirical frequencies of the N values of h provide even moderately accurate estimates of its cumulative probability distribution. An idea of the magnitude of N can be obtained by applying the well-known Kolmogorov–Smirnov* criterion of goodness of fit*. This criterion measures the maximum discrepancy D_N between the true cumulative distribution $\Pr[h \leqslant H]$ and its empirical approximation, i.e., the proportion of h values below H. It can be shown (see, e.g., Massey [1]) that

$$\Pr\left[D_N \leqslant 1.63/\sqrt{N} \right] \doteq 0.99. \qquad (9)$$

This formula shows that the error in our Monte Carlo estimates decreases with $1/\sqrt{N}$. To give an example, suppose that it is desired to compute a Monte Carlo distribution which, with 99% confidence, has three accurate decimals; then

$$1.63/\sqrt{N} = 5 \times 10^{-4} \qquad \text{or} \qquad N = 1.06 \times 10^7.$$

Numbers of samples of this magnitude may be prohibitive even on computers. It is not surprising therefore that considerable efforts were made by the Monte Carlists to modify their methods to reduce the number N of sample sequences required to obtain estimates of adequate precision. An excellent account of these methods is given by Meyer [2] as well as in numerous journal articles dealing with such methods.

COMPUTER SIMULATION IN ENGINEERING AND MANAGEMENT

The counterpart to Monte Carlo in the applications of statistics to certain engineering and management problems is the well-known technique of simulation. We confine our discussion to two examples.

Optimization of a Production Process by Computer Simulation

If we may oversimplify the situation somewhat, this technique consists of four main steps:

1. The building of a statistical model to simulate the process. For example, we may set up a model for the completion times of the various phases of a production process in a plant, including statistical generators of delay times, bottlenecks, and operation loops. Such a model would also include the parameters that we want to ultimately optimize, such as speeds of automatics, speeds of delivery belts, the number and spacing of operators and repair crews, the availability of backup equipment, etc.

2. The computer implementation of the statistical model. This is a computer programming job. There are often feedbacks from the programmer to the analyst requesting simplifications of the model.

The next step is all important but unfortunately is often omitted.

3. The validation of the model at least for certain operationally feasible parameter levels. This means that the computer outputs from the simulation model should be compared with actual operational records.

4. The optimization of the parameters in the model within the operationally feasible parameter space. This step is usually accomplished by mathematical programming, i.e., linear or (more realistically) nonlinear programming.

The high quality of the mathematical and computer techniques that are currently available for this final step of mathematical programming is liable to make us forget the maxim "garbage in, garbage out." To avoid this, it is absolutely vital that we carefully validate our model by comparison with operational data.

Example of a Simulation of an Engineering Design

At the risk of overemphasizing the discussion of a special case, let us consider a computer simulation of a neutron-scattering process in a reactor shield. Such a shield consists of a number of iron slabs with interspersed water slabs (see Fig. 1). The thicknesses of these slabs are the design parameters. The computer simulation is simplified to a one-dimensional problem using as a variable the penetration depth into the shield along its axis, which is at right angles to the shield's faces. A neutron of given energy E_0 and space angle θ_0 with the shield axis enters the first iron slab. An exponential* statistical distribution is invoked to determine the distance traveled until its first collision with an iron atom. A complex scattering law is now invoked to determine the new (reduced) energy E_1 and new direction θ_1 of the neutron, and so on for the next collision, until the reduced energy E_i is so low that it is monitored to signal neutron

capture. The statistical penetration laws and the physical scattering laws differ, of course, for the iron and water slabs. The objective of this simulation is to estimate the probability of shield penetration and the probability of neutron reflection by the shield as a function of the slab thicknesses. Somewhat sophisticated Monte Carlo procedures must be employed to obtain these as well as the "capture distribution," i.e., the frequency of neutrons captured at various penetrations through the shield.

IMPACT OF COMPUTERS ON STATISTICAL MODEL BUILDING

Finally, we turn to a development that is perhaps the most important impact of computers: their use in statistical model building. Classical statistical analysis once insisted on the use of mathematical models for which compact mathematical solutions could be found. This generated a tendency on the part of analysts to restrict their studies to such tractable models, even at the expense of making them unrealistic. With the computer's capabilities we need not be

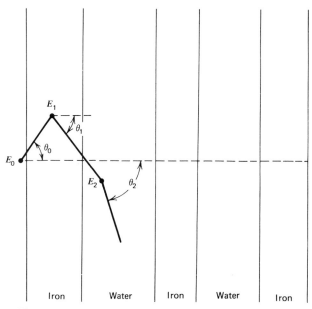

Figure 1 Simulation of neutron scattering in the reactor shield.

afraid of formulating more realistic models, thereby freeing scientists from the fetters of analytic tractability. As an example, the rate equations governing the time dependence of chemical reactions are usually assumed to be linear first-order differential equations with constant rate coefficients. These give rise to the well-known analytic solutions of mixtures of exponentials that have often been fitted to data obtained from a study of chemical reaction processes. It is well known to the chemist that a model with constant rates is often an oversimplification. Nonlinear rate equations are often more realistic. However, we usually cannot solve the resulting nonlinear rate equations analytically. By contrast, the computer has no difficulty solving more realistic rate equations by numerical integration and fitting these numerical integrals directly to the data. The parameters in the rate equations then become the unknown parameters in the nonlinear regression fit. The rapidity of numerical integration subroutines is essential for this approach to be feasible.

There are many other instances in which numerical analysis can and will replace analytic solutions. Future research will therefore be able to search more freely for information that is at the disposal of scientists. Indeed, they will use the computer as a powerful tool in trying alternative model theories, all of a complex but realistic form, to advance their theories on empirical phenomena.

References

[1] Massey, F. J. (1951). *J. Amer. Statist. Ass.*, **46**, 68–78.

[2] Meyer, H. A. ed. (1956). *Symposium on Monte Carlo Methods*. Wiley, New York.

[3] Pearson, E. S. and Hartley, H. O. (1966). *Biometrika Tables for Statisticians*, Vol. 1. Cambridge University Press, Cambridge. (Table 7, Probability integral of the χ^2-distribution and the cumulative sum of the Poisson distribution).

(EDITING STATISTICAL DATA

ERROR ANALYSIS

GRAPHICAL REPRESENTATION, COMPUTER-AIDED

RANDOM VARIABLES, COMPUTER GENERATION OF

SIMULATION)

H. O. HARTLEY

COMPUTER ASSISTED TELEPHONE SURVEYS   *See* TELEPHONE SURVEYS, COMPUTER ASSISTED

CONCENTRATION CURVE

A descriptive representation of dispersion often used for economic variables such as income or consumption (in grouped data). Let X be a positive-valued variable and T be the total of all the values of X in the population. Let $F(x)$ be the proportion of the individuals in the population with the variable values X not exceeding x. Let $T(x)$ be the total of the variate values up to (and including) x, so that $Q(x) = T(x)/T$ is the proportion of the total attributed to these individuals. The graph of $Q(x)$ against $F(x)$ (for different values of x) is called the *concentration curve*. Formally,

$$Q(x) = \frac{\int_0^x y\, dF(y)}{\int_0^\infty y\, dF(y)}$$
$$= 1 - \frac{x[1 - F(x)]}{\int_0^\infty [1 - F(y)]\, dy}.$$

Bibliography

Chakravarti, I. M., Laha, R. G., and Roy, J. (1967). *Handbook of Methods of Applied Statistics*, Vol. 1. Wiley, New York.

(CONCENTRATION RATIO

LOREZ CURVE)

CONCENTRATION INDEX   *See* DIVERSITY INDICES

CONCENTRATION PARAMETER   *See* DIRECTIONAL DISTRIBUTIONS

CONCENTRATION RATIO

A measure of dispersion* often used for economic variables. If A denotes the area between the concentration curve* [the graph of $Q(x)$ against $F(x)$; see CONCENTRATION CURVE for the definitions of $Q(x)$ and $F(x)$] and the line $Q \equiv F$, then the concentration ratio is by definition $\delta = 2A$. The larger the value of $0 \leqslant \delta \leqslant 1$, the greater is the disparity among the individuals. A computational formula for concentration ratio based on trapezoidal quadrature rule* is given, e.g., in Chakravarti et al. [1].

Reference

[1] Chakravarti, I. M., Laha, R. G., and Roy, J. (1967). *Handbook of Methods of Applied Statistics,* Vol. 1. Wiley, New York.

CONCOMITANT VARIABLES

When collecting data, several characteristics are often recorded on each observational unit. Some of these are of primary interest, whereas the others are collected because it is believed that they need to be taken into account in modeling the structure of variation. The latter are called *concomitant* ("going along with") *variables, explanatory variables, or covariables.* For example, in studying distribution of blood pressure, the age of an individual may be expected to play the role of a concomitant variable, whereas in studying lifetime distributions, the roles of age and blood pressure are exchanged.

TYPES OF CONCOMITANT VARIABLES AND THEIR USE IN STATISTICS

As with all variables, concomitant variables can be classified as *discrete** (qualitative) and *continuous** (quantitative).

Among discrete covariables, of special interest are the *indicator* variables* associated with classification of units into two subpopulations. For example, $x_j = 0$ or 1, depending whether an individual (j) belongs to the control or treatment group, respectively.

Demographic characteristics such as sex, race, social class, etc. by which the data can be stratified into more homogeneous groups, are further examples of indicator variables.

In factorial experiments*, different levels of a factor may be considered as values of a discrete variable; the corresponding analysis-of-variance* (ANOVA) models can be thought of as models with concomitant variables. On the other hand, the factors themselves can also be considered as variables of primary interest.

Continuous measurements are used as explanatory variables in a variety of problems.

In controlled experiments investigating effects of different treatments on the values of quantitative variables of primary interest, such as physiological traits of animals or plants, or strength or resistance of a mechanical device, many additional covariables, such as temperature, pressure, weight, etc., are measured. The latter are expected to contribute substantially to the observed responses. Multiple regression analysis* is a common method of allowing for the contribution of covariables—called, in this case, the *independent* variables.

Sometimes knowledge of the *initial* value(s) of the characteristic(s) of primary interest is important. For example, when comparing effects of different drugs on reduction of blood pressure, initial blood pressure is usually measured on each individual and used as one of the covariates in assessment of the drug effects.

In the factorial experiments mentioned above, the levels of factors may be considered as values of variables of primary interest, whereas other variables (often continuous) are used as concomitant variables. The resultant analysis is called the analysis of covariance* (ANCOVA).

Multiple regression ANOVA and ANCOVA models are all included in general linear models*. Other statistical methods exploiting concomitant variables in classification of units into separate, more homogeneous classes are discriminant analysis* and cluster analysis*. In these cases, the variables of primary interest are discrete (group 1, 2, 3, etc.); the covariables can be discrete or continuous.

CONCOMITANT VARIABLES IN SURVIVAL ANALYSIS

The methods mentioned above are already well established and are discussed in more detail under other headings in this encyclopedia. The use of concomitant information in the evaluation of mortality data or other age-dependent all-or-none phenomena (e.g., onset of a chronic disease) has recently attracted the attention of many researchers. As it is rather new, and so less generally known, we will devote more space to this topic. (For a review, see Elandt-Johnson and Johnson [4, Chap. 13].)

The treatment of concomitant variables starts with their introduction into a model of the hazard function*. Let $T(>0)$ denote age, or time elapsed since a certain starting point $t = 0$, and $x' = (x_1, \ldots, x_s)$ be the vector of s concomitant variables. The hazard function (HF) (intensity) is $\lambda_T(t; x)$, so the survival distribution function* (SDF) is $S_T(t; x) = \exp[-\int_0^t \lambda_T(u; x) \, du]$.

Additive and Multiplicative Models

Consider first the case when the x's are *independent of time*. Two types of hazard-rate formulas, additive and multiplicative, are currently of general interest for their mathematical tractability.

The general *additive* model* is of the form

$$\lambda_T(t; x) = \lambda(t) + \sum_{u=1}^{s} h_u(t) g_u(x_u), \quad (1)$$

where $\lambda(t)$ is the "underlying" hazard. Note that the coefficients $h_u(t)$'s are solely functions of t. The special case, commonly used, is the *linear* model*,

$$\lambda_T(t; x) = \sum_{u=0}^{s} \beta_u x_u, \quad (2)$$

where $x_0 \equiv 1$ is a "dummy" variable.

The likelihood* for N independent sets of observations $\{t_j; \lambda_T(t; x_j)\}$, $j = 1, 2, \ldots, N$ [using model (2)] is proportional to

$$\prod_{j=1}^{N} \left\{ \left(\sum_{u=0}^{s} \beta_u x_{uj} \right)^{\delta_j} \exp\left[-\left(\sum_{u=0}^{s} \beta_u x_{uj} \right) t_j \right] \right\}, \quad (3)$$

where $\delta_j = 0$ or 1, depending whether individual (j) is alive or dies at time t_j.

The general *multiplicative* model* of HF is of the form

$$\lambda_T(t; x) = \lambda(t) g(x). \quad (4)$$

Of special interest is Cox's model* [2],

$$\lambda_T(t; x) = \lambda(t) \exp\left(\sum_{u=1}^{s} \beta_u x_u \right)$$

$$= \lambda(t) \exp(\beta' x). \quad (5)$$

The (partial) likelihood for estimating β's does not depend on $\lambda(t)$ and is proportional to

$$\prod_{j=1}^{n} \left\{ \exp(\beta' x_{i(j)}) \left[\sum_{l \in \mathcal{R}_j} \exp(\beta' x_l) \right]^{-1} \right\}, \quad (6)$$

where n is the number of observed deaths, $i(j)$ denotes the individual i who is the jth to fail, and $\mathcal{R}_j$ is the set of individuals in the study alive just before the jth failure, sometimes called the risk set $\mathcal{R}_j$.

Assessment of adequacy of a model can be effected by cumulative hazard plotting* (see also Kay [5]).

Selection of appropriate sets of concomitant variables can be done by the step-up or step-down procedures used in fitting multiple regressions (*see* STEPWISE REGRESSION). As a criterion, the extended likelihood ratio principle* can be used [1].

When the concomitant variables are *time dependent*, the hazard function is $\lambda_T[t; x(t)]$. The methods of handling this problem are similar, but it is formally necessary that each of $x_{ul}(t_j')$, where t_j' is the time of the jth failure, be observed. Since this is in practice usually impossible, the most recent observations are often used.

Concomitant Variables as Random Variables

Even if the x's do not depend on time, their distribution in a population may change because of selection* in mortality.

We may consider concomitant variables as *random* variables. Let $\mathbf{X}'_0 = (X_{01}, X_{02}, \ldots, X_{0s})$ be the vector of concomitant variables at time $t = 0$, with joint *prior* CDF $F_{\mathbf{X}_0}(\mathbf{x}_0)$, and let $S_T(t; \mathbf{X}_0)$ be the SDF. The probability element of the *posterior* distribution* of $\mathbf{X}_0$ *among the survivors to age t* is

$$dF_{\mathbf{X}_0 \mid t}(\mathbf{x}_0 \mid t) = \frac{S_T(t; \mathbf{x}_0)\, dF_{\mathbf{X}_0}(\mathbf{x}_0)}{\int_{\Omega_s} S_T(t; \mathbf{x}_0)\, dF_{\mathbf{X}_0}(\mathbf{x}_0)}, \quad (7)$$

where Ω_s is the region of integration.

Suppose that the hazard rate* is of additive form $\lambda_T(t; X_0) = \alpha t + \beta X_0$ and suppose that $X_0 \sim N(\xi_0, \sigma_0^2)$. It can be shown [3] that the posterior distribution among the survivors to time t is also normal with the same variance σ_0^2, but with mean $(\xi_0 - \beta \sigma_0^2)$.

When the concomitant variables are time dependent, with a deterministic functional relation $\mathbf{X}(t) = \psi(t; X_0; \boldsymbol{\eta})$, where $\boldsymbol{\eta}$ is a vector of additional parameters, it is not difficult to derive the posterior distribution of X's. Some stochastic variation can be introduced by supposing that $\mathbf{X}_0$ and $\boldsymbol{\eta}$ are random variables. (For details, see Elandt-Johnson [3].)

Logistic Model

A popular model used in epidemiology is the *logistic-linear model*. The concomitant variables, $x_1, \ldots, x_k$, are referred to as risk factors*.

Let $q_j = q_j(\mathbf{x}_j)$ be the conditional probability that an individual j with observed values of s risk factors, $\mathbf{x}'_j = (x_{1j}, \ldots, x_{sj})$, will experience the event (e.g., onset of a disease, death) during a specific period t, and that $p_j = 1 - q_j$. The logistic-linear model is defined as

$$\log(q_j / p_j) = \sum_{u=0}^{s} \beta_u x_{uj}, \quad (8)$$

where $x_{0j} \equiv 1$ for all j. Hence

$$p_j = \left[1 + \exp\left(-\sum_{u=0}^{s} \beta_u x_{uj} \right) \right]^{-1}. \quad (9)$$

Note that q_j does not depend on the length of the period t.

GENERAL COMMENTS

The uses of concomitant variables in many fields of biological, physical, industrial, and economical disciplines are well known and proved to be important and useful. Care should be taken in their selection and in the interpretation of the results, which might be sometimes misleading if not properly analyzed and interpreted.

References

[1] Byar, D. P. and Corle, D. K. (1977). *J. Chronic. Dis.*, **30**, 445–449.

[2] Cox, D. R. (1972). *J. R. Statist. Soc. B.*, **34**, 826–838.

[3] Elandt-Johnson, R. C. (1979). *Inst. Statist. Mimeo Ser. No. 1206*. University of North Carolina at Chapel Hill, N.C., pp. 1–18.

[4] Elandt-Johnson, R. C. and Johnson, N. L. (1980). *Survival Models and Data Analysis*. Wiley, New York, Chap. 13.

[5] Kay, R. (1977). *Appl. Statist.*, **26**, 227–237.

(BIOSTATISTICS
CLINICAL TRIALS
COX'S MODEL
MULTICOLLINEARITY
REGRESSION)

REGINA C. ELANDT-JOHNSON

CONDITIONAL INFERENCE

Conditional probability* appears in a number of roles in statistical inference*. As a useful tool of probability theory, it is in particular a device used in computing distributions of many statistics used in inference. This article is not concerned with such purely probabilistic calculations but with the way in which conditioning arises in the construction of statistical methods and the assessment of their properties. Throughout this article, essentially all measurability considerations are ignored for the sake of emphasizing important concepts and for brevity. Random variables (rv's) may be thought of as having discrete case or absolutely continuous (with respect to Lebesgue measure*) case densities. Appropriate references may be consulted for general considerations.

SUFFICIENCY*

Suppose that $\{P_\theta, \theta \in \Omega\}$ is the family of possible probability laws on the sample space S (with an associated σ-field $\mathcal{Q}$). Think of the rv X under observation as the identity function on S, and $\{f_\theta, \theta \in \Omega\}$ as the corresponding family of densities of X. The usual definition of a *sufficient statistic* T on S is in terms of conditional probabilities: for all A in $\mathcal{Q}$, $P_\theta\{X \in A \mid T(X)\}$ is independent of θ. Often it is convenient to think of a sufficient partition of S, each of whose elements is a set where T is constant. The two concepts are equivalent in most common settings.

Using the definition of conditional probability and writing in the discrete case for simplicity, we have, if T is sufficient,

$$
\begin{aligned}
f_\theta(x) &= P_\theta\{X = x\} \\
&= P_\theta\{X = x, T(X) = T(x)\} \\
&= P_\theta\{X = x \mid T(X) = T(x)\} \\
&\quad \times P_\theta\{T(X) = T(x)\} \\
&= h(x)g(\theta, T(x)),
\end{aligned}
\tag{1}
$$

where $h(x) = P\{X = x \mid T(X) = T(x)\}$, independent of θ. In usual cases this development can be reversed and one has the *Fisher–Neyman decomposition* $f_\theta(x) = h(x) g(\theta, T(x))$ as necessary and sufficient for sufficiency of T. In graphic terms, Fisher's assertion that T contains all the information in X about θ is evidenced by the fact that, given that $T(X) = t$, one can conduct an experiment with outcomes in S and not depending on θ, according to the law $h(x \mid t)$

$= P\{X' = x \mid T(x) = t\}$, thereby producing a rv X' with the same unconditional law $\sum_t h(x \mid t) P_\theta\{T = t\}$ as X, for all θ; we can recover the whole sample X, probabilistically, by this randomization that yields X' from T.

This excursion into sufficiency is made both because of its relation to ancillarity discussed below (*see also* ANCILLARY STATISTICS) and also because some common developments of statistical decision theory that use the development amount formally to a conditional inference, although the usual emphasis about them is not in such terms. One such development rephrases the meaning of sufficiency by saying that, for any statistical procedure δ, there is a procedure δ^* depending only on the sufficient statistic T that has the same operating characteristic. Indeed, if δ denotes a randomized decision* function, with $\delta(\Delta \mid x)$ the probability assigned to the set Δ of decisions (a subset of the set D of all possible decisions) when $X = x$, then

$$
\delta^*(\Delta \mid t) = E\{\delta(\Delta \mid X) \mid T(X) = t\} \tag{2}
$$

defines a procedure on $T(S)$ with the desired property; δ^* and δ have the same risk function for every loss function L on $\Omega \times D$ for which the risk of δ is defined. The procedure δ^* is defined in terms of a conditioning, although the emphasis is on its unconditional properties. In particular, if $L(\theta, d)$ is convex in d on D, now assumed a convex Euclidean set, then the nonrandomized procedure d^{**} on $T(S)$ [for which δ^{**} $(d^{**}(t) \mid t) = 1$] defined by

$$
d^{**}(t) = E\left\{\int_d r\delta(dr \mid X) \mid T(X) = t\right\} \tag{3}
$$

has risk at least as small as δ; d^{**} is the *conditional expected decision* of δ, given that $T(X) = t$, and the stated improvement from δ to d^{**} is the Rao–Blackwell theorem. Thus, in unconditional decision theory, use is made of procedures defined conditionally; the emphasis in *conditional inference*, though, is usually on conditional rather than unconditional risk.

Many treatments of conditional inference use extensions of the sufficiency concept, often to settings where nuisance parameters are present. For example, if $\theta = (\phi, \tau)$ and the desired inference concerns ϕ, the statistic T is *partially sufficient* for ϕ if the law of T depends only on ϕ and if, for each τ_0, T is sufficient for ϕ in the reduced model $\Omega = \{(\phi, \tau_0)\}$. This and related concepts are discussed by Basu [5]. This topic and many other matters such as conditional asymptotics for maximum likelihood (ML) estimators* are treated by Anderson [1]; a detailed study for exponential families* is given in Barndorff-Nielsen [3]. Hájek [17] discusses some of these concepts in general terms.

ANCILLARY AND OTHER CONDITIONING STATISTICS

Fisher [14, 15] in his emphasis on ML, defined an ancillary statistic U as one that (a) has a law independent of θ and (b) together with an ML estimator $\hat{d}$ forms a sufficient statistic. Currently in the literature, and herein, we take (a) without (b) as the definition. However, whether or not we are concerned with ML estimation, Fisher's rationale for considering ancillarity is useful: U by itself contains no information about θ, and Fisher would not modify $\hat{d}$ in terms of U; however, the value of U tells us something about the precision of $\hat{d}$, e.g., in that $\text{var}_\theta(\hat{d} | U = u)$ might depend on u. If we flip a fair coin to decide whether to take $n = 10$ or $n = 100$ independent, identically distributed (i.i.d.) observations ($X = (X_1, X_2, \ldots, X_n)$ above), normally distributed on R^1 with mean θ and variance 1, and denote the sample mean by $\bar{X}_n$, then $\hat{d} = \bar{X}_n$ is ML but not sufficient, $U = n$ is ancillary, and $(\hat{d}, U)$ is minimal sufficient. The unconditional variance of $\hat{d}$ is $\frac{11}{200}$. Fisher pointed out that, knowing that the experiment with 10 observations was conducted, one would use the conditional variance $\frac{1}{10}$ as a more meaningful assessment of precision of $\hat{d}$ than $\frac{11}{200}$, and would act similarly if $n = 100$.

Much of the argumentative literature attacking or defending unconditional Ney-man–Wald assessment of a procedure's behavior in terms of examples such as this last one is perhaps due to an unclear statement of the aim of the analysis of procedures. If, before an experiment, procedures are compared in terms of some measure of their performance, that comparison must be unconditional, since there is nothing upon which to condition; even procedures whose usefulness is judged in terms of some conditional property once X is observed can only be compared before the experiment in an unconditional expectation of this conditional property. At the same time, if that conditional property is of such importance, account of its value should be taken in the unconditional comparison. An example often cited in criticism of unconditional inference is that of Welch [27], the model being that $X_1, X_2, \ldots, X_n$ are i.i.d. with uniform law* on $[\theta - \frac{1}{2}, \theta + \frac{1}{2}]$. If $W_n = \min_i X_i$, $V_n = \max_i X_i$, $Z_n = (V_n + W_n)/2$, and $U_n = V_n - W_n$, a confidence interval on θ with various classical *unconditional* optimality properties is of the form

$$\left[\max(W_n + q, V_n) - \tfrac{1}{2}, \min(W_n, V_n - q) + \tfrac{1}{2} \right]$$

for an appropriate q designed to give the desired confidence coefficient γ. Pratt [23], in a criticism from a Bayesian perspective, points out various unappealing features of this procedure; e.g., it *must* contain θ if $U_n > q$, and yet the confidence coefficient* is only γ. One may indeed find it more satisfactory to give an interval and confidence assessment *conditional* on U_n, as Welch suggests. The classical interval is what one would use if its optimum properties were criteria of chief concern, but many practitioners will not find those unconditional properties as important as conditional assessment of precision based on the value of U_n.

The last example illustrates an intuitive idea about the usefulness of conditioning. If U_n is near 1, X has been "lucky" and θ can be estimated very accurately, whereas the opposite is true if U_n is near 0. A conditional assessment is an expression of how lucky, by chance, X was in the sense of accuracy of the inference; unconditional risk or confidence averages over all possible values of X.

Many other examples of ancillarity, exhibiting various phenomena associated with the concept, occur in the literature. A famous example is that of Fisher [16] in which $X = ((Y_1, Z_1), (Y_2, Z_2), \ldots, (Y_n, Z_n))$, the vectors (Y_i, Z_i) being i.i.d. with common Lebesgue density $e^{-\theta y - \theta^{-1} z}$ for $y, z > 0$, with $\Omega = \{\theta : \theta > 0\}$. In this case

$$\hat{d} = \left(\sum_i Z_i \Big/ \sum_i Y_i \right)^{1/2}, \quad U = \left[\left(\sum_i Z_i \right) \left(\sum_i Y_i \right) \right]^{1/2},$$

and the conditional variance of $\hat{d}$ given that $U = u$ depends on u; $(\hat{d}, U)$ is minimal sufficient.

An instructive example is that of i.i.d. rv's X_i with Cauchy density $1/\{\pi[1 + (x - \theta)^2]\}$, for which the ML estimator (or other invariant estimator, such as the Pitman best invariant estimator* for quadratic loss if n is large enough) has conditional distribution depending on $U = (X_2 - X_1, X_3 - X_1, \ldots, X_n - X_1)$. For example, when $n = 2$, $\hat{d} = \bar{X}_2$ and the conditional density of $Z = \hat{d} - \theta$ given that $U = u$ is

$$2[1 + (u/2)^2] / \{\pi[1 + (u/2 + z)^2][1 + (u/2 - z)^2]\},$$

and a rough view of the spread of this density can be seen from its value $2/\{\pi[1 + (u/2)^2]\}$ at $z = 0$: large values of $|U|$ give less precise conditional accuracy of $\hat{d}$.

It is often convenient to replace (S, X) by $(T(S), T)$ for some minimal sufficient T, in these considerations. In the Cauchy example U becomes the set of *order statistic* * differences. When $X_1, \ldots, X_n$ are $\mathfrak{N}(\theta, 1)$,

$$U = (X_1 - X_2, X_1 - X_3, \ldots, X_1 - X_n)$$

is ancillary on S, but in terms of $(T(S), T)$ with $T = \bar{X}_n$ we have no nontrivial ancillary: we cannot obtain a better conditional assessment of the accuracy of $\hat{d} = \bar{X}_n$ by conditioning on an ancillary.

In all of the foregoing examples, U is a *maximal ancillary*; no ancillary U^* induces a partition of S that is a refinement of the partition induced by U. Moreover, in these examples the maximal ancillary is unique. When that is the case, a further argument along Fisherian lines would tell us that, since

a maximal ancillary gives the most detailed information regarding the (conditional) accuracy of $\hat{d}$, we should condition on such a maximal ancillary. Unfortunately, ancillary partitions do not parallel sufficient partitions in the existence of a unique finest such partition in all cases. Basu, in a number of publications (e.g., refs. 4 and 5), has considered several illustrations of this phenomenon. A simple one is a $X_1, \ldots, X_n$ i.i.d. 4-nomial with cell probabilities $(1 - \theta)/6$, $(1 + \theta)/6$, $(2 - \theta)/6$, and $(2 + \theta)/6$; the vector $T = (Y_1, Y_2, Y_3, Y_4)$ of the four observed cell frequencies is minimal sufficient, and each of $U_1 = Y_1 + Y_2$ and $U_2 = Y_1 + Y_4$ is maximal ancillary. If one adopts the *conditioning principle*, of conditioning on a maximal ancillary in assessing the accuracy of $\hat{d}$, the question arises whether to condition on U_1 or U_2.

Among the attempts to answer this are those by Cox [12] and Basu [4]. The former suggests conditioning on the ancillary U (if there is a unique one) that maximizes the variance of the conditional information. Roughly, this will give a large spread of the conditional accuracies obtained for different values of U, reflecting as much as possible the "luckiness" of X that we have mentioned; it was variability of the conditional accuracy that made conditioning worthwhile. Basu suggests that the difficulty of nonuniqueness of maximal U may lie in the difference between a real or performable experiment, such as creation of the sample size $n(10$ or $100)$ in the first example, and a conceptual or nonperformable experiment such as one from which U_1 or U_2 would result in the last example above. Basu implies that one should condition in the former case but not necessarily in the latter, and that in practice the nonuniqueness problem will not arise in terms of any ancillary representable as the result of a real experiment.

The problem of which maximal ancillary to use attracts attention in large part because of insistence on the use of an ancillary for conditioning. One may consider conditional inference based on an arbitrary conditioning variable V, and (a) require that some conditional measure of accuracy of, or confidence in, the decision, is approximately constant, given the value of V. At the same time (b) one would try, in the spirit of our comments about lucky observations and Cox's suggestion, to choose V and the decision procedure to make the variability of that conditional accuracy or confidence as large as possible. A development of Kiefer [18, 19] gives a framework in terms of which such conditional procedures can be compared. In this framework the statistician's goals are considered to be flexible so that, in an example such as that above of X_i uniformly distributed from $\theta - \frac{1}{2}$ to $\theta + \frac{1}{2}$, the length of the confidence interval* and the conditional confidence given U_n may both vary with U_n. A modification of the theory by Brown [8] includes a precise prescription for conditionings that produce most variable conditional confidence coefficients, in some settings.

The use of conditioning other than in terms of an ancillary is not new. For example, a common test of independence in 2×2 tables* conditions on the marginal totals, which are not ancillary. Similarly, inference about the difference between two Bernoulli parameters, each governing n observations, is often based on conditioning on the sum of successes in the $2n$ observations, also not an ancillary. Both of these are useful tools for which tables have been constructed.

BAYESIAN INFERENCE* AND OTHER AXIOMATICS

We have alluded to the *conditioning principle*. Various systems of foundational axioms considered by Birnbaum [6], Barnard, and others imply that inference should be based on a sufficient statistic, on the likelihood function, or conditionally on an ancillary statistic. A detailed discussion here would wander too far from the main subject. A popular axiomatic system related to conditioning is that of the Bayesian approach [25]. It is impossible to list and discuss here usual axioms of "rational behavior"* that lead to the use of Bayesian inference based on a subjective (or, infrequently, physical) prior law. Only the result of using such an approach will be described here.

If π is a prior probability law* on Ω, the element of the posterior law* of θ given that $X = x$ is given by Bayes' theorem as

$$\pi(d\theta \mid x) = f_\theta(x)\pi(d\theta) \Big/ \int_\Omega f_\theta(x)\pi(d\theta).$$

(4)

This may be thought of as the earliest basis for "conditional inference." Whatever the meaning of π (subjective or physical), $\pi(d\theta \mid x)$ updates $\pi(d\theta)$ to give probabilistic assessments in the light of the information $X = x$. Bayes procedures of statistical decision theory*, or informal "credibility intervals" that contain θ with state posterior probability, flow from (4).

The conditioning framework of Bayesian inference is conceptually quite different from that of classical conditioning in the frequentist framework (such as, in the uniform example, the assessment of a conditional confidence coefficient for an interval estimator, given that $U_n = u$). In the Bayesian context the conditioning is on the entire observation X or a sufficient $T(X)$, and the rv whose conditional law is ascertained is θ; in the conditioning of the preceding section, the conditioning was most often on a (by-itself-uninformative) ancillary U, and conditioning on X or $T(X)$ would yield nothing useful because the conditional probability assesses the accuracy of $\hat{d}$ or the coverage probability of an interval, *both functions of X*, and θ is not a rv. Thus direct comparison of the achievements of the two approaches in producing "conditional procedures" is not obvious.

Bayesians list, among the advantages of their approach, the lack of any arbitrariness in choice of point estimator or confidence interval method, or of the conditioning partition; which of two maximal ancillaries to use simply does not arise. Of course, a Bayesian credibility interval depends on the choice of π, as does the probabilistic meaning of that interval. Non-Bayesians regard the credibility intervals resulting from the use of subjective π's as appropriate for the expression of a Bayesian's subjective views, but not as meaningful for scientific discourse about θ as evidenced in X. These comments do not apply to (a) large-sample considerations in which Bayesian methods and certain frequentist methods yield essentially the same results; (b) settings in which the problem is transitive under a group on Ω that leaves the problem invariant and the Bayesian uses a (possibly improper) invariant π, yielding a procedure of frequentist invariance theory*; or (c) use of π to select a procedure which is then analyzed on a frequentist basis.

RELEVANT SUBSETS

A considerable literature, beginning with the work of Buehler [9] and Wallace [26], is concerned with questions such as the following: Given a confidence procedure with confidence coefficient γ, is there a conditioning partition $\{B, S - B\}$ such that, for some $\epsilon > 0$, the conditional confidence is $> \gamma + \epsilon$ for all θ, given that $X \in B$, and is $< \gamma - \epsilon$ for all θ, given that $X \notin B$. The set B is then called *relevant*. (The considerations have been simplified here, and a number of variants of the stated property are treated in the references.) Thus the set $B = \{U_n > c\}$, c a constant, is relevant for the classical confidence interval mentioned in the uniform example. In the example of $X_1, \ldots, X_n$ i.i.d. and $\mathfrak{N}(\mu, \sigma^2)$, the usual confidence interval $[\overline{X}_n - cS_n, \overline{X}_n + cS_n]$ on μ, with $S_n^2 = \sum_i (X_i - \overline{X}_n)^2 / (n - 1)$, was proved by Buehler and Fedderson [11] to have conditional confidence (probability of containing μ) $> \gamma + \epsilon$ for some $\epsilon > 0$, given that $X \in B$, $= \{|\overline{X}_n| / S_n < c'\}$ for some $c' > 0$. Intuitively, $E\overline{X}_n^2 = \mu^2 + \sigma^2 n^{-1}$ while $ES_n^2 = \sigma^2$, so that if $\overline{X}_n^2 / S_n^2 < n^{-1}$, there is evidence that S_n overestimates σ. This work has been extended by others, such as Brown [7] and Olshen [21]. Related work is due to Robinson [24].

Pierce [22], Buehler [10], and others have constructed a theory of "coherence"* of statistical procedures, based on the concept of relevant conditioning. If Peter takes γ confidence coefficient to mean that he will give or take $\gamma : 1 - \gamma$ odds on his interval containing θ, then Paul with a relevant B can beat him by betting for or against coverage depending on whether or not $X \in B$. The proponents of this theory regard such an "incoherent" procedure as unacceptable. Under certain assumptions they show that the only coherent procedures are obtained by using the Bayesian approach for some proper π.

These developments are interesting mathematically, and the existence of relevant sets is sometimes surprising. But a non-Bayesian response is that the confidence coefficient of the incoherent procedure is being compared unfavorably at a task for which it was not designed. All that was claimed for it was meaning as an unconditional probability γ, and the resulting frequentist interpretability of γ in terms of repeated experiments and the law of large numbers. That certain conditional probabilities differ from γ may seem startling because of our being used to unthinking unconditional employment and interpretation of such intervals, but if a finer conditional assessment is more important, the frequentist can use such an assessment with the same interval. The chance that the third toss of a fair coin is heads, given that there is one head in the first four tosses, is $\frac{1}{4}$; this does not shake one's belief in the meaning of $\frac{1}{2}$ as the unconditional probability that the third toss is a head. Which is the more useful number depends on what one is after.

OTHER CONSIDERATIONS

Among the many other topics related to conditional inference, we mention three.

By now it is well known that a test of specified level α with classical optimum properties is not necessarily obtained by using a family of conditional tests, each of conditional level α. An early paper giving a possible prescription for construction of conditional tests is Lehmann [20].

Bahadur and Raghavachari [2], in an asymptotic study of conditional tests, showed that a conditional procedure that is asymptotically optimum in Bahadur's sense of unconditional "slope" must give approximately constant conditional slope, with probability near 1.

Efron and Hinkley [13] showed that, in assessing the precision of the ML estimator $\hat{d}$ from i.i.d. $X_1, \ldots, X_n$ with density f_θ, a useful approximation to the conditional variance of $\hat{d}$, given an appropriate ancillary, is $1/I_n(X, \hat{d})$, where $I_n(X, \theta) = -\sum_i \partial^2 \log f_\theta(X_i)/\partial\theta^2$. This is Fisher's "observed information" as contrasted with the ML estimator $1/I_n(\hat{d})$ of unconditional asymptotic variance, where $I_n(\theta)$ is Fisher's information $E_\theta I_n(X, \theta)$. The observed information seems often to provide a more accurate picture.

References

[1]   Anderson, E. B. (1973). *Conditional Inference and Models for Measurement*. Mentalhygiejnisk Forlag, Copenhagen.

[2]   Bahadur, R. R. and Raghavachari, M. (1970). *Proc. 6th Berkeley Symp. Math. Statist. Prob.*, Vol. 1. University of California Press, Berkeley, Calif., pp. 129–152.

[3]   Barndorff-Nielsen, O. (1978). *Information and Exponential Families in Statistical Theory*. Wiley, New York.

[4]   Basu, D. (1964). In *Contributions to Statistics*, C. R. Rao, ed. Pergamon Press, Oxford, pp. 7–20.

[5]   Basu, D. (1977). *J. Amer. Statist. Ass.*, **72**, 355–366.

[6]   Birnbaum, A. (1962). *J. Amer. Statist. Ass.*, **57**, 269–326.

[7]   Brown, L. D. (1967). *Ann. Math. Statist.*, **38**, 1068–1075.

[8]   Brown, L. D. (1978). *Ann. Statist.*, **6**, 59–71.

[9]   Buehler, R. J. (1959). *Ann. Math. Statist.*, **30**, 845–863.

[10] Buehler, R. J. (1976). *Ann. Statist.*, **4**, 1051–1064.

[11] Buehler, R. J. and Fedderson, A. P. (1963). *Ann. Math. Statist.*, **34**, 1098–1100.

[12] Cox, D. R. (1971). *J. R. Statist. Soc. B*, **33**, 251–255.

[13] Efron, B. and Hinkley, D. V. (1978). *Biometrika*, **65**, 457–488.

[14] Fisher, R. A. (1935). *J. R. Statist. Soc. A*, **98**, 39.

[15] Fisher, R. A. (1936). *Proc. Amer. Acad. Arts Sci.*, **71**, 245.

[16] Fisher, R. A. (1956). *Statistical Methods and Scientific Inference*. Oliver & Boyd, London.

[17] Hájek, J. (1965). *Proc. 5th Berkeley Symp. Math. Statist. Prob.*, Vol. 1. University of California Press, Berkeley, Calif., pp. 139–162.

[18] Kiefer, J. (1976). *Ann Statist.*, **4**, 836–865.

[19] Kiefer, J. (1977). *J. Amer. Statist. Ass.*, **72**, 789–827.

[20] Lehmann, E. L. (1958). *Ann. Math. Statist.*, **29**, 1167–1176.

[21] Olshen, R. A. (1973). *J. Amer. Statist. Ass.*, **68**, 692–698.

[22] Pierce, D. A. (1973). *Ann. Statist.*, **1**, 241–250.

[23] Pratt, J. W. (1961). *J. Amer. Statist. Ass.*, **56**, 163–166.

[24] Robinson, G. K. (1975). *Biometrika*, **62**, 155–162.

[25] Savage, L. J. (1954). *The Foundations of Statistics*. Wiley, New York.

[26] Wallace, D. L. (1959). *Ann. Math. Statist.*, **30**, 864–876.

[27] Welch, B. L. (1939). *Ann. Math. Statist.*, **10**, 58–69.

(ANCILLARY STATISTICS
BAYESIAN INFERENCE
COHERENCE
CONDITIONAL PROBABILITY
CONFIDENCE INTERVALS
CREDIBILITY THEORY
DECISION THEORY)

J. KIEFER

CONDITIONAL INVERSE (OF A MATRIX)   See GENERALIZED INVERSE

CONDITIONAL PROBABILITY AND EXPECTATION

The fundamental ingredient that makes it possible to go beyond independence* (which dominated probability and statistics until the early part of this century) into the realm of dependence is the concept of conditioning. The notion of conditional probability, as well as that of conditional expectation, was introduced in its general form by A. N. Kolmogorov* [16], and plays a fundamental role in the theory and in the application of probability and of statistics. In probability the models are frequently described by specifying appropriate conditional probabilities or distributions; this is the case, for instance, with urn models*, Markov chains and processes* (see Feller [10, 11]) and certain random fields* (see Dobrushin [8]). In statistics the various notions of conditioning play a central role in several areas of statistical inference, such as conditional inference*, Bayesian inference*, estimation theory* and testing*, sufficiency*, and decision theory*, to mention only a few.

The core of the notion of conditioning has a great simplicity and an appealing intuitive content which transcends its occasional mathematical idiosyncracies. In this article the main concepts are introduced and discussed and their basic properties are stated. Specific applications and uses of the various notions of conditioning can be found in several other entries, such as those dealing with the topics mentioned in the preceding paragraph.

The presentation moves from the simpler notions of conditioning given an event or the value of a discrete random variable, to the more delicate and useful notions of conditioning given the value of a general random variable, and finally to the more general notions of conditioning given a σ-field of events or a random process. Regression functions*, closely related to the material of the second section, are discussed in the third section under a separate heading. Although the concepts become progressively more delicate and require more advanced mathematics, they all share a common thread and basic similarities in expressions and properties that are emphasized throughout. All mathematical aspects and prerequisites (such as the Radon–Nikodym theorem* on which the general notions of conditioning are based) are kept to a bare minimum here and all expressions are written in the simplest possible (yet sufficiently general) form. More details and further discussion and properties can be found in standard texts on probability theory, such as Ash [1], Billingsley [3], Breiman [5], Chow and Teicher [7], Feller [10, 11], Laha and Rohatgi [17], and Loève [19].

For reasons of simplicity all notions and properties are described here for (one-dimensional) random variables. Similar results hold for random vectors, and frequently appropriate analogs exist for more general random elements.

The following notation is used throughout. $(\Omega, \mathcal{F}, P)$ is a probability space*. X and Y are random variables with distribution functions* $F_X(x)$ and $F_Y(y)$ and joint distribution function $F_{X,Y}(x, y)$. Whenever functions of random variables are considered, such as $h(X)$ or $h(X, Y)$, they are assumed to be Borel-measurable functions on the real line or the plane. $\mathcal{E}$ denotes expectation, so that whenever $\mathcal{E}|h(X)| < \infty$, we have

$$\mathcal{E}[h(X)] = \int_{\Omega} h[X(\omega)] \, dP(\omega)$$

$$= \int_{-\infty}^{\infty} h(x) \, dF_X(x). \quad (1)$$

We write interchangeably $dP(\omega)$ or $P(d\omega)$ and $dF(x)$ or $F(dx)$. Finally, $1_B(x)$ is the indicator function* of the set $B(= 1$ when x is in B; $= 0$ otherwise).

CONDITIONAL PROBABILITY GIVEN AN EVENT

The conditional probability $P(F|E)$ of event F given an event E with $P(E) > 0$ is defined by

$$P(F|E) = \frac{P(F \cap E)}{P(E)} ; \quad (2)$$

i.e., given that event E has occurred, the (conditional) probability of event F is the ratio of the probability of the part of event F that lies in event E over the probability of E. It also has an appealing frequency interpretation, as the proportion of those repetitions of the experiment in which event E occurs for which event F also occurs. When (2) is written in the form $P(F \cap E) = P(F|E) \cdot P(E)$ it is called the *chain rule*. Two very useful properties are the total probability rule,

$$P(E) = \sum_{n=1}^{N} P(E|E_n) P(E_n), \quad (3)$$

and the Bayes rule,

$$P(E_k \mid E) = \frac{P(E \mid E_k)P(E_k)}{\sum_{n=1}^{N} P(E \mid E_n)P(E_n)}. \quad (4)$$

Here $\{E_n\}_{n=1}^{N}$ is a partition of the probability space Ω, i.e., $\Omega = \cup_{n=1}^{N} E_n$, and the events E_n are disjoint. More general versions of both rules are mentioned in later sections (*see also* BAYES' THEOREM).

For each fixed event E, $P(\cdot \mid E)$ is a probability measure on $\mathfrak{F}$, and when $E \mid h(X) \mid < \infty$, it is natural [cf. (1)] to define the conditional expectation of $h(X)$ given E by

$$\mathscr{E}[h(X) \mid E] = \int_{\Omega} h[X(\omega)] P(d\omega \mid E)$$

$$= \frac{1}{P(E)} \int_{E} h[X(\omega)] dP(\omega)$$

$$= \int_{-\infty}^{\infty} h(x) dF_{X \mid E}(x), \quad (5)$$

where

$$F_{X \mid E}(x) = P(X \leqslant x \mid E) \quad (6)$$

is the conditional distribution function of X given E. The analog to the total probability rule (3), the total expectation rule, now becomes

$$\mathscr{E}[h(X)] = \sum_{n=1}^{N} \mathscr{E}[h(X) \mid E_n] P(E_n). \quad (7)$$

CONDITIONAL PROBABILITY AND EXPECTATION GIVEN THE VALUE OF A RANDOM VARIABLE

When Y is a discrete random variable and $P(Y = y) > 0$, then by (2)

$$P(E \mid y = y) = \frac{P[E \cap (Y = y)]}{P(Y = y)}. \quad (8)$$

For real numbers y such that $P(Y = y) = 0$ the definition (2) is not applicable, and since such values of Y are observed with probability zero, it is in a sense irrelevant how $P(E \mid Y = y)$ is defined; it could be defined equal to 0 for such y's, which has some intuitive appeal. From (8) it follows easily that

$$P(E) = \int_{-\infty}^{\infty} P(E \mid Y = y) dF_Y(y), \quad (9)$$

which is the analog of the total probability rule (3), and also that for each real y,

$$P[E \cap (Y \leqslant y)]$$

$$= \int_{(-\infty, \, y]} P(E \mid Y = v) dF_Y(v). \quad (10)$$

If $\mathscr{E} \mid h(X, Y) \mid < \infty$, then we have as in (5),

$$\mathscr{E}[h(X, Y) \mid Y = y]$$

$$= \int_{\Omega} h[X(\omega), Y(\omega)] P(d\omega \mid Y = y)$$

$$= \frac{1}{P(Y = y)}$$

$$\times \int_{\{Y = y\}} h[X(\omega), Y(\omega)] dP(\omega)$$

$$(11)$$

when $P(Y = y) > 0$, and for all other y's it may be defined arbitrarily. Again it follows that

$$\mathscr{E}[h(X, Y)]$$

$$= \int_{-\infty}^{\infty} \mathscr{E}[h(X, Y) \mid Y = y] dF_Y(y),$$

$$(12)$$

which is the analog of the total expectation rule (7), and also that for each real y,

$$\mathscr{E}[h(X, Y) 1_{(-\infty, y]}(Y)]$$

$$= \int_{(-\infty, \, y]} \mathscr{E}[h(X, Y) \mid Y = v] dF_Y(v).$$

$$(13)$$

It is clear that, perhaps more than the defining relations (8) and (11), it is the properties (10) and (13) that capture the intuitive meaning (and desirable use) of the notions of conditional probability $P(E \mid Y = y)$ and conditional expectation $\mathscr{E}[h(X, Y) \mid Y = y]$ given the value of the random variable Y. It is therefore very pleasing that these properties (10) and (13) can be used to define the quantities $P(E \mid Y = y)$ and $\mathscr{E}[h(X, Y) \mid Y = y]$ for an arbitrary random variable Y, e.g., for a continuous random variable Y for which $P(Y = y) = 0$ for all y and the definitions (8) and (11) cannot be used. This is accomplished via the Radon–Nikodym theorem*, and the conditional probability and expectation are defined as the Radon–Nikodym derivatives of two measures on the Borel sets of the real line.

For each fixed event E, this theorem shows there exists a Borel-measurable function of y, denoted by $P(E \mid Y = y)$, which satisfies (10) for all real y, is determined uniquely almost everywhere (a.e.) (dF_Y) on the real line, and is nonnegative and dF_Y-integrable. Similarly, for each fixed function $h(x, y)$ with $\mathcal{E}|h(X, Y)| < \infty$, this theorem shows there exists a Borel-measurable function of y, denoted by $\mathcal{E}[h(X, Y) \mid Y = y]$, which satisfies (13) for all real y, is determined uniquely a.e. (dF_Y) on the real line, and is dF_Y-integrable. These two notions are naturally related by

$$P(E \mid Y = y) = \mathcal{E}[1_E \mid Y = y]$$

[just as $P(E) = \mathcal{E}(1_E)$].

Having thus defined the conditional probability and expectation $P(E \mid Y = y)$ and $\mathcal{E}[h(X, Y) \mid Y = y]$ by means of (10) and (13), we are faced with the problem of calculating them now that (8) and (11) are no longer applicable. The remainder of this section is a discussion of this problem, proceeding from the general to the more special cases.

First, one would expect that both (8) and (11) will remain valid whenever $P(Y = y) > 0$. This is indeed true and follows by noticing that (10) and (13) are valid when $(-\infty, y]$ is replaced by $(-\infty, y)$ (indeed by any Borel set) and by forming the difference of these two equations (or, which is the same, by replacing $(-\infty, y]$ by $\{y\}$). Thus if $P(Y = y) > 0$, both $P(E \mid Y = y)$ and $\mathcal{E}[h(X, Y) \mid Y = y]$ may be computed as when Y is discrete, and the remaining and main problem is how to compute them for those y's for which $P(Y = y) = 0$.

It turns out that an intuitively very satisfactory limiting form of (8) [and, of course, of (11)] is always valid. Specifically for almost every $y(dF_Y)$, we have

$$P(E \mid Y = y)$$

$$= \lim_{h \downarrow 0} \frac{P[E \cap (y - h < Y < y + h)]}{P(y - h < Y < y + h)},$$

$$\text{(14)}$$

where the ratio on the right-hand side is taken to be zero if $P(y - h < Y < y + h) = 0$ (see, e.g., Hahn and Rosenthal [12, Chap. V]).

However satisfactory (14) may be from an intuitive point of view, it does not lend itself to specific calculation, and the most efficient way of computing $\mathcal{E}[h(X, Y) \mid Y = y]$ or $P[(X, Y) \in B \mid Y = y]$ is by using the conditional distribution function of X given $Y = y$:

$$F_{X \mid Y}(x \mid y) = P(X \leqslant x \mid Y = y). \quad (15)$$

For each fixed x, it is determined uniquely a.e. (dF_Y) by (10), which with $E = (X \leqslant x)$ is written as

$$F_{X,Y}(x, y) = \int_{(-\infty, y]} F_{X \mid Y}(x \mid v) \, dF_Y(v)$$

$$\text{for all } y. \quad (16)$$

Hence, to find the conditional distribution function $F_{X \mid Y}(x \mid y)$, one starts with the joint distribution $F_{X,Y}(x, y)$ and first evaluates the distribution function of Y, $F_Y(y) = F_{X,Y}(+\infty, y)$, and then tries to write $F_{X,Y}(x, y)$ in the form

$$\int_{(-\infty, y]} \phi(x, v) \, dF_Y(v).$$

Then for each fixed y, $\phi(x, y)$ is a version of $F_{X \mid Y}(x \mid y)$ [i.e., these two are equal a.e. (dF_Y) as functions of y].

For a conditional distribution function $F_{X \mid Y}(x \mid y)$ to be useful, it should (at the least) be a distribution function in x for each fixed y. It should be recalled that, according to its definition, $F_{X \mid Y}(x \mid y)$ is determined for each fixed x—uniquely for almost every $y(dF_Y)$ but not for every y. Thus the question arises whether a version of $F_{X \mid Y}(x \mid y)$ can be found for each x, which would be a distribution function in x for each fixed y. This (and a bit more) turns out to be always possible, and a *regular* conditional distribution function $F_{X \mid Y}(x \mid y)$ always exists which has the following properties:

1. For each fixed y, $F_{X \mid Y}(x \mid y)$ is a distribution function in x; and for each fixed x, it is a Borel-measurable function of y.

2. For each fixed x, $F_{X \mid Y}(x \mid y)$ equals $P(X \leqslant x \mid Y = y)$ for almost every $y(dF_Y)$.

In fact, the procedure described at the end of the preceding paragraph, if feasible, produces a regular conditional distribution function $F_{X \mid Y}(x \mid y)$, which can then be used to evaluate conditional probabilities and expectations as follows:

$$\mathscr{E}[h(X, Y) \mid Y = y]$$
$$= \int_{-\infty}^{\infty} h(x, y) F_{X \mid Y}(dx \mid y) \quad (17)$$

$$P[(X, Y) \in B \mid Y = y] = \int_{B^y} F_{X \mid Y}(dx \mid y),$$
$$(18)$$

where $B^y = \{x : (x, y) \in B\}$. In particular, the conditional expectation of X given $Y = y$ is given by

$$\mathscr{E}[X \mid Y = y] = \int_{-\infty}^{\infty} x F_{X \mid Y}(dx \mid y). \quad (19)$$

It should be emphasized that the very useful expressions (17) to (19) hold only for regular versions of conditional distributions, which are thus very important for practical applications and computations. Combining (17) with (12), and (18) with (9), we have the useful expressions

$$\mathscr{E}[h(X, Y)]$$
$$= \int_{-\infty}^{\infty} \left\{ \int_{-\infty}^{\infty} h(x, y) F_{X \mid Y}(dx \mid y) \right\} dF_Y(y)$$
$$(20)$$

$$P[(X, Y) \in B]$$
$$= \int_{-\infty}^{\infty} \left\{ \int_{B^y} F_{X \mid Y}(dx \mid y) \right\} dF_Y(y).$$
$$(21)$$

Frequently, conditional distributions are used to describe joint distributions, via (21) and its multivariate analogs. As an example, Markov processes* are defined through their transition functions*, which are regular conditional distributions.

The computation of conditional probabilities and expectations via (17) and (18) then depends on the feasibility of computing a regular conditional distribution function of X given $Y = y$, and one such general way has been described. It should be also noted that with $E = (X \leqslant x)$, the right-hand side of (14) gives a regular conditional distribution of X given $Y = y$. [See Pfanzagl [21],

where it is in fact shown that for almost every $y(dF_Y)$, as $h \downarrow 0$, the distribution function

$$\frac{P[(X \leqslant x) \cap (y - h < Y < y + h)]}{P(y - h < Y < y + h)}$$

converges weakly to the regular conditional distribution function of X given $Y = y$.]

In certain special cases $F_{X \mid Y}$ is easily computed. The simplest are the (extreme) cases of independence or total dependence. When X and Y are independent,

$$F_{X \mid Y}(x \mid y) = F_X(x)$$

for all x and y, and when $X = \phi(Y)$, then

$$F_{X \mid Y}(x \mid y) = 1_{(-\infty, x]}[\phi(y)]$$
$$= \begin{cases} 1 & \text{if } \phi(y) \leqslant x \\ 0 & \text{otherwise.} \end{cases}$$

Finally, in the important special case, where X and Y have a joint probability density function $f_{X, Y}(x, y)$, then a conditional probability density function $f_{X \mid Y}(x \mid y)$ of X given $Y = y$ can be defined by

$$f_{X \mid Y}(x \mid y) = \frac{F_{X, Y}(x, y)}{f_Y(y)}$$
$$= \frac{f_{X, Y}(x, y)}{\int_{-\infty}^{\infty} f_{X, Y}(u, y) du}$$

when $f_Y(y) > 0$ [which is an analog of the Bayes rule (4)]. If $f_{X \mid Y}(x \mid y)$ is taken equal to an arbitrary probability density function $p(x)$ when $f_Y(y) = 0$, then

$$\int_{-\infty}^{x} f_{X \mid Y}(u \mid y) du$$

is a regular conditional distribution function of X given $Y = y$, and thus by (17) and (18),

$$\mathscr{E}[h(X, Y) \mid Y = y]$$
$$= \int_{-\infty}^{\infty} h(x, y) f_{X \mid Y}(x \mid y) dx$$

$$\mathscr{E}[(X, Y) \in B \mid Y = y]$$
$$= \int_{B^y} f_{X \mid Y}(x, y) dx.$$

Regular conditional distribution functions can be used to generate random variables with a given joint distribution function from independent uniformly distributed random variables—a very useful result. To simplify the notation, let $H(x, y)$ be the joint distribution function of X and Y, $F(y)$ the distribution function of Y, and $G_y(x)$ a regular conditional distribution function of X given $Y = y$. Denote also by $F^{-1}(\cdot)$ the usual inverse function of $F(\cdot)$, and by $G_y^{-1}(\cdot)$ the inverse function of $G_y(\cdot)$ for each y. If U and V are independent random variables, each uniformly distributed on $(0, 1)$, then the random variables

$$G_{F^{-1}(V)}^{-1}(U), F^{-1}(V)$$

have joint distribution function $H(x, y)$. Also, in the converse direction, if $F(\cdot)$ is continuous and if $G_y(\cdot)$ is continuous for each y (e.g., if X and Y have a joint probability density function*), then the random variables

$$G_Y(X), F(Y)$$

are independent and each is uniformly distributed on $(0, 1)$. These are the bivariate (and there are, of course, analogous multivariate) versions of the well-known and widely used univariate results, and they go back to Lévy [18] (see also Rosenblatt [23]).

CONDITIONAL PROBABILITY AND EXPECTATION GIVEN A RANDOM VARIABLE OR A σ-FIELD

For a fixed event E and random variable X with $\mathcal{E}|X| < \infty$, we have seen in the section "Conditional Probability Given an Event" that $P(E \mid Y = y)$ and $\mathcal{E}(X \mid Y = y)$ are (Borel-measurable) functions of y, say $a(y)$ and $b(y)$, respectively, which are determined uniquely a.e. (dF_Y) on the real line. Thus $a[Y(\omega)]$ and $b[Y(\omega)]$ are the conditional probability of E and the conditional expectation of X given that Y takes the value $Y(\omega)$. It is then natural to call $a(Y)$ the conditional probability of E given Y and denote it by $P(E \mid Y)$, and to call $b(Y)$ the conditional expectation of X given Y and denote it by $\mathcal{E}(X \mid Y)$. We also write $a[Y(\omega)] = P(E \mid Y)(\omega)$ and $b[Y(\omega)] = \mathcal{E}(X \mid Y)(\omega)$.

To summarize, we have

$$P(E \mid Y = y) = a(y) \quad \text{a.e.} \quad (dF_Y)$$
$$\text{iff } P(E \mid Y) = a(Y) \quad \text{a.s.} \quad (22)$$

and

$$\mathcal{E}(X \mid Y = y) = b(y) \quad \text{a.e.} \quad (dF_X)$$
$$\text{iff } \mathcal{E}(X \mid Y) = b(Y) \quad \text{a.s.} \quad (23)$$

We also have again $P(E \mid Y) = \mathcal{E}(1_E \mid Y)$, and we could thus restrict attention to conditional expectations only.

It is easily seen that

$$P(E) = \mathcal{E}[P(E \mid Y)],$$
$$\mathcal{E}(X) = \mathcal{E}[\mathcal{E}(X \mid Y)],$$

which are analogs of the total probability and expectation rules, and, more generally, that

$$P[E \cap (Y \leqslant y)] = \mathcal{E}[1_{(-\infty, y]}(Y)P(E \mid Y)]$$
$$(24)$$

$$\mathcal{E}[1_{(-\infty, y]}(Y)X] = \mathcal{E}[1_{(-\infty, y]}\mathcal{E}(X \mid Y)].$$
$$(25)$$

It turns out that properties (24) and (25) [which are, of course, the analogs of properties (10) and (13)] can be used as alternative definitions: $P(E \mid Y)$ and $\mathcal{E}(X \mid Y)$ are the $\sigma(Y)$-measurable random variables which are determined uniquely a.s. (again via the Radon–Nikodym theorem as Radon–Nikodym derivatives) by (24) and (25) for all y.

Properties (24) and (25) also show that $P(E \mid Y)$ and $\mathcal{E}(X \mid Y)$ depend on Y only through the σ-field $\sigma(Y)$. This suggests extending the notions of conditional probability and expectation, given a random variable Y, to those given a σ-field $\mathcal{G}(\subset \mathcal{F})$ as follows. $P(E \mid \mathcal{G})$ and $\mathcal{E}(X \mid \mathcal{G})$ are the $\mathcal{G}$-measurable random variables which are determined uniquely a.s. (via the Radon–Nikodym theorem) by the following equalities for all events G in $\mathcal{G}$:

$$P(E \cap G) = \int_G P(E \mid \mathcal{G}) dP,$$
$$\int_G X dP = \int_G \mathcal{E}(X \mid Y) dP.$$

In statistical applications the σ-field $\mathcal{G}$ is generated by a statistic, such as a sufficient statistic, or by a random function, such as a time series*. Then for any family of random variables $Y_t, t \in T$, indexed by the set T [e.g., a sequence of random variables when T is the set of (positive) integers, a random process when T is an interval on the real line], the conditional probability or expectation given the random variables $(Y_t, t \in T)$ is then defined as the conditional probability or expectation given the σ-field $\sigma(Y_t, t \in T)$ generated by the family of random variables.

In the special case where $\mathcal{G}$ is the σ-field generated by the disjoint events E_n, $n \geq 0$, with $\Omega = \cup_{n \geq 0} E_n$ and $P(E_0) = 0$, $P(E_n) > 0$, $n \geq 1$, we have a.s.

$$P(E \mid \mathcal{G}) = \sum_{n \geq 1} P(E \mid E_n) 1_{E_n},$$

$$\mathcal{E}(X \mid \mathcal{G}) = \sum_{n \geq 1} \mathcal{E}(X \mid E_n) 1_{E_n}.$$

These expressions of course agree with $P(E \mid Y)$ *and* $\mathcal{E}(X \mid Y)$ *when* Y *is discrete with values* $y_n, n \geq 1$, *and* $E_n = (Y = y_n)$.

Properties of Conditional Expectations

Conditional expectations given a σ-field have all the standard properties of ordinary expectations, such as linearity, inequalities, and convergence theorems. Some further useful properties are listed here (where all equalities hold with probability 1).

If X is $\mathcal{G}$-measurable, then $\mathcal{E}(X \mid \mathcal{G}) = X$ and more generally $\mathcal{E}(XY \mid \mathcal{G}) = X\mathcal{E}(Y \mid \mathcal{G})$, whereas if X is independent of $\mathcal{G}$, then $\mathcal{E}(X \mid \mathcal{G}) = \mathcal{E}(X)$. When $\mathcal{G}_2$ is a smaller σ-field than $\mathcal{G}_1$, then

$$\mathcal{E}[\mathcal{E}(X \mid \mathcal{G}_1) \mid \mathcal{G}_2] = \mathcal{E}(X \mid \mathcal{G}_2).$$

When Z is independent of (X, Y), then

$$\mathcal{E}(X \mid Y, Z) = \mathcal{E}(X \mid Y).$$

If $\mathcal{F}_n \uparrow \mathcal{F}_\infty$ or $\mathcal{F}_n \downarrow \mathcal{F}_\infty$, then

$$\mathcal{E}(X \mid \mathcal{F}_n) \xrightarrow[n \to \infty]{} \mathcal{E}(X \mid \mathcal{F}_\infty),$$

where the convergence is with probability 1 and in $L_1(\Omega, \mathcal{F}, P)$. Here $\mathcal{F}_n \uparrow \mathcal{F}_\infty$ means that $\mathcal{F}_n \subset \mathcal{F}_{n+1} \subset \mathcal{F}$ and $\mathcal{F}_\infty = \sigma(\cup_n \mathcal{F}_n)$, and

$\mathcal{F}_n \downarrow \mathcal{F}_\infty$ means that $\mathcal{F}_{n+1} \supset \mathcal{F}_n \supset \mathcal{F}$ and $\mathcal{F}_\infty = \cap_n \mathcal{F}_n$.

If Q is another probability measure on $(\Omega, \mathcal{F})$, which is absolutely continuous with respect to P with Radon–Nikodym derivative $Y = dQ/dP$, then conditional expectations with respect to Q, denoted by $\mathcal{E}_Q(\cdot \mid \mathcal{G})$, are expressed in terms of conditional expectations with respect to P, denoted by $\mathcal{E}_P(\cdot \mid \mathcal{G})$, by

$$\mathcal{E}_Q(X \mid \mathcal{G}) = \frac{\mathcal{E}_P(XY \mid \mathcal{G})}{\mathcal{E}_P(Y \mid \mathcal{G})}.$$

Finally, a general Bayes rule expressing the conditional expectation given $\mathcal{G}$ in terms of the conditional expectation given some other σ-field $\mathcal{H}$ has been developed by Kallianpur and Striebel [13] (and has been extensively used in filtering problems).

Conditional Independence

Conditional independence given a σ-field $\mathcal{G}$ is defined just as ordinary independence with probabilities $P(\cdot)$ replaced by conditional probabilities $P(\cdot \mid \mathcal{G})$; e.g., events E_1, E_2 are conditionally independent given $\mathcal{G}$ if $P(E_1 \cap E_2 \mid \mathcal{G}) = P(E_1 \mid \mathcal{G})P(E_2 \mid \mathcal{G})$ a.s. Conditional independence is at the heart of the notion of a Markov chain or process*; the sequence $X_n, n \geq 1$, is a Markov chain iff its past and future are conditionally independent given its present; i.e., for each $n \geq 2$ and $m \geq 1$ the random variables $X_1, \ldots, X_{n-1}$ are independent of $X_{n+1}, \ldots, X_{n+m}$ given X_n [i.e., $\sigma(X_n)$].

Conditional independence between random variables is characterized by means of conditional distributions just as independence is characterized by means of distributions.

An important theorem of de Finetti says that an infinite sequence of random variables $X_n, n \geq 1$, is exchangeable iff the random variables $X_n, n \geq 1$, are conditionally independent and identically distributed given some random variable X:

$$P(X_1 \leq x_1, \ldots, X_n \leq x_n)$$
$$= \int_{-\infty}^{\infty} F(x_1 \mid x) \cdots F(x_n \mid x) \, dF_X(x)$$

for all n, where

$$F(x_n | x) = P(X_n \leqslant x_n | X = x).$$

(The conditioning on X may be replaced by conditioning given the σ-field of exchangeable or of tail events of the sequence.) The random variables $X_1, \ldots, X_n$ are called exchangeable* when all $n!$ permutations $X_{k_1}, \ldots, X_{k_n}$ have the same joint distribution; and the infinite sequence of random variables $X_n, n \geqslant 1$, is exchangeable if $X_1, \ldots, X_n$ are exchangeable for each n. When the X_n's take only the values 0 and 1, i.e., $X_n = 1_{E_n}$, in which case the events E_n, $n \geqslant 1$, are called exchangeable, then X can be chosen to take values in $[0, 1]$ and to satisfy $P(E_n | X = x) = x$ a.e. (dF_X) on $[0, 1]$, so that

$$P\left(E_{n_1} \cap \cdots \cap E_{n_k} \cap E_{m_1}^c \cap \cdots \cap E_{m_j}^c \right)$$
$$= \int_0^1 x^k (1-x)^j \, dF_X(x).$$

(An interesting discussion of exchangeability and its applications in genetics* is given in Kingman [15]; *see also* EXCHANGEABILITY.)

Regression* and Mean Square Estimation

When $\mathcal{E}(X | Y)$ is viewed as a function of Y, it is called the regression function of X on Y [cf. (23)]. A useful general expression is

$$\mathcal{E}(X | Y = y)$$

$$= \frac{1}{2\pi f_Y(y)} \int_{-\infty}^{\infty} e^{-isy} \mathcal{E}(Xe^{isY}) \, ds$$

$$= \frac{\int_{-\infty}^{\infty} e^{-isy} \mathcal{E}(Xe^{isY}) \, ds}{\int_{-\infty}^{\infty} e^{-isy} \mathcal{E}(e^{isY}) \, ds}$$

a.e. (dF_Y), where it is assumed that Y has a probability density function $f_Y(y)$, $\mathcal{E}(Xe^{isY})$ is an integrable function of s, and for the last expression that the characteristic function $\mathcal{E}(e^{isY})$ of Y is an integrable function of s; in fact, in the latter case the version of $\mathcal{E}(X | Y = y)$ given by the second expression is continuous on the set of y's where $f_Y(y) > 0$ (see Zabell [26]). For the estimation of a regression function, *see* REGRESSION ANALYSIS.

The regression of X on Y generally produces some smoothing. For instance, the range of values of $\mathcal{E}(X | Y)$ is generally smoother than that of X. While $a \leqslant X \leqslant b$ implies that $a \leqslant \mathcal{E}(X | Y) \leqslant b$, it may happen that X takes only the values a and b while $\mathcal{E}(X | Y)$ takes all values between a and b. As an example take a binary random variable X with $P(X = -1) = \frac{1}{2} = P(X = 1)$ and a normal random variable Z, with mean 0 and variance σ^2, independent of X and let $Y = X + Z$. Then $\mathcal{E}(X | Y) = \tanh(Y/\sigma)$ takes values strictly between -1 and 1. In fact, in this example the ranges of values of X and of $\mathcal{E}(X | Y)$ are disjoint.

In certain cases the regression function is linear. Here are the standard examples. If X and Y are jointly normal with means μ_X and μ_X, variances σ_X and σ_Y, and correlation coefficient ρ_{XY}, then $\mathcal{E}(X | Y) = a + bY$, where

$$a = \mu_X - \rho_{XY} \frac{\sigma_X}{\sigma_Y} \mu_Y, \qquad b = \rho_{XY} \frac{\sigma_X}{\sigma_Y}.$$

(*See* BIVARIATE NORMAL DISTRIBUTION.) If X and Y have a joint symmetric stable* distribution, then again $\mathcal{E}(X | Y) = cY$. When regressing on a finite number of random variables, if $X, Y_1, \ldots, Y_n$ are jointly normal, then $\mathcal{E}(X | Y_1, \ldots, Y_n) = a_0 + a_1 Y_1 + \cdots + a_n Y_n$; in fact, the conditional distribution of X given $Y_1, \ldots, Y_n$ is also normal. The same is true in the more general case where $X, Y_1, \ldots, Y_n$ have a joint elliptically contoured distribution (see Cambanis et al. [6]). When $X, Y_1, \ldots, Y_n$ have a joint symmetric stable distribution, the regression of X on $Y_1, \ldots, Y_n$ is not generally linear for $n \geqslant 2$; it is linear however when $Y_1, \ldots, Y_n$ are independent (see Kanter [14]).

When regressing on an infinite sequence of random variables $Y_n, n \geqslant 1$, we have

$$\mathcal{E}(X | Y_n, n \geqslant 1) = \lim_{n \to \infty} \mathcal{E}(X | Y_1, \ldots, Y_n),$$

the convergence being both with probability 1 and in L_1. Finally when regressing on a stochastic process $Y_t, t \in T$, with T say an interval, then for each X there is an infinite sequence of points $t_n, n \geqslant 1$, in T (depending on X) such that

$$\mathcal{E}(X \mid Y_t, t \in T) = \mathcal{E}(X \mid Y_{t_n}, n \geqslant 1)$$

(see Doob [9]). These properties are very useful in computing conditional expectations given a random process. When X is jointly normally distributed with the random process $Y_t, t \in T$ (i.e., when all random variables of the form $a_0 X + a_1 Y_{t_1} + \cdots + a_n Y_{t_n}$ are normal), then the properties above can be used to show that $\mathcal{E}(X \mid Y_t, t \in T)$ is linear in $Y_t, t \in T$, in the sense that it is the (a.s. or L_2) limit as $n \to \infty$ of a sequence of the form $a_{n,1} Y_{t_1} + \cdots + a_{n,n} Y_{t_n}$. Thus regressions in Gaussian processes* are always linear. Conversely, Vershik [25] showed that the entire class of (infinite-dimensional, i.e., nondegenerate) processes with finite second moments for which regressions are linear consists of spherically invariant random processes which are scale mixtures of Gaussian processes.

When $\mathcal{E}(X^2) < \infty$, it turns out that $\mathcal{E}(X \mid \mathcal{G})$ is the unique $\mathcal{G}$-measurable random variable Z with $\mathcal{E}(Z^2) < \infty$, which minimizes the mean square error $\mathcal{E}[(X - Z)^2]$, and that the minimum is $\mathcal{E}(X^2) - \mathcal{E}[\mathcal{E}^2(X \mid \mathcal{G})]$. Thus $\mathcal{E}(X \mid \mathcal{G})$ is called the (generally nonlinear) mean square estimate of X based on $\mathcal{G}$. If $X_s, -\infty < s < \infty$, is a random process with finite second moments $\mathcal{E}(X_t^2) < \infty$, then $\mathcal{E}(X_{t+\tau} \mid X_s, s \leqslant t)$ is the mean square predictor of $X_{t+\tau}, \tau > 0$, based on observations up to present time t. When the random signal $X_s, -\infty < s < \infty$, is observed in additive noise $N_s, -\infty < s < \infty$, both having finite second moments, then $\mathcal{E}(X_{t+\tau} \mid X_s + N_s, s \leqslant t)$ is the mean square filtered estimate of $X_{t+\tau}$, based on the noisy observations up to present time t (the filtered estimate is called predictive when $\tau > 0$ and smoothed when $\tau < 0$). For the solution to the prediction and filtering problems, i.e., for the evaluation of the foregoing conditional expectations when X and N are jointly normal, see KALMAN–BUCY FILTERING, PREDICTION, TIME SERIES, and WEINER–KOLMOGOROV FILTERING.

A very useful alternative description of the minimum mean square property of the conditional expectation when $\mathcal{E} X^2 < \infty$ is that $\mathcal{E}(X \mid \mathcal{G})$ is the orthogonal projection* of X onto the space of all $\mathcal{G}$-measurable random variables with finite second moments, i.e., onto $L_2(\Omega, \mathcal{G}, P)$, so that $X - \mathcal{E}(X \mid \mathcal{G})$ is orthogonal to all such random variables. Thus when restricted to $L_2(\Omega, \mathcal{F}, P)$, i.e., for X's with $\mathcal{E} X^2 < \infty$, the conditional expectation becomes a projection operator. For the properties and characterizations of conditional expectation when restricted to $L_p(\Omega, \mathcal{F}, P), p > 1$, see, e.g., Neveu [20].

Regular Conditional Probabilities

Conditional probabilities $P(E \mid \mathcal{G})$ have all the properties of ordinary probabilities, with the important difference that these properties hold with probability 1; e.g., for disjoint events $E_n, n \geqslant 1$, with $E = \cup_n E_n$,

$$P(E \mid \mathcal{G})(\omega) = \sum_{n=1}^{\infty} P(E_n \mid \mathcal{G})(\omega) \qquad (26)$$

for all ω not in an event N with $P(N) = 0$. As the exceptional event N depends on the sequence $\{E_n\}$, it is not generally possible to find points ω for which (26) holds for all sequences of disjoint events. Thus, in general, for fixed ω, $P(E \mid \mathcal{G})(\omega)$ is not a probability measure in E. When a version of the conditional probability has this very desirable property it is called a *regular* conditional probability.

To be precise $P(E \mid \omega), E \in \mathcal{F}$ and $\omega \in \Omega$, is called a regular conditional probability on $\mathcal{F}$ given $\mathcal{G}$ if:

(a) For each fixed $\omega \in \Omega, P(E \mid \omega)$ is a probability measure on $\mathcal{F}$; and for each fixed $E \in \mathcal{F}, P(E \mid \omega)$ is a $\mathcal{G}$-measurable function of ω.

(b) For each fixed $E \in \mathcal{F}, P(E \mid \omega) = P(E \mid \mathcal{G})(\omega)$ a.s.

Regular conditional probabilities do not exist without any further assumptions on $\Omega, \mathcal{F}$, and $\mathcal{G}$. When Ω is a Borel set in a Euclidean space and $\mathcal{F}$ the σ-field of the Borel subsets of Ω, a condition satisfied in most statistical applications, then Doob [9] showed that regular conditional probabilities always exist. For more general spaces where regular conditional probabilities exist, see Blackwell [4]. When a regular conditional probability exists we can express conditional expectations as integrals with respect to it, just as ordinary expectations are expressed as integrals with respect to ordinary probabilities [cf. (1)]:

$$\mathcal{E}[h(X, Y)|\mathcal{G}](\omega)$$
$$= \int_\Omega h[X(\omega'), Y(\omega)]P(d\omega'|\omega) \text{ a.s.}$$
$$(27)$$

where Y is $\mathcal{G}$-measurable. Such an expression makes, of course, very transparent the expectation-like properties of conditional expectations. The analog to the total expectation rule is then

$$\mathcal{E}[h(X, Y)]$$
$$= \int_\Omega \left\{ \int_\Omega h[X(\omega'), Y(\omega)]P(d\omega'|\omega) \right\} P(d\omega).$$

When wishing to express $\mathcal{E}(X|\mathcal{G})$ as an integral for a fixed random variable X, we may similarly define the notion of a regular conditional probability of X given $\mathcal{G}$ by replacing $\mathcal{F}$ by $\sigma(X)$ in the previous definition. Such regular conditional distributions are denoted by $P_{X|\mathcal{G}}(E|\omega)$ and exist under very mild conditions on X, i.e., that its range of values $\{X(\omega), \omega \in \Omega\}$ is a Borel set. We can then write

$$\mathcal{E}[h(X)|\mathcal{G}](\omega)$$
$$= \int_\Omega h[X(\omega')]P_{X|\mathcal{G}}(d\omega'|\omega) \text{ a.s.}$$

The notion of a regular conditional distribution function $F_{X|\mathcal{G}}(x|\omega)$ of X given $\mathcal{G}$ is defined in analogy with the notion of the regular conditional distribution function $F_{X|Y}(x|y)$ of X given $Y = y$ introduced earlier:

(a)   For each fixed ω, $F_{X|\mathcal{G}}(x|\omega)$ is a distribution function in x; and for each

fixed x, it is a $\mathcal{G}$-measurable function of ω.

(b)   For each fixed x, $F_{X|\mathcal{G}}(x|\omega) = P(X \leqslant x|\mathcal{G})(\omega)$ a.s.

Regular conditional distribution functions always exist, and they can be used to write

$$\mathcal{E}[h(X, Y)|\mathcal{G}](\omega)$$
$$= \int_{-\infty}^\infty h[x, Y(\omega)]F_{X|\mathcal{G}}(dx|\omega) \text{ a.s.}$$

$$P[(X, Y) \in B|\mathcal{G}](\omega)$$
$$= \int_{B^{Y(\omega)}} F_{X|\mathcal{G}}(dx|\omega) \text{ a.s.}$$

[compare with (17) and (18)]. When $\mathcal{G} = \sigma(Y)$, then $\mathcal{G}$ may be replaced by Y in the expressions above, giving the analogs of (17) and (18).

Regular conditional probabilities are frequently used to define complex objects such as general Markov processes and certain random fields.

FURTHER REMARKS

The notion of conditional expectation plays a fundamental role in the definition and the study of martingales—a concept that generalizes that of a fair game.

Although conditional expectation has been defined here for random variables X with finite first absolute moment $\mathcal{E}|X| < \infty$, the notion can be extended to random variables whose positive or negative parts (but not necessarily both) have finite expectations: $\mathcal{E}X_+ < \infty$ or $\mathcal{E}X_- < \infty$, where $X_+ = \max(X, 0)$ and $X_- = \max(-X, 0)$ (e.g., positive or negative random variables). The properties of the resulting extended notion of conditional expectation are similar to those mentioned in the preceding section, but some caution is warranted when dealing with it.

An alternative approach to conditional distributions can be found in Tjur [24]. Also, an extensive and more abstract study of conditional probability measures and of conditional expectations viewed as operators can be found in Rao [22].

Finally, the notion of conditional expectation given a σ-lattice of events has been introduced and studied in connection with isotonic regression* (see Barlow et al [2]).

References

[1] Ash, R. B. (1972). *Real Analysis and Probability.* Academic Press, New York.

[2] Barlow, R. E., Bartholomew, D. J., Bremner, J. M. and Brunk, H. D. (1972). *Statistical Inference under Order Restrictions.* Wiley, New York.

[3] Billingsley, P. (1979). *Probability and Measure.* Wiley, New York.

[4] Blackwell, D. (1956). *Proc. 3rd Berkeley Symp. Math. Statist. Prob.* Vol. 2. University of California Press, Berkeley, Calif., pp. 1–6.

[5] Breiman, L. (1968). *Probability.* Addison-Wesley, Reading, Mass.

[6] Cambanis, S., Huang, S., and Simons, G. (1981). *J. Multivariate Anal.*, **11**, 368–395.

[7] Chow, Y. S. and Teicher, H. (1978). *Probability Theory.* Springer-Verlag, New York.

[8] Dobrushin, R. L. (1968). *Theor. Prob. Appl.*, **13**, 197–224.

[9] Doob, J. L. (1953). *Stochastic Processes.* Wiley, New York.

[10] Feller, W. (1968). *An Introduction to Probability Theory and Its Applications*, Vol. 1 (3rd edition), Wiley, New York.

[11] Feller, W. (1971). *ibid.*, Vol. 2.

[12] Hahn, H. and Rosenthal, A. (1948). *Set Functions.* University of New Mexico Press, Albuquerque, N.M.

[13] Kallianpur, G. and Striebel, C. (1968). *Ann. Math. Statist.*, **39**, 785–801.

[14] Kanter, M. (1972). *J. Funct. Anal.*, **9**, 441–456.

[15] Kingman, J. F. C. (1978). *Ann. Prob.*, **6**, 183–197.

[16] Kolmogorov, A. N. (1933). *Grundbegriffe der Wahrscheinlichkeitsrechnung.* Springer-Verlag, Berlin; and (1956). *Foundations of the Theory of Probability.* Chelsea, New York.

[17] Laha, R. G. and Rohatgi, V. K. (1979). *Probability Theory.* Wiley, New York.

[18] Lévy, P. (1937). *Théorie de l'addition des variables aléatoires.* Gauthier-Villars, Paris.

[19] Loève, M. (1977). *Probability Theory II.* Springer-Verlag, New York.

[20] Neveu, J. (1975). *Discrete Parameter Martingales.* North-Holland, Amsterdam.

[21] Pfanzagl, J. (1979). *Ann. Prob.*, **7**, 1046–1050.

[22] Rao, M. M. (1975). *J. Multivariate Anal.*, **5**, 330–413.

[23] Rosenblatt, M. (1952). *Ann. Math. Statist.*, **23**, 470–472.

[24] Tjur, T. (1974). *Conditional Probability Distributions.* Lecture Notes 2, Inst. Math. Statist., University of Copenhagen, Copenhagen.

[25] Vershik, A. M. (1964). *Theor. Prob. Appl.*, **9**, 353–356.

[26] Zabell, S. (1979). *Ann. Prob.*, **7**, 159–165.

(CONDITIONAL INFERENCE PROBABILITY MEASURES)

STAMATIS CAMBANIS

CONFIDENCE BAND

An extension of the concept of confidence region* to the estimation of the whole of a cumulative distribution function* $F(x)$. A confidence band $\mathscr{B}(\mathbf{X})$, calculated from random variables $\mathbf{X}$, with confidence coefficient $100(1 - \alpha)\%$ has the property that the probability that *all* the points $(x, F(x))$ (for every x) are included in $\mathscr{B}(\mathbf{X})$ is $(1 - \alpha)$. A confidence band can be regarded as a set of simultaneous confidence intervals for the (infinity of) parameters $F(x)$.

When it is not possible to construct an exact $100(1 - \alpha)\%$ bound, as is likely to be the case when the variables $\mathbf{X}$ are discrete, the requirement

$$P\big((x, F(x)) \in \mathscr{B}(x) \quad \text{for all } x\big) = 1 - \alpha$$

is often relaxed to

$$P\big((x, F(x)) \in \mathscr{B}(x) \quad \text{for all } x\big) \leqslant 1 - \alpha,$$

although (as with confidence intervals*) it is often preferred to obtain a value as *close* as possible to $(1 - \alpha)$, without excluding the possibility of exceeding this amount.

The usefulness of a confidence band depends on its width. The narrower the band, for a given confidence coefficient, the more useful it is. In general, since the width depends on x (as well as $\mathbf{X}$), "narrowness" does not always give a unique ordering, valid for all x. Indices of narrowness—such as the expected value of the width X being supposed to have some suitable distribution

—can be helpful in comparing different methods of construction of confidence bands, but they can only provide an overall comparison.

The concept extends in a straightforward fashion to confidence belts for joint cumulative distribution functions of a number of random variables.

Bibliography

Kanofsky, P. (1968). *Sankhyā A*, **30**, 369–378. (On parametric confidence bands.)

Wald, A. and Wolfowitz, J. (1939). *Ann. Math. Statist.*, **10**, 105–118. (On nonparametric confidence bands.)

(CONFIDENCE INTERVALS
ESTIMATION
KOLMOGOROV–SMIRNOV
   STATISTIC)

CONFIDENCE BELT   *See* CONFIDENCE BAND

CONFIDENCE INTERVALS AND REGIONS

Confidence intervals are used for *interval estimation*. Whether an interval estimate is required depends upon the reason for the statistical analysis.

Consider the analysis of measurements of the compressive strength of test cylinders made from a batch of concrete. If we were concerned with whether the mean strength of the batch exceeds some particular value, our problem would be one of hypothesis testing*. Our conclusion might be to accept or to reject the hypothesis, perhaps with an associated degree of confidence. If a simple indication of the strength likely to be achieved under the particular conditions of test is required, the observed mean strength might be quoted as an estimate of the true mean strength. This is called point estimation*. Interval estimation is the quoting of bounds between which it is likely (in some sense) that the real mean strength lies. This is appropriate when it is desired to give some indication of the accuracy with which the parameter is estimated. A large number of

statistical problems may be included in the classes of hypothesis testing, point estimation, or interval estimation.

It must be pointed out that there are several schools of thought concerning statistical inference. To quote confidence intervals is the interval estimation method advocated by the most widely accepted of these schools, variously referred to as the Neyman–Pearson*, Neyman–Pearson–Wald, frequentist, or classical school. There are other ways of obtaining interval estimates and we will refer to them later. (*See also* BAYESIAN INFERENCE, FIDUCIAL INFERENCE, INTERVAL ESTIMATION, LIKELIHOOD, STRUCTURAL INFERENCE.)

BASIC IDEA OF A CONFIDENCE INTERVAL

The term "confidence interval" has an intuitive meaning as well as a technical meaning. It is natural to expect it to mean "an interval in which one may be confident that a parameter lies." Its precise technical meaning differs substantially from this (see Jones [13], Cox [7], and Dempster [9]) but the intuitive idea is not entirely misleading. An example should help to explain the technical meaning.

Example 1. Suppose that some quantity is measured using a standard testing procedure. Suppose that the quantity has a well-defined true value μ, but that the measurement is subject to a normally distributed error that has known variance σ^2. Let X denote the random variable that is the result of a single measurement and let x be a particular value for X.

Now X is normally distributed with mean μ and variance σ^2. Using the properties of the normal distribution we can make probability statements about X; e.g.,

$$\Pr\left[\, \mu - 1.96\sigma \leqslant X \leqslant \mu + 1.96\sigma \,\right] = 0.95.$$

$$(1)$$

We could rewrite this as

$$\Pr\left[\, X - 1.96\sigma \leqslant \mu \leqslant X + 1.96\sigma \,\right] = 0.95$$

$$(2)$$

or

$$\Pr\left[\ \mu \in (X - 1.96\sigma, X + 1.96\sigma)\right] = 0.95.$$

$$(3)$$

Although μ may appear to be the subject of statements (2) and (3), the probability distribution referred to is that of X, as was more obvious in statement (1).

If X is observed to be x, we say that we have 95% confidence that $x - 1.96\sigma \leqslant \mu \leqslant x + 1.96\sigma$ or say that $(x - 1.96\sigma, x + 1.96\sigma)$ is a 95% confidence interval for μ. No probability statement is made about the proposition

$$x - 1.96\sigma \leqslant \mu \leqslant x + 1.96\sigma \qquad (4)$$

involving the observed value, x, since neither x nor μ has a probability distribution. The proposition (4) will be either true or false, but we do not know which. If confidence intervals with confidence coefficient p were computed on a large number of occasions, then, in the long run, the fraction p of these confidence intervals would contain the true parameter value. (This is provided that the occasions are independent and that there is no selection of cases.)

CONFIDENCE INTERVALS BASED ON A SINGLE STATISTIC

Many confidence intervals can be discussed in terms of a one-dimensional parameter θ and a one-dimensional statistic $T(X)$, which depends upon a vector of observations X. A more general formulation will be given below under the heading "Confidence Regions." Provided that $T(X)$ is a continuous random variable, given probabilities α_1 and α_2, it is possible to find $T_1(\theta)$ and $T_2(\theta)$ such that

$$\Pr\left[\ T(X) \leqslant T_1(\theta)\,|\,\theta\right] = \alpha_1 \qquad (5)$$

and

$$\Pr\left[\ T(X) \geqslant T_2(\theta)\,|\,\theta\right] = \alpha_2. \qquad (6)$$

In other words, $T_1(\theta)$ and $T_2(\theta)$ are as shown in Fig. 1.

Another diagram that can be used to illustrate the functions $T_1(\theta)$ and $T_2(\theta)$ is Fig. 2. For every particular value of θ the probability that T lies between $T_1(\theta)$ and $T_2(\theta)$ is $1 - \alpha_1 - \alpha_2$. The region between the curves $T = T_1(\theta)$ and $T = T_2(\theta)$ is referred to as a *confidence belt*. In terms of Fig. 2, the basic idea of confidence intervals is to express confidence $1 - \alpha_1 - \alpha_2$ that the point (θ, T) lies in the confidence belt after T has been observed. If T_1 and T_2 are well-behaved functions, they will have inverse functions θ_2 and θ_1, as shown in the figure, and the three propositions "$T_1(\theta) \leqslant T(X) \leqslant T_2(\theta)$," "$\theta$ lies in the confidence belt," and "$\theta_1(T) \leqslant \theta \leqslant \theta_2(T)$" will be equivalent. Thus $(\theta_1(T), \theta_2(T))$ is a $(1 - \alpha_1 - \alpha_2)$ confidence interval for θ.

Example 2. Consider eight observations from a normal distribution with known mean μ and unknown variance σ^2. Take $\theta = \sigma^2$, $X = (X_1, \ldots, X_8)$, and $T(X) = \sum_{i=1}^{8}(X_i - \mu)^2$.

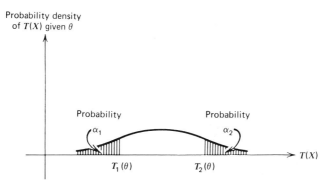

Figure 1   Illustration of the meanings of T_1 and T_2 for fixed θ.

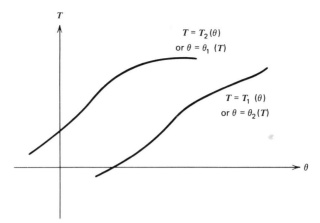

Figure 2   Confidence limits for θ based on the statistic T.

From the fact that T/θ has a χ^2 distribution with 8 degrees of freedom, we know that $\Pr[T/\theta \leqslant 2.18] = 0.025$ and that $\Pr[T/\theta \geqslant 17.53] = 0.025$. Thus we take $T_1(\theta) = 2.18\theta$, $T_2(\theta) = 17.53\theta$ and calculate $\theta_1(T) = 0.057T$, $\theta_2(T) = 0.46T$. The interval $(0.057T, 0.46T)$ may be quoted as a 95% confidence interval for σ^2.

This confidence interval may be described as a *central* confidence interval because $\alpha_1 = \alpha_2 (= 0.025)$. Noncentral confidence intervals are seldom quoted except when we are primarily concerned about large values for the parameter or about small values. In such cases it is common to quote only a single confidence limit. In Example 2

$$\Pr[T/\theta < 2.73] = 0.05$$

or, equivalently,

$$\Pr[\theta > 0.366T] = 0.05.$$

Thus $(0, 0.366T)$ is a confidence interval for $\theta = \sigma^2$ at confidence level 0.95.

DISCRETE DISTRIBUTIONS

When the statistic T is a discrete* random variable it is generally not possible to find functions T_1 and T_2 such that (5) and (6) hold precisely. Instead, we ask that $\Pr[T(X) \leqslant T_1(\theta)|\theta]$ be as large as possible but not greater than α_1 and that $\Pr[T(X) \geqslant T_2(\theta)|\theta]$ be as large as possible but not

greater than α_2. The functions T_1 and T_2 define a confidence belt which generally has a staircase-shaped perimeter. Keeping [14, p. 98] and Kendall and Stuart [15, p. 105] give examples.

Example 3.   Consider the problem of finding a 90% confidence interval for the probability, p, of success on each trial in a sequence of independent trials, if two successes are observed in 12 trials. Some calculation yields that

$$\Pr[\text{number of successes} \leqslant 2 | p]$$
$$= (1-p)^{12} + 12p(1-p)^{11}$$
$$+ 66p^2(1-p)^{10}$$
$$= 0.05 \text{ if } p = 0.438$$
$$< 0.05 \text{ if } p > 0.438$$

and

$$\Pr[\text{number of successes} \geqslant 2 | p]$$
$$= 1 - (1-p)^{12} - 12p(1-p)^{11}$$
$$= 0.05 \text{ if } p = 0.03046$$
$$< 0.05 \text{ if } p < 0.03046.$$

Thus the required 90% confidence interval is $(0.03046, 0.0348)$. (Although this method of construction does not make the probability of including the true value of p to be equal to 90%, it does ensure that this probability is not less than 90%.)

NUISANCE PARAMETERS* AND SIMILAR REGIONS*

Under some circumstances it is easy to find confidence intervals despite the presence of a nuisance parameter. Consider the following example.

Example 4. Suppose that $X_1, X_2, \ldots, X_n$ are normally distributed with mean μ and variance σ^2, both unknown. Let $\overline{X}$ and s denote the sample mean and sample standard deviation. Now $(\overline{X} - \mu)\sqrt{n}/s$ has a t-distribution with $n - 1$ degrees of freedom no matter what the value of σ^2. Therefore, letting t denote the $1\text{-}\frac{1}{2}\alpha$ quantile* of that t-distribution,

$$\Pr\left[\mu - ts/\sqrt{n} \leqslant \overline{X} \leqslant \mu + ts/\sqrt{n} \right] = 1 - \alpha;$$

or, equivalently,

$$\Pr\left[\overline{X} - ts/\sqrt{n} \leqslant \mu \leqslant \overline{X} + ts/\sqrt{n} \right] = 1 - \alpha.$$

The interval $(\overline{X} - ts/\sqrt{n}, \overline{X} + ts/\sqrt{n})$ is a confidence level for μ at confidence level $1 - \alpha$.

The parameter σ^2 is described as a *nuisance parameter* because we are not interested in estimating it, but it does affect the probability distribution of the observations. The regions of the sample space of the form $(\overline{X} - as, \overline{X} + as)$ are described as *similar regions* because the probability of each of them is independent of the parameters. Confidence regions are generally based on similar regions when they exist. However, they often do not exist.

CONFIDENCE REGIONS

Confidence regions are a generalization of confidence intervals in which the confidence set is not necessarily an interval. Let θ be a (possibly multidimensional) parameter and let Θ denote the set of possible values for θ. Let X denote a random variable, generally vector-valued. A function I that gives a subset of Θ for a value x of X is said to be a *confidence set estimator* or a *confidence region*

for θ with confidence coefficient p if

$$\Pr\left[\theta \in I(X) \right] = p. \tag{7}$$

For any such confidence region, to reject the hypothesis $\theta = \theta_0$ whenever θ_0 is not in $I(X)$ is a Neyman–Pearson hypothesis test which has probability $1 - p$ of wrongly rejecting the hypothesis $\theta = \theta_0$.

Choosing between Possible Confidence Regions

There may be many functions I such that

$$\Pr\left[\theta \in I(X) | \theta \right] = p$$

for every θ. How should we choose which to use? Within the formulation where confidence intervals are based on a single statistic T, the problem is essentially that of choosing a statistic on which to base the confidence intervals. Perhaps confidence intervals based on the sample median* would be better in some ways than confidence intervals based on the sample mean.

A number of criteria have been advanced to help decide between alternative confidence regions. We discuss some of them briefly. Standard texts on theoretical statistics may be consulted for further details.

Confidence intervals should be based on sufficient statistics (*see* SUFFICIENCY) and should be found conditional on the value of ancillary statistics (*see* ANCILLARY STATISTICS).

A confidence region I is said to be *unbiased* if

$$\Pr\left[\theta_1 \in I(X) | \theta_2 \right] \leqslant p$$

for all $\theta_1, \theta_2 \in \Theta$. This means that wrong values for the parameter are not more likely to be included in the region $I(X)$ than the correct values.

The region I is said to be *shorter, more accurate,* or *more selective* than the region J if

$$\Pr\left[\theta_1 \in I(X) | \theta_2 \right] \leqslant \Pr\left[\theta_1 \in J(x) | \theta_2 \right]$$

for all $\theta_1, \theta_2 \in \Theta$. Intuitively, this means that incorrect values for θ are more likely to be in J than in I. More selective regions correspond to more powerful tests of hypotheses and unbiased regions correspond to unbiased tests when parametric hypotheses are rejected whenever the parameter does not lie in the confidence region. The term "more selective" is preferred to "shorter" (which stems from Neyman [19]) to avoid confusion with the expected length of confidence intervals.

For complex problems it may be difficult or impossible to apply some of these and other criteria. Sometimes it may only be possible to show that a particular confidence region is optimal in some sense within a particular class of regions, such as those invariant in some way. Different criteria sometimes suggest different regions. There is no completely general way of deciding which confidence interval to use.

CRITICISMS OF THE THEORY OF CONFIDENCE INTERVALS

There have been many arguments about the foundations of statistical inference, and there will probably be many more. Three (not independent) criticisms of the theory of confidence intervals are mentioned below. Note that they are criticisms of the frequentist school of thought, not merely of confidence intervals, which are the interval estimation technique used by that school.

Likelihood Principle*

The likelihood principle states that the "force" of an experiment should depend only upon the likelihood function, which is the probability density for the results obtained as a function of the unknown parameters. Many people find this principle compelling. Pratt [20] presents a persuasive defense of it in an entertaining way. Confidence interval theory violates the likelihood principle essentially because confidence intervals are concerned with the entire sample space.

Coherence*

It has been shown in several ways (e.g., Savage [23]), using various simple coherence

conditions, that inference must be Bayesian if it is to be coherent. This means that every Neyman confidence interval procedure that is not equivalent to a Bayesian procedure violates at least one of each set of coherence properties.

Conditional Properties

For a confidence region I such that

$$\Pr[\theta \in I(X)] = \alpha \qquad \text{for all } \theta,$$

if there is a subset C of the sample space and a positive number ϵ such that either

$$\Pr[\theta \in I(X) \mid X \in C] \leq \alpha - \epsilon \qquad \text{for all } \theta$$

or

$$\Pr[\theta \in I(X) \mid X \in C] \geq \alpha + \epsilon \qquad \text{for all } \theta,$$

then the set C is a *relevant subset*. The idea stems from Fisher's use of the term "recognizable subset" [10] and was formalized by Buehler [6]. Some people argue that the existence of a relevant subset implies that the confidence coefficient α is not an appropriate measure of confidence that $\theta \in I(x)$ when it happens that x belongs to the relevant subset.

Consider the following quite artificial example, in which there are only two possible parameter values and four values for a random variable that is observed only once.

Example 5. Suppose that when $\theta = \theta_1$, $\Pr[X = 1] = 0.9$, $\Pr[X = 2] = 0.01$, $\Pr[X = 3] = 0.05$, and $\Pr[X = 4] = 0.04$, whereas when $\theta = \theta_2$, $\Pr[X = 1] = 0.02$, $\Pr[X = 2] = 0.9$, $\Pr[X = 3] = 0.03$, and $\Pr[X = 4] = 0.05$. The region

$$I(X) = \begin{cases} \{\theta_1\} & \text{if } X = 1 \text{ or } X = 3 \\ \{\theta_2\} & \text{if } X = 2 \text{ or } X = 4 \end{cases}$$

is a confidence region for θ with confidence coefficient 0.95. However,

$$\Pr[\theta \in I(X) \mid X \in \{1, 2\}, \theta] \geq 90/92$$

for both θ values   (8)

and

$$\Pr[\theta \in I(X) \mid X \in \{3, 4\}, \theta] \leq 5/8$$

for both θ values.   (9)

Thus both $\{1,2\}$ and $\{3,4\}$ are relevant subsets.

Conditional properties of confidence intervals for practical problems are seldom as poor as for this example and those of Robinson [22]. Note particularly that the complements of relevant subsets are not necessarily relevant. However, such examples do illustrate the point made by Dempster [9], Hacking [11], and others that confidence coefficients are a good measure of uncertainty before the data have been seen, but may not be afterward.

LINKS WITH BAYESIAN INFERENCE*

Bayesian confidence regions are derived by taking a prior distribution, usually considered to represent subjective belief about unknown parameters, modifying it using observed data and Bayes' theorem* to obtain a posterior distribution, and quoting a region of the parameter space which has the required probability according to the posterior distribution* (*see* BAYESIAN INFERENCE). Bayesian procedures satisfy most coherence principles, satisfy the likelihood principle, and have good conditional properties. However, their conclusions depend upon the arbitrarily or subjectively chosen prior distribution, not merely upon the data, and this is widely considered to be undesirable.

A clear distinction must be made between *proper* and *improper* Bayesian procedures. Proper Bayesian procedures are those based on prior distributions which are proper (i.e., are probability distributions) and which use bounded loss and utility functions should loss or utility functions be required. Other Bayesian procedures are called improper and sometimes lack some of the desirable properties of proper Bayesian procedures. However, they are often used because they are more tractable mathematically.

The bases of the frequentist and Bayesian schools of thought are quite different. However, many statistical procedures that are widely used in practice are both confidence interval procedures and improper Bayesian procedures. (see Bartholomew [1], Jeffreys [12], de Groot [8], and Lindley [17]). Of direct interest to people using the confidence intervals that may also be derived as improper Bayesian interval estimates is that the alternative derivation is often sufficient to ensure that these confidence intervals have most of the desirable properties of proper Bayesian procedures. An exception is that there are relevant subsets for the usual confidence intervals based on the t-distribution* for the unknown mean of a normal distribution when the variance is also unknown (see Brown [4].

RELATIONSHIP TO FIDUCIAL INFERENCE*

Fiducial inference generally proceeds by finding pivotal variables*, functions of both random variables and parameters which have a distribution that is independent of all parameters, and assuming that those pivotal variables have the same distribution after the random variables have been observed. Given the observed values of the random variables, the distribution of the pivotal variables implies a distribution for the parameters, called a fiducial distribution. To the extent that fiducial inference and confidence intervals both involve asserting faith, after seeing the data, in statements for which probabilities could be quoted before seeing the data they are similar theories. Bartlett [2] has argued that resolving the difference between these two theories is less important than resolving the difference between the pair of them and Bayesian methods. The clearest point of disagreement between them is that they support different solutions to the Behrens–Fisher problem*.

CONDITIONAL CONFIDENCE REGIONS

Brownie and Kiefer [5] consider that one of the weaknesses of Neyman–Pearson* methodology is that classical procedures generally do not give a measure of conclusiveness, which depends upon the data observed.

Most other schools of thought do vary their measure of conclusiveness with the data. Kiefer [16] has developed a theory of conditional confidence which extends Neyman–Pearson methodology to allow both a data-dependent measure of conclusiveness and a frequency interpretation of this measure. The basic idea is most easily explained by an example of testing between two hypotheses. (*See also* CONDITIONAL INFERENCE.)

Example 6. Suppose that we wish to discriminate between two simple hypotheses, $H_0: X$ has a standard normal distribution and $H_1: X$ is distributed normally with mean 3 and unit variance, on the basis of a single observation. A standard Neyman–Pearson procedure would be to accept H_0 (or fail to reject H_1) is $X \leqslant 1.5$ and to accept H_1 if $X > 1.5$, and note that the probability of being correct is 0.933 as the measure of conclusiveness. That the same conclusiveness is expressed when $X = 1.6$ and when $X = 3.6$ seems unsatisfactory to Kiefer.

Kiefer's idea is to partition the sample space and to evaluate the conclusiveness of a statistical procedure conditionally for each subset of the partition. Here the sample space might be partitioned into three sets: $(-\infty, 0] \cup (3, \infty)$, $(0, 1] \cup (2, 3]$, and $(1, 2]$. Conditionally on X being in the various sets, the probabilities of the decision to accept H_0 or H_1 being correct are 0.9973, 0.951, and 0.676. These could be considered to indicate "quite conclusive," "reasonably conclusive," and "slight" evidence, respectively.

The article and discussion of Kiefer [16] refers to most other work relevant to conditional confidence regions. Most research has addressed the problem of which partitions of the sample space to use. Until the theory is further developed, it is difficult to see whether it will escape from the known weaknesses of Neyman–Pearson inference.

CONFIDENCE INTERVALS IN PRACTICAL STATISTICS

Confidence intervals are widely used in practice, although not as widely supported by people interested in the foundations of statistics. One reason for this dominance is that the most readily available statistical computer programs are based on the methods of the Neyman–Pearson school. Another reason is that many common confidence intervals (those based on normal, t, and binomial distributions) may also be derived as improper Bayesian procedures and do not suffer from most of the possible weaknesses of confidence intervals. These common procedures have some robustness* with respect to the vagaries of inference theory and may therefore be used without worrying very much about the theory behind a particular derivation of them. Furthermore, it is fairly safe to use the intuitive notion of confidence rather than the restricted technical notion in such cases.

When interpreting confidence intervals for several comparable parameters it should be noted that for two confidence intervals to overlap does not imply that the confidence interval for the difference between the two parameters would include the point zero. Also note that comparing more than two parameters at a time requires special theory (*see* MULTIPLE COMPARISONS).

References

[1] Bartholomew, D. J. (1965). *Biometrika*, **52**, 19–35.

[2] Bartlett, M. S. (1965). *J. Amer. Statist. Ass.*, **60**, 395–409.

[3] Birnbaum, A. (1962). *J. Amer. Statist. Ass.*, **57**, 269–326. (Very difficult to read.)

[4] Brown, L. D. (1967). *Ann. Math. Statist.*, **38**, 1068–1071.

[5] Brownie, C. and Kiefer, J. (1977). *Commun. Statist. A—Theory and Methods*, **6**, 691–751.

[6] Buehler, R. J. (1959). *Ann. Math. Statist.*, **30**, 845–863. (Fundamental reference on conditional properties of statistical procedures.)

[7] Cox, D. R. (1958). *Ann. Math. Statist.*, **29**, 357–372.

[8] de Groot, M. H. (1973). *J. Amer. Statist. Ass.*, **68**, 966–969.

[9] Dempster, A. P. (1964). *J. Amer. Statist. Ass.*, **59**, 56–66.

[10] Fisher, R. A. (1956). *Statistical Methods and Scientific Inference*, Oliver and Boyd, Edinburgh. (See p. 32.)

[11] Hacking, I. (1965). *Logic of Statistical Inference*. Cambridge University Press, Cambridge.

[12] Jeffreys, H. (1940). *Annals of Eugenics*, **10**, 48–51.

[13] Jones, H. L. (1958). *J. Amer. Statist. Ass.*, **53**, 482–490.

[14] Keeping, E. S. (1962). *Introduction to Statistical Inference*. D. Van Nostrand, Princeton, N.J.

[15] Kendall, M. G. and Stuart, A. (1961). *The Advanced Theory of Statistics*, Vol. 2: *Inference and Relationship*. Charles Griffin, London.

[16] Kiefer, J. (1977). *J. Amer. Statist. Ass.*, **72**, 789–827.

[17] Lindley, D. V. (1965). *Introduction to Probability and Statistics from a Bayesian Viewpoint, Part 2, Inference*. Cambridge University Press, Cambridge, England.

[18] Neyman, J. (1934). *J. R. Statist. Soc. A*, **97**, 558–606. (Especially Note I, p. 589, and discussion by R. A. Fisher, p. 614. Mainly of historical interest.)

[19] Neyman, J. (1937). *Philos. Trans. R. Soc. Lond. A*, **236**, 333–380. (Fundamental reference on confidence intervals. These papers by Neyman are reproduced in J. Neyman, *A Selection of Early Statistical Papers of J. Neyman*, Cambridge University Press, Cambridge, 1967.)

[20] Pratt, J. W. (1962). *J. Amer. Statist. Ass.*, **57**, 314–316.

[21] Rao, C. R. (1965). *Linear Statistical Inference and Its Applications*. Wiley, New York.

[22] Robinson, G. K. (1975). *Biometrika*, **62**, 155–161. (Contrived, but reasonably simple examples.)

[23] Savage, L. J. (1954). *The Foundations of Statistics*. Wiley, New York. (Argues that Bayesian inference is the only coherent theory.)

[24] Savage, L. J. (1962). *The Foundations of Statistical Inference (a Discussion)*. Methuen, London.

(BAYESIAN INFERENCE
CONDITIONAL INFERENCE
FIDUCIAL INFERENCE
STATISTICAL INFERENCE)

G. K. ROBINSON

CONFIDENCE INTERVALS, FIXED-WIDTH AND BOUNDED LENGTH *See* FIXED-WIDTH CONFIDENCE INTERVALS

CONFIGURATION

Broadly speaking, this term is used synonymously with "pattern." However, there is a tendency to use the latter term more particularly when referring to (1) population structures or (2) specifically arranged experiment designs, and to apply the term "configuration" when discussing experimentally obtained (e.g., sample) results.

In particular, the term "configuration" is applied to the set of deviations $\{ X_i - \overline{X} \}$ of observed values $X_1, X_2, \ldots, X_n$ from their arithmetic mean $(\overline{X} = n^{-1}\sum_{i=1}^{n} X_i)$. (It is also applied to similar sets of deviations for multivariate sample values.)

(CONDITIONAL INFERENCE
DENDRITES
DESIGN OF EXPERIMENTS
PATTERN RECOGNITION)

CONFIRMATORY DATA ANALYSIS
See EXPLORATORY DATA ANALYSIS

CONFLUENCE ANALYSIS *See* REGRESSION: CONFLUENCE ANALYSIS

CONFLUENT HYPER-GEOMETRIC FUNCTION

Kummer's form of confluent hypergeometric function with argument x and parameters a and b is

$$M(a, b; x) = \sum_{j=0}^{\infty} \frac{a^{[j]}}{b^{[j]}} \frac{x^j}{j!} .$$

Another notation is $_1F_1(a; b; x)$.

The density function of the noncentral F-distribution* with ν_1 and ν_2 degrees of freedom and noncentrality parameter λ can be conveniently expressed in terms of a confluent hypergeometric distribution. In fact, the density function of $\chi_{\nu_1}'^2(\lambda)/\chi_{\nu_2}^2 = G_{\nu_1,\nu_2:\lambda}$ is

$$\left[\frac{1}{B\left(\frac{1}{2}\nu_1, \frac{1}{2}\nu_2\right)} \frac{g^{(1/2)\nu_1 - 1}}{(1 + g)^{(1/2)(\nu_1 + \nu_2)}} \right] e^{-1/2}$$

$$M\left(\tfrac{1}{2}(\nu_1 + \nu_2), \tfrac{1}{2}\nu_1; \tfrac{1}{2}\lambda g/(1 + g)\right).$$

The term in brackets is the density function of the (central) F-distribution with ν_1 and ν_2 degrees of freedom (corresponding to $\lambda = 0$).

Some other statistical distributions are conveniently expressed in terms of confluent hypergeometric functions; see, e.g., Mathai and Saxena [3, 4] and Staff [6]. For a recent statistical application in estimation in regression analysis, see Reed and Wu [5]. A detailed discussion of these and related functions is given in Erdelyi [1] and Luke [2].

References

[1] Erdelyi, A. (1954). *Transcendental Functions*, Vol. 1. McGraw-Hill, New York.

[2] Luke, Y. L. (1975). *Mathematical Functions and Their Approximation*. Academic Press, New York.

[3] Mathai, R. K., and Saxena, A. M. (1966). *Metrika*, **11**, 127–132.

[4] Mathai, R. K., and Saxena, A. M. (1969). *Ann. Math. Statist.*, **40**, 1439–1448.

[5] Reed, A. H. and Wu, G. T. (1977). *Commun. Statist.*, *A*, **6**, 405–416.

[6] Staff, P. J. (1967). *J. Amer. Statist. Ass.*, **62**, 643–654. (Application to displaced Poisson distributions.)

(*H*-FUNCTION DISTRIBUTION)

CONFOUNDING

If the plots, or experimental units, for an experiment are heterogeneous, the experimenter will usually wish to arrange the plots, schematically at least if not physically, in blocks, in such a way that the plots within a single block are fairly homogeneous. The boundaries of the blocks may be determined by the circumstances of the experiment: for example, young animals may be naturally blocked by litters, and work shifts by days of the week. Sometimes the experimenter is free to choose the block boundaries: for example, when the experimental units are plots in an agricultural field. In all cases the between-blocks variation should be substantially higher than the within-blocks variation, and the experimenter must take account of the between-blocks variation when designing* and analyzing his experiment.

If the treatments are allocated to plots in such a way that, within each block, every plot has the same treatment, then when the experiment is analyzed it will be impossible to distinguish the effect of treatment differences from the effect of block differences.

Obviously, this is undesirable. We say that treatments are *confounded* with blocks. More generally, a treatment contrast* is confounded with blocks if it takes a constant value throughout each block. If the blocks are large enough to accommodate each treatment an equal number of times, no confounding occurs.

Blocks, whether their boundaries are determined naturally or by choice, may be of equal or unequal sizes. The analysis of the experiment is considerably simpler when they are all of the same size, and we limit our discussion to this case for the rest of the article.

2^n FACTORIALS

In factorial* experiments it is usually desirable that the confounded contrasts belong to high-order interactions* rather than to main effects*. Consider a 2^3 factorial experiment, with treatment factors A, B, and C, each at two levels, which are denoted by 0 and 1. Let $q(A)$ denote the level of A, and so on, and let ab denote the treatment with $q(A) = q(B) = 1$, $q(C) = 0$, and so on. Suppose that two blocks of four plots each are available. The design shown in Table 1 confounds the main effect of A, because $q(A) = 0$ throughout block I and $q(A) = 1$ throughout block II. The design in Table 2 confounds no main effects, but now the two-factor interaction AB is confounded, for the "levels"* of AB are given by the rule

$$q(AB) = q(A) + q(B) \qquad \text{modulo 2,}$$

so $q(AB) = 0$ on block I and $q(AB) = 1$ on block II. Similarly, in the design shown in Table 3, it is the three-factor interaction ABC that is confounded. Other things being

Table 1

Block I	Block II
1	*a*
b	*ab*
c	*ac*
bc	*abc*

Table 2

Block I	Block II
1	*a*
ab	*b*
c	*ac*
abc	*bc*

Table 3

Block I	Block II
1	*abc*
ab	*c*
ac	*b*
bc	*a*

equal, the design in Table 3 would be preferred to the other two.

If the plots are arranged in four blocks of two plots each, more than one treatment effect must be confounded. Suppose that ABC and AB are confounded. Then

$$q(A) + q(B) + q(C) \qquad \text{modulo 2}$$

is constant on each block, and so is

$$q(A) + q(B) \qquad \text{modulo 2}.$$

Adding these terms together modulo 2, we find that $q(C)$ is constant on each block, so the main effect of C is also confounded. The design is shown in Table 4. Now

$$C = (ABC)(AB)$$

(canceling squares), so C is the generalized interaction* of the effects ABC and AB. Thus if two effects are confounded, so is their generalized interaction. For example, if AB and AC are confounded, so is BC, because $BC = (AB)(AC)$. A design with these three two-factor interactions confounded is shown in Table 5.

For a 2^n factorial design in 2^r blocks of size 2^{n-r} there must be 2^{r-1} confounded effects, with the property that the general-

Table 5

Block I	Block II	Block III	Block IV
1	*ab*	*ac*	*bc*
abc	*c*	*b*	*a*

ized interaction of any two is itself confounded. Block I is obtained by writing down all the treatments for which every confounded effect has level 0. All other blocks are obtained by writing down a new treatment and multiplying it by each treatment in block I (canceling squares).

p^n FACTORIALS, p, PRIME

Confounding in p^n factorial designs, where p is an odd prime, is only a little more complicated than in 2^n factorial designs. Table 6 shows a design for a 3^2 factorial experiment in three blocks of three plots each. Throughout

Block I: $\quad q(A) + 2q(B) = 0 \quad$ modulo 3;

Block II: $\quad q(A) + 2q(B) = 1 \quad$ modulo 3;

Block III: $\quad q(A) + 2q(B) = 2 \quad$ modulo 3.

Thus the 2 degrees of freedom* denoted by AB^2, which belong to the interaction AB, are confounded with blocks. The treatments in block II are obtained by multiplying each treatment in block I by a, and canceling cubes.

Table 7 shows a design for a 3^3 factorial experiment in nine blocks of three plots each. The effects AB^2 and AC^2 are both confounded, because

$$q(A) + 2q(B) \qquad \text{modulo 3}$$

and

$$q(A) + 2q(C) \qquad \text{modulo 3}$$

Table 4

Block I	Block II	Block III	Block IV
1	*bc*	*c*	*a*
ab	*ac*	*abc*	*b*

Table 6

Block I	Block II	Block III
1	*a*	*b*
ab	a^2b	ab^2
a^2b^2	b^2	a^2

Table 7

1	ab	a^2b^2	ac	a^2bc	b^2c	a^2c^2	bc^2	ab^2c^2
abc	a^2b^2c	c	a^2bc^2	b^2c^2	ac^2	b	ab^2	a^2
$a^2b^2c^2$	c^2	abc^2	b^2	a	a^2b	ab^2c	a^2c	bc

are both constant on each block. Therefore,

$$\left[q(A) + 2q(B) \right]$$
$$+ \left[q(A) + 2q(C) \right] \qquad \text{modulo } 3$$

is constant on each block; that is,

$$2\left[q(A) + q(B) + q(C) \right] \qquad \text{modulo } 3$$

is constant on each block, so ABC is confounded. Also,

$$\left[q(A) + 2q(B) \right]$$
$$+ 2\left[q(A) + 2q(C) \right] \qquad \text{modulo } 3$$

is constant on each block; that is,

$$2\left[q(B) + 2q(C) \right] \qquad \text{modulo } 3$$

is constant on each block, so BC^2 is confounded. Note that

$$(AB^2)(AC^2) = (ABC)^2$$

and

$$(AB^2)(AC^2)^2 = (BC^2)^2$$

(canceling cubes).

Thus for a p^n factorial design in p^r blocks of size p^{n-r} there must be $p^{r-1}/(p-1)$ confounded effects, each consisting of $p-1$ degrees of freedom, with the property that if D and E are confounded effects, so are DE, $DE^2, \ldots, DE^{p-1}$

FRACTIONAL FACTORIALS*

If there are two few plots to accommodate all the treatments, a fractional* design must be used. One possibility is to use one block from a single replicate blocked design. The treatment effects that were confounded with blocks in the whole design become *confounded with the mean*, or *defining contrasts*, in the fractional design, for they take the

same value on each plot and so cannot be estimated. Consider the fractional design given by block I of Table 3. The effect ABC cannot be estimated, because $q(ABC) = 0$ everywhere. Moreover, $q(AB) = q(C)$ everywhere, so the effects AB and C are *confounded with each other*, or *aliased*. Thus this design could not be used to estimate the main effect C unless the interaction AB were deemed negligible. In general, for a fraction of a p^n factorial, if the effect D is a defining contrast and E is any effect, then the effects $E, ED, ED^2, \ldots, ED^{p-1}$ are aliased.

A clear account of fractional factorial* designs constructed by this method is given by Finney [12].

Another method of construction is available when every factor has n levels (n not necessarily prime), and there are n^2 plots. When there are three factors the design is constructed from a Latin square*. Each plot corresponds to a cell of the Latin square: the level of the first factor is given by the row, of the second factor by the column, and the third factor by the letter in the cell. For example, when $n = 2$ we may use the Latin square

		Column	
		0	1
Row	0	0	1
	1	1	0

The top right-hand cell gives the treatment bc, and the fraction obtained is that shown in block I of Table 3. The Latin square properties ensure that no main effects are confounded with each other or the mean. When there are four factors, a Graeco–Latin square* is used in a similar manner.

OTHER FACTORIAL DESIGNS

Factorial designs where each treatment factor has s levels and s is not prime have not been considered above. When s is a prime power they may be dealt with by considering the finite field with s elements. The method is explained by Bose [3] and Raghavarao [31]. Chakravarti [5] explains how to extend this method to deal with asymmetrical factorial designs.

Alternatively, whatever value s has, the integers modulo s may be used to give the confounding pattern. Details are given in John and Dean [20], Dean and John [10], Bailey et al. [2], and Bailey [1]. The last three papers extend the method to deal with asymmetrical designs.

MORE COMPLICATED BLOCK STRUCTURE

Some experiments may have more complicated block structure. For example, the blocks may be grouped into superblocks. In this case some effects are confounded with blocks within superblocks, and some with superblocks. A common example of this structure is the split-plot design*.

Alternatively, the plots may form a rectangular array with rows and columns. Then some effects are confounded with rows and some with columns. Confounding can be avoided only when the number of treatments divides both the number of rows and the number of columns. A common design of this sort is the Latin square, in which there are equal numbers of rows, columns, and treatments, and each treatment occurs exactly once in each row and once in each column.

By combining the ideas of *nesting* (blocks in superblocks) and *crossing* (rows and columns), one may build up more complicated block structures. Nelder [27] described many possible block structures. Patterson [28] introduced the *design key* method for allocating treatments to plots to achieve certain confounding systems for complicated block

structures. This method is explained in more detail by Patterson and Bailey [29].

CONFOUNDING IN OTHER DESIGNS

The notion of confounding is not restricted to factorial designs. It is relevant to any design where the blocks are incomplete. For example, in the incomplete block design* shown in Table 8, the 2 treatment degrees of freedom for

123 versus 456 versus 789

are confounded with blocks.

Table 8

Block I	Block II	Block III
1	4	7
2	5	8
3	6	9

PARTIAL CONFOUNDING

A treatment effect is *orthogonal* to blocks if it takes each of its levels equally often in each block. A treatment effect that is neither confounded with blocks nor orthogonal to blocks is *partially* confounded. For example, in the design formed by the four blocks shown in Tables 2 and 3, the effects AB and ABC are partially confounded. It is often preferable, especially in nonfactorial designs, to confound several effects partially rather than to confound any one completely, because there is a certain amount of information available on partially confounded effects, but none on completely confounded effects. In the example above, it is clear that there is one-half the possible information available on AB and ABC. To find the proportion of information available in more complicated cases, see Nelder [27].

A PRACTICAL EXAMPLE

A greenhouse experiment was conducted to find the effect on the yield of tomato plants of three factors: time of sowing the seed (T), type of compost used (C), and amount of water given during growth (W). It was decided to limit each factor to two levels, so there were two times of sowing, two types of compost, and two different watering regimes, making a total of eight different combinations.

At this experimental station the stems of the growing tomato plants were intertwined, so yields (in kilograms of fruit produced per plant) could be measured unambiguously only on *groups of neighboring plants*: such groups therefore had to receive uniform treatment, and it was appropriate to consider these *groups of plants* to be the *plots* of the experiment. For reasons determined by the practical management of the greenhouse, the groups had to be fairly large, so there were only eight per greenhouse, in an array of two rows and four columns. The rows ran from east to west, and it was believed that the effect of sunlight on the southern row would be considerable. However, it was thought from past experience that there would be no appreciable difference in the environments provided by the columns. Hence the block structure of the greenhouse was considered to be two rows of four plots each.

In the absence of any other information it seemed best to confound the three-factor interaction *CTW*. Then I was told that there were *two* similarly situated greenhouses available for the experiment, and that it was believed that the time factor, *T*, would not interact with any of the others. So I recommended that the three-factor interaction *CTW* be confounded in the first greenhouse, and the two-factor interaction *CT* be confounded in the second greenhouse. Thus the layout for the experiment was that shown in Table 9 [which was, of course, randomized (*see* RANDOMIZATION) before the experiment took place]. The main effect of each of the factors, and the two-factor interaction *CW*, which might be significant, could be estimated with no loss. If the experimenter subsequently decided to check whether *T* interacted with the other factors, he or she could estimate *TW* from the whole experi-

Table 9

Greenhouse I			
Row 1	1	*ct*	*tw*	*cw*
Row 2	*t*	*c*	*w*	*ctw*

Greenhouse II			
Row 1	1	*ct*	*w*	*ctw*
Row 2	*t*	*c*	*tw*	*cw*

ment and *CT* from the second greenhouse. Here 1 denotes early sowing, first type of compost, less water; *tw* denotes late sowing, first type of compost, more water; etc.

A BRIEF HISTORY

In his work at Rothamsted in the 1920s Fisher laid down three principles for good experimentation: *replication**, *randomization**, and local control (i.e., blocking). Although he was not the first to suggest blocking, he was one of the staunchest proponents of the idea. He also recognized the advantages of the factorial experiment over the traditional one-at-a-time experiment. Blocking of factorial experiments demands confounding schemes. During the late 1920s and early 1930s, Fisher and Yates developed confounding schemes for 2^n designs and, to a lesser extent, for 3^n designs. These were presented to the general statistical world by Yates [33]. Confounding in the 3^n designs was then based on Latin squares, but later Yates [34] developed confounding schemes for 3^n designs using his "*I* and *J* diagonals," a notation that is still used in such reference works as Fisher and Yates [19] and Cochran and Cox [7]. Also during these years, Fisher defined loss of information in cases of partial confounding*, and Yates introduced his algorithm for estimating factorial effects. (*See* YATES' ALGORITHM.)

In the 1940s three separate groups of workers developed confounding schemes for p^n designs, where *p* is prime. In Great Britain, Fisher [15, 16] obtained a general scheme by using the theory of groups;

Finney [12] extended this to deal with fractional designs. In the United States, Kempthorne [22] developed the same ideas. The notation used in this article is that of Fisher, Finney, and Kempthorne. In India, Bose and Kishen [4] and Bose [3] used finite geometries to obtain general confounding schemes, which are in fact the same as those of Fisher, Finney, and Kempthorne. Their approach is described by textbooks such as Raghavarao [31].

Confounding patterns for asymmetrical factorial designs have been suggested by many workers from Fisher and Yates onward, but there is not yet any widely accepted *simple* scheme. Various complications in block structure have also been introduced from Fisher and Yates onward. Nelder [27] gave a unifying description and classification of simple orthogonal block structures, with a generalized definition of loss of information in the case of partial confounding, but the linear algebra in his paper seems to have prevented his ideas from reaching a wide audience.

Literature

The reference list contains references cited in the article, textbooks for experimenters, and textbooks for statisticians and mathematicians. The textbooks are, necessarily, only a selection of those available.

Among textbooks for experimenters, only Yates [34] assumes any previous statistical or much mathematical knowledge. Cochran and Cox [7], Cox [8], and Li [25] are written for nonspecific applications. Cochran and Cox is a standard reference: Chaps. 6 to 8 deal with confounding in 2^n designs in some detail, and in 3^n designs more briefly, using Yates' I, J notation [34]; there are a lot of plans and confounding schemes tabulated. Cox [8] has a simple discussion in Chap. 12: many confounding schemes are given for 2^n designs, and other designs are done using Latin squares. Chapter 24 of Li [25] gives a simple explanation of confounding in designs with factors at two and/or three levels, with clear examples and tables.

Chew [6], Davies [9], and Natrella [26] are aimed at industrial experimenters; Finney [14], Le Clerg et al. [24], Pearce [30], Wishart and Sanders [32], and Yates [34] at agricultural experimenters. Natrella [26] and Le Clerg et al. [24] are manuals of good practice, so they are particularly straightforward, with procedures clearly laid out. Natrella [26, Chaps. 11 to 14] gives a short clear account of blocking and confounding, with an appendix giving plans and counfounding schemes. Le Clerg et al. [24, Chap. 12] limit themselves to 2^n designs: they discuss total and partial confounding and analysis, and give numerical examples. Davies' long section [9, Chaps. 9 and 10] is an excellent coverage of 2^n and 3^n designs, with many worked examples of both the abstract schemes and numerical data. Appendices give tables, systems of confounding, and some theory for the mathematicians. Chew [6] is less elementary than the other books in this section, and uses more symbolic notation, but the discussion of real examples where confounding has been used, and its consequences, should prove interesting.

The references in Pearce [30, Chaps. 2 and 3] and Wishart and Sanders [32, Chaps. 5, 8, and 15] are brief but clear in their context for workers in the appropriate subjects. Chapters 6 and 7 of Finney [14] give a clear explanation of 2^n designs, good for workers with plants *or* animals. Yates' classic [34] assumes knowledge of analysis of variance, and uses his I, J notation for 3^n designs, but the work is a gem and merits reading by all experimenters, in whatever subject, who have the necessary background.

Of the mathematical texts, Fisher [18, Chaps. VII and VIII] does 2^n designs in the manner described in this article, but uses Latin squares for other designs. Finney [13] gives a brief but very readable account of confounding in general p^n designs in Chaps. 4 and 5. John [21] is an excellent text, Chaps. 7 to 9 covering confounding in general p^n designs and giving many confounding schemes, particularly for 2^n designs. Federer [11, Chap. IX] discusses the rationale behind blocking and total and partial confounding, and gives a good account of 2^n and 3^n designs. Kempthorne [23] gives a long, detailed treatment of the subject, including general p^n designs and partial confounding, in Chap. 14 onward: there are many examples, both abstract and numerical.

References

[1] Bailey, R. A. (1977). *Biometrika*, **64**, 597–603.

[2] Bailey, R. A., Gilchrist, F. H. L., and Patterson, H. D. (1977). *Biometrika*, **64**, 347–354.

[3] Bose, R. C. (1947). *Sankhyā*, **8**, 107–166.

[4] Bose, R. C. and Kishen, K. (1940). *Sankhyā*, **5**, 21–36.

[5] Chakravarti, I. M. (1956). *Sankhyā*, **17**, 143–164.

[6] Chew, V., ed. (1958). *Experimental Designs in Industry*. Wiley, New York.

[7] Cochran, W. G. and Cox, G. M. (1957). *Experimental Designs* (2nd ed.). Wiley, New York.

[8] Cox, D. R. (1958). *Planning of Experiments*. Wiley, New York.

[9] Davies, O. L., ed. (1956). *The Design and Analysis of Industrial Experiments* (2nd ed.). Oliver & Boyd, Edinburgh.

[10] Dean, A. M. and John, J. A. (1975). *J. R. Statist. Soc. B*, **37**, 72–76.

[11] Federer, W. T. (1955). *Experimental Design—Theory and Application*. Macmillan, New York.

[12] Finney, D. J. (1945). *Ann. Eugen.* (Lond.), **12**, 291–301.

[13] Finney, D. J. (1960). *An Introduction to the Theory of Experimental Design*. University of Chicago Press, Chicago.

[14] Finney, D. J. (1962). *An Introduction to Statistical Science in Agriculture*. Munksgaard, Copenhagen/Oliver & Boyd, Edinburgh.

[15] Fisher, R. A. (1942). *Ann. Eugen.* (Lond.), **11**, 341–353.

[16] Fisher, R. A. (1945). *Ann. Eugen.* (Lond.), **12**, 283–290.

[17] Fisher, R. A. (1958). *Statistical Methods for Research Workers* (13th ed.). Oliver & Boyd, Edinburgh. (First ed., 1925.)

[18] Fisher, R. A. (1966). *The Design of Experiments* (8th ed.). Oliver & Boyd, Edinburgh.

[19] Fisher, R. A. and Yates, F. (1963). *Statistical Tables for Biological, Agricultural and Medical Research* (6th ed.). Oliver & Boyd, Edinburgh.

[20] John, J. A. and Dean, A. M. (1975). *J. R. Statist. Soc. B*, **37**, 63–71.

[21] John, P. W. M. (1971). *Statistical Design and Analysis of Experiments*. Macmillan, London.

[22] Kempthorne, O. (1947). *Biometrika*, **34**, 255–272.

[23] Kempthorne, O. (1952). *The Design and Analysis of Experiments*. Wiley, New York.

[24] Le Clerg, E. L., Leonard, W. H., and Clark, A. G. (1962). *Field Plot Technique*. Burgess, Minneapolis, Minn.

[25] Li, C. (1964). *Introduction to Experimental Statistics*. McGraw-Hill, New York.

[26] Natrella, M. E. (1963). Experimental Statistics. *Natl. Bur. Stand.* (*U.S.*) *Handb.* 91.

[27] Nelder, J. A. (1965). *Proc. R. Soc. Lond. A*, **283**, 147–178.

[28] Patterson, H. D. (1965). *J. Agric. Sci.*, **65**, 171–182.

[29] Patterson, H. D. and Bailey, R. A. (1978). *Appl. Statist.*, **27**, 335–343.

[30] Pearce, S. C. (1976). Field Experimentation with Fruit Trees and Other Perennial Plants (2nd ed.). *Commonw. Agric. Bur. Tech. Commun.* 23.

[31] Raghavarao, D. (1971). *Constructions and Combinatorial Problems in Design of Experiments*. Wiley, New York.

[32] Wishart, J. and Saunders, H. G. (1955). Principles and Practice of Field Experimentation. *Commonw. Agric. Bur. Tech. Commun.* 18.

[33] Yates, F. (1935). *J. R. Statist. Soc. B*, **2**, 181–247.

[34] Yates, F. (1937). The Design and Analysis of Factorial Experiments. *Imp. Bur. Soil Sci. Tech. Commun.* 35.

(ANALYSIS OF VARIANCE
BLOCKS, BALANCED INCOMPLETE
CONTRAST
DEGREES OF FREEDOM
DESIGN OF EXPERIMENTS
EFFICIENCY
FACTORIAL EXPERIMENT
FRACTIONAL FACTORIAL
    EXPERIMENT
GENERAL LINEAR MODEL
GRAECO-LATIN SQUARE
INCOMPLETE BLOCK DESIGNS
INTERACTION
LATIN SQUARE
LOSS OF INFORMATION
MAIN EFFECT
NESTING
ORTHOGONALITY
PLOTS
RANDOMIZATION
REPLICATION
SPLIT-PLOT DESIGN
TREATMENTS)

R. A. BAILEY

CONFUSION MATRIX

Introduced by Massy [1], this matrix sets out the results of applying a discriminant analysis* procedure. If there are k possible categories, it is a $k \times k$ matrix with rows denoting the true category and columns the category to which assigned. "Confusion" is represented, of course, by the off-diagonal elements. In place of actual frequencies, the relative frequencies—proportions assigned to each category for a given true category, so that each row sums to 1—may be used. (This does not use the information on proportions of individuals in each true category.)

Reference

[1] Massy, W. F. (1965). *J. Advert. Res.*, **5**, 39–48.

(DISCRIMINANT ANALYSIS)

CONGESTION THEORY *See* QUEUEING THEORY; TRAFFIC FLOW THEORY

CONJOINT ANALYSIS *See* MARKETING, STATISTICS IN

CONJUGATE FAMILIES OF DISTRIBUTIONS

The idea of a parameterized family of probability distributions conjugate to a given sampling process was proposed by Raiffa and Schlaifer in their influential monograph [28]. Their work was based largely on exploring the benefits of conjugacy, which include mathematically tractable closed-form expressions for many important quantities used in Bayesian inference* and decision analysis.* When statistical data are to be analyzed in Bayesian theory, a "prior" distribution* must be chosen to model a person's belief concerning unknown parameters in the sampling process (*see* BAYES' THEOREM). The modeled uncertainty is that which exists *prior to observation* of the statistical data. There is no sense of innate "a priori" knowledge, or knowledge based on pure logical edge, or knowledge based on pure logical

reasoning free of experience. Previous statistical data, for example, may be considered when choosing a prior distribution. If a conjugate family is rich enough, a prior distribution chosen from the family can provide a useful approximation to the actual considered opinion.

DEFINITION AND PROPERTIES

In Raiffa and Schlaifer's definition [28, Sec. 3.2], a distribution is said to be natural conjugate to a given sampling process if its probability density (or mass) function is proportional to a likelihood function* corresponding to some conceivable sample from the process. For an independent, identically distributed sample, $y = (y_1, y_2, \ldots, y_n)$ with population distribution having unknown sampling parameter θ, the likelihood function is the function of θ equal (or proportional) to the joint sampling density,

$$l_{n,y}(\theta) = f_n(y; \theta)$$

$$= \prod_{i=1}^{n} f_1(y_i; \theta). \qquad (1)$$

Both the parameter θ and the observations y_i can be vector quantities. A density function, or a probability mass function, $p(\theta)$ is called *natural-conjugate* to the likelihood function $l_{n,y}(\theta)$ if the density and the likelihood are proportional as functions of θ. Hence, since $p(\theta)$ must integrate to unity,

$$p(\theta) = g(\theta; n, y)$$

$$= l_{n,y}(\theta) \Big/ \int l_{n,y}(\tilde{\theta}) d\tilde{\theta}, \qquad (2)$$

provided that the likelihood has a finite integral.

A parameterized family of natural conjugate distributions, or *natural conjugate family*, is obtained for a given sampling process by considering all probability densities of the form (2), denoted $g(\theta; n, y)$, for all conceivable sample sizes and sample values n, y. The parameters (n, y) of the family are called *hyperparameters* by Lindley [25]. It is this concept of a *family* conjugate to a sampling process, rather than a specified distribution conjugate to an individual likelihood function, which has been important in Bayesian inference*. For example, it has not been advocated that one choose the particular distribution conjugate to a given realized likelihood function as the prior distribution for use with that likelihood function. Indeed, the likelihood function depends on the data outcome, and for a distribution to represent "prior" opinion, its choice will not depend on the data.

What is, perhaps, the most notable property of a conjugate family was discussed by G. A. Barnard before its formalization by Raiffa and Schlaifer (see Wetherill [31]). A conjugate family is *closed under sampling*. That is, if sample data n, y are used to update a prior distribution in a conjugate family by Bayes' theorem, the resulting posterior distribution will also belong to the family:

$$
\begin{aligned}
p(\theta \mid n, y) &= \frac{p(\theta) l_{n,y}(\theta)}{\int p(\tilde{\theta}) l_{n,y}(\tilde{\theta}) d\tilde{\theta}} \\
&= \frac{g(\theta; n', y') l_{n,y}(\theta)}{\int g(\tilde{\theta}; n', y') l_{n,y}(\tilde{\theta}) d\tilde{\theta}} \\
&= \frac{l_{n',y'}(\theta) l_{n,y}(\theta)}{\int l_{n',y'}(\tilde{\theta}) l_{n,y}(\tilde{\theta}) d\tilde{\theta}} \\
&= \frac{l_{n'',y''}(\theta)}{\int l_{n'',y''}(\tilde{\theta}) d\tilde{\theta}} \\
&= g(\theta; n'', y'').
\end{aligned}
\tag{3}
$$

The prior hyperparameters n', y' in $g(\theta; n', y')$ are thus replaced by the posterior values n'', y'' corresponding to the "pooled sample," comprised of the observed sample n, y together with the prior parameters n', y',

$$
\begin{aligned}
n'' &= n' + n \\
y'' &= (y', y).
\end{aligned}
\tag{4}
$$

It is *as if* the prior uncertainty were somehow based on a hypothetical prior sample. This interpretation is discussed later.

Example 1: Normal Sampling. For a simple example, consider independent sampling from a normal distribution with un-

known mean θ. If, given θ, $y_i \sim N(\theta, \sigma^2)$, $i = 1, \ldots, n$, then the likelihood function of θ is proportional to a $N(\bar{y}, \sigma^2/n)$ density, centered at the sample mean $\bar{y} = (y_1 + \cdots + y_n)/n$. Thus the conjugate family consists of densities $g(\theta; n', y')$, $y' = (y'_1, \ldots, y'_{n'})$, in which $\theta \sim N(\bar{y}', \sigma^2/n')$, where $\bar{y}' = (y'_1 + \cdots + y'_{n'})/n'$. The posterior distribution from a likelihood function $l_{n,y}(\theta)$ and prior distribution $N(\bar{y}', \sigma^2/n')$ is then $N(\bar{y}''; \sigma^2/n'')$, where $n'' = n' + n$ and

$$
\begin{aligned}
\bar{y}'' &= \frac{y'_1 + \cdots + y'_{n'} + y_1 + \cdots + y_n}{n' + n} \\
&= (n'\bar{y}' + n\bar{y})/(n' + n).
\end{aligned}
$$

(Note here that the likelihood function and conjugate family could be fully parameterized merely by the sample size and sample mean, $n, \bar{y}$.)

A second important property of a conjugate family is that the Bayesian predictive density (*see* PREDICTIVE DISTRIBUTIONS) takes the form of a ratio of integrated likelihoods,

$$
\begin{aligned}
p(y) &= \int f_n(y; \theta) g(\theta; n', y') d\theta \\
&= \int l_{n'',y''}(\theta) d\theta / \int l_{n',y'}(\theta) d\theta. \tag{5}
\end{aligned}
$$

In the normal example, the predictive distribution of the sample vector y is multivariate normal with equal means $\bar{y}'$, equal variances $\sigma^2(n'^{-1} + 1)$, and equal covariances σ^2/n'. Hence for the sample mean,

$$
\bar{y} \sim N\left\{ \bar{y}', \sigma^2(n'^{-1} + n^{-1}) \right\}.
$$

The posterior predictive distribution for a future sample n^*, y^*, following a sample n, y, can be obtained merely by replacing the parameters n', y' in the prior-predictive form by the posterior parameters n'', y'',

$$
p(y^* \mid y) = \int f_{n^*}(y^*; \theta) g(\theta; n'', y'') d\theta.
$$

Thus in the normal case,

$$
\bar{y}^* \mid y \sim N\left\{ \bar{y}'', \sigma^2(n''^{-1} + n^{*-1}) \right\}.
$$

A conjugate family is particularly convenient for use with a sampling model that has tractable sufficient statistics.* Since the likelihood function for a sample (n, y) is proportional in θ to the likelihood function for a sufficient statistic (n, s), $s = s_n(y)$, the associated conjugate family is, in effect, parameterized by the sufficient statistic, (n', s'), $s' = s_{n'}(y')$. (Of course, s and s' can be vectors.) Thus for $g(\theta; n', y')$, one can write $g(\theta; n', s')$. The updating relations for the hyperparameters of the corresponding posterior distribution $g(\theta; n'', s'')$ can then be expressed in the form

$$n'' = n' + n$$
$$s'' = s_{n''}(y'')$$
$$= t_{n', n}(s', s), \qquad (6)$$

where $t_{n', n}$ is the expression for the pooled-sample sufficient statistic s'' in terms of the subsample statistics s' and s. That is, the posterior hyperparameter is related to the prior hyperparameter and the sample sufficient statistic just as the sufficient statistic for a pooled sample relates to the sufficient statistics of complementary subsamples. This phenomenon was exhibited in the normal case as the usual pooling of sample means, $t_{n', n}(\bar{y}', \bar{y}) = (n'\bar{y}' + n\bar{y})/(n' + n)$. The relation applies much more generally, as follows.

Under regularity conditions, bounded dimensionality for the sufficient statistics in a sampling process will imply that the sampling model is an exponential-type family. That is, the probability mass or density function for each observation vector in the sample, y_1 for example, will take the following form, in terms of functions h, k and vector functions $z' = (z_1, \ldots, z_J)$ and $\phi = (\phi_1, \ldots, \phi_J)$,

$$f_1(y_1; \theta) = h(y_1)k(\theta)e^{z(y_1) \cdot \phi(\theta)}, \qquad (7)$$

in which a scalar product of vectors is indicated, $z \cdot \phi = z_1\phi_1 + \cdots + z_J\phi_J$. The sampling parameter vector ϕ is called the *natural parameter* if the corresponding statistic z is minimal sufficient (no constants c_0, $c_1, \ldots, c_J$ exist for which $c_0 = \sum_{j=1}^{J} c_j z_j$ with probability 1).

The likelihood for a sample, $y = (y_1, \ldots, y_n)$, will satisfy the proportionality

in θ,

$$l_{n, y}(\theta) \propto k(\theta)^n e^{ns \cdot \phi(\theta)}, \qquad (8)$$

where the sufficient vector

$$s = s_n(y)$$
$$= \{z(y_1) + \cdots + z(y_n)\}/n. \qquad (9)$$

Hence the conjugate family of probability mass functions or densities is the class

$$g(\theta; n', s') = H(n', s')k(\theta)^{n'} e^{n's' \cdot \phi(\theta)}, \qquad (10)$$

where $1/H(n', s') = \int k(\theta)^{n'} e^{n's' \cdot \phi(\theta)} d\theta$, provided that this integral exists. This family is hyperparameterized by n' and the vector s'. The corresponding prior-to-posterior transformations again have the simple linear form reminiscent of the normal-sampling case,

$$n'' = n' + n$$
$$s'' = (n's' + ns)/(n' + n). \qquad (11)$$

Note that the posterior hyperparameter vector s'' in (11) is a weighted average with identical (scalar) weights for each coordinate.

POSTERIOR MOMENTS

In early unpublished work, Morgan [27] developed the following relations between the moments of conjugate distributions and their hyperparameters. In particular, a posterior mean is found to take a simple linear form. (See also Jewell [18–20], Diaconis and Ylvisaker [8], and Fienberg [14].) However, the population parameters that have these moments are not necessarily the "natural parameters" ϕ for the sampling model. Define the gradient vector* μ with respect to ϕ of the logarithmic normalizing constant in the exponential-type sampling model (7),

$$\mu = \nabla Q(\phi), \qquad (12)$$

with coordinates $\mu_j = \partial Q(\phi)/\partial \phi_j$, where

$$Q(\phi) = \ln \int h(y_1)e^{z(y_1) \cdot \phi} dy_1 = \ln\{1/k(\theta)\}. \qquad (13)$$

Also, define the Hessian matrix Σ of the same quantity, with entries

$$\sigma_{ij} = \frac{\partial^2}{\partial\phi_i \partial\phi_j} Q(\phi). \qquad (14)$$

Of course, both μ and Σ depend, through ϕ, on the original parameter vector θ. They are defined for interior points of the convex set of points ϕ for which the integral $Q(\phi)$ exists.

Lemma. *The parameters μ, Σ are the population first- and second-order moments of the minimal sufficient statistics $z(y_1)$,*

$$\mu = E\{z(y_1)|\phi\}$$
$$\Sigma = \mathrm{cov}\{z(y_1)|\phi\}. \qquad (15)$$

The matrix Σ is the Fisher information* for the parameter ϕ in the model (7); Σ is positive definite: hence $Q(\phi)$ is a convex function.

The following theorems require the special conjugate prior family (10) constructed by defining the parameter θ to be identical to the natural parameter ϕ,

$$\theta \equiv \phi. \qquad (16)$$

(This construction, once made, could be followed by a change of variable in θ.) The theorems hold under further mild regularity conditions [27].

Theorem 1. If θ has distribution (10) with (16), then

$$E\{\mu(\theta)\} = s'. \qquad (17)$$

In particular, the posterior expectation of μ takes the linear form (11) in terms of the sample mean s.

Theorem 2. Under the conditions of Theorem 1, the expectation of the population covariance Σ is proportional to the covariance matrix of the population mean

$$E\{\Sigma(\theta)\} = n' \mathrm{cov}\{\mu(\theta)\}. \qquad (18)$$

Theorem 3. Under the conditions of Theorem 1, the predictive expectation of the posterior covariance matrix of μ is proportional

to the corresponding prior covariance,

$$E[\mathrm{cov}\{\mu(\theta)|y\}] = \frac{n'}{n'+n} \mathrm{cov}\{\mu(\theta)\}. \qquad (19)$$

Theorem 3 has obvious importance for Bayesian experimental design [12]. Theorem 1 partially answers a call by Ericson for characterization of all situations in which the posterior mean is a linear function of the sample average. Ericson [13] showed that whenever such linearity holds, the form (11) applies, with the weights in the proportionality as indicated by Theorem 2,

$$n'/n = \frac{E\{\mathrm{cov}(s_i, s_j|\theta)\}}{\mathrm{cov}\{E(s_i|\theta), E(s_j|\theta)\}}, \qquad (20)$$

any $i, j = 1, \ldots, J$. The linearity property of a posterior mean is more general, however, than the context of exponential sampling and conjugate prior families [7, 16].

Example 2: Multivariate Normal Sampling. In the multivariate generalization of Example 1, J-dimensional vectors y_i, $i = 1, \ldots, n$, are independently sampled from a multivariate normal distribution, $N(\theta, \Omega)$. The mean vector θ is unknown, and the variance matrix Ω is assumed known. The sample average vector $\bar{y}$ is distributed as $N(\theta, n^{-1}\Omega)$. According to the theory given here so far, with $\phi = \mu = \Omega^{-1/2}\theta$, $z_1 = \Omega^{-1/2}y_1$, $\Sigma = I$, the corresponding conjugate prior distribution for the unknown mean vector θ would be

$$\theta \sim N(\bar{y}', n'^{-1}\Omega). \qquad (21)$$

The proportionality in (21) of prior and population covariance matrices is less general than the so-called "conjugate" prior distribution for multivariate normal sampling common in the literature. If one takes

$$\theta \sim N(\bar{y}', N'^{-1}), \qquad (22)$$

the posterior distribution will lie in the same family, with

$$N'' = N' + n\Omega^{-1}$$
$$\bar{y}'' = N''^{-1}(N'\bar{y}' + n^{-1}\Omega\bar{y}). \qquad (23)$$

Note the matrix-weighted-average form for the posterior mean.

To include the family (22) under the label of a conjugate prior family, one can extend the concept of conjugacy, following Jewell [19]. Let z be a vector whose J coordinates independently follow univariate exponential-type sampling models as denoted in (7). (A different exponential family is permitted for each coordinate.) Define the vector of corresponding natural parameters ϕ. Take the prior distribution in which the coordinates of ϕ are independent and each distributed conjugate to the sampling model of the corresponding coordinate of z,

$$p(\phi) = \prod_{j=1}^{J} H_j(n_j', s_j') k_j(\phi_j)^{n_j'} e^{n_j' s_j' \phi_j}. \quad (24)$$

The joint prior parameters are $N' = \mathrm{diag}(n_j')$ and $E(\phi) = s' = (s_j')$; while, following a sample of size n, the posterior parameters are $N'' = N' + nI$ and $s'' = N''^{-1}(N's' + n\bar{z})$.

Now, consider the sampling model for the transformed data,

$$y = C^{-1}z, \quad (25)$$

in which the population mean vector is

$$\theta = C^{-1}\mu. \quad (26)$$

The prior distribution for θ induced by (24) has mean $E(\theta) = C^{-1}s'$, and the corresponding induced posterior distribution has mean

$$E(\theta \mid y_1, \ldots, y_n)$$
$$= C^{-1}s'' = (C^T N' C + nC^T C)^{-1}$$
$$\times \{C^T N' C E(\theta) + nC^T C\bar{y}\}.$$
$$(27)$$

APPROXIMATE POSTERIOR DISTRIBUTIONS

In L. J. Savage's concept of "precise measurement" or "stable estimation" [10, 11], different prior distributions used with the same likelihood function tend to yield approximately the same posterior distribution for a large sample (*see* PRECISE MEASUREMENT). In particular, it has long been known that posterior distributions are asymptotically normal [2, 22, 24, 29]. Walker [30] has shown that for a large class of sampling models (including the exponential-type families), the posterior distribution of the natural parameter is asymptotically normal with mean equal to the maximum likelihood estimate and variance equal to the negative reciprocal of the second derivative of the logarithm of the likelihood function evaluated at the maximum likelihood estimate, provided that the prior density is continuous and positive. Asymptotic normality is taken to mean that with high probability the posterior distribution and the normal distribution are close as n becomes large.

Morgan [27] gives the following alternative to the standard result—"alternative" in the sense that the distribution used as an approximate posterior distribution actually takes account of the prior distribution. Again, the posterior distribution is approximately normal, but now the approximate form will depend on the prior distribution. Under the assumptions and the notation of a conjugate prior family (10), (16) with an exponential-type sampling model, the posterior distribution of the natural parameter vector approximates for large n,

$$\phi \mid y \sim N(\hat{\phi}^*, V^*), \quad (28)$$

where in the notation (15), $\hat{\phi}^*$ is the solution to

$$\mu(\hat{\phi}^*) = s'', \quad (29)$$

and

$$V^* = n''^{-1}\{\Sigma(\hat{\phi}^*)\}^{-1}. \quad (30)$$

This contrasts with a maximum likelihood approach in which the special case of (28) would be obtained with $s'' = s$ and $n'' = n$. [Note that Σ in (30) has Walker's realized likelihood function replaced by a population average.]

EXTENSIONS

The convenient properties of a conjugate family are preserved under various extensions of the concept. For example, linear transformations were introduced at the end of the preceding section. Raiffa and Schlaifer [28] proposed the use of "extended conjugate" families, in which the ranges of the hyperparameters are relaxed but the forms of the relationships are preserved. For example, the sample-size parameter n' is usually permitted to take noninteger values.

As originally proposed, a conjugate family consists of probability densities proportional to conceivable likelihood functions,

$$g(\theta; n', y') \propto l_{n', y'}(\theta). \tag{31}$$

It is also true that a constant prior pseudodensity $p(\theta) \equiv c$ (which would fail to have a finite integral over an unbounded-interval range) yields by formal use of Bayes' theorem a posterior density proportional to the likelihood function,

$$p(\theta \mid n, y) \propto l_{n,y}(\theta) \cdot p(\theta)$$
$$\propto l_{n,y}(\theta). \tag{32}$$

Hence one could interpret a conjugate prior family as consisting of posterior distributions, each coming from a suitable imaginary "prior sample" and the constant "initial-prior" density,

$$g(\theta; n', y') = p(\theta \mid n', y'), \tag{33}$$

where

$$p(\theta \mid n', y') \propto l_{n', y'}(\theta) \cdot p_0(\theta),$$

where

$$p_0(\theta) \equiv c.$$

(See, e.g., Winkler [32, p. 155].) This interpretation is appealing to some authors because of the common notion that a constant pseudodensity models "complete ignorance." However, with such an interpretation, the problem remains that a criterion must be provided to determine the choice of variable θ to receive a constant density. The problem arises more generally whenever a constant density is proposed to model prior ignorance.

Nonlinear functions of θ cannot have constant densities simultaneously with θ. Complete ignorance, in the constant-density sense, for a parameter τ implies a density proportional to θ for the change of variable to $\theta = \tau^{1/2}$, for example, and one proportional to e^{θ} for $\theta = \ln \tau$, and so forth. "Ignorance" concerning τ entails strong opinion concerning θ. There are further paradoxes, also, associated with the use of nonintegrable constant pseudodensities, particularly in high dimensions [5].

This prior sample interpretation does suggest that a conjugate family can be extended by including a further parameterized (hyperparameterized) factor in the density, a function r of θ which could take the value of a constant in θ for a particular value of its hyperparameter z'. The property of closure under sampling and the convenience of hyperparameterization by sufficient statistics could apply as well to the larger family.

$$g(\theta; n', y', z') \propto l_{n', y'}(\theta) \cdot r(\theta; z'), \tag{34}$$

Dickey [9] obtains prior independence in this way between the normal mean and variance parameters.

The difficulty encountered with the arbitrariness of the choice of variable θ and consequent nonuniqueness of the "initial-prior" density indicates that a conjugate family is not unique to a given sampling process. To see this more clearly, compare the natural conjugate family of densities constructed for a new parameter $\tau = \tau(\theta)$, followed by change of variable in the density to θ, to the family of densities $g(\theta)$ constructed directly for θ. These densities have the ratio $\{ \int l(\theta) \, d\theta / \int l(\theta(\tau)) \, d\tau \} \cdot |d\tau / d\theta|$. For example, if one first constructs the natural conjugate family for $\tau = \theta^3$ and then changes variable to θ, one obtains the family of densities proportional to

$$\theta^2 g(\theta; n', y'), \tag{35}$$

where $g(\theta; n', y')$ would be constructed directly by working with θ. This provides one way, then, of interpreting extensions (34) which append an additional factor to the conjugate density.

USE OF STRUCTURAL DISTRIBUTIONS

The approach assumed so far is to choose a particular parameter variable (vector) in a given sampling model, to consider it as a random variable, and then to construct a family of density functions of it. We have seen that in general the resulting conjugate family of distributions depends on the choice of parameter variable, and that even given such a variable, for example the "natural" parameter $\theta = \phi$ (7), or the mean parameter $\theta = \mu$ (12), the family will be subject to various useful extensions. Of course, it is also true that such an approach relies on an arbitrary choice of dominating measure for the densities. (We have tacitly assumed the usual Lebesgue measure.) We turn next to a different approach in which the distributions themselves are primary, rather than their densities.

Dawid and Guttman [4] define as conjugate to a sampling process any class of distributions constructed by starting with an arbitrary initial prior distribution (or pseudo-distribution) and including all distributions obtainable from it by sampling. Hence they do not intend a conjugate class to be unique. They then go on to propose for special interest the families of "structural" distributions interpretable as posterior distributions (*see* FIDUCIAL DISTRIBUTIONS and STRUCTURAL DISTRIBUTIONS). Fraser [15] and Hora and Buehler [17] have shown that structural distributions arise as posterior distributions from the right-Haar measure as a prior pseudo-distribution. Structural distributions and Haar measures, themselves, depend on a choice of invariant group structure in defining the inference problem. For example, in any location-parameter problem, such as our simple normal case (Example 1), the group is a translation by an additive constant, and the initial-prior pseudo-

distribution is the right-Haar measure for which the location parameter is uniform over the whole real line. In any location-scale problem, where

$$f_1(y_1; \mu, \sigma) = f_0\{(y_1 - \mu)/\sigma\}/\sigma, \quad (36)$$

the right-Haar initial-prior pseudo-distribution is such that $\theta = (\mu, \log\sigma)$ is uniform over the plane. This yields $p_0(\mu, \sigma) \equiv c/\sigma$, $\sigma > 0$.

An advantage of using structural distributions is that one need not discuss densities at all and can deal purely in terms of "pivotal quantities" whose distributions are used to define the prior and posterior distributions. A *pivotal quantity* is a function of the parameter and the data which, considered as a random variable, has a sampling distribution not depending on the value of the parameters (*see* PIVOTAL QUANTITIES). The distribution of the pivotal quantity conditional on the data is then taken to be the same as its sampling distribution. For example, in the normal location-parameter problem,

$$(\bar{y} - \mu) \mid \mu \sim N(0, \sigma^2/n), \quad (37)$$

and hence

$$(\bar{y} - \mu) \mid y \sim N(0, \sigma^2/n), \quad (38)$$

yielding $\mu \mid y \sim N(\bar{y}, \sigma^2/n)$. Applied to "prior data" n', y', this approach again gives the parameterized class of distributions, $\mu \sim N(\bar{y}', \sigma^2/n')$. Again, the prior-to-posterior transformations can be handled purely by pooling of samples,

$$\mu \mid y \sim \mu \mid (y, y') \sim N(\bar{y}'', \sigma^2/n),$$

where again $\bar{y}'' = (n'\bar{y} + ny)/(n' + n)$ and $n'' = n' + n$.

The Bayesian predictive distributions can also be obtained by referring to pivotal quantities instead of densities. For a future sample mean $\bar{y}$ in the normal example,

$$\bar{y} - \bar{y}' = (\bar{y} - \mu) - (\bar{y}' - \mu)$$

$$\sim N(0, \sigma^2/n) - N(0, \sigma^2/n')$$

$$\sim N\{0, \sigma^2(n^{-1} + n'^{-1})\},$$

or as before, $\bar{y} \sim N\{\bar{y}', \sigma^2(1/n' + 1/n)\}$. A similar trick is illustrated below for sampling from a gamma distribution (Example 4).

By an ingenious treatment of varieties of sampling processes differing only by the stopping process involved, and hence having proportional likelihoods, Dawid and Guttman found these methods to be applicable to some problems not having invariant group structure.

MIXTURES*

An important extension of a conjugate family in practice is to the family of *discrete mixtures of conjugate distributions*. Consider such a mixture,

$$p(\theta) = \sum_i g(\theta; n_i', s_i')q_i', \qquad (39)$$

where each $q_i' \geqslant 0$ and $\sum q_i' = 1$.
Note, first, that for the predictive distribution the same mixing probabilities will then apply to the conjugate predictive distributions $p(y \mid n_i', s_i')$ [from (5)],

$$p(y) = \sum_i p(y \mid n_i', s_i')q_i'. \qquad (40)$$

For the posterior distribution, one obtains a mixture of conjugate posterior distributions,

$$p(\theta \mid y) = \sum_i g(\theta; n_i'', s_i'')q_i'', \qquad (41)$$

where the posterior mixing probabilities are obtained by Bayes' theorem from the conjugate predictive densities,

$$q_i'' = p(y \mid n_i', s_i')q_i' / \sum_j p(y \mid n_j', s_j')q_j'. \qquad (42)$$

CONJUGATE UTILITIES

An optimal Bayesian decision is one that maximizes expected utility (*see* BAYES' PROCEDURE). A general class of utility functions was introduced by Lindley [26] for which the expected utility, under a conjugate distribution, will take a closed form. Define a utility function of decision d and true state θ to be

conjugate to a sampling process having likelihood functions of the form $l_{n,y}(\theta)$ if it satisfies

$$U(d, \theta) = w^*(d)l_{n^*(d), y^*(d)}(\theta), \qquad (43)$$

for suitable functions $w^*(d), n^*(d), y^*(d)$. Then for a prior density from the conjugate family, the expected utility will take the form of a predictive density,

$$E\{U(d, \tilde{\theta})\} = w^*(d) \int l_{n^*, y^*}(\theta)g(\theta; n', s') d\theta$$

$$= w^*(d)p(y^*). \qquad (44)$$

[Compare (5).] For example, a conjugate utility can be obtained for estimation in the normal case by considering the class of functions proportional to a normal density in the estimation error,

$$U(\hat{\theta}, \theta) = e^{-n^*(\theta - \hat{\theta})^2/(2\sigma^2)}. \qquad (45)$$

Then

$$E\{U(\hat{\theta}, \tilde{\theta})\} = (n^*n'^{-1} + 1)^{-1/2}$$

$$\times e^{-n'n^*(n' + n^*)^{-1}(\hat{\theta} - \bar{y}')/(2\sigma^2)}.$$

Kadane and Dickey [21] note quite generally that the optimal estimate is the same as for squared-error loss, i.e., $\hat{\theta} = E(\theta)$.

FURTHER EXAMPLES

A few examples will be given of sampling models and convenient conjugate families. Although many such examples appear in Raiffa and Schlaifer [28], the illustrations here make use of more familiar, standard notation. Further details and additional examples can be found in LaValle [23] and DeGroot [6].

Example 3: Bernoulli Sampling. The likelihood for a successes in n independent Bernoulli trials with probability parameter π is $\binom{n}{a}\pi^a(1 - \pi)^b$, where $b = n - a$. The statistic a has a binomial (π, n) sampling distribution. The family obtained by working with $\theta = \ln\{\pi/(1 - \pi)\}$ (constant initial-prior pseudo-density in θ), and then changing the variable to π and extending the range of hyperparameters to not necessarily integer values, is the beta family, $\pi \sim \text{Beta}(a', b')$, $a' > 0, b' > 0$, with density

$$g(\pi; a', b') = B(a', b')^{-1} \pi^{a'-1}(1 - \pi)^{b'-1},$$
$$0 < \pi < 1, \qquad (46)$$

where $B(a', b') = \Gamma(a')\Gamma(b')/\Gamma(a' + b')$.

Bayes' theorem yields the corresponding posterior distribution, $\text{Beta}(a'', b'')$, $a'' = a' + a, b'' = b' + b$. The moments are immediate from the gamma function property $\Gamma(a + 1) = a\Gamma(a)$. The prior mean is $m' = a'/n'$, where $n' = a' + b'$; the posterior mean is $m'' = a''/n''$, where $n'' = a'' + b''$; the weighted average

$$m'' = wm' + (1 - w)m, \qquad (47)$$

where $m = a/n$ and $w = n'/(n' + n)$. The prior and posterior variances are, respectively, $m'(1 - m')/(n' + 1)$ and $m''(1 - m'')/(n'' + 1)$. The predictive distribution for a given n is beta-binomial with probability mass function, $\binom{n}{a} B(a' + a, b' + b)/B(a', b')$, $a = 0, 1, \ldots, n, b = n - a$.

Sampling from a negative binomial distribution differs from binomial sampling merely by the stopping rule used on the Bernoulli process. The likelihood for n trials being required for a success is proportional in π to the similarly denoted binomial likelihood. Hence the same conjugate family is obtained. However, the predictive distributions for n given a has probability mass function

$$\binom{n - 1}{a - 1} B(a' + a, b' + b)/B(a', b'),$$
$$n = a, a + 1, \ldots, \qquad b = n - a.$$

In the multivariate generalization of Bernoulli sampling, π is a vector of k probabilities, identically summing to 1, and the statistic a is a vector of k cell counts, identically summing to the total sample size n. The statistic a has a multinomial$^{(k)}(\pi, n)$ sampling distribution. The conjugate prior family is Dirichlet (a'), with density (in $k - 1$ of the coordinates of π) parameterized by k-

vector a' having positive coordinates

$$g(\pi; a') = B(a')^{-1} \prod_{i=1}^{k} \pi_i^{a_i - 1}, \qquad (48)$$

where $B(a') = \Gamma(n')^{-1} \prod_{i=1}^{k} \Gamma(a_i')$, in which $n' = \sum_{i=1}^{k} a_i'$. Again the moments are immediate. The mean vector is $m' = a'/n'$. The prior-to-posterior transformation is, again, $n'' = n' + n$ and $a'' = a' + a$.

Example 4: Gamma Sampling Distribution. As is well known, a gamma random variable y, with density $(\alpha/\mu)^\alpha \Gamma(\alpha)^{-1} y^{\alpha - 1} e^{-\alpha y/\mu}$ for $y > 0$, where $\alpha > 0$, $\mu > 0$, can also be considered as a scaled chi-square random variable,

$$y/\mu \sim \chi_{2\alpha}^2/(2\alpha). \qquad (49)$$

The parameter μ is the sampling mean of y, and the variance is μ^2/α. Consider the situation of unknown mean μ and known power parameter α. (Damsleth [3] treats the case of both μ and α unknown.) The likelihood for a sample of size n is proportional (in μ) to $\mu^{-(n\alpha)} e^{-(n\alpha)m/\mu}$, where m is the sample average. Define the conjugate prior family for unknown μ as the reciprocal gamma with density in μ, $(a'm')^{a'} \Gamma(a')^{-1} \mu^{-a'-1} e^{-a'm'/\mu}$ for $\mu > 0$, where $a' > 0, m' > 0$. Synthetically, μ has the prior distribution

$$\mu/m' \sim 2a'/\chi_{2a'}^2, \qquad (50)$$

with mean $E(\mu) = [a'/(a' - 1)]m'$ and variance $[E(\mu)]^2/(a' - 2)$. (Contrast this with the sampling distribution for the sufficient-statistic sample average m, i.e., $m/\mu \sim \chi_{2n\alpha}^2/(2n\alpha)$). The prior-to-posterior transformation is

$$a'' = a' + n\alpha$$
$$m'' = a''^{-1}(a'm' + n\alpha m). \qquad (51)$$

The Bayesian predictive distribution of m can be obtained by the following trick, using pivotal quantities. Write $m/m' = (m/\mu) \cdot (\mu/m') \sim \{\chi_{2n\alpha}^2/(2n\alpha)\} \cdot \{2a'/\chi_{2a'}^2\}$. These chi-square distributions are independent, because the conditional distribution of m given

μ does not depend on μ; hence

$$m / m' \sim F_{2n\alpha, 2a'}. \qquad (52)$$

Example 5: Uniform Sampling Distribution.
For a final example, the uniform distribution for y_1 with constant density $1/(\beta - \alpha)$ on the interval (α, β) violates the regularity conditions under which the presence of bounded-dimensional sufficient statistics implies an exponential-type family. (The range, or support, of y_1 depends on the parameters α, β.) Even though this family is not of exponential type, it has the sufficient statistics $a = \min(y_1, \ldots, y_n)$ and $b = \max(y_1, \ldots, y_n)$. The likelihood vanishes for $a < \alpha$ or $\beta < b$, and

$$l_y(\alpha, \beta) = (\beta - \alpha)^{-n} \quad \text{if } \alpha < a < b < \beta. \qquad (53)$$

The family of joint conjugate prior distributions for α and β is bilateral bivariate Pareto with parameters a', b', n' [6, p. 62]. The prior density of (α, β) is proportional to (53) with a', b', n' in place of a, b, n and the normalizing constant,

$$(n' - 1)(n' - 2)(b' - a')^{n' - 2}$$
$$= 1 \Big/ \int_{b'}^{\infty} d\beta \int_{-\infty}^{a'} d\alpha \, (\beta - \alpha)^{-n'}.$$

The individual marginal and conditional distributions are essentially univariate Pareto. The associated posterior parameters are $a'' = \min(a', a)$, $b'' = \max(b', b)$, $n'' = n' + n$.

A more useful way to parameterize this problem is to define the distribution uniform$_1(\mu, \sigma)$, the same distribution as uniform (α, β) and having location and scale parameters $\mu = \frac{1}{2}(\alpha + \beta)$ and $\sigma = \frac{1}{2}(\beta - \alpha)$. Then the likelihood is proportional to σ^{-n} for $\sigma > s + |\mu - m|$, where the sample midrange and half-range are denoted $m = \frac{1}{2}(a + b)$ and $s = \frac{1}{2}(b - a)$. The same joint conjugate prior distribution then takes the form

$$g(\mu, \sigma; m', s', n') \propto \sigma^{-n'}, \qquad (54)$$

for $\sigma > s' + |\mu - m'|$. The associated conditional and marginal distributions can now be written

$$\sigma \mid \mu \sim \text{Pareto}, \qquad (55)$$

with lower-bound parameter $s' + |\mu - m'|$ and power parameter $n' - 1$ (density proportional to $\sigma^{-n'}$ for $\sigma > s' + |\mu - m'|$),

$$\mu \mid \sigma \sim \text{uniform}_1(m', \sigma - s'), \qquad (56)$$

$$|\mu - m'| / s' \sim \left(\frac{2}{2n' - 4} \right) F_{2, 2n' - 4}, \qquad (57)$$

$$\sigma / s' \sim 1 + \left(\frac{4}{2n' - 4} \right) F_{4, 2n' - 4}. \qquad (58)$$

References

[1] Bartolucci, A. A. and Dickey, J. M. (1977). *Biometrics*, **33**, 343–354. (A synthetic notation is given for the conjugate family for gamma sampling with unknown scale parameter in the presence of censoring.)

[2] Bernstein, S. (1927). *Theory of Probability* (in Russian), Gostekhizdat; Moscow-Leningrad.

[3] Damsleth, E. (1975). *Scand. J. Statist.*, **2**, 80–84. (A conjugate family of distributions is developed for gamma sampling with power parameter and scale parameter both unknown.)

[4] Dawid, A. P. and Guttman, I. (1980). *Commun. Statist. A*, **9**. (An enlightened approach in which distributions are considered as primary and densities as secondary. Pivotal quantities and structural distributions play a prominent role.)

[5] Dawid, A. P., Stone, M., and Zidek, J. V. (1973). *J. R. Statist. Soc.*, *B*, **35**, 189–233.

[6] DeGroot, M. H. (1970). *Optimal Statistical Decisions*. McGraw-Hill, New York. [Presents many examples of conjugate families. (See also Raiffa and Schlaifer [28] and Lavalle [23].)]

[7] Diaconis, P. and Ylvisaker, D. (1977). Priors with Linear Posterior Expectation. *Tech. Rep. No. 102*, Dept. of Statistics, Stanford University, Stanford, Calif.

[8] Diaconis, P. and Ylvisaker, D. (1979). *Ann. Statist.*, **7**, 269–281. (Treatment in the one-dimensional case of results similar to Morgan [27].)

[9] Dickey, J. M. (1975). In *Studies in Bayesian Econometrics and Statistics*, S. E. Fienberg and A. Zellner, eds. North-Holland, Amsterdam. (Extended conjugate family for the usual normal multiple regression sampling model, allowing prior independence of location and scale.)

[10] Dickey, J. M. (1976). *J. Amer. Statist. Ass.*, **71**, 680–689. (Generalization of Savage's concept of "precise measurement" for use of an arbitrary

approximate prior density in place of the constant pseudo-density.)

[11] Edwards, W., Lindman, H., and Savage, L. J. (1963). *Psychol. Rev.*, **70**, 193–242. (Presentation of Savage's concept of "precise measurement" in its original form. Influential general exposition of Bayesian methods for practice.)

[12] Ericson, W. A. (1968). *J. Amer. Statist. Ass.*, **61**, 964–983.

[13] Ericson, W. A. (1969). *J. R. Statist. Soc. B*, **31**, 332–334. (A posterior mean that is linear in the sample average has a special form in terms of variances.)

[14] Fienberg, S. E. (1980). In *Bayesian Analysis in Econometrics and Statistics*, Arnold Zellner, ed. North-Holland, Amsterdam. (Extensions and uses of Ericson's [13] representation for multidimensional contexts.)

[15] Fraser, D. A. S. (1961). *Biometrika*, **48**, 261–280. (Background material on group invariance and structural distributions for the approach of Dawid and Guttman [4].)

[16] Goldstein, M. (1975). *J. R. Statist. Soc. B*, **37**, 402–405. (Further work related to Ericson [13].)

[17] Hora, R. B. and Buehler, R. J. (1966). *Ann. Math. Statist.*, **37**, 643–656. (Important paper relating structural and fiducial distributions to Haar measure as a prior distribution.)

[18] Jewell, W. S. (1974). *Astin Bull.*, **8**, 77–80. (Further work related to Ericson [13] from the viewpoint of actuarial science.)

[19] Jewell, W. S. (1974). *Mitt. Ver. Schweiz. Versich.-Math.*, **74**, 193–214. (Further work related to Ericson [13] from the viewpoint of actuarial science.)

[20] Jewell, W. S. (1975). *Astin Bull.*, **8**, 336–341. (Further work related to Ericson [13] from the viewpoint of actuarial science.)

[21] Kadane, J. B. and Dickey, J. M. (1969). In *Evaluation of Econometric Models*, J. Kmenta and J. B. Ramsey, eds., Academic Press, New York. (Use of conjugate utilities.)

[22] Laplace, P. S. (1820). *Théorie analytique des probabilités*, 3rd ed., Paris, pp. 309–354. (Asymptotic forms of posterior distributions.)

[23] LaValle, I. H. (1970). *An Introduction to Probability, Decision, and Inference*. Holt, Rinehart and Winston, New York. (Complements DeGroot [6] and Raiffa and Schlaifer [28] for examples of conjugate families.)

[24] Le Cam, L. (1958). *Publ. Inst. Statist. Univ. Paris*, **7**, 17–35. (Asymptotic forms of posterior distributions.)

[25] Lindley, D. V. (1971). *Bayesian Statistics: A Review*, SIAM, Philadelphia.

[26] Lindley, D. V. (1976). *Ann. Statist.*, **4**, 1–10. (Conjugate families of utilities introduced.)

[27] Morgan, R. L. (1970). A Class of Conjugate Prior Distributions. Unpublished manuscript, Dept. of Statistics, University of Missouri, Columbia, Mo. (Elegant theory of conjugate families of distributions for exponential-type sampling models. Important theorems on the low-order posterior and predictive moments. Useful asymptotic normal form for posterior distributions. Multivariate throughout.)

[28] Raiffa, H. and Schlaifer, R. (1961). *Applied Statistical Decision Theory*. Division of Research, Harvard Business School, Boston. (Original systematization and decision-analytic exploitation of conjugate families of distributions. Useful source of examples of conjugate families, together with DeGroot [6] and LaValle [23].)

[29] von Mises, R. (1931). *Wahrscheinlichkeitsrechnung*. Springer-Verlag, Berlin.

[30] Walker, A. M. (1969). *J. R. Statist. Soc. B.*, **31**, 80–88. (Lucid and readable introduction to the asymptotic posterior theory which preceded the results of Morgan [27].)

[31] Wetherill, C. B. (1961). *Biometrika*, **48**, 281–292.

[32] Winkler, R. L. (1972). *An Introduction to Bayesian Inference and Decision*. Holt, Rinehart and Winston, New York. (View of conjugate prior distributions as corresponding to prior pseudo-samples.)

(BAYESIAN INFERENCE
BAYES' THEOREM
DECISION THEORY
EXPONENTIAL FAMILY
FIDUCIAL INFERENCE
FISHER INFORMATION
PIVOTAL QUANTITIES
POSTERIOR DISTRIBUTION
PREDICTIVE DISTRIBUTION
PRIOR DISTRIBUTION
STRUCTURAL DISTRIBUTIONS
SUFFICIENCY
UTILITY)

J. M. DICKEY

CONJUGATE RANKING

A ranking exactly opposite to a given ranking. If there are n items ranked $1, 2, \ldots, n$, the conjugate ranking assigns ranks $n, n - 1, \ldots, 1$ to these same items.

CONSERVATIVE CONFIDENCE PROCEDURES

Procedures for constructing confidence intervals, regions, belts, or bands for which the confidence coefficient* has a known minimum possible value. The term is sometimes used in contradistinction to procedures for which the confidence coefficient can be calculated exactly.

(CONFIDENCE BAND
CONFIDENCE INTERVALS
    AND REGIONS
SIMULTANEOUS CONFIDENCE
    INTERVALS)

CONSERVATIVE TESTS

Statistical tests in which the probability of the first kind of error* (significance level*) is known not to exceed a specified amount. Calling the latter the nominal significance level of the test, a conservative test is characterized by having no greater probability of incorrectly leading to rejection of the hypothesis tested than the nominal significance level.

(HYPOTHESIS TESTING)

CONSISTENT

A consistent *estimator* is one that converges in probability (*see* CONVERGENCE IN PROBABILITY) to the value of the parameter being estimated as the sample size increases.

A consistent *test* of a hypothesis* H_0 with respect to an alternative hypothesis H_1 is one for which the probability of (formal) rejection of H_0 when H_1 is valid tends to 1 as the sample size increases, the significance level* [probability of (formal) acceptance of H_0 when it is valid] being kept constant.

(POINT ESTIMATION
TESTS OF SIGNIFICANCE)

CONSONANCE INTERVAL

In many goodness-of-fit tests*, population parameter values need to be known. When they are not, values estimated from the sample may be used. (*See* BOOTSTRAPPING.) Kempthorne and Folks [2] suggested that an interval estimate—which they call a "consonance" interval—may be constructed by including all parameter values for which the goodness-of-fit* test would not result in rejection.

For example, if we have data classified according to values $X = 0, 1, \ldots, k$ giving frequencies $N_0, N_1, \ldots, N_k$ ($\sum N_j = n$), respectively, and it is supposed that each observed value has a binomial distribution* with parameters (n, p), then the chi-square* goodness-of-fit statistic is

$$X^2(p) = \sum_{j=0}^{k} \left(\left\{ N_j - n\binom{k}{j} p^j (1-p)^{k-j} \right\}^2 \right.$$

$$\left. \times \left\{ n\binom{k}{j} p^j (1-p)^{n-j} \right\}^{-1} \right)$$

and the formal test rejects the hypothesis that the parameter value is p if

$$X^2(p) > \chi^2_{k,1-\alpha}.$$

The consonance interval for p corresponding to significance level α consists of all p for which $X^2(p) < \chi^2_{k,1-\alpha}$. The probability that this consonance interval includes the true value of p is not, in general, $1 - \alpha$ (even approximately).

For additional information on this topic, see ref. 1.

References

[1] Easterling, R. G. (1976). *Technometrics*, **18**, 1–9. (Gives useful specific examples.)

[2] Kempthorne, O. and Folks, J. L. (1971). *Probability, Statistics and Data Analysis*, Iowa State University Press, Ames, Iowa.

(CONFIDENCE INTERVALS
GOODNESS OF FIT

MULTIVARIATE ANALYSIS
STATISTICAL INFERENCE)

CONSULTING, STATISTICAL

DEFINITION

Statistical consulting means quite different things to different people. In this article it will be used in a very broad sense: one or more statisticians working together with one or more persons who need statistical assistance to help solve some problem of interest. As an example, we would include in our definition the efforts of a statistician who worked for years with a chemical engineer planning experiments and analyzing data to improve the yield of a chemical process. We would also include the response of a statistician to a telephone query on how to compute a standard deviation.

Our definition of statistical consulting is by no means universally accepted. Some argue strongly that a consultant *must* take full responsibility for all statistical aspects of a cooperative venture before the effort can properly be termed consulting. Others take precisely the opposite view—that a joint enterprise in which the statistician takes a major role is by definition a *collaborative* relationship, not a consulting one. To this group, a consulting relationship is a more shallow endeavor, something undertaken rather lightly and without great responsibility. We mention these differences because there is rather strong disagreement in the profession and the reader is likely to encounter these and other views.

Our position is that both extremes are "consulting", that the difference is one of *quality*. In-depth "total involvement" is good consulting, whereas giving quick answers to casual inquiries or "sprinkling the holy water of statistical significance" on weak analyses is poor consulting. However, using the telephone to enhance continued communication is often a valuable component of a good consulting relationship.

WHAT CONSULTANTS DO

Two excellent descriptions of what good consultants do are given by Marquardt [12] and Deming [7]. In Marquardt's words, good statisticians become "totally involved" in the projects on which they are working. They learn about the subject matter, who the key people are, how the data are collected, what the goals of the project are, and what the constraints are in terms of time and resources. They then help formulate a plan of action that tries to ensure that good data will be collected and that proper analyses are carried out on the data collected. They help document the conclusions reached by the investigation. Consultants cannot really claim success until the lessons learned from the study have been accepted and put into action.

The ways in which consultants work seem endlessly varied. Here are several examples.

Example 1. Jane is the only statistician in a state agency and provides much needed statistical expertise on a wide variety of projects. One of her projects concerns the White River.

This river is heavily industrialized and many of the industries discharge wastewater into the river. Recent federal legislation has forced all of them to improve the quality of the water discharged but there is still a problem when the river is low and the water temperature is high. On such days the dissolved oxygen content of the river falls below the minimum level considered safe for aquatic life. The statistical task is to examine historical data on flow rate and temperature and seek to develop a reasonable approximation to how these two factors vary together. Data are available for the previous 45 years but the problem is complicated in several ways: the flow measurement process has been recalibrated twice; the water sampling method for temperature measurements was changed at two different times; some data are "obviously" wrong and many other values appear quite unlikely; there are a number of missing values; and the management practices for the dams that regulate flow have been changed at various times, some of which can only be guessed.

The analysis of the White River data has thus, not unexpectedly, evolved from a simple tabulation and smoothing operation into a more complicated process involving a fair amount of detective work. Key steps have included phone calls, letters, and visits to the Army Corps of Engineers (which regulates the river), officials in the city of St. Claire (where the early temperature measurements were made), and Black's Paper Mill (which took some of the later temperatures). The process has been made more difficult by the fact that some of the people who made the early measurements have died or moved away. Jane has worked closely throughout with state pollution* specialists and key personnel from other agencies. She is in the process of completing her report, which will be used as the basis for decisions involving millions of dollars of new antipollution measures. Jane finds it exhilarating to be involved in projects of such importance, but it is also a bit scary. She often wishes there were some other statistician in her group with whom she could double-check her work.

Example 2.   Fred works for a manufacturer of ceramic materials. Several months ago he was given responsibility for building a large data base to help solve some major problems in the manufacturing of an important new product. So far his work on this project has involved:

1. Gaining familiarity with complex machinery and manufacturing processes so that he can help define measurements and other data that need to be taken.
2. Designing sampling plans to measure characteristics of a two-dimensional surface, such as surface smoothness.
3. Doing exploratory analysis of pilot data.
4. Designing forms to be used in data collection.
5. Helping specify the algorithms necessary for computerized data reduction and analysis.

6. Working with computer experts to design efficient procedures for storing and accessing the data.
7. Beginning the development of a suitable report system.

Other key steps still remaining include: analyzing the data, checking results, communicating results, making recommendations for change, and following up to see that appropriate changes are made. Fred knows that this is a tough assignment; one in which much money will be invested and one in which, if he is not very careful, little useful information will be gained. However, if he succeeds, as he expects to do, his work will have been extremely useful to his company.

Example 3.   Al works for a government physical sciences research laboratory and is the leader of a group of three statisticians and two computer specialists. An important aspect of Al's approach to consulting is the emphasis he puts on in-house teaching. He tries to keep the scientists in his lab abreast of both old and new statistical developments which are relevant to their problems. As a result of his lectures, scientists in his "classes" are continually bringing him new problems. He works on each problem until he feels he understands it, then has the "student" scientist solve it with his guidance. This way both he and the scientist learn a lot.

Much of the work in his laboratory seems to fall in the areas of nonlinear* model fitting and nonstandard time-series* analysis. He has found that existing computer programs often are not satisfactory for his needs, so his group spends a substantial amount of effort developing new computer programs for analyzing data. Al and his group also do the more conventional types of consulting.

These few examples by no means exhaust the rich variety of working styles and environments experienced by statistical consultants. Some work alone, others in teams. They work in government, in industry, in universities, in banks, in other types of organizations, and as private consultants. Application areas include engineering, agriculture, medicine, biology, sociology, marketing, politics, law, economics, physical sciences, demography, meteorology, and indeed every area that attempts to learn from data. The interested reader might want to see the bibliography on consulting by Woodward and Schucany [17] and the collection of interesting examples of applications of statistics by Tanur et al. [16], as well as the works by Cameron [5] and Daniel [6].

HISTORICAL PERSPECTIVE

Modern statistical methods, together with the mathematical statistical theory that helps unify them, have largely been developed in response to the needs of consultants and others who sought to learn from data. Many of the early pioneers in statistical theory and methods were themselves scientists. For example, Francis Galton* first devised the correlation coefficient* to quantify the amount of inheritance of continuous variables in man. He later sought the assistance of a young mathematician named Karl Pearson*, who became interested in statistics and subsequently made many important contributions to statistics while consulting with Galton and others. W. S. Gosset* ("Student"), a chemist at the Guinness Brewery, similarly saw that he needed better tools to evaluate the results of his experiments and went on to develop the very widely used Student's t-test*.

R. A. Fisher*, by far the most important contributor to modern statistics, became interested in the field as an evolutionary biologist. His early mathematical training and originality enabled him to make considerable advances in statistical understanding and to an appointment at age 29 as the statistician at Rothamsted Experimental Station in England. His greatest contributions resulted from serving as statistical consultant to the diverse staff at Rothamsted. While there he served as consultant to scientists in chemistry, bacteriology, entomology, soil science, plant physiology, botany, and

agriculture. Fisher later wrote in the preface to his path-breaking *Statistical Methods for Research Workers*: "Daily contact with the statistical problems which present themselves to the laboratory worker has stimulated the purely mathematical researches upon which are based the methods here presented." Then, as now, there was strong interplay between good statistical theory and application. (*See also* Box [3].)

In the United States, Iowa State University was the first to develop a college-level program in statistics. There in 1924, Henry A. Wallace (later to become U.S. Secretary of Agriculture and Vice-President) led a group of 20 scientists in a study of correlation and regression. This soon led to the establishment of a statistical consulting center at Iowa State with George W. Snedecor* and A. E. Brandt in charge. This center was the wellspring of much of the early statistics in the United States and in the 1930s hosted extended visits from many famous statisticians, including R. A. Fisher*, John Wishart, Frank Yates*, and Jerzy Neyman*. Key faculty members included Gertrude Cox, W. G. Cochran, and Charles P. Winsor. The early learning programs at Iowa State had a strong consulting flavor. For example, when Fisher visited, local researchers took turns at presenting at seminars some of their own experimental data and associated statistical analyses. Afterward, Fisher and the others present were invited to comment on the speaker's interpretation: whether the question the experiment and analysis attempted to answer was the one the experimenter intended to ask, what additional inferences might have been drawn, and so on.

The early interests at Rothamsted and Iowa State centered on agriculture, and some of the tools developed for agriculture were readily adaptable for use in industry. But different sorts of procedures were also needed. In agriculture, time ordering within small sets of supposedly homogeneous measurements had not been a problem because measurements were not ordinarily made in close time order. But in data from the physical sciences and industry, physicist Walter

A. Shewhart found that the data sets he looked at, even those from very good laboratory scientists, almost invariably contained peculiarities when looked at with respect to *time order*. In his studies of small sets of data from supposedly stable laboratory processes, Shewhart found trends, shifts in level, and other patterns. The control chart* techniques he introduced in response were simple and effective, and soon became a vital means of monitoring manufacturing processes.

During World War II, the need to employ statistical and other quantitative methods in problem solving became apparent in a greatly expanded range of fields. The British organized operations research* teams in the armed services, and the United States quickly followed suit by employing statisticians such as A. E. Brandt of Iowa State and W. J. Youden*, originally an industrial chemist. At the same time, Harold F. Dodge and Harry G. Romig of Bell Laboratories and Hugo Hamaker of Philips (Eindhoven) were developing acceptance sampling plans*. These plans helped ensure that cartridges would fit in the rifles for which they were intended without having to be laboriously inspected one by one. Statistical analysis of survival data showed that the *number* of ships sunk in trans-Atlantic crossings was roughly independent of the number of ships in the convoy, thus implying that smaller *percentages* would be sunk in larger convoys. Other statistical analyses helped improve the accuracy of aerial gunnery. Improved test plans and analyses helped identify the median detonating power of bombs.

Soon after the end of the war, W. Edwards Deming began a series of 18 trips to Japan to teach statistical quality control* to industry. These visits and the action of Japanese management have changed the quality of Japanese goods from poor to excellent.

These and other developments meant that soon after the war there was great demand from a wide variety of sources for statistical advice. Demand rose from industry, government, agriculture, medicine, biology, education, sociology, psychology, and many other areas. Rapid growth in the demand for statistical consulting had begun. It has not abated some 35 years hence.

SKILLS NEEDED BY A CONSULTANT

A statistical consultant, to be fully effective, should have many diverse skills. Ideally, he or she should:

Have a genuine desire to solve real problems and help others to solve problems.

Be able to help investigators formulate their problem in quantifiable terms.

Be able to listen carefully and to ask probing questions.

Have a broad knowledge and true understanding of statistical and scientific methods.

Be able to adapt existing statistical procedures to novel environments.

Be able to locate or develop good statistical procedures in a timely fashion.

Be able to keep abreast of developments in statistics.

Be willing to meet deadlines, even if it requires substantial extra effort.

Be able to understand something about the clients' subject matter and speak a bit of the clients' language.

Be a good teacher—much success in consulting depends on being able to help others understand statistical tools, and their strengths and weaknesses.

Be willing to settle for a reasonably correct approximate solution, then go on to the next problem.

Be able to identify important problems (and thus avoid spending too much time on projects of little significance).

Have the confidence to use as simple a procedure as will get the job done, be it design or analysis.

Be able to convince others of the validity of a solid solution and see to it that proper action is taken.

Be able to use computers effectively and direct others in their use.

Be a good problem solver.

Be willing to meet clients regularly on their home ground, and take the responsibility to meet and communicate with all members of the working team.

Be diplomatic and know when to bend, when to stand firm, and how to help smooth over conflicts among other team members.

Be willing to get some experience in the actual collection of the data.

Be willing to take the time to check and double-check procedures and results.

Be able to communicate effectively in writing as well as orally (this often includes helping clients write *their* reports as well).

Be able to make a good estimate of how much effort will be required to solve the problem without actually having to solve the problem itself.

CONSULTING AND COMMUNICATION

Statistical consulting by its very definition implies collaboration between individuals—moreover, between individuals in different fields. Good communication is *vital* to successful consulting. Failures in communication are frequent and can lead to any number of undesirable consequences. Probably the most prevalent is what Kimball [11] termed an error of the third kind: providing a good solution to the wrong problem. Good consultants try to resist the temptation to give a quick answer. They try to make sure that they have a good understanding of the situation and that the goals of the project are clear before they make any specific proposals. Frequent continued interaction is also usually required, lest the statistician or the subject-matter specialists (or both) begin to head off in the wrong direction.

Since good communication is key to being a successful consultant, it is important that communication skills be continuously stud-

ied. The articles by Boen and Fryd [1] and Zahn and Isenberg [18] provide a good start, but since communication in statistics is in essence little different from other communication, many of the popular general-purpose treatments are also relevant. Several of these are mentioned in the Boen and Fryd references.

The critical importance to consultants of good writing skills is also emphasized by Salsburg [13] and in the important report by Snee et al. [15].

COMPUTERS AND CONSULTANTS

By far the most important development of the mid-twentieth century for statistical consultants is the widespread availability of relatively inexpensive electronic computers* and the programs that make them easy to use. With computers one can afford to try a wide variety of models and not be limited to models that are simple to compute. One can try models with nonstandard assumptions, including models whose solutions involve complicated iteration schemes (see, e.g., Efron [9]).

Probably even more important is the computer's ability to handle very large and complex data bases. In many cases computers can be used as intimate parts of the data-gathering process. For example, computers can be used to monitor household energy consumption, environmental pollution, weather, and laboratory experiments. Data bases of millions of numbers can be readily accumulated. Analyzing data sets of this size requires new ways of thinking about data. However, computers have made the analysis of data sets having 10,000 cases on 50 or more variables relatively common.

With the larger sizes come data bases of increasingly complex structures, where even the statistical procedures remain unclear. For example, one might record familial interaction patterns, digitizing for each family member each verbal or nonverbal cue and toward whom it was directed. This might be done at different times of the day, on weekdays and weekends, in different seasons, and at different ages. Some families might be given some "treatment" designed to improve their communication pattern. The research question might be: What does the treatment do to communication patterns? The data might all be available on a computer, but by and large, appropriate tools for analysis still need to be developed.

One of the most important benefits of the computer for statistical consultants is the ability to *plot** the data in many different ways with minimal effort. Plots often prove to be enormously useful in understanding what is really going on in a data set.

But computers also introduce new problems. Data that are "in the computer," but not readily accessible in a useful fashion, might almost just as well not exist. Similarly, if the only contribution of the computer is to provide stacks of tabular output, the computer and the data are likely to be of little value. An additional problem with large data bases is that of "computer error"; minor slips in programs can introduce subtle errors in results that are very difficult to detect and can lead to erroneous conclusions.

KEEPING UP WITH STATISTICS

The ideal consultant has a good general knowledge of a great many aspects of statistics. Keeping this knowledge up to date and being prepared to develop sufficient depth in a new area on a timely basis requires ongoing effort. To help in this regard, most consultants belong to one or more professional organizations, including regional, national, or international statistical societies and technical societies with a strong statistical component. Most of these societies have periodic meetings wherein members can learn of new developments in the field and mingle with colleagues who have similar interests. Many consultants also try to attend short courses, professional meetings on special topics, and in-house seminars and colloquiums.

The printed literature of statistics, like that in most professions, is growing at a rapid pace. Important developments are often summarized in new books or in encyclopedias such as this *Encyclopedia of Statistical Sciences* or the *International Encyclopedia of Statistics* (Macmillan Publishing Co., New York). The *Current Index to Statistics** (*CIS*): *Applications, Methods and Theory* is published annually by the American Statistical Association* and the Institute of Mathematical Statistics*. This index provides author and subject indexes and aspires to relatively complete coverage of the field of statistics. Each volume includes a list of related indexes and information retrieval systems.

ETHICS

"There are three kinds of lies: lies, damn lies and statistics" (Disraeli). "Statistics can prove anything." Sentiments like these represent only part of the ethical problem faced by statistical consultants. Ethical problems seem to arise more frequently for consulting statisticians than for many other professionals partly because consultants tend to work on problems where the outcome is important yet somewhat in doubt, and where the conclusions may be contrary to the immediate interests of the client who funded the consultation. Further, in most cases there may not be a single *best* mode of statistical analysis. Add to all these factors a sprinkling of human nature and some honest disagreements of opinion and it is not surprising that ethical dilemmas arise. Here are some examples.

Example 1. Salaries are being compared between two historically distinct components of a university. One group alleges that it is underpaid; central administration agrees that this group's pay is lower but attributes it to differences in experience, scholarly productivity, academic credentials, and related factors. Both groups agree that a regression analysis of salaries, adjusting for such factors as years of experience, publications, and so on, would be informative. Central administration commissions a study, but when unfavorable results begin to emerge, puts pressure on the statisticians to use the model that makes the salaries appear to be the most nearly equitable. The statisticians feel this would be unethical and insist on reporting several sets of results, explaining the strengths and weaknesses of each. They try to be very diplomatic, but realize that their sincerity may cost them future funding and support.

Example 2. In a study of a new medication a company statistician notices a strong suggestion of a possibly serious side effect. The result is not quite statistically significant and to check it would require considerable extra expense and study. Her employer points out there is no legal requirement for them to check further and company pharmacologists believe that the result is chemically unlikely. The statistician is not quite sure what to do: she knows that events with spurious statistical significance do occur all the time. She finally decides to wait and see, keeping a watchful eye on similar situations for any hint of a recurrent event.

Example 3. An analysis is done for a state department of transportation to check the effectiveness of its "driver improvement program" for problem drivers. The study shows the program has no beneficial effect whatsoever. A report is written and given to the project sponsors, who quietly file it away while keeping their multimillion-dollar program going. Should the statisticians call in the press? Tell the governor? Or trust that truth will win out in the long run? They decide to work up through channels hoping that some level of management will recognize the potential cost savings available from abolition of the program.

Examples like these are by no means everyday events, but they do occur often enough to be a legitimate concern for many consul-

tants. For further discussion, see Deming [7, 8], Bross [4], *Science* [14], and the references contained in these articles.

TEACHING CONSULTING

Many students who obtain degrees in statistics go on to become statistical consultants. Yet rarely are departments of statistics prepared to offer them a program that helps ease the transition between the classroom and the firing line of live consulting. The most eloquent statement of the problem may well be that of Box [2]:

> Swimming could be taught by lecturing the student swimmers in the classroom three times a week on the various kinds of strokes and the principles of buoyancy and so forth. Some might believe that on completing such a course of study, the graduates would all eagerly run down to the pool, jump in, and swim at once. But I think it's much more likely that they would want to stay in the classroom to teach a fresh lot of students all that they had learned.

What is thus needed is a means whereby students can work actively with good consultants and gain experience in being consultants under the watchful eye of someone who can help them see how to do it better before bad habits are developed. Being encouraged to do some consulting before leaving the academic environment also means that those who go on to teach statistics will at least have some appreciation of the actual uses of statistics. Many who have studied the problem believe that statisticians need an intern program such as that of doctors. Some schools offer these, but more are needed.

Components of a good educational program for consultants would include ways to improve interpersonal communication, how to use and keep up with statistical literature, how to analyze data, how to gather good data and recognize bad data, how to write good reports, how to use the computer, and how to develop techniques for nonstandard situations. The program should overlay all this with a heavy dose of actual analysis, design, report writing, and consulting.

A very important ingredient of such a program would be the actual conduct of a project involving data gathering and analysis. The famous consultant W. E. Deming writes: "I never lose a chance to get experience with the data; I enumerated a district in the Census of 1940; I've been out on interviews at least 40 different times on Census work, labor force, social surveys, market research; I've used the telephone; I've collected data on hundreds of physical and chemical trials and on reliability and testing and inspection in plants; to me this experience is extremely important." Hunter [10] has illustrated the usefulness of a data-gathering project in teaching even beginning students the importance of detail and the real difficulty associated with gathering good data. All consultants, and indeed all who seek to learn from data, need to be aware of the fact that many and perhaps most data sets have important errors of the sort that negate the effectiveness of any analysis that does not identify them. For example, industrial plant data often have startup effects, experiments on mice have cage-related effects, large data bases have computer processing errors, flowerpots get interchanged, and human beings make recording errors.

REWARDS OF CONSULTING

Most consultants gain enormous satisfaction from their work. Even young consultants have an opportunity to play a large and often decisive role in major decisions. They are asked to help plan the data gathering that will be used in making important decisions—then they are asked to analyze the data and help make decisions.

Statistical consultants are continually learning about new fields—from the microbiology of DNA to the relative accident rates of twin-bed to single-bed trailer trucks. Much of statistics is like detective work.

Consultants search for hidden clues in the data or the theory behind the data to find out what might have happened. Then after much hard digging, there is the joy of understanding, followed by the challenge of how to make the results clear to others.

References

[1] Boen, J. and Fryd, D. (1978). *Amer. Statist.*, **32**, 58–60.

[2] Box, G. E. P. (1979). *J. Amer. Statist. Ass.*, **74**, 1–4.

[3] Box, J. F. (1978). *R. A. Fisher, The Life of a Scientist*. Wiley, New York.

[4] Bross, I. D. J. (1974). *Amer. Statist.*, **28**, 126–127.

[5] Cameron, J. M. (1969). *Technometrics*, **11**, 247–254.

[6] Daniel, C. (1969). *Technometrics*, **11**, 241–245.

[7] Deming, W. E. (1965). *Ann. Math. Statist.*, **36**, 1883–1900. (A very careful and detailed statement of the statistician's and the client's responsibilities. Highly recommended for study. See also its references.)

[8] Deming, W. E. (1972). *Int. Statist. Rev.*, **40**, 215–219. (A leading private consultant's code of ethics.)

[9] Efron, B. (1979). *SIAM Rev.*, **21**, 460–480.

[10] Hunter, W. G. (1977). *Amer. Statist.*, **31**, 12–17. (A convincing demonstration that students can gain considerable benefit from actually doing experiments.)

[11] Kimball, A. W. (1957). *J. Amer. Statist. Ass.*, **57**, 133–142.

[12] Marquardt, D. W. (1979). *Amer. Statist.*, **33**, 102–107. (An excellent summary of the exciting role of a "totally involved" consultant.)

[13] Salsburg, D. S. (1973). *Amer. Statist.*, **27**, 152–154.

[14] *Science* (1977). **198**, 677–705. (A series of articles on the ethics of medical experimentation, including statistical aspects.)

[15] Snee, R. D., Boardman, T. J., Hahn, G. J., Hill, W. J., Hocking, R. R., Hunter, W. G., Lawton, W. H., Ott, R. L. and Strawderman, W. E., (1980). *Amer. Statist.*, **34**, 65–75. (Recommendation for graduate training of consultants.)

[16] Tanur, J. M., Mosteller, F., Kruskal, W. H., Link, R. F., Pieters, R. S., and Rising, G. R. (1978). *Statistics: A Guide to the Unknown*, 2nd ed. Holden-Day, San Francisco. (Fireside reading of exciting statistical applications.)

[17] Woodward, W. A., and Schucany, W. R. (1977). *Biometrics*, **33**, 564–565. (A nearly complete bibliography on the subject through 1977.)

[18] Zahn, D. A., and Isenberg, D. J. (1980). *1979 Proceedings of the Section on Statistical Education*, American Statistical Association, Washington, D. C., pp. 67–72.

Acknowledgments

In writing this article I have benefited enormously from the detailed and helpful comments of many. I would particularly like to thank T. A. Bancroft, Joan Fisher Box, W. Edwards Deming, Dennis Friday, Bert Gunter, Gerald J. Hahn, Ellis R. Ott, Ronald D. Snee, Douglas Zahn, and especially Alison K. Pollack.

Others who made important contributions include James R. Boen, George E. P. Box, Cathy Campbell, John Crowley, William G. Hunter, Kevin Little, Peter M. Piet, Gerald van Belle, Donald Watts, and virtually all members of the Wisconsin Statistical Laboratory. The patient and skillful typing and retyping by Debbie Dickson was also critical.

None of these people agrees completely with everything I have said.

This work was in part done at the UW Mathematics Research Center and thus was supported in part by the U. S. Army under Contract DAAG29-75-C-0024.

(BIOSTATISTICS
CLINICAL TRIALS
COLLECTION OF DATA
COMPUTERS AND STATISTICS
DEMOGRAPHY
ECOLOGICAL STATISTICS
ENGINEERING STATISTICS
EXPLORATORY DATA ANALYSIS
GEOSTATISTICS
GRAPHICAL REPRESENTATION
   OF DATA
PRINCIPLES OF PROFESSIONAL
   STATISTICAL PRACTICE
STATISTICAL EDUCATION
STATISTICS IN (various fields))

BRIAN L. JOINER

CONSUMER PRICE INDEX

The Consumer Price Index (CPI) is a measure of the *changes* in prices *paid* by *urban consumers* for the goods and services they purchase. Essentially, it measures the purchasing power of consumers' dollars by comparing what a sample "market basket" of goods and services costs today with what the same market basket would have cost at an earlier date. The CPI is compiled and released monthly by the Bureau of Labor Statistics (BLS).

UNIVERSE AND CLASSIFICATION

In 1978, the BLS began publishing two separate CPIs: (1) a new CPI for All Urban Consumers (CPI-U), which covers about 80% of the total civilian noninstitutional population, and (2) a revised CPI for Urban Wage Earners and Clerical Workers (CPI-W), which represents about half of the population covered by the CPI-U.

The CPI is based on prices of food, clothing, shelter, fuels, transportation fares, charges for doctors' services, drugs, and other goods and services that people buy for day-to-day living. Individual indexes are published for over 300 different expenditure classes. Separate indexes are also published for 28 local areas. Area indexes do not measure differences in the level of prices among cities; they measure only the average change in prices for each area since the base period.

THE INDEX

The Consumer Price Index is calculated using a modified Laspeyres index of the general form

$$I_1 = \left(\frac{\sum_i P_{1,i} Q_{0,i}}{\sum_i P_{0,i} Q_{0,i}} \right) \times 100,$$

where

I_1 = index for period 1

$P_{1,i}, P_{0,i}$ = prices for item i in periods 1 and 0, respectively

$Q_{0,i}$ = quantity of item i sold in period 0 (the base period)

These indexes may be viewed as measuring the price change of a constant set of consumption through time. Item weights are based on the Consumer Expenditure Survey (most recently for 1972–1973).

PRICES

Prices are collected in 85 urban areas across the country from over 18,000 tenants, 18,000 housing units for property taxes, and about 24,000 stores and service establishments. All taxes directly associated with the purchase and use of items are included in the index. Prices of most goods and services are obtained by personal visits of trained BLS representatives.

When an item is no longer being sold, it must be replaced in the index. The replacement must be done in such a way that only pure price change is captured by the index, and it remains as close as possible to the concept of pricing a constant set of consumption through time. This substitution procedure is generally referred to as "quality adjustment." It consists of two steps: (1) identifying all the changes in the specification of the item being priced, and (2) measuring the value of each change to the consumer.

This direct measuring of quality change is extremely difficult since measurement of the value consumers place on quality change is rarely possible. Therefore, BLS usually uses an indirect method to measure the quality change by evaluating the additional cost associated with providing the change in quality.

SAMPLING

The sample for the CPI is selected using a multistage probability proportional size (p.p.s.) sample. At the initial stage, 85 urban areas were sampled for pricing. For a majority of the items, Point of Purchase Surveys are conducted in each geographic area, identifying outlets and the amounts that consumers purchased from each. Using the survey results as a sample frame, a p.p.s. sample of outlets is then selected for each expenditure category. BLS staff then visit each outlet to make p.p.s. selections of specific items from the expenditure categories designated within each selected outlet.

HOME OWNERSHIP

One of the most difficult conceptual and measurement problems in the CPI is the home ownership component. The CPI has historically used an "asset" approach. The weights for house prices and contracted mortgage interest cost represent only those homeowners who actually purchased a home in the base period. Included are the total price paid for the home and the total amount of interest expected to be paid over half the stated life of the mortgage. Current monthly prices are used for each of these components. In effect, this approach assumes that the purchaser consumes the entire value of the house during that year.

A widely supported alternative is the "flow of services" approach, which views the resident owner as consuming housing services from the house—shelter, cooking accommodations, laundry accommodations, etc. Homeowners do not consume the entire value of a house in a single period but continue to consume it over the years they live in it. Two possible methods have been proposed for estimating home ownership price change based on the flow of services approach: (1) a *rental equivalence* technique —to measure what the owners would charge if they rented the house to themselves, (2) a *user-cost function*—to measure the major cost components (mortgage and equity financing costs, maintenance costs, taxes, etc.) that owners incur in providing themselves with housing.

Bibliography

Fisher, F. M. and Schell, K. (1972). *The Economic Theory of Price Indexes*. Academic Press, New York. (A rigorous and elegant treatment of the economic theory that underlies price indexes.)

Gillingham, R. (1980). Estimating the user cost of owner-occupied housing. *Monthly Labor Rev.*, February. (Argues for the flow-of-services approach and shows how the user-cost method cannot be used.)

Norwood, J. L. (1980). *CPI Issues*, Rep. 593, U. S. Department of Labor, Bureau of Labor Statistics, February. (Describes some experimental calculations on alternative treatments of home ownership.)

Triplett, J. E. (1971). Determining the effects of quality change on the CPI. *Monthly Labor Rev.*, May, 27–38.

U.S. Department of Labor, Bureau of Labor Statistics (1976). *BLS Handbook of Methods*, Bull. 1910, Chap. 13. (A dated but basic description of CPI. Updated editions issued periodically.)

U.S. Department of Labor, Bureau of Labor Statistics (1978). *The Consumer Price Index: Concepts and Content over the Years*, Rep. 517, May. (A basic introduction to the CPI and the 1978 revision.)

(INDEX NUMBERS
PRODUCERS PRICE INDEX
WHOLESALE PRICE INDEX)

JOHN F. EARLY

CONTAGIOUS DISTRIBUTIONS

The first systematic English-language use of *Contagious Distribution* appears to have been by Neyman [10]; it was previously used extensively by Pólya [11] and later by Feller [2]. A map that gives the domiciles of children who have suffered infection from a disease such as measles during a local epidemic often shows obvious clustering, corresponding to foci of infection. So may a map that gives the location of plants of a certain species (because of its method of propagation—e.g., along root systems, or by limited-range mechanical dispersal of seeds), or one that records the positions of insect larvae (because they hatched from clusters of eggs). Although there is no focus of infection in these cases, the analogy is clear, the parent playing the corresponding role; Neyman describes it as a case in which "the presence of one larva within an experimental plot increases the chance of there being some more larvae."

Spatial information, in one or more dimensions (two in the illustrations above), is often summarized by superimposing a grid on the region and recording the number of individuals per grid cell (quadrat), hence obtaining a discrete frequency distribution of those quadrats with 0 individuals, 1 individual, 2, 3, ... individuals. A probability distribution derived from the clustering structure above to describe the relative frequencies is called a *contagious distribution*; it

should be said immediately that all such probability distributions can be derived from assumptions other than those involving clustering, and hence it cannot be concluded from the agreement of a (theoretical) contagious distribution and an observed frequency distribution that a clustering (contagious) mechanism has operated to produce the observed distribution. In fact, Neyman's detailed derivation [10] was essentially based on mixtures*, as remarked by Feller [2]: in this context *inhomogeneity of a population* seems to be the preferred description of *mixture*, with an often derogatory comment that this is "apparent" contagion (see the following section). Alternative models can sometimes be investigated by using quadrats of different sizes for the same data sets: the theoretically predictable changes that result differ for alternative models, and these can be compared with the observational changes.

Somewhat more generally, a state of contagion can be interpreted as one in which the present state depends on earlier events (e.g., sampling without replacement from a finite population). This aspect, the one developed by Pólya [11], is not enlarged on here (see Johnson and Kotz [5]).

DEFINITION

A compact treatment follows by using probability generating functions* (PGF). If the number of clusters per quadrat is N, with PGF $\mathscr{E}(z^N) = g(z)$, and the number in the ith cluster is Y_i, independently of other clusters and N, with PGF $\mathscr{E}(z^{Y_i}) = f(z)$, then the total number of individuals per quadrat is $X = Y_1 + Y_2 + \cdots + Y_N$, with PGF

$$
\begin{aligned}
h(z) &= \mathscr{E}(z^{Y_1 + \cdots + Y_N}) \\
&= \mathscr{E}_N \big| \big\{ \mathscr{E}(z^{Y_1 + \cdots + Y_N}) \big\} | N \big| \\
&= \mathscr{E}_N \big[\{ f(z) \}^N \big] \\
&= g(f(z)).
\end{aligned}
$$

(This type of distribution is called a "generalized"* or "random sum"* distribution.)

This can now be used as a definition, although it is really too general, because with $f(z) \equiv z$ any distribution (with a PGF) can be said to be contagious. This matters little, because it is the mechanism rather than the distribution that should be looked at in a particular application. Note that there is no necessity for N to have a discrete distribution for $g(f(z))$ to be meaningful. Other names for this structure include (randomly) stopped, clustered, generalized, compound, and composed; of these, generalized (derived perhaps from a partial reading of Satterthwaite [12]) seems least suitable. When more explicit precision is needed, the structure above will be called an N-stopped Y-summed distribution, sometimes with the names of the distributions replacing N and Y, written by Gurland [3] $N \vee Y$.

If $w(z)$ is the generating function of a sequence of nonnegative normed weights $\{w_n\}$ and $\{u_n(z)\}$ is a sequence of PGFs, then $\sum w_n u_n(z)$ is a mixture PGF. A contagious, or stopped, distribution can always be interpreted as a special mixture: in the notation of the definition, take $\sum w_n z^n = g(z)$, and $u_n(z) = \{f(z)\}^n$, but it is obviously absurd, therefore, to describe clustering as "apparent" contagion.

PROPERTIES

Expanding $h(z) = g(f(z))$ in powers of z with Faà di Bruno's formula* gives explicit expressions for the probabilities in terms of those for clusters per quadrant and individuals per cluster. Write

$$h(z) = \sum h_x z^x, \quad g(z) = \sum g_x z^x, \quad f(z) = \sum f_x z^x;$$

then

$$h_x = \sum_{n=1}^{x} n! \, g_n \sum \frac{f_1^{n_1} f_2^{n_2} \cdots f_x^{n_x}}{n_1! n_2! \cdots n_x!},$$

$$x = 1, 2, \ldots,$$

where the inner sum is over the nonnegative integers $n_1, n_2, \ldots$ such that $n_1 + n_2 + \cdots + n_x = n$, $1 \cdot n_1 + 2 \cdot n_2 + \cdots + x \cdot n_x = x$, and

$$h_0 = \sum_{n=0}^{\infty} g_n f_0^n.$$

However, in applications, contagious/stopped distributions for which the stopping distribution is Poisson* (i.e., the numbers of clusters per quadrat are Poisson) have been used far more extensively than any others: for these the PGF may be taken as $h(z) = \exp[-\mu + \mu f(z)]$. Various more specific results can then be obtained: e.g.,

$$h_{x+1} = \frac{\mu}{x+1} \sum_{p=0}^{x} (x - p + 1) f_{x-p+1} h_p$$

is a useful set of recurrence relations, with $h_0 = \exp(-\mu + \mu f_0)$; and the rth (power or factorial) cumulant* of the Poisson(μ)-stopped distribution is μ times the rth (power or factorial) moment* about the origin of the distribution of $f(z)$. [It thus follows that a Poisson-stopped distribution is never a Poisson distribution unless $f(z) = z$.] These distributions also have a reproductive property: if $\{X_i\}$ is a set of independent Poisson(μ_i)-stopped f variates, then $\sum X_i$ has a Poisson $(\sum \mu_i)$-stopped f distribution, an important result for many applications. Remarkably, not only are Poisson-stopped distributions infinitely divisible*, but conversely (e.g., Feller [2, Vol. I]) *every integral-valued* infinitely divisible variate can be represented as a Poisson-stopped variate; even more remarkably, *every* infinitely divisible variate can be represented as the limit of a Poisson-stopped sequence [8].

One of the most striking features of Poisson-stopped contagious distributions is their long tails (compared with the Poisson distribution): for identical means the variance of the contagious distribution is greater and there is more probability in the zero class both absolutely and relative to the first class. It is also common for the distributions to be multimodal (not, however, the negative binomial*), with a marked half-mode at the zero class.

It should be observed that a Poisson-stopped distribution is not (except for the Neyman) a mixed Poisson distribution: the Poisson distribution plays a complementary role in these two cases.

The distributions are typically multi-parameter, those most commonly used having two parameters, and their estimation usually raises difficult questions. The estimators are frequently highly dependent, producing instability in numerical procedures and problems of interpretation (see Shenton and Bowman [13]). A number of applications are cited in Martin and Katti [9].

There are further details about specific contagious distributions under the appropriate headings: log-zero-Poisson*, negative binomial*, Neyman*, Poisson-binomial*, Poisson Pascal*, Thomas*.

All the distributions referred to above are univariate, but the ideas can be extended to produce multivariate families (although not uniquely), usually in such a way as to produce familiar marginal distributions (e.g., Holgate [4]).

Generalized power series distributions* constitute another extension; yet another is the family obtained by permitting additional probability in the zero (or any specified) class of an already introduced family (e.g., Poisson-with-zeros [6]; *see* INFLATED DISTRIBUTIONS). A review paper by Kemp [7] may be cited for an overview.

References

[1] Douglas, J. B. (1980). *Analysis with Standard Contagious Distributions*. International Co-operative Publishing House, Fairland, Md. (An extensive bibliography.)

[2] Feller, W. (1943). *Ann. Math. Statist.*, **14**, 389–400.

[3] Gurland, J. (1957). *Biometrika*, **44**, 265–268.

[4] Holgate, P. (1966). *Biometrika*, **53**, 241–244.

[5] Johnson, N. L. and Kotz, S. (1977). *Urn Models and Their Applications*: *An Approach to Modern Discrete Probability Theory*. Wiley, New York.

[6] Katti, S. K. (1966). *Biometrics*, **22**, 44–52.

[7] Kemp, A. W. (1974). *Statist. Rep. Repr. No. 15*, University of Bradford Postgraduate School of Studies in Mathematics, Bradford, England.

[8] Lukács, E. (1970). *Characteristic Functions*, 2nd ed., Hafner, New York.

[9] Martin, D. C. and Katti, S. K. (1965). *Biometrics*, **21**, 34–48.

[10] Neyman, J. (1939). *Ann. Math. Statist.*, **10**, 35–57.

[11] Pólya, G. (1930/31). *Ann. Inst. Henri Poincaré*, **1**, 117–161.

[12] Satterthwaite, F. E. (1942). *Ann. Math. Statist.*, **13**, 410–417.

[13] Shenton, L. R. and Bowman, K. O. (1977). *Maximum Likelihood Estimation in Small Samples*. Griffin's Statist. Monogr. 38. Charles Griffin, London.

(COMPOUND DISTRIBUTION
EPIDEMIOLOGICAL STATISTICS
NEYMAN TYPE A DISTRIBUTION
RANDOM-SUM DISTRIBUTION
THOMAS DISTRIBUTION)

J. B. DOUGLAS

CONTIGUITY

A standard approach for constructing tests of hypotheses suitable for a sufficiently large sample is to secure the asymptotic distribution, under the null hypothesis, of a sequence of test statistics S_n as the sample size n goes to infinity. From this asymptotic distribution, tests for individual sample sizes n can be formulated which are of approximately the desired size*. More is required if one is to (1) describe the power* of such tests under alternatives to the null hypothesis, and (2) determine the efficiency* of such tests vis-à-vis other tests of comparable size. There are several approaches for addressing these issues, but one approach, which focuses attention on "contiguous alternatives" to the null hypothesis, has considerable appeal. Roughly speaking, a contiguous alternative is one that is sufficiently close to the null hypothesis that it can reasonably be mistaken for the null hypothesis. What constitutes sufficient closeness depends on n. (Typically, its distance from the null hypothesis must be some multiple of $n^{-1/2}$ or less.) To make the notion precise, one must refer to a *sequence* of alternatives since the concept really concerns only limits as n goes to infinity.

Let $X_1, X_2, \ldots$ be a sequence of possible observations whose distribution is governed by a parameter $\theta \in \Theta$, and let P_θ denote probability and E_θ denote expectation under θ. A sequence of parameter values $\{\theta_n\}$ belonging to $\Theta - \{\theta_0\}$ is said to be *contiguous* to θ_0 in Θ (the θ-value associated with a simple null hypothesis) if for every sequence of events of the form $\{ E_n = [(X_1, \ldots, X_n) \in A_n]\}$,

$$P_{\theta_0}(E_n) \to 0 \qquad \text{implies} \quad P_{\theta_n}(E_n) \to 0$$

$$\text{as } n \to \infty.$$

(One speaks loosely of the θ_n's as being contiguous alternatives to θ_0.) This definition is essentially the same as that given by Hájek and Šidák [1] (abbreviated H-S hereafter) except for the parametric setting in which it is cast here. This concept originated with LeCam [4]. (Actually, he defines a symmetric variant of the concept described here, as does Roussas [5].)

CONDITIONS THAT GUARANTEE CONTIGUITY

Let $f_n(X_1, \ldots, X_n | \theta)$ denote the likelihood function* for $X_1, \ldots, X_n$, and $L_n = f_n(X_1, \ldots, X_n | \theta_n)/f_n(X_1, \ldots, X_n | \theta_0)$ denote the corresponding likelihood ratio* for the parameter values θ_n and θ_0. If, under θ_0, L_n converges in law to a random variable L for which $E_{\theta_0}L = 1$, then, according to what H-S refer to as "LeCam's first lemma," $\{\theta_n\}$ is contiguous to θ_0. In typical applications $\log L_n$ is asymptotically normal, in which case the condition $E_{\theta_0}L = 1$ is satisfied if, and only if, L is log-normally* distributed with parameters $(-\sigma^2/2, \sigma^2)$ for some $\sigma^2 \geqslant 0$ (see ASYMPTOTIC NORMALITY).

Example. If $X_1, X_2, \ldots$ are independent Poisson* random variables with unknown common mean θ, then

$$\log L_n = (X_1 + \ldots + X_n)\log(\theta_n/\theta_0)$$
$$- n(\theta_n - \theta_0).$$

If, in addition, $\theta_n = \theta_0(1 + cn^{-1/2})$ for some real constant c (so that the distance between θ_n and θ_0 is a multiple of $n^{-1/2}$), then, under θ_0, $\log L_n$ is asymptotically normal* $(-c^2\theta_0/2, c^2\theta_0)$ (the asymptotic mean and variance, respectively). Thus $\{\theta_n\}$ is contiguous to θ_0. (Hence the distribution of L, i.e., the limit distribution of the L_n's, depends upon θ_0 and upon the θ_n's through c.)

Hall and Loynes [3] describe other conditions that are equivalent to or imply contiguity. For instance, if L_n converges in law under θ_n, i.e., if

$$\lim_{n\to\infty} P_{\theta_n}(L_n \leqslant u) = F(u)$$

for all points of continuity u of a distribution function F, then $\{\theta_n\}$ is contiguous to θ_0. For the example above, such a limit exists; F is a lognormal* distribution function with parameters $(c^2\theta_0/2, c^2\theta_0)$. (The first component changes sign from that for L. As with the distribution of L, F depends upon θ_0 and one's choice of the θ_n's.)

When $X_1, X_2 \ldots$ are independent under each θ, H-S describe (under the heading "LeCam's second lemma") another situation that guarantees contiguity. A certain statistic W_n (in their notation) must be asymptotically normal $(-\sigma^2/4, \sigma^2)$ under θ_0.

ASYMPTOTIC POWER UNDER CONTIGUOUS ALTERNATIVES

It is frequently possible to obtain the asymptotic distribution of a test statistic S_n (based upon $X_1, \ldots, X_n$) under the contiguous alternatives $\{\theta_n\}$, and thereby approximate, for each n, the power of a test based upon S_n at the alternative θ_n. H-S describe this under the heading "LeCam's third lemma"; it may be expressed as follows:

Lemma. If the pair $(S_n, \log L_n)$ is asymptotically bivariate normal* $(\mu_1, \mu_2; \sigma_1^2, \sigma_2^2; \sigma_{12})$ under θ_0 with $\mu_2 = -\sigma_2^2/2$, then the same holds under θ_n with μ_1 replaced by $\mu_1 + \sigma_{12}$ and the sign of μ_2 changed. Thus S_n is asymptotically normal $(\mu_1 + \sigma_{12}, \sigma_1^2)$ under θ_n.

(Here the parameters $(\mu_1, \mu_2; \sigma_1^2, \sigma_2^2; \sigma_{12})$ have their standard meanings: the asymptotic means, variances, and covariance, respectively.) The lemma asserts, among other things, that

$$\lim_{n\to\infty} P_{\theta_n}\big(S_n \leqslant \sigma_1 u + (\mu_1 + \sigma_{12})\big)$$

$$= \int_{-\infty}^{u} \frac{1}{\sqrt{2\pi}} e^{-v^2/2} \, dv, \qquad -\infty < u < \infty.$$

The condition $\mu_2 = -\sigma_2^2/2$ guarantees that $\{\theta_n\}$ is contiguous to θ_0.

Example (Continued).   Consider the (known to be inefficient) test statistics

$$S_n = n^{-1/2}\left[\sum_1^n (X_i - \bar{X}_n)^2 - n\theta_0\right], \quad n \geqslant 1,$$

where $\bar{X}_n = n^{-1}(X_1 + \cdots + X_n)$. ($S_n$ is recommended by the fact that Poisson random variables have mean and variance equal.) Under $\theta_0, (S_n, \log L_n)$ is asymptotically jointly normal $(0, -c^2\theta_0/2, \theta_0 + 2\theta_0^2, c^2\theta_0, c\theta_0)$. The lemma asserts that, under θ_n, S_n is asymptotically normal $(c\theta_0, \theta_0 + 2\theta_0^2)$.

PITMAN EFFICIENCY*

There is a simple way to compute the Pitman efficiency for tests based upon the test statistics $\{S_n\}$. It is equal to ρ^2, where $\rho = \sigma_{12}/\sigma_1\sigma_2$ is the correlation coefficient* for the asymptotic bivariate normal distributions referred to in the lemma. For the example above, $\rho^2 = (1 + 2\theta_0)^{-1}$.

REMARKS

The notion of contiguity may be generalized to include composite null hypotheses*. This is done in H-S. (A contiguous alternative must only be near to *some* θ satisfying the null hypothesis.) The notion of contiguity has been described as "asymptotic absolute continuity": For $\{\theta_n\}$ to be contiguous to θ_0, $P_{\theta_n}(X_1, \ldots, X_n)^{-1}$ does not necessarily have to be absolutely continuous* to $P_{\theta_0}(X_1, \ldots, X_n)^{-1}$. But the singular part of the former must go to zero, i.e., $P_{\theta_n}(f_n(X_1, \ldots, X_n | \theta_0) = 0) \to 0$ as $n \to \infty$ (a trivial consequence of the definition of contiguity) (see Hall and Loynes [3]). Roussas' book [5] provides the most complete description of the subject of contiguity, including applications; H-S provide one of the more readable ones, as well as some interesting applications involving linear rank statistics. Hall and Loynes [2] extend the theory given

in H-S to a setting involving likelihood ratio processes ("time" related to sample size), thereby permitting application to sequential analysis*.

References

[1]   Hájek, J. and Šidák, Z. (1967). *Theory of Rank Tests*. Academic Press, New York.

[2]   Hall, W. J. and Loynes, R. (1977). *Ann. Statist.*, **5**, 330–341.

[3]   Hall, W. J. and Loynes, R. (1977). *Ann. Prob.*, **5**, 278–282.

[4]   LeCam, L. (1960). *Univ. Calif. Publ. Statist.*, **3**, 37–98.

[5]   Roussas, G. G. (1972). *Contiguity of Probability Measures: Some Applications in Statistics*. Cambridge University Press, Cambridge.

(ASYMPTOTIC NORMALITY
HYPOTHESIS TESTING)

Gordon Simons

CONTINGENCY TABLES

HISTORICAL REMARKS

Multivariate statistical analysis has occupied a prominent place in the classical development of statistical theory and methodology. The analysis of cross-classified *categorical data**, or contingency-table analysis as it is often referred to, represents the *discrete* multivariate analog of *analysis* of *variance** for continuous response variables, and now plays an important role in statistical practice. This presentation is intended as an introduction to some of the more widely used techniques for the analysis of contingency-table data and to the statistical theory that underlies them.

The term *contingency*, used in connection with tables of cross-classified categorical data, seems to have originated with Karl Pearson* [50], who for an $s \times t$-table defined contingency to be any measure of the total deviation from "independent probability." The term is now used to refer to the table of counts itself. Prior to this formal use of the term, statisticians going back at least to Quetelet* [53] worked with cross-classifications of counts to summarize the association between variables. Pearson [48] had laid the groundwork for his approach to contingency tables, when he developed his X^2 test* for comparing observed and expected (theoretical) frequencies. Yet Pearson preferred to view contingency tables involving the cross-classification of two or more polytomies as arising from a partition of a set of multivariate, normal data, with an underlying continuum for each polytomy*. This view led Pearson [49] to develop his tetrachoric correlation coefficient* for 2×2 tables, and this work in turn spawned an extensive literature well chronicled by Lancaster [10].

The most serious problems with Pearson's approach were (a) the complicated infinite series linking the tetrachoric correlation coefficient with the frequencies in a 2×2 table, and (b) his insistence that it always made sense to assume an underlying continuum, even when the dichotomy of interest was dead–alive or employed–unemployed, and that it was reasonable to assume that the probability distribution over such a continuum was normal. In contradistinction, Yule* [57] chose to view the categories of a cross-classification as fixed, and he set out to consider the structural relationship between or among the discrete variables represented by the cross-classification, via various functions of the cross-product ratio. Especially impressive in this, Yule's first paper on the topic, is his notational structure for n attributes or 2^n tables, and his attention to the concept of partial and joint association of dichotomous variables.

The debate between Pearson and Yule over whose approach was more appropriate for contingency-table analysis raged for many years (see, e.g., Pearson and Heron [52]) and the acrimony it engendered was exceeded only by that associated with Pearson's dispute with R. A. Fisher* over the adjustment in the degrees of freedom (d.f.) for the *chi-square test** of independence in

the $s \times t$-table. [In this latter case Pearson was simply incorrect; as Fisher [33] first noted, d.f. $= (s - 1)(t - 1)$.]

Although much work on two-dimensional contingency tables followed the pioneering efforts by Pearson and Yule, it was not until 1935 that Bartlett, as a result of a suggestion by Fisher, utilized Yule's cross-product ratio to define the notion of second-order interaction in a $2 \times 2 \times 2$ table, and to develop an appropriate test for the absence of such an interaction [21]. The multivariate generalizations of Bartlett's work, beginning with the work of Roy and Kastenbaum [55], form the basis of the log-linear model approach to contingency tables, which is described in detail below.

The past 25 years has seen a burgeoning literature on the analysis of contingency tables, stemming in large part from work by S. N. Roy and his students at North Carolina, and from that of David Cox on binary *regression**. Some of this literature emphasizes the use of the minimum modified chi-square* approach (e.g., Grizzle et al. [42]) or the use of the minimum discrimination information approach (e.g., Ku and Kullback [46] and Gokhale and Kullback [5], but the bulk of it follows Fisher in the use of *maximum likelihood**. For most contingency-table problems, the minimum discrimination information approach yields maximum likelihood estimates.

Except for a few attempts at the use of additive (linear) models* (see, e.g., Bhapkar and Koch [22]), almost all the papers written on the topic emphasize the use of log-linear or logistic models. Key papers by Birch [23], Darroch [27], Good [34], and Goodman [35, 36], plus the availability of high-speed computers, served to spur renewed interest in the problems of categorical data analysis. This, in turn, led to many articles by Leo Goodman (e.g., Goodman [37–39]) and others, and finally culminated in books by Bishop et al.[1], Cox [2], Gokhale and Kullback [5], Haberman [7], and Plackett [11], all of which focus in large part on the use of log-linear models* for both two-dimensional and multidimensional tables. A detailed bibliography for the statistical literature on contingency tables through 1974 is given by Killion and Zahn [45].

The subsequent sections of this presentation are concerned primarily with the use of log-linear models for the analysis of contingency-table data. For details on some related methods, see the book by Lancaster [10], and the series of papers on *measures of association** by Goodman and Kruskal, which have been recently reprinted as ref. 6. Several book-length but elementary presentations on log-linear models are now available, including Everitt [3], Fienberg [4], Haberman [8, 9], and Upton [12].

The next section describes two examples that will serve to illustrate some of the methods of analysis. The third section discusses briefly some alternative methods for estimation of parameters used in conjunction with categorical data* analysis, and the fourth section outlines the basic statistical theory associated with maximum likelihood estimation* and log-linear models. These theoretical results are then illustrated in the fifth section. The final section presents a guide to some recent applications of log-linear and contingency-table modeling, and computer programs for contingency-table analysis.

TWO CLASSIC EXAMPLES

The data reported by Bartlett [21] in his pioneering article, and included here in Table 1, are from an *experiment* giving the response (alive or dead) of 240 plants for each combination of two explanatory variables, time of planting (early or late) and length of cutting (high or low).

Table 1   2 × 2 × 2 Table

Time of planting:		Early		Late	
Length of cutting:		High	Low	High	Low
Response:	Alive	156	107	84	31
	Dead	84	133	156	209
Total		240	240	240	240

The questions to be answered are: (a) What are the effects of time of planting and length of cutting on survival? (b) Do they interact in their effect on survival?

The data in Table 2, from Waite [56], give the cross-classification or right-hand fingerprints according to the number of whorls and small loops. The total number of whorls and small loops is at most 5, and the resulting table is triangular. Here the question of interest is more complicated because, as a result of the constraint forcing the data into the triangular structure, the number of whorls is "related to" to the number of small loops. Such an array of counts is referred to as an *incomplete contingency table**, and the incomplete structure, in the case of the Waite data, was the cause of yet another controversy involving Karl Pearson [51], this time with J. A. Harris (see Harris and Treloar [43]). In the section "Contingency-Table Analyses," the fit of a relatively simple model to these data is explored.

Table 2   Fingerprints of the Right Hand Classified by the Number of Whorls and Small Loops

Whorls	Small Loops						Total
	0	1	2	3	4	5	
0	78	144	204	211	179	45	861
1	106	153	126	80	32		497
2	130	92	55	15			292
3	125	38	7				170
4	104	26					130
5	50						50
Total	593	453	392	306	211	45	2000

ESTIMATING PARAMETERS IN CONTINGENCY-TABLE MODELS

Let $\mathbf{x}' = (x_1, x_2, \ldots, x_t)$ be a vector of observed counts for t cells, structured in the form of a cross-classification such as in Tables 1 and 2, where $t = 2^3 = 8$ and $t = 21$, respectively. Now let $\mathbf{m}' = (m_1, m_2, \ldots, m_t)$ be the vector of expected values that are assumed to be functions of unknown parameters $\boldsymbol{\theta}' = (\theta_1, \theta_2, \ldots, \theta_s)$, where $s < t$. Thus one can write $\mathbf{m} = \mathbf{m}(\boldsymbol{\theta})$.

There are three standard sampling models for the observed counts in contingency tables.

POISSON* MODEL.   The $\{\mathbf{x}_i\}$ are observations from independent Poisson random variables with means $\{m_i\}$ and likelihood function

$$\prod_{i=1}^{t} \left[m_i^{x_i} \exp(-m_i) / x_i! \right]. \tag{1}$$

MULTINOMIAL* MODEL.   The total count $N = \sum_{i=1}^{t} x_i$ is a random sample from an infinite population where the underlying cell probabilities are $\{m_i / N\}$, and the likelihood is

$$N! \cdot N^{-N} \prod_{i=1}^{t} (m_i^{x_i} / x_i!). \tag{2}$$

PRODUCT-MULTINOMIAL MODEL.   The cells are partitioned into sets, and each set has an independent multinomial structure, as in the multinomial model.

For the Bartlett data in the preceding section, the sampling model is product-multinomial—there are actually four independent binomials, one for each of the four experimental conditions corresponding to the two factors, time of planting and length of cutting. (*See* the discussion of factors and responses in CATEGORICAL DATA.)

For each of these sampling models the estimation problem can typically be structured in terms of a "distance" function, $K(\mathbf{x}, \mathbf{m})$, where parameter estimates $\hat{\boldsymbol{\theta}}$ are chosen so that the distance between $\mathbf{x}$ and $\mathbf{m} = \mathbf{m}(\boldsymbol{\theta})$, as measured by $K(\mathbf{x}, \mathbf{m})$, is minimized. The minimum chi-square method uses the distance function,

$$X^2(\mathbf{x}, \mathbf{m}) = \sum_{i=1}^{t} (x_i - m_i)^2 / m_i, \tag{3}$$

the *minimum modified chi-square method** uses the function

$$Y^2(\mathbf{x}, \mathbf{m}) = \sum_{i=1}^{t} (x_i - m_i)^2 / x_i, \tag{4}$$

and the *minimum discrimination information method** uses either

$$G^2(\mathbf{x}, \mathbf{m}) = 2 \sum_{i=1}^{t} x_i \log(x_i / m_i), \tag{5}$$

or

$$G^2(\mathbf{m}, \mathbf{x}) = 2 \sum_{i=1}^{t} m_i \log(m_i / x_i). \tag{6}$$

Rao [54] studies these and other choices of "distance"* functions.

For the three basic sampling models for contingency tables, choosing $\hat{\boldsymbol{\theta}}$ to minimize $g^2(\mathbf{x}, \mathbf{m})$ in (5) is equivalent to maximizing the likelihood function provided that

$$\sum_{i=1}^{t} m_i(\hat{\boldsymbol{\theta}}) = \sum_{i=1}^{t} x_i \tag{7}$$

[and that constraints similar to (7) hold for each of the sets of cells under product-multinomial sampling]. Moreover, the estimators that minimize each of (3) to (6) in such circumstances belong to the class of *best asymptotic normal* (BAN) *estimates* (*see* ASYMPTOTIC NORMALITY) for $\mathbf{m}$ (see Bishop et al. [1] and Neyman [47] for further discussion of asymptotic equivalence). Because of various additional asymptotic properties, and because of the smoothness of maximum likelihood estimates in relatively sparse tables, many authors have preferred to work with maximum likelihood estimates (MLEs), which minimize (5).

BASIC THEORY FOR LOG-LINEAR MODELS

For expected values $\{m_{ij}\}$ for a 2×2 table,

		B	
		1	2
A	1	m_{11}	m_{12}
	2	m_{21}	m_{22}

a standard measure of association for the row and column variables, A and B, respectively, is the cross-product ratio proposed by Yule [57]:

$$\alpha = \frac{m_{11} m_{22}}{m_{12} m_{21}} \tag{8}$$

(for a discussion of the properties of α, see Bishop et al. [1], or Fienberg [4]). Independence of A and B is equivalent to setting $\alpha = 1$, and can also be expressed in log-linear form:

$$\log m_{ij} = u + u_{1(i)} + u_{2(j)}, \qquad (9)$$

where

$$\sum_{i=1}^{2} u_{1(i)} = \sum_{j=1}^{2} u_{2(j)} = 0. \qquad (10)$$

Note that the choice of notation here parallels that for analysis-of-variance models. (*See* CATEGORICAL DATA for a related discussion, using somewhat different notation.)

Bartlett's [21] no-second-order interaction model for the expected values in a $2 \times 2 \times 2$ table

m_{111}	m_{121}		m_{112}	m_{122}
m_{211}	m_{221}		m_{212}	m_{222}

is based on equating the values of α in each layer of the table, i.e.,

$$\frac{m_{111}m_{221}}{m_{121}m_{211}} = \frac{m_{112}m_{222}}{m_{122}m_{212}}. \qquad (11)$$

Expression (11) can be represented in log-linear form as

$$\log m_{ijk} = u + u_{1(i)} + u_{2(j)} + u_{3(k)} + u_{12(ij)}$$
$$+ u_{13(ik)} + u_{23(jk)}, \qquad (12)$$

where, as in (10), each subscripted u-term sums to zero over any subscript, e.g.,

$$\sum_{i} u_{12(ij)} = \sum_{j} u_{12(ij)} = 0. \qquad (13)$$

All of the parameters in (12) can be written as functions of cross-product ratios (see Bishop et al. [1]).

For the sampling schemes described in the preceding section, the *minimal sufficient statistics** (MSSs) are the two-dimensional marginal totals, $\{x_{ij+}\}$, $\{x_{i+k}\}$, and $\{x_{+jk}\}$ (except for linearly redundant statistics included for purposes of symmetry), where a "+" indicates summation over the corresponding subscript. The MLEs of the $\{m_{ijk}\}$ under model (12) must satisfy the likelihood

equations,

$$\hat{m}_{ij+} = x_{ij+}, \qquad i,j = 1,2,$$
$$\hat{m}_{i+k} = x_{i+k}, \qquad i,k = 1,2, \qquad (14)$$
$$\hat{m}_{+jk} = x_{+jk}, \qquad j,k = 1,2,$$

usually solved by some form of iterative procedure. For the Bartlett data the third set of equations in (14) corresponds to the binomial sampling constraints.

More generally, for a vector of expected values $\mathbf{m}$, if the log expectations $\boldsymbol{\lambda}' = (\log m_1, \ldots, \log m_t)$ are representable as a linear combination of the parameters $\boldsymbol{\theta}$, the following results hold under the Poisson and multinomial sampling schemes:

1. Corresponding to each parameter in $\boldsymbol{\theta}$ is a MSS that is expressible as a linear combination of the $\{x_i\}$. (More formally, if $\mathfrak{M}$ is used to denote the log-linear model specified by $\mathbf{m} = \mathbf{m}(\boldsymbol{\theta})$, then the MSSs are given by the projection of $\mathbf{x}$ onto $\mathfrak{M}$, $P_{\mathfrak{M}}\mathbf{x}$. For a more detailed discussion, see Haberman [7].)

2. The MLE, $\hat{\mathbf{m}}$, of $\mathbf{m}$, if it exists, is unique and satisfies the likelihood equations

$$P_{\mathfrak{M}} \hat{\mathbf{m}} = P_{\mathfrak{M}} \mathbf{x}. \qquad (15)$$

[Note that the equations in (14) are a special case of those given by expression (15).]

Necessary and sufficient conditions for the existence of a solution to the likelihood equations, (15), are relatively complex (see Haberman [7]). A sufficient condition is that all cell counts be positive, i.e., $\mathbf{x} > 0$, but MLEs for log-linear models exist in many sparse situations where a large fraction of the cells have zero counts.

For product-multinomial sampling situations, the basic multinomial constraints (i.e., that the counts must add up to the multinomial sample sizes) must be taken into account. Typically, some of the parameters in $\boldsymbol{\theta}$ which specify the log-linear model $\mathfrak{M}$, i.e., $\mathbf{m} = \mathbf{m}(\boldsymbol{\theta})$, are fixed by these constraints.

More formally, let $\mathfrak{M}^*$ be a log-linear model for $\mathbf{m}$ under product-multinomial sampling which corresponds to a log-linear model $\mathfrak{M}$ under Poisson sampling such that the multinomial constraints "fix" a subset of the parameters, θ, used to specify $\mathfrak{M}$. Then:

3. The MLE of $\mathbf{m}$ under product-multinomial sampling for the model $\mathfrak{M}^*$ is the same as the MLE of $\mathbf{m}$ under Poisson sampling for the model $\mathfrak{M}$.

As a consequence of result 3, equations (14) are the likelihood equations for the $2 \times 2 \times 2$ table under the no-second-order interaction model for Poisson or multinomial sampling, as well as for product-multinomial sampling when any set of one-way or two-way marginal totals are fixed (i.e., these correspond to the multinomial constraints).

A final result, which is used to assess the fit of log-linear models, can be stated in the following informal manner:

4. If $\hat{\mathbf{m}}$ is the MLE of $\mathbf{m}$ under a log-linear model, and if the model is correct, then the statistics

$$X^2 = \sum_{i=1}^{t} (x_i - \hat{m}_i)^2 / \hat{m}_i \qquad (16)$$

and

$$G^2 = 2 \sum_{i=1}^{t} x_i \log(x_i / \hat{m}_i) \qquad (17)$$

have asymptotic χ^2 distributions with $t - s$ degrees of freedom, where s is the total number of independent constraints implied by the log-linear model and the multinomial sampling constraints (if any). If the model is not correct, then X^2 and G^2, in (16) and (17), are stochastically larger than χ^2_{t-s}. (See CHI-SQUARE TESTS.) Expression (17) is the minimizing value of the distance function (5), but (16) is not the minimizing chi-square value for the function (3).

In the next section these basic results are applied in the context of the Bartlett and Waite data sets of the section "Two Classic Examples."

Many authors have devised techniques for selecting among the class of log-linear models applicable for contingency table structures. These typically (although not always) resemble corresponding model selection procedures for analysis of variance* and regression* models. See, for example, Goodman [39] and Aitken [20], as well as the discussions in Bishop et al. [1], and Fienberg [4].

CONTINGENCY-TABLE ANALYSES

Illustrative Analyses

For the 2^3 table of Bartlett, variables 2 and 3 are fixed by design, so that $\hat{m}_{+jk} = 240$, and the estimated expected values under the no-second-order interaction model of expressions (12) are given in Table 3. These values were computed by Bishop et al. [1] using the method of iterative proportional fitting. Bartlett originally found the solution to equations (14), by noting that the constraints in his specification, (11), reduced (14) to a single cubic equation for the discrepancy $\Delta = \hat{m}_{111} - x_{111}$. Note that the expected values satisfy expression (12), e.g., $\hat{m}_{12+} = 78.9 + 36.1 = 115 = 84 + 31 = x_{12+}$. The goodness-of-fit* statistics for this model are $X^2 = 2.27$ and $G^2 = 2.29$. Using result 4 of the preceding section, one compares these values to tail values of the chi-square distri-

Table 3 Observed and Expected Values for the Bartlett Data, Including the No-Second-Order Interaction Model

Cell	Observed, x	Estimated Expected, $\hat{m}$
1, 1, 1	156	161.1
2, 1, 1	84	78.9
1, 2, 1	84	78.9
2, 2, 1	156	161.1
1, 1, 2	107	101.9
2, 1, 2	133	138.1
1, 2, 2	31	36.1
2, 2, 2	209	203.9

bution with 1 d.f., e.g., $\chi_1^2(0.10) = 2.71$, and this suggests that the no-second-order interaction model provides an acceptable fit to the data.

Since the parameters u, $\{u_{2(j)}\}$, $\{u_{3(k)}\}$, and $\{u_{23(jk)}\}$ are fixed by the binomial sampling constraints for these data, model (12) is often rewritten as

$$\log\left(\frac{m_{1jk}}{m_{2jk}}\right) = 2\left[u_{1(1)} + u_{12(1j)} + u_{13(2k)}\right]$$

$$= w + w_{2(j)} + w_{3(k)}, \qquad (18)$$

where

$$\sum_j w_{2(j)} = \sum_k w_{3(k)} = 0.$$

Expression (18) is referred to as a *logit** model for the log odds for alive versus dead. The simple additive structure corresponds to Bartlett's notion of no-second-order interaction.

For the Waite fingerprint data of Table 2, one model that has been considered is the simple additive log-linear model of expression (9), but only for those cells where positive counts are possible, i.e., in the upper triangular section. For cells with $i > j$, $m_{ij} = 0$ a priori. This restricted version of the independence model is referred to as quasi-independence, and the results of the preceding section can be used in connection with it. The MSSs are still the row and column totals (result 1). The likelihood equations under multinomial sampling are (applying

results 1 and 2):

$$\hat{m}_{i+} = x_{i+}, \qquad i = 0, 1, 2, \ldots, 5,$$
$$\hat{m}_{+j} = x_{+j}, \qquad j = 0, 1, 2, \ldots, 5, \qquad (19)$$

where $m_{ij} = 0$ for $i > j$. A solution of equations (19) satisfying the model can be found directly (see Goodman [37] or Bishop and Fienberg [24]), or by using a standard iterative procedure. The estimated expected values for the fingerprint data under the model of quasi-independence are given in Table 4, and they satisfy the marginal constraints in expression (19).

The goodness-of-fit statistics for this model are $X^2 = 399.8$ and $G^2 = 450.4$, which correspond to values in the very extreme right-hand tail of the χ_{10}^2 distribution. Thus the model of quasi-independence seems inappropriate. Darroch [28] describes the log-linear model of F-independence (with more parameters than the quasi-independence model), which takes in account the way in which the constraint—that the number of small loops plus the number of whorls cannot exceed 5—makes the usual definition of independence inappropriate. This model in log-linear form is

$$\log m_{ij} = u + u_{1(i)} + u_{2(j)} + u_{3(5-i-j)}, \qquad (20)$$

where the u_3-parameters correspond to diagonals where the sum of the numbers of whorls and small loops is constant. Darroch and Ratcliff [29] illustrate the fit of the F-

Table 4 Estimated Expected Values for Fingerprint Data under Quasi-Independence

Whorls	Small Loops						Total
	0	1	2	3	4	5	
0	200.6	167.4	166.6	150.3	131.1	45.0	861
1	122.2	101.9	101.4	91.6	79.9		497
2	85.5	71.4	71.0	64.1			292
3	63.8	53.2	53.0				170
4	70.9	59.1					130
5	50.0						50
Total	593	453	392	306	211	45	2000

independence model to a related set of fingerprint data involving large rather than small loops.

Multidimensional Contingency-Table Analyses

Not all applications of log-linear models involve such simple structures as 2^3 tables, or even incomplete 6×6 arrays. Indeed, much of the methodology was developed in the mid-1960s to deal with very large, highly multidimensional tables. For example, in the National Halothane Study [26], investigators considered data on the use of (a) 5 anesthetic agents in operations involving (b) 4 levels of risk, and patients of (c) 2 sexes, (d) 10 age groups, with (e) 7 differing physical statuses (levels of anesthetic risk) and (f) previous operations (yes, no) for (9) 3 different years, from (h) 34 different institutions. Two sets of data were collected, the first consisting of all deaths within six weeks of surgery, and the second consisting of a sample (of comparable size) of all those exposed to surgery. Thus the data consisted of two very sparse $5 \times 4 \times 2 \times 10 \times 7 \times 2 \times 3 \times 34$ tables, each containing in excess of 57,000 cells. One of the more successful approaches used in the analysis of the data in these tables was based on log-linear models and the generalizations of the methods illustrated in this section.

One of the key reasons why log-linear models have become so popular in such analyses is that they lead to a simplified description of the data in terms of marginal totals—the minimal sufficient statistics of result 1 of the section "Basic Theory for Log-Linear Models." This is especially important when the table of data is large and sparse. For more details on the halothane study analyses, as well as examples of other applications involving four-way and higher-dimensional tables of counts, see Bishop et al. [1].

A second reason for the popularity of log-linear models relates to their interpretation. A large subset of these models can be interpreted in terms of independence or the

*conditional independence** of several discrete random variables given the values of other discrete variables, thus generalizing the simple ideas for 2×2 tables outlined in the section on log-linear models. For further details, see any of the books cited in the first section.

BRIEF GUIDE TO ADDITIONAL APPLICATIONS AND COMPUTING PROGRAMS

Novel Applications Involving Contingency Tables

Many data sets can profitably be structured to appear in the form of a cross-classification of counts, and then analyzed using methods related to those described in this entry. Some examples of applications where this has been done include the following:

CAPTURE (MULTIPLE)–RECAPTURE ANALYSIS*. This type of analysis estimates the size of a nonchanging population [1, 30]. If the members of nonchanging populations are sampled k successive times (possibly dependent), the resulting recapture history data can be displayed in the form of a 2^k table with one missing cell, corresponding to those never sampled. Such an array is amenable to log-linear analysis, the results of which can be used to project a value for the missing cell.

GUTTMAN SCALING*. Guttman scaling is performed on a sequence of p dichotomous items [41]. The items form a perfect Guttman Scale if they have an order such that a positive response to any item implies a positive response to those items lower in the ordering. Goodman describes an application of techniques for incomplete multidimensional contingency tables in which he measures departures from perfect Guttman scales.

LATENT STRUCTURE ANALYSIS*. In latent structure analysis, unobservable categorical variables are included as part of the analysis of categorical data structures, and the observable variables are taken to be conditionally independent given the unobservable latent variables (ref. 40; *see also* CATEGORICAL DATA).

PAIRED COMPARISONS*. Paired comparisons of several objects are made by a set of judges, the outcome being the preference of one object over the other. A well-known model for paired comparisons, first proposed by Bradley and Terry [25], and several extensions to it, can be viewed as log-linear models. Then relatively standard contingency table methods can be used to analyze paired comparisons data (see Imrey et al. [44], Fienberg and Larntz [32], and Fienberg [31]).

Computer Programs for Log-Linear Model Analysis

As with other forms of multivariate analysis, the analysis of multidimensional contingency tables relies heavily on computer programs. A large number of these have been written to compute estimated parameter values for log-linear models and associated test statistics, and most computer installations at major universities have one or more programs available for users.

The most widely used numerical procedure for the calculation of maximum likelihood estimates for log-linear models is the method of iterative proportional fitting (IPF), which iteratively adjusts the entries of a contingency table to have marginal totals equal to those used in specifying the likelihood equations*. Detailed FORTRAN listings for this method are available in Haberman [17, 18], and they have been implemented in the BMDP Programs distributed by the UCLA Health Sciences Computing Facility [15], as well as in a variety of other forms. IPF programs also exist in other languages, such as APL (see, e.g., Fox [16]). The major advantage of the IPF method is that it requires limited computer memory capabilities since it does not require matrix

inversion or equivalent computations, and thus can be used in connection with the analysis of very high dimensional tables. Its major disadvantage is that it does not provide, in an easily accessible form, estimates of the basic log-linear model parameters (and an estimate of their asymptotic covariance matrix)*; it provides only estimated expected values.

The other numerical approaches suggested for the computation of maximum likelihood estimates are typically based on classical procedures for solving nonlinear equations, such as modifications of Newton's method or the Newton–Raphson method* (see, e.g., the listing in Haberman [9]). Currently, the most widely used such program is GLIM*, distributed by the Numerical Algorithms Group of the United Kingdom [13], which fits a class of generalized linear models, of which log-linear and logit models are special cases. The virtue of these programs is that they produce both estimated expected values and estimated parameter values, and an estimate of the asymptotic covariance matrix*. Unfortunately, such output comes at the expense of added storage, and these programs cannot handle analyses for very large contingency tables. Several groups of researchers are currently at work adapting variants of Newton's method using numerical techniques that will allow for increased storage capacity and thus the analysis of larger tables than is currently possible.

Computation problems remain as a major stumbling block to the widespread application of log-linear-model methods to the analysis of large data sets structured in the form of multidimensional cross-classifications of counts.

References

BOOKS ON THE ANALYSIS OF CONTINGENCY TABLES

[1] Bishop, Y. M. M., Fienberg, S. E., and Holland, P. (1975). *Discrete Multivariate Analysis: Theory and Practice*. MIT Press, Cambridge, Mass. (A systematic exposition and development of the log-

linear model for the analysis of contingency tables, primarily using maximum likelihood estimation, and focusing on the use of iterative proportional fitting. Includes chapters on measures of association, and others on special related topics. Contains both theory and numerous examples from many disciplines with detailed analyses.)

[2] Cox, D. R. (1970). *Analysis of Binary Data*. Methuen, London. (A concise treatment of log-linear and logistic response models, primarily for binary data. Emphasis on statistical theory, especially related to exact tests; includes several examples.)

[3] Everitt, B. S. (1977). *The Analysis of Contingency Tables*. Chapman & Hall, London. (A brief and very elementary introduction to contingency-table analysis, with primary emphasis on two-dimensional tables.)

[4] Fienberg, S. E. (1980). *The Analysis of Cross-Classified Categorical Data*, 2nd ed. MIT Press, Cambridge, Mass. (A comprehensive introduction, for those with some training in statistical methodology, to the analysis of categorical data using log-linear models and maximum likelihood estimation. Emphasis on methodology, with numerous examples and problems.)

[5] Gokhale, D. V. and Kullback, S. (1978). *The Information in Contingency Tables*. Marcel Dekker, New York. (A development of minimum discrimination information procedures for linear and log-linear models. Contains a succinct theoretical presentation, followed by numerous examples.)

[6] Goodman, L. A. and Kruskal, W. (1979). *Measures of Association for Cross Classifications*. Springer-Verlag, New York. (A reprinting of four classical papers, written between 1954 and 1972, on the construction of measures of association for two-way tables, historical references, sample estimates, and related asymptotic calculations.)

[7] Haberman, S. J. (1974). *The Analysis of Frequency Data*. University of Chicago Press, Chicago. (A highly mathematical, advanced presentation of statistical theory associated with log-linear models and of related statistical and computational methods. Contains examples, but is suitable only for mathematical statisticians who are familiar with the topic.)

[8] Haberman, S. J. (1978). *Analysis of Qualitative Data, Vol. 1: Introductory Topics*. Academic Press, New York.

[9] Haberman, S. J. (1979). *Analysis of Qualitative Data, Vol. 2: New Developments*. Academic Press, New York. (An intermediate-level, two-volume introduction to the analysis of categorical data via log-linear models, emphasizing maximum likelihood estimates computed via the Newton–Raphson algorithm. Volume 1 examines complete cross-classifications, and Vol. 2 considers multino-

mial response models, incomplete tables, and related topics. Contains many examples, problems, and solutions, and a computer program listing (for two-way tables) in Vol. 2.)

[10] Lancaster, H. O. (1969). *The Chi-Squared Distribution*. Wiley, New York, Chaps. 11 and 12. (A mathematical statistics monograph developing ideas on the chi-square distribution and quadratic forms for both discrete and continuous random variables, with several chapters related to the analysis of contingency tables. Emphasis is on topics other than log-linear models.)

[11] Plackett, R. L. (1974). *The Analysis of Categorical Data*. Charles Griffin, London. (A concise introduction to statistical theory and methods for the analysis of categorical data. Assumes a thorough grasp of basic principles of statistical inference. Considerable emphasis on two-way tables. Contains many examples and exercises.)

[12] Upton, G. J. G. (1978). *The Analysis of Cross-Tabulated Data*. Wiley, New York. (A brief introduction to the analysis of contingency tables via log-linear models and measures of association for those with some training in statistical methodology. Contains several examples.)

COMPUTER PROGRAM DESCRIPTIONS AND DOCUMENTATION

[13] Baker, R. J. and Nelder, J. A. (1978). *The GLIM System, Release 3, Manual*. Numerical Algorithms Group, Oxford.

[14] Bock, R. D. and Yates, G. (1973). *MULTIQUAL: Log-Linear Analysis of Nominal or Ordinal Qualitative Data by the Method of Maximum Likelihood*. International Education Services, Chicago. (A manual for a log-linear model program that uses a modified Newton–Raphson algorithm.)

[15] Dixon, W. J. and Brown, M. B., eds. (1979). *BMPD, Biomedical Computer Programs, P-Series*. University of California Press, Berkeley, Calif. (See Chap. 11 on frequency tables and Sec. 14.LR on logistic regression.)

[16] Fox, J. (1979). *Amer. Statist.* **33**, 159–160. (Contains a program description, but no listing.)

[17] Haberman, S. J. (1972). *Appl. Statist.* **21**, 218–225. (Contains FORTRAN listing of program that uses iterative proportional fitting.)

[18] Haberman, S. J. (1973). *Appl. Statist.* **22**, 118–126. (Contains FORTRAN listing of program.)

[19] SAS Institute (1979). *SAS User's Guide*, 1979 ed. SAS Institute, Raleigh, N.C. (See pp. 298–301 for instructions on the use of general programs for nonlinear equations for computing minimum modified chi-square estimates, and maximum likelihood estimations using a Newton–Raphson algorithm.)

[*Editors' note.* Two further computer programs:
(i) Goodman, L. A. (1973). *ECTA* (*Everyman's Contingency Table Analysis*). Dept. of Statistics, University of Chicago, Chicago, Ill. Estimates parameters and tests hypotheses for log-linear models in ordered or unordered classes; multidimensional tables.
(ii) *TRICHI*. Statistical Laboratory, Southern Methodist University, Dallas, TX. Simultaneous estimates and tests of basic models in three-dimensional tables; includes the partition of chi-square.]

OTHER REFERENCES CITED

[20] Aitken, M. (1978). *J. R. Statist. Soc. A*, **141**, 195–223.

[21] Bartlett, M. S. (1935). *J. R. Statist. Soc. B*, **2**, 248–252.

[22] Bhapkar, V. P. and Koch, G. (1968). *Biometrics*, **24**, 567–594.

[23] Birch, M. W. (1963). *J. R. Statist. Soc. B*, **25**, 229–233.

[24] Bishop, Y. M. M. and Fienberg, S. E. (1969). *Biometrics*, **25**, 119–128.

[25] Bradley, R. A. and Terry, M. E. (1952). *Biometrika*, **39**, 324–345.

[26] Bunker, J. P., Forrest, W. H., Jr., Mosteller, F., and Vandam, L. (1969). *The National Halothane Study*. Report of the Subcommittee on the National Halothane Study of the Committee on Anesthesia, Division of Medical Sciences, National Academy of Sciences–National Research Council, National Institutes of Health, National Institute of General Medical Sciences, Bethesda, Md. U.S. Government Printing Office, Washington, D. C.

[27] Darroch, J. N. (1962). *J. R. Statist. Soc. B*, **24**, 251–263.

[28] Darroch, J. N. (1971). *Biometrika*, **58**, 357–368.

[29] Darroch, J. N. and Ratcliff, D. (1973). *Biometrika*, **60**, 395–402.

[30] Fienberg, S. E. (1972). *Biometrika*, **59**, 591–603.

[31] Fienberg, S. E. (1979). *Biometrics*, **35**, 479–481.

[32] Fienberg, S. E. and Larntz, K. (1976). *Biometrika*, **63**, 245–254.

[33] Fisher, R. A. (1922). *J. R. Statist. Soc. A*, **85**, 87–94.

[34] Good, I. J. (1963). *Ann. Math. Statist.*, **34**, 911–934.

[35] Goodman, L. A. (1963). *J. R. Statist. Soc. A*, **126**, 94–108.

[36] Goodman, L. A. (1964). *J. R. Statist. Soc. B*, **26**, 86–102.

[37] Goodman, L. A. (1968). *J. Amer. Statist. Ass.*, **63**, 1091–1131.

[38] Goodman, L. A. (1969). *J. R. Statist. Soc. B*, **31**, 486–498.

[39] Goodman, L. A. (1971). *Technometrics*, **13**, 33–61.

[40] Goodman, L. A. (1974). *Biometrika*, **61**, 215–231.

[41] Goodman, L. A. (1975). *J. Amer. Statist. Ass.*, **70**, 755–768.

[42] Grizzle, J. E., Starmer, C. F., and Koch, G. G. (1969). *Biometrics*, **25**, 489–504.

[43] Harris, J. A. and Treloar, A. E. (1927). *J. Amer. Statist. Ass.*, **22**, 460–472.

[44] Imrey, P. B., Johnson, W. D. and Koch, G. G. (1976). *J. Amer. Statist. Ass.*, **71**, 614–623.

[45] Killion, R. A. and Zahn, D. A. (1976). *Int. Statist. Rev.*, **44**, 71–112.

[46] Ku, H. H. and Kullback, S. (1968). *J. Res. Bur. Stand.*, **72B**, 159–199.

[47] Neyman, J. (1949). *Proc. Berkeley Symp. Math. Statist. Prob.* University of California Press, Berkeley, Calif., pp. 239–273.

[48] Pearson, K. (1900). *Philos. Mag. 5th Ser.*, **50**, 157–175.

[49] Pearson, K. (1900). *Philos. Trans. R. Soc. Lond. A*, **195**, 79–150.

[50] Pearson, K. (1904). *Draper's Co. Res. Mem. Biom. Ser. I*, 1–35.

[51] Pearson, K. (1930). *J. Amer. Statist. Ass.*, **25**, 320–323.

[52] Pearson, K. and Heron, D. (1913). *Biometrika*, **9**, 159–315.

[53] Quetelet, M. A. (1849). *Letters Addressed to H.R. H. the Grand Duke of Saxe Coburg and Gotha on the Theory of Probabilities as Applied to the Moral and Political Sciences* (translated from the French by Olinthus Gregory Downs). Charles and Edwin Layton, London.

[54] Rao, C. R. (1962). *J. R. Statist. Soc. B*, **24**, 46–72.

[55] Roy, S. N. and Kastenbaum, M. A. (1956). *Ann. Math. Statist.*, **27**, 749–757.

[56] Waite, H. (1915). *Biometrika*, **10**, 421–478.

[57] Yule, G. U. (1900). *Philos. Trans. R. Soc. Lond. A*, **194**, 257–319.

Acknowledgment

The preparation of this entry was supported in part by Contract N00014-7-C-0600 from the Office of Naval Research, Statistics and Probability Program, to the School of Statistics, University of Minnesota.

(ASSOCIATION, MEASURES OF
CATEGORICAL DATA
CHI-SQUARE TESTS
ITERATIVE PROPORTIONAL
FITTING)

S. E. FIENBERG

CONTINUITY CORRECTIONS

In an article in *The American Statistician* entitled "What Is the Continuity Correction?", Mantel and Greenhouse [9] have provided a description of a continuity correction as being "simply a device for evaluating probabilities when a discrete distribution is being approximated by a continuous one."

In most applications, the approximating continuous distribution is the normal distribution* and the procedure consists of approximating a desired point probability, $\Pr[X = x]$, from the discrete distribution with a corresponding interval probability, $\Pr[x - \frac{1}{2} \leqslant X \leqslant x + \frac{1}{2}]$, from the appropriate normal distribution. Similarly, an interval probability such as $\Pr[X \leqslant x]$ from the discrete distribution would be approximated by $\Pr[X \leqslant x + \frac{1}{2}]$ from the corresponding normal distribution.

The adjustment of x by the addition and/or subtraction of $\frac{1}{2}$ is the *continuity correction*. It serves to transform the point value x in a discrete set to an interval $[x - \frac{1}{2}, \ x + \frac{1}{2}]$ in a continuum so that the approximate value for the discrete probability, $\Pr[X = x]$, is found as the area under the approximating continuous probability density function over the unit interval centered on the value x.

The three most common distributions for which this form of continuity correction is applied are the binomial*, hypergeometric*, and Poisson* distributions. In each case, the approximating normal distribution is the one with the same mean and variance as the discrete distribution of interest.

Another common correction, often referred to as *Yates' correction for continuity*, involves the use of the continuous χ^2 distribution* to approximate the discrete exact probabilities related to 2×2 contingency tables*. In this case, the correction consists of reducing the magnitude of the difference between each observed frequency and the corresponding expected frequency by $\frac{1}{2}$. This correction derives its name from its first introduction by Yates [19], although Pearson [13], "without wishing to detract from the

value of Yates' suggestion," has pointed out that the correction is based on the same adjustment as the first type of continuity correction and that this adjustment "was used by statisticians well before 1934," and perhaps for several years prior to 1921.

Since the continuity corrections represent very simple adjustments and have generally been found to provide improvements over the same approximations applied without the corrections, they have gained general acceptance and are widely used. The corrections applied to the binomial distribution and applied to the calculation of the χ^2 statistic in 2×2 contingency tables are especially prominent in texts on statistical methodology.

It is interesting to note the popularity of Yates' correction in the analysis of 2×2 tables since its application represents an exception to the general rule of improvement. Pearson [13] expressed doubts about the usefulness of Yates' correction and his doubts have been reinforced in the works of several other researchers, many of which are referenced in the articles by Conover [1] (including discussion) and Maxwell [10].

In the investigation of the behavior of outcomes in 2×2 tables in which individual observed configurations may be represented as

$$\frac{a \mid b}{c \mid d}$$

with the total frequency n, the uncorrected measure

$$\frac{n(ad - bc)^2}{(a + b)(c + d)(a + c)(b + d)}$$

and the Yates' corrected measure

$$\frac{n(|ad - bc| - n/2)^2}{(a + b)(c + d)(a + c)(b + c)}$$

have been subject to several comparisons with regard to their appropriateness as approximate χ^2 variables with 1 degree of freedom.

With the assumption of fixed marginal totals, comparisons have been based on assessing the closeness of the approximating χ^2

distribution to the accumulated exact probabilities for individual configurations given (e.g., in Fisher [3, Sec. 21.02]) by

$$\frac{(a+b)!(c+d)!(a+c)!(b+d)!}{n!a!b!c!d!}.$$

(*See* FISHER'S EXACT TEST.)

For cases involving random marginal totals, conclusions have been based on general results demonstrated in cases of fixed totals and on assessments based on randomly generated tables under given hypotheses.

These comparisons have indicated that Yates' correction is better in cases involving at least one set of equal marginal totals than it is in other cases; however, the uncorrected measure, although itself subject to weakness, is generally preferable in all cases in which a particular level of significance* is to be achieved in a test based on a 2×2 table.

Advances in computer technology have eliminated many difficulties of computation in cases for which Yates' correction was meant to apply and for which the χ^2 approximation is poor. As illustrated by Robertson [18], with the assumption of fixed marginal totals, computation of the exact probabilities for 2×2 tables can be programmed quite readily and the need for approximations avoided.

The correction applied in the normal approximation to the binomial has not been subject to the same criticism of inappropriateness, although other approximations have been considered which provide a closer fit but are considerably more complex.

Plackett [15] refers to an early proof by Laplace* [8, Livre 2, Chap. 3] of a result equivalent to using

$$\Phi\left\{\frac{x-np}{\sqrt{npq}}\right\} + \frac{1}{2\sqrt{npq}}\phi\left\{\frac{x-np}{\sqrt{npq}}\right\}$$

as an approximation to $\Pr[X \leq x]$, where X is the outcome from a binomial distribution with n trials and a probability p of success and $q = 1 - p$ of failure, and where $\phi(\cdot)$ denotes the standard normal density function and $\Phi(\cdot)$ denotes its distribution function.

He indicates further that it is this result which has led to the more usual approximation with the continuity correction in which $\Pr[X \leq x]$ is approximated by

$$\Phi\left\{\frac{x+\frac{1}{2}-np}{\sqrt{npq}}\right\},$$

noting further that the result has been shown to be valid asymptotically if $n \to \infty$ and $x \to \infty$ in such a way that $(x-np)^3/(npq)^2 \to 0$.

Feller [2, Chap. 7] considers the accuracy of the approximation, including an illustrative example of the limits of its usefulness. In a discussion on the accuracy of the foregoing approximation with the continuity correction, Raff [17] notes that the maximum absolute error is less than $0.140/\sqrt{npq}$. Johnson and Kotz [7, Chap. 3] discuss other approximations with an indication of source references for these approximations and, when appropriate, degree of accuracy of the approximations. Other approximations are also discussed by Peizer and Pratt [14], Pratt [16], Gebhardt [6], and Molenaar [11].

The normal approximation to the hypergeometric distribution with continuity correction is not included in texts as often as the approximation to the binomial; however, the fact that some improvement is provided by the continuity correction is generally accepted. The approximation incorporating the continuity correction is similar to that for the binomial and (in the cumulative probability form) involves approximating $\Pr[X \leq x]$, with

$$\Phi\left\{\frac{x+\frac{1}{2}-np}{\sqrt{npq(N-n)/(N-1)}}\right\},$$

where X is the number of successes in n draws without replacement from a finite population of N items of which S represent success, $N - S$ represent failure, where $p = S/N$ and $q = 1 - p = (N - S)/N$.

Further improvements to this approximation are discussed by Nicholson [12], Johnson and Kotz [7], and Molenaar [11].

The Poisson distribution with parameter λ

is generally approximated by the normal distribution with mean λ and variance λ and, once again, it is generally accepted that the approximation is improved by incorporating the continuity correction so that the approximation consists of approximating $\Pr[X = x]$ with

$$\Phi\left\{ \frac{x + \frac{1}{2} - \lambda}{\sqrt{\lambda}} \right\} - \Phi\left\{ \frac{x - \frac{1}{2} - \lambda}{\sqrt{\lambda}} \right\}.$$

As with the binomial and hypergeometric distributions, other, more complicated, approximations exist for the Poisson and are discussed by Johnson and Kotz [7] and Molenaar [11]. One simple approximation involves considering $2(\sqrt{x} - \sqrt{\lambda})$ as a standard normal variable. This approximation is compared to the normal approximation with continuity correction by Fraser [4, Chap. 6; 5, Chap. 6].

References

[1] Conover, W. J. (1974). *J. Amer. Statist. Ass.*, **69**, 374–382.

[2] Feller, W. (1968). *An Introduction to Probability Theory and Its Applications*, Vol. 1, 3rd ed. Wiley, New York.

[3] Fisher, R. A. (1963). *Statistical Methods for Research Workers*, 13th ed. Oliver & Boyd, Edinburgh.

[4] Fraser, D. A. S. (1968). *The Structure of Inference*. Wiley, New York.

[5] Fraser, D. A. S. (1976). *Probability and Statistics: Theory and Applications*. Duxbury Press, North Scituate, Mass.

[6] Gebhardt, F. (1969). *J. Amer. Statist. Ass.*, **64**, 1638–1646.

[7] Johnson, N. L. and Kotz, S. (1969). *Distributions in Statistics: Discrete Distributions*. Wiley, New York.

[8] Laplace, P. S. (1820). *Théorie analytique des probabilités*, 3rd ed. Courcier, Paris.

[9] Mantel, N. and Greenhouse, S.W. (1968). *Amer. Statist.*, **22**, 27–30.

[10] Maxwell, E. A. (1976). *Canad. J. Statist.*, **4**, 277–290.

[11] Molenaar, W. (1970). *Approximations to the Poisson, Binomial and Hypergeometric Distribution Fractions*, Mathematische Centrum, Amsterdam.

[12] Nicholson, W. L. (1956). *Ann. Math. Statist.*, **27**, 471–483.

[13] Pearson, E. S. (1947). *Biometrika*, **34**, 139–167.

[14] Peizer, D. B. and Pratt, J. W. (1968). *J. Amer. Statist. Ass.*, **63**, 1416–1456.

[15] Plackett, R. L. (1964). *Biometrika*, **51**, 327–337.

[16] Pratt, J. W. (1968). *J. Amer. Statist. Ass.*, **63**, 1457–1483.

[17] Raff, M. S. (1956). *J. Amer. Statist. Ass.*, **51**, 293–303.

[18] Robertson, W. H. (1960). *Technometrics*, **2**, 103–107.

[19] Yates, F. (1934). *J. R. Statist. Soc. B*, **1**, 217–235.

(CONTINGENCY TABLES)

E. A. MAXWELL

CONTINUOUS SAMPLING PLAN(S) (CSP)

The "official" definition of CSP as suggested by the Standards Committee of the ASQC [10] states:

> A plan intended for application to a continuous flow of individual units of product that (1) involves acceptance or nonacceptance on a unit-by-unit basis and (2) uses alternate periods of 100% inspection and sampling depending on the quality of the observed product.

The first continuous sampling plan was CSP-1, proposed by Dodge [4] for the purpose of screening mass-produced items. The idea is to reduce sampling, yet still ensure certain long-term quality requirements. Since 1943, many modifications and extensions of Dodge's original plan have been proposed (e.g., Derman et al. [2,3], Dodge and Torrey [5], Endres [6], Lieberman and Solomon [8]).

A formal definition of CSP-1, proposed by Dodge [4], reads as follows:

> (a) At the outset, inspect 100% of the units consecutively until i units in succession are found clear of defects. (b) When i units in succession are found clear of defects, discontinue 100% inspection, and inspect only a fraction $1/k$ of the units, selecting individual sample units one at a time from the flow of product, in such a manner as to assure an unbiased sample. (c) If a sample unit is found defective, revert immediately

to 100% inspection of succeeding units and continue until again i units in succession are found clear of defects, as in paragraph (a). (d) Correct or replace with good units all defective units found.

Derman et al. [2] observed that (b) does not specify how the sampling should be conducted when on partial inspection. In fact, either systematic sampling* (inspect every kth item) or random sampling* (sample only a fraction $1/k$ of the units by choosing one item at random from each segment of size k (until a defective unit is observed, then begin 100% inspection with the first item following the segment in which the defective item was observed) or even probability sampling* (with probability $1/k$) on partial inspection all lead to the same (Dodge's) operating characteristic formulas. Moreover, CSP-1 as well as other Dodge-type plans guarantees some average outgoing quality limit (AOQL)*, provided that the assumption of statistical control is imposed on the production process. In general, outgoing quality will be higher than incoming quality, because any defectives found on inspection will be replaced by nondefectives. Lieberman [7] has shown that even without the assumption of statistical control, the CSP-1 plan still guarantees an unrestricted AOQL. The problem of partial inspection in CSP-1 was studied more recently by Sackrowitz [9]. Previously, investigations of partial sampling in CSP-1 for particular cases of various sampling schemes (without the assumption of statistical control) had been carried out by Derman et al. [2].

The CSP-2 plan developed by Dodge and Torrey [5] differs from plan CSP-1 in that once sampling inspection is started, 100% inspection is not invoked every time a defect is found, but only if a *second* defect occurs in the next k or fewer sample units. In other words, if two defects observed during sampling are separated by k, or fewer, good inspected units, 100% inspection is invoked. Otherwise, sampling is continued. CSP-2 plans have $k = i$ (where i is the *clearance number*—the number of units in succession found clear of defects at the outset of sampling).

Plan CSP-3 is a refinement of CSP-2 to provide greater protection against a sudden run of bad quality. When one sample defective is found, the next four units from the production line are always inspected. If none of these is defective, the sampling procedure is continued as in CSP-2. If one of the four units is defective, 100% inspection is resumed at once and continued under the rules of CSP-2.

Graphical methods for finding quick and accurate approximations to AOQL and to i values for CSP-1 and CSP-2 plans have been developed by Abraham [1].

References

[1] Abraham, F. L. (1971). *J. Quality Tech.*, **3**, 2–5.

[2] Derman, C., Johns, M. V., and Lieberman, G. J. (1959). *Ann. Math. Statist.*, **30**, 1175–1191.

[3] Derman, C., Littauer, S., and Solomon, H. (1957). *Ann. Math. Statist.*, **28**, 395–404.

[4] Dodge, H. F. (1943). *Ann. Math. Statist.*, **14**, 264–279. (Basic paper.)

[5] Dodge, H. F. and Torrey, M. N. (1951). *Ind. Quality Control*, 7(5), 7–12.

[6] Endres, A. C. (1969). *J. Amer. Statist. Ass.*, **64**, 665–668.

[7] Lieberman, G. J. (1953). *Ann. Math. Statist.*, **24**, 480–484.

[8] Lieberman, G. J. and Solomon, H. (1955). *Ann. Math. Statist.*, **26**, 686–704.

[9] Sackrowitz, H. (1975). *J. Quality Tech.*, **7**, 77–80.

[10] Standards Committee of ASQC (1978). Terms, Symbols and Definitions. *ASQC Standard*, American Society for Quality Control, Milwaukee, Wis.

[11] Stephens, K. S. (1979). *How to Perform Continuous Sampling (CSP)*. American Society for Quality Control, Milwaukee, Wis.

(ACCEPTANCE SAMPLING
AVERAGE OUTGOING QUALITY
 LIMIT
DODGE–ROMIG TABLES
DOUBLE SAMPLING
QUALITY CONTROL)

CONTRAST

This term is used with a very specialized meaning in connection with general linear models*. It means a linear function of the parameters with the *sum of the coefficients equal to zero*. Many of the linear hypotheses tested in standard analyses of variance* specify that the values of one or more contrasts be zero.

(INTERACTION)

CONTRAST ANALYSIS

Contrast analysis is a technique closely related to regression analysis*. The essence of contrast analysis is to divide the data from the experiment into two groups and to examine the difference (the contrast*) between them. In the simplest case, when a single factor is examined at each of two levels, the contrast is the difference between the average at the first level and the average at the second level.

In general, the effect of a factor is measured by contrasting the responses of its low and high levels (or some weighted combination of responses for more levels) and then evaluated against a measure of variability* due to causes not explained by deliberate changes introduced into the experiment. At the second stage, regression coefficients are calculated in order to construct a predictive model*. The mechanics of contrast analysis involve a plus and minus table*, which tells us which response is to be added or subtracted to form the contrast for each assignable source of variation to be considered in the model; the purpose of this table is to define how the responses are to be combined for any assignable source. The contrast for each assignable source is found by algebraically summing the product of the response value by the corresponding plus and minus table coefficient ($X = \pm 1$ or 0). The regression coefficient estimate* is the contrast divided by the sum of squares of the plus and minus coefficients ($\sum X^2$) used to determine that contrast.

Both regression analysis and contrast analysis examine the questions of whether a factor affects the response patterns and pro-

duce the regression coefficients necessary for modeling. Computationally, contrast analysis uses the orthogonalization method* (orthogonalized plus and minus values), whereas regression analysis uses matrix algebra involving matrix inversion*. However, contrast analysis is less general and is useful only for reasonably well balanced experiments*. Its advantage is that it works with the data in original units and lends itself readily to data plotting*.

Bibliography

Daniel, C. and Wood, F. S. (1971). *Fitting Equations to Data*. Wiley, New York.

Freund, R. A. (1974). *J. Quality Tech.*, **6**(1), 2–21. (A useful didactic and expository paper.)

(ANALYSIS OF VARIANCE GENERAL LINEAR MODEL)

CONTROL, STATISTICAL   *See* CONTROL CHART

CONTROL CHARTS

A statistical control chart is a graphical device for monitoring a measurable characteristic of a process for the purpose of showing whether the process is operating within its limits of expected variation. The most commonly used form is referred to as the *Shewhart control chart*, named after Walter A. Shewhart (1891–1967), who invented it in 1924 and used it as the basis for laying the foundation of modern quality control* in his seminal 1931 book, *Economic Control of Quality of Manufactured Product* [24].

In this form the control chart provides information on (1) shorter-term variability estimated from a measure of the variation within a small sample (a subgroup) of from two to a dozen or more observations (usually, four or five) taken within a relatively short time, and (2) longer-term variation estimated from sample-to-sample changes in some measure of location*. Most commonly, the range* and the arithmetic mean* are used as measures of variability and location, respectively. However, the standard deviation estimate is sometimes used to measure variability.

In considering the operation of a control chart, it is useful to think of variation as arising from two sources. The first is the inherent variability which is characteristic of a process and cannot readily (economically) be reduced by isolating and eliminating its major components. This is sometimes referred to as the "noise"* in the system. The second is variation arising from identifiable sources (called assignable causes) which can be found and eliminated. These can be thought of as "signals" which can be detected above the background "noise."

The fundamental idea of the Shewhart control chart involves first removing, to the extent that it is economically feasible, the assignable causes of variation from the subgroup (short-term) variability, and then using the resulting estimate of variability to set limits for judging the long-term variation. In this regard, an important concept introduced by Shewhart is that of a rational subgroup which refers to the fact that "short-term" should be meaningfully related to the process to be controlled. Thus a subgroup might be an operator, the heads on a machine, or simply measurements closely spaced in time. The general principle is that subgroups should be as homogeneous as possible, thereby maximizing the opportunity for variation among them. In this way the assignable cause of a significant variation can be more readily identified.

A process is said to have reached a state of statistical control when changes in measures of variability *and* location from one sampling period to the next are no greater than statistical theory would predict. That is, assignable causes of variation have been detected, identified, and eliminated.

A Shewhart chart using the range to measure variability and the arithmetic mean to detect changes in the process average is shown in Fig. 1. In this illustration, each subgroup consists of four observations. It can be imagined that prior study, accompanied by elimination of assignable causes of variation, yielded an estimate of the process average of 16.08. This is taken as the center line (₵) for the $\bar{x}$-chart. Similarly, an estimate of the average range is 8.74. The aver-

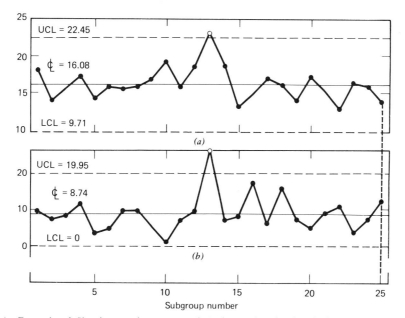

Figure 1 Example of Shewhart $\bar{x}$-charts (*a*) and *R*-charts (*b*), showing both out of control at the thirteenth subgroup.

age range is used with factors (discussed later) from Table 1 to give lower and upper control limits (LCL* and UCL*) for both charts. These limits on the $\bar{x}$-chart are $16.08 \pm (0.729)(8.74)$ or 9.71, 22.45, and on the range chart are zero and $2.282(8.74) = 19.94$.

The usual practice is to plot the charts one above the other as shown in Fig. 1, using a common horizontal (time) scale. Whether the points are connected is a matter of personal choice. If connecting lines are drawn, it is solely as a visual aid; no statistical interpretation involving the lines is appropriate. Points lying outside the control limits indicate potential assignable causes that should be investigated and eliminated. The out-of-control points, shown as open circles, on both the $\bar{x}$-chart and the R-chart for subgroup number 13 in Fig. 1 indicate that a significant effect was probably operating and that remedial action should be taken to prevent its recurrence.

In the following sections additional information on Shewhart charts is provided and brief discussions are given of acceptance control charts, cumulative sum charts*, moving-average charts, and multivariate charts. Finally, various uses of control charts are considered and references to pertinent literature are provided.

TYPES OF CONTROL CHARTS

Shewhart Charts

As indicated, the purpose of Shewhart control charts is to identify the time at which a significant deviation in a process occurs. Knowledge of the time of such an occurrence is the best evidence for searching out and eliminating a process disturbance.

In quality control work, characteristics that can be measured on a continuous scale are called *variables*. Those that are classified as being either present or absent are referred to as *attributes**. There are control charts for each type of characteristic.

CHARTS FOR VARIABLES. In addition to the range and arithmetic mean charts already discussed, constants have been determined (see Table 1) for setting up charts based on the standard deviation. If variation is normally distributed, the sample standard deviation provides a more precise estimate of the population standard deviation than does the range (for $n > 2$), but the difference is not great for sample sizes less than about a dozen; and the simplicity of calculation of the range provides an overriding practical advantage. In Table 1, central-line factors c_4 and d_2 are unbiasing factors which, when multiplied by the known or given standard deviation, yield the mean of the distribution. Factors for the control chart limits are multipliers of $\bar{R}$, $\bar{s}$, or σ_0 to give the distance from the central line to the 3σ limits for averages, or the upper and lower 3σ limits directly for ranges and standard deviations.

Several statistics alternative to the arithmetic mean* have been suggested to reduce the calculations for control charts. These include the median* [4, 8] and the midrange* [8]. It is, of course, also possible to plot individual observations (subgroups of size 1) on a chart bearing control limits based on previous knowledge of variability. However, such a chart has less power than an arithmetic mean chart, and is more vulnerable to the effects of deviations from normality which are usually assumed.

CONTROL LIMITS. An examination of Table 1 will show two sets of conditions associated with establishing control limits. These are "no standard given" and "standards given." The "standards" refer to values of the true process mean μ_0 standard deviation σ_0. When these are known, they can be used to calculate control limits. When they are not known it is the usual practice to estimate them, or their analogs, from some two dozen initial subgroups. The problem of setting control limits on small numbers of means and ranges which are themselves used to calculate the limits has been solved by Hillier [13] for subgroup samples of size 5.

Table 1 Factors for Limits on Shewhart Control Charts

Number of Observations in Sample	Chart for Averages			Chart for Standard Deviations					Chart for Ranges				
	Factors for Control Limits			Factor for Central Line	Factors for Control Limits				Factor for Central Line	Factors for Control Limits			
n	A	A_2	A_3	c_4	B_3	B_4	B_5	B_6	d_2	D_1	D_2	D_3	D_4
2	2.121	1.880	2.659	0.7979	0	3.267	0	2.606	1.128	0	3.686	0	3.267
3	1.732	1.023	1.954	0.8862	0	2.568	0	2.276	1.693	0	4.358	0	2.575
4	1.500	0.729	1.628	0.9213	0	2.266	0	2.088	2.059	0	4.698	0	2.282
5	1.342	0.577	1.427	0.9400	0	2.089	0	1.964	2.326	0	4.918	0	2.115
6	1.225	0.483	1.287	0.9515	0.030	1.970	0.029	1.874	2.534	0	5.078	0	2.004
7	1.134	0.419	1.182	0.9594	0.118	1.882	0.113	1.806	2.704	0.205	5.203	0.076	1.924
8	1.061	0.373	1.099	0.9650	0.185	1.815	0.179	1.751	2.847	0.387	5.307	0.136	1.864
9	1.000	0.337	1.032	0.9693	0.239	1.761	0.232	1.707	2.970	0.546	5.394	0.184	1.816
10	0.949	0.308	0.975	0.9727	0.284	1.716	0.276	1.669	3.078	0.687	5.469	0.223	1.777
11	0.905	0.285	0.927	0.9754	0.321	1.679	0.313	1.637	3.173	0.812	5.534	0.256	1.744
12	0.866	0.266	0.886	0.9776	0.354	1.646	0.346	1.610	3.258	0.924	5.592	0.284	1.716
13	0.832	0.249	0.850	0.9794	0.382	1.618	0.374	1.585	3.336	1.026	5.646	0.308	1.692
14	0.802	0.235	0.817	0.9810	0.406	1.594	0.399	1.563	3.407	1.121	5.693	0.329	1.671
15	0.775	0.223	0.789	0.9823	0.428	1.572	0.421	1.544	3.472	1.207	5.737	0.348	1.652

The Control Limit Formulas for Shewhart Control Charts are as follows:

Statistic	No Standard Given		Standard Given	
	Central Line	Control Limits	Central Line	Control Limits
Averages, $\bar{X}$	$\bar{X}$	$\bar{X} \pm A_2 \bar{R}$ or $\bar{X} \pm A_3 \bar{s}$	μ_0	$\mu_0 \pm A\sigma_0$
Ranges, R	$\bar{R}$	$D_3 \bar{R}, D_4 \bar{R}$	$d_2\sigma_0$	$D_1\sigma_0, D_2\sigma_0$
Standard deviations, s	$\bar{s}$	$B_3\bar{s}, B_4\bar{s}$	$c_4\sigma_0$	$B_5\sigma_0, B_6\sigma_0$

The limits resulting from the factors given in Table 1 are 3σ limits. That is, they lie at a distance from the expected value of the statistic equal to three times the standard error of that statistic. The multiplier 3 was chosen by Shewhart [24] partly on the theoretical basis that Chebyshev's inequality ensures that no less than 89% of the distribution will lie within the control limits. On the other hand, with normality, which is approached by virtue of the central limit theorem, 99.73% of the distribution will lie within the control limits for the $\bar{x}$-chart. Shewhart found from experience that the multiplier 3 gave a good economic balance between having the control chart fail to indicate a significant effect that was real (a type II error*) and having it indicate trouble when none existed (a type I error*). Reference to Table 1 shows that the lower 3σ limit for the range control chart does not become positive until the sample size reaches seven. For smaller sample sizes it is taken to be zero, as in the example shown in Fig. 1.

The effect of nonnormality on Shewhart control charts has been found to be minimal [1, 23]. The effect of serial correlation* has been found to be substantial [27].

It might be noted that for statistics not normally distributed, the chance of a point lying beyond the upper 3σ limit generally is not 0.00135. For the extreme case of subsamples of size 2, for example, the chance of finding a value of the range beyond $\bar{R} + 3\sigma_R$ (when control exists) is about 0.009. In practice, this variation in type I error is of little consequence.

The sensitivity of control charts can be enhanced by applying supplementary rules. One such rule establishes "warning limits" at $\pm 2\sigma$ with an out-of-control declaration being made if a run of two points falls between these warning limits and the "action limits" at $\pm 3\sigma$. The effects of choosing runs of various lengths have been studied [28]. Additionally, a run of 9 or 10 successive points on one side of the centerline of an $\bar{X}$-chart might be taken as indicating the presence of an assignable cause. The same rule for a range or standard deviation chart would

have to relate to the median rather than the centerline, which is the expected value. Eisenhart [7] gives median values for these statistics.

CHARTS FOR ATTRIBUTES

p-Chart. When items are classified only as being either good or defective, a control chart of the proportions found defective in subgroup samples can be set up based on the binomial model. This p-chart is analogous to an $\bar{x}$-chart, but because the variance of a proportion is $np(1 - p)$, control limits for this chart are derived from the mean value of p. These limits are taken as $\bar{p} \pm 3\sqrt{\bar{p}(1 - \bar{p})/n}$. When sample sizes are not constant, it is the usual practice to compute control limits using an average sample size, provided that the extreme sample sizes do not differ from $\bar{n}$ by more than a factor of 2. Otherwise, limits are computed for individual subgroups.

np-Chart. When samples for a p-chart are all the same size, it is simpler (and mathematically equivalent) to plot the numbers of defectives found in the samples. Such a chart is called an np-chart and has control limits set at $n\bar{p} \pm 3\sqrt{n\bar{p}(1 - \bar{p})}$.

c-Chart. An item can be defective because it contains one or more defects. If, instead of counting the number of defective items in a sample, one counts the number of defects in an item, a c-chart may be the appropriate control chart to use. The c-chart is based on the Poisson model. For this to be appropriate, it is necessary that the following two requirements be met (or at least approximated): (1) an indefinitely large number of defects can occur in any item, and (2) the probability of getting a defect in any given place is very small. The space within which defects can occur must be kept the same from item to item. For example, if blankets were being inspected, separate control charts would be required for blankets of different sizes. Control limits are set at $\bar{c} \pm 3\sqrt{\bar{c}}$, where $\bar{c}$ is the average number of defects.

u-Chart. The *u*-chart is a variation of the *c*-chart and is also based on the Poisson model. Each point is the average number of defects per unit in a sample of *n* units. Control limits are set at $\bar{u} \pm 3\sqrt{\bar{u}/n}$, where

$$u = \frac{\text{number of defects per sample}}{\text{number of units per sample}}.$$

For example, if *u* represents "defects per 5280 feet of tubing," a "unit" of the product is 5280 feet of tubing. Then for a number of samples of, say 26,400 feet each, *u'* would represent the standard number of defects per 5280 feet and the sample size *n* = 5. It is not required that *n* be an integer. If the samples each contained 7 kilometers of tubing, then *n* = 4.35.

Acceptance Control Chart

The Shewhart $\bar{x}$-chart has a single acceptable process level located at the grand mean $\bar{\bar{x}}$. It may be more reasonable to have a band of acceptable process levels. Such a situation could exist, for example, in a continuous process that was subject to a large number of acceptably small shifts in level. The acceptance control chart was devised by Freund [9] to deal with this by combining the ideas of acceptance sampling* with those of the $\bar{x}$-chart. This is reviewed by Woods [29].

CUMULATIVE SUM CONTROL CHART*. Although the cumulative sum chart can be applied to either attributes or variables, it is most commonly used with variables to detect changes in the level of a process. For such a purpose it is assumed that the variability is in statistical control. This technique is very sensitive to a shift in level by virtue of using cumulative deviations from the target value which are plotted in the usual time sequence. In the graphical procedure a mask with a cutout *V* is used to judge when a significant shift in level has occurred. The mask is placed so that the cutout *V* is horizontal with its apex pointed to the right and located a given number of units ahead of the latest plotted point. A significant shift is signaled when some previously plotted point falls beyond the boundaries of the mask. The relationship between the dimensions of the mask and the statistical properties of the chart, as well as general directions for setting up these charts, is given by Johnson and Leone [17]. A detailed description is given by van Dobben de Bruyn [26].

It is, of course, possible to apply the cumulative sum chart technique to estimates of variability, as illustrated by Johnson and Leone [17], and, for example, to estimates of parameters of the Weibull distribution* [16]. Cumulative sum tests are not robust with respect to departures from normality [19] or from independence [18]. Distributions having slightly longer tails than normal distributions can greatly reduce the average run length (ARL)* between out-of-control signals occurring at an acceptable quality level (AQL)*. Serial correlation when positive (negative) serves to decrease (increase) the ARL.

MOVING-AVERAGE CHARTS. As with the cumulative sum chart, moving-average charts continuously make use of past data. However, these charts either limit how far back the sample extends (the "moving-average" chart) or apply successively decreasing weights to the points averaged (the "geometric moving-average" chart). Such tests are described and compared with many other control procedures by Roberts [22].

MULTIVARIATE CHARTS. When it is required to control the levels of two or more characteristics simultaneously, the most exact procedure involves the use of Hotelling's T^2-distribution* [14, 15]. The observed averages and standard deviation estimates are combined into a statistic that has the T^2-distribution. A plot of this is monitored.

APPLICATIONS OF CONTROL CHARTS

In the manufacturing area, control charts are used to examine historical data, provide the basis for process capability studies, and serve as tools for monitoring the characteristics of processes to detect deviations from target values and from statistical control.

In research and engineering*, control charts provide a method for tracking the stability of measurement processes, and a way of examining the homogeneity of sets of data. It is a useful tool for data analysis and provides an easily understood way of displaying results.

In forecasting, a control-chart approach to examining errors (residuals) can provide not only information concerning correction but also insight into improved model formulation.

Several models have been developed for the economics associated with Shewhart control charts for attributes [3, 11, 20] and for variables [2, 5, 6, 10]. Similarly, the economics of cumulative sum charts have also been treated [12, 21, 25]. Cost functions to be minimized have included the costs associated with sampling, maintaining the charts, process adjustment, defective items, downtime, repair, false alarms, and so on.

Bibliography

The following books are recommended as general references. The first four contain collections of statistical quality control techniques well illustrated with case histories. A ranking of the level of presentation, beginning through intermediate, is Grant and Leavenworth, Ott, Burr, and Duncan. Juran's handbook serves primarily as a reference work in which a large number of quality control procedures are succinctly discussed and exemplified.

Burr, I. W. (1976). *Statistical Quality Control Methods*. Marcel Dekker, New York.

Duncan, A. J. (1974). *Quality Control and Industrial Statistics*, 4th ed. Richard D. Irwin, Homewood, Ill.

Grant, E. L. and Leavenworth, R. S. (1972). *Statistical Quality Control*, 4th ed. McGraw-Hill, New York.

Juran, J. M., ed. (1974). *Quality Control Handbook*, 3rd ed. McGraw-Hill, New York.

Ott, E. R. (1975). *Process Quality Control*, McGraw-Hill, New York.

References

[1] Burr, I. W. (1967). *Ind. Quality Control*, **23**, 563, 566–568.

[2] Chiu, W. K. (1974). *J. Quality Tech.*, **6**, 63–69.

[3] Chiu, W. K. (1975). *Technometrics*, **17**, 81–87.

[4] Clifford, P. C. (1959). *Ind. Quality Control*, **15**(11), 40–44.

[5] Duncan, A. J. (1956). *J. Amer. Statist. Ass.*, **51**, 228–242.

[6] Duncan, A. J. (1971). *J. Amer. Statist. Ass.*, **66**, 107–121.

[7] Eisenhart, C. (1949). *Ind. Quality Control*, **6**(1), 24–26.

[8] Ferrell, E. B. (1964). *Ind. Quality Control*, **20**(10), 22–25.

[9] Freund, R. A. (1957). *Ind. Quality Control*, **14**(4), 13–19, 22.

[10] Gibra, I. N. (1971). *Manag. Sci.*, **17**, 635–646.

[11] Gibra, I. N. (1978). *J. Quality Tech.*, **10**, 12–19.

[12] Goel, A. L., and Wu, S. M. (1973). *Manag. Sci.*, **19**, 1271–1282.

[13] Hillier, F. S. (1969). *J. Quality Tech.*, **1**, 17–26.

[14] Jackson, J. E. (1956). *Ind. Quality Control*, **12**(7), 4–8.

[15] Jackson, J. E. (1959). *Technometrics*, **1**, 359–377.

[16] Johnson, N. L. (1966). *Technometrics*, **8**, 481–491.

[17] Johnson, N. L. and Leone, F. C. (1962). *Ind. Quality Control*, **18**(12), 15–21; **19**(1), 29–36; **19**(2), 22–28 (1963).

[18] Johnson, R. A. and Bagshaw, M. (1974). *Technometrics*, **16**, 103–112.

[19] Kemp, K. W. (1967). *Technometrics*, **9**, 457–464.

[20] Ladany, S. P. (1973). *Manag. Sci.*, **19**, 763–772.

[21] Montgomery, D. C. and Klatt, P. J. (1972). *Manag. Sci.*, **19**, 76–89.

[22] Roberts, S. W. (1966). *Technometrics*, **8**, 411–430.

[23] Schilling, E. G. and Nelson, P. R. (1976). *J. Quality Tech.*, **8**, 183–188.

[24] Shewhart, W. A. (1931). *Economic Control of Quality of Manufactured Product*. D. Van Nostrand, New York.

[25] Taylor, H. M. (1968). *Technometrics*, **10**, 479–488.

[26] van Dobben de Bruyn, C. S. (1968). *Cumulative Sum Tests: Theory and Practice*. Hafner, New York.

[27] Vasilopoulos, A. V. and Stamboulis, A. P. (1978). *J. Quality Tech.*, **10**, 20–30.

[28] Weindling, J. I., Littauer, S. B., and de Oliveira, J. T. (1970). *J. Quality Tech.*, **2**, 79–85.

[29] Woods, R. F. (1976). *J. Quality Tech.*, **8**, 81–85.

(ACCEPTANCE SAMPLING
CUMULATIVE SUM (CUSUM)
  CONTROL CHARTS
DOUBLE SAMPLING
QUALITY CONTROL)

LLOYD S. NELSON

CONVERGENCE, STRONG *See* CONVERGENCE OF SEQUENCES OF DISTRIBUTIONS

CONVERGENCE, WEAK *See* CONVERGENCE OF SEQUENCES OF DISTRIBUTIONS

CONVERGENCE IN LAW *See* CONVERGENCE OF SEQUENCES OF DISTRIBUTIONS

CONVERGENCE IN *r*TH MEAN *See* CONVERGENCE OF SEQUENCES OF DISTRIBUTIONS

CONVERGENCE OF DISTRIBUTIONS
See CONVERGENCE OF SEQUENCES OF RANDOM VARIABLES

CONVERGENCE OF SEQUENCES OF RANDOM VARIABLES

This article is concerned with the convergence of infinite sequences of random variables and with the convergence of their distributions. The concept of convergence was originally introduced in connection with sequences of numbers and then extended to sequences of functions. A sequence of random variables $\{X_n\}$ is a sequence of measurable functions $\{X_n(\omega)\}$ defined on a space Ω. The most obvious approach would therefore be the requirement that the sequence $X_n = X_n(\omega)$ should converge for all $\omega \in \Omega$. It is easy to give an example: Let X be an arbitrary random variable and put $X_n = X + 1/n$; then the sequence $\{X_n\}$ converges to X as $n \to \infty$. This kind of conver- gence is not appropriate since it does not take probabilities into account. As an example we consider a sequence of coin tossings. Let S_n be the number of heads observed in n independent trials. Then X_n/n is the relative frequency of heads in n trials. Let p be the probability of heads in a single trial. Bernoulli's law of large numbers asserts that the difference between the relative frequency S_n/n and the probability p will be small if n is large. However, the sequence S_n/n of relative frequencies does not converge[1] to p. It is therefore not advisable to insist on the convergence of a sequence of random variables at all points of Ω, but it is desirable to introduce more suitable convergence concepts.

One can define such convergence concepts in several ways and one speaks about stochastic convergence to distinguish this concept from convergence in the sense of classical analysis. Stochastic convergence concepts take into account the existence of a probability measure defined on Ω. This can be done in a variety of ways; hence one can give a number of definitions for stochastic convergence. Since the functions that we consider are random variables, we assume that all random variables are given on the same probability space* $(\Omega, \mathfrak{A}, \mathrm{Pr})$.

The following result is crucial for making the definition of stochastic convergence possible.

Let $\{X_n\}$ be a sequence of random variables and let c be a real number. Then the set $\{\omega : X_n(\omega) = c\} \in \mathfrak{A}$, so that the probability $\mathrm{Pr}[\lim_{n \to \infty} X_n = c]$ is always defined.

DEFINITION OF THE MOST IMPORTANT MODES OF STOCHASTIC CONVERGENCE

1. (a) Let $\{X_n\}$ be a sequence of random variables. We say that the sequence $\{X_n\}$ converges almost certainly (almost surely, with probability 1) to zero if $\mathrm{Pr}[\lim_{n \to \infty} X_n = 0] = 1$.

    (b) Let $\{X_n\}$ be a sequence of random variables. We say that the sequence

$\{X_n\}$ converges almost certainly (almost surely, with probability 1) to a random variable X, if the sequence $\{X_n - X\}$ converges almost certainly to zero. We write that a.c. $\text{l.}_{n\to\infty} X_n = X$. (a.c.l.$\equiv$"almost certain limit.")

For instance, let $X_n = X_n(\omega)$ be a sequence of random variables defined on the interval $\Omega = [0, 1]$ such that

$$X_n = X_n(\omega) = \begin{cases} n & \text{if } 0 \leqslant \omega \leqslant 1/n \\ 0 & \text{if } 1/n < \omega \leqslant 1; \end{cases}$$

then a.c.l.$_{n\to\infty} X_n = 0$.

2. A sequence $\{X_n\}$ of random variables is said to converge in probability to a random variable X if

$$\lim_{n\to\infty} \Pr[|X_n - X| > \epsilon] = 0$$

for any $\epsilon > 0$. We then write

$$\plim_{n\to\infty} X_n = X.$$

Definitions 1 and 2 are applicable to any sequence of random variables. This is not true for the next two modes of stochastic convergence since these definitions require the existence of certain moments.

3. (a) Let $\{X_n\}$ be a sequence of random variables and let X be a random variable. Suppose that all X_n and also X have finite second-order moments. We say that the sequence $\{X_n\}$ converges in the (quadratic) mean to X if

$$\lim_{n\to\infty} \mathcal{E}(|X_n - X|^2) = 0.$$

We write then l.i.m.$_{n\to\infty} X_n = X$.

(b) Let $\{X_n\}$ be a sequence of random variables and let X be a random variable. Suppose that all X_n and also X have finite moments of order r and that

$$\lim_{n\to\infty} \mathcal{E}(|X_n - X|^r) = 0.$$

We say then that $\{X_n\}$ converges in

the rth mean to X and write

$$L_r - \lim_{n\to\infty} X_n = X.$$

For example, let $\{X_n\}$ be a sequence of random variables such that

$$\Pr[X_n = -1/n] = \Pr[X_n = 1/n] = \tfrac{1}{2}.$$

Then

$$\mathcal{E}(|X_n|^r) = \frac{1}{n^r}$$

and

$$L_r - \lim_{n\to\infty} X_n = 0.$$

4. Let $\{X_n\}$ be a sequence of random variables with distribution functions $\{F_n(x)\}$. Suppose that the $F_n(x)$ converge weakly to $F(x)$ [which means that they converge in all continuity points of the limiting function $F(x)$]. Let X be a random variable that has the distribution function $F(x)$; we say, then, that the sequence $\{X_n\}$ converges in law to X and write $\text{Lim}_{n\to\infty} X_n = X$.

Example. Let X be a random variable such that $\Pr[X = 0] = \Pr[X = 1] = \tfrac{1}{2}$ and let $Y = 1 - X$. Then X and Y have the same distribution, $F(x) = \tfrac{1}{2}[\epsilon(x) + \epsilon(x - 1)]$, while $|Y - X| = 1$ no matter which value X assumes. Define a sequence $\{X_n\}$ by $X_n = Y$; then the X_n converge in law to X.

RELATIONS AMONG THE VARIOUS CONVERGENCE CONCEPTS

In this section we list the connections between the various convergence concepts introduced above.

1. Almost certain convergence implies convergence in probability.

2. Convergence in the mean implies convergence in probability.

3. Convergence in probability implies convergence in law.

4. Let $r > s$; then L_r-convergence implies L_s-convergence.

The converse of these four statements is in general not true. To show this, one can construct examples; these may be found in Lukács [2].

We also note that there exist no implications between almost certain convergence and convergence in the mean (of any order). These modes are, however, compatible.

Under certain additional assumptions connection 3 can be equivalent to connection 1 [respectively to connection 2 or 3]. These additional assumptions are also discussed in Lukács [2].

BEHAVIOR OF CERTAIN CHARACTERISTICS OF CONVERGENT SEQUENCES

We treat here the question of whether the stochastic convergence of a sequence of random variables implies the convergence of their mean values or of other characteristics.

CONVERGENCE IN PROBABILITY.   The assumption that

$$\operatorname*{plim}_{n \to \infty} X_n = X$$

does not imply that

$$\lim_{n \to \infty} \mathscr{E}(X_n) = \mathscr{E}(X).$$

The operator $\mathscr{E}$ denotes the expected value.

CONVERGENCE IN THE rTH MEAN ($r \geqslant 2$ AN INTEGER).   Let $\{X_n\}$ be a sequence of random variables and suppose that $L_r - \lim_{n \to \infty} X_n = X$; then $\lim_{n \to \infty} \mathscr{E}(|X_n|^k) = \mathscr{E}(|X|^k)$ for $k \leqslant r$.

An important particular case is obtained for $r = 2$. If

$$\operatorname*{l.i.m.}_{n \to \infty} X_n = X,$$

then

$$\lim_{n \to \infty} \mathscr{E}(X_n) = \mathscr{E}(X)$$

and

$$\lim_{n \to \infty} \mathscr{E}(X_n^2) = \mathscr{E}(X^2).$$

Let $\{X_n\}$ and $\{Y_n\}$ be two sequences of random variables and suppose that

$$\operatorname*{l.i.m.}_{n \to \infty} X_n = X$$

while

$$\operatorname*{l.i.m.}_{n \to \infty} Y_n = Y;$$

then

$$\lim_{n \to \infty} \mathscr{E}(X_n Y_n) = \mathscr{E}(XY).$$

CONVERGENCE IN LAW.   The convergence of a sequence in law does not imply the convergence of the expectations of the sequence.

Remark.   The sequence of medians behaves differently. Let $\{X_n\}$ be a sequence of random variables and let m_n be the median of X_n. Suppose that $\operatorname{Lim}_{n \to \infty} X_n = X$; then every accumulation point of the m_n is a median of X.

One can also consider continuous functions of random variables. Let $\{X_n\}$ be a sequence of random variables such that

$$\operatorname*{plim}_{n \to \infty} X_n = X.$$

Let $g(y)$ be a continuous function; then $\operatorname{plim} g(X_n) = g(X)$.

Similarly, if $\{X_n\}$, $\{Y_n\}$ are two sequences of random variables and if

$$\operatorname*{plim}_{n \to \infty} X_n = X$$

$$\operatorname*{plim}_{n \to \infty} Y_n = Y$$

and if $g(x, y)$ is continuous in x and y, then

$$\operatorname*{plim}_{n \to \infty} g(X_n, Y_n) = g(X, Y).$$

CRITERION FOR STOCHASTIC CONVERGENCE

The similarity between the concepts of ordinary and of stochastic convergence becomes more noticeable by the fact that stochastic convergence admits Cauchy-type convergence criteria.

CONVERGENCE IN PROBABILITY.   Let $\{X_n\}$ be a sequence of random variables and let X be a random variable. The necessary and sufficient condition for the validity of the relation

$$\operatorname*{plim}_{n \to \infty} X_n = X$$

is that for every $\epsilon > 0$, $\delta > 0$, there exists an $N = N(\epsilon, \delta)$ such that $P(|X_n - X_m| > \epsilon) < \delta$ for $n, m > N$.

Let $\{X_n\}$ be a sequence of random variables such that

$$\operatorname*{plim}_{n \to \infty} X_n = X,$$

then there exists a subsequence $\{X_{n_j}\}$ of the $\{X_n\}$ such that

$$\operatorname*{a.c.l.}_{j \to \infty} X_{n_j} = X.$$

It can be shown that a squence $\{X_n\}$ converges in probability to a random variable X if and only if every subsequence $\{X_{n_k}\}$ of $\{X_n\}$ contains a subsequence that converges to X with probability 1.

CONVERGENCE IN THE rTH MEAN.   This mode of stochastic convergence admits the following convergence criterion. The necessary and sufficient condition for the convergence of a sequence $\{X_n\}$ in the rth mean is that for any $\epsilon > 0$ there exists an $N = N(\epsilon)$ such that

$$(|X_m - X_n|^r) \leqslant \epsilon \qquad \text{for } m, n \geqslant N.$$

A Cauchy-type convergence condition exists also for almost certain convergence. Its formulation is more complicated and is therefore not presented here. It can be found in Lukács [2].

OTHER MODES OF STOCHASTIC CONVERGENCE

We discussed the most important modes of stochastic convergence in the first section. In the present section we mention a few further modes of stochastic convergence.

A sequence $\{X_n\}$ of random variables is said to be completely convergent to zero if

$$\lim_{n \to \infty} \sum_{j=n}^{\infty} P\big[|X_j| > \epsilon\big] = 0$$

for any $\epsilon > 0$. Some authors use the term "almost completely convergent" instead of "completely convergent."

A sequence $\{X_n\}$ converges almost certainly uniformly to X if there exists a set A of probability measure zero such that the sequence $X_n = X_n(\omega)$ converges uniformly (in ω) to the random variable X on the complement of the set A.

We mention next conditions between these concepts.

Almost certain uniform convergence and almost certain convergence are equivalent.

Almost certain uniform convergence implies almost uniform convergence. The converse is not true.

Complete convergence implies almost certain convergence. The converse is not true.

Almost certain uniform convergence implies almost certain convergence. The converse is not true.

CONCLUDING REMARKS

Using the concepts of stochastic convergence, it is possible to study the convergence of infinite series of random variables. Such series can either be divergent or convergent to a limit that is either a proper or a degenerate random variable. As an example we mention the following result.

Let $\{X_n\}$ be a sequence of random variables and let $\{\epsilon_n\}$ be a sequence of positive numbers such that the following two conditions are satisfied:

(a)
$$\sum_{n=1}^{\infty} \epsilon_n < \infty$$

and

(b)
$$\sum_{n=1}^{\infty} P\big[|X_n| \geqslant \epsilon_n\big] < \infty.$$

Then the series

$$\sum_{n=1}^{\infty} X_n$$

is almost certainly convergent; i.e., the partial sums

$$S_n = \sum_{j=1}^{n} X_j$$

of the series converge to a (possible degenerate) random variable.

Another interesting development, which was made possible by the introduction of stochastic convergence concepts, is the study of random power series, that is, of series whose coefficients are random variables. Let $\{a_n\}$ be an infinite sequence of complex-valued random variables defined on the same probability space. Then

$$f(z, \omega) = \sum_{k=0}^{\infty} a_k(\omega) z^k \qquad (z \text{ complex})$$

is a random power series. The study of random power series was first suggested by E. Borel [1]. Random power series have a radius of convergence

$$r(\omega) = \left[\limsup_{n \to \infty} \sqrt[n]{|a_n(\omega)|} \right]^{-1},$$

the convergence of a random power series can be defined in any mode of stochastic convergence. If the coefficients of a random power series are independent random variables, then the series converges almost certainly if and only if it converges in probability.

Instead of sequences of random variables, one can also study random functions. This means that one considers instead of sequences $\{X_n\}$ of random variables depending on a discrete parameter n, random functions $X(\omega, t)$ which depend on a continuous parameter t, where t belongs to an arbitrary (finite or infinite) interval T.

Then one extends the idea of stochastic convergence to random functions. This makes it possible to introduce a "random calculus" by defining stochastic integration* and stochastic differentiation*.

The study of random power series and of random calculus is definitely of considerable mathematical interest and is discussed in detail in Lukács [2]. However, it is as yet of only limited interest to statisticians and is therefore not treated in this article.

Concepts of stochastic convergence are essential for the study of limit theorems. Limit theorems* constitute a very important chapter of probability theory. Convergence concepts also play a certain role in the theory of stochastic processes. Limit theorems, as well as stochastic processes, can be of great importance to mathematical statisticians.

For further results on stochastic convergence with detailed proofs, we refer to Lukács [2], where an extensive bibliography can be found.

NOTE. That is, the difference $|S_n/n - p|$ cannot be made arbitrarily small by taking n sufficiently large.

References

[1] Borel, E. (1896). *C. R. Acad. Sci. Paris*, **123**, 1051–1052.

[2] Lukács, E. (1975). *Stochastic Convergence*, 2nd ed. Academic Press, New York.

EUGENE LUKÁCS

CONVEXITY *See* GEOMETRY IN STATISTICS: CONVEXITY

CONVOLUTION

Given two independent random variables X and Y, the probability distribution of the sum $Z = (X + Y)$ is called the convolution of the distributions of X and Y. It is often denoted by $F_X * F_Y(\cdot)$, where $F_X(F_Y)$ is the cumulative distribution function* of $X(Y)$. Explicitly,

$$F_X * F_Y(z) = \int_{-\infty}^{\infty} F_X(z - y) dF_Y(y).$$

If the PDF* of Z exists, it is

$$f_Z(z) = \int_{-\infty}^{\infty} f_X(z - y) f_Y(y) dy.$$

For discrete random variables we have

$$p_Z(z) = \sum_j p_X(z - y) p_Y(y),$$

where $p_X(x) = P[X = x]$. The operation denoted by $*$ is associative and commutative. It has some properties like multiplication but there is not a complete analogy, since "division" is not always possible and, when possible, is not always unique. The class of all distributions is made into a semigroup* (but not a group) by the operation of convolution. The nth convolution of a distribution $F(x)$—the distribution of the sum of n independent random variables all having this distribution—is $F(x) * F(x) * \cdots * F(x)$, which is sometimes written as $F(x)^{n*}$ or $F(x)^{*n}$ (the latter is preferable).

Bibliography

Gnedenko, B. V. (1962). *The Theory of Probability*, Chelsea, New York.

Harris, B. (1966). *Theory of Probability*. Addison-Wesley, Reading, Mass.

CORNISH-FISHER AND EDGEWORTH EXPANSIONS

The representation of one distribution function in terms of another, as well as the representation of a quantile* (percentile) of one distribution in terms of the corresponding quantile (percentile) of another, is widely used as a technique for obtaining approximations* of distribution functions and percentage points. One of the most popular of such quantile representations was introduced by Cornish and Fisher [10] and later reformulated by Fisher and Cornish [18] and is referred to as the Cornish-Fisher expansion. Essentially, this expansion may be obtained from the distribution function representation introduced by Edgeworth [15], and now referred to as the Edgeworth expansion, for the purpose of calculating values of a distribution function F in terms of values of the normal distribution*. The Edgeworth expansion, although arrived at independently, is formally equivalent to the Gram-Charlier* expansion. From a practical point of view, however, it is distinct since it represents a rearrangement of that series to one that has better asymptotic convergence properties. A historically interesting commentary on this

property has been given by Edgeworth [16]. For a more current statement of this asymptotic behavior see Hill and Davis [25].

An introduction to Cornish-Fisher and Edgeworth expansions is presented in Kendall and Stuart [29] and Johnson and Kotz [28]. For survey papers see Wallace [44] and Bickel [4].

We now formulate these concepts for the univariate case. Later we discuss extensions of the Edgeworth expansion to the multivariate representation.

The Expansions

Let $F(\cdot; \lambda)$ and Φ be probability distribution functions with cumulants* k_i and α_i, respectively, where we assume for convenience that $k_1 - \alpha_1 = k_2 - \alpha_2 = 0$. In addition assume that $k_i - \alpha_i = 0(\lambda^{1-(i/2)})$ for $i = 3, 4, \dots$.

We begin with the formal expression of $F(\cdot; \lambda)$ in terms of Φ, where Φ is the standard unit normal distribution function. It can be shown that

$$F(x; \lambda) = \exp\left\{ \sum_{i=1}^{\infty} (k_i - \alpha_i)\left[(-D)^i / i!\right] \right\} \Phi(x),$$

$$(1)$$

where

$$D^i \Phi(x) = \frac{d^i}{dx^i} \Phi(x) = \Phi^i(x). \qquad (2)$$

Now (2) can be expressed as

$$\Phi^i(x) = (-1)^{i-1} H_{i-1}(x) Z(x) \text{ for } i \geqslant 1,$$

$$(3)$$

where $z(x) = \Phi'(x)$ and the H_i are Chebyshev-Hermite polynomials* defined by the recursive relations

$$H_0(x) = 1,$$

$$H_1(x) = x, \quad \text{and}$$

$$H_i(x) = x H_{i-1}(x) - (i-1) H_{i-2}(x)$$

$$\text{for } i \geqslant 2.$$

Since Φ is the unit normal distribution, we have $\alpha_i = 0$ for $i \geqslant 3$. Now, if we formally expand (1), collect terms of equal order in $\lambda^{-1/2}$ and arrange in ascending order, we obtain the formal *Edgeworth expansion*

$$F(x;\lambda) = \Phi(x) - \frac{k_3}{6}(x^2 - 1)Z(x)$$

$$- \frac{k_4}{24}(x^3 - 3x)Z(x)$$

$$- \frac{k_3^2}{72}(x^5 - 10x^3 + 15x)Z(x) + \dots$$

$$(4)$$

For more terms of this expansion see Draper and Tierney [13]. In many instances (see Wallace [44] and Draper and Tierney), expansion (4) is at least an asymptotic expansion and can be expressed in the form

$$F(x;\lambda) = F_n(x;\lambda) + 0(\lambda^{-(n+1)/2}) \quad (5)$$

where

$$F_n(x,\lambda) = \Phi(x) + \sum_{i=1}^{s} h_i(\lambda)\Phi^{(m_i)}(x) \quad (6)$$

and the h's are functions of λ determined by the cumulants.

Now let x_p and u_p be $100p$ percentiles (quantiles*) of $F(\ ;\lambda)$ and Φ respectively, that is

$$F(x_p;\lambda) = \Phi(u_p) = p \quad (7)$$

Using Taylor's formula, Φ can be expressed as

$$\Phi(u_p) = \Phi\big[x_p + (u_p - x_p)\big]$$

$$= \Phi(x_p) - \sum_{i=1}^{\infty} \frac{(x_p - u_p)^i}{i!}$$

$$\times H_{i-1}(x_p)Z(x_p) \quad (8)$$

Substituting (1) and (8) into (7), and after a considerable amount of algebraic manipulation and rearranging of terms into ascending orders of $\lambda^{-1/2}$ (as in the Edgeworth expansion), we obtain the *Cornish-Fisher expansion* of u_p as a function of x_p (see Johnson and Kotz [28]) given by

$$u_p = u(x_p) = x_p - \frac{1}{6}(x_p^2 - 1)k_3 - \frac{1}{24}(x_p^3 - 3x_p)k_4$$

$$+ \frac{1}{36}(4x_p^3 - 7x_p)k_3^2 - \frac{1}{120}(x_p^4 - 6x_p^2 + 3)k_5$$

$$+ \frac{1}{144}(11x_p^4 - 42x_p^2 + 15)k_3k_4$$

$$- \frac{1}{648}(69x_9^4 - 187x_p^2 + 52)k_3^3$$

$$- \frac{1}{720}(x_p^5 - 10x_p^3 + 15x_p)k_6$$

$$+ \frac{1}{360}(7x_p^5 - 48x_p^3 + 51x_p)k_3k_5$$

$$+ \frac{1}{384}(5x_p^5 - 32x_p^3 + 35x_p)k_4^2$$

$$- \frac{1}{864}(111x_p^5 - 547x_p^3 + 456x_p)k_3^2k_4$$

$$+ \frac{1}{7776}(948x_p^5 - 3628x_p^3 + 2473x_p)k_3^4 + \dots$$

$$(9)$$

and the inverse Cornish-Fisher expansion of x_p as a function of u_p given by

$$x_p = x(u_p) = u_p + \frac{1}{6}(u_p - 1)k_3$$

$$+ \frac{1}{24}(u_p^3 - 3u_p)k_4$$

$$- \frac{1}{36}(2u_p^3 - 5u_p)k_3^2$$

$$+ \frac{1}{120}(u_p^4 - 6u_p^2 + 3)k_5 - \frac{1}{24}(u_p^4 - 5u_p^2 + 2)k_3k_4$$

$$+ \frac{1}{324}(12u_p^4 - 53u_p^2 + 17)k_3^3$$

$$+ \frac{1}{720}(u_p^5 - 10u_p^3 + 15u_p)k_6$$

$$- \frac{1}{180}(2u_p^5 - 17u_p^3 + 21u_p)k_3k_5$$

$$- \frac{1}{384}(3u_p^5 - 24u_p^3 + 29u_p)k_4^2$$

$$+ \frac{1}{288}(14u_p^5 - 103u_p^3 + 107u_p)k_3^2k_4$$

$$- \frac{1}{7776}(252u_p^5 - 1688u_p^3 + 1511u_p)k_3^4 + \dots$$

$$(10)$$

Note that the expansions given by Cornish and Fisher [10] and Fisher and Cornish [18] do not require zero mean and unit variance, that is, the data need not be standardized. In (9) and (10), we assumed, however, that $k_1 - \alpha_1 = k_2 - \alpha_2 = 0$. That is, we assumed $F(\cdot;\lambda)$ had been standardized and hence in making use of the expansions here the data should first be standardized to zero mean and unit variance. Note also that the functional forms $u(\cdot)$ and $x(\cdot)$ do not depend upon the value of p. For more terms of expansions (9) or (10) see Draper and Tierney [13].

As is the case for the Edgeworth expansion, (9) and (10) are asymptotic expansions in many instances (see Draper and Tierney

[13]) and can be expressed in the form

$$u_p = u_{p,n} + O(\lambda^{-(n+1)/2}) \qquad (11)$$

with

$$u_{p,n} = x_p + \sum_{i=1}^{s} h_i(\lambda) P_i(x_p) \qquad (12)$$

and

$$x_p = x_{p,n} + O(\lambda^{-(n+1)/2}) \qquad (13)$$

with

$$x_{p,n} = u_p + \sum_{i=1}^{s} h_i(\lambda) W_i(u_p) \qquad (14)$$

where the P_i's and W_i's are polynomial functions independent of λ.

Applications of the Edgeworth Expansion

The primary application of the Edgeworth expansion has been in the approximation of a distribution function in terms of its cumulants and the normal distribution. An immediate consequence of such an application is in the study of the robustness* and power of tests done by obtaining approximations to the distributions of test statistics under fixed alternatives. General applications of the Edgeworth expansion have been discussed by Wallace [44] and a survey of Edgeworth expansions of linear rank statistics* (including the sign, Wilcoxon, and normal scores tests) and other statistics in non-parametric* problems was presented by Bickel [4]. Other applications consist of an important paper by Sargan [37] in which he established the validity of the Edgeworth expansion for the sampling distributions of quite general estimators and test statistics, including simultaneous equations estimators and t ratio test statistics. Developments of Edgeworth expansions for Kendall's rank correlation* coefficient were presented by Práškova-Vízková [36] and Albers [1]. The expansions for the distribution function of quadratic forms* were studied by Gideon and Gurland [21]. The Edgeworth expansion was also used by Assaf and Zirkle [2] to develop a method to approximate statistical characteristics of the response of nonlinear stochas-

tic systems* Phillips [35] obtained an Edgeworth expansion for the distribution of the least squares estimator of the autoregressive coefficient for the first order noncircular autoregressive model*. For an application of the Edgeworth expansion to sample data, we reference a study by Sethuraman and Tichler [39] on some micrometeorological observations.

Generalizations and Modifications of Edgeworth Expansions

Very little has been done concerning Edgeworth expansions with limiting distributions other than the normal. Bickel [4] references some work in nonparametric statistics using nonnormal limiting distributions. Edgeworth-type asymptotic expansions using the chi-square* as a limiting distribution have been presented by Fujikoshi [19], Han [24], and Tan and Wong [42].

Gray, Coberly, and Lewis [23] utilized the general Edgeworth expansion in such a way as to eliminate the requirement for knowing the cumulants without affecting the order of the error of the approximation. Their expansion makes use of the derivatives rather than the cumulants of the distribution functions. Takeuchi and Akahira [41] presented Edgeworth expansions for the case when moments do not necessarily exist but when the density can be approximated by rational functions.

Hipp [26] obtained theoretical results on Edgeworth expansions of integrals of smooth functions. The validity of the formal Edgeworth expansion for a class of statistics including all appropriately smooth functions of sample moments has been examined by Bhattacharya and Ghosh [3].

Multivariate Edgeworth Expansions

Chambers [8] developed the Edgeworth expansion for the distribution of multivariate statistics in a fairly general setting. He also presented conditions for validity along with computational algorithms. Theory and applications of multivariate Edgeworth expansions in nonparametric statistics are discussed by Bickel [4]. Further extensions of Chambers results have been obtained by Sargan [38] and Phillips [35]. These expansions have been applied frequently in the economic literature.

Applications of Cornish-Fisher Expansions

In applications only a finite number of terms of expansion (9) or (10) are used. The function $u(x)$ defined by (9) is applied in two basic ways. It is sometimes regarded as a normalizing transformation of the random variable X (see Bol'shev [5]). It can also be used in the same manner as the Edgeworth expansion, that is, by evaluating the normal distribution Φ at the value $u(x)$ one obtains an approximation for F at x. The most utilized of the Cornish-Fisher expansions is $x(u)$ defined by (10) and generally referred to as the inverse Cornish-Fisher expansion; it enables one to express the percentiles of fairly complicated distributions in terms of the percentiles of the normal distribution. An obvious application of such expansions is to incorporate them into calculator or computer programs, hence avoiding computer storage of tables (see Zar [46]).

Much study has been directed toward expressing the percentile of a distribution in terms of the corresponding normal percentile. These efforts include percentile expansions of the chi-square and t distributions* by Goldberg and Levine [22], Fisher and Cornish [18] and Zar [46]; rectangular, double-exponential, sech, and $sech^2$ distributions by Chand [9]; the skewness statistics D'Agostino and Tietjen [11] and Bowman and Shenton [6, 7]; k-sample Behrens-Fisher* distributions by Davis and Scott [12]; distribution of sums of differences of independent random variables by Howe [27] and Ghosh [20]; and for the distribution of the moment estimator of the shape parameter of the gamma distribution* by Dusenberry and Bowman [14]. Cornish-Fisher expansions were used by Waternaux [45] to study the asymptotic distribution of the characteristic roots of the sample covariance

matrix drawn from a nonnormal multivariate population and to study robustness properties. Venables [43] used Cornish-Fisher expansions to calculate the end points of a confidence interval* for the noncentrality parameter of the noncentral chi-square* and F distributions*.

Generalizations and Modifications of the Cornish-Fisher Expansions

Some work has been done with Cornish-Fisher expansions using a kernel or limiting distribution other than the normal. Finney [17] suggests the possible use of a limiting distribution in the Gamma family.

Hill and Davis [25] obtained formal expansions which generalized the Cornish-Fisher relations to an arbitrary analytic limiting distribution Φ. The formulas of Hill and Davis were used by Nagao [34] to obtain Cornish-Fisher type expansions for distributions of certain test statistics concerning covariance matrices where the limiting distribution used was the chi-square. Shenton and Bowman [40] obtained Cornish-Fisher expansions of the distribution of skewness* and kurtosis* using Johnson's S_U distribution* as a kernel or limiting distribution.

McCune and Gray [30] and McCune [31] utilized the generalized Cornish-Fisher expansions to eliminate the requirement for knowing the cumulants without affecting the order of the error of the approximations. Their expansions make use of the derivatives of the distribution functions instead. Further studies on these expansions have been conducted by McCune [32] in which the derivatives were estimated from sample data. McCune and Adams [33] investigated transformations on Cornish-Fisher expansions for improvement in accuracy.

Concluding Remarks

Although the theory of Cornish-Fisher and Edgeworth expansions has had a rich history, it is doubtful that these expansions will play a large role in the future of statistics. This is probably due to the digital computer and to expansions like those of this paper being complicated and not necessarily computationally efficient. As a result, many users find the alternatives more satisfying.

References

[1] Albers, W. (1978). *Ann. Statist.*, **6**, 923–925.

[2] Assaf, Sh. A. and Zirkle, L. D. (1976). *Int. J. Control.*, **23**, 477–492.

[3] Battacharya, R. N. and Ghosh, J. K. (1978). *Ann. Statist.*, **6**, 434–451. Settles a conjecture by Wallace [44]. The class of statistics considered includes all appropriately smooth functions of sample moments—a theoretical presentation.

[4] Bickel, P. J. (1974). *Ann. Statist.*, **2**, 1–20. Special invited paper, a review of results obtained since the general review paper by Wallace [44]. The bibliography contains 59 references.

[5] Bol'shev, L. W. (1959). *Theor. Probab. Appl.*, **IV**, 129–141.

[6] Bowman, K. O. and Shenton, L. R. (1973a). *Biometrika*, **60**, 155–167.

[7] Bowman, K. O. and Shenton, L. R. (1973b). *J. Amer. Statist. Ass.*, **68**, 998–1002.

[8] Chambers, J. M. (1967). *Biometrika*, **54**, 367–383. An extension of the Edgeworth expansion to the multivariate case.

[9] Chand, U. (1949). *J. Res. Natl. Bur. Stand.*, **43**, 79–80.

[10] Cornish, E. A. and Fisher, R. A. (1937). *Rev. Inst. Int. Statist.*, **4**, 307–320. The original work in which Cornish-Fisher expansions are introduced.

[11] D'Agostino, R. B. and Tietjen, G. L. (1973). *Biometrika*, **60**, 169–173.

[12] Davis, A. W. and Scott, A. J. (1973). *Sankhyā Ser. B*, **35**, 45–50. An approximation to the *k*-sample Behrens-Fisher distribution.

[13] Draper, N. R. and Tierney, D. E. (1973). *Commun. Statist.*, **1**, 495–524. Provides exact formulas for the first eight terms in the Edgeworth and Cornish-Fisher expansions with the standard normal as the limiting distribution.

[14] Dusenberry, W. E. and Bowman, K. O. (1977). *Commun. Statist. B*, **6**, 1–19.

[15] Edgeworth, F. Y. (1905). *Cambridge Philos. Trans.*, **20**, 36–66, 113–141. One of the early works giving a formal expansion with the normal as the limiting distribution.

[16] Edgeworth, F. Y. (1907). *J. R. Statist. Soc. Ser. A*, **70**, 102–106.

[17] Finney, D. J. (1963). *Technometrics*, **5**, 63–69. A generalization of the Cornish-Fisher expansion to include limiting distributions other than the standard normal.

[18] Fisher, R. A. and Cornish, E. A. (1960). *Technometrics*, **2**, 209–225. Exact expressions for more terms are presented along with examples of types of problems to which the expansions have been applied.

[19] Fujikoshi, Y. (1973). *Ann. Inst. Statist. Math.*, **25**, 423–437.

[20] Ghosh, B. K. (1975). *J. Amer. Statist. Ass.*, **70**, 350, 463–467.

[21] Gideon, R. A. and Gurland, J. (1976). *J. Amer. Statist. Ass.*, **71**, 353, 227–232.

[22] Goldberg, H. and Levine, H. (1946). *Ann. Math. Statist.*, **17**, 216–225.

[23] Gray, H. L., Coberly, W. A., and Lewis, T. O. (1975). *Ann. Statist.*, **3**, 741–746. An expression in terms of derivatives instead of cumulants.

[24] Han, C. P. (1975). *Ann. Inst. Statist. Math.*, **27**, 349–356.

[25] Hill, G. W. and Davis, A. W. (1968). *Ann. Math. Statist.*, **39**, 1264–1273. Techniques for obtaining generalized Cornish-Fisher expansions with arbitrary analytic limiting distribution functions are presented.

[26] Hipp, C. (1977). *Ann. Probab.* **5**, 1004–1011.

[27] Howe, W. G. (1974). *J. Amer. Statist. Ass.*, **69**, 347, 789–794.

[28] Johnson, N. L. and Kotz, S. (1970). *Distributions in Statistics. Continuous Univariate Distributions − 1*. Houghton Mifflin, Boston. See especially pages 15–22 and 33–35.

[29] Kendall, M. G. and Stuart, A. (1977). *The Advanced Theory of Statistics, Vol. I*, 4th ed. MacMillan, New York. See especially pages 169–179.

[30] McCune, E. D. and Gray, H. L. (1975). *Commun. Statist.*, **4**, 1043–1055.

[31] McCune, E. D. (1977). *Commun. Statist.*, **6**, 243–250. An expression in terms of derivatives instead of cumulants.

[32] McCune, E. D. (1978). *Proc. 1978 Amer. Statist. Ass. Statist. Comput. Sec.*, pp. 268–270.

[33] McCune, E. D. and Adams, J. E. (1978). *Texas J. Sci.*, **XXX**, 301–307.

[34] Nagao, H. (1973). *Ann. Statist.*, **1**, 700–709.

[35] Phillips, P. C. B. (1977). *Econometrica*, **45**, 1517–1534.

[36] Prášková-Vizková, Zuzana (1976). *Ann. Statist.* **4**, 597–606.

[37] Sargan, J. D. (1975). *Econometrica*, **43**, 327–346.

[38] Sargan, J. D. (1976). *Econometrica*, **44**, 421–448.

[39] Sethuraman, S. and Tichler, J. (1977). *J. Appl. Meteor.*, **16**, 455–461.

[40] Shenton, L. R. and Bowman, K. D. (1975). *J. Am. Statist. Assoc.*, **70**, 349, 220–228.

[41] Takeuchi, K. and Akahira, M. (1977). *Ann. Inst. Statist. Math.*, **29**, Part A, 397–406.

[42] Tan, W. Y. and Wong, S. P. (1977). *J. Am. Statist. Assoc.*, **72**, 360, 875–885.

[43] Venables, W. (1975). *J. R. Statist. Soc. B*, **37**, 406–412.

[44] Wallace, D. L. (1958). *Ann. Math. Statist.*, **29**, 635–654. A general review paper concerning theoretical properties of expansions of distribution functions and quantiles including the Edgeworth and Cornish-Fisher expansions. Usage of these expansions in theoretical statistics as well as in applications are discussed. The bibliography contains 76 references.

[45] Waternaux, C. M. (1976). *Biometrika*, **63**, 639–645.

[46] Zar, J. H. (1978). *Appl. Statist.*, **27**, 280–290.

E. D. McCune
H. L. Gray

CORRECTION FOR GROUPING

When data are grouped, information on individual values is lost. Considering the simple case of groups of fixed width h, with centers at points $x_0 + jh$ ($j = \ldots -1, 0, 1, \ldots$), this means that the information that there are n_j values in the interval $(x_0 + (j - \frac{1}{2})h, x_0 + (j + \frac{1}{2})h]$ replaces the n_j individual values in this interval.

Conventional calculation of moments from grouped data proceeds on the basis of assigning the value $x_0 + jh$ to all n_j values in the interval. Thus the rth crude sample moment* is calculated as

$$(\sum n_j)^{-1}(\sum n_j(x_0 + jh)^r).$$

The error so introduced cannot be determined in any particular case, but if certain assumptions are introduced, its average amount can be estimated and corrections introduced to reduce its effect. The commonest corrections, obtained by Sheppard [4], are based on the assumptions that the individual values are from a population with a continuous density function* which is smooth* and has high contact at each end of its range of variation. (The latter means that its derivatives* of all orders tend to zero at these points.) The corrections are:

1. To the second crude moment: *subtract* $\frac{1}{12}h^2$.
2. To the third crude moment: *subtract* $\frac{1}{4}h^2 \times$ first crude moment.
3. To the fourth crude moment: *subtract* $\frac{1}{2}h^2 \times$ second crude moment and *add* $\frac{7}{240}h^4$.

There is an interesting, partly historical discussion in Carver [1]. Further discussion in Dwyer [2] includes allowances for grouping arising from recording values "correct to the nearest unit (of some kind)" as well as from grouping of recorded values. *See also* Kendall and Stuart [3, Section 3, 18] and GROUPED DATA.

References

[1] Carver, H. C. (1936). *Ann. Math. Statist.*, **7**, 154–163.

[2] Dwyer, P. S. (1942). *Ann. Math. Statist.*, **13**, 138–155.

[3] Kendall, M. G. and Stuart, A. (1977). *The Advanced Theory of Statistics, Vol. 1* (4th edition). Macmillan, New York.

[4] Sheppard, W. F. (1898). *Proc. Lond. Math. Soc.*, **29**, 353–380.

CORRELATION

Correlation methods for determining the strength of the linear relationship between two or more variables are among the most widely applied statistical techniques. Theoretically, the concept of correlation has been a starting point or a building block in the development of a number of areas of statistical research. This article summarizes the theory of correlation for two variables, outlines the historical development of correlation methods, and describes some of the problems of interpreting correlation analysis. For discussions of correlation in situations involving more than two variables, the reader should consult the articles on general topics, such as CANONICAL ANALYSIS, FACTOR ANALYSIS, MULTIPLE REGRESSION, MULTIVARIATE ANALYSIS, PATH ANALYSIS, and TIME SERIES, as well as entries on specific forms of correlation, such as INTRACLASS CORRELATION, MULTIPLE CORRELATION, PARTIAL CORRELATION, and RANK CORRELATION.

BASIC THEORY OF CORRELATION

Correlation is a measure of the strength of the linear relationship between two random variables. Theoretically, the correlation between X and Y is defined as

$$\text{corr}(X, Y) = \frac{\text{cov}(X, Y)}{\left[\text{var}(X)\,\text{var}(Y)\right]^{1/2}},$$

where

$$\text{cov}(X, Y) = E\left[(X - E(X))(Y - E(Y))\right]$$

is the covariance* of X and Y, and where $\text{var}(X)$ and $\text{var}(Y)$ denote the variances* of X and Y, respectively. (It is assumed that the second moments* of X and Y are both finite.) If one or both of the variables are constant, the correlation is undefined.

The absolute value of the correlation is bounded by 1. The correlation is equal to $+1$ or -1 if and only if X and Y are linearly related with probability 1; in other words, there exist constants α and $\beta \neq 0$ such that

$$Y = \alpha + \beta X \qquad \text{a.e.,}$$

with $\text{corr}(X, Y) = -1$ for $\beta < 0$ and $\text{corr}(X, Y) = +1$ for $\beta > 0$. If X and Y are independent, then $\text{corr}(X, Y) = 0$, but the converse is not necessarily true. The addition of constants to X and Y does not alter the value of $\text{corr}(X, Y)$. Similarly, $\text{corr}(X, Y)$ is unchanged if X and Y are multiplied by constant factors with the same sign; if the signs of the factors differ, then the sign of $\text{corr}(X, Y)$ is reversed.

The term "correlation" also refers to the sample correlation of a set of N bivariate observations, $(x_1, y_1), \ldots, (x_N, y_N)$. In this context the sample correlation most often used is the Pearson product-moment correlation,

$$r = \frac{1}{N} \sum_{i=1}^{N} \left(\frac{x_i - \bar{x}}{s_x}\right)\left(\frac{y_i - \bar{y}}{s_y}\right);$$

$$\bar{x} = \frac{1}{N} \sum_{i=1}^{N} x_i, \qquad \bar{y} = \frac{1}{N} \sum_{i=1}^{N} y_i,$$

$$s_x^2 = \frac{1}{N} \sum_{i=1}^{N} (x_i - \bar{x})^2,$$

$$s_y^2 = \frac{1}{N} \sum_{i=1}^{N} (y_i - \bar{y})^2.$$

The sample correlation r satisfies the inequality $-1 \leqslant r \leqslant 1$, and equality is achieved if and only if the data are distributed along a perfect line in a scatter plot.

Numerically, r can be interpreted as the average product of the x- and y-coordinates for a scatter plot of the *standardized* data. If points in the two "positive coordinate product" quadrants predominate, then r is positive. If points in the two "negative coordinate product" quadrants predominate, then r is negative. For a scatter plot of the *original* data (where the y-data correspond to the ordinate), the product-moment correlation r measures clustering about a "standard deviation line" which passes through the point $(\bar{x}, \bar{y})$ and has slope s_y/s_x. (See the example in Fig. 1.)

If it is assumed that $(x_1, y_1), \ldots, (x_N, y_N)$ are N independent pairs of observations with the same bivariate distribution, r can be used to estimate the population correlation $\text{corr}(X, Y)$, which will be designated throughout this article by the symbol ρ. The expected value of r is

$$\rho + (1/N)\left[-\rho(1 - \rho^2)/2 \right.$$
$$+ 3\rho(\gamma_{40} + \gamma_{04})/8$$
$$\left. - (\gamma_{31} + \gamma_{13})/2 + \rho\gamma_{22}/4\right],$$

γ_{ij} denoting the cumulant* ratio

$$\kappa_{ij}\kappa_{20}^{-i/2}\kappa_{02}^{-j/2}.$$

Thus r is an approximately unbiased estimator* of ρ. (See Quensel [41], Gayen [19], and Chap. 32 of Johnson and Kotz [26] for de-

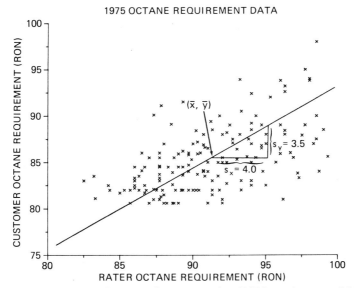

Figure 1   Scatter plot for the gasoline research octane number (RON) requirements of 169 vehicles, as determined by their owners (customer) and experts (rater). The sample correlation ($r = 0.556$) measures clustering about a "standard deviation line" which passes through $(\bar{x}, \bar{y}) = (91.3 \text{ RON}, 85.3 \text{ RON})$ and has slope $s_y/s_x = (3.5 \text{ RON})/(4.0 \text{ RON})$. (Data from Coordinating Research Council [5].)

tails concerning the distribution of r in non-normal distributions.) It can also be shown that r is a consistent estimator* of ρ.

Under the additional assumption that the data have a bivariate normal distribution*, r is a maximum likelihood estimator* of ρ. The distribution of r was derived by Fisher [12] via a geometrical argument, and its density function* can be expressed as

$$p(r) = \frac{(1 - \rho^2)^{(N-1)/2}(1 - r^2)^{(N-4)/2}}{\sqrt{\pi}\,\Gamma[(N-1)/2]\Gamma[(N-2)/2]}$$

$$\times \sum_{j=0}^{\infty} \frac{\{\Gamma[(N-1+j)/2]\}^2}{j!}(2\rho r)^j$$

for $-1 \leqslant r \leqslant 1$. See Chap. 16 of Kendall and Stuart [29] and Chap. 32 of Johnson and Kotz [26] for alternate forms of the density; the distribution has been tabulated by David [7]. Moments of the distribution of r can be expressed in terms of hypergeometric functions, and these representations yield asymptotic expansions for the moments. In particular (see Hotelling [24] and Ghosh

[20]),

$$E(r) = \rho \left[1 - \frac{1}{2N}(1 - \rho^2) + O(N^{-2}) \right]$$

and

$$\text{var}(r) = \frac{(1 - \rho^2)^2}{N - 1}\left(1 + \frac{11\rho^2}{2N} \right) + O(N^{-3}).$$

Olkin and Pratt [33] showed that an unbiased estimator of ρ is the hypergeometric function $rF(\frac{1}{2}, \frac{1}{2}, (N-2)/2, 1 - r^2)$, and they proposed the estimator

$$r \left[1 + \frac{1 - r^2}{2(N - 4)} \right]$$

for ρ. If $\rho = 0$, the statistic

$$(N - 2)^{1/2} r/(1 - r^2)^{1/2}$$

has a t distribution* with $N - 2$ degrees of freedom; this result is useful for testing the hypothesis* that $\rho = 0$.

For large samples, the asymptotic distribution of r (appropriately centered and scaled) is normal. However, the quantity

$$z = \tanh^{-1} r = \tfrac{1}{2} \log[(1 + r)/(1 - r)],$$

referred to as Fisher's variance-stabilizing transformation* of r, has a distribution which approaches normality much faster than that of r, particularly when $\rho \neq 0$ (see Fisher [13, 14]). Asymptotically, the expectation of z is $\tanh^{-1} \rho$, and the standard deviation is approximately $(N - 3)^{-1/2}$. This result is highly useful for constructing confidence limits* and tests of hypothesis* for ρ; see, for example, Chap. 10 of Snedecor and Cochran [43].

Improvements on the z transformation were derived by Hotelling [24], who developed a new approach for obtaining the moments of r and z. One of Hotelling's variance-stabilizing transformations · is defined as

$$z^* = z - (3z + r)/4(N - 1),$$

which has variance

$$\frac{1}{N - 1} + O(N^{-3}),$$

whereas z has variance

$$\frac{1}{N - 1} + \frac{4 - \rho^2}{2(N - 1)^2} + O(N^{-3}).$$

It can be shown that for small samples, z^* is more nearly normally distributed than z.

A different type of function of r was studied independently by Samiuddin [41a] and Kraemer [29a], both of whom assumed bivariate normality. Samiuddin showed that for $\rho \neq 0$ the statistic

$$t = \frac{\sqrt{N - 2}\,(r - \rho)}{\sqrt{(1 - r^2)(1 - \rho^2)}}$$

has, approximately, a t distribution* with $N - 2$ degrees of freedom, and he concluded that the approximation is sufficiently close that t may be used to construct two-tailed tests and confidence intervals for ρ. Kraemer showed that

$$T(r|\rho, N) = \frac{\sqrt{N - 2}\,(r - \rho')}{\sqrt{(1 - r^2)(1 - \rho'^2)}}$$

has approximately a t distribution with $N - 2$ degrees of freedom when $\rho' = \rho'(\rho, N)$ is a function of ρ and N such that $|\rho'(\rho, N)| \geqslant \rho$, $\rho'(\rho, N) = \rho$ when $\rho = 0, \pm 1$, $\rho'(\rho, N) = -\rho'(-\rho, N)$, and $\lim_{N \to \infty} \rho'(\rho, N) = \rho$. She suggested setting $\rho'(\rho, N)$ equal to the median of the distribution of r (given by David [7]) and derived and compared various approximations to the non-null distribution of r. Other approximations to the distribution of r and its transformations are summarized in Chap. 10 of Patel and Read [34a].

Various procedures are available for estimating the population correlation ρ when the assumption of bivariate normality is not appropriate for the data. In some cases it is possible to make a preliminary transformation of the data to bivariate normality; however, it can be difficult to assess the effect of the transformation on subsequent procedures involving a correlation computed from transformed data. An alternative approach is to apply a nonparametric correlation method to the ranks of the x_i and the y_i values.

The rank correlation* coefficients commonly used include Spearman's ρ, Kendall's τ, and the Fisher–Yates coefficient r_F. Whereas r measures the strength of the *linear* relationship between the x data and the y data, the rank correlation coefficients measure the strength of the *monotone* relationship (which may not be linear). Spearman's ρ (see Spearman [44]) is the Pearson product-moment correlation of the ranks of x_i (within the x data) and the ranks of y_i (within the y data). (N.B. The ranks can be assigned to the data arranged in increasing or decreasing order, provided the same type of ordering is used with the x data and the y data.) A formula for Spearman's ρ is

$$1 - 6\left(\sum_{i=1}^{N} d_i^2\right) \Big/ \{N(N^2 - 1)\},$$

where d_i is the difference between the rank of x_i and the rank of y_i. The value of Spearman's ρ is $+1$ if there is complete agreement between the two sets of ranks, and it is -1 if there is complete disagreement between the two sets of ranks. Kendall's τ (see Kendall [27]) is defined as

$$1 - 4Q/\{N(N-1)\},$$

where Q is the number of inversions of order in the ranks of the x_i values corresponding to the y_i values listed in their natural order. The value of Kendall's τ also ranges from -1 to $+1$. The Fisher–Yates coefficient r_F is the Pearson product-moment correlation of the normal scores of x_i and y_i. For discussions of inference procedures based on rank correlation, see Fieller et al. [11], Kendall [27], Owen [34], and Chap. 31 of Kendall and Stuart [28]. The interpretation of rank correlation measures is discussed by Kruskal [30].

Measures of correlation referred to as *biserial correlations** have been developed for situations in which the variable Y is measured quantitatively, whereas the variable X is dichotomized so that it takes only two values (typically, 0, if X is less than some threshold, and 1, otherwise). These measures are frequently applied in the social sciences, marketing, and public opinion research. Generally, it is assumed that the underlying distribution of X and Y is bivariate normal. Procedures for estimating the correlation between the original X and Y, as well as the correlation between the dichotomized X and Y, are discussed by Tate [45]. A *tetrachoric correlation** r_t is used to measure the correlation between X and Y, assuming bivariate normality, when both variables are dichotomized. Correlation procedures for bivariate normal data collapsed into categories are not well developed, although they were studied by Karl Pearson* and others in the early 1900s; see, for example, Pearson [38].

HISTORY OF CORRELATION

Francis Galton* (1822–1911), an English anthropologist and eugenist, is generally regarded as the founder of correlation analysis. Although a number of writers, including Carl Friedrich Gauss* (1777–1855), Auguste

Bravais (1811–1863), and Francis Edgeworth (1845–1926), dealt with multivariate normal distributions* as the basis for a theory of measurement error*, Galton was the first to recognize the need for a measure of correlation in bivariate data.

On Februrary 9, 1877, Galton presented a lecture at the Royal Institution of Great Britain entitled "Typical Laws of Heredity in Man" [15], which introduced the concepts of regression (termed "reversion") and correlation. Galton was interested in quantifying relationships between physical characteristics of parent and offspring for human beings, but because suitable anthropological data were difficult to obtain at that time, he discussed instead the analysis of measurements of the sizes of sweet pea seeds from mother and daughter plants. (See K. Pearson [40] for details of the early history of correlation.)

In subsequent work, having managed to collect anthropological data, Galton tabulated and smoothed the bivariate frequencies of characteristics such as height for parents and their adult children sampled from human populations. From his tables, Galton observed that the marginal distributions of the data were normal, the array (conditional) means occurred on a nearly straight line, the array variances were constant, and the contours of equal frequency were concentric ellipses. On September 10, 1885, Galton described his "normal correlation surface" in a lecture to the British Association for the Advancement of Science, and his paper on "Regression Towards Mediocrity in Hereditary Stature" was published in 1885 in the *Journal of the Anthropological Institute* [16]. The diagram that Galton used to discover the "normal correlation surface" is reproduced in Fig. 2. It is interesting to note that Galton cited an 1874 population density map of the city of Paris as an inspiration for his correlation surface (see Beniger and Robyn [1]).

In 1886, Galton referred the problem of formulating such surfaces analytically to Hamilton Dickson, a Cambridge mathematician. Dickson's solution was the equation

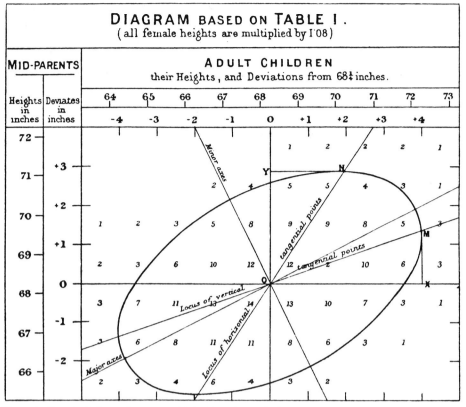

Figure 2 Francis Galton's diagram, published in 1885, of the bivariate frequency distribution of heights of parents and their adult children. From this "correlation surface," Galton observed various properties of the bivariate normal density surface.

now known as the density function for the bivariate normal distribution. This work appeared as an appendix to a paper by Galton, "Family Likeness in Stature," published in the *Royal Society Proceedings* in 1886 [17].

During these years Galton became interested in the problem of measuring the linear association between two observed variables independently of their location and scale of measurement. In this context, the term "correlation" (from "co-relation") first appeared in a paper entitled "Correlations and Their Measurement Chiefly from Anthropometric Data," which he presented to the Royal Society on December 5, 1888 [18]. Galton determined correlation as the slope of the least-squares regression line* for the data in standardized form. Karl Pearson [40] pointed out that Galton standardized his variables by subtracting their medians* and dividing by their semi-interquartile ranges.

The symbol r, which currently denotes the product-moment correlation, was used by Galton in 1888 (and in 1877) for correlations and presumably stood for "reversion." At this point, Galton realized that r is symmetric in both variables, that r is less than 1, that r measures the degree of linear relationship between variables, and that r is the slope of the fitted regression line when the x data and the y data have equal variability. More significantly, Galton was aware that correlation analysis could be applied to a wide variety of problems.

Galton's correlation method was soon adopted and modified by other researchers. In an 1892 paper on shrimp [47], Walter F. R. Weldon (1860–1906) introduced the use of means, rather than medians, for standardizing data before computing r; Weldon also published the first negative correlation coefficients. Having obtained similar values of correlations between 22 pairs of organs for five subspecies of shrimp, Weldon believed that these values of r should be the same for all members of the species. Later, Karl Pearson [40] wrote that "it was this very series of values which led to the investigation of the probable error of r."

Galton's work greatly influenced the career of Karl Pearson (1857–1936), who systematized the application of correlation analysis. In 1896, Pearson developed the version of the product-moment correlation r now used, and he stated the theory of correlation for three variables [37]. Subsequently, he and his associates at University College, London, pioneered the use of multiple regression* and correlation as generally applicable tools for analyzing data (see E. S. Pearson [35] and F. N. David [7]). Among those who contributed to the development of correlation analysis during this period were Edgeworth and G. Udny Yule* (1871–1951). In two basic papers, Yule [50, 51] dealt with the theory and application of partial correlation for more than two variables.

Methods for correlating ranked data were of interest early in the development of correlation analysis. Karl Pearson [39] wrote that Galton "dealt with the correlation of ranks before he even reached the correlation of variates." However, the first published work on rank correlation appeared in a 1904 psychological study of intelligence [44] by Charles E. Spearman (1863–1945).

In 1908, William S. Gosset* (1876–1937) discovered that the Pearson product-moment correlation r is symmetrically distributed about zero according to a Pearson type II distribution*, assuming a bivariate normal distribution with $\rho = 0$ for the data. He also showed that when $\rho \neq 0$, the distribution of r is skew and does not belong to the Pearson system (see Eisenhart [10]). The exact distribution of r was derived with a geometric argument in 1915 by Ronald A. Fisher* (1890–1962). This result, together with Fisher's z-transformation of r, made possible statistical tests of the significance of observed correlations.

A lesser known development in correlation analysis at that time was the introduction of path analysis* by Sewall Wright [48]. Path analysis determines whether the intercorrelations between a set of variables are consistent with assumptions concerning the causal relationships between the variables. Although path analysis cannot "prove" causation, it does allow the researcher to reject linear causal models that are not consistent with the correlations between variables. Thus unrealistic causal assumptions can be detected and rejected. During the last 20 years there has been a resurgence of interest in this area among social scientists; see Blalock [2] and Duncan [9].

The practical problem of interpreting correlation analysis became an issue with the increasing application of correlation methods in the early 1900s. Karl Pearson was aware that the high correlation between two variables may be due to their correlations with a third variable. However, this phenomenon was not generally recognized until it was pointed out by Yule [52] in a paper entitled "Why Do We Sometimes Get Nonsense-Correlation Between Time-Series?"

Since the 1930s, correlation analysis has played a major role in the development of various branches of modern statistics. For example, the concept of correlation was a starting point for the theory of canonical correlation*, pioneered in 1936 [23] by Harold Hotelling (1895–1973). Time-series analysis* is an example of an area in which specialized versions of correlation measures are applied.

Recently, attention has been drawn to the fact that inference procedures based on the use of the product-moment correlation r are heavily dependent on the assumption of bivariate normality. This has resulted in the development of robust measures of correlation and techniques for identifying outliers* in bivariate data. A discussion of robust correlation* is presented in Chap. 10 of Mosteller and Tukey [31]. New techniques for diagnosing and dealing with unusual data points utilize the influence curve* for correlation coefficients and the jackknife*. See Devlin et al. [8] and Hinkley [21, 28] for applications of these tools to correlation analysis.

INTERPRETATION OF CORRELATION

During the early development of correlation analysis, the major obstacles encountered by investigators were scarcity of appropriate data, lack of computational facilities, and difficulty in interpreting the values of correlation measures. One hundred years later, the first two obstacles are relatively minor. However, the problem of interpreting correlation persists, and it is further compounded by the plethora of correlations generated routinely by statistical computing packages.

Statistically, the value of a correlation measure such as r cannot be interpreted properly unless it is accompanied by a probability model for the chance variation in the data. The two models most commonly used are the bivariate normal distribution and the simple linear regression model.

When the observations, $(x_1, y_1), \ldots, (x_N, y_N)$, are assumed to have an underlying bivariate normal distribution, the sample correlation r estimates the population correlation ρ, which is a parameter* of the bivariate normal density function*:

$$f(x, y) = \left(2\pi\sigma_x\sigma_y\sqrt{1 - \rho^2}\right)^{-1}$$

$$\times \exp\left[-\frac{1}{2(1 - \rho^2)}\left(x'^2 - 2\rho x'y' + y'^2\right)\right],$$

where $x' = (x - \mu_x)/\sigma_x$, $y' = (y - \mu_y)/\sigma_y$, $-\infty < x < \infty$, and $-\infty < y < \infty$. Examples of bivariate normal densities with different values of ρ are illustrated in Fig. 3. An advantage of assuming bivariate normality, provided that it is a valid assumption, is that

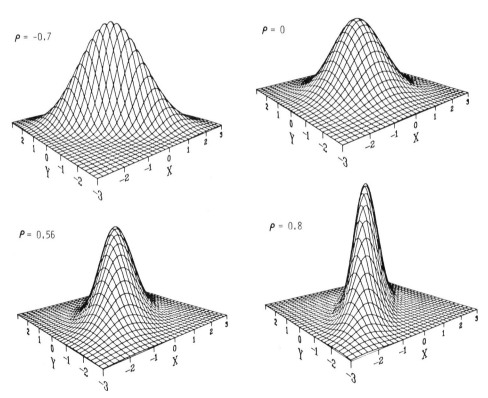

Figure 3   Examples of standard bivariate normal density surfaces with different values of correlation ρ.

one can compute confidence intervals and carry out hypothesis tests for ρ.

In the context of the simple linear regression model,

$$y_i = \beta_1 + \beta_2 x_i + \text{(random error)},$$

the slope of the least-squares fitted line is $\hat{\beta}_2 = rs_y/s_x$, and r^2 is the proportion of total variability in the y data which can be explained by the linear regression model:

$$r^2 = \left[\sum_{i=1}^{N} (\hat{y}_i - \bar{y})^2 \right] \bigg/ \left[\sum_{i=1}^{N} (y_i - \bar{y})^2 \right],$$

where $\hat{y}_i = \hat{\beta}_1 + \hat{\beta}_2 x_i$. This interpretation, which presumes that $\text{var}(Y|X)$ is a constant, can be expressed in terms of the theoretical correlation as

$$\rho^2 = 1 - \text{var}(Y|X)/\text{var}(Y).$$

More generally, the quantity

$$\eta^2 = 1 - E(Y - E(Y|X))^2/\text{var}(Y)$$

is referred to as the "correlation ratio of Y on X"; see Chap. 26 of Kendall and Stuart [28] for a discussion of correlation ratios* and linearity of regression.

Another interpretation of r^2 in simple linear regression follows from the fact that r equals the sample correlation of $(y_1, \widehat{y_1}), \ldots, (y_N, \widehat{Y_N})$. Thus the quantity r^2 is sometimes referred to as an " index of determination." (*See* COEFFICIENT OF DETERMINATION.) Some statistical computing packages currently in use fit nonlinear functions to bivariate data by transforming the y data with the inverse function and fitting a least squares line to the x data and the transformed y data. The "index of determination" generated is the squared correlation of the transformed y data and the fitted transformed y data. Consequently, it should be used with caution in judging the explanatory power of the nonlinear fit for the original data.

Researchers applying correlation techniques to data analysis are sometimes led to wrong or meaningless conclusions by incorrect interpretation of correlation measures. Although the interpretation of a correlation value such as $r = 0.98$ is usually apparent, it is not so easy to draw conclusions from lower values of r. For example, $r = 0.56$ does not mean "about half as much clustering as perfect linearity." Similarly, $r = 0.0$ does not mean that there is no relationship between the x data and the y data; indeed, a scatter plot might reveal a definite nonlinear relationship for data whose sample correlation is numerically near zero, as illustrated in Fig. 4.

Although it is valid to state that one set of data is more highly correlated than another set on the basis of their sample correlations, one should not attempt to interpret the numerical difference of two r values without doing a significance test.

A number of authors have discussed the problem of "false" or "spurious" correlation between two observed variables, which is induced artificially by their dependent relationships to a third variable. This problem can occur, for instance, if one computes correlations with scaled variables (such as rates and percentages) rather than the original variables. Neyman [32] illuminated this phenomenon with several examples, and he pointed out that the label "spurious" is more descriptive of the method of correlating scaled variables than of the correlations themselves. According to Neyman, a more appropriate method is to compute the partial correlation between the two original variables with the effect of the third variable removed.

It is also possible for two variables to have a high observed correlation purely by chance, particularly when dealing with small samples. If the bivariate normal distribution (or some other probability model) can be assumed, a test of significance can be used to decide whether the value of r represents a real population correlation.

Often, the problem of artificially high correlation is compounded by the confusion of correlation with causation. With some techniques, such as path analysis, one can determine whether the correlations between a number of variables are consistent with causal assumptions (see Simon [42]). How-

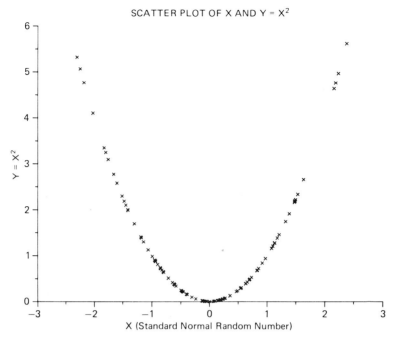

Figure 4 Scatter plot for 100 standard normally distributed random numbers and their squares. The plot reveals a strong nonlinear relationship, although the sample correlation is only $r = -0.06$, and the theoretical correlation is $\rho = 0$.

ever, it is not possible to *prove* causal relationship on the basis of observed correlation. For didactic examples of conceptual problems in the interpretation of correlation, see Campbell [4], Box et al. [3], Huff [25], Wallis and Roberts [46], and Neyman [32].

The numerical value of r does not, in itself, identify or assess the influence of unusual data points. Consequently, the computation of r should be accompanied by the use of diagnostic tools such as scatter plots. In situations where the data are not bivariate

1975 OCTANE NUMBER REQUIREMENT DATA
(169 CARS)

1978 OCTANE NUMBER REQUIREMENT DATA
(229 CARS)

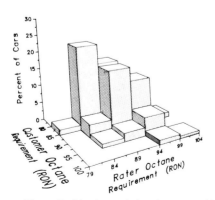

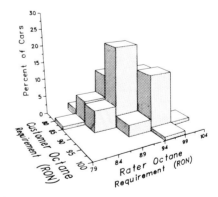

Figure 5 Bivariate relative frequency histograms for the gasoline research octane number (RON) requirements of 169 vehicles in 1975 and 229 vehicles in 1978, as determined by their owners (customer) and experts (rater). The two bivariate distributions are different, although the sample correlations for both sets of data coincidentally equal 0.556. (Data from Coordinating Research Council [5, 6].)

normally distributed, the value of r can provide very limited information about the actual distribution of the observations (see the examples illustrated in Fig. 5). Whenever possible, the interpretation of correlation should be based on a combination of careful assumptions, diagnostic analysis, and inference, in addition to the computed measure of correlation.

References

[1] Beniger, J. R. and Robyn, D. L. (1978). *Amer. Statist.*, **32**, 1–11.

[2] Blalock, H. W., Jr. (1964). *Causal Inferences in Nonexperimental Research.* University of North Carolina Press, Chapel Hill, N.C.

[3] Box, G. E. P., Hunter, W. G., and Hunter, J. S. (1978). *Statistics for Experimenters.* Wiley, New York.

[4] Campbell, S. K. (1974). *Flaws and Fallacies in Statistical Thinking.* Prentice-Hall, Englewood Cliffs, N.J.

[5] Coordinating Research Council (1977). 1975 CRC Customer/Rater Knock Perception Study. *Rep. No. 492*, Coordinating Research Council, Inc., Atlanta, Ga.

[6] Coordinating Research Council (1980). 1978 CRC Customer Versus Rater Octane Number Requirement Program. *Rep. No. 514*, Coordinating Research Council, Inc., Atlanta, Ga.

[7] David, F. N. (1938). *Tables of the Correlation Coefficient.* Cambridge University Press, Cambridge.

[8] Devlin, S., Gnanadesikan, R., and Kettenring, J. R. (1975). *Biometrika*, **62**, 531–45.

[9] Duncan, O. D. (1975). *Introduction to Structural Equation Models.* Academic Press, New York.

[10] Eisenhart, C. (1979). *Amer. Statist.*, **33**, 6–10.

[11] Fieller, E. C., Hartley, H. O., and Pearson, E. S. (1957). *Biometrika*, **44**, 470–481.

[12] Fisher, R. A. (1915). *Biometrika*, **10**, 507–521.

[13] Fisher, R. A. (1921). *Metron*, **1**, 3–32.

[14] Fisher, R. A. (1925). *Statistical Methods for Research Workers*, 14th ed. Hafner, New York.

[15] Galton, F. (1877). *Proc. R. Inst. G. Brit.*, **8**, 282–301.

[16] Galton, F. (1885). *J. Anthropol. Inst.*, **15**, 246–263.

[17] Galton, F. (1886). *Proc. R. Soc. Lond.*, **29**, 42–63. (An appendix by J. D. H. Dickson appears on pp. 63–72.)

[18] Galton, F. (1888). *Proc. R. Soc. Lond.*, **45**, 135–145.

[19] Gayen, A. K. (1951). *Biometrika*, **38**, 219–247.

[20] Ghosh, B. K. (1966). *Biometrika*, **53**, 258–262.

[21] Hinkley, D. V. (1977). *Biometrika*, **64**, 21–27.

[22] Hinkley, D. V. (1978). *Biometrika*, **65**, 13–21.

[23] Hotelling, H. (1936). *Biometrika*, **28**, 321–377.

[24] Hotelling, H. (1953). *J. R. Statist. Soc. B*, **15**, 193–224. (Discussion on pp. 225–232.)

[25] Huff, D. (1954). *How to Lie with Statistics.* W. W. Norton, New York.

[26] Johnson, N. L. and Kotz, S. (1970). *Continuous Univariate Distributions*, Vol. 2. Wiley, New York.

[27] Kendall, M. G. (1970). *Rank Correlation Methods*, 4th ed. Charles Griffin, London.

[28] Kendall, M. G. and Stuart, A. (1973). *The Advanced Theory of Statistics* Vol. 2: *Inference and Relationship*, 3rd ed. Hafner, New York.

[29] Kendall, M. G. and Stuart, A. (1977). *The Advanced Theory of Statistics*, Vol. 1: *Distribution Theory*, 4th ed. Macmillan, New York.

[29a] Kraemer, H. C. (1973). *J. Amer. Statist. Ass.*, **68**, 1004–1008.

[30] Kruskal, W. H. (1958). *J. Amer. Statist. Ass.*, **53**, 814–861.

[31] Mosteller, F. and Tukey, J. W. (1977). *Data Analysis and Regression.* Addison-Wesley, Reading, Mass.

[32] Neyman, J. (1952). *Lectures and Conferences on Mathematical Statistics and Probability*, 2nd ed. U.S. Department of Agriculture, Washington, D.C.

[33] Olkin, I. and Pratt, J. W. (1958). *Ann. Math. Statist.*, **29**, 201–211.

[34] Owen, D. B. (1962). *Handbook of Statistical Tables.* Addison-Wesley, Reading, Mass.

[34a] Patel, J. K. and Read, C. B. (1982). *Handbook of the Normal Distribution.* Dekker, New York.

[35] Pearson, E. S. (1967). *Biometrika*, **54**, 341–355. Reproduced in *Studies in the History of Statistics and Probability*, Vol. 1, E. S. Pearson and M. Kendall, eds. Charles Griffin, London.

[36] Pearson, E. S. (1968). *Biometrika*, **55**, 445–457. Reproduced in *Studies in the History of Statistics and Probability*, Vol. 1, E. S. Pearson and M. Kendall, eds. Charles Griffin, London.

[37] Pearson, K. (1896). *Philos. Trans. R. Soc. Lond. A*, **187**, 253–318.

[38] Pearson, K. (1913). *Biometrika*, **9**, 116–139.

[39] Pearson, K. (1914–1930). *The Life, Letters and Labours of Francis Galton.* Cambridge University Press, Cambridge. (In three volumes.)

[40] Pearson, K. (1920). *Biometrika*, **13**, 25–45. Reproduced in *Studies in the History of Statistics and Probability*, Vol. 1, E. S. Pearson and M. Kendall, eds. Charles Griffin, London.

[41] Quensel, C. E. (1938). *Lunds Univ. Arsskr. N. F. Afd.*, (2) **34**(4), 1–111.

[41a] Samiuddin, M. (1970). *Biometrika*, **57**, 461–464.

[42] Simon, H. A. (1971). In *Causal Models in the Social Sciences*, H. W. Blalock, Jr., ed. Aldine-Atherton, Chicago.

[43] Snedecor, G. W. and Cochran, W. G. (1980). *Statistical Methods*, 7th ed. Iowa State University Press, Ames, Iowa.

[44] Spearman, C. E. (1904). *Amer. J. Psychol.*, **15**, 201–293.

[45] Tate, R. F. (1955). *Biometrika*, **9**, 116–139.

[46] Wallis, W. A. and Roberts, H. V. (1956). *Statistics: A New Approach*. Free Press, Glencoe, Ill.

[47] Weldon, W. F. R. (1892). *Proc. R. Soc. Lond.*, **51**, 2–21.

[48] Wright, S. (1918). *Genetics*, **3**, 367–374.

[49] Wright, S. (1921). *J. Agric. Res.*, **20**, 557–585.

[50] Yule, G. U. (1897). *J. R. Statist. Soc.*, **60**, 1–44.

[51] Yule, G. U. (1907). *Proc. R. Soc. Lond. A*, **79**, 182–193.

[52] Yule, G. U. (1926). *J. R. Statist. Soc. A*, **89**, 1–64.

(BIVARIATE NORMAL DISTRIBUTION
DEPENDENCE, CONCEPTS OF
DEPENDENCE, MEASURES AND
   INDICES OF
FISHER'S z-TRANSFORMATION
LINEAR REGRESSION
REGRESSION
TRANSFORMATIONS)

R. N. RODRIGUEZ

CORRELATION RATIO

This is a name once given to the ratio

$$E^2 = \frac{\text{between-groups sum of squares}}{\text{total sum of squares}}$$

of the quantities occurring in analysis of variance* for a one-way classification. It is particularly relevant when the "groups" are, in fact, arrays defined by values of an independent variable. In such cases it provides a general measure of correlation not specifically related to linearity of regression. The squared correlation coefficient* (r^2) cannot exceed E^2 and ($E^2 - r^2$) may be regarded as the "nonlinear" part of the correlation. The statistic

$$(E^2 - r^2)/(1 - r^2)$$

is used in the standard analysis of variance test for departure from linearity of regression*.

(ASSOCIATION, MEASURES OF
CORRELATION)

CORRELOGRAM

A plot of the values of the *serial* or *lag correlation** against the lag, calculated from observed values in a time series. The term is also used to describe a similar plot for values obtained from a model.

The correlogram reflects the nature of variation in the time series, and has been used to assist in deciding which type of model (e.g., autoregressive or moving average*) is likely to be more appropriate to represent the generation of an observed time series. At present, correlograms are not used to so great an extent, as the methods of spectral analysis* have tended to supplant them.

(ARMA MODEL
AUTOREGRESSIVE MODEL
CORRELATION COEFFICIENT
FOURIER SERIES
MOVING-AVERAGE MODEL
SPECTRAL ANALYSIS
TIME SERIES)

CORRESPONDENCE ANALYSIS

Correspondence analysis is an algebraic technique analogous to principal components* analysis, but appropriate to categorical rather than to continuous variates. It originated simultaneously in the United States (Richardson in ref. 14; [6]) and in Britain [3, 13]. During the next 20 years it was rediscovered independently by numerous workers (for a list of these, see Guttman [7]), but never acquired a permanent name. Richardson (in Horst [14]) had called it the "method of reciprocal averages," and Hill [9] independently coined the name "reciprocal averaging." This name has subsequently

Table 1   Relationship of Hair Color and Eye Color

Eye Color	Hair Color					Total
	Fair	Red	Medium	Dark	Black	
Blue	326	38	241	110	3	718
Light	688	116	584	188	4	1580
Medium	343	84	909	412	26	1774
Dark	98	48	403	681	85	1315
Total	1455	286	2137	1391	118	5387

become popular with plant ecologists. In the meantime, however, Benzécri and his collaborators had already established the popularity of the method in France, calling it at first "analyse factorielle des correspondances" and later simply "analyse des correspondances" [1]. Benzécri's name for the method was soon adopted also by Hill [10], and is accordingly used here.

Correspondence analysis arises in several different statistical contexts. One of its commonest uses in practice is to scale a set of objects on the basis of the attributes that they possess (see the upcoming section on applications). However, the simplest derivation is due to Fisher [3], who considered data relating hair and eye color in a sample of schoolchildren from Caithness, Scotland, as shown in Table 1.

Clearly, darker hair is positively correlated with darker eyes. The aim of Fisher's analysis was to quantify the correlation. He did this by representing the categories of pigmentation by numbers ("scores"), and then looking for those scores for which hair and eye color were as highly correlated as possible. The scores in Table 2 maximize the correlation, and have been standardized so

Table 2   Score Functions Maximizing Correlation for Hair Color and Eye Color

Eye Color	Score	Hair Color	Score
Blue	−0.897	Fair	−1.219
Light	−0.987	Red	−0.523
Medium	0.075	Medium	−0.094
Dark	1.574	Dark	1.319
		Black	2.452

that the mean score for eye color is 0, and the mean square deviation is 1. The hair-color scores are similarly standardized.

The correlation coefficient* $r = 0.4464$. Fisher showed that these scores and the correlation coefficient can be derived as the maximal nontrivial solution of the correspondence analysis problem $C_0(\mathbf{A})$ of the contingency table $\mathbf{A}$, where $C_0(\mathbf{A})$ is defined as follows.

DEFINITION

The notation of Hill [10] is used. Let $\mathbf{A} = [a_{ij}](i = 1, \ldots, m; j = 1, \ldots, n)$ be a matrix of nonnegative real numbers, and let $\mathbf{x} = [x_i](i = 1, \ldots, m)$ and $\mathbf{y} = [y_j](j = 1, \ldots, n)$ be column vectors. Then a triple $(r, \mathbf{x}, \mathbf{y})$ is said to be a solution of the correspondence analysis problem $C_0(\mathbf{A})$ if

$$rx_i = \sum_j a_{ij} y_j / a_{i.} \qquad (i = 1, \ldots, m)$$

$$ry_j = \sum_i a_{ij} x_i / a_{.j} \qquad (j = 1, \ldots, n).$$

Note that the row scores x_i are proportional to weighted means* of the column scores y_j, and that the column scores are proportional to weighted means* of the row scores x_i. The number r is the correlation between the row and column scores, in the sense of Fisher's problem considered above, and also in the sense of the equivalent canonical correlation problem considered below.

BASIC PROPERTIES

Solutions of correspondence analysis can be obtained as the eigenvalues* and eigenvectors* of a square symmetric matrix. Defining $\mathbf{R} = \mathbf{diag}(a_{i\cdot})$, $\mathbf{C} = \mathbf{diag}(a_{\cdot j})$, $\mathbf{R}^{1/2} = \mathbf{diag}(a_{i\cdot}^{1/2})$, etc., the condition for $\mathbf{x}$ to be a solution is

$$r^2(\mathbf{R}^{1/2}\mathbf{x}) = (\mathbf{R}^{-1/2}\mathbf{A}\mathbf{C}^{-1/2})(\mathbf{R}^{-1/2}\mathbf{A}\mathbf{C}^{-1/2})$$
$$\times (\mathbf{R}^{1/2}\mathbf{x}). \qquad (1)$$

From this it follows that differing solutions of the eigenvalue problem are orthogonal in the sense that

$$(\mathbf{R}^{1/2}\mathbf{x})'(\mathbf{R}^{1/2}\mathbf{z}) = \mathbf{x}'\mathbf{R}\mathbf{z} = \sum\sum a_{ij}x_i z_i = 0.$$
$$(2)$$

The definition of correspondence analysis is completely symmetrical between $\mathbf{x}$ and $\mathbf{y}$, so an analogous result holds for the column scores y_j. For the solution defined in (1), the eigenvalue is r^2.

The maximal eigenvalue of C_0 is always 1, corresponding to the solution $(1, \mathbf{1}_m, \mathbf{1}_n)$, where $\mathbf{1}_m, \mathbf{1}_n$ are m- and n-vectors of 1's. This is referred to as the trivial solution of C_0. The condition of orthogonality to the trivial solution is

$$\sum_i a_{i\cdot}x_i = \sum_i a_{\cdot j}y_j = 0.$$

When $\mathbf{A}$ is a contingency table, this is simply the condition for $\mathbf{x}$ and $\mathbf{y}$ to be centered to zero mean.

If the matrix $\mathbf{A}$ is irreducible, there are no other solutions corresponding to the eigenvalue 1. If the matrix $\mathbf{A}$ is not irreducible, each partition of the matrix into noncommunicating blocks

$$\mathbf{A} = \begin{bmatrix} \mathbf{A}_1 & \mathbf{0} \\ \mathbf{0} & \mathbf{A}_2 \end{bmatrix}$$

corresponds to a nontrivial solution $(1, \mathbf{x}, \mathbf{y})$.

STANDARDIZATION OF THE SOLUTIONS

The approach to correspondence analysis developed by Benzécri [1] is essentially geometric, and consists of looking for directions of maximum inertia for a cloud of points in a metric space. (For an English-language exposition, see Greenacre and Degos [5].)

The "chi-square metric" [6] is used. In this case the appropriate standardization is to demand that the length of an eigenvector is proportional to the square root of the corresponding eigenvalue. Referring to (1), this condition is

$$\|\mathbf{R}^{1/2}\mathbf{x}\|^2 = \sum a_{i\cdot}x_i^2 = \sum\sum a_{ij}x_i^2 = r^2.$$

However, there is no necessity to use this standardization, which is not appropriate to the contingency-table* rationale considered next.

RELATION TO CONTINGENCY-TABLE CHI-SQUARE

In the case where $\mathbf{A}$ is a contingency table, the analysis $C_0(\mathbf{A})$ defines a factorial partition of the contingency-table* chi-square*. Consider the problem of minimizing

$$X^2 = \sum\sum \left\{ (a_{ij} - u_i v_j)^2 / (\bar{a}_{i\cdot}\bar{a}_{\cdot j}/\bar{a}_{\cdot\cdot}) \right\}.$$

Writing

$$u_i = a_{i\cdot}x_i; \qquad v_j = a_{\cdot j}y_j,$$

the stationary values of X^2 are found when (in the obvious notation) $(r, \mathbf{x}, \mathbf{y})$ is a solution of $C_0(\mathbf{A})$. The maximal (trivial) solution corresponds to fitting the expected values. Subsequent solutions partition the chi-square. Denoting the various solutions by a superscript $s(s = 1, \ldots, k)$, the contingency table is factored by correspondence analysis, according to the equation

$$a_{ij} = \sum_{s=1}^{k} r^{(s)}a_{i\cdot}x_i^{(s)}a_{\cdot j}y_j^{(s)}.$$

Here the eigenvectors $\mathbf{x}^{(s)}$, $\mathbf{y}^{(s)}$ are standardized to unit length, in the sense that

$$\sum\sum a_{ij}x_i^{(s)2} = \sum\sum a_{ij}y_j^{(s)2} = 1.$$

RELATION TO CANONICAL CORRELATION ANALYSIS

Correspondence analysis can be viewed as a special case of canonical correlation analysis (*see* CANONICAL ANALYSIS). Consider, for example, the matrix

$$A = \begin{bmatrix} 2 & 0 & 1 \\ 0 & 1 & 2 \end{bmatrix}.$$

Assuming that A is a contingency table, it will have been assembled from a raw data matrix

$$T(A) = \begin{bmatrix} 1 & 1 & 1 & 0 & 0 & 0 \\ 0 & 0 & 0 & 1 & 1 & 1 \\ 1 & 1 & 0 & 0 & 0 & 0 \\ 0 & 0 & 0 & 1 & 0 & 0 \\ 0 & 0 & 1 & 0 & 1 & 1 \end{bmatrix}.$$

The columns of $T(A)$ represent the individual observations from which the contingency table was assembled. The first two rows represent the two rows of A, and the last three rows represent the three columns of A.

In general, $T(A)$ is defined only for matrices A with integer elements. $T(A)$ has $(m + n)$ rows and $a_{..}$ columns. All its elements are 0 or 1. The matrix element $[T(A)]_{k,l}$ is nonzero if and only if there exist numbers r, s for which either $k = r$ or $k = m + s$, and

$$\sum_{i=1}^{r-1} a_{i.} + \sum_{j=1}^{s-1} a_{rj} < l \leqslant \sum_{i=1}^{r-1} a_{i.} + \sum_{j=1}^{s} a_{rj}.$$

It is easily shown (see Hill [10]) that the correspondence analysis problem $C_0(A)$ is essentially identical to a canonical correlation analysis problem, in which it is sought to maximize the correlation between a linear combination of those rows of $T(A)$ that correspond to the rows of A and a linear combination of the rows of $T(A)$ corresponding to the columns of A. If (r, x, y) is a solution of $C_0(A)$, then r is the correlation coefficient derived from the canonical correlation*.

RELATION TO PRINCIPAL COMPONENTS ANALYSIS*

A curious feature of correspondence analysis is that the solutions of $C_0(T(A))$ are the same as those of $C_0(A)$, in the sense that if (r, u, v) is a solution of $C_0(T(A))$, and if the

vector u is partitioned into two parts,

$$u = \begin{bmatrix} x \\ y \end{bmatrix},$$

corresponding to the rows and columns of A, then $(2r^2 - 1, x, y)$ is a solution of $C_0(A)$. Using the notation of Hill [10], $C_0(T(A))$ is referred to as $C_1(A)$.

The advantage of considering the problem $C_1(A)$ rather than $C_0(A)$ is that it can be applied to general p-way contingency tables, and allows correspondence analysis to be related to principal components* analysis. Consider the principal components problem for the matrix $B = [b_{ij}]$ ($i = 1, \ldots, m$; $j = 1, \ldots, n$). Rows represent attributes, columns individuals. Without loss of generality, assume that the attributes are centered to zero mean. Then principal components analysis of the correlation matrix of the attributes is equivalent to finding loadings c_i such that

$$\sum_i \sum_{i'} \sum_j (c_i b_{ij} - c_{i'} b_{i'j})^2$$

is minimized [2].

This approach to principal components analysis can be generalized by allowing the rows of B to be linear combinations of the rows of another matrix, say $T(A)$. If each row of B is constrained to be an arbitrary linear combination of a group of rows in $T(A)$ corresponding to an attribute, then the generalized principal components problem for B turns out to be equivalent to $C_1(A)$ [10]. In the case where a raw data matrix has some categorical variates and some continuous variates, it is possible to combine a principal components analysis of the continuous variates with a correspondence analysis of the categorical variates [12]. In symbols, let x be a canonical variate derived in this manner, and let $\hat{x}_{(i)}$ be the least-squares estimate of x provided by a linear combination of the ith group of rows. Then

$$x = \lambda \sum_i \hat{x}_{(i)},$$

where λ is a scalar multiplier. This approach is pursued in great generality and with a very elegant notation by Healy and Goldstein [8].

SAMPLING THEORY

Historically, the two most influential papers were those of Guttman [6] and Fisher [3]. Both attempted to develop a sampling theory. Their work appears superficially different because Fisher considered the problem C_0 and Guttman the problem C_1. Recent users of correspondence analysis have generally regarded it as an exploratory technique for which a sampling theory is not necessary. The topic has therefore received little attention.

The analytical form of the solutions for several bivariate distributions has been examined by Naouri [15], but the question of how these might be approximated by solutions for finite samples has scarcely been considered. This problem is not too intractable, however, as the row scores are weighted means of the column scores, and vice versa, so that standard errors can be calculated, at least to a first approximation.

APPLICATIONS

Correspondence analysis has two main types of application: (a) as a method of scaling, and (b) to provide a basis for automatic classification. For both these purposes its advantages are its simplicity and the low computational cost of solving large problems.

As a method of scaling, correspondence analysis compares favorably with principal components analysis when applied to problems in plant ecology (Gauch et al. [4]). In general, the maximal nontrivial eigenvector provides a good ordering of data in one dimension, but the later eigenvectors can be hard to interpret. They are constrained to be orthogonal to the maximal eigenvector in the sense of (2), but this is far from guaranteeing their independence. They may or may not be approximately independent, and

there is no way of telling in advance whether they will be. A less important difficulty is that even the maximal canonical variate is distorted, in the sense that points near the ends of the range of variation are placed closer together than points similarly different but near the middle (see Hill [10, Figs. 1, 2]).

The other major use of the method is as a basis for automatic classification [11]. The reason why it is suitable for this purpose has been indicated above: if the data can be partitioned into noncommunicating blocks, correspondence analysis will detect them. Indeed, the method is potentially a very powerful way of detecting reducibility and near-reducibility in matrices.

As an example of the use of correspondence analysis both to scale and to classify, consider the incidence structure in Table 3. This is a common type of structure in many applications, with two groups of attributes (columns 1–5 and columns 6–9) that rarely occur together, and a group of attributes (columns 10–12) whose occurrence is unrelated to the others. Two scalings of the rows and columns are indicated along the margins. Both scalings were obtained as the maximal nontrivial solution of a correspondence analysis problem, and both were standardized according to the formula used by Fisher,

$$\sum \sum a_{ij} x_i^2 = \sum \sum a_{ij} y_j^2 = \sum \sum a_{ij}.$$

The weighted analysis was obtained merely by multiplying the columns of the matrix by weights $(n_+ - n_-)^2/(n_+ + n_-)^2$, where n_+ is the number of occurrences of the attribute in individuals with a positive score in the unweighted analysis, and n_- is defined similarly. The effect of this attribute weighting is to polarize the scaling of the individuals so that it approximates a classification. The correlation coefficients for the two scalings are $r = 0.755$ and $r = 0.981$ respectively.

The intermediate nature of row 6 in Table 3 is indicated clearly by its score in the unweighted scaling. The weighted scaling shows that if the individuals must be dichotomized, then row 6 belongs with rows

Table 3 Incidence Matrix, Showing Occurrence of a Set of Attributes (Columns) in a Set of Individuals (Rows)

												Unweighted	Weighted
1	1	1	.	.	.	.	.	.	.	1	1	−1.10	−1.18
1	.	1	1	.	.	.	.	.	1	.		−1.22	−1.12
1	1	1	1	.	.	.	.	.	1	1	1	−0.95	−1.13
.	1	1	.	1	.	.	.	.	1	1	.	−0.84	−1.12
.	1	.	1	1	.	.	.	.	1	1		−0.78	−1.02
.	.	.	1	1	1	1	.	.	1	1	1	0.13	0.65
.	.	.	.	.	1	1	1	1	1	.	1	1.24	0.92
.	.	.	.	.	1	1	.	1	1	1	1	0.86	0.91
.	.	.	.	.	.	1	1	1	.	1	.	1.52	0.94
.	.	.	.	.	.	1	1	1	.	1	1	1.10	0.91
−1.4	−1.2	−1.4	−0.9	−0.7	1.1	1.3	1.7	1.6	0.1	−0.0	−0.1	Unweighted	
−1.2	−1.1	−1.2	−0.7	−0.5	0.9	0.9	0.9	0.9	0.0	−0.1	−0.1	Weighted	
1.0	1.0	1.0	0.25	0.11	1.0	1.0	1.0	1.0	0.02	0.00	0.00	Weights	

7–10 rather than rows 1–5. The weighting used to obtain the polarized scaling is designed to diminish the influence of attributes whose number of occurrences in individuals with a positive score is close to the number in individuals with a negative score.

COMPUTATION

Small and medium-sized problems can be solved using standard eigenanalysis programs for square symmetric matrices. Large problems with sparse data matrices are better tackled by direct iteration techniques, for which square matrices of cross-products are never formed. The following "two-way iteration algorithm" [10] is fast enough for use in some applications.

Let $\mathbf{x} = [x_i]$ be a trial m-vector of row scores. Form a new trial vector $\mathbf{x}' = [x_i']$, according to the equations

$$y_j = \sum_i a_{ij} x_i / a_{.j} \qquad (j = 1, \ldots, n)$$

$$x_i' = \sum_j a_{ij} y_j / a_{i.} \qquad (i = 1, \ldots, m).$$

If $\mathbf{x}, \mathbf{x}'$ are orthogonalized with respect to the first $k - 1$ eigenvectors and scaled to unit length, the iteration converges to the kth eigenvector. For very large problems this algorithm can be greatly speeded up by the following device. Start with a trial vector and make a few iterations of the two-way algorithm. Approximate the linear transformation of the two-way iteration in a low-dimensional space spanned by the trial vector and its immediate successors. Solve the eigenvector problem for the low-dimensional space, and use this solution as the next trial vector.

OUTLOOK

Correspondence analysis is not perfect for scaling (see the preceding applications section) nor does it directly provide a classification. On the other hand, it is very rapid and robust, so that it provides an excellent starting point for further refinement. In the future it will probably be used more often to obtain a first approximation to a desired structure (whether a scaling or a classification) than as the final answer.

References

[1] Benzécri, J.-P. (1973). *Analyse des données*, Tome 2: *Analyse des correspondances*. Dunod, Paris. (In French. Extended exposition of the geometric approach; many examples.)

[2] Edgerton, H. A. and Kolbe, L. E. (1936). *Psychometrika*, **1**, 183–187.

[3] Fisher, R. A. (1940). *Ann. Eugen. (Lond.)*, **10**, 422–429. (Succinct introduction, with estimate of precision of the scores; bivariate case only.)

[4] Gauch, H. G., Whittaker, R. H., and Wentworth, T. R. (1977). *J. Ecol.*, **65**, 157–174. (Empirical comparison with principal components analysis in applications to plant ecology.)

[5] Greenacre, M. J. and Degos, L. (1977). *Amer. J. Hum. Gen.*, **29**, 60–75. (English-language exposition of geometric approach.)

[6] Guttman, L. (1941). In *The Prediction of Personal Adjustment*, P. Horst, ed. Social Science Research Council, New York, pp. 319–348. (Application to multivariate categorical data; good explanation of rationale.)

[7] Guttman, L. (1959). *Sankhyā*, **21**, 257–268. (Good bibliography of the early literature.)

[8] Healy, M. J. R. and Goldstein, H. (1976). *Biometrika*, **63**, 219–229. (Generalized approach, of which correspondence analysis is a special case; good notation.)

[9] Hill, M. O. (1973). *J. Ecol.*, **61**, 237–249. (Explanation of rationale for applications in plant ecology.)

[10] Hill, M. O. (1974). *Appl. Statist.*, **23**, 340–354. (Up-to-date review with a substantial bibliography.)

[11] Hill, M. O. (1977). In *First International Symposium on Data Analysis and Informatics, Versailles, 7–9 Sept. 1977*, Vol. 1, E. Diday et al. eds., Institut de Recherche d'Informatique et d'Automatique, Rocquencourt, France, pp. 181–199. (Application to a problem in automatic classification.)

[12] Hill, M. O. and Smith, A. J. E. (1976). *Taxon*, **25**, 249–255. (Application to data with mixed categorical and continuous variates.)

[13] Hirschfeld, H. O. (1935). *Proc. Camb. Philos. Soc.*, **31**, 520–524. (First full explanation of the method.)

[14] Horst, P. (1935). *J. Social Psychol.*, **6**, 369–374. (Allusion to what must be this method, developed by M. W. Richardson; explanation in words only; no mathematics.)

[15] Naouri, J. C. (1970). *Publ. Inst. Statist. Univ. Paris*, **19**, 1–100. (In French. Analytical solutions are obtained for several continuous bivariate distributions.)

[16] Nishisato, S. (1980). *Analysis of Categorical Data: Dual Scaling and Its Applications*. University of Toronto Press, Toronto. (A good general text.)

(CLASSIFICATION
ECOLOGICAL STATISTICS
FACTOR ANALYSIS
MULTIDIMENSIONAL SCALING

PRINCIPAL COMPONENTS
PSYCHOLOGY, STATISTICS IN)

M. O. HILL

COS THETA (Cos θ)

An alternative designation of the coefficient of correlation* r:

$$\cos\theta = r = \frac{\sum X_i Y_i}{\sqrt{(\sum X_i)^2 (\sum Y_i)^2}},$$

which is also called the *coefficient of proportional similarity* in statistical methods in geology.

θ can be viewed as the angle between two samples $\{X_i\}$ and $\{Y_i\}$. For $\theta = 90° = \pi/2$, $\cos\theta = 0$ and the samples are interpreted as having nothing in common; for $\theta = 0$, $\cos\theta = 1$ and the samples are interpreted—in the case of mapping geological data—as being identical in composition.

Bibliography

Krumbein, W. C. and Graybill, F. A. (1965). *An Introduction to Statistical Models in Geology*. McGraw-Hill, New York.

(GEOLOGY, STATISTICS IN)

COUNTABLE ADDITIVITY

A property assigned to probability measures* in Kolmogorov's axiomatic theory of probability*.

If $M(\mathcal{Q})$ denote the measure of a set $\mathcal{Q}$, then this measure is countably additive with respect to sets $\mathcal{Q}_1, \mathcal{Q}_2, \ldots$ if

$$M\left(\bigcup_{j=1}^{\infty} \mathcal{Q}_j\right) = \sum_{j=1}^{\infty} M(\mathcal{Q}_j).$$

In applications to probability theory $M(\cdot)$ is a probability measure* and the $\mathcal{Q}_j$'s are disjoint sets of events*.

COURNOT, ANTOINE AUGUSTIN

Born: August 28, 1801, in Gray (Haute-Saône), France.

Died: March 31, 1877.

Contributed to: mathematical economics, philosophy of science.

French educator, mathematician, economist, and philosopher. Cournot attended the *collèges* of Gray (1809–1916) and Besançon (1820–1821), the École Normale Supérieure in Paris (1821–1822), and received the *licence* in mathematics after studying under Lacroix* at the Sorbonne (1822–1823). A series of mathematical papers during the next decade culminated in a doctorate (1829) and brought him to the attention of Poisson, who secured him the chair in analysis at Lyons (1834). Several distinguished administrative posts quickly followed (rector at Grenoble, 1835–1838; inspector general of studies, 1838–1854; rector at Dijon, 1854–1862).

A competent pure mathematician, Cournot is best known as a founder of mathematical economics and for his contributions to the philosophy of science. Noteworthy among the latter is his distinction among objective, subjective, and philosophical senses of probability, and his theory of chance as the conjunctions of events in two causally independent series. Cournot's *Exposition de la théorie des chances et des probabilités* (1843) was a highly regarded mathematical text, cited by Boole and Todhunter. Among its many interesting features are a clear formulation of a frequency theory of probability (pp. iii, 437–439), and perhaps the earliest precise statement of the confidence interval property of interval estimates (pp. 185–186). In the tradition of Condorcet, Laplace, and Poisson, Cournot also wrote a lengthy article on legal applications of probability, "Sur les Applications du calcul des chances à la statistique judiciaire" [*Journal de mathématiques pures et appliquées*, **3** (1838), pp. 257–334].

Interest in Cournot arose in France at the turn of the century and has continued there since, but he has suffered (undeserved) neglect elsewhere. Not surprisingly, however, Edgeworth* and Keynes*, both economists interested in the foundations of probability and statistics, were familiar with his work. See also the entries under Bienaymé* (a contemporary and friend of Cournot's) and Chuprov* (who was influenced by Cournot in his conception of statistical regularity).

Literature

An annotated edition of Cournot's works is currently in preparation: *Oeuvres complètes* (1973–, J. Vrin); the 1843 *Exposition* has been reprinted by Edizioni Bizzarri (1968). The entry on Cournot in the *Dictionary of Scientific Biography* gives a balanced account of his life and work, that in the *Encyclopedia of Philosophy* emphasizes his philosophical efforts, and that in the *International Encyclopedia of Statistics* highlights his contributions to economics. All three contain useful select bibliographies. E. P. Bottinelli, *A. Cournot, métaphysicien de la connaissance* (1913, Librarie Hachette), contains an extensive list of works on Cournot, virtually complete up to the date of its publication. For further details of Cournot's life, see his posthumous *Souvenirs, 1760–1860* (1913, Librarie Hachette).

Virtually all serious work on Cournot is in French; the most important among these discussing his philosophy of probability are F. Mentré, *Cournot et la renaissance du probabilisme au XIXᵉ siècle* (1908, Marcel Rivière); A. Darbon, *Le Concept du hasard dans la philosophie de Cournot* (1911, Felix Alcan); and G. Milhaud, *Études sur Cournot* (1927, J. Vrin). The *Revue de mètaphysique et de morale*, Vol. 13 (1905), No. 3, is a special number devoted to all aspects of Cournot's work, as is *A. Cournot: Études pour le centenaire de sa mort, 1877–1977* (1978, Economica), a collection of essays useful as a guide to the recent French literature up to 1978. The annual *Répertoire bibliographique de la philosophie* may be consulted for subsequent work.

For an English view of Cournot, see J. M. Keynes, *A Treatise on Probability* (1921, Macmillan), pp. 166, 283–284. C. C. Heyde and E. Seneta, *I. J. Bienaymé: Statistical Theory Anticipated* (1977, Springer-Verlag), although not directly concerned with Cournot, touch on the mathematical aspects of his work on probability, as does Oskar Anderson, *Probleme der statistischen methodenlehre in den Sozialwissenschaften* (1954, Physica-Verlag), pp. 131–133.

SANDY L. ZABELL

COVARIANCE

The covariance of two random variables X and Y is

$$\text{cov}(X, Y) = E\big[\{X - E[X]\}\{Y - E[Y]\}\big]$$
$$= E[XY] - E[X]E[Y].$$

(ANALYSIS OF COVARIANCE
CORRELATION)

COVERAGE

Consider a tree that is beginning to lose its leaves in autumn. If we focus our attention on a small area of ground under the tree, it is reasonable to suppose that the leaves fall independently of each other and are equally likely to land anywhere within the region and with any orientation. Broadening our perspective, we would notice that the leaves fall more densely near the trunk and become sparser as we move farther away. It is more tractable mathematically if the tree is radially symmetric and there is no wind deflecting the leaves as they fall, in which case a model such as a circular normal* can be adopted to describe the pattern of dispersion.

There are many questions which we could ask about the coverage of the ground below the tree, which correspond to the types of two-dimensional coverage problems that have been discussed in the literature of the subject. First, consider a small convex region of area A_0 and B_0 containing a fixed point $\mathbf{x}$, and wait until the first leaf falls on this area. The probability that the leaf covers $\mathbf{x}$, assuming a convex leaf of area A_1 and perimeter B_1, is found from integral geometry (see ref. 8) to be

$$\frac{A_1}{A_0 + A_1 + B_0 B_1 / 2\pi}. \qquad (1)$$

For nonconvex or nonisotropic leaves the third term in the denominator of (1) must be modified. If, instead of a fixed point $\mathbf{x}$, we are concerned with the probability that the leaf will cover an ant who wanders randomly over the region according to a probability density* $g(\mathbf{x})$, formula (1) is still valid as a consequence of the uniform random position assumed for the leaf.

If we suppose the center of the leaf to have an arbitrary probability density $f(\mathbf{y})$, then the coverage probability of a fixed point $\mathbf{x}$ is

$$\frac{1}{2\pi} \int_0^{2\pi} \int_{C(\theta, \mathbf{x})} f(\mathbf{y}) \, d\mathbf{y} \, d\theta, \qquad (2)$$

where $C(\theta, \mathbf{x})$ is the set of positions of the center of the leaf such that for orientation θ, the leaf covers $\mathbf{x}$. When $\mathbf{x}$ is a random ant instead of a fixed point, the probability is

$$\frac{1}{2\pi} \int \int_0^{2\pi} \int_{C(\theta, \mathbf{x})} f(\mathbf{y}) g(\mathbf{x}) \, d\mathbf{y} \, d\theta \, d\mathbf{x}. \qquad (3)$$

Evaluation of the integrals in (2) and (3) is usually complicated. Various tables have been constructed for particular cases such as circular leaves and circular normal f and g. See the review paper by Guenther and Terragno [3] for further details.

So far we have discussed only the impact of a single leaf. As more leaves fall (see Fig. 1), overlapping each other, more and more of the ground becomes covered. It is of interest to know the distribution of area covered, or of the proportion of area covered within a certain region, after either a fixed time or a fixed number of leaves. This problem, although easy to pose, is largely

A_0

Figure 1 Coverage of a region by randomly located leaves.

unsolved, except for finding first moments and asymptotic distributions. If the leaves are uniformly and isotropically distributed, with independent sizes and shapes, and expected area EA_1 and perimeter EB_1, then the expected area that is covered after n leaves have fallen onto the region is

$$A_0\left[1 - \frac{(A_0 + B_0 \cdot EB_1/2\pi)^n}{(A_0 + EA_1 + B_0 \cdot EB_1/2\pi)^n}\right]. \quad (4)$$

This formula takes into account the fact that larger leaves are more likely to overlap the observed region than are smaller ones.

If the centers of the newly fallen leaves are governed by a Poisson process* of intensity λ per unit time per unit area, and the other assumptions are as for (4), then the expected area covered after a time t is

$$A_0\left[1 - \exp(-\lambda t EA_1)\right]. \quad (5)$$

A somewhat more realistic model is obtained by retaining the homogeneity of the leaf process with respect to time, but adopting a circular normal distribution with variance σ^2 to describe the spatial dispersion. In this case one can determine the total expected covered area instead of being restricted to a bounded observation area. This total expected area for circular leaves of radius R is

$$2\pi\int_0^\infty r\left\{1 - \exp\left[-\lambda t H(r)\right]\right\} dr. \quad (6)$$

Here $H(r)$ is the integral of the circular normal density function with unit variance

over a disk of radius R/σ whose center is r/σ units from the origin.

Under suitable conditions, the area of uncovered ground is asymptotically normal as either the observed region becomes large, the observation time becomes large, or the fixed number of leaves becomes large (see, e.g., Ailam [1]).

The sequential coverage problem of finding the distribution of the number of leaves required to completely cover a region seems once again to be intractable. However, asymptotic bounds have been given by Cooke [2]. Instead of considering coverage on the plane, some authors have considered coverage of the surface of a sphere, including Moran and Fazekas de St. Groth [7], Miles [4], and Wendel [10]. This has certain biological applications and has the mathematical advantage of avoiding edge effects. The probability of covering a sphere by n uniformly randomly distributed hemispheres, for example, is

$$1 - (n^2 - n + 2)/2^n. \quad (7)$$

For independent uniformly distributed sets of area A and perimeter B on the surface of the unit sphere, the asymptotic coverage probability is

$$1 - (n^2 - n)B^2(4\pi - A)^{n-2}/(4\pi)^n. \quad (8)$$

Coverage problems also exist in lower and higher dimensions. In one dimension, there are the problems of coverage of an interval by overlapping line segments and coverage of the circumference of a circle by overlapping arcs. The probability of covering a circle of unit circumference by n independent uniformly distributed arcs of length x was derived by Stevens [9]:

$$1 - \binom{n}{1}(1 - x)^{n-1} + \binom{n}{2}(1 - 2x)^{n-1}$$
$$+ \cdots + \binom{n}{j}(1 - jx)^{n-1}, \quad (9)$$

where j is the integral part of $1/x$.

In three dimensions, the volume occupied by uniformly or normally distributed balls of fixed radius has been investigated by Moran [5, 6]. Asymptotic normality applies in both cases, although for uniformly distributed balls the asymptotic variance differs according to whether the number of balls is fixed or Poisson* distributed. The expected covered (occupied) volume in the case of normally distributed balls with unit variance is asymptotically equivalent to

$$\tfrac{4}{3}\pi\left[2\log\lambda - 2\log(2\log\lambda)\right]^{3/2}, \quad (10)$$

where λ is the mean number of balls.

References

[1] Ailam, G. (1970). *Ann. Math. Statist.*, **41**, 427–439.

[2] Cooke, P. J. (1974). *J. Appl. Prob.*, **11**, 281–293.

[3] Guenther, W. C. and Terragno, P. J. (1964). *Ann. Math. Statist.*, **35**, 232–260.

[4] Miles, R. E. (1969). *Biometrika*, **56**, 661–680.

[5] Moran, P. A. P. (1973). *J. Appl. Prob.*, **10**, 483–490.

[6] Moran, P. A. P. (1974). *Acta Math.*, **133**, 273–286.

[7] Moran, P. A. P. and Fazekas de St. Groth, S. (1962). *Biometrika*, **49**, 389–396.

[8] Santaló, L. A. (1976). *Integral Geometry and Geometric Probability*. Addison-Wesley, Reading, Mass.

[9] Stevens, W. L. (1939). *Ann. Eugen. (Lond.)*, **9**, 315–320.

[10] Wendel, J. G. (1962). *Math. Scand.*, **11**, 109–111.

(GEOMETRIC PROBABILITY
INCLUSION–EXCLUSION PRINCIPLE
OCCUPANCY PROBLEMS)

Pamela J. Davy

COVERAGE PROBLEMS *See* Target
Coverage

COX'S REGRESSION MODEL

THE MODEL

In medical and industrial statistics, methods for evaluating the dependence of survival* time or response time T on independent variables or covariates $\mathbf{x} = (x_1, \ldots, x_p)'$ have received considerable attention. The independent variables could include, e.g.,

state of disease, duration of symptoms prior to treatment, or a binary (0/1) variable representing control/treatment group. Cox's regression model [1] assumes that $\lambda(t; \mathbf{x})$, the hazard function of the continuous random variable T, is given by

$$\lambda(t; \mathbf{x}) = \lambda_0(t)e^{\boldsymbol{\beta}'\mathbf{x}}, \quad (1)$$

where $\boldsymbol{\beta}$ is a $p \times 1$ vector of unknown parameters reflecting the effects of $\mathbf{x}$ on survival and $\lambda_0(t)$ is an unspecified function of time. An extension of model (1) introduced by Kalbfleisch [7] allows strata* within the population to have distinct underlying hazard functions $\lambda_{0j}(\cdot)$.

PARAMETER AND FUNCTION ESTIMATION

Usually, survival data of the type envisaged here will be subject to right censoring* since, e.g., some individuals will not have failed on termination of the study. It will be assumed throughout that the censoring and failure mechanisms are independent [14].

For the n individuals in the study with independent variable vectors $\mathbf{x}_i$, $i = 1, \ldots, n$, let $t_{(1)} < t_{(2)} < \cdots < t_{(k)}$ denote the ordered uncensored failure times with corresponding independent variables $x_{(1)}, x_{(2)}, \ldots, x_{(n)}$ and denote by $R(t_{(i)})$ the collection of individuals with censored or uncensored failure times $\geq t_{(i)}$. Following Prentice and Kalbfleisch [14], let A_i be the event that individual (i) fails at $t_{(i)}$ and B_i be the event that describes the pattern of failure and censorings to $t_{(i)}$ and the information that a failure occurs at $t_{(i)}$. The likelihood factors to produce

$$\prod_{i=1}^{k} p(A_i \mid B_i)\left\{ \prod_{i=2}^{k} p(B_i \mid B_{i-1}, A_{i-1}) \right.$$

$$\left. \times p(B_1)p(B_{k+1}) \right\}$$

(with $t_{(k+1)} = \infty$). The first of these terms does not involve $\lambda_0(t)$ and reduces to

$$L(\boldsymbol{\beta}) = \prod_{i=1}^{k} \left\{ \frac{e^{\boldsymbol{\beta}'\mathbf{x}_{(i)}}}{\sum_{j \in R(t_{(i)})} e^{\boldsymbol{\beta}'\mathbf{x}_j}} \right\}. \quad (2)$$

Cox [2], through the concept of partial likelihood*, indicates that inferences concerning β, using the usual large-sample likelihood methods, can be based on (2). In the absence of censoring, (2) is the marginal distribution of the ranks of the failure times.

Estimation of $\lambda_0(t)$ can be achieved [9] by approximating this "underlying hazard function" as a step function with

$$\lambda_0(t) = \lambda_i, \quad t \in [b_{i-1}, b_i), \quad i = 1, \ldots, r,$$

where $b_0 = 0 < b_1 < \ldots < b_{r-1} < b_r = \infty$ is a subdivision of the time scale. Conditional on $\beta = \hat{\beta}$ (the partial likelihood estimator of β), the maximum likelihood estimator* of λ_i is given by

$$\hat{\lambda}_i = d_i \left\{ \sum_{j=1}^n e^{\hat{\beta}'x_j} D_{ij} \right\}^{-1}, \quad i = 1, \ldots, r,$$

where D_{ij} is the length of time spent in the interval $[b_{i-1}, b_i)$ by individual j and d_i is the number of uncensored failures in that interval. For estimation in the stratified form of the model, see Kay [10].

In practice data may involve ties. Prentice and Gloeckler [13] have presented some results on the use of grouped form of model (1) to handle ties and have discussed other approaches to this problem.

EFFICIENCY CALCULATIONS

The efficiency* of $\hat{\beta}$ compared to the corresponding estimator based on parametric forms of (1) is clearly of interest. Efron [5] obtains the general element of the Cox model information matrix* from (2) for $k, l = 1, \ldots, p$, as

$$I_{kl}(\beta) = E\left\{ \frac{-\partial^2 \log L(\beta)}{\partial \beta_k \partial \beta_l} \right\}$$

$$= \int_0^\infty E_{R(t)} \left\{ \sum_{j \in R(t)} y_{jk} y_{jl} \right\} \lambda_0(t) \, dt$$

$$(3)$$

where

$$y_{jr} = x_{jr} - \left\{ \sum_{i \in R(t)} x_{ik} e^{\beta'x_i} \right\} \left\{ \sum_{i \in R(t)} e^{\beta'x_i} \right\}^{-1}$$

and $E_{R(t)}$ denotes expectation over the probability distribution of rankings in $R(t)$.

A censoring model of interest in the clinical trial* setting assumes that individuals enter the trial at random in the interval $(0, \tau)$, termination of the trial at τ giving a collection of censored and uncensored survival times. Choosing τ to ensure a specified proportion, c_p are censored. Kay [11] shows, e.g., in the two-group case ($p = 1, x_i = 0/1$) that the Fisher information* from the exponential form of model (1) $[\lambda_0(t) \equiv 1]$ is given by

$$I_E(\beta) = \frac{n[1 - a_1(\beta, \tau)][1 - a_2(\beta, \tau)]}{1 - \frac{1}{2}[a_1(\beta, \tau) + a_2(\beta, \tau)]},$$

where

$$a_1(\beta, \tau) = \frac{e^{\beta}(1 - e^{-\tau e^{-\beta}})}{\tau}$$

and

$$a_2(\beta, \tau) = \frac{e^{-\beta}(1 - e^{-\tau e^{\beta}})}{\tau}$$

and (3) reduces to

$$I_C(\beta) = 2n \int_{e^{-\tau e^{\beta}}}^1 \frac{\tau + e^{-\beta}\log u}{\tau \{1 + e^{2\beta} u^{(2\beta - 1)e^{-2\beta}}\}} \, du.$$

The asymptotic efficiency of $\hat{\beta}$ under the exponential assumption is then $R(\beta) = I_C(\beta)/I_E(\beta)$. At $\beta = 0$, $R(\beta) = 1$. In the uncensored case $R(0.2) = 0.963$ and $R(0.4) = 0.874$ while for $C_p = 0.3$, $R(0.2) = 0.988$ and $R(0.4) = 0.950$. Again for $p = 1$ and in the uncensored case Kalbfleisch [8] shows that in the neighborhood of $\beta = 0$, $R(\beta) \simeq e^{-u_2\beta^2}$, where μ_2 is the second central moment of the x-values.

MODEL CHECKING

Some informal methods, based on residuals* and hazard plotting*, have been suggested for assessing model (1) assumptions by Kay

[10], Prentice and Kalbfleisch [14], and Cox [3], among others. Defining for $r = 1, 2, \ldots$

$$B_r^{(m)} = \sum_{j \notin R(t_{(rm)})} t_j e^{\hat{\beta}' x_j} + \sum_{j \in R(t_{(rm)})} t_{(rm)} e^{\hat{\beta}' x_j}$$

as the total estimated "operational" time at risk up to the rmth failure, Cox shows that approximately $Z_r^{(m)} = d - \log[B_r^{(m)} - B_{(r-1)}^{(m)}]$, where d is a constant independent of r and $B_0^{(m)} = 0$, has mean $\log \bar{\lambda}_r$, where $\bar{\lambda}_r$ is the average underlying hazard between the failures at $t_{((r-1)m)}$ and $t_{(rm)}$, and variance $(m - \frac{1}{2})^{-1}$. In addition, the Z's are independent and, for chosen m, plots of $Z_r^{(m)}$ against the midpoint of that interval will display the form of $\lambda_0(t)$. To check that a particular independent variable x affects the hazard in the assumed multiplicative way, the sample may be divided into a small number of subsamples on the basis of the x values and the stratified form of model (1) fitted. Plots as above, separately within each stratum, can then be viewed for evidence of trends. These would indicate violation of the multiplicative assumptions with regard to the effect of x.

TIME-DEPENDENT COVARIATES

A generalization of (1) allows the independent variables to be time dependent and Crowley and Hu [4] and Farewell [6] provide interesting examples of their use. Prentice and Kalbfleisch [14] give a thorough discussion of their application and show how they may be used with model (1) in the analysis of competing risks*.

AN APPLICATION

The data set to be used as an illustration of the methods described above is from the Manchester (U.K.) Regional Breast Study. Further details are given in Lythgoe et al. [12]. The independent variables, recorded at

entry into the study, are:

$$x_1 = \log \text{age} \qquad x_2 = \begin{cases} 1 & \text{tumor size} > 2 \text{ cm} \\ 0 & \text{tumor size} \leqslant 2 \text{ cm} \end{cases}$$

$$x_3 = \begin{cases} 1 & \text{tumor site lateral} \\ 0 & \text{otherwise} \end{cases}$$

$$x_4 = \begin{cases} 1 & \text{tumor site central} \\ 0 & \text{otherwise} \end{cases}$$

$$x_5 = \begin{cases} 1 & \text{postmenopausal} < 3 \text{ years} \\ 0 & \text{otherwise} \end{cases}$$

$$x_6 = \begin{cases} 1 & \text{postmenopausal} \geqslant 3 \text{ years} \\ 0 & \text{otherwise} \end{cases}$$

$$x_7 = \begin{cases} 1 & \text{clinical state II} \\ 0 & \text{clinical stage I} \end{cases}$$

$$y_1 = \begin{cases} 1 & \text{local} + \text{XRT, stage I} \\ 0 & \text{local, stage I} \end{cases}$$

$$y_2 = \begin{cases} 1 & \text{local} + \text{XRT, stage II} \\ 0 & \text{radical, stage II} \end{cases}$$

Of the 881 patients, there were 205 deaths, of which 185 were distinct. The 10 pairs of tied values were broken at random for the purposes of analysis. Table 1 presents Cox model β estimates with estimated standard errors obtained from the inverse of the negative of the matrix of second partial derivatives of (2) evaluated at $\beta = \hat{\beta}$. Chi-square statistics were obtained by fitting reduced models with each variable omitted in turn and employing the large-sample likelihood ratio test*. Stage and menopausal status are the only significant factors and in particular the treatment variables y_1 and y_2 produce nonsignificant effects. The model checking ideas of §4 are illustrated by investigating the effect of clinical stage in the model. Defining two strata according to the value of x_7, Fig. 1 plots the $Z_r^{(m)}$ values, for $m = 12$, having fitted the stratified form of the Cox model with x_7 omitted from the variables listed above. The pattern of the two sets of Z values, particularly in view of the associated standard deviations, are similar, suggesting that clinical stage affects the underlying hazard function in the assumed multiplicative way.

Table 1   Cox Model Fit Giving Estimates, Estimated Standard Errors, and χ_1^2 Statistics for Testing Effects

Independent Variable		Estimated Coefficient	Estimated Standard Error	χ_1^2 Statistic
Age,	x_1	0.004	0.672	0
Tumor size,	x_2	0.236	0.169	2.01
Tumor site,	$\begin{cases} x_3 \\ x_4 \end{cases}$	0.068 0.015	0.184 0.218	0.14 0
Menopausal status,	$\begin{cases} x_5 \\ x_6 \end{cases}$	0.573 0.323	0.293 0.283	3.56^a 1.31
Stage,	x_7	0.553	0.201	6.86^b
Treatment,	$\begin{cases} y_1 \\ y_2 \end{cases}$	-0.229 0.159	0.195 0.205	1.39 0.61

$^a 0.05 < p < 0.10$
$^b p < 0.001.$

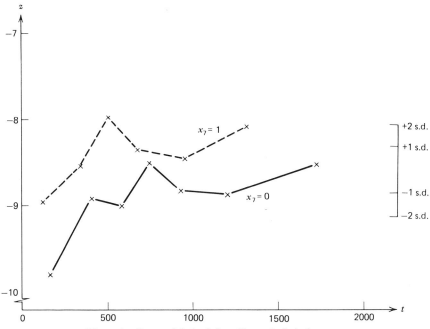

Figure 1   Cox model check for effects of clinical stage.

References

[1] Cox, D. R. (1972). *J. R. Statist. Soc. B*, **34**, 187–220.

[2] Cox, D. R. (1975). *Biometrika*, **62**, 269–276.

[3] Cox, D. R. (1979). *Biometrika*, **66**, 188–190.

[4] Crowley, J. and Hu, M. (1977). *J. Amer. Statist. Ass.*, **72**, 27–36.

[5] Efron, B. (1977). *J. Amer. Statist. Ass.*, **72**, 557–565.

[6] Farewell, V. T. (1979). *Appl. Statist.*, **28**, 136–143.

[7] Kalbfleisch, J. D. (1974). *Biometrics*, **30**, 561.

[8] Kalbfleisch, J. D. (1974). *Biometrika*, **61**, 31–38.

[9] Kalbfleisch, J. D. and Prentice, R. L. (1973). *Biometrika*, **60**, 267–278.

[10] Kay, R. (1977). *Appl. Statist.*, **26**, 227–237.

[11] Kay, R. (1979). *Biometrika*, **66**, 91–96.

[12] Lythgoe, J. P., Leck, I., and Swindell, R. (1978). *Lancet*, 744–747.

[13] Prentice, R. L. and Gloeckler, L. A. (1978). *Biometrics*, **34**, 57–67.

[14] Prentice, R. L. and Kalbfleisch, J. D. (1979). *Biometrics*, **35**, 25–39.

(CLINICAL TRIALS
CONCOMITANT VARIABLES
HAZARD PLOTTING
SURVIVAL ANALYSIS)

RICHARD KAY

COX'S TEST OF RANDOMNESS

It is often necessary to test whether a sequence of events is random in time, the alternative being that there exists a trend in the rate of occurrence. Let n events occur at times $t_1, \ldots, t_n$ during the time interval $(0, T)$ and let $P[\text{event occurs in } (t, t + \Delta t)] = \lambda(t)\Delta t + O(\Delta t)$. Assuming that the rate $\lambda(t) = Ce^{Bt}$, Cox's test of randomness [2] is testing $H_0: B = 0$ against the alternative $H_1: B \neq 0$. The test statistic is

$$m = \sum_{i=1}^{n} t_i / nT$$

and the distribution of this statistic under H_0 is the Irwin–Hall distribution* with mean $\frac{1}{2}$ and variance $1/(12n)$. As n increases the distribution approaches normality very rapidly and the normal approximation can be used safely for $n \geqslant 20$. The power function* of the test for values of $n = 5$ was computed by Bartholomew [1] and for $n = 20(10)$, 80, 100, 200 by Mansfield [3] for values of $M = BT = \pm 0.10, \pm 0.20, \pm 0.40, \pm 0.80, \pm 1.2, \pm 1.6, \pm 3.0$ at the two significance levels* $\alpha = 0.05$ and $\alpha = 0.010$ for both one-tailed and two-tailed tests.

References

[1] Bartholomew, D. J. (1956). *J. R. Statist. Soc. B*, **18**, 234–239.

[2] Cox, D. R. (1955). *J. R. Statist. Soc. B*, **17**, 129–157. [Followed by discussion.]

[3] Mansfield, E. (1962). *Technometrics*, **4**, 430–432.

C_P STATISTICS

Introduced by Mallows [3] for evaluating the adequacy of (multiple) regression functions* of different orders.

Consider a multiple linear regression model* with n observations on k independent variables $x_1, \ldots, x_k$ and dependent variable y. The model for the expected value of y given $\mathbf{x} = (x_1, \ldots, x_k)$ is of the form

$$\eta_P(\mathbf{x}) = \beta_0 + \sum_{j \in P} \beta_j x_j, \tag{1}$$

where P is a subset of $(p - 1)$ of the elements of $\{1, \ldots, k\}$. There are 2^k regression functions of the form (1). Let $\hat{\beta}_P$ be the $(p - 1) \times 1$ least-squares* estimator of the elements $\beta_j (j \in P)$ when using regression function (1) and let $\mathbf{x}_P$ be the corresponding $(p - 1) \times 1$ vector of elements of $\mathbf{x}$. The C_P statistic is defined by

$$C_P = \frac{\sum_{i=1}^{n} (y_i - \hat{\beta}_0 \hat{\beta}'_P \mathbf{x}_{Pi})^2}{\hat{\sigma}^2} - n + 2p.$$

The numerator is the residual* sum of squares (RRS_P), taken over the n sets of observed values of $(y, x_1, \ldots, x_k)$ and is an estimate of $(n - p)\sigma^2$, which is the *common* variance of the residual variables. Intuitively, for an adequate regression, the function C_P should be close to p. If the regression is not adequate, it is likely to be much larger than p. Hence a plot of C_P against p is used to indicate whether the regression is adequate.

An upper bound for an "acceptable" C_p is

$$C_P < (2p - k - 1)k \times F_\alpha; \tag{2}$$

F_α is the upper $100\alpha\%$ percentile of the F-distribution* with k, $n - k - 1$ degrees of freedom. The bound is such that the *set of* regression functions (1) satisfying (2) contains all the adequate functions with probability at least $1 - \alpha$. Detailed discussion of these plots, with a comprehensive bibliography, is given by Mallows [4]. Some alternative plots for the same purpose have been suggested by Spjøtvoll [5].

For additional information on this topic, see refs. 1 and 2.

References

[1] Daniel, C. and Wood, F. S. (1971). *Fitting Equations to Data*. Wiley, New York, Chap. 6.

[2] Gorman, J. W. and Toman, R. J. (1966). *Technometrics*, **8**, 27–51.

[3] Mallows, C. L. (1964). *Choosing Variables in a Linear Regression: A Graphical Aid*. Presented at the Central Regional Meeting of the IMS, Manhattan, Kansas, May 7–9, 1964.

[4] Mallows, C. L. (1973). *Technometrics*, **15**, 661–675.

[5] Spjøtvoll, E. (1977). *Biometrika*, **64**, 1–8.

(ANALYSIS OF COVARIANCE
ANALYSIS OF VARIANCE
MULTIPLE REGRESSION)

CRAMÉR–RAO INEQUALITY   *See* CRAMÉR–RAO LOWER BOUND; MINIMUM VARIANCE UNBIASED ESTIMATION

CRAMÉR–RAO LOWER BOUND

If the joint likelihood* of a vector of random variables $\mathbf{X}$ is $L(\mathbf{X};\theta)$ depending on a single parameter *θ, and the expected value of a statistic* $T = T(\mathbf{X})$, given θ, is $\theta + b(\theta)$ [$b(\theta)$ is the bias* of T as an estimator of θ], then under certain regularity conditions the mean-square error of T as an estimator of θ, $E[(T - \theta)^2|\theta]$, cannot be less than

$$\left(1 + \frac{db(\theta)}{d\theta}\right)^2 \bigg/ \left\{ -E\left[\frac{\partial^2 \log L}{\partial \theta^2} |\theta \right] \right\}.$$

In particular, no unbiased estimator of θ (i.e., with zero bias) can have variance less than

$$\left\{ -E\left[\frac{\partial^2 \log L}{\partial \theta^2} |\theta \right] \right\}^{-1}.$$

This is the reciprocal of the Fisher information*.

If the X's are independent and identically distributed, $L(\mathbf{X};\theta) = \prod_{j=1}^{n} L^*(X_j;\theta)$ and

$$-E\left[\frac{\partial^2 \log L}{\partial \theta^2} |\theta \right] = -nE\left[\frac{\partial^2 \log L^*}{\partial \theta^2} |\theta \right],$$

where $L^*(X;\theta)$ is the common likelihood function of the X's. In this case, for any unbiased estimator T, of θ we have

$$n \, \text{var}(T|\theta) \geqslant -E\left[\frac{\partial^2 \log L^*}{\partial \theta^2} |\theta \right].$$

If there are several parameters $\boldsymbol{\theta}' = (\theta_1, \ldots, \theta_s)$ and $E[T|\boldsymbol{\theta}] = \tau(\boldsymbol{\theta})$, then

$$\text{var}(\tau|\boldsymbol{\theta}) \geqslant \text{trace}\left(\left(\frac{\partial \tau}{\partial \boldsymbol{\theta}} \right)\left(\frac{\partial \tau}{\partial \boldsymbol{\theta}} \right)' \mathbf{J}^{-1} \right)$$

where $(\partial \tau / \partial \boldsymbol{\theta})' = (\partial \tau / \partial \theta_1, \ldots, \partial \tau / \partial \theta_s)$ and $\mathbf{J}$ has (g,h)th element $E[\partial \log L / \partial \theta_g \cdot \partial \log L / \partial \theta_h | \boldsymbol{\theta}]$.

In some cases it is possible to improve (i.e., increase) the Cramér–Rao lower bound.

(EFFICIENCY
ESTIMATION
FISHER INFORMATION
HOEFFDING INEQUALITIES)

CRAMÉR('S) SERIES

This is a tool used in the analysis of large deviations* in the distribution of standard sums of large numbers of independent, identically distributed (i.i.d.) random variables.

The series is defined in terms of the cumulants of a distribution, which are supposed to be all finite.

Denoting the characteristic function* of the distribution of $\phi(z)$ and the cumulant generating function* by $\psi(z) = \ln \phi(z)$, we define a function $s(z)$ by

$$\psi'(s) = z \times (\text{variance of the distribution})^{1/2}$$

$$= z\sqrt{\kappa_2} = z\sigma \qquad (1)$$

(using κ_r to denote the rth cumulant of the distribution). Since $\psi(\cdot)$ is an analytic function in the neighborhood of zero, (1) defines $s(z)$ as an analytic function of z, for sufficiently small $|z|$. Hence $s(z)$ can be expanded as a power series for $|z| < \zeta$ for sufficiently small ζ.

The power series $\lambda(z) = \sum_{j=1}^{\infty} \lambda_j z^{j-1}$, defined by

$$z^3 \lambda(z) = \psi(s) - s\psi'(s) + \tfrac{1}{2}\psi''(s),$$

is called the *Cramér series* of the distribution. It is uniquely defined by the distribution, and defines the distribution uniquely if the distribution is determined by its moments.

If the expected value of the distribution is zero, then

$$\lambda_1 = \frac{\kappa_3}{3!\sigma^3} \; ; \quad \lambda_2 = \frac{\kappa_4\sigma^2 - 3\kappa_3^2}{4!\sigma^6} \; ;$$

$$\lambda_3 = \frac{\kappa_5\sigma^2 - 10\kappa_2\kappa_3\sigma + 15\kappa_3^3}{5!\sigma^9} \; .$$

(ASYMPTOTIC NORMALITY
LIMIT THEOREMS
LIMIT THEOREMS, CENTRAL)

CRAMÉR–VON MISES STATISTIC

The Cramér–von Mises statistic, W_n^2, is generally defined to be the statistic

$$W_n^2 = n \int_{-\infty}^{\infty} \left[F_n(x) - F_0(x) \right]^2 dF_0(x),$$

where $F_0(x)$ is the hypothesized CDF, and $F_n(x)$ is the sample or empirical CDF based on the sample $x_1, \ldots, x_n$. It is used to test the null hypothesis $H_0 : F(x) = F_0(x)$, where it is assumed the sample comes from a population with CDF $F(x)$.

The statistic was suggested independently by Cramér [1] and von Mises [10] (actually they defined it with $dF_0(x)$ replaced by $dG(x)$, $G(x)$ being some suitable positive weight function), and Smirnov [7] found the asymptotic null distribution of W_n^2. The statistic was introduced as an alternative to the chi-squared* goodness-of-fit* statistic, which requires the data to be grouped before calculation. For evaluation of the statistic, let $t_i = F_0(x_i')$, where $x_1' \leqslant x_2' \leqslant \cdots \leqslant x_n'$ are the original ordered observations, then W_n^2 is given by

$$W_n^2 = \frac{1}{12n} + \sum_{i=1}^{n} \left(t_i - \frac{2i-1}{2n} \right)^2. \quad (1)$$

The null hypothesis H_0 is rejected in favor of $H_1 : F(x) \neq F_0(x)$ for large values of W_n^2. If F_0 is continuous, the null distribution of W_n^2

is independent of F_0. Small-sample percentage points of W_n^2 have been found and are tabulated in Pearson and Hartley [5, Table 54].

As an example, suppose that we have five observations 0.22, 2.53, 1.16, 0.55, 0.14 and it is wished to test the null hypothesis that these observations constitute a random sample from a negative exponential distribution* with mean equal to 1. In this case $F_0(x) = 1 - e^{-x}$, $x > 0$. The ordered observations are 0.14, 0.22, 0.55, 1.16, 2.53; the t_i's given by 0.131, 0.197, 0.423, 0.687, 0.920; this gives 0.347 as the value of W_n^2. Referring to Table 54 of Pearson and Hartley, we see that this value of W_n^2 is certainly not significantly large.

The test statistic can also be used to test for normality; that is, the null hypothesis is that the sample comes from a normal population with mean, μ, and variance, σ^2, unspecified. In this case the empirical CDF is compared with $\Phi\{(x - \hat{\mu})/\hat{\sigma}\}$ [$\Phi(\cdot)$ is the standard normal CDF], where

$$\hat{\mu} = \bar{x}, \qquad \hat{\sigma}^2 = \sum (x_i - \bar{x})^2/(n - 1).$$

This procedure gives W_n^2 as before, equation (1), but with t_i replaced by $\hat{t}_i = \Phi\{(x_i' - \hat{\mu}) /\hat{\sigma}\}$. Small-sample percentage points of this statistic are again given in Pearson and Hartley [5].

The asymptotic distribution theory of W_n^2, when parameters are estimated from the data, was pioneered in Darling [2]. Later important papers, dealing with the test of normality, are those of Stephens [9] and Durbin [4].

Other modifications have been made to the statistic to make tests of fit for discrete data, censored data* for distributions apart from the normal, and multivariate distributions. Extensions to two samples have also been made.

Further modifications of the original W_n^2 statistic have included the class of statistics defined by

$$\omega_n^2 = n \int_{-\infty}^{\infty} \{ F_n(x) - F_0(x) \}^2$$

$$\times \psi(F_0(x)) \, dF_0(x),$$

where $\psi(\cdot)$ is a positive weight function. The most important of these statistics is the statistic with $\psi(t) = \{t(1 - t)\}^{-1}$, and this statistic is known as the Anderson–Darling statistic* and usually denoted by A_n^2 or A_n. Pearson and Hartley [5] give computing formulas for A_n and small-sample percentage points. The statistic A_n provides a very good omnibus test of normality and compares favorably, in terms of power, ease of computation and available percentage points, with other tests for normality (e.g., the Shapiro–Wilk* statistic); see Stephens [8] for results on power and small-sample comparisons of A_n and W_n^2 with other tests of fit.

Pettitt [6], by a particular choice of $\psi(t)$, shows how ω_n^2 is related to the goodness-of-fit statistics $\sqrt{b_1}$ (sample skewness) and b_2 (sample kurtosis).

The study of the asymptotic distributions of Cramér–von Mises statistics has resulted in much work in the literature. The basic results involve the weak convergence of the empirical process

$$y_n(t) = \sqrt{n}\left\{ F_n \cdot F_0^{-1}(t) - t \right\}$$

to the Brownian* bridge. The statistics ω_n^2 can be expressed as functionals of the process $y_n(t)$ and under certain regularity conditions the asymptotic distribution of ω_n^2 is the same as the equivalent functional of the Brownian bridge. In many cases ω_n^2 has the same distribution as $\sum_{j=0}^{\infty} Z_j^2 \lambda_j$, where the Z_j are i.i.d. standard normal and the λ_j are eigenvalues of a particular integral equation. Durbin [3] gives extensive details of this theory.

References

[1] Cramér, H. (1928). *Skand. Aktuar.*, **11**, 141–180.

[2] Darling, D. A. (1955). *Ann. Math. Statist.*, **26**, 1–20. (Pioneering theoretical paper.)

[3] Durbin, J. (1973). *Regional Conference Series in Applied Mathematics*, Vol. 9. SIAM, Philadelphia. (An excellent theoretical review.)

[4] Durbin, J., Knott, M., and Taylor, C. C. (1975). II. *J. R. Statist. Soc. B*, **37**, 216–237.

[5] Pearson, E. S. and Hartley, H. O. (1972). *Biometrika Tables for Statisticians*, Vol. 2. Cambridge University Press, Cambridge.

[6] Pettitt, A. N. (1977). *J. R. Statist. Soc. B*, **39**, 364–370.

[7] Smirnov, N. (1936). *C. R. Acad. Sci. Paris*, **202**, 449–452.

[8] Stephens, M. A. (1974). *J. Amer. Statist. Ass.*, **69**, 730–737. (An excellent practical review.)

[9] Stephens, M. A. (1976). *Ann. Statist.*, **4**, 357–369.

[10] von Mises, N. (1931). *Wahrscheinlichkeitsrechnung*. Deuticke, Leipzig.

(ANDERSON–DARLING TEST DISTRIBUTION-FREE METHODS EDF STATISTICS KOLMOGOROV–SMIRNOV TESTS)

A. N. PETTITT

CRAMÉR–WOLD THEOREM *See* LIMIT THEOREMS, CENTRAL

CRAUSE AND LEMMER DISPERSION TEST

A test based on the index of dispersion, c, defined by $P[|X - \mu| \geqslant c] = 0.10$. Given a random sample $X_1, \ldots, X_n$ from an unknown distribution F with mean μ and variance σ^2, let ξ_p satisfy $F(\xi_p) = p$ for fixed p, $0 < p < 1$. A measure of spread (or dispersion) c is defined by $P[|X - \mu| \geqslant c] = p_0 = 0.10$. (In case of a symmetric distribution, $\xi_{0.95} - \xi_{0.05} = 2c$.) To test the hypothesis $H_0: c = c_0$ against the alternative of greater dispersion $H_1: c > c_0$, Crause and Lemmer [1] propose the following test for a symmetric distribution with known mean.

Count the number of observations lying outside the interval $(\mu - c_0, \mu + c_0)$ and reject H_1 if too many observations lie outside this interval. Specifically, the test statistic is

$$T = n^{-1} \sum_{i=1}^{n} I(|X_i - \mu| \geqslant c_0),$$

where $I(\cdot)$ is the indicator function. The rejection rule is reject H_0 if $nT > k$ and reject it with probability γ if $nT = k$. The values of k and γ for $\alpha = 0.01, 0.025, 0.05$, and 0.10 are given in Table 1 for $n = 2(1)10$. The choice of $p_0 = 0.10$ was made by Crause and Lemmer [1] after an extensive power

Table 1 Values of k and γ for Different n and α ($p_0 = 0.10$)

		α						
		0.01		0.025		0.05		0.10
n	k	γ	k	γ	k	γ	k	γ
2	1	0	1	0.0833	1	0.2222	1	0.5000
3	2	0.3333	2	0.8889	1	0.0905	1	0.2963
4	2	0.1296	2	0.4383	2	0.9527	1	0.1636
5	2	0.0206	2	0.2263	2	0.5684	1	0.0567
6	3	0.5988	2	0.0930	2	0.3470	2	0.8551
7	3	0.3130	3	0.9652	2	0.1960	2	0.5984
8	3	0.1505	3	0.6041	2	0.0800	2	0.4160
9	3	0.0381	3	0.3734	3	0.9334	2	0.2731
10	4	0.7500	3	0.2125	3	0.6482	2	0.1538

study of the test of samples of various sizes from the variate of distributions.

The asymptotic relative efficiency* of the test based on T relative to the χ^2-test in the case of normality is 64%. The test is robust to deviations from normality and compares favorably with the jackknife test*. An extension of this statistic when μ is unknown and the distribution is not necessarily symmetric was proposed and studied by Lemmer [2]. This test is a randomized decision* test and so subject to severe practical disadvantages.

References

[1] Crause, C. F. and Lemmer, R. H. (1975). A Distribution-Free One-Sample Test for Dispersion for Symmetrical Distributions. Manuscript, Dept. of Statistics, Rand Afrikaans University, South Africa.

[2] Lemmer, R. H. (1978). *J. Amer. Statist. Ass.*, **73**, 419–421.

(RANDOMIZED DECISIONS SCALE TESTS)

CREDIBILITY

As a branch of actuarial* science, credibility theory seeks to improve estimates used for premium rate making in order to reflect developing claim experience. More narrowly, credibility may be defined as a linear estimate of the true expectation, which estimate represents a comprise between hypoth-

esis and observation. See Hickman [5] and Hewitt [4] in *Credibility: Theory and Applications*, edited by P. M. Kahn [8]. This book is a principal reference on this subject and contains an extensive bibliography. Hickman and Miller [6] describe credibility in a more general sense as "the adaptive estimation of the parameters of risk theory models," where risk theory* is the study of the fluctuations of an insurance or risk enterprise.

Originally developed by American actuaries for setting premium rates for workmen's compensation insurance, credibility theory in its classic form studies models with fixed parameters and uses sampling theory to estimate them. Modern developments focus on models with parameters that are random variables and on Bayesian estimation techniques. See Hickman [5] and Mayerson [9].

CLASSICAL MODELS

If $Z(t)$ is the credibility factor depending upon a measure of the risk class of size t, if y is a function of actual claims, and m' is a prior estimate of expected claims, the adjusted claim estimate may be expressed as

$$Z(t)y + \left[1 - Z(t)\right]m'.$$

Classical credibility theory first must fix the minimum size of a risk class, t_0, say, to be assigned full credibility, i.e.,

$$Z(t_0) = 1;$$

then the claim estimate for a class of this size or larger is a function only of the experience of the particular class. As an example, consider the group of employees of a single employer covered under group life insurance. If the group is large enough, i.e., if its experience has full credibility, the premium will be based on the actual claim history of this group alone rather than partly on its own experience and partly on statistics representing experience of a collection of similar groups. The full credibility level is usually determined by assuming that the claim process can be adequately described by the Poisson distribution and then by approximating it with the normal. Hence a class of at least 1084 is required to provide full credibility with a 90% confidence level. Next, partial credibility is then assigned to smaller classes, usually by various formulas, such as

$$Z(t) = \sqrt{t/t_o}$$
$$Z(t) = tc/(t + h),$$
$$\text{where } c = (t_0 + h)/t_0, \quad t \geqslant t_0.$$

BAYESIAN MODELS

Although Bayesian concepts were first applied to credibility problems by Bailey in 1945 [1], they were not put into modern Bayesian language until 1964 by Mayerson [9].

In this model, past experience and our subjective interpretation of it can be adequately described by a parameter θ (or set of parameters) with a prior distribution. If recent claim experience has a likelihood distribution, then Bayes' theorem gives the posterior distribution of θ conditioned on the occurrence of x representing claim experience. Thus we have

$$p''(\theta|x) = l(x|\theta)p'(\theta)N(\theta),$$

where $p'(\theta)$ is the prior density of θ, $p''(\theta|x)$ is the posterior density of θ conditioned on x, $l(x|\theta)$ is the likelihood density of claims, and $N(x)$ is a normalizing constant.

The derivation of the posterior distribution is made tractable if the prior and the likelihood distributions are conjugate. When this is so, the prior and the posterior belong to the same family, but with different parameters. For example, if the prior is a beta and the likelihood is binomial, the posterior is beta also. Again, if the prior is gamma and the likelihood is Poisson, the posterior is gamma.

Bailey [1] showed that in these two special cases there is a credibility factor Z of the form

$$Z(t) = t/(t + k),$$

such that

$$E''(\theta|x) = Z(t)\bar{x} + [1 - Z(t)]E'(\theta),$$

where E' and E'' are expectations with respect to the prior and posterior distributions, respectively. Bühlmann [2, 3] and Mayerson [9] have noted that the best linear approximation to $E''(\theta|x)$ in these cases is

$$\rho^2 x + (1 - \rho^2)E'(\theta),$$

where the credibility factor ρ^2 is the square of the correlation coefficient between x and θ.

In the beta binomial and gamma Poisson cases, k in the formulas above has the following form:

for the beta binomial:
$$\{E'(\theta)[1 - E'(\theta)] - \text{var}'(\theta)\}/\text{var}'(\theta),$$
and for the gamma Poisson:
$$E'(\theta)/\text{var}'(\theta).$$

In more general cases, Bühlmann [3] uses least-squares lines to give credibility factors of the form

$$Z(t) = \text{var}'(\theta)/[\text{var}'(\theta) + E'(\theta)\text{var}(x|\theta)].$$

GENERALIZATIONS

Current literature is rich in generalizations of these models. Hickman and Miller [6] consider models with and without sufficient statistics for estimating parameters of the distributions involved. Jewell [7] has introduced numerous variations in credibility models using least-squares* theory. He gives results for a time-varying, multidimensional risk and for multidimensional data assumed to arise from different but related units.

Bühlmann [3] has applied minimax* concepts to credibility. He has also sought to include credibility theory within a comprehensive premium determination system consistent with decision theory* (see ref. 2).

References

[1] Bailey, A. (1945). *Proc. Casualty Actuarial Soc.*

[2] Bühlmann, H. (1970). *Mathematical Methods in Risk Theory*. Springer-Verlag, New York.

[3] Bühlmann, H. (1975). In *Credibility: Theory and Applications*, P. M. Kahn, ed. Academic Press, New York, pp. 1–18.

[4] Hewitt, C. C. (1975). In *Credibility: Theory and Applications*, P. M. Kahn, ed. Academic Press, New York, pp. 171–179.

[5] Hickman, J. (1975). In *Credibility: Theory and Applications*, P. M. Kahn, ed. Academic Press, New York, pp. 181–192.

[6] Hickman, J. and Miller, R. (1975). In *Credibility: Theory and Applications*, P. M. Kahn, ed. Academic Press, New York, pp. 249–270.

[7] Jewell, W. (1975). In *Credibility: Theory and Applications*, P. M. Kahn, ed. Academic Press, New York, pp. 193–244.

[8] Kahn, P. M., ed. (1975). *Credibility: Theory and Applications*, Academic Press, New York.

[9] Mayerson, A. L. (1964). *Proc. Casualty Actuarial Soc.*

(ACTUARIAL STATISTICS
BAYESIAN INFERENCE
BAYES' THEOREM
POSTERIOR DISTRIBUTION
PRIOR DISTRIBUTION
RISK THEORY)

P. M. KAHN

CRISSCROSS DESIGNS   *See* STRIP-PLOT
DESIGNS

CRITICALITY THEOREM

The criticality theorem of branching process* theory delineates the conditions under which the descendants of a single initial ancestor will ultimately become extinct with probability 1, or will, with positive probability, survive. In its fundamental form it involves the prior assumptions that all individuals produce offspring independently, and that the probability distribution of number of offspring of any one individual is identical with that of the initial ancestor. Denoting by p_r, $r = 0, 1, 2, \ldots$, the probability that any one individual has r offspring, then, if Z_n, $n = 0, 1, 2, \ldots$ is the total number of individuals in the nth generation ($Z_0 = 1$), $p_r = \Pr[Z_1 = r]$, and $G(s) = \sum_{r=0}^{\infty} p_r s^r$, $s \in [0, 1]$, is the probability generating function (PGF)* of Z_1. The mean number of offspring per individual is, thus, $m = \sum_{r=0}^{\infty} r p_r = G'(1)$. If we exclude the triv-

ial case $p_1 = 1$, the criticality theorem asserts that if $m \leq 1$ extinction will occur with probability $q = 1$; but if $m > 1$, the probability q of ultimate extinction is the unique root of the equation $G(s) = s$ in the interval $0 \leq s < 1$, and thus $q < 1$. (Thus the "critical value" of m at which there is a transition of behavior structure, is $m = 1$.) The proof (see, e.g., refs. 3 and 5) depends first on noticing that if $G'(1-) \leq 1$, then the equation $G(s) = s$, involving the convex function $G(s)$, has only the root $q = 1$ in $s \in [0, 1]$; but if $G'(1-) > 1$, it has one additional root denoted by q in this closed interval. The second crucial aspect is that the PGF

$$G_n(s) = \sum_{r=0}^{\infty} \Pr[Z_n = r]s^r, \quad s \in [0, 1],$$

is the nth functional iterate of $G(s)$ (i.e., $G_n(s) = G(G_{n-1}(s))$, where $G(s) = G_1(s)$), and that, consequently, $\Pr[Z_n = 0] = G_n(0) \to q$ as $n \to \infty$, irrespective of the value of $G'(1-)$.

The theorem is important partly because of the breadth of the practical applicability of the underlying description (or model) [3, 5]. The individuals in the stochastic process* $\{Z_n\}$ may for example be (as in the original applications) direct male descendants of a single ancestor (the situation of extinction of surnames), or carriers of copies of a mutant gene, electrons in an electron multiplier, neutrons in a nuclear chain reaction, branch units in a polymer molecule [4], or branches emanating from a point of propagation in crack growth [13]. There are also (e.g., in ref. 3) subtle applications to the theory of queues (*see* QUEUEING THEORY), specifically to the $M/G/1$ system. Even though the independence assumptions will tend to break down in practice if numbers become large, particularly in biological applications, the value q calculated under these assumptions in the case $m > 1$ will nevertheless often provide a good approximation to the true situation [5].

From a theoretical aspect, the process $\{Z_n\}$, $n \geq 0$, usually called the Galton–Watson process*, is the most extensively studied example of both an absorbing Markov* chain on a countable state space, and of a branching process*, as a result of the functional-iteration property. Its structure is easily generalized to several types of particles (the multitype Galton–Watson process) with an accompanying criticality theorem; this generalization is of very great applicability, for example, in population genetics and polymer chemistry. There are generalizations to other settings, which have resulted in extensive theories developed on the basis of corresponding criticality theorems [1, 5, 8, 11, 12].

The criticality theorem has a fascinating history. I. J. Bienaymé* gave a completely correct statement of it in 1845 [2], but his contribution has passed unnoticed till recently [6]. An excellent account of historical developments, dating from the partly correct statement in 1873–1874 of F. Galton* and H. W. Watson, in which scholars such as R. A. Fisher*, J. B. S. Haldane, W. P. Elderton, A. K. Erlang, and J. F. Steffensen (who was the first to give a detailed proof with a clear statement in 1930) also figure, is given in ref. 9.

References

[1] Athreya, K. B. and Ney, P. (1972). *Branching Process.* Springer-Verlag Berlin. (A standard reference on the theory of branching processes.)

[2] Bienaymé, I. J. (1845). *Société Philomatique de Paris—Extraits*, Ser. 5, pp. 37–39. (Also in *L'Institut, Paris*, **589**, 131–132; and reprinted in ref. 10.)

[3] Feller, W. (1968). *An Introduction to Probability Theory and Its Applications*, 3rd ed., Vol. 1. Wiley, New York. (See especially Secs. XII.4 and XII.5.)

[4] Flory, P. J. (1953). *Principles of Polymer Chemistry.* Cornell University Press, Ithaca, N.Y. (Chapter IX, especially pp. 352–353.)

[5] Harris, T. E. (1963). *The Theory of Branching Processes.* Springer-Verlag, Berlin. (A standard reference on the theory of branching processes. Chapter 1, Secs. 1 to 7, deal with the criticality theorem and applications.)

[6] Heyde, C. C. and Seneta, E. (1972). *Biometrika*, **59**, 680–683.

[7] Heyde, C. C. and Seneta, E. (1977). *I. J. Bienaymé: Statistical Theory Anticipated.* Springer-Verlag, New York. (Section 5.9 expands on ref. 6 in relation to the criticality theorem with references and discussion.)

[8] Jagers, P. (1975). *Branching Processes with Biological Applications.* Wiley, New York.

[9] Kendall, D. G. (1966). *J. Lond. Math. Soc.*, **41**, 385–406.

[10] Kendall, D. G. (1975). *Bull. Lond. Math. Soc.*, **7**, 225–253. (An attempt to reconstruct Bienaymé's thinking, in a sequel to refs. 6 and 9.)

[11] Mode, C. J. (1971). *Multitype Branching Processes.* American Elsevier, New York.

[12] Sevastyanov, B. A. (1971). *Vetviashchiesia Protsessi* [*Branching Processes*]. Nauka, Moscow. (Standard Russian-language reference.)

[13] Vere-Jones, D. (1977). *Math. Geol.*, **9**, 455–481. (Pages 460–461 give references to earlier manifestations of the criticality theorem in this setting.)

(BRANCHING PROCESSES
MARKOV PROCESSES
QUEUEING THEORY
STOCHASTIC PROCESSES)

E. Seneta

CRITICAL PHENOMENA

The term "critical phenomena" as used in statistical physics* refers to the properties of certain statistical-mechanical models which imitate the behavior of a fluid near its liquid–vapor critical point, where the coexisting liquid and vapor phases become identical, or a ferromagnet near its Curie or critical temperature, where the spontaneous magnetization disappears. The best known model of this type is the Ising model in two dimensions, whose solution by Onsager [7] initiated the modern work on the subject. A brief description of this model will serve to illustrate the major points of interest.

Let l and m be integers labeling points on a square lattice. With each point is associated a variable σ_{lm} taking the values $+1$ and -1, and $\boldsymbol{\sigma}$ denotes the collection of these variables. On a finite square $-L \leqslant l \leqslant L$, $-L \leqslant m \leqslant L$, the joint probability distribution is of the Gibbs* or Boltzmann type:

$$\Pr[\boldsymbol{\sigma}] = Z^{-1}\exp[H(\boldsymbol{\sigma})], \qquad (1)$$

where

$$H(\boldsymbol{\sigma}) = K\sum_l \sum_m (\sigma_{lm}\sigma_{l+1,m} + \sigma_{lm}\sigma_{lm+1})$$

$$(2)$$

represents a *coupling* or *interaction* between σ's on neighboring sites, and the partition sum

$$Z = \sum_{\sigma} \exp\left[H(\sigma) \right] \qquad (3)$$

serves to normalize the distribution in (1). The parameter K in (2) is a positive number (which in physical applications is inversely proportional to the temperature).

This model exhibits various types of behavior in the "thermodynamic limit" in which L tends to infinity. In particular, the free energy (omitting factors involving the temperature)

$$f = \lim_{L \to \infty} L^{-2} \ln Z \qquad (4)$$

is an analytic function of $K \geqslant 0$ except at the critical point,

$$K = K_c = 0.4406 \ldots = 0.5 \sinh^{-1}(1),$$

where its second derivative diverges as $|\ln|K - K_c||$. The correlation* functions for two points l, m and l', m' on the lattice,

$$E\left[\sigma_{lm} \sigma_{l'm'} \right], \qquad (5)$$

evaluated using the distribution (1) and taking the limit $L \to \infty$, have the property that their dominant behavior when the separation

$$r = \left[(l' - l)^2 + (m' - m)^2 \right]^{1/2} \qquad (6)$$

between sites is large is exponential, i.e., $\exp(-r/\xi)$ times a power of r, for $K < K_c$, and algebraic, as $r^{-1/4}$, for $K = K_c$. For $K > K_c$ there is an exponential decrease to a finite positive value, denoted by M^2 (the square of the spontaneous magnetization) as r becomes infinite.

As one might expect, the changeover from exponential to algebraic decay of correlations is signaled by an increase of ξ, the correlation length, to infinity as K approaches K_c from above or below; in fact,

$$\xi \propto |K - K_c|^{\nu} \qquad (7)$$

with $\nu = 1$. Also, M goes to zero as K decreases to K_c,

$$M \propto (K - K_c)^{\beta} \qquad (8)$$

with $\beta = 1/8$. Note that (7) and (8) hold only for K close to K_c; i.e., they give the dominant behavior upon approaching the critical point. The quantities β and ν are typical examples of "critical exponents."

The divergence of the correlation length, the appearance of an algebraic decay with distance in certain correlation functions, and the nonanalytic behavior of f are the principal characteristics of critical phenomena in Ising and similar statistical models. All of these have their counterparts in experimentally measurable effects in physical systems such as carbon dioxide near its liquid–vapor critical point.

Recent studies of critical phenomena (see the References) have focused on the following sorts of questions:

1. To what extent do other model systems exhibit the same qualitative critical behavior as the two-dimensional Ising model?

2. What are the values of the critical exponents, of which β and ν in (6) and (7) are examples, in various models, and how do they depend on the dimensionality d?

3. What is the effect of introducing variations in the function $H(\sigma)$ in (2)?

4. What is the behavior of other correlation functions analogous to (5) but involving only one σ, or products of three σ's or four σ's, etc.?

The generalization of (2) to $d = 3$ is obvious: the lattice is a cubic lattice and each σ interacts with six nearest neighbors. Unfortunately, it has thus far not been possible to find an exact solution for this or any of the models discussed below for $d \geqslant 3$. Even for $d = 2$ it is only in exceptional cases that one has exact solutions for various generalizations of the Ising model. Hence the available answers to the questions posed above depend largely on approximation methods, such as series expansions or renormalization-group transformations, whose errors cannot be estimated rigorously. Nevertheless, the variety of different approaches makes it possible to provide at least some answers, with varying degrees of reliability, to all of these questions.

One of the conclusions to emerge from such studies is that while the $d = 2$ Onsager solution provides a good qualitative guide to critical behavior in three dimensions, its quantitative results must be modified. Experiments and theory indicate that $\beta \simeq 0.32$ and $\nu \simeq 0.67$ for the $d = 3$ Ising model, quite different from the $d = 2$ values, and suggest that

$$d^2f/dK^2 \propto |K - K_c|^{-\alpha} \qquad (9)$$

with $\alpha \simeq 0.1$, in contrast to the logarithm for $d = 2$. For $d \geqslant 5$ it is believed that the critical behavior is qualitatively the same as found in some of the older approximate theories in which $\beta = \nu = 1/2$, while for $d = 4$ there are small (logarithmic) corrections to the older results.

It is possible to modify (2) in various ways so as to introduce additional parameters. The case where the interaction K has a different value for sites in the same column as for sites in the same row was studied by Onsager, and the results are similar to those already discussed. If one adds to (2) a term

$$h \sum_l \sum_m \sigma_{lm} \qquad (10)$$

with h (proportional to the magnetic field divided by the temperature) a real number not equal to zero, it is known from rigorous arguments that no critical point occurs for $K > 0$ and $h \neq 0$, even though the model has not been solved exactly. However, the presence of a critical point at $h = 0$ manifests itself in the fact that various quantities, including f and ξ, exhibit singularities as h tends to zero while $K = K_c$. For any $K > K_c$, all the correlations involving a product of an odd number of σ's, including $E[\sigma_{lm}]$, are discontinuous functions of h at $h = 0$. Such a discontinuity is referred to as a "first-order" phase transition.

Other modifications of (2) that have been considered include: the addition of interactions between pairs of σ's which are nearby, but not nearest neighbors; the possibility of interactions between all pairs of sites, with a strength decreasing with some inverse power of r; interactions involving products of three, four, or more σ's; etc. All of these except the first are believed capable of giving rise to significant changes in critical behavior, both for $d = 2$ Ising models and for $d > 2$. Another interesting class of models arises when K in (2) is negative, or when some of the additional pair interactions just discussed are negative. Some of these "antiferromagnets" can give rise to correlation functions which, in the thermodynamic limit, lack the translational symmetry of the lattice.

There are also many interesting models which involve more fundamental modifications of (2). In the n-vector model each σ_{lm} is an n-component (real) vector of unit length, and the products in (2) and (5) are to be interpreted as vector dot products. The sum in (3) is replaced by integrals over the unit spheres at each site.

No exact solutions (except for $d = 1$, where there is no critical point) are available for $n \geqslant 2$. For $d = 2$ it is believed that for $n = 2$ the correlations decay algebraically, rather than exponentially, for all K larger than some K_c, whereas for $n \geqslant 3$ the decay is exponential for all K. Rigorous arguments show that if $d = 2$ the correlations (5) always decay to 0 as $r \to \infty$ for $n \geqslant 2$; the decay to a positive M^2 as found in the Ising model is impossible. (These arguments do not apply for $d \geqslant 3$.)

Still other models have been studied in which the "object" located on a lattice site takes a finite number of values greater than 2, or in which it is a quantum mechanical operator (i.e., a matrix) in a suitable tensor product space. The number of possible models is obviously extremely large, and hence the need for some organizing or classification principle. The one that is currently most popular is the proposal that critical points fall into certain "universality classes," all points in the same class having identical critical exponents, and possibly other characteristics in common. One hopes that in the space of parameters such as K in (2), h in (10), and strengths of further-neighbor interactions, etc., critical points in the same universality class will lie on smooth manifolds. Points that lie at the limits of manifolds of different universality classes will then have a special behavior; such "multicritical" points are the focus of much current research. The main support for the universality hypothesis just discussed comes from renormalization-group calculations, which also provide a mathematical mechanism for explaining the breakdown of universality in certain two-dimensional models.

Literature

The three 1967 review articles [2, 4, 5], although now somewhat out of date, still contain much valuable material. Domb and Green [1] is a standard reference work for the subject of critical phenomena. Stanley's book [9] is introductory. The other two books [6, 8] stress the renormalization-group approach to the subject, and Fisher's review [3] provides a good introduction to this approach.   .

References

[1] Domb, C. and Green, M. S., eds. (1972). *Phase Transitions and Critical Phenomena*. Academic Press, London. (Several volumes.)
[2] Fisher, M. E. (1967). *Rep. Prog. Phys.*, **30**, 615–730.
[3] Fisher, M. E. (1974). *Rev. Mod. Phys.*, **46**, 597–616.
[4] Heller, P. (1967). *Rep. Prog. Phys.*, **30**, 731–826.
[5] Kadanoff, L. P., Götze, W., Hamblen, D., Hecht, R., Lewis, E. A. S., Palciauskas, V. V., Rayl, M., Swift, J., Aspnes, D., and Kane, J. (1967). *Rev. Mod. Phys.*, **39**, 395–431.
[6] Ma, S.-K. (1976). *Modern Theory of Critical Phenomena*. W. A. Benjamin, Reading, Mass.
[7] Onsager, L. (1944). *Phys. Rev.*, **65**, 117.
[8] Pfeuty, P., and Toulouse, G. (1975). *Introduction to the Renormalization Group and to Critical Phenomena*. Wiley, New York.
[9] Stanley, H. E. (1971). *Phase Transitions and Critical Phenomena*. Oxford University Press, New York.

(GIBBS DISTRIBUTION
LATTICE THEORY
STATISTICAL PHYSICS)

ROBERT B. GRIFFITHS

CRITICAL REGION

A test of significance* is defined by a rule for deciding to " accept" or "reject" as specified statistical hypothesis* on the basis of a set of observed values of random variables. Those sets of values that lead to (formal) rejection are said to constitute the "critical region" of the test.

(HYPOTHESIS TESTING
POWER
SIGNIFICANCE LEVEL
SIGNIFICANCE LIMIT
TESTS OF SIGNIFICANCE)

CRONBACH α *See* PSYCHOLOGICAL TESTING THEORY

CROSS DATING   *See* ARCHAEOLOGY, STATISTICS IN; DENDROCHRONOLOGY

CROSS VALIDATION   *See* JACKKNIFE

CROSSINGS

A term used in time-series* analysis. The crossing of a level (L) consists of two successive observed values, one greater than and the other less than L. The term *level crossing* is also used.

(STOCHASTIC PROCESSES
TIME SERIES)

CROSSOVER DESIGNS   *See* CHANGEOVER DESIGNS

CROSS-PRODUCT RATIO   *See* CONTINGENCY TABLES; ODDS RATIO

CROSS-SECTIONAL DATA

Data obtained at (more or less) the same time, as opposed to *longitudinal data** obtained by observation over a relatively extended period of time.

(CLINICAL TRIALS
FOLLOW UP)

CRUDE MOMENT

The hth crude moment of a random variable X is the expected value* of the hth power of the variable. Symbolically, $\mu_h' = E[X^h]$. The term "hth crude moment about (a fixed value) A" is sometimes applied to the quantity $E[(X - A)^h]$. When A is zero, the quantity is called simply "hth crude moment," the words "about zero" being omitted.

(ABSOLUTE MOMENTS
CENTRAL MOMENTS
FACTORIAL MOMENTS)

CRUDE SUM OF SQUARES

The statistic $\sum_{i=1}^{n} X_i^2$ as distinct from the "sum of squares" $\sum_{i=1}^{n} (X_i - \bar{X})^2 = \sum_{i=1}^{n} X_i^2 - (\sum_{i=1}^{n} X_i)^2/n$. The term "crude" is used especially in connection with analysis-of-variance* calculations, to distinguish from the sums of squares appearing in the ANOVA tables.

CUBE LAW *See* ELECTION FORECASTING IN THE U.K.

CUBE-ROOT TEST OF HOMOGENEITY

This test of variances was introduced by Samiuddin [2] for testing the null hypothesis $H_0: \sigma_1^2 = \cdots = \sigma_k^2$ for the standard k normal population samples situation. It serves as an alternative to Bartlett's M test* [1]. A modified version motivated by the Wilson–Hilferty transformation* [4] was proposed by Samiuddin et al. [3]. The test statistic is

$$X_2^2 = \tfrac{9}{2} \sum_{i}^{k} \nu_i \{ T_i - (1 - b_i) T \}^2 / T^2,$$

where $\nu_i = n_i - 1$, n_i being the sample size of the ith sample, $b_i = (2/9)\nu_i^{-1}$, $T_i = (s_i^2)^{1/3}$, and $T = (\sum_{i=1}^{k} \nu_i s_i^2 / \nu)^{1/3}$, $\sum_{i=1}^{k} \nu_i = \nu$. For large ν_i, $1 - b_i$ is close to 1, which yields the simplified formula $X_3^2 = \tfrac{9}{2} \sum^k \nu_i (t_i - T)^2 / T^2$. Under H_0, X_2^2 and X_3^2 asymptotically (for large ν_i) have the χ_{k-1}^2 distribution. At the 1% significance level, X_3^2 seems to be more powerful than X_2^2.

References

[1]   Bartlett, M. S. (1937). *Proc. R. Soc. Lond. A*, **160**, 268–282.

[2]   Samiuddin, M. (1976). *J. Amer. Statist. Ass.*, **71**, 515–517.

[3]   Samiuddin, M., Hanif, M. and Asad, H. (1978). *Biometrika*, **65**, 218–221.

[4]   Wilson, E. B. and Hilferty, M. M. (1931). *Proc. Natl. Acad. Sci. USA*, **17**, 684–688.

(BARTLETT'S TEST OF
   HOMOGENEITY OF VARIANCES
WILSON–HILFERTY
   TRANSFORMATION)

CUBIC EFFECT *See* LINEAR REGRESSION; POLYNOMIAL REGRESSION

CUMULANTS

The coefficients $\{\kappa_j\}$ of $\{(it)j/j!\}$ in the expansion in powers of t of $\chi(t) = \log \phi(t)$, where $\phi(t)$ is the characteristic function* are called cumulants (or semi-invariants). The name "semi-invariant" is due to the fact that these coefficients—except for κ_1—are invariant for translation* of the corresponding random variable. The relation between cumulants κ_j, the moments about the origin μ_j', and the central moments μ_j is given below.

$$\kappa_1 = \mu_1' = \mu$$

  (the mathematical expectation)

$$\kappa_2 = \mu_2' - \mu_1'^2 = \sigma^2$$

  (the variance)

$$\kappa_3 = \mu_3' - 3\mu_1' \mu_2' + 2\mu_1'^3 = \mu_3$$

  (the third central moment)

$$\kappa_4 = \mu_4' - 3\mu_2'^2 - 4\mu_1' \mu_3' + 12\mu_1'^2\mu_2 - 6\mu_1'^4$$

$$= \mu_4 - 3\mu_2^2,$$

and conversely

$$\mu_1' = \kappa_1$$

$$\mu_2' = \kappa_2 + \kappa_1^2$$

$$\mu_3' = \kappa_3 + 3\kappa_2\kappa_1 + \kappa_1^3$$

$$\mu_4' = \kappa_4 + 3\kappa_2^2 + 4\kappa_1\kappa_3 + 6\kappa_1^2\kappa_2 + \kappa_1^4.$$

The normal distribution has the characterizing property that all its cumulants of order 3 and higher are zero. (*See also* FACTORIAL CUMULANTS, MOMENT GENERATING FUNCTION, SEMI-INVARIANTS, and entries for specific distributions.)

Bibliography

Johnson, N. L. and Kotz, S. (1969). *Distributions in Statistics, Discrete Distributions*. Wiley, New York, Chap. 1.

Kendall, M. G. and Stuart, A. (1978). *The Advanced Theory of Statistics*, 3rd ed., Vol. 1. Charles Griffin, London.

CUMULATIVE DAMAGE MODELS

Cumulative damage models have been applied principally in three different areas. These are (a) to predict the times of failure of mechanical systems for reasons of maintenance or safety; (b) to calculate the safety of structures, or of structural components, for duration of load effects on members bearing dynamic loads over time; and (c) to determine the tolerance levels, for health or safety, to persons occasionally exposed to toxic or latently injurious materials.

The first utilization of a theory of cumulative damage was in engineering, by Palmgren [17], who sought to calculate the life length of ball bearings. He used a deterministic formula giving the life of metallic components sustaining repetitions of combinations of different stresses, to wit: If under repetitions of the ith load the component will last μ_i cyles for $i = 1, \ldots, k$, then under repetitions of a spectrum of service loads each of which contains n_i applications of the ith load, the number N of such spectra be-

fore failure is

$$N = \left[\sum_{i=1}^{k} (n_i/\mu_i) \right]^{-1}. \qquad (1)$$

This result gives the average damage rate per cycle as the harmonic mean* of the rates of accumulated damage per spectrum. This result, rederived by Miner [16], is often called Miner's cumulative damage rule and it is still used in certain applications.

With increasing demands of technology a more precise calculation of the fatigue life of metals became a major scientific problem. Many deterministic models, alternate to (1), were advanced to overcome its failure to account for either scatter or influence of load order. See, e.g., Freudenthal and Heller [9] and Impellizzeri [12]. For an overview, see Madayag [14]. In all of these models damage was accumulated in a deterministic manner as a function of each duty cycle.

During this period some competing models, using statistical distributions to account for observed variability in fatigue life, were proposed. Among these distributions were the log-normal* and the gamma*, but the most successful was an extreme value distribution advocated by Weibull* [24]. Various damage models relating stress to life were proposed for these distributions (see Weibull [25, Sec. 85]).

Several authors suggested stochastic cumulative damage models in which the incremental damage for the ith work cycle was a random variable X_i. If we denote cumulative damage for $k \geqslant 1$ by

$$Z_0 = 0, \qquad Z_k = X_1 + \ldots + X_k, \qquad (2)$$

then the number of such cycles before cumulative damage exceeds a critical level x_t, which may decrease with age t, is

$$\overline{H}(t) = P[Z_t \leqslant x_t] \qquad \text{for} \quad t = 1, 2, \ldots.$$

Parzen [18] introduced renewal theory* to evaluate this probability. Sweet and Kozin [23] related cumulative damage from a harmonic stress cycle to the stress–strain hysteresis loop thus taking into account order of load applications.

Birnbaum and Saunders [2] gave a stochastic interpretation to Miner's rule assuming incremental damage was a variate "new better than used in expectation" (N.B.U.E.)*. Later [3] they proposed a cumulative damage model also based on renewal theory. Recently, Bogdanoff [4, 5] has advanced a model using Markov chain theory, which is history dependent and attempts to relate the parameters of the distribution of fatigue life to the physical constants governing the failure mechanism. This paper also contains a survey of some realistic engineering aspects of the problem.

Some textbooks that provide expository accounts of statistical theory as related to cumulative damage are refs. 1, 7, 10, and 15. Some reference to fatigue models that are related more to physical concepts and require more knowledge of fracture mechanics and engineering are refs. 8, 11, and 13.

The fatigue of metals and composite materials has provided the first stimulus for the development of mathematical models of cumulative damage, which, as a consequence, are among the most mathematically advanced and most closely related to the actual physical process of fatigue.

One of the general models of cumulative damage with simple (not to say inapplicable) mathematical assumptions results when shocks occur randomly in time accordingly to a Poisson process* with a specified intensity rate, say λ. (See Barlow and Proschan [1, pp. 91 ff.].)

Assume that the ith shock will case a random nonnegative amount of damage, say X_i, for $i = 1, 2, \ldots$. It is again postulated that the device fails when a critical amount of damage, call it $x > 0$, has been accumulated from the repeated shocks. The survival probability of such a device, i.e., the probability that it does not fail in the interval $[0, t]$, is, for $0 \leqslant t < \infty$, given by

$$\bar{H}(t) = \sum_{k=0}^{\infty} \frac{e^{-\lambda t}(\lambda t)^k}{k!} \bar{P}_k(x).$$

The word "shock" could be replaced by "dosage" in another application.

Here $e^{-\lambda t}(\lambda t)^k / k!$ is the Poisson* probability of exactly k shocks in $[0, t]$, while $\bar{P}_k(x)$ is the probability that the cumulative damage due to the k shocks has not exceeded the critical level x. From (2) we have for $k = 1, 2, \ldots$,

$$\bar{P}_0(x) = 1 \quad \text{and} \quad \bar{P}_k(x) = P[Z_k \leqslant x].$$

There are three separate assumptions concerning the distribution of damage which may obtain. These assumptions say, equivalently, that (a) the damage due to the kth shock, given the previous shocks, depends only upon the accumulated damage and not, say, upon the order of occurrence of prior shocks; (b) higher accumulated damage lowers resistance to further damage; and (c) for any given accumulation of damage, successive shocks become more severe. It is known that the distribution $H = 1 - \bar{H}$ has a failure rate that increases on the average (IFRA); *see* HAZARD RATE CLASSIFICATION OF DISTRIBUTIONS.

One may further specialize the model above by assuming that the damage generated at each shock is independent of all others. In this case, letting $X_i \sim F_i$ for $i \geqslant 1$ and $*$ denote convolution*,

$$\bar{P}_k = F_1 * \cdots * F_k \quad \text{for} \quad k \geqslant 1.$$

But further, the damages may also be identically distributed, in which case $\bar{P}_k = F^{k*}$ is the k-fold convolution of the common damage distribution F having support on $[0, \infty)$.

With the addition of physical identification of the parameters of the distribution F, this model is useful in specific applications when the hypotheses can be satisfied. Two important, but restrictive, assumptions were that the order of the occurrence of shocks is not important and that there was an exponential distribution of time between shocks. In some instances these are unrealistic and other assumptions must be made.

If a renewal process, rather than just a Poisson process, governs the time between shocks, say $N(t)$ is the renewal random variable counting the number of shocks in $[9, t]$, then using the notation previously introduced,

$$\overline{H}(t|x) = P[Z_{N(t)} \leqslant x]$$

$$= \sum_{k=0}^{\infty} P[N(t) = k]\overline{P}_k(x)$$

is the probability of cumulative damage not exceeding x by time t. In this form cumulative damage is related to storage theory* for reservoirs and can be treated using the same mathematical methods; see refs. 1 and 7.

In certain applications the cumulative damage does not result from sporadic shocks but from the gradual wear, or accretion of damage, which occurs in use at different rates, depending upon variations in both the work environment and the strength of the component. In such cases one may often describe the amount of cumulative damage S_t at time $t \geqslant 0$ as a stochastic process, which in practice, is often modeled as the solution of a differential equation having random coefficients. From its probabilistic behavior one must calculate the distribution H of the random time T until failure by

$$\overline{H}(t) = P[T \geqslant t] = P[S_t \leqslant x],$$

where x is the critical amount of damage that can be tolerated. In many specific instances this is so laborious that Monte Carlo methods (*see* SIMULATION) are used to determine the distribution. General classes of life distributions that show stochastic wear are studied by Hanes and Singpurwalla in Proschan and Serfling [20] and by Bryson and Siddiqui [6].

Under certain assumptions S_t may be a Gaussian process*, in which case this time of failure T has an "inverse Gaussian distribution"* and exact formulas can be obtained.

In specific damage models, for duration of load effects, the order in which loads occur is important. (It is also frequently true in health applications that the order of stress plays a role.) In such cases cumulative damage is not just a sum but a specific function of all loads encountered up to time t, of which there are a random number $N(t)$. This cumulative damage, call it $S_t = f(X_1, \ldots, X_{N(t)})$, governs life at any time

t with failure occurring when for the first time it exceeds the residual strength (or health) (see ref. 26 and the references there).

Another damage model, which has been studied for calculating the strength of cables or ropes, called bundles, as well as for structural components formed from composite materials, involves the redistribution of the imposed load upon the surviving components within a multicomponent structure. In a bundle the load is distributed equally to all surviving components. In a structure the load is redistributed to the surviving components in a manner determined by the design of the structure.

Some general results are known when the failure rate of each component is proportional to some nonanticipatory functional of the imposed load. The asymptotic strength properties for a large number of fibers have been determined (see ref. 19 and the references given there).

Another so-called "damage model" occurs when a random number of objects, such as eggs, has a Poisson distribution* with parameter λ, each one of which is independently subjected to damage with probability p. By letting X and Y denote the resulting number of damaged and undamaged objects, respectively, it follows that X has a Poisson distribution with parameter λp, and

$$P[X = r] = P[X = r|Y = 0]$$

$$= P[X = r|Y \neq 0].$$

This property is a characterization of the Poisson distribution shown by Rao and Rubin [21], and also of the Poisson process (see ref. 22).

References

[1] Barlow, R. E. and Proschan, F. (1975). *Statistical Theory of Reliability and Life Testing*. Holt, Rinehart and Winston, New York.

[2] Birnbaum, Z. W. and Saunders, S. C. (1968). *SIAM J. Appl. Math.*, **16**, 637–652.

[3] Birnbaum, Z. W. and Saunders, S. C. (1969). *J. Appl. Prob.*, **6**, 328–337.

[4] Bogdanoff, J. L. (1978). *J. Appl. Mech.*, **45**, 246–257.

[5] Bogdanoff, J. L. (1978). *J. Appl. Mech.*, **45**, 733–739.

[6] Bryson, M. C. and Siddiqui, M. M. (1969). *J. Amer. Statist. Ass.*, **64**, 1472–1483.

[7] Cox, D. R. (1969). *Renewal Theory*. Wiley, New York.

[8] Freudenthal, A. M. (1974). *Eng. Fract. Mech.*, **6**, 775–793.

[9] Freudenthal, A. M. and Heller, R. A. (1959). *J. Aerosp. Sci.*, **26**, 431–442.

[10] Gertsbakh, I. B. and Kordanskiy, Kh. B. (1962). *Models of Failure*. Engineering Science Library. Springer-Verlag, New York.

[11] Heller, R. A., ed. (1972). Probabilistic Aspects of Fatigue. *Amer. Soc. Test. Mater. Spec. Tech. Publ.* 511.

[12] Impellizzeri, L. F. (1968). Effects of Environment and Complex Load History on Fatigue Life. *Amer. Soc. Test. Mater. Spec. Tech. Publ.* 462, pp. 40–68.

[13] Liebowitz, H., ed. (1976). *Progress in Fatigue and Fracture*. Pergamon Press, Elmsford, N.Y.

[14] Madayag, A. F. (1969). *Metal Fatigue: Theory and Design*. Wiley, New York.

[15] Mann, N. R., Schafer, R. E., and Singpurwalla, N. D. (1974). *Methods for Statistical Analysis of Reliability and Life Data*. Wiley, New York.

[16] Miner, M. A. (1945). *J. Appl. Mech.*, **12**, A159–A164.

[17] Palmgren, A. (1924). *Z. Ver. Dtsch. Ing.*, **68**, 339–341.

[18] Parzen, E. (1959). On Models for the Probability of Fatigue Failure of a Structure. *NATO 245*.

[19] Phoenix, S. L. and Taylor, H. N. (1973). *Adv. Appl. Prob.*, **5**, 200–216.

[20] Proschan, F. and Serfling, R. J. (1974). *Reliability and Biometry, Statistical Analyses of Life Length*. SIAM, Philadelphia.

[21] Rao, C. R. and Rubin, H. (1964). *Sankhyā A*, **26**, 295–298.

[22] Srivastava, R. C. (1971). *J. Appl. Prob.*, **8**, 615–616.

[23] Sweet, A. L. and Kozin, F. (1968). *J. Mater.*, **3**, 802–823.

[24] Weibull, W. (1951). *ASME J. Appl. Mech.*, **18**, 293–297.

[25] Weibull, W. (1961). *Fatigue Testing and Analysis of Results*. Pergamon Press, Elmsford, N.Y., Sec. 85.

[26] Whittemore, A. S., ed. (1977). *Environmental Health: Quantitative Methods*. SIAM, Philadelphia.

(LIFE TESTING
MIXTURE DISTRIBUTIONS
WEIBULL PROCESS)

Sam C. Saunders

CUMULATIVE DISTRIBUTION FUNCTION (CDF)

The probability that a random variable X does not exceed a value x, regarded as a function of x, is called the cumulative distribution function of x. A common notation is

$$F_X(x) = \Pr[X \leqslant x].$$

From the definition, it follows that $0 \leqslant F_X(x) \leqslant 1$, and $F_X(x)$ is a nondecreasing function of x.

The notation $F_X(x)$ is intended to indicate that the (mathematical) function of x represents properties of the random variable X. The symbol $F_X(X)$, for example, represents a random variable: the *probability integral transform** of X.

If $F_X(x) \to 0$ as $x \to -\infty$ and $F_X(x) \to 1$ as $x \to \infty$, the distribution is called *proper*; otherwise, it is called *improper*.

CUMULATIVE SUM CONTROL CHARTS

The classical control chart* procedures as proposed by Shewhart [39] are based on a one-point rule where the process is said to be out of control if the last plotted point falls outside the control limits*. Such procedures are equivalent to the repeated application of the fixed sample-size test. The advantages of these charts are their simplicity and their ability to detect large changes quickly. The main disadvantage is that they are slow in signaling small or moderate changes. Several modifications to the one-point rule have been proposed to overcome this disadvantage and to improve the performance of these procedures.

A fundamental change in the classical procedure was proposed by Page [30], who suggested constructing control charts based on sums of observations rather than individual observations. This system of charting takes full advantage of the historical record and provides a rapid means of detecting shifts in the process level. Starting from a given point, all subsequent plots contain information from all the points up to and including the plotted point. The ordinate at the rth point equals the ordinate at the $(r - 1)$st point plus the value of a statistic computed from the current sample. Thus the ordinate is the sum of the cumulated values of a statistic and hence the name cumulative sum control or cusum chart in brief.

During the 25 years since their introduction, much work has been done on the theoretical as well as the practical aspects of cusum charts. They have been used extensively in the chemical industry and for control of manufacturing processes. The cusum techniques have also been found useful in the control of sales forecasts, determination of restructuring points in a computerized data base [43], etc.

This article discussed the basic theory and the main characteristics of the cusum charts.

OPERATION

Consider the use of a cusum chart to detect deviations in the mean level of a process characteristic, X, with $E[x] = \mu$ and $\text{var}[X] = \sigma^2$. Let μ_a be the acceptable quality level (AQL) or the in-control state and μ_r be the rejectable quality level (RQL) or the out-of-control state. Sometimes, μ_a is also called the "target" value. Samples of size n are taken at regular intervals and cumulative sums of statistics are plotted against the sample number. The method of computing a statistic from the sample and the decision criterion depend on whether the chart is to detect deviations in only one direction, positive or negative (one-sided chart), or in both directions (two-sided chart).

The original procedure suggested by Page [30] to detect one-sided deviations is to assign a score d_j to the jth sample and plot the sums $\sum_{j=1}^{r} d_j$ against r. For a scheme to detect positive deviations, the system of scoring is so chosen that the mean path of the plotted points is downward when the process is in control and upwards with it is out of control. Action is taken at the first rth sample for which

$$\sum_{j=1}^{r} d_j - \min_{0 \leqslant j < r} \sum_{j=1}^{r} d_j \geqslant h, \quad (1)$$

where h is called the decision limit. In other words, the decision that the process mean has increased is taken after $\sum_{j=1}^{r} d_j$ has changed direction and has exceeded the previous lowest result by a value h or more. A similar procedure is used for detecting a decrease in the process mean.

A commonly used scoring scheme consists of subtracting a constant k from each sample mean $\bar{x}_j = \sum_{t=1}^{n} x_{jt}/n$ and plotting the cumulative sums $\sum_{j=1}^{r}(\bar{x}_j - k)$ against r. Such a scheme is called a cusum chart with reference value k. For detecting positive deviations (μ_a to μ_{r_1}) the reference value is denoted by k_1 and by k_2 for detecting negative deviations (μ_a to μ_{r_2}). It is concluded that a positive change has occurred whenever

$$\sum_{j=1}^{r} (\bar{x}_j - k_1) - \min_{0 \leqslant j < r} \sum_{j=1}^{r} (\bar{x}_j - k_1) \geqslant h^+ \quad (2)$$

and a negative change has occurred whenever

$$\max_{0 \leqslant j < r} \sum_{j=1}^{r} (\bar{x}_j - k_2) - \sum_{j=1}^{r} (\bar{x}_j - k_2) \geqslant h^-. \quad (3)$$

The quantities h^+ and h^- are the decision limits for the two one-sided charts.

An abbreviated version of the procedure above is commonly employed when the chart is to be used only to check deviations and not for record keeping. For checking positive deviations, say, the results are cumulated starting with a value $\bar{x}_j > k_1$ and continued until the sum of $(\bar{x}_j - k_1)$'s either returns below zero or exceeds h^+. The plotting is started afresh at zero after either occurrence. This method of plotting is identical to the repeated application of a Wald sequential test* with boundaries at 0 and h^+ with the starting point always at the acceptance boundary, or it can be interpreted as a random walk* between a reflecting barrier and an absorbing barrier. This method can also be used to set up tables of successive results without any plotting if no graph is desired.

Two methods of plotting have been suggested if the cusum chart is to be employed to detect both negative and positive shifts in the process mean. The first is to run concurrently two one-sided charts: one each for positive and negative shifts with parameters (k_1, h^+) and (k_2, h^-), respectively.

The second method proposed by Barnard [3] is to subtract the target value μ_a from each $\bar{x}_j$ and plot the sums

$$S_r = \sum_{j=1}^{r} (\bar{x}_j - \mu_a)$$

against r. If the process mean remains at the target value, the plots will not deviate too much from the horizontal. To check this, a V-mask is placed at the last plotted point (r, S_r) with its vertex pointing horizontally forward at a distance d, the lead distance (Fig. 1). A shift in the process mean is indicated if the arms of the V-mask obscure any previously plotted point; the origin ($S_0 = 0$) counts as a previous point for this purpose. A positive shift is indicated if the lower arm obscures the points and a negative shift is indicated when the points are obscured by the upper arm. For simplicity only, symmetric deviations in μ are considered in this article; i.e., it is assumed that $\mu_{r_1} = \mu_a + \delta\sigma$ and $\mu_{r_2} = \mu_a - \delta\sigma$, where $\delta\sigma$ is the deviation from the target mean. Then the V-mask is symmetric with half-angle ϕ and the control procedure is determined by d and ϕ. As a modification, Lucas [26] has

proposed the inclusion of a parabolic section inside and tangential to the V. This modification provides a quicker detection of large changes in the process mean.

The visual picture of the chart will obviously depend on the choice of the scale factor, w, i.e., on the ratio of the plotting scales for S_r and r. A recommended choice is to take $w = 2\sigma/\sqrt{n}$ so that a shift of 2σ units in the process mean, or equivalently a shift of $2\sigma/\sqrt{n}$ units in the sample mean, will change the average slope of the plotted points from 0 to 45°.

A V-mask for monitoring a process mean is shown in Fig. 1. For this illustration the target value μ_a is 100 units the process standard deviation σ is known to be 10 units, the sample size is 4, and shifts of 1 standard deviation are to be detected, i.e., $\delta\sigma = 10$. Specifications are $d = 2.6$ in units of the horizontal scale and $\tan\phi = \frac{1}{2}$ (methods for determining d and $\tan\phi$ are described later in the article). The scale factor for this chart is $w = 2(10)/\sqrt{4} = 10$. Samples of size 4 are taken from the process and cumulative sums S_r are computed as shown in Table 1. These S_r values are plotted and the status of the process is judged by placing the V-mask after each sample as shown (for sample numbers 4 and 9) in Fig. 1. For this illustration the first indication that the process is out of control is given by the chart at sample

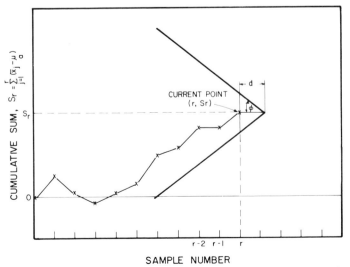

Figure 1   Use of a V-mask.

Table 1   Computations for a V-Mask Chart

Sample Number	$\bar{x}_j$	$\bar{x}_j - 100$	$S_r = \sum\limits_{j=1}^{r} (\bar{x}_j - 100)$
1	110	10	10
2	102	2	12
3	102	2	14
4	106	6	20
5	109	9	29
6	99	−1	28
7	110	10	38
8	109	9	47
9	110	10	57

9 and the plot suggests an upward shift in the mean.

The V-mask scheme is equivalent to the simultaneous operation of two one-sided charts with

$$k_1 = (\mu_a + \mu_{r_1})/2,$$
$$k_2 = (\mu_a + \mu_{r_2})/2,$$
$$|h^+| = |h^-| = h.$$

The equivalence relationships as shown by Kemp [23] are

$$\tan \phi = \frac{k_1 - k_2}{2w} \tag{4}$$

and

$$d = \frac{h}{w \tan \phi} . \tag{5}$$

In other words, the results will be exactly equivalent if a process is monitored by either of the schemes defined by the parameters in (4) and (5).

RUN-LENGTH DISTRIBUTION AND AVERAGE RUN LENGTH (ARL)

The run length is the number of samples taken at a given process level before the chart indicates a lack of control. The average run length (ARL) is simply its expected value and has been commonly used as a measure of the effectiveness of a cusum chart.

For a one-sided chart, let $L(0)$ represent the run length when the plotting starts at 0.

Page [30] and Ewan and Kemp [9] obtained the approximate distribution of $L(0)$ when the observations are normally distributed. At AQL the distribution of $L(0)$ can be approximated by a geometric distribution* while at RQL the distribution is not geometric but has the desirable property that the probability of obtaining a run length less than the ARL is large and the probability of getting a run length greater than, say three times the ARL, is small.

Brook and Evans [5] and Shah [38] studied the distribution of $L(0)$ using a Markov chain approach which gives exact results for a discrete distribution and accurate approximation when the distribution is continuous.

The ARL of a cusum chart is given by Page [30]:

$$\mathrm{ARL} = \frac{N(0)}{1 - P(0)} . \tag{6}$$

$P(0)$ and $N(0)$ are the special cases $z = 0$ of $P(z)$ and $N(z)$, where

$$P(z) = \int_{-\infty}^{-z} f(x)\,dx + \int_0^h P(x) \cdot f(x - z)\,dx,$$
$$0 \leqslant z \leqslant h, \tag{7}$$

$$N(z) = 1 + \int_0^h N(x) \cdot f(x - z)\,dx,$$
$$0 \leqslant z \leqslant h, \tag{8}$$

and $f(x)$ is the probability density function of the increments in the cumulative sum. Equations (7) and (8) are Fredholm integral equations of the second kind. For the case when x is normally distributed, these equations have been solved either numerically or by approximation to obtain the ARL values [9, 13, 22, 30, 41]. Tables of $P(0)$, $N(0)$, and ARL for $(\mu - k) = -5(0.2)5$ and $h = 0.5(0.5)10$ along with bounds on the computed values are also available [12].

Approximations to the average run length have been developed using the analogy between cusum and Wald's sequential probability ratio test* (SPRT) and by using a Brownian motion* approximation to the cumulative sum [36]. The first approximation assumes that the $\bar{x}$'s are normally distributed while the Browinan motion approximation does not require this assumption. The latter, however, underestimates ARL.

To get the ARL of a two-sided scheme or the V-mask, the ARL values L_1 and L_2 for the two equivalent one-sided charts are obtained as above. The relationship [23]

$$(L_v)^{-1} = (L_1)^{-1} + (L_2)^{-1} \qquad (9)$$

is then used to calculate L_v, the ARL of the equivalent one-sided chart or the V-mask.

DESIGN

The design consists of finding n, h, and k (k_1 or k_2) for a one-sided scheme or n, d, and ϕ for a V-mask scheme such that a desired criterion is satisfied. A commonly used criterion is to select those values of the parameters that yield desired ARLs at AQL and RQL. Another method [20] is to regard the cusum as a sequential sampling* procedure "in reverse" [17] and obtain the parameters that satisfy the type I and type II risks. The third method is based on minimizing the long-run average cost for a given set of cost and risk factors [14, 40]. The cost criterion also enables the determination of the sampling interval. The first two methods are relatively easy, while the third requires a computer solution. A simplified method that approximately satisfies the cost criterion is given by Chiu [6]. A brief description of the first two methods is given below.

ARL Criterion

For a one-sided scheme, the design consists of determining n, h, and k for specified ARL values L_a and L_r at AQL and RQL, respectively. It is assumed that the process standard deviation σ is known or can be reliability estimated. The reference value k is usually taken to be between AQL and RQL, i.e., $k = (\mu_a + \mu_r)/2$. Such a choice provides maximum discrimination between the two values of μ. However, it has not been conclusively shown that such a choice is optimum. The values of h and n can then be determined from a nomogram [13] which gives contours of L_a and L_r in the $(h\sqrt{n}/\sigma, |\mu - k|\sqrt{n}/\sigma)$ plane.

To design a two-sided scheme using a symmetric V-mask, the values of n, d, and ϕ can be obtained from the nomogram in Fig. 2. This nomogram gives the contours of the ARLs of the V-mask in the $(d, (|\mu_a - \mu_r|\sqrt{n})/(2\sigma))$ plane. To illustrate the use of this nomogram, consider a process that is to be maintained at $\mu_a = 50$. The standard deviation (σ) of the process is 5. A cusum chart with a V-mask is to be used to detect deviations of $\delta\sigma = 5$ in either direction, i.e., $\mu_{r_1} = 55$ and $\mu_{r_2} = 45$. The desired ARLs for the V-mask are $L_{va} = 250$ and $L_{vr} = 5$. In other words, the chart is expected to give a false alarm 0.4% of the times and indicate that the process is in control when it is not about 20% of the times.

The procedure for designing the chart is as follows. Locate the point of intersection of the contours $L_{va} = 250$ and $L_{vr} = 5$. The ordinate of this point is $|\mu_a - \mu_r|\sqrt{n}/(2\sigma) = 0.73$ (Fig. 2). For given μ_a, μ_r, and σ, calculate $n = \{0.73(2)(5)/5\}^2 = 2.132$. Round off n to, say, 2, and compute $(|\mu_a - \mu_r|\sqrt{n})/(2\sigma) = 5\sqrt{2}/2 \times 5 = 0.707$ as the new ordinate. Draw a horizontal line at 0.707 and note its intersection with either $L_{va} = 250$ or $L_{vr} = 5$, depending upon which ARL is intended to be fixed. The abscissa of the point of intersection with $L_{va} = 250$ gives a value $d = 4.5$. If w is taken to be 10, the tangent of the half-angle ϕ can now be computed from $(w\sqrt{n}/\sigma) \tan \phi = 0.707$ as

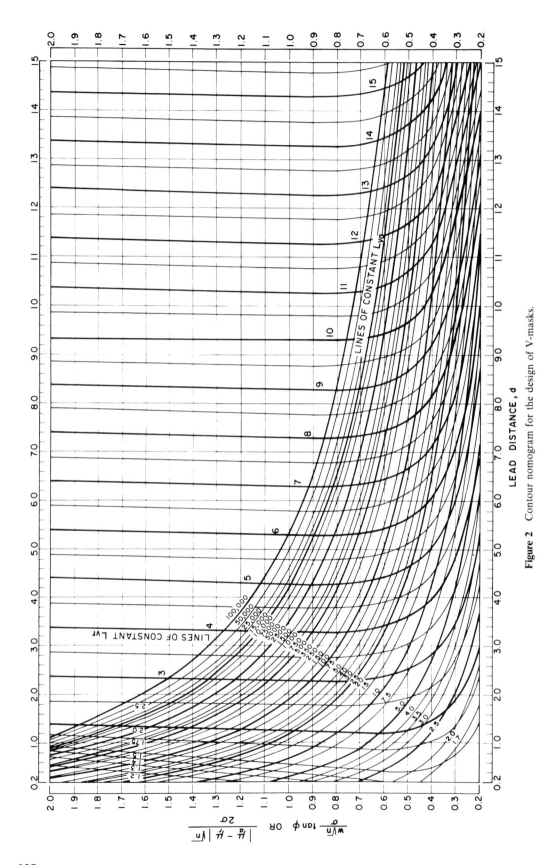

Figure 2 Contour nomogram for the design of V-masks.

$\tan \phi = (0.707)(5)/10(\sqrt{2}) = 0.25$. Thus a cusum scheme satisfying the requirements will consist of taking samples of size 2 and plotting $\sum_{j=1}^{r}(\bar{x}_j - 50)$ against r on a chart where the distance between 10 units on the vertical scale is the same as a unit distance on the horizontal scale. A V-mask with $d = 4.5$ horizontal units and $\phi = \arctan(0.25)$ is to be employed to detect deviations in the mean.

To obtain the dimensions of a V-mask based on its equivalence to sequential sampling "in reverse," [17], the values of $\delta\sigma$, α, and β must be specified. Here α is the probability of wrongly indicating a shift in the process level from μ_a to $\mu_a + \delta\sigma$ or $\mu_a - \delta\sigma$ and β is the probability of failure to detect such a shift. In other words, α represents the type I error for a two-sided test while β is the type II error. For β small, the dimensions of the V-mask are approximated by the following equations [21, p. 368]:

$$d = -\frac{2}{\delta^2 n}\ln\left(\frac{\alpha}{2}\right) \qquad (10)$$

and

$$\phi = \arctan\left(\frac{\delta\sigma}{2w}\right), \qquad (11)$$

where w, as defined earlier, is the scale factor. It should be noted that the charts based on (10) and (11) will have the same ARL when the process is in control even though the proportion of false alarms will not be exactly equal to α.

PERFORMANCE COMPARISON WITH $\bar{X}$-CHARTS

An $\bar{X}$-chart* is used for process control or quality control and consists of three horizontal lines drawn at μ_a (centerline), $\mu_a + B\sigma/\sqrt{n}$ (upper control limit), and $\mu_a - B\sigma/\sqrt{n}$ (lower control limit), where B is the control limit factor and is usually taken to be 3 (see CONTROL CHARTS). Samples of size n are taken at specified intervals and sample averages $\bar{x}_j$ are plotted against the sample number. If a plotted point falls outside the control limits, the process is said to be *out of control*.

Even though several modifications to this

procedure have been proposed in the literature [7, 21], this basic form of the Shewhart chart* is still extensively used because of its simplicity and ease of operation. For this reason it is worthwhile to compare the performance of the cusum chart with that of the $\bar{X}$-chart.

Three criteria can be used for this purpose: ARL, sample size, and cost. For the ARL criterion, both charts are designed for a specified $\delta\sigma$ and a fixed L_a. The ARLs are then compared for various shifts in the process level. For example, if $\delta\sigma = 1$ and $L_a = 125$, the L_r value for the $\bar{X}$-chart is 8 and for the cusum chart is 20.3 (Fig. 3). If both charts are designed to detect the same shift in the process level with equal risks, both the sample size and the cost criteria favor the cusum chart [11], even though the savings in cost may not be significant in all cases.

A cusum chart will be better than the corresponding $\bar{X}$-chart based on any of the three criteria if the shift in the process mean is equal to the design value. This is so because in a cusum procedure information from previous samples is explicitly used in detecting a lack of control while for an $\bar{X}$-chart the decision is based solely on the most recent sample. However, the superiority of cusum is diminished for $\delta\sigma$ greater

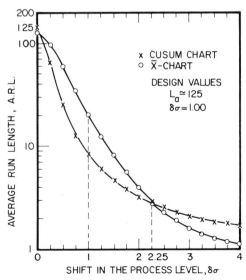

Figure 3 Comparison of $\bar{X}$ and cusum charts based on ARL criterion.

than the design value. In fact, in Fig. 3, for $\delta\sigma > 2.25$, the $\bar{X}$-chart catches a shift faster than the cusum chart.

CORRELATED OBSERVATIONS

A few studies have been reported to investigate the effect of serial correlation on the run-length distribution and ARL of a cusum scheme. Using a Monte Carlo simulation approach, Goldsmith and Whitfield [15] found that positive correlation reduces the ARL of a V-mask while negative correlation increases it. Bagshaw and Johnson [1] through a Wiener process approximation for the cumulative sums noted that serial correlation has a major influence on the run-length distribution. Some exact results have also been reported by Barasia [2]. Mikhailov [28] has given conditions under which a sequence of processes, formed by sequences of cumulative sums of dependent variables, converges in some sense to a Poisson process.

DEVIATIONS FROM NORMALITY

Kemp [24] pointed out the dangers of erroneously using the normal theory results for the nonnormal case by considering a generalized gamma variate. Bissell [4] showed that if a distribution is positively (negatively) skewed, L_a decreases (increases) and L_r is little affected except in the extreme cases. Using Monte Carlo simulation, Bissell also developed a simple nomogram to account for skewness in designing a cusum chart.

DISTRIBUTIONS OTHER THAN NORMAL

Cusum charts for binomial and Poisson variates to control fractions defective and number of defects, respectively, have been discussed by Page [30], Ewan and Kemp [9], Bissell [4], and Johnson and Leone [21]. They also provide tables and formulae for designing the appropriate charts. Cusum

schemes for controlling range, standard deviation, etc., have also been discussed by these authors. Charts for the folded normal and Weibull distributions are discussed by Johnson [18, 19], and cusum procedures to detect changes in distribution for the gamma family have been investigated by Regula [35].

OTHER PERTINENT ASPECTS

Models for finite horizon and long-run time-average costs of a finite jump process have been studied by Shah [38] for the Markovian case and by Barasia [2] for the semi-Markovian case. The plot of the cumulative sums $\sum_{j=1}^{r}(\bar{x}_j - \mu_a)$ can be used for estimating the point of an apparent significant change in the process level. Hinkley [16] has derived the asymptotic distribution of such an estimate and associated test statistics, McGilchrist and Woodyer [27] and Pettitt [34] have discussed nonparametric* approaches to the change-point problem, and Schweder [37] has discussed the use of cusum for detecting structural shifts in regression problems.

Literature

Page [33], Ewan [8], and the monograph by Woodward and Goldsmith [44] are recommended for introductory material. Chapters in Duncan [7] and Wetherill [42] describe the underlying theory and methodology, while Johnson and Leone [21] present the subject matter based upon its equivalence to sequential sampling "in reverse." Page [30], Barnard [3], Ewan and Kemp [9], and van Dobben de Bruyn [41] are good references for the basic theory. Survey papers by Gibra [10] and Montgomery [29] are useful references for a perspective of cusum and other control charts.

References

[1] Bagshaw, M. and Johnson, R. A. (1975). *Technometrics*, **17**, 73–80.

[2] Barasia, R. K. (1975). Economic Design and ARL of Cusum and Generalized $\bar{x}$-Charts for a

Process Mean under Semi-Markovian Deterioration in the Presence of Correlation and Nonnormality. Ph.D. dissertation, Syracuse University.

[3] Barnard, G. A. (1959). *J. R. Statist. Soc. B*, **21**, 239–271.

[4] Bissell, A. F. (1969). *Appl. Statist.*, **18**, 1–30.

[5] Brook, D. and Evans, D. A. (1972). *Biometrika*, **59**, 539–549.

[6] Chiu, W. K. (1974). *Appl. Statist.*, **23**, 420–433.

[7] Duncan, A. J. (1974). *Quality Control and Industrial Statistics*, 4th ed. Richard D. Irwin, Homewood, Ill.

[8] Ewan, W. D. (1963). *Technometrics*, **5**, 1–22.

[9] Ewan, W. D. and Kemp, K. W. (1960). *Biometrika*, **47**, 369–380.

[10] Gibra, I. N. (1975). *J. Quality Tech.*, **7**, 183–192.

[11] Goel, A. L. (1968). A Comparative and Economic Investigation of $\bar{x}$ and Cumulative Sum Control Charts. Ph.D. dissertation, University of Wisconsin.

[12] Goel, A. L. (1979). Tables of $P(0)$, $N(0)$ and ARL for Cusum Charts Based on the Nystrom Method. *Tech. Rep. No. 79-14*, Dep. of Industrial Engineering and Operations Research, Syracuse University, Syracuse, N.Y.

[13] Goel, A. L. and Wu, S. M. (1971). *Technometrics*, **13**, 221–230.

[14] Goel, A. L. and Wu, S. M. (1973). *Manag. Sci.*, **19**, 1271–1282.

[15] Goldsmith, P. L. and Whitfield, H. (1961). *Technometrics*, **3**, 11–20.

[16] Hinkley, D. V. (1971). *Biometrika*, **58**, 509–523.

[17] Johnson, N. L. (1961). *J. Amer. Statist. Ass.*, **56**, 835–840.

[18] Johnson, N. L. (1963). *Technometrics*, **5**, 451–458.

[19] Johnson, N. L. (1966). *Technometrics*, **8**, 481–491.

[20] Johnson, N. L. and Leone, F. C. (1962). *Ind. Quality Control*, **18, 19** (three parts).

[21] Johnson, N. L. and Leone, F. C. (1976). *Statistics and Experimental Design in Engineering and the Physical Sciences*, Vol. 2, 2nd ed. Wiley, New York.

[22] Kemp, K. W. (1958). *J. R. Statist. Soc. B*, **20**, 379–386.

[23] Kemp, K. W. (1961). *J. R. Statist. Soc. B*, **23**, 149–153.

[24] Kemp, K. W. (1967). *Technometrics*, **9**, 457–464.

[25] Kemp, K. W. (1971). *J. R. Statist. Soc. B*, **33**, 331–360.

[26] Lucas, J. M. (1973). *Technometrics*, **15**, 833–847.

[27] McGilchrist, C. A. and Woodyer, K. D. (1975). *Technometrics*, **17**, 321–325.

[28] Mikhailov, V. G. (1974). *Theory Prob. Appl.*, **19**, 403–407.

[29] Montgomery, D. G. (1980). *J. Quality Tech.*, **12**, 75–87.

[30] Page, E. S. (1954). *Biometrika*, **41**, 100–115.

[31] Page, E. S. (1954). *J. R. Statist. Soc. B*, **16**, 136–139.

[32] Page, E. S. (1957). *Biometrika*, **44**, 248–252.

[33] Page, E. S. (1961). *Technometrics*, **3**, 1–9.

[34] Pettitt, A. N. (1979). *Appl. Statist.*, **28**, 126–135.

[35] Regula, G. A. (1976). Optimal Cusum Procedures to Detect a Change in Distribution for the Gamma Family. Ph.D. dissertation, Case Western Reserve University.

[36] Reynolds, M. R. (1975). *Technometrics*, **17**, 65–71.

[37] Schweder, T. (1976). *J. Amer. Statist. Ass.*, **71**, 491–501.

[38] Shah, D. S. (1973). Models for the Economic Design of Some Markovian Control Chart Schemes for a Finite Jump Process. Ph.D. dissertation, Syracuse University.

[39] Shewhart, W. A. (1931). *Economic Control of Quality of Manufactured Product*, Van Nostrand, New York.

[40] Taylor, H. M. (1968). *Technometrics*, **10**, 479–488.

[41] van Dobben de Bruyn, C. S. (1968). *Cumulative Sum Tests: Theory and Practice*. Charles Griffin, London.

[42] Wetherill, G. B. (1977). *Sampling Inspection and Quality Control*, 2nd ed. Halsted Press, New York.

[43] Winslow, L. E. and Lee, J. C. (1975). *Proc. Int. Conf. Very Large Data Bases, Ass. Computing Mach.*, pp. 353–363.

[44] Woodward, R. H. and Goldsmith, P. L. (1964). *Cumulative Sum Techniques*. Oliver & Boyd, Edinburgh.

(ACCEPTANCE SAMPLING
CONTROL CHARTS
HYPOTHESIS TESTING
QUALITY CONTROL
SEQUENTIAL ANALYSIS)

A. L. GOEL

CURRENT INDEX TO STATISTICS

The *Current Index to Statistics: Applications, Methods and Theory* is an annual publication that provides relatively comprehensive indexing coverage for the field of statistics. It views statistics in a very broad sense and the articles range from probability to such topics in applied statistics as how to increase the response rate of mail surveys.

The index includes an author index and a subject index. The subject index is based on all important words in the title of the article and supplemental key words. All articles from a list of over 40 core journals are included as are articles from a wide range of related journals and other publications which are selected by the editor and a network of contributing editors and editorial collaborators. Other helpful information is included, such as a listing of related indexes and information retrieval systems and a full listing of all index sources with addresses.

The ordering of entries in the subject index is based on an adaption of an algorithm developed by Ian Ross and John Tukey (1974–1975) which makes much better use of the information to both the left and right of the "gutter" word. This is a particularly useful enhancement when there is a long string of titles all having the same gutter word. For example, under "distribution," the sorting order for the title "The Uniform Distribution in Signal Detection Theory" is determined by the following "sort key": "Distribution Uniform Signal Detection" instead of "Distribution in Signal Detection Theory," as it would have been under a more naive sorting order. This procedure, called "chunk sorting," is fully explained in the *Index to Statistics and Probability*. by Ross and Tukey [1].

Another novel feature of the subject index is the use of boldface type to isolate the substantive portion of the title being used for sorting. Enough words always appear in boldface to ensure that the boldface portion of any entry differs in some way from the boldface portions of the entries above and below it.

The Current Index is sponsored by the American Statistical Association and the Institute of Mathematical Statistics. The Index commenced publication in 1975. Each volume contains about 550 pages and refers to some 7500 papers. These are derived from complete coverage of about 50 "core" jour-

nals, regular scanning of about 300 others and occasional reference to about a further 250 journals; together with the contents of proceedigs of conferences, symposia, etc., numbering about 15 per year. The price (in 1981) is $13 and the circulation about 3000. Brian L. Joiner served as editor and Jessie Gwynne as associate editor from 1975 to 1979, and James E. Gentle became editor in 1980. His address is IMSL, Inc., 7500 Bellaire Blvd., Houston, Texas 77036.

Reference

[1]   Ross, I. C. and Tukey, J. W. (1974–1975). *Index to Statistics and Probability*, 4 vols. R & D Press, Los Altos, Calif.

CURRENT POPULATION SURVEY   *See*
BUREAU OF LABOR STATISTICS

CURVED EXPONENTIAL FAMILY   *See*
EXPONENTIAL FAMILIES

CURVE FITTING

The problem of curve fitting arises when we look for a mathematical expression relating one set of observed values, say y_i, $i = 1, \ldots, n$, either to a second set of observations z_{ij}, or to a suitable set of functions $Z_{ij} = f_i(z_{i1}, \ldots, z_{im})$. The first case requires the finding of a set of coefficients a_j in the equation

$$\eta_i = \sum_{j=0}^{m} a_j z_{ij}. \qquad (1a)$$

In the second case an appropriate function Z_{ij} must be selected and the coefficients b_j for

$$\eta_i = \sum_{j=0}^{m} b_j Z_{ij} \qquad (1b)$$

in each case be determined, $m < n$, and η_i is an appropriate estimate for y_i. Since (1a) is a special case of (1b), the subsequent solutions will be presented largely for formula (1b).

Before a particular solution is found, a decision on the properties of Z_{ij} must be made. We must select either an orthogonal or a nonorthogonal function Z_{ij}. Solutions for a nonorthogonal system will be discussed first.

THE NONORTHOGONAL CASE

Let us assume that Z_{ij} or z_{ij} is represented by a series of nonorthogonal terms, e.g., X_i^j, where $j = 0, 1, \ldots, m$. We must find the coefficients for

$$\eta_i = \sum_{j=0}^{m} b_j X_i^j \tag{2}$$

Let X_i be a linear function of equal spacing, e.g., $X_k = X_0 + k$, with $k = 0, 1, 2, \ldots, n - 1$, and X_0 being an initial value (in this case the minimum). It is customary to determine the coefficients so that the squared difference between the calculated analytical value η_i and the observed value y_i becomes a minimum for the number of assumed terms in (2). Thus

$$\sum (y_i - \eta_i)^2 \rightarrow \text{minimum}. \tag{3}$$

The solution is well known in mathematics; in matrix notation

$$b_j = (-1)^j D_j / D, \tag{4}$$

where D and D_j are the determinants of the matrix which is defined and written below for Z_{ij}:

$\mathbf{M}_Z =$

$$\begin{bmatrix} 1 & \sum_{i=1}^{n} y_i & \sum y_i Z_{i1} & \sum y_i Z_{i2} & \cdots & \sum y_i Z_{im} \\ b_0 & n & \sum Z_{i1} & \sum Z_{i2} & & \sum Z_{im} \\ b_1 & \sum Z_{i1} & \sum Z_{i1}^2 & \sum Z_{i1} Z_{i2} & \cdots & \sum Z_{i2} Z_{im} \\ \vdots & \vdots & \vdots & \vdots & & \vdots \\ b_m & \sum Z_{im} & \sum Z_{im} Z_{i1} & \sum Z_{im} Z_{i2} & \cdots & \sum Z_{im}^2 \end{bmatrix} \tag{5}$$

(The summation over i is spelled out for the first term and is not repeated.) D denotes the determinant of the minor matrix $\mathbf{M}_{Z_{11}}$ (eliminate row 1 and column 1), and D_j is the determinant of the "cofactor" matrix $\mathbf{M}_{Z_j}$. (Eliminate the row and column in which b_j appears.)

The substitution $Z_{ij} = X_i^j$ leads to a matrix $\mathbf{M}_X$.

Linear and Quadratic Equations

Let us assume that we need

$$\eta_i = b_0 + b_1 X_i + b_2 X_i^2 \tag{6a}$$

or without the quadratic term

$$\eta_i = b_0 + b_i X_i. \tag{6b}$$

The coefficient matrix with $\bar{X} = 0$ can be stated as

$$\mathbf{M}_X = \begin{bmatrix} 1 & \sum y_i & \sum y_i X_i & \sum y_i X_i^2 \\ b_0 & n & 0 & \sum X_i^2 \\ b_1 & 0 & \sum X_i^2 & 0 \\ b_2 & \sum X_i^2 & 0 & \sum X_i^4 \end{bmatrix}. \tag{7}$$

For the quadratic case we derive after some arithmetic:

$$b_0 = \frac{\sum y_i \sum X_i^4 - \sum X_i^2 \sum y_i X_i^2}{n \sum X_i^4 - \left(\sum X_i^2 \right)^2} \tag{8a}$$

$$b_1 = \sum y_i X_i / \sum X_i^2 \tag{8b}$$

$$b_2 = \frac{n \sum y_i X_i^2 - \sum y_i \sum X_i^2}{n \sum X_i^4 - \left(\sum X_i^2 \right)^2}. \tag{8c}$$

In the linear case we find that

$$b_0 = \sum y_i \sum X_i^2 / \left(n \sum X_i^2 \right) = \sum y_i / n \tag{9a}$$

$$b_1 = \sum y_i X_i / \sum X_i^2 \tag{9b}$$

or for $\tilde{y} = \sum y_i / n = 0$:

$$b_0 = 0. \tag{9c}$$

We learn from a comparison of (8a) and (8b) with (9a) and (9b) that the coefficient b_1 remains the same while b_0 changes. This reflects a general rule. When an even-numbered coefficient is added, all even-numbered coefficients change. This rule also applies to odd-numbered coefficients.

The coefficient b_1 is related to the correlation coefficient:

$$b_1 = \rho(y, X) \left[\text{var}(y) / \text{var}(X) \right]^{1/2}. \tag{10}$$

Higher-Order Equations

The addition of higher-order terms makes the use of (4) more complex; it becomes difficult to derive analytical terms for the coefficients. Matrix diagonalization reduces the computational efforts. A short explanation follows. We can derive a set of equations

$$b_{0i}Z_{i0} + b_{1i}Z_{i2} + b_{2i}Z_{i2} + \cdots$$
$$+ b_{mi}Z_{im} = y_i, \quad (11)$$

where $i = 1, 2, \ldots, n$. Because $m < n$ we have more equations than coefficients (oversaturated system). For h coefficients we need only h equations. Consequently, any permutation or selection of h equations leads to a solution. We select a "least-squares" solution. We sum (11) over all $i = 1, \ldots n$, which would leave only one equation. More equations are added by multiplication of (11) by Z_{ik} with $k = 0, 1, \ldots, m$, e.g., $Z_{ik} = X_i^k$. In summary:

$$b_{0k}\sum_{i=1}^{n} Z_{ij}Z_{ik} + b_{1k}\sum Z_{ij}Z_{ik} + b_{2k}\sum Z_{ij}Z_{ik}$$
$$+ \cdots + b_{mk}\sum_{i=1}^{n} Z_{ij}Z_{ik} = \sum y_i Z_{ik}. \quad (12)$$

The expanded writing in (12) is identical with the minor matrix of (5) without the coefficients. The left side forms a matrix and the right side a column vector.

Diagonalization makes the terms of the matrix below the diagonal equal to zero. We present an example for b_0, b_1, and b_2 with $Z_{ij} = X_i^j$ and $\overline{X} = 0$. We derive from (12) or (7):

$$\begin{bmatrix} n & 0 & \sum X_i^2 \\ 0 & \sum X_i^2 & 0 \\ \sum X_i^2 & 0 & \sum X_i^4 \end{bmatrix} \cdot \begin{bmatrix} b_0 \\ b_1 \\ b_2 \end{bmatrix} = \begin{bmatrix} \sum y_i \\ \sum y_i X_i \\ \sum y_i X_i^2 \end{bmatrix}. \quad (12a)$$

The procedure of diagonalization is explained in mathematical texts, e.g., Boas [4], Guest [11], Daniel et al. [5], or Essenwanger [7].

First, we divide row one by n. Since the second row already shows zero in the first column, it is left alone. Row one is now multiplied by $\sum X_i^2$, and this product row is subtracted from the third row, etc. The end product:

$$\begin{bmatrix} 1 & 0 & \sum X_i^2/n \\ 0 & \sum X_i^2 & 0 \\ 0 & 0 & \sum X_i^4 - \left(\sum X_i^2\right)^2/n \end{bmatrix} \cdot \begin{bmatrix} b_0 \\ b_1 \\ b_2 \end{bmatrix}$$
$$= \begin{bmatrix} \sum y_i/n \\ \sum y_i X_i \\ \sum y_i X_i^2 - \sum y_i/n \end{bmatrix}. \quad (12b)$$

Now we reintroduce the b's and write the system in equation form:

$$b_0 + 0 + b_2\sum X_i^2/n = \sum y_i/n \quad (13a)$$

$$b_i\sum X_i^2 = \sum y_i X_i \quad (13b)$$

$$b_2\left[\sum X_i^4 - \left(\sum X_i^2\right)^2/n\right]$$
$$= \sum y_i X_i^2 - \sum y_i/n. \quad (13c)$$

The values of b_2 and b_1 can be determined immediately, then b_0. The sets of (13a) to (13c) are identical with (8a) to (8c). Although mathematical expressions for more than three coefficients are lengthy, the calculation of numerical values for the coefficients is a simple arithmetic problem.

Numerical Example

Given in Table 1 are pairs of observations y_i and $z_i = X_i$. The following matrix $\mathbf{M}_X$ is derived from the data given [minor matrix of (5)]. Analogously to (12a), we write

$\mathbf{M}_X, yX$:

$$\begin{bmatrix} 7 & 0 & 28 & 0 & 196 \\ 0 & 28 & 0 & 196 & 0 \\ 28 & 0 & 196 & 0 & 1588 \\ 0 & 196 & 0 & 1588 & 0 \\ 196 & 0 & 1588 & 0 & 13636 \end{bmatrix}, \begin{bmatrix} 16.8 \\ 18.3 \\ 119.3 \\ 133.5 \\ 1012.7 \end{bmatrix}$$
$$(14a)$$

Table 1   Polynomial Curve Fitting with Nonorthogonal Functions

i	y_i	X_i	η_i (14c)	η_i (14d)
1	3.9	-3	3.42	3.86
2	0.1	-2	1.24	0.23
3	0.2	-1	0.01	0.15
4	1.1	0	-0.11	0.76
5	0.6	1	1.03	1.17
6	2.9	2	3.56	2.55
7	8.0	3	7.64	8.08
Mean	2.4	0	2.4	2.4

With the diagonalized matrix, we have

$$
\begin{bmatrix}
1 & 0 & 4 & 0 & 28 \\
0 & 1 & 0 & 7 & 0 \\
0 & 0 & 1 & 0 & 9.57 \\
0 & 0 & 0 & 1 & 0 \\
0 & 0 & 0 & 0 & 1
\end{bmatrix},
\begin{bmatrix}
2.4 \\
0.65 \\
0.63 \\
0.025 \\
0.084
\end{bmatrix}. \quad (14b)
$$

The result is

$$
\eta_i = -0.108 + 0.478X_i + 0.627X_i^2
$$
$$
+ 0.025X_i^3 \quad (14c)
$$
$$
\eta_i = 0.756 + 0.478X_i - 0.177X_i^2
$$
$$
+ 0.025X_i^3 + 0.084X_i^4. \quad (14d)
$$

Notice the change of even-numbered coefficients from (14c) to (14d).

We obtain from Table 1 the squared difference $\sum_{i=1}^{n}(y_i - \eta_i)^2 = 3.78$ and 0.59 for η_i by (14c) and (14d), respectively. If we assume that $\eta_i = \bar{y} = b_0$, the squared difference $\sum(y_i - \bar{y})^2 = 48.92$. For (14c) and (14d) the residual variance (error) is reduced by 92.3% and 98.8%, respectively. See the section "Residual Variance."

ORTHOGONAL SYSTEMS

The systems of (1a) or (1b) require elaborate arithmetic unless the number (order) of terms is known a priori. An orthogonal system has the advantage that terms can be added without affecting previous coefficients. By definition the Z_{ij}'s are orthogonal

if

$$
\sum_{i=1}^{n} Z_{ij} Z_{ik} = 0 \quad \text{for} \quad j \neq k. \quad (15)
$$

The reader recognizes that only the diagonal term is left in (5) or in (12). Thus the coefficients are

$$
b_j = \sum_{i=1}^{n} y_i Z_{ij} / \sum Z_{ij}^2. \quad (16)
$$

In mathematics and statistics numerous functions have been derived that fulfill (15). Some examples are presented next.

Chebyshev Polynomials

Let us fit the data y_i of Table 1 using orthogonal functions — "$\{\phi_{ij}\}$," where ϕ_{ij} is a polynomial in X_i^k of jth order with coefficients not depending on i but in general depending on the set of values $X_1, \ldots, X_n$. We write

$$
\eta_i = c_0 + c_1\phi_{i1} + c_2\phi_{i2} + c_3\phi_{i3} + c_4\phi_{i4}. \quad (17)
$$

Now $Z_{ij} = \phi_{ij}$. Then

$$
c_j = \sum_{i=1}^{n} y_i\phi_{ij} \Big/ \sum_{i=1}^{n} \phi_{ij}^2 \quad \text{with} \quad c_0 = \bar{y}. \quad (18)
$$

ϕ_{ij} are called Chebyshev polynomials.

Chebyshev polynomials for sets $\{X_i\}$ at equal intervals are listed in various publications, e.g., Pearson and Hartley [15], Beyer [3], etc. They give values of ϕ_{ij} for each of the X_i's in the set. Since the property of orthogonality is not affected by multiplying each ϕ_{ij} by a constant, θ_j, say (depending on j, not on i), the latter can be chosen arbitrarily. It is convenient to make this choice so that the values of the ϕ_{ij}'s are integers as in Table 2. Also, whatever the original values of the X_i's, provided that they are at equal intervals, they can be transformed linearly to make $X_i = \phi_{i1}$, so standard sets of values X_i are used. Usually, these are centered at zero (as in Tables 1 and 2)—for even n they are often taken at 2 units apart (e.g., with $n = 4$, the values $X_1 = -3$, $x_2 = -1$, $X_3 = 1$, $X_4 = 3$ can be used).

The use of continuous orthogonal polynomials* (e.g., Laguerre polynomials, Legendre

Table 2   Curve Fitting with Orthogonal Polynomials

| | Chebyshev Polynomials | | | | | η_i | |
i	ϕ_{i1}	ϕ_{i2}	ϕ_{i3}	ϕ_{i4}	y_i	3rd Order	4th Order
1	-3	5	-1	3	3.9	3.42	3.86
2	-2	0	1	-7	0.1	1.24	0.24
3	-1	-3	1	.1	0.2	0.02	0.16
4	0	-4	0	6	1.1	-0.11	0.75
5	1	-3	-1	1	0.6	1.02	1.16
6	2	0	-1	-7	2.9	3.56	2.55
7	3	5	1	3	8.0	7.65	8.08
$\sum_{i=1}^{n} \phi_{ij}^2$	28	84	6	154	—	—	—
$\sum \eta_i$ or $\sum y_i$					16.8	16.80	16.80

polynomials) for curve fitting with discrete observations may lead to difficulties. Deficiencies and how to solve them have been treated in Essenwanger [7].

We now fit a polynomial to the data of Table 1, using Chebyshev polynomials. The coefficients are

$$
\begin{aligned}
c_0 &= \bar{y} = 16.8/7 &&= 2.4 \\
c_1 &= 18.3/28 &&= 0.654 \\
c_2 &= 52.7/84 &&= 0.627 \quad (19) \\
c_3 &= 0.9/6 &&= 0.15 \\
c_4 &= 22.1/154 &&= 0.144.
\end{aligned}
$$

The fitted fourth-order polynomial is now

$$
\eta_i = 2.4 + 0.654\phi_{i1} + 0.627\phi_{i2}
$$
$$
+ 0.15\phi_{i3} + 0.144\phi_{i4}. \quad (20)
$$

The third-order polynomial (values shown in column 7 of Table 2) is obtained by omitting $0.144\phi_{i4}$. A comparison of the values η_i in Tables 1 and 2 reveals that the result is identical except for rounding differences. The computational effort is considerably reduced, however, by the use of orthogonal polynomials. A curve-fitting process by orthogonal function can be applied only if we are permitted to select the functions. A curve-fitting task relating y_i and z_{ij} will have to follow the procedure for nonorthogonal functions unless z_{ij} is orthogonal or an orthogonal system can be derived for z_{ij}. The process of deriving orthogonal functions from a set of nonorthogonal functions is

described in Essenwanger [7]. (*See also* GRAM–SCHMIDT ORTHOGONALIZATION.)

Residual Variance

In general, some judgment of the goodness of fit is made by the calculation of the residual variance. We can define a residual variance by

$$
\mathrm{var}(y,\eta) = \sum_{i=1}^{n} (y_i - \eta_i)^2 \Big/ n, \quad (21)
$$

although it is usual to decrease the divisor (n) by the number (k, say) of parameters (c's) fitted or the residual sum of squares:

$$
R_k^2 = \sum_{i=1}^{n} (y_i - \eta_i)^2. \quad (22)
$$

When the fit is perfect, $R_k^2 = 0$. The simplest case is a fitting by the mean, i.e., $\eta_i = \bar{y}$, giving

$$
R_0^2 = \sum_{i=1}^{n} (y_i - \bar{y})^2. \quad (23)
$$

Thus the proportional reduction is

$$
P_k = 1 - R_k^2 / R_0^2. \quad (24)
$$

Numerical values have been presented in the numerical example above.

R_k^2, $\mathrm{var}(y,\eta)$, and P_k are not readily obtainable for a nonorthogonal system. A detailed description of the necessary matrix can be found in Essenwanger [7].

In turn, a simple relationship for R_k^2 may be derived in an orthogonal system. Let us define

$$R_{\phi_j}^2 = c_j^2 \sum_{i=1}^{n} \phi_{ij}^2. \qquad (25)$$

Then $R_k^2 - R_{k-1}^2 = R_{\phi_k}^2$ and

$$R_k^2 = R_0^2 - \sum_{j=1}^{k} R_{\phi_j}^2. \qquad (26)$$

$R_0^2 = \sum (y_i - \bar{y})^2 = 48.92$ was given in the example (Table 1). $R_{\phi_1}^2 = (0.654)^2 \cdot 28 = 11.96$, $R_{\phi_2}^2 = 33.06$, $R_{\phi_3}^2 = 0.14$. and $R_{\phi_4}^2 = 3.17$. In summary, $R_4^2 = 48.92 - 48.33 = 0.59$. This residual sum of squares is identical with the calculated value given in the numerical example. In an orthogonal system in the contribution of any individual term is readily evaluated from the coefficients and $\sum \phi_{ik}^2$ [see (18) and (25)]. This sum is already needed for the calculation of the coefficient [see (18)] and does not require new computations (see Table 2). Determination of R_k^2 from $\sum_{i=1}^{n}(y_i - \eta_i)^2$ in the numerical example may not appear to be an elaborate arithmetic task. It becomes elaborate whenever n becomes larger. For a nonorthogonal system it may be easier to calculate η_i and utilize (22) rather than find a solution by matrices. In the orthogonal case the calculation of η_i can be skipped [see (25) and (26)].

Fourier Series

Another orthogonal system in widespread use is the Fourier series, which is based on trigonometric functions. The Fourier series is more appropriately fitted to data sets with periodicity. As a mathematical tool of curve fitting, however, it can also be utilized for the approximation of any arbitrary set of discrete data points. In this case it is advisable to evaluate whether polynomial functions need fewer terms to approximate y_i. For the Fourier series we set

$$b_j Z_{ij} = A_j \sin(jt_i + \alpha_j) \qquad (27)$$

with

$$t_i = 2\pi i/p, = 360 \cdot i/p, \qquad (28)$$

where $i = 1, \ldots, p$. The first expression in (28) is applicable for radians, the second expression for angular degrees, and p denotes the "basic period" as a reference cycle. In general, $p \equiv n$.

By substituting (27) into (1b), we cast

$$\eta_i = A_0 + \sum_{j=1}^{k} A_j \sin(jt_i + \alpha_j). \qquad (29)$$

The coefficients A_j and α_j, called *amplitude* and *phase angle*, respectively, are determined from

$$A_0 \equiv \bar{y} \qquad (30a)$$

$$A_j = \left(a_j^2 + b_j^2\right)^{1/2}, \qquad (30b)$$

where

$$a_j = (2/p) \sum_{i=1}^{p} y_i \sin jt_i \qquad (31a)$$

$$b_j = (2/p) \sum_{i=1}^{p} y_i \cos jt_i \qquad (31b)$$

$$\tan \alpha_j = b_j/a_j. \qquad (31c)$$

The reader should note that a_j and b_j are not identical with the notation used in (1a) and (1b). The angular association of α_j can be found from

	b_j	a_j		b_j	a_j
2nd	+	−	1st	+	+
3rd	−	−	4th	−	+

where the ordinal number refers to the quadrant (i.e., 1st = from 0 to 90°, etc.).

Because every Fourier term in (27) has two unknown parameters (A_j and α_j), the summation over k stops at $k = n/2 = p/2$ for p even and $(n - 1)/2 = (p - 1)/2$ for p odd. For p even, the last term is

$$a_k = 0 \qquad (32a)$$

$$b_k = (1/p) \sum_{i=1}^{p} (-1)^i y_i. \qquad (32b)$$

For example, for $p = 7$ we find $k = 3$, and (32a) and (32b) do not apply; a_3 and b_3 are calculated from (31a) and (31b). Because the variance of a Fourier term (wave) is known a priori, the residual variance simplifies:

$$R_{F_j}^2 = A_j^2/2 \qquad (33a)$$

and

$$R_k^2/n = \text{var}(y) - \sum_{j=1}^{k} A_j^2 \Big/ 2. \qquad (33b)$$

Although the data set of Table 1 does not include a periodicity, it was decided to exemplify the Fourier analysis for this data set to illustrate to the reader that the Fourier series is a mathematical tool of curve fitting, and that a display of periodicity is not necessary.

The following coefficients have been computed:

$$A_0 = 2.40$$

$$A_1 = 3.12, \qquad \alpha_1 = 90.5°$$

$$A_2 = 1.98, \qquad \alpha_2 = 74.2°$$

$$A_3 = 0.58, \qquad \alpha_3 = 91.7°.$$

This leads to the fitted equation

$$\eta_i = 2.4 + 3.12\sin(t_i + 90.5)$$
$$+ 1.98\sin(2t_i + 74.2)$$
$$+ 0.58\sin(3t_i + 91.7). \qquad (29a)$$

Table 3 shows the result of computing η_i by (29a) with one, two, and three trigonometric terms. It is obvious from Table 3 that we have an exact fit for $k = 3$. (The differences are due to rounding errors.) The variance of the individual term and percentage reduction is

$$R_{F_1}^2 = 9.74/2 = 4.87 \qquad (P_{F_1} = 69.7\%),$$

$$R_{F_2}^2 = 3.90/2 = 1.95 \qquad (P_{F_2} = 27.9\%),$$

$$R_{F_3}^2 = 0.34/2 = 0.17 \qquad (P_{F_3} = 2.4\%);$$

$$\sum_{j=1}^{3} R_{F_j}^2 = 13.98/2 = 6.99.$$

Earlier we calculated $\sum (y_i - \bar{y})^2 = 48.92$. This value compares to $n\sum_{j=1}^{3} R_{F_j}^2$, i.e., $7 \times 6.99 = 48.93$, which is an excellent agreement. The first term provides 69.7% of the variance. This corresponds to $45.02/48.92 = 92.0\%$ from the polynomial representation (i.e., two polynomial terms, or up to the

Table 3 Result of Computing η_i by (29a)

		Number of Terms		
i	y_i	One	Two	Three
1	3.9	4.33	4.43	3.90
2	0.1	1.68	− 0.27	0.11
3	0.2	− 0.42	0.34	0.19
4	1.1	− 0.40	1.21	1.10
5	0.6	1.73	0.25	0.60
6	2.9	4.37	3.41	2.89
7	8.0	5.52	7.42	8.00

second order, because of two parameters, A_1 and α_1). In this case curve fitting by polynomials leads to a good approximation faster. After the second Fourier term we have 2.4% of the variance left; the corresponding fitting by polynomials after the fourth term is $0.59/48.92 = 1.2\%$.

SMOOTHING AND CURVE FITTING

Various techniques are available for smoothing of existing data or sets of data. From the previous discussions in curve fitting we may recognize that curve fitting can also be a tool for smoothing. Because $m < n$ in (1a) or (1b), the calculated η_i is an approximation of y_i. If the data set y_i has many irregular, random or small-scale fluctuations in its sequence, limiting $m \ll n$ will suppress these undesirable fluctuations. Consequently, curve-fitting techniques may serve as smoothing tools (see GRADUATION.)

Polynomials

Let us assume that a set of data y_i is given with $i = 1, \ldots, n$. A primary technique of smoothing is overlapping means:

$$\bar{\eta}_i = \sum_{k=i-s}^{i+s} y_k \Big/ s, \qquad (35)$$

where s defines the range k of the temporal or spatial smoothing of the data set. For example, if $s = 2$ and $i = 3$, $k = 1, \ldots 5$. For $i = 4$, $k = 2, \ldots, 6$, etc. Thus five data points of the set would be utilized for overlapping averaging. Consequently, $s < i <$

$n - s$, and the number of smoothed points is reduced to $n - 2s$.

Smoothing by overlapping means generates a bias if the data set y_i has maxima or minima because these are blunted. Although this bias can be reduced by introducing weighting functions,

$$\bar{\eta}_i = \sum_{k=1-s}^{i+s} w_k y_k \bigg/ \sum w_k, \qquad (35a)$$

it may sometimes be more advantageous to smooth by overlapping polynomials:

$$\eta_i = \sum_{j=1}^{m} b_j Z_{ij}, \qquad (36)$$

where $m \leqslant 2s + 1$. Then polynomial fitting for b_j takes place over $2s + 1$ data points equivalently to the overlapping mean case. We need only the center point η_i from (36) except for the margins. For these we obtain data points from $i = 1, \ldots, s$ and $i = n - s$ to n in contrast to the overlapping means, because the first data set provides a fitting from y_i for $i = 1, \ldots, 2s + 1$. The last set includes $i = n - (2s + 1)$ to n. If a weighting of the data is intended, then we could expand the right-hand expression:

$$\eta_i = \sum_{j=1}^{m} b_j w_i Z_{ij} \qquad (36a)$$

(see later unequal spacing). Other details can be found in a forthcoming text by Essenwanger [8].

Smoothing by overlapping polynomials is illustrated with Table 4. The fitting by overlapping means and polynomials is based on five data points. The first two columns list i and y_i. Following next is the overlapping mean $\bar{\eta}_i$. The adjacent column provides the addition to (or subtraction from) the mean $\bar{\eta}$ by adding the fitting of a second-order polynomial from five data points (see ϕ_{ji} in the last two columns of Table 4). Finally, η_i from the polynomial fitting is given under η_i. As expected, $\bar{\eta}_i$ and η_i differ because the concepts of smoothing in both cases are different.

The selection of the smoothing method, by and large, is a subjective decision, as is the number of polynomial terms. We could

Table 4   Smoothing by Overlapping Means and Polynomials

i	y_i	$\bar{\eta}_i$	$\Delta\eta_i^a$	η_i	ϕ_{1i}	ϕ_{2i}
1	4.1	—	0.34	5.76	-2	2
2	9.2	—	0.13	5.55	-1	-1
3	4.4	5.42	-0.04	5.38	0	-2
4	2.9	5.30	-1.24	4.06	1	-1
5	6.5	5.60	-1.54	4.06	2	2
6	3.5	6.38	0.26	6.64	$\sum\phi^2 = 10$	14
7	10.7	6.60	1.74	8.34		
8	8.3	—	0.85	7.45		
9	4.0	—	-1.78	2.22		

$^a \Delta\eta_i = b_1\phi_{1i} + b_2\phi_{2i}$.

add a third polynomial term, but for an odd number of points the central $\phi_{3i} = 0$. Thus a change of η_i for $i = 3$ to 7 will occur only by adding a fourth-order term. In our case, however, $2s + 1 \equiv m \equiv 5$ and $\eta_i \equiv y_i$ for the four-term polynomial solution. The following equations have been determined for the given set of y_i:

$i = 1, \ldots, 5:$ $\eta_i = 5.42 - 0.15\phi_{1i} + 0.021\phi_{2i},$

$i = 2, \ldots, 6:$ $\eta_i = 5.30 + 5.37\phi_{1i} + 0.621\phi_{2i},$

$i = 3, \ldots, 7:$ $\eta_i = 5.60 + 0.25\phi_{1i} + 0.771\phi_{2i},$

$i = 4, \ldots, 8:$ $\eta_i = 6.38 + 0.15\phi_{1i} - 0.129\phi_{2i},$

$i = 5, \ldots, 9:$ $\eta_i = 6.60 - 0.02\phi_{1i} - 0.871\phi_{2i}.$

Curve Fitting by Spline Functions*

Let us assume that a set of data y_i is given as in Table 4. If these discrete data were plotted into a diagram, they would be given only at the discrete points. We could arbitrarily decide to connect them by drawing a straight line from y_i to y_{i+1}, etc. If the task of drawing a connection line is given to a draftsman, he may very likely prefer to draw smooth lines between these points by using a draftman's spline. This process resembles the task of curve fitting, and recently various authors have developed a mathematical formulation for computer application [2, 9, 10]. Thus spline functions as a tool of curve fitting provide a special sort of connection line between discrete points y_i and y_{i+1}.

We select a solution function $\eta = F(y)$ which is defined by two conditions:

(a). Over each interval (y_i, y_{i+1}), $F(y)$ is a polynomial of degree m or less.

(b). These polynomials are such that the derivatives of $F(y)$ of order $(m - 1)$ or less are continuous.

Under these conditions the spline function $F(\eta_i)$ corresponds to a piecewise fitting of the data set η_i from $i = 1, \ldots, n$. Very frequently a cubic spline is used. Then, for $y_i \leqslant y \leqslant y_{i+1}$,

$$\eta = F(y) = c_{1,i}(y_{i+1} - y)^3 + c_{2,i}(y - y_i)^3$$
$$+ c_{3,i}(y_{i+1} - y) + c_{4,1}(y - y_i). \quad (37)$$

The determination of the coefficients is a lengthy arithmetic process and the listing of the mathematical formulas exceeds the frame of this contribution. For details, see Jupp [12] or Pennington [16]. Further information on spline functions can be found in Essenwanger [8] and in Greville [9, 10] or Ahlberg et al. [2]. The practical application is stressed in Pennington [16]. Notice, $F(\eta_i)$ is an interpolation function and $\eta_i \equiv y_i$ at the discrete data points. Because of this identity no numerical example with discrete points is given. Other recent texts on spline functions are Karlin et al. [13] or de Boor [6].

UNEQUAL INTERVALS

The procedures that were presented in the second through the fourth sections were limited by the postulation that the sequence of the observations of the data set y_i follows equal temporal or spatial intervals; i.e., the related variate z_i, Z_i, or X_i is defined so that Δz_i, ΔZ_i, or ΔX_i is constant for all i values where $\Delta z_i = z_{i+1} - z_i$, etc. For the Fourier series it means that $\Delta t_i = t_{i+1} - t_i = $ constant. In some measurement or observational programs it may be impractical to fulfill this condition. For example, if x_i is a space coordinate, it may not be possible to

obtain observations y_i at equal intervals of x_i. How does this affect curve fitting? Three basic principles will be discussed about how to resolve the problem.

Conversion of the Data to Equal Intervals

In this case we convert a data set u_k related with v_k, $k = 1, \ldots, n_k$, into a data set y_i related with z_i, Z_i or X_i, $i = 1, \ldots, n$, where the latter sequence is equally spaced. The simplest scheme is a linear interpolation, but more sophisticated methods, including interpolation by spline functions, can be utilized. In most cases the choice between several procedures is a subjective decision. Nonstatistical background of the data, such as physical properties, precision of the final result, conveniences of the arithmetic, and other factors may enter into this decision process. Statistical significance and the size of the residual error are other considerations.

It would lead too far to describe all possible interpolation* schemes. The reader is referred to Essenwanger [8] or Abramovitz and Stegun [1], etc. However, linear interpolation is briefly discussed. Let us assume that $u_k \leqslant y_i \leqslant u_{k+1}$ and $v_k \leqslant X_i \leqslant v_{k+1}$. Then

$$y_i = u_k + (u_{k+1} - u_k)$$
$$\times (X_i - v_{k+1})/(v_k - v_{k+L}). \quad (38)$$

(*See also* INTERPOLATION.)

Adjustment by Weighting

A second widely used technique is the adjustment of z_i, etc., by introducing a weighting function, ω_i. For example, instead of (8b) or (9b), we would obtain b_1 from

$$b_1 = \sum \omega_i y_i X_i / \sum \omega_i X_i. \quad (39)$$

This scheme can be generalized as

$$\eta_i = \sum_{j=1}^{m} b_j \omega_{ji} Z_{ji}. \quad (40)$$

Suitable weighting functions ω_{ji} may be determined independently from z_i, etc.

Fitting of Discrete Points

A third technique is applicable when the interest is only in a mathematical description at the given discrete points v_k. No emphasis is placed on what the function does between these intervals. In this case the unequal spacing can be neglected and the set u_i could be treated as though it were a set of equal intervals. The reader must be cautioned, however, that a conflict with the principle of least-squares* solution may exist because some observations may exert any unduly heavy weight in the balance of the least-squares sum. This can be corrected by weighting, but then we are back at the preceding section.

SPECIAL FUNCTIONS

Up to now only curve fitting by polynomial terms or trigonometric functions has been treated. If other functions were selected, e.g., $\ln U_i$, we could set:

$$y_i = \ln U_i \qquad (41a)$$

or

$$y_i = \ln(U_i - c). \qquad (41b)$$

Subsequently, all formulas would be modified by the substitution. This replacement has already been demonstrated in the case of the Fourier analysis, although later the coefficients have been calculated by special formulas. It must be cautioned that substitutions as in (41a) and (41b) will usually nullify an originally orthogonal system.

Another method is the transformation into equations whose solutions are known. Let us assume that

$$u_i = a_0 + a_1 \ln(z_i + a_2). \qquad (42a)$$

We could substitute

$$v_i = \ln(z_i + a_2) \qquad (42b)$$

and derive

$$u_i = a_0 + a_1 v_i. \qquad (42c)$$

This is a curve-fitting problem which has been discussed previously. Other transformations can be found in Essenwanger [7].

MULTIDIMENSIONAL CURVE FITTING

Up to now curve fitting has been discussed in terms of one dimension only, i.e., relating y_i to z_i, etc. In a two-dimensional case we would relate y_{ik} to u_i and v_k, $i = 1, \ldots, n_i$, $k = 1, \ldots, n_k$. A simple linear relationship could then be written

$$\eta(u_i, v_k) = \bar{\eta} + a_{1u}u_i + a_{1v}v_k + a_{1uv}u_i v_k. \qquad (43)$$

Besides the linear fitting by u_i and v_k a "cross-product" term uv is involved. The curve-fitting process can be converted to

$$\eta_i = c_0 + c_1 z_{1i} + c_2 z_{2i} + c_3 z_{3i}, \qquad (43a)$$

which is an equation discussed earlier.

The system can be expanded by adding dimensions and higher-order terms. The more dimensions and order terms are augmented, the lengthier is the arihmetic process involved in the determination of the coefficients and basic functions. Simplification can be achieved by switching to orthogonal systems: e.g.,

$$\eta_{hk} = \bar{\eta} + \sum_{s=1}^{m_s} a_s U_{sh} + \sum_{t=1}^{m_t} a_t V_{tk}$$

$$+ \sum_{s=1}^{m_s} \sum_{t=1}^{m_t} a_{st} U_{sh} V_{tk}, \qquad (44)$$

where $\bar{\eta}$ is the mean, U_{sh}, V_{tk} are orthogonal functions, and a_s, a_t, and a_{st} are coefficients. In the orthogonal case the residual variance calculation simplifies because

$$\text{var}(\eta) = S_s^2 + S_t^2 + S_{st}^2, \qquad (45)$$

where S_s^2 designates the contribution from the dimension s, etc. A residual variance results whenever $m_s < n_s$, $m_t < n_s$, or $m_s \cdot m_t < n_s n_t$, where the subscript denotes the dimensions and n stands for the total possible number in the respective dimension of the data set y_{st}.

Multidimensional problems do not receive widespread treatment in the literature, but among the best sources are the texts by Daniel et al. [5] or Rice [18].

References

[1] Abramowitz, M. and Stegun, I. A. (1971). Handbook of Mathematical Functions with Formulae, Graphs, and Mathematical Tables. *Nat. Bur. Stand. (U.S.) Appl. Math. Ser. 55*, (Washington, D.C.). (The text is a comprehensive reference of mathematical functions that arise in physics and engineering. It contains many tables for these functions and is of great assistance, primarily to persons without any access to electronic computers. Twenty-nine different topics are covered. A large subject index is added. The text is sometimes very brief and requires a thorough background in mathematics for many sections.)

[2] Ahlberg, J. H., Nilson, E. N., and Walsh, J. L. (1967). *The Theory of Splines and Their Application.* Academic Press, New York. (The book treats cubic, polynomial, and generalized splines. Doubly cubic splines and generalized splines in two dimensions close out the topics. The book is highly theoretical, and the applications are not very simple to find, although the text is an excellent comprehensive treatment of spline functions.)

[3] Beyer, W. H. (1966). *Handbook of Tables for Probability and Statistics.* Chemical Rubber Co., Cleveland, Ohio. (This very popular book of statistical tables covers a wide range of topics. The text is very brief and explanations of the tables are kept to a minimum. Thus a knowledge of basic statistical theory is a prerequisite. Persons without easy access to electronic computers benefit most from the book.)

[4] Boas, M. L. (1966). *Mathematical Methods in the Physical Sciences.* Wiley, New York. (Selected topics in calculus, vector analysis, and mathematical functions are treated in 15 chapters. Although sufficient theoretical background is given, the primary goal is practical application. Many examples are provided. Knowledge of calculus is required, but the text provides an excellent explanation of the topics.)

[5] Daniel, C., Wood, F. S., and Gorman, J. W. (1971). *Fitting Equations to Data.* Wiley-Interscience, New York. (This text is written primarily for the fitting of equations, but many examples apply to curve fitting. The examples for multidimensional variates are selected for digital computer usage with many computer listings included. Although written for the practitioner, some sections are not very easy to understand. The text is an excellent reference source, however, and includes a glossary and computer programs for linear and nonlinear curve fitting.)

[6] de Boor, C. (1978). *A Practical Guide to Splines.* Springer-Verlag, New York. (The text is part of an applied mathematics series. It covers various kinds of spline functions. In 17 chapters a wide variety of spline applications is discussed. The text is difficult in part and requires a thorough mathematical background. However, a list of FORTRAN programs, a bibliography, and a subject index are included.)

[7] Essenwanger, O. M. (1976). *Applied Statistics in Atmospheric Science,* Part A: *Frequencies and Curve Fitting.* Elsevier, Amsterdam. (The text treats nonelementary frequency distributions with examples from atmospheric science in Chapters 1 and 2, and problems related to curve fitting in Chapters 3 and 4. Chapter 3 includes factor analysis and transformation, and Chapter 4 provides tools for the practitioner in matrix analysis and eigenvector computation. Many examples are given. This advanced text requires knowledge of basic statistical theory but may prove more useful to the practitioner than the theoretician.)

[8] Essenwanger, O. M. (1981). Elements of Statistical Analysis. Elsevier, Amsterdam. In *World Survey of Climatology*, Vol 1a, O. M. Essenwanger, ed. (Editor-in-Chief, H. E. Lanbsberg). (This text covers the more common frequency distributions, regression, polynomials, Fourier analysis, smoothing and filtering, and basic tests applied to atmospheric data. Numerous examples are given for practical application. Publication is expected in late 1981.)

[9] Greville, T. N. E. (1967). *Spline Functions, Interpolation and Numerical Quadrature in Mathematical Methods for Digital Computers*, Vol. 2, A. Ralston and H. W. Wilf, eds. Wiley, New York. (This text is one section of a book on mathematical methods for digital computers. The article starts with a theoretical background. It also includes a program for digital computers on spline functions, interpolation, and numerical quadrature.)

[10] Greville, T. N. E. (1969). *Theory and Applications of Spline Functions.* Academic Press, New York. (This text is written more for the theoretician than for the practitioner and is an excellent but concise treatment of the theory of spline functions and their historical development.)

[11] Guest, P. G. (1961). *Numerical Methods of Curve Fitting.* Cambridge University Press, Cambridge. (This text covers a wide variety of topics in 12 chapters, starting with single variables and Gaussian distributions and extending to the problems of general regression with several variables. Numerous examples illustrate the applications. The book is directed toward practical work, although requiring good mathematical background.)

[12] Jupp, D. L. B. (1972). In *Optimization*, R. S. Anderssen, L. S. Jennings, and D. M. Ryan, eds. University of Queensland Press, Saint Lucia, Brisbane, Queensland. (This article covers the theory of the location of the knots of best linear square cubic splines and is part of the proceedings of a seminar on optimization.)

[13] Karlin, S., Micchelli, C. A., Pinkus, A., and Schoenberg, I. J. (1976). *Studies in Spline Functions and Approximation Theory*. Academic Press, New York. (This text is a comprehensive discussion of spline functions, Part I treats approximations, optimal quadrature, and mono splines. In Part II cardinal splines are analyzed. Interpolation with splines and miscellaneous applications are presented in Parts III and IV. The book is an excellent reference of the theory but is sometimes difficult to read for the practitioner.)

[14] Korn, G. A. and Korn, T. M. (1961). *Mathematical Handbook for Scientists and Engineers*. McGraw-Hill, New York. (This handbook is a comprehensive survey of definitions, theorems, and mathematical formulas in science and engineering, and covers the entire field of mathematics in 21 chapters. Appendix F contains numerical tables ranging from simple squares, logarithms, etc., to t, F, and χ^2 distributions of statistical analysis. Some of the text is very brief and requires background knowledge in the special topic. It is an excellent reference source. The main text is presented in large print; advanced topics are added in fine print.)

[15] Pearson, E. S. and Hartley, H. O. (1974). *Biometrika Tables for Statisticians*, Vol. I, 3rd ed. Cambridge University Press, Cambridge. (This set of tables is primarily designed for statisticians, but general mathematical tables on polynomials, squares, logarithms, etc., are added. Volumes II and III list the incomplete beta and gamma functions, which are of little interest here. The tables, first published in 1914, have been consistently updated. Although the tables are explained in the introductory text, knowledge of statistical analysis is required.)

[16] Pennington, R. H. (1970). *Introductory Computer Methods and Numerical Analysis*, 2nd ed. Macmillan, London. (This text is written for the practitioner. The first five chapters cover digital programming. Chapters on functions, quadratures, equations, curve fitting, and spline functions follow. Knowledge of some advanced mathematics is required. Numerous examples illustrate digital programming. Thus the text is an excellent source of digital programs.)

[17] Ralston, A. and Wilf, H. S. (1967). *Mathematical Methods for Digital Computers*, Vols. 1 and 2. Wiley, New York. (These two volumes cover a variety of mathematical topics: functions, matrices, differential equations, statistics, linear algebra, numerical quadrature, and numerical solutions for equations; see Greville [9]. The text is an excellent source for digital programming because flowcharts are included for every topic.)

[18] Rice, G. R. (1969). *The Approximation of Functions*, Vol. 2. Addison-Wesley, Reading, Mass. (This text deals largely with functions and treats

spline functions in Chapter 10. In Chapter 12 multivariate approximations are presented. The text requires advanced knowledge of mathematics and is not always easy to read for the practitioner.)

(ESTIMATION
GOODNESS OF FIT
GRADUATION
INTERPOLATION
SPLINE FUNCTIONS)

OSKAR ESSENWANGER

CYCLES

Time series*, or variables observed over time, are studied in the physical sciences (geophysics, meteorology, oceanography, atmospheric physics), engineering sciences (mechanics, acoustics, speech), biological sciences (bio-rhythms), medicine (EEK and EKG analysis), social sciences (history), economics (business cycles); and management science (forecasting). All time series exhibit oscillatory behavior, and it is natural to regard the aim of time-series analysis to be the identification of cycles, rhythms, and periodicities in the data. Modern time-series analysis regards its aim to be fitting models to the data that "best" predict future values of the time series and help develop scientific theories to explain the time series. Cyclic components found by statistical techniques in past values of a time series should be expected to continue in the future, and thus make the time series predictable, only if they are explained by a scientific model. Scientists seek to detect and measure cycles in time series because they can be interpreted as manifestations of regulatory mechanisms in the system the variable is measuring.

Time-series* analysis seems to be the easiest field of statistics in which to reach spurious conclusions; spurious correlations and spurious periodicities are obtained by failing to take into account the phenomenon of autocorrelation* (or statistical dependence) in the observations. Another source of spurious periodicites is the autocorrelation induced in time series by moving-average operations used to smooth them (this result is often called Slutsky's theorem*). If one seeks to determine a relationship between two time series, it does not suffice to "establish" the existence of some common cycles in the two time series; the detection and measurement of relations between time series is a part of the modern theory of multiple time series, while the study of cycles in a single time series is a part of the theory of univariate time series.

Modern statistical methods for defining and modeling cycles involve consideration of various models to be fitted to the data, including the following models: (a) strict periodicity in white noise*, (b) stationary time series, (c) strict periodicities in stationary noise, (d) nonstationary seasonal models, and (e) nonlinear models.

The *strict periodicity in white noise* approach to searching for periodicities or cycles in a time series $Y(t)$, $t = 1, 2, \ldots$ assumes a model for it as a sum

$$Y(t) = \mu + S_1(t) + \cdots + S_k(t) + N(t),$$

$$(1)$$

where μ is a constant representing the mean level of the time series, $N(t)$ is a white noise* time series (sequence of independent random variables with zero means and constant variance σ^2), and for $j = 1, 2, \ldots, k$ the component time series $S_j(t)$ is a sine wave of a single fixed frequency $\omega_j = 2\pi/p_j$. One represents $S_j(t)$ by

$$S_j(t) = A_j \cos \frac{2\pi}{p_j} t + B_j \sin \frac{2\pi}{p_j} t,$$

where p_j is the period of the cycle; it satisfies the periodicity condition $S_j(t + p_j) = S_j(t)$ for all t.

The parameters of a strict periodicity model for a time series can be divided into two categories according to the relative difficulty involved in estimating them. The difficult problem of determining k, $p_1, \ldots, p_k$ is called *model identification**. The simpler problem of estimating $\mu, \sigma^2, A_1, B_1, \ldots, A_k$, B_k corresponding to assumed values of k and $p_1, \ldots, p_k$ is called *parameter estima-*

*tion**. The most difficult problem is that of testing whether model (1) provides an adequate fit to the data.

To avoid "spurious periodicities," or identifying periods p_j whose reality is suspect, one should consider replacing the assumption that the residual or error series $N(t)$ is white noise by a more realistic assumption that it is a zero mean covariance stationary* time series. Its statistical parameters are its covariance function

$$R(v) = E[N(t)N(t + v)],$$

$$v = 0, \pm 1, \pm 2, \ldots$$

and its correlation function

$$\rho(v) = \frac{R(v)}{R(0)}, \qquad v = 0, \pm 1, \pm 2, \ldots .$$

White noise corresponds to $\rho(v) = 0$ for $v \neq 0$.

The *disturbed periodicity* or stationary time-series approach to searching for cycles in a time series assumes the model

$$Y(t) = \mu + Z(t),$$

where $Z(t)$ is a zero mean covariance stationary time series whose correlation function satisfies $\sum_{v=-\infty}^{\infty} |\rho(v)| < \infty$. From a sample $\{Y(t), t = 1, 2, \ldots, T\}$, one usually estimates μ and $\rho(v)$ by

$$\hat{\mu} = \overline{Y} = \frac{1}{T} \sum_{t=1}^{T} Y(t)$$

$$\hat{\rho}(v) = \frac{\sum_{t=1}^{T-v} \{Y(t) - \overline{Y}\}\{Y(t + v) - \overline{Y}\}}{\sum_{t=1}^{T} \{Y(t) - \overline{Y}\}^2}$$

for $v = 0, 1, \ldots, T - 1$. One calls $\hat{\rho}(v)$ the *sample correlation function* or *correlogram**. If the graph of $\hat{\rho}(v)$ is sinusoidal or periodic, one regards the time series as having cycles or rhythms with a "disturbed period." To measure this disturbed period quantitatively one estimates the Fourier transforms of $R(v)$ and $\rho(v)$ and determines the frequencies at which they have relative maxima. There are almost as many notations for the Fourier transforms of covariance and correlation functions as there are authors, and the reader must learn to cope with the lack of a standard notation. The following definitions will be adopted here.

When $\rho(v)$ is summable, its Fourier transform is denoted

$$f(u) = \sum_{v=-\infty}^{\infty} e^{2\pi i u v} \rho(v), \qquad 0 \leqslant u \leqslant 1,$$

and called the *spectral density** of the time series $Y(\cdot)$. The Fourier transform of $R(v)$, equal to $R(0)f(u)$, is called the *power spectral density* of the time series. In the physical sciences and engineering, the argument of the power spectral density function is often denoted by "f" to denote frequency, and the function itself is denoted P or S. Time-series analysts often use ω as the argument of f.

The function $f(u)$ is actually defined for all u in $-\infty < u < \infty$. However, it is periodic with period 1, and its domain can be taken to be either $-0.5 \leqslant u \leqslant 0.5$ or $0 \leqslant u \leqslant 1$. The interval $-0.5 \leqslant u \leqslant 0.5$ is customary in the engineering literature, but only the subinterval $0 \leqslant u \leqslant 0.5$ has physical significance. The deterministic time series $Y(t) = \cos 2\pi u t$ or $Y(t) = \sin 2\pi u t$ have period $1/u$; thus frequency u varies from 0 to 0.5, period varies from 2 to ∞. One plots $f(u)$ on the interval $0 \leqslant u \leqslant 0.5$, since it is an even function. The spectral density has two important mathematical properties; it is even, $f(u) = f(-u)$, and nonnegative, $f(u) \geqslant 0$.

The basic building block of modern statistical techniques for modeling cycles in an observed time series $\{ Y(t), t = 1, 2, \ldots, T \}$ of length T is the function

$$\tilde{f}(u) = \left| \sum_{t=1}^{T} Y(t) \exp(2\pi i u t) \right|^2 \div \sum_{t=1}^{T} Y^2(t)$$

$$= \sum_{|v| < T} \exp(2\pi i u v) \hat{\rho}(v)$$

defined for $-0.5 \leqslant u \leqslant 0.5$ and called the sample spectral density. When defined for $u = k/T$, $k = 0, 1, \ldots, T/2$, it is called the *periodogram** or discrete Fourier spectrum. If the time series $Y(t)$ has a strictly sinusoidal component $S(t)$ of period p,

$$S(t) = A \cos \frac{2\pi}{p} + B \sin \frac{2\pi}{p} t,$$

the sample spectral density $\tilde{f}(u)$ would have peaks at $u = 1/p$. To detect a strictly periodic component $S(t)$ of period p in a sample $\{ Y(t), t = 1, 2, \ldots, T \}$ of a time series, classical techniques for searching for periodicities would look for a relative maximum at $u = 1/p$ in $\tilde{f}(u)$.

When the time series $Y(t)$ is normal zero mean, stationary $\tilde{f}(u)$ has a probability distribution which is approximately exponential* with mean $f(u)$. Thus, approximately, $\tilde{f}(u)$ has mean $f(u)$ and variance $f^2(u)$; since the variance of $\tilde{f}(u)$ does not decrease to 0 as T tends to infinity, it is not a consistent estimator of $f(u)$. In fact, the graph of $\tilde{f}(u)$ as a function of u is usually very wiggly, with many relative maxima, which could be spuriously interpreted as indicating strict periodicities. To estimate the function $f(u)$, one uses an estimator $\hat{f}(u)$ which is a smooth curve passed through the wiggly curve $\tilde{f}(u)$. The theory of time-series analysis since 1945 has developed many techniques for forming estimators $\hat{f}(u)$ of the spectral density function of a stationary time series and has inspired techniques for the estimation of a probability density function*. To form an estimator $\hat{f}(u)$ of $f(u)$, two principal methods are available: the *window** method, which forms an average of $\tilde{f}(u')$ for u' is a suitable neighborhood of u, and the *filter* method, which determines approximately a transformation of the time series $Y(t)$ to a white noise using a time domain filter, especially autoregressive schemes.

The mixed spectral approach to searching for cycles assumes a model of the form of (1), but with the assumption that $N(t)$ is a stationary time series with spectral density $f(u)$; one calls this a model of *strict periodicities in stationary noise*. Nonstationary seasonal time-series models correspond to a model that can be transformed to a stationary time series by a differencing operation; for example one might assume that $Y(t + p) - Y(t)$ is a stationary time series where the "differencing" period p has to be determined by the researcher. Space is not available here to describe methods for identifying the form of model appropriate for a particular time series. In addition to the linear models discussed in the foregoing, statisticians are currently developing nonlinear models for time series which possess strong evidence of cycles.

The foregoing discussion has aimed to show how strict or disturbed periodicities in time series can be found by suitable interpretations of the models fitted to time series. It is to be emphasized that such cycles are regarded as being scientifically significant only when they can be explained in terms of scientific theory.

An important research area concerned with the existence of cycles, and measurement of their characteristics, is biological rhythms. This field is surveyed by Sollberger [7]; current research is reported in the *Journal of Interdisciplinary Cycle Research.*

Popular books, and newspaper and magazine articles, are constantly being published about the cycles present in the time series of prices of stocks and commodities, the sizes of animal populations, weather (rainfall and temperature), sunspots, ice ages, earthquakes, etc. (see, e.g., Dewey [3] and the journal *Cycles* founded by Dewey, which advocates that "cycles" in diverse time series be cataloged according to their periods in order to discern an underlying regulatory mechanism). Statisticians commenting on the writings of 'cyclists' are likely to warn of the need for caution. The intuitive notion of cycles as sinusoidal or periodic waveforms present in a time series seem to be best understood when related to models fitted to a time series in terms of mathematical operations on the time series which transforms it to a white noise time series.

Literature

Current books by statisticians concerned with building models for time-series data do not usually discuss methods for testing for hidden periodicities or cycles. Good discussions are in Bloomfield [1], Fuller [4], Hannan [5], and Nerlove et al. [6]. Discussions of cycles that may exist in various physical time series are to be found in the *Annals of*

the New York Academy of Sciences (1961) **95** ("Solar Variations, Climatic Change, and Related Geophysical Problems") and (1967) **130** ("Interdisciplinary Perspectives of Time").

Two famous time series exhibiting disturbed periodicities which have been extensively analyzed are the sunspot series and the Canadian lynx series, references to the statistical literature on the analysis of these series may be found in Woodward and Gray [9] for sunspots and Campbell and Walker [2] and Tong [8] for lynx.

References

[1] Bloomfield, P. (1976). *Fourier Analysis of Time Series: An Introduction.* Wiley, New York.

[2] Campbell, M. J. and Walker, A. M. (1977). *J. R. Statist. Soc. A*, **140**, 411–431, 448–468 (discussion).

[3] Dewey, E. R. (1971). *Cycles, The Mysterious Forces That Trigger Events.* Hawthorn Books, New York.

[4] Fuller, A. (1976). *Introduction to Statistical Time Series.* Wiley, New York.

[5] Hannan, E. J. (1970). *Multiple Time Series.* Wiley, New York.

[6] Nerlove, M., Grether, D. and Carvalho, J. (1979). *Analysis of Economic Time Series.* Academic Press, New York.

[7] Sollberger, A. (1976). *CRC Clin. Rev. Clin. Lab. Sci.*, 247–285.

[8] Tong, H. (1977). *J. R. Statist. Soc. A.*, **140**, 432–436.

[9] Woodward, W. A. and Gray, H. L. (1978). *Commun. Statist. B*, **7**, 97–115.

(AUTOREGRESSIVE MODELS
CORRELOGRAM
FOURIER ANALYSIS
PERIODOGRAM
PREDICTION
TIME SERIES)

EMANUEL PARZEN

CYCLIC DESIGNS

Cyclic designs are incomplete block designs* consisting in the simplest case of a set of blocks obtained by cyclic development of an initial block. More generally, a cyclic design consists of a combination of such sets and is said to be of size (t, k, r), where t is the number of treatments, k the block size, and r the number of replications.

Label the treatments $0, 1, \ldots, t - 1$. To fix ideas consider the arrangement of $t = 7$ treatments in blocks of size $k = 3$. The complete design of $\binom{7}{3} = 35$ distinct blocks may be set out as follows:

$$
\begin{array}{lccccccc}
\{012\}: & 012 & 123 & 234 & 345 & 456 & 560 & 601 \\
\{013\}: & 013 & 124 & 235 & 346 & 450 & 561 & 602 \\
\{014\}: & 014 & 125 & 236 & 340 & 451 & 562 & 603 \\
\{015\}: & 015 & 126 & 230 & 341 & 452 & 563 & 604 \\
\{024\}: & 024 & 135 & 246 & 350 & 461 & 502 & 613
\end{array}
\tag{1}
$$

From any block its neighbor to the right may be obtained by increasing each treatment label by 1 and reducing modulo 7. Performing this operation on the last column of (1) takes us back to the first column. The rows have been arranged to start with the block of lowest numerical value and are designated by the initial block placed in braces. We call each row a cyclic set. Note that the order effects in each set are balanced out, giving two-way elimination of heterogeneity.

A block may also be conveniently represented by identical beads placed at k of t equispaced positions on a circular necklace. Figure 1 shows the blocks 013 and 124. The set $\{013\}$ is then generated by successive unit rotations.

The same procedure can be used for any t and k except that when t and k are not relative primes, fractional sets arise consisting of t/d blocks, where d is any common divisor of t and k. In terms of Fig. 1, such sets correspond to arrangements of beads

that can be reproduced in fewer than t rotations of the necklace.

SYSTEMATIC CONSTRUCTION

We begin with the systematic enumeration of cyclic sets, which for this purpose are conveniently characterized by a circular partition of t. Thus we replace $\{0, x_1, x_2, \ldots, x_{k-2}, x_{k-1}\}$ by $(x_1, x_2 - x_1, \ldots, x_{k-1} - x_{k-2}, t - x_{k-1})$.

Example. For $t = 8$, $k = 4$, the set $\{0123\}$ becomes (1115). The cyclic sets may now be written down in increasing order of the numerical value of the corresponding partition: (1115), (1124), (1133), (1142), (1214), (1223), (1232), (1313), (1322), (2222). After (1142) we omit (1151), this being identical with (1115), etc. As the repetition of digits indicates, the set (1313) consists of the four blocks

$$
0145 \quad 1256 \quad 2367 \quad 3470 \quad (r = 2)
$$

and (2222) of the two (disconnected) blocks 0246, 1357 ($r = 1$). As a check, note that all $\binom{8}{4}$ blocks are accounted for, since $8 \times 8 + 4 + 2 = 70$.

For any t and k, the total number of sets, being equal to the number of distinct arrangements of k white beads and $n - k$ black beads on a necklace of n beads (which may not be turned over), is given by

$$
\begin{aligned}
N(k, t - k) \\
= \frac{1}{t} \sum \phi(d) \frac{(t/d)!}{(k/d)! \left[(t - k)/d \right]!},
\end{aligned}
\tag{2}
$$

where the summation is over all integers d (including unity) which are divisors of both k and $t - k$, and where $\phi(d)$ is Euler's function, the number of positive integers less than and prime to d, with $\phi(d) = 1$. Thus

$$
N(4, 4) = \frac{1}{8} \left(\frac{8!}{4!4!} + \frac{4!}{2!2!} + 2\frac{2!}{1!1!} \right) = 10,
$$

in agreement with the example.

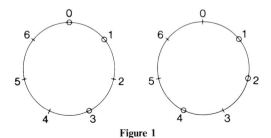

Figure 1

For the cyclic set $\{012\}$ of $t = 7$ and $k = 3$, let us now apply the renumbering permutation

$$R(7,3) = \begin{pmatrix} 0 & 1 & 2 & 3 & 4 & 5 & 6 \\ 0 & 3 & 6 & 2 & 5 & 1 & 4 \end{pmatrix}$$

obtained by multiplying each of the seven treatment labels by 3(modulo 7). Then $\{012\}$ becomes

036 362 625 251 514 140 403,

which is merely a rearrangement of $\{014\}$. Thus $\{012\}$ and $\{014\}$ are isomorphic*. Two further applications of $R(7,3)$ give $\{024\}$ and the original $\{012\}$. We have therefore established the equivalence class $\{012\}$ $\sim \{014\} \sim \{024\}$; also, $\{013\} \sim \{015\}$. Of these two classes the latter is to be preferred because it represents a balanced incomplete block* (BIB) design.

More generally, whatever one's efficiency criterion, the choice of design is greatly reduced by the establishment of equivalence classes. The procedure above can be used for any prime t and any k. To see this, note that the permutations $R(t,1)$ (the identity permutation), $R(t,2), \ldots, R(t,t-1)$ form a group under "multiplication" $*$ defined by

$$R(t,i) * R(t,j) = R(t, ij \bmod t), \quad (3)$$

which is isomorphic with the multiplicative group of residues mod t. Hence all elements $R(t,i)$ are generated by powers of $R(t,g)$, where g is a primitive root of t (i.e., $g^x \not\equiv 1$ mod t for $x = 1, 2, \ldots, t - 2$ but $g^{t-1} \equiv 1$ mod t). All possible isomorphisms between cyclic sets can be established conveniently by repeated application of $R(t,g)$.

When t is not prime the $R(t,i)$ continue to form a group under $*$ of (3) provided that i and j are restricted to integers relatively prime to t. The group is now of order $\phi(t)$ and is clearly isomorphic with the multiplicative group of the reduced set of residues. g is said to be the primitive root of t if $\phi(t)$ is the smallest power making $g^{\phi(t)} \equiv 1 \bmod t$. Primitive roots exist only if t equals 2, 4, p^n, or $2p^n$, where p is any prime greater than 2 and n any integer. For values of t admitting a primitive root, we proceed as before; oth-

erwise, multiplication by each member of the reduced set of residues will establish most isomorphisms.

Cyclic sets for given t may now be combined to produce a wide variety of cyclic designs. Construction of a design of size (t,k,r) is always possible if r is a multiple of k and may be possible for certain other values of r if fractional sets exist. Equivalence classes may again be established. Further details, references, and an extensive tabulation of designs are given in John et al. [12].

ANALYSIS

The analysis of cyclic designs can be carried out by the usual methods applicable to any incomplete block design. Advantage may be taken, however, of the circulant nature of the coefficient matrix $\mathbf{A}$ in the normal equations, $\mathbf{A}\hat{\boldsymbol{\tau}} = \mathbf{Q}$, where $\hat{\boldsymbol{\tau}}$ is the vector $(\hat{\tau}_1, \ldots, \hat{\tau}_t)'$ of (intrablock) estimates of treatment effects and $\mathbf{Q}$ is the vector of adjusted treatment totals. From a method described by Kempthorne [14] we find that $\hat{\boldsymbol{\tau}} = \mathbf{CQ}$, where $\mathbf{C} = \{c_{ij}\}$ is any generalized inverse* of $\mathbf{A}$ (i.e., $\mathbf{ACA} = \mathbf{A}$) and is also circulant. See John et al. [12], where the interblock analysis* is also presented together with analyses for resolvable designs and for two-way elimination of heterogeneity.

EFFICIENCY*

The standard measure of the efficiency of an incomplete block design is the mean variance of treatment differences divided into $2\sigma^2/r$, the corresponding variance for a randomized block* design with r replications. For the cyclic designs,

$$\text{var}(\hat{\tau}_i - \hat{\tau}_j) = \sigma^2(c_{ii} + c_{jj} - 2c_{ij})$$

$$= 2\sigma^2(c_{ii} - c_{ij})$$

$$= 2\sigma^2(c_{11} - c_{1,j-i+1}), \quad j > i.$$

Summing over all possible $\frac{1}{2}t(t-1)$ differences, we get the mean variance $2\sigma^2 tc_{11}/(t-1)$. Hence the efficiency* becomes $E = (t-1)/rtc_{11}$. We can also define the efficiency of $\hat{\tau}_i - \hat{\tau}_j$ as

$$E_{ij} = \frac{2\sigma^2/r}{2\sigma^2(c_{ii} - c_{ij})} = \frac{1}{r(c_{11} - c_{1,j-i+1})}$$

$$= E_{1,j-i+1}$$

From the cyclic nature of the design it follows moreover that $c_{1,j-i+1} = c_{1,i-j+t}$, so that there are at most m different values among the c_{ij} (and E_{ij}), where $m = \frac{1}{2}t + 1$ or $\frac{1}{2}(t+1)$, according as t is even or odd.

TABLES OF CYCLIC DESIGNS

A catalog of 460 cyclic designs is given in John et al. [12]. A very concise representation is possible because it is one of the great advantages of a cyclic design that no experimental plan is needed beyond a statement of initial block(s). When more than one cyclic design of a given size exists, only the one with maximum overall efficiency E is included. Each entry contains numerical values of E and of E_{ij}, c_{ij} for $j = 2, \ldots, m$.

HISTORY AND RELATION TO PARTIALLY BALANCED INCOMPLETE BLOCK (PBIB)* DESIGNS

In a major paper Bose and Nair [2] apply Bose's [1] method of differences to the construction of various types of PBIB designs. One special case (Sec. 6.3) leads directly to what would now be called cyclic designs. The term "cyclic" was apparently first used in the classification scheme for PBIB (2) designs developed by Bose and Shimamoto [3], where it was confined to those cyclic designs not classified as group divisible. This restriction is dropped in current usage. Cyclic designs may, in fact, be regarded as a convenient cross section of PBIB designs, with up to m associate classes (David and Wolock [6]; John [10]).

Cyclic designs as a class in their own right

(but without name) were introduced for $k = 2$ by Kempthorne [14]. Methods of systematic construction and enumeration for $k = 2$ are given in David [4, 5].

APPLICATIONS

Because of their convenience, cyclic designs have served to supplement the limited number of available BIB and PBIB(2) designs. The need for additional designs is especially strong for $k = 2$, a case of particular interest because of applications to paired-comparison and diallel cross experiments. Apart from their use in field experiments, for any k, cyclic designs have also been employed in preference testing, usually for small k. See John et al. [12] for further details.

GENERALIZATIONS AND EXTENSIONS

As noted in the introduction, cyclic sets, and hence combinations thereof, provide two-way elimination of heterogeneity. This Youden type design property is necessarily lost in fractional sets (see the example in the section on systematic construction). Nevertheless, when t is large, a single fractional set may be all the experimenter can afford. Such fractional designs are included in John et al. [12]. Additional possibilities arise when t is composite, say $t = uv$. Jarrett and Hall [9] study designs in which treatment labels are increased by $u \bmod t$ rather than by $1 \bmod t$. This results in many new designs particularly useful when t is very large and sometimes markedly superior to any fractional designs of the same size.

A different generalization for $t = uv$ had been considered earlier by John [11]: to express each label as $a_1 a_2$, where $a_1 = 0, 1, \ldots, u - 1, a_2 = 0, 1, \ldots, v - 1$. Increments are in steps of 1, but reduction is $\bmod u$ for a_1 and $\bmod v$ for a_2. Such designs can be used for partial confounding* in factorial experiments*, so as to provide an orthogonal analysis of main effects and interactions*. See also Dean and John [8].

Cyclic changeover designs* have been investigated by Davis and Hall [7].

Cyclic designs are also useful in the construction of resolvable designs, i.e., designs that may be arranged in groups such that each group is a complete replicate. For $k = 2$ resolvable designs provide tournament schedules, especially useful when time does not permit every team to meet every other team (see Williams [16]). Resolvable designs for general k are treated in Patterson and Williams [15].

Literature

A general exposition, including a fairly complete bibliography of earlier work, is given in a monograph by J. A. John et al. [12]. The book by P. W. M. John [13], a general account of the design and analysis of experiments, contains substantial entries on cyclic designs. For additional information, see the other entries in the reference list.

References

[1] Bose, R. C. (1939). *Ann. Eugen.* (*Lond.*), **9**, 353–399.

[2] Bose, R. C. and Nair, K. R. (1939). *Sankhyā*, **4**, 337–372.

[3] Bose, R. C. and Shimamoto, T. (1952). *J. Amer. Statist. Ass.*, **47**, 151–190.

[4] David, H. A. (1963). *J. Aust. Math. Soc.*, **3**, 117–127.

[5] David, H. A. (1972). *J. Comb. Theory*, **13**, 303–308.

[6] David, H. A. and Wolock, F. W. (1965). *Ann. Math. Statist.*, **36**, 1526–1534.

[7] Davis, A. W. and Hall, W. B. (1969). *Biometrika*, **56**, 283–293.

[8] Dean, A. M. and John, J. A. (1975). *J. R. Statist. Soc. B*, **37**, 72–76.

[9] Jarrett, R. G. and Hall, W. B. (1978). *Biometrika*, **65**, 397–401.

[10] John, J. A. (1969). *Sankhyā B*, **31**, 535–540.

[11] John, J. A. (1973). *Biometrika*, **60**, 55–63.

[12] John, J. A., Wolock, F. W., and David, H. 'A. (1972). Cyclic Designs, *Nat. Bur. Stand.* (*U.S.*) *Appl. Math. Ser.* 62 (Washington, D.C.).

[13] John, P. W. M. (1971). *Statistical Design and Analysis of Experiments*. Macmillan, New York.

[14] Kempthorne, O. (1953). *Ann. Math. Statist.*, **24**, 76–84.

[15] Patterson, H. D. and Williams, E. R. (1976). *Biometrika*, **63**, 83–92.

[16] Williams, E. R. (1976). *J. R. Statist. Soc. B*, **38**, 171–174.

(ANALYSIS OF VARIANCE
CONFOUNDING
DESIGN OF EXPERIMENTS
FACTORIAL EXPERIMENTS
FRACTIONAL FACTORIAL DESIGNS
INCOMPLETE BLOCK DESIGNS
PAIRED COMPARISONS
PARTIALLY BALANCED
   INCOMPLETE BLOCKS
PBIB DESIGNS
RANDOMIZED BLOCK DESIGNS
RESOLVABLE DESIGNS)

H. A. David

CYCLIC SAMPLING

Selection of a sample by taking every kth individual in the population to be chosen. This results in very nearly a sample of size $(1/k)$th of the population. When the number of individuals in the population is a multiple of k, there is exactly $(1/k)$th of the population in the sample.

This is a convenient and simple way of selecting a sample provided that an accurate and exhaustive list of the members of the population is available. Care is needed to ensure that the values of characters to be measured do not depend on position in the list.

(SAMPLING
SURVEY SAMPLING)

D

D'AGOSTINO TEST OF NORMALITY

A number of tests of normality* use criteria of the form

$$\frac{\text{linear estimates of } \sigma \text{ from order statistics}}{s}.$$

The D'Agostino test [1] is obtained by taking the numerator to be $\sum_{i=1}^{n}\{i - \frac{1}{2}(n + 1)\} X_{(i)}$, where $X_{(1)} \leqslant X_{(2)} \leqslant \cdots \leqslant X(n)$ is an ordered sample of size n with the mean $\bar{X}$. His test statistic is thus given by the formula

$$D(n) = \frac{\sum_{i=1}^{n}\{i - \frac{1}{2}(n + 1)\} X_{(i)}}{n\sqrt{n\sum_{i=1}^{n}\left(X_{(i)} - \bar{X}\right)^2}}.$$

The test, originally proposed for moderate-size and large samples possesses the omnibus property*, i.e., is appropriate to detect deviations from normality due either to skewness* or kurtosis*. (The test statistic equals up to a constant to the ratio of *Downton's linear unbiased estimator* [3] of the population standard deviation to the sample standard deviation.) Tables of percentage points of

$$Y(n) = \frac{\sqrt{n} D(n) - 0.28209479}{0.02998598}$$

[the approximate standardized $D(n)$] were published by D'Agostino [2] for $n = 10(2)50$ and for $n = 50(10)100(5)1000$ in D'Agostino [1]. Investigations by Theune [5] indicate that Shapiro and Wilk's [4] test of normality* (which is computationally more involved) is preferable over the D'Agostino test for sample sizes of 50 or smaller for four alternatives [log normal* ($\mu = 0$, $\sigma^2 = 1$), $\chi^2(1)$, uniform, and U-shaped distributions]. For a sample size of $n = 100$, the Shapiro–Wilk test is superior only for the U-shaped alternative. For more details on comparisons among tests of normality, including the D'Agostino test, *see* DEPARTURES FROM NORMALITY, TESTS FOR.

References

[1] D'Agostino, R. B. (1971). *Biometrika*, **58**, 341–348.
[2] D'Agostino, R. B. (1972). *Biometrika*, **59**, 219–221.
[3] Downton, F. (1966). *Biometrika*, **53**, 129–141.
[4] Shapiro, S. S. and Wilk, M. B. (1965). *Biometrika*, **52**, 591–611.
[5] Theune, J. A. (1973). *Statist. Neerlandica*, **27**, 163–169.

(ANDERSON–DARLING TEST
DEPARTURES FROM NORMALITY,
     TESTS FOR

EDF STATISTICS
GINI'S MEAN DIFFERENCE
KOLMOGOROV–SMIRNOV TEST
SHAPIRO–WILK TEST)

DAMAGE FUNCTION   *See* TARGET COV-
ERAGE

DAMAGE MODELS

The concept of damage model arose quite
naturally in probabilistic modeling of a certain type of random phenomenon which
may not be observable as a whole undamaged. To be specific, let X be a nonnegative integer-valued random variable with
probability law $g_n, n = 0, 1, \ldots$, i.e., $\Pr[X
= n] = g_n, n = 0, 1, 2, \ldots$, representing as a
probability model for some random phenomenon under observation. For example, it
may be of interest to provide a probability
model for the total number of eggs laid by
an insect selected at random from a given
species. In this case, X will denote the number of eggs laid by an insect and this is
indeed a random variable. Specification of
the probability model, in such a case, is
tantamount to the specification of the probabilities $g_n, n = 0, 1, 2, \ldots$. A finite number
of independent observations on X would
then provide information on g_n's. But in
reality, it could happen that the observations
on X are damaged, and in a probabilistic
sense, this means that we observe a related
random variable Y less than or equal to X.
While counting the eggs of an insect, some
of the eggs might have been lost or squashed
into one lump beyond recognition and we
note down only the total number of eggs
that are intact. Such an observation could be
regarded as a realization of Y. It is natural
to call Y the *undamaged part* of X and
$X - Y$ the *damaged part* of X.

A variety of phenomena fit the description
above and problems relating to this area
come under the pervasive label "damage
models." One of the central problems in this
area is the following. Knowing the condi-

tional probability law of Y given X, how
much information can we glean about the
probabilities $g_n, n = 0, 1, 2, \ldots$?

Rao and Rubin [14] initiated research in
this area and after the publication of their
fundamental paper, many papers appeared
in the literature in the same area. The aim of
this note is to provide an insight into the
salient features of this area and indicate
current trends of research. Before this, we
formulate one of the problems in this area in
simplistic mathematical terms.

The damage model can be typified by a
random vector (X, Y) of nonnegative integer-valued components, with the joint probability law of X and Y having the following
structure:

$$\Pr[X = n, Y = r] = g_n S(r \mid n),$$
$$r = 0, 1, \ldots, n; \quad n = 0, 1, \ldots,$$

where $S(r \mid n), r = 0, 1, 2, \ldots, n$, is a discrete
probability law for each $n = 0, 1, 2, \ldots$,
and g_n's are nonnegative and add up to
unity. Observe that the probability law of X
is $g_n, n = 0, 1, 2, \ldots$, and that of $Y \mid X = n$
is $S(r \mid n), r = 0, 1, 2, \ldots, n$. In the context
of damage models, the conditional probability law of $Y \mid X$ is called the "survival" probability law. Once we know the survival probability law, under what circumstances will it
tell all about the probability law of X or
characterize the probability law of X?

Rao and Rubin [14] proved the following
result. If $g_0 < 1$ and

$$S(r \mid n) = \binom{n}{r} p^r (1 - p)^{n-r},$$
$$r = 0, 1, 2, \ldots, n; n = 0, 1, 2, \ldots,$$

where p is a fixed number in $(0, 1)$, then

$$\Pr[Y = r] = \Pr[Y = r \mid X = Y],$$
$$r = 0, 1, 2, \ldots \quad (1)$$

if and only if X has a Poisson* probability
law. (*See* CHARACTERIZATION THEOREMS.) Intuitively, if the conditional probability law
of $(Y \mid X = n)$ is binomial* with index parameter n and success probability parameter
p for every $n \geqslant 0$ with fixed p in $(0, 1)$, then
the probability law of the undamaged part Y
of X and the conditional probability law of

(Y | no damage) are identical if and only if X has a Poisson probability law. This is called the Rao–Rubin characterization of the Poisson probability law. Rao and Rubin proved this result using a deep theorem in analysis, namely, Bernstein's theorem on absolutely monotonic functions. Using essentially the same argument, Talwalker [21] obtained a bivariate version of the Rao–Rubin characterization of the Poisson probability law. Shanbhag [15] has arrived at the results of Rao–Rubin and Talwalker using elementary techniques. Another derivation of the Rao–Rubin characterization has been given by Srivastava and Singh [19].

Let us call condition (1) the Rao–Rubin condition. Variants of the Rao–Rubin result are sought and they are essentially of the following type. The probability laws $S(r \mid n)$, $r = 0, 1, 2, \ldots, n$; $n = 0, 1, \ldots$, are specified. Does the fulfillment of the Rao–Rubin condition determine the probability law of X? In this spirit, characterizations of the negative binomial* and the Lagrangian Poisson* probability laws have been given by Patil and Ratnaparkhi [10] and Consul [4], respectively. The state of the literature in this area up to 1975 was reviewed by Patil and Ratnaparkhi [10]. Another reference is Kagan et al. [6].

It was felt that unification of all the foregoing characterizations in the framework of damage models was possible and accordingly, Shanbhag [15] proved the following result. Suppose that $g_0 < 1$ and there exist nonnegative real sequences $a_n, n \geq 0$, and b_n, $n \geq 0$, satisfying $a_n > 0$ for every n and b_0, $b_1 > 0$ such that whenever $g_n > 0$,

$$S(r \mid n) = a_r b_{n-r}/c_n, \qquad r = 0, 1, 2, \ldots, n,$$

where $c_n, n = 0, 1, 2, \ldots$, is the convolution* of $a_n, n = 0, 1, 2, \ldots$, and $b_n, n = 0, 1, 2, \ldots$. Then the Rao–Rubin condition holds if and only if $g_n/c_n = (g_0/c_0)\theta^n$ for some $\theta > 0$. This result yields the results of Rao and Rubin [14], Patil and Ratnaparkhi [10], and Consul [4] as special cases. One could also derive countless characterizations of other discrete distributions trivially using the result above. This result does not remain

valid if the condition $a_n > 0$ for every $n = 0$, $1, 2, \ldots$ is dropped. For a counterexample, see Shanbhag and Panaretos [17]. Multivariate extensions have also been derived. For example, see Shanbhag [16].

A forerunner of the Rao–Rubin result is a result of Moran [8]. If $g_0 < 1, S(r \mid n), r = 0$, $1, 2, \ldots, n$, is binomial with index parameter n and success probability parameter p_n for every $n = 0, 1, 2, \ldots$, for which g_n is positive and Y and $X - Y$ are independent random variables satisfying the condition that $\Pr[Y = X - Y]$ is positive, then X has a Poisson probability law. Moreover, $p_n = a$ constant. Chatterji [3] has given a slightly stronger version of this result. Haight [5] reviewed some of these results.

The damage model theory has an unusually large number of proponents who committed *faux pas* in their investigations into characterizations. Some of these aspects have been discussed in Shanbhag [15] and Shanbhag and Panaretos [17].

The Rao–Rubin condition can be rewritten in the form

$$\Pr[Y = r] = \Pr[Y = r \mid X - Y = 0],$$
$$r = 0, 1, 2, \ldots.$$

It has been conjectured that the condition

$$\Pr[Y = r] = \Pr[Y = r \mid X - Y = m],$$
$$r = 0, 1, 2, \ldots,$$

for some fixed support point m of the random variable $X - Y$ in addition to the condition that $Y \mid X = n$ follows the binomial probability law with parameters n and $0 < p < 1$ for all $n = 0, 1, 2, \ldots$ for which g_n is positive would uniquely determine the probability law of X. This is false. However, if for two distinct support points m_1 and m_2 of the nondegenerate random variable $X - Y$,

$$\Pr[Y = r] = \Pr[Y = r \mid X - Y = m_1]$$
$$= \Pr[Y = r \mid X - Y = m_2],$$
$$r = 0, 1, 2, \ldots,$$

holds and the additional condition mentioned above is valid, then X has a Poisson probability law (see Patil and Taillie [12]).

Shanbhag's [16] general result can be extended to cover this modified Rao–Rubin condition (see Shanbhag and Taillie [18]).

In the spirit of Moran's [8] result, Patil and Seshadri [11] have characterized the probability law of X as belonging to a certain power series family when Y and $X - Y$ are independent and for each $n = 0, 1, 2, \ldots$

$$S(r \mid n) \propto a_r b_{n-r}, \qquad r = 0, 1, 2, \ldots, n,$$

with $a_n, n = 0, 1, 2, \ldots,$ and $b_n, n = 0, 1, 2, \ldots,$ as real sequences satisfying certain conditions. They also gave analogs of this result when Y and $X - Y$ are absolutely continuous. Some of the conditions and the arguments in this paper were obscure, and to a certain extent, Menon [7] and Berk [2] ameliorate these results.

Now, we look into the problem of characterizing the survival probability law in damage models. Srivastava and Srivastava [20] proved the following. $(X_\lambda, Y_\lambda) : \lambda \in (a, b)$ with $a \geqslant 0$ and $a < b$ is a family of random vectors of nonnegative integer-valued components such that for all values of the parameter λ,

$$\Pr_\lambda[X_\lambda = n, Y_\lambda = r]$$
$$= \exp(-\lambda)(\lambda^n / n!) S(r \mid n),$$
$$r = 0, 1, 2, \ldots, n; \; n = 0, 1, 2, \ldots$$

where $S(r \mid n), r = 0, 1, 2, \ldots, n$ is a discrete probability law independent of λ for each $n = 0, 1, 2, \ldots$ and $0 < S(n \mid n) < 1$ for some n, then

$$\Pr_\lambda[Y_\lambda = r] = \Pr_\lambda[Y_\lambda = r \mid X_\lambda = Y_\lambda],$$
$$r = 0, 1, 2, \ldots,$$

for every $\lambda \in (a, b)$ if and only if

$$S(r \mid n) = \binom{n}{r} p^r (1 - p)^{n-r},$$
$$r = 0, 1, 2, \ldots, n; \; n = 0, 1, 2, \ldots$$

for some fixed p in $(0, 1)$. Recently, Alzaid [1] has shown that if, in the above setup, we have

$$\Pr_\lambda[X_\lambda = n, Y_\lambda = r] = \Pr_\lambda[X_\lambda = n] S(r \mid n),$$
$$r = 0, 1, 2, \ldots, n; \; n = 0, 1, 2, \ldots,$$

with X_λ having a general power series distri-

bution* instead of the specific Poisson distribution for every λ in (a, b), then

$$\Pr_\lambda[Y_\lambda = r] = \Pr_\lambda[Y_\lambda = r \mid X_\lambda = Y_\lambda],$$
$$r = 0, 1, 2, \ldots,$$

for every λ in (a, b) if and only if Y_λ and $X_\lambda - Y_\lambda$ are independent for each λ in (a, b). In view of the well-known Raikov theorem*, Srivastava and Srivastava's [20] result follows as a special case. Alzaid [1] has also succeeded in characterizing a certain class of infinitely divisible distributions via Rao–Rubin-type conditions.

On a practical note, one should mention Rao's [13] paper on applications of some of the results discussed above. The *Proceedings of the International Symposium on Classical and Contagious Discrete Distributions* edited by Patil [9] is another source for practical applications.

References

[1] Alzaid, A. H. (1979). Some Characterizations in Rao–Rubin Damage Models. M.Sc. dissertation, University of Sheffield.

[2] Berk, R. H. (1977). *J. Appl. Prob.*, **14**, 806–816.

[3] Chatterji, S. D. (1963). *Amer. Math. Monthly*, **70**, 958–964.

[4] Consul, P. C. (1975). In *Statistical Distributions in Scientific Work*, Vol. 3, G. P. Patil, S. Kotz, and J. K. Ord, eds. D. Reidel, Dordrecht-Holland, pp. 279–290.

[5] Haight, H. A. (1967). *Handbook of the Poisson Distribution*. Wiley, New York.

[6] Kagan, A. M., Linnik, Y. V., and Rao, C. R. (1973). *Characterization Problems in Mathematical Statistics*. Wiley, New York.

[7] Menon, M. V. (1966), *J. R. Statist. Soc. B*, **28**, 143–145.

[8] Moran, P. A. P. (1952), *Proc. Camb. Philos. Soc.*, **48**, 206–207.

[9] Patil, G. P., ed. (1963), *Proc. Int. Symp. on Classical and Contagious Discrete Distributions*, Statistical Publishing Society, Calcutta.

[10] Patil, G. P. and Ratnaparkhi, M. V. (1975). In *Statistical Distributions in Scientific Work*, Vol. 3, G. P. Patil, S. Kotz, and J. K. Ord, eds. D. Reidel, Dordrecht-Holland, pp. 255–270.

[11] Patil, G. P. and Seshadri, V. (1964), *J. R. Statist. Soc. B*, **26**, 286–292.

[12] Patil, G. P. and Taillie, C. (1980), To appear in *Sankhyā*.

[13] Rao, C. R. (1963). In *Proc. Int. Symp. on Classical and Contagious Distributions*. Statistical Publishing Society, Calcutta, pp. 320–332.

[14] Rao, C. R. and Rubin, H. (1964). *Sankhyā A*, **26**, 295–298.

[15] Shanbhag, D. N. (1974). *J. Appl. Prob.*, **11**, 211–215.

[16] Shanbhag, D. N. (1977). *J. Appl. Prob.*, **14**, 640–646.

[17] Shanbhag, D. N. and Panaretos, J. (1977). *Aust. J. Statist.*, **21**, 78–83.

[18] Shanbhag, D. N. and Taillie, C. (1980). To appear in *Sankhyā*.

[19] Srivastava, R. C. and Singh, J. (1975). In *Statistical Distributions in Scientific Work*, Vol. 3. D. Reidel, Dordrecht-Holland, pp. 271–277.

[20] Srivastava, R. C. and Srivastava, A. B. L. (1970). *J. Appl. Prob.*, **7**, 495–501.

[21] Talwalker, S. (1970). *Sankhyā* A, **32**, 265–270.

(CUMULATIVE DAMAGE MODELS)

M. Bhaskara Rao
D. N. Shanbhag

DAM THEORY

Dam theory is concerned with data analysis and stochastic modeling with regard to the operation and control of water storage in reservoirs. The probabilistic structure of dam models is similar to that of queues (*see* QUEUEING THEORY) in telephone traffic, inventory stock (*see* INVENTORY THEORY) in a warehouse, and financial capital in an insurance company (*see* RISK THEORY). In realistic terms, dam theory must be considered a subfield of stochastic hydrology (*see* HYDROLOGY, STATISTICS IN), which is itself a small if important part of the broad field of hydrology. This includes diverse areas such as infiltration and basin loss, overland flow and runoff, flood waves in channels and reservoirs, conceptual models of basins, frequency and control of floods, and man-made changes, for which there is an enormous literature (see, e.g., the book by Buras [4] and that edited by Ciriani et al. [5]).

The earliest probability treatment of stream flow and overflow regulation is due to Savarenskiy [25]. This Russian engineer developed methods for calculating the flow coefficients of rivers and their reliability, and used these for the Belaja, Volga, and Don, for which 50-year records of streamflows were available. His results did not become well known, even in the USSR, until two decades later; meanwhile, Moran [17] had developed his stochastic theory of storage (*see* STORAGE THEORY), which has since become widely used by engineers and hydrologists. The principles of Moran's model are described below.

THE BASIC DAM MODEL OF MORAN

Let $Z_t (0 \leqslant Z_t \leqslant K - M)$ to be the integer-valued content of a dam at yearly epochs $t = 0, 1, 2, \ldots$, its initial content being $Z_0 = u > 0$. Suppose that inputs $X_t = 0, 1, 2, \ldots$, flow into the dam annually in the intervals $(t, t + 1)$, and that a release of the smaller of $M(0 < M \leqslant K - M)$ or the total dam content is allowed just before the end of the year at $t + 1$. Then the dam content Z_{t+1} at time $t + 1$, after the release, will be

$$Z_{t+1} = \min\{Z_t + X_t, K\} - \min\{Z_t + X_t, M\},$$

where an overflow $Z_t + X_t - K$ occurs during $(t, t + 1)$ if the content of the dam exceeds its capacity K.

Moran [17] regarded this discrete model as a rough approximation to the realistic dam problem, and assumed that the annual inputs $\{X_t\}$ were independently and identically distributed random variables with probability distribution $p_i = \Pr[X_t = i]$, $i = 0, 1, 2, \ldots$. In this case, the dam content Z_t forms a homogeneous Markov chain; *see* MARKOV PROCESSES. Most of the early work on the Moran dam as reviewed by Gani [7], Moran [20], and Prabhu [23] was concerned with deriving the time-dependent content distributions

$$P_{uj}(t) = \Pr[Z_t = j \mid Z_0 = u]$$

and, historically earlier, the stationary distributions*

$$P_j = \lim_{t \to \infty} P_{uj}(t)$$

for particular types of independent inputs, when the dam capacity K was either finite or infinite. It was these stationary distributions, and more particularly the probabilities $P_{u0}(t)$ and P_0 of emptiness of the dam which proved to be of greatest interest to hydrologists.

Perhaps the most interesting of Moran's models remains that for which the input is continuous with an infinitely divisible gamma-type distribution (*see* INFINITELY DIVISIBLE DISTRIBUTIONS) in any time interval T. This is obtained by a limiting operation from the discrete model for which the $\{p_i\}$ form a negative binomial distribution*. Again using limiting methods, Moran [18] was able to derive the stationary distribution of this continuous dam content for infinite K; Gani and Prabhu [9] later found its time-dependent distribution. While the discrete dam is closely related in structure to queueing models, Moran's continuous dam has no queueing analog and thus represents a distinct innovation in the theory of stochastic processes*.

THE LLOYD DAM AND AUTOREGRESSIVE MARKOV MODELS

As Moran himself was aware, the assumption of independence of inputs $\{X_t\}$ is hardly tenable, except as an approximation. Many annual river flows exhibit a significant serial correlation*, and it is a distinct advantage of the Lloyd dam that such flow dependence can be taken into account. In Lloyd's [15] model, the assumption is made that the discrete inputs $\{X_t\}$ themselves form a homogeneous Markov chain. While the content $\{Z_t\}$ is no longer a Markov chain, the joint process $\{U_t\} = \{Z_t, X_t\}$ forms a bivariate Markov chain.

Using methods similar to those for the simpler Moran dam, Odoom and Lloyd [21] were able to derive formulas for the stationary distributions of the dam content $\{Z_t\}$ for finite and infinite K. The time-dependent results for the case of infinite K were later obtained by Ali Khan and Gani [1].

Meanwhile, Fiering [6], in his book on streamflow, explored Markov autoregressive schemes (*see* AUTOREGRESSIVE MODELS) for modeling the flows of rivers, and also suggested simulation* as a method of producing long data series. These developments were to have significant effects on stochastic hydrology.

FLOOD FLOWS AND CONTROL

The prediction of floods and their control by dams is an area of great importance. Moran [19] considered the problem of estimating the distribution of largest monthly floods (during a year) from runs of river flow data. The difficulty lies in estimating the tail of the largest monthly river flow distribution from a restricted number of available annual observations. If, for example, these distributions are assumed to be either (1) lognormal*, or (2) gamma*, then estimates of flood flows for Australian rivers based on them are somewhat different.

As an example, estimates based on the two assumed distributions for the River Murray at Jingellic, Victoria, are given in Table 1. These estimates, although different, show some measure of agreement and would allow engineers to work out the capacity of dams needed to contain 100-year floods.

Also in the area of flood control, some simple mathematical modeling for the regulation of two confluents was carried out by Anis and El-Naggar [2]. Gani [8] later compared two models for the damming of a river system; in the first, dams were built on tributaries before their junction with the main river, while in the second they were

Table 1   River Murray: Estimated Flood Flows for 100-Year Floods (Thousands of Acre-Feet)

Distribution	Flood Estimate	95% Confidence Interval
Log-normal	1108	852–1442
Gamma	941	768–1114

constructed on the main river after the confluence of a tributary.

The optimal operation and control of dams, with a view to satisfying water demand as well as minimizing floods, has been studied by Rozanov [24] and Klemeš [13] among others. Both authors consider a dynamic programming* approach; Klemeš also compares this with the older graphical mass-curve technique used by engineers.

RECENT WORK IN THE THEORY OF DAMS

During the past decade dam theory has developed in two main directions: (1) applications in stochastic hydrology*, and (2) the development of new probabilistic models.

The recent paper of Lawrance and Kottegoda [14] reviews some of the early history of stochastic hydrology, and discusses, among other topics, the Hurst* effect for long-term flow persistence and the beginnings of streamflow simulation methods. A variety of autoregressive and other models with both short and long memories are analyzed. Further contributions in this area are numerous, among them the book of Hipel and McLeod [10] and the review of Klemeš [12]. Developments in this area have been reasonably realistic and have taken account of observed data, or used simulated data for testing models.

Meanwhile, probabilistic models have become increasingly complex and sophisticated (*see* MODEL BUILDING). Phatarfod [22] has reviewed some aspects of dam theory and suggested that analytic results should complement the simulation techniques of stochastic hydrology. So far, little progress has been made on these lines. The recent work of Brockwell and Chung [3], Yeo [27], Tin and Phatarfod [26], Kennedy [11], and Lloyd and Saleem [16], among many others, while contributing to the theory of stochastic processes, is perhaps too complex to be of immediate practical use to engineers and hydrologists. It may in future, like Moran's and Lloyd's earlier work, become useful in

obtaining numerical solutions for realistic dam models.

The fact remains, however, that dam theory has already contributed considerably to stochastic hydrology and to the mathematical analysis of the operation and control of water reservoirs. There seems every reason to believe that probabilistic modeling will continue to prove its value in the management of water resources.

References

In this list, T indicates publications that are technical in nature, S denotes surveys, and G identifies books and papers of general interest.

[1] Ali Khan, M. S. and Gani, J. (1968). *J. Appl. Prob.*, **5**, 72–84. (T)

[2] Anis, A. A. and El-Naggar, A. S. T. (1968). *J. Inst. Math. Appl.*, **4**, 223–231. (T)

[3] Brockwell, P. J. and Chung, K. L. (1975). *J. Appl. Prob.*, **12**, 212–217. (T)

[4] Buras, N. (1972). *Scientific Allocation of Water Resources*. Elsevier, New York. (G)

[5] Ciriani, T. A., Maione, U., and Wallis, J. R., eds. (1977). *Mathematical Models for Surface Water Hydrology*. Wiley, New York. (G)

[6] Fiering, M. B. (1967). *Streamflow Synthesis*. Harvard University Press, Cambridge, Mass. (T)

[7] Gani, J. (1957). *J. R. Statist. Soc. B*, **19**, 181–206. (S)

[8] Gani, J. (1969). *Adv. Appl. Prob.*, **1**, 90–110. (S)

[9] Gani, J. and Prabhu, N. U. (1963). *Proc. Camb. Philos. Soc.*, **59**, 417–429. (T)

[10] Hipel, K. W. and McLeod, A. I. (1978). *Applied Box–Jenkins Modelling for Water Resources Engineers*. Elsevier, Amsterdam. (T)

[11] Kennedy, D. P. (1978). *J. Appl. Prob.*, **15**, 171–178. (T)

[12] Klemeš, V. (1978). *Adv. Hydrosci.*, **11**, 285–356. Academic Press, New York. (G)

[13] Klemeš, V. (1979). *Water Resour. Res.*, **15**, 359–370. (G)

[14] Lawrance, A. J. and Kottegoda, N. T. (1977). *J. R. Statist. Soc. A*, **140**, 1–47. (S)

[15] Lloyd, E. H. (1963). *Technometrics*, **5**, 85–93. (T)

[16] Lloyd, E. H. and Saleem, S. D. (1979). *J. Appl. Prob.*, **16**, 117–128. (T)

[17] Moran, P. A. P. (1954). *Aust. J. Appl Sci.*, **5**, 116–124. (T)

[18] Moran, P. A. P. (1956). *Quart. J. Math.*, **2**(7), 130–137. (T)

[19] Moran, P. A. P. (1957). *Trans. Amer. Geophys. Union*, **38**, 519–523. (G)

[20] Moran, P. A. P. (1959). *The Theory of Storage*. Methuen, London. (T)

[21] Odoom, S. and Lloyd, E. H. (1965). *J. Appl. Prob.*, **2**, 215–222. (T)

[22] Phatarfod, R. M. (1976). *J. Hydrol.*, **30**, 199–217. (T)

[23] Prabhu, N. U. (1964). *J. Appl. Prob.*, **1**, 1–46. (S)

[24] Rozanov, Yu. A. (1977). In *Multivariate Analysis*, Vol. 4, P. R. Krishnaiah, ed. North-Holland, Amsterdam, pp. 431–444. (T)

[25] Savarenskiy, A. D. (1940). *Gidrotekh. stroit.*, No. 2, 24–28. (T)

[26] Tin, P. and Phatarfod, R. M. (1976). *J. Appl. Prob.*, **13**, 329–337. (T)

[27] Yeo, G. F. (1975). *J. Appl. Prob.*, **12**, 205–211. (T)

(AUTOREGRESSIVE-INTEGRATED
 MOVING AVERAGE (ARIMA)
 MODELS
AUTOREGRESSIVE-MOVING
 AVERAGE (ARMA) MODELS
HYDROLOGY, STATISTICS IN
QUEUEING THEORY
STOCHASTIC PROCESSES
TIME SERIES)

J. Gani

DANDEKAR (V.M.) CONTINUITY CORRECTION

(Quoted by Rao [1] for calculating the value of χ^2 in a 2×2 contingency table*.) "Standard" χ^2 values (here denoted χ_0^2) are calculated for the observed configuration* and also values obtained by decreasing and increasing the smallest frequency in the table by unity (χ_{-1}^2 and χ_1^2, respectively). The continuity-corrected χ^2 is

$$\chi_c^2 = \chi_0^2 - \frac{\chi_0^2 - \chi_{-1}^2}{\chi_1^2 - \chi_{-1}^2} \left(\chi_1^2 - \chi_0^2 \right).$$

For example, in the following 2×2 contingency table

	R	$\bar{R}$	Totals
A	13	4	17
B	6	14	20
Totals	19	18	37

the calculated values are

$$\chi_0^2 = \frac{37(13 \times 14 - 6 \times 4)^2}{17 \times 20 \times 19 \times 18} = 7.9435,$$

$$\chi_{-1}^2 = \frac{37(13 \times 14 - 6 \times 3)^2}{17 \times 20 \times 19 \times 18} = 12.0995,$$

$$\chi_1^2 = \frac{37(13 \times 14 - 6 \times 5)^2}{17 \times 20 \times 19 \times 18} = 4.6587,$$

and the corrected value is

$$\chi_c^2 = 7.9435 - \frac{7.9435 - 4.6587}{12.0995 - 4.6587}$$
$$\times (12.0995 - 7.9435)$$
$$= 6.1068.$$

Comparisons with exact values (based on multinomial* probabilities) show that Dandekar's correction is slightly better than Yates' continuity correction*, although it involves more elaborate calculations.

Reference

[1] Rao, C. R. (1973). *Linear Statistical Inference and Its Applications*, 2nd ed. Wiley, New York, pp. 414–415.

(CONTINGENCY TABLES
CONTINUITY CORRECTIONS
YATES' CORRECTION)

DARMOIS–KOOPMAN FAMILY   *See*
KOOPMAN-DARMOIS-PITMAN FAMILY

DARMOIS–SKITOVICH THEOREM

This theorem asserts the normality of *independent* random variables $X_1, X_2, \ldots, X_n$ if there exist any two linear functions

$$a_1 X_1 + \cdots + a_n X_n$$
$$b_1 X_n + \cdots + b_n X_n$$

with the conditions $a_i b_i \neq 0$, $i = 1, \ldots, n$, which are independently distributed. (This is one of the earliest characterization theorems for the normal distribution.)

A generalization of the Darmois–Skitovich theorem, provided by Rao [4], is based on the weaker condition of the conditional expectation of one linear function, given another, being zero. Thus

$$E(a_1 X_1 + \cdots + a_n X_n \mid b_1 X_1$$
$$+ \cdots + b_n X_n) = 0$$

for some $\{a_i\}$, $\{b_i\}$ with

(a) $E[X_1] = 0$, var (X_1) finite
(b) $\text{var}(X_1) \neq 0$ or $\sum_{i=1}^{n} a_i b_i = 0$
(c) $|b_n| > \max(|b_1|, \ldots, b_{n-1}|)$; $a_n \neq 0$, $a_i b_i a_n b_n < 0$, for $i = 1, 2, \ldots, n-1$

ensures that the X_i's follow the normal law.

For more information on this topic, see the works by Darmois [1], Kagan et al. [2], Mathai and Pederzoli [3], and Skitovich [5].

References

[1] Darmois, G. (1953). *Rev. Inst. Int. Statist.*, **21**, 2–8.

[2] Kagan, A. M., Linnik, Yu. V., and Rao, C. R. (1973). *Characterization Problems in Mathematical Statistics*. Wiley, New York.

[3] Mathai, A. M. and Pederzoli, G. (1978). *Characterizations of the Normal Probability Law*. Halsted Press, New York.

[4] Rao, C. R. (1967). *Sankhyā A*, **29**, 1–14.

[5] Skitovich, V. P. (1954). *Izv. Acad. Nauk SSSR ser. mat.*, **18**, 185–200.

(CHARACTERIZATION OF DISTRIBUTIONS)

DATA COLLECTION

A distinction needs to be made between numbers, adjectives, and other forms of description and data. Numbers and adjectives can be available simply as entities in themselves or as data. The existence of numbers and adjectives does not imply the existence of a set of data, but the existence of a set of data implies the existence of some descriptive forms such as numbers, adjectives, phrases, pictures, graphs, etc. For example, the set of numbers $\{3, 1, 0, 4, 9, 6\}$ and the set of adjectives {small, pretty, personable, harsh, susceptible, resistant} do not in themselves convey information about any phenomenon. When numbers and adjectives

convey information about some entity such as 0, 1, 3, and 4 worms in four particular apples and small, pretty, heat-resistant spores of a bacterium, these facts are denoted as data. A *datum* is defined to be a fact (numerical or otherwise) from which a conclusion may be drawn such as, for example, that none of the four apples have the same number of worms. A datum contains information whereas a number, adjective, or other form of description may not.

The information in a set of data may be contaminated (mixed up or confounded) with other kinds of information. For example, the particular four apples may have been the only ones in a basket of apples that contained worms and the apples were from a part of the orchard that had a very light infestation of the codling moth. Thus we do not know if 4 out of 120 wormy apples per basket represents the proportion of wormy apples in the entire population of apples or only for this variety in lightly infested orchards. The various factors affecting variation and response are denoted as *sources of variation** in the data. When the various effects in a set of data are inseparable, they are said to be completely confounded. Partial confounding* of effects occurs when the effects are partly mixed together, but it is possible to obtain estimates of each of the various sources of variation. No confounding of effects (orthogonality) implies maximum separability of sources of variation. For example, suppose that there is technician-to-technician variation, day-to-day variation, and method-to-method variation. If technician I conducts method 1 on day 1, technician II method 2 on day 2, technician III method 3 on day 3, etc., one cannot separate the technician effect from the method effect or from the day effect. If, on the other hand, all technicians used all methods on each of several days, the different effects would be completely separable. The preceding plan would be an orthogonal (completely unconfounded) one, whereas the former would be completely confounded. Thus the plan or the design of the investigation has a large influence on the confounding aspects in data and hence on the amount of information derivable from the data. With adequate planning and foresight, data can be obtained that have little or no confounding of sources of variation of interest to the investigator. On the other hand, unplanned and/or haphazard collection of data, no matter how voluminous, most frequently results in highly confounded data sets which may be of little or no use in studying effects of various sources of variation. There is a tendency for people to believe that largeness of a data set removes the effects of confounding. This is not necessarily true. The closer one gets to obtaining responses for 100% of the population, the nearer one is to the values of the population parameters. In this sense, largeness does remove the difficulties of haphazard or selective sampling. However, the complete enumeration of a subpopulation provides no information on the other subpopulation parameters. In this sense, largeness of a set of data does nothing to remove the biases.

ASPECTS OF DATA COLLECTION

Whenever data are to be collected, several aspects require consideration. The first one to consider is *why* these data should be collected and what uses these data will serve. If there is no purpose or no use for these data, why collect them? The ready accessibility of microcomputers, minicomputers, and macrocomputers, together with recording equipment, makes it a simple matter to record voluminous sets of data with relatively little effort. As an illustration, a number of instruments for recording temperature, pulse rate, blood pressure, changes in blood pressure, etc., are attached to a dog undergoing surgery. A minicomputer is used to determine the frequency with which temperature, pulse rate, blood pressure measurements, etc., are recorded. As a second example, suppose that temperature in one-hundredths of a degree celsius and humidity in one-hundredths of a percent are to be recorded every second over a 20-year period in 1000 locations of a field; a minicomputer is used

to determine the frequency with which observations are recorded and the observations may be recorded on tape and/or paper. This results in $60 \times 60 \times 24 \times 365.25 \times 1000 \times 20 = 631,150,000,000$ measurements on temperature and the same number on humidity. The data set would consist of more than 1.25 trillion observations. Before embarking on such a large data collection venture, one should seriously consider taking fewer measurements (e.g., one every hour rather than 3600 per hour), the uses for data of this nature, and which individuals would actually use these data. Unfortunately, large data sets have been and are being collected, merely on the premise that they might be useful to someone at some time. A computer is available, and data are collected to utilize the capabilities of the computer.

A second aspect of data collection is *what* data will be collected. All pertinent data for studying a phenomenon should be collected. Thus it is essential to determine what data to collect in light of why the data are collected. Sufficient data should be collected to achieve the goals and purposes of the investigation. Provision should also be made to obtain pertinent data that surface during the course of an investigation. The investigator must be aware of evidence that comes to his or her attention, and obtain the necessary data to explain the phenomenon. An example was given above illustrating how a large data set might be obtained. In some investigations, such as sending missiles to the moon, one can obtain only a few observations on selected entities. In both these cases, it is essential to determine what data to collect.

A third aspect of data collection is *how* and *where* the data are to be collected. The statistician can be very helpful in designing and planning the investigation and in determining how and where the data are to be collected. The how and where of data collection are intimately entwined with the plan and type of the investigation. Also involved are how to measure and quantify information on the various phenomena in an investigation. This is especially true of social phenomena. Methods of measuring responses must be carefully considered, and it should be ascertained if a measuring instrument is measuring what it is supposed to, or if it is actually measuring something else.

Two other aspects to be considered are—*who* is to collect the data and *when* are they to be obtained. Unless these aspects are firmly regulated, the data may not be collected or only a portion of the data may be obtained. In many investigations it is imperative to have unbiased and highly trained technicians in order to carry out the investigation. The timing of an investigation may be important to its success, and/or it may be necessary to carry out the investigation in a specified time period. If these aspects are ignored, the data may be so unreliable as to make them useless.

In addition to the *why*, *what*, *how*, *where*, *who*, and *when* aspects of data collection, it is imperative to have a complete, written description of all data obtained. If this information is recorded only in the investigator's mind, it may soon be lost and be unavailable to other investigators desiring to use the data. Plans need to be made for storage and/or disposal of all data collected. Discarding valuable data that are needed in future research or the storage of useless data result in economic losses that can be avoided with proper planning. Data may be stored in notebooks, on tapes, or other forms; precautions need to be taken to avoid loss or damage due to fire, water, tape erasures, etc.

TYPES OF DATA COLLECTION

Three main types of data collection are observational investigations or studies, sample surveys and censuses, and experimental investigations. In the observational types of investigation, records are kept on whatever observations are available with little or no thought of making them representative of the population. In censuses*, an attempt is made to obtain observations on every member of the population, whereas in sample surveys* only a portion of the population is surveyed with an idea of the sample being representative of the population. In an experiment, conditions are often introduced which do not appear in the population, and there is a considerable degree of control over the conditions of the experiment. The controlled conditions in an experiment may not be available in observational* and survey investigations.

Observational Investigations

An important source of observational data sets is record keeping. In all societies of the world, records are kept on a wide variety of phenomena. For example, there are records of births, of deaths, of marriages, of church members, of traffic tickets issued, of convictions, of diplomas, of fraternal and social organizations, of gem collectors, of daily maximum and minimum temperatures, of daily rainfall, of auto sales by dealerships, of treatments administered to patients by doctors, and on and on. In totality, records of events comprise a very voluminous set of data. These records, although possibly useful for the purpose for which they were collected, may be useless for another purpose because of the method for determining whether or not a record would be kept. For example, in the issuance of traffic tickets, no record may be kept of traffic tickets that were issued and then destroyed. Thus the records available on traffic tickets issued may be only those tickets issued which are eventually presented to a judge, with no mention being made of the number of tickets issued and destroyed before presentation. The omission of a segment of the population would make the remaining data useless in determining ratio of convictions to number of traffic tickets issued. One of the major faults of observational data is the omission of subsets of the data without knowing or having a description of the subsets omitted. In many cases, no valid conclusions are possible about the entire population from a portion of the subpopulations in the observational data set.

Observational data sets are often used in studies simply because of their availability. Some statistical studies have been made for utilizing observational data, but no general methods exist for drawing valid conclusions from observational data sets. Each set has to be considered on its own merits, and the investigator must exercise caution in making inferences from the data to the population.

To illustrate an observational investigation, consider the records on patients coming to a specified general medical practitioner's office (*what* and *where*). The M.D. has decided to record only those observations pertinent to a patient's physical condition. He or she may record one word (e.g., "healthy") or 10 pages of notes, together with weight, height, temperature, two blood pressure readings, and specified laboratory tests on urine and blood (*what*). The M.D. personally records the observations, measurements, and test results (*who*). A folder is made up for each patient and the observations are recorded on standardized $8\frac{1}{2} \times 11\frac{1}{2}$ medical forms (*how*). The forms are completed as soon as the observations are available (*when*). Patients' folders are stored in a central file in alphabetical order (*how*). The purpose for obtaining these observational data is to keep a readily accessible account of a patient's medical history for future reference by the M.D. and/or nurses (*why*). Note that the sample of patients coming to the M.D.'s office may be a very biased sample of the general population of patients, and that observations are not uniformly recorded. That is, for each patient the M.D. makes a decision as to what is to be recorded. Another M.D. would record different observations in a different manner, and it would be difficult, if not impossible, for a researcher to use these data in a medical study and draw valid conclusions from them.

Censuses* and Sample Surveys*

The population to be surveyed is defined and then either a strictly 100% sample (a census) or a smaller percentage of the population is surveyed. Interest centers on the conditions and events that occur within the specified population. Censuses are usually not 100% samples because some sampling units are inaccessible and others may appear more than once. However, an attempt is made to obtain a 100% sample with no repetitions. For example, the U.S. Department of Commerce, through its Census Bureau*, is bound by law to carry out a population census for the entire United States every 10 years. This is an enormous and costly task and there are omissions and duplications among the results despite elaborate precautions. It is an even more formidable task to take censuses in countries with very large populations, as in India and China. Instead, these countries must obtain population estimates through sample survey methods. The sample percentage may be small, say 0.1 to 1%, in order that the work may be within the capacity of the available personnel.

Sample surveys may be grouped into two broad categories: probability sample* surveys and nonprobability sample* surveys. In the former type, the probability of selecting sampling units from the population is known, whereas it is not in the latter type. Whether or not nonprobability sample estimates can be regarded as representative of the population parameters is unknown. Despite this, a large number of surveys are of the nonprobability type. The lower cost and convenience of such surveys is attractive to investigators even if they are unable to check on the biases in the survey.

Some types of nonprobability sample surveys are:

1. **Judgment.** The investigator limits the selection to the sampling units to those he or she judges to be representative of the population.

2. **Convenience.** The investigator selects sampling units that are convenient or readily accessible.

3. **Quota.** The only restriction on the selection of the sampling units is that there be a specified number in each of the specified groups. A quota in each group is thus maintained by a variety of unspecified sampling procedures.

4. **Purposely biased.** The investigator devises a sampling procedure that eliminates all sampling units of an undesirable class, or he or she selects only the sampling units that give the desired result, say two to one, nine out of ten, etc.

5. **Haphazard.** The investigator selects the sampling units in such a manner as to leave the reader with the impression that the selection involved randomization* and a probability survey.

Some types of probability sample surveys are:

1. **Simple random sample.** The population does not contain subpopulations and each sampling unit in the population has an equal and independent chance of being selected. An equivalent definition is that every possible sample of size n has an equal chance of being selected.

2. **Stratified-simple random sample.** The population is composed of subpopulations and a simple random sample of sampling units is selected within *each* of the subpopulations (strata).

3. **Cluster-simple random sample.** The population is composed of subpopulations (clusters) but a simple random sample of subpopulations (clusters) is made; then a simple random sample is made within each of the selected clusters. When the clusters are areas, the sample survey design is known as an area-simple random sample.

4. **Every kth sampling unit with a random start.** In many cases, a list of sampling units or a serial ordering of sampling units is available. The list is partitioned in subgroups of k items each. A number between 1 and k, say t, is randomly selected. Then the tth item, the $(t + k)$th item, the $(t + 2k)$th item, the $(t + 3k)$th item, etc., form the sample. This survey design has also been called systematic or cyclic sampling*.

5. **More complex designs.** There are many types of sample survey designs involving more complexity than the above. Several of the types described above may be included in these designs.

Probability sample survey designs are recommended for use in data collection because their properties are known. Representative and unbiased estimates of population parameters are possible with these designs, but not with the others described above.

As an example of a judgment sampling procedure for predicting voter's preference for a presidential candidate in the United States, suppose that a survey organization (*who*) decided to use 100 voting precincts and that these 100 were those that had most nearly mirrored the national percentages in each of the last eight elections (*how*). It was judged that these precincts had been, and would continue to be, most representative of the national percentages of an election. To predict the outcome of the current national election (*why* and *what*), an enumerator (*who*) was assigned to each precinct (*where*) for the purpose of seeking out the voting results as soon as the polls closed (*what* and *when*). The results were to be phoned to a central place (*how* and *where*) and were recorded in a computer that was programmed to predict the national percentages for each of the candidates (*how* and *what*). These predictions were to be released to the press and national television as soon as a winner was determined (*what* and *how*).

Experiment Designs*

Every experiment involves data collection and has a plan of procedure, some involving randomization and some not. Because of the statistical properties of the randomized designs, they are recommended over systematic or nonrandomized designs. In order to have higher precision (repeatability*), blocking is used to group the units (experimental units) used in the experiment into blocks, or groups, which are relatively homogeneous within blocks. The blocks are the subpopulation (clusters) in the population. A simple random sample of blocks is made and a simple random sample of units in each block is made. The entities of interest are called treatments. These are randomly allocated to the experimental units within each block in the randomized design and selectively placed in the systematic ones. Then the experiment is put into operation and data are collected on each experimental unit in order to obtain information as a basis for comparing treatments and/or of describing the action of treatments on the responses for each experimental unit, and responses may be obtained for a large number of variables.

Experiment designs (plans for arrangement of treatments in an experiment) are used for data collection in research, developmental, and investigational studies. They are used under relatively controlled conditions as compared to survey designs. A selected set of phenomena are studied in any particular experiment, and these may or may not be present in the general population. (*See* DE-SIGN OF EXPERIMENTS.)

DATA COLLECTION AGENCIES

Data collection agencies appear in every part of a community. The judicial, legislative, and executive branches of village, city, county, state, national, and international governments have a variety of data collection agencies. Some of the better known U.S. federal data collection agencies are the Census Bureau*, the Bureau of Labor Statistics*, the Statistical Reporting Service*, the Central Intelligence Agency, the Internal Revenue Service, the Food and Drug Administration*, the Bureau of the Budget, and the General Services Administration. In addition to the agencies located in Washington, D.C., branches are located throughout the United States; *see* FEDERAL STATISTICS.

The United Nations has several large data collection agencies, as do all the governments of the world. The more developed a country becomes, the more extensive and sophisticated are the data collection agencies. The data are often utilized in planning the economic, business, and social policies of a country. To have intelligent planning, it is necessary to have data on the phenomenon at hand.

In the private sector of the United States, there are many national and state survey organizations collecting data for specific purposes. The Roper, Gallup, and Harris survey organizations provide the polls frequently reported in newspapers. In addition, volumes of data are collected by pharmaceutical, industrial, business, and educational organizations. The *Wall Street Journal* is devoted to reporting data on business transactions on stocks, bonds, mutuals, futures, etc. Large quantities of data are accumulated and reported every weekday.

Research organizations of colleges, universities, industries, corporations, institutes, etc., conduct and report a vast number of research, development, and investigational studies and surveys. Data from these studies are reported in scientific journals of which there is a very large number. The reporting of results from these studies is an effort to make the results of data collections and the resulting conclusions available to the general scientific population.

From the above it should be apparent that a vast amount of data collection is made in all countries of the world and that individual, corporate, and political relations are affected to a considerable degree by the data collected. Data collection is an inescapable and essential part of the lives of all individuals in a community. Hence there should be considerable desire to obtain the best and most complete data possible on the desired variables of interest. Statisticians can be of considerable aid in the planning of all data collection programs.

Bibliography

Anastasi, A. (1967). *Amer. Psychol.*, **22**, 297–306. (An intriguing article on psychological testing and measurement.)

Campbell, N. R. (1928). *An Account of the Principles of Measurements and Calculations*. Longmans, Green, London. [A basic book in the philosophical aspects of measurements; this was the leading book in this area until 1962 (see Ellis (1962).]

Campbell, S. K. (1974). *Flaws and Fallacies in Statistical Thinking*. Prentice-Hall, Englewood Cliffs, N.J. (A humorous and enlightening account of many misconceptions about and of statistics.)

Cochran, W. G. (1977). *Sampling Techniques*, 3rd ed. Wiley, New York. (A lucid and clear account of sample survey techniques, theory, and application is presented. The text is an introductory book in sampling with prerequisites of a course in statistical methods and some mathematical statistics.)

Deming, W. E. (1950). *Some Theory of Sampling*. Wiley, New York.

Ellis, B. (1962). *Basic Concepts of Measurement*. Cambridge University Press, Cambridge. (The first book on philosophical concepts of measurement since Campbell's 1928 book.)

Federer, W. T. (1973). *Statistics and Society—Data Collection and Interpretation*. Marcel Dekker, New York. (An introductory, second-year college-level text covering introductory concepts of statistical design, analysis, and inference, but with little space being devoted to hypothesis testing. It stresses concepts rather than techniques and emphasizes the population, sampling aspects, and data collection aspects that are missing from most texts.)

Haack, D. G. (1979). *Statistical Literacy. A Guide to Interpretation*. Duxbury Press, North Scituate, Mass. (An elementary and introductory book that deals with statistical doublespeak, which is defined as the "inflated, involved, and often deliberately ambiguous use of numbers" in many areas of our life.)

Hansen, M. H., Hurwitz, W. N., and Madow, M. G. (1953). *Sample Survey Methods and Theory*, Vols. 1 and 2. Wiley, New York. (An advanced text on the theory and methods of sample survey design.)

Kish, L. (1965). *Survey Sampling*. Wiley, New York. (An intermediate-level text of survey design and analysis.)

McCarthy, P. J. (1957). *Introduction to Statistical Reasoning*. McGraw-Hill, New York. (Chapters 6 and 10 are recommended reading. The basic ideas of sampling design are presented in these two chapters.)

McCarthy, P. J. (1958). Sampling: Elementary Principles. *Bull. No. 15*, New York State School of Industrial and Labor Relations, Cornell University, Ithaca, N.Y. (Somewhat more technical than the book by Slonim; the presentation is elementary and is directed more toward survey analysis aspects.)

Moe, E. O. (1952). New York Farmers' Opinion on Agricultural Programs. *Cornell Univ. Ext. Bull.* 864. (A detailed account of a sample survey of farmers in the 55 agricultural counties of New York.)

Moore, D. S. (1979). *Statistics: Concepts and Controversies*. W. H. Freeman, San Francisco. [This book, like Federer (1973), places its emphasis on how the data are collected rather than assuming the statistical design, population structure, and sampling procedures as most texts do. Both books tend to focus on ideas and their impact on everyday life.]

Raj, D. (1972). *The Design of Sample Surveys*. McGraw-

Hill, New York. (A nontechnical discussion of sample surveys.)

Slonim, M. J. (1960). *Sampling* (original title: *Sampling in a Nutshell*). Simon and Schuster, New York. (An elementary and easily understood presentation of the ideas, concepts, and techniques of sample surveys.)

Sukhatme, P. V. (1954). *Sampling Theory of Surveys with Applications*. Iowa State College Press, Ames, Iowa. (An advanced-level book in sample survey design.)

Tanur, J. M., et al. (1972). *Statistics: A Guide to the Unknown*. Holden-Day, San Francisco. (A collection of 44 essays on human beings in their biologic, political, social, and physical works. The level is elementary to intermediate.)

Wallis, W. A., et al. (1971). *Federal Statistics—A Report of the President's Commission*, Vols. 1 and 2. U.S. Government Printing Office, Washington, D.C.

Wilson, E. B., Jr. (1952). *An Introduction to Scientific Research*. McGraw-Hill, New York. (An easy-to-understand and comprehensive account of the principles of scientific investigation.)

Yates, F. (1960). *Sampling Methods for Censuses and Surveys*. 3rd ed., rev. and enl. Charles Griffin, London/Hafner, New York.

(ANALYSIS OF VARIANCE
CENSUS
COMPUTERS IN STATISTICS
CONSULTING, STATISTICAL
DESIGN OF EXPERIMENTS
MEASUREMENT ERROR
STRATIFICATION
SURVEY SAMPLING)

WALTER T. FEDERER

DATA IMPUTATION *See* EDITING STATISTICAL DATA

DATUM *See* DATA COLLECTION

DAVID, PEARSON, AND HARTLEY'S TEST FOR OUTLIERS (STUDENTIZED RANGE)

This test is devised to assist in judging whether either or both of the two extreme observed values [$y_{(1)}$ (least) and $y_{(n)}$ (greatest)] are from different sources than the bulk of the sample. The statistic used is the ratio of the sample range to the sample standard deviation

$$(y_{(n)} - y_{(1)})/s,$$

where $s^2 = \sum_{i=1}^{n}(y_i - \bar{y})^2/(n-1)$. Large values are regarded as significant of difference of source. Upper percentage points of the statistic are given by David et al. [2]. The test is particularly sensitive to an outlier on each side of the sample mean, but there is a pronounced masking effect with outliers on the same side of the mean.

For more information on this topic, see David and Paulson [1] and Prescott [3].

References

[1]   David, H. A. and Paulson, A. S. (1965). *Biometrika*, **52**, 429–436.

[2]   David, H. A., Hartley, H. O., and Pearson, E. S. (1954). *Biometrika*, **41**, 482–483.

[3]   Prescott, P. (1978). *Ann. Statist.*, **27**, 10–25.

(CHAUVENET'S CRITERION
OUTLIERS
STUDENTIZED RANGE)

DEBYE ENERGY FUNCTION

Originates from the Debye heat-capacity equation [3]. The function is

$$D(x) = 3x^{-3}\int_0^x y^2 P(y)\,dy,$$

where

$$P(x) = x/(e^x - 1)$$

is the Planck function*.

For computational purposes the integral $\int_0^x y^3\,dy/(e^y - 1)$ can be represented by the following infinite series:

$$\int_0^x \frac{y^3\,dy}{e^y - 1} = \frac{\pi^4}{15} - x^4 \sum_{n=1}^{\infty} e^{-nx}$$

$$\times \left(\frac{1}{nx} + \frac{3}{n^2x^2} + \frac{6}{n^3x^2} + \frac{6}{n^4x^4} \right).$$

In statistical work this function appears in Planck distributions*. Beattie [2] presents values of this function to six significant figures for $x = 0.00(0.01)24.00$. There are

shorter tables [for $x = 0.1(0.1)10$] in Abramowitz and Stegun [1].

For more information on this topic, see Johnson and Kotz [4].

References

[1]   Abramowitz, M. and Stegun, I. A. (1964). Handbook of Mathematical Functions. *Natl. Bur. Stand. (U.S.) Appl. Math. Ser. 55* (Washington, D.C.).

[2]   Beattie, J. A. (1926). *J. Math. Phys.*, **6**, 1–32.

[3]   Debye, P. (1912). *Ann. Phys. (IV)*, **39**, (12), 789–839.

[4]   Johnson, N. L. and Kotz, S. (1970). *Distributions in Statistics*, Vol. 2: *Continuous Univariate Distributions*. Wiley, New York, p. 273.

DECAPITATED DISTRIBUTIONS

A term applied to discrete distributions truncated in a special way—by omission of the value zero for the variable. Thus the decapitated Poisson variable corresponding to the Poisson distribution*

$$\Pr[X = x] = e^{-\theta}\theta^x/x! \qquad (x = 0, 1, \dots)$$

is

$$\Pr[X = x] = (1 - e^{-\theta})^{-1}e^{-\theta}\theta^x/x!$$

$$(x = 1, 2, \dots).$$

An equivalent name is zero-truncated distribution. The term is sometimes extended to include truncation by omission of more than one variable value. If values $x = 0$ and $x = 1$, the term "doubly decapitated" is used occasionally.

Decapitated variables are sometimes used to approximate expected values of reciprocals of variables when the original distribution would formally give an infinite expected value (because $\Pr[X = 0]$ is not zero). One might interpret such results as applying to sampling in which records are kept only when the original variable is not zero. If $\Pr[X = 0]$ is small, this may give useful approximations: for example, in calculating probabilities based on central limit results.

(TRUNCATED DISTRIBUTIONS)

DECILE

A value that is exceeded by $10, 20, \ldots, 80$ or 90% of the values in a distribution. Each distribution can have nine deciles; that corresponding to 10% is usually called the first, to 20% the second decile, etc. Often, the first and ninth deciles are called the *lower decile* and *upper decile*, respectively.

In terms of the cumulative distribution function* $F(x)$, the pth decile $\xi_{p/10}$ satisfies the equation

$$F(\xi_{p/10}) = \tfrac{1}{10}\, p.$$

It is possible that this equation may be satisfied by any value in a certain interval. The decile may then be defined by some more or less arbitrary rule.

(MEDIAN
PERCENTILE
QUARTILE)

DECISION ANALYSIS   *See* DECISION THEORY

DECISION THEORY

Every day, in our professional work and our personal lives, each of us must make a multitude of decisions, both major and minor, under various conditions of uncertainty and partial ignorance. Decision theory deals with the development of methods and techniques that are appropriate for making these decisions in an optimal fashion. In fact, statistics itself is sometimes described as the science of decision making under uncertainty. Although decision theory may not be tantamount to the entire field of statistics, the importance of decision theory has steadily grown during the past 30 years as virtually all the classical problems of statistical inference, and many new problems as well, have been formulated in decision-theoretic terms.

The mathematical basis of statistical decision theory was developed mainly by Abraham Wald* during the 1940s. In many re-

spects, this theory was an outgrowth, and a special case, of the *theory of games* (*see* GAME THEORY) as developed by von Neumann and others during the 1920s and 1930s. The central difference is that in the theory of *zero-sum two-person games**, the decision maker must act against an intelligent opponent whose interests are diametrically opposed to his own, whereas in a statistical decision problem there is usually no such opponent. For this reason, the theory of *minimax decision rules**, which play a central part in the theory of games, play at best a very minor part in modern decision theory.

PARAMETERS, DECISIONS, AND CONSEQUENCES

Consider a problem in which a decision maker (DM) must choose a decision from some class of available decisions, and suppose that the consequences of this decision depend on the unknown value θ of some *parameter* Θ. We use the term "parameter" here in a very general sense, to represent any variable or quantity whose value is unknown to the DM but is relevant to his or her decision. Some authors refer to Θ as the "unknown state of nature" or "state of the world." The set Ω of all possible values of Θ is called the *parameter space*.

The set D of all possible decisions d that the DM might make in the given problem is called the *decision space*.

For each value of $\theta \in \Omega$ and each possible decision $d \in D$, let $\gamma(\theta, d)$ denote the consequence to the DM if he or she chooses decision d when the parameter has value θ. Let $\mathcal{C}$ denote the set of all consequences that might result from all possible pairings of θ and d. If Θ has a specified probability distribution, then the choice of any particular decision d will induce a probability distribution of $\gamma(\Theta, d)$ on the set $\mathcal{C}$ of possible consequences. Hence the DM's choice among the decisions in D is tantamount to a choice among various probability distributions on the set $\mathcal{C}$.

UTILITY*

The DM will typically have preferences among the consequences in $\mathcal{C}$. In some problems, these consequences might be monetary gains or losses; in others they might be much more complicated and abstract quantities. In general, the DM's preferences among the consequences in $\mathcal{C}$ will result in his or her having preferences among the different possible probability distributions on $\mathcal{C}$. In other words, if the DM could have a consequence from $\mathcal{C}$ generated by a random process in accordance with some specified probability distribution, he or she would generally have a preference as to which distribution was used.

Now let U denote a real-valued function on the set $\mathcal{C}$, i.e., a function that assigns a real number to each consequence in $\mathcal{C}$. Also, for any probability distribution P on the set $\mathcal{C}$, let $E(U \mid P)$ denote the expectation* of U with respect to the distribution P. Then under certain conditions regarding the coherence of the DM's preferences among probability distributions, it can be shown that there exists such a function U with the following property: For any two distributions P_1 and P_2, P_1 is not preferred to P_2 if and only if $E(U \mid P_1) \leqslant E(U \mid P_2)$.

A function U with this property is called a *utility function*, and the value that U assigns to any particular consequence is called the *utility* of that consequence. The *expected utility hypothesis*, as we have just described, states that the DM will prefer a probability distribution P for which $E(U \mid P)$ is as large as possible. In other words, the DM will prefer a distribution for which the expected utility of the resulting consequence is a maximum.

It should be noted that there is more than one utility function that could be used in a given problem. If U is a utility function, then $V = aU + b$, where a and b are constants ($a > 0$, $-\infty < b < \infty$) is also a utility function. The reason is that for any two distributions P_1 and P_2, $E(U \mid P_1) \leqslant E(U \mid P_2)$ if and only if $E(V \mid P_1) \leqslant E(V \mid P_2)$. Hence both U and V represent the DM's preferences equally well. In practice, this arbitrariness is exploited and removed by choosing two particular consequences and assigning them the utilities 0 and 1, or 0 and 100, or some other convenient pair of reference values.

COMPONENTS OF A DECISION PROBLEM

We now return to the original decision problem. For each value of $\theta \in \Omega$ and each decision $d \in D$, let $U(\theta, d)$ denote the utility of the consequence $\gamma(\theta, d)$. We may think of $U(\theta, d)$ as the utility of choosing decision d when the parameter Θ has the value θ. Suppose that Θ has a specified probability distribution ξ. Then in accordance with the expected utility hypothesis, the DM will choose a decision d for which the expected utility $E(U \mid \xi, d)$ is a maximum. Such a decision is called an *optimal decision* or a *Bayes decision* with respect to the distribution ξ.

In many decision problems, it has become standard to specify the negative of the utility function, rather than the utility function itself, and to call this function the *loss function*. Thus the *loss* $L(\theta, d)$ is the *disutility* to the DM of choosing decision d when the parameter has the value θ. An optimal or *Bayes decision* with respect to the distribution ξ will be a decision d for which the expected loss $E(L \mid \xi, d)$ is a minimum.

Thus the components of a decision problem are a parameter space Ω, a decision space D, and a loss function $L(\theta, d)$. For any given distribution ξ of Θ, the expected loss $E(L \mid \xi, d)$ is called the *risk* $R(\xi, d)$ of the decision d. The risk of the Bayes decision, i.e., the minimum $R_0(\xi)$ of $R(\xi, d)$ over all decisions $d \in D$, is called the *Bayes risk*.

SUBJECTIVE PROBABILITY

In some decision problems, the probability distribution ξ that the DM assigns to Θ will be based on a large amount of historical data or on theoretical frequency considerations. In such problems, the distribution ξ

will be "objective" in the sense that any other DM who faced the same problem would assign the same distribution. In most decision problems, however, the distribution ξ will be a "subjective" distribution that is based, at least in part, on the DM's personal information and beliefs about what the value of Θ is likely to be.

The existence of *subjective probabilities** is based on the assumption that certain conditions are satisfied regarding the coherence of the DM's judgments about the relative likelihoods of various subsets of values of Θ. When these conditions are satisfied, it can be shown that there exists a unique probability distribution P on the set Ω that satisfies all the mathematical properties of probability and has the additional property that, for any two subsets $A \subset \Omega$ and $B \subset \Omega$, $P(A) \leqslant P(B)$ if and only if the DM does not believe that the value of Θ is more likely to lie in A than in B.

Some statisticians feel that there are different types of probability and that subjective probabilities are of a different type from *logical*, *frequency*, or *physical probabilities*. On the other hand, it can be argued that subjective probability is the only type of probability that can be put on a sound foundation and the only type of probability that exists. In this view, all probabilities are subjective; some are more "objective" than others only because larger groups of DM's would all assign the same values for these probabilities based on their experience.

Together, the concepts of subjective probability and utility provide a unified theory of decision making. The DM's subjective probabilities represent his or her knowledge and beliefs, and the DM's utilities represent his or her tastes and preferences. The expert DM is careful to maintain the distinction between these concepts, and does not confuse the value that he or she wishes Θ would have with the value that he or she thinks Θ is likely to have. In other words, the DM does not let utilities influence his or her subjective assignment of probabilities, and vice versa. The DM then chooses a decision that maximizes his or her subjective expected utility or, equivalently, minimizes his or her subjective expected loss.

DECISION ANALYSIS

Many problems of decision making, such as deciding where to locate a new airport, are extremely complicated, and it is often not immediately clear how to apply the concepts of decision theory that have just been described. The process of aiding the DM in applying these concepts in a particular problem is called *decision analysis*. In recent years techniques of decision analysis have been developed which are intended to aid the DM in (a) identifying all the relevant dimensions of the parameter Θ, (b) specifying the spaces Ω and D of all possible parameter values θ and decisions d, and especially (c) specifying the DM's probabilities and utilities.

Various procedures are available, including some computer programs, for the elicitation of a DM's subjective probabilities. A probability distribution on Ω must be determined on the basis of the DM's responses when questioned about the relative likelihoods of different events. Some type of fitting procedure is typically needed because few, if any, persons exhibit the perfect coherence* necessary for the existence of a unique distribution. Similarly, procedures are available for fitting a utility function on the basis of the DM's responses when questioned about his or her preferences among different probability distributions that might yield a consequence from the set $\mathcal{C}$.

STATISTICAL DECISION PROBLEMS

In many decision problems, the DM has the opportunity, before choosing a decision in D, of observing the value of a random variable or random vector X that is related to the parameter Θ. The observation of X provides information that may be helpful to the DM in choosing a good decision. We shall assume that the conditional distribution of X

given $\Theta = \theta$ can be specified for every value of $\theta \in \Omega$. A problem of this type is called a *statistical decision problem.*

Thus the components of a statistical decision problem are a parameter space Ω, a decision space D, a loss function L, and a family of conditional densities $f(x \mid \theta)$ of an observation X whose value will be available to the DM when he or she chooses a decision. The conditional densities $f(x \mid \theta)$ might be of either the discrete or the continuous type. In a statistical decision problem, the probability distribution of Θ is called its *prior distribution** because it is the distribution of Θ before X has been observed. The conditional distribution of Θ given the observed value $X = x$ is then called the *posterior distribution** of Θ.

Suppose that Θ has the prior density $\xi(\theta)$, which again could be of either the discrete or the continuous type. Then for any given value x of X, the posterior density $\xi(\theta \mid x)$ of Θ given $X = x$ can be found from the following result (Bayes' theorem*):

$$\xi(\theta \mid x) = \frac{f(x \mid \theta)\xi(\theta)}{\int_\Omega f(x \mid \theta')\xi(\theta')\,d\theta'}. \qquad (1)$$

This expression for $\xi(\theta \mid x)$ is written using an integral, as though $\xi(\theta)$ is a PDF. If it is actually a discrete probability function, the integral should be replaced by a sum. Throughout this article we write expressions with integral signs, as if $\xi(\theta)$ or $f(x \mid \theta)$ were PDFs, but all these expressions will apply equally well to discrete distributions by replacing the integrals with sums.

A *decision function* is a rule that specifies a decision $\delta(x) \in D$ that will be chosen for each possible observed value x of X. The *risk* $R(\xi, \delta)$ of a decision function with respect to the prior distribution ξ is the expected loss

$$R(\xi, \delta) = E\{L[\theta, \delta(X)]\}$$

$$= \int_\Omega \int_X L[\theta, \delta(x)] f(x \mid \theta)\xi(\theta)\,dx\,d\theta.$$

$$(2)$$

We recall that a decision function $\delta^*(x)$ for which the risk $R(\xi, \delta)$ is a minimum is called

a *Bayes decision function* with respect to ξ, and its risk $R(\xi, \delta^*)$ is called the *Bayes risk.**

A Bayes decision function is easily described. After the value x has been observed, the DM simply chooses a Bayes decision $\delta^*(x)$ with respect to the posterior distribution $\xi(\theta \mid x)$ rather than the prior distribution $\xi(\theta)$. Thus if the DM has no control over whether or not the observation X will be obtained, he or she need not specify the entire Bayes decision function in advance. The DM can wait until the observed value x is known, and then simply choose a Bayes decision $\delta^*(x)$ with respect to the posterior distribution. At this stage of the decision-making process, the DM is not interested in the risk $R(\xi, \delta)$, which is an average over all possible observations, but in the posterior risk

$$\int_\Omega L[\theta, \delta^*(x)]\xi(\theta \mid x)\,d\theta, \qquad (3)$$

which is the risk from the decision he or she is actually making.

On the other hand, if the DM must decide whether or not to obtain the observation X at some specified cost, or must choose among different possible observations or different possible sample sizes, at varying costs, then he or she must consider the Bayes risk $R(\xi, \delta^*)$. In these situations, the DM must compare the reduction in risk that can be obtained from an observation X with the cost of observing X.

It can be shown that information is never harmful in the following sense. In every statistical decision problem, for every possible observation X, the Bayes risk $R(\xi, \delta^*)$ is never greater than the Bayes risk $R_0(\xi)$ that could be obtained by immediately choosing a decision $d \in D$ without observing X. Thus if an observation X is offered without cost, it can never hurt the DM to observe its value, and it will typically help him or her in the sense that it will reduce the Bayes risk.

It should be emphasized that the reduction in Bayes risk from the observation X, as represented by the inequality

$$R(\xi, \delta^*) \leqslant R_0(\xi),$$

means that the overall average risk $R(\xi, \delta^*)$ from observing X is not greater than the risk

$R_0(\xi)$ that could be attained by an immediate decision. It is possible that for certain values of x, the posterior risk given by (3) is actually larger than $R_0(\xi)$, even if the observation is costless. If one of these values of x is observed, the DM will find that the posterior risk from a Bayes decision $\delta^*(x)$ is larger than the risk $R_0(\xi)$ he or she would have suffered by making a decision without observing X. Should the DM regret making the observation or even "forget" that it was made and just revert to what his or her original decision would have been?

The answer to this question is, of course, no. If the posterior risk is larger than $R_0(\xi)$, the reason is usually that the posterior density $\xi(\theta \mid x)$ is more spread out and less concentrated than the prior density $\xi(\theta)$. This means that the DM was relatively certain that he or she knew the value of Θ and concentrated the prior density around that value. The Bayes risk from an immediate decision was, therefore, small *with respect to that prior density*. If the observed value x indicates that Θ may be far from the favored value, the posterior density will be more spread out and the DM will be less certain than originally about the value of Θ. Although the posterior risk of the Bayes decision $\delta^*(x)$ will be relatively large, the posterior risk from the original Bayes decision would be even larger with respect to the posterior density, which represents the DM's current state of knowledge. Thus the DM should be glad that he or she obtained the observation. Even though it may have left the DM more confused than before, it saved him or her from making what might have been a bad decision and indicated that the DM might have concentrated the prior density around an incorrect value of Θ. Further observations, when available, would ultimately lead the posterior density to become more and more concentrated around the correct value.

CONJUGATE FAMILIES* OF PRIOR DISTRIBUTIONS

Consider a problem in which the observation vector $X = (X_1, \ldots, X_n)$ is a random sample from one of the standard distributions that are used in statistics such as a Bernoulli*, Poisson*, normal*, exponential*, or uniform* distribution. If the distribution depends on a parameter Θ, it is convenient to have another standard family of prior distributions that can be used to represent the DM's prior information. One convenient property of such a family is that it be *closed under sampling*; i.e., if the prior distribution of Θ belongs to the family, then for any sample size n and any values of the observations in the sample, the posterior distribution will also belong to the family. A family with this property is also called a *conjugate family of (prior) distributions*.

For a conjugate family of distributions to be useful it must be small enough so that it can itself be described by just a few parameters (or *hyperparameters*, as they are called, in order to distinguish these parameters of the conjugate family from the parameter Θ). On the other hand, it must be rich enough to permit the DM to find within the family a distribution that will adequately represent the prior distribution. Thus both the family of *all* probability distributions on Ω and the family containing just a single degenerate distribution that concentrates all its probability on one particular value in Ω are always conjugate families in every problem, but one family is much too large and the other is much too small to be useful.

The conjugate families that have proven to be useful for sampling from a normal distribution are as follows:

1. Suppose that $X_1, \ldots, X_n$ form a random sample from a normal distribution* with unknown mean Θ and known precision h. (The *precision* of a normal distribution is the reciprocal of its variance.) Suppose also that the prior distribution of Θ is a normal distribution with mean μ and precision τ. Then the posterior distribution of Θ given the observed values $x_1, \ldots, x_n$ is again a normal distribution with mean

$$\frac{\tau\mu + h\sum_{i=1}^{n} x_i}{\tau + nh} \tag{4}$$

and precision $\tau + nh$.

2. Suppose that $X_1, \ldots, X_n$ form a random sample from a normal distribution with both unknown mean Θ and unknown precision H. Suppose also that the joint prior distribution of Θ and H is as follows. The conditional distribution of Θ given $H = h$ is a normal distribution with mean μ and precision τh, and the marginal distribution of H is a gamma distribution* with hyperparameters α and β; i.e., the marginal density of H is

$$\xi(h) = \frac{\Gamma(\alpha + \beta)}{\Gamma(\alpha)\Gamma(\beta)} h^{\alpha - 1}(1 - h)^{\beta - 1}$$

$$\text{for } h > 0. \quad (5)$$

Then the posterior joint distribution of Θ and H given the observed values $x_1, \ldots, x_n$ is as follows. The conditional distribution of Θ given $H = h$ is again a normal distribution with mean

$$\frac{\tau\mu + \sum_{i=1}^{n} x_i}{\tau + n} \quad (6)$$

and precision $(\tau + n)h$, and the marginal distribution of H is again a gamma distribution with hyperparameters $\alpha + \frac{1}{2}n$ and

$$\beta + \frac{1}{2} \sum_{i=1}^{n} (x_i - \bar{x})^2 + \frac{\tau n(\bar{x} - \mu)^2}{2(\tau + n)} . \quad (7)$$

IMPROPER PRIOR DISTRIBUTIONS

In some problems, the prior information of the DM about the parameter Θ is small and vague relative to the information about Θ that he or she will obtain by observing X. Because of the vagueness of the prior information, it may be difficult for the DM to select a particular distribution from the conjugate family to represent the prior distribution. Also, because the DM will soon acquire relatively precise information about Θ from the observation X, it will typically not be worthwhile for the DM to spend much time making a very careful determination of the prior distribution. In a situation like this it may be convenient to use a standard prior distribution that is suitable as an approximate representation of vague prior information.

The standard prior distribution that is used is often an *improper distribution** in the sense that it is represented by a nonnegative density $\xi(\theta)$ for which the integral over the entire space Ω is infinite rather than 1. For example, suppose that the DM is going to draw a random sample $X_1, \ldots, X_n$ from a normal distribution with unknown mean Θ and known precision h, and that the DM has only vague prior information about Θ. To use a prior distribution for Θ from the conjugate family, the DM must specify a normal distribution for Θ with a very small precision (a very large variance). Rather than try to select particular values for the mean and precision of the prior normal distribution, it is common practice to represent this prior distribution by a uniform, or constant, density over the entire real line, even though this density does not represent a proper probability distribution. In a sense that will now be described, this uniform density can be thought of as the limit of a proper normal prior distribution as the precision of that normal distribution approaches zero.

Although a uniform density* over the entire real line represents an improper distribution, it is often possible to develop a proper posterior distribution for Θ by formally inserting the uniform density into Bayes' theorem, as given by (1), and carrying out the calculation for the posterior density. When Bayes' theorem is applied in this way for the observed values $x_1, \ldots, x_n$ of the normal random sample, it is found that the posterior distribution of Θ is a proper normal distribution with mean $\bar{x}$ and precision nh.

This result can be compared with the result (1) that was obtained when a conjugate normal prior distribution with mean μ and precision τ was used. It can be seen that the posterior distribution derived from the uniform prior density is the limit of the posterior distribution derived from the normal prior distribution as the precision τ of that prior distribution approaches zero.

When a random sample $X_1, \ldots, X_n$ is to be drawn from a normal distribution for which both the mean Θ and the precision H are unknown, and about which the DM has only vague information, it is common to represent the joint prior distribution of Θ and H by the joint density

$$\xi(\theta, h) = \frac{1}{h}; \quad -\infty < \theta < \infty, h > 0. \quad (8)$$

This density again represents an improper prior distribution in which Θ and H are independent and Θ and $\log H$ each have uniform densities over the entire real line. For $n \geq 2$, a proper posterior distribution is obtained from the formal application of Bayes' theorem that is the same as the limiting posterior distribution obtained from a conjugate prior distribution (result 2) when we let $\tau \to 0$, $\alpha \to -\frac{1}{2}$, and $\beta \to 0$.

Improper prior distributions must always be used with the utmost caution, since they do not satisfy the rules of probability. Although the posterior density $\xi(\theta \mid x)$ may represent a proper probability distribution, the conditional densities $f(x \mid \theta)$ and $\xi(\theta \mid x)$ will typically be mutually incompatible because they did not arise from a proper probability model. In certain circumstances, especially when the parameter Θ is a vector of high dimension, decisions based on posterior distributions derived from improper prior distributions can be unreasonable.

ESTIMATION* AND TESTS OF HYPOTHESES*

The standard estimation problems of statistical inference can be formulated as decision problems. For example, consider a real-valued parameter Θ that is to be estimated. The parameter space Ω in a problem like this is usually an interval of the real line, possibly an unbounded interval or the entire line. Since the decision to be chosen is an estimate of Θ, the decision space D is the same interval as Ω itself.

The most widely used loss function* in problems involving the estimation of a real parameter is the *squared error loss*

$$L(\theta, d) = (\theta - d)^2. \quad (9)$$

A Bayes estimate of Θ for this loss function is a number d for which the *mean squared error** $E[(\Theta - d)^2]$ is a minimum. It follows that the Bayes estimate is the mean of the posterior distribution of Θ, and the posterior risk is the variance of that posterior distribution (assuming that the mean and variance exist).

Another loss function that is often used in the estimation of a real parameter is the *absolute error loss*

$$L(\theta, d) = |\theta - d|. \quad (10)$$

A Bayes estimate for this loss function is a number d for which the *mean absolute error** $E(|\Theta - d|)$ is a minimum. It follows that the Bayes estimate is the median of the posterior distribution of Θ (or, more precisely, *any* median of the posterior distribution of Θ, since the median need not be unique).

Both the loss functions (9) and (10) are of the general form

$$L(\theta, d) = \lambda(\theta)|\theta - d|^p, \quad (11)$$

where p is a positive number and $\lambda(\theta)$ is a positive function of θ. If $\lambda(\theta)$ is not constant, the loss in (11) depends not only on the magnitude of the error $|\theta - d|$ but on the value of θ as well. However, every loss function of the form (11) is still rather special, because overestimates and underestimates of the same magnitude yield the same loss. Loss functions that do not have this symmetric feature can also be introduced.

The standard problems of testing hypotheses can also be formulated as decision problems. In fact, every test of hypotheses is, at least theoretically, a problem with exactly two decisions: accept the null hypothesis H_0, which we shall call decision d_0, and accept the alternative hypothesis H_1 (or, equivalently, reject H_0), which we shall call decision d_1.

Loss functions appropriate to testing hypotheses can easily be developed. For example, suppose that Θ is a real-valued parameter and it is desired to test the hypotheses

$H_0 : \Theta \leqslant \theta_0$ and $H_1 : \Theta > \theta_0$, where θ_0 is a specified number. A typical loss function for this problem would have the following form:

$$
\begin{aligned}
L(\theta, d_0) &= 0 && \text{for } \theta \leqslant \theta_0, \\
L(\theta, d_0) &= \lambda_0(\theta) && \text{for } \theta > \theta_0, \\
L(\theta, d_1) &= \lambda_1(\theta) && \text{for } \theta \leqslant \theta_0, \\
L(\theta, d_1) &= 0 && \text{for } \theta > \theta_0,
\end{aligned}
\tag{12}
$$

where $\lambda_0(\theta)$ is positive and nondecreasing for $\theta > \theta_0$ and $\lambda_1(\theta)$ is positive and nonincreasing for $\theta < \theta_0$. The posterior PDF of Θ can be calculated from any specified prior PDF. The Bayes test procedure would then choose the decision with the smaller posterior risk.

Special consideration is needed for testing hypotheses in which the dimension of the set of values of Θ that satisfy H_0 is smaller than the dimension of the whole parameter space Ω. For example, suppose again that Θ is a real-valued parameter and it is desired to test the hypotheses $H_0 : \Theta = \theta_0$ and $H_1 : \Theta \neq \theta_0$. Under any prior PDF that is assigned to Θ, the probability of H_0 will be zero since the probability of any single value of Θ is zero. Hence the posterior probability of H_0 after any observations will again be zero, and the Bayes decision is always to choose d_1 and reject H_0.

This decision is not entirely unreasonable in many problems, since the DM knows to begin with that the continuous parameter Θ is not *exactly* equal to θ_0, even though it may be very close. Therefore, H_0 cannot strictly be true, and it is appropriate to reject it immediately. Often, however, the DM wishes to accept H_0 if Θ is sufficiently close to θ_0. In such problems, the hypothesis H_0 should be widened to include a suitable interval around θ_0. The hypothesis H_0 would then have positive probability under the prior distribution, and a Bayes test procedure can be developed in the usual way.

Finally, there are some problems in which the DM actually does believe that the null hypothesis $H_0 : \Theta = \theta_0$ might be true. For example, there might be a physical theory that leads to this particular value. One way to proceed in such problems is to assign an atom of positive probability p_0 to the point θ_0 and to spread the remaining probability $1 - p_0$ over the values $\theta \neq \theta_0$ in accordance with some PDF $\xi(\theta)$. Test procedures based on this type of prior distribution require careful study, because they can exhibit some unusual features.

SEQUENTIAL DECISION PROBLEMS

In many statistical decision problems, the DM can obtain the observations X_1, $X_2, \ldots$ in a random sample one at a time. After each observation X_n the DM can calculate the posterior distribution for Θ based on the observed values of $X_1, \ldots, X_n$ and can decide whether to terminate the sampling process and choose a decision from D or to continue the sampling process and observe X_{n+1}. A problem of this type is called a *sequential decision problem* (*see* SEQUENTIAL ANALYSIS).

In most sequential decision problems there is either an explicit or an implicit cost associated with each observation. A procedure for deciding when to stop sampling and when to continue is called a stopping rule or a sampling plan*. The fundamental task of the DM in a sequential decision problem is to determine a stopping rule that will minimize some overall combination of loss from choosing a decision in D and sampling cost.

If $X_1, X_2, \ldots$ form a sequential random sample from some distribution that depends on the parameter Θ, it is relatively easy for the DM to update the posterior distribution after each observation. This updating can be done one observation at a time: the posterior distribution after X_n has been observed serves as the prior distribution of Θ for X_{n+1}. The DM can simply use this current prior density for Θ, together with the conditional density of X_{n+1} given Θ, in Bayes' theorem (1) to obtain the posterior distribution after X_{n+1} has been observed.

The standard problems of estimation and tests of hypotheses can be treated sequentially. Once the DM decides to stop sampling, the choice of decision from D is clear:

the DM will simply choose the Bayes estimate or Bayes test procedure with respect to the posterior distribution of Θ. It is often assumed that there is constant cost per observation (although a varying cost is more realistic in most problems) and the DM must find a stopping rule* that minimizes a linear combination of risk and expected sampling cost.

There are a wide variety of other kinds of sequential decision problems that have been discussed in the statistics literature, including gambling* problems, inventory* problems, control problems, clinical trials*, and search* problems. The following problems of optimal stopping are illustrative of a different class of sequential decision problems. Suppose as before that a sequential random sample can be taken from some distribution involving an unknown parameter at a constant cost of c per observation. Suppose that sampling is terminated by the DM after $X_1, \ldots, X_N$ have been observed. The problem of *sampling without recall* is to determine a stopping rule that maximizes $E(X_N - cN)$. The problem of *sampling with recall* is to determine a stopping rule that maximizes $E[\max(X_1, \ldots, X_N) - cN]$.

Bibliography

Blackwell, D. and Girschick, M. A. (1954). *Theory of Games and Statistical Decisions.* Wiley, New York. (An early and influential graduate-level textbook on statistical decision theory. Although the presentation is restricted to discrete distributions, the coverage is comprehensive and the treatment is highly technical.)

Box, G. E. P. and Tiao, G. C. (1973). *Bayesian Inference in Statistical Analysis.* Addison-Wesley, Reading, Mass. (An extensive account of Bayesian methods in statistics, but with the emphasis on statistical analysis rather than decision theory. The authors develop posterior distributions based on noninformative or improper prior distributions in a wide variety of useful statistical models.)

Decision Analysis Group, SRI International (1977). *Readings in Decision Analysis.* Stanford Research Institute, Menlo Park, Calif. (A collection of articles by members of the Stanford Research Institute on a wide variety of aspects of decision analysis. Problems of modeling and the elicitation of subjective probabilities are discussed, and applications in several different areas are presented.)

DeGroot, M. H. (1970). *Optimal Statistical Decisions.* McGraw-Hill, New York. (A comprehensive textbook

at an intermediate level that covers all of the topics discussed in this article, with a large bibliography.)

DeGroot, M. H. (1975). *Probability and Statistics.* Addison-Wesley, Reading, Mass. (An introduction to probability and statistics, based on elementary calculus, that discusses Bayesian methods as part of an integrated approach to statistical inference and decision making.)

Fienberg, S. E. and Zellner, A., eds. (1975). *Studies in Bayesian and Econometrics and Statistics.* North-Holland, New York. (A valuable collection of articles, many by widely known econometricians and statisticians. The papers are written at an advanced level and cover the theory of Bayesian statistics and its application to various aspects of economics.)

Keeney, R. L. and Raiffa, H. (1976). *Decisions with Multiple Objectives: Preferences and Value Tradeoffs.* Wiley, New York. (An intensive study of decision analysis. The authors study the development of preferences and utility functions in the presence of multiple conflicting objectives, with an emphasis on applications.)

Lindley, D. V. (1971). *Bayesian Statistics, A Review.* SIAM, Philadelphia. (An excellent survey of Bayesian statistics based on a sequence of lectures given by the author. This monograph is only 83 pages long, and includes a lengthy bibliography.)

Marschak, J. (1974). *Economic Information, Decision, and Prediction; Selected Essays, Vol. 1.* D. Reidel, Dordrecht, Holland. (A collection of papers on the economic aspects of decision making by one of the leading contributors to this field. The book contains 18 articles published by the author and his co-workers between 1950 and 1974.)

Pratt, J. W., Raiffa, H., and Schlaifer, R. (1965). *Introduction to Statistical Decision Theory.* McGraw-Hill, New York. (An authoritative introduction to many of the concepts mentioned in this article. Unfortunately, only a preliminary edition of this book has ever been published.)

Raiffa, H. (1968). *Decision Analysis.* Addison-Wesley, Reading, Mass. (A useful introduction to aspects of decision analysis. The book is written at an elementary level and includes subjective probability, utility, risk sharing, and group decisions.)

Raiffa, H. and Schlaifer, R. (1961). *Applied Statistical Decision Theory.* (A ground-breaking study of Bayesian methods in decision theory with extensive technical details. Conjugate prior distributions are heavily utilized.)

Savage, L. J. (1972). *The Foundations of Statistics*, 2nd ed. Dover, New York. (One of the original explications and derivations of the Bayesian approach to statistics and decision theory. Although the book uses only elementary mathematics, it is thorough and rigorous in its development.)

von Neumann, J. and Morgenstern, O. (1974). *Theory of Games and Economic Behavior*, 2nd ed. Princeton University Press, Princeton, N. J. (A path-breaking book that developed the theory of games and utility functions in the context of economic theory, and opened the door to the development of statistical decision theory.)

Wald, A. (1950). *Statistical Decision Functions.* Wiley, New York. (The first book to present the mathematical basis of statistical decision theory, written by the man who had invented and developed most of the theory himself. The book is written at an advanced mathematical level.)

Winkler, R. L. (1972). *Introduction to Bayesian Inference and Decision.* Holt, Rinehart and Winston, New York. (A fine introductory text covering many of the topics described in this article. The book is written at an elementary level and contains an extensive bibliography.)

(BAYESIAN INFERENCE
CONJUGATE FAMILIES OF
    DISTRIBUTIONS
LOSS FUNCTION
STATISTICAL INFERENCE)

MORRIS H. DeGROOT

DECOMPOSITION THEOREM

There are a number of theorems of this kind. The basic theorem for probability distributions is due to Khinchine [1]. It asserts that every probability law (distribution) can be represented as a composition of at most two laws: one consisting of a countable number of indecomposable (nondegenerate) components and the other having no indecomposable components (the so-called I_0-laws*). This decomposition is not necessarily unique; for example, the uniform distribution* on $[-1, 1]$ has infinitely many distinct decompositions into indecomposable components.

For more information on this topic, see Linnik and Ostrovskii [2].

References

[1] Khinchine, A. Ja. (1937–1938). *Bull. Univ. Moscow*, Sec. A, No. 5, 1–6.

[2] Linnik, Ju. V. and Ostrovskii, I. V. (1972). Decomposition of Random Variables and Vectors. *Transl. Math. Monogr.* **48**, American Mathematical Society, Providence, R. I.

(INFINITE DIVISIBILITY)

DEFF   *See* DESIGN EFFECT

DEFICIENCY   *See* SECOND-ORDER EFFICIENCY

DEFLATED DISTRIBUTIONS

Discrete distributions constructed by reducing a few (usually only one or two) of the probabilities, while the remainder are increased proportionately are termed "deflated." The most common form of deflation is reduction of the probability of observing a zero value.

Thus if $\Pr[X = j] = p_j$ for $j = 0, 1, \ldots$ with $\sum_j p_j = 1$, then the distribution with

$$\Pr[X = 0] = \theta p_0;$$

$$\Pr[X = j] = \frac{1 - \theta p_0}{1 - p_0} p_j$$

$$(j = 1, 2, \ldots)$$

with $0 < \theta < 1$ is a deflated distribution.

If $\theta = 0$, we would obtain a *truncated distribution**, which in this case would also be a *decapitated distribution**.

(INFLATED DISTRIBUTIONS)

DEFT   *See* DESIGN EFFECT

DEGENERATE DISTRIBUTION

The joint distribution of n random variables $X_1, \ldots, X_n$ is said to be *degenerate* if there is at least one relationship among the variables $g(X_1, \ldots, X_n) = 0$ which holds with probability 1, the function $g(\cdot)$ itself not being identically constant for all $X_1, \ldots, X_n$.

In the case of a single random variable X with

$$P(X = a) = 1$$

the corresponding CDF is

$$P(X \leqslant x) = F(x) = \begin{cases} 0, & x < a \\ 1, & x \geqslant a \end{cases}$$

and the characteristic function* is $\phi(t) = e^{ita}$. The moments of this distribution are $\mu_k' = E(X^k) = a^k, k = 1, 2, \ldots$, and var$(X) = 0$. It is sometimes said, loosely, that this degenerate distribution describes a "non-random variable." The converse is also valid. If a random variable X possesses finite expectation and zero variance, then

$$P(X = E(X)) = 1.$$

DEGREE OF CONFIRMATION

A fundamental concept of Carnap's theory of probability [1] based on inductive logic*. This concept can be best explained using the following example due to Magnes.

1. The premise is formed of two data propositions called e: For example:

   This pack P contains 52 playing cards, 13 of which are hearts (first data proposition).

   This card X comes from the pack P (second data proposition).

2. The hypothesis h: X is hearts (conclusion).

3. The *degree of confirmation* $c(\cdot)$ of the hypothesis h, given the premise, is

$$c(h \mid e) = r, = \tfrac{13}{52} = \tfrac{1}{4}$$

in this example, with the assumption that no further evidence of the kind e with respect to hypothesis h is available.

$r(r \in [0, 1])$ is called the "quantitative explicative" of the degree of confirmation or probability 1 in Carnap's sense. (This is an a priori probability which, for the sake of symmetry, a priori assigns the same probability measure to all noncontradictory propositions of the language system.)

References

[1] Carnap, R. (1962). *Logical Foundations of Probability*. University of Chicago Press, Chicago.

[2] Magnes, G. (1974). In *Information, Inference and Decision*, G. Magnes, ed. D. Reidel, Dordrecht, Holland, pp. 3–49.

(INDUCTIVE LOGIC
STATISTICAL INFERENCE)

DEGREES OF BELIEF

The main controversy in statistics is between "Bayesian" or neo-Bayesian methods and non-Bayesian or sampling theory methods. If you, as a statistician, wish to attach a probability to a hypothesis or to a prediction, or if you wish to use statistics for decision making, you are usually forced to use Bayesian methods. (*See* BAYESIAN STATISTICS; DECISION THEORY.) The kinds of probability that are basic to Bayesian methods are known as (1) subjective (personal) probabilities, here taken as a synonym for "degrees of belief," and (2) logical probabilities (credibilities) or the degrees of belief of a hypothetical perfectly rational being. Sometimes the expressions "intuitive probability" or "epistemic probability" are used to cover both subjective probability and credibility. Although "Each man's belief is right in his own eyes" (William Cowper: Hope), you would presumably accept "credibilities" as your own probabilities if you knew what they were. Unfortunately, the question of the existence of credibilities is controversial even within the Bayesian camp.

Sometimes "degree of belief" is interpreted as based on a snap judgment of a probability, but it will here be assumed to be based on a well-considered mature judgment. This perhaps justifies dignifying a degree of belief with the alternative name "subjective probability." We shall see that a theory of subjective probability provides help in arriving at mature judgments of degrees of belief. It does this by providing criteria by which inconsistencies in a body of beliefs can be detected. We shall return to this matter in more detail.

The simplest way to give an operational meaning to degrees of belief is by combining them with payoffs as in a bet. For example, suppose that your subjective probability of an event is $\frac{3}{4}$, and that you know that your opponent has the same judgment. If you are forced to bet, the unit stake being $\pm \$100$ and the sign being chosen by your opponent, then you would offer odds of 3 to 1. Any other odds will give you a negative expected gain. More generally, the basic recommendation of Bayesian decision theory is "maximize the mathematical expectation of the utility" where "utility" is a measure of value, not necessarily monetary or even material. This "principle of rationality" is used informally even in matters of ordinary life, as when we decide whether to carry an umbrella. In fact, it is a characteristic of the Bayesian philosophy that its basic principles are the same in all circumstances, in ordinary life, in business decisions, in scientific research, and in professional statistics. For other applications of the principle of rationality, see Good [12] and DECISION THEORY. There are those who believe that Bayesian decision theory is the true basis for the whole of statistics, and others who consider that utilities should be avoided in purely scientific reporting.

Because of the subjectivity of degrees of belief, most statisticians still try to avoid their formal use though much less so than in the second quarter of the twentieth century. For this avoidance the concept of a confidence interval was introduced for the estimation of a parameter. (Confidence intervals have a long history; see Wilks [44, p. 366].) It is possible that clients of statisticians usually incorrectly interpret confidence intervals as if they were Bayesian estimation intervals. As far as the present writer knows, a survey has not been made to determine whether this is so.

Although subjective probabilities vary from one person to another, and even from time to time for a single person, they are not arbitrary because they are influenced by common sense, by observations and experiments, and by theories of probability. Sampling is just as important in Bayesian statistics as in non-Bayesian or sampling theory methodology, and the staunch Bayesian

would claim that whatever is of value in the definition of probability by long-run proportional frequency can be captured by theorems involving subjective probabilities. For the sake of emphasis we repeat here that "subjective probability" is used in this article interchangeably with "degree of belief." But in the article BELIEF FUNCTIONS a "degree of belief" is for some reason identified with "lower probability" in a sense that will be explained later.

According to the *Oxford English Dictionary*, "belief" is an all-or-none affair: you either believe something or you do not. It was this sense of belief that Bernard Shaw had in mind when he said: "It is not disbelief that is dangerous to our society; it is belief" (preface to *Androcles and the Lion*). But the extension of language that is captured by the expression "degrees of belief" is justified because we all know that beliefs can be more or less strong; in fact, the O.E.D. itself refers to "strong belief" in one definition of "conviction"! Clearly, Shaw was attacking dogmatic belief, not degrees of belief.

Beliefs, then, are quantitative, but this does not mean that they are necessarily numerical; indeed, it would be absurd to say that your degree of belief that it will rain tomorrow is 0.491336230. Although this is obvious, Keynes [26, p. 28n] pointed out that "there are very few writers on probability who have explicitly admitted that probabilities, although in some sense quantitative, may be incapable of numerical comparison," and what he said was still true 40 years later, but is appreciably less true in 1980. He quotes Edgeworth as saying in 1884 that "there may well be important quantitative, although not numerical, estimates." In fact, sometimes, but not always, one can conscientiously say that one degree of belief exceeds or "subceeds" another one. In other words, degrees of belief are only partially ordered, as emphasized by Keynes [26], Koopman [27, 28], and in multitudinous publications by Good, beginning with Good [11]. Keynes's emphasis was on the degrees of belief of a perfectly rational person, in other words on logical probabilities or credibilities, whereas Good's is on those of a real person, in other words on subjective or personal probabilities. (For "kinds of probability" see, e.g., Fréchet [10], Kemble [25], Good [13], and Fine [9].)

One reason why degrees of belief are not entirely numerical is that statements often do not have precise meanings. For example, it is often unclear whether a man has a beard or whether he has not shaved for a few days. But in spite of the fuzziness of faces it is often adequate to ignore the fuzziness of language; *see* FUZZY SETS. Even when language is regarded as precise, it must still be recognized that degrees of belief are only partially ordered. They may be called comparative probabilities.

Some degrees of belief are very sharp, such as those occurring in many games of chance, where well-balanced roulette wheels and well-shuffled packs of cards are used. These sharp degrees of belief are fairly uncontroversial and are sometimes called "chances." Every rational number (the ratio of two integers, such as p/q) between 0 and 1 can thus be an adequately uncontroversial sharp degree of belief because a pack of cards could contain just q cards of which p are of a specialized class. With only a little idealization, of the kind that is conventional in all applied mathematics, we can include the irrational numbers by imagining a sharp rotating pointer instead of a roulette wheel. Such degrees of belief provide sharp landmarks. Every judgment of a subjective probability is in the form of an inequality of the form $P(A \mid B) \leqslant (P(C \mid D)$. By allowing for the sharp landmark probabilities provided by games of chance, each subjective probability can be enclosed in an interval of values; so we may regard subjective probabilities as "interval valued." This leads one to talk about "upper" and "lower" degrees of belief, also often called upper and lower (subjective) probabilities. These are defined as the right-hand and left-hand end points of the shortest interval that is regarded as definitely containing a specified degree of belief. (The deduction of upper and lower

probabilities from other sharp probabilities, as was done by Boole [2, Chap. XIX], and by Dempster [8] by means of multivalued mappings, is logically distinct from recognizing that degrees of belief can only be *judged* to be "interval valued." The terminology of upper and lower probabilities has caused some people to confuse a "philosophy" with a "technique.") Thus a theory of partially ordered probabilities is essentially the same as a theory of probabilities that are "interval valued." An extreme anti-Bayesian would regard these intervals for subjective probabilities to be $(0, 1)$, whereas they reduce to points for "sharp" Bayesians. Otherwise, the partially ordered or comparative theory constitutes a form of Bayes/non-Bayes compromise.

In a theory of comparative degrees of belief the upper and lower degrees of belief are themselves liable to be fuzzy, but such a theory is at any rate more realistic than a theory of sharp (precise) degrees of belief if no allowance is made for the greater simplicity of the latter.

This greater simplicity is important and much of the practical formal applications and theory of subjective probability have so far depended on sharp probabilities [3, 30, 37, 38]. This is true also for writings on logical probability [24, 29].

Without loss of generality, a degree of belief can be regarded as depending on propositions that might describe events or hypotheses. A proposition can be defined as the meaning of a statement, and a subjective probability always depends upon *two* propositions, say A and B. We then denote the "subjective probability of A given or assuming B" by $P(A \mid B)$. Upper and lower probabilities are denoted by P^* and P_*. The axioms are all expressed in terms of these notations in the article on axiomatics. (*See* AXIOMS OF PROBABILITY.)

Theories of degrees of belief can be either descriptive or prescriptive (= normative). Descriptive theories belong strictly to psychology; but they are useful also in the application of a normative theory because the understanding of your own psychology enables you to introspect more effectively. See Hogarth [23] for a review of the psychological experiments. It is perhaps fair to say that these experiments show that in some respects people are good at judging probabilities and in other respects are not good even when they are uninfluenced by gambler's superstitions and wishful thinking.

The function of a normative theory of degrees of belief is similar to that of logic; it removes some of your freedom because it constrains your beliefs and tends to make them more rational, in other words, it encourages your subjective judgments to be more objective. We then have the apparent paradox that those people who refuse to use a good theory of subjective probability might be more subjective than those who do use the theory. The freedom to be irrational is a freedom that some of us can do without. It is because the beliefs of real people are not entirely rational that it is useful to have a normative theory of degrees of belief: at least this is the opinion of those people who make use of such theories.

It is convenient to distinguish between (1) psychological probabilities which depend on your snap judgments, and (2) subjective probabilities that depend on your mature judgments made in an unemotional disinterested manner and with some attempt to make your body of beliefs more consistent with a good theory. There is, of course, no sharp line of demarcation between psychological and subjective probabilities. Research on the nature of psychological probability is more a problem for psychology and artificial intelligence than for statistics.

It is not known in detail how people make judgments; if it were we could build an intelligent but possibly not entirely rational machine. Obviously, the neural circuitry of the brain gradually adapts, both by conscious and unconscious thought, and not necessarily by formal Bayesian methods, to some degree of rationality. One difficulty in the analysis of judgments is that thought is not purely linguistic; as Balth van der Pol commented at the First London Symposium on Information Theory in 1951, you usually know in some sense what you are going to say before you know what words you are going to use. It is as if thought depends on propositions more than on statements. This might give some justification for expressing the axioms of probability in terms of propositions. (*See* AXIOMS OF PROBABILITY.)

The two primary methods for measuring subjective probabilities are linguistic and behavioral. In the linguistic method a person is asked to estimate a probability, and in the behavioral method he has to make bets, although the stakes need not be monetary. He has to put his decisions "where his mouth is." Or in the words of James 2:26: "Faith without works is dead." This principle constitutes a threat to the use of confidence intervals. A statistician who provides confidence intervals would not necessarily in the long run "break even" if his client, taking independent advice from a Bayesian of good judgment, had the choice of whether or not to bet. A Bayesian of bad judgment might suffer even more.

The approach in terms of betting can be used to arrive at the usual axioms of sharp subjective probability: see, e.g., Ramsey [36], de Finetti [6], and Savage [38]. (*See* AXIOMS OF PROBABILITY.) De Finetti's definition of $P(A \mid B)$ is the price you are just willing to pay for a small unit amount of money conditional on A's being true, B being regarded as certain all along (e.g., de Finetti [7, p. 747]). By means of a modification of Savage's approach, axioms for partially ordered subjective probability can be derived [43]. See also Shimony [42] and AXIOMS OF PROBABILITY.

The simplest form for a theory of subjective probability can be regarded as a "black box" theory. The black box or mathematical theory contains the axioms of a theory of *sharp* probabilities. Its input consists of judgments of inequalities between subjective probabilities, and its output consists of *discernments* of the same kind. The purposes of the theory are to increase your body of beliefs and to detect inconsistencies in it [11]. This description of a theory applies *mutatis mutandis* to other scientific theories, not just to a theory of subjective probability.

For example, it applies to the theory of rational decisions [12]. In a more complete description of the black box theory of probability, the probabilities within the black box are represented by a different notation from the ones outside. A typical judgment, $P(A \mid B) < P(C \mid D)$, can be regarded as represented by $P'(A \mid B) < P'(C \mid D)$ in the black box, where the prime indicates that the probability is a real number and is not interval valued. This double notation, P and P', can be avoided only by using somewhat complicated axioms relating to upper and lower probabilities, such as are mentioned in the article AXIOMS OF PROBABILITY.

The plugging in of the input judgments and the reading out of the discernments can be regarded as the simplest rules of application of the theory. But for formulating your subjective probabilities you need more than just the axioms and rules of application. You also need a variety of "suggestions," and these are perhaps almost as important as the axioms, although many mathematical statisticians might not agree. In an attempt to codify the suggestions, 27 of them were collected together by Good [15] and called "priggish principles," with citations of earlier writings. For example, number 5 was: "The input and output to the abstract theories of probability and rationality are judgments (and discernments) of inequalities of probabilities, odds, Bayesian factors (ratios of final to initial odds), log-factors or weights of evidence, information, surprise indices, utilities, *and any other functions of probabilities and utilities* . . . . It is often convenient to forget about the inequalities for the sake of simplicity and to use precise estimates (see Principle 6)."

Principle 6 was: "When the expected time and effort taken to think and to do calculations is allowed for in the costs, then one is using the principle of *rationality of type II*. This is more important than the ordinary principle of rationality, but is seldom mentioned because it contains a veiled threat to conventional logic by incorporating a time element. It often justifies *ad hoc* procedures such as confidence methods and this helps to decrease controversy." (For an extended account of this "time element" effect, see Good [19].)

Principle 12 dealt with the Device of Imaginary Results, which is the recommendation that you can derive information about an initial or prior distribution by an imaginary (*Gedanken*) experiment. Perhaps this principle is overlooked so often because of the "tyranny of words," that is, because a "prior distribution" sounds as if it must be thought of before the posterior distribution. (The expression "tyranny of words" is apparently due to Stuart Chase, but the concept was emphasized by Francis Bacon, who called it the Idol of the Market Place.) To take a simple example, suppose that you would be convinced that a man had extra-sensory perception if he could correctly "guess" 40 consecutive random decimal digits, fraud having been ruled out. Then your prior probability must, for consistency, be at least 10^{-40} and not strictly zero. This can be shown by using Bayes' theorem backwards. The Device of Imaginary Results can be used in more standard statistical problems, such as in Bayesian testing for equiprobability of the cells of a multinomial distribution, or for "independence" in contingency tables [14, 17, 21]. Consider, for example, the estimation of a binomial parameter p. A class of priors that seems broad enough in some cases is that of symmetric beta priors, proportional to $p^{k-1}(1-p)^{k-1}$. If you have difficulty in selecting a value for k, you can consider implications of a few values of k when used in connection with various imaginary samples of the form "r successes in n trials." You can ask yourself, for example, whether you would in a bet give odds of 3 to 1 that $p > 0.6$ if $n = 100$ and $r = 80$. As an aid to making such judgments you can even make use of non-Bayesian methods. See also the comment below concerning hyperpriors.

Other priggish principles dealt with the Bayes/non-Bayes compromise, compromises between subjective and logical probability, quasi-utilities, such as amounts of information or weights of evidence whose expectations can be maximized in the design of experiments, the sharpening of the Ockham–Duns razor (expounded more fully by Good [18]), and the hierarchical Bayes technique. The latter is related to the fuzziness of upper and lower probabilities, and attempts, with some success, to cope with it by distributions of higher types. These deal with subjective probabilities concerning the values of rational subjective probabilities. For example, a binomial parameter p might be assumed provisionally to have a beta prior proportional to $p^{k-1}(1-p)^{k-1}$, where the "hyperparameter" k is not fixed, but is assumed to have a distribution or "hyperprior." This comes to the same as assuming a prior that is a *mixture* of beta distributions and the extension to mixtures of Dirichlet distributions is also useful. Some of the history and applications of the hierarchical Bayes technique is reviewed by Good [20]. The technique may well be necessary for the sound application of neo-Bayesian theory to multiparameter problems. Experience so far seems to support the following comments from Good [12]: "It might be objected that the higher the type the woollier (fuzzier) the probabilities. It will be found, however, that the higher the type the less the woolliness matters, provided the calculations do not become too complicated."

In short, "suggestions" for handling degrees of belief are probably essential for the practical implementation of neo-Bayesian theory.

Introspection can be aided by working with appropriate interactive computer programs. See, e.g., Novick and Jackson [33], Novick et al. [34], and Schlaifer [40]. These programs are designed to return deductions from your judgments and to allow you to change them. In other words, the computer is more or less the embodiment of the abstract black box mentioned before. The programs provide an excellent example of human–machine synergy.

A problem that has received much attention since 1950 is that of eliciting accurate degrees of belief by means of suitable rewards. The assessor specifies sharp degrees of belief $p_1, p_2, \ldots, p_n$ to n mutually exclusive events, and the reward depends on these probabilities and on the event that occurs. The reward scheme is known to the assessor in advance and is chosen in such a manner that the assessor's expected payoff is maximized, in his own opinion, if he is honest. The concept was independently suggested by Brier [4], without explicit reference to expected payoffs, and by Good [12]. Marschak [31, p. 95] says that the problem "opens up a new field in the economics of information."

According to McArthy [32], it was pointed out by Andrew Gleason (see also Aczél and Daróczy, [1, p. 115]) that if the reward is a function only of the asserted probability p of the event that occurs, and if $n > 2$, then the reward must be of the form $a + b \log p$, a formula proposed by Good [12]. This is therefore an especially pertinent form for eliciting degrees of belief, provided that the assessor has a good "Bayesian" (or neo-Bayesian) judgment. Such judgment, like any other, can presumably be improved by education and experience. The logarithmic payoff formula has an intimate relationship with weight of evidence in the sense of Peirce [35] and with the communication theory of Shannon [41] and these relationships do not depend on the condition $n > 2$.

The topic of eliciting honest degrees of belief has generated a large literature, and a very good review of the topic was written by Savage [39]. For some later work, see Buehler [5], Hendrickson and Buehler [22], and Good [16].

References

[1] Aczél, J. and Daróczy, Z. (1975). *On Measures of Information and Their Characterizations*. Academic Press, New York.

[2] Boole, G. (1854). *An Investigation of the Laws of Thought.* (Reprinted by Dover, New York, undated.)

[3] Box, G. E. P. and Tiao, G. C. (1973). *Bayesian Inference in Statistical Analysis*. Addison-Wesley, Reading, Mass.

[4] Brier, G. W. (1950). *Monthly Weather Rev.,* **78**, 1–3.

[5] Buehler, R. J. (1971). In *Foundations of Statistical Inference,* V. P. Godambe and D. A. Sprott, eds.

Holt, Rinehart and Winston, Toronto, pp. 330–341. (With discussion.)

[6] de Finetti, B. (1937). *Ann. Inst. Henri Poincaré*, **7**, 1–68. English translation in *Studies in Subjective Probability*, H. E. Kyburg and H. E. Smokler, eds. Wiley, New York.

[7] de Finetti, B. (1978). *International Encyclopedia of Statistics*, Vol. 2, W. H. Kruskal and J. M. Tanur, eds. Free Press, New York, pp. 744–754.

[8] Dempster, A. P. (1967). *Ann. Math. Statist.*, **38**, 325–329. [*Math. Rev.*, **34** (1967), Rev. No. 6817.]

[9] Fine, T. (1973). *Theories of Probability*. Academic Press, New York.

[10] Fréchet, M. (1938). *J. Unified Sci.*, **8**, 7–23.

[11] Good, I. J. (1950). *Probability and the Weighing of Evidence*. Charles Griffin, London/Hafner, New York.

[12] Good, I. J. (1952). *J. R. Statist. Soc. B*, **14**, 107–114.

[13] Good, I. J. (1959). *Science*, **129**, 443–447.

[14] Good, I. J. (1967). *J. R. Statist. Soc. B*, **29**, 399–431. (With discussion.)

[15] Good, I. J. (1971). In *Foundations of Statistical Inference*, V. P. Godambe and D. A. Sprott, eds. Holt, Rinehart and Winston, Toronto, pp. 124–127.

[16] Good, I. J. (1973). In *Science, Decision, and Value*, J. J. Leach, R. Butts, and G. Pearce, eds. D. Reidel, Dordrecht, Holland, pp. 115–127.

[17] Good, I. J. (1976). *Ann. Statist.* **4**, 1159–1189.

[18] Good, I. J. (1977). *Proc. R. Soc. Lond. A*, **354**, 303–330.

[19] Good, I. J. (1977). In *Machine Intelligence*, Vol. 8, E. W. Elcock and D. Michie, eds. Ellis Horwood, /Wiley, New York, 1976.

[20] Good, I. J. (1980). International Meeting on Bayesian Statistics, 1979, Valencia, Spain. *Trab. Estadíst. Invest. Oper.* (in press). Also in *Bayesian Statistics*, J. M. Bernardo, M. H. DeGroot, D. V. Lindley, and A. F. M. Smith, eds. Univ. of Valencia Press, 1981, pp. 489–519 (with discussion).

[21] Good, I. J. and Crook, J. F. (1974). *J. Amer. Statist. Ass.*, **69**, 711–720.

[22] Hendrickson, A. and Buehler, R. J. (1972). *J. Amer. Statist. Ass.*, **67**, 880–883.

[23] Hogarth, R. M. (1975). *J. Amer. Statist. Ass.*, **70**, 271–294. (With discussion.)

[24] Jeffreys, H. (1939/1961). *Theory of Probability*. Clarendon Press, Oxford.

[25] Kemble, E. C. (1942). *Amer. J. Phys.*, **10**, 6–16.

[26] Keynes, J. M. (1921). *A Treatise on Probability*. Macmillan, London.

[27] Koopman, B. O. (1940). *Bull. Amer. Math. Soc.*, **46**, 763–774.

[28] Koopman, B. O. (1940). *Ann. Math.*, **41**, 269–292.

[29] Lindley, D. V. (1965). *Introduction to Probability and Statistics from a Bayesian Viewpoint*, 2 vols. Cambridge University Press, Cambridge.

[30] Lindley, D. V. (1971). *Bayesian Statistics: A Review*. SIAM, Philadelphia.

[31] Marschak, J. (1959). In *Contributions to Scientific Research in Management*, Western Data Processing Center, University of California at Los Angeles, ed. University of California, Berkeley, Calif., pp. 79–98.

[32] McArthy, J. (1956). *Proc. Natl. Acad. Sci. USA*, **42**, 654–655.

[33] Novick, M. R. and Jackson, P. H. (1974). *Statistical Methods for Educational and Psychological Research*. McGraw-Hill, New York.

[34] Novick, M. R., Isaacs, G. L., and DeKeyrel, D. F. (1977). In Manual for the CADA Monitor. University of Iowa, Iowa City, Iowa.

[35] Peirce, C. S. (1878). *Popular Sci. Monthly*; reprinted in *The World of Mathematics*, Vol. 2, J. R. Newman, ed. Simon and Schuster, New York, 1956, pp. 1341–1354.

[36] Ramsey, F. P. (1931). *The Foundations of Mathematics and Other Logical Essays*. Kegan Paul, London.

[37] Rosenkrantz, R. (1977). *Inference, Method and Decision*. D. Reidel, Dordrecht, Holland.

[38] Savage, L. J. (1954). *The Foundations of Statistics*. Wiley, New York.

[39] Savage, L. J. (1971). *J. Amer. Statist. Ass.*, **66**, 783–801.

[40] Schlaifer, R. (1971). Computer Programs for Elementary Decision Analysis. Graduate School of Business Administration, Harvard University.

[41] Shannon, C. E. (1948). *Bell Syst. Tech. J.*, **27**, 379–423, 623–656.

[42] Shimony, A. (1955). *J. Symb. Logic*, **20**, 1–28.

[43] Smith, C. A. B. (1961). *J. R. Statist. Soc. B*, **23**, 1–37. (With discussion.)

[44] Wilks, S. S. (1962). *Mathematical Statistics*, Wiley, New York.

I. J. GOOD

DEGREES OF FREEDOM

This term is used in a number of different fields of science and has a variety of meanings. In statistics, it is most commonly thought of as a parameter in the chi-square* distribution and in distributions related thereto, such as central and noncentral t^* and F^*. Although the number of degrees of freedom is usually a positive integer, fractional numbers occur in some approximations, and one can, for example, have a noncentral chi-square distribution with zero degrees of freedom, obtained by taking this value for the degrees of freedom parameter. (It is a mixture of (a) a degenerate distribution $P(X = 0) = 1$, with weight $\exp(-\frac{1}{2}\lambda)$, and (b) central chi-square distributions with degrees of freedom $2, 4, \ldots$ and weights equal to the Poisson* probabilities $(\frac{1}{2}\lambda)$ $\exp(-\frac{1}{2}\lambda), (1/2!)(\frac{1}{2}\lambda)^2\exp(-\frac{1}{2}\lambda)$, where λ is the noncentrality parameter [1].)

In another connotation, the degrees of freedom of a model for expected values of random variables is the excess of the number of variables over the number of parameters in the model. In the case of the general linear model* with independent, homoscedastic residuals, the expected value of the sum of squares of differences between observed values and corresponding expected values, fitted by least squares*, is

(number of degrees of freedom)

$\times$ (residual variance).

Reference

[1] Siegel, A. F. (1979). *Biometrika*, **66**, 381–386.

DELTA FUNCTION *See* DIRAC DELTA FUNCTION

DELTA METHOD *See* STATISTICAL DIFFERENTIALS, METHODS OF

DEMOGRAPHY

HISTORY AND GENERAL CHARACTERISTICS

The study of human populations by statistical methods began in London in 1666 with the work of John Graunt. He analyzed the only available data, lists of deaths classified by cause, and he attempted, on the basis of general observation, to assess the average

sizes of families and households, the extent of migratory movements, and other elements of population structure and change. Graunt and his contemporary Sir William Petty made recommendations for the closer study of populations and for the establishment of a central organization for the collection of statistics. Late in the seventeenth century, Edmund Halley, the astronomer, examined the church records of Breslau and added the numbers of deaths at various ages to construct a life table* (see below). During the eighteenth century, statistics of this kind were assembled elsewhere and improved analyses were made, notably by Richard Price, leading to the development of life assurance. Johann Süssmilch collected large masses of data from parish records in Germany. Later, there was much discussion of "population philosophies"—speculation about the likely effects of population growth on general prosperity.

These early advances illustrate some of the main characteristics of demography, for instance, the possibilities for introspective thought; the dependence of the direction of that thought on actual records; the importance of the collection of relevant data, which tend, however, to be imperfect; and the consequent need for mathematical models (of which the life table is an example) as an aid to analysis.

During the nineteenth century, national population censuses and the civil registration of births, deaths, and marriages began in the United States and in many European countries. Although demographic studies had pointed to the need for such information, it was not so much for scientific as for legal and other reasons that it was collected. Nevertheless, the circumstances of demographic work were profoundly affected by its availability. In particular, a much more comprehensive assessment of the chances of dying became possible, showing how they varied by region and by occupation. This helped to pave the way for improvements in public health and reductions in hazards at work. The nature and characteristics of population change could be measured in some detail, and from this it became possible to attempt forecasts of future development (*see* "Population Projections" below). Much of the change in the earlier part of the century consisted of the rapid increase in total numbers without much variation in birth rate, death rate, and age distribution. A powerful mathematical technique for calculating some important parameters in these circumstances was developed by Lotka.

In a *stable population*, as it is called, the annual rate of growth r is linked to the rate of fertility* at age x (ϕ_x)—that is, the number of female births to women aged x expressed as a ratio of the total number of women aged x—and the chance of a female infant surviving from birth to age x($_xp_0$) as follows:

$$\int_0^\infty e^{-rx} {}_xp_0\phi_x \, dx = 1$$

(*see* "Population Mathematics" below).

In countries peopled by Europeans a decline in the birth rate began in the nineteenth century, and this continued with increasing effect in the first part of the twentieth century. Interest accordingly began to be focused on fertility (defined by demographers in terms of actual live births, to distinguish it from fecundity, the capacity to reproduce). Provision was made for the collection of data showing the length of time married or number of existing children of couples enumerated at a census* or to whom a child was born. The interpretation of such statistics gave rise to new analytical techniques, notably the study of family building as a cumulative process related to time elapsed (since puberty or marriage). This can be more meaningful than to compare the experience of different calendar periods. The big fall in fertility gave much impetus to population forecasting, but it also became apparent that the results are often unreliable. Consequently, calculations of this kind are usually made on series of alternative hypotheses and regarded only as "projections" illustrating what could happen on stated assumptions.

The major developments in mathematical statistical techniques that took place after 1900 have not found much place in demography. This is because population data and vital statistics* rarely satisfy the conditions of consistency, randomness, or independence which are necessary for the useful application of those techniques. Even so, demography has greatly expanded and diversified; it has also developed close links with other disciplines such as sociology*. One consequence of these advances has been the emergence of demographic journals and periodicals, for example *Demography* (United States), *Population* (France), and *Population Studies* (Britain). Many countries now have such journals, and in addition the United Nations Organization issues a *Demographic Year Book* and many manuals of demographic practice.

Toward the middle of the twentieth century, demographers began to study the population of the world as a whole rather than just those economically developed countries for which statistical data were available. This new interest received a considerable impetus after World War II, when it became apparent that successful efforts to combat tropical and other infectious diseases were leading to rapid population growth in Africa, Asia, and America south of the United States. Attention has been paid principally to (a) the collection of new or better census and vital registration data, mainly under the auspices of the United Nations Organization; (b) as such data are still unreliable and incomplete in many developing areas, the specification of mathematical and other statistical models that may provide intelligent guesses about untrustworthy or nonexistent information; (c) the study of the effect of population growth on resources and of the limits that scarce resources can impose on good demographic development; and (d) the formulation of social policies in relation to marriage, the family, children, and other factors involved in population growth and the study of such policies and their effects. It also began to be realized that certain restricted data still existed in relation to the past history of Western European countries, notably in the seventeenth and eighteenth centuries. These are now studied not only for themselves but for the light they may throw on the demography of the less well-developed countries today.

Those less-developed countries have, in general, fertility rates twice as high as those in Western Europe and the United States. Their populations are rapidly increasing in size as mortality* falls, so, in consequence, their prosperity is retarded and efforts are now being made to encourage reduction in the number of births.

SOCIAL AND ECONOMIC ASPECTS

The work of the demographer is associated to some extent with that of economists, agronomists, and nutritionists in multidisciplinary studies, relating particularly to the population of the countries of the Third World, where there is a good deal of malnourishment and some starvation. The well-being of people in such countries depends largely upon agricultural production, and where this is scanty the calorie and protein requirements for normal health are studied. Other relevant resources are land, fertlizers, farming implements and techniques, and also water, minerals, fuel, and other sources of energy. Education is necessary to enable improvements in production to be made, and injections of financial and other capital are essential for the construction of buildings, roads, and harbors. In analyses relating to the interactions between population characteristics and such nondemographic elements, the demographer contributes his expert knowledge of such basic features as age distribution, family size, regional and occupational distribution, sickness, and mortality. On this foundation he can assess dependency ratios, household size, and degrees of social stratification and urbanization, which are important in many forms of economic and other research. With the aid of large-scale electronic computers, it is now possible to construct mathematical models incorpo-

rating such demographic elements along with other social and economic statistics or hypotheses. These models can be employed to demonstrate the likely future outcome of current trends and the consequences of various possible policies, and to show how mankind may adapt to changing circumstances in order to maintain or improve its standards or reduce the risk of catastrophe.

This is just one of the ways in which the development of electronic computers* has influenced the course of demographic analysis. In general, elaborate tabulations and calculations, which before would have been too laborious to undertake, are now readily possible, and this applies notably to popluation projections (see below).

Population prospects depend very much on how many children couples will have and how soon they are born. Success in projection, and also in the correct interpretation of current happenings, must therefore depend on the couples' family-building intentions and on how closely they achieve them. Sample surveys* are therefore conducted into the attitudes and aspirations of husbands and wives and, as views can change quickly, are regularly repeated. Other surveys investigate how accurately the performance matches the expectation. If it does not, it may be because economic or personal circumstances have changed, the contraceptive method used was unreliable (possibly because of ignorance or lack of motivation on the parents' part), or biological factors may have intervened. Demographers therefore try to collect statistics of all such elements of uncertainty, including the availability, cost, and effectiveness of birth prevention devices. They also study governmental policy in regard to such matters—whether states ban some forms of contraception, encourage or discourage abortion, prevent, allow, or promote the spread of knowledge about birth control. Some demographers specialize in one or other of these branches, or—as wide differences exist, notably between well-developed and less-developed countries—concentrate on particular regions.

POPULATION MATHEMATICS

There are still many areas in which population growth is steady, and so stable population mathematics in its original form, or some modification of it, may well be useful. Solutions have been found to the equation quoted above. By a variety of methods, it has been found that, approximately,

$$r = \frac{1}{G} \log_e \int {}_x p_0 \phi_x \, dx,$$

where G is the length of a generation or the average time between the birth of parents and children. The expression $\int {}_x p_0 \phi_x \, dx$ is called the "net reproduction rate", and it gives a measure of the intrinsic speed of growth of a population. This may well not be the same as its actual current rate of growth, but it indicates the rate that would eventually be experienced if circumstances did not change for several decades. Although the formulae are normally applied to women, they can be used for men also; a disadvantage is that the net reproduction rates often differ between men and women, and reconciliation is difficult.

This form of mathematics can still be of value for the estimation of parameters for populations about to develop economically, but, as one of the earliest departures from the "stable" state is a decline in mortality, an adjusted form has been devised that allows for this. It is known as *quasi-stable* population mathematics. But both forms are deterministic and do not allow for stochastic variations. More complex formulae have been devised to incorporate such variations but their application in practice is very limited. With the development of electronic computers it has proved more rewarding to make projections starting from a combination of actual data and hypotheses. The outcome of such models can be studied and the reasonableness or unreasonableness of the reslut can cast light on the validity of the assumptions made. Among the demographic subjects studied in this manner are the biology of reproductive processes and of family building, and the effectiveness of expenditures on birth control promotion programmes.

MORTALITY ANALYSIS

The statistical study of mortality began with the life table*, which shows the numbers living and dying at each age out of a group of people all born at the same moment, for example:

Age	Number Alive	Number dying
(x)	(l_x)	(d_x)
0	1000	27
1	973	10
2	963	5
3	958	4

and so on up to the oldest age of, say, 110. The normal method of constructing a life table is first to calculate probabilities of death, $d_x \div l_x$. *Rates of mortality*, as these probabilities are often called, vary considerably with increasing age, falling from infancy to childhood, and later rising as youth leads to middle life and old age. They differ between men and women. Their size and age pattern also vary with time and between countries, areas, climates, seasons, occupations, and social systems.

Rates of mortality can be usefully calculated for specific causes of death, and they are naturally higher among groups of people suffering from illness than among those who are well. For such sufferers it is often meaningful to construct life tables on the basis of the time elapsed since the onset of the illness. This form of analysis is called *cohort study* or, where the cohort begins at birth, *generation study*. Normally, however, whole life tables are not constructed on the basis of generations because a 100-year span nowadays encompasses so many social and economic changes affecting mortality that it would present a confusing picture. Instead, life tables are of necessity made from "secular" mortality rates, those experienced by people of all ages in a given short period of years.

Demographers study mortality for a variety of purposes, for instance, as a basis for population projection. Some aim to show how to reduce the incidence of death, and much of their work is comparative, for example, showing differences between countries. The pattern of mortality in a country depends on the age and sex distribution of its population, so the making of valid comparisons is a complex business. *See also* VITAL STATISTICS.

"Crude" death rates, or ratios of total deaths to total population in given years, are often misleading, and detailed studies by separate age and cause can be tedious. Much attention has therefore been devoted to processes of "standardization*", designed to eliminate the influence of age, sex, and other factors. Among the standardized rates adopted are the following:

1. The *crude death rate* that would be experienced if the population in question had the age and sex distribution of a selected standard population. When such death rates are calculated for two populations using the same selected standard, their ratio is called the "Comparative Mortality Factor."

2. The *Standardized Mortality Ratio*, which is a weighted average of the ratios of the mortality rates of two populations at individual ages.

3. *Life Table Death Rates*, or $\sum d_x / \sum l_x$ over all or part of the age range, based on the mortality experiences of the populations to be compared.

MARRIAGE* AND FERTILITY*

Marriage is of interest to demographers mainly because of its influence on fertility. Comparative analysis is complicated by the differing nature of the marriage bond and the degree of ease with which marriage can be ended in various parts of the world. The most significant features are the age at which marriage takes place and the proportion that remains celibate. Techniques similar to those described above with respect to mortality can be used for marriage experience; the proportion of people married and unmarried at a given time can also be compiled.

Greater analytical scope is provided by the demographic study of fertility. The chances of the birth of a child to a parent can be investigated according to (a) the age and sex of the parent; (b) the time elapsed since marriage; or (c) the number of children the parent already has (sometimes called "parity"). Couples usually have an idea of the size of family they want, and everyone can see that two children are needed for replacement. Nevertheless, plans accord to varying circumstances: They can be speeded up or delayed for different reasons during the 20 or so childbearing years. Marriage can also be hurried (sometimes for reasons connected with the arrival of a child) or delayed, so that assessment of the respective merits of the three methods is complex. Method (a) permits the compilation of net reproduction rates (see above), though this is not an important advantage.

Fertility can be analyzed, not only as a ratio of birth occurrences per unit population in a given period of time, but also on a census basis: Married people can be asked, at a particular moment, to state how many living children they have or how many have been born to them in the past. The second method eliminates the effect of child mortality but introduces errors dependent on the possibly faulty memories of the respondents. Fertility varies significantly with the occupations of the parents. For the wife, there may be a negative association between the number of children she has and the extent of her work outside the home. Attitudes toward childbearing may depend on urban or rural domicile, and on whether the parents' occupations are managerial, skilled, or unskilled.

"Social class" can be an important factor affecting fertility studies. Because there are a number of fertility analyses possible, and no one has been established as the best, many methods of summary and presentation are possible. Those used for mortality and marriage—life tables, rates, proportions, and standardization—are all useful. But fertility standardization can be for age, for duration of marriage, or for parity. A commonsense approach to the interpretation of basic statistics, or of rates based on them, is valuable. "Internal" analysis, that is, the examination of fertility rates at increasing ages or marriage durations, or of successive marriage cohorts, often reveal errors or special features.

MIGRATION

International migration is normally a minor contributor to population change. It is studied by those with a special interest in political and economic influences on population. Internal migration (within the borders of a country) is more highly significant in relation to regional distribution. Migration statistics are collected at ports and airports where this movement is distinguished from temporary holiday trips. In census taking, questions are asked about place of birth and earlier residence. Variations between countries and at different times make careful appraisal of the data more important than close technical analysis.

POPULATION PROJECTIONS

As the population cannot continue to grow forever, owing to space and other resource limitations, estimates of future numbers have been based on diminishing growth rates, such as

$$r - k({}^{t}P),$$

where r and k are constants and ${}^{t}P$ is the total population at time t. Forecasts made by such methods have not been accurate, and they do not yield useful information about age distribution and other demographic features. Most projections today are made by the "component" method: In this,

rates or numbers of births and deaths are first specified and then normally assumed to vary as the years unfold. The corresponding population size and age distribution at all future times can then be calculated quite readily. The work is iterative and so very suitable for an electronic computer. Mathematically, the process consists of multiplying a vector* by a matrix*; the vector represents the starting population classified by age and the matrix represents the fertility and mortality factors. Migration is usually accounted for separately, as subsequent births and deaths among net immigrants may well have a different pattern from that of the resident population. This is an example of the type of formula used:

$$ {}^{n}P_{x} = {}^{n-1}P_{x-1}\left(1 - {}^{n-1}q_{x-(1/2)}\right) \pm {}^{n}M_{x}, $$

where ${}^{n}P_{x}$ is the number aged x in year n, q_{x} represents the mortality rate at age x and ${}^{n}M_{x}$ adjusts for migration. For age 0, the corresponding formula is based on the number of live births in the 12-month period $n - 1$ to n.

The method illustrated is one applicable to projections of the whole population of a country for a period of up to 50 years. For a much shorter period, for regions or smaller areas within a country, or for special attention to the aged or the very young, the technique can readily be altered. Projections made in the past are tested against subsequent populations, and such tests help to improve the assumptions made in future projections.

An illustration of the kind of picture a projection may present is given in the diagram that appears later in POPULATION PYRAMID showing by its width the numbers of males and females in successive 10-year age groups, starting with children at the bottom. The outline with thin lines refers to the current population, in which a deficiency of men compared to women may be noted at ages 30–39, and a surplus of women over men at the oldest ages. The outline with thick lines shows the population forecast for 20 years later. In most age groups this is generally larger than the starting population but it is smaller under age 10, as a fall in the number of births is expected. The shortfall in males now appears at ages 50–59, the result perhaps of a war or big emigratory movement 30 years before.

Bibliography

Benjamin, B., Cox, P. R., and Peel, J., eds. (1973). *Resources and Population*. Academic Press, London.

Bowen, I. (1976). *Economics and Demography*. Allen and Unwin, London.

Cox, P. R. (1976). *Demography*, 5th ed. Cambridge University Press, Cambridge. (A work of wide range, including a discussion of current population situations and prospects.)

Hauser, P. M. and Duncan, O. D. (1959). *The Study of Population: An Inventory and Appraisal*. Chicago University Press, Chicago. (A very thorough discussion of the nature of demographic work.)

Henry, L. (1976). *Population—Analysis and Models*. Edward Arnold, London.

Human Populations. (1973). Cambridge University Press, Cambridge.

Matras, J. (1973). *Populations and Societies*. Prentice-Hall, Englewood Cliffs, N.J.

Parsons, J. (1977). *Population Fallacies*. Elek Books, London.

Pollard, J. H. (1973). *Mathematical Models for the Growth of Human Populations*. Cambridge University Press, Cambridge.

Population Projections 1976–2016. (1978). Her Majesty's Stationery Office, London.

Pressat, R. (1974). *A Workbook in Demography*. Methuen, London. (This shows detailed examples of how to make specific demographic analyses.)

Shryock, H. S. and Siegel, J. S. (1976). *The Methods and Materials of Demography* (condensed ed.). Academic Press, New York. (There is also a full edition.)

United Nations. (1973) *The Determinants and Consequences of Population Trends*. United Nations, New York. (A highly valuable social and economic analysis.)

(ACTUARIAL STATISTICS, LIFE
FERTILITY
INFANT MORTALITY
LIFE TABLES
MARRIAGE
MATHEMATICAL THEORY OF
   POPULATION
STOCHASTIC DEMOGRAPHY
VITAL STATISTICS)

PETER R. COX

DEMOGRAPHY, STOCHASTIC *See* STOCHASTIC DEMOGRAPHY

DE MOIVRE, ABRAHAM

Born: May 26, 1667, in Vitry (in Champagne), France.

Died: November 27, 1754, in London, England.

Contributed to: mathematical analysis, actuarial science, probability.

De Moivre's early education was in the humanities, at the Protestant University of Sedan where he was sent at age 11, and the University of Saumur, but he soon showed a flair for mathematics which was encouraged by his father, a poor surgeon, and by an early age had read, *interalia*, the work of C. Huygens* on games of chance (*De Ratiociniis in Ludo Aleae*), which was the kernel of his own later work on chance. At the Sorbonne he studied mathematics and physics with the famous Jacques Ozanam. In 1688 he emigrated to London to avoid further religious persecution as a Protestant, after the repeal in 1685 of the Edict of Nantes. Here he was forced to earn his living first as a traveling teacher of mathematics, and then by coffeehouse advice to gamblers, underwriters, and annuity brokers, in spite of his eminence as a mathematician. He became acquainted with Halley in 1692, was elected Fellow of the Royal Society of London in 1697, and to the Berlin and Paris Academies. His contact with the work of Newton at an early stage in England had important consequences for his mathematical growth; the two men became very close friends. Todhunter [9] writes of De Moivre: "In the long list of men ennobled by genius, virtue and misfortune, who have found an asylum in England, it would be difficult to name one who has conferred more honour on his adopted country than De Moivre."

His principal contributions to probability theory are contained in the book [4] *The*

Doctrine of Chances (dedicated to Newton), the two later editions of which contain English versions of the rare second supplement {as bound with his *Miscellanea Analytica de Seriebus et Quadraturis*}, which is actually a seven-page privately printed pamphlet [3] dated November 12, 1733, discovered by K. Pearson* [6] in which the density function of the normal distribution first appears. In the supplement of the *Miscellanea Analytica*, De Moivre obtained the result $n! \sim B n^{n+1/2} e^{-n}$, now called Stirling's formula* (and of extensive use in the asymptotics of combinatorial probability)—the contribution of James Stirling was the determination of B as $\sqrt{2\pi}$. Using this formula, De Moivre [3] initially investigated the behavior of the modal term of the symmetric binomial distribution $\binom{n}{k} 2^{-n}$, $k = 0, \ldots, n$, and the term t terms distant, for large n. If we call these α_0 and α_t, respectively, he concluded that $\alpha_0 \simeq 2/\sqrt{2\pi n}$, $\alpha_t \simeq \alpha_0 \exp(-2t^2/n)$, and determined a series approximation to $\alpha_0 + \alpha_1 + \cdots + \alpha_t$, which is a series expansion of the integral $(2/\sqrt{2\pi n}) \int_0^t \exp(-2t^2/n)\, dt$. He also deduced similar results for the general case of a binomial distribution* with probability of success p, $(0 < p < 1)$, giving $\alpha_0 \simeq 1 / \sqrt{2\pi n p(1-p)}$, $\alpha_t \simeq \alpha_0 \exp\{-t^2/[2np(1-p)]\}$, although again in different notation. We may therefore attribute to him the local and global central limit theorem* in the case of sums of $(0, 1)$ random variables, now known in probability collectively as "De Moivre's theorem." He also had a very clear notion of the significance of *independence*.

De Moivre's work is well explored [1]. A detailed account of his other probabilistic writings is given in refs. 2 and 9; it is, however, worthwhile to comment on his treatment of the simple random walk* on $\{0, 1, 2, \ldots, a + b\}$ with starting point a, probability of a step to the right p, and 0 and $a + b$ absorbing barriers. It is desired to find the probabilities of ultimate absorption into each of the two absorbing barriers, the expected time to absorption, and the probability of absorption into a specific barrier in

n steps. De Moivre treats it in the framework of "gambler's ruin." The problem and its solution in the simple case $a = b = 12$ originates with Pascal* and Fermat* [5], whence it is to be found in Huygens. De Moivre's *method* of solution in the general case is ingenious and shorter than most modern demonstrations [8], although there is some doubt on *priority* because of solutions by Montmort, and N. Bernoulli (possibly due to James Bernoulli; *see* BERNOULLIS, THE) in the same year (1711).

References

[1] Adams, W. J. (1974). *The Life and Times of the Central Limit Theorem*. Kaedmon, New York. (Chapter 2 contains a careful bibliographical analysis of the origins of De Moivre's theorem, and a picture of De Moivre. There is also a useful bibliography of secondary sources.)

[2] Czuber, E. (1899). *Jahresber. Dtsch. Math.-Ver.*, **7** (2nd part), 1–279.

[3] De Moivre, A. (1733). Approximatio ad Summam terminorum Binomii $\overline{a + b}\backslash^n$ in Seriem expansi. [A facsimile may be found in R. C. Archibald, *Isis*, **8**, 671–683 (1926).]

[4] De Moivre, A. (1738). *The Doctrine of Chances; or a Method of Calculating the Probability of Events in Play*, 2nd ed. H. Woodfall, London. (First ed., 1718; 3rd ed., 1756. The first edition is an enlarged English version of De Moivre's first published work on probability: De Mensura Sortis, seu de Probabilitate Eventuum in Ludis a Casu Fortuito Pendentibus. *Philos. Trans. R. Soc. (London)*, **27** (1711) 213–226. Reprinted by Kraus, New York, 1963.)

[5] Ore, O. (1960). *Amer. Math. Monthly*, **67**, 409–419.

[6] Pearson, K. (1978). *The History of Statistics in the 17th and 18th Centuries*. Charles Griffin, London. (Lectures by Karl Pearson 1921–1933, edited by E. S. Pearson. Pages 141–166 contain details of De Moivre's life and work on annuities as well as on De Moivre's theorem.)

[7] Schneider, I. (1968). *Arch. History Exact Sci.*, **5**, 177–317.

[8] Thatcher, A. R. (1957). *Biometrika*, **44**, 515–518.

[9] Todhunter, I. (1865). *A History of the Mathematical Theory of Probability*. Macmillan, London. (Reprinted by Chelsea, New York, 1949 and 1965.)

(BINOMIAL DISTRIBUTION
COMBINATORICS
GAMBLING, STATISTICS IN

LIMIT THEOREMS, CENTRAL
MARKOV PROCESSES
RANDOM WALK
STIRLING'S FORMULA)

E. SENETA

DE MOIVRE NUMBERS   *See* BALLOT
PROBLEM

DENDRITES

The term "dendrite" is associated with early applications of graph theory* to numerical taxonomy. It was introduced by H. Steinhaus and his co-workers from the Wrocław school of applied mathematics, and has been used in various fields of research, at the beginning in phytosociology and anthropology (see Florek et al. [5,6], Perkal [12], Kowal and Kuźniewski [9]; also see Hubac [8] and Sneath and Sokal [16, p. 14]). From the graph theoretical point of view, "dendrite" is a synonym of "tree" (Gk. *dendron*). A brief definition is as follows: A dendrite (or a tree) is a connected graph that has no circuits (see, e.g., Ore [11]). Some introduction to the graph-theoretic terminology may be helpful here. This will be followed by indicating the role played by dendrites in the application of multivariate analysis*, with special reference to taxonomic and classificatory studies. (*See* CLASSIFICATION.)

Let V be a nonempty set of points called vertices. If $u, v \in V$, then an undirected segment $e = \{u, v\}$ joining u and v is called an edge. A pair $G = (V, E)$, where E is a finite family of edges, is called a graph. Two vertices are said to be adjacent (or nonadjacent) if there is an edge (or no edge) joining them; the vertices are then said to be incident to such an edge. Similarly, two distinct edges are said to be adjacent if they have at least one vertex in common. An edge sequence in a graph is a finite sequence of edges of the form $\{v_0, v_1\}, \{v_1, v_2\}, \ldots, \{v_{m-1}, v_m\}$; vertex v_0 may be called the initial vertex and v_m the final vertex of the edge sequence, the

edge sequence being then called from v_0 to v_m. An edge sequence in which all the edges are distinct and, moreover, all the vertices $v_0, v_1, \ldots, v_m$ are different (except, possibly, $v_0 = v_m$) is called a chain. A chain is said to be closed if $v_0 = v_m$. If a closed chain contains at least one edge it is called a circuit. A graph G is said to be connected if for any pair of vertices $v, w \in G$ there is a chain from v to w. A graph can be split up into disjoint connected subgraphs called (connected) components. A connected graph has, clearly, only one component; if a graph has more than one component it is called disconnected (see also Wilson [17]). Note that if for a pair of vertices $v, w \in G$ there are more chains from v to w, at least one circuit occurs in G. On the contrary, if for every pair $v, w \in G$ there is at most one chain from v to w, G has no circuits. Finally, if for every pair $v, w \in G$ there is exactly one chain from v to w, the graph is connected and has no circuits; i.e., it is a tree, a dendrite. It will be denoted by T (to be consistent with most of the literature). If T has n vertices, it has $n - 1$ edges, the smallest possible number in a connected graph. (This is the reason for sometimes calling it a minimally connected graph, in contrast to a complete or maximally connected graph, in which every pair of distinct vertices are adjacent.) A removal of one edge from a dendrite disconnects it into two subgraphs. Each of the components obtained is again a dendrite. If two edges are removed, three component dendrites are obtained, etc.

In some applications of graph theory a nonnegative real number $\mu(e)$ may be assigned to each edge e of G and called its measure or its length. If G is a complete graph of n vertices the problem of interest may be to find as its subgraph a dendrite T connecting (spanning) all n vertices of G in such a way that its measure sum or overall length, $M(T) = \sum \mu(e_i)$, is as small as possible, the sum being taken over all $n - 1$ edges of T. This problem, known as the minimal connector problem, originates from Borůvka [2]. It seems that the first rigorous solution of the problem was given by Florek et al. [5], although it is usually ascribed to Kruskal [10]. The resulting dendrite is called the shortest dendrite or the shortest spanning subtree or the most economical spanning tree or the minimum spanning tree (MST). The latter term seems to be most common among statisticians (see, e.g., Gower and Ross [7]) and taxonomists (see, e.g., Rohlf [14]).

There are several algorithms now available for finding the MST (see the references in Gower and Ross [7]). They all operate iteratively. At each stage one of the $n(n - 1)/2$ edges of the complete graph is chosen and assigned to the searched subgraph in such a way that after a completion of the choice of $n - 1$ edges the subgraph becomes the shortest dendrite (the MST). The algorithms differ only in the order of the sequence of choices. One possibility (algorithm I in Gower and Ross [7]) is to choose at each stage the shortest edge among those not previously chosen and not giving rise to any circuit when added to the subgraph already constructed. Another possibility (algorithm II, originally given by Prim [13]) is to initiate the subgraph with any one vertex and to add the shortest edge incident to it, and then continue by adding at each stage the shortest edge adjacent to any of the edges already in the subgraph but not forming a circuit with them. If in any of the algorithms a choice of several edges of the same minimum length occurs, any one of them may be chosen. In such cases the MST may not be unique, but otherwise each algorithm must give the same result.

Applications of shortest dendrites (MSTs) can be found in various fields of endeavor, most naturally in problems of operations research* (e.g., in some communication problems). Berge and Ghouila-Houri [1] and Gower and Ross [7] give accounts of different inventions and uses of the idea of the MST. As far as statistical sciences are concerned, the most common applications are in multivariate analysis*, particularly in connection with numerical taxonomy. This type of application seems to have been stimulated, as already mentioned, mainly by

phytosociologists and anthropologists, under the impact of the precursory work by Czekanowski [3, 4].

From a taxonomical point of view, vertices of a graph represent operational taxonomic units (OTUs; for a definition, see Sneath and Sokal [16, p. 69]), i.e., the lowest-ranking taxa employed in a given classificatory study, while edges of the graph indicate inter-OTU relationships. The length of an edge is then a measure of the relationship between two adjacent OTUs. Most frequently, the relationships are some kind of dissimilarities or distances between the OTUs. If connections between not only the individual OTUs but also between them and their sets (clusters) and between the sets themselves are to be established on the basis of single edges (links) between two most similar OTUs, then the shortest dendrite (the MST) is the answer. A method utilizing this principle of connecting OTUs is known in numerical taxonomy as single linkage cluster analysis. In fact, the single linkage clusters (SLCs) can be obtained from the MST by successively removing its edges, largest first, the seeond largest next, etc. At each step the resulting components (each being a shortest dendrite for itself) represent some SLCs. If all edges of the length exceeding certain chosen threshold d are removed, the components obtained are the d-level SLCs (see Ross [15], Gower and Ross [7], Sneath and Sokal [16, pp. 216ff.]).

An obvious feature of the shortest dendrite (MST) is that it provides no information on the inter-OTU relationships other than those indicated by the edges of the dendrite. This is a disadvantage when drawing the MST, since nothing is then known about how the various branches of the dendrite should lie relative to each other. If the number of vertices is not too large, this can be overcome by drawing the MST on a vector diagram provided by a suitable ordination method, such as principal component analysis*, principal coordinate analysis, or canonical variate analysis. This approach is indicated, e.g., by Gower and Ross [7] or Sneath and Sokal [16, p. 257].

Table 1   Mahalanobis Generalized Distances Between the Means of Eleven Sunflower Strains

	B	C	D	E	F	G	H	I	J	K
A	9.70	10.06	9.38	9.71	1.53	13.08	9.01	9.04	0.73	12.73
B		2.16	2.00	1.00	9.70	6.47	1.74	2.59	10.26	13.21
C			1.85	1.23	9.90	6.54	1.97	1.76	10.60	12.34
D				1.41	9.22	6.73	2.55	1.70	9.93	13.37
E					9.65	6.59	1.49	1.79	10.27	12.77
F						12.42	9.10	9.02	1.32	13.33
G							7.53	7.98	13.36	17.65
H								1.92	9.60	11.49
I									9.63	11.76
J										13.15

The application of the shortest dendrite as an ancillary method accompanying the multivariate analysis of experimental data is illustrated here by an example taken from plant breeding research. Multivariate analysis of variance* together with the canonical variate analysis was applied to data consisting of measurements on four characteristics of 11 sunflower strains compared in a field experiment with six replications. The Mahalanobis generalized distances* between strains are given in Table 1. A plot of the first two canonical variate means is shown in Fig. 1. The distances between means in this diagram do not necessarily reproduce those

of the table exactly. Although the two-dimensional approximation given by the first two canonical variates accounts for 94% of the between-strains variation, there is still some distortion. This is readily apparent from examining the shortest dendrite (MST) superimposed upon the means. The MST shows that there are two distinct clusters of strains, {B, C, D, E, H, I} (strain I appears to be closer to D than to H) and {A, F, J}, and two outlying strains, {G} and {K}, each well separated from all the others.

References

[1] Berge, C. and Ghouila-Houri, A. (1965). *Programming, Games and Transportation Networks.* Methuen, London.

[2] Borůvka, O. (1926). *Acta. Soc. Sci. Math. Moravicae*, **3**, 37–58.

[3] Czekanowski, J. (1909). *Korespondenzbl. Dtsch. Ges. Anthropol. Ethnol. Urgesch.*, **40**, 44–47.

[4] Czekanowski, J. (1932). *Anthropol. Anz.*, **9**, 227–249.

[5] Florek, K., Łukaszewicz, J., Perkal, J., Steinhaus, H. and Zubrzycki, S. (1951). *Colloq. Math.*, **2**, 282–285.

[6] Florek, K., Łukaszewicz, J., Perkal, J., Steinhaus, H. and Zubrzycki, S. (1951). *Przegląd. Antropol.*, **17**, 193–211.

[7] Gower, J. C. and Ross, G. J. S. (1969). *Appl. Statist.*, **18**, 54–64.

[8] Hubac, J. M. (1964). *Bull. Soc. Bot. Fr.*, **111**, 331–346.

[9] Kowal, T. and Kuźniewski, E. (1959). *Acta. Soc. Bot. Pol.*, **28**, 249–262.

[10] Kruskal, J. B. (1956). *Proc. Amer. Math. Soc.*, **7**, 48–50.

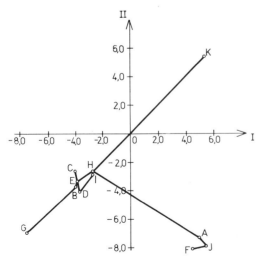

Figure 1   The means of the first two canonical variates for 11 strains of sunflower, with the shortest dendrite (MST) superimposed upon the means.

[11] Ore, O. (1963). *Graphs and Their Uses*. Random House, New York.

[12] Perkal, J. (1953). *Przegląd Antropol.*, **19**, 82–96.

[13] Prim, R. C. (1957). *Bell Syst. Tech. J.*, **36**, 1389–1401.

[14] Rohlf, F. J. (1973). *Computer J.*, **16**, 93–95.

[15] Ross, G. J. S. (1969). In *Numerical Taxonomy*, A. J. Cole, ed. Proc. Colloq. Numer. Taxon., University St. Andrews, Sept. 1968. Academic Press, London, pp. 224–233.

[16] Sneath, P. H. A. and Sokal, R. R. (1973). *Numerical Taxonomy*. W. H. Freeman, San Francisco.

[17] Wilson, R. J. (1972). *Introduction to Graph Theory*. Academic Press, New York.

(CANONICAL ANALYSIS
CLASSIFICATION
DENDROGRAMS
GRAPH THEORY
MULTIDIMENSIONAL SCALING
MULTIVARIATE ANALYSIS
MULTIVARIATE ANALYSIS
   OF VARIANCE
PATTERN RECOGNITION)

T. CALIŃSKI

DENDROCHRONOLOGY

Dendrochronology is the science of dating historical and environmental events, utilizing variations in the width of annual growth rings that are characteristic of many tree species growing in temperate and subpolar regions. Tree-ring data may also be used to reconstruct past climates (dendroclimatology) and hydrologic events (dendrohydrology).

Ring widths generally vary in response to moisture and temperature or other climatic factors that may affect growth. Moisture is usually the dominant factor in arid regions and temperature in cold areas. Wide rings are formed when growth-limiting factors are abundant and narrow rings when they are in short supply. Narrow rings are best for diagnostic purposes and field sampling is deliberately stratified toward environmentally sensitive species or individuals.

Adequate sampling is necessary to ensure the reliability of measurements and dating. Typically, samples from arid sites should consist of two cores from each of 10 to 20 trees. Larger samples may be required in regions where climate is less limiting to ensure an adequate climatic signal over random variations or noise.

Absolute dating of each ring is essential regardless of the intended use of the data. The dating procedure, known as cross-dating, requires the matching of narrow and wide ring-width patterns among trees from the same region so that any discrepancy due to mistakes, double rings, or missing rings can be identified and corrected. This *cross-dating* procedure is perhaps the most tedious and critical aspect of tree-ring analysis. The relative pattern of wide and narrow rings of appropriately dated *tree-ring series* from a given site constitutes a *chronology* covering a specific time span. Chronologies so developed may be cross-dated with older, overlapping material from dead or living trees, or wood from archaeological sites, to extend the chronology further back in time.

Ring widths are affected by both the age of the tree and the productivity of its site. Rings tend to become narrower as trees grow older and measured ring widths cannot be used directly to develop statistical relationships. This systematic change resembles, especially for coniferous tree species, a decreasing exponential curve when ring widths are plotted against time. This trend is easily removed by fitting an exponential curve with a positive asymptote k. When the growth curve* is more complicated, orthogonal polynomials* or some other flexible functions may be used. A *ring-width index* is then developed by dividing the width of each dated ring by the value of the fitted curve for the same year. This procedure provides an essentially stationary time series of ring-width indices, referred to as a *standardized ring-width chronology*. The indices for individual trees on a site are averaged by year to form a *mean chronology*. This chronology is a mean value function in which the variance common to all trees is considered the climatic signal and the random variation, the nonclimatic noise. The larger the number of sample trees and the more limiting the climatic factors, the greater is the climatic signal compared to noise.

In general the averaged indices of successive growth rings are related due to the conditioning and carryover effects of growth-related environmental and physiological factors. This variation is both an autoregressive and a moving-average process in that the values are both interdependent and ordered in a time series.

Ring-width indices are routinely characterized by several statistical parameters including the mean*, variance*, standard deviation*, standard error*, correlation coefficient* and mean sensitivity. The latter statistic measures the relative difference in width from one ring to the next. It is unique to tree-ring analyses and may be defined as

$$\frac{2}{n-1} \sum_{t=1}^{t=n-1} \frac{|x_{t+1} - x_t|}{x_{t+1} + x_t}$$

where x_t is the ring-width index for year t and n is the number of successive points in an index series. Values of mean sensitivity can range from zero to two. A value of zero is obtained from series with no variation, while series that alternate indices of zero with those of measurable size give values up to two. Beyond these simple facts the properties of mean sensitivity remain largely uninvestigated.

High-speed computers and advanced statistical techniques have enabled dendrochronologists to reconstruct climatic data and hydrologic events as far back as 500 to 1000 years. Such reconstructions have significantly increased our knowledge of the seasonal to century-long variations in past climates. These include the occurrence, duration, and severity of drought periods and major fluctuations in annual streamflow for some important river basins in the western United States.

Large arrays of tree-ring chronologies can be used by averaging the climatic data by regions, or by using principal components* to reduce the number of chronologies to a more tractable number of orthogonal components. Regression analysis* is used on the principal components in two ways. First, when climate is the statistical predictor of tree growth and the coefficients interpreted as to which climatic factor affects growth, they are termed *response functions*. In contrast, when tree growth is the statistical predictor, they are termed *transfer functions** and transform growth data into estimates of climate (reconstructions).

A model of the climatic response includes the effect of climate on growth for the year t; for the year $t + 1$ and for the year $t + k$. Such influences can occur in any season of the year prior to the cessation of growth in the summer, and can extend up to a period of 14 or more months prior to actual ring growth. This results in a significant moving average in the ring-width response. In addition, the growth in year t is also related to growth in the previous year, $t - 1$ through autocorrelated phenomena. Climatic factors can also be directly related to growth in one season but inversely related in other seasons, such as summer. Multiple linear regression* or canonical regression (*see* CANONICAL ANALYSIS) can be used to statistically predict ring width from the amplitude of the eigenvector principal components* of precipitation and temperature. Similarly, climate can be predicted from the eigenvector principal components of growth together with other variables of prior growth representing the autoregressive persistence within the ring-width series.

When the complicated tree growth response is properly modeled, one is able to reconstruct both large-scale, long-term variation as well as certain small-scale and seasonal variations in climate. Since many reconstructions emphasize long-term climatic variations, low-pass filters are often applied to the tree-ring indices or to the climatic reconstructions themselves.

The development of the climate–tree growth relationships, used in the various reconstructions, requires an overlapping period between the ring-width series and instrumented meteorological records. A portion of the overlapping record is used to develop the statistical relationship between ring widths and climatic factors. Once developed, the model is tested and verified against the remaining part of the instrumented data.

Reconstructed climatic histories are in themselves weakly stationary time series*. Quasi-periodicities, patterns, and trends in those series can be investigated by applying well-known techniques of time-series analysis. Power and cross-spectral analysis* have, for instance, been used to investigate relationships between reconstructions of drought severity and variations in the double sunspot record. Autoregressive-integrated moving-average* modeling techniques have been used in Monte Carlo simulations* of drought series.

Currently available tree-ring series uniquely provide an accurately dated time series that extends backward in time from a few hundred to about 8500 years B. P. (before present). These chronologies are useful outside the generally established interests of dendrochronologists. For instance, several investigators have used tree-ring data to examine various aspects of the Hurst statistic (or Hurst coefficient*). This statistic, developed by Hurst [1] as a measure of long-term fluctuations in geophysical time series, has been the subject of considerable controversy, especially in the field of hydrology. Others have used tree rings to improve estimates and develop long-term model parameters for autoregressive moving-average models*.

The Hurst coefficient itself has motivated much of Mandelbrot's theory of self-affine and self-similar processes* (see Mandelbrot [2]). It seems likely that the study of tree-ring reconstructions might well suggest new applications for the theory of such processes.

Bibliography

Dean, J. S. (1968). Tree-Ring Dating in Archaeology. *Misc. Archaelogical Paper No.* 99, University of Utah, Logan, Utah. (Provides good background material on dendrochronology with applications to archaeology.)

Dewitt, E. and Ames, M., eds. (1978). Tree-Ring Chronologies of Eastern North America. *Lab. Tree-Ring Res. Chronol. Ser. IV*, **1**. University of Arizona, Tucson, Ariz. (Listing of tree-ring indices for sample trees on each of 39 sites in the eastern United States. Also provides standard deviation, serial correlation, and mean sensitivity for each site chronology.)

Douglass, A. E. (1928). Climatic Cycles and Tree Growth. *Carnegie Inst. Wash. Publ.* 289, II. (A key historical reference by the "father" of dendrochronology in the United States and founder of the Laboratory of Tree-Ring Research at the University of Arizona.)

Drew, L. G. (1972, 1974–1976). Tree-Ring Chronologies of Western North America. *Lab. Tree-Ring Res. Chronol. Ser. I*, **2–6**; *Ser. II*, **1**. University of Arizona, Tucson, Ariz. (Provides the same information for western U.S. sites as listed under reference by Dewitt and Ames. All sites listed are considered adequate for dendrochronologic analyses.)

Ferguson, C. W. (1970). In *Radiocarbon Variation and Absolute Chronology*, I. U. Olsson, ed. Nobel Symp., Vol. 12. Almqvist & Wiksell, Stockholm/Wiley, New York. (A lucid description of the longest chronology yet developed and its use in checking radiocarbon dating procedures. Note: This chronology has been further lengthened to 8,580 years or to 6600 B.C.)

Fritts, H. C. (1976). *Tree Rings and Climate*. Academic Press, London. (A comprehensive treatment of dendrochronology. Detailed discussions with many numerical examples of tree-growth processes, model development, and statistical procedures used in preparing and processing data for climatic reconstructions. A key reference.)

LaMarche, V. C., Jr. (1978). *Nature* (*Lond.*), **276**, 334–338. (A good summary of tree-ring and climatic relationships.)

Schulman, E. (1956). *Dendroclimatic Changes in Semiarid America*. University of Arizona Press, Tucson, Ariz. (A pioneering effort in the application of tree-ring data to hydrologic reconstructions in the precomputer era.)

Stockton, C. W. (1975). *Long-Term Streamflow Records Reconstructed from Tree Rings*. University of Arizona Press, Tucson, Ariz. (A key reference emphasizing the application of tree-ring data to hydrologic problems; good discussion of statistical methods.)

Stokes, M. A. and Smiley, T. L. (1968). *An Introduction to Tree-Ring Dating*. University of Chicago Press, Chicago. (An excellent general treatment of principles, field methods, and laboratory techniques in dendrochronology. A key reference.)

References

[1] Hurst, H. E. (1951). *Trans. Amer. Soc. Civil Eng.*, **116**, 770.

[2] Mandelbrot, B. B. (1977). *Fractals: Form, Chance, and Dimension*. W. H. Freeman, San Francisco.

(AUTOREGRESSIVE-INTEGRATED
  MOVING AVERAGE MODELS
AUTOREGRESSIVE MOVING
  AVERAGE MODELS
FRACTALS
HURST COEFFICIENT
HYDROLOGY, STOCHASTIC)

WILLIAM R. BOGGESS

DENDROGRAMS

The term "dendrogram" (Gk. *dendron,* a tree) is used in numerical taxonomy for any graphical drawing or diagram giving a tree-like description of a taxonomic system. More generally, a dendrogram is a diagram representing a tree of relationships, whatever their nature.

Some of the earliest examples of dendrograms are the customary phylogenetic trees used by systematists. It seems that the term dendrogram was first used by Mayr et al. [6]. (See also Sneath and Sokal [8, p. 58].)

Depending on the nature of relationships described by the diagram, the term "dendrogram" is sometimes replaced by another, such as phenogram or cladogram. The former is used for a dendrogram representing phenetic, and the latter for that representing cladistic relationships (see Mayr [5] and Camin and Sokal [1]).

The representation of a taxonomic system by a dendrogram is particularly suitable in connection with a cluster analysis applied to investigate the structure of the corresponding operational taxonomic units (OTUs), i.e., the lowest-ranking taxa within the system. This becomes apparent when it is desirable to interpret the results of the analysis in terms of a natural nonoverlapping taxonomic hierarchy.

There are various ways of drawing a tree diagram. The early practice of drawing dendrograms tended to have the branches of the treelike diagram pointing upward or downward. But later, with ever-increasing numbers of OTUs, it has become more conve-

nient to place the dendrograms, and particularly the phenograms, almost uniformly on their side, with branches running horizontally across the page. The abscissa is then scaled in the resemblance measure on which the clustering has been based, and the points of furcation between stems along the scale imply that the resemblance between two stems is at the similarity coefficient value shown on the abscissa. It is customary to place code numbers or names of the OTUs to the right of the tips of the dendrogram. It should be realized, however, that the order in which the branches of a dendrogram are presented has no special significance and can be changed within wide limits without actually changing the taxonomic relationships implied by the dendrogram. The multiplicity of ways in which the same relationships can be represented in a dendrogram may be regarded as a disadvantage. Several methods have been suggested to overcome this.

Examples of dendrograms and methods of their presentation are described in many textbooks on numerical taxonomy (see, particularly, Sokal and Sneath [9, pp. 197–201] and Sneath and Sokal [8, pp. 207–295]).

Two different clustering methods may lead to different dendrogram representations of the results, even if both methods are based on the same resemblance measure. Among the various clustering methods, one is particularly relevant to a dendrogram representation. It is the single linkage cluster analysis (SLCA), also known as the nearest-neighbor technique, introduced by Florek et al. [2, 3] and Sneath [7]. As shown by Gower and Ross [4], the most efficient procedure for the SLCA is based on producing the shortest dendrite (the minimum spanning tree). The shortest dendrite itself also gives an alternative graphical representation of the SLCA results. This may appear more convenient than the usual application of a dendrogram, particularly when superimposing the dendrite on the OTUs scattered in an ordination plot of two or three dimensions obtained, e.g., from the principal component analysis (see Sneath and Sokal [8, p. 257]; *see also* DENDRITES).

References

[1] Camin, J. H. and Sokal, R. R. (1965). *Evolution*, **19**, 311–326.

[2] Florek, K., Łukaszewicz, J., Perkal, J., Steinhaus, H., and Zubrzycki, S. (1951). *Colloq. Math.*, **2**, 282–285.

[3] Florek, K., Łukaszewicz, J., Perkal, J., Steinhaus, H., and Zubrzycki, S. (1951). *Przegląd Antropol.*, **17**, 193–211.

[4] Gower, J. C. and Ross, G. J. S. (1969). *Appl. Statist.*, **18**, 54–64.

[5] Mayr, E. (1965). *Syst. Zool.*, **14**, 73–97.

[6] Mayr, E., Linsley, E. G. and Usinger, R. L. (1953). *Methods and Principles of Systematic Zoology*. McGraw-Hill, New York.

[7] Sneath, P. H. A. (1957). *J. Gen. Microbiol.*, **17**, 201–226.

[8] Sneath, P. H. A. and Sokal, R. R. (1973). *Numerical Taxonomy*, W. H. Freeman, San Francisco.

[9] Sokal, R. R. and Sneath, P. H. A. (1963). *Principles of Numerical Taxonomy*, W. H. Freeman, San Francisco.

(DENDRITES)

T. CALIŃSKI

DENSITY ESTIMATION

Since the mid-1950s, density estimation has been an extremely popular though somewhat controversial subject. At least part of the controversy stems from the fact that a density may be specified only up to a set of measure zero; hence it is, in effect, meaningless to estimate a density at a point. Even admitting sufficient regularity conditions so that the density in question is uniquely specified, many argue that the density carries no more information about the probability structure of a sample than the distribution. Nevertheless, there appear to be situations other than estimating probabilities in which knowledge of the density function is necessary. Among these are empirical Bayes* procedures, robust procedures (*see* ROBUST-

NESS), and cluster analysis. Moreover, densities are much more easily interpreted than distributions, so that there is also the aesthetic appeal of a density as motivation.

Various parametric approaches to density estimation have existed since as early as 1890. All these schemes involve systems of distributions or frequency curves which are intended to represent as wide a variety of observed density as is possible. Parameters are estimated and the resulting density is taken as the estimated density. Perhaps the best known system is the Pearson system* originated by Pearson between 1890 and 1900. A Pearson density satisfies a differential equation of the form

$$\frac{df(x)}{dx} = -\frac{(x-a)f(x)}{b_0 + b_1 x + b_2 x^2}.$$

If f is unimodal, the mode is a and f has smooth contact with the x-axis as f tends to 0. A thorough treatment of this system is given in Elderton and Johnson [12] or in Johnson and Kotz [16]. Other approaches include systems based on expansions of densities in orthogonal series (Gram–Charlier)*, based on translations (Johnson)*, and based on representations of the distribution (Burr)*. See Johnson and Kotz for a treatment of all of these.

Receiving far more attention since 1956 have been the nonparametric approaches. See Cover [8], Rosenblatt [25], and Wegman [33] for review papers and Tapia and Thompson [29] for a review monograph on the subject. One may distinguish four main traditions of nonparametric density estimation:

1. Kernel methods*
2. Orthogonal* series methods
3. Maximum likelihood* methods
4. Spline* methods

The smoothing kernel approach has been the most thoroughly developed theoretically and has an extensive literature. If we let $X_1, \ldots, X_n$ be a sequence of independent, identically distributed (i.i.d.) random variables with probability density f, the general kernel estimate has the form

$$\hat{f}_n(x) = \int_{-\infty}^{\infty} K_n(x, y)\, dF_n(y)$$

$$= \frac{1}{n} \sum_{i=1}^{n} K_n(x, X_i). \qquad (1)$$

Here F_n is the empirical distribution function based on the first n observations.

The idea of these estimates is the following. The empirical distribution function is a discrete distribution with mass $1/n$ placed at each of the observations. The formula (1) smears this probability out continuously, smoothing according to the choice of $K_n(x, y)$. Thus the choice of $K_n(x, y)$ is very important and to a large extent determines the properties of $\hat{f}_n(x)$. The first published work on estimates of this type was that of Rosenblatt [24]. Rosenblatt considers a "naive estimator,"

$$\hat{f}_n(x) = \frac{F_n(x + h) - F_n(x - h)}{2h}.$$

This estimate is the special case of (1) when $K_n(x, y)$ is $1/2h$ for $|x - y| \leqslant h$ and 0 elsewhere. Of course, no assumptions are necessary on f to form this estimate. However, Rosenblatt goes on to consider asymptotic mean square error* (MSE) as well as some other asymptotic properties. In general, the problem is to choose the sequence of $h = h(n)$ converging to zero at an appropriate rate. If $h = kn^{-\alpha}, \alpha > 0$, the choice of α minimizing MSE is $\frac{1}{5}$ and the optimum value of k is

$$\frac{9}{2} \frac{f(x)}{|f''(x)|^2}.$$

Since we are attempting to estimate f, it is unlikely that we will know enough to choose optimum k. Nonetheless, with a satisfactory choice of the constant $k > 0$, we should still have either pointwise consistency in quadratic mean or integrated consistency in quadratic mean. Rosenblatt also proposes estimates with $K_n(x, y) = [1/h(n)]K((x - y)/h(n))$, where K is a nonnegative function

such that

$$\int_{-\infty}^{\infty} K(u)\, du = 1$$

$$\int_{-\infty}^{\infty} [K(u)]^2\, du < \infty$$

$$\int_{-\infty}^{\infty} [K(u)]|u|^3\, du < \infty$$

$K(u)$ is symmetric about 0.

Under these conditions and the condition that f have derivatives of the first three orders, the optimum choice of h leads to mean square error* $E(\hat{f}_n(x) - f(x))^2$ no smaller than $O(n^{-4/5})$.

Parzen [19] considers these estimates and requires that the nonnegative, even, Borel function $K(\cdot)$ satisfy the foregoing conditions and

$$\sup_{-\infty < x < \infty} |K(x)| < \infty$$

$$\int_{-\infty}^{\infty} |K(x)|\, dx < \infty$$

$$\lim_{|x| \to \infty} |xK(x)| = 0.$$

He finds expressions for asymptotic variance, bias, and MSE. With added conditions, Parzen shows that the sequence of estimates is asymptotically normal. While Parzen obtained most of the weak convergence* results, additional strong consistency results were obtained by Nadaraya [18] and Van Ryzin [30], as well as others. Kernel estimates have been generalized in many directions. Bhattacharyya [2] and Schuster [26] give necessary and sufficient conditions for strong consistency. Whittle [35] gives a Bayesian approach. Epanechnikov [13] finds a choice of the kernel, K, that is optimal but shows that this choice is not too critical. The choice of $h(n)$, however, is critical and is addressed by Woodroofe [37] and Silverman [28]. Other interesting variations include the recursive/sequential approach of Wolverton and Wagner [36], Yamato [38], Davies [9], Wegman and Davies [34] and Deheuvels [10].

Estimates of the form (1) have received a majority of the attention in the literature. An alternative approach is to represent the density by means of an orthogonal series in the following formulation. Let X be a random variable and R the real line. Let f be the density function of X. Let r be some arbitrary but fixed function and define an inner product by

$$(g, h) = \int_R g(x)h(x)r(x)\,dx.$$

The set L of functions g such that $\int_R g^2(x)r(x)\,dx$ is finite, together with the inner product, form a Hilbert space. In the special case $r(x) = 1$, we have the ordinary set of square integrable functions.

Let us further consider a subspace E spanned by the orthonormal basis* $\{g_k\}, k \in I$, where I is some index set. Finally, let us consider the projection of the density f onto E. This is given by

$$f^*(x) = \sum_{k \in I} (g_k, f)g_k(x),$$

where $(g_k, f) = \int_R g_k(x)f(x)r(x)\,dx$. Let $a_k = (g_k, f)$. To estimate f^* (hence f), one must estimate a_k, $k \in I$. In general, of course, if $X_1, \ldots, X_n$ is a set of observations,

$$\hat{a}_k = (1/n) \sum_{i=1}^n g_k(X_j)r(X_j)$$

is a strongly consistent estimate of a_k. Thus

$$\hat{f}_n(x) = \sum_{k \in I} \hat{a}_k g_k(x) \tag{2}$$

is an estimate of the probability density f. This general description is due largely to Cencov [6]. Cencov considers cases where the index set I is finite and chooses square error as error criterion. By choosing a sufficiently good approximating space, Cencov argues that a good estimate can be made. Of course, if the set I is fixed, no consistency results can be obtained. In fact, if the g_k are chosen as indicators of disjoint intervals, this estimator is essentially a histogram*.

More interesting cases exist when the $g_k(x)$ are chosen as an infinite orthonormal series. Schwartz [27] considers the sequence of normalized Hermite functions

$$g_k(x) = (2^k k \pi^{1/2})^{-1/2} e^{-x^2/2} H_k(x),$$
$$k = 0, 1, \ldots,$$

where (*see* CHEBYSHEV-HERMITE POLYNOMIALS)

$$H_k(x) = (-1)^k e^{x^2} (d^k/dx^k)(e^{-x^2}).$$

If $r(x) = 1$, and the density f is continuous, bounded variation, absolutely and square integrable, then $f(x) = \sum_{i=0}^\infty a_i g_i(x)$, where $a_i = \int f(x)g_i(x)\,dx$. Schwartz considers as an estimate of the density,

$$\hat{f}_n(x) = \sum_{i=0}^{q(n)} \hat{a}_i g_i(x),$$

where $a_j = (1/n)\sum_{i=1}^n g_j(X_i)$. If $q(n)$ is chosen so that $q(n) \to 0$ as $n \to \infty$, the estimate is integratedly consistent in quadratic mean. In addition, if $q^2(n)/n$ converges to zero as n diverges to infinity, then the estimate is uniformly consistent in quadratic mean. Schwartz gives convergence rates. We mention here that the Schwartz approach is closely connected to the Gram-Charlier parametric system.

Kronmal and Tarter [17] consider an estimate based on trigonometric functions rather than the Hermite function as in Schwartz. The form of their estimate is (2), where $\{g_k\}$ are chosen as one of the orthogonal systems $\{\sin \pi kx\}$, or $\{\cos \pi kx\}$ or $\{e^{i\pi kx}\}$. Kronmal and Tarter show that if we have a density f, with finite support $\{x : f(x) > 0\}$, and if the density f may be represented by a Fourier cosine series, then for a choice of $q(n) = O(\sqrt{n})$, the mean square error (MSE) and the mean integrated square error* (MISE) converge to zero. An interesting point of contact with estimates of form (1) exists. One may express the Fourier cosine estimates as estimates of the form (1) where the weighting function is chosen as $K(y) = \frac{1}{2}\pi[\sin(y/2)/(y/2)]^2$. Kronmal and Tarter discuss quite thoroughly the choice of $q(n)$. They conclude that the optimal number of terms, $q(n)$, depends on $(b - a)/B$ and "the nature of the distribution which one wishes to estimate." Here a, b, and B are, respectively, the lower and upper limits on the range on which one is estimating the density, and a scale parameter for the density of interest. They do give "stopping rules" in several cases based on the fact that when too many terms are included in the series, the MISE begins to increase. Expressions are given for the MISE which are free of the density function f. At least part of the problem with orthogonal series estimators is that the choice of $q(n)$, in analogy to the choice of $h(n)$ for the kernel estimators, is fairly critical. Very interesting recent work on this problem has been done by Brunk [4], who espouses a Bayesian approach to the estimation of $q(n)$.

A third rather different approach is the maximum likelihood* approach. In principle, if $\hat{f}$ is the density based on $X_1, \ldots, X_n$, the sample, one wants to choose $\hat{f}_n$ so that it maximizes

$$L(\hat{f}) = \prod_{i=1}^{n} \hat{f}(X_i).$$

However, this cannot be done nonparametrically without additional constraints on $\hat{f}$. This follows since by merely putting a spike of arbitrarily large magnitude and correspondingly narrow width at each observation, one can create a likelihood that is as large as one would like. The problem then is to find a class of candidate estimators that are suitably restricted. One restriction considered is that the estimates are measurable with respect to a σ-lattice. A σ-lattice L is a collection of subsets (of R) which is closed under countable unions and intersections and contains both the whole set if $\{x : f(x) > a\} \in L$ for all real a. Under nominal restrictions, the set of densities that are measurable with respect to a σ-lattice will form a closed convex cone in L_2. Hence just as we projected the density onto the subspace spanned by the orthogonal functions, we may also project the density onto the cone. Robertson [23] considers estimates that are measurable with respect to a σ-lattice L. An interesting case, considered by Robertson, is the case of a unimodal density* with known mode. A unimodal density with known mode may be characterized as measurable with respect to the σ-lattice L of intervals containing the mode. The estimate

of Robertson has the form

$$\hat{f}_n(x) = \sum_{i=1}^{k} \frac{n_i}{n} \cdot \frac{I_{A_i}(x)}{\lambda(A_i)}.$$

Here A_i is an interval determined from the lattice L and the particular set of observations, $X_1, \ldots, X_n$. The function I_{A_i} is the indicator of A_i, and n_i is the number of observations in A_i. Notice that by letting $g_i = I_{A_i}/\lambda(A_i)$, we may define an orthonormal basis and n_i/n is an estimate of $P(A_i)$, which is an orthogonal series estimator. Robertson shows pointwise consistency with probability 1, and under the assumption of continuity of f, he shows almost uniform consistency with probability 1. (Grenander [15] first considered estimating monotone densities in connection with mortality studies*.)

Prakasa Rao [22] considers the same unimodal case as Robertson. He derives, rather elegantly but tediously, the asymptotic distribution theory. In practice, the mode is rarely known. Wegman [32] considers maximum likelihood estimation of a unimodal density with unknown mode and obtains consistency and distribution results similar to Robertson and Prakasa Rao.

An alternative to restricting the class of candidate estimators is to penalize an estimator that is too "rough," an approach pioneered by Good and Gaskins [14]. One considers in general a manifold H contained in L_2 (or L_1) and a real-valued functional on H. The ϕ-penalized likelihood is

$$L(\hat{f}) = \prod_{i=1}^{n} \hat{f}(x_i) \exp\left[-\phi(\hat{f})\right].$$

Depending on the choice of H and ϕ, a variety of estimators can be generated in this form. Good and Gaskins offer two choices of ϕ, one of the form

$$\phi(f) = \alpha \int_{-\infty}^{\infty} \frac{f'(t)^2}{f(t)} \, dt, \qquad \alpha > 0,$$

and the other

$$\phi(f) = \alpha \int_{-\infty}^{\infty} f'(t)^2 \, dt + \beta \int_{-\infty}^{\infty} f''(t)^2 \, dt,$$

$$\alpha \geqslant 0, \beta > 0.$$

A considerable effort is spent showing the existence and uniqueness of such estimators. Later, de Montrichier et al. [11] show that if H is chosen as a suitable Sobolev space* and ϕ to correspond to an appropriate inner product, then the maximum penalized likelihood estimator becomes simply a polynomial spline, which brings us to the last major tradition, the spline approach.

The inaugural paper in the use of splines to estimate densities appears to be that of Boneva et al. [3]. Their histosplines are empirical densities in the nature of a smooth analog of a histogram, with pleasant mathematical features. Perhaps the most pleasant of these is the fact that, with suitable regularity conditions, the empirical spline-fitted density is the derivative of the spline-fitted distribution. Thus there is a natural relationship between distribution and density. To make these analysis feasible, the authors allow the possibility of negative densities. The work of Boneva et al. also contains considerable empirical material on histospline behavior. Perhaps the most interesting work on the spline approach, aside from the above-mentioned work of de Montrichier et al. [11], is that of Wahba [31]. She defines a variant of the histospline and shows that the expected mean square error at a point has the same order of magnitude for several estimators, including Parzen's kernel estimates, Wahba's spline, Kronmal–Tarter's orthogonal series, and the ordinary histogram. Several other pieces of work on the spline approach have been excellently documented in Tapia and Thompson [29].

It is worthwhile to point out some connections between these methods. We have already alluded to the connection between kernel estimates and the Kronmal–Tarter estimates. Similarly, we have pointed out the connection of the maximum likelihood estimates and the orthogonal series. In fact, this connection is much more profound. Cencov [6] describes a subspace of L_2 spanned by a finite orthonormal basis. The estimator becomes a projection onto this subspace using the usual L_2 norm. For the maximum likelihood estimates measurable with respect to a σ-lattice, the subspace is replaced by a closed convex cone and the MLE is again a projection with respect to the usual L_2 norm. For the spline estimators (and some of the penalized likelihood estimators), the subspace is a very special kind of Sobolev subspace of smooth functions and the norm is typically modified to something based on $\int (f''(x))^2 \, dx$, but once again the spline estimator is a projection onto a subspace. Thus the orthogonal series, maximum likelihood, and spline estimators share a very close conceptual framework. Still another connection exists. Cogburn and Davis [7] show that the spline estimate (of spectral densities) is in effect a kernel estimate with the kernel chosen in some optimal sense. Thus there is a close but largely unexplored connection between all these estimators.

We mentioned only tangentially in the preceding paragraph the problem of spectral density* estimation. Again there is a vast literature in this area which we shall not attempt to address fully. Kernel estimation procedures exist for spectral densities and are summarized by Anderson [1]. The development of kernel density and kernel spectral density estimates took place in parallel over the last 15 to 20 years and the theory of spectral densities is quite similar. A more recent approach to spectral density estimation is a parametric approach (parametric in the sense of a parametric model of the time series), which is called the autoregressive (AR) approach by statisticians and the maximum entropy method (MEM) by electrical engineers. Essentially, in this approach, an autoregressive scheme is fit to the time series in question; then parameters are estimated and plugged into the formula for an autoregressive spectral density*. In general, one can comment that the AR-MEM estimator seems very good at identifying the discrete components (lines, resonances) of the spectrum, while the classical kernel–spline approach seems superior for the absolutely continuous part. In general qualitative terms, the AR-MEM appears to be superior on short data sets, the kernel–spline method stronger on large data sets. The AR approach was developed by Parzen [20]; the MEM appears to have been developed by Burg [5] independently.

Finally, we note that Parzen has developed a rather extensive inference system based on quantile* density estimation. This scheme, which is described elsewhere (*see* QUANTILES), is based in part on the AR approach described above.

References

[1] Anderson, T. W. (1971). *The Statistical Analysis of Time Series*. Wiley, New York. (Contains a very complete treatment of a kernel approach to spectral density estimation.)

[2] Bhattacharyya, P. K. (1967). *Sankhyā A*, **29**, 373–383. (Addresses the questions of necessary and sufficient conditions for consistency and also introduces the idea of estimating the derivatives of a density. Related to Schuster [26]).

[3] Boneva, L., Kendall, D., and Stefanov, I. (1971). *J. R. Statist. Soc. B*, **33**, 1–70. (One of the earliest papers on the use of splines or density estimation. Suggests the use of interpolating splines to construct estimators called histosplines based on histograms.)

[4] Brunk, H. D. (1978). *Biometrika*, **65**, 521–528. (A Bayesian approach to estimating the upper limit of summation in an orthogonal series approach. Other Bayesian approach by Whittle [35]).

[5] Burg, J. P. (1975). Maximum Entropy Spectral Analysis. Ph.D. dissertation, Dept. of Geophysics, Stanford University. (Suggests the method now known as the maximum entropy method for estimating spectral densities. Burg's idea was first set forth in a paper entitled "Maximum entropy power spectral analysis," presented at the 37th Annual Int. SEG Meeting, Oklahoma City, OK, Oct. 31, 1967. It is closely related to Parzen [20].)

[6] Cencov, N. N. (1962). *Sov. Math.*, **3**, 1559–1562. (Suggests the orthogonal series method for density estimation but with a finite basis.)

[7] Cogburn, R. and Davis, H. T. (1974). *Ann. Statist.*, **2**, 1108–1126. (A general treatment of periodic splines with an application to spectral density estimation. The authors show that the spline estimate can be regarded as a kernel estimator.)

[8] Cover, T. (1972). In *Frontiers of Pattern Recognition*, S. Watanabe, ed. Academic Press, New York. (An expository paper on density estimation from a more engineering/pattern recognition point of view. A good treatment.)

[9] Davies, H. I. (1973). *Bull. Math. Statist.*, **15**, 49–54. (Establishes strong consistency of Yamato's estimator. Has priority over Deheuvels [10].)

[10] Deheuvels, P. (1974). *C. R. Acad. Sci. Paris A*, **278**, 1217–1220. (Establishes strong consistency properties for a class of recursive probability density estimators. In French. Priority is to Davies [9].)

[11] de Montrichier, G. M., Tapia, R. A., and Thompson, J. R. (1975). *Ann. Statist.*, **3**, 1319–1348. (A sophisticated mathematical treatment of the penalized likelihood approach which discusses the relationship of this approach to smoothing with splines.)

[12] Elderton, W. P., and Johnson, N. L. (1969). *Systems of Frequency Curves*. Cambridge University Press, Cambridge. (The definitive work on systems of frequency curves including the Pearson system and the Johnson translation system.)

[13] Epanechnikov, V. A. (1969). *Theory Prob. Appl.*, **14**, 153–158. (Establishes optimal kernels among a limited class for density estimation.)

[14] Good, I. J. and Gaskins, R. A. (1971). *Biometrika*, **58**, 255–277. (Establishes the concept of penalized likelihoods for density estimation. Related to de Montrichier et al. [11].)

[15] Grenander, U. (1956). *Skand. Aktuarietidskr.*, **39**, 125–153. (Establishes the isotonic approach to density estimation. See also Robertson [23] and Wegman [32].)

[16] Johnson, N. L. and Kotz, S. (1969–1972). *Distributions in Statistics*, Vols. 1–4. Wiley, New York. (An encyclopedic work on distributions, including the Pearson, Johnson, and Burr families. An important reference work for any statistician. The mathematical level is intermediate.)

[17] Kronmal, R. and Tarter, M. (1968). *J. Amer. Statist. Ass.*, **63**, 925–952. (Establishes the orthogonal series approach to density estimation based on Fourier series.)

[18] Nadaraya, E. A. (1965). *Theory Prob. Appl.*, **10**, 186–190. (First results on strong consistency for kernel estimators. Preceded in 1963 by a Russian-language version.)

[19] Parzen, E. (1962). *Ann. Math. Statist.*, **33**, 1065–1076. (The classic work on kernel estimators. It establishes most of the important weak convergence results. There have been many derivative papers with minor variations.)

[20] Parzen, E. (1969). In *Multivariate Analysis II*, P. R. Krishnaiah, ed. Academic Press, New York. (The paper that establishes the AR approach to spectral density estimation. See also Burg [5]. Extremely important work in time series.)

[21] Pearson, K. (1895, 1901, 1916). *Philos. Trans. R. Soc. Lond. A*, **186**, 343–414, **197**, 443–459, **216**, 429–457. (Earliest work on Personian curves. See Elderton and Johnson [12] for fuller details.)

[22] Prakasa Rao, B. L. S. (1969). *Sankhyā A*, **31**, 26–36. (Establishes the asymptotic distribution for isotonic (unimodal) estimators with fixed mode. See also Robertson [23].)

[23] Robertson, T. (1967). *Ann. Math. Statist.*, **38**, 482–493. (Establishes the form of maximum likelihood isotonic estimators as well as some of the asymptotic convergence properties. Related to Prakasa Rao [22], Grenander [15], and Wegman [32].)

[24] Rosenblatt, M. (1956). *Ann. Math. Statist.*, **27**, 832–837. (The first published work on kernel estimators. This paper establishes most of the weak convergence results for the so-called "naive" kernel estimator.)

[25] Rosenblatt, M. (1971). *Ann. Math. Statist.*, **42**, 1815–1842. (An excellent expository on the general problem of curve estimation, including density estimation.)

[26] Schuster, E. F. (1969). *Ann. Math. Statist.*, **40**, 1187–1195. (Gives necessary and sufficient conditions for mean square consistency. Closely related to Bhattacharyya [2].)

[27] Schwartz, S. C. (1967). *Ann. Math. Statist.*, **38**, 1261–1265. (Takes the orthogonal series approach using Hermitian polynomials.)

[28] Silverman, B. (1978). *Biometrika*, **65**, 1–12. (Contains a heuristic approach to estimating *h* in the kernel estimator. An alternative to Woodroofe [37].)

[29] Tapia, R. A. and Thompson, J. R. (1978). *Nonparametric Probability Density Estimation*. Johns Hopkins University Press, Baltimore, Md. (An excellent expository book on the penalized likelihood approach to density estimation. Somewhat weaker on other types of estimators. Intermediate mathematical level.)

[30] Van Ryzin, J. (1969). *Ann. Math. Statist.*, **40**, 1765–1772. (Contains early strong consistency results. Alternative conditions to those of Nadaraya [18].)

[31] Wahba, G. (1975). *Ann. Statist.*, **3**, 15–29. (Representative of a number of excellent papers exploring the spline approach to density estimation. This one compares convergence rates for a variety of estimators and concludes they are essentially identical.)

[32] Wegman, E. J. (1970). *Ann. Math. Statist.*, **41**, 457–471. (Representative of a series of papers using the isotonic approach to arrive at a maximum likelihood unimodal estimator. Related closely to work of Robertson and Prakasa Rao [22], this particular paper establishes the maximum likelihood property and strong consistency.)

[33] Wegman, E. J. (1972). *Technometrics*, **14**, 533–546; *J. Statist. Comp. Simul.*, **1**, 225–245. (Gives an expository overview of density estimation circa 1972. Has an extensive pre-1972 bibliography and a Monte Carlo study of a variety of estimators.)

[34] Wegman, E. J. and Davies. H. I. (1979). *Ann. Statist.*, **7**, 316–327. (Gives the law of the iterated logarithm for the recursive density estimates of Wolverton and Wagner and Yamato. Also suggests the sequential approach to density estimation.)

[35] Whittle, P. (1958). *J. R. Statist. Soc. B*, **20**, 334–343. (One of the early papers on density estimations giving a Bayesian approach to kernel estimates. See also Brunk [4].)

[36] Wolverton, C. T. and Wagner, T. J. (1969). *IEEE, Trans. Inf. Theory*, **IT-15**, 258–265. (Contains the first work on recursive density estimators apparently independent of Yamato [38].)

[37] Woodroofe, M. (1970). *Ann. Math. Statist.*, **41**, 1665–1671. (One of the early approaches to estimating *h* in the kernel estimators. It is a two-step procedure.)

[38] Yamato, H. (1971). *Bull. Math. Statist.*, **14**, 1–12. (Introduces independently of Wolverton and Wagner the recursive density estimator, but erroneously calls it a sequential estimator. This paper, following the model of Parzen [19], establishes most of the weak convergence results.)

(ESTIMATION, POINT
ISOTONIC INFERENCE
KERNEL ESTIMATORS
SPECTRAL ANALYSIS)

E. J. Wegman

DEPARTURES FROM NORMALITY, TESTS FOR

The normal distribution* first appeared in 1733 in the works of De Moivre* dealing with the large-sample properties of the binomial distribution. This aspect of De Moivre's work appears to have gone unnoticed and the normal distribution was rediscovered in 1809 by Gauss* and in 1812 by Laplace*. Laplace actually touched upon the subject around 1789, but it was not until 1812 that he presented an in-depth treatment of it. Both Gauss and Laplace developed the normal distribution in connection with their work on the theory of errors of observations. The works of Gauss and Laplace were extremely influential and for a long time it was assumed that frequency distributions of most measurable quantities would be normally distributed if a sufficient number of accurate observations were removed.

Although empirical investigations often did produce distributions that were normal or at least approximately so, it was obvious that the appropriate distributions for many phenomena were not normal. In response to this the techniques for judging or testing for departures from normality came into being.

FIRST TESTS FOR DEPARTURES FROM NORMALITY

Moment Tests: $\sqrt{b_1}, b_2$

The modern theory of testing for departures from normality can be regarded as having been initiated by Karl Pearson [30], who recognized that deviations from normality could be characterized by the standard third and fourth moments of a distribution. To be more explicit, the normal distribution with density

$$f(x; \mu, \sigma) = \frac{1}{\sqrt{2\pi}\,\sigma} e^{-(1/2)[(x-\mu)/\sigma]^2};$$

$$-\infty < x < \infty,$$
$$-\infty < \mu < \infty, \quad \sigma > 0, \quad (1)$$

has as its standardized third and fourth moments

$$\sqrt{\beta_1} = \frac{\mu_3}{\sigma^3} = \frac{E(X-\mu)^3}{\left[E(X-\mu)^2\right]^{3/2}} = 0 \quad (2)$$

and

$$\beta_2 = \frac{\mu_4}{\sigma^4} = \frac{E(X-\mu)^4}{\left[E(X-\mu)^2\right]^2} = 3. \quad (3)$$

The third standardized moment $\sqrt{\beta_1}$ characterizes the skewness* of a distribution. If a distribution is symmetric about its mean μ, as is the normal distribution, $\sqrt{\beta_1} = 0$. Values of $\sqrt{\beta_1} \neq 0$ indicate skewness and so nonnormality. The fourth standardized moment β_2 characterizes the kurtosis* (or peak-

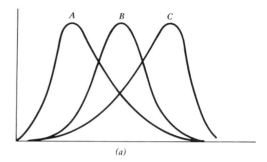

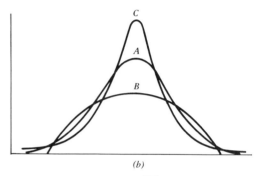

Figure 1 Distributions with $\sqrt{\beta_1} \neq 0$ and $\beta_2 \neq 3$. (a) Distributions differing in slowness: A, $\sqrt{\beta_1} > 0$; B, $\sqrt{\beta_1} = 0$; C, $\sqrt{\beta_1} < 0$. (b) Distributions differing in kurtosis: A, $\beta_2 = 3$; B, $\beta_2 < 3$; C, $\beta_2 > 3$.

edness) of a distribution. For the normal distribution, $\beta_2 = 3$. Values of $\beta_2 \neq 3$ indicate nonnormality. See Fig. 1 for illustrations of distributions with $\sqrt{\beta_1} \neq 0$ and $\beta_2 \neq 3$.

Pearson suggested that in the sample, the standardized third and fourth moments given by

$$\sqrt{b_1} = m_3/m_2^{3/2} \quad (4)$$

and

$$b_2 = m_4/m_2^2, \quad (5)$$

where

$$m_k = \sum(X - \bar{X})^k/n, \quad k > 1, \quad (6)$$

and

$$\bar{X} = \sum X/n. \quad (7)$$

could be used to judge departures from normality. He found the first approximation (i.e., to n^{-1}) to the variances and covariances of $\sqrt{b_1}$ and b_2 for samples drawn at random from any population, and assuming that $\sqrt{b_1}$ and b_2 were distributed jointly with bivariate normal probability, constructed equal probability ellipses*. From these the probability of the same values as observed in a sample occurring, if in fact the underlying population were normal, could be inferred approximately. For situations where $\sqrt{b_1}$ and b_2 deviated substantially from expectation under normality, K. Pearson* developed his elaborate system of Pearson curves* as possible alternative distributions for the populations under investigation (see Elderton and Johnson [14] for a full discus-

sion of Pearson curves; also *see* FREQUENCY CURVES, SYSTEMS OF).

Later, Egon S. Pearson [25, 26] tabulated more precise percentage points of $\sqrt{b_1}$ and b_2, Geary [16, 17] suggested the ratio of the sample mean deviation to the sample standard deviation as a possible alternative test to b_2 and David et al. [13] replaced the mean deviation in this latter ratio by the sample range to produce a general test of nonnormality. Excellent references for these tests are Pearson and Hartley [27, 28]. Table 1 contains numerical examples of these statistics

Table 1   Numerical Examples for Moment and Related Tests (Ten Observations per Sample/Five Samples)

		Distribution			
	Uniform	Johnson Unbounded (0, 2)	Johnson Unbounded (1, 2)	Negative Exponential, Mean 5	Normal: $\mu = 100$, $\sigma = 10$
	8.10	0.10	− 0.41	8.15	92.55
	2.06	− 0.31	− 0.91	4.69	96.20
	1.60	− 0.09	− 0.63	2.17	84.27
	8.87	− 0.58	− 1.25	0.37	90.87
Data[a]	9.90	1.15	0.50	16.69	101.58
	6.58	0.17	− 0.34	0.06	106.82
	8.68	− 1.39	− 2.46	6.48	98.70
	7.31	− 0.14	− 0.68	2.63	113.75
	2.85	− 0.31	− 0.97	0.44	98.98
	6.09	0.68	0.13	0.89	100.42
$\sqrt{\beta_1}$	0	0	− 0.87	2.	0
β_2	1.80	4.51	5.59	9.	3.
$\bar{x}$	6.204	− 0.078	− 0.701	4.257	98.414
s	3.009	0.691	0.807	5.169	8.277
$\sqrt{b_1}$	− 0.49	− 0.04	− 0.73	1.49[b]	0.16
b_2	1.75	3.06	3.57	4.33[b]	2.76
a	0.86	0.74	0.73	0.77	0.76
u	2.76	3.68	3.67	3.22	3.56

[a] Note that

$$\sqrt{b_1} = \frac{m_3}{m_2^{3/2}} \; ; \quad b_2 = \frac{m_4}{m_2^2} \; ; \quad m_k = \frac{\sum (X - \bar{X})^k}{n} \; ;$$

$$a = \text{Geary's statistic} = \frac{\sum |X - \bar{X}|/n}{m_2^{1/2}} \; ;$$

$$u = \text{David et al. [13] statistic} = (\text{sample range})/s,$$

$$s = [n/(n - 1)]^{1/2} m_2^{1/2}.$$

[b] Reject null hypothesis of normality at 0.10 level of significance.

from simulations for a variety of distributions with different $\sqrt{\beta_1}$ and β_2 values.

Chi-Square Test*

Karl Pearson [31] developed the chi-square test, which can also be used to test for deviations from normality. In this procedure the distribution is discretized or categorized into k categories. If the population from which the data are drawn is normal with a particular mean μ_0 and standard deviation σ_0, then the proportion of the population in the ith category can be computed as

$$p_i = \int_{y_i}^{y_{i+1}} f(x; \mu_0, \sigma_0)\, dx$$
$$\text{for } i = 1, \dots, k, \quad (8)$$

where y_i and y_{i+1} are the boundaries of the ith class ($y_{k+1} = -y_1 = \infty$). Further, for a sample of size n the expected number of observations in each category is

$$e_i = np_i \qquad \text{for } i = 1, \dots, k. \quad (9)$$

Pearson's chi-square statistic contrasts the observed values o_i (i.e., the observed number of observations for the ith category) with the expectations e_i as

$$X^2 = \sum_i \frac{(o_i - e_i)^2}{e_i}. \quad (10)$$

If n is large and the underlying distribution is normal, X^2 of (10) is distributed as a chi-square distribution with $k - 1$ degrees of freedom*. So large values of X^2 indicate nonnormality. If μ_0 and σ_0 were not known Pearson suggested replacing them with estimates based on the data and using $k - 3$ degrees of freedom for the chi-square distribution. Later it was shown that the null distribution of X^2 is chi-square with $k - 3$ degrees of freedom if these estimates are the maximum likelihood estimates* based on the grouped data. If the sample mean and standard deviation based on the ungrouped data are used to replace μ_0 and σ_0, the actual degrees of freedom are between $k - 1$ and $k - 3$ [4]. Use of $k - 3$ produces a conservative test—i.e., the actual level of significance is smaller than the nominal or stated level.

The chi-square test of normality is often presented in textbooks as the preferred test. Actually, it is not very sensitive for this use and should not be used. Other tests described in this entry are more appropriate.

The reader is referred to the entry on the chi-square test for numerical examples.

Empirical Cumulative Distribution Function: The Kolmogorov–Smirnov Test*

Still another general procedure was developed which could be used to test for departures from normality. This was the Kolmogorov–Smirnov test [21, 37]. In this technique the theoretical cumulative distribution function (CDF), $F(x; \mu_0, \sigma_0)$, for the normal distribution defined by

$$F(x; \mu_0, \sigma_0) = \int_{-\infty}^{x} f(y; \mu_0, \sigma_0)\, dy \quad (11)$$

is contrasted with the empirical cumulative distribution function (ECDF), $F_n(x)$,

$$F_n(x) = \frac{\#(\lambda \leqslant x)}{n} \quad (12)$$

in the statistic

$$D = \sup_{-\infty < x < \infty} |F_n(x) - F(x; \mu_0, \sigma_0)|. \quad (13)$$

Large values of D indicate nonnormality. For μ_0 and σ_0 known, critical values of D are readily available [28]. Use of these for μ_0 and/or σ_0 unknown produce very conservative tests. Stephens [38] developed adjustments to the significance values for these latter situations. (*See* KOLMOGOROV–SMIRNOV TESTS for numerical examples.)

Probability Plots*

The techniques described above lead to formal statistical tests in which statistical significance is declared. A graphical procedure, called normal probability plotting or probit* plotting, was developed as an informal technique for judging deviations from normality. The objective of the normal probability plot is to graph the data in such a way that if the underlying population is normally distributed, the graph will be a straight line. The deviation indicates the degree and type of nonnormality. Figures 2 and 3 contain normal plots of simulated data. Figure 2 is a plot of data from a normal distribution. Figure 3 contains four plots of nonnormal data, illustrating how deviations from normality $(\sqrt{\beta_1} \neq 0$ and/or $\beta_2 \neq 3)$ are reflected in deviations of the plots from linearity.

In Figs. 2 and 3 the horizontal axis contains the ordered observations $X_1' \leqslant X_2' \leqslant \cdots \leqslant X_n'$ and the vertical axis contains the inverse of the cumulative of the standard normal distribution

$$\Phi^{-1}(q_i). \qquad (14)$$

Much discussion has centered over what is the appropriate q_i in (14) to relate for plotting purposes to the order statistics X_i'. In Figs. 2 and 3 the plotting positions are selected so that

$$q_i = \frac{i - 0.5}{n}. \qquad (15)$$

For other suggestions and further discussion, see Chernoff and Lieberman [5] and Tukey [39]. Excellent textbook references for probability plotting are Hald [19], Bliss [2], and D'Agostino and Stephens [11]. *See* also GRAPHICAL REPRESENTATION OF DATA.

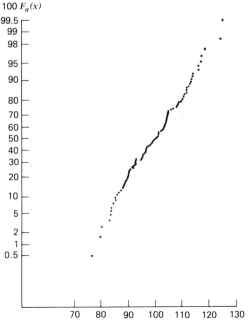

Figure 2 Normal probability plot of data from a normal distribution ($\mu = 100$, $\sigma = 10$).

LATER DEVELOPMENTS

Over the years a large number of other techniques have been suggested for testing for departures from normality. Most, if not all, are extensions and elaborations of the four classes of procedures given above. The following highlights the major developments.

Moment Tests

D'Agostino and Tietjen [12] presented approximations to the null distribution of $\sqrt{b_1}$ valid for samples of size $n \geqslant 7$. D'Agostino and Pearson [10] presented results of computer simulations of the null distribution of b_2 for samples up to $n = 200$. Further, they suggested various ways of combining $\sqrt{b_1}$ and b_2 to produce one final test statistic that is sensitive to departures from normality due either to skewness $\sqrt{\beta_1} \neq 0$ and/or kurtosis ($\beta_2 \neq 3$). One such statistic is

$$K^2 = X^2\left(\sqrt{b_1}\right) + X^2(b_2), \qquad (16)$$

where $X(\sqrt{b_1})$ and $X(b_2)$ are the standardized normal deviates equivalent (in probability) to the observed $\sqrt{b_1}$ and b_2. Bowman and Shenton [3] presented graphs that make possible the performance of the K^2 test for samples of size $n < 1000$ and levels of significance $100\alpha = 10, 5, 1\%$. Thus this test is available for many if not all practical applications.

Empirical Cumulative Distribution Function Tests

A large number of goodness-of-fit techniques have been developed which employ the ECDF. Most were primarily designed to deal with simple null hypotheses (i.e., for the normal distribution μ and σ being completely specified) and so are not of much practical use. Stephens [38] considered five of these and adapted them so that they are applicable for testing composite null hypoth-

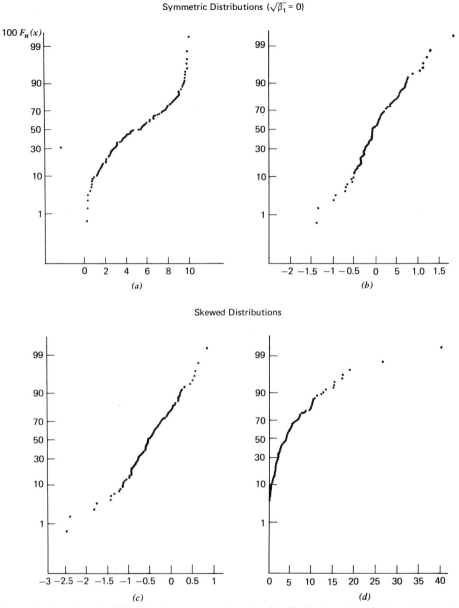

Figure 3 Normal probability plots for nonnormal unimodal distributions. (a) Uniform distribution: $\beta_2 = 1.80$. (b) Johnson unbounded $(0, 2)$ distribution: $\beta_2 = 4.51$. (c) Johnson unbounded $(1, 2)$ distribution: $\sqrt{\beta_1} = -0.87$, $\beta_2 = 5.59$. (d) Negative exponential distribution: $\beta_1 = 2$, $\beta_2 = 9$.

eses (i.e., normal distribution with μ and/or σ unknown). His method of attack was to employ large-scale Monte Carlo simulations* plus smoothing of these to obtain the null distributions of the test statistics. He presented his results so that all that is needed to perform the test is a computation of the test statistic, a simple further computation employing this value and a table look-

up to determine significance of the observed results. The five ECDF tests that Stephens adapted are the Kolmogorov–Smirnov test, the Cramér–von Mises W^2-test* [6], the Kuiper V-test [22], Watson's U^2-test* [41], and the Anderson–Darling A-test* [1]. Green and Hegazy [18] further modified the Kolmogorov–Smirnov, the Cramér–von Mises, and the Anderson–Darling tests.

Csörgö et al. [7] reviewed another approach for applying ECDF tests to composite null hypotheses. In this approach the data are first transformed into independent observations free of unknown parameters and then ECDF tests (e.g., W^2 or A) are applied to these transformed observations. O'Reilly and Quesenberry [24] and Quesenberry [32] present a general theory for obtaining transformations such that the transformed variables are independent uniform random variables. With use of these it is now possible to apply any ECDF test as a test for deviations from normality. However, these transformation procedures require randomization* of the data, and to many users this is considered an undesirable feature.

Regression Techniques

Clearly, the recent interest in tests for departures from normality is due mainly to the exciting work of S. S. Shapiro and M. B. Wilk [35]. Their work represents the first true innovation in the field since the 1930s. Basically, it consisted of returning to the normal probability plots and quantifying the information contained in them. The Shapiro–Wilk statistic is

$$W = \frac{\left\{ \sum_{i=1}^{n} w_i X_i' \right\}^2}{\sum_{i=1}^{n} \left(X_i - \bar{X} \right)^2}, \qquad (17)$$

where $X_1' \leqslant \cdots \leqslant X_n'$ are the ordered observations and w_i are the optimal weights for the weighted least-squares estimator of the population standard deviation given that the population is normally distributed. This W statistic can be viewed as the square of the ratio of the best linear estimator of σ (i.e., obtained from the probability plot) to the sample standard deviation. Further, W can also be viewed as the R^2 (square of the correlation coefficient) obtained from the normal probability plots.

Small values of W indicate nonnormality. Shapiro and Wilk [35] give a table of critical values for tests of level $100\alpha = 1, 2, 5,$ and 10% and for samples up to size $n = 50$. Shapiro and Francia [34] extended W up to samples of size $n = 99$ by replacing w_i of (17) by coefficients, say a_i, which are simple functions of the so-called expected normal order statistics*. However, there are questions concerning the adequacy of the probability points presented by them (see Pearson et al. [29].)

D'Agostino [8, 9] modified the W-test further by basically replacing the w_i of (17) with a simple linear function c_i, where

$$c_i = i - \tfrac{1}{2}(n + 1). \qquad (18)$$

So no tables of coefficients are needed to compute it. Probability points are available for this test for samples of size up to 1000. Table 2 contains numerical examples of the Shapiro–Wilk and D'Agostino tests.

Others, notably Filliben [15], Hegazy and Green [20], and La Brecque [23], have extended this regression idea further for developing tests for nonnormality.

POWER STUDIES: WHICH TESTS ARE BEST?

As the above demonstrates, there are a plethora of tests for judging departures from normality. There is no one test that is optimal for all possible deviations from normality. The procedure usually adopted to investigate the sensitivity of these tests is to perform power studies where the tests are applied to a wide range of nonnormal populations for a variety of sample sizes. A number of such studies have been undertaken. The major ones, in order of completeness and importance, are Pearson et al. [29], Shapiro et al. [36], Saniga and Miles [33], and Stephens [38]. Although it is not possible to obtain definitive answers from these studies, it does appear appropriate to make some general summary statements.

1. $\sqrt{b_1}$ and b_2 have excellent sensitivity over a range of alternative distributions which deviate from normality with re-

Table 2   Numerical Examples of Shapiro–Wilk and D'Agostino Tests (Samples of Size 10)[a]

| Weight, | Shapiro–Wilk | | | |
| | Normal, $\mu = 100$, $\sigma = 10$ | | Negative Exponential, $\mu = 5$ | |
w_i	X_i'	$w_i X_i'$	X_i'	$w_i X_i'$
− 0.5739	84.27	− 48.363	0.06	− 0.034
− 0.3291	90.87	− 29.905	0.37	− 0.122
− 0.2141	92.55	− 19.815	0.44	− 0.094
− 0.1224	96.20	− 11.775	0.89	− 0.109
− 0.0399	98.70	− 3.938	2.17	− 0.087
0.0399	98.98	3.949	2.63	0.105
0.1224	100.42	12.291	4.69	0.574
0.2141	101.58	21.748	6.48	1.387
0.3291	106.82	35.154	8.15	2.682
0.5739	113.75	65.281	16.69	9.578
		24.627		13.880
$\sum (X - \bar{X})^2$		616.554		240.498
$W = \dfrac{(\sum wX')^2}{\sum (X - \bar{X})^2}$		0.984		0.801[b]

| Weight, | D'Agostino | | | |
| | Normal, $\mu = 100$, $\sigma = 10$ | | Negative Exponential, $\mu = 5$ | |
c_i	X_i'	$c_i X_i'$	X_i'	$c_i X_i'$
− 4.5	84.27	− 379.215	0.06	− 0.270
− 3.5	90.87	− 318.045	0.37	− 1.295
− 2.5	92.55	− 231.375	0.44	− 1.100
− 1.5	96.20	− 144.300	0.89	− 1.335
− 0.5	98.70	− 49.350	2.17	− 1.085
0.5	98.98	49.490	2.63	1.315
1.5	100.42	150.630	4.69	7.035
2.5	101.58	253.950	6.48	16.200
3.5	106.82	373.870	8.15	28.525
4.5	113.75	511.875	16.69	75.105
		217.53		123.095
$\sum (X - \bar{X})^2$		616.554		240.498
$D = \dfrac{\sum c_i X_i'}{n\sqrt{n \sum (X - \bar{X})^2}}$		0.27703		0.25101[b]

[a] For the Shapiro–Wilk test we reject the null hypothesis of normality at the 0.10 level of significance if $W \leqslant 0.869$. For the D'Agostino test we reject the null hypothesis of normality at the 0.10 level of significance if $D \leqslant 0.2573$ or $D \geqslant 0.2843$.

[b] Reject null of hypothesis of normality at the 0.10 level of significance.

spect to skewness* and kurtosis*, respectively.

2. K^2 of (16) and similar tests are sensitive to a wide range of nonnormal populations. They can be considered omnibus tests.

3. The most powerful omnibus ECDF test appears to be the Anderson–Darling A-test as modified by Stephens.

4. The Shapiro–Wilk W-test and the Shapiro–Francia extension are very sensitive omnibus tests. For many skewed

populations they are clearly the most powerful.

5. While the D'Agostino test using the weights given by (18) is an omnibus test, it has best power for distributions with $\beta_2 > 3$.

The following references deal with different procedures and aspects of testing for deviations from normality. To aid the reader in sorting the various topics, the following symbols are used. The appropriate topic or topics are indicated after each reference. The symbols refer to the major topics of the references with regard to tests of nonnormality.

C: Chi-square tests

E: empirical cumulative distribution function techniques

G: general; treatment of all techniques

M: moment or related techniques

P: normal probability plotting

PS: power studies

R: regression techniques

T: transformation techniques

Most of the articles and texts in this reference list require only a good background in mathematical statistics for a complete understanding. In addition, most require only a background in statistical methods for a working understanding of the techniques and comparisons contained in them. To aid the reader further in using the reference list we identify the mathematical content level of each article and textbook by the symbols

e: elementary, noncalculus

i: intermediate, mathematical statistics needed

a: advanced level

Many articles identified as elementary (e) do contain more advanced material. We identify these as elementary because the major contribution in them is at that level.

References

[1] Anderson, T. W. and Darling, D. A. (1954). *J. Amer. Statist. Ass.*, **49**, 765–769. (E) (i)

[2] Bliss, C. I. (1967). *Statistics in Biology*, Vol. 1. McGraw-Hill, New York. (M, P) (e to i)

[3] Bowman, K. O. and Shenton, L. R. (1975). *Biometrika*, **62**, 243–250. (M) (e)

[4] Chernoff, H. and Lehmann, E. L. (1954). *Ann. Math. Statist.*, **25**, 579–586. (C) (i)

[5] Chernoff, H. and Lieberman, G. J. (1956). *Ann. Math. Statist.*, **27**, 806–818. (P) (i)

[6] Cramer, H. (1928). *Skand. Aktuarietidskr.*, **11**, 141–180. (E) (i)

[7] Csörgö, M., Seshadri, V., and Yalovsky, N. (1973). *J. R. Statist. Soc. B*, **35**, 507–522. (T, E) (a)

[8] D'Agostino, R. B. (1971). *Biometrika*, **58**, 341–348. (R) (e)

[9] D'Agostino, R. B. (1972). *Biometrika*, **59**, 219–221. (R) (e)

[10] D'Agostino, R. B. and Pearson, E. S. (1973). *Biometrika*, **60**, 613–622. (M) (e)

[11] D'Agostino, R. B. and Stephens, M. A. (1980). *Handbook of Goodness-of-Fit Techniques*. Marcel Dekker, New York. (G) (e to a)

[12] D'Agostino, R. B. and Tietjen, G. L. (1973). *Biometrika*, **60**, 169–173. (M) (e)

[13] David, H. A., Hartley, H. O., and Pearson, E. S. (1954). *Biometrika*, **41**, 482–493. (M) (e)

[14] Elderton, W. P. and Johnson, N. L. (1969). *Systems of Frequency Curves*. Cambridge University Press, Cambridge. (M) (i)

[15] Filliben, J. J. (1975). *Technometrics*, **17**, 111–117. (R) (e)

[16] Geary, R. C. (1935). *Biometrika*, **27**, 310–332. (M) (i)

[17] Geary, R. C. (1947). *Biometrika*, **34**, 209–242. (M) (i)

[18] Green, J. R. and Hegazy, Y. A. S. (1976). *J. Amer. Statist. Ass.*, **71**, 204–209. (E, PS) (e)

[19] Hald, A. (1952). *Statistical Theory with Engineering Applications*. Wiley, New York. (P) (e to i)

[20] Hegazy, Y. A. S. and Green, J. R. (1975). *Appl. Statist.*, **24**, 299–308. (R, PS) (e)

[21] Kolmogorov, A. (1933). *G. Ist. Ital. Attuari*, **4**, 83–91. (E) (a)

[22] Kuiper, N. H. (1960). *Proc. Kon. Nederlandse Akad. Wet. A*, **63**, 38–47. (E) (a)

[23] La Brecque, J. (1977). *Technometrics*, **19**, 293–306. (R) (e)

[24] O'Reilly, F. and Quesenberry, C. P. (1973). *Ann. Statist.*, **1**, 74–83. (T) (a).

[25] Pearson, E. S. (1930). *Biometrika*, **21**, 239–249. (M) (e to i)

[26] Pearson, E. S. (1931). *Biometrika*, **22**, 423–424. (M) (e to i)

[27] Pearson, E. S. and Hartley, H. O. (1966). *Biometrika Tables for Statisticians*, Vol. 1. Cambridge University Press, Cambridge. (G) (e to i)

[28] Pearson, E. S. and Hartley, H. O. (1972). *Biometrika Tables for Statisticians*, Vol. 2. Cambridge University Press, Cambridge. (G) (e to i)

[29] Pearson, E. S., D'Agostino, R. B., and Bowman, K. O. (1977). *Biometrika*, **64**, 231–246. (PS) (e)

[30] Pearson, K. (1895). *Philos. Trans. R. Soc. Lond.*, **191**, 343. (M) (i)

[31] Pearson, K. (1900). *Philos. Mag. 5th Ser.*, **50**, 157. (C) (i)

[32] Quesenberry, C. P. (1975). *Commun. Statist.*, **4**, 1149–1155. (T) (a)

[33] Saniga, E. M. and Miles, J. A. (1980). *J. Amer. Statist. Ass.*, **74**, 861–865. (PS) (e)

[34] Shapiro, S. S. and Francia, R. S. (1972). *J. Amer. Statist. Ass.*, **67**, 215–216. (R) (e)

[35] Shapiro, S. S. and Wilk, M. B. (1965). *Biometrika*, **52**, 591–611. (R) (e)

[36] Shapiro, S. S., Wilk, M. B., and Chen, H. J. (1968). *J. Amer. Statist. Ass.*, **63**, 1343–1372. (PS) (e)

[37] Smirnov, N. V. (1939). *Proc. Math.*, **6**, 3–26. (E) (a)

[38] Stephens, M. A. (1974). *J. Amer. Statist. Ass.*, **69**, 730–737. (E, PS) (e)

[39] Tukey, J. W. (1962). *Ann. Math. Statist.*, **33**, 1–67. (P) (e to i)

[40] Watson, G. S. (1961). *Biometrika*, **48**, 109–114. (E) (i)

(EDF STATISTICS
GOODNESS OF FIT
LIMIT THEOREMS, CENTRAL
NORMAL DISTRIBUTION
ORDER STATISTICS
OUTLIERS)

R. B. D'AGOSTINO

DEPENDENCE, CONCEPTS OF

Dependence relations between random variables is one of the most widely studied subjects in probability and statistics. The nature of the dependence can take a variety of forms and unless some specific assumptions are made about the dependence, no meaningful statistical model can be contemplated.

The study of the notion of dependence may be broadly classified into two categories. The first one can be described as "temporal"; it arises when observations are made on a process. Thus the conditions that define Markov*, martingale*, mixing*, and similar notions of dependence usually pay attention to the *order* or time at which the random variables are generated so that notions such as "future," "past," etc., play an important role. The second category of conditions, on the other hand, treat the random variables in a somewhat symmetric fashion, and this category is of principal interest in this article. As an illustration, suppose that an *n*-vector $\mathbf{X} = (X_1, \ldots, X_n)$ satisfies

$$\Pr[X_i \leqslant a_i, i = 1, \ldots, n] \geqslant \prod_{i=1}^{n} \Pr[X_i \leqslant a_i]$$

for every $\mathbf{a} \in R^n$, then dependence could be called "(lower) quadrant dependence"*. This condition also illustrates a concept of *positive* dependence which is of primary importance for this exposition.

First, we will consider briefly some measures of dependence*. It should be noted that although it is customary to compute a correlation coefficient of some sort, one number alone cannot reveal the nature of dependence adequately. As will be seen, most of these measures can be expressed in terms of a covariance*. According to properties desired for the measure of dependence, one chooses a particular pair of functions among a suitable class and the covariance of this pair is used for the construction of the measure. On the other hand, a *concept* of dependence typically requires the covariance of every pair of functions chosen from a suitable class to have some particular behavior (positive, nonincreasing, etc.).

After discussing measures of dependence, various concepts of dependence will be introduced for the special, but important bivariate case, which provides a theoretical stepping stone toward the concepts of dependence in the multivariate situation.

The first section deals with the notions that are based on the properties of covariance between certain functions. These notions are then utilized to obtain dependence orderings within families of bivariate distributions. Applications of these concepts are given.

In the second section, some of the notions are generalized to multivariate distributions. As one may expect, some concepts are "bivariate bound," while others apply under more restrictive assumptions. The notion of "association"* is developed in more detail while some other such as "exchangeability"* are discussed briefly. As we will find out, the area of "dependence" is presently active in research and has potential in applications.

BIVARIATE DISTRIBUTIONS

Let (X, Y) be a pair of real random variables. One of the most popular measures of dependence is the correlation coefficient* between (X, Y),

$$\rho(X, Y) = \frac{\text{cov}(X, Y)}{\left[\text{var} X \text{ var} Y\right]^{1/2}}.$$

The basic reason for its popularity is that it is an important parameter for the bivariate normal* family plus the fact that it gives the extent to which X and Y are linearly dependent. It is well known that zero correlation does not imply independence. In the following we consider some measures where independence *is* implied whenever the measure is 0. Rényi [21] considered

$$S(X, Y) = \sup_{f, g} \rho(f(X), g(Y)),$$

where f, g varied over the class of Borel-measurable functions. It is easy to see that this measure satisfies the following properties.

1. $0 \leqslant S(X, Y) \leqslant 1$.
2. $S(X, Y) = S(Y, X)$.
3. $S(X, Y) = 0$ if and only if X and Y are independent. (This can be seen by considering indicator functions* of $X \leqslant a$, $Y \leqslant b$, where a, b are varied.)
4. If X, Y are functionally dependent, then $S(X, Y) = 1$.

Although such a measure has some good properties, it is usually difficult to compute and a "sample" version of this is even harder to deal with.

Inspired by an information theoretic viewpoint, Linfoot [15] considered the following measure. Suppose that (X, Y) has a discrete distribution with $P[X = x_i, Y = y_j] = P_{ij}$; $i = 1, \ldots, n; j = 1, \ldots, m$. Then

$$r_0(X, Y) = \sum_{i, j} \left[P_{ij} \log p_{ij} - p_i q_j \log p_i q_j \right].$$

It can be shown that $r_0(X, Y) \geqslant 0$ and satisfies properties 2 to 4. In case (X, Y) has a density function $p(x, y)$, say, then, the sum on the right side of the defining relation of r_0 is replaced by an integral and thus

$$r_0 = E \log \left\{ \frac{p(x, y)}{p_1(x) p_2(y)} \right\},$$

where p_1, p_2 represent the marginal densities.

Another approach is to make monotone transformations so as to make the marginals uniform*. Thus, if F_X, F_Y are distribution functions of X and Y, then $U = F_X(X)$ and $V = F_Y(Y)$ are uniformly distributed on $[0, 1]$ and their joint distribution function $c(u, v)$, say, indicates the dependence between X and Y. Schweizer and Wolff [23] propose and study the measure

$$\sigma(X, Y) = 12 \int_0^1 \int_0^1 |uv - c(u, v)| \, du \, dv.$$

This measure is in the same spirit as the Cramér–von Mises* analog,

$$\gamma(X, Y) = \int\int_{-\infty}^{\infty} \{ F_{X, Y}(x, y) - F_X(x) F_Y(y) \}^2 \cdot dF_X(x) \, dF_Y(y),$$

or the Kolmogorov–Smirnov* analog,

$$\delta(X, Y) = \sup_{x, y} |F_{X, Y}(x, y) - F_X(x) F_Y(y)|.$$

It should be noted that all these measures (except the correlation coefficient) are nonnegative and do not reveal whether the dependence is "positive" or "negative." In the following we first state four basic conditions describing positive dependence with increasing order of stringency.

(a) $\text{cov}(X, Y) \geqslant 0$.

(b)   For every pair of nondecreasing functions f and g, defined on R,

$$\text{cov}[f(X), g(Y)] \geqslant 0. \qquad (1)$$

(c)   For every pair of functions f, g defined on R^2 which are nondecreasing in each of the arguments (separately),

$$\text{cov}[f(X, Y), g(X, Y)] \geqslant 0.$$

(This condition was introduced by Esary et al. [5] and is called "association.")

(d)   Either for every y

$$P[Y > y | X = x]$$

is nondecreasing in x, or, for every x,

$$P[X > x | Y = y]$$

is nondecreasing in y.

This dependence can be described as *positive regression dependence*.

The fact that (c)$\Rightarrow$(b)$\Rightarrow$(a) is obvious. To see that (d)$\Rightarrow$(c) as well as to express (b) in another useful form, the following simple device is helpful. We will assume that the distribution functions are continuous. Let I denote the indicator function; then

$$x - x' \equiv \int_{-\infty}^{\infty} \{ I[x' \leqslant u] - I[x \leqslant u] \} \, du. \qquad (2)$$

From this identity it follows that:

1.   If X, X' is a pair of random variables, then

$$E[X - X'] = \int_{-\infty}^{\infty} \{ F_{X'}(u) - F_X(u) \} \, du.$$

Thus whenever X is *stochastically* larger* than $X' (X \overset{s}{>} X'$ for short); i.e., for every u,

$$\Pr[X > u] \geqslant \Pr[X' > u],$$

then

$$EX \geqslant EX',$$

equality holding *only if* X, X' are identically distributed. Note that if $X \overset{s}{>} X'$, so is $g(X) \overset{s}{>} g(X')$, for every nondecreas-

ing function g, and it follows that

$$Eg(X) \geqslant Eg(X'). \qquad (3)$$

2.   Writing X as the difference between its positive part X^+ and negative part X^-, it is easy to see from (1) that

$$EX = E(X^+ - X^-)$$

$$= -\int_{-\infty}^{0} F_X(u) \, du + \int_0^{\infty} [1 - F_X(u)] \, du.$$

3.   Let (X, Y) and (X', Y') be independent identically distributed pairs. Then from (2) it follows that

$$2\,\text{cov}(X, Y) = E(X - X')(Y - Y')$$

$$= 2 \int_{-\infty}^{\infty} \int_{-\infty}^{\infty} \{ F_{X, Y}(u, v)$$

$$- F_X(u) F_Y(v) \} \, du \, dv. \qquad (4)$$

Consequences **1** and **2**, although well known, are usually proved by lengthy methods. Consequence **3** is due to Hoeffding [6], who used the simple trick given by (2). From (4) it follows immediately that if a pair (X, Y) is positive quadrant dependent (PQD), then $\text{cov}(X, Y) \geqslant 0$, equality holding only if X, Y are independent. Further, if f, g is a pair of nondecreasing real functions, then (X, Y) is PQD $\Rightarrow (f(X), g(Y))$ is PQD; thus $\text{cov}[f(X), g(Y)] \geqslant 0$. Consequently, conditions (b) and PQD are equivalent. This equivalence and its ramifications were studied by Lehmann [14]. A simple special case of the above is that for f, g as defined above,

$$\text{cov}[f(X), g(X)] \geqslant 0. \qquad (5)$$

This inequality is due to Chebyshev* (inequality not to be confused with the one giving a probability upper bound) and follows from (b) since the pair (X, X) is PQD.

Condition (d) requires a stochastically ordered family of conditional distributions*. To see that it implies (c), consider the following identity, which is a useful tool in its own right. Let X, Y be real random variables and Z be an arbitrary random vector; then

$$\text{cov}(X, Y) = \text{cov}\{ E(X|Z), E(Y|Z) \}$$

$$+ E\{ \text{cov}[(X, Y)|Z] \}. \qquad (6)$$

If f, g are nondecreasing functions on R^2,

then

$$\text{cov}\big[\,f(X,Y),g(X,Y)\,\big]$$

$$= \text{cov}\big\{E\big[\,f(X,Y)|Y\big],E\big[\,g(X,Y)|Y\big]\big\}$$

$$+ E\big\{\text{cov}\big[\,f(X,Y),g(X,Y)|Y\big]\big\}.$$

(7)

Suppose now that (d) holds. Then the expected values in the first term on the right side of (7) are nondecreasing in Y (almost surely) so that in view of (5), this term is nonnegative. Further, monotonicity of f and g makes the conditional covariance in the second term to be nonnegative and the same holds for its expected value. Condition (c) has a natural multivariate analog, although its verification is somewhat difficult.

Suppose that $P[X \leqslant x|Y \leqslant y]$ is nonincreasing in y for every x. This may be described as X being *left tail decreasing* in Y. Then it is easy to see that this condition is weaker than X, being positively regression dependent on Y. Esary and Proschan [4] show that this weaker condition implies (c). Similarly, if X is right tail increasing in Y, then also (X,Y) satisfy (c).

When the inequality signs in (a), (b), and (d) are reversed, analogs for negative dependence are obtained. However, when $f \equiv g$ and $f(X,Y)$ does not have a degenerate distribution, $\text{cov}[\,f(X,Y),f(X,Y)] > 0$, no matter how X,Y are related. Thus no negative analog of (c) is possible. It is also easy to see that the negative analog of (b) is equivalent to negative quadrant dependence (NQD).

We will now state the basic theorem due to Lehmann [14] derived from condition (b), using the following notation. Let r and s be a pair of real functions defined on R^n which are monotone in each of their n arguments. The functions r and s are said to be *concordant* in ith argument if the directions of the monotonicity for ith arguments are the same (i.e., both functions are simultaneously either nondecreasing or nonincreasing in the ith argument while others are kept fixed) and *discordant* if the directions are opposite.

Theorem 1 (Lehmann). Let (X_i, Y_i), $i = 1, \ldots, n$, be n independent pairs each satisfying either PQD or NQD. Suppose that r, s is a pair of real functions on R^n such that if (X_i, Y_i) is PQD (NQD), then r, s are concordant (discordant) in the ith argument, $i = 1, \ldots, n$. Then

$$\text{cov}\big[\,r(X),s(X)\,\big] \geqslant 0.$$

This theorem is a direct consequence of (6), as seen by conditioning on (X_n, Y_n), and then using the method of induction. This theorem can be used to show that when a random sample (X_i, Y_i), $i = 1, \ldots, n$, is taken from a bivariate population satisfying (b), several of the measures of dependence commonly used have their expected values nonnegative.

1. Kendall's τ^* is a measure based on the sample covariance of terms such as $U = \text{sign}(X_2 - X_1)$ and $V = \text{sign}(Y_2 - Y_1)$. From Theorem 1 it follows that with (b), the measure $\tau \geqslant 0$.

2. Spearman's ρ_s is based on the covariance between $U = \text{sign}(X_2 - X_1)$ and $V = \text{sign}(Y_3 - Y_1)$. Treating U, V as $r(X_1, X_2, X_3)$ and $s(Y_1, Y_2, Y_3)$, it is seen that the pair r, s is concordant and Theorem 1 implies that under (b) the measure $\rho_s \geqslant 0$.

3. Let μ and ν denote the medians of marginals of X_i and Y_i. Divide the plane into four quadrants and find the number of pairs in positive and negative quadrants, that is, those pairs where either $X_i \geqslant \mu$, $Y_i \geqslant \nu$ holds or $X_i \leqslant \mu$, $Y_i \leqslant \nu$ holds. If p_n is the proportion of pairs in the union of positive and negative quadrants, then $(2p_n - 1)$ is a measure of dependence proposed by Blomquist [2]. Under (b) it follows that

$$E(2p_n - 1)$$

$$= 2\big\{\text{cov}\big[\,I(X_i \geqslant \mu),I(Y_1 \geqslant \nu)\big]$$

$$+ \text{cov}\big[\,I(X_i \leqslant \mu),I(Y_1 \leqslant \mu)\big]\big\}$$

$$\geqslant 0.$$

4. A class of multiple decision* problems, called slippage problems, typically consist of the following kinds of statistical

tests. Under hypothesis H all k populations parametrized by $\theta_1, \ldots, \theta_k$ are assumed to be same, while under the alternative K_i, $i = 1, \ldots, k$, the ith population is supposed to have *slipped* to the right. A typical test procedure is to reject H in favor of K_i when a statistic $T_i \geqslant c_i$. It can be shown that if α_i denotes the probability of falsely rejecting H due to the event $T_i \geqslant c_i$, then negative quadrant dependence among T_i yields

$$\sum_i \alpha_i \geqslant \alpha \geqslant \sum_i \alpha_i - \sum_{i<j} \alpha_i \alpha_j,$$

where α denotes the total probability of falsely rejecting H. (For more details, *see* SLIPPAGE TESTS.)

The applications cited above are qualitative in nature, that is, one can infer that under positive quadrant dependence, certain proposed measures of dependence are positive, a desirable property of any reasonable measure. There are situations where one has a family of distributions which exhibit ordering in dependence; for example, bivariate normal distributions where the correlation coefficient ρ varies from -1 to $+1$ while other parameters are fixed. Under these situations it would be desirable that the proposed measure is nondecreasing in ρ. Such behavior would have an important consequence in testing of hypotheses of independence. The measures considered before are "nonparametric"* in nature and the tests based on those would not only be unbiased but would have nondecreasing power* against those ordered alternatives. This improvement was made by Yanagimoto and Okamoto [29]. In the following we summarize the related results.

Suppose that $G(x, y)$ and $F(x, y)$ have the same marginals which are continuous. Distribution G is said to have *larger quadrant dependence* than F if for every (x, y),

$$G(x, y) \geqslant F(x, y). \qquad (8)$$

With this ordering it follows that for every pair of nondecreasing functions f, g defined

on R,

$$\mathrm{cov}_G[\, f(X), g(Y)] \geqslant \mathrm{cov}_F[\, f(X), g(Y)]. \qquad (9)$$

Note that (8) implies that the joint distributions of $[\,f(X), g(Y)]$ obey the same ordering, so that (9) becomes an immediate consequence of (4), Hoeffding's formula for the covariance. Further extension similar to Theorem 1 can also be made.

Instead of two distributions, one can consider a family of bivariate distributions that can be ordered by the notion of larger quadrant dependence. For example, the family of bivariate normal distributions* having the same marginals and parametrized by the correlation coefficient is such a family. This fact was first established by Slepian [27] by a neat trick which is worth mentioning. The multivariate normal density* $\varphi(x, \mu, \Sigma)$ has the rather special property

$$\frac{\partial}{\partial \sigma_{ij}} \varphi(\mathbf{x} : \mu, \Sigma) = \frac{\partial}{\partial x_i \partial x_j} \varphi(\mathbf{x}, \mu, \Sigma),$$
$$i \neq j, \quad (10)$$

which was observed by Plackett [19]. If $\mathbf{X}$ has density φ, then

$$\frac{\partial}{\partial \sigma_{12}} \Pr[X_i \leqslant c_i, i = 1, \ldots, n]$$

$$= \frac{\partial}{\partial \sigma_{12}} \int_{-\infty}^{c_1} \int_{-\infty}^{c_2} \cdots \int_{-\infty}^{c_n} \varphi(\mathbf{x}; \mu, \Sigma)\, d\mathbf{x}$$

$$= \int_{-\infty}^{c_1} \int_{-\infty}^{c_2} \cdots \int_{-\infty}^{c_n} \frac{\partial^2}{\partial x_1 \partial x_2} \varphi(\mathbf{x}; \mu, \Sigma)\, d\mathbf{x}$$

$$= \int_{-\infty}^{c_3} \cdots \int_{-\infty}^{c_n} \varphi(c_1, c_2, x_3, \ldots, x_n; \mu, \Sigma)$$
$$\cdot dx_3 \ldots dx_n$$

$$> 0.$$

As a result, if the variances and the means are kept fixed, the probability content of an orthant* under multinormal distribution is a *strictly* increasing function of the correlation coefficients. In particular, the bivariate normal family with fixed marginals is ordered by quadrant dependence through the correlation coefficient. Further, if (X, Y) has a bivariate normal distribution and for *some*

c_1, c_2,

$$\Pr[X \leqslant c_1, Y \leqslant c_2]$$
$$= \Pr[X \leqslant c_1]\Pr[Y \leqslant c_2]$$

then X, Y are independent!

In general, a stronger ordering can be used to check the quadrant dependence ordering. The stronger one (also studied by Yanagimoto and Okamoto [29]) corresponds to the condition (d). Thus a distribution G is said to have a larger regression dependence on x than F on x when G and F have continuous marginals with a common x marginal and for $x' > x$,

$$F^{-1}(u|x') \geqslant F^{-1}(v|x) \Rightarrow G^{-1}(u|x')$$
$$\geqslant G^{-1}(v|x). \quad (11)$$

The motivation behind such a definition and its relation to stochastic ordering* is as follows (see Jogdeo [10] for more details). It can be seen that G is stochastically larger than F if $G^{-1}(u) \geqslant F^{-1}(u)$, for every $u \in (0, 1)$ and due to the nondecreasing nature of F^{-1},

$$G^{-1}(u) \geqslant F^{-1}(v)$$
$$\text{for every } 0 \leqslant v \leqslant u \leqslant 1.$$

Let $S_{GF} = \{(u,v) : G^{-1}(u) \geqslant F^{-1}(v)\}$, where $S_{G,F}$ is a subset of the unit square. $S_{G,F}$ describes the *stochastic* difference between G and F. If $S_{G,F}$ is larger than the triangular region $\{(u,v) : 0 \leqslant v \leqslant u \leqslant i\}$, then G is stochastically larger than F. Condition given by (11) says that for $x_2 > x_1$,

$$S_{F_{x_2}, F_{x_1}} \subset S_{G_{x_2}, G_{x_1}}.$$

In particular, if the conditional distributions under x_2, x_1 are stochastically ordered under F, so are those under G.

It is easy to check that the bivariate normal family parametrized by ρ is ordered by such regression dependence. So are bivariate distributions of (X, Y) generated by linear models

$$Y = \alpha X + Z,$$

where X and Z are independent.

The concept of regression dependence can be strengthened further. Instead of stochastic ordering induced by conditioning, one may require the *monotone likelihood ratio** property (which implies stochastic ordering) for the conditional distributions. (*See* TOTAL POSITIVITY.) This concept of dependence was formulated and studied by Lehmann [14]. It was generalized further by Yanagimoto [28] and applied to some concepts in reliability theory by Shaked [24].

Finally, we state an application of negative quadrant dependence in conjunction with Bonferroni inequalities* to obtain confidence intervals*. Suppose that

$$q_i = \Pr[|X_i - \mu_i| > c_i].$$

Then Bonferroni inequalities give

$$1 - \sum_1^k q_i$$
$$\leqslant \Pr[|X_i - \mu_i| \leqslant c_i, i = 1, \ldots, k]$$
$$\leqslant 1 - \sum_1^k q_i$$
$$+ \sum_{i<j} \Pr[|X_i - \mu_i| > c_i, |X_j - \mu_j| > c_j]. \quad (12)$$

If $|X_i - \mu_i|$, $i = 1, \ldots, k$, are pairwise negatively quadrant dependent, then the last term on the right side of (12) may be replaced with $\sum_{i<j} q_i q_j$. In fact, Dykstra et al. [3] show that

$$\left| \Pr[|X_i - \mu_i| \leqslant c_i, i = 1, \ldots, k] \right.$$
$$\left. - \prod_{i=1}^k \Pr[|X_i - \mu_i| \leqslant c_i] \right|$$
$$\leqslant \frac{k-1}{2k} \left\{ \log \prod_{i=1}^k \Pr[|X_i - \mu_i| \leqslant c_i] \right\}^2 \quad (13)$$

Thus rectangular-shaped confidence intervals could be set up by utilizing the knowledge about q_i, $i = 1, \ldots, k$, alone under the assumption of pairwise negative quadrant dependence. The bound in (13) gives the extent of the error in the confidence coefficient*. Note that the bound is small when the $\Pr[|X_i - \mu_i| \leqslant c_i]$ are large.

MULTIVARIATE GENERALIZATIONS

A straightforward generalization of quadrant dependence can be made by requiring, for every $c_1, \ldots, c_k$,

$$\Pr[X_i \leqslant c_i, i = 1, \ldots, k]$$

$$\geqslant \prod_{i=1}^{k} \Pr[X_i \leqslant c_i]. \tag{14}$$

Unfortunately, this does not lead to consequences similar to those in the bivariate case. In fact, it *does not imply* that

$$\Pr[X_i \geqslant c_i, i = 1, \ldots, k]$$

$$\geqslant \prod_{i=1}^{k} \Pr[X_i \geqslant c_i]. \tag{15}$$

Nor does it set up an inequality between an expected value of the product of nondecreasing functions and the product of their expected values. The most fruitful concepts of positive dependence related to those considered in the first section are (a) association and (b) mixtures of *product measures**. First we will develop the notion of association and consider some applications. (The following is a brief summary of the results of Esary et al. [5].)

Definition 1. A random vector $\mathbf{X} = (X_1, \ldots, X_k)$ or its distribution is said to be associated if for every pair of nondecreasing real functions f, g defined on R^k,

$$\mathrm{cov}(f(\mathbf{X}), g(\mathbf{X})) \geqslant 0.$$

Association has the following useful properties which can be verified easily from the definition. Suppose that $(X_1, \ldots, X_k)$ is associated; then

1. $\mathrm{cov}(X_i, X_j) \geqslant 0$, $i, j = 1, \ldots, k$.
2. Any subset of $(X_1, \ldots, X_k)$ is associated.
3. The union of two independent associated sets is associated. [Follows from (6).]
4. A set consisting of a single real random variable is associated. [Assertion equivalent to (5).]
5. Nondecreasing (nonincreasing) functions of associated random variables are associated.
6. A set of independent random variables is associated. [Follows from assertions 3 and 4.]

The last assertion sounds somewhat odd, since we are attributing a concept of positive dependence to independent random variables! However, the use of assertion 5 will show that even for the independent case, nontrivial inequalities can be obtained rather easily with this notion.

7. For every $\mathbf{c}$,

$$P[X_i \leqslant c_i, i = 1, \ldots, k] \geqslant \prod_{i=1}^{k} P[X_i \leqslant c_i], \tag{16}$$

$$P[X_i \geqslant c_i, i = 1, \ldots, k] \geqslant \prod_{i=1}^{k} P[X_i \geqslant c_i]. \tag{17}$$

More generally, if J_1 and J_2 partition $\{1, 2, \ldots, k\}$,

$$P\left[X_i \underset{(\geqslant)}{\leqslant} c_i, i = 1, \ldots, k\right]$$

$$\geqslant P\left[X_i \underset{(\geqslant)}{\leqslant} c_i, i \in J_1\right] P\left[X_i \underset{(\geqslant)}{\leqslant} c_i, i \in J_2\right]. \tag{18}$$

Applications

PARTIAL SUMS. Let $X_i, \ldots, X_n$ be associated and $S_i = \sum_{j=1}^{i} X_j$; then $S_1, \ldots, S_n$ are associated. In particular, for every $c_1, \ldots, c_k$,

$$\Pr\left[S_i \underset{(\geqslant)}{\leqslant} c_i, i = 1, \ldots, k\right] \geqslant \prod_{i=1}^{k} \Pr\left[S_i \underset{(\geqslant)}{\leqslant} c_i\right],$$

a result proved by Robbins [22] using a more complex method.

ORDER STATISTICS*. Let $X^{(i)}$ be the ith-order statistic corresponding to a random sample $X_1, \ldots, X_n$. Then $(X^{(1)}, \ldots, X^{(n)})$ is associated. In particular, $\mathrm{corr}(X^{(i)}, X^{(j)}) \geqslant 0$. With the concept of association this inequality is a triviality, although more complex proofs had been used in the previous literature.

Although some special conditions on correlations yield association, independence plays a rather crucial role in generating association. In many applications the statistics are based on absolute values of the random variables. In order to apply the notion of association to various statistics, it is important to study the operations under which either the property of association or positive orthant dependence is preserved. The concept of concordance can be generalized for the real functions defined on R^m, where $m = n \times k$ and the arguments are n independent k-dimensional vectors. Such a generalization is given in Jogdeo [11]. As an application, suppose that $\mathbf{X}$ has a multinormal distribution with mean vector $\mathbf{0}$ and that $|\mathbf{X}|$ is associated. Let $\mathbf{t}$ be the vector of t-statistics based on a random sample of observations from such a population. Then $|\mathbf{t}|$ is associated. Such a property allows us to construct *conservative* confidence intervals via inequalities of the form (18). For example, if

$$\Pr\left[|u_i - \mu_i|/s_i \leqslant c_i, i = 1, \ldots, k\right]$$
$$\geqslant \prod_{i=1}^{k} \Pr\left[|u_i - \mu_i|/s_i \leqslant c_i\right], \quad (19)$$

then the simultaneous confidence intervals* may be chosen so that the right side of (19) has a certain value $1 - \alpha$, say. The actual confidence coefficient associated with the intervals is *at least* $1 - \alpha$. [Note that the error bound such as (13) is not applicable here since association implies PQD and not NQD.]

Another important operation is taking mixtures of distributions possessing association. The condition under which the association is preserved can be expressed through

conditional expectation* (see Jogdeo [12] for details). The distribution of a k-vector X is said to be a *monotone mixture* with an m-vector W if for every nondecreasing f,

$$h(W) = E[f(X)|W]$$

is nondecreasing in W. With this assumption, if X is conditionally associated given W, which itself is associated, then the vector (X, W) is associated. This has some applications in reliability theory* for obtaining probability bounds (see Marshall and Olkin [17]).

So far we studied the hierarchy of definitions which considered the behavior of the joint density or distribution functions. An alternative approach may be to study those models where independence is *contaminated* by an extraneous random variable(s). For example, an infinite sequence of random variables $\{X_m\}$ may be exchangeable (*see* EXCHANGEABILITY). Loosely speaking, exchangeable variables are *mixtures* of independent, identically distributed (i.i.d.) random variables. It should be noted that the joint distribution of $X_1, \ldots, X_n$ may be permutation invariant and still may not be a subset of an exchangeable sequence. Dykstra et al. [3] and Shaked [25] studied the concept of *positive dependence* by *mixture* (PDM), which amounts to having conditionally i.i.d. random variables. One of the immediate consequences is that if $X_1, \ldots, X_m$ are PDM and h is an arbitrary (measurable) real function, then

$$\mathrm{cov}\left[h(X_1, \ldots, X_k), h(X_{k+1}, \ldots, X_{2k})\right] \geqslant 0. \quad (20)$$

A simple example of PDM is that of a multinormal vector whose components have a common marginal and every pair has the same *nonnegative* correlation coefficient. From (20) it can be seen that under PDM the χ^2 statistics, say $V_1, \ldots, V_m$, constructed from a random sample have the property that for $0 < a < b$,

$$\Pr\left[a \leqslant V_i \leqslant b, i = 1, \ldots, m\right]$$
$$\geqslant \prod_{i=1}^{m} \Pr\left[a \leqslant V_i \leqslant b\right]. \quad (21)$$

This was first stated by Jensen [7] for the bivariate case. The result above has applications to the construction of conservative confidence interval* estimation. Jensen [8] showed that if the bivariate density can be represented as a series of a certain kind, then the inequality (21) holds. In the same article, several examples of such distributions have been given. It should be noted, however, that (21) may hold for every pair (a, b) and still the covariance could be negative. To illustrate this by a simple example, let (X, Y) take values $(1, 1)$ and $(-1, -1)$, each with probability 0.02, $(-1, 1)$ and $(1, -1)$ with probability 0.1, and $(0, 0)$ with probability 0.76. It is easy to check that (X, Y) satisfies inequality (21) for every a and b (there are only two distinct cases to verify). However, $\operatorname{cov}(X, Y) = -0.16$. Thus (21) cannot be a defining relation for a concept of positive dependence. In fact, in the example above, $(-X, Y)$ also satisfies this relation and a reasonable concept of dependence would imply independence in such a situation.

The permutation symmetry in conjunction with a condition of unimodality along certain sections of the density gives rise to the notion of Schur concavity. This condition gives rise to a host of inequalities and is the subject of a book by Marshall and Olkin [18]. Although Schur concavity (see MAJORIZATION) has some common ground with PDM, we will discuss other notions based on unimodality* and symmetry* which have applications to study of the order within distributions in a family exhibiting positive dependence.

The notion of multivariate unimodality was first introduced by Anderson [1]: a density function f defined on $R^n \to R$ is unimodal if for every $c > 0$, the set $\{x : f(x) \geq c\}$ is convex. An important result, basically motivated by the multinormal density, was that if f is unimodal and centrally symmetric [e.g., $f(x) = f(-x)$], then

$$h(\lambda) = \Pr[X \in C + \lambda z]$$

is nonincreasing in $\lambda > 0$ for every z and every centrally symmetric convex set C. A simple interpretation is that when a centrally symmetric convex set is "moved" along in any direction, the probability assigned to it decreases. Among numerous applications of this result, the following one was given by Šidák [26]. Let X be multinormal with mean vector 0 and consider

$$h(\rho_{12}, \rho_{13}, \ldots, \rho_{k-1}, k)$$
$$= P[|X_i| \leq c_i, i = 1, \ldots, k]. \quad (22)$$

If all the correlation coefficients of one component say X_1, with other components are increased by a common factor, then the probability increases. More precisely, suppose that $\rho_{1j} = \lambda \xi_{1j}, j > 1$, where ξ_{1j} is fixed. Then as long as $|\rho_{1j}| \leq 1$, the probability above, considered as a function of λ, is nondecreasing for $\lambda > 0$. A simpler proof of this result is given in Jogdeo [9]. Note that this sets up a partial ordering among multinormal distributions and implies that

$$P[|X_i| \leq c_i, i = 1, \ldots, k]$$
$$\geq P[|X_1| \leq c_1] P[|X_j| \leq c_j, j = 2, \ldots, k]$$
$$\geq \prod_{i=1}^{k} P[|X_i| \leq c_i]. \quad (23)$$

It should be noted that the function h defined by (22) may actually decrease if only one of the correlation coefficients is increased. Second, if $|X_i| \leq c_i$ is replaced by $|X_i| \geq c_i$, the inequality (23) may fail. In other words, $|X_i|$ are not necessarily orthant dependent when positive orthants are considered and thus not associated. A natural question would be whether there are some simple conditions (besides independence) on correlation coefficients implying association for $|X_i|, i = 1, \ldots, k$. Such conditions are given by Jogdeo [11], and apply to the situation of *linear contaminated independence*. Briefly stated, suppose that

$$X = Z + U,$$

where (a) Z and U are independent, (b) $|U|$ is associated, and (3) the components of Z are independent and have symmetric unimodal distributions. Then $|X|$ is associated. As a particular case, consider

$$\operatorname{cov}(X_i, X_j) = \lambda_i \lambda_j \quad \text{where } \operatorname{var} X_i \geq \lambda_i^2.$$

Then it can be shown that **X** having a multinormal distribution centered at **0** satisfies the foregoing conditions. These results have applications to multivariate t-distributions* and to the elements of matrices with certain Wishart distributions*.

Finally, some results regarding negative dependence can be stated. Many of these can be obtained by the following simple result, which is a consequence of *negative regression dependence*. Let A be some event and X be a real random variable. If $\Pr[A|X = x]$ is a *nonincreasing* function of x, then for every c,

$$\Pr[A, X \geq c] \leq \Pr[A]\Pr[X \geq c].$$

Several discrete distributions that were considered by Mallows [16] and Proschan and Sethuraman [20] can be shown to obey *negative orthant dependence* by this simple device (see Jogdeo and Patil [13] for details). The following are two examples.

MULTINOMIAL DISTRIBUTION*. Suppose that

$$\Pr[X_i = x_i, i = 1, \ldots, m]$$
$$= \binom{N}{x_0, x_1, \ldots, x_m} \prod_{i=0}^{m} p_i^{x_i},$$

where $x_0 = N - \sum_1^m x_i$ and $P_0 = 1 - \sum_1^m p_i$. It is easy to verify that

$$\Pr[X_i \geq c_i, i = 1, \ldots, m - 1 | X_m = t]$$

is nonincreasing in t, so that

$$\Pr[X_i \geq c_i, i = 1, \ldots, m]$$
$$\leq \Pr[X_1 \geq c_1]\Pr[X_i \geq c_i, i = 1, \ldots, m - 1]$$
$$\leq \prod_{i=1}^{m} \Pr[X_i \geq c_i].$$

Note that $X_i \geq c_i$ may be replaced throughout by $X_i > c_i$. The multivariate hypergeometric*, Dirichlet*, and negative multinomial* distributions may be considered in the same fashion.

CONTINUOUS DISTRIBUTION. Let X_i, $i = 1, \ldots n$, be a random sample from a continuous distribution and R_i, $i = 1, \ldots, n$, be their ranks. Let $c_1 \leq c_2 \leq \cdots \leq c_m \leq n$

be m positive integers. It can be shown that

$$\Pr[R_i \leq c_i, i = 1, \ldots, m]$$
$$= \frac{c_1}{n} \cdot \frac{c_2 - 1}{n} \cdots \frac{c_m - m + 1}{n - m + 1}.$$

Using this it can be shown that

$$\Pr[R_i \leq c_i, i = 1, \ldots, m | R_{m+1} = t]$$

is nondecreasing in t, so that

$$\Pr[R_i \leq c_i, i = 1, \ldots, m + 1]$$
$$\leq \Pr[R_i \leq c_i, i = 1, \ldots, m]$$
$$\times \Pr[R_{m+1} \leq c_{m+1}].$$

Thus the ranks exhibit negative orthant dependence.

Further Reading

The references listed below can be broadly classified in the following fashion. References 5, 8, 10, 14, and 24 describe several applications of the concepts of positive dependence. References 2–4, 9, 11–13, 15, 16, 20, 23, 25, 26, 28, and 29 develop theory and are devoted to the concepts of dependence. A beginning graduate level of mathematical statistics is adequate for understanding the proofs.

References

[1] Anderson, T. W. (1955). *Proc. Amer. Math. Soc.*, **6**, 170–176.

[2] Blomquist, N. (1950). *Ann. Math. Statist.*, **21**, 593–600.

[3] Dykstra, R. L., Hewett, J. E., and Thompson, W. A. (1973). *Ann. Statist.*, **1**, 647–681.

[4] Esary, J. D. and Proschan, F. (1972). *Ann. Math. Statist.*, **43**, 651–655.

[5] Esary, J., Proschan, F., and Walkup, W. (1967). *Ann. Math. Statist.*, **38**, 1466–1474.

[6] Hoeffding, W. (1940). *Schr. Math. Inst. Univ. Berlin*, **5**, 181–233.

[7] Jensen, D. R. (1969). *J. Amer. Statist. Ass.*, **64**, 333–336.

[8] Jensen, D. R. (1971). *SIAM J. Appl. Math.*, **20**, 749–753.

[9] Jogdeo, K. (1970). *Ann. Math. Statist.*, **41**, 1357–1359.

[10] Jogdeo, K. (1974). In *Statistical Distributions in Scientific Work*, Vol. 1, G. P. Patil, S. Kotz, and J.

K. Ord, eds. D. Reidel, Dordrecht, Holand, pp. 271–279.

[11] Jogdeo, K. (1977). *Ann. Statist.*, **5**, 495–504.

[12] Jogdeo, K. (1978). *Ann. Statist.*, **6**, 232–234.

[13] Jogdeo, K. and Patil, G. P. (1975). *Sankhyā B*, **37**, 158–164.

[14] Lehmann, E. L. (1966). *Ann. Math. Statist.*, **37**, 1137–1153.

[15] Linfoot, E. H. (1957). *Inf. Control*, **1**, 85–89.

[16] Mallows, C. L. (1968). *Biometrika*, **55**, 422–424.

[17] Marshall, A. W. and Olkin, I. (1974). *Ann. Statist.*, **2**, 1189–1200.

[18] Marshall, A. W. and Olkin, I. (1979). *Inequalities: Theory of Majorization and Its Applications*. Academic Press, New York.

[19] Plackett, R. L. (1954). *Biometrika*, **41**, 351–360.

[20] Proschan, F. and Sethuraman, J. (1975). *Teor. veroyatn. ee primen.*, **20**, 197–198.

[21] Rényi, A. (1959). *Acta Math.*, **10**, 217–226.

[22] Robbins, H. (1954). *Ann. Math. Statist.*, **25**, 614–616.

[23] Schweizer, B. and Wolff, F. F. (1976). *C. R. Acad. Sci. Paris*, **283**, 609–611.

[24] Shaked, M. (1975). A Family of Concepts of Dependence for Bivariate Distributions. *Tech. Rep.*, Dept. of Statistics, University of Rochester, Rochester, N.Y.

[25] Shaked, M. (1977). *Ann. Statist.*, **5**, 505–515.

[26] Šidák, Z. (1968). *Ann. Math. Statist.*, **39**, 1425–1434.

[27] Slepian, D. (1962). *Bell Syst. Tech. J.*, **41**, 463–501.

[28] Yanagimoto, T. (1972). *Ann. Inst. Statist. Math. Tokyo*, **24**, 559–573.

[29] Yanagimoto, T. and Okamoto, M. (1969). *Ann. Inst. Statist. Math. Tokyo*, **21**, 489–506.

Acknowledgment

This work was partially supported by NSF Grant 7902581.

KUMAR JOGDEO

DEPENDENCE, MEASURES AND INDICES OF

NOTATION

Let X and Y be random variables with distribution functions G and H, respectively and joint distribution function F. If it can be defined, the likelihood ratio or Radon–Nikodym derivative of bivariate or F-measure with respect to product or $G \times H$-measure is written as

$$\Omega \equiv \Omega(x, y) = dF(x, y)/\left[dG(x) dH(y) \right].$$

(1)

$\{x^{(i)}\}$ and $\{y^{(j)}\}$ are always complete orthonormal systems on G and H, respectively, with $x^{(0)} = 1 = y^{(0)}$. X is said to be completely dependent on Y or determined by Y if $X = f(Y)$, where f is a measurable function on the space of Y.

MEASURES AND INDICES

A measure of dependence indicates, in some defined way, how closely X and Y are related, with extremes at mutual independence and complete mutual dependence. If the measure can be expressed as a scalar, it is convenient to refer to it as an index. Such indices were first devised by Karl Pearson*. However, many later authors such as C. Gini and M. Fréchet have made refinements to the theory.

CONDITIONS ON AN INDEX OF DEPENDENCE

The conditions for an index $\delta(X, Y)$ to be useful can be stated after Rényi [25], although they are not original with him (e.g., see Pearson [21], Pearson and Heron [24], Fréchet [5], and Gini [6, 7]). The conditions are followed by short explanations.

1. $\delta(X, Y)$ is defined for any pair of random variables, neither of them being constant with probability 1. This is to avoid trivialities.

2. $\delta(X, Y) = \delta(Y, X)$. Although independence is a symmetrical property, complete dependence is not, e.g., $Y = \text{sign } X$, where X is standard normal.

3. $0 \leqslant \delta(X, Y) \leqslant 1$. This is an obvious choice.

4. $\delta(X, Y) = 0$ if and only if X and Y are mutually independent. This is the "strong" condition of A. N. Kolmogorov.

5. If the Borel-measurable functions $\xi(\cdot)$ and $\eta(\cdot)$ map the spaces of X and Y respectively onto themselves biuniquely, then $\delta(\xi(X), \eta(Y)) = \delta(X, Y)$. This condition requires that the index remain invariant under measure-preserving transformations of either or both marginal variables, so δ is a functional of Ω.

6. $\delta(X, Y) = 1$ if and only if X and Y are mutually determined. Equivalently:

6'. $\delta(X, Y) = 1$ if and only if measure-preserving transformations can be made $X \rightarrow \xi(X)$, $Y \rightarrow \eta(Y)$, where ξ and η are standardized and such that $\text{corr}(\xi, \eta) = 1$. Note that this implies that $\xi = \eta$ a.e.

7. If X and Y are jointly normal, then $\delta(X, Y) = |\rho|$. If δ is to have general validity, it must hold for the joint normal distribution.

8. In any family of distributions defined by a vector parameter $\boldsymbol{\theta}$, $\delta(X, Y)$ must be a function of $\boldsymbol{\theta}$. This is reasonable.

9. For a given F and for any sequence of partitions generating the σ-rings of the marginal spaces, let $\delta_k \equiv \delta(x_k, Y_k)$ be calculated at the kth partition; then the sequence $\{\delta_k\}$ must tend to a unique limit δ. This is necessary to avoid inconsistencies.

MATRIX OF CORRELATIONS

Let $\rho_{ij} = E(x^{(i)}y^{(j)})$ and $\mathbf{R} = (\rho_{ij})$, $i, j = 1, 2, \ldots$. For given G and H, $\mathbf{R}$ completely determines F [17], so that $\mathbf{R}$ can be said to be a matrix measure of dependence. In particular, $\mathbf{R} = \mathbf{0}$ if and only if X is independent of Y. $\mathbf{R}$ is orthogonal if and only if X and Y are mutually determined [18], for the orthogonal condition implies that $L_G^2 \subseteq L_H^2$ and $L_H^2 \subseteq L_G^2$. Let ξ be $X/(1 + |X|)$ in standardized form. Then $\xi \in L_G^2$, so is in L_H^2, and is determined by Y. Since the transformation is

strictly monotonic, X is determined by Y. Similarly, Y is determined by X. $\{x^{(i)}\}$ and $\{Y^{(j)}\}$ can be specialized in some cases to be polynomials or other functions of the marginal variables. Special interest holds if the two series possess the biorthogonal property $E(x^{(i)}y^{(j)}) = \delta_{ij}\rho_j$, and so $\mathbf{R}$ is diagonal. Some properties of $\mathbf{R}$ can be summarized in a single scalar, e.g., $\rho_{11} = \text{corr}(X, Y)$ with a suitable choice of $x^{(1)}$ and $y^{(1)}$; $\rho_{11} = 1$ in this case forces $\mathbf{R}$ to be orthogonal. $\phi^2 = \text{tr}\,\mathbf{R}\mathbf{R}^T$ is an important scalar property. If X and Y have been transformed to rectangular variables on the unit interval, perhaps after a randomized partition to give rectangular distributions, and if $\{x^{(i)}\}$ and $\{y^{(j)}\}$ are taken to be the standardized Legendre polynomials*, ρ_{ij} is termed a quasi-moment.

MEAN SQUARE CONTINGENCY

In the joint normal distribution with the orthonormal sets taken to be the standardized Hermite polynomials, $\rho_{ij} = \delta_{ij}\rho^i$ and $1 + \phi^2 = \text{tr}\,\mathbf{R}\mathbf{R}^T + 1 = (1 - \rho^2)^{-1}$ and $\rho = \phi(1 + \phi^2)^{-1/2}$. Pearson [23] suggested an estimate from an $m \times n$ contingency table

$$\hat{\phi}^2 = \{\chi^2 - (m-1)(n-1)\}/N, \quad (2)$$

where χ^2 is the usual Pearson χ^2 calculated from a sample of size N. This estimate can be improved by equating this expression to a series in $\hat{\rho}^2$ to obtain a variation on Pearson's polychoric correlation [19]. ϕ^2 is a strong index, but it can be infinite without there being mutual determination, so that no monotonic transformation can satisfy condition 6.

PRODUCT MOMENT CORRELATION

$|\rho|$, where ρ is the common coefficient of correlation, is not a strong index. It can be zero when X and Y are mutually determined. However, if $|\rho| = 1$, X is a linear form in Y. Moreover, $|\rho|$ completely determines the joint distribution for the random elements in the common model,

$X = \sum_{i \in A} U_i$, $Y = \sum_{j \in B} U_j$, where $\{U_i\}$ is a set of independent, identically distributed (i.i.d.) variables with finite variance and A and B are sets of the positive integers, for $\rho = n(A \cap B)[n(A)n(B)]^{-1/2}$, where $n(\cdot)$ is the cardinality of the set.

CORRELATION RATIOS

The correlation ratio η_{XY}^2 was defined by Pearson [22] for variables X of finite variance. None of η_{XY}^2, η_{YX}^2, and $\frac{1}{2}(\eta_{XY}^2 + \eta_{YX}^2)$ is strong. Each is equal to ρ^2 in the normal distribution, and in the random elements in common model for any Meixner class of variables. $\eta_{XY}^2 = 1$ if and only if X is determined by Y.

MAXIMAL CORRELATION

The maximal [13] or first canonical correlation is a strong index since it vanishes if and only if $\mathbf{R} = \mathbf{0}$. It always exists in ϕ^2-bounded distributions. Examples show that even an infinity of canonical correlations of unity does not imply mutual determination or agree otherwise with intuition; *see* CANONICAL ANALYSIS.

HOEFFDING'S Φ^2

Hoeffding [14] defined

$$\Phi^2 = 90 \int_0^1 \int_0^1 (F(x, y) - xy)^2 \, dx \, dy \quad (3)$$

for rectangularly distributed X and Y. If either is not so distributed, it is to be made so by a strictly monotonic transformation, perhaps after a randomized partition; this partition does not affect the distribution of Ω. This is a a strong measure of dependence and $\Phi^2 = 1$ if and only if the variables are mutually determined. It can be proved that $8\Phi^2 \geqslant 5\rho^2$, where ρ is the correlation between the first pair of Legendre polynomials. Hoeffding [15] also defined c-dependence if the points of increase of F were all included in a set M which had zero product measure.

This is a weaker condition than mutual determination, although it implies an infinite Pearson ϕ^2.

POSITIVE CONTINGENCY

The integral of $(\Omega - 1)$ with respect to product measure is zero; the corresponding integral, ψ, taken over the positive values of $(\Omega - 1)$ is thus equal to half the value of the integral $|\Omega - 1|$ over the whole product space. Pearson [21] defined ψ as the positive contingency and this was obtained in a slightly different manner by Hoeffding [15]. The line delineating areas of positive contingency, namely $\Omega = 1$, is a hyperbola in the joint normal distribution, and Pearson [21] showed that ψ is a monotonically increasing function of $|\rho|$. Pearson's ψ is related to Steffensen's ω by

$$\omega = 2\psi/(1 + \psi) \quad (4)$$

in the limit. All three authors hoped to estimate ω or ψ from discrete data, by

$$\hat{\psi} = N^{-1}\sum a_{ij} \quad \text{for } a_{ij} > a_{i.} a_{.j}/N. \quad (5)$$

ψ is strong and $\psi = 1$ if and only if the joint measure is singular with respect to product measure but $\psi = 1$ does not imply mutual determination.

TETRACHORIC CORRELATION

The standard normal density $f(x, y)$ can be written

$$f(x, y) = g(x)h(y)\left(1 + \sum_1^\infty \rho^n x^{(n)} y^{(n)}\right),$$
$$|\rho| < 1, \quad (6)$$

where g and h are standard normal densities and $x^{(n)}$ and $y^{(n)}$ are standardized Hermite polynomials. Convergence is absolute a.e. with respect to product measure. The value of the normal measure in a quadrant defined by a double dichotomy at (x_0, y_0) can thus be determined as a sum of products $\rho^n \int_{-\infty}^{x_0} x^{(n)} g(x) \, dx \int_{-\infty}^{y_0} y^{(n)} h(y) \, dy$. Pearson [20] assumed that an underlying normal distribution had generated a given fourfold or

2×2 contingency table and that x_0 and y_0 could be accurately estimated by setting the probability in the first row or column of the table equal to the cumulative probability of the standard normal distribution. The sum obtained by integrating (6) was thus a series in the powers of ρ and this was truncated and equated to the probability in the appropriate cell of the fourfold table. The resultant solution $\hat{\rho}$ is the tetrachoric correlation. K. Pearson and his school spent much time in elaborating other indices, such as the polychoric correlation mentioned above. These methods could be extended to distributions in other nonnormal Meixner variables.

YULE'S COEFFICIENTS

Yule [28] reviewed measures of dependence particularly in the fourfold table, to which we now turn. His κ is the cross ratio, $f_{12}f_{21}/(f_{11}f_{22})$. Q is $(1 - \kappa)/(1 + \kappa)$ and r is the coefficient of correlation. These three are related biuniquely in pairs. Q is a strong measure of association, but $Q = 1$ does not imply mutual determination, except when there is symmetry, so that $f_{12} = 0$ if and only if $f_{21} = 0$. Although F is determined by the ensemble of correlations r, for every fourfold table made by a double dichotomy, there seems no way of applying these measures conveniently except by the use of the correlation matrix $\mathbf{R}$, already defined.

STEFFENSEN'S INDICES OF DEPENDENCE

Steffensen [26, 27] defined

$$\psi^2 = \frac{\sum_{i,j} f_{ij}(f_{ij} - f_{i.}f_{.j})^2}{\left[f_{i.}(1 - f_{i.})f_{.j}(1 - f_{.j}) \right]} \qquad (7)$$

and

$$\omega = \frac{\sum |f_{ij} - f_{i.}f_{.j}|}{\left[\sum' (f_{ij} - f_{ij}^2) + \sum'' f_{i.}f_{.j} \right]}, \qquad (8)$$

where $\sum'$ and $\sum''$ are summation operators for i, j such that $f_{ij} > f_{i.}f_{.j}$ and $f_{ij} \leqslant f_{i.}f_{.j}$. For variables assuming only a finite set of values, both indices are strong and take a value

of unity if and only if X and Y are mutually determined. However, if the underlying distributions are continuous, ψ^2 can be unity for some finite partition but can tend to zero for increasingly fine partitions. ω is related to the positive contingency by (4).

CRAMER'S INDEX OF DEPENDENCE

Cramér [1] defined

$$\mu = \sigma_X \sigma_Y \int \int \left[f(x, y) - r(x)c(y) \right]^2 dx \, dy,$$

$$(9)$$

where $r(x)$ and $c(y)$ are continuous functions chosen to minimize μ. The integral can be evaluated in joint normal distribution. This index has been little used. Condition 5 is not satisfied.

GUTTMAN'S INDEX OF DEPENDENCE

Guttman [12] gave an index of dependence for $m \times n$ distributions in which asymmetry is supposed to exist so that Y is dependent on X in some causal or chronological sense. The problem is: Given $X = a$, what is the best estimate of the value of Y? If there is independence, the value of y, corresponding to max $p_{.j}$, would be chosen, M say.

$$\lambda_f = 1 - \frac{(\text{Pr of error conditional on } X)}{(\text{Pr of unconditional error})}$$

$$= \left(\sum_x f_{x, \max_x} - f_{.M} \right) / (1 - f_{.M}). \quad (10)$$

If symmetry is supposed to hold, λ_a can be similarly defined. λ can be defined as the sum of numerators of λ_a and λ_b divided by the sum of the denominators. λ_a, λ_b, and λ are not strong. The index depends on the fineness of the partition.

FRÉCHET'S DISTANCE INDEX

Fréchet [3] proposed to use an average of the distances of the distributions of Y conditional on X from some typical value such as the conditional mean or median. Fréchet [4] proposed that the Lévy metric be used. The two indices are strong and take the unit value when Y is determined by X. Neither would be invariant as required by condition 5.

INFORMATION-THEORETICAL INDICES

After Csiszár [2], we define

$$I_f(F, GH) = \int f(\Omega) \, dG \, dH.$$

f may be given particular functional forms and if we write u for Ω, I_f takes values as follows:

1. The directed divergence when $f = u \log u$
2. The directed divergence of order α, when $f = u^\alpha / (\alpha - 1)$
3. Pearson's ϕ^2 when $f = (u - 1)^2$
4. Pearson's positive contingency when $f = |u - 1|$
5. The symmetrized divergence of order 1 when $f = (u - 1) \log u$

All these are strictly convex functions of u except $|u - 1|$, which is only strictly convex at $u = 1$. Items 3 and 4 have been treated above. An informational theory index of dependence can be formed from any of the other three by making a strictly monotonic transformation to a function on the unit interval strictly convex at the origin. There results in each case a strong index, but there is difficulty at the upper end of the scale.

CONCLUSIONS

A general index of dependence, whereby joint distributions can be arranged in order of the degree of dependence, does not exist. For some defined classes of distribution, such as the joint normal and the random elements in common models, the product moment correlation is the index of choice. In other classes, there may be indices useful for some purposes. There is inevitably loss of

information* in passing from, say, the matrix of correlations to a single index.

The sampling errors of the indices are often determined only with difficulty [11, 16]. Extensive bibliographies have been provided by Goodman and Kruskal [8, 9, 10].

References

[1] Cramér, H. (1924). *Skand. Aktuarietidskr.*, **7**, 220–240.

[2] Csiszár, I. (1967). *Stud. Sci. Math. Hung.*, **2**, 299–318.

[3] Fréchet, M. (1946). *Proc. Math. Phys. Soc. Egypt*, **3**(2), 13–20.

[4] Fréchet, M. (1948). *Proc. Math. Phys. Soc. Egypt*, **3**, 73–74.

[5] Fréchet, M. (1958–1959). Trans. from the French by C. de La Menardière, *Math. Mag.*, **32**, 265–268.

[6] Gini, C. (1914). *Atti. R. Ist. Veneto Sci. Lett. Arti*, (8) **74**, 185–213.

[7] Gini, C. (1914). *Atti R. Ist. Veneto Sci. Lett. Arti*, (8) **74**, 583–610.

[8] Goodman, L. A. and Kruskal, W. H. (1954). *J. Amer. Statist. Ass.*, **49**, 732–764; Correction: **52**, 578.

[9] Goodman, L. A. and Kruskal, W. H. (1959). *J. Amer. Statist. Ass.*, **54**, 123–163.

[10] Goodman, L. A. and Kruskal, W. H. (1963). *J. Amer. Statist. Ass.*, **58**, 310–364.

[11] Goodman, L. A. and Kruskal, W. H. (1972). *J. Amer. Statist. Ass.*, **67**, 415–421.

[12] Guttman, L. (1941). In The Prediction of Personal Adjustment, P. Horst, et al., eds. *Bull. No. 48*, Social Science Research Council, New York, pp. 253–318.

[13] Hirschfeld, H. O. (1935). *Proc. Camb. Philos. Soc.*, **31**, 520–524.

[14] Hoeffding, W. (1940). *Schr. Math. Inst. Angew. Math. Univ. Berlin*, **5**(3), 181–233.

[15] Hoeffding, W. (1942). *Skand. Aktuarietidskr.*, **25**, 200–227.

[16] Kruskal, W. H. (1958). *J. Amer. Statist. Ass.*, **53**, 814–861.

[17] Lancaster, H. O. (1963). *Ann. Math. Statist.*, **34**, 532–538.

[18] Lancaster, H. O. (1963). *Ann. Math. Statist.*, **34**, 1315–1321.

[19] Lancaster, H. O. and Hamdan, M. A. (1964). *Psychometrika*, **29**, 383–391.

[20] Pearson, K. (1900). *Philos. Trans. R. Soc. A*, **195**, 1–47.

[21] Pearson, K. (1904). Mathematical Contributions

to the Theory of Evolution. XIII. On the Theory of Contingency and Its Relation to Association and Normal Correlation. *Drapers' Co. Res. Mem. Biom. Ser. I.*

[22] Pearson, K. (1905). Mathematical Contributions to the Theory of Evolution. XIV. On the General Theory of Skew Correlation and Non-linear Regression. *Drapers' Co. Res. Mem. Biom. Ser. II.*

[23] Pearson, K. (1913). *Biometrika*, **9**, 116–139.

[24] Pearson, K. and Heron, D. (1913). *Biometrika*, **9**, 159–315.

[25] Rényi, A. (1959). *Acta Math. Acad. Sci. Hung.*, **10**, 441–451.

[26] Steffensen, J. F. (1934). *Biometrika*, **26**, 251–255.

[27] Steffensen, J. F. (1941). *Skand. Aktuarietidskr.*, **24**, 13–33.

[28] Yule, G. U. (1912). *J. R. Statist. Soc.*, **75**, 579–652.

(ASSOCIATION, MEASURES OF
CANONICAL ANALYSIS
DEPENDENCE, CONCEPTS OF)

H. O. Lancaster

DEPENDENCE, TESTS FOR

ABSTRACT

This paper is designed to provide a sound introduction for a reasonably well informed reader who is, however, not a specialist in tests for dependence. The paper contains references to many tests but emphasizes the parametric test of independence based on Pearson's sample correlation coefficient* r and certain nonparametric tests based on ranks. The ranks tests are generally preferable to the test based on r in that they have wider applicability, are much less sensitive to outlying observations, are exact under mild assumptions that do not require an underlying bivariate normal* population, and have good efficiency (power) properties.

INTRODUCTION

Many studies are designed to explore the relationship between two random variables X and Y, say, and specifically to determine whether X and Y are independent or dependent. Some particular examples are:

1. **Obesity and blood pressure.** Are obesity and blood pressure independent or, for example, do men who are overweight also tend to have high blood pressure? Here X could be the degree of overweight as measured by the ratio of actual body weight to ideal body weight as given in certain standard tables, and Y could be systolic blood pressure.

2. **Color and taste of tuna.** Are color and quality of canned tuna independent or perhaps do consumers tend to prefer light tuna? Here X could be a measure of lightness and Y could be a quality score determined by a consumer panel.

3. **Infants walking and their IQ.** Is the time until it takes an infant to walk alone independent of the infant's IQ at a later age, or do children who learn to walk early tend to have higher IQs? Here X could be the number of days measured from birth until the infant walks alone, and Y could be the infant's IQ score at age 5.

4. **System reliability and the environment.** Is the life length X (say) of a specific system independent of a certain characteristic of the environment, for example, the temperature Y, within which the system operates, or do high temperatures tend to shorten the life length?

One can test the null hypothesis that the two variables X and Y are independent, against alternatives of dependence, using a random sample from the underlying bivariate population. We suppose that such a sample of size n is available, and we denote the sample by $(X_1, Y_1), (X_2, Y_2), \ldots, (X_n, Y_n)$. Our assumptions are:

1. The n bivariate observations $(X_1, Y_1), \ldots, (X_n, Y_n)$ are mutually independent.

2. Each (X_i, Y_i) comes from the same bivariate population with continuous dis-

tribution function $H(x, y) = P(X \leqslant x, Y \leqslant y)$ and continuous marginal distributions $F(x) = P(X \leqslant x)$ and $G(y) = P(Y \leqslant y)$.

The *hypothesis of independence* asserts that

$$H_0 : H(x, y) = F(x)G(y) \quad \text{for all } (x, y); \tag{1}$$

that is, the variables X and Y are independent. Under H_0, all $2n$ random variables are mutually independent; that is,

$$P(X_1 \leqslant x_1, Y_1 \leqslant y_1, \ldots, X_n \leqslant x_n, Y_n \leqslant y_n)$$
$$= \prod_{i=1}^{n} F(x_i)G(y_i).$$

When we discuss alternatives to H_0, we will be assuming that X and Y are dependent, so that (1) fails to hold, but we still insist that the independence *between* the n pairs is preserved.

TEST BASED ON PEARSON'S CORRELATION COEFFICIENT

The concept of correlation* is due to Francis Galton* in a series of papers in the 1880s; see Galton [9]. The Pearson correlation coefficient r was derived by the eminent statistician Karl Pearson* [26] in further developments and studies of Galton's methods. The statistic r, often called Pearson's product-moment correlation coefficient, is

$$r = S_{XY} / \{ S_{XX} S_{YY} \}^{1/2}; \tag{2}$$
$$S_{XY} = \textstyle\sum_{i=1}^{n} X_i Y_i - (\sum_{i=1}^{n} X_i)(\sum_{i=1}^{n} Y_i)/n,$$
$$S_{XX} = \textstyle\sum_{i=1}^{n} X_i^2 - (\sum_{i=1}^{n} X_i)^2/n,$$
$$S_{YY} = \textstyle\sum_{i=1}^{n} Y_i^2 - (\sum_{i=1}^{n} Y_i)^2/n,$$

The statistic r is the sample correlation coefficient and is an estimator of the corresponding population parameter ρ, the correlation coefficient of the bivariate population defined by $H(x, y)$. Specifically,

$$\rho = \frac{E(XY) - E(X)E(Y)}{\sigma_X \sigma_Y}, \tag{3}$$

where E denotes expectation, σ_X is the standard deviation of the X population, and σ_Y is the standard deviation of the Y population. It can be shown that for all samples $-1 \leqslant r \leqslant 1$, and for all bivariate populations $-1 \leqslant \rho \leqslant 1$. When $\rho > 0$, this may be interpreted as X and Y being positively associated (as measured by ρ) and $\rho < 0$ may be interpreted as X and Y being negatively associated (as measured by ρ). Assuming that $H(x, y)$ is a bivariate normal cumulative distribution function with correlation ρ, an exact α level test of H_0 versus $\rho \neq 0$ is

reject H_0 in favor of $\rho \neq 0$ if $|T| \geqslant t_{\alpha/2, n-2}$, accept H_0 if $|T| < t_{\alpha/2, n-2}$, $\tag{4}$

where $t_{\alpha/2, n-2}$ is the upper $\alpha/2$ percentile point of Student's t-distribution with $n - 2$ degrees of freedom, and

$$T = (n - 2)^{1/2} r / (1 - r^2)^{1/2}. \tag{5}$$

Since $|T|$ is an increasing function of $|r|$, the test defined by (4) is equivalent to the test that rejects for large values of $|r|$, and the latter is easily derived to be the likelihood ratio test of H_0 versus $\rho \neq 0$ in the model that assumes bivariate normality. (Of course, under the bivariate normality assumption, X and Y are independent if and only if $\rho = 0$.)

One-sided tests based on T are readily defined. To test H_0 versus $\rho > 0$, at the α level, reject H_0 if $T \geqslant t_{\alpha, n-2}$ and accept H_0 if $T < t_{\alpha, n-2}$. To test H_0 versus $\rho < 0$, at the α level, reject H_0 if $T \leqslant -t_{\alpha, n-2}$ and accept H_0 if $T > -t_{\alpha, n-2}$.

The two-sided test defined by (4), and the corresponding one-sided tests, are exact (i.e., have true type I error probability equal to the nominal value α) if at least one of the two variables X, Y has a (marginal) normal distribution. Of course, assuming that the joint distribution of (X, Y) is bivariate normal is enough to ensure this condition.

Normal approximations to the distribution of r are summarized in Bickel and Doksum [4, Sec. 6.5.A]. One such approximation treats T as an approximate standard normal random variable under H_0.

The test based on T can also be developed within a linear regression model. Assume that the conditional distribution of one variable (Y, say) given the other ($X = x$, say) is normal with linear regression $E(Y|x) = \alpha + \beta x$ and constant variance so that $\rho = \beta \sigma_X / \sigma_Y$. The least-squares estimator b, of the slope β, is

$$b = \frac{n\sum_{i=1}^{n} X_i Y_i - (\sum_{i=1}^{n} X_i)(\sum_{i=1}^{n} Y_i)}{n\sum_{i=1}^{n} X_i^2 - (\sum_{i=1}^{n} X_i)^2},$$

and r can be rewritten as $r = bs_X / s_Y$, where s_X, s_Y are the sample standard deviations for the X and Y samples, respectively. Thus the test of $\rho = 0$ can be viewed as a test of the hypotheses that the slope β is zero.

Of course, in many cases regressions are not linear. Kendall and Stuart [20, Secs. 26.21–26.24] discuss the population parameter $\eta_1^2 - \rho^2$, where $\eta_1^2 = \text{var}\{E(X|Y)\}/\sigma_X^2$, as an indicator of nonlinearity of regression. (η_1 is called the correlation ratio* of X on Y.) Kendall and Stuart describe how, under certain conditions, the sample analog of $\eta_1^2 - \rho^2$ can be used for a test of linearity of regression.

For more information on testing independence in a parametric context where an assumption of normality is made, see Bickel and Doksum [4, Sec. 6.5.A] and Kendall and Stuart [20, Chap. 26].

In the following section we present nonparametric tests of H_0 which are exact without requiring an assumption of normality.

RANK TESTS OF INDEPENDENCE

Let R_i be the rank of X_i in the joint ranking from least to greatest of $X_1, \ldots, X_n$, and let S_i be the rank of Y_i in the (separate) joint ranking from least to greatest of $Y_1, \ldots, Y_n$.

Under assumptions 1 and 2 and H_0, the vector of X ranks $R = (R_1, \ldots, R_n)$ is independent of the vector of Y ranks $S = (S_1, \ldots, S_n)$, and both R and S have uniform distributions over the space $\mathcal{P}$ of the $n!$ permutations $(i_1, \ldots, i_n)$ of the integers $(1, \ldots, n)$. That is, for each permutation $(i_1, \ldots, i_n)$,

$$P_0\{(R_1, \ldots, R_n) = (i_1, \ldots, i_n)\} = 1/n!,$$

with the same result holding for $(S_1, \ldots, S_n)$. (The subscript 0 indicates the probability is computed under H_0.) It follows that rank statistics (i.e., statistics that are solely based on R and S) are distribution-free under H_0.

One important class of rank statistics for testing H_0 are the *linear* rank statistics of the form

$$L = \sum_{i=1}^{n} a(R_i) b(S_i), \qquad (6)$$

where the "scores" $a(R_i)$, $b(S_i)$ satisfy $a(1) \leqslant \cdots a(n)$, $b(1) \leqslant \cdots \leqslant b(n)$.

Test Based on Spearman's Rank Correlation Coefficient*

Making the choice $a(i) = b(i) = i$ in (6), L reduces to

$$M = \sum_{i=1}^{n} R_i S_i. \qquad (7)$$

Then if M is linearly transformed so that the minimum and maximum values are -1 and 1, we obtain Spearman's rank-order correlation coefficient

$$r_s = \frac{12\sum_{i=1}^{n} [R_i - (n+1)/2][S_i - (n+1)/2]}{n(n^2 - 1)}. \qquad (8)$$

An even simpler formula for computational purposes is

$$r_s = 1 - \frac{6\sum_{i=1}^{n} D_i^2}{n^3 - n}, \qquad (9)$$

where $D_i = R_i - S_i$. Note also that r_s is obtainable from $r(2)$ by replacing X_i with its X-rank R_i and Y_i with its Y-rank S_i.

The statistic r_s does not estimate ρ as given in (3); rather, it estimates the population parameter

$$\rho_s = 6P\{(X_1 - X_2)(Y_1 - Y_3) > 0\} - 3. \qquad (10)$$

It can be shown that for all samples $-1 \leqslant r_s \leqslant 1$, and for all bivariate populations $-1 \leqslant \rho_s \leqslant 1$. Note that

$$P\{(X_1 - X_2)(Y_1 - Y_3) > 0\}$$
$$= P(X_1 > X_2, Y_1 > Y_3)$$
$$+ P(X_1 < X_2, Y_1 < Y_3)$$

and when H_0 is true,

$$P\{(X_1 - X_2)(Y_1 - Y_3) > 0\}$$
$$= P(X_1 > X_2)P(Y_1 > Y_3)$$
$$+ P(X_1 < X_2)P(Y_1 < Y_3)$$
$$= \tfrac{1}{4} + \tfrac{1}{4} = \tfrac{1}{2},$$

so that when H_0 is true, $\rho_s = 0$. In addition, $\rho_s > 0$ may be interpreted as X and Y being positively associated (as measured by ρ_s), and $\rho_s < 0$ may be interpreted as X and Y being negatively associated (as measured by ρ_s). (For further information and interpretation of the parameter ρ_s as a measure of association, see Kruskal [22].)

Under assumptions **1** and **2**, an exact α level test of H_0 versus $\rho_s \neq 0$ is

$$\text{reject } H_0 \text{ in favor of } \rho_s \neq 0$$
$$\text{if } |r_s| \geq r_s(\alpha/2, n), \qquad (11)$$
$$\text{accept } H_0 \text{ if } |r_s| < r_s(\alpha/2, n),$$

where $r_s(\alpha/2, n)$ is the upper $\alpha/2$ percentile point of the null distribution of r_s. To test H_0 versus the one-sided alternative $\rho_s > 0$, at the α level, reject H_0 if $r_s \geq r(\alpha, n)$ and accept H_0 otherwise. To test H_0 versus $\rho < 0$, at the α level, reject H_0 if $r_s \leq -r(\alpha, n)$ and accept H_0 otherwise.

From (9) we see that tests based on r_s are equivalent to tests based on the statistic $\sum D_i^2$. Glasser and Winter [10] give critical values of r_s and $\sum D_i^2$ for $n = 4(1)30$. Tables of the complete null distribution of r_s and $\sum D_i^2$ are given for $n = 4(1)11$ in Kraft and van Eeden [21].

Under H_0, $E(r_s) = 0$, $\text{var}(r_s) = 1/(n-1)$, and as n gets large, the distribution of $(n-1)^{1/2}r_s$ tends to the standard normal distribution. Thus approximate (for large n) tests of H_0 can be obtained by treating

$$r_s^* = (n-1)^{1/2}r_s \qquad (12)$$

as a standard normal variable under H_0.

The eminent psychologist C. Spearman [30] suggested r_s as a measure of the degree of dependence between X and Y. The nonparametric test of H_0 based on r_s is due to Hotelling and Pabst [15].

Test Based on Kendall's Rank Correlation Coefficient

Kendall's rank correlation coefficient can be written as

$$r_k = 2 \sum_{i=1}^{n-1} \sum_{j=i+1}^{n} \xi(X_i, X_j, Y_i, Y_j)/\{n(n-1)\},$$
$$(13)$$

where $\xi(a, b, c, d) = 1$ if $(a - b)(c - d) > 0$, and $= -1$ if $(a - b)(c - d) < 0$. When $(X_i - X_j)(Y_i - Y_j) > 0$, we say that the pairs $(X_i, Y_i), (X_j, Y_j)$ are concordant and when $(X_i - X_j)(Y_i - Y_j) < 0$, we say that the pairs are discordant. Note that r_k is a rank statistic* ($\xi(X_i, X_j, Y_i, Y_j) = \xi(R_i, R_j, S_i, S_j)$, so that one only needs the ranks to compute r_k), but it is not a linear rank statistic. However, it can be shown (see Hájek and Šidák [13, Sec. II.3.1]) that, up to a multiplicative constant, Spearman's r_s is the "projection" of Kendall's r_k into the family of linear rank statistics. The statistic r_k estimates the parameter $\tau = 2P\{(X_1 - X_2)(Y_1 - Y_2) > 0\} - 1$. It can be shown that for all samples $-1 \leq r_k \leq 1$, and for all bivariate populations $-1 \leq \tau \leq 1$. When H_0 is true, $\tau = 0$. In addition, $\tau > 0$ may be interpreted as X and Y being positively associated (as measured by τ), and $\tau < 0$ may be interpreted as X and Y being negatively associated (as measured by τ). The reader should note that τ is analogous to the parameter $\rho_s(10)$ estimated by Spearman's r_s. For details of the relationship between ρ_s and τ, see Kruskal [22].

From (13) we see that tests based on r_k are equivalent to tests based on

$$K = \sum_{i=1}^{n-1} \sum_{j=i+1}^{n} \xi(X_i, X_j, Y_i, Y_j). \qquad (14)$$

Under assumptions **1** and **2**, an exact α level test of H_0 versus $\tau \neq 0$ is

reject H_0 in favor of $\tau \neq 0$ if $|K| \geq k(\alpha/2, n)$,

accept H_0 if $|K| < k(\alpha/2, n)$, $\qquad (15)$

where $k(\alpha/2, n)$ is the upper $\alpha/2$ percentile point of the null distribution of K. To test H_0 versus $\tau > 0$, at the α level, reject H_0 if $K \geq k(\alpha, n)$ and accept H_0 otherwise. To test H_0 versus $\tau < 0$, at the α level, reject H_0 if $K \leqslant -k(\alpha, n)$ and accept H_0 otherwise. Kaarsemaker and van Wijngaarden [17] give tables of the null distribution of K for $n = 4(1)40$. See also Table A.21 of Hollander and Wolfe [14]. Extended tables up to $n = 100$ are made available on request by D. J. Best [2].

Under H_0,

$$E(K) = 0, \qquad \text{var}(K) = n(n-1)(2n+5)/18,$$

and as n gets large, the standardized distribution of K tends to the standard normal distribution. Thus approximate (for large n) tests of H_0 can be obtained by treating

$$K^* = K / \left[n(n-1)(2n+5)/18 \right]^{1/2}$$

as a standard normal variable under H_0.

Kendall [18] considered K in detail, although the statistic has been in the literature since the nineteenth century. Kruskal [22] gives the history of certain independent discoveries of K.

Ties

Although assumption **2** precludes the possibility of ties, ties may occur in practice. One method of treating ties, when dealing with rank statistics, is to replace R_i by R_i^* (the average of the ranks that X_i is tied for), S_i by S_i^* (the average of the ranks that Y_i is tied for), compute the rank statistic using the R^*'s and S^*'s, and refer it to the appropriate null distribution tables derived under the assumption of continuity. This, however, yields only an approximate, rather than an exact test.

Exact conditional tests, in the presence of ties, can be performed but they are computationally tedious. See, e.g., Lehmann [24, Sec. 7.3]. For more information on ties, see Hájek [12, Chap. VII].

Advantages of Rank Tests

Advantages of rank tests, as compared to the parametric test based on r, include:

1. *Wider applicability.* To compute a rank statistic, we need only know the ranks, rather than the actual observations.

2. *Outlier* insensitivity.* Rank statistics are less sensitive than r to wildly outlying observations.

3. *Exactness.* Tests based on rank statistics are exact under the mild assumptions 1 and 2, whereas the significance test based on r is exact when at least one of the two marginal distributions is normal.

4. *Good efficiency properties.* Rank tests of H_0 are only slightly less efficient than the normal theory test based on r when the underlying bivariate population is normal (the home court of r), and they can be mildly and wildly more efficient than r when the underlying bivariate population is not normal. Of course, the efficiency question is complicated, as it depends both on the specific rank test under consideration and the specific measure of efficiency used. Roughly speaking, for large n and dependency alternatives "close" to the null hypothesis, the tests based on r_s and r_k sacrifice 9% of the information in the sample, as compared to the test based on r, when the underlying population is bivariate normal, and can be much more efficient for certain nonnormal populations. For more details on efficiency and power, see Lehmann [24, Sec. 7.5E] and Hájek and Šidák [13, Sec. VII.2.4], and the references therein.

Other Rank Tests

A "normal scores" rank test studied by Fieller and Pearson [8] and Bhuchongkul [3] is particularly noteworthy. The normal scores* test statistic for independence is a linear rank statistic of the form (6) with $a(i) = b(i) = EV_n^{(i)}$, where $V_n^{(1)} < \cdots < V_n^{(n)}$ is an ordered sample of n observations from the standard normal distribution. For a suitable choice of the definition of efficiency and a suitable choice of the nature of dependency alternatives, the normal scores test of independence and the test based on r are equally efficient under "normality" and Srivastava [31] has shown that the normal scores test is more efficient than the test based on r for "all" (i.e., subject to mild regularity) other cases.

References to other nonparametric tests of independence can be found in Secs. 8.1 and 10.2 of Hollander and Wolfe [14] and in Sec. 7.5D of Lehmann [24].

RECENT DEVELOPMENTS

The relatively new areas of robustness, censoring, and multivariate generalizations are currently receiving much attention. It is reasonable to expect that future developments will focus, to some degree, on these areas.

Robustness

Pearson's r is not robust. In fact, one sufficiently bad outlying observation (X_i, Y_i), say, can shift r to any value in $(-1, 1)$. Gnanadesikan and Kettenring [11] and Devlin et al. [7] consider the problem of robust estimation and outlier detection with correlation coefficients. Huber [16] considers robust regression and robust covariances within a general treatment of robust statistical procedures.

Censoring

In clinical trials*, where the observation may be the time to the occurrence of an end-point event, the data usually are analyzed before all patients have experienced the event. For example, in a study of the length of post-transplant survival for patients receiving a heart transplant, many patients in the study may be alive at the time of data analysis. For such a survivor, all that is known is that his or her survival time is at least as great as the survival time thus far. Incomplete information also occurs when a patient is lost to the study by, for example, moving or dying in an automobile accident.

These situations yield incomplete observations called *censored* observations. Brown et al. [5] have developed some generalizations of Kendall's K to test for independence when one or both of the variables X, Y are censored. Similar generalizations of Spearman's r_s have been advanced by Latta [23]. Miller [25] considered the independence problem in the presence of censored data and in the framework of the usual linear regression model. See also Cox [6] for regression methods with censored data.

Multivariate Generalizations

Suppose that, instead of a random sample of n bivariate observations $(X_1, Y_1), \ldots,$ (X_n, Y_n), we observe a random sample of n p-variate $(p \geqslant 2)$ observations $(X_{11}, \ldots, X_{p1}), \ldots, (X_{1n}, \ldots, X_{pn})$, each with continuous distribution function $F(\mathbf{x}) = P(X_{11} \leqslant x_1, \ldots, X_{p1} \leqslant x_p)$ for $\mathbf{x} = (x_1, \ldots, x_p)$, an arbitrary element in p-dimensional Euclidean space R_p. Denoting the marginal distribution function of X_{1i} by $F_i(x)$, $i = 1, \ldots, p$, the hypothesis H_0^p (say) of total independence is

$$H_0^p : F(\mathbf{x}) = \prod_{i=1}^{p} F_i(x_i) \qquad \text{for all } \mathbf{x} \in R_p.$$
(17)

One can also consider hypotheses that assert that certain subvectors are independent. For example, let $\mathbf{X}_i = (X_{1i}, \ldots, X_{pi})$ be partitioned into $q(q \geqslant 2)$ subvectors,

$$\mathbf{X}_i = (\mathbf{X}_i^{(1)}, \ldots, \mathbf{X}_i^{(q)}), \quad i = 1, \ldots, n,$$

where $\mathbf{X}_i^{(j)}$ is a subvector of i_j components, where $i_j \geqslant 1$, $j = 1, \ldots, q$, and $\sum_{j=1}^{q} i_j = p$. The hypothesis that these subvectors are independent is

$$H_0^q : F(\mathbf{x}) = \prod_{j=1}^{q} F^j(\mathbf{x}^{(j)}) \qquad \text{for all } \mathbf{x} \in R_p,$$

where $F^j(\mathbf{x}^{(j)})$ denotes the marginal cumulative distribution function of the subvector $\mathbf{X}_i^{(j)}$. When $q = p$, H_0^q and H_0^p are equivalent. When $F(\mathbf{x})$ is a multivariate normal cumulative distribution function, H_0^p and H_0^q reduce to hypotheses about the structure of the covariance matrix. Anderson [1, Chap. 9] considers this multivariate independence problem under assumptions of multivariate normality. Puri and Sen [27, Chap. 8] consider the multivariate independence problem in a nonparametric framework. Simon [29], with a multivariate nonparametric viewpoint, suggests the computation of all $\binom{p}{2}$ Kendall rank correlation coefficients. Beyond this, Simon suggests computation of more complicated Kendall-type correlations with the aim of investigating dimensionality of the data and developing techniques for selecting variables that are redundant (and could therefore be dropped to reduce the dimension).

CONSUMER PREFERENCE EXAMPLE

The following example is based on data of Rasekh et al. [28] in a study designed to ascertain the relative importance of the various factors contributing to tuna quality and to find objective methods for determining quality parameters and consumer preference. Table 1 gives values of the Hunter L measure of lightness, together with panel scores for nine lots of canned tuna. The original consumer panel scores of excellent, very good, good, fair, poor, and unacceptable were converted to the numerical values of 6, 5, 4, 3, 2, and 1, respectively. The panel scores in Table 1 are averages of 80 such values. The Y random variable is thus dis-

Table 1   Hunter L Values and Consumer Panel Scores for Nine Lots of Canned Tuna

Lot	Hunter L Value, X	Panel Score, Y
1	44.4	2.6
2	45.9	3.1
3	41.9	2.5
4	53.3	5.0
5	44.7	3.6
6	44.1	4.0
7	50.7	5.2
8	45.2	2.8
9	60.1	3.8

Source.   Rasekh et al. [28].

crete, and hence the continuity portion of assumption 2 is not satisfied. Nevertheless, since each Y is an average of 80 values, we need not be too nervous about this departure from assumption 2.

It is suspected that the Hunter L value is positively associated with the panel score. Thus we will illustrate the one-sided tests of H_0 versus positive association, based on r, r_s, and r_k. The reader will soon see that all three tests reach the same conclusion; i.e., there is positive association between the Hunter L value and the panel score.

Test Based on r

From Table 1 we can easily calculate $\sum X_i Y_i = 1584.88$, $\sum X_i = 430.3$, $\sum Y_i = 32.6$, $(\sum X_i)^2 = 185,158.09$, $(\sum Y_i)^2 = 1062.76$, $\sum X_i^2 = 20,843.11$, $\sum Y_i^2 = 125.90$, and from (2) and (5) with $n = 9$, we obtain $r = 0.57$ and $T = 1.84$. Referring $T = 1.84$ to a t-distribution with 7 degrees of freedom yields a one-sided P value of 0.054. Thus the test based on r leads to the conclusion that the Hunter L lightness variable and the panel score variable are positively associated.

The large-sample approximation refers $T = 1.84$ to the standard normal distribution yielding an approximate P value of 0.034.

Test based on r_s

We use Table 2 to illustrate the computation of r_s. From (9) with $n = 9$, we obtain

$$r_s = 1 - \frac{6(48)}{(9)^3 - 9} = 0.60.$$

Referring $r_s = 0.60$ to Table J of Kraft and van Eeden [21] yields a one-sided P value of 0.048. Thus the test based on r_s leads to the conclusion that the Hunter L lightness variable and the panel score variable are positively associated.

From (12) we see that the large-sample approximation refers $r_s^* = (8)^{1/2}(0.6) = 1.70$ to the standard normal distribution, yielding an approximate P value of 0.045. This is in good agreement with the exact P value of 0.048 based on r_s.

Table 2 Computation of r_s for the Canned Tuna Data

Lot	R	S	D	D^2
1	3	2	1	1
2	6	4	2	4
3	1	1	0	0
4	8	8	0	0
5	4	5	−1	1
6	2	7	−5	25
7	7	9	−2	4
8	5	3	2	4
9	9	6	3	9
				$\sum D^2 = 48$

Test Based on r_k

Table 3 contains the $\xi(X_i, X_j, Y_i, Y_j)$ values used to compute r_k. For example, the $i = 2$, $j = 5$ entry in Table 3 is a "-1" because $X_2 > X_5$ and $Y_2 < Y_5$, yielding $(X_2 - X_5)$ $(Y_2 - Y_5) < 0$ and thus $\xi(X_2, X_5, Y_2, Y_5)$ $= -1$. Summing the 1's and -1's of Table 3 yields $K = 16$ and from (13), $r_k = 0.44$. Referring $K = 16$ to Table A.21 of Hollander and Wolfe [14] yields a one-sided P value of 0.060. Thus there is evidence that the Hunter L lightness variable and the panel score variable are positively associated.

To apply the large-sample approximation we compute, from (16), $K^* = 1.67$, yielding an approximate P value of 0.048. This is in good agreement with the exact P value of 0.060 based on K.

Literature

Bickel and Doksum [4] present an introduction to mathematical statistics for students with a mathematics background that includes linear algebra, matrix theory, and advanced calculus. In our list of references, articles and books that require mathematics beyond this level are indicated with an asterisk (*).

In the body of the text, we have indicated that Bickel and Doksum [4], Kendall and Stuart [20], and Anderson [1] are good books for learning about tests of independence in a parametric setting. Kendall and Stuart cover both parametric and nonparametric tests. Anderson specializes in the multivariate situation.

Nonparametric texts referenced are Hájek [12], Hájek and Šidák [13], Hollander and Wolfe [14], Kendall [19], Kraft and van Eeden [21], Lehmann [24], and Puri and Sen [27]. Hájek [12] gives an introduction to nonparametric theory, whereas Hájek and Šidák [13] present an elegant, advanced treatment of nonparametric theory that requires an asterisk according to our system of indicating level. Hollander and Wolfe [14] is a modern methods text that features actual applications and is geared to users. Both Hollander and Wolfe, and Kraft and van Eeden [21], are introductory and have excellent tables for implementing the procedures. Kendall [19] features rank correlation methods. Lehmann [24] presents both applications and an introduction to nonparametric theory. Puri and Sen [27] feature an advanced presentation of multivariate nonparametric methods, but the material is difficult to apply. Kendall's [19] book, devoted to rank correlation methods, is a classic in its fourth edition. The first edition, which appeared in 1948, marked the appearance of the first textbook solely devoted to ranking methods.

Huber [16] presents a modern summary of robust statistical procedures, with suggestions and directions for future research.

Table 3 $\xi(X_i, X_j, Y_i, Y_j)$ Values for Canned Tuna Data

j \ i	1	2	3	4	5	6	7	8
2	1							
3	1	1						
4	1	1	1					
5	1	−1	1	1				
6	−1	−1	1	1	−1			
7	1	1	1	−1	1	1		
8	1	1	1	1	−1	−1	1	
9	1	1	1	−1	1	−1	−1	1

References

An asterisk denotes that the level is advanced.

[1] *Anderson, T. W. (1958). *An Introduction to Multivariate Statistical Analysis.* Wiley, New York.

[2] Best, D. J. (1973). *Biometrika*, **60**, 429–430.

[3] *Bhuchongkul, S. (1964). *Ann. Math. Statist.*, **35**, 138–149.

[4] Bickel, P. J. and Doksum, K. A. (1977). *Mathematical Statistics: Basic Ideas and Selected Topics.* Holden-Day, San Francisco.

[5] Brown, B. W., Jr., Hollander, M., and Korwar, R. M. (1974). In *Reliability and Biometry*, F. Proschan and R. J. Serfling, eds. SIAM, Philadelphia, pp. 327–354.

[6] Cox, D. R. (1972). *J. R. Statist. Soc. B*, **34**, 187–202.

[7] Devlin, S. J., Gnanadesikan, R., and Kettenring, J. R. (1975). *Biometrika*, **62**, 531–545.

[8] Fieller, E. C. and Pearson, E. S. (1961). *Biometrika*, **48**, 29–40.

[9] Galton, F. (1889). *Natural Inheritance.* Macmillan, London.

[10] Glasser, G. J. and Winter, R. F. (1961). *Biometrika*, **48**, 444–448.

[11] Gnanadesikan, R. and Kettenring, J. R. (1972). *Biometrics*, **28**, 81–124.

[12] Hájek, J. (1969). *Nonparametric Statistics.* Holden-Day, San Francisco.

[13] *Hájek, J. and Šidák, Z. (1967). *Theory of Rank Tests.* Academic Press, New York.

[14] Hollander, M. and Wolfe, D. A. (1973). *Nonparametric Statistical Methods.* Wiley, New York.

[15] Hotelling, H. and Pabst, M. R. (1936). *Ann. Math. Statist.*, **7**, 29–43.

[16] *Huber, P. (1977). *Robust Statistical Procedures.* SIAM, Philadelphia.

[17] Kaarsemaker, L. and van Wijngaarden, A. (1953). *Statist. Neerlandica*, **7**, 41–54.

[18] Kendall, M. G. (1938). *Biometrika*, **30**, 81–93.

[19] Kendall, M. G. (1970). *Rank Correlation Methods*, 4th ed. Charles Griffin, London.

[20] Kendall, M. G. and Stuart, A. (1973). *The Advanced Theory of Statistics*, Vol. 2, 3rd ed. Charles Griffin, London.

[21] Kraft, C. H. and van Eeden, C. (1968). *A Nonparametric Introduction to Statistics.* Macmillan, New York.

[22] Kruskal, W. H. (1958). *J. Amer. Statist. Ass.*, **53**, 814–861.

[23] Latta, R. B. (1977). Rank Tests for Censored Data. *Tech. Rep. No. 112*, Dept. of Statistics, University of Kentucky, Lexington, Ky.

[24] Lehmann, E. L. (1975). *Nonparametrics: Statistical Methods Based on Ranks.* Holden-Day, San Francisco.

[25] Miller, R. G., Jr. (1977). *Biometrika*, **64**, 449–464.

[26] Pearson, K. (1896). "1948 Mathematical contributions to the theory of evolution. III. Regression, heredity and panmixia." In Karl Pearson, *Karl Pearson's Early Statistical Papers.* Cambridge University Press, Cambridge, pp. 113–178. First published in *Philos. Trans. R. Soc. Lond. Ser. A*, **187**.

[27] *Puri, M. L. and Sen, P. K. (1971). *Nonparametric Methods in Multivariate Analysis.* Wiley, New York.

[28] Rasekh, J., Kramer, A., and Finch, R. (1970). *J. Food Sci.*, **35**, 417–423.

[29] Simon, G. (1977). *J. Amer. Statist. Ass.*, **72**, 367–376.

[30] Spearman, C. (1904). *Amer. J. Psychol.*, **15**, 72–101.

[31] *Srivastava, M. S. (1973). *Canad. Math. Bull.*, **16**, 337–342.

Acknowledgment

Research sponsored by the Air Force Office of Scientific Research, AFSC, USAF, under Grant AFOSR-78-3678. The U.S. government is authorized to reproduce and distribute reprints for governmental purposes.

(CORRELATION
DEPENDENCE, CONCEPTS OF
DEPENDENCE, MEASURES OF
DISTRIBUTION-FREE METHODS
HOLLANDER BIVARIATE SYMMETRY
   TEST
RANK TESTS
REGRESSION)

MYLES HOLLANDER

DESIGN EFFECT

"Design effect" denotes the ratio (v^2/v_0^2) of the actual variance of a statistic to that for a simple random sample (srs) with the same number (n) of elements. From complex survey samples v^2 is computed in accord with the complexity of the sample design; v_0^2 is computed with standard formulas based on assumptions of n independent selections. It has been frequently computed for means with $v_0 = s^2/n$, sometimes under other names: variance ratio, variance factor, efficiency ratio, etc. (historically, also as the Lexis ratio* and Poisson ratio* for special cases).

There is a slight difference between two symbolic definitions: $deff = v^2/[(1 - f) s^2/n]$ and $deft^2 = v^2/[s^2/n]$. Thus an srs of size n has deff $= 1$, but deft$^2 = (1 - f)$; $(1 - f)$ is commonly negligible, and *deft* is a convenient symbol for confidence intervals. The s^2 computed simply from complex samples is a robust, consistent estimate of the population value S^2. The other population values are also usually denoted with capitals. The sample values of deft^2 are often subject to great sampling variations, especially because of few degrees of freedom for v^2.

Deft^2 are computed to assess the overall effects of the sample design, free of the disturbing factors of scale and of distribution, present in v^2 but removed by v_0^2. They facilitate generalizations for inferences from a set of computations (1) to other variables from the same sample, (2) to subclasses partitioned from the same sample, and (3) to other similar samples.

For simplicity and clarity we now discuss means based on equal selection probabilities. For proportionate element sampling the gains of stratification* are reflected in reduced variances s_w^2 within strata when deft^2 $= s_w^2/s^2 < 1$. More important are the typical losses due to cluster sampling* when deft^2 > 1. These increases are often large, with great differences for diverse variables from the same sample. However, the "portability" of deft^2 is limited for the three purposes named above and deft$^2 = [1 + \text{roh}(\bar{b} - 1)]$ can take us further. Here $\bar{b}$ is the average size of sample clusters, and roh is an overall complex measure of homogeneity of elements in the clusters; in random sampling of equal-sized (b) clusters, roh = rho, the intraclass correlation. Note that deft^2 depends on the sizes of sample clusters; hence for subclasses that cut across the design (cross classes) deft^2 has been found to decline roughly and nearly proportionately with subclass size. Even greater declines of deft^2 have been noted for differences of subclass means.

Values of deft$^2 > 1$ have also been computed for complicated (analytical) statistics from complex samples [3]. For weighted computations of v^2, deft^2 may be computed with weighted v_0^2 for external use, but with unweighted v_0^2 for uses internal to the sample.

Details, data, and references are given in refs. 1, 2, and 4, and also in publications of the U.S. Bureau of the Census* and of Statistics Canada.

References

[1] Kish, L. (1965). *Survey Sampling*. Wiley, New York.

[2] Kish, L. et al. (1976). *Sampling Errors for Fertility Surveys, Occasional Survey Paper 17.*, World Fertility Surveys, London.

[3] Kish, L. and Frankel, R. M. (1974). *J. R. Statist. Soc. B*, **36**, 1–37.

[4] Verma, V. et al. (1980). *J. R. Statist. Soc. A*, **143**, 431–473.

L. Kish

DESIGN MATRIX

A matrix representing assignment of factor levels to blocks in a factorial experiment*. If there are b blocks and t factor levels ("treatments"), the design matrix is a $b \times t$ matrix; the entry in the ith row and jth column is the number of times the jth factor level appears in the ith block. For example, the design matrix for a complete randomized block* experiment is a matrix with each entry 1. In most designs all entries are either 0 or 1, but this need not be the case.

The term "design matrix" is also applied to the sets of values of controlled variables used in an experiment. These are regarded as coordinates in a space of dimension p equal to the number of controlled variables. If there are n observations in the experiment, the design matrix is an $n \times p$ matrix. If there are repeated measurements for some sets of values, the corresponding row is repeated in the design matrix.

For example, if a quadratic polynomial regression of the form

$$E[Y|X_1, X_2] = \beta_0 + \beta_1 X_1 + \beta_2 X_2 + \beta_{11} X_1^2 + \beta_{12} X_1 X_2 + \beta_{22} X_2^2$$

is to be fitted, and there are observations for each of the nine points $X_1 = -1$, 0, 1; $X_2 = -1$, 0, 1 with three additional points at the origin ($X_1 = X_2 = 0$), the design matrix is

$$
\begin{array}{cccccc}
1 & 0 & 0 & 0 & 0 & 0 \\
1 & 0 & 0 & 0 & 0 & 0 \\
1 & 0 & 0 & 0 & 0 & 0 \\
1 & 0 & 0 & 0 & 0 & 0 \\
1 & -1 & -1 & 1 & 1 & 1 \\
1 & -1 & 0 & 1 & 0 & 0 \\
1 & -1 & 1 & 1 & -1 & 1 \\
1 & 0 & -1 & 0 & 0 & 1 \\
1 & 0 & 0 & 0 & 0 & 0 \\
1 & 0 & 1 & 1 & -1 & 1 \\
1 & 1 & -1 & 1 & -1 & 1 \\
1 & 1 & 0 & 1 & 0 & 0 \\
1 & 1 & 1 & 1 & 1 & 1
\end{array}
$$

Kempthorne [1] regards the term "design matrix" as inappropriate in this last case, and suggests "model matrix" as a preferable alternative.

Reference

[1] Kempthorne, O. (1980). *Amer. Statist.*, **34**, 249 (letter).

(DESIGN OF EXPERIMENTS
RESPONSE SURFACES)

DESIGN OF EXPERIMENTS: INDUSTRIAL AND SCIENTIFIC APPLICATIONS

Obtaining valid results from a test program calls for commitment to sound statistical design. In fact, proper experimental design is more important than sophisticated statistical analysis. Results of a well-planned experiment are often evident from simple graphical analyses (*see* GRAPHICAL REPRESENTATION OF DATA). However, the world's best statistical analysis cannot rescue a poorly *planned* experimental program.

The main reason for designing an experiment statistically is to obtain unambiguous results at minimum cost. The need to learn about interactions* among variables and to measure experimental error are some of the added reasons.

Many think of experimental design solely in terms of standard plans for assigning treatments to experimental units. Such designs are described in various books, such as those summarized in the annotated bibliography* in the accompanying article and cataloged in various reports and papers. Important as such formal plans are, the final selection of test points represents only the proverbial tip of the iceberg, the culmination of a careful planning process. Often the desired plan is not a standard textbook design, but one that is tailormade to the needs and peculiarities of the specific problem and testing constraints.

This article discusses basic considerations in designing an experiment. A brief introduction to some formal test plans is also provided. The analysis of the results involves statistical methods such as regression analysis* and the analysis of variance*. The methods for analyzing the results of a planned experiment are not discussed in this article.

Statistically planned experiments were first used prior to World War II, primarily in argicultural applications. Their use has rapidly expanded to science, industry, and the social sciences. The emphasis here is on industrial and scientific applications.

Much of this article previously appeared in refs. 2 and 3.

DEFINING THE PURPOSE AND SCOPE OF THE EXPERIMENT

Designing an experiment is like designing a product. Every product serves a purpose; so should every experiment. This purpose must be clearly defined at the outset. It may, for example, be to optimize a process, to estimate the probability that a component will operate properly under a given stress for a specified number of years, to evaluate the relative effects on product performance of different sources of variability, or to determine whether a new process is superior to an existing one.

In addition to defining the purpose of a program, one must decide upon its scope. An experiment is generally a vehicle for drawing inferences about the real world. Since it is highly risky to draw inferences about situations beyond the scope of the experiment, care must be exercised to make this scope sufficiently broad. This must especially be kept in mind if laboratory findings are to apply to the manufacturing line. Thus if the results are material-dependent, the material for fabricating experimental units must be representative of what one might expect to encounter in production. If the test program were limited to a single heat of steel or a single batch of raw material, the conclusions might be applicable only to that heat or batch, irrespective of the sample size. Similarly, in deciding whether or not temperature should be included as an experimental variable to compare different preparations, one must decide whether the possible result that one preparation outperforms another at a constant temperature would also apply for other temperatures of practical interest. If this is not expected to be the case, one would generally wish to include temperature as an experimental variable.

WAYS OF HANDLING VARIABLES

An important part of planning an experimental program is the identification of the variables that affect the response and deciding what to do about them. The decision as to how to deal with each of the candidate variables can be made jointly by the experimenter and the statistician. However, *identifying* the variables is the experimenter's responsibility. Controllable or independent variables in a statistical experiment can be dealt with in four different ways. To which category a particular variable is assigned often involves a trade-off among information, cost, and time.

Primary Variables

The most obvious variables are those whose effects upon performance are to be evaluated directly; these are the variables that, most likely, created the need for the investigation in the first place. Such variables may be quantitative, such as temperature, pressure, or concentration of catalyst, or they may be qualitative, such as preparation method, catalyst type, or batch of material.

The results of experiments involving quantitative variables are frequently analyzed using regression analysis*. Experiments with qualitative variables are often analyzed by the analysis of variance*. This, however, is not a hard and fast rule. For example, experiments involving qualitative variables can be analyzed by regression analysis methods using dummy and indicator variable techniques. In any case, graphical representation of the results is generally enlightening.

Quantitative controllable variables are frequently related to the performance variable by some assumed statistical relationship or model. The minimum number of conditions or levels per variable is determined by the form of the assumed model. For example, if a straight-line relationship can be assumed, two levels (or conditions) may be sufficient; for a quadratic relationship a minimum of three levels is required.

Qualitative variables can be broken down into two categories. The first consists of those variables whose specific effects are to be compared directly; for example, comparison of the effect on performance of two proposed preparation methods or of three specific types of catalysts. The required number of conditions for such variables is generally evident. Such variables are sometimes referred to as fixed-effects* or Type I variables.

The second type of qualitative variables are those whose individual contributions to performance variability are to be evaluated. The specific conditions of such variables are randomly determined. Material batch would be such a variable if one is not interested in the behavior of specific batches per se, but instead needs information concerning the magnitude of variation in performance due to differences between batches. In this case, batches would be selected randomly from a large population of batches. For variables of this type, it is generally desirable to have a reasonably large sample of conditions (e.g., five or more) so as to obtain an adequate degree of precision in estimating perfor-

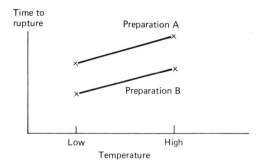

Figure 1   Situation with no interaction.

mance variability attributable to the variable. Such variables are sometimes referred to as random-effects or Type II variables.

When dealing with fixed-effects or Type I variables, one is generally interested in obtaining point and interval estimates for the performance or response variable or in comparing treatment averages. For random-effects or Type II variables, one frequently wishes to obtain a point or interval estimate of the variability in the response as measured by a variance component*, due to each of the controllable variables.

When there are two or more variables, they might *interact* with one another; that is, the effect of one variable upon the response depends on the condition of the other variable. Figure 1 shows a situation where two noninteracting variables, preparation type and temperature, independently affect time to rupture; i.e., the effect of temperature on time to rupture is the same for both preparation types. In contrast, Fig. 2 shows two examples of interactions between prepara-

tion and temperature. In the first example, an increase in temperature is beneficial for preparation A, but does not make any difference for preparation B. In the second example, an increase in temperature raises time to rupture for preparation A, but decreases it for preparation B.

An important purpose of a designed experiment is to obtain information about interactions among the primary variables. This is accomplished by varying factors simultaneously rather than one at a time. Thus in the preceding example, each of the two preparations would be run at both low and high temperatures, using, for example, a full factorial* or a fractional factorial experiment* (see later discussion).

Background Variables and Blocking

In addition to the primary controllable variables there are those variables which, although not of primary interest, cannot, and perhaps should not, be held constant. Such variables arise, for example, when an experiment is to be run over a number of days, or when different machines and/or operators are to be used. It is of crucial importance that such background variables are not varied in complete conjunction with the primary variables. For example, if preparation A were run only on day 1 and preparation B only on day 2, one cannot determine how much of any observed difference in performance between the two preparations is due to normal day-to-day process variation (such mixing of effects is called confounding). If

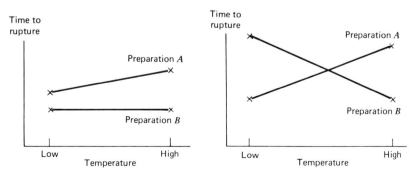

Figure 2   Situations with interactions.

the background variables do not interact with the primary variables, they may be introduced into the experiment in the form of experimental blocks*. An experimental block represents a relatively homogenous set of conditions within which different conditions of the primary variables are compared.

A well-known example of blocking arises in the comparison of wear for different types of automobile tires. Tire wear may vary from one automobile to the next, irrespective of the tire type, due to differences among automobiles, variability among drivers, etc. Say, for example, that one wishes to compare four tire types (A, B, C, and D) and four automobiles are available for the comparison. A *poor* procedure would be to use the same type of tire on each of the four wheels of an automobile and vary the tire type between automobiles, as in the following tabulation:

Automobile			
1	2	3	4
A	B	C	D
A	B	C	D
A	B	C	D
A	B	C	D

Such an assignment would be undesirable because the differences among tires cannot be separated from the differences among automobiles in the subsequent analysis. Separation of these effects can be obtained by treating automobiles as experimental blocks and randomly assigning tires of each of the four types to each automobile as follows:

Automobile			
1	2	3	4
A	A	A	A
B	B	B	B
C	C	C	C
D	D	D	D

The arrangement above is known as a *randomized block design*.

The symmetry of the preceding example is not always found in practice. For example, there may be six tire types under comparison and 15 available automobiles. Tires are then assigned to automobiles to obtain the most precise comparison among tire types, using a so-called *incomplete block design*. Similar concepts apply if there are two or more primary variables, rather than tire type alone.

A main reason for running an experiment in blocks is to ensure that the effect of a background variable does not contaminate evaluation of the primary variables. However, blocking also permits the effect of the blocked variables to be removed from the experimental error, thus providing more precise evaluations of the primary variables. This is especially important when the background variable has an important effect since, in that case, without blocking the effect of the primary variable may be hidden. Finally, in many situations, the effect of the blocking variables on performance can also be readily evaluated in the statistical analysis—a further advantage of blocking.

In some situations, there may be more than one background variable whose possible contaminating effect need be removed by blocking. Thus in the automobile tire comparison, differences between wheel positions, as well as differences between automobiles, may be of concern. In this case, wheel position might be introduced into the experiment as a second blocking variable. If there are four tire types to be compared, this might, for example, be done by randomly assigning the tires of each of the four types according to the following plan, known as a *Latin square design*:

Wheel Position	Automobile			
	1	2	3	4
1	A	D	C	B
2	B	A	D	C
3	C	B	A	D
4	D	C	B	A

In this plan the effects of both automobile and wheel position are removed by blocking. It should, however, be kept in mind that for the Latin square design, as for other blocking plans, it is generally assumed that the blocking variables do not interact with the main variable to be evaluated.

Uncontrolled Variables and Randomization

A number of further variables, such as ambient conditions, can be identified, but not controlled, or are only hazily identified or not identified at all, but affect the results of the experiment. To ensure that such uncontrolled variables do not bias the results, randomization is introduced, in various ways, into the experiment to the extent that this is practical.

Randomization means that the sequence of preparing experimental units, assigning treatments, running tests, taking measurements, etc., is randomly determined, based, for example, on numbers selected from a random number table. The total effect of the uncontrolled variables is thus lumped together into experimental error as unaccounted variability. The more influential the effect of such uncontrolled variables, the larger will be the resulting experimental error and the more imprecise will be the evaluations of the effects of the primary variables. (Sometimes, when the uncontrolled variables can be measured, their effect can be removed from the experimental error statistically; see below.)

Background variables could also be introduced into the experiment by randomization rather than by blocking techniques. Thus in the previous example, the four tires of each type could have been assigned to automobiles and wheel positions completely randomly instead of treating automobiles and wheel positions as experimental blocks. This could have resulted in an assignment such as the following:

Wheel Position	Automobile			
	1	2	3	4
1	B	C	B	D
2	A	D	D	A
3	C	C	A	D
4	B	B	C	A

Both blocking and randomization generally ensure that the background variables do not contaminate the evaluation of the primary variables. Randomization sometimes has the advantage of greater simplicity compared to blocking. However, under blocking the effect of a background variable is removed from the experimental error, whereas under randomization it usually is not. Thus one might aim to remove the effects of the one or two most important background variables by blocking while counteracting the possible contaminating effect of others by randomization.

Variables Held Constant

Finally, there are some variables that one wishes to hold constant in the experiment. Holding a variable constant limits the size and complexity of the experiment but, as previously noted, can also limit the scope of the resulting inferences. The variables to be held constant in the experiment must be identified and the mechanisms for keeping them constant defined. The experimental technique should be clearly specified at the outset of the experiment and closely followed.

DEFINING THE EXPERIMENTAL ENVIRONMENT AND CONSTRAINTS

The operational conditions under which the experiment is to be conducted and the manner in which each of the factors is varied must be clearly spelled out.

All variables are *not* created equal; some can be varied more easily than others. For example, a change in pressure may be implemented by a simple dial adjustment; on the other hand, stabilization requirements might make it difficult to change the temperature. In such situations, completely randomizing the testing sequence is impractical. However, basing the experimental plan on convenience alone can lead to ambiguous and unanalyzable results. For example, if the first half of the experiment were run at one temperature and the second half at another temperature, there is no way of knowing whether the observed difference in the results is due to the difference in temperature or to some other factors that varied during the course of the experiment, such as raw material, ambient conditions, or operator technique. Thus the final experimental plan must be a compromise between cost and information. The experiment must be practical to run, yet still yield statistically valid results (see refs. 4 and 5).

Practical considerations enter into the experimental plan in various other ways. In many programs, variables are introduced at different operational levels. For example, in evaluating the effect on tensile strength of alloy composition, oven temperature, and varnish coat, it may be convenient to make a number of master alloys with each composition, split the alloys into separate parts to be subjected to different heat treatments, and then cut the treated samples into subsamples to which different coatings are applied. Tensile-strength measurements are then obtained on all coated subsamples.

Situations such as the preceding arise frequently in practice and are referred to as *split-plot** experiments. (The terminology is due to the agricultural origins of experimental design; e.g., Farmer Jones needed to compare different fertilizer types on a plot of land with varying normal fertility.) A characteristic of split-plot plans is that more precise information is obtained on the low-level variable(s) (varnish coats in the preceding example) than on the high-level variable(s) (alloy composition). The split-plot nature of the experimental environment, if present, is an important input both in the planning and in the analysis of the experiment.

USING PRIOR KNOWLEDGE

Prior knowledge concerning the expected outcome at certain experimental conditions is sometimes available. For example, some combinations of conditions might be known to yield poor results or might not be attainable with the available equipment, or worse yet, could result in blowing up the plant (and the experiment with it). Also, all proposed conditions in the experiment need make sense.

Clearly, unreasonable conditions must be omitted from the experimental design, irrespective of whether they happen to fall in with a standard statistical pattern. Thus the experiment must be adjusted to accommodate the real-world situation and not the reverse.

DEFINING THE PERFORMANCE VARIABLE

A clear statement is required of the performance characteristics or dependent variables to be evaluated. Even a well-designed experiment will fail if the response cannot be measured properly. Frequently, there may be a number of response variables: for example, tensile strength, yield strength, or percent elongation. It is important that standard procedures for measuring each variable be established and documented. Sometimes, the performance variable is on a semiquantitative scale; for example, material appearance may be graded into one of five categories, such as outstanding, superior, good, fair, and poor. In this case it is particularly important that standards be developed initially, especially so if judgments are to be made at different times and perhaps by different observers.

UNDERSTANDING DIFFERENT TYPES OF REPEAT INFORMATION

The various ways of obtaining repeat results in the experiment need to be specified. Different information about repeatability is obtained by (1) taking replicate measurements on the same experimental unit, (2) cutting a sample in half at the end of the experiment and obtaining a reading on each half, and (3) taking readings on two samples prepared independently of one another (e.g., on different runs) at the same aimed-at conditions. One would expect greater homogeneity* among replicate measurements on the same sample than among measurements on different samples. The latter reflect the random unexplained variation on repeat runs conducted under identical conditions; the former do not. A skillfully planned experiment imparts information about each component of variability if such information is not initially available, and uses a mixture of replication and repeat runs so as to yield the most precise information for the available testing budget. The ways in which such information is obtained must also be known to allow one to perform a valid analysis of the results.

OBTAINING PRELIMINARY ESTIMATES OF REPEATABILITY

Initial estimates of overall repeatability should be obtained before embarking on any major test program. Such information may be available from previous testing; if it is not, it may be wise to conduct some valid preliminary runs at different times under supposedly identical conditions. If these runs result in large variability in performance, the important variables that affect the results have not been identified, and further research may be needed before the experiment can commence. As suggested in the previous discussion, experimental error is a fact of life which all experimental programs need to take into account.

ESTABLISHING CONSISTENT DATA RECORDING PROCEDURES

Clear procedures for recording all pertinent data from the experiment must be developed and documented and unambiguous data sheets must be established. These should include provisions not only for recording the values of the measured responses and the aimed-at experimental conditions, but also the conditions that actually resulted if these differ from those planned. It is usually preferable to use the values of the actual condi-

tions in the statistical analysis of the experimental results. For example, if a test was supposed to have been conducted at 150°C but was actually run at 148.3°C, the actual temperature should be used in a subsequent regression analysis* of the data. In experimentation with industrial processes, the process equilibrium should be reached before the responses are measured. This is particularly important when complex chemical reactions are involved.

The values of any other variables that might affect the responses should also be recorded, if possible. For example, although it may not be possible to control the ambient humidity, its value should be measured if it might affect the results. In addition, variations in the factors to be held constant, special "happenings," or other unplanned events should also be recorded. The values of such "covariates" can be factored into the evaluation using regression analysis* or other techniques, thereby reducing the unexplained variability or experimental error. If the covariates do indeed have an effect, this leads to more precise evaluations of the primary variables. Alternatively, such covariates may be related to the unexplained variation that remains after the analysis, using residual plotting* or other techniques.

RUNNING THE EXPERIMENT IN STAGES

A statistically planned experiment does not require all testing to be conducted at one fell swoop. Instead, the program can be conducted in stages of, say, 8 to 20 runs; this permits changes to be made in later tests based on early results and provides the practical benefit of allowing preliminary findings to be reported. A recent experiment to improve the properties of a plastic material involved such variables as mold temperature, cylinder temperature, pressure, ram speed, and material aging. The experiment was conducted in three stages. After the first stage, the overall or main effects of each of the variables were evaluated; after the second stage, interactions between pairs of variables were analyzed; and after the third stage, nonlinear effects were assessed. Each stage involved about a month of elapsed time, and management was apprised of progress each month. If unexpected results had been obtained at an early stage—for example, poor results at one of the selected ram speeds—the later stages of the experiment could have been changed.

Whether or not a particular experiment should be run in stages depends upon the program objectives and the specific experimental situation. A stagewise approach makes sense when units are made in groups or one at a time, and a rapid feedback of results is possible. Running the experiment in stages is also attractive in searching for an optimum response, because it might allow one to move closer to the optimum from stage to stage. On the other hand, a one-shot experiment may be desirable if there are large startup costs at each stage or if there is a long waiting time between fabricating the units and measuring their performance. This is the case in many agricultural experiments and also when the measured variable is product life.

If the experiment is run in stages, precautions must be taken to ensure that possible differences between the stages do not invalidate the results. Appropriate procedures to compare the stages must be included, both in the test plan and in the statistical analysis. For example, some standard test conditions, known as controls, may be included in each stage of the experiment.

OTHER CONSIDERATIONS

Many other questions must be considered in planning the experiment. In particular:

1. What is the most meaningful way to express the controllable or independent variables? For example, should current density and time be taken as the experimental variables, or are time and the product of current density and time the real variables affecting the response? Judicious selection of the independent variables often reduces or eliminates interactions between variables, thereby leading to a simpler experiment and analysis.

2. What is a proper experimental range for the selected quantitative controllable variables? *Assuming* a linear relationship between these variables and performance, the wider the range of conditions or settings, the better are the chances of detecting the effects of the variable. (However, the wider the range, the less reasonable is the assumption of a linear relationship.) Also, one usually would not want to go beyond the range of physically or practically useful conditions. The selection of the range of the variables depends in part upon the ultimate purpose of the experiment: Is it to learn about performance over a broad region or to search for an optimum condition? A wider range of experimentation would be more appropriate in the first case than in the second.

3. What is a reasonable statistical model, or equation, to approximate the relationship between the independent variables and each response variable? Can the relationship be approximated by an equation involving linear terms for the quantitative independent variables and two-factor interaction terms only, or is a more complex model, involving quadratic and perhaps even multifactor interaction terms, necessary? As indicated, a more sophisticated statistical model may be required to describe adequately relationships over a relatively large experimental range than over a limited range. A linear relationship may thus be appropriate over a narrow range, but not over a wide one. The more complex the assumed model, the more runs are usually required.

4. What is the desired degree of precision of the final results and conclusions? The greater the desired precision, the larger

is the required number of experimental runs.

5. Are there any previous benchmarks of performance? If so, it might be judicious to include the benchmark conditions in the experiment to check the results.

6. What statistical techniques are required for the analysis of the resulting data, and can these tools be rapidly brought to bear after the experiment has been conducted?

FORMAL EXPERIMENTAL PLANS: AN INTRODUCTION

After the preceding considerations have been attended to, a formal statistical test plan is evolved. This might involve the use of one of the standard plans developed by statisticians. Such plans are described in various texts (see annotated bibliography) and separate articles and are considered only briefly here. Some type plans are:

BLOCKING DESIGNS. Such designs use blocking techniques to remove the effect of extraneous variables from experimental error (see above). Well-known blocking designs include randomized and balanced incomplete block designs* to remove the effects of a single extraneous variable, Latin square* and Youden square* designs to remove the effects of two extraneous variables, and Graeco-Latin square* and hyper-Latin square* plans to remove the effects of three or more extraneous variables.

COMPLETE FACTORIAL AND FRACTIONAL FACTORIAL* DESIGNS. Such designs apply for two or more primary independent variables. Factors are varied simultaneously, rather than one at a time, so as to obtain information about interactions among variables and to obtain a maximum degree of precision in the resulting estimates. In complete (or full) factorial plans all combinations of conditions of the independent variables are run. For example, a $3 \times 3 \times 2 \times 2$ factorial de-

sign is one that requires running all 36 combinations of two variables with three conditins each and two variables with two conditions each. A fractional factorial design is often used when there is a large number of combinations of possible test points, arising from many variables or conditions per variable, and it is not possible, or practical, to run all combinations. Instead, a specially selected fraction is run. For example, a 2^{6-1} fractional factorial plan is one where there are six variables each at two conditions, resulting in a total of 64 possible combinations, but only a specially selected one half, or 32, of these combinations are actually run. Reference 6 provides a comprehensive catalog of fractional factorial designs.

A bonus associated with full factorial and fractional factorial designs is that by providing a comprehensive scanning of the experimental region, they can often identify, without any further analyses, one or two test conditions which are better than any others seen previously.

RESPONSE SURFACE DESIGNS*. These are special multivariable designs for quantitative independent variables (temperature, pressure, etc.). The relationship between the independent variables and the performance variable is fitted to develop prediction equations using the technique of least-squares regression analysis*. Popular response surface designs include the orthogonal central composite* (or Box) designs and the rotatable designs*.

Frequently, combinations of the preceding plans are encountered, such as a factorial experiment conducted in blocks or a central composite design using a fractional factorial base. There are also plans to accommodate special situations. For example, extensive work has been conducted recently in the development of designs for so-called "mixture experiments"*; these apply for the situation where the experimental variables consist of the percentages of the ingredients that make up a material and must therefore add up to 100% (see ref. 1).

CONCLUDING REMARKS

Statistically planned experiments are characterized by:

1. The proper consideration of extraneous variables.
2. The fact that primary variables are changed together, rather than one at a time, in order to obtain information about the magnitude and nature of the interactions of interest and to gain improved precision in the final estimates.
3. Built-in procedures for measuring the various sources of random variation and for obtaining a valid measure of experimental error.

A well-planned experiment is often tailormade to meet specific objectives and to satisfy practical constraints. The final plan may or may not involve a standard textbook design. If possible, a statistician knowledgeable in the design of experiments should be called in early and made a full-fledged team member. He or she should be told the objectives of the program and the practical considerations and constraints, and not merely what is felt need be known to evolve a matrix of test points. One of the statistician's greatest contributions may be in asking embarrassing questions. After the problem and the constraints have been clearly defined, the statistician can evolve an experimental layout to minimize the required testing effort to obtain the desired information. This may involve one of the formal statistical designs. However, designing an experiment is often an iterative process, requiring rework as new information and preliminary data become available. With a full understanding of the problem, the statistician is in an improved position to respond rapidly if last-minute changes are required and to provide meaningful analyses of the subsequent experimental results.

Literature

The books discussed in the annotated bibliography* in the accompanying article provide additional information concerning the technical and practical aspects of experimental design. An important recent development has been the development of computer programs to generate, assess, and compare alternative experimental designs. For example, ref. 7 describes a computer program for evaluating the precision of an experiment before running it. Reference 8 describes a case study that illustrates many of the considerations discussed here and introduces a few others (e.g., the desirability of forming "a committee of the most knowledgeable individuals in the area of interest to select the variables to be examined, their levels and the measurement criteria"). Finally, refs. 9 and 10 provide two handbook surveys dealing with the design of experiments and provide more detailed discussions of specific designs and their analysis.

References

[1] Cornell, J. A. (1979). *Technometrics*, **21**, 95–106.
[2] Hahn, G. J. (1975). *Chem. Tech.*, **5**, 496–498, 561–562.
[3] Hahn, G. J. (1977). *J. Quality Tech.*, **9**, 13–20.
[4] Hahn, G. J. (1977). *Chem. Tech.*, **7**, 630–632.
[5] Hahn, G. J. (1978). *Chem. Tech.*, **8**, 164–168.
[6] Hahn, G. J. and Shapiro, S. S. (1966). A Catalog and Computer Program for the Design and Analysis of Orthogonal Symmetric Fractional Factorial Experiments. *General Electric TIS Rep. 66C165.*
[7] Hahn, G. J., Meeker, W. Q., Jr., and Feder, P. I. (1976). *J. Quality Tech.*, **8**, 140–150.
[8] Mueller, F. X. and Olsson, D. M. (1971). *J. Paint Tech.*, **43**, 54–62.
[9] Natrella, M. G. (1974). In *Quality Control Handbook*, 3rd ed., J. M. Juran, ed. McGraw-Hill, New York, Chap. 27.
[10] Tingey, F. G. (1972). In *Treatise on Analytical Chemistry*, Part 1, I. M. Kolthoff and P. J. Elving, eds. Wiley-Interscience, New York, Chap. 106.

(ANALYSIS OF COVARIANCE
ANALYSIS OF VARIANCE
BLOCKS, RANDOMIZED COMPLETE
CHANGEOVER DESIGNS
CONFOUNDING
FRACTIONAL FACTORIAL DESIGNS
GENERAL BALANCE

INTERACTIONS
MIXTURE EXPERIMENTS
SPLIT-PLOT DESIGNS
STRIP-PLOT DESIGNS)

GERALD J. HAHN

DESIGN OF EXPERIMENTS: AN ANNOTATED BIBLIOGRAPHY

There are now a large number of books on experimental design on the market. This entry provides an annotated bibliography of 44 such texts, drawing in part upon the information given by the authors. A summary table indicates the major topics covered, the technical level, the emphasized field of application, and other pertinent information about each book.

The determination of which books to include and exclude is necessarily a somewhat arbitrary one. In general, the guide is limited to books that deal principally with the design of experiments. Thus books that are mainly concerned with the analysis of experimental data and linear models have been excluded. Similarly, judgment was required in describing the books and in preparing the summary tabulation.

For an extensive bibliography on the literature of experimental design prior to 1968, see W. T. Federer, and L. N. Balaam's *Bibliography on Experiment and Treatment Design Pre-1969* (Hafner, New York, 1973). This lists 8378 works under 44 subheadings, including one on bibliographies, which references about 150 additional works. A list of the complete table of contents of 28 books on experiment and treatment design (most of which are included in the present bibliography) is also included.

Comments on Individual Books

Anderson, V. L. and McLean, R. A. (1974). *Design of Experiments—A Realistic Approach*. Marcel Dekker, New York. (This book provides an extensive exposition of experimental design at a relatively elementary level. It includes most of the standard material, as well as detailed discussions of such subjects as nested and split-plot experiments. Restrictions on randomization receive special emphasis.)

Bannerjee, K. S. (1975). *Weighing Designs: For Chemistry, Medicine, Operations Research, Statistics*. Marcel Dekker, New York. (This short specialized volume is devoted to a single type of experimental design: designs for "determining with maximum possible precision the weights of light objects by weighing the objects in suitable combinations." The book's four chapters are devoted to formulation of the weighing problem, the chemical balance problem, the spring balance problem, and miscellaneous issues concerning the weighing problem.)

Box, G. E. P., Hunter, W. G., and Hunter, J. S. (1978). *Statistics for Experimenters*. Wiley, New York. [This book, by three eminent practitioners, "is an introduction to the philosophy of experimentation and the part that statistics plays in experimentation." Subtitled "An Introduction to Design, Data Analysis, and Model Building," it provides a practically motivated introduction to basic concepts and methods of experimental design. It "is written for those who collect data and try to make sense of it," and gives an "introduction to those ideas and techniques" that the authors have found especially useful. Statistical theory is introduced as it becomes necessary. Readers are assumed to have no previous knowledge of the subject; "the mathematics needed is elementary." The book includes numerous examples and case studies and provides appreciable detail about many elementary and a few more advanced designs. It is, however, an introductory treatment; therefore, more advanced situations are not discussed. Heavy emphasis is placed on the use of graphical methods for analyzing experimental results. In contrast, the more computationally involved analysis-of-variance tools receive relatively little attention. The four parts of the book deal with (1) comparing two treatments, (2) comparing more than two treatments, (3) measuring the effects of variables, and (4) building models and using them.]

Campbell, D. T. and Stanley, J. C. (1966). *Experimental and Quasi-Experimental Designs for Research*. Rand McNally, Chicago. (This short manual originally appeared in the *Handbook of Research on Teaching*. Although "many of the illustrations come from educational research ... the survey draws from the social sciences in general." The major chapters are devoted to three preexperimental designs, three true experimental designs, quasi-experimental designs, and correlational and ex post facto designs.)

Chakrabarti, M. C. (1962). *Mathematics of Designing and Analysis of Experiments*. Asia Publishing House, New York. [Not shown in 1978–1979 *Books in Print*. This book deals principally with the theory of experimental design. Its six chapters deal with theory of linear estimation, general structure of analysis of designs, standard designs (two chapters), applications of Galois fields and finite geometry in the construction of designs, and some selected topics in design of experiments.]

Chapin, F. S. (1955). *Experimental Designs in Sociological Research*, rev. ed. Reprinted in 1974 by Greenwood Press, Westport, Conn. [The author defines the concept of experimental design in sociological research as the "systematic study of human relations by making observations under conditions of control." He differentiates between the following three types of designs: (1) a cross-sectional design that makes controlled observations for a single date, (2) a projected experimental design that makes "before" and "after" comparisons, and (3) an ex post facto design in which one traces some present effect backward to an assumed past causal complex. The book "is an account of nine experimental studies . . . conducted in normal community situations . . . outside of the schoolroom or laboratory situations." Chapters are also devoted to sociometric scales available for control and the measurement of effects, some fundamental problems and limitations to study by experimental designs, and some problems in psychosocial measurement. Discussion of formal statistical techniques is limited principally to two chapters dealing with the analysis of variance and the *t*-statistic and with nonparametric methods. An interesting feature is an appendix which responds to reviews in the literature to an earlier edition of the book.]

Chew, Victor, ed. (1958). *Experimental Designs in Industry*. Wiley, New York. (Not shown in 1978–1979 *Books in Print*. This book is a compilation of a series of papers presented by well-known practitioners at a symposium on industrial applications of experimental design. It is written with the realization that the "basic designs originally constructed for agricultural experiments [and treated in most of the standard texts on experimental design] . . . may not take into account conditions peculiar to industry." Included are papers on basic experimental designs, complete factorials, fractional factorials, and confounding; simple and multiple regression analysis; experimental designs for exploring response surfaces; experiences with incomplete block designs; experiences with fractional factorials; application of fractional factorials in a food research laboratory; experiences with response surface designs; and experiences and needs for design in ordnance experimentation.)

Chilton, N. W. (1967). *Design and Analysis in Dental and Oral Research*. J. B. Lippincott, Philadelphia. (This book is designed "for classroom teaching of graduate and undergraduate students in dental schools. The use of illustrative examples from studies in oral biology should make the exposition much more applicable to the problems faced by the dental investigator." This book is devoted principally to methods for analyzing experimental data. Only a single chapter is devoted to the topic of experimental design per se.)

Cochran, W. G. and Cox, G. M. (1957). *Experimental Designs*, 2nd ed. Wiley, New York. (This is one of the earliest, best known, and most detailed books in the field. The classical experimental designs are described in a relatively simple manner. The treatment is oriented toward agricultural and biological applications. Exten-

sive catalogs of designs are included, making the book a useful reference guide.)

Cornell, J. A. (1981). *Experiments with Mixtures*. Wiley, New York. (This book deals with experimental designs for situations where the experimental variables involve mixing together two or more ingredients of a product. Its "primary purpose . . . is to present the fundamental concepts in the design and analysis . . . " of such experiments. "The book is written for anyone who is engaged in planning or performing experiments with mixtures. In particular, research scientists and technicians in the chemical industries, whether or not trained in statistical methods, should benefit from the many examples that are chemical in nature . . . . The mathematical prerequisites have been kept to a minimum." However, background in mathematics and statistics is probably required for most readers to derive full benefit from the book.)

Cox, D. R. (1958). *Planning of Experiments*. Wiley, New York. (This book provides a simple survey of the principles of experimental design and of some of the most useful experimental schemes. It tries "as far as possible, to avoid statistical and mathematical technicalities and to concentrate on a treatment that will be intuitively acceptable to the experimental worker, for whom the book is primarily intended." As a result, the book emphasizes basic concepts rather than calculations or technical details. Chapters are devoted to such topics as "Some Key Assumptions," "Randomization," and "Choice of Units, Treatments, and Observations."

Daniel, C. (1976). *Applications of Statistics to Industrial Experimentation*. Wiley, New York. [This book is based upon the personal experiences and insights of the author, an eminent practitioner of industrial applications of experimental design. It provides extensive discussions and new concepts, especially in the areas of factorial and fractional factorial designs. "The book should be of use to experimenters who have some knowledge of elementary statistics and to statisticians who want simple explanations, detailed examples, and a documentation of the variety of outcomes that may be encountered." Some of the unusual features are chapters on "Sequences of Fractional Replicates" and "Trend-Robust Plans," and sections entitled "What Is the Answer? (What Is the Question?)," and "Conclusions and Apologies."]

Das, M. N. and Giri, N. C. (1979). *Design and Analysis of Experiments*. Wiley, New York. (This book "can serve as a textbook for both the graduate and postgraduate students in addition to being a reference book for applied workers and research workers and students in statistics." Chapters dealing with designs for bio-assays and weighing designs are included.)

Davies, O. L., et al. (1956). *The Design and Analysis of Industrial Experiments*, 2nd ed. Hafner, New York. Reprinted by Longman, New York, 1978. [This book, which is a sequel to the authors' basic text *Statistical Methods in Research and Production*, is directed specifically at industrial situations and chemical applications.

Three chapters are devoted to factorial experiments and one chapter to fractional factorial plans. A lengthy chapter (84 pages) discusses the determination of optimum conditions and response surface designs, which are associated with the name of George Box, one of the seven co-authors. Theoretical material is presented in chapter appendices.]

Dayton, C. M. (1970). *The Design of Educational Experiments*. McGraw-Hill, New York. (This book represents in detail "those experimental designs which have proved to be widely usable in education and related research fields. . . . " No mathematical training beyond algebra is required. The treatment includes chapters devoted to "Nested Designs" and "Repeated Measures Designs.")

Diamond, W. J. (1981). *Practical Experimental Designs*. Lifetime Learning Publications, Belmont, Calif. ["This book is for engineers and scientists with little or no statistical backgound who want to learn to design efficient experiments and to analyze data correctly . . . . The emphasis is on practical methods, rather than on statistical theory." Sample size requirements for simple comparitive experiments, two-level fractional factorial experiments (based upon Hamard matrices and John's three-quarter designs), and a computer program for generating such designs are emphasized.]

Edwards, A. L. (1972). *Experimental Design in Psychological Research*, 4th ed. Holt, Rinehart and Winston, New York. (The first third of this book deals with statistical analysis methods, while the remainder provides a fairly detailed discussion of standard blocking and factorial plans. Designs with repeated measures on the same subject received special emphasis.)

Federer, W. T. (1955). *Experimental Design: Theory and Application*. Macmillan, New York. (Not shown in 1978–1979 *Books in Print*. This book describes in detail classical plans of experimentation with emphasis on blocking and factorial designs. Consideration is also given to covariance analysis, missing data, transformations, and to multiple range and multiple *F*-tests. The orientation is toward biological and agricultural applications.)

Federov, V. V. (1972). *Theory of Optimal Experiments*. Academic Press, New York. Translated from the Russian by W. J. Studden and E. M. Klimko. (This book attempts "to present the mathematical apparatus of regression experimental design" and includes much research done by the author. The treatment is mainly in terms of theorems.)

Finney, D. J. (1955, 1974). *Experimental Design and Its Statistical Basis*. University of Chicago Press, Chicago. (This small introductory book is intended to be "in a form that will be intelligible to students and research workers in most fields of biology . . . . Even a reader who lacks both mathematical ability and acquaintance with standard methods of statistical analysis ought to be able to understand the relevance of these principles to his work . . . ." A chapter on biological assay is also included.)

Finney, D. J. (1975). *An Introduction to the Theory of Experimental Design*. University of Chicago Press, Chicago. (This book is written "with the intention of interesting the mathematician who is beginning to specialize in statistics, by giving him a broad survey of the mathematical techniques involved in design, yet not immersing him too deeply in details of analysis or of particular application." Actually, this book is not highly mathematical. It provides fairly detailed expositions of various blocking and factorial design schemes.)

Fisher, R. A. (1966). *Design of Experiments*, 8th ed. Hafner, New York; first ed., 1935. (An early book on the subject and a classic in the field, *Design of Experiments* includes detailed and readable discussions of experimental principles and confounding in experimentation. The treatment is directed principally at agricultural situations. The author is regarded as the father of modern statistics.)

Gill, J. L. (1978). *Design and Analysis of Experiments in the Animal and Medical Sciences*, 3 Vols., Iowa State University Press, Ames, Iowa. (This book "has been designed in two volumes for a one-year course for research students in the animal, medical, and related sciences . . . . The two volumes also will serve as a reference for experimenters in many areas of zoology and animal agriculture and in the food, health, and behavioral sciences. Also statisticians who consult with biologists may find some of the topics and references useful." The first volume concentrates on the analysis of experimental data and consists of chapters on introduction and review of basic statistical principles and procedures, completely randomized designs and analysis of variance, single covariate in experimental design, and analysis of nonorthogonal and multivariate data. The second volume emphasizes the design of experiments and has chapters on randomized complete block designs, incomplete block designs and fractional replication, double blocking, Latin squares and other designs, repeated measurement split-plot designs, and response surface designs. The discussion is generally quite detailed, thus making the book suitable as a reference work. The third volume includes tables and charts.)

Hersen, M. and Barlow, D. H. (1975). *Single-Case Experimental Designs: Strategies for Studying Behavior Change*. Pergamon Press, Elmsford, N.Y. (This book is devoted to a specialized subject—that where an individual is exposed to two or more treatments or conditions over time. Typical chapters are on "Repeated Measurement Techniques" and "Basic A–B–A Designs.")

Hicks, C. R. (1973). *Fundamental Concepts of Design of Experiments*, 2nd ed., Holt, Rinehart and Winston, New York; first ed., 1964. ("It is the primary purpose of this book to present the fundamental concepts in the design of experiments using simple numerical problems, many from actual research work . . . . The book is written for anyone engaged in experimental work who has a good background in statistical inference. It will be most profitable reading to those with a background in statistical methods including analysis of variance." This work provides an intermediate-level coverage of most of the basic experimental designs.)

John, A. and Quenouille, M. H. (1977). *Experiments: Design and Analysis*, 2nd ed. Hafner, New York. [This is an update of a 1953 book by the second author (previously entitled *The Design and Analysis of Experiment*), which was one of the earliest detailed expositions of the subject. "The emphasis is on the principles and concepts involved in conducting experiments, and on illustrating the different methods of design and analysis by numerical example rather than by mathematical and statistical theory. It is hoped that ... the book will be of value to those statisticians, researchers and experimenters who wish to acquire a working knowledge of experimental design and an understanding of the principles governing it."]

John, P. W. (1971). *Statistical Design and Analysis of Experiments*. Macmillan, New York. ("This book ... is for the mathematically oriented reader." The author has, however, "endeavored throughout to include enough examples from both engineering and agricultural experiments to give the reader the flavor of the practical aspects of the subject." Most of the standard experimental plans are treated in some detail.)

Johnson, N. L. and Leone, F. C. (1977). *Statistics and Experimental Design in Engineering and the Physical Sciences, Vol. 2*, 2nd ed. Wiley, New York. (Although this book is the second of two volumes, it stands reasonably well on its own and provides an intermediate-level treatment of various experimental plans. The calculational aspects of the analysis of variance associated with different designs are emphasized and some relatively complex situations are illustrated.)

Kempthorne, O. (1952). *Design and Analysis of Experiments*. Wiley, New York. Reprinted by Krieger, Huntington, N.Y., 1975. (This book is directed principally at readers with some background and interest in statistical theory. The early chapters deal with the general linear hypothesis and related subjects using matrix notation. Factorial experiments are discussed in detail. Chapters are devoted to fractional replication and split-plot experiments. Examples are generally taken from biology or agriculture.)

Keppel, G. (1973) *Design and Analysis—A Researcher's Handbook*, Prentice-Hall, Englewood Cliffs, N.J. (This book is directed principally to researchers and students in psychology. It is "intended for students who have aquired some degree of sophistication in statistics, such as they would obtain from an undergraduate introductory course. . . . Mathematical arguments are held to a minimum, and emphasis is placed on an intuitive understanding of the mathematical operations involved." Following an introductory section, the book has sections devoted to single-factor experiments, factorial experiments with two factors, higher order factorial experiments, designs with repeated measurements, designs intended to decrease error variance, and sensitivity of experimental designs and controversial topics.)

Kimmel, H. D. (1970). *Experimental Principles and Design in Pyschology*, Ronald Press, New York. ["This text for advanced courses in experimental psychology describes the major methods that have been developed for planning and conducting experiments on behavior and analyzing their results. It deals with (1) experimental principles in relation to the general background in which experiments are done and what they achieve; (2) quantitative matters dealing with measurement in psychology and psychological scaling; (3) experimental design principles and methods." Only four of the 12 chapters are devoted to the discussion of experimental designs per se. Others chapters deal with such subjects as "The Nature of Experiments," "The Dependent Variable," "Psychophysics and Scaling," and "Choosing the Independent Variables."]

Kirk, R. E. (1968). *Experimental Design: Procedures for the Behavioral Sciences*. Brooks-Cole, Belmont, Calif. (This work reflects the "belief that there is a need for a book requiring a minimum of mathematical sophistication on the part of the reader but including a detailed coverage of the more complex designs and techniques available to behavioral scientists." Thus it provides a fairly detailed exposition of certain designs at a reasonably elementary level. Most chapters conclude with a discussion of the advantages and disadvantages of the plans considered in that chapter.)

LeClerg, E. L., Leonard, W. H., and Clark, A. G. (1962). *Field Plot Technique*, 2nd ed. Burgess, Minneapolis, Minn. (Not shown in 1978–1979 *Books in Print*. "The book is an attempt to bring together and explain, as simply as possible, the general principles, techniques, assumptions, and guides to procedures applicable to experimentation in agriculture and biology. An effort was made to furnish the reader some of the practical aspects of statistical procedures, with detailed examples of analysis, rather than a general exposition of the mathematical theory underlying them." Three chapters are devoted to various types of lattice designs.)

Lee, W. (1975). *Experimental Design and Analysis*. W. H. Freeman, San Francisco. ("This book is an introductory treatment of multifactor experimental design and analysis of variance. It is aimed at students and researchers in the behavioral sciences, particularly psychology. The only prerequisite ... is a one-semester course in introductory statistics. . . . The main reason ... [for] the book is to present comprehensive and general methods ... for creating and analyzing the designs having factors related by crossing and nesting." Latin square designs are also discussed in detail.)

Li, C. C. (1964). *Introduction to Experimental Statistics*. McGraw-Hill, New York. (This book is directed at "the student whose field is not mathematics—the biological or medical research worker, for example—[who] is in genuine need of a short, nonmathematical course on the design and analysis of experiments, written in a rather informal style." The book consists of three parts devoted, respectively, to basic mechanics and theory, experimental designs, and some related topics.)

Lindquist, E. F. (1956). *Design and Analysis of Experiments in Psychology and Education.* Houghton Mifflin, London. (Not shown in 1978–1979 *Books in Print.* The purpose of this book "is to help students and research workers in psychology and education to learn how to select or devise appropriate designs for the experiments they may have occasion to perform, and to analyze and interpret properly the results obtained through the use of those designs." Chapters include ones devoted to "Treatments × Levels Designs," "Treatments × Subject Designs," and "Groups-Within-Treatment Designs.")

Lipson, C. and Sheth, N. J. (1973). *Statistical Design and Analysis of Engineering Experiments.* McGraw-Hill, New York. ["This book is written in a relatively simple style so that a reader with a moderate knowledge of mathematics may follow the subject matter. No prior knowledge of statistics is necessary." Appreciably more discussion is devoted to statistical analysis than to the planning of experiments. Such relatively nonstandard subjects (for an introductory text) as accelerated experiments, fatigue experiments, and renewal analysis are also included.]

Mendenhall, W. (1968). *Introduction to Linear Models and the Design and Analysis of Experiments.* Duxbury Press, Belmont, Calif. (This book provides an introduction to basic concepts and the most popular experimental designs without going into extensive detail. In contrast to most other books on the design of experiments, the emphasis in the development of many of the stated models and analysis methods is principally on a regression, rather than an analysis-of-variance viewpoint, thus providing a more modern outlook.)

Montgomery, D. C. (1976). *Design and Analysis of Experiments.* Wiley, New York. (This "introductory textbook dealing with the statistical design and analysis of experiments ... is intended for readers who have completed a first course in statistical methods." It provides a basic treatment of standard experimental plans and techniques for analyzing the resulting data.)

Myers, J. L. (1966). *Fundamentals of Experimental Design,* 2nd ed. Allyn and Bacon, Boston; first ed. 1966. (This book is intended to "provide a reasonably sound foundation in experimental design and analysis .... The reader should be familiar with material usually covered in a one semester introductory statistics course." Although the treatment is quite general, many of the examples deal with applications from psychology.)

Myers, R. H. (1971). *Response Surface Methodology.* Allyn and Bacon, Boston. (Not shown in 1978–1979 *Books in Print.* This specialized book is devoted to a detailed exposition of one type of experimental design —response-surface plans. Such designs are appropriate when independent variables on a quantitative scale, such as temperature, pressure, etc., are to be related to

the dependent variable using regression analysis. It is frequently desired to determine the combination of the independent variables which leads to the optimum response. "The primary purpose of *Response Surface Methodology* is to aid the statistician and other users of statistics in applying response surface procedures to appropriate problems in many technical fields .... It is assumed that the reader has some background in matrix algebra, elementary experimental design, and the method of least squares ....")

Namboodiri, N. K., Carter, L. F., and Bialock, H. M., Jr. (1975). *Applied Multivariate Analysis and Experimental Designs.* McGraw-Hill, New York. ("This book is intended primarily for sociologists and political scientists and secondarily for social scientists in other fields. It is directed toward advanced graduate students and professionals who are actively engaged in quantitative research ...." The book contains four parts: one, comprising approximately one-third of the volume, deals directly with experimental design. The other three deal with "Multiple Regression and Related Techniques," "Simultaneous-Equation Models," and "Models Involving Measurement Errors." The use of the term "Multivariate Analysis" in the book's title might mislead some statisticians; the term "multivariable analysis" might have been more appropriate.)

Ogawa, J. (1974). *Statistical Theory of the Analysis of Experimental Designs.* Marcel Dekker, New York. (This book "covers the fundamental portions of the statistical theory of the analysis of experimental designs .... The author relies heavily on linear algebraic methods as his mathematical tools." The six chapters are devoted to analysis of variance, design of experiments, factorial design, theory of block designs—intrablock and interblock analysis (two chapters), and a (mathematically rigorous) treatment of randomization of partially balanced incomplete block designs.)

Peng, K. C. (1967). *The Design and Analysis of Scientific Experiments.* Addison-Wesley, Reading, Mass. (Not shown in 1978–1979 *Books in Print.* This book is subtitled "An Introduction with Some Emphasis on Computations." It is written primarily for statisticians, computer programmers, and persons engaged in experimental work who have some background in mathematics and statistics. The mathematical background should include calculus and elementary matrix theory. The statistical background should be equivalent to a one-year course in statistics.)

Raghavarao, D. (1971). *Constructions and Combinatorial Problems in Design of Experiments.* Wiley, New York. (This book provides the mathematical development for the construction of experimental designs. It is intended "as a secondary and primary reference for statisticians and mathematicians doing research on the design of experiments.")

Raktoe, B. L., Hedayat, A., and Federer, W. T. (1981). *Factorial Designs.* Wiley, New York. (This book attempts to provide "a systematic treatise" on factorial

Table 1 Summary of Characteristics of Experimental Design Texts[a]

| | Characteristics | | | | | | | | | | | |
| | Subject Coverage | | | | | | | | | | | |
	Blocking Designs	Factorial Designs	Fractional Factorial Designs	Response-Surface Designs	Split-Plot Situations	Nested Situations	Number of Pages	Emphasis on Calculations and Analysis	References (Approximate Number)	Examples	Technical Level	Applications emphasis
Anderson and McLean	M-H	M-H	M-H	M-H	H	M-H	418	L-M	150	M	I	G/I
Bannerjee	N-L	N-L	N-L	N	N	N	141	L-M	83	L	A	G
Box et al.	M	M	L-M	M-H	N	L	653	L	150	M-H	E	G/I
Campbell and Stanley	L	L	N	N	N	L	84	N	150	M	E	P
Chakrabarti	M	M	L-M	N	M	N	120	N	37	N	A	G
Chapin	N	N	N	N	N	N	297	L	100	M	E	P
Chew	L-M	M	M-H	M-H	L	L	268	L-M	350	M	I	I
Chilton	L	N	N	N	N	N	365	M-H	200	M	E	O
Cochran and Cox	H	H	M-H	M-H	M-H	N	611	L-M	150	M	I	A/B
Cox	L-M	L-M	L-M	L	M	N	308	N	100	M	E	G
Daniel	N-L	H	H	L	L-M	N	294	L-M	100	M	I	I
Davies	M-H	M-H	M-H	M-H	N	L-M	656	L-M	50	M	I	I
Dayton	M	M	L-M	N-L	L	M	441	M	50	M	E	P
Edwards	M	M-H	L	N	L-M	N	488	M	125	M	E-I	P
Federer	M-H	M-H	L	N	M-H	L-M	591	M	340	L-M	I	A/B
Federov	N	N	N	M	N	N	292	L	62	L	A	G
Finney (1974)	L-M	L	L	N	L	N	169	L	70	M	E	B
Finney (1975)	M-H	M-H	M-H	L	L-M	N	223	L	150	L	I	A/B
Fisher	M	M	M	N	N	N	245	L	50	M	E-I	A/B
Gill	M-H	M-H	M	M	M	M-H	409(1) 301 182(3)	M-H	500	L-M	I	A/B

Hersen and Barlow	N	N	N	N	N	N	374	L	100	M	E	P
Hicks	M	M-H	L-M	L-M	M	M	349	M	20	M	I	I
John and Quenouille	M-H	M-H	L	N	M-H	N	356	L-M	150	M	I	A/B
John	M-H	M-H	M-H	M-H	M	M	356	L-M	250	L-M	A	G
Johnson and Leone	M	M	L-M	M-H	M-H	M-H	449	M-H	75	M	I	I
Kempthorne	M-H	M-H	M-H	N	N	N	631	L-M	150	L-M	A	A/B
Kimmel	M	L-M	N	N	N	N	280	L	100	M	E	P
Kirk	M-H	M-H	M-H	N	M-H	L	577	M	450	M	I	P
LeClerg et al.	M-H	M	N	N	M-H	N	373	M	300	M	E	A/B
Lee	L-M	M	N	N	N-L	M-H	353	L-M	15	L-M	I	P
Li	M	M	N	N	M	M	460	M	35	M	E-I	B/G
Lindquist	M	M	M	N	N	N	393	M	50	L-M	I	P
Lipson and Sheth	N	L	N	N	N	N	518	M	100	M	E	I
Mendenhall	L-M	L-M	L-M	L-M	L	L-M	464	L-M	100	M	I	G/I
Montgomery	M	M	M	M	M	M	418	M-H	85	L-M	E-I	I
J. L. Myers	M	N	N	N	N	M	465	M	100	L-M	I	G
R. H. Myers	N	L	L	H	N	N	246	M	50	L-M	E-I	G
Namboodiri et al.	M	L	L	N	M	M	688	M	250	L-M	I	P
Ogawa	M-H	M	M	N	N	N	465	M	220	L	A	G
Peng	M	M	M-H	M	L-M	M	252	M-H	90	L-M	I	G
Raghavarao	M	M-H	M-H	N	N	N	386	N	500	L	A	G
Vajda	N	N	N	N	N	N	120	N	109	L	A	A
Winer	M-H	M-H	L	L-M	L-M	M	907	M-H	100	M	I	B/P
Yates	M-H	M-H	N	L	M	N	296	L-M	100	M	I	A

[a]Includes publications through 1978.

design. It is directed at "the person interested in the pursuit of knowledge about the *concepts* of factorial design ... at a level suitable for persons with knowledge of the theory of linear models .... This book is also a place to which statisticians and mathematicians may turn in order to become acquainted with the subject." Thus this book provides a rigorous development of the concepts of factorial designs.)

Vajda, S. (1967). *The Mathematics of Experimental Design—Incomplete Block Designs and Latin Squares.* Hafner, New York. (This small volume is "concerned with the mathematical aspect of designs which have been developed for the purpose of analyzing first agricultural, and then also other statistical experiments. The analysis of the results of such experiments is outside the scope of the book." The treatment is on a relatively advanced mathematical level.)

Winer, B. J. (1971). *Statistical Principles in Experimental Design,* 2nd ed. McGraw-Hill, New York. [This book, "written primarily for students and research workers in the area of the behaviorial and biological sciences ... is meant to provide a text as well as a comprehensive reference on statistical principles underlying experimental design. Particular emphasis is given to those designs that are likely to prove useful in research in the behavioral and biological sciences." Chapters are devoted to single-factor and multi-factor experiments with repeated measures on the same elements (or subjects).]

Yates, F. (1970). *Experimental Design: Selected Papers of Frank Yates.* Hafner, Darien, Conn. (This book consists of 12 papers previously published in the years 1933 to 1965 by one of the giants in the development of experimental design. It "is intended to give a representative selection of Yate's publications on experimental design, with particular attention to some of the earlier papers now not readily accessible to statisticians. The reader will not find here a comprehensive textbook, but he can learn much about the practice of good design that is as true today as when it was first written. Although developed in connection with agricultural research, Yate's ideas are equally relevant to any investigations in which comparative experiments are used ....")

A summary of various characteristics of these experimental design texts is presented in Table 1. Abbreviations used in the table are as follows:

General:
  N, none
  L, little
  M, moderate
  H, heavy
Technical Level:
  E, elementary (no background in statistics needed)

  I, intermediate (assumes one or two elementary statistics courses)
  A, advanced (assumes more than one or two statistics courses)
Applications Emphasis:
  A, agriculture
  B, biology/medicine
  G, general
  I, industrial/engineering/scientific
  M, management/business/economics
  P, psychology/education/sociology/ behavioral sciences/social sciences
  O, other

GERALD J. HAHN

DESTRUCTIVE SAMPLING

This term does not apply, strictly, to the sampling process but to the effect of measurement operations on the individuals in a sample. Tests of lifetime of equipment in which time to failure are recorded, result in "destruction" (as viable items) of the elements in the sample. Tests of strength of materials also result in destruction of many of the items on test. The same is true of much experimentation on the effects of drugs on animals.

When sampling is "destructive" there are several important consequences. Such sampling tends to be costly, but also sampling (rather than 100% inspection) is necessary. In control of quality, it would clearly be of no use to find the lifetime of every lamp bulb in a lot of electric lamp bulbs, as there would be no usable lamp bulb left to use.

(ACCEPTANCE SAMPLING
COST FUNCTIONS
QUALITY CONTROL
SAMPLING)

DETERMINISTIC TESTS *See* EDITING STATISTICAL DATA

DEVIANCE

Another name for sum of squares, especially as used in analysis-of-variance (ANOVA)* tables. *See also* GENERALIZED LINEAR MODEL.

DEVIATE

Usually, the difference between a variable and its expected value*. (Sometimes, the difference between the value of an estimator and the parameter being estimated is called a "deviate," but a more usual term is "error" or "residual.")

DEVIATE, STANDARDIZED

The difference between a variable (X) and its expected value* (ξ) divided by its standard deviation* (σ). Thus the standardized deviate corresponding to X is

$$X' = \frac{X - \xi}{\sigma}.$$

The expected value of X' is zero; the standard deviation of X' is 1.

If X does not possess a mean and a standard deviation (e.g., if X has a Cauchy distribution*), then there is no standardized deviate corresponding to X.

The *shape** of the distribution of X' is the same as that of X. In particular, X' has the same moment ratios* and cumulant* ratios as X.

Standardized deviates are useful in connection with tabulation because they can depend on two fewer parameters than the corresponding unstandardized deviates. This substantially reduces the amount of tables needed. For example, tables of the cumulative distribution function* of the unit normal can be used to calculate the CDF of any normal distribution by means of the relationship

$$\Pr[X < x] = P(\xi + \sigma X' < x)$$
$$= P\left(X' < \frac{x - \xi}{\sigma}\right).$$

(STANDARDIZED DISTRIBUTION TABLES, STATISTICAL)

DICHOTOMY

The term is used to describe the results of dividing a population into two parts, usually according to some known criterion. For example, when the values of a variable are recorded only as being greater or less than some specific value, the resulting observations are *dichotomized.*

By obvious extensions we have the terms trichotomy, tetrachotomy, pentachotomy, . . . , polychotomy or polytomy (splitting into three, four, five, . . . , many parts). Variables taking only two, three, four, five, . . . or any finite number of possible values can be called dichotomous, trichotomous, tetrachotomous, pentachotomous, . . . , polychotomous. A more common term for dichotomous variables is *binary variables*.*

(BINARY DATA)

DIFFERENCE EQUATION

Constituting an important topic in *finite difference** *calculus*, difference equations are a finite difference analog of differential equations*. A difference equation is simply a relationship

$$g(u_r, r_{r+1}, \ldots, u_{r+p-1}) = 0$$

among p successive terms of a sequence u_1, $u_2, \ldots$.

Linear difference equations are of form

$$a_p u_{r+p} + a_{p-1} u_{r+p-1} + \cdots + a_1 u_{r+1}$$
$$+ a_0 u_r = b. \qquad (1)$$

They are analogs of linear differential equations* and are solved in a similar manner, using roots $\theta_1, \ldots, \theta_p$ of the auxiliary equation

$$a_p \theta^p + a_{p-1} \theta^{p-1} + \cdots + a_1 \theta + a_0 = 0$$

to obtain a general solution of form $u_r = \sum_{j=1}^p A_j \theta_j^r$, to which must be added a special (particular) solution of (1).

Bibliography

Milne-Thomson, L. M. (1933). *Calculus of Finite Differences*. Macmillan, London.

(AUTOREGRESSIVE-INTEGRATED
    MOVING AVERAGE MODELS
AUTOREGRESSIVE-MOVING
    AVERAGE MODELS

FINITE DIFFERENCES, CALCULUS OF)

DIFFERENCE OF ZERO

The r-th forward difference* of x^s is

$$\Delta^r x^s = \sum_{j=0}^{r} (-1)^j \binom{r}{j}(x+j)^s.$$

Putting $x = 0$, we obtain the *difference of zero*,

$$\Delta^r 0^s = \sum_{j=0}^{r} (-1)^j \binom{r}{j} j^s.$$

Since $\Delta^r x^s = 0$ for $r > s$ (*see* FORWARD DIFFERENCES),

$$\Delta^r 0^s = 0 \qquad \text{for } r > s$$

and similarly,

$$\Delta^s 0^s = s! \ .$$

Differences of zero provide convenient ways of representing certain probabilities.

Some useful approximate formulas for the values of differences of zero are given in Good [3].

As r and n increase, the values of the differences of zero increase very rapidly. It is convenient to use the *reduced difference of zero* $\Delta^r 0^s / s!$ (a Stirling number of the second kind). Tables of this quantity are available [1, 2].

The recurrence formula

$$\Delta^r 0^{s+1} = r(\Delta^r 0^s + \Delta^{r-1} 0^s)$$

is useful if it is necessary to compute numerical values; it can also be useful in analytical work.

References

[1] Abramovitz, M. and Stegun, I. A. eds. (1964). *Handbook of Mathematical Functions, Natl. Bur. Stand. (U.S.) Appl. Math. Ser.* **55** (Washington, D.C.).

[2] Goldberg, K., Leighton, F. I., Newman, M., and Zuckerman, S. L. (1976). *J. Res. Natl. Bur. Stand.*, **80B**, 99–171.

[3] Good, I. J. (1961). *Ann. Math. Statist.*, **32**, 249–256.

(FINITE DIFFERENCES, CALCULUS OF
FORWARD DIFFERENCES)

DIFFUSION PROCESSES

A *diffusion process* is defined to be a Markov process* whose sample processes are continuous. The state space S may be the real line R, the r-dimensional space R^r or more generally a topological space.

Let $\{X_t, t \in [0, \infty)\}$ be a diffusion process on S. Since it is a Markov process, the probability law governing the process $\{X_t\}$ is determined by the *transition probability**:

$$P_{s,t}(x, E) = \Pr[X_t \in E \mid X_s = x]$$
$$(s \leqslant t) \qquad (1)$$

and the initial distribution $\mu(E) = \Pr[X_0 \in E]$. The transition probability $P_{s,t}(x, E)$ satisfies the *Chapman–Kolmogorov equation*:

$$P_{s,u}(x, E) = \int_S P_{s,t}(x, dy) P_{t,u}(y, E)$$
$$(s \leqslant t \leqslant u). \qquad (2)$$

The *generator* A_s defined by

$$(A_s f)(x) = \lim_{h \downarrow 0} h^{-1} E[f(X_{s+h})$$
$$- f(X_s) \mid X_s = x]$$
$$= \lim_{h \downarrow 0} h^{-1}\left[\int_S P_{s,t}(x, dy) f(y)$$
$$- f(x)\right] \qquad (3)$$

plays an important role in the theory of diffusion processes.

Because of continuity of sample processes the transition probability must satisfy an additional condition besides the Chapman–Kolmogorov equation*. A conceivable one, although neither necessary nor sufficient in general, is

$$P_{s,s+h}(x, U(x)^c) = o(h) \qquad (h \downarrow 0) \quad (4)$$

for every neighborhood $U(x)$. This condition implies that A_s is a local operator, namely, that $(A_s f)(x)$ is determined by the behavior of f near x. This explains why A_s

turns out to be a differential operator in many practical cases.

KOLMOGOROV'S DIFFUSION PROCESSES

The analytical theory of diffusion processes was initiated by A. Kolmogorov in 1931. Let X_t be a diffusion process on R. Kolmogorov assumed the following conditions:

$$
\left.
\begin{aligned}
&E[X_{s+h} - X_s \mid X_s = x]\\
&\quad = a(s,x)h + o(h)\\
&E[(X_{s+h} - X_s)^2 \mid X_s = x]\\
&\quad = b(s,x)h + o(h)
\end{aligned}
\right\}
\quad (h\downarrow 0) \quad (5)
$$

in addition to (4). Such a diffusion process is called *Kolmogorov's diffusion process*. The generator A_s of this process is given by

$$
A_s = a(s,x)\frac{d}{dx} + \frac{1}{2}b(s,x)\frac{d^2}{dx^2}. \quad (6)
$$

Using the Chapman–Kolmogorov equation, we can check that the function

$$
U(s,x) = P_{s,t}(x, E), \qquad s \leqslant t, \quad (7)
$$

satisfies the following parabolic equation, called the *Kolmogorov backward equation*:

$$
\frac{\partial U}{\partial s} = -A_s U, \qquad s < t, \quad (8)
$$

and the terminal condition

$$
U(t,x) = \delta(x, E), \quad (9)
$$

where $\delta(x, E)$ is 1 or 0 according to whether x lies in E or not.

Suppose that we are given a differential operator A_s of the form (6). We want to construct a diffusion process whose generator is A_s. In 1936, W. Feller proved that the backward equation (8) together with the terminal condition (9) has a unique solution under the assumption "$b(s,x) > 0$" and some regularity conditions on $a(s,x)$ and $b(s,x)$. Denote the solution by $P_{s,t}(x, E)$, because it depends not only on (s,x) but also on (t, E). Then we can check that $P_{s,t}(x, E)$ is a probability measure in E and satisfies the Chapman–Kolmogorov equation. Also, $P_{s,t}(x, E)$ has density $p_{s,t}(x, y)$ relative to the Lebesgue measure. The func-

tion

$$
u(s,x) = p_{s,t}(x, y), \qquad s \leqslant t,
$$

satisfies the backward equation (8) and the terminal condition

$$
u(t,x) = \delta(x - y)
$$

(δ = the Dirac delta function*).

Fix s and x and consider the function

$$
v(t, y) = p_{s,t}(x, y), \qquad t \geqslant s.
$$

Then $v(t, y)$ satisfies the following parabolic equation, called the Kolmogorov forward equation:

$$
\frac{\partial v}{\partial t} = A_t^* v, \quad (10)
$$

where A_t^* is the adjoint operator of A_t, i.e.,

$$
(A_t^* g)(y) = -\frac{d}{dy}\big[a(t, y)g(y)\big]
$$

$$
+ \frac{1}{2}\frac{d^2}{dy^2}\big[b(t, y)g(y)\big] \quad (11)
$$

Let $p(t, y)$ be the density of the probability distribution of X_t. Then we have

$$
p(t, y) = \int_R \Pr[X_0 \in dx]\, p_{0,t}(x, y),
$$

so $p(t, y)$ also satisfies the forward equation (10), i.e.,

$$
\frac{\partial p}{\partial t} = A_t^* p. \quad (12)
$$

This equation is called the *Fokker–Planck equation**.

We can easily extend this theory to the r-dimensional case where A_s is an elliptic operator:

$$
A_s = \sum_{i=1}^{r} a^i(s,x)\frac{\partial}{\partial x^i}
$$

$$
+ \frac{1}{2}\sum_{i,j=1}^{r} b^{ij}(s,x)\frac{\partial^2}{\partial x^i \partial x^j}. \quad (13)
$$

Corresponding to the assumption "$b(s,x) > 0$" we impose the assumption that the matrix (b^{ij}) is strictly positive-definite.

The special case where $a^i(s,x) = 0$ and (b^{ij}) = the identity matrix I is called the r-dimensional *Wiener process** or *Wiener's Brownian motion**. The generator of this process is $\Delta/2$ and the transition probability

$P_{s,t}(x, E)$ is Gauss measure with mean vector x and variance matrix $(t - s)I$.

In the discussion above we have been concerned only with the transition probability. From the transition probability obtained above we can construct a Markov process*. Now we have to check that this process has a version with continuous sample processes. R. Fortet did this by estimating $p_{s,t}(x, y)$ elaborately. The method of stochastic differential equations due to K. Itô will provide a more direct way.

STOCHASTIC DIFFERENTIAL EQUATIONS*

Consider a stochastic differential equation

$$dX_t = a(t, X_t) dt + \sigma(t, X_t) dB_t, \qquad X_0 = E,$$

$$(14)$$

where B_t is a one-dimensional Wiener process. If $a(t, x)$ and $\sigma(t, x)$ are continuous in t and Lipschitz continuous in x uniformly in t and if E is independent of the Wiener process B_t, $0 \leqslant t < \infty$, then the solution is a diffusion process whose generator A_s is given by

$$A_s = a(s, x) \frac{d}{dx} + \frac{1}{2} \sigma(s, x)^2 \frac{d^2}{dx^2}. \quad (15)$$

If we take a vector $(a^1, a^2, \ldots, a^r)$ for a, an $r \times r$-matrix (σ_k^i) for σ and an r-dimensional Wiener process for B in the stochastic differential equation (14), we obtain an r-dimensional diffusion process X_t whose generator is given by (13) with

$$b^{ij} = \sum_{k=1}^{r} \sigma_k^i \sigma_k^j.$$

The stochastic differential equation

$$dX_t = -\alpha X_t dt + \beta dB_t \qquad (16)$$

is the *Langevin equation* in physics. This determines a diffusion process with the generator

$$-\alpha x \frac{d}{dx} + \frac{\beta^2}{2} \frac{d^2}{dx^2},$$

which is called the *Ornstein–Uhlenbeck process* (*see* GAUSSIAN PROCESSES). The solution

of this equation is

$$X_t = X_0 + \beta \int_0^t e^{-\alpha(t-u)} dB_u,$$

so

$$X_t = X_s + \beta \int_s^t e^{-\alpha(t-u)} dB_u \qquad (s < t).$$

Hence the transition probability $P_{x,t}(x, E)$ is the Gauss measure* with mean x and variance

$$\beta^2 \int_s^t e^{-2\alpha(t-u)} du = \frac{\beta^2}{2\alpha} \left[1 - e^{-2\alpha(t-s)} \right].$$

This fact can also be proved by solving the Kolmogorov backward equation for the process:

$$\frac{\partial U}{\partial s} = \alpha x \frac{\partial U}{\partial x} - \frac{1}{2} \beta^2 \frac{\partial^2 U}{\partial x^2} \qquad (s < t).$$

TIME-HOMOGENEOUS DIFFUSION PROCESSES

A diffusion process $\{X_t\}$ is called *time homogeneous* if its transition probability is invariant under time shift. In this case the transition probability $P_{s,t}(x, E)$ can be written as $P_{t-s}(x, E)$, and hence the generator A_t is independent of t, so it can be written as A. The operator T_t defined by

$$(T_t f)(x) = \int_S P_{s,t}(x, dy) f(y)$$

$$= E[f(X_t) | X_s = x] \quad (17)$$

is called the *transition operator* at t. Since the Chapman–Kolmogorov equation implies that

$$T_{t+s} = T_t T_s,$$

the transition operators T_t, $0 \leqslant t < \infty$, form a semi-group*. Applying the Hille–Yosida theory of semigroups, W. Feller initiated the modern theory of diffusion processes, which was extensively developed by E. B. Dynkin, K. Itô, H. P. Mckean, and many others. Here we will restrict our discussion to the case considered by Feller, called *Feller's diffusion process*. As a matter of fact, practically all diffusion processes useful in application can be reduced to this case.

Following Feller we assume the following:

1. The state space S is a compact metric space.
2. The transition operator T_t carries the space $C(S)$, the continuous functions on S, into itself.

Then $\{T_t, \ 0 \leqslant t < \infty\}$ is a semigroup of operators on the separable Banach space $C(S)$ in the Hille–Yosida sense and the generator of this semi-group is the generator of our diffusion process. The most important property of this diffusion process is the *strong Markov property*, which means that the process starts afresh at every *stopping time* (sometimes called *Markov time*), in particular at every *first passage time* (sometimes called *hitting time*). Using this property Dynkin derived the following formula, called *Dynkin's representation* of the generator A:

$$(Af)(x) = \lim_{r \downarrow 0} \frac{E(f(X_{\tau(r)}) \mid X_0 = x) - f(x)}{E(\tau(r) \mid X_0 = x)},$$

$$(18)$$

where $\tau(r)$ is the hitting time for the sphere with center x and radius r. Since $X_{\tau(r)}$ lies on the sphere because of continuity of the sample processes of X_t, Dynkin's representation assures that A is a local operator.

If $Au = 0$, u is called *harmonic* relative to X_t. If X_t is Wiener process, this definition coincides with the classical definition of harmonicity. Thus we can generalize the theory of harmonic functions and potentials from this viewpoint. This idea was developed by J. L. Doob, G. Hunt, R. K. Getoor, R. M. Blumenthal, and many others not only for diffusion processes but also for general Markov processes.

Let D be an open subset of S and $\tau = \tau_D$ denote the hitting time for the boundary ∂D. Then we have the following:

1.

$$u(x) = E\left[f(X_\tau) \mid X_0 = x\right] \quad (19)$$

is the solution of the generalized Dirich-

let problem:

$$Au = 0 \text{ in } D \qquad \text{and} \qquad u = f \text{ on } \partial D. \quad (20)$$

2.

$$u(x) = E\left[\int_0^\tau \exp\left[-\int_0^t q(X_s)\,ds\right] f(X_t)\,dt\right]$$

$$(21)$$

is the solution of

$$Au - qu = -f \text{ in } D \quad \text{and} \quad u = 0, \text{ on } \partial D.$$

$$(22)$$

The representation (21) is called the *Kac formula*. In case $q \equiv 0$, (22) is the generalized *Poisson equation*.

Observing Kolmogorov's time homogeneous diffusion processes in such a way, we can see that the formula (21) gives the solution of the elliptic differential equation $(x \in D)$:

$$\left[\frac{1}{2} \sum_{i, j=1}^r b^{ij}(x) \frac{\partial^2}{\partial x^i \partial x^j}\right.$$

$$\left. + \sum a^i(x) \frac{\partial}{\partial x^i} - q(x)\right]u(x) = -f(x)$$

$$(23)$$

with the boundary condition $u = 0$ on ∂D. The case where b and a are not necessarily smooth but only continuous or measurable has recently been investigated by S. R. Varadhan, D. W. Stroock, and N. V. Krylov.

Assumption 1 does not hold even when $\{X_t\}$ is the Wiener process. Adding the point at infinity, denoted by ∞, to its state space R^r and assuming that $P_t(\infty, \infty) = 1$, we obtain a diffusion process satisfying assumptions 1 and 2. Thus we can apply Feller's theory to the Wiener process.

We can also apply Feller's theory to the time nonhomogeneous case by considering the *space–time process* introduced by J. L. Doob. Let $\{X_t\}$ be a diffusion process on S which may not be time homogeneous. The space–time process $\{\tilde{X}_t\}$ defined by

$$\tilde{X}_t = (X_t, t) \quad (24)$$

is a time-homogeneous diffusion process on

$S \times [0, \infty)$. If X_t has generator A_s, then the generator $\tilde{A}$ of $\{\tilde{X}_t\}$ is given by

$$(\tilde{A}f)(x,s) = A_{s(x)}f(x,s) + \frac{\partial}{\partial s} f(x,s), \quad (25)$$

where the suffix (x) suggests that A_s acts on $f(x,s)$ viewed as a function of x for each s fixed. Replacing X_t, D and f in item 1 by $\tilde{X}_t$, $\tilde{D} = S \times [0,t)$, and $\tilde{f}(x,t) = \delta(x,E)$, respectively, we can easily see that the transition probability $u(s,x) = P_{s,t}(x,E)$ of the original process X_t satisfies the Kolmogorov backward equation (8) and the terminal condition (9). Let τ be the hitting time of the original process X_t for the boundary of an open subset D. Replacing X_t, D and f in item 1 by $\tilde{X}_t$, $\tilde{D} = S \times [0,t)$, and $\tilde{f}(x,s) = \delta(x, \partial D)\delta(s,[0,t))$, respectively, we can see that

$$u(x,s) = \Pr[\tau < t \mid X_s = x] \quad (26)$$

satisfies the Kolmogorov backward equation

$$A_{s(x)}u + \frac{\partial u}{\partial s} = 0 \quad (s < t, x \in D), \quad (27)$$

with the boundary conditions

$$u(x,s) = \begin{cases} 0 & (s = t, x \in D) \\ 1 & (s < t, x \in \partial D). \end{cases}$$

The fundamental problem in the theory of diffusion processes is to determine all possible types of time-homogeneous diffusion processes with the strong Markov property. This problem was completely solved for one-dimensional diffusions by Feller in the analytical sense. The probabilistic interpretation of Feller's theory was given by Itô, McKean, and Dynkin. Let X_t be a time-homogeneous strong Markov diffusion process on $S = [a,b]$. A point $x \in S$ is called *regular* if

$$\Pr[X_t > x \mid X_0 = x] > 0 \quad \text{and}$$
$$\Pr[X_t < x \mid X_0 = x] > 0 \quad (t > 0).$$

We assume that every interior point of S is regular. Then the generator A of X_t is given by

$$A = \frac{d}{dm}\frac{d}{ds}, \quad (28)$$

where m and s are strictly increasing functions and s is continuous. This is much more general than the Kolmogorov case, where

$$A = \frac{1}{2} b(x) \frac{d^2}{dx^2} + a(x) \frac{d}{dx}.$$

In fact, this can be written in the form (28) with

$$dm = \frac{2e^B}{b} dx \text{ and } ds = e^{-B}dx$$

$$\left(B = \int \frac{2a}{b} dx\right),$$

so $m(x)$ and $s(x)$ are absolutely continuous in x. Feller determined all possible boundary conditions to discuss the behavior of X_t at the boundary points. If we remove the regularity assumption, the situation is more complicated.

Feller also treated the case where *absorption* (or *killing*) occurs. If $k(x)$ is the absorption rate, then the generator A should be replaced by $A - k$. The most general absorption is given by absorption measure, as McKean showed.

Bibliography

Here we will only list several important monographs where the reader can find the original papers.

Blumenthal, R. M. and Getoor, R. K. (1968). *Markov Processes and Potential Theory*, Academic Press, New York.

Doob, J. L. (1953). *Stochastic Processes*. Wiley, New York.

Dynkin, E. B. (1965). *Markov Processes*, Vols. 1 and 2. Springer-Verlag, New York.

Friedman, A. (1975, 1976). *Stochastic Differential Equations and Applications*, Vols. 1 and 2. Academic Press, New York.

Itô K. and McKean, H. P., Jr. (1965). *Diffusion Processes and Their Sample Paths*. Springer-Verlag, New York.

McKean, H. P. (1960). *Stochastic Integrals*, Academic Press, New York.

Stroock, D. W. and Varadhan, S. R. S. (1979). *Multidimensional Diffusion Processes*. Springer-Verlag, New York.

KIYOSI ITÔ

DIGAMMA FUNCTION

Also called the psi function, it is defined by the equation

$$\psi(x) = d\{\log \Gamma(x)\}/dx = \frac{\Gamma'(x)}{\Gamma(x)},$$

for a real positive x, where $\Gamma(x)$ is the gamma function. This function often occurs in statistical practice, particularly when beta or gamma densities are involved.

For real positive x, $\psi(x)$ is a concave increasing function which satisfies the relations

$$\psi(1) = -\gamma = 0.57721566\ldots,$$

where γ is the Euler constant*,

$$\psi(1 + x) = \psi(x) + \frac{1}{x},$$

$$\psi(x) = \log x - \frac{1}{2x} - \frac{1}{12x^2} + \frac{1}{120x^4}$$
$$- \frac{1}{252x^6} + O\left(\frac{1}{x}\right) \qquad \text{as } x \to \infty.$$

Also,

$$\psi(x) = -\gamma - \frac{1}{x} + O(x) \qquad \text{as } x \to 0.$$

Additional properties are discussed in Abramowitz and Stegun [1, pp. 258–259]; for values, see Davis [3].

A computer routine in ISO FORTRAN language for calculating $\psi(x)(x > 0)$ was developed by Bernardo [2].

References

[1] Abramowitz, M. and Stegun, I. A., eds. (1964). *Handbook of Mathematical Functions*. Dover, New York.

[2] Bernardo, J. M. (1976). *Appl. Statist.*, **25**, 315–317.

[3] Davis, H. T. (1933). *Tables of the Higher Mathematical Functions*, Vol. 1. Principia Press, Bloomington, Ind.

(POLYGAMMA FUNCTION)

DIGRESSION ANALYSIS

Digression analysis is a method for clustering of observations and for estimation of regression models in heterogeneous data. Digression analysis can be considered as a generalization of normal regression analysis*; the ordinary least-squares method is replaced by a *selective least-squares* (SLS) *method**. Digression analysis is also closely related to *switching regression**. In digression analysis, however, no extraneous information, as time or other support variables, is used for classification of observations.

Let there be n observations $y_j, x_{1j}, \ldots, x_{mj}, j = 1, \ldots, n$, on the variables $Y, X_1, \ldots, X_m$ and assume that these observations belong in an unknown way to two (or more) groups. In group i $(i = 1, 2)$ we have $E(Y \mid x_1, \ldots, x_m) = f_i(x_1, \ldots, x_m, \alpha_i)$, where the form of the regression function f_i is known but possibly different for each group and α_i is the vector of the parameters.

The parameters are estimated by generalizing the least-squares criterion to a selective form

$$S(\alpha_1, \alpha_2) = \sum_{j=1}^{n} \min\{(y_j - f_1(x_{1j}, \ldots, x_{mj}, \alpha_1))^2,$$
$$(y_j - f_2(x_{1j}, \ldots, x_{mj}, \alpha_2))^2\}$$

and this SLS criterion is minimized with respect to α_1, α_2. Thus each observation will be attributed to the nearest regression curve. This set up and the SLS criterion can be extended to more than two submodels. In applications it is natural to expect that the submodels have a similar form and the parameters may also be partially common.

The estimates obtained by the SLS method may be biased particularly if the subgroups overlap strongly. This digression bias diminishes rapidly when the heterogeneity increases. The magnitude of the bias depends also on the nature of the parameter.

Minimization of the SLS criterion is usually a nonlinear optimization problem that must be solved by iterative methods. It is also difficult to study the SLS principle theoretically.

In digression analysis it is necessary to check that the data are really heterogeneous in the intended manner. One possibility to do this is to fit also an ordinary regression model $E(Y) = f(x_1, \ldots, x_m, \alpha)$ and compare the residual sum of squares S_R of regression analysis and $S_D = \min S(\alpha_1, \alpha_2)$ of digression analysis. If $f = f_1$, we have $S_D \leqslant S_R$, since the digression model is always more flexible than its submodels. Typically, $2S_D < S_R$ even in homogeneous samples.

Testing for heterogeneity may be based on the ratio S_D/S_R, which is asymptotically normal in homogeneous samples when the error terms are normal. In the decomposition of two univariate normal distributions S_D/S_R is asymptotically

$$N\big(1 - 2/\pi, 8(1 - 3/\pi)/(\pi n)\big)$$

when the population is homogeneous, i.e., $N(\mu, \sigma^2)$. For $n \geqslant 8$ a good approximation for the mean is $1 - 2\pi^{-1} - n^{-1}$.

In Fig. 1, an artificial heterogeneous sample of 200 observations

$$Y = \begin{cases} \sigma_1 x + \beta_1 + \epsilon_1 & \text{with probability } p_1 \\ \alpha_2 x + \beta_2 + \epsilon_2 & \text{with probability } p_2 \end{cases}$$

$[\alpha_1 = 1.2, \ \beta_1 = 1, \ \epsilon_2 \sim N(0, 0.8^2), \ p_1 = 0.4, \ \alpha_2 = 0.5, \ \beta_2 = 2.5, \ \epsilon_2 \sim N(0, 0.4^2), \ p_2 = 0.6]$ is displayed. In Fig. 2 the estimated digression lines $y = 1.344x + 0.507$ and $y = 0.504x + 2.525$ are plotted together with the theoretical digression lines (dashed). By simulation

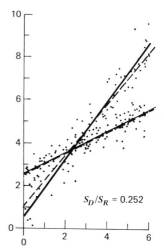

Figure 2   Estimated and theoretical digression lines.

it has been found that in this case the digression estimates have the following properties:

Parameter	Mean of Estimate	Bias	Mean Square Error
$\alpha_1 : 1.2$	1.324	0.124	0.0226
$\beta_1 : 1$	0.645	-0.355	0.2180
$\alpha_2 : 0.5$	0.497	-0.003	0.0008
$\beta_2 : 2.5$	2.532	0.032	0.0133

Bibliography

Mustonen, S. In *COMPSTAT 1978, Proc. Comp. Statist.* Physica-Verlag, Vienna, 1978, pp. 95–101.

(HOMOGENEITY, TESTS OF
MIXTURES
REGRESSION ANALYSIS
SELECTIVE LEAST SQUARES
SWITCHING REGRESSION)

SEPPO MUSTONEN

DILATION   *See* GEOMETRY IN STATISTICS—
CONVEXITY; MAJORIZATION AND SCHUR CONVEXITY

DILUTION EXPERIMENTS

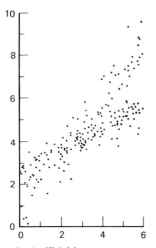

Figure 1   Artificial heterogeneous sample.

A form of experiment design especially associated with bioassay*. In its simplest form, a drug is given at successively greater dilutions, until the response under study appears very rarely or not at all. There may be an objective rule for deciding when this stage has been reached—for example, application to a specified number of individuals without response being observed.

Commonly, dilution is in the same ratio at each stage, so that if at each stage active solution and inactive medium are mixed in ratios $1:k$, the doses per unit volume of solution at successive stages are in ratio $1:k^{-1}:k^{-2}:\ldots$.

(BINARY DATA
BIOASSAY, STATISTICAL METHODS IN
QUANTAL DATA
TOLERANCE DISTRIBUTION)

DIRAC DELTA FUNCTION

The δ function was introduced and systematically employed by P. A. M. Dirac in his book, *The Principles of Quantum Mechanics* [4]. Dirac formally defined the δ-function as a real-valued function on $(-\infty, \infty)$ having the property that $\int_{-\infty}^{\infty} \delta(x)\,dx = 1$, $\delta(x) = 0$ for $x \neq 0$. Such a definition cannot be justified in terms of Lebesgue integration theory. However, Dirac realized that this definition does not define a function in the usual sense; he actually used the δ-function to obtain certain useful linear operations on continuous functions through formal integration. For example, if f is a real-valued continuous function on $(-\infty, \infty)$, then

$$\int_{-\infty}^{\infty} f(x)\delta(x-a)\,dx \equiv f(a).$$

Mathematicians sought for many years to develop a rigorous mathematical structure that would encompass the δ-function. This was brought to fruition by Schwartz in his book, *Théorie des Distributions* [6].

In Schwartz's framework, the δ-function is a family of elements belonging to a linear space of distributions. A "distribution" in this sense is defined as follows. The space of distributions is the linear space of all continuous linear functionals on a specified topological linear space of test functions. The space of test functions is frequently taken as the set $\mathfrak{D}$ of all complex-valued infinitely differentiable functions on $(-\infty, \infty)$ that vanish outside some closed and bounded set (which can vary with f in $\mathfrak{D}$) endowed with an appropriate topology [2, 3, 5, 6]. Another important space of test functions is the "Schwartz space" $\mathfrak{S}$ of infinitely differentiable functions f that are rapidly decreasing, i.e., for every p and q, $x^q(d^p/dx^p)f(x)$ is bounded on $(-\infty, \infty)$. The space of distributions $\mathfrak{S}'$ on $\mathfrak{S}$ is called the space of *tempered distributions*. For further discussion of $\mathfrak{D}$, $\mathfrak{D}'$, $\mathfrak{S}$, and $\mathfrak{S}'$, see refs. 2, 3, 5, 6, and 7.

A distribution cannot always be defined as a function on $(-\infty, \infty)$, although locally integrable functions do define distributions. Instead, a distribution is defined by its set of values on the space of test functions. Thus if ϕ is a distribution, then ϕ is defined by $\{\langle \phi, f \rangle$, all test functions $f\}$, where $\langle \phi, f \rangle$ is the value of ϕ at the point f. The Dirac δ-function is defined by the family of distributions $\{\phi_a, a$ in $(-\infty, \infty)\}$ having the property $\langle \phi_a, f \rangle = f(a)$ for each test function f in $\mathfrak{S}$ (or $\mathfrak{D}$). This definition is consistent with Dirac's definition. The definitions above can be readily extended to test functions defined on $\mathbb{R}^n$.

The space of tempered distributions is important in Fourier analysis*. The usual L_1 Fourier transform is a continuous linear map of $\mathfrak{S}$ into $\mathfrak{S}$; from this, one defines the Fourier transform $F\phi$ of a tempered distribution ϕ in $\mathfrak{S}'$ by $\langle F\phi, f \rangle = \langle \phi, Ff \rangle$ for all f in $\mathfrak{S}$. F is a continuous linear map of $\mathfrak{S}'$ into $\mathfrak{S}'$ [6].

The generality of the structure developed by Schwartz permits not only a mathematically rigorous definition of the δ-function and its various operations, but also of other ill-defined "functions" that have been very useful to physicists and engineers. Other approaches to justifying these formal operations have been developed [1, 8]. In all these approaches, the δ-function is not defined as an ordinary function. Thus distributions are often called generalized functions.

In applications involving stochastic processes*, engineers have for many years employed the δ-function to represent the covariance function of "white noise." White noise represents a stationary stochastic process having equal energy at all frequencies; such a phenomenon is not observed in real physical systems (which necessarily have finite energy). Thus white noise is an idealization. Nevertheless, formal use of white noise and the δ-function have been found to be very useful in many practical problems (*see* COMMUNICATION THEORY, STATISTICAL). For applications of the white noise* model, see ref. 9.

References

[1] Antosik, P., Mikusinski, J., and Sikorski, R. (1973). *Theory of Distributions—The Sequential Approach.* Elsevier, New York.

[2] Barros-Neto, J. (1973). *Introduction to the Theory of Distributions.* Marcel Dekker, New York.

[3] Challifour, J. L. (1972). *Generalized Functions and Fourier Analysis.* W. A. Benjamin, Menlo Park, Calif.

[4] Dirac, P. A. M. (1930, 1958). *The Principles of Quantum Mechanics,* 1st and 4th eds. Oxford University Press, London.

[5] Gel'fand, I. M. and Shilov, G. E. (1964, 1968). *Generalized Functions.* Vol. 1: *Properties and Operations*: Vol. 2: *Spaces of Fundamental and Generalized Functions.* Academic Press, New York.

[6] Schwartz, L. (1950, 1951). *Théorie des Distributions,* Vols. 1 and 2. Hermann, Paris.

[7] Schwartz, L. (1963). In *Lectures in Modern Mathematics.* Vol. 1, T. L. Saaty, ed. Wiley, New York, pp. 23–58.

[8] Temple, G. (1953). *J. Lond. Math. Soc.,* **28**, 134–148.

[9] Wozencraft, J. M. and Jacobs, I. M. (1965). *Principles of Communication Engineering.* Wiley, New York.

CHARLES R. BAKER

DIRECTIONAL DATA ANALYSIS

The current methods of analyzing directional data were motivated by measurements of (1) the direction of the long axis of pebbles [8], (2) the direction (not strength) of magnetization of rocks [5], and (3) the vanishing bearings of homing pigeons. In case (1) we have axial, not directional, data, and the axes could be oriented in two or three dimensions. The data could be displayed by marking the two points where the axis cuts a unit circle or sphere. In case (2), each measurement can be thought of as a unit vector or as a point on a sphere of unit radius. In case (3), each measurement can be thought of as an angle, a point on a circle of unit radius, or a unit vector in the plane. In each of these cases, the sample of axes or directions has a fairly symmetric cluster about some "mean" direction, so that some scalar might be sought to describe the "dispersion" of the data.

Thus we may seek for directions, analogs of the mean and variance of data on the real line—and even of the normal distribution. The distribution used is known by the names *von Mises* on the circle and *Fisher* on the sphere and higher dimensions. This density is proportional to $\exp(K\cos\theta)$, where θ is the angle between the population mean direction and the direction of an observation. $K \geqslant 0$ is an accuracy parameter. For some axial data, the density proportional to $\exp(K\cos^2\theta)$ is helpful. The scatter of data in some applications suggests a generalization proportional to

$$\exp\left\{ K_1(\mathbf{r}'\boldsymbol{\mu}_1)^2 + K_2(\mathbf{r}'\boldsymbol{\mu}_2)^2 + K_3(\mathbf{r}'\boldsymbol{\mu}_3)^2 \right\},$$

where the terms $\mathbf{r}'\boldsymbol{\mu}_i$ are the scalar products of the observed direction $\mathbf{r}$ with three mutually orthogonal directions. (*See* DIRECTIONAL DISTRIBUTIONS.)

Assuming that our data are fitted by one of these distributions, we may find maximum likelihood estimates and make likelihood ratio tests of various hypotheses in the usual way. We show that if the data are not too dispersed, these methods can be reduced to analogs of the familiar analysis of variance*. These tests are appropriate when the K's are large and due to Watson [14, 15, 18]. If the sample size is large, there are the usual simplified methods.

It could well be that there is no preferred direction—for example, the pigeons may be unable to use any navigational clues and leave in random directions. A test for the stability of magnetization of rocks that left a formation (which would be magnetized uniformly) to be part of a conglomerate is that the direction of magnetization of pebbles in the latter is uniform. Thus tests here for uniformity are perhaps of more practical importance than on the line.

The book by Mardia [10] provides references to all the pre-1971 original papers and tables of significance points. Extensive references to earth science applications are given in Watson [20]. The references and discussion in Kendall [7] open up other related areas, practical and theoretical.

EXPLORATORY ANALYSIS

Data in two dimensions could be grouped by angle and sectors drawn with radii proportional to frequency. This "rose diagram" is the analog of the histogram. One might, use (frequency)$^{1/2}$. In three dimensions, it is hard to view points on a sphere, so that projections are used. The equal area or Lambert projection makes the density of points easy to interpret. One may see only one hemisphere, so one tries to position it conveniently. Indeed, the ability to rotate the data freely and view projections is invaluable in practice. The programs are easy to write. Such plots will reveal the general shape of the data—one or more clusters, points clustered around great circles, etc. On the sphere, "histograms" are rarely used, but contouring methods are often used; e.g., one might compute a density estimate at $\mathbf{r}$ from

$$\hat{f}(\mathbf{r}) = 1/N \sum_{i=1}^{N} w_N(\mathbf{r}'\mathbf{r}_i),$$

where $w_N(z)$ is a probability density on $(-1, 1)$ suitably peaked at $z = 1$. As an example, we could use $w(\mathbf{r}'\mathbf{r}_i)$ proportional to $\exp(K\mathbf{r}'\mathbf{r}_i)$.

The position of a single cluster is clearly suggested by $\hat{\boldsymbol{\mu}} = \mathbf{R}/R$, where $\mathbf{R} = \sum \mathbf{r}_i$ and

R = length of $\mathbf{R}$. If the cluster is very dispersed, R will be much smaller than N, so that $N - R$ is a measure of total dispersion of the sample. This suggests that $(N - R)/N$ should be an analog of the sample variance or dispersion. For example, if all the vectors are identical, $R = N$, and this quantity is zero.

For axial data, data with diametrically opposed modes or clusters, and data around a great circle, one might look at

$$\sum \cos^2 \theta_i = \sum (\mathbf{r}'_i \boldsymbol{\nu})^2$$

as $\boldsymbol{\nu}$ varies over the sphere. Its stationary values are the eigenvalues* of the matrix $\sum \mathbf{r}_i \mathbf{r}'_i$ and the eigenvectors interesting directions of $\boldsymbol{\nu}$. One large and two small eigenvalues suggest a single cluster or an axial distribution since there is a $\boldsymbol{\nu}$ that is nearly parallel or antiparallel to all the observations. A single small and two nearly equal large roots suggest a uniform distribution around a great circle whose normal is the eigenvector for the small root, etc.

There is no severe problem here with "wild" observations, although they may affect measures of dispersion.

To illustrate the suggestions above, we consider data on the orbits of comets given in Marsden [11]. The orientation of the orbital planes and directions of motion may give clues on their origin. The normal to the orbit plane in the direction suggested by the right-hand rule (fingers in the direction of motion, thumb indicating the normal) is a unit vector. Looking down onto the plane of the ecliptic, and using an equal-area projection, the vectors associated with all periodic comets are shown in Fig. 1. The clumping of 658 points in the center (or pole of the hemisphere) indicates that many comets move like the planets in orbits near the plane of the earth's orbit. Cometary orbits change, so we have used orbits associated with their last apparition, or sighting. If only the 505 comets with periods greater than 1000 years are plotted (see Fig. 2), their normals appear to be uniformly distributed. The distribution of the 153 normals to the orbital planes of comets with periods of less

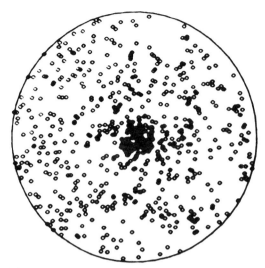

Figure 1 Equal-area plot of the normals to the last seen orbit of 658 comets, as seen looking vertically down onto the ecliptic.

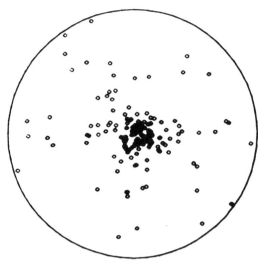

Figure 3 Equal-area plot of the normals to the last seen orbit of the 153 comets with periods less than 1000 years, as seen looking vertically down onto the ecliptic.

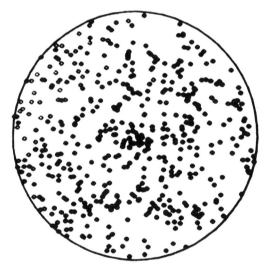

Figure 2 Equal-area plot of the normals to the last seen orbit of all 505 comets with periods greater than 1000 years, as seen looking vertically down onto the ecliptic.

than 1000 years (see Fig. 3) is concentrated. The superposition of Figs. 2 and 3 is of course Fig. 1. The eigenvectors of the matrix $\sum r_i r_i'$ for the data in Fig. 2 are, as should be expected, fairly equal. They are 195, 167, 144. But for Fig. 3 they are 119, 20, 13. The sums of these numbers, but for rounding errors, are 505 and 153, respectively. The

vector resultant $\mathbf{R}$ of the vectors in Fig. 3 is (5.13, 1.36, 99.23), so its length $(5.13^2 + 1.36^2 + 99.23^2)^{1/2} = 99.37 = R$. The point where $\mathbf{R}$ meets the hemisphere is an estimate of the mean direction of these normals. $(153 - 99.37)/153 = 0.35$ is a measure of the dispersion of the data about this mean direction. The direction cosines of the mean direction are $(5.13, 1.36, 99.23)/99.37 = (0.052, 0.014, 0.998)$. The eigenvector associated with the eigenvalue 119 is very similar $(0.071, 0.018, 0.997)$.

PARAMETRIC ANALYSES

For a single cluster on the *sphere*, it is reasonable to assume the Fisher distribution, which yields a likelihood

$$\prod_{i=1}^{N} \frac{K}{4\pi \sinh K} \left\{ \exp(K\mathbf{r}_i'\boldsymbol{\mu}) \right\}$$

$$\propto \left(\frac{K}{\sinh K} \right)^N \exp(K\mathbf{R}'\boldsymbol{\mu}).$$

Thus the m.l. estimates are $\hat{\boldsymbol{\mu}} = \mathbf{R}/R$ and $\hat{K}$ such that $\coth \hat{K} - 1/\hat{K} = R/N$ or, approximately,

$$\hat{K} \simeq k = (N-1)/(N-R).$$

Thus for the data in Fig. 3, which seem (this

could be examined more carefully) to follow the Fisher distribution, $\hat{\mu} = (0.052, 0.014, 0.998)$ and $k = (153 - 1)/(153 - 99.37) = 2.87$. These are the intuitive estimators derived above. Writing $r_i'\mu = \cos\theta$, and letting K be large so that $2\sinh K = \exp K$, one may show that $2K(1 - \cos\theta)$ is distributed like χ_2^2. Hence if μ is known,

$$\sum 2K(1 - \cos\theta_i) = 2K(N - R)$$

is distributed like χ_{2N}^2. One might guess that when μ is fitted to the data, 2 d.f. will be lost, so that $2K(N - R)$ is approximately $\chi_{2(N-1)}^2$. Hence we may write, setting $\mathbf{R}'\mu = X$, the identity

$$N - X = N - R + R - X$$

and give it the interpretation

$$\begin{pmatrix} \text{dispersion about} \\ \text{true } \mu \end{pmatrix} = \begin{pmatrix} \text{dispersion about} \\ \text{estimate } \hat{\mu} \end{pmatrix}$$
$$+ \text{ dispersion of } \hat{\mu} \text{ about } \mu.$$

Continuing this analog to the analysis of variance, we have

$$2K(N - X) = 2K(N - R) + 2K(R - X),$$

$$\chi_{2N}^2 = \chi_{2(N-1)}^2 + \chi_2^2,$$

so that the test of a prescribed mean μ is provided by

$$\frac{\text{dispersion of } \hat{\mu} \text{ about } \mu}{\text{dispersion about } \hat{\mu}} = \frac{2K(R - X)}{2K(N - R)}$$

$$= \frac{X_2^2}{X_{2(N-1)}^2}.$$

Thus one may use $F_{2,2(N-1)}$ to make the test. The reader should examine Fig. 4 to see the common sense of the test.

The data in Fig. 3 have a k of only 2.87, about the minimum for which the approximation above makes sense. To test the null hypothesis that the true mean normal is perpendicular to the ecliptic, i.e., direction $\mu = (0, 0.1)$, $\mathbf{R}'\mu = 99.23 = X$. Thus $(N - 1)(R - X)/(N - R) = 0.04$ is very small compared to $F_{2,306}$—clearly, the null hypothesis is strongly supported.

Similarly, to test that two populations (with the same large K) have the same mean .direction given samples sizes N_1, N_2 and re-

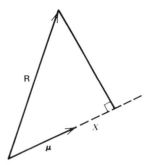

Figure 4 Sample vector resultant $\mathbf{R}$ whose projection down onto the hypothetical mean direction μ has length X. If $\mathbf{R}$ and μ are nearly parallel (orthogonal), $R - X$ will be small (large).

sultants $\mathbf{R}_1$ and $\mathbf{R}_2$, Fig. 5 suggests the statistic

$$\frac{R_1 + R_2 - R}{(N_1 - R_1) + (N_1 - R_1)}$$

and the identities ($N = N_1 + N_2$)

$$2K(N - R) = 2K(N_1 - R_1) + K(N_2 - R_2)$$
$$+ 2K(R_1 + R_2 - R),$$

$$\chi_{2(N-1)}^2 = \chi_{2(N_1 - 1)}^2 + \chi_{2(N_2 - 1)}^2 + \chi_2^2$$

$$= \chi_{2(N-2)}^2 + \chi_2^2$$

suggest that $F_{2,2(N-2)}$ may be used.

Similar tests (with tables) are available on the circle for the von Mises distribution*. Details of these, some exact and further approximate tests for all the distributions above, are given in Mardia's book [10].

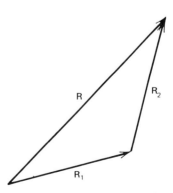

Figure 5 Sum of the lengths vector resultants $\mathbf{R}_1$ and $\mathbf{R}_2$ of two samples will be only slightly (much) larger than the length of $\mathbf{R}$, the resultant of all the data, if $\mathbf{R}_1$ and $\mathbf{R}_2$ are nearly parallel (greatly inclined).

Much of this work is due to M. A. Stephens. The result is a fairly complete set of analogs of normal tests for independent observations. Conspicuously lacking so far are satisfactory analogs for correlated directions (but see Stephens [13]) and time series* or spatial fields of directions. Fortunately, such problems seem rare in practice. Wellner [22] extends the two-sample theory—see his references to related work. Bingham [4] gives methods for his distribution.

TESTS OF UNIFORMITY

As mentioned earlier with reference to the homing directions of disoriented pigeons, the direction of magnetization of pebbles in a conglomerate, and the normals to the orbits of comets with periods over 1000 years (see Fig. 2), the problem of testing uniformity* arises more often here than on the line. One has an intuitive feeling whether a set of points on a circle, or sphere, or an equal-area projection of a sphere suggest nonuniformity in the population they come from, and intuition suggests test statistics.

If the population is unimodal, the length R of the vector resultant should be longer than it would be when sampling from a uniform parent. The Fisher and von Mises unimodal distributions become uniform when $K = 0$. R does not depend upon the coordinate system—it is invariant. Hypothesis testing* theory shows that in this case, the best invariant test of $K = 0$ is Rayleigh's test: reject uniformity if R is significantly large. Now

$$R^2 = \left(\sum x_i\right)^2 + \left(\sum y_i\right)^2 + \left(\sum z_i\right)^2,$$

where $(x_i, y_i, z_i) = \mathbf{r}_i$. When the $\mathbf{r}_i$ are independently uniformly distributed,

$$Ex_i = Ey_i = Ez_i = 0$$

$$E(x_i y_i) = E(y_i z_i) = E(z_i x_i) = 0$$

$$E(x_i^2) = E(y_i^2) = E(z_i^2) = \tfrac{1}{3}.$$

Then Ex_i, Ey_i, and Ez_i become independently Gaussian, means zero, variances

$N/3$, and R^2 becomes, on the null hypothesis, $N\chi_3^2/3$. On the circle, $R^2 = (\sum x_i)^2 + (\sum y_i)^2$ is, by a similar argument, asymptotically $N\chi_2^2/2$. These are also likelihood ratio tests.

If the distribution is antipodally symmetric, R is clearly not powerful. The likelihood ratio test* for a Bingham alternative might then be used—see Bingham [4]. When this test is used on the data in Fig. 2, it shows that the points are significantly nonrandom. The eye is a poor judge in the other direction, too—one often thinks one sees features in purely random data.

One may ask for tests that are in the spirit of the Kolmogorov and Cramér–von Mises tests. Although one may choose a starting point on the circle and form the sample distribution, the resulting tests depend upon the starting point; i.e., they are not invariant. Invariant tests for the circle were first constructed intuitively; Kuiper gave an invariant form of the Kolmogorov–Smirnov tests* and Watson [16, 17] gave U^2, an invariant form of the Cramér–von Mises tests*.

Beran [1] discovered a very general theory to derive statistics of the U^2 type on homogeneous spaces as locally optimal tests. See also Giné [6] and Prentice [12] for further generalizations. Wellner's [23, 24] two-sample (asymptotic) tests arise by applying the permutation idea to the Fourier coefficients used since Watson [16], Beran [1] in this literature on uniformity testing. (*See also* GOODNESS-OF-FIT TESTS.) Watson [21] produced Kuiper–Kolmogorov-type tests as optimal tests for distant, not local, alternatives.

MISCELLANEOUS REMARKS

These topics flow naturally into more general orientation problems—we have dealt with the orientation of a line or arrow, but a solid body is oriented by an orthogonal matrix. They also raise particular cases of the fascinating problem of finding suitable definitions on new manifolds of familiar quantities such as means, dispersions, and correlations.

More references to modern work can be traced through Wellner [23] and Beran [2], who exploits the exponential family simplification; Beran avoids the complex maximum likelihood estimation by using a nonparametric estimator of the logarithm of the density.

The probability and statistics of directed quantities has some very early history. Buffon solved his needle problem in 1733. Daniel Bernoulli tried in 1734 to show that it is very unlikely that the near-coincidence of the planetary orbits is an accident. (See, e.g., Watson [20].)

References

[1] Beran, R. J. (1968). *J. Appl. Prob.*, **5**, 177–195.

[2] Beran, R. J. (1969). *Biometrika*, **56**, 561–570.

[3] Beran, R. J. (1974). *Ann. Math. Statist.*, **7**, 1162–1179.

[4] Bingham, C. (1974). *Ann. Math. Statist.*, **6**, 292–313.

[5] Fisher, R. A. (1953). *Proc. R. Soc. Lond. A*, **217**, 195–305.

[6] Giné, E. M. (1975). *Ann. Statist.*, **3**, 1243–1266.

[7] Kendall, D. G. (1974). *J. R. Statist. Soc. B*, **36**, 365–417.

[8] Krumbein, W. C. (1939). *J. Geol.*, **47**, 673–706.

[9] Kuiper, N. H. (1960). *Ned. Akad. Wet. Proc. Ser A*, **63**, 38–47.

[10] Mardia, K. V. (1972). *Statistics of Directional Data*. Academic Press, New York.

[11] Marsden, B. G. (1979). *Catalogue of Cometary Orbits*. Smithsonian Astrophysical Observatory, Cambridge, Mass.

[12] Prentice, M. J. (1978). *Ann. Statist.*, **6**, 169–176.

[13] Stephens, M. A. (1979). *Biometrika*, **66**, 41–48.

[14] Watson, G. S. (1956). *Monthly Notices R. Astron. Soc., Geophys. Suppl.*, **7**, 153–159.

[15] Watson, G. S. (1956). *Monthly Notices R. Astron. Soc., Geophys. Suppl.*, **7**, 160–161.

[16] Watson, G. S. (1961). *Biometrika*, **48**, 109–114.

[17] Watson, G. S. (1962). *Biometrika*, **49**, 57–63.

[18] Watson, G. S. (1965). *Biometrika*, **52**, 193–201.

[19] Watson, G. S. (1967). *Biometrika*, **54**, 675–677.

[20] Watson, G. S. (1970). *Bull. Geol. Inst., Univ. Upps.*, **2**, 73–89.

[21] Watson, G. S. (1974). In *Studies in Probability and Statistics*, Jerusalem Academic Press, Jerusalem, pp. 121–128.

[22] Wellner, J. A. (1978). Two-Sample Tests for a Class of Distributions on the Sphere. Unpublished manuscript.

[23] Wellner, J. A. (1979). *Ann. Statist.*, **7**, 929–943.

(DIRECTIONAL DISTRIBUTIONS)

G. WATSON

DIRECTIONAL DISTRIBUTIONS

There are various statistical problems that arise in the analysis of data when the observations are directions. Directional data* are often met in astronomy, biology, medicine, and meteorology; for example, in investigating the origins of comets, solving bird navigation problems, interpreting paleomagnetic currents, assessing variation in the onset of leukemia, and analyzing wind directions.

The subject has recently been receiving increasing attention, but the field is as old as mathematical statistics itself. Indeed, the theory of errors was developed by Gauss primarily to analyze certain directional measurements in astronomy. It is a historical accident that the observational errors involved were sufficiently small to allow Gauss to make a linear approximation and, as a result, he developed a linear rather than a directional theory of errors (see Gauss [7]; *see also* GAUSS'S CONTRIBUTIONS TO STATISTICS and LAW OF ERRORS).

CIRCULAR DISTRIBUTIONS

The direction of a unit random vector in two dimensions can be represented by an angle X. The distribution of X is called circular since the unit random vector, which can be written as $(\cos X, \sin X)$, lies on the unit circle.

A given function f is the probability density* function (P.D.F.) of an absolutely continuous circular distribution if and only if

$$f(x) \geqslant 0, \qquad \int_0^{2\pi} f(x)\,dx = 1.$$

We can define the pth *trigonometric moments*[*] about the origin as $\alpha_p = E(\cos px)$ and $\beta_p = E(\sin px)$.

Let ϕ_p be the characteristic function[*] of the random variable X. This is defined by

$$\phi_p = E(e^{ipx}) = \alpha_p + i\beta_p,$$
$$p = 0, \pm 1, \pm 2, \ldots.$$

A circular distribution is always uniquely defined by its moments. For $p = 1$, we write

$$\phi_1 = \rho e^{i\mu_0},$$

where μ_0 is the *mean direction* and ρ is the *resultant length*. We define the sample counterparts of μ_0 and ρ as $\bar{x}_0$ and $\bar{R}$, respectively.

Let $x_1, x_2, \ldots, x_n$ be n observations on the random angle X. Let

$$\bar{C} = (1/n) \sum_{i=1}^{n} \cos x_i, \qquad \bar{S} = (1/n) \sum_{i=1}^{n} \sin x_i.$$

Then $\bar{x}_0$ and $\bar{R}$ are defined by

$$\bar{C} = \bar{R}\cos\bar{x}_0, \qquad \bar{S} = \bar{R}\sin\bar{x}_0.$$

We have $\bar{R} = (\bar{C}^2 + \bar{S}^2)^{1/2}, 0 \leqslant \bar{R} \leqslant 1$. These statistics play the same role as $\bar{x}$ and s^2 on the line except that $\bar{R}$ is a measure of precision rather than variance.

SPECIFIC CIRCULAR DISTRIBUTIONS

Uniform Distribution[*]

A random variable X is uniformly distributed on the circle if its probability density function is given by

$$f(x) = 1/2\pi, \qquad 0 < x \leqslant 2\pi.$$

Its characteristic function is given by

$$\phi_p = \begin{cases} 1 & \text{if } p = 0, \\ 0 & \text{if } p \neq 0. \end{cases}$$

von Mises Distribution

A circular random variable X is said to have a *von Mises distribution* if its probability density function is given by

$$f(x) = \frac{1}{2\pi I_0(\kappa)} e^{\kappa\cos(x-\mu_0)}$$
$$0 < x \leqslant 2\pi; \kappa > 0; 0 < \mu_0 < 2\pi,$$

where $I_r(\kappa)$ is the modified Bessel function[*] of the first kind and order r.

This distribution was introduced by von Mises [24] who wished to test the hypothesis that atomic weights are integers subject to error.

The parameter μ_0 is the mean direction and κ is called the *concentration* parameter. The von Mises distribution can be considered as the circular analog to the normal distribution[*] on the line. The distribution is unimodal and symmetrical about $x = \mu_0$. For large κ, the random variable X is distributed as $N(\mu_0, 1/\kappa^{1/2})$, while for $\kappa = 0$, the von Mises distribution reduces to the uniform distribution. (See Fig. 1.)

The von Mises distribution has the maximum likelihood[*] and maximum entropy[*] characterizations, both of which produce the normal distribution on the line (see Mardia [16]). Kent [10] gives a diffusion process[*] leading to the von Mises distribution and also shows that the distribution is infinitely divisible (*see* INFINITE DIVISIBILITY).

The trigonometric moments are given by

$$\alpha_p = A_p(\kappa)\cos p\mu_0, \qquad p = 1, 2, \ldots,$$
$$\beta_p = A_p(\kappa)\sin p\mu_0, \qquad p = 1, 2, \ldots,$$

where $A_p(\kappa) = I_p(\kappa)/I_0(\kappa)$. Maximum likelihood estimates (m.l.e.) $\hat{\mu}_0, \hat{\kappa}$ of μ_0, κ are given by

$$\hat{\mu}_0 = \bar{x}_0$$

and

$$A(\hat{\kappa}) = \bar{R},$$

where $A(\hat{\kappa}) = A_1(\hat{\kappa})$. A test of the null hypothesis of uniformity against the alternative of a von Mises distribution, with unknown

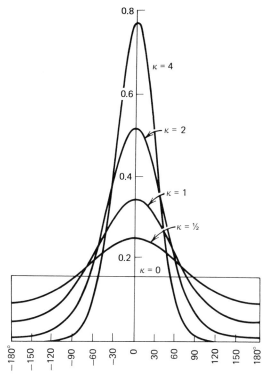

Figure 1 Density of the von Mises distribution for $\mu_0 = 0°$ and $\kappa = 0, \frac{1}{2}, 1, 2, 4$.

μ_0 and κ, is given by the Rayleigh test (see Rayleigh [20]), where the null hypothesis is rejected if

$$\bar{R} > K,$$

where K is a constant depending on n and the significance level. For large n, we have the χ^2 approximation that under H_0,

$$2n\bar{R}^2 \sim \chi_2^2.$$

Wrapped Distributions

Given a distribution on the line, we can wrap it around the circumference of the circle of unit radius. If Y is the random variable on the line, the random variable X of the wrapped distribution is given by

$$X = Y(\text{mod } 2\pi).$$

In particular, we can wrap the normal distribution around the unit circle. Let Y be $N(0, \sigma^2)$; then the probability density function of the *wrapped normal distribution* is given by

$$f(x) = \frac{1}{\sigma\sqrt{2\pi}} \sum_{k=-\infty}^{\infty} \exp\left\{-\frac{1}{2} \frac{(x + 2\pi k)^2}{\sigma^2}\right\},$$

$$0 < x \leq 2\pi.$$

The wrapped normal and the von Mises distributions can be made to approximate each other very closely by equating their first trigonometric moments to give [11, 22]

$$e^{-1/2\sigma^2} = A(\kappa).$$

Although the two distributions are very similar, for inference purposes it is much more convenient to work with the von Mises distribution. We can also wrap other distributions on the line.

Let the random vector (Z_1, Z_2) have a bivariate normal distribution with mean vector $\boldsymbol{\mu}$ and covariance matrix $\boldsymbol{\Sigma}$. Then the distribution of the random angle X defined by

$$Z_1 = R \cos X, \qquad Z_2 = R \sin X$$

is called the *offset normal distribution*.

AXIAL DISTRIBUTIONS

In some situations, we have random axes (lines) rather than random angles; i.e., X and $(X + \pi) \bmod 2\pi$ represent the same line. Such data can be modeled by using antipodally symmetric distributions; i.e., the P.D.F. satisfies $f(x) = f(x + \pi)$. This procedure is identical to doubling the angle x.

SPHERICAL DATA

Let θ and ϕ be colatitude and longitude, respectively, on the unit sphere. Define the direction cosines (l, m, n) of the point (θ, ϕ) by

$$l = \cos \phi \sin \theta, \quad m = \sin \phi \sin \theta, \quad n = \cos \theta.$$

Let $(l_i, m_i, n_i), i = 1, \ldots, n$, be n observations from a continuous distribution on the sphere. The direction cosines of the mean direction $(\bar{l}_0, \bar{m}_0, \bar{n}_0)$ are therefore

$$\bar{l}_0 = \sum_{i=1}^{n} l_i / R, \qquad \bar{m}_0 = \sum_{i=1}^{n} m_i / R,$$

$$\bar{n}_0 = \sum_{i=1}^{n} n_i / R,$$

where R is the length of the resultant given by

$$R = \left\{ \left(\sum l_i \right)^2 + \left(\sum m_i \right)^2 + \left(\sum n_i \right)^2 \right\}^{1/2}.$$

It is useful to gain some insight into the configuration of sample points on the sphere as to whether the distribution is uniform, unimodal, bimodal, or girdle. This can be studied quite effectively by considering the *moment of inertia* of the sample points, considering each point to have unit mass. This is equivalent to considering $\mathbf{T}$, the 3×3 matrix of sums of squares and products of $(l_i, m_i,$

$n_i)$, where

$$\mathbf{T} = \begin{bmatrix} \sum l_i^2 & \sum l_i m_i & \sum l_i n_i \\ \sum l_i m_i & \sum m_i^2 & \sum m_i n_i \\ \sum l_i n_i & \sum m_i n_i & \sum n_i^2 \end{bmatrix}.$$

Let $t_1 \leqslant t_2 \leqslant t_3$ be the eigenvalues of $\mathbf{T}$.

1. If $t_1 \doteq t_2 \doteq t_3$, the configuration is uniform.
2. If t_3 is large, t_1 and t_2 small, the configuration is unimodal if R is large, bimodal otherwise.
3. If t_1 is small, t_2 and t_3 large, the configuration is girdle.

This intuitive interpretation of the moment of inertia as a general diagnostic tool is due to Watson [26].

The extension of the von Mises distribution is given by the Fisher distribution, which has the probability density function

$$f(l, m, n) = c \exp \left\{ \kappa (l\lambda + m\mu + n\nu) \right\},$$

$$\kappa \geqslant 0,$$

where (λ, μ, ν) are the direction cosines of the mean direction, and κ is the *concentration* parameter. Note that (l, m, n) is a unit random vector.

The distribution was studied by Fisher [6] to investigate certain statistical problems in paleomagnetism. The distribution first appeared in statistical mechanics in 1905, [12], in considering a statistical assemblage of weakly interacting dipoles, subject to an external electric field.

In polar coordinates with (μ_0, ν_0) as the mean direction for (θ, ϕ), the probability density function becomes

$$f(\theta, \phi) = c \exp \left[\kappa \left\{ \cos \mu_0 \cos \theta + \sin \mu_0 \sin \theta \right. \right.$$
$$\left. \left. \times \cos (\phi - \nu_0) \right\} \right] \sin \theta,$$

where

$$0 < \theta < \pi, \qquad 0 < \phi \leqslant 2\pi, \qquad \kappa \geqslant 0,$$

and

$$c = \kappa / 4\pi \sinh \kappa.$$

For $\kappa = 0$, the Fisher distribution reduces to

the uniform distribution on the sphere. Like the von Mises distribution on the circle, the Fisher distribution has the maximum likelihood and maximum entropy characterizations on the sphere (see Mardia [16]). The Brownian motion* distribution on the sphere can be closely approximated by the Fisher distribution (see Roberts and Ursell [21]). The distribution is unimodal with mode at (λ, μ, ν) and antimode at $(-\lambda, -\mu, -\nu)$. Further, the distribution is rotationally symmetric about the mean direction.

The sample mean direction is the m.l.e. of (λ, μ, ν), while the m.l.e. of κ is the solution $\hat{\kappa}$ of

$$\coth \hat{\kappa} - 1/\hat{\kappa} = \overline{R}.$$

Now consider a situation where one observes not *directions* but *axes*, which are random variables on a projective hemisphere. It is convenient to represent such a random variable by an antipodally symmetric distribution on the sphere; i.e., the probability density function of $\mathbf{l} = (l, m, n)'$ has antipodal symmetry, i.e.,

$$f(\mathbf{l}) = f(-\mathbf{l}).$$

Note that the procedure of doubling the angles on the circle has no analog here.

The unit random vector $\mathbf{l}$ is said to have the *Bingham distribution** [2] if its probability density function has the form

$$f(\mathbf{l}) = d(\boldsymbol{\kappa}) \exp\{\operatorname{tr}(\boldsymbol{\kappa}\boldsymbol{\mu}'\mathbf{l}\mathbf{l}'\boldsymbol{\mu})\},$$

where $\boldsymbol{\mu}$ now denotes an orthogonal matrix, $\boldsymbol{\kappa} = \operatorname{diag}(\kappa_1, \kappa_2, \kappa_3)$ is a matrix of constants, and $d(\boldsymbol{\kappa})$ is the normalizing constant, depending only on $\boldsymbol{\kappa}$. Since $\operatorname{tr}(\boldsymbol{\mu}'\mathbf{l}\mathbf{l}'\boldsymbol{\mu}) = 1$, the sum of the κ_i is arbitrary and it is usual to take $\kappa_3 = 0$.

The distribution contains a number of different forms, such as symmetric axial and girdle distributions of Watson [25] and Dimroth [4], as well as asymmetric axial and girdle distributions. For its various characterizations, see Mardia [16]. A small circle distribution on the sphere is given by Bingham and Mardia [3].

EXTENSIONS

Generalizations to p dimensions can readily be made from the preceding discussion. Let S_p denote the unit sphere in R^p. A unit random vector $\mathbf{l}$ is said to have a *p-variate von Mises–Fisher* distribution if its probability density function is given by

$$f(\mathbf{l}) = c_p(\kappa) e^{\kappa \boldsymbol{\mu}' \mathbf{l}},$$

$$\kappa \geqslant 0, \qquad \boldsymbol{\mu}'\boldsymbol{\mu} = 1, \qquad \mathbf{l} \in S_p,$$

where

$$c_p(\kappa) = \kappa^{(1/2)p-1} / \left\{ (2\pi)^{(1/2)p} I_{(1/2)p-1}(\kappa) \right\}.$$

This distribution was first introduced by Watson and Williams [28]. For a discussion of the distribution, see Mardia [15].

Mackenzie [13] put forward the general concept of spherical dependence and correlation by introducing the idea of rotational dependence. For some recent developments on directional correlation, see Downs [5], Mardia and Puri [17], and Stephens [23]. For distributions on a cylinder (x, θ), $-\infty < x < \infty, 0 < \theta \leqslant 2\pi$, see Mardia and Sutton [18] and Johnson and Wehrly [8]. For an important process involving orientation, see Kendall [9].

Various applications of directional data analysis can be found in Watson [27], Batschelet [1], and Mardia [14]. Pearson and Hartley [19, Chap. 9] contain a useful set of tables and introductory material.

References

[1] Batschelet, E. (1965). *Statistical Methods for the Analysis of Problems in Animal Orientation and Certain Biological Rhythms*. American Institute of Biological Sciences, Washington, D.C. (Monograph written for biologists who deal with directional data. The text is "prepared at a fairly low mathematical level" and contains good introductory material. It is very readable with a number of practical examples from the biological sciences.)

[2] Bingham, C. (1974). *Ann. Statist.*, **2**, 1201–1225.

[3] Bingham, C. and Mardia, K. V. (1978). *Biometrika*, **65**, 379–389.

[4] Dimroth, E. (1962). *Tschermaks. Mineral. Petrogn. Milt.*, **8**, 248–274.

[5] Downs, T. D. (1974). In *Biorhythms and Human Reproduction*, M. Ferrin, F. Halberg, R. M. Richart, and L. van der Wiele, eds. Wiley, New York, pp. 97–104.

[6] Fisher, R. A. (1953). *Proc. R. Soc. Lond. A*, **125**, 54–59.

[7] Gauss, C. F. (1809). *Theoria motus corporum coelestium in sectionibus conicus solem ambientium.* Perthes et Besser, Hamburg.

[8] Johnson, R. A. and Wehrly, T. E. (1978). *J. Amer. Statist. Ass.*, **73**, 602–606.

[9] Kendall, D. G. (1974). *J. R. Statist. Soc. B*, **36**, 365–417.

[10] Kent, J. T. (1975). *J. R. Statist. Soc. B*, **37**, 349–393.

[11] Kent, J. T. (1978). *Math. Proc. Camb. Philos. Soc.*, **84**, 531–536.

[12] Langevin, P. (1905). *Ann. Chim. Phys.*, **5**, 70–127.

[13] Mackenzie, J. K. (1957). *Acta. Cryst. Camb.*, **10**, 61–62.

[14] Mardia, K. V. (1972). *Statistics of Directional Data.* Academic Press, London. Russian translation: Nauka, Moscow. (The book assumes a basic knowledge of mathematical statistics at the undergraduate level and incorporates many practical examples as well as an extensive bibliography and tables. It treats the topics described here in Secs. 1–5.)

[15] Mardia, K. V. (1975). *J. R. Statist. Soc. B.* **37**, 349–393. (Review article of developments since Mardia [14] with useful discussion.)

[16] Mardia, K. V. (1975). In *A Modern Course on Statistical Distributions in Scientific Work*, Vol. 3, G. P. Patil, S. Kotz, and J. K. Ord, eds. D. Reidel, Dordrecht, Holland, pp. 365–385.

[17] Mardia, K. V. and Puri, M. L. (1978). *Biometrika*, **65**, 391–395.

[18] Mardia, K. V. and Sutton, T. W. (1978). *J. R. Statistic. Soc. B*, **40**, 229–233.

[19] Pearson, E. S. and Hartley, H. O. (1972). *Biometrika Tables for Statisticians*, Vol. 2. Cambridge University Press, Cambridge. (Contains useful tables and introductory material.)

[20] Rayleigh, Lord. (1919). *Philos. Mag.*, (6), **37**, 321–347.

[21] Roberts, P. H. and Ursell, H. D. (1960). *Phil. Trans. R. Soc. Lond. A*, **252**, 317–356.

[22] Stephens, M. A. (1963). *Biometrika*, **50**, 385–390.

[23] Stephens, M. A. (1979). *Biometrika*, **66**, 41–48.

[24] von Mises, R. (1918). *Phys. Z.*, **19**, 490–500.

[25] Watson, G. S. (1965). *Biometrika*, **52**, 193–201.

[26] Watson, G. S. (1966). *J. Geol.*, **74**, 786–797.

[27] Watson, G. S. (1970). Orientation Statistics in the Earth Sciences. *Bull. Geol. Inst., Univ. Upps.* (Contains a description of various applications of directional statistics in the earth sciences.)

[28] Watson, G. S. and Williams, E. J. (1956). *Biometrika*, **43**, 344–352.

(BROWNIAN MOTION

DIRECTIONAL DATA ANALYSIS)

K. V. Mardia

DIRICHLET DISTRIBUTION

The $(k-1)$-dimensional Dirichlet distribution with parameters $(\alpha_1, \alpha_2, \ldots, \alpha_k)$, $\alpha_j > 0$, is denoted by $\mathcal{D}(\alpha_1, \alpha_2, \ldots, \alpha_k)$. The standard form is defined as the distribution of $(Y_1, Y_2, \ldots, Y_k)$, where $Y_j = Z_j / \sum_{i=1}^{k} Z_i$ and $Z_j, j = 1, 2, \ldots, k$, are independent, standard gamma-distributed* random variables with shape parameters $\alpha_j, Z_j \sim \mathcal{G}(\alpha_j, 1)$. Since $\sum_{i=1}^{k} Y_i = 1$, the distribution $\mathcal{D}(\alpha_1, \alpha_2, \ldots, \alpha_k)$ can be described by the joint density of $(k-1)$ of the variables

$$f(y_1, y_2, \ldots, y_{k-1}) = \frac{\Gamma(\alpha_1 + \alpha_2 + \cdots + \alpha_k)}{\Gamma(\alpha_1)\Gamma(\alpha_2) \cdots \Gamma(\alpha_k)}$$

$$\times \prod_{i=1}^{k-1} y_i^{\alpha_i - 1} \left(1 - \sum_{j=1}^{k-1} y_j\right)^{\alpha_k - 1}$$

for $y_i \geq 0$, $\sum_{i=1}^{k-1} y_i \leq 1$, and $f(y_1, y_2, \ldots, y_{k-1}) = 0$, otherwise. The joint distribution of $cY_1, \ldots, cY_{k-1}(c > 0)$ is also Dirichlet (but not *standard* Dirichlet). The main properties of the distribution are as follows:

1. When $k = 2$, $\mathcal{D}(\alpha_1 \alpha_2), \alpha_i > 0$, is the beta distribution* $Be(\alpha_1, \alpha_2)$.

2. If $(Y_1, Y_2, \ldots, Y_k) \stackrel{d}{=} \mathcal{D}(\alpha_1 \alpha_2, \ldots, \alpha_k)$ and $\gamma_1, \gamma_2, \ldots, \gamma_l$ are integers such that $0 < \gamma_1 < \gamma_2 < \cdots < \gamma_l = k$, then

$$\left(\sum_{i=1}^{\gamma_1} Y_i, \sum_{i=\gamma_1+1}^{\gamma_2} Y_i, \ldots, \sum_{i=\gamma_{l-1}+1}^{\gamma_l} Y_i\right)$$

$$\stackrel{d}{=} \mathcal{D}\left(\sum_{i=1}^{\gamma_1} \alpha_i, \sum_{i=\gamma_1+1}^{\gamma_2} \alpha_i, \ldots, \sum_{i=\gamma_{l-1}+1}^{\gamma_l} \alpha_i\right).$$

This property follows from the definition and the additivity property of the gamma distribution*.

3. If $(Y_1, Y_2, \ldots, Y_k) \overset{d}{=} \mathcal{D}(\alpha_1, \alpha_2, \ldots, \alpha_k)$, then the marginal moments are

$$EY_i = \alpha_i/\alpha,$$

$$EY_i^2 = \alpha_i(\alpha_i + 1)/\left[\alpha(\alpha + 1)\right]$$

$$EY_i Y_j = \alpha_i\alpha_j/\left[\alpha(\alpha + 1)\right], \qquad i \neq j,$$

where $\alpha = \sum_{j=1}^{k}\alpha_j$.

4. The conditional distribution of any subset of the Y's given any other subset is also (nonstandard) Dirichlet.

5. The Dirichlet distribution has the following "Bayesian property" useful in applications to nonparametric statistics [1]: If the prior distribution* of $(Y_1, Y_2, \ldots, Y_k)$ is $\mathcal{D}(\alpha_1, \alpha_2, \ldots, \alpha_k)$ and if $P(X = j \mid Y_1, Y_2, \ldots, Y_k) = Y_j$ almost surely for $j = 1, 2, \ldots, k$ (where X is an integer-valued random variable distributed on $\{1, 2, \ldots, k\}$), then the posterior distribution of $(Y_1, Y_2, \ldots, Y_k)$ given $X = j$ is also a Dirichlet distribution $\mathcal{D}(\alpha_1^{(j)}, \alpha_2^{(j)}, \ldots, \alpha_k^{(j)})$, where

$$\alpha^{(j)} = \begin{cases} \alpha_i, & i \neq j, \\ \alpha_i + 1, & i = j \end{cases}$$

(*See* DAMAGE MODELS*.) Other properties and applications are found in Johnson and Kotz [2, Chap. 40, Sec. 5].

The properties described above were used by Ferguson [1] to define the Dirichlet process, which is a major tool in studying some Bayesian nonparametric problems concerning estimation of population distributions, means, medians*, quantiles*, variances and covariances, etc.

For more information on this topic, see Wilks [3].

References

[1] Ferguson, T. S. (1973). *Ann. Statist.*, **1**, 209–230.

[2] Johnson, N. L. and Kotz, S. (1972). *Distributions in Statistics. Continuous Multivariate Distributions*. Wiley, New York.

[3] Wilks, S. S. (1962). *Mathematical Statistics*. Wiley, New York.

DIRICHLET PROBLEM *See* DIFFUSION PROCESSES

DISCOUNT FACTOR *See* DYNAMIC PROGRAMMING

DISCOUNTING

An amount of money P, accumulated at compound interest at rate $100i\%$ per unit of time, will amount to $P(1 + i)^t$ after t units of time. (It is convenient to think of t as being an integer.) In order for the accumulated amount of be A after t units of time, we must have $P(1 + i)^t = A$; that is, the present value of A due after t time units is $P = A(1 + i)^{-t}$.

The operation of correcting A back to its present value is *discounting*. It is useful when payments have to be made at different points of time, and it is desired to establish the rate of interest implicit in a group of such transactions. This is done by equating values of payments out and in at a convenient time (possibly, but not necessarily the present) and solving the resulting equation for i.

DISCRETE DISTRIBUTION

The distribution of any random variable that can take only some one of a countably infinite number of possible values is said to be discrete. If $\Pr[X = x_j] = p_j$ for all j, then the CDF of X is $F_X(x) = \Pr[X \leqslant x] = \sum_{x_j \leqslant x} p_j$.

By far the most commonly used discrete distributions are those for which the x_j's are the nonnegative integers. They are used in models for "count data," which include variables representing the results of counts (of defective items, apples on a tree, etc.). However, it is not necessary that the variable takes only integer values (an observed proportion is a simple counterexample); it can even take an infinity of values in any finite interval and still have a discrete distribution. As a simple example, consider the distribution of a variable X which can take only rational values (with $|m|$ and $|n|$ mutually prime integers)

$$\Pr[X = m/n] = \tfrac{1}{2}(\theta - 1)^2(\theta^{|m|+|n|} - 1)^{-1}.$$

Of course, such examples are not likely to be of wide practical use.

(ABSOLUTE CONTINUITY)

DISCRETE FOURIER TRANSFORM

A sequence $\{X = X_0, X_1, \ldots, X_{N-1}\}$ is defined as

$$Y_n = N^{-1} \sum_{k=0}^{N-1} X_k \exp\left(-j \frac{2\pi kn}{N}\right),$$

$$j = \sqrt{-1}, \; n = 0, 1, \ldots, N - 1.$$

The sequence $\{X\}$ can be recovered from the inverse transform:

$$X_k = \sum_{n=0}^{N-1} Y_n \exp\left(j \frac{2\pi kn}{N}\right),$$

$$k = 0, 1, \ldots, N - 1.$$

Alternatively, Y_n is expressed as

$$Y_n = N^{-1} \sum_{k=0}^{N-1} X_k \exp\left(\frac{-j\pi n^2}{N}\right)$$

$$\times \exp\left(\frac{-j\pi k^2}{N}\right) \exp\left(\frac{j\pi(k-n)^2}{N}\right)$$

$$= N^{-1} \exp\left(\frac{-j\pi n^2}{N}\right)$$

$$\times \left\{ X_k \exp\left(\frac{-j\pi k^2}{N}\right) \odot \exp\left(\frac{j\pi k^2}{N}\right) \right\},$$

where $\odot$ denotes the convolution* of two series in brackets.

An algorithm in ISO FORTRAN language for the discrete Fourier transform of a sequence of a general length was developed by Monro and Branch [2]. For the case when $\{X\}$ consists of real numbers, a more efficient algorithm in the same language was developed earlier by Monro [1].

References

[1] Monro, D. M. (1976). *Appl. Statist.*, **25**, 166–172.

[2] Monro, D. M. and Branch, J. L. (1977). *Appl. Statist.*, **26**, 351–361.

(FOURIER COEFFICIENTS
PERIODOGRAM ANALYSIS
TIME SERIES)

DISCRETE PROCESSES   See PROCESSES, DISCRETE

DISCRETE RECTANGULAR DISTRIBUTION

A distribution for which any one of a finite set of equally spaced values is equally likely —that is,

$$\Pr[X = a + jh] = (n + 1)^{-1}$$

$$(j = 0, 1, \ldots, n),$$

so that X can take any one of the equally spaced values $a, a + h, \ldots, a + nh$. As $n \to \infty$ and $h \to 0$, with $nh = b - a$, the distribution tends to a continuous rectangular (or uniform*) distribution over $a \leqslant X \leqslant b$.

The *standard* discrete rectangular distribution has $a = 0$ and $h = n^{-1}$, so X takes the values $0, n^{-1}, 2n^{-1}, \ldots, 1 - n^{-1}, 1$. This distribution has expected value zero, and variance $\frac{1}{12}\{(n + 1)^2 - 1\}n^{-2}$.

The distribution with $a = 0$, $h = 1$, $n = 9$ has been used as the null distribution of numbers in a random decimal sequence.

Distributions with $a = 0$, $h = 1$ can arise as mixtures of binomial (n, p) distributions with p distributed uniformly ("rectangularly") over $0 \leqslant p \leqslant 1$. Formally,

discrete rectangular

$$\equiv \text{binomial } (n, p) \underset{p}{\wedge} \text{rectangular } (0, 1).$$

(BINOMIAL DISTRIBUTION
COMPOUND DISTRIBUTIONS
MIXTURES
UNIFORM DISTRIBUTION)

DISCRIMINANT ANALYSIS

BASIC PROBLEM AND
FURTHER SOURCES

The basic problem of discriminant analysis is to assign an observation, **x**, to one of two or more groups on the basis of its value **x**. It is to be distinguished from the problem of testing equality of means or covariance matrices. Many of the techniques and computations are similar but the problem of discriminant analysis is that of allocation of new observations rather than one of testing equality of distributions.

We give three examples of problems that may be studied using discriminant analysis.

1. A woman who has a history of hemophilia in her family comes to a physician for genetic counseling. Her family history gives an indication of an a priori probability* of her being a carrier. On the basis of this a priori probability and blood tests, she can be assigned to a carrier or noncarrier group. In this case it may be more appropriate to indicate her posterior probability* of being a carrier [6].

2. In the last year of secondary school a student is given three tests. On the basis of scores on an arithmetic test, an English test, and a foreign relations test, the student is to be advised on a course of future study. There are four choices available, and the student is told to which profession his test profile is most similar (Porebski, cited in Lachenbruch [13].

3. On the basis of a number of predictors available at five weather stations it is desired to predict the ceiling at an airfield in 2 hours. The choices are closed, low instrument, high instrument, low open, or high open. This is an example of a multiple group discriminant problem (Miller, cited in Lachenbruch [13]).

In many problems the underlying distributions are assumed to be normal, although this is not required in a theoretical development. Other possibilities include discrete models, such as the multinomial distribution* and nonparametric density estimation* techniques. A Bayesian approach may be taken with any of these procedures, although thus far the procedures have only been worked out for the multivariate normal* case.

A general source for material on applied linear discriminant analysis is Lachenbruch [13]. This monograph concentrates on normal theory discrimination, the robustness of the linear discriminant function, and briefly considers nonnormal and multiple-group problems. The important class of problems of discrete discriminant analysis is reviewed in the book by Goldstein and Dillon [10]. A number of discrete classification models are reviewed and several computer programs are provided. Recent reviews are Lachenbruch and Goldstein [15] and Geisser [8]. In engineering, discrimination is often referred to as pattern recognition* and a large, sophisticated literature has arisen. Some recent texts are Fukanaga [7] and Young and Calvert [23]. Two important symposium proceedings are Cacoullos (cited in Lachenbruch [13] and Van Ryzin [21]). These include papers on theoretical and applied aspects of the discrimination problems. The remainder of this article will be concerned primarily with discriminant analysis under the assumption of normality which leads to a linear or quadratic rule.

BASIC THEORY OF DISCRIMINANT
ANALYSIS

We shall use the following notation:

x	$k \times 1$ observation
π_i	population i
$\bar{\mathbf{x}}_i$	ith sample mean
p_i	a priori probability

P_i probability of misclassifying an observation from $\prod_i$

R_i region in which an individual is allocated to $\prod_i$

$\boldsymbol{\mu}_i$ $k \times 1$ mean

$\boldsymbol{\Sigma}$ $k \times k$ covariance matrix in $\prod_i$

$\mathbf{S}_i$ $k \times k$ sample covariance matrix

$f_i(x)$ PDF of $\mathbf{X}$ in $\prod_i$

$\Phi(\cdot)$ cumulative normal distribution function

Welch (cited in Lachenbruch [13]) suggested that an appropriate approach to the problem of obtaining meaningful allocation rules was to minimize the total probability of misclassification, $T(R, f)$, for the two-group case. This is given by

$$T(R, f) = p_1 \int_{R_2} f_1(\mathbf{x}) \, d\mathbf{x} + p_2 \int_{R_1} f_2(\mathbf{x}) \, d\mathbf{x}$$

$$= p_1 + \int_{R_1} \left[p_2 f_2(\mathbf{x}) - p_1 f_1(\mathbf{x}) \right] d\mathbf{x}.$$

$$(1)$$

It is clear that this quantity is minimized if R_1 is chosen so that the integrand is negative for all points in R_1. Thus the allocation rule is: assign $\mathbf{x}$ to π_1 if $f_1(\mathbf{x})/f_2(\mathbf{x}) > p_2/p_1$ and π_2 otherwise. The probabilities of misclassification are given by

$$P_1 = \int_{R_2} f_1(\mathbf{x}) \, d\mathbf{x}$$

$$P_2 = \int_{R_1} f_2(\mathbf{x}) \, d\mathbf{x}.$$

$$(2)$$

The demonstration of the optimal allocation regions as given by (1) is formally identical to the proof of the Neyman–Pearson lemma*. The most important special case arises when it is assumed that π_1 and π_2 are multivariate normal distributions with means $\boldsymbol{\mu}_1$, and $\boldsymbol{\mu}_2$ and common covariance $\boldsymbol{\Sigma}$. It is a simple exercise to show that the optimal rule is to allocate to π_1 if

$$D_T(\mathbf{x}) = \left[\mathbf{x} - \tfrac{1}{2}(\boldsymbol{\mu}_1 + \boldsymbol{\mu}_2) \right]' \boldsymbol{\Sigma}^{-1}(\boldsymbol{\mu}_1 - \boldsymbol{\mu}_2)$$

$$> \ln p_2/p_1.$$

$$(3)$$

$D_T(\mathbf{x})$ may be referred to as the theoretical discriminant function. Its sample analog may be obtained by replacing $\boldsymbol{\mu}_i$ by $\bar{\mathbf{x}}_i$ and $\boldsymbol{\Sigma}$ by $\mathbf{S}$. This is the well-known linear discriminant function. If $\boldsymbol{\Sigma}_1$ is not equal to $\boldsymbol{\Sigma}_2$, one has the quadratic discriminant function, whose theoretical form is

$$Q_T(\mathbf{x}) = \frac{1}{2} \ln \frac{|\boldsymbol{\Sigma}_2|}{|\boldsymbol{\Sigma}_1|} - \tfrac{1}{2}(\mathbf{x} - \boldsymbol{\mu}_1)'\boldsymbol{\Sigma}_1^{-1}(\mathbf{x} - \boldsymbol{\mu}_1)$$

$$+ \tfrac{1}{2}(\mathbf{x} - \boldsymbol{\mu}_2)'\boldsymbol{\Sigma}_2^{-1}(\mathbf{x} - \boldsymbol{\mu}_2).$$

$$(4)$$

An important special case of quadratic discrimination occurs when the mean vectors are equal. This could arise in the context of discriminating monozygotic and dizygotic twins on the basis of a series of measurements (Bartlett and Please, cited in Lachenbruch [13]). A recent study by Wahl and Kronmal [22] has examined the behavior of the quadratic discriminant function for moderate-size samples. They state: "When the dimension and covariance differences are small, it makes little difference whether Fisher's linear or the quadratic function is used. However when the dimension and the covariance differences are large, the quadratic performance is much better than Fisher's linear, provided the sample size is sufficient." The problem of assigning an unknown observation to one of several groups has not been studied to the extent that the two-group problem has been. It is easy to see that the rule which minimizes the total probability of misclassification is given by assigning the unknown observation to π_i if

$$p_i f_i(\mathbf{x}) = \max_j p_j f_j(\mathbf{x}).$$

$$(5)$$

If the underlying observations are multivariate normal*, with common covariances, this rule is equivalent to assigning $\mathbf{x}$ to π_i if

$$\ln p_i + (\mathbf{x} - \tfrac{1}{2} \boldsymbol{\mu}_i)' \boldsymbol{\Sigma}^{-1} \boldsymbol{\mu}_i$$

$$= \max \left\{ \ln p_j + (\mathbf{x} - \tfrac{1}{2} \boldsymbol{\mu}_j)' \boldsymbol{\Sigma}^{-1} \boldsymbol{\mu}_j \right\}.$$

$$(6)$$

Some results are available in the papers of Lachenbruch and of Michaelis in Cacoullos (cited in Lachenbruch [13]).

If the underlying distributions are multinomial*, $f_i(\mathbf{x})$ refers to the probability that an observation from the ith population falls into the state defined by **x**. **x** may be a single multinomial variable or it may refer to a set of categorical variables. In any event, the allocation rule has the identical form given earlier. If the parameters are known, the allocation problem is trivial. However, if the parameters are unknown, they must be estimated and generally require very large sample sizes. The text of Goldstein and Dillon [10] suggests some alternatives to this problem by assuming structure to the multinomial model, as in a log-linear model.

Nonparametric rules were initially proposed by Fix and Hodges (cited in Lachenbruch [13]), who suggested a nearest-neighbor* rule. One chooses an integer K, and a distance function, and lets K_i be the number of observations from π_i in the K closest observations to X. Then one assigns X to π_i if

$$p_i \frac{K_i}{n_i} = \max_j p_j \frac{K_j}{n_j} . \qquad (7)$$

More recently, various nonparametric density estimation* procedures have been proposed (e.g., Parzen, cited in Lachenbruch [13]). One problem with these seems to be the large number of parameters that need to be estimated. Recent references include Habbema and Hermans, [11, 12] and Van Ness and Simpson [20].

Thus far we have been considering the problem of minimizing the total probability of misclassification. In many contexts this leads to rules that assign the vast majority of the observations to one population. For example, if one is using a discriminant procedure to screen for breast cancer, the annual age-adjusted incidence rate is 75/100,000 in the the U. S. white population. Any rule constructed on the basis of minimizing a total probability of misclassification would assign the vast majority of observations to the normal group and miss most if not all of the breast cancers. One solution that has been proposed is to assign costs to the misclassification errors and to minimize the total cost of the procedure. Unfortunately, it may be difficult to assign costs to misalloca-

tions. Another alternative is to fix an acceptable error rate in one of the groups (in this example it might be the breast cancer group), and determine the error rate in the other group. Usually, the fixed error rate is in the group that is more serious to misclassify. It may be difficult to do this because the number of observations available in this group may be small and the estimated error rate may be quite variable. It is also possible to constrain both error rates to be less than certain values, but such rules may lead to a large number of points remaining unassigned (see, e.g., Anderson, cited in Lachenbruch [13]). As an example, let us consider the hemophilia-carrier detection problem. After considerable preliminary analysis, the variable set was reduced to two variables called AC and R. The means are given by

	AC	R	n
π_1 carrier	57.13	1.89	16
π_1 noncarrier	118.04	0.81	27

The pooled covariance matrix is

$$\mathbf{S} = \begin{pmatrix} 499.82 & -2.692 \\ -2.692 & 0.1686 \end{pmatrix}.$$

The discriminant coefficients are

$$\mathbf{b} = \mathbf{S}^{-1}(\bar{\mathbf{x}}_1 - \bar{\mathbf{x}}_2) = \begin{pmatrix} -0.0956 \\ 4.880 \end{pmatrix}.$$

The constant is

$$0.5(\bar{\mathbf{x}}_1 + \bar{\mathbf{x}}_2)'\mathbf{b} = -1.785.$$

Thus the final rule is assign to carrier if

$$D(x) = -0.0956AC + 4.880R + 1.785$$

$$> \ln \frac{1-p}{p} .$$

In estimating the error rates, this rule missed one carrier and no noncarriers using the resubstitution method.

By checking the individual covariance matrices, it was noted that they are quite different. Thus a quadratic rule would be appropriate. Such a rule was calculated, and no clinically useful changes were noted. In the region of interest, the linear and quadratic rules perform identically.

ESTIMATION OF ERROR RATES

After determining a discriminant function, a question of major importance is how the discriminant function will perform on future samples. This suggests that we should estimate the error rates of the proposed discriminant function. If the underlying distributions are normal and the parameters are known, the total error rate is that given by (1) and may be easily calculated. It turns out to be a function of the Mahalanobis distance* δ^2 between the two groups and the a priori probability p_1 of population 1, where

$$\delta^2 = (\boldsymbol{\mu}_1 - \boldsymbol{\mu}_2)'\boldsymbol{\Sigma}^{-1}(\boldsymbol{\mu}_1 - \boldsymbol{\mu}_2).$$

Table 1 [13] gives these probabilities for the two-group problem for a variety of p_1 and δ^2. It should be noted that this total error rate is a weighted sum of two error rates, namely, those given by (2). These may be considerably different. In practice, of course, we will not know the form and parameters of the distribution.

Table 1   Total Error Rates
$$p_1 P_1 + (1 - p_1)P_2$$

δ^2	p_1				
	0.1	0.2	0.3	0.4	0.5
1	0.098	0.186	0.253	0.295	0.309
4	0.074	0.112	0.139	0.154	0.159
9	0.034	0.050	0.060	0.065	0.067

It is nevertheless of great interest to evaluate the population specific error rates and the total error rates. There are at least five error rates of interest that have been considered:

1. The optimum error rate.
2. The error rate for the sample discriminant function as it will perform in future samples (the actual error rate).
3. The expected error rate for discriminant functions based on samples of N_1 from Π_1 and N_2 from Π_2 (the expected actual error rate).
4. The plug-in estimate of the error rate obtained by using the estimated parameters of f_1 and f_2 as if they were the true parameters.
5. The apparent error rate, the fraction of observations in the initial sample which are misclassified by the sample discriminant. This error rate is sometimes referred to as the resubstitution error rate.

If we assume that the underlying distributions are multivariate normal, the plug-in error rates are a function of the sample Mahalanobis distance and the a priori probability. They have the property that they are consistent estimators if the underlying assumptions are true, but they do underestimate the optimum error rate, possibly by a considerable amount, if sample sizes are small. The resubstitution error rate was proposed by Smith (cited in Lachenbruch [13]). In this method the observations in the training sample are classified. This method also has a optimistic bias and can be bad for very small samples. For large samples the resubstitution error rate is quite satisfactory and is the one used in many computer programs. The problems of bias with the resubstitution estimator for small to moderately sized samples led to the development of the leaving-one-out or jackknife* estimators. In this estimator one observation is deleted from the training sample. A discriminant function is calculated using the remaining observation and the deleted observation is allocated. This procedure is followed for each of the observations in the training sample and one is able to obtain an estimate of the expected actual error rate as if it were based upon samples of size $(N_1 + N_2 - 1)$. This is almost unbiased for the error rate in which one is interested [13].

Recently, Glick [9] has investigated additive estimators for probabilities of correct classification and has indicated that for relatively small samples the jackknife method, while almost unbiased, appears to have a large variance in situations where one would be especially interested in using it. He proposes some smooth estimators, which show great promise. He concludes his article by noting the following points:

1. Parametric estimators of error or nonerror probability are not robust and hence should not be used in practical problems.

2. The leave-one-out estimator is robust and nearly unbiased . . . [it] is an "anti-smooth" modification of counting; so its standard deviation exceeds that of the sample success proportion or of "smooth" modification.

3. Sample success proportion (the simple counting or "resubstitution" estimator) is intuitive and robust, but optimistically biased.

4. The counting method can be modified to reduce bias magnitude and standard deviation simultaneously without sacrificing robustness.

ROBUSTNESS* OF DISCRIMINANT FUNCTIONS AND THE EFFECTS OF THE FAILURE OF ASSUMPTIONS TO HOLD

Linear Discriminant Functions (LDF)

The basic assumptions of the linear discriminant function include multivariate normality in the populations, equal covariance matrices, uncontaminated samples, and correct classification of the training samples. The effects of nonnormality on the LDF have been studied by several authors. The failure of normality assumptions to hold may arise either from continuous nonnormal distributions or from discrete nonnormal distribution. The first discrete studies were performed, using multivariate Bernouilli distributions, by Gilbert and by Moore (both cited in Lachenbruch [13]). They compared

the performance of the LDF in the full multinomial model and a first-order independence model. Two other models were different in the two studies. They both found that the full multinomial model tended to do poorly for a large number of Bernouilli variables. Gilbert concluded that if the parameters of the distribution are "moderate," the linear discriminant and the optimal procedure are highly correlated. Moore's conclusions generally agreed with Gilbert's. If there are reversals in the log likelihood ratio, the LDF can suffer considerably.

Lachenbruch et al. (cited in Lachenbruch [13]) considered the robustness to certain types of continuous nonnormality. They transformed normal variables to distributions that were members of the Johnson family* of distributions: the log-normal*, the logit-normal, and the $\sinh^{-1}$ normal. Because of these nonlinear transformations, the underlying observations no longer have the same covariance matrices. In this study the authors found that the total error rate was often greatly increased and the individual error rates were distorted in such a way that one error rate was increased and the other decreased. The smallest effect was for the logit-normal* transformed distributions, presumably because they are bounded. The worst cases arose for log-normal distributions, which may be highly skewed.

The effects of unequal covariance matrices on the linear discriminant function was studied by Gilbert for the large-sample case and by Marks and Dunn for small samples (both works are cited in Lachenbruch [13]). Gilbert assumed that one covariance matrix was the identity and the other was a scalar multiple of the identity. If this multiple is much greater or much less than 1, the LDF can be considerably affected and the quadratic discriminant is preferable to the linear. Marks and Dunn compared the sample linear discriminant to the sample quadratic discriminant function in a wide variety of situations. They concluded, as did Gilbert, that the linear function is quite satisfactory if the covariance matrices are not too different. If covariances are considerably different and sample sizes are sufficiently large, they recommend the quadratic discriminant function.

The third question of robustness is related to contamination of the initial samples. Ahmed and Lachenbruch [1] show that with scale contamination and equal prior probabilities, the LDF is unaffected by this contamination. If the prior probabilities are unequal, scale contamination can hurt discrimination ability. With mild contamination, say the contaminating covariance matrices between four and nine times as large as the uncontaminated matrix, only mild effects are observed. Various types of location contamination can induce severe effects. However, recently, Broffitt et al. [3] have studied this problem for a kind of location contamination, namely a variablewise shift in which the shifts are independent of each other and found only small effects. Ahmed and Lachenbruch [1] also studied a number of robust location estimators for protecting against contamination. They found statistically significant improvement over the ordinary LDF for a variety of trims and hubers among others. However, these improvements are measured in the third decimal place of the error rates and so are unlikely to be of major practical importance unless there is heavy contamination.

A final problem in constructing a discriminant function is the assumption that the training samples have been correctly classified. If they have not been, Lachenbruch (cited in Lachenbruch [13]) showed, for large samples, if the initial misclassification rates were the same and the samples were randomly misallocated, then there was no effect on the performance of the linear discriminant. If the initial samples are not randomly misallocated, but are the borderline cases, the previous results do not apply. Lachenbruch (cited in Lachenbruch [13]) investigated this problem and found that the actual error rates were relatively unaffected by nonrandom initial misclassification. However, the apparent error rates were grossly distorted and unreliable for any sample size. McLachlan (cited in Lachenbruch [13]) provided asymptotic results on the initial misclassification problem.

Quadratic Discriminant Function (QDF)

The robustness of quadratic discriminant function has been studied less than the linear discriminant. The underlying assumptions to be concerned with are normality, contamination*, and initial misclassification. Clarke et al. [4] have studied the effects of nonnormality on QDF's using the Johnson system of transformations.

The misclassification rates for the QDF are most seriously affected when the underlying distributions are quite skewed as in the log-normal system. In both the S_L and S_U systems, the total error rate is always less affected than the individual error rates. There were only small effects on the total error rate in the S_U system when comparing the observed rates in the nonnormal data to the observed rate in the normal data. Huberizing and trimming* transformations have been studied and found to provide only slight gains, even in highly skewed distributions. The huberized QDF performs nearly as well as the ordinary QDF for normal and in the less skewed situations, and thus has some potential as a robust procedure if contamination is a major concern. In general, using robust measures such as huberizing or trimming on the QDF do not produce sufficient gains to make them satisfactory procedures for nonnormal distributions. It is better to transform skewed data into a more normal form.

The second failure of assumptions in quadratic discrimination may be contaminated observations. At the present time, no results are available regarding either location or scale contamination, but it may be conjectured that the results will be similar to that found in the linear situation.

A third area of robustness studies is correct initial classification [14]. If the initial misclassification is random, there are serious effects on the individual error rates, the total error rate. Equality of initial misclassification rates does not neutralize the effects of initial misclassification as it does with the linear discriminant. Therefore, it is of extreme importance that the initial samples be correctly classified if one is using a QDF.

Multiple Discriminant Function (MDF)

The multiple discriminant function has been studied in far less detail than the linear or the quadratic discriminant function. See, e.g., the works by Lachenbruch and by Michaelis (both cited in Lachenbruch [13]). The optimal rule is easily obtained if one specifies the form of the distribution and the parameters. The asymptotically optimal rule is also easily obtained by estimating parameters by maximum likelihood. One of the problems involved with multiple-group discriminant analysis is the complexity of the mathematics. The problems are often studied by simulation methods. Unfortunately, there is no canonical form for the MDF as in the two-group case. However, special configurations of means may be considered such as a collinear regular simplex pattern.

Little has been done in studying the robustness of the MDF to failure of assumptions. At present, Lin [16] is examining the effects of nonnormality on the MDF. His results indicate, as in the linear and quadratic cases, that if the underlying observations are badly skewed, the MDF can be considerably affected as the probability of correct classification is substantially reduced. If the nonnormal distributions are fairly symmetric, there is only a small decrement in performance. There have been no large simulation studies of the behavior of the MDF when the underlying covariances are not equal. However, Michaelis (cited in Lachenbruch [13]) considered a special case of the multiple-group discrimination problem when the underlying population covariance matrices were unequal. He chose means and covariances based on some data sets and generated normal random samples with these parameters. This leads to a multi-ple quadratic discriminant function. The results indicated that if the covariance matrices are unequal, the multiple quadratic discriminant performs considerably better than the linear form. The apparent error rate is badly biased and is more so for the multiple quadratic discriminant than the linear. He also noted that larger sample sizes are needed for good convergence to the optimal correct classification rates for the quadratic function approach than for the linear function approach in the multiple-group situation. At the present time there are no studies available indicating the effects of contamination or initial misclassification on the MDF.

Selection of Variables

A wide variety of methods for selecting a "best" set of variables for discrimination have been proposed. These include, for example, selecting those variables with the largest F between groups, selecting the variables in a stepwise manner based on the conditional F statistics, choosing the variable that minimizes the total probability of misclassification at any given step, examining all possible subsets of variables of a given size, choosing that subset which yields the minimum total error rate, and using a simultaneous test procedure* for selecting subsets of variables [17]. Weiner and Dunn (cited in Lachenbruch [13]) compared a number of methods of variable selection, in particular the t and stepwise selection method. They found that use of the t is quite satisfactory for constructing discriminant functions based on a small number of variables. For a large number of variables it appears that the t did better than the stepwise and sometimes did poorer than the stepwise. One might use Bonferroni* inequalities on the t-tests* to adjust for the potentially large number of tests that one may do. Using all possible combinations of variables guarantees a minimum error rate for a given number of variables, but may not be worth the cost in computing time if the subset size is large.

Packaged Computer Programs and Some of Their Features

The three major computer program packages available in the United States are BMDP, SAS, and SPSS. All three of these provide computer programs for performing discriminant analysis. BMD's discriminant procedure is BMDP7M and may be used for two-group or multiple-group problems [5]. This provides for stepwise selection or removal of variables based on an F criterion or Wilks' λ and variables may be forced into the discriminant function. Prior probabilities may be user specified. BMDP also provides for the leaving-one-out method of error rate estimation and the apparent error rate. BMDP does not perform quadratic discrimination directly, although by using cross-products and square terms, one can obtain a quadratic discriminant. This is less than satisfactory if one's heart is set on doing quadratic discrimination. BMDP includes an option for plotting these points.

SAS Institute [19] does not provide a stepwise selection program nor a leaving-one-out method of error rate estimation. It offers the user a choice of linear or quadratic rules. It can test if the covariance matrices are equal and if not will use the quadratic function. However, in the error rate evaluation, the apparent error rate is used and thus the closest point to the point being classified is always the point itself. If one considers nearest neighbors of the order 1, no points are misclassified.

The SPSS [18] subprogram performs discriminant analysis for two- or multiple-group problems. Stepwise selection using Wilks' λ*, minimum Mahalanobis distance, F statistics, multiple correlation*, or Rao's V are available. Priors may be specified.

There are no widely distributed programs for doing density estimation discriminant analysis; however, the program ALLOC is available [11, 12], which performs this using Parzen windows* and has many desirable features, including maximum likelihood estimates* of the smoothing parameters and jackknife estimation of error rates.

References

[1] Ahmed, S. and Lachenbruch, P. A. (1975). *EDV Med. Biol.* **6**, 34–42.

[2] Ahmed, S. and Lachenbruch, P. A. (1977). In *Classification and Clustering*, J. Van Ryzin, ed. Academic Press, New York, pp. 331–354.

[3] Broffitt, B., Clarke, W. R., and Lachenbruch, P. A. (1981). *Commun. Statist.—Simulation Comput.* **B10** (2), 129–141.

[4] Clarke, W. R., Broffitt, B., and Lachenbruch, P. A. (1979). *Commun. Statist. A*, **8**, 1285–1301.

[5] Dixon, W. J. and Brown, M. B. (1979). *BMDP Biomedical Computer Programs P. Series 1979.* University of California Press, Berkeley, Calif., (pp. 711–733).

[6] Elston, R. C., Graham, J. B., Mitler, C. H., Reisner, H. M., and Bouma, B. M. (1976). *Thromb. Res.* **8**, 683–695.

[7] Fukanaga, K. (1972). *Introduction to Statistical Pattern Recognition.* Academic, New York.

[8] Geisser, S. (1977). In *Classification and Clustering*, J. Van Ryzin, ed. Academic Press, New York, pp. 301–330.

[9] Glick, N. (1978). *Pattern Recognition*, **19**, 211–222.

[10] Goldstein, M. and Dillon, W. (1978). *Discrete Discriminant Analysis.* Wiley, New York.

[11] Habbema, J. D. F. and Hermans, J. J. (1977). *Technometrics*, **19**, 487–493.

[12] Hermans, J. and Habbema, J. D. F. (1976). Manual for the ALLOC Discriminant Analysis Program. *Tech. Rep.*, Department of Medical Statistics, University of Leiden.

[13] Lachenbruch, P. A. (1975). *Discriminant Analysis.* Hafner Press, New York.

[14] Lachenbruch, P. A. (1979). *Technometrics*, **21**, 129–132.

[15] Lachenbruch, P. A. and Goldstein, M. (1979). *Biometrics*, **35**, 69–85.

[16] Lin, L. (1979). Personal communication.

[17] McKay, R. (1976). *Technometrics*, **18**, 47–54.

[18] Nil, N. H., Hull, C. H., Jenkins, J. G., Steinbrenner, K., and Bent, D. H. (1975). *SPSS: Statistical Package for the Social Sciences.* McGraw-Hill, New York, pp. 434–467.

[19] SAS Institute, Inc. (1979). *SAS User's Guide.* SAS Institute, Raleigh, N.C., pp. 183–190, 307–311.

[20] Van Ness, J. and Simpson, G. (1976). *Technometrics*, **18**, 175–187.

[21] Van Ryzin, J., ed. (1977). *Classification and Clustering*. Academic Press, New York.

[22] Wahl, P. W. and Kronmal, R. A. (1977). *Biometrics*, **33**, 479–484.

[23] Young, F. Y. and Calvert, T. W. (1974). *Classification, Estimation and Pattern Recognition*. American Elsevier, New York.

Bibliography

Goldstein, M., and Dillon, W. (1978). *Discrete Discriminant Analysis*. Wiley, New York.

Lachenbruch, P. A. (1975). *Discriminant Analysis*. Hafner Press, New York.

(CLASSIFICATION
DENDRITES
MULTIPLE COMPARISONS)

PETER A. LACHENBRUCH

DISJOINT SETS

Sets that have no element in common. When the sets represent events (as in probability theory*), disjoint sets correspond to mutually exclusive events.

(SAMPLE SPACE)

DISPERSION THEORY

The term "dispersion theory" in its historical setting refers to sequences of binomial trials, generally taken to be independent with fixed probability of success p_i in the ith trial. If X is the number of successes in n such trials, one of the problems studied in the area was the *stability* of X/n. Bernoulli's theorem* asserts stability under the assumption of Bernoulli trials (independent trials with p_i = const.); Poisson's law of large numbers*, on the other hand, merely asserts the loss of variability of X/n in independent binomial trials. The major problem considered, however, was the classification of heterogeneous (non-Bernoulli) situations; and the construction of statistical tests for *homogeneity** (the Bernoulli trials case).

To investigate homogeneity in a long series of binomial trials, W. Lexis* suggested (ca. 1876) considering m disjoint sets (blocks) of independent trials with n trials in each, where n may be small, although $N = mn$ is large; it is assumed the probability of success in the ith set, p_i, $i = 1, \ldots, m$, is constant. If P_i is the actual proportion in the ith set, and P the proportion of successes in all N trials, so that $P = \sum_{i=1}^{m} P_i/m$, then an empirical criterion for the investigation is the Lexis quotient (sometimes Lexis ratio*, or empirical dispersion coefficient):

$$D = \sum_{i=1}^{m} \left[(P_i - P)^2/m \right] / \left[P(1 - P)/n \right].$$

In order to make simpler the consideration of the distributional properties, D was replaced in the early work of Lexis and his disciple L. J. Bortkiewicz* by a variable that replaces P in D by $\bar{p} = \sum_{i=1}^{m} p_i/m$. If we denote the new random variable by D', the quantity $Q = \sqrt{ED'}$ is called the (theoretical) dispersion coefficient or divergence coefficient (in German: *Fehlerrelation*). It is then easy to show that $Q \geqslant 1$, in which case the dispersion may be *normale* ($Q = 1$; as in the case of Bernoulli trials) or *übernormale* ($Q > 1$).

The random variables D, D' may, more generally, be considered in m sets of n binomial trials, where the probabilities of success in individual trials may fluctuate in an arbitrary manner. In particular, we may take the m sets to be replications of each other with the trials still independent, in which case the scheme of Poisson* produces an *unternormale* ($Q < 1$) situation; while that of Bienaymé* produces an *übernormale* one (see ref. 1, Chap. 3, for details). A further conceptual extension by A. A. Chuprov* to dependent trials in each of m sets, but with the sets still independent of each other, achieves the same ends.

The work of Chuprov and A. A. Markov* in the period 1913–1922 reverted to a consideration of the statistic D and its sampling distribution under the (null) hypothesis that the $N = mn$ trials are all Bernoulli. Chuprov proved that $EL = 1$, where

$$L = \left[(N - 1)D/\{n(m - 1)\}\right]$$

and is defined as 1 if $P = 0$ or 1, and extended the concept of this *dispersion coefficient* L to general variables by defining it as

$$L = (m - 1)^{-1}S_b^2/\left((N - 1)^{-1}S^2\right),$$

where

$$N = mn, \qquad S_b^2 = n\sum_{i=1}^{m}\left(\overline{X}_i - \overline{X}\right)^2,$$

$$S^2 = \sum_{i=1}^{m}\sum_{j=1}^{n}\left(X_{ij} - \overline{X}\right)^2,$$

$$\overline{X} = N^{-1}\sum_i\sum_j X_{ij}, \qquad \overline{X}_i = n^{-1}\sum_{j=1}^{n}X_{ij},$$

X_{ij} being the jth observation ($j = 1, \ldots, n$) in the ith set of observations. He showed that in the case of general i.i.d. variables X_{ij}, it is still true (just as in the Bernoulli case where $X_{ij} = 0$ or 1) that $EL = 1$. Markov investigated var L; and determined the asymptotic distribution of $(m - 1)L$ as $n \to \infty$ to be the $\chi^2(m - 1)$ distribution (and even in the more general case where the sample size may not be the same for each set, but each is made to approach infinity). This work, under the guiding influence of Markov, thus produced one of the first instances of a rather complete sampling theory of a test statistic L, which could in particular be used to test for homogeneity of statistical trials.

The more general significance of all these results in the light of the subsequent flowering of statistical theory lies in the following facts:

1. mD may be recognized as the *chi-square statistic** for testing the homogeneity of m populations whose elements take on only the values 0 and 1. Thus Lexis's ratio anticipates the various chi-square statistics and Markov's limit result appropriately gives its limit distribution as $\chi^2(m - 1)$.

2. If we define $S_w^2 = S^2 - S_b^2$, and note that in the Bernoulli trials case $S^2 = NP(1 - P)$, $S_b^2 = n\sum_{i=1}^{m}(P_i - P)^2$, and $S_w^2 = n\sum_{i=1}^{m}P_i(1 - P_i)$, we see that Markov's result is also tantamount to showing that the asymptotic distribution

of $(m - 1)Y(m, n)$ as $n \to \infty$, where

$$Y(m, n) = \left(S_b^2/(m - 1)\right)/\left(S_w^2/(m(n - 1))\right)$$

is $\chi^2(m - 1)$. $Y(m, n)$ has the form of an *F*-statistic in single-factor analysis of variance* which Lexis's ratio thus also anticipates. Although the usual normality assumptions of the analysis of variance do not apply, Markov's result is consistent with the limit distribution in that case.

Dispersion theory is often claimed as being the framework of the "Continental direction" of statistics, typified especially by the names of E. Dormoy, Lexis, Bortkiewicz, Chuprov, Markov, and Chuprov's disciple O. N. Anderson*, as opposed to the English biometric school*, founded by K. Pearson*. In contrast to the English school, it is said of the Continental direction that it sought to give test criteria of more universal applicability by avoiding precise distributional assumptions. The main area of present-day statistical theory that embraces this view is that of nonparametric statistics*, where the contributions of more recent Russian researchers such as Kolmogorov and Smirnov are well known.

Bibliography

(i) Bauer, R. K. (1955). *Mitteilungsbl. math. Statist. Anwend.*, **7**, 25–45. (Also in Russian translation as pp. 225–238 of ref. (ii). The influence of Lexis's ratio is traced through to the theory of analysis of variance, experimental design, C. R. Rao's dispersion analysis, and multivariate analysis; and the philosophy of the Continental direction as seen by O. N. Anderson is presented.)

(ii) Chetverikov, N. S., ed. (1968). *O Teorii Dispersii*. Statistika, Moscow. (Presents evolution of dispersion theory through a series of key papers, in Russian versions, including Lexis's, Bortkiewicz's, and Chuprov's.)

(iii) Heyde, C. C. and Seneta, E. (1977). *I. J. Bienaymé: Statistical Theory Anticipated*. Springer-Verlag, New York. (Chapter 3 gives an account of dispersion theory from a modern standpoint.)

(iv) Ondar, Kh. O., ed. (1977). *O teorii veroiatnostei i matematicheskoi statistike*. Nauka, Moscow. (Annotated correspondence between Chuprov and Markov, Nov. 1910–Feb. 1917.)

(v)   Uspensky, J. V. (1937). *Introduction to Mathematical Probability*. McGraw-Hill, New York. (Gives an extensive but incomplete account of Markov's contributions to dispersion theory.)

Reference

[1]   Heyde, C. C. and Seneta, E. (1977). *I. J. Bienaymé: Statistical Theory Anticipated*. Springer-Verlag, New York.

(ANALYSIS OF VARIANCE
CHI-SQUARE DISTRIBUTION
CHI-SQUARE TESTS
LAWS OF LARGE NUMBERS
NONPARAMETRIC STATISTICS
SAMPLING DISTRIBUTIONS)

E. SENETA

DISPLACEMENT OPERATOR

An operator used in finite difference calculus*. It is usually denoted E and has the effect of increasing the argument of a function by 1. Thus

$$Ef(x) = f(x + 1).$$

Similarly,

$$E^2f(x) = E[Ef(x)] = Ef(x + 1)$$
$$= Ef(x + 2).$$

Generally, $E^nf(x) = f(x + n)$, for integer n, and by extension

$$E^hf(x) = f(x + h)$$

for any real value h.

(FINITE DIFFERENCES,
   CALCULUS OF)

DISSOCIATED RANDOM VARIABLES

A concept related to a set of multiply subscripted random variables introduced by McGinley and Sibson [2]. A precise definition is as follows.

Suppose that we have a family $\{X_r\}$, of random variables subscripted by j-tuples of integers in the range 1 to n: i.e., the index $\mathbf{r} \in \{(r_1, \ldots, r_j)$ with $1 < r_i \leqslant n$ for all $i = 1, \ldots, j\}$. This family is *dissociated* if, for every pair of subfamilies A and B which have the property that the set of integers occurring in the subscripts of elements of A is disjoint from the corresponding set for B, the joint distributions of A and B are independent of one another.

The families, in which the subscripts $r_1, \ldots, r_j$ are all distinct, i.e., $\mathbf{r} \in R_j^n = \{(r_1, \ldots, r_j),\ 1 < r_i \leqslant n$ for all $i = 1, \ldots, j$ and $r_j \neq r_k$ for $i \neq k\}$ are called an off-diagonal disassociated family.

McGinley [1] used this notion to prove the almost sure convergence of some convex combinations of order statistics*, including Gini's* coefficient of mean difference. (Gini's coefficient is just the mean of an off-diagonal dissociated family.)

References

[1]   McGinley, W. G. (1979). *J. R. Statist. Soc. B*, **41**, 111–112.

[2]   McGinley, W. G. and Sibson, R. (1974). *Math. Proc. Camb. Philos. Soc.*, **1**, 185–188.

DISTANCE FUNCTIONS

The term "distance function" is used in a variety of contexts. Usually, it does not denote any real distance, but rather a measure of difference or distinctiveness between two rather complicated (and often abstract) entities. For example, we can have "distance" between

1.   Two CDFs*
2.   Two individuals, based on values of a number of characters measured on each individual

"Distance" can be defined arbitrarily, but there are certain basic conditions that most distances satisfy. These conditions are natural ones, if "distance" is to retain much of its intuitive meaning.

The conditions are:

1. The distance of an individual from itself is zero.
2. The distance from A to B equals the distance from B to A.
3. For any three entities A, B, C,

   (distance A to B) + (distance B to C)

   $\geqslant$ (distance A to C).

Condition 3 is called the *triangle* (or *triangular*) inequality.

A rule for calculating distances is often termed a "metric"—that is, a "sort of measurement."

(CLASSIFICATION
DENDRITES
LEVY DISTANCE
MAHALANOBIS DISTANCE)

DISTRIBUTED LAG MODELS *See* LAG MODELS, DISTRIBUTED

DISTRIBUTION-FREE METHODS

Distribution-free methods include statistical techniques of estimation* and inference* that are based on a function of the sample observations whose corresponding random variable has a distribution that does not depend on complete specification of the distribution of the population from which the sample was drawn. As a result, the techniques are valid under only very general assumptions about the underlying population. Strictly speaking, the term "nonparametric statistical techniques" implies an estimation or inference statement that is not directly concerned with parameters. These terms are therefore not exactly synonymous, yet the body of distribution-free techniques is perhaps more commonly known as nonparametric* statistical methods than as distribution-free methods.

Nonparametric methods provide valuable alternative techniques to classical parametric

methods for many reasons. Nonparametric methods are sometimes called "weak assumption" statistics because the assumptions required for validity usually are quite general and minimal, at least much less stringent than those required for classical parametric methods. In many cases the procedures are inherently robust and even these weak assumptions can be relaxed. This property implies that conclusions reached by a nonparametric inference procedure generally need not be tempered by qualifying statements. Some other advantages are as follows: nonparametric methods (1) are easy to understand and apply, (2) are especially appropriate for small samples, (3) frequently require data measured only on an ordinal scale (inference procedures based on ranks), (4) may frequently be applied to "dirty data" (i.e., incomplete or imprecise data), and (5) have a very wide scope of applications.

HISTORY AND LITERATURE

It is generally accepted that the history of nonparametric statistics dates back to Arbuthnott in 1710 [1] and his introduction of the sign test. However, most methods were not developed before the middle 1940s and early 1950s. The original articles that introduced those distribution-free tests that are best known and in most common use today are Kolmogorov [16], Smirnov [29], Mood [22, 23] Wald and Wolfowitz [30], Westenberg [34], Brown and Mood [3], Cochran [4], Friedman [8], Kruskal and Wallis [18], Mann and Whitney [20], and Wilcoxon [35, 36, 37]. Since these pioneer articles appeared, the learned journals in statistics have published countless articles in the field. Savage [27] published a bibliography listing approximately 3000 papers that had appeared in the statistical literature up to 1962.

A very large number of papers have appeared each year since 1962. As a result, it is not possible to describe the literature in an encyclopedia entry of this length. Suffice it to say that this literature has dealt with refinements, properties, and extensions of procedures already introduced as well as new distribution-free procedures that can be applied in a variety of new problem situations as well as new procedures that compete with old procedures for the same problem. The most recent literature has been concerned with distribution-free confidence interval estimation, density estimation*, scale problems, general linear models*, regression* problems, multiple comparisons*, sequential procedures*, two-stage procedures*, Bayesian* methods, performance (power and efficiency) studies, multivariate* problems, and selection problems*, to name only the most predominant problems.

Govindarajulu [11] gives a brief survey of nonparametric statistics in the case of univariate fixed-sample-size problems. Other surveys are found primarily in books. The first books in the nonparametrics area were those of Siegel [28] and Fraser [7]. Walsh [31, 32, 33] prepared a three-volume handbook giving a complete compendium of nonparametric statistical methods up to 1958 in a concise but complete manner. At present there are also numerous text and reference books on the general subject; these include Noether [24], Bradley [2], Conover [5], Gibbons [9, 10], Hollander and Wolfe [14], Noether [24], Marascuilo and McSweeney [21], and Daniel [6]. Hájek and Šidák [13], Hájek [12], and Lehmann [19] cover the subject of rank tests. Kendall [15] gives a complete treatment of rank correlation* methods. The most recent book is Pratt and Gibbons [26]. All of these sources have a bibliography, and most include tables that give the sampling distributions of many nonparametric test statistics or the appropriate rejection regions.

NONPARAMETRIC HYPOTHESIS TEST PROCEDURES

Nonparametric methods are applicable to a wide variety of statistical problems, which may be most easily classified according to type of inference problem and type of data. The following outline lists the best known

nonparametric hypothesis-testing procedures classified in this manner. A location problem is to be interpreted as a null hypothesis concerning the value of one or more location or central tendency parameters (such as the mean or median), and a scale problem means a null hypothesis concerning the value of one or more dispersion or spread parameters (such as the standard deviation or variance). Association problems refer to tests concerning the statistical independence of two or more variables. Many of the test procedures listed here have corresponding procedures for finding confidence interval estimates of the relevant parameter or for making multiple comparisons among the relevant parameters.

I.   Location Problems
   A.   Data on one sample or two related samples (paired sample data)
       1.   Sign test*
       2.   Wilcoxon signed rank test*
   B.   Data on two independent samples
       1.   Mann–Whitney U-test* or Wilcoxon rank-sum test
       2.   Median test (*see* BROWN–MOOD MEDIAN TEST)
       3.   Tukey's quick test*
       4.   Normal scores* tests (e.g., Terry–Hoeffding and van der Waerden*; *see also* FISHER–YATES TESTS)
       5.   Wald–Wolfowitz runs test
   C.   Data on k independent samples ($k \geqslant 3$)
       1.   Kruskal–Wallis* (one-way analysis-of-variance) test
       2.   Extended median test (*see* BROWN–MOOD MEDIAN TEST)
   D.   Data on k related samples ($k \geqslant 3$)
       1.   Friedman (two-way analysis-of-variance) test*

2. Durbin test for balanced incomplete block designs

II. Scale Problems
   A. Data on two independent samples
      1. Siegel–Tukey test*
      2. Mood test*
      3. Tests of the Freund, Ansari*, Bradley, David, or Barton type
      4. Normal scores tests (e.g., Klotz)
      5. Moses ranklike tests
III. Goodness-of-Fit* Problems (*See* GOODNESS-OF-FIT-TESTS)
   A. Data on one sample
      1. Chi-square* goodness-of-fit test
      2. Kolmogorov–Smirnov test
   B. Data on k independent samples ($k \geqslant 2$)
      1. Chi-square test
      2. Kolmogorov–Smirnov test*
IV. Association Problems
   A. Data on two related samples
      1. Spearman rank correlation*
      2. Kendall's tau coefficient*
      3. Corner test (Olmstead–Tukey)*
   B. Data on k related samples ($k \geqslant 3$)
      1. Kendall coefficient of concordance* for complete rankings
      2. Kendall coefficient of concordance* for balanced incomplete rankings
   C. Contingency table data
      1. Chi-square test* of independence
V. Randomness* Problems
   A. Data on one sample
      1. Runs test*
      2. Runs above and below the median
      3. Runs up-and-down test

VI. Trend* Problems
   A. Data on one sample
      1. Rank correlation test* (Daniels test)
      2. Kendall tau test* (Mann test)
      3. Runs test
      4. Runs up-and-down test
      5. Cox–Stuart test
VII. Linear Regression* Problems
   A. Data on two related samples
      1. Theil test
      2. Brown–Mood test*
   B. Data on two independent samples
      1. Hollander and Wolfe test
VIII. Proportion Problems
   A. Data on one sample
      1. Binomial test (*see* BINOMIAL DISTRIBUTION)
   B. Data on two related samples
      1. McNemar test*
   C. Data on two independent samples
      1. Fisher's exact test*
      2. Chi-square test*
   D. Data on k independent samples ($k \geqslant 3$)
      1. Chi-square test
   E. Data on k related samples ($k \geqslant 3$)
      1. Cochran's Q test*

The specific methodology for the most important and useful procedures in this listing are now outlined.

Wilcoxon Signed Rank Procedures

ASSUMPTIONS. $X_1, X_2, \ldots, X_n$ are a random sample of n observations that are measured on at least an ordinal scale and drawn from a population that is continuous and symmetric about its median M.

TEST PROCEDURE. The null hypothesis is $M = M_0$. Calculate the absolute differences $|X_i - M_0|$ for $i = 1, 2, \ldots, n$, arrange them from smallest to largest, and assign rank 1 to the smallest, $\ldots$, n to the largest, while keeping track of the sign of $X_i - M_0$. Define a *positive rank* as a rank assigned to a positive $X_i - M_0$ and a *negative rank* as a rank assigned to a negative $X_i - M_0$. The test statistic is

$$T_+ = \text{sum of positive ranks.}$$

The appropriate rejection regions are as follows:

Alternative	Rejection Region
$M > M_0$	T_+ too large
$M < M_0$	T_+ too small
$M \neq M_0$	T_+ either too small or too large

Tables of the null distribution of T_+ for small samples that can be used to find P-values or rejection regions are given in the text and reference books cited earlier. Since the asymptotic sampling distribution of T_+ is the normal distribution, the test statistic (without a continuity correction*) for large samples is

$$z = \frac{T_+ - n(n+1)/4}{\sqrt{n(n+1)(2n+1)/24}}.$$

The distribution of z is approximately the standard normal distribution, and the direction of the appropriate rejection region for z is the same as the corresponding rejection region for T_+. The accuracy of the approximation may be improved by using a continuity correction in the numerator of z.

A zero occurs if any $X_i = M_0$ and a tie occurs if $|X_i - M_0| = |X_j - M_0|$ for any $i \neq j$. The probability of a zero or a tie is zero, but either may occur in practice. The recommended procedure when a tie or zero occurs is to ignore the zeros and reduce n accordingly, and to assign the midrank to each member of any set of tied observations. The midrank for each set of tied observations is calculated as the average of the smallest and largest ranks that the tied observations would be assigned if they were not tied. Then T_+ is calculated as before and the same table can be used for an approximate test. The correction for ties

with large samples is to replace the denominator of z by

$$\sqrt{\frac{n(n+1)(2n+1)}{24} - \frac{\Sigma u^3 - \Sigma u}{48}},$$

where u is the number of $|X_i - M_0|$ that are tied for a given rank and the sum is over all different sets of tied ranks.

CONFIDENCE INTERVAL PROCEDURES.   A confidence interval for M can be found by computing the $n(n+1)/2$ Walsh averages defined by

$$U_{ij} = \frac{X_i + X_j}{2} \quad \text{for } 1 \leqslant i \leqslant j \leqslant n.$$

The confidence interval end points are the kth smallest and kth largest Walsh averages. The value of k depends on the confidence coefficient $1 - \alpha$; for small samples it is determined from tables of the sampling distribution of T_+ and for large samples it is calculated from

$$k = 0.5 + \tfrac{1}{4}n(n+1)$$
$$- z\sqrt{\tfrac{1}{24}n(n+1)(2n+1)},$$

where z is a standard normal deviate that satisfies $\Phi(z) = (1 - \alpha)/2$ for Φ the standard normal CDF. Any noninteger value of k should be rounded downward to the next smaller integer to obtain a conservative result.

PAIRED SAMPLES.   For data on paired samples $(X_1, Y_1), (X_2, Y_2), \ldots, (X_n, Y_n)$, the test and confidence interval procedures are the same as those described earlier, with X_i replaced by $D_i = X_i - Y_i$. Then M refers to the median of the population of differences $X - Y$ that is assumed to be continuous and symmetric.

Mann–Whitney–Wilcoxon Procedures

ASSUMPTIONS. $X_1, X_2, \ldots, X_m$ and $Y_1, Y_2, \ldots, Y_n$ are mutually independent random samples of observations that are each measured on at least an ordinal scale and drawn from continuous populations F_X and F_Y with medians M_X and M_Y, respectively.

TEST PROCEDURE. The null hypothesis is $F_X(u) = F_Y(u)$ for all u, or $M_X = M_Y$ under the additional assumption that $F_X(u) = F_X(u - M_X + M_Y)$.

The $m + n = N$ observations are pooled and arranged from smallest to largest and assigned the corresponding integer ranks 1, 2, . . . , N according to their relative magnitude, while keeping track of the sample from which each observation comes. Define an X rank as a rank assigned to an observation from the X set. The test statistic is

$$T_X = \text{sum of the } X \text{ ranks.}$$

The appropriate rejection regions are as follows:

Alternative	Rejection Region
$M_X > M_Y$	T_X too large
$M_X < M_Y$	T_X too small
$M_X \neq M_Y$	T_X either too small or too large

Tables of the null distribution of T_X or rejection regions for small samples are given in the textbooks cited earlier. For large samples we can treat

$$z = \frac{T_X - m(N + 1)/2}{\sqrt{mn(N + 1)/12}}$$

as an approximately standard normal variable. A continuity correction of ± 0.5 can be added to the numerator of z. The appropriate rejection region for z has the same direction as the corresponding region for T_X.

Ties between or within samples are handled by the midrank method. The correction for ties in large samples is to replace the denominator of z by

$$\sqrt{\frac{mn(N + 1)}{12} - \frac{mn(\Sigma u^3 - \Sigma u)}{12N(N - 1)}},$$

where u is the total number of observations from either sample that are tied for a given

rank and the sum is over all different sets of tied ranks.

CONFIDENCE INTERVAL PROCEDURES. A confidence interval for $M_X - M_Y$ can be found under the assumption that $F_X(u) = F_Y(u - M_X + M_Y)$. The procedure is to calculate the mn differences

$$U_{ij} = X_i - Y_j$$

for all $1 \leq i \leq m$ and $1 \leq j \leq n$.

The confidence interval end points are the kth smallest and kth largest values of U_{ij}. The value of k depends on the confidence coefficient $1 - \alpha$; for small samples it is determined from tables of the sampling distribution of T_X and for large samples it is calculated from

$$k = 0.5 + \frac{mn}{2} - z\sqrt{\frac{mn(N + 1)}{12}},$$

where z is a standard normal deviate that satisfies $\Phi(z) = (1 - \alpha)/2$. Any noninteger value of k should be rounded downward to the next smaller integer to obtain a conservative result.

Kruskal–Wallis (One-Way Analysis-of-Variance) Procedures

ASSUMPTIONS. The data are k mutually independent random samples of observations measured on at least an ordinal scale and drawn from populations $F_1, F_2, \ldots, F_k$ that are continuous and have medians $M_1, M_2, \ldots, M_k$, respectively.

TEST PROCEDURE. The null hypothesis is $F_1(u) = F_2(u) = \cdots = F_k(u)$ for all u, or $M_1 = M_2 = \cdots = M_k$ under the additional assumption that if the k populations differ, they differ only in location.

Let n_j denote the sample size for the data from population F_j for $j = 1, 2, \ldots, k$ and $N = \sum_{j=1}^{k} n_j$. The procedure is to pool the N observations, arrange them from smallest to largest, and assign integer ranks according to their relative magnitude, while keeping track of the population from which each observation was drawn. Let R_j denote the sum of the ranks assigned to the observations from population F_j. The test statistic compares each R_j with its null expected value $n_j(N + 1)/2$ by using the expression

$$H = \frac{12}{N(N + 1)} \sum_{j=1}^{k} \frac{\left[R_j - n_j(N + 1)/2\right]^2}{n_j}$$

$$= \frac{12}{N(N + 1)} \sum_{j=1}^{k} \frac{R_j^2}{n_j} - 3(N + 1).$$

The appropriate rejection region for the alternative that at least two F's (or M's) differ is large values of H. Tables of the null distribution of H for small sample sizes and small k are given in many textbooks. For large samples the chi-square distribution with $k - 1$ degrees of freedom is used for an approximate test.

Midranks should be assigned to all tied observations. The correction for ties is to divide H by

$$1 - \frac{\sum u^3 - \sum u}{N(N^2 - 1)},$$

where u is the total number of observations from any sample that are tied for a given rank and the sum is over all different sets of tied ranks.

MULTIPLE COMPARISONS*. Instead of, or in addition to, a hypothesis test under the assumption that if the populations differ, they differ only in location, a multiple-comparisons procedure can be used to determine which pairs of populations differ significantly from each other with regard to their medians. In a multiple-comparisons procedure at level α, α is the simultaneous or overall error rate in the sense that α is the null probability that at least one statement is incorrect, or, equivalently, $1 - \alpha$ is the null probability that all statements are correct. Any pair of populations i and j are declared to have significantly different medians if

$$\left| \frac{R_i}{n_i} - \frac{R_j}{n_j} \right| \geq z \sqrt{\frac{N(N + 1)}{12} \left(\frac{1}{n_i} + \frac{1}{n_j} \right)},$$

where z is the standard normal deviate that satisfies $\Phi(z) = 1 - [\alpha/k(k - 1)]$.

Friedman (Two-Way Analysis-of-Variance) Procedures

ASSUMPTIONS. The data are k related random samples of observations measured on at least an ordinal scale and drawn from continuous populations with treatment effects $M_1, M_2, \ldots, M_k$, respectively. Each of k treatments is observed once in each of n blocks (groups or matching conditions) so that there are n observations for each treatment. The blocks are formed in such a way that the units within each block are homogeneous or matched in some way and there is no interaction between blocks and treatments. Then the k treatments are assigned randomly to the k units in each block.

TEST PROCEDURE. The null hypothesis is that the treatment effects are all the same, or $M_1 = M_2 = \cdots = M_k$.

The procedure is to assign ranks 1, 2, ..., k to the treatment observations within each block and let R_j denote the sum over all blocks of the ranks assigned to the jth treatment. The test statistic compares each R_j with its null expected value $n(k + 1)/2$ by using the expression

$$S = \sum_{j=1}^{k} \left[R_j - n(k + 1)/2 \right]^2$$

$$= \sum_{j=1}^{k} R_j^2 - \frac{n^2 k(k + 1)^2}{4}.$$

The appropriate rejection region for the alternative that at least two treatment effects differ is large values of S. Tables of the exact null distribution of S (or a monotonic function of S) are given in many textbooks (sometimes under the heading Kendall's coefficient of concordance with n and k reversed). For large samples the chi-square distribution with $k - 1$ degrees of freedom can be used for the value of the statistic $Q = 12S/nk(k + 1)$.

Midranks should be assigned to all sets of tied observations. The correction for ties is to divide Q by

$$1 - \frac{\Sigma u^3 - \Sigma u}{nk(k^2 - 1)},$$

where u is the total number of observations in any block that are tied for a given rank and the sum is over all sets of tied ranks.

MULTIPLE COMPARISONS.   Instead of, or in addition to, a hypothesis test of homogeneity of treatment effects, a multiple comparisons procedure can be used to determine which pairs of treatment effects differ significantly from each other. At overall level α, any pair of treatments i and j are declared to have significantly different effects if

$$|R_i - R_j| \geqslant z\sqrt{\frac{nk(k+1)}{6}},$$

where z is the standard normal deviate that satisfies $\Phi(z) = 1 - [\alpha/k(k-1)]$.

Kendall's Tau* Coefficient

ASSUMPTIONS. $(X_1, Y_1), (X_2, Y_2), \ldots,$ (X_n, Y_n) are a random sample of paired observations that are measured on at least an ordinal scale and drawn from a population that is continuous.

CALCULATION OF THE TAU COEFFICIENT.   Assume without loss of generality that the data have been arranged in natural order of the X observations so that $X_1 < X_2 < \cdots < X_n$. For a particular Y_j, compare it with each Y_i that follows it and score a $+1$ each time that $Y_j < Y_i$ and a -1 each time that $Y_j > Y_i$, and repeat this for each $j = 1, 2, \ldots, n$. Let P be the number of positive scores and Q the number of negative scores and define $S = P - Q$. The Kendall tau coefficient τ is calculated as the ratio of S to the number of distinguishable pairs of Y values, or

$$\tau = \frac{S}{n(n-1)/2}.$$

Note that P is the number of pairs of Y values that occur in natural order and Q is the number of pairs of Y values that occur in reverse natural order. A pair in natural order is called concordant and a pair in reverse natural order is called discordant. Hence τ measures the relative agreement between pairs with respect to order.

τ ranges between -1 and $+1$; $\tau = 1$ indicates perfect agreement between the rankings and $\tau = -1$ indicates perfect disagreement between the rankings. Increasing values of τ indicate increasing agreement.

TEST PROCEDURE.   The coefficient τ or, equivalently, the value of S, can be used to test the null hypothesis that no association exists between the variables X and Y, i.e., that X and Y are statistically independent. The appropriate rejection regions are as follows:

Alternative	Rejection Region
Direct (positive) association	S (or τ) too large
Indirect (negative) association	S (or τ) too small
Association exists	S (or τ) either too small or too large

Tables of the null distribution of S (or τ) for small samples that can be used to find P-values or rejection regions are given in the textbooks cited earlier. For large samples, the approximate test procedure is to treat

$$z = \frac{S}{\sqrt{n(n-1)(2n+5)/15}}$$

as a standard normal deviate with rejection region in the same direction as that given for S.

If a tie occurs in either an X pair or a Y pair, zero is scored and hence there is no contribution to S. The value of τ can then be adjusted to take the ties into account by calculating

$$\tau = \frac{S}{\sqrt{\binom{n}{2} - \Sigma\binom{u}{2}}\sqrt{\Sigma\binom{v}{2} - \binom{v}{2}}}$$

where u is the number of observations in the X sample that are tied for a given rank and the sum is over all sets of u tied ranks, and v is defined similarly for the Y sample. To test the null hypothesis of no association, the correction for ties in large samples is to replace the denominator of z by

$$\left\{ \tfrac{1}{18}\left[n(n-1)(2n+5) - \sum u(u-1)(2u+5) - \sum v(v-1)(2v+5) \right] \right\}^{1/2}.$$

References

[1] Arbuthnott, J. (1710). *Philos. Trans. R. Soc. Lond.*, **27**, 186–190.

[2] Bradley, J. V. (1968). *Distribution-Free Statistical Tests*. Prentice-Hall, Englewood Cliffs, N.J. (Elementary; organized according to type of distribution of test statistic; tables of critical values; references.)

[3] Brown, G. W. and Mood, A. M. (1951). In *Proc. 2nd Berkeley Symp.* University of California Press, Berkeley, Calif., pp. 159–166.

[4] Cochran, W. J. (1950). *Biometrika*, **37**, 256–266.

[5] Conover, W. J. (1971). *Practical Nonparametric Statistics*. Wiley, New York. (Elementary; extensive tables of critical values; extensive references.)

[6] Daniel, W. (1978). *Applied Nonparametric Statistics*. Houghton Mifflin, Boston. (Elementary; extensive tables of critical values; exhaustive references.)

[7] Fraser, D. A. S. (1957). *Nonparametric Methods in Statistics*. Wiley, New York. (Theoretical.)

[8] Friedman, M. (1937). *J. Amer. Statist. Ass.*, **32**, 675–701.

[9] Gibbons, J. D. (1971). *Nonparametric Statistical Inference*. McGraw-Hill, New York. (Intermediate level; mostly theory.)

[10] Gibbons, J. D. (1976). *Nonparametric Methods for Quantitative Analysis*. Holt, Rinehart and Winston, New York. (Elementary; uses *P*-values; emphasizes applications; good examples from all fields of application; extensive tables of exact null distributions.)

[11] Govindarajulu, Z. (1976). *Commun. Statist. A*, **5**, 429–453. (Few recent references.)

[12] Hájek, J. (1969). *Nonparametric Statistics*. Holden-Day, New York. (Theoretical treatment of rank tests.)

[13] Hájek, J. and Šidák, Z. (1967). *Theory of Rank Tests*. Academic Press, New York. (Theoretical.)

[14] Hollander, M. and Wolfe, D. A. (1973). *Nonparametric Statistical Methods*. Wiley, New York. (Large coverage with cryptic descriptions of procedures; extensive tables and bibliography.)

[15] Kendall, M. G. (1962). *Rank Correlation Methods*, 3rd ed. Hafner, New York. (Methods and theory; references and tables.)

[16] Kolmogorov, A. N. (1933). *G. Ist. Ital. Attuari*, **4**, 83–91.

[17] Kruskal, W. H. (1957). *J. Amer. Statist. Ass.*, **52**, 356–360.

[18] Kruskal, W. H. and Wallis, W. A. (1952). *J. Amer. Statist. Ass.*, **47**, 583–621; errata, *ibid.*, **48**, 907–911.

[19] Lehmann, E. L. (1975). *Nonparametrics: Statistical Methods Based on Ranks* Holden-Day, San Francisco. (Treatment of rank tests at introductory to advanced level; covers power; theory in appendix; tables and references.)

[20] Mann, H. B. and Whitney, D. R. (1947). *Ann. Math. Statist.*, **18**, 50–60.

[21] Marascuilo, L. A. and McSweeney, M. (1977). *Nonparametric and Distribution-Free Methods for the Social Sciences*. Brooks/Cole, Monterey, Calif. (Cookbook approach.)

[22] Mood, A. M. (1940). *Ann. Math. Statist.*, **11**, 367–392.

[23] Mood, A. M. (1950). *Introduction to the Theory of Statistics*. McGraw-Hill, New York, Chap. 16.

[24] Noether, G. E. (1967). *Elements of Nonparametric Statistics*. Wiley, New York. (Theoretical; no examples or exercises.)

[25] Noether, G. E. (1973). *Introduction to Statistics: A Nonparametric Approach*. Houghton Mifflin, Boston. (Elementary; integrates nonparametric and parametric techniques.)

[26] Pratt, J. W. and Gibbons, J. D. (1981). *Concepts of Nonparametric Theory*. Springer-Verlag, New York: (Mostly theory; extensive problems and references; tables.)

[27] Savage, I. R. (1962). *Bibliography of Nonparametric Statistics*. Harvard University Press, Cambridge, Mass. (References only.)

[28] Siegel, S. (1956). *Nonparametric Statistics for the Behavioral Sciences*. McGraw-Hill, New York. (A landmark for its time.)

[29] Smirnov, N. V. (1939). *Bull. Mosc. Univ.*, **2**, 3–16 (in Russian).

[30] Wald, A. and Wolfowitz, J. (1940). *Ann. Math. Statist.*, **11**, 147–162.

[31] Walsh, J. E. (1962). *Handbook of Parametric Statistics*, Vol. 1. D. Van Nostrand, Princeton, N.J. (Handbook only.)

[32] Walsh, J. E. (1965). *Handbook of Nonparametric Statistics*, Vol. 2. D. Van Nostrand, Princeton, N.J. (Handbook only.)

[33] Walsh, J. E. (1968). *Handbook of Nonparametric Statistics*, Vol. 3. D. Van Nostrand, Princeton, N.J. (Handbook only.)

[34] Westenberg, J. (1948). *Proc. Kon. Ned. Akad. Wet.*, **51**, 252–261.

[35] Wilcoxon, F. (1945). *Biometrics*, **1**, 80–83.

[36] Wilcoxon, F. (1947). *Biometrics*, **3**, 119–122.

[37] Wilcoxon, F. (1949). *Some Rapid Approximate Statistical Procedures*. American Cyanamid Company, Stanford Research Laboratories, Stanford, Calif.

JEAN DICKINSON GIBBONS

DISTRIBUTION-FREE TOLERANCE LIMITS (INTERVALS)

A statistical tolerance interval* is based on a random sample and has a preassigned confidence of γ, including at least proportion P of the population. The underlying distribution is, of course, a basic factor in the derivation of this interval, but if the underlying distribution is unknown, recourse has to be made to *distribution-free tolerance intervals*. Statements associated with the latter are valid for any distribution, provided that it is continuous. The problem is usually to determine the sample size n needed to be γ confident that at least P of the population lies below (above) an upper (lower) limit or between two limits. For a one-sided distribution-free tolerance limit, taking the least value, n should satisfy $P^n > 1 - \gamma$ or $n > \log(1 - \gamma)/\log P$.

Example 1. If $\gamma = 0.90$, $P = 0.95$, then $n \geqslant \log 0.10/\log 0.095 \approx 45$. For two-sided distribution-free tolerance limits, based on least and greatest observed value we need $nP^{n-1} - (n - 1)P^n > 1 - \gamma$. A nomograph for obtaining two-sided distribution-free tolerance intervals has been devised by Nelson [2, 3].

Example 2. If $\gamma = 0.90$ and $n = 45$, the corresponding value of P in the two-sided case is 0.916 (as compared with $P = 0.95$ in the one-sided case).

For more information on this topic, see Lindgren [1].

References

[1] Lindgren, B. W. (1976). *Statistical Theory*, 3rd ed. Macmillan, New York.

[2] Nelson, L. S. (1963). *Ind. Quality Control*, **19** (12), 11–13.

[3] Nelson, L. S. (1974). *J. Quality Tech.*, **6**, 163–164.

(DISTRIBUTION-FREE METHODS TOLERANCE INTERVALS)

DISTURBANCES

In an autocorrelation model

$$E_t = E_{t-1} + u_t \tag{1}$$

the variables u_t are called disturbances. In regression analysis any error term E_t given by (1) is the sum of the previous error term E_{t-1} and a new disturbance term u_t, which are usually assumed to be independently normally distributed with mean 0 and variance 1.

(AUTOREGRESSIVE–INTEGRATED MOVING AVERAGE MODELS AUTOREGRESSIVE–MOVING AVERAGE MODELS)

DISUTILITY   *See* DECISION THEORY

DIVERSITY INDICES

A diversity index is a measure of the "qualitative dispersion" of a population of individuals belonging to several qualitatively different categories. In the same way that such statistics as variance*, standard deviation*, mean deviation*, and range* serve to measure quantitative variability, diversity indices measure qualitative variability. The topic of diversity is of especial concern to biologists, particularly ecologists and biogeographers. An ecological community is, by definition, composed of co-occurring populations of several (two or more) different species of plants and/or animals. A community is said to have high diversity if it has many rather than few different species;

and for a given number of species, the more nearly equal their relative abundances, the higher the diversity.

MOST COMMONLY USED DIVERSITY INDICES

Numerous indices that could serve as measures of diversity have been proposed. The best known are the following:

SHANNON–WIENER INDEX, H'. Suppose that the community (treated as indefinitely large) contains s species. Let p_j be the proportion of individuals in the jth species $(j = 1, \ldots, s)$. Then the index is defined as

$$H' = - \sum_{j=1}^{s} p_j \log p_j.$$

The index is derived from information theory*. The units are bits, decits, or nats, depending on whether logs to base 2, 10, or e are used; usage is not standardized.

BRILLOUIN INDEX, H. Suppose that a finite community consists of N individuals belonging to s species with N_j individuals in the jth species $(j = 1, \ldots, s)$. Then

$$H = \frac{1}{N} \log \frac{N!}{\prod_j N_j!}.$$

This index is the finite-community equivalent of the Shannon–Wiener index. If N and all the N_j become large, so that $\log N! \rightarrow N \log N - N \log e$, then $H \rightarrow H'$ with $p_j = N_j/N$.

SIMPSON INDEX, D OR D'. For an indefinitely large community,

$$D' = - \log \sum_j p_j^2 \equiv - \log \lambda', \qquad \text{say.}$$

For a finite community

$$D = - \log \left[\frac{\sum_j N_j(N_j - 1)}{N(N - 1)} \right]$$

$$\equiv - \log \lambda, \qquad \text{say.}$$

The index is derived from Simpson's [17] "index of concentration" (see CONCENTRA-

TION INDEX), which is the probability that, if two individuals are picked independently and at random from a many-species community, they will belong to the same species. Clearly, the probability is

$$\lambda = \sum_j N_j(N_j - 1)/\{N(N - 1)\}$$

for a finite community and tends to $\lambda' = \sum p_j^2$ as all the N_j become indefinitely large. "Concentration" is the opposite of diversity and an index of diversity based on Simpson's measure of concentration must therefore be a function of the reciprocal of concentration. The logarithm of the reciprocal is the most appropriate function, as may be seen from the following consideration.

In communication theory*, a general function, H_α, called the entropy* of order α of a code, has been defined [14]. For a code composed of s different kinds of symbols, and with symbols of the jth kind forming a proportion p_j of all the symbols, H_α is defined as

$$H_\alpha = \left(\log \sum p_j^\alpha \right)/(1 - \alpha).$$

Putting $\alpha = 1, 2$, it is found that

$$H_1 = - \sum p_j \log p_j = H',$$

the Shannon–Wiener index;

$$H_2 = - \log \sum p_j^2 = D',$$

the Simpson index* (see SHANNON INFORMATION).

Thus H' and D' are both special cases of a more general function. The Simpson index has the merit of intuitive reasonableness: its numerical value is high if the probability is low that two randomly drawn individuals are conspecific. The Shannon–Wiener index (and likewise its finite community equivalent, the Brillouin index) has the merit that it is decomposable into additive components corresponding to the different levels in a hierarchical classification*, as will now be shown.

HIERARCHICAL DIVERSITY

The individual members of biological communities are always classified hierarchically. The chief divisions, in descending order of rank, are phylum, class, order, family, genus, and species. Considering (for conciseness) only the lowest two levels, genus and species, it is clear that a community has generic diversity and that each genus treated by itself has within-genus specific diversity. Thus suppose that a community has g genera and a proportion q_i of the individuals belong to the ith genus ($i = 1, \ldots, g$). Then, using the Shannon–Wiener index, the community's generic diversity, say $H'(G)$, is

$$H'(G) = -\sum_{i=1}^{g} q_i \log q_i.$$

Now suppose that genus i contains s_i species and that the proportion, within genus i, of individuals belonging to species j of the genus is p_{ij}. Then the specific diversity within genus i is

$$H_i'(S) = -\sum_{j=1}^{s_i} p_{ij} \log p_{ij}.$$

The weighted mean, over all genera in the community, of the specific diversities within the genera is then

$$H_G'(S) = -\sum_{i=1}^{g} q_i H_i'(S).$$

Now note that the individuals in species j of genus i form a proportion $q_i p_{ij}$ of the whole community. Therefore, the overall specific diversity of the whole community, $H'(SG)$, say, is

$$H'(SG) = -\sum_i \sum_j q_i p_{ij} \log q_i p_{ij}$$
$$= H'(G) + H_G'(S).$$

The decomposition of overall diversity into additive components representing diversities at the different, nested hierarchical levels is possible only if the Shannon–Wiener index (or its finite-community equivalent, the Brillouin index) is used as a measure of diversity. The method can be generalized to as many hierarchical levels as desired.

SPECIES RICHNESS AS A MEASURE OF DIVERSITY

Many biologists regard the diversity indices described above, and others like them that take account of the relative abundances of the different species in a community, as needlessly elaborate. They consider s, the number of species present, as by far the most important aspect of a community's diversity and the best and simplest measure of it; s is often called the species richness of a community. If a descriptive statistic measuring the variability of the relative abundances of the several species is required as well, *evenness* may be used (see below).

To use s as a measure of diversity is analogous to using the range as a measure of the dispersion of a quantitative variate. Like range, s is subject to large sampling error. Also, its value in a sample obviously depends on sample size; clearly, the expected value of s in a sample is less than the community s and the bias is often serious.

To compensate for this defect in sample s as a descriptive statistic, an alternative measure of species richness known as the *rarefaction diversity*, $E(s_n)$, is useful. It is the expected number of species that would be found in a subsample of size n drawn from a larger, N-member sample in which s species are present with N_j individuals in the jth species ($j = 1, \ldots, s$). It is easy to show that

$$E(s_n) = s - \sum_{j=1}^{s} \binom{N - N_j}{N} \bigg/ \binom{N}{n}.$$

Further, if $n \ll N$,

$$E(s_n) \simeq s - \sum_{j=1}^{s} (1 - N_j/N)^n.$$

Exact and approximate values of var(s_n) have been given by Heck et al. [5]. Use of rarefaction diversities permits estimates of species richness to be compared even when the samples on which they are based are unequal in size.

EVENNESS

In diversity indices such as H', H, D', and D, the number of species and the evenness of their representation are confounded. Thus a community with a few, evenly represented species can have the same diversity index as one with many, unevenly represented species. It is often desirable to keep these two aspects of diversity separate. An index is therefore needed for measuring evenness independently of species richness. Two ways of doing this are the following.

1. One may use the fact that, for given s, diversity is a maximum when all s species are present in equal proportions, that is, when the proportion of individuals in the jth species is $1/s$ for all j. The Shannon–Wiener index is then $H'_{max} = \log s$. As an index of evenness, say J', we may therefore use the ratio of the observed H' to the maximum value it could have given the same number of species. Thus

$$J' = \frac{H'}{H'_{max}} = \frac{H'}{\log s}.$$

It follows that $J' = -\sum_{j=1}^{s} p_j \log_s p_j$; in other words, J' is numerically identical to the diversity (H') calculated using logs to base s.

An analogous measure of evenness based on Brillouin's index is

$$J = \frac{H_{obs} - H_{min}}{H_{max} - H_{min}},$$

where H_{max} and H_{min} are the maximum and minimum possible Brillouin indexes for communities with the given, observed values of N and s (see Pielou [13]).

2. Another way of measuring the evenness of a community whose Shannon–Wiener index is H' is to take $\exp[H']$.

Clearly, $\exp[H'_{max}] = s$. Thus $\exp[H']$ is the number of species that would be found in a hypothetical community of perfect evenness having the same H' as the community whose evenness is to be measured [2]. The index $\exp[H']$ has obvious intuitive appeal.

SAMPLING PROBLEMS

The estimation of indices of diversity, species richness, and evenness is fraught with difficulties, both practical and conceptual.

The greatest practical difficulty consists in the fact that, for the majority of ecological communities, it is very difficult to obtain a random sample of their member individuals. This is especially true for communities of sessile or sedentary organisms. Although there may be no difficulty in placing sample plots at random on the ground, the contents of the plots hardly ever constitute a random sample of the individuals present. This is because most species have patchy patterns. Methods of coping with this problem have been suggested by Pielou [12, 13] and Zahl [18].

Another serious difficulty arises from the fact that many ecological communities lack well-defined boundaries and are not internally homogeneous. Delimitation of the sampling universe must often be to some degree arbitrary. It may be arguable whether a biological collection (e.g., the contents of an insect light trap, or a plankton haul, either of which may contain millions of individuals, belonging to hundreds of species) should be treated as a universe in its own right or as a sample (nonrandom) from a larger universe of unspecified extent.

A third difficulty is revealed by the fact that both the Shannon–Wiener and the Simpson indices exist in versions suitable for conceptually infinite communities (H' and D') or for finite communities (Brillouin's H, and D). Problems arise, in sampling and definition, from the fact that in any collection, large or small, the numbers of individuals in the species-samples comprising the collection nearly always have enormous range. Thus even when a big collection has large samples of the common species, it will usually have extremely small samples of the rare species; often, indeed, many specimens in the collection are "singletons," that is, samples of size 1 of the species to which they belong. Therefore, large-sample approximations (such as that for log factorials), which may be used for the abundant species, cannot be used for the rare ones. In ordinary statistical practice, members of rare species would be disregarded as outliers. In ecological contexts, they may be important.

HISTORICAL OUTLINE AND LITERATURE

Diversity as a concept was first introduced by Williams in Fisher et al. [3]. On the assumption that in all (or nearly all) biological collections, the frequencies of the species abundances have a log-series distribution*, he proposed that the single parameter of that distribution, denoted by α, be treated as measuring the diversity of a collection. This index has fallen into disuse with the realization that species abundance distributions can have many forms, and that a distribution-free index is therefore desirable.

The Simpson index [17] was the next diversity index to be proposed.

The idea of an index derived from information theory* is due to Margalef [8]. The underlying mathematical theory is given in Shannon and Weaver [16], Goldman [4], and Khinchin [7] for the infinite-collection case; and by Brillouin [1] for the finite-collection case.

Rarefaction diversity was first introduced by Sanders [15].

Recent summaries of ecological diversity and its measurement will be found in Pielou [11–13], Hurlbert [6], and Peet [9]. Hierarchical diversity was introduced by Pielou [10].

References

[1] Brillouin, L. (1962). *Science and Information Theory*, 2nd ed. Academic Press, New York.

[2] Buzas, M. A. and Gibson, T. G. (1969). *Science* **163**, 72–75.

[3] Fisher, R. A., Corbet, A. S., and Williams, C. B. (1943). *J. Anim. Ecol.*, **12**, 42–58.

[4] Goldman, S. (1953). In *Information Theory in Biology*, H. Quastler, ed. University of Illinois Press, Champaign, Ill.

[5] Heck, K. L., Jr., van Belle, G. and Simberloff, D. (1975). *Ecology*, **56**, 1459–1461.

[6] Hurlbert, S. H. (1971). *Ecology*, **52**, 577–586.

[7] Khinchin, A. I. (1957). *Mathematical Foundations of Information Theory*. Dover, New York.

[8] Margalef, D. R. (1958). *Gen. Syst.*, **3**, 36–71.

[9] Peet, R. K. (1974). *Ann. Rev. Ecol. Syst.*, **5**, 285–307.

[10] Pielou, E. C. (1967). *Proc. 5th Berkeley Symp. Math. Stat. Prob.*, Vol. 4, University of California Press, Berkeley, Calif., pp. 163–177.

[11] Pielou, E. C. (1969). *Introduction to Mathematical Ecology*. Wiley, New York.

[12] Pielou, E. C. (1975). *Ecological Diversity*. Wiley, New York.

[13] Pielou, E. C. (1977). *Mathematical Ecology*. Wiley, New York.

[14] Renyi, A. (1961). *Proc. 4th Berkeley Symp. Math. Stat. Prob.*, Vol. 1, University of California Press, Berkeley, Calif., pp. 547–561.

[15] Sanders, H. L. (1968). *Amer. Nat.*, **102**, 243–282.

[16] Shannon, C. E. and Weaver, W. (1949). *The Mathematical Theory of Communication*. University of Illinois Press, Champaign, Ill.

[17] Simpson, E. H. (1949). *Nature* (*Lond.*), **163**, 688.

[18] Zahl, S. (1977). *Ecology*, **58**, 907–913.

E. C. Pielou

(GINI–SIMPSON INDEX)

DIVISIA INDICES

The Divisia index is a highly regarded continuous-time statistical index number. In addition, discrete-time approximations to the Divisia index are among the best available discrete-time statistical index numbers. Index numbers* acquire their link with economic theory through the aggregator functions of aggregation theory, since index numbers can be viewed as approximations to aggregator functions.

AGGREGATION THEORY

Structure of the Economy

Aggregation* theory is a branch of economic theory. In economic theory, the structure of an economy is defined by the tastes (utility functions) of consumers and the technology (production functions) of firms. Let there be n consumers and N firms. Let f_i, $i = 1, \ldots, n$, be the utility function of the ith consumer, and let f_i, $i = n + 1, \ldots, n + N$, be the production function (possibly vector-valued) of the ith firm. Then the structure of the economy is defined by $\{ f_i : i = 1, \ldots, n + N \}$.

Aggregator Functions

Let $\mathbf{x}_i$ be the vector of arguments of f_i for economic agent (firm or consumer) i. Then $\mathbf{x}_i$ consists of consumer goods, if i is a consumer, or factors of production, if i is a firm. For some economic agent, i, let there exist functions, $g_{ij}(j = 1, \ldots, k)$ and F_i and a partition of $\mathbf{x}_i$, $\mathbf{x}_i = (\mathbf{x}'_{i1}, \ldots, \mathbf{x}'_{ik})'$, such that $f_i(\mathbf{x}_i) = F_i(g_{i1}(\mathbf{x}_{i1}), \ldots, g_{ik}(\mathbf{x}_{ik}))$ for all feasible $\mathbf{x}_i$. Then for economic agent i, g_{ij} is defined to be the "quantity aggregator function" over the components $\mathbf{x}_{ij}$, and $g_{ij}(\mathbf{x}_{ij})$ is the economic quantity aggregate over the components $\mathbf{x}_{ij}$.

Exact Aggregator Functions

An aggregator function or its corresponding economic aggregate is called "exact" or "consistent" if the aggregator function is linearly homogeneous. It can be shown (see, e.g., Green [5, Theorem 4] that economic agents behave as if exact economic aggregates were elementary goods. The specification of an aggregator function is defined to be "flexible" if it can provide a second-order approximation to any arbitrary aggregator function. Examples include the quadratic and translog specifications.

INDEX NUMBER THEORY

Functional Index Numbers

To approximate an aggregator function, a specification commonly is selected having unknown parameters. If the parameters are estimated empirically and the unknown parameters are replaced by their estimates, the resulting estimated function (normalized to equal 1 in a fixed base year) is called a "functional index number." Although such index numbers* are valuable in research, their dependence upon estimated parameters commonly discourages their publication as data by governmental agencies.

Statistical Index Numbers

Statistical index numbers provide parameter-free and specification-free approximations to economic aggregates. An economic quantity aggregate, $g_{ij}(\mathbf{x}_{ij})$, depends upon the component quantities, $\mathbf{x}_{ij}$, and upon the unknown function, g_{ij}. A statistical quantity index depends upon the component quantities, $\mathbf{x}_{ij}$, and also upon the corresponding prices, $\mathbf{p}_j$, but not upon any unknown parameters or functions. The inclusion of prices in statistical quantity indices permits the index (under conventional economic behavioral assumptions) to reveal information regarding the current point, $g_{ij}(\mathbf{x}_{ij})$, on the aggregator function, i.e., the current value of the aggregate. Statistical index numbers cannot provide information regarding the form or properties (such as substitution elasticities) of the aggregator function, g_{ij}, itself.

Statistical index numbers are characterized by their statistical properties and by their economic properties. The classical source on the statistical properties of statistical index numbers is Fisher [4]. See also Theil [8]. The economic properties of statistical index numbers are defined in terms of the indices' abilities to approximate economic aggregates.

A particularly desirable economic property of a statistical index number is the ability of the index always to attain the current value of a "flexible" aggregator function. A statistical index number possessing that property is called "superlative." (See Diewert [2].)

DIVISIA INDEX

Continuous Time

Let us fix (i, j). To simplify the notation, now drop the subscripts i and j from $\mathbf{x}_{ij}$ and $\mathbf{p}_j$. Our data then are $(\mathbf{x}, \mathbf{p})$. In continuous time, $(\mathbf{x}, \mathbf{p}) = (\mathbf{x}(t), \mathbf{p}(t))$ is a continuous function of time.

The Divisia [3] quantity index, $Q(t)$, is the line integral defined by the differential

$$d \log Q(t) = \sum_{k=1}^{M} s_k(t) d \log x_k(t),$$

where M is the dimension of the vector $\mathbf{x}$, and $s_k = p_k x_k / \mathbf{p}'\mathbf{x}$ is the expenditure share of good k in total expenditure, $\mathbf{p}'\mathbf{x}$, on the M goods. In short, the growth rate of the Divisia index is defined to be the weighted average of the growth rates of the components, where the weights are the corresponding expenditure value shares, $\mathbf{s} = (s_1, \ldots, s_M)'$, of the components.

The statistical properties of the Divisia index are provided by Richter [7]. Hulten [6] has shown that under the conditions sufficient for the existence of a consistent aggregator function, the Divisia index line integral is path independent and the resulting Divisia index always exactly attains the current value of the economic aggregate.

Discrete Time

In discrete time Törnquist [9] and Theil [8] have proposed the following approximation to the Divisia index:

$$\log Q_t - \log Q_{t-1}$$

$$= \sum_{k=1}^{M} s_k^* (\log x_{kt} - \log x_{k,t-1}),$$

where $s_k^* = \frac{1}{2}(s_{k,t} + s_{k,t-1})$ is the average of

the current expenditure value share, $s_{kt} = p_{kt} x_{kt} / \mathbf{p}_t'\mathbf{x}_t$, and the lagged share, $s_{k,t-1}$.

The Törnquist–Theil approximation to the Divisia index does not possess all of the properties proved for the continuous-time Divisia index by Hulten [6]. In fact, no known discrete-time statistical index is capable of attaining the current value of every economic aggregate without error. However, the Törnquist–Theil Divisia index does fall within Diewert's class of "superlative" index numbers. The magnitude of the error of the approximation generally is very small. Applications of the Törnquist–Theil Divisia index are discussed in Theil [8] and Barnett [1].

DUAL PRICE INDICES

In economic theory, price and quantity aggregator functions are duals. Hence the discussion on quantity aggregation above applies directly to price aggregation. While the price aggregator function dual to a quantity aggregator function is unique, two price index numbers can be acquired from a statistical quantity index. One price index can be acquired by interchanging prices with quantities in the quantity index. In the case of the Divisia index, we get

$$d \log P(t) = \sum_{k=1}^{M} s_k(t) d \log p_k(t)$$

in the continuous-time case, and

$$\log P_t - \log P_{t-1}$$

$$= \sum_{k=1}^{M} s_k^* (\log p_{kt} - \log p_{k,t-1})$$

in the Törnquist–Theil discrete-time case. Another price index can be acquired by dividing actual total expenditure on the components by the quantity index number.

If the two methods result in the same price index, the index number formula is said to satisfy Fisher's factor reversal test. The Divisia index in discrete time fails the factor reversal test*, although the magnitude of the discrepancy commonly is very small.

References

[1] Barnett, W. A. (1980). *J. Econometrics*, **14**, 11–48.

[2] Diewert, W. E. (1976). *J. Econometrics*, **4**, 115–146.

[3] Divisia, F. (1925–1926). *Rev. Econ. Polit.*, **29**, 842–861, 980–1008, 1121–1151; **30**, 49–81.

[4] Fisher, I. (1922). *The Making of Index Numbers*. Cambridge University Press, Cambridge.

[5] Green, H. A. J. (1964). *Aggregation in Economic Analysis*. Princeton University Press, Princeton, N.J.

[6] Hulten, C. R. (1973). *Econometrica*, 1017–1026.

[7] Richter, M. K. (1966). *Econometrica*, **34**, 739–755.

[8] Theil, H. (1967). *Economics and Information Theory*. North-Holland, Amsterdam.

[9] Törnquist, L. (1936). *Bank Finland Bull.* No. 10, 1–8.

(INDEX NUMBERS)

WILLIAM A. BARNETT

DIXON TEST

A distribution-free test for testing the hypothesis H_0 that two samples giving observed values $x_1, x_2, \ldots, x_{n_1}$ and $y_1, y_2, \ldots, y_{n_2}$ come from the same population (having a continuous CDF). Let $n_1 \leqslant n_2$. Consider the order statistics* $X_{(1)} < X_{(2)} < \cdots < X_{(n_1)}$. Let f_1 be the number of y's less than $X_{(1)}$, f_i the number of y's between $X_{(i-1)}$ and $X_{(i)}$ for $i = 1, 2, \ldots, n_1$, and f_{n_1+1} the number of y's greater than $X_{(n_1)}$. Dixon's [2] test statistic is

$$D^2 = \sum_{i=1}^{n_1+1} \left(\frac{f_i}{n_2} - \frac{1}{n_1+1} \right)^2.$$

The hypothesis H_0 is rejected if D^2 exceeds a preassigned constant. If H_0 is true, the distribution of kD^2 in large samples is approximated by the chi-square distribution* with ν d.f., where

$$k = \frac{n_2^2(n_1+2)(n_1+3)(n_1+4)}{2(n_2-1)(n_1+n_2+2)(n_1+1)};$$

$$\nu = \frac{n_1 n_2(n_1+n_2+1)(n_1+3)(n_1+4)}{2(n_2-1)(n_1+n_2+2)(n_1+1)^2}.$$

For n_1 and n_2 large ($\geqslant 20$) approximate formulas for k and ν are

$$k \doteq \frac{(n_1+2)^2 n_2}{2(n_1+n_2)} \left(\underset{\approx}{\doteq} \frac{n_1^2 n_2}{2(n_1+n_2)} \right),$$

$$\nu \underset{\approx}{\doteq} \frac{n_1+5}{2}.$$

The hypothesis H_0 is rejected at the approximate level of significance α if $T \equiv kD^2 > \chi_\nu^2(\alpha)$.

For more information on this topic, see Chakravarti et al. [1].

References

[1] Chakravarti, I. M., Laha, R. G., and Roy, J. (1967). *Handbook of Methods of Applied Statistics*, Vol. 1. Wiley, New York.

[2] Dixon, W. J. (1940). *Ann. Math. Statist.*, **11**, 199–204.

(DISTRIBUTION-FREE METHODS)

D-LINE *See* ADMISSIBILITY

DODGE–ROMIG LOT TOLERANCE TABLES

These are sampling inspection tables which specify *consumer's risk** ($\beta_2 = 0.10$) and also minimize the total amount of inspection, including detailing* for lots of expected quality. Dodge–Romig tables also show AOQL* for each plan so that the consumer can ascertain his or her long-run protection (*see* AVERAGE OUTGOING QUALITY LIMIT).

A portion of this table is presented as Table 1.

Lot sizes and process averages are indicated by class limits only; hence the consumer risk is only approximate. For example, if the process average is 0.05%, the *lot tolerance percent defective* is taken as 1%, and the lot size is 250, we find that the sample size $n = 165$ and $c = 0$; i.e., we reject the lot if there are any defectives in the sample. The AOQL of this plan is shown to be 0.1%.

Table 1   Dodge–Romig Lot Tolerance Single Sampling Inspection Table: Table SL-1, Lot Tolerance Percent Defective = 1.0[a]

Process Average %	0–0.010			0.011–0.10			0.11–0.20		
Lot Size	n	c	AOQL %	n	c	AOQL %	n	c	AOQL %
1–120	all	0	0.00	all	0	0.00	all	0	0.00
121–150	120	0	0.06	120	0	0.06	120	0	0.06
151–200	140	0	0.08	140	0	0.08	140	0	0.08
201–300	165	0	0.10	165	0	0.10	165	0	0.10
301–400	175	0	0.12	175	0	0.12	175	0	0.12
401–500	180	0	0.13	180	0	0.13	180	0	0.13
501–600	190	0	0.13	190	0	0.13	190	0	0.13
601–800	200	0	0.14	200	0	0.14	200	0	0.14
800–1000	205	0	0.14	205	0	0.14	205	0	0.14
1001–2000	220	0	0.15	220	0	0.15	220	0	0.15
2001–3000	220	0	0.15	375	1	0.20	505	2	0.23
3001–4000	225	0	0.15	380	1	0.20	510	2	0.24
4001–5000	225	0	0.16	380	1	0.20	520	2	0.24
5001–7000	230	0	0.15	385	1	0.21	655	3	0.27
7001–10,000	230	0	0.16	520	2	0.25	660	3	0.28
10,001–20,000	390	1	0.21	525	2	0.26	785	4	0.31
20,001–50,000	390	1	0.21	530	2	0.26	920	5	0.34
50,001–100,000	390	1	0.21	670	3	0.29	1040	6	0.36

Source. Reprinted from H. F. Dodge and H. G. Romig. *Sampling Inspection Tables*, Wiley, New York, 1944.

[a]n, size of sample (entry of "all" indicates that each piece in lot is to be inspected); c, allowable defect number for sample; AOQL, average outgoing quality limit.

Bibliography

Cowden, D. J. (1957). *Statistical Methods in Quality Control*. Prentice-Hall, Englewood Cliffs, N.J.

Dodge, H. F. and Romig, H. G. (1944). *Sampling Inspection Tables*. Wiley, New York.

(ACCEPTANCE SAMPLING
AVERAGE OUTGOING QUALITY
  LIMIT
CONSUMER'S RISK
QUALITY CONTROL
SAMPLING)

DOMAIN OF ATTRACTION

Let $X_1, X_2, \ldots$ be independent identically distributed random variables. Let X be a random variable. Then if there exist constants $b(n)$ and $g(n)$ such that

$$\frac{\sum_{i=1}^{n} X_i}{b(n)} - g(n) \to X \qquad (1)$$

in law, then X_1 is said to be attracted to X. Let X_1 and X have distribution functions F and G, respectively. Alternatively, then, we may say that F is attracted to G. The totality of all random variables (alternatively distribution functions) attracted to X is said to be the domain of attraction of X (alternatively, G).

Introduction of the concept of stable laws* allows us to characterize those distributions that have nonempty domains of attraction: A random variable X is said to be stable* if for each n there are independent random variables $X_1, X_2, \ldots, X_n$ with common distribution that of X, centering constants $e(n)$, and scaling constants $a(n) > 0$ such that

$$\frac{\sum_{i=1}^{n} X_i}{a(n)} - e(n) \qquad (2)$$

has the distribution of X. The basic result about stable random variables is that a random variable X has a nonempty domain of attraction if and only if X is stable. The stable distributions form essentially a two-parameter family (see ref. 4, p. 164) and include the familiar normal* and Cauchy* distributions.

We know of course by the central limit theorem* that any random variable X_1 with finite variance is in the domain of attraction to the normal law. Moreover, referring to (1), we may choose $b(n) = n^{1/2}$. Supposing normalization by $b(n) = n^{1/2}$ in (1), one can ask which random variables are attracted to a normal random variable. It turns out that only those random variables with finite variance are in the domain of attraction of a normal random variable. However, if more general choices of $b(n)$ are permitted, the class of random variables attracted to the normal is wider: X_1 belongs to the domain of attraction to a (nonzero variance) normal distribution if and only if, as $x \to \infty$,

$$\frac{x^2 \Pr[|X_1| > x]}{\int_{|y| < x} y^2 \, dF_{X_1}(y)} \to 0. \qquad (3)$$

Referring to the definition of a stable random variable given by (2), it turns out that $a(n)$ may only be of the form $a(n) = cn^\alpha$ for $0 < \alpha \leq 2$, $c > 0$. Here α is said to be the index of the stable law. This suggests that $b(n) = n^\alpha$ is an appropriate normalization for a stable random variable of index α in (1). When (1) holds for such a $b(n)$, then X_1 is said to be in the domain of *normal* attraction to a stable law. We have already seen in the case of attraction to a normal distribution that a random variable X_1 can be in the domain of attraction to a random variable X

without being in the domain of normal attraction to X.

For the case $0 < \alpha < 2$ a random variable X_1 is in the domain of normal attraction to X if and only if, as $x \to \infty$,

$$x^\alpha \Pr[X_1 > x] \qquad \text{converges} \qquad (4)$$

and

$$x^\alpha \Pr[X_1 < -x] \qquad \text{converges} \qquad (5)$$

with at least one of these limits being non-zero. More generally, for the case $0 < \alpha < 2$, a random variable X_1 is in the domain of attraction to X if and only if

$$\frac{\Pr[X_1 < -x]}{\Pr[X_1 > x]} \qquad \text{converges} \qquad (6)$$

(possibly to ∞) as $x \to \infty$ and

$$x^\alpha \Pr[|X_1| > x] = L(x), \qquad (7)$$

where $L(x) > 0$ for all x and $L(x)$ is slowly varying $[L(tx)/L(x) \to 1$ as $x \to \infty$ for each $t > 0]$. This last result is the fundamental result concerning the topic "domain of attraction." Neither of the results above, as we have seen, applies when $\alpha = 2$.

Some really amazing results hold if one considers the topic "domain of partial attraction." We say that X_1 is in the domain of partial attraction to X if (1) holds along some subsequence of the integers. It turns out that every infinitely divisible distribution (a random variable X has an infinitely divisible distribution* if there is a sum of n independent identically distributed random variables with the distribution of X for each n) has a nonempty domain of partial attraction. Further, the domain of partial attraction of a particular stable distribution is wider than its domain of attraction. Third, remarkably, there exist distribution functions belonging simultaneously to the domain of partial attraction to every infinitely divisible law. There are random variables X that belong to no domain of partial attraction. For example,

$$\lim_{t \to \infty} \lim_{x \to \infty} \frac{\Pr[X > tx]}{\Pr[X > x]} \neq 0 \qquad (8)$$

is a sufficient condition for this.

Techniques of proof of results concerning domains of attraction are based on characteristic functions and thus are analytic in character. However, recently in ref. 6, a probabilistic proof of the sufficiency of the conditions for being in the domain of attraction to a stable has been given.

The theory of domains of attraction has important implications for applied work: Recall that a random phenomenon is often modeled by a normal random variable, the justification being that the phenomenon results from the summation of many independent and roughly identically distributed random quantities. However, this argument is invalid if we are unwilling to assume that the summed random quantities have finite variances. Because of the theory of domains of attraction, one is forced in the infinite variances case to consider that a random phenomenon may be modeled by a nonnormal stable random variable. Further, if one is willing to assume as a reasonable approximation that the summed random quantities are identically distributed, then the only allowable distributions for a random phenomenon resulting from summation are the stable random variables. There is a vigorous literature on the suitability of modeling random phenomena by stable random variables in economics. The pioneering article is that of Mandelbrot [5]. A very readable article written in the spirit of the comments above is that of Fama [1]. For a specific application to the stock market, see ref. 2.

Although written at a high mathematical level, Chapter 7 of ref. 4 is a very readable and complete reference to the subject of domains of attractions. Reference 3 has a more terse style but is also an excellent reference. The three applied articles [1, 2, 5] are all very readable.

References

[1] Fama, E. F. (1963). *J. Bus. Univ. Chicago*, **36**, 420–429.

[2] Fama, E. F. (1965). *J. Bus. Univ. Chicago*, **38**, 34–105.

[3] Feller, W. (1971). *An Introduction to Probability Theory and Its Applications*, Vol. 2. Wiley, New York.

[4] Gnedenko, B. V. and Kolmogorov, A. N. (1954). *Limit Theorems for Sums of Independent Random Variables*. Addison-Wesley, Reading, Mass.

[5] Mandelbrot, B. (1963). *J. Bus. Univ. Chicago*, **36**, 394–419.

[6] Simons, G. and Stout, W. (1978). *Ann. Prob.*, **6**, 294–315.

(STABLE DISTRIBUTIONS)

W. Stout

DONSKER'S THEOREM   *See* LIMIT THEOREMS, CENTRAL

DOOB'S INEQUALITY   *See* MARTINGALES

DOOLITTLE METHOD (TECHNIQUE)

The forward Doolittle technique is a numerical method for solving the system of normal equations* associated with estimating parameters in a multiple linear regression*. Hald [3] presents a detailed description of the method and a computer program for this method has been recently published by Nelson [4]. The Doolittle method is also used for inverting a matrix. A shortcut method for inverting a symmetric matrix is known as the *abbreviated* Doolittle method and is described by Dwyer [1, 2].

As a consequence of the increased use of electronic computers, the Doolittle method is now used only rarely.

References

[1] Dwyer, P. S. (1941). *Psychometrika*, **6**, 101–129.

[2] Dwyer, P. S. (1951). *Linear Computations*. Wiley, New York.

[3] Hald, A. (1952). *Statistical Theory with Engineering Applications*. Wiley, New York.

[4] Nelson, P. R. (1974). *J. Quality Tech.*, **6**, 160–162.

D-OPTIMALITY   *See* OPTIMAL DESIGNS

DOSAGE–RESPONSE CURVE

The regression* of response on stimulus may be represented graphically as a curve. When the stimulus is in the form of a "dose" (e.g., of a drug, or possibly of an applied force or some other source), this may be called a "dose–response curve." The term "dosage" rather than "dose" is used to allow for the possibility that a dose metameter* may be used rather than the directly measured dose.

In most applications the response is binary* (0, 1)—in particular death or survival, or more generally failure or nonfailure. The dosage–response curve then is obtained by plotting the probability of obtaining a nonzero response against the dosage (dose metameter).

(BINARY DATA
BIOASSAY, STATISTICAL METHODS IN
LOGIT
NORMIT
PROBIT
QUANTAL RESPONSE
RANKIT
TOLERANCE DISTRIBUTION)

DOSE METAMETER

A function of the directly observable dose, used in dosage–response* curves for mathematical convenience. By far the most common form of dosage metameter is the logarithm of the dose, but other functions are sometimes encountered.

DOT DIAGRAM

When data consist of a small set of numbers (usually fewer than 20 to 25), they can be represented conveniently by drawing a line with a scale covering the range of values of the measurements and plotting the individual measurements above this line as prominent dots. This forms a dot diagram.

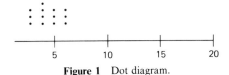

Figure 1 Dot diagram.

Example.   Given the data

6, 3, 5, 6, 4, 3,   5,
4, 6, 3, 4, 5, 4, 18,

the corresponding dot diagram is as shown in Fig. 1. This type of graphical representation is especially useful in drawing attention to outlying observations (as, e.g., 18 in the set above).

(GRAPHICAL REPRESENTATION OF DATA
HISTOGRAMS
OUTLIERS)

DOUBLE SAMPLING

The technique of double sampling is used mainly in industrial sampling inspection*, where it is necessary to decide on acceptance or rejection of a lot of articles on the basis of tests on one or more samples. In the double sampling method, a first sample of a given size is taken at random from the lot and inspected. If the results are good enough, the lot is accepted without further inspection. If the results are bad enough, the lot is rejected without further inspection. If, however, the results are intermediate, a second sample is taken at random and inspected. The final decision is then taken using the combined results of the first and second samples.

When inspection is by attributes*, the decisions are taken on the basis of the number of defectives* found in each sample. When inspection is by variables*, the decisions are taken by checking the distances of the sample (arithmetic) mean* from upper and lower limits, in units of the sample standard deviation*. Apart from this difference the methods are the same in principle, and only inspection by attributes will be dealt with in detail here.

Tables of double sampling plans for variables inspection are to be found in Bowker and Goode ([2]; a useful textbook, but beware of its use of an outdated definition of acceptable quality level*), and in Bravo and Wetherill [3]. The latter reference contains a useful discussion of the difficulties of matching plans for attributes and for variables with particular reference to the double sampling case.

INSPECTION BY ATTRIBUTES

In the sort of plan in common use there are five adjustable constants: the first sample size* (n_1), the first acceptance number* (c_1), the first rejection number* (r_1), the second sample size (n_2), and the second acceptance number (c_2). The rules are:

1. Inspect a random sample* of size n_1.
2. If c_1 defectives or fewer were found, accept the lot, and finish.
3. If r_1 defectives or more were found, reject the lot, and finish.
4. If more than c_1 but fewer than r_1 defectives were found, inspect a further random sample of size n_2.
5. If the total number of defectives in the two samples combined was c_2 or fewer, accept the lot; otherwise, reject the lot. It will be noted that the second rejection number (r_2) is not a separate constant, always being $c_2 + 1$.

The official definition, as given by the Standards Committee of the American Society for Quality Control [9], is:

Double Sampling: Sampling inspection in which the inspection of the first sample of size n_1 leads to a decision to accept a lot; not to accept it; or to take a second sample of size n_2, and the inspection of the second sample then leads to a decision to accept or not to accept the lot.

This definition might perhaps be criticized on the grounds that it does not make clear that decision, after inspecting a second sample, should be based on *all* the results and not on the second sample results alone.

PLANS IN PUBLISHED TABLES

Because five constants are rather a large number to handle, and were particularly so before the days of electronic computation, somewhat arbitrary restrictions have often been put on the plans in published double sampling tables. Thus Dodge and Romig* (1944 [6], 1959 [7]; the 1944 edition is the pioneering book in this field, although articles in journals had appeared earlier) always took $r_1 = r_2$, and also aimed to make the probability of acceptance on the first sample take a value of about 0.06, at the quality for which the entire procedure was to give a probability of 0.10. The first of these rules leads to some very peculiar results; for example, in the plan $n_1 = 200$, $c_1 = 2$, $r_1 = 16$, $n_2 = 555$, $c_2 = 15$, the rules require that if 15 defectives are found in the first sample of 200, a second sample of 555 must be inspected to see whether 1 defective (or more) is found. (It is worth noting that if there are 15 defectives altogether in 755, the probability that they all occur in the first 200 is as low as 1.5×10^{-9}). Dodge and Romig's reply to this criticism would, presumably, have been that their tables were designed for rectifying inspection*, in which rejection of a lot calls for 100% inspection* of the entire lot, and that therefore no extra inspection was called for by not rejecting on the first sample. Although this is true, it remains a fact that to draw a random sample of 555 items is in itself a substantial task, all of which is wasted when 100% inspection of the remainder of the lot could have started at once without the necessity of further sampling.

Dodge and Romig's description of what is being done also seems peculiar now (although it should be noted that this in no way devalues the importance of their pioneering work). They say that "a lot is given a second chance of acceptance if the first sample results are unfavourable" and refer to the probability of acceptance after the second sample "if the first fails." Nowadays it is usual to regard the first sample as failing only if immediate rejection is called for, and to say that a second sample is required if the result of the first is inconclusive (or some such wording).

The later Military Standard 105A tables [4] retained the $r_1 = r_2$ rule, and additionally introduced the further arbitrary rules $n_2 = 2n_1$ and $n_1 = 2n/3$, where n was the sample size of an equivalent single sampling plan—equivalent plans* being taken as those that give approximately the same operating characteristic* (or power*) curve. The last rule was necessary if equivalents were to be found to single sampling plans with acceptance number 1, rejection number 2, because even $c_1 = 0$ will lead to too high an OC curve at the lower end otherwise. (The OC curves for sample size n, acceptance number 1 and sample size m, acceptance number 0 cross at a probability higher than 0.05 unless $m > 0.6315n$, assuming the Poisson* formula for the probabilities.) Together with the $n_2 = 2n_1$ rule, however, it led to some peculiar results:

1. The second sample, if needed, was greater than the single sample that could have replaced both first and second samples.

2. In most parts of the table, an observed proportion defective* that would lead to acceptance on the first sample could lead to rejection on the combined samples. For example, $n_1 = 75$, $c_1 = 5$, $r_1 = 12$, $n_2 = 150$, $c_2 = 11$ would lead to acceptance if four or five defectives were observed in a sample of 75, but to rejection if three times as many (12 or 15 defectives) were observed in a total sample of three times the size (225).

3. If the quality were not particularly good or particularly bad, the average sample size (usually, and regrettably, called average sample number or ASN*) for double sampling was considerably greater than the equivalent single-sample size.

Indeed, some quality control* textbooks of the era took this to be a general rule, and warned that double sampling was not an efficient method for intermediate quality, whereas in fact it was merely that these particular tables had chosen to use some arbitrary rules that necessarily resulted in inefficiency.

Hamaker and van Strik [8] pointed out the inconsistency and inefficiency of these tables, and recommended keeping the $n_2 = 2n_1$ rule, but using $n_1 = 2n/5$ instead of $n_1 = 2n/3$. They showed that much more satisfactory tables could be produced in this way, but unfortunately their recommendation for an equivalent plan to a single sample with acceptance number 1 was unsatisfactory. As noted above, a good match is not possible with so small a first sample.

The current Military Standard 105D [5] sought a way out of the difficulties by taking $n_2 = n_1$ instead of $n_2 = 2n_1$, and $n_1 = 0.631n$ approximately. This last choice resulted from the fact that the sample sizes in these tables are planned to be in an approximate geometric progression with common ratio $\sqrt[5]{10}$. The reciprocal of this value is 0.631, and thus the same set of sample sizes could be used for double sampling as for single sampling. Within this framework, efficient plans were found, in the sense that the ASN for a double plan is nearly always less than the equivalent single-sample size.

ADVANTAGES AND DISADVANTAGES

Compared with single sampling, double sampling has the advantage of giving a smaller average sample size for a given degree of discrimination between qualities. However, it has the disadvantages of being more complicated to operate, and of giving variability in the amount of inspection that needs to be done, which may lead to difficulties in such matters as estimating the amount of labor needed. Whether the advantages outweigh the disadvantages, or vice versa, will depend upon the circumstances of each particular case.

EXTENSIONS

The extension to multiple sampling*, in which a decision is not necessarily taken after the second sample, but third, fourth, etc., samples may be called for, is self-evident.

Sequential sampling* may be regarded as a particular case of multiple sampling in which successive samples are of size one unit only.

The advantages and disadvantages of these extensions, as compared with double sampling, are precisely similar to those of double sampling as compared with single sampling.

OPERATING CHARACTERISTIC (OC) FUNCTION

Let p be the independent probability* that each item in the lot is defective, and let $q = 1 - p$. Let each acceptance condition be denoted by a sample size m_i and number of defectives in the sample d_i. Then the probability of acceptance at that particular acceptance condition is

$$\text{constant} \times p^{d_i} q^{m_i - d_i}$$

where the constant is the number of ways in which that acceptance condition can be reached. The overall probability of acceptance (P) is the sum of all such probabilities.

For example, consider the plan $n_1 = 6$, $c_1 = 1$, $r_1 = 4$, $n_2 = 4$, $c_2 = 4$. There are five possible acceptance conditions, as follows:

i	m_i	d_i
1	6	0
2	6	1
3	10	2
4	10	3
5	10	4

(The conditions 10, 0 and 10, 1 are impossi-

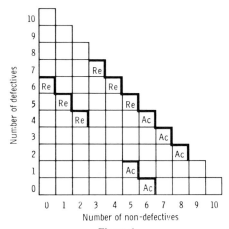

Figure 1

ble, as they would have led to acceptance on the first sample.) We therefore have

$$P = Aq^6 + Bpq^5 + Cp^2q^8 + Dp^3q^7 + Ep^4q^6,$$

where the constants A, B, C, D, and E have to be determined. The easiest way to determine them is from a diagram such as Fig. 1. It is clear that the number of ways of reaching any square can be obtained as the sum of the two numbers in (1) the square below, and (2) the square to the left, except when the square below or to the left has already led to acceptance or rejection. (For further information on this "lattice diagram"* approach, see, e.g., Section 3 of Barnard [1].)

The numbers for this particular case are shown in Fig. 2, and lead to

$$P = q^6 + 6pq^5 + 15p^2q^8 + 80p^3q^7 + 170p^4q^6.$$

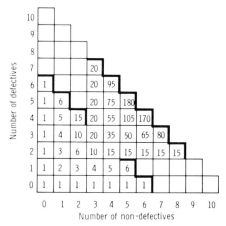

Figure 2

The process of counting the number of ways, for larger and more realistic examples, would be tedious, and likely to lead to error, if done by hand but is easy to program for a computer. In such a program it is well to use "real" rather than "integer" representation for the count, as the numbers can become large enough to lead to integer overflow on some machines.

AVERAGE-SAMPLE-SIZE (ASN) FUNCTION

The ASN function can be calculated in a very similar way, but it is necessary to take account of the individual probabilities of rejection as well as those of acceptance, and to multiply each probability by the appropriate sample size. Thus, using Fig. 2 again, a precisely similar argument leads to

$$\text{ASN} = 6(q^6 + 6pq^5 + 15p^4q^2 + 6p^5q + p^6)$$
$$+ 10(15p^2q^8 + 80p^3q^7 + 170p^4q^6$$
$$+ 180p^5q^5 + 95p^6q^4 + 20p^7q^3).$$

References

[1] Barnard, G. A. (1954). *J. R. Statist. Soc. B*, **16**, 151–174.

[2] Bowker, A. H. and Goode, H. P. (1952). *Sampling Inspection by Variables*. McGraw-Hill, New York.

[3] Bravo, P. C. and Wetherill, G. B. (1980). *J. R. Statist. Soc. A*, **143**, 49–67.

[4] Department of Defense (1950). *Military Standard 105A: Sampling Procedures and Tables for Inspection by Attributes*. U.S. Government Printing Office, Washington, D.C.

[5] Department of Defense (1963). *Military Standard 105D: Sampling Procedures and Tables for Inspection by Attributes*. U.S. Government Printing Office, Washington, D.C.

[6] Dodge, H. F. and Romig, H. G. (1944). *Sampling Inspection Tables*. Wiley, New York.

[7] Dodge, H. F. and Romig, H. G. (1959). *Sampling Inspection Tables*, 2nd ed. Wiley, New York.

[8] Hamaker, H. C. and van Strik, R. (1955). *J. Amer. Statist. Ass.*, **50**, 830–849.

[9] Standards Committee of ASQC (1979). Terms, Symbols and Definitions. *ASQC Standard*, American Society for Quality Control, Milwaukee, Wis.

(ACCEPTABLE QUALITY LEVEL
ACCEPTANCE NUMBER
ACCEPTANCE SAMPLING
AVERAGE SAMPLE NUMBER
INDEPENDENCE
MULTIPLE SAMPLING
OPERATING CHARACTERISTIC
POISSON DISTRIBUTION
POWER
QUALITY CONTROL, STATISTICAL
RANDOM SAMPLING
RECTIFYING INSPECTION
REJECTION NUMBER
SAMPLING INSPECTION
SEQUENTIAL PROCEDURES)

I. D. HILL

DOUBLE-TAILED TEST

A significance test* in which the critical region* is the union* of two disjoint sets of values $T_{(-)} < T_-$ and $T_{(-)} > T_+$ of values of the test statistic T. The name "double-tailed" comes from the fact that it is the "tails" of the distribution of T that provide significant results for the test.

Decision to use a double-tailed test is closely associated with the classes of alternative hypotheses* deemed to be relevant.

Synonymous terms are *two-tailed*, *double (or two)-sided*, and *bilateral*.

(HYPOTHESIS TESTING
POWER)

DOUBLY NONCENTRAL DISTRIBUTIONS

Distributions may depend on two or more noncentrality parameters*. The most common are the doubly noncentral F distribution (which is the distribution of the ratio of independent variables, each distributed as noncentral χ^2's, divided by their degrees of freedom) and the doubly noncentral t. Formally, a noncentral F with ν_1, ν_2 degrees of freedom and noncentrality parameters λ_1, λ_2

is the distribution of

$$\left[\chi_{\nu_1}'^2(\lambda_1)/\nu_1 \right] \left[\chi_{\nu_2}'^2(\lambda_2)/\nu_2 \right]^{-1}.$$

Similarly, doubly noncentral t is the distribution of $(U + \delta)[\chi_\nu'(\lambda)/\sqrt{\nu}]$ with U unit normal, and independent of $\chi'\sqrt{(\cdot)}$.

The doubly noncentral beta is related to the doubly noncentral F in the same way as the (central) beta to the (central) F. It is the distribution of $\chi_{\nu_1}^2(\lambda_1)[\chi_{\nu_1}^2(\lambda_1) + \chi_{\nu_2}^2(\lambda_2)]^{-1}$.

Bibliography

Bulgren, W. G. (1971). *J. Amer. Statist. Ass.*, **66**, 184–186. (Doubly noncentral *F*.)

Bulgren, W. G. and Amos, D. E. (1968). *J. Amer. Statist. Ass.*, **63**, 1013–1019.

Johnson, N. L. and Kotz, S. (1970). *Continuous Univariate Distributions*, Vol. 2. Wiley, New York, pp. 197–198. (Doubly noncentral beta.)

DUAL SCALING *See* CORRESPONDENCE ANALYSIS

DUNCAN'S MULTIPLE RANGE TEST

A test proposed by D. B. Duncan in 1953 for studying details of differences among sample means, usually applied after an analysis of variance* F-test* gives a significant result.

The procedure is as follows: All k (say) sample means are arranged in increasing order of magnitude; the differences between adjacent sample means are calculated. Next, the existence of significant variation within each of the two groups of $k - 1$ adjacent ordered means is checked. If a set does not give a significant result, it is concluded that the variability of means within that group of $k - 1$ is random and no further testing for differences within that group of $k - 1$ means is conducted. To indicate this, the means in the groups are underlined by a common line. If there is significance, two sets of $(k - 2)$ adjacent ordered means in the block are examined, and so on.

At each stage, the test consists of comparing the range of the group of g (say) adjacent means under study with a critical limit

$$R_g = C(g, \nu, \alpha) \left[\frac{\text{(residual mean square)}}{\text{(number of observations in group)}} \right]^{1/2}$$

The constants $C(g, \nu, \alpha)$ depend on the number of degrees of freedom* (ν) in the residual mean square* in the complete analysis of variance, and also on the (nominal) significance level* (α) being used. These constants are, in fact, the upper $100[1 - (1 - \alpha)^{g-1}]\%$ points of the appropriate studentized range* distribution. Reasons for choosing these critical values are given in Duncan [1]. Values of $C(g, \nu, \alpha)$ are available in Harter [2] for $\alpha = 0.001, 0.005, 0.01, 0.05, 0.10$. These strictly apply only if the numbers of observations are the same in each group; slight differences among these numbers are, however, of little practical importance.

If the range of the g adjacent ordered means is greater than R_g, the test is regarded as giving a significant result.

Klugh [3] has suggested that the test should be applied in a conservative manner —that is, that a low value of α be used. (Harter's [2] tables do give $\alpha = 0.001$ and 0.005, which should be as low as is needed.)

Example. Suppose that we have five groups $A-E$, with sample values as shown:

A	B	C	D	E
7	14	9	22	24
9	15	8	21	27
13	19	11	23	30
17	19	9	27	15
23	24	16	14	18
14	22	15	19	22
15	21	16	26	17

The ANOVA* table is

Source	d.f.	Sum of Squares	Mean Square
Between groups	4	571.54	142.88
Residual (within groups)	30	619.20	20.64

The F-ratio is $142.88/20.64 = 6.92$, which is significant at the 1% level ($F_{4,30,0.99} = 4.02$); *see* F-TESTS.

Ordering the group means, we have

Group	C	A	B	D	E
Sample Mean	12	14	$19\frac{1}{7}$	$21\frac{5}{7}$	$21\frac{6}{7}$

For this pattern of data we have $k = 5$ groups, 7 observations per group, ν (degrees of freedom of residual mean square) $= 5(7 - 1) = 30$, so

$$R_g = C(g, 30, \alpha)\sqrt{(20.64/7)}$$
$$= 1.72C(g, 30, \alpha).$$

From Harter [2] with $g = 4$, $C(4, 30, 0.01) = 4.168$, whence

$$R_4 = 1.72 \times 4.168 = 7.17.$$

For block $CABD$, we have range $= 21\frac{5}{7} - 12 = 9\frac{5}{7} > 7.17$. For block $ABDE$, we have range $= 21\frac{6}{7} - 14 = 7\frac{6}{7} > 7.17$.

Both tests give significant results, so we proceed to consider the three-component blocks:

CAB: range $= 7\frac{1}{7}$ ⎫ $C(3, 30, 0.01) = 4.056$,
ABD: range $= 7\frac{5}{7}$ ⎬ $R_3 = 1.72 \times 4.056$
BDE: range $= 2\frac{5}{7}$ ⎭ $\quad = 6.98$

Since there is a nonsignificant result for BDE, we draw a line under the numbers of this block and need only test the one two-component blocks CA and AB.

CA: range $= 2$ ⎫ $C(2, 30, 0.01) = 3.889$,
AB: range $= 5\frac{1}{7}$ ⎭ $R_3 = 6.69$.

Neither of these is significant, so we draw lines under CA and AB (but *not* under the block CAB). [If a 5% level had been used, we would have had $R_2 = 1.72 \times 2.888 = 4.97$ and the range $AB(= 5\frac{1}{7})$ would have been significant.]

There are a number of competing procedures—some are included in the other entries listed below. It is not easy to assess relative merits—indeed, it is difficult even to calculate the expected properties of even a single procedure, and, further there is a wide variety of possible situations [actual differ- ences among population (true) means of the different groups] to be considered.

References

[1] Duncan, D. B. (1955). *Biometrics*, **11**, 1–42. (Gives details of the test procedure, and explains the reasons for using modified significance levels.)
[2] Harter, H. L. (1960). *Biometrics*, **16**, 671–685. (Tables of critical values.)
[3] Klugh, H. E. (1970). *Statistics: The Essentials for Research*. Wiley, New York. (An elementary discussion of the test, with examples.)

(*F*-TESTS
MULTIPLE COMPARISONS
NEWMAN–KEULS PROCEDURE
SCHEFFÉ'S SIMULTANEOUS
    COMPARISON PROCEDURE
STUDENTIZED RANGE
TUKEY'S SIMULTANEOUS
    COMPARISON PROCEDURE)

DURATION OF PLAY See GAMBLING, STATISTICS IN

DURBIN'S DISTRIBUTION-FREE RANK TEST

Durbin's distribution-free rank test is a test for differences among treatments in a BIBD (*see* BLOCKS, BALANCED INCOMPLETE). The assumptions are that the blocks are mutually independent of each other; the underlying distribution functions are continuous, so that no ties occur. Denote by t the total number of treatments, b the total number of blocks, k the number of experimental units per block, r the number of times each treatment appears, and λ the number of blocks in which the rth treatment and the jth treatment appear together and is the same for all pairs (r, j) of treatments. (*See* FISHER'S INEQUALITY.)

The test can be described as follows. Let x_{ij} be the observation corresponding to the jth treatment in the ith block. Rank the observations x_{ij} within block i in increasing order and denote the rank of x_{ij} by R_{ij} (whenever such x_{ij} exists). We test H_0: the rankings are chosen at random from the collection of all the permutations of numbers $1, 2, \ldots, k$. The alternative hypothesis, H_1, is that at least one treatment tends to yield larger observed value than at least one other treatment.

Compute the sum of ranks corresponding to the jth treatment and denote this sum by R_j ($R_j = \sum_i R_{ij}$, where the summation is over all i for which R_{ij} exists). Durbin's test statistic [2] is defined as

$$D = \frac{12(t-1)}{rt(k-1)(k+1)} \sum_{j=1}^{t} \left\{ R_j - \frac{r(k+1)}{2} \right\}^2$$

$$= c_1 S - c_2,$$

where the positive constants c_1 and c_2 are

$$c_1 = \frac{12(t-1)}{rt(k-1)(k+1)},$$

$$c_2 = 3 \frac{r(t-1)(k+1)}{k-1},$$

and $S = \sum_{j=1}^{t} R_j^2$.

H_0 is rejected at significance level α if the test statistics are larger than or equal to $D_{1-\alpha}$, where $D_{1-\alpha}$ is the smallest value satisfying $\Pr[D \geqslant D_{1-\alpha} \mid H_0] \leqslant \alpha$.

The exact distribution of D under H_0 was calculated by van der Laan and Prakken [3] for $t \leqslant 7$, $b \leqslant 12$ and the case $t = 3$ and $b = 15$ (excluding four exceptional cases that require large amounts of computer time). Tables of the exact probabilities related to $\alpha = 0.10$, 0.05, 0.025, 0.01 for $t \leqslant 7$, $\alpha \leqslant 5$ (corresponding to $r \leqslant 10$) and the approximations used in practice are also given by these authors. If the number of repetitions r is large, one approximates the distribution of D by $\chi^2_{(t-1)}$. Another approximation suggested by Durbin [2] is to use the $F_{2p,2q}$-distribution for

$$W = \left\{ \frac{\lambda(t+1)}{k+1} - 1 \right\} \frac{W^*}{1 - W^*},$$

where

$$p = \frac{b\left\{ 1 - \dfrac{k+1}{t+1} \right\}}{2\left(\dfrac{b}{t-1} - \dfrac{1}{k-1} \right)} - \frac{k+1}{\lambda(t+1)},$$

$$q = \left\{ \frac{(t+1)}{k+1} - 1 \right\} p,$$

and

$$W^* = \frac{k+1}{r(t+1)(k-1)} D.$$

Often a continuity correction* for this approximation is used, by taking for the value of S the average of the (current) value S and its preceding (smaller) value. (No correction is used for the smallest value of S.)

Investigations of van der Laan and Prakken [3] show that the F-approximation with continuity correction is far superior to the χ^2-approximations (for the cases indicated above). (The χ^2-approximation tends to overestimate the probabilities substantially, whereas the F-approximation without continuity correction underestimates them.)

For more information on this topic, see Conover [1] and Walsh [4].

References

[1] Conover, J. (1971). *Practical Non-parametric Statistics*. Wiley, New York.

[2] Durbin, J. (1951). *Brit. J. Psychol.* **4**, 85–90.

[3] van der Laan, P. and Prakken, J. (1972). *Statist. Neerlandica*, **26**, 1554–1644.

[4] Walsh, J. E. (1968). *Handbook of Non-parametric Statistics*, Vol. 3. D. Van Nostrand, Princeton, N.J.

(BLOCKS, BALANCED INCOMPLETE FISHER INEQUALITY)

DURBIN–WATSON TEST

Durbin and Watson [2] introduced a statistic, d, to test for serial correlation* of the error term in the linear model using residuals calculated from an ordinary least-squares* fit. The problem frequently arises in time-series* analysis when one is fitting a linear model and one suspects that the error term is not independent and follows a first-order autoregressive model*.

In more detail suppose that we are regressing y on k independent variables x_1, $x_2, \ldots, x_k$. The model for a sample of n observations, in the usual matrix form, is given by

$$\mathbf{y} = \mathbf{X}\boldsymbol{\beta} + \boldsymbol{\epsilon}.$$

The residuals using the ordinary least-squares* estimate, $\mathbf{b} = (\mathbf{X}'\mathbf{X})^{-1}\mathbf{X}'\mathbf{y}$ of $\boldsymbol{\beta}$ are given by

$$\mathbf{z} = \mathbf{y} - \mathbf{X}\mathbf{b} = \left\{\mathbf{I} - \mathbf{X}(\mathbf{X}'\mathbf{X})^{-1}\mathbf{X}'\right\}\mathbf{y}.$$

The d statistic is defined by

$$d = \frac{\sum_{i=2}^{n}(z_i - z_{i-1})^2}{\sum_{i=1}^{n}z^2_i}.$$

It is algebraically equivalent to von Neumann's statistic.

The statistic d is introduced to test the null hypothesis of independence for the ϵ_i's against the alternative that they follow the Markov (first-order autoregressive) process*

$$\epsilon_i = \rho\epsilon_{i-1} + u_i,$$

where $|\rho| < 1$ and the u_i are normal with mean zero variance σ^2 and independent. In terms of ρ the null hypothesis is that $\rho = 0$ and the alternative is $\rho \neq 0$.

Durbin and Watson [2, Part I] show that d is bounded by two variables d_L and d_u so that $d_L \leqslant d \leqslant d_u$. The variables d_L and d_u depend only on the number of regressor variables k and the sample size n and not on the regressor matrix X. The bounds are the best in the sense that d_L and d_u can both be attained for particular choices of X. On the null hypothesis of independence ($\rho = 0$), the distributions of d_L and d_u are the same as those of

$$\left\{\frac{\left(\sum_{i=1}^{n-k}a_iV_i^2\right)}{\left(\sum_{i=1}^{n-k}V_i^2\right)}\right\}, \quad \left\{\frac{\left(\sum_{i=1}^{n-k}b_iV_i^2\right)}{\left(\sum_{i=1}^{n-k}V_i^2\right)}\right\},$$

respectively, where the V_i are independent standard normal random variables, the a_i, b_i satisfy

$$\left.\begin{array}{l} a_1 \leqslant a_2 \leqslant \cdots \leqslant a_{n-k}, \\ b_1 \leqslant b_2 \leqslant \cdots \leqslant b_{n-k}, \end{array}\right\} \quad a_i \leqslant b_i.$$

Durbin and Watson [2, Part II] give approximate percentage points of d_L and d_u for the number of independent variables, k, less than or equal to 5. These approximations, in a later paper [2, Part III], are found to be fairly accurate. For testing H_0: independence ($\rho = 0$) against H_1: positive serial correlation ($\rho > 0$), the null hypothesis is rejected for significantly small values of d. An

approximate test can be carried out as follows. Let $d_{L,\alpha}$ and $d_{u,\alpha}$ be the lower α 100% points of the statistics d_L and d_u on the null hypothesis. If $d < d_{L,\alpha}$, reject H_0 at the α level; if $d > d_{u,\alpha}$, accept H_0 at the α level; if $d_{L,\alpha} < d < d_{u,\alpha}$, the test is inconclusive. This procedure is known as the "bounds test." For tests of independence against negative serial correlation ($\rho < 0$), the procedure is repeated as above, but $(4 - d)$ is considered in place of d.

Durbin and Watson [2, Part III] consider the exact distribution of d and approximations to d. These results are needed especially when the results of the bounds test are inconclusive. They suggest a reasonable approximation (the "$a + bd_u$ approximation") to the distribution of d, which uses the distribution d_u and the exact mean and variance of d, for which they give formulas. Savin and White [8] extend the tables of Durbin and Watson for $d_{L,\alpha}$, $d_{u,\alpha}$ with $\alpha = 0.1, 0.05$, $6 \leqslant n \leqslant 200$, and $k \leqslant 20$. Computer programs (FORTRAN) can be found in Koerts and Abrahamse [5] to compute the exact distribution of d.

To illustrate the use of the Durbin–Watson statistic, we consider the following example. The amounts of wheat (in bushels per acre) produced in a given country in 15 successive years are given by

14.3, 16.5, 15.7, 15.4, 19.1, 19.4,
22.1, 23.7, 19.9, 22.2, 24.8,
25.9, 24.5, 26.1, 28.3.

It appears reasonable to fit the linear model*, amount = $a + (b \times \text{time})$. The fitted least-squares* line is amount = $14.64 + 0.94t$, $t = 0, 1, \ldots, 14$. The residuals are -0.34, 0.92, -0.81, -2.05, 0.71, 0.08, 1.84, 2.51, -2.23, -0.87, 0.80, 0.96, -1.37, -0.71, 0.55. The value of d is $(0.92 + 0.34)^2 + \cdots + (0.55 + 0.71)^2$, or 52.34 divided by $(0.34)^2 + \cdots + (0.55)^2$, or 25.99, giving 2.01. To carry out the test of no serial correlation, $H_0 : \rho = 0$ against $H_1 : \rho > 0$, we find the values of $d_{L,0.05}$ and $d_{u,0.05}$ from Durbin and Watson [2, Part II, Table 4] for one independent regressor variable. These are, respectively, 1.08 and 1.36. Since the value of the statistic, 2.01, is greater than 1.36, we accept H_0 at the 5% level.

Durbin and Watson [2, Part II] show that the d-test is the locally most powerful invariant test* in the neighborhood of the null hypothesis. Using simulation, L'Esperance and Taylor [7] compare the d-test with tests based on BLUS* residuals and other statistics based on standardized residuals independent of the regressor matrix **X**: the d-test is found generally to be the most powerful test.

Epps and Epps [3] and Harrison and McCabe [4] consider the robustness of the d-test against heteroscedasticity*; both pairs of authors find the d-test to be generally robust and the former find the power of the d-test to be superior to that of Geary's test.

Further developments of the d-test include the following: the problem of fitting the regression line through the origin [6]; Wallis's [10] modified fourth-order statistic for quarterly data; missing observations [9]; and lagged regressor variables [1].

Bibliography

Durbin, J. (1969). *Biometrika*, **56**, 1–15. (Alternative method to d-test.)

Habibahati, J. and Pratschbe, J. L. (1972). *Rev. Econ. Statist.*, **54**, 179–185.

Henshaw, R. C. (1966). *Econometrica*, **34**, 646–660. (Early review plus beta distribution approximation to d-statistic.)

Richardson, S. M. and White, K. J. (1979). *Econometrica*, **47**, 785–788. (A modified d-test is considered.)

Tillman, J. A. (1975). *Econometrica*, **43**, 959–974.

References

[1]   Durbin, J. (1970). *Econometrica*, **38**, 410–421.

[2]   Durbin, J. and Watson, G. S. (1950). *Biometrika*, **37**, 409–428; *ibid.*, **38**, 159–178 (1951); *ibid.*, **58**, 1–19 (1971). (5% and 1% points of d_L and d_u in **38**.)

[3]   Epps, T. W. and Epps, M. L. (1977). *Econometrica*, **45**, 745–753. (Compares d-test with Geary's test.)

[4]   Harrison, M. J. and McCabe, B. P. M. (1975). *Biometrika*, **62**, 214–216.

[5]   Koerts, J. and Abramhamse, A. P. J. (1969). *On the Theory and Application of the General Linear Model*. Rotterdam University Press, Rotterdam.

[6]   Kramer, G. (1971). *Jb. Nationalökon. Statist.*, **185**, 345–358.

[7]   L'Esperance, W. L. and Taylor, D. (1975). *J. Econometrics*, **3**, 1–21.

[8]   Savin, N. E. and White, K. J. (1977). *Econometrica*, **45**, 1989–1996.

[9]   Savin, N. E. and White, K. S. (1978). *Econometrica*, **46**, 59–67.

[10] Wallis, K. F. (1972). *Econometrica*, **40**, 617–636. (Bounds of 5% significance points for modified d-test for $H_1 : \rho > 0$.)

(AUTOREGRESSIVE—MOVING AVERAGE MODELS TIME SERIES)

A. N. Pettitt

DYNAMIC PROGRAMMING

Dynamic programming can be viewed as the mathematics of sequential decision* making. If, as is assumed here, decisions must be made in the face of uncertainty, the natural mathematical setting is probabilistic. More or less similar theories of such decision problems are known by a number of names, including gambling* theory, Markov decision theory, and stochastic control*. The multitude of names is in part a consequence of the many fields of application, which include statistics, operations research, engineering, and probability theory itself.

DEFINITIONS AND EXAMPLES

Starting with an initial state s_0 in the state space S, you select a first action a_1 from the set $A(s_0)$ of actions available at s_0. The pair (s_0, a_1) determine the conditional distribution $q(\cdot \mid s_0, a_1)$ of the next state s_1 and thereby the conditional distribution of your reward $r(s_0, a_1, s_1)$ for the first day of play. Next, select an action a_2 from $A(s_1)$ to determine the conditional distribution $q(\cdot \mid s_1, a_2)$ of s_2 and of the reward $r(s_1, a_2, s_2)$. You continue in this fashion to select actions and accumulate rewards. The sequence of actions (which may be chosen at random and conditional on what has transpired prior to their selection) is called a *plan* and is denoted by π. Your *return* from the plan π starting from s_0 consists of the expected total reward and is written

$$I(\pi)(s_0) = E\left(\sum_{k=0}^{\infty} \beta^k r(s_k, a_{k+1}, s_{k+1}) \right). \quad (1)$$

Various conditions can be imposed to ensure that the expectation in this definition is well defined. For example, if r is bounded and the *discount factor* β is in the open interval $(0, 1)$, then the sum is bounded. Such problems are called *discounted* [5]. If $\beta = 1$ and $r \geqslant 0$ ($r \leqslant 0$), the problem is *positive* (*negative*) [6, 19]. It is useful to consider also the case when play terminates after n days. The *n-day return* from π starting at s_0 is

$$I_n(\pi)(s_0) = E\left(\sum_{k=0}^{n} r(s_k, a_{k+1}, s_{k+1}) \right).$$

For a rigorous treatment, measurability conditions must be imposed, but I will just assume that all the integrals that occur are meaningful.

To get an idea of the scope of dynamic programming, consider the following three examples taken from optimal stopping*, gambling*, and sequential analysis*, respectively.

Example 1: A Stopping Problem. Suppose that a nonnegative utility function* u is specified on the set $I = \{0, 1, \ldots, n\}$ and that the process $X_0, X_1, X_2, \ldots$ is a simple symmetric random walk* starting from $X_0 = x \in I$ and with absorbing barriers at 0 and n. You may stop the process at any time t and receive $Eu(X_t)$, the expected utility at the time of stopping. The time t may be random, but cannot employ knowledge of future states.

To formulate the problem as a dynamic programming problem, let y be an element not in I to be thought of as a terminal state. Set $S = I \cup \{y\}$ and let $A(s) = \{a, b\}$ for every $s \in S$. Whenever $s \in I$ and action a is selected, motion to the next state is governed by the transition probabilities of the random walk*. If $s \in I$ and action b is selected, the next state is y, with probability 1. Thus action a corresponds to letting the process

continue and b to stopping it. Once at the terminal state y, all future states are also y, regardless of the choice of actions. The reward function r is equal to zero except when a transition is made from a state $s \in I$ to the terminal state y, in which case the reward equals $u(s)$.

There is a well-developed general theory of optimal stopping* problems [7] and such problems can always be viewed as gambling or dynamic programming problems [10].

Example 2: Discrete Red-and-Black. A gambler makes bets at even odds on independent repetitions of a given event such as the appearance of red or black on each spin of a roulette wheel. The gambler's object is to attain a certain amount of cash, say n dollars, and anything less is of no value to him.

This problem is also easily formulated as a positive dynamic programming problem. Take S to be the set $\{0, 1, \ldots, n\}$ and, for each $s \in S$, let $A(s) = \{a_0, \ldots, a_l\}$, where $l = l(s)$ is the minimum of s and $n - s$. The action a_i corresponds to a bet of i dollars. The gambler wins each bet with fixed probability w and loses with probability $\overline{w} = 1 - w$. Thus

$$q(s + i \mid s, a_i) = w = 1 - q(s - i \mid s, a_i).$$

A unit reward is gained when and only when the goal is achieved so that $r(s, a, s')$ is 1 if $s' = n$ and $s \neq n$, and $r(s, a, s')$ is zero, otherwise.

Two extreme plans are of special interest. The first, which is called *bold play*, is always to make the largest possible bet. The second, known as *timid play*, always bets 1 except at the states 0 and n, where the only available bet is zero.

Example 3: The Sequential Choice of Experiments: A Two-Armed Bandit. Consider a sequence of trials on each of which an experimenter (gambler) has a choice of two possible treatments (pulling the right or the left arm). The outcome of each trial is either a success (worth 1 dollar) or a failure (worth nothing), and the objective is to maximize the expectation of the total payoff, this expectation being calculated from a prior* joint distribution $F(p_1, p_2)$ for the success probabilities p_1 and p_2 for the two experiments.

For a dynamic programming formulation, take S to be the collection of all distributions F on the square $\{(p_1, p_2): 0 \leqslant p_1 \leqslant 1, 0 \leqslant p_2 \leqslant 1\}$ and let $A(F) = \{L, R\}$ for every F in S. A choice of action R leads to a success with probability

$$E(p_1) = \int p_1 dF(p_1, p_2)$$

and, with the same probability, the next state F' is the (posterior*) conditional distribution of (p_1, p_2) given that an observation on the right arm resulted in a success. Similarly, action R leads to failure with probability $1 - E(p_1)$ and, with this probability, the next state is the conditional distribution of (p_1, p_2) given a failure on the right arm. Thus the distribution $q(\cdot \mid F, R)$ has been specified and $q(\cdot \mid F, L)$ is defined similarly. The reward function is, for simplicity, taken to be the expected payoff, which is just the probability of success. For example,

$$r(F, R, F') = E(p_1).$$

So that the expected total reward will be finite, the problem is either discounted or play is limited to a finite number of trials.

The principles introduced in the next section make possible the solutions of the problems of the first two examples and are helpful also for problems, such as the two-armed bandit, which defy an explicit solution in general.

SOME FUNDAMENTAL PRINCIPLES

Let $U(s)$ be the supremum over all plans π of $I(\pi)(s)$. Thus $U(s)$ is the *optimal return* at s. The two basic problems associated with any dynamic programming problem are the calculation of the optimal return function U and the determination of a plan π whose return function $I(\pi)$ equals or nearly equals U.

Suppose that π is a plan whose possible optimality is under investigation. Let $Q = I(\pi)$ be the return function for π and, for each state s and action a in $A(s)$, let

$$(T_a Q)(s)$$
$$= \int [r(s, a, s') + \beta Q(s')] dq(s' \mid s, a).$$

$$(2)$$

Then $T_a Q$ is the return from a plan that uses action a the first day and then follows plan π. If π is optimal, then no plan has a greater return and so, in particular,

$$(T_a Q)(s) \leqslant Q(s) \qquad (3)$$

for all s and all a in $A(s)$. A function Q that satisfies (3) is called *excessive*. Thus the return function of an optimal plan is excessive. Somewhat surprisingly, the converse is also true for many problems.

Theorem 1. A plan is optimal for a discounted or a positive dynamic programming problem if and only if its return function is excessive.

To see how this theorem is applied, consider Example 2 again and assume that the win probability w is greater than $\frac{1}{2}$. Let π be timid play and set $Q = I(\pi)$. Then $Q(s)$ is the probability that a simple random walk* which moves to the right with probability w and starts from s will reach n before 0. This probability is that of the classical gambler's ruin problem [13] and is given by

$$Q(s) = \frac{1 - (\overline{w}/w)^s}{1 - (\overline{w}/w)^n}. \qquad (4)$$

For $a = a_i$ in $A(s)$, the inequality (3) becomes

$$wQ(s + i) + \overline{w}Q(s - i) \leqslant Q(s),$$

and this inequality is not difficult to verify using (4). So, by Theorem 1, timid play is optimal and the optimal return function U is given by (4). If $w < \frac{1}{2}$, then bold play is optimal, but the application of Theorem 1 is more intricate [9].

It can easily happen that no optimal plan is available for a problem. Thus it is useful to have the following characterization of the optimal return function U in addition to the characterization of optimal plans given by Theorem 1.

Theorem 2. For a discounted (positive) dynamic programming problem, the optimal return function is the least bounded (nonnegative) excessive function.

This theorem has a particularly simple interpretation in the case of Example 1. For in that problem, an excessive function Q is one that satisfies

$$\tfrac{1}{2} Q(k+1) + \tfrac{1}{2} Q(k-1) \leqslant Q(k)$$

for $k = 1, \ldots, n-1$, and

$$u(k) \leqslant Q(k)$$

for $k = 0, \ldots, n$. That is, Q is excessive if and only if Q is concave and dominates u on $\{0, \ldots, n\}$. So by Theorem 2, U is the concave envelope over u. It is now easy to argue that an optimal plan is to stop (i.e., use action b) at k if $u(k) = U(k)$ and to go (i.e., use action a) if $u(k) < U(k)$.

Ordinarily, the determination of the optimal return function is more difficult than in the example just considered. However, if S is finite and $A(s)$ is finite for every s in S, there are algorithms that permit the calculation of U. For example, Theorem 2 can be restated for a positive problem as follows. U is that function Q on S which maximizes $\sum_s Q(s)$ subject to the restrictions $Q(s) \geqslant 0$ and $(T_a Q)(s) \leqslant Q(s)$ for all s and all a in $A(s)$. But this problem is a linear programming* problem whose solution can be calculated using the simplex algorithm*. (Additional information about the relationship between linear and dynamic programming can be found in the book by Mine and Osaki [16].) Another algorithm will be described below.

THE OPTIMALITY EQUATION AND BACKWARD INDUCTION

It is true in great generality that

$$\sup_a (T_a U)(s) = U(s). \tag{5}$$

Intuitively, the left-hand side is no greater

than the right because it is for every a, at least as good to play optimally from the beginning as it is to use action a the first day and then plan optimally. On the other hand, if a is a nearly optimal action for the first play, then $(T_a U)(s)$ must be nearly as large as $U(s)$. The same reasoning applied to an n-day problem leads to the equation

$$\sup_a (T_a U_{n-1})(s) = U_n(s), \tag{6}$$

which, in a rough translation to English, says that, to play well for n days, one must choose a good first action and then play well for $n-1$ days. The formula works even for $n = 1$ if the convention is made that $U_0 = 0$. Thus U_n can be calculated inductively, and if the values of a, which achieve (or nearly achieve) the supremum at each stage, are retained, they can be used to make up an optimal n-day plan. This procedure (due to Arrow et al. [1]) is called backward induction, because, for example, the actions calculated at the beginning to get U_1 will be used by the n-day plan at the end when there is one day left to play.

For a simple illustration of this method, consider Example 2 in the special case when the win probability is $w = \tfrac{1}{2}$ and $S = \{0, \ldots, 4\}$. So the goal is to reach 4 and $U_n(s)$ is the maximum probability of reaching 4 in no more than n plays starting from s. Here is a table showing U_1 and the optimal actions (or bets) for the one-day problem, which are also the optimal last actions for an n-day problem.

s	0	1	2	3	4
$U_1(s)$	0	0	$\tfrac{1}{2}$	$\tfrac{1}{2}$	1
a	0	0 or 1	2	1	0

Using this table together with (6), one easily calculates U_2 and the optimal first actions for a two-day problem, which are the optimal next-to-last actions for an n-day problem when $n \geqslant 2$. For example,

$$\begin{aligned}
U_2(1) &= \max \{ (T_0 U_1)(1), (T_1 U_1)(1) \} \\
&= \max \{ U_1(1), \tfrac{1}{2} U_1(0) + \tfrac{1}{2} U_1(2) \} \\
&= \max \{ 0, \tfrac{1}{4} \} \\
&= \tfrac{1}{4} .
\end{aligned}$$

Notice that $(T_1 U_1)(1) > (T_0 U_1)(1)$, so that the optimal bet at 1 is 1. Here is the complete table:

s	0	1	2	3	4
$U_2(s)$	0	$\frac{1}{4}$	$\frac{1}{2}$	$\frac{3}{4}$	1
a	0	1	2	1	0

Further calculation shows that $U_3 = U_2$ and hence, by induction and (6), $U_n = U_2$ for all $n \geq 2$. In this problem, as in many others, U_n converges to U as n approaches infinity; so it follows that $U = U_2$ as well.

For more complicated problems, calculations can often be done using electronic computers. However, calculations can become unwieldy even when machines are used, in which case backward induction may become a useful theoretical tool. For example, it has been used to find solutions for interesting special cases of the two-armed bandit problem [3, 12].

STATIONARY PLANS

Let f be a function that assigns to each state s an action $f(s)$ in $A(s)$. Then f determines the plan π which uses action $f(s)$ whenever the current state is s. Such a plan is called stationary because under it the process of states becomes a stationary Markov chain. All the specific plans considered above, such as bold play for red-and-black, have been stationary (*see* STATIONARY PROCESSES).

A basic theoretical problem is to find necessary and sufficient conditions that guarantee the existence of optimal or nearly optimal stationary plans. Many sufficient conditions are known [11], but the theorem below gives the simplest.

Theorem 3. Suppose that S is finite and $A(s)$ is finite for every s in S. Then there is an optimal stationary plan regardless of whether the problem is discounted, positive, or negative.

The theorem allows the search for an optimal plan to be made in the relatively small collection of stationary plans. In the discounted case, there is an efficient algorithm for finding an optimal stationary plan [15].

Literature

The name and an appreciable part of the theory of dynamic programming are due to Richard Bellman (see ref. 2), but most of the specific results of this article are to be found in the works of David Blackwell, Lester Dubins and Leonard Savage, and Ronald Howard. The book by Howard [15] remains an excellent introduction to the subject and is not mathematically demanding. A short, useful, and fairly elementary treatment is given in Chapter 6 of the book by Ross [18]. The book by Mine and Osaki [16] is at roughly the same level but goes into much greater detail. For the mathematically prepared reader, the papers by Blackwell [5, 6] and Strauch [19] and the book by Dubins and Savage [9] are highly recommended. The recent book by Bertsekas and Shreve [4] is also at a high mathematical level and is a useful guide to the literature, as is Mine and Osaki [16] and the 1976 edition of Dubins and Savage [9].

This article has neglected deterministic dynamic programming to concentrate on those aspects of the subject more closely connected to statistics. Bellman's book provides a nice introduction to the deterministic theory and its many applications to problems of allocation, scheduling, inventory control*, etc. The more recent book by Dreyfus and Law [8] is also rich in examples and contains a large collection of problems and their solutions.

References

[1]   Arrow, K. J., Blackwell, D., and Girshick, M. A. (1949). *Econometrica*, **17**, 213–244.

[2]   Bellman, R. (1957). *Dynamic Programming*. Princeton University Press, Princeton, N.J.

[3]   Berry, D. A. (1972). *Ann. Math. Statist.*, **43**, 871–897.

[4]   Bertsekas, D. P. and Shreve, S. (1978). *Stochastic Optimal Control: The Discrete Time Case*. Academic Press, New York.

[5] Blackwell, D. (1965). *Ann. Math. Statist.*, **36**, 226–235.

[6] Blackwell, D. (1966). *Proc. 5th Berkeley Symp. Math. Statist. Prob.*, Vol. 1. University of California Press, Berkeley, Calif. pp. 415–418.

[7] Chow, Y. S., Robbins, H., and Siegmund, D. (1971). *Great Expectations: The Theory of Optimal Stopping*. Houghton Mifflin, Boston.

[8] Dreyfus, S. E. and Law, A. M. (1977). *The Art and Theory of Dynamic Programming*. Academic Press, New York.

[9] Dubins, L. E. and Savage, L. J. (1965). *Inequalities for Stochastic Processes: How to Gamble If You Must*. McGraw-Hill, New York (Dover ed., 1976).

[10] Dubins, L. E. and Sudderth, W. D. (1977). *Z. Wahrscheinlichkeitsth.*, **41**, 59–72.

[11] Dubins, L. E. and Sudderth, W. D. (1979). *Ann. Prob.*, **7**, 461–476.

[12] Feldman, D. (1962). *Ann. Math. Statist.*, **33**, 847–856.

[13] Feller, W. (1950). *An Introduction to Probability Theory and Its Applications*, Vol. 1. Wiley, New York.

[14] Hinderer, K. (1970). *Foundations of Non-stationary Dynamic Programming with Discrete Time Parameter*, Lect. Notes Operat. Res. 33. Springer-Verlag, New York.

[15] Howard, R. A. (1961). *Dynamic Programming and Markov Processes*. Wiley, New York.

[16] Mine, H. and Osaki, S. (1970). *Markovian Decision Processes*. Elsevier, New York.

[17] Puterman, M. L., ed. (1979). *Dynamic Programming and Its Applications*. Academic Press, New York.

[18] Ross, S. M. (1970). *Applied Probability Models with Optimization Applications*. Holden-Day, San Francisco.

[19] Strauch, R. E. (1966). *Ann. Math. Statist.*, **37**, 871–890.

(DECISION THEORY DISCOUNTING GAMBLING, STATISTICS IN SEQUENTIAL ANALYSIS)

WILLIAM D. SUDDERTH

DYNKIN REPRESENTATION *See* DIFFUSION PROCESSES

E

ECOLOGICAL STATISTICS

Studies of the dynamics of natural communities and their relation to environmental variables must rely largely upon field data, and hence upon statistical techniques for observational rather than experimental design and analysis. Sampling procedures thus play a critical role in ecological investigations, and a number of specialized sampling methods, as well as an immense variety of sampling devices, have been developed in attempts to collect probability samples and thereby obtain meaningful data on relatively inaccessible, cryptic, and often elusive organisms in their natural environment. Methods for reducing the spatial and temporal multispecies data to comprehensive and more comprehensible lower-dimensional measures of community structure and function is the other major component of ecological statistics.

SAMPLING PROCEDURES

Conventional sample survey methods are generally applicable to the probability sample selection of physical sites for biological sampling, but biological measurement at these sites then commonly entails further sampling stages which attempt to enumerate the target organisms present. Size and shape of the organisms, spatial patterns and density of the local distribution, and in the case of animals, degree of mobility, largely determine the nature of a biological sampling procedure and the tractability of the problem of quantifying either absolute or relative sample selection probabilities. The principal sampling methods employed at this stage may be categorized as quadrat, point and line intercept, point and line intersect, nearest neighbor*, capture–recapture*, and survey removal methods.

Quadrat Sampling

Communities of sessile or highly sedentary organisms present relatively tractable sampling problems once the statistical universe of interest has been delimited. Square plots or quadrats at sites distributed over the delimited area in a random, stratified random or systematic manner represent commonly used sampling designs in plant ecology. Measures of vegetation in a quadrat may consist simply of a species list for the quadrat as a whole, or species presence or absence in each of n^2 square subdivisions of the plot, or more refined quantitative measures for each species, such as total number of individuals, total yield, or measures of ground area coverage such as basal area, ground projected area of all aerial parts (cover) or its convex hull (canopy). Except for the species list, such data are relatively unambiguous and amenable to interpretation and statistical analysis.

Species presence or absence in a quadrat, defined either in terms of basal presence or aerial cover, is the easiest of these measures to determine but somewhat awkward to analyze and interpret. Such species lists form a species × quadrat table with zero or 1 entries and appear amenable to categorical data* analysis for such purposes, say, as comparing environmental classes of quadrats. Difficulties of interpretation arise due to the unknown relationship between probability of occurrence in a quadrat and such factors as quadrat size, spatial pattern and density of species distributions, and variation in size and shape of plant cover within and between species. When the objective is to estimate species abundance in terms of area coverage, however, the ease and simplicity of presence or absence measurements in a quadrat may be exploited and difficulties of interpretation largely circumvented by reducing the dimensions of the quadrat to a line or a point.

Point and Line Sampling

Contact between the aerial part of a plant and a vertical pin or vertical line of sighting at a selected point quadrat is taken as an indicator of presence of the species at that point. The proportion of sample points so covered then estimates the proportion of the delimited study area covered by that species; although when pins rather than sighting lines are used the "point" becomes a tiny circular quadrat, and the estimate of species area coverage becomes inflated by a correspondingly tiny proportional factor of the perimeter of the coverage of that species, which is again dependent on growth form.

Vertical-line interceptions with basal or crown parts serve to determine not only the presence of a species at that ground point but also to identify the one or more individual plants of that species whose projected area contains the sample point. For the added cost of measuring the projected areas $a_1, \ldots, a_m$ of those $m \geq 1$ individuals of a species so identified, the sample point then provides an estimate $(1/a_1) + \cdots + (1/a_m)$ of the density (individuals per unit area) of that species. More generally, if another plant characteristic, such as a yield component x_j, is also measured, then $(x_1/a_1) + \cdots + (x_m/a_m)$ estimates yield per unit area for that species; this follows from the fact that a randomly placed point falls in a_j with probability a_j/A, where A denotes the area of the population or the stratum in question. Each sample point provides such an estimate.

Applied to basal area sampling of trees, where a point on the ground would ordinarily intercept at most one tree base, the areas a_j may be taken to represent the (circular cross-sectional) basal area of the tree expanded by some fixed and known factor c; thus $a_j = c\pi d_j^2/4$, where d_j is the breast height diameter of the tree. This technique, known as Bitterlich point sampling, is implemented with the aid of an angle gauge which subtends a fixed horizontal angle

$$\alpha = 2\sin^{-1}(1/\sqrt{c}).$$

Any tree that at breast height subtends an angle greater than α when viewed through the gauge from the sample point is then included in the sample, and its d_j and x_j are measured.

Such point-intersect methods substantially increase the amounts of information gained and effort expended at each sample point, thereby giving greater justification for the travel costs required to reach the point. Other less substantial modifications of point-intercept sampling which retain the simplicity of presence or absence observations while expanding the effort at a sample site include the use of a battery of pins or discrete points at the site, or a continuum of points in the form of a line segment of fixed length. In either case the proportion of points covered by a species estimates the corresponding proportional area coverage in the population. In the line intercept method, the n line segments in the sample might be portions of parallel transects. With stratified random spacing of parallel transects, for example, and the conceptual latitudinal subdivision of each into equal-length, nonoverlapping segment sampling units, conventional stratified two-stage sampling and estimation procedures apply.

The line intercept design may be extended to a line intersect design by including supplementary measurement of the projected crown of those plants intercepted by a line segment within a stratum. Maximum width w_j of a crown, measured in a direction perpendicular to the transect, determines the geometric probability* of intersecting this individual, and in transect sampling this supplementary measurement is therefore sufficient for the purpose of estimating species density. Any other measurements x_j weighted inversely to the now known intersection probability would, when summed over the transect, provide unbiased estimates of the population total for x.

Spatial randomization* in the design of quadrat, point, or line sampling systems is sufficient to ensure the existence of unbiased estimators of species abundance in a study area, regardless of the pattern in the spatial distribution of the sessile organisms. Pattern does effect the sampling variability of such estimators, but their unbiasedness is a spatial distribution-free property. One consequence of this property is the absence of any rigid control over sample size as measured by the number of specimens in the sample, and hence lack of control over costs and precision in estimating frequency distributions for organism attributes. Size-biased (a_j and w_j) selection probabilities in point and line methods do, in fact, result in severely distorted distributions within the sample. A point sampling procedure that selects the nearest individual, regardless of its size, overcomes such distortion and results in a fixed sample size. The measured distance X

from sampling point to the nearest individual provides a potential basis for estimating density, since density D is a decreasing function of mean distance, but the exact nature of this relationship is dependent upon the spatial pattern, and hence only biased density estimates are achievable by distance methods. In the case of a "Poisson forest" (a completely random spatial pattern of trees), e.g., $2D\pi X^2$ is distributed as chi-square* on 2 degrees of freedom; thus $1/(\pi X^2)$ is a parametric estimator of D. Similarly, in the case of "T-square" sampling, where X again measures the distance from a random point P to the location Q of the nearest individual and Z is the distance from Q to the nearest individual falling in the half-plane on the other side of Q from P, then $2D\pi X^2$ and $D\pi Z^2$ are independent χ_2^2-variables. The ratio $2\sum X^2/\sum Z^2$, where the sum extends over n sample points, is $F_{2n,2n}$-distributed in a Poisson forest, and tends to be larger in a clumpy forest and smaller in a forest plantation.

Distance methods applied to individuals of the same species or to individuals irrespective of species are thus found to provide useful test of randomness*, and in this respect are competitive with quadrat sampling and its associated Poisson variance test. In other respects, however, distance methods remain tantalizingly appealing to ecologists but restrictively parametric and ill adapted to the multispecies needs of ecological sampling.

A review of all these methods of sampling and estimation is given by Cormack et al. [2].

The claim of unbiasedness for quadrat, point, and line sampling methods is largely restricted to the sampling of plants and other sessile and relatively conspicuous organisms, posing no serious problem of detectability or taxonomic classification. Although used also in sampling animal populations, these methods then acquire bias due to avoidance behavior patterns peculiar to the species and dependent on such factors as age, sex, and social groupings within the species. In cases relying on animal sightings by observers, learning and behavioral characteristics of the observer as well as the observed may become factors influencing the biological measurement process. Mechanical collecting devices, whether active or passive, operate on species at efficiencies which also vary differentially with environmental conditions. These virtually self-evident complications, readily demonstrated in calibration* and comparative sampling experiments, represent the grain of salt which must always be taken along with animal abundance data. Some of these biases may sometimes be circumvented, however, by expending the added cost and effort required for removal or capture–recapture sampling.

SURVEY REMOVAL AND CAPTURE–RECAPTURE SAMPLING

Repeated sampling and removal of organisms at the same site results in a sequence of decreasing catches for each species as their numbers become depleted. If the operation can be completed over a short time under the same set of environmental conditions so that the population remains closed and capture efficiency remains constant on k successive trials, the capture probability p_j for the jth species may be estimated from the (decreasing) sequence of catch sizes x_{1j}, $x_{2j}, \ldots, x_{kj}$. An example might be the repeated electrofishing of a stream section that has been temporarily blocked off with nets to prohibit escape, the fish captured on any pass through the section being recorded and held aside until k passes have been made. For the jth species, the chance of avoidance on any one pass is $q_j = 1 - p_j$ and the chance of avoiding all k passes is then q_j^k. The number actually caught, $x_{.j}$, may then be adjusted for this escapement rate to give an estimate

$$\hat{N}_j = x_{.j} / \left(1 - \hat{q}_j^k \right)$$

of the number actually present, where

$$\frac{\hat{q}_j}{\hat{p}_j} - \frac{k\hat{q}_j^k}{1 - \hat{q}_j^k} = \frac{\sum_{i=1}^k (i-1)x_{ij}}{\sum_{i=1}^k x_{ij}}.$$

This "survey removal" method of estimation has also been generalized to the case where the units of effort f_i expended on the ith trial is variable but known, and catch per unit of effort for the jth species, x_{ij}/f_i, is expected to be proportional to the number still at large,

$$E\left(X_{ij} \mid x_{1j}, \ldots, x_{i-1, j}\right) = K_j \left(N_j - \sum_{\nu=1}^{i-1} x_{\nu j} \right) f_i.$$

Conditional regression may be used to estimate the assumed constant capture efficiency K_j for "species" j, and thereby estimate the initial number N_j. Note that the classification "species" may, for these purposes, also include a partition into size classes or other identifiable categories that might influence catchability within a species.

Removal disallows recapture in subsequent samples, but if marking and remarking the animals at each capture and recapture is practicable, some of the restrictive assumptions required in the removal method may be relaxed. Marking and releasing on every occasion renders the capture history known for every individual in a sample, and thereby allows relaxation of the assumption that the population is closed to mortality and recruitment or (permanent) emigration and immigration, and allows capture probability to change arbitrarily from one sampling occasion to the next.

Species $\times$ environmental interaction* effects on capture probabilities do not invalidate capture–recapture methods for estimating vital statistics of the community. These estimation procedures are collectively known as the Jolly–Seber method and are presented by Seber [9] together with various generalizations to accommodate such complications as transient effects of capture and marking on subsequent capture probability and on survival rate, age-specific survival rates for identifiable age groups, and destructive recapture sampling. Specializations are also presented to allow improved efficiency in both the conduct of the experiment and the statistical estimation when simplifying assumptions can be made concerning the degree of closure of the population.

STATISTICAL ANALYSES OF COMMUNITY STRUCTURE

An ecological community, subjectively defined in terms of spatial extent and taxonomic group under consideration, is commonly described by an $N \times s$ data matrix $\mathbf{X}$ giving the amount x_{ij} of species j in sample i, where "sample" is generically referred to below as a quadrat. Map location of the quadrat will ordinarily be known, together with some environmental data descriptive of that location. In this ecometrics context, as in analogous sociometric and psychometric contexts, the objective of the analysis is to identify intrinsic structure in the community and to relate this to extrinsic variables.

Classification* and Ordination

Motivation for classification or cluster analysis of the quadrats may be the identification of subcommunities representing distinct sociological entities existing in a homogeneous environment or, more commonly, may be an attempt to use the revealed discontinuities in taxonomic composition as a guide in the search for causative, abiotic environmental discontinuities. Quadrat clusters so identified are then often treated as sample collections representing distinct populations, and either separately subjected to further multispecies analyses, such as the ordination of quadrats within each cluster, or the quadrat clusters may then be compared with respect to environmental variables measured on each quadrat. Discriminant analysis*, for example, may be used as a means of studying the dimensionality and specific nature of cluster differences with respect to environmental variables.

Principal components* defined by the first few eigenvectors of the species (taxa) covariance matrix also offer potential aid in determining the dimensionality and nature of environmental gradients. If, in every quadrat i, the amount (e.g., density) x_{ij} of each species j were uniquely determined by the level v_i of a single environmental factor, then a Taylor

expansion of $x_{ij} = f_j(v_i)$ would give

$$f_j(v_i) = f_j(\bar{v}) + (v_i - \bar{v})f_j'(\bar{v})$$
$$+ \tfrac{1}{2}(v_i - \bar{v})^2 f_j''(\bar{v}) + \cdots .$$

To the extent that each of these response functions could be approximated by a second- or third-degree polynomial in the (unknown) variable v, the first two or three principal components would correspondingly "explain" the between-quadrat variance and covariance of species amounts. In a case where each species obeyed its own, exactly quadratic response

$$f_j(v_i) = a_j + b_j v + c_j v^2,$$

the species covariance matrix would have only two nonzero eigenvalues. The first two principal components,

$$y_{1i} = \sum_{j=1}^{s} u_{1j} x_{ij}$$

$$= \left(\sum u_{1j} a_j \right) + v_i \left(\sum u_{1j} b_j \right)$$

$$+ v_i^2 \left(\sum u_{1j} c_j \right)$$

$$= A_1 + B_1 v_i + C_1 v_i^2$$

$$y_{2i} = \sum_{j=1}^{s} u_{2j} x_{ij} = A_2 + B_2 v_i + C_2 v_i^2,$$

when ordinated (plotted one against the other) would reveal their structural relation and the unidimensionality of the environmental factor space. Quadratic and cubic trends in structural relations among the first three principal components are often noted in ordinations of plant communities (e.g., Hill [5]). A Taylor series* expansion in a response function of several (unknown) variables also offers a conceptual basis for analysis of higher-dimensional nonlinear ordinations.

An alternative ordination method now frequently used in the analysis of binary* (species presence/absence) data, but not restricted to this case, confers symmetrical roles to the rows and columns of $\mathbf{X}$ by first standardizing x_{ij} to $z_{ij} = x_{ij}/\sqrt{x_i. x_{.j}}$ and then diagonalizing $\mathbf{ZZ}'$ or, equivalently, diagonalizing $\mathbf{Z}'\mathbf{Z}$. An $N \times 1$ eigenvector* $\mathbf{a}$ of $\mathbf{ZZ}'$ (excluding the vector corresponding to the eigenvalue 1) then plays the role of a principal component to ordinate quadrats, but now the dual eigenvector $\mathbf{b} = \mathbf{Z}'\mathbf{a}$ serves symmetrically to ordinate species. Analogous symmetry would be achieved in a principal component analysis of $z_{ij} = x_{ij} - \bar{x}_{i.} - \bar{x}_{.j} + \bar{x}_{..}$.

These methods have application also in agricultural* experiments, as in mixed cropping experiments where several crop species are grown competitively in the same field plot, and in crop variety trials where "species" become the potential new varieties of a single species in a number of "quadrats" = locations or blocks. Locations, like quadrats, differ with respect to a number of observable environmental factors, and the problem is then to discern the relationship between these observable factors and the unobserved factors revealed in the ordination analysis. More generally, and divorced from any particular subject matter context, ordination methods are seen to be applicable to the problem analyzing nonadditivity in a two-way analysis of variance. In the latter context Mandel [6] derived yet another dual ordination method. Green [3] presents a useful source listing of computer programs for implementing a variety of classification and ordination methods.

Index of Diversity

An ordination variable

$$y_i = u_1 x_{i1} + u_2 x_{i2} + \cdots + u_s x_{is},$$

calculated at two quadrats or sites containing equal numbers of individuals but in a different permutation with respect to species labels, will, in general, receive different ordination scores. In one sense, however, these two communities are equivalent; they show equal degrees of species diversity if all species are considered equal. Diversity in a community (i) is understood to reflect both the number of species (s) present and the evenness of their relative abundances

$p_{ij} = x_{ij}/x_{i.}$. A score function of ($p_1, \ldots, p_s$) that is symmetric in its arguments, is maximum when $p_1 = \cdots = p_s = 1/s$ for a fixed s, and is then an increasing function of s is called a diversity* index. A great variety of such indices have been devised and are used to supplement other multivariate measures that, perforce, lack these qualifications. The two most commonly used are Shannon's* measure of information

$$H = - \sum_{j=1}^{s} p_j \log p_j$$

and Simpson's index* $1 - C$ where

$$C = \sum_{j=1}^{s} \frac{x_j(x_j - 1)}{x_{.}(x_{.} - 1)} \doteq \sum_{j=1}^{s} p_j^2.$$

Simpson's index is the probability that two individuals, selected randomly and without replacement at this site, will belong to different species. The number of species (s) in a community is a measure of species richness and is also commonly used as an index of diversity. A parametric class of diversity indices which includes all three of the above is given by Patil and Taillie (1979) as

$$\Delta_\beta = \left(1 - \sum_{j=1}^{s} p_j^{\beta+1} \right) \Big/ \beta$$

which produces a diversity profile when graphed as a function of β for $\beta \geqslant -1$.

Spatial Pattern

Heterogeneity in the spatial pattern of individuals ordinarily exists on several scales, consisting of patches within patches. The size of these scales is of both intrinsic interest as a structural property of a community, and practical interest as a determinant of efficient plot size for sampling the community. Greig-Smith [4] devised an experimental sampling procedure to determine these scales empirically for any individual species in a plant community. A grid of contiguous quadrats in which plant density has been measured is subjected to a within-blocks and among-blocks analysis of variance for blocks consisting of 2^s grid units. If this species were randomly distributed, the "among blocks of size *s*" mean square should remain constant for all *s*. If the patches-within-patches phenomenon is occurring, however, this mean square should show a small peak at the average size of the finest patch and higher peaks at the successively coarser patch sizes. Modifications of this design and statistical tests for peaks have been developed in subsequent papers; for references to this and other spatial analysis methods, see Cormack and Ord [1].

Models for regression* and autoregression* analysis of a species response measured on such a regular lattice of cells are also reviewed by these authors. The simplest such method called *trend surface analysis*, does not require the lattice configuration of sites and consists simply of fitting a polynomial regression to the map coordinates of the sites on the assumption of no correlation between sites. The simplest autoregressive model for a square lattice configuration is a first-order model for the response X_{ij} at the (i, j)th lattice point,

$$X_{ij} = \beta\left(X_{i-1, j} + X_{i+1, j}\right.$$
$$\left. + X_{i, j-1} + X_{i, j+1}\right) + \epsilon_{ij},$$

where the residuals ϵ_{ij} are assumed uncorrelated and homoscedastic with mean zero. Such models, modified to incorporate the effects of measured environmental variables at the sample sites and extended to multispecies analysis, are expected to play a key role in the future developments of statistical analysis of ecological communities.

For more information on this topic, see Pielou [7, 8].

References

[1] Cormack, R. M. and Ord, J. K., eds. (1979). *Spatial and Temporal Analysis in Ecology*. Satellite Program in Statistical Ecology. International Co-operative Publishing House, Fairland, Md.

[2] Cormack, R. M., Patil, G. P., and Robson, D. S., eds. (1979). *Sampling Biological Populations*. Satellite Program in Statistical Ecology. International Co-operative Publishing House, Fairland, Md.

[3] Green, R. H. (1979). *Sampling Design and Statistical Methods for Environmental Biologists*. Wiley-Interscience, New York.

[4] Greig-Smith, P. (1952). *Ann. Bot.*, **16**, 293–316.

[5] Hill, M. O. (1973). *J. Ecol.*, **61**, 237–249.

[6] Mandel, J. (1961). *J. Amer. Statist. Ass.*, **56**, 878–888.

[7] Pielou, E. C. (1975). *Ecological Diversity*. Wiley-Interscience, New York.

[8] Pielou, E. C. (1977). *Mathematical Ecology*. Wiley, New York.

[9] Seber, G. A. F. (1980). *Estimation of Animal Abundance and Related Parameters*. Macmillan, New York.

(AGRICULTURE, STATISTICS IN
CAPTURE–RECAPTURE METHODS
DIVERSITY, MEASURES OF
FISHERIES, STATISTICS IN
FORESTRY, STATISTICS IN
STRATIFICATION
STRATIFIED SAMPLING
SURVEY SAMPLING)

D. S. ROBSON

ECONOMETRICA

Econometrica is the official journal of the Econometric Society. The Society was founded on December 29, 1930, in Cleveland, Ohio, by a small group of economists, statisticians, and mathematicians that included Ragnar Frisch, Harold Hotelling*, Karl Menger, C. F. Roos, Joseph Schumpeter, Henry Schultz, W. A. Shewhart, and Norbert Wiener. Irving Fisher served as the first president of the society and *Econometrica* appeared in January 1933 with Ragnar Frisch as Editor.

Econometrica is published bimonthly and mailed to approximately 5200 subscribers; each yearly volume contains approximately 1650 pages. The goal of the journal is manifest in the constitution of the Society: "The Econometric Society is an international society for the advancement of economic theory in its relation to statistics and mathematics. . . . Its main object is to promote studies that aim at the unification of the theoretical-quantitative and the empirical-quantitative approach to economic problems and that are penetrated by constructive and rigorous thinking. . . . " This special focus on unification was emphasized by Frisch in an editorial that appeared in the first issue: "There are several aspects of the quantitative approach to economics, and no single one of these aspects, taken by itself, should be confounded with econometrics. Thus, econometrics is by no means the same as economic statistics. Nor is it identical with . . . general economic theory. . . . Nor should econometrics be taken as synonymous with the application of mathematics to economics. Experience has shown that each of these three view-points, that of statistics, economic theory, and mathematics, is a necessary, but not by itself a sufficient, condition for a real understanding of the quantitative relations of modern economic life. It is the *unification* of all three that is powerful. And it is this unification that constitutes econometrics."

Econometrica has no tightly controlled policy toward subject matter. Manuscripts published during the past several years range from highly theoretical to very applied. An example of recent papers in the statistics area include:

Deaton, A., "The Analysis of Consumer Demand in the United Kingdom, 1900–1970," March 1974.

Dreze, J. H., "Bayesian Limited Information Analysis of the Simultaneous Equation Model," September 1976.

Goldberger, A. S., "Structural Equation Methods in the Social Sciences," November 1972.

Heckman, J. J., "Shadow Prices, Market Wages and Labor Supply," July 1974.

Sargan, J. D., "Econometric Estimators and the Edgeworth Approximation," May 1976.

Hugo Sonnenschein is the Editor of *Econometrica*. Angus Deaton, Eytan Sheshinski, Christopher Sims, and Kenneth Wallis serve as Co-Editors, and Dorothy Hodges is the Managing Editor. Details concerning the submission of manuscripts are available from *Econometrica*, Department of Economics, Princeton University, Princeton, New Jersey 08544, U.S.A. Subscription and membership information may be obtained by writing to The Econometric Society, Northwestern University, Department of Economics, Evanston, Illinois, 60201.

HUGO SONNENSCHEIN

ECONOMETRICS

Econometrics is concerned with the application of statistical methods to economic data. Economists often apply statistical methods to data in order to quantify or test their theories or to make forecasts. However, traditional statistical methods are not always appropriate for application to economic data, in the sense that the assumptions underlying these methods may fail to be satisfied. Basically, this is so because much of traditional statistics has been developed with an eye toward application in the natural sciences, where data are generated by experimentation. In economics, data are virtually always nonexperimental. (This is, of course, also the case in other social sciences; not surprisingly, there is substantial overlap in the statistical methodologies of economics, sociology, political science, etc.) Furthermore, the nature of the economist's view of the world is such that the mechanism viewed as generating the data creates some statistical problems which are distinctly "econometric," and whose solution constitutes a large portion of econometric theory.

SINGLE-EQUATION LINEAR REGRESSION MODELS

The usual assumptions underlying the linear regression* model are that the regressors have fixed (nonrandom) values and are linearly independent, and that the disturbances are uncorrelated and have zero mean and constant variance (*see* GENERAL LINEAR

MODEL). Under these assumptions the least-squares estimator is best linear unbiased (*see* GAUSS–MARKOV THEOREM). Furthermore, if the disturbances are assumed to be normal, likelihood ratio tests* of linear hypotheses concerning the regression coefficients, or concerning forecasts of the dependent variable outside the sample period, are possible using the *F*-distribution*.

Of course, when these assumptions are not satisfied, the least-squares estimator does not have such nice properties. Accordingly, for each of the assumptions above, it is reasonable to ask what damage is done by its violation, and what cure (if any) exists for this damage. This line of inquiry is by no means peculiar to econometrics. Nevertheless, the consequences (and cures thereof) of the violations of the assumptions of the general linear model do receive considerable attention in all econometrics texts and in current econometric research.

The assumption that the regressors are nonrandom will be maintained throughout this section; its violation will be discussed in the next two sections. In this section we discuss briefly the consequences of violations of the other assumptions of the general linear model.

First, consider the assumption that the regressors are linearly independent. Its violation is a condition called *multicollinearity*, under which the regression coefficients are not estimable (*see* ESTIMABILITY). The term "multicollinearity" is also applied to the case in which this assumption "almost" fails, due to one of the regressors being highly (although not perfectly) correlated with a linear combination of the other regressors. In this case, the coefficients are estimable, but only imprecisely. The "solution" that is most commonly advanced is to attempt to reduce mean square error by shrinking the least-squares estimator toward zero, through the use of *ridge regression** or *Stein-rule** estimators. Good surveys (by econometricians) include Vinod [34] and Judge and Bock [17].

Next, consider the assumption that the mean of the disturbances is zero. Basically, this is the assumption that the model is correctly specified, and the study of its violation is the study of *specification error**. This has been a standard part of econometric theory at least since the work of Theil [30], and appears to have received somewhat more attention from econometricians than from other statisticians.

Specification error is a serious problem because it potentially invalidates all the results of a regression; it causes biased and inconsistent estimators and invalid tests of hypotheses. There is no cure except to make sure that one's model is (more or less) correctly specified. On the other hand, there are ways to test the hypothesis of correct specification. Besides such heuristic (but useful) methods as looking for outliers* or patterns in the residuals, a number of more formal specification error tests have been developed. Good sources for these include Ramsey [25] and Hausman [13].

Finally, we consider the assumptions that the disturbances are uncorrelated, and that their variance is constant. Failure of the first assumption is *serial correlation**—sometimes called autocorrelation—while failure of the second assumption is called *heteroscedasticity**. In either case the least-squares estimator is unbiased but inefficient, and the usual inferences about it are invalidated; the cure is to use the *generalized least-squares** estimator.

The *random coefficient model** is an example of a rationale for heteroscedasticity; another popular scheme is discussed by Amemiya [2]. Tests for heteroscedasticity are surveyed by Goldfeld and Quandt [9, Chap. 3]. In the case of autocorrelation, the standard test is the *Durbin–Watson test**, and the disturbances are typically modeled by an autoregressive, moving-average, or ARMA* scheme (*see* TIME SERIES).

REGRESSION MODELS WITH STOCHASTIC REGRESSORS

We now return to the assumption that the regressors are fixed, nonrandom variables. This assumption will often be appropriate in the analysis of experimental data, since the explanatory variables will generally represent conditions of the experiment that were fixed by the experimenter. However, it is generally an unreasonable assumption when one is dealing with nonexperimental data. Thus it is necessary to consider the *mixed model*, in which the randomness of the regressors is explicitly recognized (*see* FIXED, RANDOM, AND MIXED EFFECTS MODELS).

As an example, suppose that one has cross-sectional data on individuals, and is trying to explain income as a function of the individual's age, education, sex, and other demographic variables. Clearly, although individuals are endowed at birth with a given birth date and sex (and even the latter is not as permanently fixed as it used to be!), they are not so endowed at birth with either education or income. Both are subject to random influences over the course of the individual's lifetime. In this sense, education is no more "fixed" than income is.

Given a set of regressors, at least some of which are random, it should not be surprising that the properties of the least-squares estimator depend on the relationship between these regressors and the disturbances. To consider the simplest case first, suppose that the regressors and disturbances are independent. In such a case one is justified in treating the regressors as if they were nonrandom, in the sense that the least-squares estimates remain unbiased and consistent, and the usual tests remain valid. As a result, the assumption that the regressors are independent of the disturbance is the random-regressor case counterpart to the assumption that the regressors are nonrandom.

Philosophically, the assumption that the regressors are independent of the disturbances is tied to the notions of exogeneity and unidirectional causality. Clearly, given the nature of a regression equation, any random effect on a regressor must cause an effect on the dependent variable. However, the assumption of independence of regressors and disturbances implies that the converse is not true—random effects on the dependent variable, as captured by the disturbance, do not affect the regressors. In other words, the assumption that the distur-

bances and regressors are independent is roughly equivalent to the notion that the regressors cause the dependent variable, but not vice versa. Attempts have been made to make this statement more precise, but they have not been entirely successful because it is hard to get agreement on a definition of causality. For one fairly rigorous such attempt, see Granger [10] and Sims [29]. For present purposes it is sufficient to simply use the word *exogenous* for explanatory variables that are independent of the disturbances. Exogenous variables are determined apart from the model under consideration.

To carry through with the previous example, it may be plausible to assume that education is exogenous in an earnings function. This assumes that random effects on one's earnings do not affect one's educational level. Of course, to argue the other way, it is conceivable that earnings do affect education. This could happen, for example, if unexpectedly high earnings increased one's ability to afford higher education. Models allowing for this type of feedback are of considerable importance and will be discussed in the next section.

A second case worth considering is one in which any observation on the regressors is independent of the corresponding observation on the disturbance, although it may not be independent of all observations on the disturbance. The typical example of this occurs in a time-series* context, when one or more of the regressors is a lagged value of the dependent variable (*see* AUTOREGRESSION). In such a case the desirable large-sample properties of least squares (consistency, asymptotic normality*, and asymptotic efficiency) still hold, although its desirable small-sample properties are lost. The proof of this assertion is complicated since it requires establishing a *central limit theorem** for a sum of dependent random variables; this problem was first solved by Mann and Wald [22].

Models with lagged dependent variables* as regressors are quite common in a time-series context, expecially in dealing with aggregate economic data. Indeed, such models had routinely been fitted by least squares by economic forecasters for some time prior to the Mann and Wald article just cited; this article is a significant one in the history of econometrics because it was one of the first to identify a violation of the usual assumptions of the general linear model* (which, furthermore, is distinctly due to the nonexperimental nature of the data), and to consider its consequences.

The third case to be considered in this section is one in which corresponding observations on the regressors and disturbance are correlated. This is a serious problem whenever it occurs, since it causes the least-squares estimator to be biased and inconsistent, and invalidates the usual tests. The nature of the solution to this problem depends on the context, but a general method for obtaining consistent estimates in this case is the method of instrumental variables*, if the necessary instruments can be found.

Correlation between corresponding observations on regressors and disturbance can occur in several ways. One is to have feedback from the dependent variable to the regressors, as discussed previously. The statistical implications of this are the subject of the next section.

Another way in which this may happen is to have a model with lagged dependent variables among the regressors *and* a serially correlated disturbance. If the pattern of serial correlation is such that the current disturbance is correlated with the random component of the lagged dependent variable regressor, least squares will be biased and inconsistent.

The conjunction of lagged dependent variables and autocorrelated errors causes other substantial difficulties worth mentioning. For one thing, the usual tests for autocorrelation (e.g., the Durbin–Watson test*) are invalidated with lagged dependent variables among the regressors. Asymptotically valid tests are given by Durbin [5]. Another problem, closely related to the testing problem, is that the usual estimates of the serial correlation pattern of the disturbances (e.g., sample autocorrelations of the least-squares residu-

als) are inconsistent. Consistent estimates of the serial correlation pattern of the disturbances can be obtained from instrumental variables residuals, where reasonable instruments for the lagged dependent variables might be the lagged values of other regressors. Finally, generalized least squares based on a consistent estimate of the disturbance covariance matrix is asymptotically inefficient (relative to maximum likelihood) in this case. An asymptotically efficient two-step estimator has been suggested by Hatanaka [12], however.

A third way in which correlation between regressors and disturbance can be generated is by *errors in measurement* of the regressors (*see* MEASUREMENT ERRORS). As an example, consider the simple regression model

$$y_i = \alpha + \beta X_i + \epsilon_i, \qquad (1)$$

which would satisfy the usual assumptions except that X_i is not observed. Suppose that what is observed is $X_i^* = X_i + v_i$, where v_i is a measurement error. Then in terms of observables the equation becomes

$$y_i = \alpha + \beta X_i^* + (\epsilon_i - \beta v_i). \qquad (2)$$

Since X_i^* and v_i will be correlated, OLS will be inconsistent here. No simple solution exists since, if v and ϵ are both normally distributed, β is not identified—see Reiersøl [26]. Any consistent estimator of β would have to exploit the nonnormality of v or ϵ, which some allegedly consistent estimators (e.g., those based on grouping of the data) do not. However, there is an interesting (and more hopeful) literature on measurement errors in systems of equations, to which we return in a later section.

SIMULTANEOUS EQUATION MODELS

For reasons based in economic theory, economists tend to view the world as determining the values of economic variables by the solution of sets of equations, each of which holds simultaneously. For example, no one can escape a first course in economics with-

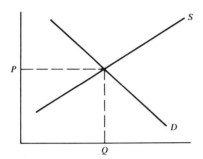

Figure 1 Determination of price and quantity sold by the intersection of supply and demand curves.

out seeing a graph like that of Fig. 1. It depicts the determination of the price and quantity sold of some commodity by the intersection of supply and demand curves. Quantity supplied depends positively on price, as given by S. Quantity demanded depends negatively on price, as given by D. Price and quantity are determined at the point where quantity supplied equals quantity demanded.

Algebraically, the example above can be represented by the set of equations

$$Q = a + bP + \epsilon_1 \tag{3a}$$

$$Q = \alpha + \beta P + \epsilon_2, \tag{3b}$$

where (3a) is the demand curve and (3b) is the supply curve. These two equations determine the variables P and Q, the price and quantity sold of the commodity in question. Variables determined by the model are called *endogenous**. We can express the solution for the endogenous variables P and Q explicitly:

$$P = \left[(\alpha - a) + (\epsilon_2 - \epsilon_1) \right] / (b - \beta) \tag{4a}$$

$$Q = \left[(\alpha b - a\beta) + (b\epsilon_2 - \beta\epsilon_1) \right] / (b - \beta). \tag{4b}$$

We will call the original equations of the model, such as (3), *structural equations**. On the other hand, the equations giving the solutions for the endogenous variables, such as (4), are called the *reduced form** equations.

The system given by (3) is an unusually simple one in many ways. One is that it

contains no exogenous variables except for a constant term. As another example, if we consider the market for wheat as being represented by a system of equations such as (3), we might argue that weather affects supply. Thus we might enlarge the model to allow for this effect:

$$Q = a + bP + \epsilon_1 \tag{5a}$$

$$Q = \alpha + \beta P + \gamma W + \epsilon_2, \tag{5b}$$

where W is a measure of weather (e.g., rainfall). Clearly, weather can be regarded as exogenous, since it is hard to think of a mechanism by which the behavior of wheat suppliers or demanders will affect it. Note that when we write the reduced form,

$$P = \left[(\alpha - a) + \gamma W + (\epsilon_2 - \epsilon_1) \right] / (b - \beta) \tag{6a}$$

$$Q = \left[(\alpha b - a\beta) + b\gamma W + (b\epsilon_2 - \beta\epsilon_1) \right] / (b - \beta), \tag{6b}$$

the solution for the endogenous variables depends on the exogenous variables and disturbances.

It should be noted that the decision of which variables in a system are endogenous and which are exogenous is basically a theoretical (i.e., economic rather than statistical) one. However, the internal consistency of the model requires that the number of endogenous variables equal the number of equations in the system. For example, if we decided that weather were endogenous in (5), then to complete the system we would need to add an equation explaining it.

Note that the reduced form gives the solution for each endogenous variable, and that this solution will in general depend on every exogenous variable and on every disturbance. Thus every endogenous variable will in general be correlated with every disturbance. The implication of this is that least squares will give biased and inconsistent estimates when applied to structural equations that have endogenous variables as right-hand-side variables. This phenomenon is referred to as the simultaneous equations bias of least squares, and was first systematically identified by Haavelmo [11]. Its obvious implication is that least squares is not an appropriate way to estimate structural equations. However, it should be noted explicitly that least squares can be used to estimate reduced-form equations consistently, since the explanatory variables in reduced-form equations are exogenous.

Before turning to the problem of estimation of structural parameters, it should be noted that there is a problem of identification of the structural parameters (*see* IDENTIFICATION PROBLEMS). The reduced-form parameters are always identified, but the structural parameters are identified if and only if it is possible to solve for them uniquely from the reduced-form parameters. In general, this is not possible; many different sets of structural parameters would imply the same reduced-form parameters. (For example, in Fig. 1, many different supply and demand curves could yield the same intersection point.)

If there are sufficient a priori restrictions on the structural parameters, they may be identified. Usually, these restrictions take the form of the exclusion of some variables from some equations. For example, the variable W (weather) appears in (5b) but not in (5a), and this suffices to identify the structural parameters of (5a). The structural parameters of (5b) are not identified, however, without further restrictions. A very complete treatment of the identification of structural parameters can be found in Fisher [6], which also treats identification under other kinds of theoretical restrictions than exclusions of variables from particular equations.

We now return to the question of finding consistent estimates of structural parameters (assuming, of course, that they are identified). As we saw earlier, this is motivated by the inconsistency of least squares.

The oldest, and at first glance simplest, method of estimating structural parameters consistently is indirect least squares. The procedure is first to estimate the reduced form by least squares, and then to solve for estimates of the structural parameters in terms of the estimated reduced-form param-

eters. Such a solution ought to be possible if the structural parameters are identified. The consistency of the indirect least-squares estimates follows directly from the consistency of the least-squares estimates of the reduced-form parameters.

As an example, consider the model (5). The supply curve (5b) is not identified and cannot be estimated consistently by any method. However, the demand curve is identified, and we would like to estimate its parameters. If we estimate the reduced form (6) by least squares, we get consistent estimates of the reduced-form parameters, from which we can solve for consistent estimates of the structural parameters in the demand equation (5a). For example, the parameter b can be consistently estimated by the ratio of the estimated coefficient of W in (6b) to the estimated coefficient of W in (6a).

One serious problem with indirect least squares arises in cases of *overidentification**. A structural parameter is overidentified if there is more than one (distinct) way to solve for it in terms of the reduced-form parameters. If this is the case, indirect least squares is not uniquely defined; different sets of indirect least-squares estimates are possible.

The simplest procedure for estimating structural parameters consistently, but which handles the overidentified case reasonably, is two-stage least squares*. If the equation being estimated is exactly identified, two-stage least squares and indirect least squares are identical. If the equation is overidentified, two-stage least squares is more efficient than indirect least squares. (It can also be expressed as a weighted average of the possible indirect least-squares solutions.) There are numerous other estimators which are asymptotically equivalent to two-stage least squares, but two-stage least squares is most widely used because of its simplicity.

It is still possible to find more efficient estimates than the two-stage least-squares estimates if (as seems reasonable) the disturbances in the different equations are correlated. These more efficient techniques estimate the parameters of all (identified) equations jointly, and are thus somewhat burdensome computationally. One such technique is three-stage least squares, which is beyond the scope of this survey, but which can be found in any econometric text. Another technique that is conceptually straightforward is maximum likelihood (sometimes called "full information" maximum likelihood), which involves maximizing the likelihood function of the system numerically with respect to all the structural parameters, usually by some iterative procedure. This method is feasible only for fairly small systems.

Finally, it should be admitted that the notion of a simultaneous set of structural equations, although by now deeply rooted in the intuition of most economists, is not accepted by all. There is still a dispute, touched off by the work of Wold [35], over whether the world is, or could be, simultaneous, or whether things merely seem that way (e.g., because the time lag in people's actions is small relative to the period of observation of the data). This dispute is rather philosophical, and has really never been resolved; however, most practicing econometricians appear to have revealed a preference for use of simultaneous models. (For Wold's alternative, *see* RECURSIVE MODEL*.)

Another objection to the structural systems discussed above is that they rely on theoretical restrictions (exclusions of certain variables from certain equations) for identification. Some argue that one rarely has a strong theoretical basis for such restrictions, and that identification may therefore be illusory. The alternative is some sort of "unrestricted" model, usually based on time-series methods. For an example, see Sargent and Sims [27]. The comments by Klein [18] are also of interest since they give the traditional defense of structural models, which is really to argue the strength and relevance of economic theory.

LINKS TO OTHER SOCIAL SCIENCE METHODOLOGIES

Much of the theory of simultaneous structural systems appears to have been rediscov-

ered by sociologists, who discuss such systems under the label *path analysis**. Many path models are recursive, and tied to the notion of causality. However, by now econometricians and sociologists have both come to understand the correspondences between their models, and are working together on similar problems.

The best example of this is recent work on unobservable variables. This is tied to earlier statistical work on *measurement error*, since for most unobservables there exist one or more proxy measures, of varying accuracy. (For example, "intelligence" is unobservable, but it has various observable measures, such as various test scores.) Now, in a single equation there is not much that can be done about measurement error, since measurement error on an explanatory variable makes the model underidentified. However, it turns out that in structural models the overidentification due to exclusions of variables can be used to compensate for the underidentification due to measurement error. In this way certain classes of simultaneous models with measurement error can be identified and estimated consistently.

One such model is the multiple indicator–multiple cause (MIMIC)* model surveyed by Goldberger [8]. Suppose that there is an unobservable variable (such as intelligence), which we denote x^*. Suppose that it is related to K observable variables:

$$y_1 = \beta_1 x^* + \epsilon_1$$
$$y_2 = \beta_2 x^* + \epsilon_2$$
$$\vdots$$
$$y_K = \beta_K x^* + \epsilon_K. \qquad (7)$$

The y's are called *indicators*. The variable y_i may be a measure of x^*, in which case $\beta_i = 1$, or it may be some variable that x^* affects in some other way. [There may be other exogenous variables as regressors in (7).] With $K > 1$ we have multiple indicators, hence half of the name used above. Under the assumption that the ϵ's are independent, the multiple-indicators model (without multiple causes) is identified for $K \geqslant 3$. To add the notion of multiple causes, suppose that we add to (7) the specification that

$$x^* = \sum_{j=1}^{Q} \alpha_j x_j, \quad (8)$$

where the x_j are the *causes* of x^*. The resulting MIMIC model is identified for $Q \geqslant 2$ (and any $K \geqslant 1$) and can be estimated in a number of ways, including maximum likelihood.

It is also the case that models with multiple indicators are related to *factor analysis**, which is much discussed in the psychometric literature. Factor analysis has not gotten much attention in econometrics, largely because of the severe identification (in psychometric terms, "rotation") problems one encounters. However, these can be solved by theoretical restrictions (exclusions of variables from equations) in the usual way.

An excellent source on this general topic is the volume edited by Aigner and Goldberger [1].

TIME-SERIES MODELS

For many years *time-series** methods were largely ignored by econometricians. They were considered relevant mainly to the correction of autocorrelation in the disturbance of a model, and even this was typically done under the rather simplistic assumption of a first-order autoregressive process. Despite the fact that structural systems were being used on time series data in the 1940s, most of the development of techniques for correcting the autocorrelation in such systems did not take place until the late 1960s.

However, in the last 10 years or so there has been a spectacular rise in the use of time-series methodology in econometrics. To some extent, this is a reflection of the influence of the work of Box and Jenkins [4]. As their ARIMA models were applied to economic data, a striking thing occurred. It was quickly discovered that very simple, univariate ARIMA models provided forecasts of economic time series such as gross national product which were about as good as those provided by elaborate structural models. This was a bit of a blow to forecasters who used large models, and illustrated at the least that they might wish to pay more attention to the time-series aspects of their models. On the other hand, the model builders have argued (somewhat convincingly, in my view) that at least with a structural model the source of forecasting errors is more easily identified, so that it is easier to learn from one's mistakes and, hopefully, avoid them in the future. (With ARIMA models forecast errors are just random events, and that is not very informative.) Also, ARIMA models are sometimes criticized for being "mechanical," and for not making use of economic theory. This criticism assumes that economic theory is worth using, of course, and the relatively good performance of ARIMA models may bring this assumption into some question; *see* AUTOREGRESSIVE-INTEGRATED MOVING AVERAGE MODELS.

Since both structural models and time-series methods appear useful, it is reasonable to try to combine them. This is the aim of recent work originating with Zellner and Palm [37]. Suppose that one sets up a structural model, and also identifies ARIMA processes for the exogenous variables. These, plus the form of the structural equations, imply very specific ARIMA processes for the endogenous variables. These implied ARIMA processes can be compared to the ARIMA processes actually found by analyzing the endogenous variables separately. Such a comparison can be viewed as a test of the structural specification. From this point of view, one reason why structural models do not forecast better may be that the structure and the time-series properties of the data are not compatible.

MODELS FOR QUALITATIVE AND LIMITED VARIABLES

With the increasing availability of large cross-sectional data bases, there has also been a dramatic increase in interest in the analysis of variables that are distinctly incompatible with the assumptions of the linear regression model. For example, many demographic variables are discrete rather than continuous, and this is incompatible with a linear model with a continuously distributed disturbance.

We will first discuss variables that are *qualitative* in nature. Qualitative variables have a finite number of possible values (rather than a continuous range of values), and no intrinsic numerical scale. For example, a variable indicating whether or not an individual belongs to a union has two possible values and no natural numerical values. It might well be coded as a 0–1 dummy variable (member $= 1$, nonmember $= 0$), but this choice is completely arbitrary.

Suppose that we try to explain such a variable with a linear regression model. Let y be the 0–1 dummy variable and X be a vector of explanatory variables (which may be either continuous or qualitative). Then a regression specification is that

$$y = \beta'X + \epsilon. \tag{9}$$

But if y must equal 0 or 1, ϵ must equal $1 - \beta'X$ or $-\beta'X$. If $E(\epsilon) = 0$, then $\beta'X = P(y = 1) \equiv P$ and $\text{var}(\epsilon) = P(1 - P)$. So ϵ is nonnormal and heteroscedastic. Furthermore, there is no guarantee that P is in the range $[0, 1]$ for every observation. As a result, a linear regression model must be considered unsuitable for analysis of qualitative variables.

More suitable models for analysis of dichotomous (two-valued) variables include the *logit** and *probit** models (*see* BIOASSAY). The logit specification is that

$$\ln \frac{P(y = 1)}{P(y = 0)} = \beta'X. \tag{10}$$

This implies that

$$P(y = 1) = e^{\beta'X}/(1 + e^{\beta'X}), \tag{11}$$

which has the advantage of being restricted to the range $[0, 1]$. It is of the form $P(y = 1) = G(\beta'X)$, where G is the logistic* CDF. A similar model is the *probit* model, which is as follows:

$$y^* = \beta'X + \epsilon, \tag{12}$$

with $\epsilon \sim N(0, 1)$, y^* *unobservable*, and with y determined as

$$y = \begin{cases} 1 & \text{if } y^* \geqslant 0 \\ 0 & \text{if } y^* < 0. \end{cases} \tag{13}$$

This implies that $P(y = 1) = F(\beta'X)$, F being the standard normal CDF. Clearly, this is also restricted to the range $[0, 1]$. Since the logistic and normal CDFs are very similar, the logit and probit models give very similar results for most binary variables. Both are usually estimated by maximum likelihood, with the logit model being slightly easier to estimate.

Things are slightly more complicated when the variable to be analyzed is polytomous (has three or more possible values). Here the logit and probit specifications diverge in a fundamental way. The logit model assumes a purely qualitative variable; that is, it is unordered in the sense that there is no numerical comparison whatever between the values of the dependent variable. An equation such as (10) is specified for each of the $K - 1$ distinct comparisons possible, where K is the number of possible values of the dependent variable. Probabilities such as those in (11) can be expressed using nothing more complicated than exponentiation, so estimation is fairly easy. Good surveys, from rather different points of view, are McFadden [23] and Nerlove and Press [24].

The most natural polytomous probit specification assumes an ordered response. For example, if we know only that individuals are poor, middle class, or rich, we do not know how to assign numbers to these classes, but we do know in what order the numbers would have to be. The specification is basically the same as (12), but (13) is replaced by an equation that splits the range of y^* into K possible subsets, with $K - 1$ dividing points to be estimated. Probabilities are given by a univariate normal integral, so estimation is again not too difficult. See Amemiya [3] for more details.

We now turn to variables that are limited in their possible range. Here there are (at least) two cases of interest. One is the case in which we have a truncated variable. For example, certain variables, such as expenditure on some commodity, are nonnegative and have a finite probability of zero. The most popular model for such variables is the *tobit* specification of Tobin [33]. This assumes a standard linear regression model

$$y^* = \beta'X + \epsilon, \qquad (14)$$

but with y^* unobservable and with the observable y given by $y = \max(0, y^*)$. (*See* RANKIT.)

The tobit model assumes that one has drawn a random sample of individuals, and has data both on those individuals for whom $y > 0$ and for those for whom $y = 0$. An alternative case is one in which one has data only on the random sample of those for whom $y > 0$. In this case we have a standard regression model $y = \beta'X + \epsilon$, but with ϵ constrained to be $\geq -\beta'X$. This case is treated by Hausman and Wise [14].

CONCLUDING REMARKS

The development of simultaneous equation models to use in forecasting economic time series was the historical genesis of econometrics as a distinct field, and it remains an important part of econometrics today. However, the current trend seems to be in the direction of a less distinctively "econometric" methodology in economics. Partly this is the result of the increasing influence of so-called time-series methods in problems of forecasting and control, and partly it is the result of the increasing availability of good cross-sectional data, the analysis of which has created bridges to the methodologies of the other social sciences. This broadening of the field will no doubt continue.

There are available a good many textbooks in econometrics to which the interested reader can be referred. At the introductory level (i.e., supposing basic knowledge of statistics, calculus, and sometimes matrix algebra) some examples are Intriligator [15], Johnston [16], Kmenta [19], Maddala [20], and Theil [32]. At a more advanced level, examples are Malinvaud [21], Schmidt [28], and Theil [31]. The primary text from a Bayesian point of view is Zellner [36].

References

[1] Aigner, D. J. and Goldberger, A. S., eds. (1977). *Latent Variables in Socio-economic Models.* Elsevier/North-Holland, New York.

[2] Amemiya, T. (1973). *J. Amer. Statist. Ass.*, **68**, 928–934.

[3] Amemiya, T. (1975). *Ann. Econ. Social Meas.*, **5**, 363–372.

[4] Box, G. E. P. and Jenkins, G. M. (1970). *Time Series Analysis, Forecasting and Control*, Holden-Day, San Francisco.

[5] Durbin, J. (1970). *Econometrica*, **38**, 410–421.

[6] Fisher, F. M. (1966). *The Identification Problem in Econometrics.* McGraw-Hill, New York.

[7] Goldberger, A. S. (1964). *Econometric Theory.* Wiley, New York.

[8] Goldberger, A. S. (1974). In *Frontiers in Econometrics*, Paul Zarembka, ed. Academic Press, New York.

[9] Goldfeld, S. M. and Quandt, R. E. (1972). *Nonlinear Methods in Econometrics.* North-Holland, Amsterdam.

[10] Granger, C. W. J. (1969). *Econometrica*, **37**, 424–438.

[11] Haavelmo, T. (1943). *Econometrica*, **11**, 1–12.

[12] Hatanaka, M. (1974). *J. Econometrics*, **2**, 199–220.

[13] Hausman, J. A. (1978). *Econometrica*, **46**, 1251–1272.

[14] Hausman, J. A. and Wise, D. A. (1977). *Econometrica*, **45**, 919–938.

[15] Intriligator, M. D. (1978). *Econometric Models, Techniques and Applications.* Prentice-Hall, Englewood Cliffs, N. J.

[16] Johnston, J. (1972). *Econometrica Methods*, 2nd ed. McGraw-Hill, New York.

[17] Judge, G. G. and Bock, M. E. (1978). *The Statistical Implications of Pre-test and Stein-Rule Estimators in Econometrics.* Elsevier/North-Holland, New York.

[18] Klein, L. R. (1977). In *New Methods of Business Cycle Research: Proceedings from a Conference.* Federal Reserve Bank of Minneapolis, Minneapolis, Minn.

[19] Kmenta, J. (1971). *Elements of Econometrics.* Macmillan, New York.

[20] Maddala, G. S. (1977). *Econometrics*. McGraw-Hill, New York.

[21] Malinvaud, E. (1970). *Statistical Methods of Econometrics* North-Holland, Amsterdam.

[22] Mann, H. B. and Wald, A. (1943). *Econometrica*, **11**, 173–220.

[23] McFadden, D. (1974). In *Frontiers in Econometrics*, Paul Zarembka, ed. Academic Press, New York.

[24] Nerlove, M. and Press, S. J. (1973). Univariate and Multivariate Log-Linear and Logistic Models. *Report R-1306-EDA/NIH*, The Rand Corporation, Santa Monica, Calif.

[25] Ramsey, J. B. (1969). *J. R. Statist. Soc. B*, **31**, 350–371.

[26] Reiersøl, O. (1950). *Econometrica*, **18**, 375–389.

[27] Sargent, T. J. and Sims, C. A. (1977). In *New Methods in Business Cycle Research: Proceedings from a Conference*. Federal Reserve Bank of Minneapolis, Minneapolis, Minn.

[28] Schmidt, P. (1976). *Econometrics*, Marcel Dekker, New York.

[29] Sims, C. A. (1972). *Amer. Econ. Rev.*, **62**, 540–552.

[30] Theil, H. (1957). *Rev. Int. Statist. Inst.*, **25**, 41–51.

[31] Theil, H. (1971). *Principles of Econometrics*. Wiley, New York.

[32] Theil, H. (1978). *Introduction to Econometrics*. Prentice-Hall, Englewood Cliffs, N.J.

[33] Tobin, J. (1958). *Econometrica*, **26**, 24–36.

[34] Vinod, H. D. (1977). *Rev. Econ. Statist.*, **60**, 121–131.

[35] Wold, H. (1953). *Demand Analysis*. Wiley, New York (in association with L. Jurien).

[36] Zellner, A. (1971). *An Introduction to Bayesian Inference in Econometrics*. Wiley, New York.

[37] Zellner, A. S. and Palm, F. (1974). *J. Econometrics*, **2**, 17–54.

(AUTOREGRESSIVE-INTEGRATED
  MOVING AVERAGE MODELS
AUTOREGRESSIVE-MOVING
  AVERAGE MODELS
FIX POINT METHOD
GENERAL LINEAR MODEL
LEAST SQUARES
MULTIPLE LINEAR REGRESSION
TIME SERIES)

P. SCHMIDT

ECONOMIC INDEX NUMBERS   *See* IN-
DEX-NUMBERS

EDF STATISTICS

The empirical distribution function (EDF) of a random sample $X_1, X_2, \ldots, X_n$ is a function $F_n(x)$ defined by (number of $X_i \leq x)/n$; it is the proportion of values less than or equal to x, and is an estimate of $F(x)$, the distribution function of x. EDF statistics measure the discrepancy between $F_n(x)$ and $F(x)$, usually for making tests of fit. Many such measures of discrepancy have been proposed. Some of the major ones are defined as follows, with the names often associated with them. Let

$$w_n(x) = \{ F_n(x) - F(x) \},$$

$$y_n(x) = \sqrt{n}\, w_n(x);$$

then

$$
\left.
\begin{aligned}
D^+ &= \sup\{ w_n(x) \}, \\
D^- &= \sup\{ -w_n(x) \}, \\
D &= \max(D^+, D^-)
\end{aligned}
\right\}
\quad
\begin{aligned}
&\text{(Kolmogorov-} \\
&\text{Smirnov*);}
\end{aligned}
$$

$$V = D^+ + D^- \quad \text{(Kuiper)}.$$

The Cramér–von Mises* family is

$$Y^2 = \int \{ y_n(x) \}^2 \psi(x)\, dF(x),$$

with $\psi(x)$ a weight function; when $\psi(x) = 1$, $Y^2 = W^2$, the Cramér–von Mises statistic itself, and when $\psi(x) = \{ F(x)(1 - F(x)) \}^{-1}$, $Y^2 = A^2$, the Anderson–Darling* statistic. A modification of W^2 is Watson's statistic

$$U^2 = \int \{ y_n(x) - \bar{y}_n \}^2 dF(x),$$

where $\bar{y}_n = \int y_n(x)\, dF(x)$. In all these expressions the integrals and suprema are taken over the range of x. We shall call D^+, D^-, D, and V the *supremum* statistics, and W^2, U^2, and A^2 the *quadratic* statistics. The foregoing distributions can be turned into straightforward computing formulas, when $F(x)$ is continuous, as follows. Suppose that the sample is placed in ascending order $X_{(1)} < X_{(2)} < \cdots < X_{(n)}$, and let $z_i = F(X_{(i)})$.

Then

$$D^+ = \max_i\{(i/n) - z_i\},$$

$$D^- = \max_i\{z_i - (i-1)/n\},$$

$$D = \max(D^+, D^-),$$

$$V = D^+ + D^-,$$

$$W^2 = \sum_i \{z_i - (2i-1)/(2n)\}^2 + (12n)^{-1},$$

$$U^2 = W^2 - (\bar{z} - 0.5)^2,$$

$$A^2 = -n - n^{-1}\sum_i (2i-1)$$
$$\{\ln z_i + \ln(1 - z_{n+1-i})\}.$$

Another formula for A^2 is

$$A^2 = -n - n^{-1}\sum_i \{(2i-1)\ln z_i$$
$$+ (2n+1-2i)$$
$$\ln(1 - z_i)\}.$$

DISTRIBUTIONS

The statistics are now functions of the z_i (which will be in ascending order); when they are not ordered, it is well known that the z_i have a uniform distribution* between 0 and 1, written $U(0,1)$. Thus the distributions of EDF statistics do not depend on $F(x)$, but on a set of n ordered uniforms. Distribution theory of these statistics has been much studied, and for practical purposes the distributions are exactly known or well approximated. In particular, asymptotic distributions are known exactly, and percentage points have been provided for all n. Results of these calculations have been condensed in Table 1.

Suppose that T_n is a typical EDF statistic, calculated from a sample of size n. Table 1 gives modifications T of T_n, calculated from T_n and n, which can then be referred to the percentage points given; these are the asymptotic points of T_n or of $T_n\sqrt{n}$. The quadratic statistics converge rapidly to their asymptotic distributions, and the modifications are relatively minor; in particular, no modification is needed for the upper tail of A^2, for $n \geqslant 5$. The supremum statistics converge more slowly. See also Durbin (1971).

GOODNESS OF FIT*: CASE 0

A major use of EDF statistics is to test for fit, i.e., to test H_0: the distribution of X is $F_0(x)$, where $F_0(x)$ is a continuous distribution, which may contain a vector θ of pa-

Table 1   Modifications and Percentage Points for EDF Statistics for Testing a Completely Specified Distribution (Case 0)

Statistic T	Modified Form T^*	0.25	0.15	0.10	0.05	0.025	0.01	0.005	0.001
		\multicolumn{8}{c}{Upper-Tail Percentage Points}							
$D^+(D^-)$	$D^+(\sqrt{n} + 0.12 + 0.11/\sqrt{n})$	0.828	0.973	1.073	1.224	1.358	1.518	1.628	1.859
D	$D(\sqrt{n} + 0.12 + 0.11/\sqrt{n})$	1.019	1.138	1.224	1.358	1.480	1.628	1.731	1.950
V	$V(\sqrt{n} + 0.155 + 0.24/\sqrt{n})$	1.420	1.537	1.620	1.747	1.862	2.001	2.098	2.303
W^2	$(W^2 - 0.4/n + 0.6/n^2)(1.0 + 1.0/n)$	0.209	0.284	0.347	0.461	0.581	0.743	0.869	1.167
U^2	$(U^2 - 0.1/n + 0.1/n^2)(1.0 + 0.8/n)$	0.105	0.131	0.152	0.187	0.222	0.268	0.304	0.385
A^2	For all $n \geqslant 5$:	1.248	1.610	1.933	2.492	3.070	3.857	4.500	6.000
		\multicolumn{8}{c}{Lower-Tail Percentage Points.}							
D	$D(\sqrt{n} + 0.275 - 0.04/\sqrt{n})$	—	0.610	0.571	0.520	0.481	0.441	—	—
V	$V(\sqrt{n} + 0.41 - 0.26/\sqrt{n})$	—	0.976	0.928	0.861	0.810	0.755	—	—
W^2	$(W^2 - 0.03/n)(1.0 + 0.5/n)$	—	0.054	0.046	0.037	0.030	0.025	—	—
U^2	$(U^2 - 0.02/n)(1 + 0.35/n)$	—	0.038	0.033	0.028	0.024	0.020	—	—

rameters with possibly one or more components unknown. When all components of θ are known, we describe the situation as case 0; then $F_0(x)$ is completely specified, and we can put $z_i = F_0(X_{(i)})$ in the definitions above. In general, H_0 will be rejected for large values of the test statistic used; a small value of the statistic will indicate that the z_i are superuniform, i.e., more regularly spaced than expected for an ordered uniform sample. Thus the test of fit to a fully specified continuous distribution $F_0(x)$ is given by the following steps:

1. Calculate $z_i = F_0(X_{(i)})$ for all i.
2. Calculate the statistic desired from the formulas presented in the first section.
3. Modify the statistic as in Table 1 and compare with the appropriate table of upper-tail percentage points. If the modified statistic exceeds a point at level α, reject H_0, that the observations came from $F_0(x)$, at significance level α. Alternatively, the significance level of the test statistic can be found from tables in Stephens (1980).

The transformation $z_i = F(X_{(i)})$ or $z_i = F(X_i)$, often called the probability integral transformation, is not the only way to produce an ordered uniform sample in testing for goodness of fit. Several other methods exist and, when the distribution tested is not correct, some of these may easily produce superuniform observations; then the lower tail of the appropriate distribution should be used. Seshadri et al. (1969) have demonstrated this possibility in the context of tests for the exponential distribution.

A special case to notice is when observations are given on a circle; then only V or U^2 should be used, as, in contrast with the other statistics, they are invariant with respect to the origin for x.

POWER: CASE 0

EDF statistics are usually much more powerful, especially for relatively small samples, than Pearson's X^2, certainly if the Cochran (1952) rules are followed for X^2. This is because of the loss of information when the continuous observations are grouped into cells. Apart from this, the statistics have different powers against different types of alternative. Statistic D^+, for example, becomes significant if the z-values tend toward zero; this could arise if the hypothesized $F_0(x)$ had a mean greater than $F(x)$, the true distribution, and in particular if $F(x) \geqslant F_0(x)$ everywhere. Similarly, D^- is significant when z-values approach 1. Then D, and also W^2, detect a shift in mean in either direction; V and U^2 detect a change in variance (a clustering of z-values, or two clusters near 0 and 1), and A^2 gives weight to observations in the tails and so tends to detect alternatives where more such observations will arise.

GOODNESS OF FIT: CONTINUOUS DISTRIBUTIONS WITH UNKNOWN LOCATION OR SCALE PARAMETERS

It may be that, in a test of fit, the distribution under test is defined by a family $F_0(x)$ of distributions, e.g., the normal or the exponential, containing a vector θ of parameters, and with one or more of the parameters in θ unknown. Let $\hat{\theta}$ refer to θ with unknown components replaced by estimates, and let $\widehat{F_0(x)}$ denote $F_0(x)$ with $\hat{\theta}$ instead of θ. We can still transform by $z_i = \widehat{F_0(X_{(i)})}$ and calculate EDF statistics as above, but their null distributions [i.e., when $F_0(x)$ is the correct family] will be much changed, even asymptotically. The z_i are no longer ordered uniforms, because θ has been replaced by $\hat{\theta}$. Nevertheless, for many important situations, the distributions of EDF statistics will depend only on the sample size and on the family tested, not on the true value of θ. This will be so, in general, when θ contains unknown location and/or scale parameters and when estimates $\hat{\theta}$ are obtained by maximum likelihood. Further, when $\hat{\theta}$ is estimated by efficient estimators, asymptotic theory is available for the quadratic statistics W^2, U^2, and A^2. As also for case 0, this theory possesses considerable mathematical elegance. It was encouraged by the heuristic argument of Doob, later made rigorous by Donsker, that $y_n(x)$ approaches a Gaussian process* as $n \to \infty$; the covariance of this process for different tested distributions with efficiently estimated parameters can be found, and asymptotic distributions of W^2, U^2, and A^2 can be examined. These distributions are sums of weighted χ_1^2 variables; the calculation of the weights and then the percentage points requires for each situation some analysis and much computation. Asymptotic theory is not yet available for the supremum statistics, except for case 0. For small samples, distribution theory when parameters must be estimated is in general very difficult; many authors have provided small-sample percentage points by means of Monte Carlo sampling.

In general, the quadratic statistics W^2, U^2, and A^2 converge more rapidly to their asymptotic distributions than do the supremum statistics D^+, D^-, D, and V. A survey of distribution theory is given by Durbin (1971); Stephens (1980) gives more recent references. Stephens has also given modifications to the quadratic statistics, on the lines of those in Table 1, so that they can be used with only the asymptotic percentage points. EDF statistics are now available for tests of fit to the following distributions, with unknown parameters estimated by maximum likelihood: normal*, exponential*, extreme-value*, logistic*, and gamma* with unknown scale parameter. These tests are set out in detail in Stephens (1980); *see also* ANDERSON–DARLING STATISTIC and KOLMOGOROV–SMIRNOV TYPE TESTS OF FIT.

POWER: COMPARISONS WITH OTHER METHODS

EDF statistics, for the test situations described above, are highly competitive with other techniques for goodness of fit. In particular, they are much more powerful than

Pearson's X^2, and although there have been recent developments with X^2 tests* which can be expected to give improved results, the loss of information involved in grouping into cells, noted in case 0, will still work to the detriment of X^2 when used with continuous data. Among the EDF statistics, the differences in performance noted above tend to disappear when parameters are estimated from the data; the estimation allows a fit of one member of the family tested, in a fairly even fashion. Nevertheless, D^+, and D^- will be more effective than other statistics against one-sided alternatives, and A^2 is good as an overall statistic when the alternative to the family tested is not clearly specified. EDF statistics compare well with regression and correlation methods, which can also be used for unknown location and scale parameters, where the order statistics are regressed against expected values of order statistics from a standard member of the family tested.

MISCELLANEOUS TOPICS

EDF Statistics for Censored* Data and for Discrete Data

Asymptotic theory has been given for the quadratic statistics when the sample has been censored or the parent distribution truncated; see Pettitt and Stephens (1976) and Pettitt (1977) for case 0 and tests for exponentiality. Analogs of EDF statistics have also been proposed for discrete data, based on a comparison of the observed cumulated histogram with its expected value (*see* KOLMOGOROV–SMIRNOV STATISTICS).

Confidence Intervals

The Kolmogorov–Smirnov statistics can be used to give confidence intervals* for an unknown $F(x)$, derived from $F_n(x)$. EDF statistics can also be used for confidence intervals for unknown parameters of a distribution, thus incorporating goodness-of-fit and parameter estimation; see Easterling (1976) and Littell and Rao (1978).

EDF Tests When Unknown Parameters Are Not Location or Scale

If a tested distribution has an unknown parameter that is not location or scale, EDF statistics, after estimating the parameter, will depend in distribution on its true value and, since this is unknown, the tests cannot be applied accurately. The half-sample* method is an interesting technique for overcoming this difficulty when samples are large. Half of the sample is selected at random and used to estimate the parameters; these estimates are then inserted to transform the entire sample by $z_i = \widehat{F}(X_{(i)})$ as in the section on goodness of fit. Asymptotically, EDF statistics will then have case 0 distributions and the percentage points in Table 1 can be used, although the modifications will not still apply. Stephens (1978) has examined this method; see also Braun (1980) for an alternative technique. The half-sample method appears to compare well with X^2, which is of course still available for this situation: *see* CHI-SQUARE TESTS.

Bibliography

To avoid a long list, two references are given from which the others may be obtained:

Durbin, J. (1971). *Distribution Theory for Tests Based on the Sample Distribution Function*. SIAM Regional Conference Series in Applied Mathematics No. 9. SIAM, Philadelphia. (This reference contains a comprehensive bibliography of work to that date.)

Stephens, M. A. (1980). EDF Statistics: A Review with Tables. *Stanford Univ. Tech. Rep.*, Dept. of Statistics, Stanford University, Stanford, Calif. (The present article is an abbreviated version of this technical report.)

M. A. STEPHENS

(GLIVENKO-CANTELLI THEOREMS
GOODNESS OF FIT
KOLMOGOROV-SMIRNOV TYPE
  TESTS)

EDGEWORTH EXPANSIONS *See* CORNISH-FISHER/EDGEWORTH EXPANSIONS

EDGEWORTH-MARSHALL-BOWLEY INDEX *See* INDEX NUMBERS

EDITING STATISTICAL DATA

Many statistical and data-processing techniques have been developed to monitor, detect, and correct errors in data. These techniques are referred to as data editing, screening, laundering, validation, or input accuracy control.

Much of the early statistical research into data editing concentrated on probabilistic parametric procedures to detect extreme values. Rider [28] surveys this early research, most of which assumes normally distributed data. Sheppard [32] studies, for the normal model, rejection limits based on the sample size n which are such that the probability that an individual items falls within the limits is $(1 - 1/(cn))$, and therefore the probability that at least one item in the sample falls outside the limits is for large sample sizes approximately $1 - \exp(-1/c)$. Irwin [17] studies a rejection criterion based on the ratio of the difference between successive order statistics* from a normal population to the standard deviation. Dixon [8] studies the performance characteristics of various rejection criteria on normally distributed data contaminated with either scalar or location errors. (*See* OUTLIERS.)

Subsequently, the research into probabilistic outlier procedures was extended in various directions: nonparametric approaches [37], Bayesian procedures [4, 6], and approaches for multiple classification [1, 34]. The probabilistic procedures were typically tailored to specific analyses to be performed on the data and in light of particular distributional assumptions.

With the advent of computers with the capacity to handle huge quantities of data, a different editing task emerged. Large quantities of data did not even pass simple format or elementary logical checks, with the result that processing programs would not run. As an idea of the magnitude of the task, a census* of population and housing involved 50 million inconsistencies and omissions (out of a total of 2 billion fields that were processed). The data typically involved many variates with large numbers of interrelationships between variates. In addition, there were many potential users and uses of the data. This made it difficult to choose from among methods designed for specific applications and for particular combinations of variates. At the same time there was both external and internal pressure on the producers of the data to edit and launder the data as much as possible. In this context a variety of general procedures were developed and used to organize logical and probabilistic tests and to "correct" data that were inconsistent or in error. Many of the procedures seemed intuitively reasonable but little was known either theoretically or empirically about their operating characteristics*. Recent research by Fellegi and Holt, Minton, Nordbotten, O'Reagan, Scott, Tukey, Varley, and others has improved this situation and has provided structure to modern data editing.

DETERMINISTIC TESTS

Errors can enter data in a variety of ways. In a survey or census, errors can enter due to nonresponse, incomplete or biased frames, selection, and interviewer and questionnaire bias. Errors can and typically do arise at the selection, measurement, recording, transcribing, and keypunching stages of data collection and processing. Dalenius [5] gives an extensive bibliography on nonsampling errors in surveys.

For a given set of observations there are a variety of consistency and empirical checks that help detect some of the gross errors. There are known constraints on the values or combinations of values of variates. There is historical experience on the probable range of values for variates.

The consistency checks are sometimes referred to as logical or *deterministic tests*. Deterministic tests include *range tests** on individual variates and *if–then tests* for acceptable combinations of values for several variates. An example of a range test is the check of whether the variate "day of month" assumes an integer value between 1 and 31.

An example of an if–then test is a check of whether a respondent whose position in family is daughter is also recorded as being female.

A variety of procedures have been developed to efficiently summarize deterministic information. The code control method gives allowable combinations of codes. For example, the method would check whether individuals who had the code for daughter in the field designated position in family also had the code for female in the field for sex. Various techniques are used to reduce storage space. The binary string technique assigns a string of zeros and ones together with a starting and ending number to indicate allowable and nonallowable values for variates. For example, suppose that the allowable diameters for a part are 2, 3, 5, 7, 8, 9, 11, 12 mm. The binary string would list the starting and ending numbers 2, 12 together with the string 101011101. To check whether a recorded entry of 6 was in error, one would compute 6 minus the starting number of $2 = 4$, and then go to fourth position in the string of zeros and ones. Since the fourth position in the string is zero, this indicates that the recorded entry of 6 is in error.

Other techniques to summarize deterministic information are based on functional relationships between variates. The ratio control method is based on the ratio of variates falling between certain limits. The *zero control method* is based on certain variates summing to another variate. For example, a company sells three different products and reports the dollar sales S_1, S_2, S_3 of each. The company also reports total sales S. The zero control method would check whether $S - S_1 - S_2 - S_3 = 0$. Deterministic tests are also based on more complex functional relations between variates, such as the relation between speed, distance, and time traveled.

When a deterministic test is violated, one knows that an error has been made, but does not necessarily know which variate is in error. Sometimes several deterministic tests involving some of the same variates are violated and this gives clues as to where the errors lie. In data systems with large numbers of variates with complex interrelationships, it was found useful to organize the many deterministic tests in a dictionary. Naus [20] and Fellegi and Holt [10] describe ways to develop and check such a dictionary for completeness, redundancy, and consistency and to use the tests to locate records and variates in error. The following example illustrates the use of deterministic tests to locate errors.

A survey gathered information from 12,000 households on the use of neighborhood health care facilities. 132 variates were measured for each family and 78 variates were measured for each individual in selected families. For each variate a range check was constructed, and an additional 60 bivariate and higher variate if–then tests were developed and applied to values that passed the range tests. For example, three of the variates were:

V_1: date of first family use of facility
V_2: date of first use of facility by head of household
V_3: date of last use of facility by head of household

Two obvious deterministic tests are:

$$\text{test 1: } V_1 \leqslant V_2; \quad \text{test 2: } V_2 \leqslant V_3.$$

Data violating either test was flagged. For example, one questionnaire had the entries: $V_1 =$ September 1979; $V_2 =$ January 1980; $V_3 =$ February 1979. Test 1 was not violated, whereas test 2 was. This indicates that either V_2 or V_3 is in error. Observe that test 1 and test 2 together imply a third test: test 3: $V_1 \leqslant V_3$. test 3 is redundant for detecting errors since it can only be violated if either test 1 or test 2 is violated. However, test 3 is useful in helping to locate errors. For the recorded data, tests 2 and 3 are both violated. Since V_3 appears in both violated tests, we tend to be more suspicious of it. Naus et al. [21] give a probabilistic approach to formalizing a measure of suspicion. Fellegi and Holt [10] formalize the generation of the implied tests.

PROBABILISTIC TESTS

In addition to deterministic tests, many data-editing systems use a variety of empirical or *probabilistic tests* to check whether extreme values are sufficiently unusual to warrant investigation. The extreme values are called *outliers*, stragglers, or mavericks, and there is a large statistical literature on approaches to detect and handle outliers. (*See* OUTLIERS.)

When a probabilistic test tags an observation as unusual, this does not necessarily mean that an error has been made. In large data sets one expects to see a certain number of extremely large or small (correct) observations as well as unusual combinations of observations. An individual can be 8 feet tall. A 90-year-old woman can have a natural child 1 year old. The tagging of such observations by a probabilistic test is an indication of suspicion and provides one basis for allocating costly verification effort.

Because in data editing for large censuses or surveys there are many potential users and uses for the data, and because it is often not practical for users to edit the data, producers of data have tried various types of general-purpose outlier procedures.

One set of procedures assumes that the variates are normally distributed either with known or unknown means, variances, and correlations. In the case where the parameters are not known, they can be estimated from the data. Unfortunately, large errors affect the estimates of the parameters and make it hard to compare the observations with the fitted model. Several approaches around this problem have been used. The gross control method checks the internal consistency in small portions of the data that appear in a sequence. Let x_i, x_{i+1}, x_{i+2} denote three consecutive observations in the sequence and let $y_1 < y_2 < y_3$ denote the ordered values of these x's. The test rejects the observations as being unusual if the ratios

$$(y_2 - y_1)/(y_3 - y_1) \text{ or } (y_3 - y_2)/(y_3 - y_1)$$

exceed certain bounds calculated by assuming that the x's are normally distributed. Tukey [35] describes a more sensitive approach called full normal rejection (FUNOR)* that orders the observations and takes deviations about the median*. The deviations are then divided by the expected deviations for ordered standard normal variates. These ratios are computed only for the largest and smallest thirds of the original observations. The sample median of these ratios is computed and compared with the individual ratios. An observation is considered unusual if it has both a large deviation from the median of observations and a ratio that has a large deviation from the sample median of ratios computed. The deviations of the original observations are considered large if they are more than A times the median of ratios computed. The deviation of the ratio is considered large if the ratio is more than B times the median of ratios computed. Scott [31] suggests that a practical operating range for data screening is for A between zero and one, and B between 1.5 and 2.0. For higher variate data Scott describes a regression* approach and computer program designed for data editing. Hawkins [15] details a principal component* approach to screening for multivariate data.

O'Reagan [26] discusses the application of outlier* approaches that do not require normality assumptions. One group of approaches fit the data by various distributions chosen from a system of models. O'Reagan noted that a system that emphasized good fit in the tails was particularly appropriate for data screening, and suggested the Elderton–Pearson system. A second group of outlier approaches makes few assumptions about the distributional patterns of the underlying variates. The approaches based on Chebyshev-type inequalities makes assumptions only about moments but tend to be extremely conservative and typically rely on substantial amounts of correct historical experience (see Godwin [13] and CHEBYSHEV INEQUALITIES). The order statistics* approaches assume that the data come in a random order (at least within various strata). Given this assumption one can say, for example, that it is highly unlikely that the 1001st observation will fall outside the range given by the first 1000 observations. (See Gumbel [14] and ORDER STATISTICS and DISTRIBUTION-FREE TESTS.)

DATA IMPUTATION

Data editing deals with both the detection and handling of errors. Given that an error is detected by a deterministic test that involves an individual variate, one would check to see that the response on the questionnaire was coded and keypunched correctly. If the answer on the questionnaire is clearly in error, one might go back to the source to clear it up. In many surveys and censuses it is not possible to recontact the respondent or to reobserve variates. In the case of a deterministic test that indicates that one of several variates is in error, the data producer might flag the record and treat all the variates involved as missing. The user of the data, given enough information, could develop specific procedures for handling the missing observations, where procedures are tailored for specific applications. Often the data collector seeks to put the data in as complete a form as possible for many different users and therefore applies a general correction procedure. The new values used to replace the old values are called *imputations*, *allocations*, or *assignments*. Automatic correction of data is an important area of statistical investigation and various approaches have been developed.

Several of the approaches for handling the correction of data are tied to the corresponding approaches for detecting errors. Tukey's *full normal rejection* (FUNOR) has a corresponding correction procedure called full normal modification (FUNOM). This correction procedure changes rejected values to a median value, and winsorizes extreme nonrejected values down to the next most extreme values. Fellegi and Holt's [10] approach for setting up a dictionary of tests has associated with it procedures for making imputations in a way to be consistent with the set of tests. A variety of regression approaches are used for estimating missing values (*see* REGRESSION). Fellegi [9] indicates how Hawkins' principal component technique for detecting errors in data can be adapted for imputation. Rubin [30] develops a phenomenological Bayesian approach to multiple imputations. Dempster et al. [7] describe a converging iterative procedure based on maximum likelihood estimation for handling incomplete data. Missing data are estimated given initial estimation of parameters; then parameters are reestimated and the procedure is repeated. (*See* ESTIMATION; MAXIMUM LIKELIHOOD; MISSING VALUES.)

Several different imputation rules replace variates or records that contain errors by "similar" records that do not seem to contain errors. The cold deck method replaces a record that contains errors in an individual variate by another record that is similar on many of the other variates and has been chosen to be representative for that class of records. The particular class of records is sometimes referred to as the adjustment cell or poststrata. In the hot deck method a rejected record is replaced by the last previously accepted record in the same adjustment cell. In the modified hot deck method only the values for variates found to be in error are replaced by the values from the last previously accepted record in the same adjustment cell. The *Monte Carlo method** replaces an erroneous value on one variate by a value chosen at random from the conditional distribution of that variate given the observed values on some of the other variates. Ford [11] describes variations on the procedures noted above, such as a modified (modified) hot deck procedure that replaces values on specific variates from an adjustment cell mean, or from a "close" record. Nordbotten [23] empirically and thru modeling investigated the efficiency of automatic correction techniques, and found that sometimes such correction can do more harm than good. Bailar and Bailar [3] study some of the operating characteristics of the hot deck method.

In censuses and large surveys there are typically many cross classifications of variates summarized in contingency tables*. Missing values on certain variates within a record cause there to be fewer observations within the cells of the tables than there are in the marginals. Rather than impute the missing values on a record-by-record basis, one can try to fit the cell entries in such a way that they are consistent with the marginal ratios. Raking* is an iterative procedure for such fitting. For example, in a two-way table let n_{ij} denote the observed entry in the ijth cell, and let $n_{i.} = \sum_j n_{ij}$ and $n_{.j} = \sum_i n_{ij}$. Let $m_{i.}$ denote the number of individuals observed to be at the ith level of the row variate, and $m_{.j}$ be the number at the jth level of the column variate. Because some individuals have been observed on only one variate, $m_{i.}$ can exceed $n_{i.}$ and $m_{.j}$ can exceed $n_{.j}$. Raking seeks to find adjusted cell counts $\hat{n}_{ij}$ that here satisfy the constraints $\sum_i \hat{n}_{ij} = m_{.j}$ and $\sum_j \hat{n}_{ij} = m_{i.}$. One approach is to first adjust by rows to find a first-stage estimate $\hat{n}_{ij}^{(1)} = (m_{i.}/n_{i.})n_{ij}$; then the procedure adjusts by columns to find $\hat{n}_{ij}^{(2)} = (m_{.j}/\hat{n}_{.j}^{(1)})\hat{n}_{ij}^{(1)}$, and then continues this cycle. Oh and Scheuren [24, 25] illustrate this procedure for a three-way table, detail the application of raking to a large data study, and give an extensive bibliography on raking.

A variety of other statistical procedures are applied to monitor and control the quality of data. Minton [19] describes the combination of acceptance sampling* and quality control* procedures to check the editing and reverification activities. (See ACCEPTANCE SAMPLING and QUALITY CONTROL.) Varley [36] and O'Reagan [27] and others develop models to determine the cost effectiveness of verification, reverification, setting of levels of probabilistic tests, and automatic correction under certain assumptions. Empirical and theoretical studies investigate the efficiency of inspection and verification activities (see Minton [18]). Other studies deal with the error-making process and observe the types and frequencies of errors of various types (for references, see Naus [20, pp.

132–136]. Recently, a large number of papers on data editing were given in the 1978 American Statistical Association* annual meeting (summarized in the *Proceedings of the Section on Survey Research Methods*; see ref. 2) and in the 1979 Symposium on Incomplete Data [33]. These papers indicate both the current interest and the opportunities for research in many aspects of data editing.

References

[1] Anscombe, F. J. (1960). *Technometrics*, **2**, 123–147.

[2] Aziz, F. and Scheuren, F., eds. (1978). *Imputation and Editing of Faulty or Missing Survey Data*. Selected papers, primarily from the Section on Survey Research Methods, Proceedings of American Statistical Association 1978 meeting. Compiled and edited by the Social Security Administration, Office of Research and Statistics. Printed by U.S. Department of Commerce, Bureau of the Census, Oct. 2, 1978.

[3] Bailar, J. C., III and Bailar, B. A. (1978). In Aziz and Scheuren [2, pp. 65–75]. (Subsequent research in ref. 33.)

[4] Box, G. E. P. and Tiao, C. C. (1968). *Biometrika*, **55**, 119–129. (Estimates regression coefficients in linear model with normally distributed data contaminated with scalar errors.)

[5] Dalenius, T. (1977). *Int. Statist. Rev.*, **45**, Pt. I (A to G), 71–89; Pt. II (H to Q), 181–197; Pt. III (R to Z), 303–314.

[6] de Finetti, B. (1961). *Proc. 4th Berkeley Symp. Math. Statist. Prob.*, Vol. 1. University of California Press, Berkely, Calif., pp. 199–210. (General paper outlining Bayesian view that instead of rejecting outlying observations, one should use all the observations to get the posterior distributions of the statistics of interest; suspicious observations might be assigned low weights.)

[7] Dempster, A. P., Laird, N. M., and Rubin, D. B. (1977). *J. R. Statist. Soc. B*, **38**, 1–11.

[8] Dixon, W. J. (1950). *Ann. Math. Statist.*, **21**, 488–506. (Rejection criteria evaluated include chi-square test, extreme deviation or range, modified *F*-test, and ratio of ranges and subranges. Deals with single or multiple outliers and cases where variance is either known or unknown.)

[9] Fellegi, I. P. (1975). *Proc. Int. Statist. Inst.* (40th Meet.), **3**, 249–253. (Synthesizes principal component approach of Hawkins with that of Fellegi and Holt [10].)

[10] Fellegi, I. P. and Holt, D. (1976). *J. Amer. Statist. Ass.*, **71**, 17–35. (Develops a complex but usable formal structure for organizing a complete dictionary of deterministic tests for coded data. Gives a method for identifying fields in error.)

[11] Ford, B. (1976). *Missing Data Procedures: A Comparative Study*. U.S. Department of Agriculture, Washington, D.C., August.

[12] Frane, J. W. (1976). "A New BMDP Program for the Identification and Parsimonious Description of Multivariate Outliers." *1976 Proc. Statist. Computing Sect. Amer. Statist. Ass.*, pp. 160–162.

[13] Godwin, H. J. (1955). *J. Amer. Statist. Ass.*, **50**, 923–945.

[14] Gumbel, E. J. (1954). Statistical Theory of Extreme Values and Some Practical Applications. *Natl. Bur. Stand. (U.S.) Appl. Math. Ser. 33* (Washington, D.C.).

[15] Hawkins, D. M. (1974). *J. Amer. Statist. Ass.*, **69**, 340–344.

[16] Hewitt, J. M. (1977). *Concor Editing and Imputation System for Censuses and Surveys: System Reference Manual*. Int. Statist. Prog. Center, U.S. Bureau of the Census, Washington, D. C.

[17] Irwin, J. O. (1925). *Biometrika*, **17**, 238–250.

[18] Minton, G. (1969). *J. Amer. Statist. Ass.*, **64**, 1256–1275.

[19] Minton, G. (1970). *J. Quality Tech.*, **2**, 86–98.

[20] Naus, J. I. (1975). *Data Quality Control and Editing*. Marcel Dekker, New York. (Provides a nonmathematical integrated approach to and survey of data editing. Chapters are given on setting up deterministic tests and dictionary of tests, probabilistic tests, quality control and acceptance sampling procedures, automatic data correction, locating errors, and cost effectiveness of procedures.)

[21] Naus, J. I., Johnson, T., and Montalvo, R. (1972). *J. Amer. Statist. Ass.*, **67**, 943–950.

[22] Nordbotten, S. (1963). *Conf. Eur. Statist., Statist. Stand. Stud. No. 2*. United Nations, New York (U.N. Sales Number 64 II E/MIM 4).

[23] Nordbotten, S. (1965). *Bull. Int. Statist. Inst.* (Proc. 35th session, Belgrade), **41**, 417–441.

[24] Oh, H. L. and Scheuren, F. (1978). In Aziz and Scheuren [2, pp. 120–127].

[25] Oh, H. L. and Scheuren, F. (1978). In Aziz and Scheuren [2, pp. 128–135]. (49 references.)

[26] O'Reagan, R. T. (1965). *Possible Techniques for Computer Editing of Data*. Gen. Res. Rep. No. 1, U.S. Bureau of Census, Economics Operations Division, Operations Research.

[27] O'Reagan, R. T. (1969). *J. Amer. Statist. Ass.*, **64**, 1245–1255.

[28] Rider, P. R. (1933). "Criteria for Rejection of Observations." Wash. Univ. Stud., New Ser., *Science and Technology*, **8**, St. Louis.

[29] Rockwell, R. (1975). *J. Amer. Statist. Ass.*, **70**, 39–42.

[30] Rubin, D. R. (1978). In Aziz and Scheuren [2, pp. 1–18, with discussion and rejoinder, pp. 19–23]. (29 references. Approach generates several imputations for each missing value and uses these to estimate distribution of desired statistics under various models.)

[31] Scott, A. J. (1964). *Optimizing Statistical Analysis: Data Screening and Preconditioning*. Task 389 135, Contract No. 1228 26. Northwestern University, Evanston, Ill. (U.S.G.R.R. Document Number AD-433-551).

[32] Sheppard, W. F. (1899). *Proc. Lond. Math. Soc.*, **31**, 70–99.

[33] Symposium on Incomplete Data (1979). Sponsored by National Research Council, National Academy of Science, Washington, D.C., August 10–11, 1979.

[34] Tukey, J. W. (1949). *Biometrics*, **5**, 99–114.

[35] Tukey, J. W. (1962). *Ann. Math. Statist.*, **33**, 1–67.

[36] Varley, T. C. (1969). Data Input Error Detection and Correction Procedures. Ph.D. dissertation, George Washington University. (A readable compendium of organizational procedures, economic and computational approaches, and insights based on practical experience.)

[37] Walsh, J. E. (1959). *Ann. Inst. Statist. Math. (Tokyo)*, **10**, 223–232.

(CENSUS
COLLECTION OF DATA
FUNOP, FUNOR–FUNOM
OUTLIERS
SURVEY SAMPLING)

JOSEPH I. NAUS

EDUCATIONAL STATISTICS

Educational statistics overlaps almost completely with psychological statistics* and may be broadly categorized into techniques that are used in the analysis of experimental data and those that utilize statistical principles for modeling phenomena that occur in the discipline. The first category, data analysis methods, is shared in common not only with psychology but with a variety of other disciplines such as agriculture (where many of the techniques originated), economics, and sociology. The second category, statistical models, again contains many things common to psychology, but it also includes some that are unique to education.

One reason for the large intersection of methods with psychology is that, in this country, statistics was introduced (c. 1900) to both disciplines by the same people—primarily J. McK. Cattell and E. L. Thorndike—in the context of courses in mental measurement. Since these early common origins, the expanding applications of statistics in the two disciplines have proceeded hand in hand because the main area of education that uses sophisticated statistical methods in research is educational psychology, whose boundary with psychology is indeed a fine line.

DATA ANALYSIS METHODS

Within the rubric of experimental design and analysis, techniques used in educational research range from the simplest *t*-test* to the most complicated analysis of variance (ANOVA)* and analysis of covariance (ANCOVA)*. A typical use of the *t*-test would be in comparing the mean achievement scores of groups taught by two different methods. A one-way ANOVA would extend the comparisons to groups taught by more than two methods.

Factorial* designs in which independent variables besides teaching method, (such as distribution of practice, "social climate" of the classroom, and so forth) are considered, are probably more prevalent than simple *t*-tests and one-way ANOVA. Proceeding up the ladder of complexity of design, mixed designs are also frequently used. The between-subjects factor would typically be teaching method again, or some other variable hypothesized to affect learning, and the within-subjects factor might simply be testing occasions or could represent different conditions under which tests are given.

In all the ANOVA-type studies cited above, one may frequently identify a relevant variable that cannot be controlled or manipulated, such as general intelligence or some specific aptitude. In such situations ANCOVA may be used, which achieves a statistical control of sorts where experimental control is infeasible. Then the presence or absence of treatment differences over and above what may be expected on the basis of, say, IQ differences may be investigated.

Examples of the foregoing methods are too numerous to cite, and the reader may scan a few issues of such journals as *American Educational Research Journal*, the *Journal of Educational Psychology*, and the *Journal of Educational Research* to get an idea of the extent and level of statistical usage. Of course, simpler techniques like chi-square tests* and product-moment correlations are also used.

Aptitude Treatment Interaction

In most applications of ANOVA, interaction* effects usually constitute a nuisance that beclouds the main effects and makes their interpretation difficult. In some educational applications, however, they may signal a phenomenon that was dubbed "Aptitude Treatment Interaction" (ATI) by Cronbach [11] and subsequently popularized by Cronbach and Snow [12]. This refers to a situation in which two treatments are differentially effective for individuals at different levels of aptitude (or any characteristic that affects one's response to treatments). In extreme cases, the regression lines of outcome on aptitude in the two treatment groups may intersect, indicating that Treatment A is more effective for individuals with aptitude level greater than some value X_A, say, while Treatment B is more effective for individuals whose aptitude is less than $X_B(< X_A)$. The interval $[X_B, X_A]$ constitutes an indifference region, such that for individuals with aptitudes in this range it makes no difference whether Treatment A or B is given. This interval is determined by the Johnson-Neyman technique [17]; the method was extended to two aptitude variables by Potthoff [26].

Quasi-Experimental Designs

It is often impossible to adhere strictly to the requirements of experimental design in edu-

cational research in the classroom setting because students are not assigned at random to classes, class schedules cannot be disrupted, and for other reasons beyond the researcher's control. Mindful of this fact, Campbell and Stanley [6] discuss a number of "quasi-experimental designs" in which some of the requirements are relaxed, but which nevertheless permit valid probing of a hypothesis and elimination of rival hypotheses. The designs they discuss include such things as time-series, equivalent time sample designs, nonequivalent control group designs, and so forth. Appropriate statistical analyses are described for each design, as are actual studies reported in the literature that employed that design.

Space forbids our even summarizing the discussions of the ten quasi-experimental designs presented in that article. However, the time-series experiment is briefly described since, due to its popularization by Glass, Willson, and Gottman [16], it is receiving increased attention by educational researchers.

Long used in economics, time-series analysis*, as applied to educational research, seeks to detect and test for significance the effect of a treatment (such as a social intervention) on a series of observations made before and after the intervention. The observations may either be made repeatedly on a single group of subjects—in which case the study is longitudinal or "unit-repetitive"—or on a series of separate but conceptually equivalent groups, such as the set of third graders in a certain school district in successive years; the study is then sequentially cross-sectional or "unit-replicative."

The first step is to decide whether a stationary* or nonstationary process is involved. In educational research, especially in longitudinal studies where learning takes place, the process will most likely be nonstationary. Then the appropriate time-series model is set up, or, rather, two such models are employed—one for the series of observations before the intervention, the other for the post-intervention observations. In the simplest case where the effect of the intervention is assumed to be immediate and constant for all post-intervention outcome data, the two models differ only in that the second contains an extra additive parameter δ representing the effect. Then the parameters for the "collapsed" model (i.e., the two combined without the δ) are estimated by solving the Yule-Walker equations*. Finally, suitable algebraic tansformations of the observations lead to a matrix equation analogous to a multiple regression equation, and a least-squares estimation procedure is used to estimate the magnitude of the intervention effect δ and the "level" of the system at the initial time-point.

The effect may be more complicated than the "immediate and constant" type assumed above, with a time lag before it appears and a waxing or waning thereafter. All these cases are discussed in Glass et al. The one matter that is not addressed in that book, however, is the problem of appropriate parameter-estimation procedures when time-series models are applied to longitudinal data. A partial solution to this problem was given by M. M. Tatsuoka [32], but much more work needs to be done on it.

Multivariate Methods

By their very nature, data in educational research are more often multivariate than not. Traditionally, multiple regression* analysis has been the most widely used multivariate method in education—both for research and for administrative decisions such as screening candidates for college admission. This trend was given a boost recently by articles by Cohen [7] and Darlington [13] and books by Kerlinger and Pedhazur [21] and Cohen and Cohen [8], which describe multiple regression analysis as a "general data-analytic system"—that is, not just a system for predicting a criterion variable from a set of predictors in the usual sense, but in a role customarily played by ANOVA.

A good example of the last-mentioned use for multiple regression analysis is G. J. Anderson's [1] study of the effects of classroom social climate on learning. The social

climates of some 110 high school physics classes were measured by the Learning Environment Inventory (LEI) and used as a predictor along with IQ, IQ $\times$ LEI, $(IQ)^2$ and $(LEI)^2$ in four multiple regression equations using different criterion variables. The entering of IQ $\times$ LEI led to a significant incremental multiple-R by the stepdown procedure, thus indicating a significant interaction effect of intelligence and classroom social climate on the understanding of physics.

In contrast to the stepdown (or sequential) method of assessing the contributions of the various predictors in the multiple regression equation, a simultaneous partitioning-of-variance method called commonality analysis* was developed by Mood [24]. For example suppose that achievement is to be predicted by two sets of variables, one describing student background and the other school characteristics. Then the unique contribution of each set of variables and the joint contribution (or "commonality") of both are each estimated by a suitable incremental squared multiple-R.

Multivariate analysis of variance (MANOVA*), discriminant analysis*, and canonical correlation analysis* were relatively late arrivals to the field of educational research. In particular, the acceptance of multigroup discriminant analysis as a research tool was due, in no small part, to the publicizing efforts of Rulon [28], Tatsuoka and Tiedeman [33], and Cooley and Lohnes [9]. The last-mentioned was the first book on multivariate statistical analysis addressed specifically to behavioral scientists, and it contained a large assortment of computer programs. [It has now been replaced by a revised edition under a new title (Cooley and Lohnes, [10]).] These works were quickly followed by others such as those by Tatsuoka and Tiedeman [34], Bock [3], L. V. Jones [18], Bock and Haggard [5], and Rulon and Brooks [29], most of which contain reanalyses of actual educational studies using MANOVA and discriminant analysis.

Use of all varieties of multivariate analysis* in a large-scale educational study was first made in Project TALENT. Initiated by Flanagan in 1959 under the sponsorship of the U.S. Office of Education, this study continues to this day in the form of periodic follow-up* studies on parts of the sample, which originally comprised some 440,000 students in over 1300 high schools. A report titled *Project TALENT: One-Year Follow-Up Studies* (Flanagan et al., [15]) includes one by Cooley on the prediction of career-plan changes, using canonical correlation analysis, and another by Cooley and Lohnes using discriminant analysis and classification* procedures to draw implications of patterns of career-plan changes for educational and vocational guidance. Other examples of the application of multivariate analysis abound in subsequent reports, all published by American Institutes of Research.

More recent examples of educational research employing multivariate methods may be found in Cooley and Lohnes [10a], which deals with the evaluation of educational programs, and in Kerlinger [20], which, although it is addressed mainly to beginning students, also contains material that is informative to the mature researcher.

Finally, the treatment of all species of multivariate analysis in the framework of the general linear model* has been expounded on by a number of educational psychologists, notably Finn [14] and Bock [4], who also present many research examples. Recent advances in this as well as other areas, such as Bayesian inference*, (many of which are made by educational researchers themselves) are frequently reported in the relatively new *Journal of Educational Statistics*, whose founding editor, Novick, has been influential in publicizing Bayesian statistics among educational researchers (e.g., Novick and Jackson, [25]).

STATISTICAL MODELS

By far the best known statistical model for psychological and educational phenomena is factor analysis*, in which observable variables such as test scores are expressed as linear combinations of hypothetical constructs known as factors. The latter may represent distinct ability dimensions if the tests are in the cognitive domain, distinct personality traits if they are in the affective domain, and so forth. Relatively recent offshoots of this time-honored field are: (1) Tucker's [35] three-mode factor analysis, which starts with a three-dimensional data matrix showing individual scores on several tests administered under different conditions; and (2) Jöreskog's [19] covariance structure analysis, which compares observed and theoretically-dictated covariance matrices to decide which of several competing theories or hypotheses about the factorial structure of the variables is the most tenable.

Latent trait theory seeks to "explain" a person's test score in terms of an underlying latent ability or some other relevant latent characteristic. For simplicity, the test items have traditionally been taken to be dichotomously scored, but extensions to polychotomous items have recently been made (cf. Samejima, [30]).

The first step is to *assume* that the regression of item response on the latent trait has some specific form. The most commonly assumed regression function is the logistic* cumulative distribution function (cdf) in its one-, two-, or three-parameter forms (Rasch, [27]; Birnbaum, [2]; and Lord, [22]). The basic form of this cdf is

$$\Psi(X) = 1/(1 + e^{-X}), \qquad -\infty < X < \infty,$$

but when it is used as the regression of an item score ($U = 1$ or 0) on a latent trait, the X is replaced by a linear function of the latent-trait variable θ involving one or two parameters, a and/or b. In the dichotomous, cognitive item case, then, the regression function represents the probability that a person with latent ability θ will respond correctly to the item. The function is called the item characteristic curve (ICC).

The parameter a stands for the discrimination power of the item, and b for the difficulty. In the one-parameter case (Rasch model) a is equal to unity. The third parameter, c, if used, represents the guessing probability. It does not occur in the argument of Ψ, but is used as a mixing probability, so the

ICC in this case takes the form

$$P(\theta) = c + (1 - c)\Psi\left[a(\theta - b)\right].$$

With an ICC postulated for each item (with separate parameters), the response matrix for a large number of examinees enables estimation of the item parameters and the latent-ability θ_i for each person via the maximum-likelihood method, provided the assumption of local independence holds (i.e., when θ is fixed, correctly answering the various items are statistically independent). Other estimation procedures were recently proposed by Lord [23], K. Tatsuoka [31], and Wainer [36].

Once the parameters have been estimated, various properties of the test and individual ability can be parsimoniously explained. This is one of the most elegant psychometric theories, but it has several unsolved problems, such as the practical need for unidimensionality of the ability underlying the test. When these are solved, latent trait theory promises to become a very powerful one with a host of applications in many fields of behavioral and social science besides education.

CONCLUDING REMARKS

The brief survey above shows statistics in education to be as vigorous an endeavor as that in any field. Contributions from within the discipline are visible and will continue to proliferate, especially in the category of statistical models, which is an interface between the substantive and methodological domains. "Built-in" statistical theories which model the structure of educational data and render external statistical analysis all but unnecessary (as in statistical mechanics), may eventually become the order of the day.

References

[1] Anderson, G. J. *Amer. Educ. Res. J.*, 1970, **7**, 135–152.

[2] Birnbaum, A. (1968). In *Statistical Theories of Mental Test Scores*, by F. N. Lord and M. R.

Novick, Addison-Wesley, Reading, Mass. These chapters are considered to be the most definitive treatment of latent trait theory in print. The entire book is perhaps the most authoritative one in test theory. Birnbaum wrote Chapters 17 and 18.

[3] Bock, R. D. (1966). In *Handbook of Multivariate Experimental Psychology*, R. B. Cattell, ed. Rand McNally, Chicago. Another excellent chapter in another excellent—albeit somewhat dated—book.

[4] Bock, R. D. (1975). *Multivariate Statistical Methods in Behavioral Research*. McGraw-Hill, New York. Probably the most mathematical of the multivariate analysis texts for behavioral scientists. Difficult reading, but contains many real-data examples.

[5] Bock, R. D. and Haggard, E. A. In *Handbook of Measurement and Assessment in Behavioral Sciences*, D. K. Whitla, ed. Addison-Wesley, Reading, Mass.

[6] Campbell, D. T. and Stanley, J. C. (1963). In *Handbook of Research on Teaching*. N. L. Gage, ed. Rand-McNally, Chicago. Considered a "must" on the reading list of educational researchers-to-be. Also available as a separate monograph, with the phrase "on teaching" omitted from the title.

[7] Cohen, J. *Psychol. Bull.*, **70**, 426–443.

[8] Cohen, J. and Cohen, P. (1975). *Applied Multiple Regression/Correlation Analysis for the Behavioral Sciences*. Lawrence Erlbaum Associates, Hillsdale, N. J. Perhaps the most comprehensive treatment of multiple regression analysis as a general data-analytic tool: an " informal" general linear model approach.

[9] Cooley, W. W. and Lohnes, P. R. (1962). *Multivariate Procedures for the Behavioral Sciences*. Wiley, New York. The first multivariate analysis textbook for behavioral-science students. Relatively easy reading, and contains many computer programs. Now superseded by the next title.

[10] Cooley, W. W. and Lohnes P. R. (1971). *Multivariate Data Analysis*. Wiley, New York. See remarks under preceding title.

[10a] Cooley, W. W. and Lohnes, P. R. (1976). *Evaluation Research in Education*. Irvington, New York. Perhaps the most statistically oriented treatment of evaluation methods in education.

[11] Cronbach, L. J. *Amer. Psychol.*, **12**, 671–684.

[12] Cronbach, L. J. and Snow, R. E. (1977). *Aptitudes and Instructional Methods*. Irvington, New York. A definitive, thorough treatise on "aptitude treatment interactions" and related topics. A must for the serious quantitative educational researcher.

[13] Darlington, R. B. *Psychol. Bull.*, **69**, 161–182.

[14] Finn, J. D. (1974). *A General Model for Multivariate Analysis*. Holt, Rinehart & Winston, New York. One of the first texts on multivariate analy-

sis for the behavioral scientist written in the general linear model framework.

[15] Flanagan, J. C., Cooley, W. W., Lohnes, P. R., Schoenfeldt, L. F., Holdeman, R. W., Combs, J., and Becker, S. J. (1966). *Project TALENT: One-Year Follow-Up Studies*. University of Pittsburgh Press, Pittsburgh, Penn.

[16] Glass, G. V., Willson, V. L., and Gottman, J. M. (1975). *Design and Analysis of Time-Series Experiments*. Colorado Associated University Press, Boulder Colo. The only book on time-series analysis, to date, addressed to behavioral scientists.

[17] Johnson, P. O. and Neyman, J. (1936). *Statist. Res. Mem.*, Vol. 1.

[18] Jones, L. V. (1966). In *Handbook of Multivariate Experimental Psychology*, R. B. Cattell, ed. Rand McNally, Chicago, Ill.

[19] Jöreskog, K. G. (1974). In *Contemporary Developments in Mathematical Psychology (Vol. II): Measurement, Psychophysics, and Information Processing*. D. H. Krantz, R. C. Atkinson, R. D. Luce, and P. Suppes, eds. Freeman, San Francisco, Calif. A definitive account of one of the recent breakthroughs in statistical models of data in the factor analysis tradition.

[20] Kerlinger, F. N. (1979). *Behavioral Research: A Conceptual Approach*. Holt, Rinehart & Winston, New York. An introductory survey of quantitative behavioral research methods. Easy reading.

[21] Kerlinger, F. N. and Pedhazur, E. J. (1973). *Multiple Regression in Behavioral Research*. Holt, Rinehart & Winston, New York. Similar to the Cohen and Cohen book but somewhat less comprehensive.

[22] Lord, F. N. (1968). In *Statistical Theories of Mental Tests Scores*. F. N. Lord and M. R. Novick, eds. Addison-Wesley, Reading, Mass. Ch. 16. See remarks under Birnbaum [2].

[23] Lord, F. M. (1980). In *The 1979 Computerized Adaptive Testing Conference*, D. J. Weiss, ed. University of Minnesota Press, Minneapolis, Minn.

[24] Mood, A. M. *Amer. Educ. Res. J.*, **8**, 191–202.

[25] Novick, M. R. and Jackson, P. H. (1974). *Statistical Methods for Educational and Psychological Research*. McGraw-Hill, New York. An introductory text on Bayesian inference for behavioral science students and researchers.

[26] Potthoff, R. F. *Psychometrika*, **29**, 241–256.

[27] Rasch, G. (1960). *Probabilistic Models for Some Intelligence and Attainment Tests*. Nielson and Lydiche, Copenhagen. Definitive treatment of Rasch models. Unfortunately out of print.

[28] Rulon, P. J. *Harvard Educ. Rev.*, **21**, 80–90.

[29] Rulon, P. J. and Brooks, W. D. (1968). In *Handbook of Measurement and Assessment in Behavioral Sciences*. D. K. Whitla, ed. Addison-Wesley, Reading, Mass.

[30] Samejima, F. (1969). Estimation of Latent Ability Using a Response Pattern of Graded Scores. *Psychometrika Monogr. No. 17*.

[31] Tatsuoka, K. K. (1979). The Least-Squares Estimation of Latent Trait Variables by a Hilbert Space Approach. *Tech. Rep. No. 1*, ONR Contract No. N00014–78–C–0159.

[32] Tatsuoka, M. M. (1976). Investigation of Methodological Problems in Educational Research: Longitudinal Methodology. *Final Report for NIE Project No. 4–1114*, Contract No. NIE–C–74–0124, Ch. 8.

[33] Tatsuoka, M. M. and Tiedeman, D. V. *Rev. Educ. Res.*, **24**, 402–420.

[34] Tatsuoka, M. M. and Tiedeman, D. V. (1963). In *Handbook of Research on Teaching*, N. L. Gage, ed. Rand McNally, Chicago, Ill.

[35] Tucker, L. R. (1964). In *Contributions to Mathematical Psychology*, N. Frederiksen and H. Gulliksen, eds. Holt, Rinehart & Winston, New York.

[36] Wainer, H. (1979). In *The 1979 Computerized Adaptive Testing Conference*. D. J. Weiss, ed. University of Minnesota Press, Minneapolis, Minn.

(FACTOR ANALYSIS

MULTIVARIATE ANALYSIS

PSYCHOLOGICAL TESTING THEORY

PSYCHOLOGY, STATISTICS IN

SOCIOLOGY, STATISTICS IN)

MAURICE M. TATSUOKA

EDWARDS' TEST FOR SEASONALITY

Edwards' test for seasonality [1] has been a standard technique in epidemiology. It applies to data that fall into categories having a natural cyclic order, e.g., months of the year or seasons. The null hypothesis is that individuals are equally likely to be allocated to each of the categories. The alternative is that the frequencies in the categories have some periodic (e.g., seasonal) variation. (When an alternative hypothesis is envisioned which includes *any* type of deviation from the null hypothesis or equiprobable categories, the standard χ^2 goodness-of-fit test* might be used.)

Denote by N_i the number of events observed in the ith category, $i = 1, \ldots, k$. A sequence of k weights, $\sqrt{N_i}$, is placed at equally spaced points on the unit circle and the test statistic is derived based on the distance of the center of gravity of these weights from the center of the unit circle. The square of this distance is

$$T = \frac{8n\left\{\left(\sum_{i=1}^{k} s_i\sqrt{N_i}\right)^2 + \left(\sum_{i=1}^{k} c_i\sqrt{N_i}\right)^2\right\}}{\left(\sum_{i=1}^{k}\sqrt{N_i}\right)^2},$$

where $c_i = \cos(2\pi i/k)$, $s_i = \sin(2\pi i/k)$, $n = \sum_{i=1}^{k} N_i$. It is asymptotically distributed as χ_2^2 if there is no variation in population frequencies.

A major drawback is that this asymptotic approximation is not satisfactory in small, or even in medium-sized samples. It has the effect of making the type I errors* in the test too large, leading to too many spurious significant results. A modification of this test statistic of the form

$$R = 2\left[\{\sum N_i s_i\}^2 = \{\sum N_i c_i\}^2\right]/n$$

suggested by Roger [2] has the same asymptotic distribution and enjoys superior small-sample properties. A generalization of Edwards' test for the case of unequal expected numbers in each category and/or when there are apparently unequal spacings of the categories in a cyclic order was devised by Walter and Elwood [4]. St. Leger [3] provides modified percentage points for sample sizes less than 200. He also derives a corresponding likelihood ratio* test for seasonality based on a set of 12 consecutive monthly observations. [Although the distribution of the test statistic (against the same alternative of simple sinusoidal form) approximates moderately well to the χ_2^2 distribution for as few as 20 observations and remarkably well for as few as 50 observations and seems to have (marginally) higher power, it is computationally much more complicated.]

References

[1] Edwards, J. H. (1961). *Ann. Hum. Gen.*, **25**, 83–85.
[2] Roger, J. H. (1977). *Biometrika*, **64**, 152–155.
[3] St. Leger, A. S. (1976). *Appl. Statist.*, **25**, 280–286.
[4] Walter, S. D. and Elwood, J. M. (1975). *Brit. J. Prev. Soc. Med.*, **29**, 18–21.

(LIKELIHOOD RATIO TESTS
PERIODOGRAM ANALYSIS
TIME SERIES)

EFFECTIVE DEGREES OF FREEDOM

If $X_1, \ldots, X_n$ denote a sample of normal independent random variables with mean μ and variance σ^2, then the quadratic form $Z = \sum(X_i - \bar{X}.)^2$ is distributed as σ^2 times a χ^2 variate with $n - 1$ degrees of freedom*. In this context, degrees of freedom represents a parameter identifying the particular member of the χ^2 family. In applied statistics it is often necessary to obtain approximate distributions of more general quadratic forms of normal variables than just those that can be written as sums of squares. The nature of the problem typically dictates that the quadratic form be chosen to have a specific expected value. The distribution of this quadratic form is then approximated by a χ^2 distribution*. The degrees of freedom of this approximating distribution, the effective degrees of freedom of the quadratic form, are chosen so that the variances of the quadratic form at hand and the approximating χ^2 distribution are equal. A general discussion of this can be found in Box [2].

One of the early applications of this technique is given by Welch [9] for testing the difference between means for two samples with unequal variances. A subsequent application is given by Satterthwaite [5] to the analysis-of-variance case, where it is necessary to synthesize one or more mean squares in order to test hypotheses of interest. The techniques are discussed in standard statistical methods books, such as Snedecor and Cochran [7] and Steel and Torrie [8]. In general, if $Z_1, \ldots, Z_m$ denote m independent mean squares with $v_1, \ldots, v_m$ degrees of freedom, respectively, and $a_1, \ldots, a_m$ are constants, then the synthesized mean square $Z = \sum a_i Z_i$ has effective degrees of freedom $(\sum a_i Z_i)^2 / (\sum a_i^2 Z_i^2 / v_i)$. Common practice is to insist that all $a_i > 0$ [1]. However, Gaylor and Hooper [4] provide guidelines for relaxing this rule.

The same general formula for assigning effective degrees of freedom to the estimated variances of means and totals in stratified random samples* is given by Cochran [3]. A third application of the technique is to construct approximate confidence intervals for estimates of variance components*. The estimate is a linear function of mean squares or more generally quadratic forms with a particular expected value. The variance of this function is used to compute effective degrees of freedom and consequently specify an approximating distribution. This application was first proposed by Satterthwaite [6].

References

[1] Anderson, R. L. and Bancroft, T. W. (1952). *Statistical Theory in Research*. McGraw-Hill, New York.

[2] Box, G. E. P. (1954). *Ann. Math. Statist.*, **25**, 290–302.

[3] Cochran, G. W. (1977). *Sampling Techniques*, 3rd ed. Wiley, New York.

[4] Gaylor, D. W. and Hooper, F. N. (1969). *Technometrics*, **11**, 691–706.

[5] Satterthwaite, F. E. (1941). *Psychometrika*, **6**, 309–316.

[6] Satterthwaite, F. E. (1946). *Biometrics*, **2**, 110–114.

[7] Snedecor, G. W. and Cochran, W. G. (1967). *Statistical Methods*, 6th ed. Iowa State University Press, Ames, Iowa.

[8] Steel, R. G. D. and Torrie, J. H. (1979). *Principles and Procedures of Statistics*, 2nd ed. McGraw-Hill, New York.

[9] Welch, B. L. (1937). *Biometrika*, **29**, 350–362.

(ANALYSIS OF VARIANCE
BEHRENS–FISHER PROBLEM
DEGREES OF FREEDOM)

Francis G. Giesbrecht

EFFICACY

If the critical region* of a test of a statistical hypothesis specifying the value of a parameter θ—$\theta = \theta_0$, say—can be expressed as $T > K$ or $T < K$, or more generally as T outside a certain interval, its *efficacy* is defined as

$$\frac{\left(\dfrac{\partial E(T|\theta)}{\partial \theta} \right)^2 \Big|_{\theta = \theta_0}}{\operatorname{var}(T|\theta = \theta_0)}$$

This quantity is sometimes referred to as the efficacy of the test statistic T. It is related to the efficiency* of the test, but not in a deterministic way. The efficacy is defined in terms of behavior of T when θ is close to θ_0, so it does not necessarily reflect properties of the test when θ differs substantially from θ_0. Note also that, unlike efficiency, efficacy depends on sample size (usually it is a multiple of sample size).

Asymptotic relative efficiency (ARE), however, is more closely related to efficacy, since ARE also depends essentially on properties of the test for small departures of θ from θ_0. (ARE is also used to describe properties of estimators.)

(EFFICIENCY, ASYMPTOTIC
   RELATIVE)

EFFICIENCY, ASYMPTOTIC RELATIVE (ARE) OF ESTIMATORS

Let $\hat{\theta}_1$ and $\hat{\theta}_2$ be two consistent* estimators of θ. Denote the variances of $\hat{\theta}_i (i = 1, 2)$ by $\sigma_i^2 (i = 1, 2)$, respectively. (These variances usually depend on the sample size n, but this dependence is not exhibited here.) The asymptotic relative efficiency of $\hat{\theta}_1$ with respect to $\hat{\theta}_2$ is the limit

$$\lim_{n \to \infty} \frac{\sigma_2^2}{\sigma_1^2}.$$

If this limit (for a particular underlying population) is greater (less) than 1, we say that $\hat{\theta}_1$ is more (less) efficient than $\hat{\theta}_2$. If the limit is 1, the two estimators are asymptotically *equally* efficient (for the population under consideration).

In general, asymptotic relative efficiency

is the limiting value of relative efficiency* as sample size (n) increases without limit. 'Efficiency' is also used in describing properties of tests.

Bibliography

Johnson, N. L. and Leone, F. C. (1977). *Statistics and Experimental Design*, 2nd ed., Vol. 1. Wiley, New York.

Kendall, M. G. and Stuart, A. (1975). *The Advanced Theory of Statistics*, 3rd ed., Vol. 2. Charles Griffin, London.

Zacks, S. (1971). *The Theory of Statistical Inference*, Wiley, New York.

(EFFICACY)

EFFICIENCY, SECOND-ORDER

INTUITIVE INTRODUCTION FOR NONMATHEMATICAL READERS

In problems of statistical inference concerning an unknown parameter θ, a practicing statistician has an ample choice of criteria to compare the performance of possible statistics T_n based on a moderate finite sample of size n. If there is no clear choice and the sample size is sufficiently large, one can consider an asymptotic comparison. Under some regularity assumptions, the first-order asymptotic efficiency of an unbiased estimator T_n, as compared to the maximum likelihood* estimator $\hat{\theta}_n^*$, $e_1(T; \hat{\theta}^*)$, can be defined as

$$e_1(T; \hat{\theta}^*) = \lim_{n \to \infty} \frac{\operatorname{var} \hat{\theta}_n^*}{\operatorname{var} T_n}. \qquad (1)$$

It is clear that $0 \leqslant e_1 \leqslant 1$, and if the first-order efficiency e_1 is less than 1, then T_n is inferior to $\hat{\theta}_n^*$ for large n. If, however, $e_1(T; \hat{\theta}^*) = 1$, the first-order efficiency fails to distinguish between the relative performance of the two sequences of estimators. There are many cases when this happens (*see* ASYMPTOTIC NORMALITY). The natural question arises of the speed of convergence of the variance ratio to 1. Although there are several different formal definitions of second-order efficiency not always equivalent, in simple very regular cases the typical speed

is inversely proportional to n. One can therefore say that

$$1 - \frac{\operatorname{var} \hat{\theta}_n^*}{\operatorname{var} T_n} \simeq \frac{D}{n}$$

or

$$\lim_{n \to \infty} n \left[1 - \frac{\operatorname{var} \hat{\theta}_n^*}{\operatorname{var} T_n} \right] = D.$$

Some formal definitions reduce in regular cases to the proportionality constant D in a sense that the smaller D the better is the second-order efficiency. A letter D was used here to indicate that it represents the second-order "deficiency," and in fact Fisher [4] observed that D/n represents the proportion of independent observations "lost" because of using T_n rather than $\hat{\theta}_n^*$. Therefore, $(1 - D/n)$ would represent the proportion of observations effectively used when employing T_n.

DEFINITIONS

The concept of second-order efficiency is used in estimation theory and other statistical inference problems. There are several definitions discussed in the literature, each having its merits and disadvantages. The ones that are more intuitively appealing often involve difficult calculations. Consider a single-parameter-point estimation problem. Let $X_1, X_2, \ldots, X_n, \ldots$ be independent, identically distributed (i.i.d.) random variables each with density $f = dP^{X; \theta}/d\mu$ (for some μ) with θ varying in an open set Θ. It will be assumed that f is regular "enough" to admit at least two differentiations with respect to θ under the integral ($d\mu$). Although assuming this type of regularity is usually not too restrictive, to justify the use of the Cramér–Rao inequality* some regularity of the statistic T_n under consideration is required. That would put unnecessary restrictions on a possible choice of estimators; hence T_n will be quite general. To consider the second-order efficiency one assumes that T_n is a consistent sequence and first order efficient. The last condition should be made more precise in view of possible "irregularity" of T_n. Following the approach in ref. 10, we shall say that T_n is first order efficient if

$$\lim_{n \to \infty} (p) \left[n^{-1/2} \frac{d \log L}{d\theta} \right.$$
$$\left. - A - Bn^{1/2}(T_n - \theta) \right] = 0 \quad (2)$$

for some constants A and B possibly depending on θ, where L denotes the likelihood function and $\lim(p)$ is the limit in probability. Some definitions of the second-order efficiency will be listed below. Suppose that the sequence of statistics T_n possesses density ϕ_{T_n} with respect to some measure ν. Let I_f denote the Fisher information associated with density f in a sample of size 1 and let

$$I_{\phi_n} = \frac{1}{n} E \left(\frac{d \log \phi_{T_n}}{d\theta} \right)^2,$$

i.e., Fisher's information* per unit observation using statistic T_n.

Definition 1. $D_1 = \lim_{n \to \infty} n[I_f - I_{\phi_n}]$.

Definition 2. $D_2 = \lim_{n \to \infty} \text{var}((d \log L / d\theta) - d \log \phi_{T_n} / d\theta)$.

Definition 3.

$$D_3 = \lim_{n \to \infty} \inf_{h \in H} E \left[(d \log L / d\theta) \right.$$
$$\left. - h(T_n - \theta) \right]^2,$$

where H is the class of all measurable functions possibly depending on θ.

Definition 4.

$$D_4 = \lim_{n \to \infty} \inf_{h \in H_2} E \left[(d \log L / d\theta) \right.$$
$$\left. - h(T_n - \theta) \right]^2,$$

where H_2 is the class of second-degree polynomials.

Definition 5.

$$D_5 = \lim_{n \to \infty} \inf_{h \in H_1} E \left[(d \log L / d\theta) \right.$$
$$\left. - h(T_n - \theta) \right]^2,$$

where H_1 is the class of linear functions.

Definition 6. $D_6 = \lim_{n \to \infty} n^2[\text{var } T_n - (nI_f)^{-1}]$ (with some regularity restrictions on T_n).

Note that D_6 reduces to the variance expansion var $T_n = 1/nI_f + D_6/n^2 + o(1/n^2)$. One could also use this definition with suitable correction for bias.

One relationship among the D_i, $i = 1$, $2, \ldots, 6$, follows directly, i.e., $D_3 \leqslant D_4 \leqslant D_5$. Upon closer inspection it turns out that $D_1 = D_2 = D_3$ (see refs. 3 and 10). It has been established in ref. 3 that in some cases $D_1 < D_4$. The question of general conditions under which $D_1 = D_4$ was posed by C. R. Rao [10] and appears to be open as of this writing.

HISTORICAL NOTE

First studies of this concept appear in the literature in the work of R. A. Fisher [4] in connection with multinomial distributions*. The name "second-order efficiency" was introduced by C. R. Rao (see ref. 10), who discussed D_i, $i = 2, 3, 4, 5$, and later D_6. The pioneering work of C. R. Rao in this subject was followed later by several leading researchers. Efron studied the concept in connection with statistical curvature [3]. Ghosh and K. Subrahmaniam considered extension to a larger class of distributions [6]. The concept of second-order efficiency under quadratic loss structure was studied by Rao [12], who also posed a question of the second-order efficiency under more general loss function. This was studied by Ghosh et al. [7]. As the concept is closely related to properties of maximum likelihood estimators, one should mention the fundamental work of LeCam (e.g., ref. 9) in this connection. Approaches based on concepts other than Fisher's information have also been considered, e.g., by Torgersen [13] in connection with test performance. Finally, one should mention a most recent paper by Ghosh and Sinha [5] on a closely related subject of second-order admissibility. No recent review paper on the subject of second-order efficiency with extensive bibliography exists. The latest paper containing essential ideas is that of Efron and discussants [3].

For additional information, see Amemiya [1], Berkson [2], and Rao [11].

AN EXAMPLE AND RELATED CONCEPTS

The following example is a simplified version of a problem discussed in ref. 3 whose solution appeared before in ref. 12 in a more complex version.

Consider $X_1, X_2, \ldots, X_n$ i.i.d. random variables with bivariate components whose density forms a "curved" exponential family

$$f^{X_1; X_2}_{\eta_1; \eta_2}(x_{1i}; x_{2i}) = c(\eta_1; \eta_2)h(x_{1i}; x_{2i})$$
$$\times \exp(\eta_1 x_{1i} + \eta_2 x_{2i}),$$
$$i = 1, 2, \ldots, n. \quad (3)$$

Suppose that the parameter space is restricted by an equation $\eta_2 = g(\eta_1)$, where g is a smooth twice-differentiable function. Suppose that $\theta = \theta(\eta_1; \eta_2)$ is to be estimated with quadratic loss structure. To evaluate the performance of estimators, the risk function* is considered. Let $\overline{X} = (\overline{X}_1; \overline{X}_2)$ denote the sufficient statistic*. Any "regular" estimator $\hat{\theta}(\overline{X})$ can be modified to assure that the bias function and its derivative can be absorbed in the last term of the expression below. It can be shown that for any "regular" estimator $\hat{\theta}$,

$$\operatorname{var} \hat{\theta} = \frac{1}{nI_f} + \frac{1}{n^2 I_f}\left\{\gamma^2 + 4\frac{\Gamma^2}{I_f} + \Delta(\hat{\theta})\right\}$$
$$+ o(1/n^2), \quad (4)$$

where $\Delta(\hat{\theta}) \geqslant 0$ and $= 0$ when $\hat{\theta}$ is the maximum likelihood estimator. The quantities γ^2 and Γ^2 defined below depend on f and θ only and are independent of the choice of $\hat{\theta}$. (This type of variance expansion for the maximum likelihood estimator was first obtained by Haldane and Smith [8].) It can be observed that the maximum likelihood estimator has the smallest variance up to order n^{-2} and hence is second order most efficient according to Definition 6, with D_6 being the coefficient of n^{-2} in (4).

To complete the notation we have

$$\gamma^2 = \frac{d^2 g(\eta_1)/d\eta_1^2}{\left[1 + (dg(\eta_1)/d\eta_1)^2\right]^3} \quad (5)$$

evaluated at (η_1, η_2) such that $\theta(\eta_1; \eta_2) = \theta$, and

$$\Gamma^2 = \frac{d^2\psi/d\theta'^2\big|_{\theta' = \theta}}{\left[1 + (d\psi/d\theta')^2_{\theta' = \theta}\right]^3}, \quad (6)$$

where

$$\psi = \theta' + E\left[\frac{d\log L}{d\theta}\frac{1}{nI_f}\bigg|\theta = \theta'\right].$$

All calculations as well as regularity conditions on $\hat{\theta}$ are omitted here. One should only mention that they are based on large deviations probabilities, which provide necessary tools in this type of calculations.

References

[1] Amemiya, T. (1978). The n^{-2} Order Mean Squared Error of the Maximum Likelihood and Minimum Logit Chi-Square Estimates. *Tech. Rep. No. 267, Econ. Ser.*, Stanford University, Stanford, Calif.

[2] Berkson, J. (1980). *Ann. Statist.*, **8**, 457–478.

[3] Efron, B. (and discussants) (1975). *Ann. Statist.*, **3**, 1189–1242.

[4] Fisher, R. A. (1925). *Proc. Camb. Philos. Soc.*, **22**, 700–725.

[5] Ghosh, J. K. And Sinha, B. K. (1979). A Necessary and Sufficient Condition for Second Order Admissibility with Applications to Berkson's Problem. Unpublished manuscript.

[6] Ghosh, J. K. and Subrahmaniam, K. (1974). *Sankhyā A*, **36**, 325–358.

[7] Ghosh, J. K., Sinha, B. K., and Wieand, H. S. (1980). *Ann. Statist.*, **8**, 506–521.

[8] Haldane, J. B. S. and Smith, S. M. (1956). *Biometrika*, **43**, 96–103.

[9] LeCam, L. (1974). Notes on Asymptotic Methods in Statistical Decision Theory. Centre de Recherches Mathématiques, Université de Montréal.

[10] Rao, C. R. (1961). *Proc. 4th Berkeley Symp. Math. Statist. Prob.*, Vol. 1. University of California Press, Berkeley, Calif., pp. 531–545.

[11] Rao, C. R. (1962). *J. R. Statist. Soc. B*, **24**, 46–72.

[12] Rao, C. R. (1963). *Sankhyā*, **25**, 189–206.

[13] Torgersen, E. N. (1970). *Z. Wahrscheinlichkeitsth. verwend. Geb.*, **16**, 219–249.

**(ASYMPTOTIC NORMALITY
MAXIMUM LIKELIHOOD
ESTIMATION)**

Piotr W. Mikulski

EFFICIENT SCORE

Let X_j be a random variable with density $f_j(x; \theta)$ of known form, θ in some convex parameter space such as the real line. The *efficient score* ([5]; see also Cox and Hinkley [4]) for X_j evaluated at θ_0 is defined to be the random variable

$$U_j(\theta_0) = \frac{\partial}{\partial \theta} \log f_j(X_j; \theta_0).$$

If $X = (X_1, \ldots, X_n)$ with all X_j independent, then the *efficient score* for X is $U(\theta_0) = \sum U_j(\theta_0)$, summing over all j from 1 to n.

Implicit in the following discussion are certain regularity conditions (see LeCam [6]). We note that these conditions are met by any density from the exponential family*. In such cases,

$$E\left[U(\theta_0) \mid \theta_0 \right] = 0$$

and the univariate variance or multivariate covariance matrix* of $U(\theta_0)$ given θ_0 is

$$i(\theta_0) = E\left[-\frac{\partial}{\partial \theta} U(\theta_0) \mid \theta_0 \right],$$

the *Fisher information** about θ_0 from X. Also the asymptotic distribution of $U(\theta_0)$ is normal with these parameters.

To test the one-dimensional hypothesis $H_0 : \theta = \theta_0$ against the alternative $H_\delta : \theta = \theta_0 + \delta$, we apply a Taylor expansion* to the log likelihood ratio

$$\log f(X; \theta_0 + \delta) - \log f(X; \theta_0)$$
$$= \delta U(\theta_0) + o(\delta)$$

and if $\delta > 0$, reject H_0 if $U(\theta_0)$ is too large, for $\delta < 0$, if too small. Hence the efficient score is the locally optimum test statistic. Uniformly most powerful test statistics for $H_0 : \theta = \theta_0$ against $H_+ : \theta > \theta_0$ or $H_- : \theta < \theta_0$ are efficient score statistics. This concept can be extended to asymptoti-

cally equivalent forms of the likelihood ratio statistic; see, for example, Cox and Hinkley [4, Sec. 9.3]. In similar manner, the efficient score leads to good confidence interval* estimates for the parameter θ when the sample size is reasonably large [1–3].

The efficient score plays an interesting role in point estimation. Let T be any one-dimensional statistic with finite second moment. Let $g(\theta) = E[T; \theta]$. Then T achieves the *Cramér–Rao** lower bound—implying that T is a uniformly minimum variance unbiased estimate of $g(\theta)$—if and only if T is a linear function of $U(\theta)$. On the other hand, $\hat{\theta}$ is a maximum likelihood estimate (MLE) of θ for a regular distribution only if $U(\hat{\theta}) = E[U(\theta); \theta] = 0$. The approximation

$$\sqrt{n}\, i(\theta)(\hat{\theta} - \theta) \doteq U(\theta)/\sqrt{n}$$

serves to establish both the consistency and the asymptotic normality for a unique MLE.

References

[1] Bartlett, M. S. (1953). *Biometrika*, **40**, 12–19.

[2] Bartlett, M. S. (1953). *Biometrika*, **40**, 306–317.

[3] Bartlett, M. S. (1955). *Biometrika*, **42**, 201–204.

[4] Cox, D. R. and Hinkley, D. V. (1974). *Theoretical Statistics*. Chapman & Hall, London.

[5] Fisher, R. A. (1925). *Proc. Camb. Philos. Soc.*, **22**, 700–725.

[6] LeCam, L. (1970). *Ann. Math. Statist.*, **41**, 802–828.

**(FISHER INFORMATION
MAXIMUM LIKELIHOOD
ESTIMATION)**

Gerald A. Shea

EIGENVALUE; EIGENVECTOR

A scalar λ is called an eigenvalue and a nonzero $(1 \times n)$ vector $\mathbf{v}$ is called an eigenvector of the $n \times n$ matrix $\mathbf{T}$ if $\mathbf{Tv} = \lambda \mathbf{v}$. λ must be a root of the polynomial equation $|\mathbf{T} - \lambda \mathbf{I}| = 0$. A linear transformation $\mathbf{Y} = \mathbf{TX}$ on a vector space of dimension n

has at most n distinct eigenvalues. If there exists a basis* consisting of eigenvectors, the linear transformation is *diagonalizable*. Eigenvalue is also called characteristic root, latent root, proper value, or spectral value; similarly, eigenvector is referred to as characteristic vector, latent vector, proper vector, or spectral vector.

ELASTICITY

The elasticity of a variable Y with respect to a variable X is the ratio of the proportional change in Y to the proportional change in X associated with it. Formally,

$$\text{elasticity} = \frac{d(\log Y)}{d(\log X)} = \frac{X}{Y}\frac{dY}{dX}.$$

The concept originated in econometrics*, with particular reference to relations between demand (Y) and price (X), but it is now more widely used. As defined above, the elasticity of demand relative to price is nearly always negative. However, in such cases—when the sign is well understood—it is usual to omit the sign and just refer to the magnitude of the elasticity (often as a percentage).

When there are a number of variables $X_1, X_2, \ldots, X_k$ which may affect the value of Y, the elasticity of Y with respect to X_j can be defined as

$$\frac{\partial(\log Y)}{\partial(\log X_j)} = \frac{X_j}{Y}\cdot\frac{\partial Y}{\partial X_j}.$$

Generally, this will depend on the values of the other X's as well as on X_j. This is not so, however, if $\log Y$ is a linear function of $\log X_1, \ldots, \log X_k$. If we have a multiple linear regression*

$$E\big[\log Y \mid \log X_1, \ldots, \log X_k\big]$$
$$= \beta_0 + \beta_1 \log X_1 + \ldots + \beta_k \log X_k,$$

then the elasticity of y with respect to X_j will be approximately β_j, for all values of the X's.

(ECONOMETRICS)

ELECTION FORECASTING TRANSFERABLE VOTE SYSTEM

Election to a parliament by single transferable vote ensures the proportionate representation of minority groups in the parliament and allows the voter freedom to choose between candidates within a group [4]. The system is used in the Republic of Ireland, Northern Ireland, Malta, and Tasmania and for elections to the upper houses in Australia and South Africa. It is also used in many local government elections (including the City Council of Cambridge, Mass.) and for the election of many other bodies, such as the General Synod of the Church of England and the New York School Board. The system is relatively complicated and it was only in 1975 that the first successful election forecast by computer was made.

THE TRANSFERABLE VOTE SYSTEM

Election by single transferable vote is described in detail by Knight and Baxter-Moore [3]. Below we give an abbreviated and simplified description of the system.

The constituency is always a multimember constituency, i.e., one that elects m members in a ballot, where m is usually between 3 and 8. Each voter casts a single vote, listing the candidates in the order of his or her preference by marking 1, 2, 3, etc., opposite the names on the ballot paper.

During the count the ballot papers are first arranged in parcels, according to the first preferences recorded for each candidate on the ballot papers. A quota, q, is calculated as

$$q = \frac{N}{m+1} + 1, \tag{1}$$

where N is the total number of valid votes.

The count then proceeds in stages. At each stage if any candidate's vote, v_k, equals or exceeds q, the candidate is elected; (1) ensures that no more than m candidates can be elected. When $v_k > q$ the candidate has more votes than he or she needs for election. The surplus votes, $s_k = v_k - q$, are therefore distributed to the other candidates by transferring a fraction of a vote, $f_k = s_k/v_k$, to the next eligible candidate listed on each of his or her ballot papers. An eligible candidate is one who is not already elected or excluded (see below). Thus if the candidate with a surplus has v_{ki} ballots showing next preferences for candidate i, then the transferred vote to i is

$$t_{ki} = v_{ki}f_k = \left[v_{ki}/v_k\right]s_k. \qquad (2)$$

All candidates $\{i\}$ then have their total votes increased by $\{t_{ki}\}$. This completes one stage in the count.

At any stage in the count when no surplus is to be distributed and when fewer than m candidates have been elected, the candidate with the lowest vote is excluded from the count and his or her ballots distributed to the other candidates by the same process used for the distribution of a surplus. In (2), s_k is replaced by v_k and the transferred vote from excluded candidate k to candidate i is given by

$$t_{ki} = \left[v_{ki}/v_k\right]v_k = v_{ki}. \qquad (3)$$

The count continues in a sequence of stages, each distributing either a surplus or an excluding candidate's total vote, until m candidates are elected.

FORECASTING UNDER THE TRANSFERABLE VOTE SYSTEM

This relatively complex process complicates forecasting considerably. A few hours after the count begins the early results come in, but they give only the first preference votes at stage 1. The forecaster must first use these early results to predict the total number of first preference votes in each of the remaining constituencies by the same methods used for forecasting election results for first-past-the-post elections (*see* ELECTION FORECASTING IN THE U.K.). However, this forecast gives only a small part of the final result. For example, in the 1973 Assembly Election

for Northern Ireland [2] only 17 members had been elected out of 78 from the votes counted at the first stage in each constituency. The election of the other 61 candidates depended on transfers indicated by second, third, or later preferences on the ballot papers. An average of 14 stages per constituency was needed to complete the count over a period of 2 days.

The forecaster, therefore, must be able to predict how votes will be transferred in order to forecast the later stages in the count. Such forecasting, without a digital computer, is almost impossible.

THE COMPUTER PROGRAM

It is straightforward for a computer program to calculate the total number of votes, s_k or v_k, to be transferred at any stage. The votes transferred to each candidate t_{ki} in (2) and (3) are not so easily calculated. These equations can be written in matrix form as

$$\mathbf{t} = \mathbf{Cv} \qquad \text{or} \qquad \mathbf{t} = \mathbf{Cs}, \qquad (4)$$

where matrix $\mathbf{C}$ is defined by the relation

$$C_{ki} = v_{ki}/v_k. \qquad (5)$$

We call this the *candidate transfer probability matrix*. It has the property that

$$\sum_i C_{ki} = 1. \qquad (6)$$

The candidates can be grouped according to their party and to a first approximation, neglecting the influence of individual characteristics of candidates and their position on the ballot paper, the transfers to all candidates of a particular party, r, are equal. Thus $C_{ki} = C_{kj}$ if i and j belong to the same party r.

We now define a new matrix, $\mathbf{P}$, called the *party transfer probability matrix*, such that

$$P_{qr} = \sum_i C_{ki}, \qquad (7)$$

where candidates k and i belong to parties q and r, respectively, and the sum is over all candidates i belonging to party r. The element P_{qr} is the probability that a voter marking a particular preference for a candidate of party q will transfer his or her next preference to a candidate of party r. A detailed analysis of transfers in a local government election [1] has shown that the distribution of values of this probability over almost 100 different constituencies has a small variance and that we can therefore use the mean value of this probability for election forecasting.

A difficulty occurs because candidates for all parties are not present in all constituencies and at all counts. We consider, therefore, the position when no candidate of party s is present. Then the probability of transfer from party q to party r, with s absent, is

$$P_{qr}^s = P_{qr} + P_{qs}P_{sr}/(1 - P_{ss}) \qquad \text{all } r, r \neq s,$$

$$(8)$$

since the probability of a vote transferring from p to s and then from s to r is the product of the two probabilities. The factor $1 - P_{ss}$ must be included since transfers from s to itself are not possible and the equation must be normalized to satisfy (6). If candidates for two parties s and t are not present, then the probability is

$$P_{qr}^{s,t} = P_{qr}^s + P_{qt}^s P_{tr}^s/(1 - P_{tt}^s). \qquad (9)$$

These equations allow recursive calculations of the party transfer probability matrix associated with any subset of the parties left at any stage in the count. From this the appropriate candidate transfer probability matrix $\mathbf{C}$ is computed using (7) and the forecast transferred votes from (4).

In this way, the computation of all stages in the count are made. As the results for each subsequent stage are made known they can be read into the computer for a recalculation of the later stages and a more accurate forecast.

1975 NORTHERN IRELAND ELECTION PREDICTION

This method was first used by the BBC to forecast the election result for the N. Ireland Convention in 1975, with the aid of a 32K-store Alpha 2 minicomputer. It was very successful. At 3:00 P.M. on the first day of the count it predicted the final result within two seats, although it took another 24 hours for the last stage in all counts to be completed.

References

[1] Elliott, S. and Smith, F. J. (1977). *The Northern Ireland Local Government Elections of 1977*. The Queen's University of Belfast, Belfast.

[2] Knight, J. (1974). *Northern Ireland: The Elections of 1973*. The Arthur McDougall Fund, London.

[3] Knight, J. and Baxter-Moore, N. (1972). *Northern Ireland: The Elections of the Twenties*. The Arthur McDougall Fund, London.

[4] Lakeman, E. (1974). *How Democracies Vote*, 4th ed. Faber and Faber, London.

F. J. SMITH

ELECTION FORECASTING IN THE UNITED KINGDOM

Forecasting the composition of the next parliament has long been regarded by political scientists and the media as an interesting and integral part of the coverage of national elections in the U.K. To this end statistical techniques have been used ranging from crude extrapolation* to sophisticated model building.

The U.K. electoral system has some important distinguishing features. It is based on a number of single-member constituencies (currently 635) in each of which a political party may field one candidate, a voter has one vote to cast, and the winner is the candidate who receives the greatest number of votes (the "first past the post" system). The aim of election forecasting is to predict the winning candidate in each constituency, and from this the number of seats won by each party. Most postwar elections have been fought between the two major parties, Conservative and Labour, although in recent years some minor parties have had an important effect on the outcome. This has led to a fundamental change in the nature of the contest, from a mainly two-party system to a three-, and in some places, four-party contest. There has also been an increasingly more complex pattern of movement of voters between the parties and the closeness of most recent elections has set a premium on accurate predictions. For these reasons the problems of forecasting have become more complicated and the statistical techniques used more sophisticated.

The boundaries of U.K. constituencies usually do not coincide with the boundaries of administrative areas so that there is a paucity of up-to-date information about their socioeconomic characteristics. Accordingly, the main data used for prediction purposes are the results of previous elections (or the estimated results when boundaries have changed).

It is important to distinguish two forecasting contexts which raise different problems. First, in the period between elections and leading up to the election itself, regular opinion polls are carried out and the aim is to translate voting intentions into a preelection forecast of the number of constituencies that would be won by each party if an election were held on the date of the poll. The second context is on the election night itself when results start to come in and the aim is to forecast the final composition of parliament from the results received to date. Such forecasts are broadcast throughout the television and radio programmes covering the election.

The main technique used for preelection forecasts is based on "swinging," a measure of the change in relative shares of the vote of the two major parties. If C_1 and L_1 were the percentages of the vote for Conservative and Labour respectively at the previous election and C_2 and L_2 the percentages forecast by the poll, the swing, S, is defined as $S = \frac{1}{2}[(C_2 - C_1) + (L_1 - L_2)]$. Berrington [1] describes and evaluates some alternative measures of swing. Given the assumption that all electoral movements involving other parties and abstention cancel out, swing

measures the net transfer of votes between Conservative and Labour. A forecast is obtained by applying the swing S to each constituency in turn. If $C\hat{n}$ and $L\hat{n}$ were the Conservative and Labour percentages in the previous election in constituency n, then the predicted percentage of the Conservative vote is $C\hat{n} + S$, and of the Labour vote $L\hat{n} - S$, the larger of these two quantities giving the predicted major party winner. When there are significant minor parties the measure is calculated for the top two parties and applied similarly.

Recent work has improved estimates of the shares of the vote. Whiteley [9] has used a Box–Jenkins forecasting model* to produce a better estimate of the shares of the vote using a time series* of poll predictions. Budge and Farlie [3] describe a new and completely different approach. They have developed a statistical model of rational voting behavior using survey evidence about voting intentions, which can be used to forecast the shares of the vote. The model is based on a data-dependent application of Bayes' theorem*. The proportion of respondents who are in category j of an attribute A (e.g., social class), given that the respondents intended to vote for party i, $\Pr[Aj \mid Pi]$, is estimated from the survey data. Bayes' theorem is then used to calculate the posterior probability* of a respondent choosing party i, given that he or she is in category j of A as

$$\Pr[Pi \mid Aj] = \frac{\Pr[Aj \mid Pi]\Pr[Pi]}{\sum_j \Pr[Aj \mid Pi]\Pr[Pi]},$$

where $\Pr[Pi]$ is the prior probability* that the respondent would vote for party i. This formula is applied sequentially to a set of attributes to obtain the posterior probability $\Pr[Pi \mid Aj, Bk, \ldots]$ for each voter profile, i.e., combination of the values of the attributes $A, B, \ldots$. Typical attributes used here are age, sex, class, religion and region. At each stage $\Pr[Pi]$ is replaced by the posterior probability for the attributes incorporated in the previous step. Initially $\Pr[Pi]$ is set to the tie value of $1/I$, where I is the number of parties. The attributes used may be characteristics such as sex or age, positions on salient issues in the election, and combinations of these where significant interactions are identified. A forecast of the number of constituencies won by each party is obtained by applying the final posterior probabilities to each constituency whose distribution of voter profiles has been estimated from the survey data.

Election night forecasting presents rather different problems since there is a premium on obtaining accurate forecasts as the outcome will soon be known. Here statistical methods are used for essentially entertainment purposes and there is a healthy competition between the organizations sponsoring the forecasts. In the United Kingdom results are declared on a constituency basis so that the total number of votes for each candidate is given; this means that results are received as discrete quanta and not as a sample of electoral districts as in many other countries. The main problem encountered is that the first constituency results to be declared do not form a representative sample* of all constituencies. In particular, constituencies in the predominantly Labour urban areas tend to declare early. These constituencies have shown rather different behavior in terms of movement between parties than the mainly Conservative rural constituencies, where minor parties also have their strongest support. Thus the problem is to use a biased sample of results to predict the outcome in the remaining undeclared constituencies.

A variety of statistical methods have been used for election night forecasting. In the early 1950s the method used was based on Kendall and Stuart's cube law [4]. This asserted that the number of constituencies won by a party was proportional to the cube of its share of the vote. This law was based on the empirical fact that the proportions of the vote obtained by a party followed a normal distribution* with particular properties. Laakso [5] has refined the cube law and demonstrates that a proportionality of $2\frac{1}{2}$ rather than 3 has, on average, given rather better forecasts in successive U.K. elections.

Tufte [8] gives a cross-national comparison of the operation of the cube law.

The main defects of the cube law approach were that it only worked satisfactorily in an election where there were two major parties predominant, it did not deal with bias in the declaration order, and it did not give predictions for individual constituencies. The last deficiency was met by the use of swing, which has provided the basis for most prediction algorithms since 1951. The mean swing, $\bar{S}$, from the previous election for the sample of results declared was calculated and then applied to each undeclared constituency in turn as described above. This method was extended in various ways, primarily by grouping constituencies (e.g., by pattern of candidacy) so that adjustments were made to the mean swing for each such type. Milledge and Mills [6] give an example of this approach. The swing-based approaches worked well in elections fought by two predominant major parties, but have been found to be seriously deficient in recent elections where there have been significant minor parties.

A more sophisticated method used by the BBC since 1974 has given good predictions and is described in Brown and Payne [2]. Multiple regression* was used to predict the change in share of electorate in each constituency for all significant parties, using independent variables such as the party shares at the previous election, candidate pattern and socioeconomic characteristics of the constituencies. For the top three parties in each undeclared constituency a vector of predicted shares of the electorate, $Zi = (Z_1, Z_2, Z_3)$, assumed to be trivariate normal with a variance–covariance* structure, V, was obtained from the regressions on the set of declared results. The probability, $Pr[Pi]$, that party i would win the constituency was then calculated as

$$Pr[Pi] = Pr\{[(Zi - Zi') > 0] \cap [(Zi - Zi'') > 0]\},$$

$i \neq i' \neq i'' = 1, 2, 3$, where $(Zi - Zi', Zi - Zi'')$ is bivariate normal* with mean and variance–covariance matrix obtained directly from Zi and V. The calculation thus involved a bivariate normal integration. Any remaining parties were assumed to have zero probability of a win.

A similar but simpler approach was used for constituencies that were considered to be contests between the two major parties only. Here the regression model was used to predict the swing from Labour to Conservative, s, in each undeclared constituency to give a predicted Conservative share of the two-party vote of $c + s$, where c is the Conservative share of the two party vote at the previous election. Table 1 lists the independent variables in inclusion order and with their estimated coefficients for this regression obtained when 5% of the results had been declared in the May 1979 General Election.

The probability of a Conservative win was then estimated as

$$\Pr[\text{con}] = \Phi\{(c + s - 0.5)/a)\}$$

where Φ is the cumulative standard normal distribution function*, 0.5 is the tie value, and a is the prediction standard error for the constituency. The probability of a Labour win is $\Pr[\text{lab}] = 1 - \Pr[\text{con}]$.

The forecast of the number of seats won by each party was obtained by summing the probabilities over constituencies. Confidence intervals* for party totals were also calculated. A modified form of ridge regression* was used; this performed better for estima-tion purposes than the least-squares method* and had the important practical advantage that the problem of multi-collinearity* is dealt with automatically. This problem often arises in the early stages of the prediction when some of the independent variables are so highly correlated that with ordinary least squares it becomes very difficult or perhaps impossible to obtain reasonably precise estimates of the relative effects of the variables. The ridge modification drives parameter estimates toward zero in ill-conditioned directions.

Consideration has been given in recent elections to the use of log-linear models* for forecasting. Here the proportion of the constituency vote going to a party is assumed to have an expectation whose logarithm is linear in the set of independent variables. A log-linear model was used for the prediction of the 1975 Referendum on entry to the Common Market, where the voting alternatives where either for or against entry.

The problem of bias* in the declaration order, although dealt with to some extent by the regression approach, remained. Most forecasters in the U.K. have attacked this problem by producing prior predictions based on such information as regional opinion polls available immediately before the election and expert judgment of what would happen in unusual constituencies. The television companies have commissioned surveys

Table 1   Regression Model for Dependent Variable: Labour to Conservative Swing (May 1979 General Election after 32 Results Received)

Order	Variable	Coefficient
	Constant	6.51
1	Liberal party withdraws = 1; otherwise, 0	− 0.32
2	Labour percent of electorate in previous election	− 1.36
3	National Front party stood in previous election = 1; otherwise, 0	0.00
4	Marginal seat held by Labour = 1; otherwise, 0	− 0.13
5	Liberal percent of electorate in previous election	− 1.56
6	Constituency in SE region = 1; otherwise, 0	0.09
7	Marginal seat held by Conservatives = 1; otherwise, 0	0.45
8	Constituency in NE region = 1; otherwise, 0	0.08
9	Percent of housing in public ownership	1.55

on the election day itself at polling stations in selected marginal constituencies where the respondents, who had just voted, were asked to declare their vote to the interviewer. Predictions of changes in party shares of the vote were then produced for these constituencies and were applied to other constituencies of a similar type to produce a prior forecast. The actual predictions made when the results began to come in were a weighting of the prior forecast and that obtained from the statistical analysis of the sample of results declared.

The sophisticated forecasting techniques have consistently given better forecasts on election night than those based on the swing approach, and a high degree of accuracy has usually been obtained. In the most recent national election (May 1979), for example, the forecasts obtained using the method of Brown and Payne [2] were within ± 5 seats of the final outcome for the two major parties in the very early stages and were within ± 2 seats after a third of the results had been declared. In the European Direct Elections of June 1979 when the U.K. had 81 constituencies formed from aggregates of from 8 to 13 national constituencies, the early forecasts of the final party totals were all within two seats of the outcome (see Payne and Brown [7]).

References

[1] Berrington, H. B. (1965). *J. R. Statist. Soc. A*, **128**, 17–66. (A good description of the various measures of swing.)

[2] Brown, P. and Payne, C. (1975). *J. R. Statist. Soc. A*, **138**, 463–498. (A description of the use of multivariate ridge regression models in election night forecasting. A good reference source.)

[3] Budge, F. and Farlie, D. (1977). *Voting and Party Competition*. Wiley, New York. (A lengthy but important account of a new approach to forecasting based on survey data.)

[4] Kendall, M. G. and Stuart, A. (1950). *Brit. J. Sociol.*, **1**, 183–196. (The seminal article describing the first application of statistical methods to election forecasting.)

[5] Laakso, M. (1979). *Brit. J. Polit. Sci.*, **9**, 355–361.

[6] Milledge, D. and Mills, M. J. (1960). *Computer J.*, **2**, 195–198.

[7] Payne, C. and Brown, P. (1981). *Brit. J. Polit. Sci.*, **11**, 235–248.

[8] Tufte, E. R. (1973). *Amer. Polit. Sci. Rev.*, **68**, 540–554.

[9] Whiteley, P. (1979). *Brit. J. Polit. Sci.*, **9**, 219–236.

CLIVE PAYNE

ELECTION PROJECTIONS

The purpose of this paper is to discuss briefly the various statistical techniques that are used in television and radio election projection and analysis efforts. The discussion will cover four broad topics: the types of election projections that are made, the kinds of information available to the persons making the projections, the various statistical considerations that underlie this information, and a brief description of the nature of the decision process.

TYPES OF PROJECTIONS

Election projections are made for a specific political unit, covering the outcome of the race, i.e., designating a "winner," as well as estimating the percentages that the various candidates will receive. The political unit may be a city in the case of a mayoralty contest, a congressional district where House seats are at issue, or an entire state of gubernatorial, senatorial, and presidential elections.

For congressional elections there is interest, as well, in the total composition of the House of Representatives, i.e., in the results of all 435 races being decided that night; and in the case of the presidential match the main interest is in the outcome nationwide, specifically which, if any, candidate has received more than 270 electoral votes, and therefore has been elected president.

These are the most common sorts of races that are projected, but sometimes other contests merit consideration, such as statewide propositions (Proposition 13 in California is a good example of this) or even minor city or state offices which have generated great interest.

It is perhaps worthwhile stating just what a projection is. Basically, it is an estimate of the outcome of a political race based on partial information, and made as early as possible in order to provide a reliable estimate of what has happened and what the final result will be after all the votes are counted. The reason there is great interest in such an early estimate is that a more or less complete unofficial vote count is usually not available until after the day of the election, or even later; and the official results, or complete vote count, may not be determined for months after an election.

WHAT INFORMATION IS AVAILABLE

Information about the probable outcome of an election begins to become available in many instances well before the election is held. NBC News/AP, Gallup, Harris, Yankelovich, Skelly, and White—just to mention a few—routinely announce estimates of presidential standings before every presidential race. Sometimes information of state or city elections is published before the fact. The *Daily News* publishes a Straw Poll before each New York City mayoralty election and before each state gubernatorial, senatorial, and presidential election. Various other polls from a variety of sources are available which give indications of how the election may go.

On election day itself, organizations such as NBC and CBS conduct election day exit polls, asking the people who have voted and have left the polls questions about their attitude toward various issues and also how they voted in selected races. This information forms the basis of the analyses that are given on the networks and adds more information about the possible outcomes of many races.

After the polls have closed the electronic media—ABC, CBS, and NBC—collect votes from so-called key precincts that have been chosen to give an early indication as to what is happening in the contest and to furnish information about the probable outcome.

At the same time the entire vote down to the county level is collected by an organization called News Election Service (NES). This organization is a joint venture supported by AP and UPI, the two wire services, and by ABC, CBS, and NBC, the three major networks. It has been collecting the vote for these media since the general election of 1964. NES provides the votes to the various subscribers both at the largest relevant political unit level (city, congressional district, state, or nation) and also at lower levels, i.e., by towns in New England and by county in the rest of the country. It is generally the source of all the vote seen on television screens and in newspapers.

NES information usually comes in somewhat later than the data from key precincts.

Thus the information available to the person making projections comes from several sources of varying quality and timeliness. Generally speaking, however, after the polls close the quantity and precision of the available information increases until a judgment can be made which allows the projection of a winner before all the votes are counted.

STATISTICAL TECHNIQUES USED

A variety of statistical techniques are employed in the information stream that reaches the decision makers. We shall describe the techniques employed in the same order as the time sequence that the information generally becomes available.

Preelection polls employ sampling techniques that involve stratification of the electorate, usually on the factors of geography, urbanity, and history (i.e., past voting behavior). The population distribution is available at the census tract level from census* information and is often obtainable in an updated version from various commercial services. The voting behavior is available at the county level from published sources, the most notable of which is "America Votes," published by the Elections Research Center. Actual historical precinct data (sometimes called election district data) are available from appropriate governmental sources. All of these sources provide material for constructing samples.

It should be noted that there is a severe sampled versus target population problem in this activity. The target population is those people who will vote election day; this population is unknowable to preelection pollsters. Many devices are used to attempt to overcome this problem, including the use of various screening questions about registration and voting history or intent.

The polling is administered by in-person interviews, telephone interviews, or by self-administered questionnaires. (The *Daily News* Straw Poll is an example of the latter technique). Telephone interviewing has become increasingly popular as a technique because of its high speed, ease of control, and low cost per interview. In-person interviewing, although theoretically preferable, has many practical difficulties and is quite expensive. The advantages of telephone interviewing tend to overcome the new problems induced by the fact that not all voters have telephones. Fortunately, the percentage of voters with telephones is very high and empirically it is apparent that this does not exacerbate the sample versus target population problem to any significant degree.

Election day exit polling again depends on the application of sampling techniques. The principal considerations in the sampling development are geography, urbanity, and past voting behavior, and stratification is accomplished utilizing these factors.

Election day polls do not have the same degree of sampled versus target population problems as do preelection polls, as only voters are interviewed. There are modest snags in that the precinct sample frame may be incomplete or outdated, but these problems do not exist to any troublesome extent. Generally, the election day polls are self-administered and secret, i.e., the person collecting the filled-in questionnaire does not know what has been written thereon.

There are many areas of judgment in the execution of the election day polls, such as what distribution of polling hours will be appropriate to represent the actual election day vote distribution over time, what should be done about the multiple exit problem, etc. There is a strong statistical component in making these judgments but other operational considerations also loom large.

Precincts to serve as keys may be selected in a variety of ways. The most popular methods appear to be stratified random selection*, augmenting this selection with key history so that the results from the keys can be compared with previous results, and using barometric keys (keys that reflect the results of an appropriately chosen race). If there is to be a street or election day poll, the first of the two methods is usually chosen for reasons of economy. If there is to be no polling effort, the use of barometric keys can be very advantageous. Of course, operational considerations lead to the selection of precincts that report early. Thus precincts that have voting machines or historically have fast counts are often preferred. Again, a large element of judgment enters into the selection, this judgment being most relevant in the choice of the race that is to be considered the best model for the upcoming election.

Finally, the NES vote in its various forms can be combined utilizing a model to provide estimates about the results. The county or town vote can be stratified into appropriate pieces on the basis of geography, urbanity, or other political factors to form more or less homogeneous groups. Estimates can be made for each of the subgroups by a variety of techniques, and the estimates of the subgroups can be weighted together (based on historical relative contributions) to provide estimates for the total political unit.

Error estimates are available for all of these sources of information. They can be prior estimates or can be calculated on the bases of internal evidence. Whatever technique is employed, the error estimate for the projection will be a weighted average of the variances of the several pieces that go into the estimate (it being assumed that the various pieces will behave in an essentially statistically independent fashion). This error estimate can then be used to provide interval estimates for the projection.

The precise details of the key and county

projection models tend to be proprietary. However, in general they do exploit the advantage of utilizing swing calculations. That is to say, basic estimates are of the form

$$\text{estimate} = \text{base value} + \text{swing}$$

where the base value is the result for the unit in some previous race, and the swing compares the vote coming in this time in those reporting subunits with the vote in the base race for the subunits.

THE DECISION PROCESS

Election projections do highlight one very important problem that is often obscured in other statistical situations, namely, those situations where large amounts of data are available from a variety of sources. These data need to be sorted out and evaluated for the purpose of establishing their meaning. In the case of projection, the evaluation results in the choice of a winner at an appropriate point in time.

This evaluation is carried out in at least one network by the combined judgments of several disciplines. The evaluation is aided by having nominal interval estimates as well as having point estimates. However, no hard and fast rules are used; each case is judged by itself, taking into account all available information.

In many other statistical applications this evaluation step tends to be left out of the process. For example, in a maintenance data collection system, there is often no central organization which continually tries to extract meaning from the data; rather frequently, the data are merely accumulated and stored. In general, it is the lack of a formal evaluation procedure that limits the use of statistics in making practical decisions.

RICHARD F. LINK

ELEMENTS ANALYSIS *See* COMMONAL-ITY ANALYSIS

ELIMINATION OF VARIABLES

Elimination of variables refers to the situation where various properties of the individuals being studied have been, or can be, recorded as variables, and one has to decide which to eliminate from the analysis. Alternatively, the problem can be regarded as one of selecting a subset of the variables to be used in the analysis. Cox and Snell [3] provide a lucid introduction to this topic. This article concentrates on the more technical problem of how computation can help.

The first section discusses multiple linear regression*, which is the main field of application for elimination of variables. The second section discusses other multivariate analyses more briefly.

MULTIPLE LINEAR REGRESSION

If some regression coefficients* in a multiple linear regression* analysis are not statistically significantly different from zero, it may seem natural to assume that they are zero and to eliminate the corresponding variable from the analysis. This is sensible if:

1. The analysis is intended to suggest which variables apparently affect the response, the prediction of future numerical values of the response being of secondary importance; or

2. The equation with zero coefficients is more convenient, perhaps because the corresponding variables need not be measured; or

3. One believes that the true values of these coefficients may be small, and therefore hopes for more accurate predictions if nonsignificant coefficients are set to zero than if all coefficients are estimated by least squares*.

If none of these reasons applies, it may be wrong to eliminate variables. And if only the third reason applies, one might prefer to keep all the variables and use ridge regression*.

The relative merits of different algorithms

for automatic elimination of variables are discussed by Beale [1]. They all use the process variously known as Gauss–Jordan elimination* or Pivoting (by students of linear programming) or Sweeping. When applied to the matrix of sums of squares and products of the independent and dependent variables, this represents the effect of adding one variable to the selected subset, or eliminating one variable from it.

A popular approach is backward elimination (or step-down). This starts by estimating regression coefficients for all linearly independent variables. It then finds the variable whose elimination makes the smallest increase in the residual sum of squares, and eliminates it if this increase is not significantly larger than the residual mean square. The process is repeated until no further variable can be eliminated in this way. This seems a natural approach, but it works awkwardly in practice if the variables are highly correlated in the data; if each is a nearly linear function of the others, it is largely a matter of chance which is eliminated first. A generally more successful approach is forward selection (or step-up). This starts with all variables excluded. It then finds the variable whose selection makes the largest decrease in the residual sum of squares, and selects it if this decrease is significantly larger than the residual mean square. The process is repeated until no further variable can be selected in this way. Stepwise regression, due to Efroymson [4], combines these operations. Given any trial subset of variables, it attempts a backward elimination step. If this is impossible, it attempts a forward selection step. If this is also impossible, it stops.

Unless the independent (or regressor) variables are statistically independent, there is no guarantee that any of these algorithms will yield a subset that is optimum in any objective sense. We may define that optimum subset as one that yields a regression equation that minimizes the residual sum of squares (*see* ANOVA TABLE) for any given number r of nonzero regression coefficients. If n, the total number of independent variables, is not much more than 10, optimum subsets for each value of r can be found by complete enumeration. For somewhat larger values of n, optimum solutions can be found by partial enumeration or tree-search methods (*see* DENDRITES). This saves work by using the fact that if a subset S of size greater than r gives a larger residual sum of squares than some other subset of size r, then no subset of S can be optimum. Furnival and Wilson [5] describe an implementation of this approach, and quote reasonable solution times for problems with n as large as 35. But, like others, they found that the computing task roughly doubles when n increases by 2. So approximate methods must be used for larger values of n. McHenry [9] suggests finding a solution that cannot be improved by removing any single variable and replacing it with the best alternative, and then repeating the exercise with each variable in turn omitted.

Two objections to the automatic generation of "optimum" subsets are worth noting. One is that negative values for certain regression coefficients may be physically meaningless. Subsets giving such coefficients can be automatically rejected, although this extension of the algorithm prevents the use of some computational simplifications due to Furnival and Wilson [5]. Another objection is that alternative subsets may give almost as small a residual sum of squares as the theoretical optimum, and be preferable on physical or other grounds.

So computer programs for optimum regression have been extended to produce the m best equations for each value of r, where m is a piece of input data with a typical value of, say, 10. In practice this option is often not very illuminating, the subsets produced being selections from the optimum subsets of size $r + 1$ or $r + 2$. But this is not always so.

The choice of the optimum value of r brings us back from computing to statistical theory. If σ^2, the variance of the random error, is known, it is natural to choose r to minimize

$$R + \alpha\sigma^2,$$

where R denotes the residual sum of squares* and α is a parameter. The rejection of all regression coefficients that are not significant at the 5% level implies setting $\alpha = 4$. But a more fundamental approach is to seek the equation that minimizes the expected mean square error* in prediction. This means square error consists of a bias (squared) caused by the omission of any terms assumed to be zero plus a variance caused by the random errors in estimating the other terms. Mallows ([7], and earlier unpublished papers) shows that this approach produces a definite answer if we assume that the sets of values of independent variables for which predictions are required have the same first and second moments as the observed data set. And this answer is that we should use the value $\alpha = 2$, although Mallows warns against using this criterion blindly. Similar answers are given by the similar (but more general) principles of Akaike's information criterion* and of cross-validation. (See Stone [12].) Sprevak [11] reports a detailed computational study of the performance of such methods when $n = 2$. This shows that, at least in this case, the statistical properties of the methods are not unduly sensitive to correlation* between the independent variables.

Further discussion and references can be found in Chapter 12 of Seber [10].

OTHER MULTIVARIATE ANALYSES

It is not so easy to find convincing criteria for eliminating variables from a multivariate analysis* with no dependent variable. One motivation for eliminating in these circumstances comes from principal component analysis*. If the points representing the values of the n variables recorded on each individual all lie very near a hyperplane in $r(< n)$ dimensions, one can plausibly argue that $n - r$ of the variables are redundant, since their values can be reconstructed as linear functions of the remaining r variables. One might think of trying to associate a variable with each of the eigenvectors asso-

ciated with the $n - r$ smallest eigenvalues of the sample correlation matrix. But this is an arbitrary and unsatisfactory process. Beale et al. [2] therefore introduced the process known as "interdependence analysis," which selects r variables out of n to maximize the smallest of the multiple correlation coefficients of the rejected variables with the set of selected variables.

One practical difficulty with interdependence analysis is that the original variables to be considered must be chosen carefully. In multiple regression analysis there is no great harm in considering a variable that later proves irrelevant: it will simply be eliminated. But in interdependence analysis a variable that is useless for predicting the other variables will often have to be selected because it cannot itself be predicted from the values of the other variables. It is therefore useful to extend interdependence analysis to allow each variable to be classified as either dependent (i.e., requiring prediction) or independent (i.e., available for selection) or both. The analysis then selects r of the independent variables to maximize the smallest of the multiple correlation coefficients of any of the dependent variables with the selected independent variables. The original interdependence analysis is then the special case where all variables are both dependent and independent.

Other analyses are possible when there is more information than just the values of the independent variables. McCabe [8] considers discriminant analysis*. If only two populations are involved, the problem can be analyzed as multiple regression with a binary dependent variable equal to 1 for one population and 0 for the other. With more populations it is natural to choose the set of variables that minimizes Λ defined by

$$\Lambda = |\mathbf{W}|/|\mathbf{T}|,$$

where $\mathbf{W}$ denotes the sum of squares and products matrix for the deviations of the selected variables from their sample means within each population, and $\mathbf{T}$ denotes the corresponding matrix for the deviations of the selected variables from their overall sample means. McHenry [9] proposes an approximate solution to this problem when there are too many variables for an enumerative approach. And he suggests that the same approach is applicable when the set of discrete alternatives is replaced by a set of values of some other variables that may affect the original variables either linearly or not at all. The matrix **W** is then the matrix of sums of squares and products of the residuals when each selected variable is regressed on these other variables. Hawkins [6] suggests a similar approach to multivariate analysis of variance*

References

[1] Beale, E. M. L. (1970). *Technometrics*, **12**, 909–914.

[2] Beale, E. M. L., Kendall, M. G., and Mann, D. W. (1967). *Biometrika*, **54**, 356–366.

[3] Cox, D. R. and Snell, E. J. (1974). *Appl. Statist.*, **23**, 51–59.

[4] Efroymson, M. A. (1960). In *Mathematical Methods for Digital Computers*, Vol. 1, A. Ralston and H. S. Wilf, eds. Wiley, New York, pp. 191–203.

[5] Furnival, G. M. and Wilson, R. W. (1974). *Technometrics*, **16**, 499–511.

[6] Hawkins, D. M. (1976). *J. R. Statist. Soc. B*, **38**, 132–139.

[7] Mallows, C. L. (1973). *Technometrics*, **15**, 661–675.

[8] McCabe, G. P., Jr. (1975). *Technometrics*, **17**, 103–109.

[9] McHenry, C. E. (1978). *Appl. Statist.*, **27**, 291–296.

[10] Seber, G. A. F. (1977). *Linear Regression Analysis*. Wiley, New York.

[11] Sprevak, D. (1976). *Technometrics*, **18**, 283–289.

[12] Stone, M. (1977). *J. R. Statist. Soc. B*, **39**, 44–47.

(MULTIVARIATE ANALYSIS
REGRESSION)

E. M. L. Beale

ELLIPSOID OF CONCENTRATION

In the bivariate case, given a random variable (X, Y) with bivariate CDF $F_{X, Y}(x, y)$,

marginal moments μ_X and μ_Y, marginal variances σ_X^2 and σ_Y^2, and the correlation coefficient $\rho(\rho < 1)$, a *uniform distribution* of probability mass over the area enclosed by the ellipse

$$\frac{1}{1 - \rho^2} \left[\frac{(x - \mu_X)^2}{\sigma_X^2} - \frac{2\rho(x - \mu_X)(y - \mu_Y)}{\sigma_X \sigma_Y} + \frac{(y - \mu_Y)^2}{\sigma_Y^2} \right] = 4$$

has the same first- and second-order moments as the given distribution. This ellipse is called the "ellipse of concentration" corresponding to the given distribution. When two bivariate distributions with the same center of gravity are such that one of the concentration ellipses lies wholly within the other, the first distribution is said to have a greater concentration than the second. This concept is of importance in statistical estimation theory. Extension of this concept to the multivariate case is straightforward.

Bibliography

Cramér, H. (1945). *Mathematical Methods of Statistics*. Princeton University Press, Princeton, N.J.

(BIVARIATE NORMAL DISTRIBUTION
CHI SQUARE DISTRIBUTION
MULTINORMAL DISTRIBUTION)

ELLIPTIC OPERATOR *See* DIFFUSION PROCESSES

EMBEDDED PROCESSES

The method of embedded processes is a basic tool utilized in the study of Markov branching processes* and queueing processes (*see* QUEUES, THEORY OF). The reader is therefore advised to consult these entries for notation, definition, etc. The terms "embedded" and "imbedded" are used interchangeably in the theory of branching processes*, whereas the term "imbedded" is invariably used in the theory of queueing processes. In topology, complex analysis, functional equations*, and other areas of mathematics, these two terms are used synonymously. Different areas use the term "embedding" differently. But the general idea is to suitably "map" or include a given class of mathematical objects into a large and/or well-behaved class of objects. The fundamental reason behind embedding is to simplify analysis.

The study of continuous-time stochastic processes* that arise in applications is not always easy to analyze for reasons such as that they need not be Markovian (the queue $M/M/s$ is Markovian, whereas queues $E_k/G/1$ and $GI/M/s$ are not). In such cases appropriate discrete (time) processes* are introduced. Roughly, these processes are the so-called embedded processes. A general recipe for extracting an embedded process Y_n from a given process X_t is as follows. Let $\{T_n, n \geqslant 1\}$ be a sequence of stopping times* with respect to (w.r.t.) X_t. Define $Y_n = X(T_n)$. Then Y_n is called an *embedded process* as long as it is a useful Markov chain with discrete or continuous state space or any other suitable, well-studied and/or easy-to-analyze process. Before considering examples and properties of embedded processes, some remarks are in order. Instead of the process X_t we can sometimes find another process Z_t associated with X_t such that $Y_n = Z(T_n)$ is Markov. (See Example 5.) The process Y_n need not always be Markov. There are situations in branching processes where $\{Y_n\}$ is obtained as a martingale* rather than a Markov chain and the martingale properties are used in the study of X_t (see ref. 1). Even in the general study of continuous-time Markov processes* X_t, a useful device is to reduce the problem to an embedded Markov chain. For example, in classifying the states of a continuous-time Markov chain, one can reduce the problem to an embedded Markov chain, and then establish that if a given state is of one type (recurrent*/transient*) in the embedded case, then it is of the same type in the original case (see Foster's criteria in ref. 6). Therefore, the general idea behind an embedded process is to extract a well-

behaved/studied discrete-time process Y_n from a hard-to-analyze process X_t and then to draw conclusions about X_t from the properties of Y_n. In this article we shall concentrate on the method of embedded processes in branching and queueing processes.

Example 1. Let $\{X(t), t \geqslant 0\}$ be a continuous-time Markov branching process with probability generating function* $F_t(x) = \sum_{k \geqslant 0} P\{X(t) = k | X(0) = 1\} x^k$, state space Z^+—the set of nonnegative integers—and transition probabilities* $p_{ij}(t)$. Arbitrarily fix a $\delta > 0$. The fixed time points $\{n\delta\}$, $n \geqslant 0$, form an increasing sequence of stopping times. Define $Y_n(\omega) = X_{n\delta}(\omega)$, $n \geqslant 0$, $\omega \in \Omega$, the basic sample space*. Then $\{Y_n\}$ is a Markov chain known as the embedded *Galton–Watson process** whose probability generating function* is given by $f(x) = F_\delta(x)$.

Example 2. Let $\xi_0 = 0$, and ξ_n, $n \geqslant 1$, be a sequence of nonnegative independent random variables with a common nonlattice distribution* $F(t)$ with $F(0 +) = 0$ and $m = \int_0^\infty t\, dF(t)$. Define $S_n = \sum_{k=0}^n \xi_k$, $n \geqslant 0$, $Z_0 = 0$, and $Z_t = \max\{n : S_n < t\}$, $t > 0$. Then Z_t, $t \geqslant 0$ is a renewal process*. Let $\{X(t), t \geqslant 0\}$ be the regenerative process* w.r.t. $\{Z_t\}$ with $p_{ij}(t - s) = P\{X(t) = j | X(s) = i, S_n = s, S_{n+1} = u\}$, for $0 \leqslant s \leqslant t \leqslant u$ and $p_{ij}(t - s) = 0$ for $t < s$. Define $Y_0 = X_0$, $Y_n = X(S_n)$. Then $\{Y_n, n \geqslant 0\}$ is a stationary Markov chain known as the *embedded chain of the regenerative process* X. The transition probabilities p_{ij} of $\{Y_n\}$ are given by $p_{ij} = \int_0^\infty p_{ij}(t)\, dF(t)$.

Example 3. Let $\{X(t), t \geqslant 0\}$ be the queue size at time t of a queue of type $G/M/1$, and T_n, $n \geqslant 1$, be the sequence of successive arrival times of customers. In the queueing examples we will assume that nobody is waiting but that someone may arrive at time $t = 0$. For this system $G/M/1$, the interarrival times $\sigma_n = T_n - T_{n-1}$ are independent and exponentially distributed with same intensity. As a consequence, $\{X(t)\}$ is a Markovian regenerative process. Define Y_0

$= X_0$, $Y_n = X(T_n)$, $n > 1$. Then $\{Y_n\}$ is an *embedded chain* of $X(t)$. Similarly, consider a queue with s servers, Poisson* service times, and the interarrival time forming a sequence of independent and identically distributed random variables with an arbitrary common distribution. Using arrival epochs one can define, as above, an embedded chain.

Example 4. Consider an $M/G/1$ queue with Poisson arrivals and arbitrary service-time distribution $G(t)$. Let $X(t)$ denote the queue length at time t, and D_n, $n \geqslant 1$, be the random times of departure of customers from the system. Assuming that a new service started at $t = 0$, we set $D_0 = 0$, $Y_0 = X(0 +)$, and $Y_n = X(D_n +)$. Whereas $X(t)$ is not a regenerative process w.r.t. D_n, $n \geqslant 0$, $\{Y_n\}$ is an *embedded Markov chain* of $\{X(t)\}$.

Example 5. Let $X(t)$ be a queueing process of type $G/G/1$ with successive arrival and departure times denoted by T_n and D_n, respectively. In the present case, neither the process $\{X(T_n)\}$ nor the process $\{X(D_n)\}$, $n \geqslant 1$, is Markovian. There is an embedded Markov process used in the study of the $G/G/1$ queue. Let Z_t, $t \geqslant 0$, be the process of virtual waiting time * at time t. Then the waiting time of the nth arriving customer $W_n = Z(T_n)$, $n \geqslant 1$, is an embedded Markov chain of $Z(t)$ with state space R^+. Several books mentioned in the references contain more examples. Next we look at some of the properties and uses of embedded processes.

We first take up, for further analysis, the embedded Galton–Watson process Y_n of a Markov branching process $X(t)$. The strength of the result that the embedded process Y_n is a Galton–Watson process lies in the fact that we could use our knowledge about the Galton–Watson process to derive properties of our embedded Galton–Watson process Y_n, for every $\delta > 0$, and in turn apply them to study $X(t)$. For example, the extinction of the branching process $X(t)$ is equivalent to the extinction of the embedded Galton–Watson process Y_n, for each $\delta > 0$.

In similar and appropriate occasions, the following result of Kingman (cf. ref. 1) is very useful in passing to the continuous-time case. If a continuous function $g(t)$ defined on $(0, \infty)$ satisfies the condition that the limit $\lim_{n \to \infty} g(n\delta) = c(\delta)$ exists for each $\delta > 0$, then $\lim_{t \to \infty} g(t) = c$ exists and $c(\delta) = c$ for all $\delta > 0$. The asymptotic behavior of $X(t)$ as $t \uparrow \infty$ is similar to that of the discrete time Galton–Watson process. Therefore, we use the asymptotic properties of the embedded Galton–Watson process Y_n to establish the limiting behavior of X_t that is *continuous in probability**. Here we have to verify whether or not certain hypotheses hold for each $\delta > 0$ and then apply the results of Galton–Watson processes to Y_n.

Let Z_n, $n > 0$, be an arbitrary Galton–Watson process with $E(Z_1) = m$. Then $E(Z_n) = m^n$. Define $W_n = m^{-n}Z_n$, and $\mathcal{Q}_n$ = all sensible events (= σ-algebra*) generated by $Z_0, \ldots, Z_n$. Let $0 < m < \infty$. Then $\{W_n, \mathcal{Q}_n, n > 0\}$ is a nonnegative martingale*, and hence the limit $\lim_{n \to \infty} W_n = W$ exists a.s. (see Doob's theorem in ref. 6). For each $\delta > 0$, the embedded Galton–Watson process has similar properties. Noting that $E[X(t)|X(0) = 1] = e^{\lambda t}$, we similarly see that $W(t) = e^{-\lambda t}X(t)$ is a nonnegative martingale and consequently that $\lim_{t \to \infty} W(t) = W$ exists a.s. The cases $m < 1$, $= 1$, and > 1 (correspondingly, $\lambda < 0$, $= 0$, and > 0) are called the *subcritical, critical, and supercritical cases*.

1. If $m < 1$, Yaglom has shown that the probability $P\{Z_n = k | Z_n > 0\}$ that the nth generation size is k given that the population is not yet extinct converges to a (proper) probability function. Using this result as applied to the embedded Galton–Watson process Y_n, for every $\delta > 0$, and utilizing Kingman's result, one can arrive, in the case $\lambda < 0$, at similar conclusions in the continuous case.

2. The critical case $m = 1$ (resp., $\lambda = 0$) is slightly cumbersome in notation and mathematics. In this case, one can show that the embedded process has, for every

$\delta > 0$, an exponential distribution* as its limiting distribution. Applying Kingman's result to this property of an embedded process, one sees that the original process $X(t)$ has an exponential distribution as its limiting distribution.

3. Now let $m > 1$ (resp., $\lambda > 0$), and $Z_0 \equiv 1$. A result of Levinson says that $E[W] = 1$ if $E[Z_1 \log Z_1] < \infty$, and $E[W] = 0$ if $E[Z_1 \log Z_1] = \infty$. Using this for our embedded Galton–Watson process we see that, if $W(\delta) = \lim_{n\to\infty} e^{-n\lambda\delta} X(n\delta)$, $\delta > 0$, then $P\{W(\delta) = 0\}$ or $E[W(\delta)] = 1$, and $W(\delta) = W$ a.s. where $W =$ a.s. $\lim_{t\to\infty} W(t)$. Furthermore, $E[W] = 1$ if and only if $\sum k p_k \log k < \infty$.

We have seen above some uses of the embedded Galton–Watson process Y_n in studying the limiting behavior of $X(t)$. Fortunately, the present continuous-time case $X(t)$ lends itself more easily to mathematical analysis. Consequently, one can now say more about the embedded Galton–Watson process than what we could derive from applying properties of an arbitrary Galton–Watson process. The reverse problem, called the *embeddability problem*, becomes more interesting once we note that not all (discrete-time) Galton–Watson processes are embeddable. Let $F_t(x)$ be the probability generating function of a continuous-time Markov BP $X(t)$ and Y_n be an associated embedded GWP corresponding to an arbitrary $\delta > 0$. Then $F_{t+s}(x) = F_t(F_s(x))$, $t, s > 0$, $|x| < 1$, and the PGF of Y_n is given by $f(x) = F_\delta(x)$. The embeddability problem is that, given a GWP $\{Z_n\}$ with PGF $f(x)$, can one find a BP $X(t)$, $t > 0$, with PGF $F_t(x)$ such that, for some $\delta > 0$, the embedded GWP Y_n of $X(t)$ is equivalent to Z_n? Since the problem is analytical in nature (and is a particular case of a more general problem arising, e.g., in the theory of conformal mapping), we omit any further details, and refer the reader to the paper of Karlin and McGregor [7] and the references given there.

We shall now consider the embedded Markov chain Y_n associated with the regenerative process* X_t that is introduced in Example 2. If σ_{ij} is the entrance time of the embedded chain Y_n to state j, that is, $\sigma_{ij} = \min\{n > 0 : Y_n = j \mid Y_0 = i\}$, then from the independence of ξ_k's and Wald's equation*, we get $m_{jj} = E[\xi_1 + \cdots + \xi_{\sigma_{jj}}] = E[\xi_1] E(\sigma_{jj}) = m E[\sigma_{jj}]$. Let the embedded chain Y_n of the regenerative process X_t be irreducible*, aperiodic, and ergodic*. Then:

1. The limiting distribution is

$$a_i = \lim_{n\to\infty} P\{Y_n = i\}$$

$$= \lim_{t\to\infty} P\{X(S_{Z(t)}) = i\} = m / m_{ii},$$

and it satisfies

$$a_j = \sum_i a_i \int_0^\infty P_{ij}(t) \, dF(t),$$

$$\sum_i a_i = 1, \text{ and}$$

2.

$$\lim_{t\to\infty} P\{X_t = k\}$$

$$= \sum_j m^{-1} a_j \int_0^\infty P_{jk}(t)[1 - F(t)] \, dt.$$

3. Let now X_t be a Markovian regenerative process, and η_t, $t \geqslant 0$, be the past life time at time t of the renewal process Z_t. Set $f_j(t) = m^{-1} \sum_i a_i p_{ij}(t)$. Then the density $g_j(u)$ of $\lim_{t\to\infty} P\{X_t = j, \; \eta_t < u\}$ exists and is given by 0 for $u \leqslant 0$ and by $f_j(u)[1 - F(u)]$ for $u > 0$.

4. Let $E[\sigma_{ii}] < \infty$, for every i. Then the classification of states for the regenerative process X_t follows the corresponding classification of states for the embedded process.

Let Y_n, $n \geqslant 1$, be the embedded chain of the Markovian regenerative process X_t that was introduced in Example 3 w.r.t. the queue $G/M/1$. Then Y_n represents the number of customers in the queue just before the arrival of the nth customer after $t = 0$. Let $F(t)$, $t > 0$, be the interarrival time distribution with $m = \int_0^\infty t \, dF(t)$, and λ be the intensity of the exponential service time distribution. Then:

1. The embedded chain Y_n is an irreducible aperiodic Markov chain with transition probabilities given by

$$p_{ij} = \int_0^\infty \int_0^t \frac{(s/\lambda)^i e^{-s/\lambda}}{i!} \, ds \, dF(t)$$

$$\text{for} \quad j = 0,$$

$p_{ij} = 0 \quad \text{for} \quad j = i + 2, i + 3, \ldots,$

and

$$p_{ij} = \int_0^\infty \frac{(t/\lambda)^{i+1-j}}{(i+1-j)!} e^{-t/\lambda} dF(t)$$

$$\text{for} \quad j = 1, 2, \ldots, i + 1.$$

2. The Markov chain $\{Y_n\}$ is recurrent if $\lambda < m$, and transient if $\lambda > m$. More precisely, it is ergodic if $\lambda < m$, null recurrent* if $\lambda = m$, and transient if $\lambda > m$.

3. If $\lambda < m$, the embedded chain Y_n is ergodic and $\lim_{n\to\infty} P\{Y_n = i \mid Y_0 = k\} = \alpha^i(1 - \alpha)$, $i = 0, 1, \ldots$, where α is the zero inside the unit circle of $z - m[(1 - z)/\lambda]$.

4. Using now the results 1 and 2 of the preceding list, one can show, if $\lambda < m$ and F is not a lattice distribution*, that $\lim_{t\to\infty} P\{X_t = j \mid Y_0 = X_0 = k\} = 1 - (\lambda/m)$ for $j = 0$, and is $= \lambda(1 - \alpha)\alpha^{j-1}/m$ for $j = 1, 2, \ldots$. If $\lambda \geq m$, then the limit above is zero.

5. When $F(t)$ is not a lattice distribution, we can show that

$\lim_{n\to\infty} E[Y_n \mid X_0 = k] = \alpha/(1 - \alpha)$,

$\lim_{t\to\infty} E[X_t \mid X_0 = k] = m/(1 - \alpha)$,

$\lim_{n\to\infty} \operatorname{var}[Y_n \mid X_0 = k] = \alpha/(1 - \alpha)^2$,

$\lim_{t\to\infty} \operatorname{var}[X_t \mid X_0 = k]$

$\quad = (1 - \alpha)^{-2}(m - m^2 - m\alpha)$.

As a final example from queueing theory, let us consider the embedded Markov chain of the queueing system $M/G/1$. Here the interarrival times are independent and follow the exponential distribution with intensity λ. We shall assume that a new service starts at

$t = 0$. Let the general service time distribution be denoted by $G(t)$ with finite $m = \int_0^\infty t \, dG(t)$ and $m_2 = \int_0^\infty t^2 \, dG(t)$.

1. The embedded process Y_n, $n \geq 0$, of the queue X_t is an irreducible and aperiodic Markov chain with transition probabilities given by

$$p_{0j} = \int_0^\infty \frac{(t/\lambda)^j}{j!} e^{-t/\lambda} dG(t) \quad \text{for} \quad j \geq 0,$$

$p_{ij} = 0 \quad \text{for} \quad j = 0, 1, \ldots, i - 2;$

$$i = 2, 3, \ldots,$$

and

$$p_{ij} = \int_0^\infty \frac{(t/\lambda)^{j+1-i}}{(j+1-i)!} e^{-t/\lambda} dG(t)$$

$$\text{for} \quad i \geq 1, j \geq i - 1.$$

2. The embedded process Y_n, $n > 0$, is ergodic if $m < \lambda$, is null recurrent if $m = \lambda$, and is transient if $m > \lambda$.

3. Let $m < \lambda$ and $\sigma_j = \lim_{n\to\infty} p_{ij}^{(n)}$. If $S(x)$ is the probability generating function of σ_j, then $\sigma_0 = (1 - a)$ and

$$S(x) = (1 - a)(1 - x) \frac{m[(1 - x)/\lambda]}{m[(1 - x)/\lambda] - x},$$

$|x| \leq 1$, where $a = m/\lambda$. Moreover, the embedded process Y_n and the queue X_t have identical limit distribution given by $\lim_{n\to\infty} P\{Y_n = j \mid Y_0 = k\} = \lim_{t\to\infty} P\{X_t = j \mid Y_0 = k\} = \sigma_j$, provided that $m < \lambda$, and by 0, otherwise.

4.

$\lim_{t\to\infty} E[X_t \mid Y_0 = 0]$

$\quad = a + [a^2 m_2/(2(1 - a)m^2)]$,

provided that $a = m/\lambda < 1$. If $a > 1$,

$\lim_{t\to\infty} t^{-1} E[X_t \mid Y_0 = 0] = (a - 1)/m$.

Below we give a short list of selected references. Further examples and properties can be found there. All references except refs. 7 and 10 are books. Reference 6 is a first course in stochastic processes introducing various main classes of stochastic processes. References 1 and 5 are on branching processes and assume knowledge of measure theory. The remaining references [2–4, 8, 9] are first-year graduate-level books on queues (measure theory is not needed).

References

[1] Athreya, K. B. and Ney, P. E. (1972). *Branching Processes.* Springer-Verlag, New York.

[2] Benes, V. E. (1963). *General Stochastic Processes in the Theory of Queues.* Addison-Wesley, Reading, Mass.

[3] Cohen, J. W. (1969). *The Single Server Queue.* North-Holland, New York.

[4] Cox, D. R. (1963). *Renewal Theory.* Methuen, London.

[5] Harris, T. E. (1963). *The Theory of Branching Processes.* Springer-Verlag, New York.

[6] Kannan, D. (1979). *An Introduction to Stochastic Processes.* North-Holland, New York.

[7] Karlin, S. and McGregor, J. (1968). *Trans. Amer. Math. Soc.*, **132**, 115–136.

[8] Prabhu, N. U. (1965). *Queues and Inventories.* Wiley, New York.

[9] Saaty, T. L. (1965). *Elements of Queueing Theory.* McGraw-Hill, New York.

[10] Yaglom, A. M. (1947). *Dokl. Akad. Nauk SSSR*, **56**, 795–798.

(BRANCHING PROCESSES GALTON–WATSON PROCESS QUEUES, THEORY OF STOCHASTIC PROCESSES)

D. KANNAN

EMPIRICAL BAYES THEORY

The empirical Bayes approach to statistical decision theory* is applicable when one is confronted with an independent sequence of Bayes decision problems having similar structure. The structural similarity includes the assumption that there is a prior distribution G on the parameter space, and that G is unknown. Such an empirical Bayes approach, in the words of Robbins [21], "offers certain advantages over any approach which ignores the fact that the parameter is itself a random variable as well as over any approach which assumes a personal or conventional distribution of the parameter not subject to change with experience." Further arguments in support of the empirical Bayes approach to statistical decision problems can be found in the paper by Neyman [16], in the book by Maritz [15], and in the monograph by Gilliland et al. [8].

In this expository article, we present some of the elements of the theory. It is organized as follows. In the first section we introduce the empirical Bayes problem, and state some definitions and two general theorems. A couple of brief examples illustrating the construction of empirical Bayes procedures are also given in the second section. In the third section we develop some general theory applicable to the estimation of the prior distribution* function G mentioned above. The fourth section deals with Stein's estimator constructed in the context of estimating the mean of a multivariate normal distribution*. It is shown that this estimator can be approached also from an empirical Bayes point of view. The fifth section describes some of the recent developments of the empirical Bayes theory of Robbins. The sixth section gives a description of the compound decision problem and its relation to the empirical Bayes problem. The seventh section gives an application of empirical Bayes theory to the estimation of a survival probability function using a set of randomly right censored data*.

The results provided here are neither necessarily the best available in the literature (mostly due to the lack of measure theoretic formalism) nor exhaustive, but they do present some insight into how empirical Bayes theory can be (or has been) developed.

A brief statement about the notation. Integrals without limits are on the entire range of the variables of integration. Unless otherwise stated, all limits are as $n \to \infty$. E stands for expectation with respect to all the random variables involved. $\sim$ abbreviates the phrase "distributed according to."

THE EMPIRICAL BAYES PROBLEM

As already pointed out, an empirical Bayes decision problem is composed of a sequence of repetitions of the same decision problem. So we begin with a description of the usual Bayes statistical decision problem, which is composed of the following elements:

1. A parameter space Θ, with a generic point θ representing the true state of nature.

2. An action space A, with generic point a.

3. A loss function L on $A \times \theta$ to $[0, \infty)$ with $L(a, \theta)$ representing the loss of taking action a when θ is the true state of nature.

4. A prior distribution* G on Θ.

5. A random variable X taking values in a space $\mathfrak{X}$, and for a given realization θ of a random variable having distribution G, X has a specified probability density $f_\theta(\cdot)$ with measure μ on a σ-field in $\mathfrak{X}$.

The statistical decision problem is to choose a decision function t defined on X with range contained in A. Assuming that $L(t(\cdot), \theta)$ is a measurable function on $\mathfrak{X} \times \Theta$, the average (or expected) loss of t is given by

$$R(t, \theta) = \int L(t(x), \theta) f_\theta(x) \, d\mu(x). \quad (1)$$

The overall expected loss with $\theta \sim G$ is then given by

$$R(t, G) = \int R(t, \theta) \, dG(\theta). \quad (2)$$

In the statistical decision problem composed of the five elements described above, the object is to choose, if possible, a decision function t_G for which $R(t, G)$ is a minimum among all decision functions t. This choice may not always be possible, but we shall bypass this difficulty and assume throughout that there exists a decision function t_G such that

$$\int L(t_G(x), \theta) f_\theta(x) \, dG(\theta)$$

$$= \min_{a \in A} \int L(a, \theta) f_\theta(x) \, dG(\theta) \quad (3)$$

for each x in $\mathfrak{X}$. It is then obvious that for any decision function t,

$$R(t, G) \geqslant R(t_G, G)[= R(G)]; \quad (4)$$

t_G is called a Bayes decision function with respect to G, and $R(G) = R(t_G, G)$ is the minimum Bayes risk attainable by any decision function relative to G.

If G is known, we can use t_G, and attain the minimum possible risk $R(G)$. But in practical situations, G is rarely known even if it is believed to exist. Suppose now that the decision problem described occurs repeatedly and independently with the same unknown prior distribution G in each repetition of the decision problem. That is, let

$$(\theta_1, X_1), \ldots, (\theta_n, X_n), \ldots$$

be a sequence of independent, identically distributed (i.i.d.) random pairs where θ_i are i.i.d. G and where X_n has density $f_{\theta_n}(\cdot)$ w.r.t. μ given θ_n. $X_1, \ldots, X_n, \ldots$ are observable, whereas $\theta_1, \ldots, \theta_n, \ldots$ are not. Viewing this setup at stage $(n + 1)$ with G unknown, we have already accumulated the $x_1, \ldots, x_n$ and x_{n+1}, and we want to make a decision about θ_{n+1} with loss L. Since G is assumed to be unknown, and since $x_1, \ldots, x_n$ is a random sample from the population with density

$$f_G(x) = \int f_\theta(x) \, dG(\theta) \quad (5)$$

w.r.t. μ, it is reasonable to expect that $x_1, \ldots, x_n$ do contain some information about G. Eliciting this information about G from $x_1, \ldots, x_n$ and then using it to define $t_n(\cdot) = t_n(x_1, \ldots, x_n, \cdot)$ a decision rule for use in the $(n + 1)$th decision problem to decide about θ_{n+1}, we incur an expected loss at stage $n + 1$ given by

$$R_n(T, G) = E[R(t_n(\cdot), G)]$$

$$= \int \int E[L(t_n(x), \theta)]$$

$$\times f_\theta(x) \, d\mu(x) \, dG(\theta) \quad (6)$$

with $T = \{t_n\}$. It is obvious from the equality above and the definition of $R(G)$ that

$$R_n(T, G) \geqslant R(G) \quad (7)$$

for all $T = \{t_n\}$. Inequality (7) leads one in an intuitive way to the following definition [21].

Definition 1. $T = \{t_n\}$ is said to be asymptotically optimal (abbreviated a.o.) relative to G if

$$\lim_{n \to \infty} R_n(T, G) = R(G). \qquad (8)$$

Definition 2. $T = \{t_n\}$ is said to be asymptotically optimal of order α_n relative to G if

$$R_n(T, G) - R(G) = 0(\alpha_n),$$
$$\text{where } \lim \alpha_n = 0 \qquad (9)$$

The "consequence of T being a.o." is that even while G is assumed to be unknown, one can still achieve risk very close to the optimal Bayes risk $R(G)$ by using the accumulated observations "to improve the decision rule t_n" at each stage n. Although Definition 1 provides the a.o. of T, it is Definition 2, with the constant recovered in the $0(\alpha_n)$ order, that is more important from the point of view of application.

Two general theorems concerning the asymptotic optimality of an empirical Bayes testing procedure and an empirical Bayes estimator are given below.

We first consider the testing situation in which $A = \{a_0, \dots, a_k\}$. For $j = 0, \dots, k$, let

$$\Delta_G(a_j, x) = \int \{ L(a_0, \theta) - L(a_0, \theta) \}$$
$$\times f_\theta(x) \, dG(\theta) \qquad (10)$$

and $\{\Delta_{j,n}(X_1, \dots, X_n; x)\}$ be such that

$$\sup_{0 \le j \le k} |\Delta_{j,n}(X_1, \dots, X_n; x) - \Delta_G(a_j, x)|$$
$$\to 0 \text{ in probability for each } x. \qquad (11)$$

Then the following theorem can be obtained.

Theorem 1. Let $\int L(a_j, \theta) \, dG(\theta) < \infty$ for $j = 0, 1, \dots, k$.

1. Define $T = \{t_n\}$ by $t_n(X_1, \dots, X_n; X_{n+1}) = a_l$ where l is any integer such

that $\Delta_{l,n}(X_1, \dots, X_n; X_{n+1}) = \min\{\Delta_{j,n}(X_1, \dots, X_n, X_{n+1}) | 0 \le j \le k\}$. Then $T = \{t_n\}$ is a.o.

2. Let

$$\Delta_{j,n}(X_1, \dots, X_n; x)$$
$$= n^{-1} \sum_{i=1}^{n} h_j(x, X_i), \qquad (12)$$

where $E[h_j(x, X_1)] = \Delta_G(a_j, x)$ for $j = 0, 1, \dots, k$. If

$$\int |\Delta_G(a_j, x)|^{1-\delta} \{\text{var}(h_j(x, X_1))\}^{\delta/2} d\mu(x) < \infty$$

for some $0 < \delta < 2$, and all $j = 0, 1, \dots, k$, then $T = \{t_n\}$ defined in **1** with $\Delta_{j,n}$ defined by (12) [note that $\Delta_{0,n}(x) \equiv 0$] is a.o. of order $n^{-\delta/2}$ relative to G.

3. If unbiased estimators* $\Delta_{j,n}$ of $\Delta_G(a_j, \cdot)$ do not exist (see Step 2), the following result holds. Let $\{h_{j,n}(\cdot, \cdot)\}$ be k real-valued sequences of measurable functions on $\mathfrak{X} \times \mathfrak{X}$. Define

$$\Delta_{j,n}(X_1, \dots, X_n; x)$$
$$= n^{-1} \sum_{i=1}^{n} h_{j,n}(x, X_i) \qquad (13)$$

for $j = 1, \dots, k$, and $\Delta_{0,n}(x) \equiv 0$. Assume that the following inequalities hold for $j = 0, 1, \dots, k$:

$$\int |\Delta_G(a_j, x)|^{1-\delta} |E[h_{j,n}(x, X_1)] - \Delta_G(a_j, x)|^\delta d\mu(x)$$
$$\le c_G a_n$$

and

$$\int |\Delta_G(a_j, x)|^{1-\delta} \{\text{var}(h_{j,n}(x, X_1))\}^{\delta/2} d\mu(x)$$
$$\le c'_G a'_n$$

for some constants c_G and c'_G. Then $T = \{t_n\}$ defined in step 1 with $\Delta_{j,n}$ defined by (13) is a.o. of order $\alpha_n = \max\{n^{-\delta/2} a'_n, a_n\}$.

Let us stop for a moment and interpret what Theorem 1 says. In the Bayes problem with G known, the Bayes rule in the $(n + 1)$th decision problem, for a given realization x of X_{n+1}, will be to choose an a_j for which the left-hand side of (10) is minimum. Since G is unknown in the empirical Bayes setup, Δ_G is unknown. If we can somehow find a good estimator $\Delta_{j,n}(X_1, \ldots, X_n; x)$ of $\Delta_G(a_j, x)$ for all $j = 0, 1, \ldots, k$ and all x, then the obvious rule for use in the $(n + 1)$th problem is simply t_n given in step 1 of Theorem 1. The proof of step 1 of the theorem uses the dominated convergence theorem and the condition $\max\{\int L(a_j, \theta) \, dG(\theta) | j = 0, 1, \ldots, k\} < \infty$. Part 2 of the theorem provides sufficient conditions for $T = \{t_n\}$ to be a.o. with rate $O(n^{-\delta/2})$ whenever unbiased estimators* for $\Delta_G(a_j, \cdot)$ exist for $j = 0, 1, \ldots, k$. The assumption of the existence of unbiased estimators $h_j(\cdot)$ of $\Delta_G(a_j, \cdot)$ for $j = 0, 1, \ldots, k$ is relaxed in step 3. Applications of Theorem 1 (or results similar to Theorem 1) to empirical Bayes (a) multiple decision problems*, (b) slippage tests*, (c) classification problems, and (d) nonparametric decision problems can be found in several references, including Gilliland and Hannan [7], Johns and Van Ryzin [10, 11], Robbins [20, 21], Van Ryzin [25, 26], and Van Ryzin and Susarla [27]. (*See also* the upcoming section "Monotonizing Empirical Bayes Tests.") A discussion of the best obtainable rates $O(n^{-1})$ in an empirical Bayes linear loss two-action problem involving Poisson and other distributions is given in [10].

The next general theorem involves the usual statistical squared error loss estimation problem: $L(\theta, a) = (\theta - a)^2$. For the sake of ease of exposition, assume that $\Theta = A$ is an interval contained in $(-\infty, \infty)$. In this setup, the Bayes rule t_G is given by

$$t_G(x) = \frac{\int \theta f_\theta(x) \, dG(\theta)}{\int f_\theta(x) \, dG(\theta)}. \quad (14)$$

$T = \{t_n\}$ satisfies (8) iff

$$R_n(T, G) - R(G)$$
$$= E\left[\{t_n(X_1, \ldots, X_n; X_{n+1}) - \theta_{n+1}\}^2 \right]$$
$$- R(G) \to 0. \quad (15)$$

A sufficient condition for (15) to hold is given by

Theorem 2. If $\phi(n) = E[\{t_n(X_1, \ldots, X_n; X_{n+1}) - t_G(X_{n+1})\}^2 \to 0$, then $T = \{t_n\}$ is

a.o. relative to G. If $\phi(n) = O(\alpha_n)$, then T is a.o. of order α_n relative to G.

The interpretation of Theorem 2, as in Theorem 1, is that $T = \{t_n\}$ is a.o. whenever $t_n(X_1, \ldots, X_n; X_{n+1})$ is a "good" estimator of $t_G(X_{n+1})$. It is interesting to note that neither in Theorem 1 nor in Theorem 2 did we need to estimate G. We only needed estimates of certain functionals of G such as $\int f_\theta(x) \, dG(\theta)$, etc. That this observation is extremely useful in the construction of empirical Bayes procedure can be seen from the following two examples.

Example 1: Case Involving Poisson Distribution*. Let $\Theta = (a, b) \subseteq (0, \infty)$ and $f_\theta(x) = e^{-\theta} \theta^x / x!$ for $x = 0, 1, \ldots$ (the measure μ here is the counting measure on $\{0, 1, 2, \ldots\}$). The statistical decision problem is to test the hypothesis $H_0: \theta \leq c$ against the alternative $H_1: \theta > c$ for a known c in (a, b). The loss of deciding in favor of H_i (denoted by a_i) is given by $L(\theta, a_0) = (\theta - c)^+$ and $L(\theta, a_1) = (\theta - c)^-$. If $\theta \in (a, c]$, a_0 is the correct action to be taken. Then, in the notation of Theorem 1, $\Delta_G(a_0, x) \equiv 0$, and

$$\Delta_G(a_1, x) = \int \{(\theta - c)e^{-\theta} \theta^x / x!\} \, dG(\theta)$$
$$= (x + 1) f_G(x + 1) - c f_G(x),$$

where f_G is the common density of the independent discrete random variables $X_1, \ldots, X_n$. Obvious estimators for $\Delta_G(a_0, x)$ and $\Delta_G(a_1, x)$ are given respectively by $\Delta_{1,n}(x) = (x + 1)\hat{f}_G(x + 1) - c\hat{f}_G(x)$, and $\Delta_{0,n}(x) \equiv 0$, where $n\hat{f}_G(x) = \sum_{i=1}^n I_{[X_i = x]}$. If $\int \theta \, dG(\theta) < \infty$, then $T = \{t_n\}$ with $t_n = a_0$ or a_1 as $\Delta_{1,n}(X_{n+1}) \geq$ or < 0, is a.o. by step 1 of Theorem 1. Applying part 2 of the same theorem, since $\hat{f}_G$ is an unbiased estimator of f_G, we can recover a rate like $O(n^{-\gamma})$ with $\gamma (< 1)$, depending on the moment conditions on G.

The Bayes estimator under squared error loss function is given by

$$t_G(x) = \int \theta f_\theta(x) \, dG(\theta) \Big/ \int f_\theta(x) \, dG(\theta)$$
$$= (x + 1) f_G(x + 1) / f_G(x).$$

Hence the empirical Bayes estimator would be $T = \{t_n\}$, where

$$t_n(X_1, \ldots, X_{n+1}) = (X_{n+1} + 1)\frac{\hat{f}_G(X_{n+1} + 1)}{\hat{f}_G(X_{n+1})}$$

with an appropriate modification when the right-hand side is either not defined or $= \infty$. As in the testing problem, we can recover rates like $O(n^{-\gamma})$ with $\gamma(\leqslant 1)$ depending on the moment conditions on G. (We implicitly assumed here that $E[\theta^2] < \infty$.)

Example 2: Case Involving Normal Distributions*. Let $\Theta = (a, b) \subseteq (-\infty, \infty)$, and $f_\theta(x) = (2\pi)^{-1/2} e^{-(x-\theta)^2/2}$, $-\infty < x < \infty$. [The measure μ is the Lebesgue measure on the Borel σ-field in $(-\infty, \infty)$.] The Bayes estimator under squared error loss function is given by

$$t_G(x) = x - \frac{f'_G(x)}{f_G(x)},$$

where f'_G is the derivative of f_G and $f_G = \left(\sqrt{2\pi}\right)^{-1} \int e^{-(x-\theta)^2/2} dG(\theta)$. In the empirical Bayes version of the situation above, it is not easy to see a good estimator for t_G since, unlike Example 1, we do not have here an unbiased estimator for the common density f_G based on $X_1, \ldots, X_n$. In this situation, we can use the so-called kernel estimators* $\hat{f}_G$ (introduced by Rosenblatt) to estimate f_G, and an appropriate modification $\hat{f}'_G$ of $\hat{f}_G$ to estimate f'_G arriving at the empirical Bayes estimator $T = \{t_n\}$ with

$$t_n(X_1, \ldots, X_n; X_{n+1}) = X_{n+1} - \frac{\hat{f}'_G(X_{n+1})}{\hat{f}_G(X_{n+1})},$$

with suitable modifications if the right-hand side of the equality above is equal to ∞ or not defined. $T = \{t_n\}$ can be shown to be a.o. of order $n^{-\gamma}$ with $\gamma(\leqslant 1)$ under certain moment conditions on G. It is worth mentioning here (could be applicable in other examples as well) that James Hannan noted that an appropriate truncation* of t_n to the usual estimator X_{n+1} (for θ_{n+1}) will give an empirical Bayes estimator that is a.o. with minimal moment conditions on G.

ESTIMATION OF A PRIOR DISTRIBUTION FUNCTION

One of the assumptions involved in the empirical Bayes problem described in detail in the preceding section is that there is an unknown prior distribution G on Θ associated with each repetition of the statistical decision problem. In this section we provide a general method of estimating G using $X_1, \ldots, X_n$. A motivation for estimating G comes from the following observation. Generally, if $t_n = t_n(X_1, \ldots, X_n; X_{n+1})$ is a Bayes rule (using X_{n+1}) about θ_{n+1} when the prior distribution is $\hat{G}_n$, an estimator of G based on $X_1, \ldots, X_n$, then $T = \{t_n\}$ will be a.o. and for each n, and t_n is admissible under fairly general conditions on Θ, A, and L. As in some parts of the preceding section, the results are best presented in the case when $\Theta = \mathfrak{X} = (-\infty, \infty)$ and μ is the Lebesgue measure. Extensions to other cases can be obtained with appropriate changes.

To estimate G, we need to have the following condition, called the identifiability condition. To state this, let $\mathcal{Q} = \{f_\theta(\cdot) : \theta \in \Theta\}$. Recall that $f_G(\cdot) = \int f_\theta(\cdot) dG(\theta)$ is the common marginal density of independent variables $X_1, \ldots, X_n$.

Definition 3. $\mathcal{Q}$ is said to be identifiable if the following condition is satisfied.

(C1) $f_G(\cdot) = f_H(\cdot) \Leftrightarrow G = H.$

Identifiability condition, which is assumed true in the rest of the section, simply means that there is a one-to-one correspondence between G and f_G. Without this condition, it is impossible to estimate G using $X_1, \ldots, X_n$.

Example 3: Mixtures* of Normal $N(\theta, 1)$ Distributions. Let $\theta \sim G$, and X, given θ, $\sim N(\theta, 1)$. Then the identifiability condition holds since $f_G(\cdot) = f_H(\cdot)$ implies that G and H have the same characteristic function.

Example 4: A Nonidentifiable Situation.
Let G and H be any two distribution functions (with support in [0, 1]) having the same first two moments, but distinct third moments. Define $f_\theta(x) = \binom{2}{x}\theta^x(1-\theta)^{2-x}$ for $x = 0$, 1, and 2. It can then be verified that $f_G(\cdot) = f_H(\cdot)$; but $G \neq H$. Hence $\{f_\theta(\cdot) : 0 \leq \theta \leq 1\}$ is not identifiable.

There are several methods (all are difficult in our opinion) for constructing estimators $\hat{G}_n$ of G. In the case when Θ is a finite set of known states, it is possible to obtain unbiased estimators for G (for details, see Gilliland et al. [8], Robbins [21], and Van Ryzin [26]). We describe below a general method for estimating G. The intuitive idea behind the procedure is as follows. The idea involves discretizing the problem in the following sense. For each n, form a grid $\theta_{n,-1}(= -\infty) < -n = \theta_{n,0} < \theta_{n,1} < \cdots < \theta_{n,m(n)} = n < \theta_{n,m(n)+1} = \infty$ with maximum length in the grid $\to 0$. Then to estimate $G(x)$ for $\theta_{n,l} \leq x \leq \theta_{n,l+1}$ at stage n, find estimators $\hat{p}_{n,0}, \ldots, \hat{p}_{n,l}$ for the unknown probabilities $G(\theta_{n,0})$, $G(\theta_{n1}) - G(\theta_{n0})$, $\ldots$, $G(\theta_{nl}) - G(\theta_{nl-1})$. Then $\sum_{i=0}^{l} p_{n,i}$ is our estimator for $G(x)$. Since the maximum length of the grid goes to zero, it is reasonable to expect that $\hat{G}_n(x) = (\sum_{i=0}^{l} \hat{p}_{n,i})I_{[\theta_{n,l} \leq x < \theta_{n,l+1}]} \to G(x)$ under certain general conditions.

We now describe formally the estimator $\hat{G}_n$. Assume that there exists an estimator $\hat{f}_G(\cdot)(= \hat{f}_G(X_1, \ldots, X_n; \cdot))$ such that

(C2) $\sup\{|\hat{f}_G(x) - f_G(x)| : -\infty < x < \infty\}$
$$\to 0 \quad \text{almost surely.}$$

Define

$$M_{n,l}(x) = \sup\{f_\theta(x) : \theta_{n,l} \leq \theta \leq \theta_{n,l+1}\},$$
$$m_{n,l}(x) = \inf\{f_\theta(x) : \theta_{n,l} \leq \theta \leq \theta_{n,l+1}\}.$$

Let $\hat{p}_n = \{\hat{p}_{n,-1}, \ldots, \hat{p}_{n,m(n)}\}$ be such that $\hat{p}_{n,l} \geq 0$ and $\sum_{l=-1}^{m(n)} \hat{p}_{n,l} = 1$, and for which there exists an $\epsilon_n > 0$ such that $\sum_{l=-1}^{m(n)} \hat{p}_{n,l}M_{n,l}(x) \geq \hat{f}_G(x) - \epsilon_n$ and $\sum_{l=-1}^{m(n)} \hat{p}_{n,l}m_{n,l}(x) \leq \hat{f}_G(x) + \epsilon_n$ for $x = \theta_{n,0}, \ldots, \theta_{n,m(n)}$. Then an estimator for G is

given by

$$\hat{G}_n(x) = \begin{cases} 0, & y < \theta_{n,0} \\ \hat{p}_{n,-1} + \hat{p}_{n,0}, & \theta_{n,0} \leq y < \theta_{n,1} \\ \sum_{i=1}^{l} \hat{p}_{n,i}, & \theta_{n,l} \leq x < \theta_{n,l+1}, \\ & l = 1, 2, \ldots, m(n). \end{cases}$$

(16)

Clearly, $\hat{G}_n$ is a discrete distribution function. The problem of solving the inequalities above for obtaining $\hat{p}_n$ is a simple linear programming problem and there are efficient computational algorithms available for the solution of such inequalities.

$\hat{G}_n$ given by (16) satisfies the following theorem. Let $\mathcal{Q} = \{f.(x) : x \in (-\infty, \infty)\}$.

Theorem 3. Let $\mathcal{Q} \subset C_0((-\infty, \infty))$, (C1), and (C2) hold. Let the following two conditions hold for each x.

(C3) $\sup\{|f_\theta(x) - f_\theta(x')| : -\infty < \theta < \infty\}$
$$\to 0 \quad \text{as } x' \to x.$$

(C4)   For each $\gamma > 0, \exists \delta, \delta' > 0$ such that
$$|x' - x| < \delta \quad \text{and}$$
$$|\theta' - \theta| < \delta' \to |f(x',\theta') - f(x',\theta)| < \gamma.$$

Then $G_n \xrightarrow{\text{weakly}} G$ almost surely.

Example 5: Location Parameter Case. Let h be a density such that $h(x) \to 0$ as $|x| \to \infty$, and $\sup\{|h'(x)| : -\infty < x < \infty\} < \infty$ [and hence h satisfies (C3) and (C4)]. Define $f_\theta(x) = h(x - \theta)$. Assume that $f.(\cdot)$ satisfies (C1). [For $f.(\cdot)$ to satisfy (C1), it is sufficient that the characteristic function of h does not vanish.] Let

$$na_n\sqrt{2\pi}\,\hat{f}_G(x) = \sum_{i=1}^{n} \exp\left[-(x - X_i)^2/(2a_n^2)\right]$$

with $a_n = n^\gamma$ and $0 < 4\gamma < 1$. Then $\hat{G}_n$ defined by (16) $\xrightarrow{\text{weakly}} G$ almost surely. A similar result holds for the case of scale parameter densities $f_\theta(x) = \theta h(\theta x)$ with $\theta > 0$.

We conclude this section with an example in which an explicit representation of G in terms of f_G is possible.

Example 6: Uniform Distribution* $U(0, \theta)$ Case. Let $\theta \sim G$, and let

$$f_\theta(x) = \begin{cases} mx^{m-1}/\theta^m, & 0 < x < \theta \\ 0, & \text{otherwise} \end{cases} \quad (17)$$

[$m = 1$ is the $U(0, \theta)$ case]. It is then easy to verify that

$$G(x) = F_G(x) - \frac{x}{m} f_G(x), \quad (18)$$

where $F_G(\cdot) = \int_0^\cdot f_G(t) \, dt$. In view of the foregoing representation for G, a "good" estimator of G based on $X_1, \dots, X_n$ is given by

$$\hat{G}_n(x) = n^{-1} \sum_{i=1}^{n} I_{[X_i < x]}$$

$$- \frac{x}{m} \frac{1}{n\epsilon_n} \sum_{i=1}^{n} I_{[x < X_i \le x + \epsilon_n]}, \quad (19)$$

where $\epsilon_n \to 0$. $\hat{G}_n(x)$ can be shown to be asymptotically unbiased, mean square consistent, and asymptotically normal under certain conditions on G and on ϵ_n. It is also worth noting that if (18) holds for a large class $\mathcal{G}$ of distributions G, then $f_\theta(\cdot)$ has to have the form given by (17).

STEIN'S ESTIMATOR VIEWED FROM AN EMPIRICAL BAYES FORMULATION

Let $(\xi_1, X_1), \dots, (\xi_p, X_p)$ be bivariate random vectors where $\xi_1, \dots, \xi_p$ are i.i.d. normal random variables with $E[\xi_1] = 0$ and $\text{var}(\xi_1) = \tau^2$ [denoted hereafter by $N(0, \tau^2)$], and given ξ_i, $X_i \sim N(\xi_i, \sigma^2)$. Stein considered the problem of squared error loss estimation of $\boldsymbol{\xi} = (\xi_1, \dots, \xi_p)'$ using $X_1, \dots, X_p$. The structure described above is similar to the formulation described in the second section with prior distribution $G = N(0, \tau^2)$ and with $n = p$. The Bayes solution (under squared distance loss) of $\boldsymbol{\xi}$ is given by

$$E[\boldsymbol{\xi}|\mathbf{X}] = \frac{\tau^2}{\tau^2 + \sigma^2} \mathbf{X}. \quad (20)$$

If we do not know τ^2, G is partially un-

known (or partially known), an assumption made in the empirical Bayes formulation. Since we do not know τ^2, it is reasonable to replace τ^2 in (20) by a good estimator of τ^2, given here by

$$\hat{\tau}^2 = p^{-1} \sum_{i=1}^{p} X_i^2 - \sigma^2. \quad (21)$$

Thus an estimator for $E[\boldsymbol{\xi}|\mathbf{X}]$ is

$$\hat{E}[\boldsymbol{\xi}|\mathbf{X}] = \frac{p^{-1}\sum_{i=1}^{p} X_i^2 - \sigma^2}{p^{-1}\sum_{i=1}^{p} X_i^2} \mathbf{X}. \quad (22)$$

If σ^2 is also assumed to be unknown, but an estimator S/N of σ^2 is available, then $\hat{E}[\boldsymbol{\xi}|\mathbf{X}]$ can be further modified to give the estimator

$$\hat{E}[\boldsymbol{\xi}|\mathbf{X}] = \left(1 - \frac{p}{N} \frac{S}{\|\mathbf{X}\|^2}\right) \mathbf{X}, \quad (23)$$

where $\|\mathbf{X}\|^2 = \sum_{i=1}^{p} X_i^2$. Stein points out that $(p - 2)/(N + 2)$ turns out to be a better constant in place of p/N, giving the famous Stein's estimator

$$\hat{\boldsymbol{\xi}} = \left(1 - \frac{p-2}{N+2} \frac{S}{\|X\|^2}\right) \mathbf{X}. \quad (24)$$

Stein shows that $E[\|\hat{\boldsymbol{\xi}} - \boldsymbol{\xi}\|^2] \le E[\|\mathbf{X} - \boldsymbol{\xi}\|^2]$ for $p \ge 3$ giving the inadmissibility of $\mathbf{X}$.

For the following discussion, let $\sigma^2 = 1$. In the empirical Bayes situation of Robbins [19, 21], we take a slightly different route from (20) onward. In the empirical Bayes setup, an estimator of ξ_{p+1} based on $x_1, \dots, x_{p+1}$ [here we have the extra vector (ξ_{p+1}, X_{p+1})] is desired when the prior $G = N(0, \tau^2)$. In view of (21), the appropriate empirical Bayes estimator [restricted to $N(0, \tau^2)$ priors] will be

$$\xi^*(X_1, \dots, X; X_{p+1}) = \left(1 - \frac{p}{\|\mathbf{X}\|^2}\right) X_{p+1}, \quad (25)$$

which is similar to (23) with S/N now taken to be unity since $\sigma^2 = 1$. The empirical Bayes estimator (25) is much less complicated than the empirical Bayes estimator pointed out in Example 2, where G is assumed to have no particular form.

We end this section by pointing out that if one were to assume that the ξ_i are indeed i.i.d. G, then the average risk of ξ, namely, $p^{-1}E[\|\hat{\xi} - \xi\|^2] \neq R(G)$, a goal that can be attained by using the empirical Bayes estimator suggested in Example 2. Connections between Stein's estimators and restricted empirical Bayes estimators have been considered in great detail by Cogburn [2] and in the recent papers of Efron and Morris (see, e.g., Efron and Morris [5] and references in that paper).

SOME RECENT DEVELOPMENTS IN EMPIRICAL BAYES THEORY

In this section, we provide a brief survey of the recent results concerning the original empirical Bayes approach of Robbins. In the standard empirical Bayes problem of Robbins, empirical Bayes solutions (i.e., finding $T = \{t_n\}$) are obtained utilizing the fact that the random variables $X_1, \ldots, X_n, \ldots$ associated with the repetitions of the decision problem are i.i.d. Such is not the case in the following subsection. For notation, please refer to the earlier section on the empirical Bayes problem.

The Case of Varying Sample Sizes

In the modified version of the empirical Bayes problem to be described here, the decision problems occurring in the sequence are not identical in that the sample sizes associated with each repetition of the decision problem could vary from problem to problem. Thus similar to the situation in the earlier section, there is a sequence of independent random pairs $\{(\theta_n, X_n)\}$, where $\theta_1, \ldots, \theta_n, \ldots$ are i.i.d. G, and conditional on $\theta_n = \theta$, X_n has density $f_{\theta, m(n)}(\cdot)$ where $m(n)$ is the sample size in the nth decision problem. If $R^m(G)$ denotes the minimum Bayes rule (i.e., the Bayes risk of a Bayes risk $t_G^m(\cdot)$ in the decision problem described by the five elements of the earlier section, with m observations instead of one observation), then a reasonable goal to be achieved at the $(n + 1)$th stage is $R^{m(n+1)}(G)$. If $t_n(X_1, \ldots, X_n; X_{n+1})$ is any decision rule for use in the $(n + 1)$th

decision problem, it is then obvious that $R_n(T, G) \geq R^{m(n+1)}(G)$, where $T = \{t_n\}$. This inequality motivates two definitions that are parallel to Definitions 1 and 2. For example, we say that $T = \{t_n\}$ is a.o. relative to G if $R_n(T, G) - R^{m(n+1)}(G) \to 0$. The other parallel definition is obvious. *Just as in the standard empirical Bayes problem, $T = \{t_n\}$ is a.o. in the modified empirical Bayes problem above whenever $t_n(X_1, \ldots, X_n; X_{n+1})$ is a "good" estimator of $t_G^{m(n+1)}(X_{n+1})$* (see the paragraphs following Theorems 1 and 2). It is possible to state analogs of Theorems 1 and 2; instead, we explain the modification that becomes necessary by means of the following example.

Example 7: Case Involving Poisson Distribution and Varying Sample Sizes. We reconsider here Example 1 with sample size at stage n taken to be $m(n)$. We first consider the two-action problem. It is obvious that X_{n+1}, the sum of all the $m(n+1)$ observations in the $(n + 1)$th problem, is a sufficient statistic, and

$$P[X_{n+1} = x] = f_{G, m(n+1)}(x)$$
$$= (x!)^{-1} \int e^{-m(n+1)\theta} (m(n+1)\theta)^x dG(\theta)$$
$$\text{for } x = 0, 1, 2, \ldots.$$

Then the Bayes rule is to choose a_0 or a_1 as

$$\Delta_{G, m(n+1)}(a_1, X_{n+1})$$
$$= (X_{n+1} + 1) f_{G, m(n+1)}(X_{n+1})$$
$$- c f_{G, m(n+1)}(X_{n+1})$$

≥ 0 or < 0. Since it can be verified that

$$E\left[(m(n+1))^x (m(i))^{-X_i} (m(n+1) - m(i))^{X_i - x} \binom{X_i}{X_i - x} I_{[X_i \geq x]} \right]$$
$$= f_{G, m(n+1)}(x)$$

for all $x = 0, 1, 2, \ldots$ and $i = 1, 2, \ldots$, we can obtain an unbiased estimator $\hat{f}_{G, m(n+1)}(\cdot)$ of $f_{G, m(n+1)}(\cdot)$ and hence an unbiased estimator $\Delta_n(a_1, X_{n+1})$ of $\Delta_{G, m(n+1)}(a_1, X_{n+1})$, both based on independent $X_1, \ldots, X_n$. Then $T = \{t_n\}$, with $t_n = a_0$ or a_1 as $\Delta_n(a_1, X_{n+1}) \geq 0$ or < 0, is a.o. with order $O(n^{-(1-2\epsilon)/2})$ for any $0 < 2\epsilon < 1$ relative to G under certain conditions, including (a) $m(n) \leqslant \bar{m} < \infty$, and (b) the support of $G \subseteq [\alpha, \beta] \subset (0, \infty)$. In the case of empirical Bayes squared error loss estimation problem, the empirical Bayes estimator $T^* = \{t_n^*\}$, with

$$t_n^*(X_1, \ldots, X_n; X_{n+1})$$
$$= \frac{(X_{n+1} + 1)\hat{f}_{G, m(n+1)}(X_{n+1} + 1)}{\hat{f}_{G, m(n+1)}(X_{n+1})},$$

with an appropriate modification when the right-hand side is either not defined or infinity, can be shown to be a.o. with $O(n^{-(1-\epsilon)})$ for any $0 < \epsilon < 1$ under certain conditions, including the two mentioned above.

Other such results are possible for decision problems involving estimation of the normal means with variances changing from problem to problem, and the location parameter exponential distributions with scale changing from problem to problem. A lot of work needs to be done in this direction of modified empirical approach to decision problems.

Empirical Bayes Interval Estimation

In this subsection we introduce the problem of empirical Bayes interval estimation, and suggest an intuitive method for finding a.o. procedures in such a problem. For ease of exposition, we consider the empirical Bayes problem formulated in the second section with Θ an interval, $A = \{(\epsilon, \delta) : \delta \geqslant 0; \epsilon, \epsilon + \delta \in \theta\}$ and the loss function given by

$$L(\theta, (\epsilon, \delta))$$
$$= C_1\{1 - I_{[\epsilon \leqslant \theta \leqslant \epsilon + \delta]}\} + C_2(\epsilon - \theta)^2$$
$$+ C_3(\epsilon + \delta - \theta)^2 + C_4\delta,$$

where C_1, C_2, C_3, and C_4 are known positive constants. Such a loss function accounts for the coverage of θ, the length of the interval, and the closeness of θ to the end points of the interval. By taking $C_2 = C_3 = C_4 = 0$, and $C_1 = 1$, we obtain the $0 - 1$ loss structure associated with the coverage of θ by an interval.

Obviously, to solve the empirical Bayes problem above, we first need to solve the Bayes problem. The Bayes rule is simply given by $(\epsilon_G(\cdot), \delta_G(\cdot))$ where for any x, $(\epsilon_G(x), \delta_G(x))(\epsilon A)$ is a minimizer of $\int L(\theta, (\epsilon, \delta))f_\theta(x)\,dG(\theta)$. Provided that G satisfies certain conditions, it is reasonable to expect that the empirical Bayes procedure $T = \{t_n\}$, with $t_n(X_1, \ldots, X_n; X_{n+1}) = (\epsilon_{G_n}(X_{n+1}), \delta_{G_n}(X_{n+1}))$, is a.o. whenever G_n is a "good" estimator of G. (Here the problem of estimation of G considered earlier is extremely useful.) There are no general results in this area except in a simple case (given below) in which the explicit representation of G in terms of the marginal distribution F_G of X_1 is possible.

Example 8: Uniform Distribution $U(0, \theta)$ Case. We revisit Example 6. Equation (18) of Example 6 gives a representation of G in terms of F_G. Let $\hat{G}_n(x) = F_n(x) - m^{-1}x(n\epsilon)^{-1}(F_n(x + \epsilon) - F_n(x))$, where

$$F_n(x) = \sum_{i=1}^n \left[(i-1)n^{-1} + (x - X_{(i-1)}) \right.$$
$$\left. \times n^{-1}(X_{(i)} - X_{(i-1)})^{-1} \right]$$
$$\times I_{[X_{(i-1)} \leqslant x \leqslant X_{(i)}]} + I_{[x > X_{(n)}]}$$

with $X_{(1)} < \cdots < X_{(n)}$ denoting the order statistic corresponding to $X_1, \ldots, X_n$. (Note that F_n is continuous, and that its distance from the empirical distribution of $X_1, \ldots, X_n$ is at most n^{-1}.) Then $T = \{t_n\}$, with

$$t_n(X_1, \ldots, X_n; X_{n+1})$$
$$= (\epsilon_{\hat{G}_n}(X_{n+1}), \delta_{\hat{G}_n}(X_{n+1})),$$

is a.o. with order $O(n^{-1/2})$ provided that G has no point masses, has bounded support, and $\int \theta^{-m}\,dG(\theta) < \infty$.

Lwin and Maritz [14] consider the restricted empirical Bayes approach to the problem of interval estimation in the following sense. Let $(\theta, \mathbf{X}) = (\theta, X_1, \ldots, X_m)$ be an $m + 1$ variate random vector with $\mathbf{X} \sim \times_1^m f_\theta(x_j)$. If the form of G is known, the posterior distribution of θ given $\mathbf{X}$ can be calculated and hence one can find a region R_α such that $P[\theta \in R_\alpha | \mathbf{X} = \mathbf{x}] = 1 - \alpha$ (R_α could be a one-sided interval like $[\theta_L(\alpha, G, x), \infty,)$ or $(-\infty, \theta_U(\alpha, G, x)]$, etc.) Lwin and Maritz [14] assume that G is known except for a parameter η and consider the empirical Bayes situation, as described earlier, which results in n independent, identically distributed (as $\mathbf{X}$) random vectors $\mathbf{X}_1, \ldots, \mathbf{X}_n$ with $\mathbf{X}_i = (X_{i1}, \ldots, X_{im})$, corresponding to the realization θ_i of θ in the ith repetition. Then one can hopefully estimate η by $\hat{\eta}$ $(\mathbf{X}_1, \ldots, \mathbf{X}_n)$ and replace η in R_α by $\hat{\eta}$ giving the *restricted empirical Bayes* confidence region $\hat{R}_\alpha$. They discuss the goodness of the region $\hat{R}_\alpha$ by two criteria which they labeled the expected cover criterion and the β-*content region*. They explain their procedures when R_α is $[\theta_1(\eta, x), \infty)$ or $(-\infty, \theta_v(\eta, x)]$ and when $\sqrt{2\pi}\, \sigma f_\theta(x) = \exp[-(x - \theta)^2/2\sigma^2]$ with σ^2 known. For full details, see Lwin and Maritz [14].

Lord and Cressie [13] obtain simultaneous confidence intervals for $E[\theta | X = x]$ for $x = 0, 1, \ldots, m$ using the empirical Bayes approach when $(\theta, X), (\theta_1, X_1), \ldots, (\theta_n, X_n)$ are i.i.d. with $\theta_i \sim G$ and given $\theta_i = \theta_0$, X_i is binomial with parameters and θ_0. They describe their procedure using a set of data consisting of the scores of $n = 12,990$ persons on a psychological test composed of $m = 20$ questions.

A recent paper by Cox [3] discusses yet another method by which one can obtain confidence intervals in *restricted empirical Bayes* situations. As in Lwin and Maritz [14], Cox illustrates his method using a normal prior with known variance and unknown mean, and conditional normal distributions with known variance. Cox's method depends on a general solution of the prediction intervals based on a large set of data.

Empirical Bayes Estimation of a Distribution Function

Let $(\theta_i, \mathbf{X}_i)$, $i = 1, \ldots, n + 1$, be a sequence of i.i.d. stochastic processes with the following probability structure: $\theta_1, \ldots, \theta_{n+1}$ are Dirichlet processes (for basic definitions and

the properties of Dirichlet processes, see the fundamental paper by Ferguson [6]) with parameter α, a σ-additive nonnull measure on the Borel σ-field in $(-\infty, \infty)$ with $\alpha((-\infty, \infty))$ known. (With probability 1, θ_i are discrete probability measures.) Given that $\theta_i = \theta$, $\mathbf{X}_i = (X_{i1}, \ldots, X_{im}) \sim \times_1^m \theta$. In the empirical Bayes formulation, Korwar and Hollander [12] consider the problem of estimating $F_{n+1}(\cdot)(= \theta_{n+1}(-\infty, \cdot])$ using the weighted squared error loss function $L(F, \hat{F})$ $= \int_{-\infty}^{\infty} (F(u) - \hat{F}(u))^2 \, dw(u)$ for a weight function w on $(-\infty, \infty)$. Korwar and Hollander [12] suggest the empirical Bayes estimator for $F_{n+1}(t)$

$$F_{n+1}(X_1, \ldots, X_n, t)$$
$$= (nm)^{-1} \sum_{i=1}^{n} \sum_{j=1}^{m} I_{[X_{ij} \leqslant t]}$$
$$+ m^{-1}(1 - P_m) \sum_{j=1}^{m} I_{[X_{n+1j} \leqslant t]}$$

for $-\infty < t < \infty$ where $(m + \alpha < (-\infty, \infty)$ $P_m = \alpha((-\infty, \infty))$ and show that it is a.o. in the sense of Definition 2 with rate of risk convergence $= 0(n^{-1})$. Since it is easy to calculate the moments of X_{ij}, and the correlation between X_{ij} and X_{ik}, they also obtain the exact expression for the risk r_{n+1} of $\hat{F}_{n+1}$. The rate $O(n^{-1})$ of risk convergence result can be extended to the following case. Let $Y_1, \ldots, Y_n, Y_{n+1}$ be independent identically G-distributed random variables and independent also of $(\theta_1, X_1), \ldots, (\theta_n, X_{n+1})$. Define $\delta_i = 1$ if $X_i \leqslant Y_i$ and $= 0$ otherwise, and $Z_i = \min\{X_i, Y_i\}$. The rate $O(n^{-1})$ of risk convergence has been extended in [22] to the case when we observe $(\delta_1, Z_1), \ldots, (\delta_n, Z_n)$ instead of $X_1, \ldots, X_n$ under certain conditions α and G. An application of this result is given in a later section.

Monotonizing Empirical Bayes Tests

Here we describe a recent idea of Van Houwelingen [24]. Consider the empirical Bayes setup described earlier with the following identification. Let μ = Lebesgue measure, and $f_\theta(x) = m(x)h(\theta)\exp(\theta x)$, where $m(x) > 0$ if and only if $a < x < b$ ($-\infty \leqslant a$

$< b \leqslant \infty$). Let $\Omega = \{ \theta : \int m(x)e^{\theta x} \, dx < \infty \}$, the natural parameter space of the exponential family, and assume without loss of generality that $0 \in \Omega$ (if not, we can reparametrize the exponential family). Consider testing $H_0 : \theta \leqslant 0$ versus $H_1 : \theta > 0$ when the loss, defined on $\Omega \times \{a_0, a_1\}$, $L(\theta, a_0) = \theta^+$, and $L(\theta, a_1) = \theta^-$ (this loss function is the one given in Example 1 with $c = 0$). If $\theta \sim G$, then the Bayes rule ϕ_G is given by $\phi_G(x) = a_1$ if $x > c_G$ and $= a_0$ if $x < c_G$ by the monotonicity of the Bayes rule. Van Houwelingen uses this cut-type property for ϕ_G to develop an empirical Bayes cut-type rule $\phi_n(x)$ based on $X_1, \ldots, X_n$ as follows. Define $M(x) = \int_a^x m(t) \, dt$. Let $\alpha_n(X_1, \ldots, X_n)$ be an estimator of $M(C_G) = \int_a^{c_G} m(t) \, dt$ and define the empirical Bayes rule by $\phi_n(X_1, \ldots, X_n, X_{n+1}) = a_1$ if $X_{n+1} > M^{-1}(\alpha_n)$ and a_0 otherwise. [Observe that for each realization $(x_1, \ldots, x_n)$ of $(X_1, \ldots, X_n)$, ϕ_n is a cut-type rule just as ϕ_G is.] He shows that ϕ_n is a.o. with rate $O(n^{-2r/(2r+3)}(\ln n)^2)$ under certain conditions, including:

$$(1) \int_{-\infty}^{\infty} |\theta|^{r+1} \, dG(\theta) < \infty,$$

(2) m is r times differentiable.

For each n and all $x_1, \ldots, x_n$, note that $\phi_n(x_1, \ldots, x_n, \cdot)$ is an admissible test in the $(n+1)$st decision problem. He refers to this property as a weak admissibility property. A lemma, useful in obtaining the foregoing rate of convergence result along with weak admissibility, has been generalized by Gilliland and Hannan [7] to k-arbitrary action, monotone loss, monotone likelihood ratio family $\{ f_\theta : \theta \epsilon \Omega \}$ multiple decision problems.

COMPOUND DECISION PROBLEMS

In this section we describe a compound decision problem, since it has been observed that a solution in the compound decision problem generally leads to an a.o. procedure in the corresponding empirical Bayes decision problem (for such a connection, see Gilliland et al. [8] and see also Neyman [16] for

Robbins' [18] introduction of this idea), and mention a few results available for compound decision problems. We also describe some of the recent results of Gilliland and Hannan involving the novel idea of restricting the risk in the component problems involved in the setup of a compound decision problem.

Simultaneous consideration of n decision problems each having the same general structure constitutes a compound decision problem. The general structure of the decision problem (see also the beginning of the section on the empirical Bayes function) involves (a) a set θ of states, (b) a set $\mathcal{C}$ of acts, and (c) a loss function $L(a, \theta) \geqslant 0$ defined on $\mathcal{C} \times \Theta$. The action chosen depends on an observation of a random variable X which is $\sim P_\theta$, $\theta \in \Theta$. Let $R(t, \theta)$ be the risk of the decision rule t when θ is the true state of nature. For $\theta \sim G$, let $R(G)$ denote the minimum attainable Bayes risk. Thus in the compound decision problem, we have vector $\boldsymbol{\theta}_n = (\theta_1, \ldots, \theta_n)'$ of states of nature, and $\mathbf{X}_n = (X_1, \ldots, X_n)' \sim \times_1^n P_{\theta_i}$. The loss in the compound decision problem is taken to be the average of the losses in the n decision problems. We say that $\mathbf{t}_n = (t_1, \ldots, t_n)'$ is a *compound decision rule for* $\boldsymbol{\theta}_n$ *if, for each* $i = 1, \ldots, n$, t_i *is a decision rule based on* $\mathbf{X}_n$. (If, for $i = 1, \ldots, n$, t_i is allowed to depend only on $X_1, \ldots, X_i$, then $t = \{t_n\}$ is said to be a sequence compound decision rule.) As an example, it can be seen that Stein's estimator (24) is a compound decision rule.

A goal that one tries to achieve in the setup of a compound decision problem is $R(G_n)$, where G_n is the empirical distribution of $\theta_1, \ldots, \theta_n$. Thus we generally look for compound decision rules $\mathbf{t}_n$ such that $D_n = R(\mathbf{t}_n, \boldsymbol{\theta}_n) - R(G_n) = n^{-1} \sum_{j=1}^{n} R(t_i, \theta_i) - R(G_n) \to 0$ uniformly for every sequence $(\theta_1, \theta_2, \ldots)$ in Θ^∞, preferably with a rate of convergence. Generally, solutions for compound decision problems have been attained by seeking $\mathbf{T}_n = (T_1, \ldots, T_n)$, where T_i is a "good" estimator of the Bayes rule ψ_{G_n} versus G_n in the ith decision problem. This idea for the construction of T_n is similar to the idea behind the construction of empirical Bayes rules. [In the case of a sequence-compound decision problem, Hannan's [9] inequality

$$\sum_{j=1}^{n} R(\psi_{G_i}, \theta_i) \leqslant nR(G_n) \leqslant \sum_{j=1}^{n} R(\psi_{G_{i-1}}, \theta_i)$$

can be used to show that the sequence compound decision rule $\{t_n\}$, with t_n a "good" estimator of the Bayes rule ψ_{G_n} in the nth decision problem, is such that $D_n \to 0$, again uniformly in all $(\theta_1, \theta_2, \dots)$ in Θ^∞.] Gilliland, Hannan, Oaten, Robbins, Van Ryzin, and others have exhibited set compound and/or sequence compound procedures for which $D_n = o(1)$, $O(n^{-(1-2\epsilon)}/2)$ with $0 < 2\epsilon < 1$, $O(n^{-1/2})$, $O(n^{-1}\log n)$, and $O(n^{-1})$, depending on the conditions on Θ, $\mathcal{C}$, L, and the class of distributions $\{P_\theta : \theta \in \Theta\}$. In the case of finite Θ and $\mathcal{C}$, some of the rate results mentioned above are obtained by using the fact that unbiased estimators of G_n based on $\mathbf{X}_n$ can be constructed.

We close this section by describing a recent idea of Gilliland and Hannan. Let the state space consist of $m + 1$ elements and thus let the possible (distinct) distributions of X be $P_0, P_1, \dots, P_m$. Let S denote the risk set in the general decision problem described earlier. We assume that S is a bounded subset of $[0, \infty)^{m+1}$. (If S is equal to the largest possible risk set for given $\mathcal{C}$ and L, then this is the usual decision problem.) Consider now the compound decision problem involving n independent repetitions of the decision problem above. Let $X_1, \dots, X_n$ be independent random variables with distributions $F_1, \dots, F_n$, respectively, where $F_i \in \{P_0, P_1, \dots, P_m\}$. Let $\mathbf{F} = \times_1^n F_i$. Let $\mathbf{S} = \{(\mathbf{s}_1, \dots, \mathbf{s}_n) : \mathbf{s}_\alpha = (s_{\alpha 0}, \dots, s_{\alpha m})$ is a (measurable) mapping from $\mathcal{X}^{n-1}$ to $S\}$. Let $s_{\alpha i}(X_1, \dots, X_{\alpha-1}, X_{\alpha+1}, \dots, X_n)$ denote the risk, conditional on $(X_1, \dots, X_{\alpha-1}, X_{\alpha+1}, \dots, X_n)$, incurred by $\mathbf{S}$ in component α when $F_\alpha = P_i$. Then $n\mathbf{R}(\mathbf{F}, \mathbf{S}) = \sum_{i=1}^{n} R_\alpha(\mathbf{F}, \mathbf{S})$ denotes the compound risk of $\mathbf{s}$ evaluated at $\mathbf{F}$. (If S is the largest possible risk set for given $\mathcal{C}$ and L, then this is the usual compound decision problem.

The formulation above provides a structure in which there is control over the component risk behavior of compound decision rules through the choice of S.) Let $\mathbf{E} = \{\mathbf{s} : \mathbf{s} = (\mathbf{s}_1, \dots, \mathbf{s}_n)$, where $\mathbf{s}_\alpha$ is constant with respect to α, and is a symmetric function of its remaining arguments$\}$. Let $\tilde{\psi}(\mathbf{n}) = \inf\{\mathbf{R}(\mathbf{n}, \mathbf{s}) : \mathbf{s} \in \mathbf{E}\}$, where $\mathbf{n} = (n_0, \dots, n_m)$ with $n_\alpha = $ number of $i \ni F_i = P_\alpha$; $i = 1, \dots, n$. A rule $\mathbf{s} \in \mathbf{E}$ is said to be *simple* if $\mathbf{s}_\alpha$ is a constant function with respect to α. Let $\psi(\mathbf{n}) = \inf\{\mathbf{R}(\mathbf{n}, s^n) : S \in \mathcal{S}\}$, where $s^n = (s, \dots, s)$. Gilliland and Hannan show that the difference between $\tilde{\psi}(\mathbf{n}) - \psi(\mathbf{n})$, is $O(n^{-1/2})$.

They also relate a useful upper bound for $\mathbf{R}(\mathbf{n}, \mathbf{s}) - \psi(\mathbf{n})(= D_n \geqslant 0)$ to an L_1-error estimation of $\mathbf{N}$ using $X_1, \dots, X_n$. Thus this last result provides a strengthening of some of the results stated earlier, and as Efron and Morris [4] have done in the case of normal distributions, the compound procedures of Hannan and Gilliland provide restricted risk component compound procedures for any finite state space Θ. For details, see [8].

A BRIEF APPLICATION

We consider an application of the result mentioned in subsection "Empirical Bayes Estimation of a Distribution Function" when the data are randomly right censored. This application was chosen because of the ease of computations involved in calculating the empirical Bayes estimator. The application is to a set of data consisting of the survival times (in weeks) of 81 $(= n + 1)$ patients participating in a melanoma study. This data were obtained at the central oncology headquarters located at the University of Wisconsin, Madison. The data reported and analyzed in ref. 22 were also analyzed by Phadia [17], whose analysis we describe below. The data, which were listed sequentially in order of entry into study, consisted of 46 uncensored observations and 35 censored observations. The censored observation 16 was treated as the 81 $(n + 1)$th observation. So we have 80 observations for use in the 81st problem. Phadia considered the following empirical Bayes distribution function estimator:

$$F_{n+1}(u) = \begin{cases} 1 - 2^{-1}(1 + \hat{\alpha}(u)), & u < Z_{n+1} \\ 1 - \hat{\alpha}(u)/2 \\ \qquad \text{if } \delta_{n+1} = 1, u \geqslant Z_{n+1}, \\ 1 - \{(1 + \hat{\alpha}(Z_{n+1})\hat{\alpha}(u)/2\hat{\alpha}(Z_{n+1}))\} \\ \qquad \text{if } \delta_{n+1} = 0, \quad u \geqslant Z_{n+1}, \end{cases}$$

where

$$\hat{\alpha}(u) = \{N^+(u)/n\}$$
$$\times \prod_{j=1}^{n} \left\{ \frac{C + 1 + N^+(Z_j)}{C + N^+(Z_j)} \right\}^{(1 - \delta_j)I_{[Z_j \leqslant u]}}$$

with $N^+(\cdot)$ denoting the number of $Z_i(i = 1, \ldots, N) > \cdot$ and C denoting a positive constant. For each realization of $(\delta_1, Z_1), \ldots, (\delta_{n+1}, Z_{n+1}), \hat{F}_{n+1}(\cdot)$ is a distribution function on $(0, \infty)$, a desirable characteristic since we are estimating the (random) distribution function F_{n+1}. $\hat{F}_{n+1}$ is very easy to compute (in fact, can be calculated using a hand calculator). We list below $\hat{F}_{n+1}(u)$ for various values of u when $C = 1$.

u	< 13	$14 \leqslant u < 16$	16	20
$\hat{F}_{n+1}(u)$	1	0.998	0.988	0.962

u	40	60	80	100
$\hat{F}_{n+1}(u)$	0.744	0.650	0.593	0.552

u	120	140	160	180
$\hat{F}_{n+1}(u)$	0.479	0.380	0.302	0.302

u	200	220	233	$\geqslant 234$
$\hat{F}_{n+1}(u)$	0.259	0.173	0.173	0

Other applications have been pointed out in the literature, for example to pattern recognition*, pricing insurance policies, multiple-lease auctions, and multiple-choice tests. In some of these applications, the authors have assumed that the prior distribution function is known except for a parameter η and they estimate η using the available data.

CONCLUDING REMARKS

In conclusion, we believe that a lot of work needs to be done in the problem of empirical Bayes confidence interval estimation* discussed previously. Here, results concerning estimation of a prior distribution should prove to be extremely useful. There are very few papers dealing with the simulation* of empirical Bayes procedures, especially for small sample sizes. Van Houwelingen [23] and some of the papers [1] provide some numerical results in some empirical Bayes decision problems. Although rates of convergence results in an empirical Bayes setup are important, attention should be paid to the two problems described above and also to estimation of $g(\theta_1, \ldots, \theta_n)$ of the first n parameters based on the data gathered in the first n repetitions of the experiment and search for risk convergent estimators of g. [For example, $g(\theta_1, \ldots, \theta_n) = \sum_{i=1}^{n} \theta_i$.]

This expository article presents only some of the basic results in empirical Bayes theory and it is hoped that the reader will consult other sources for further interesting aspects.

References

[1] Atchison, T. A. and Martz, H. F., Jr., eds., *Proc. Symp. Empirical Bayes Estimation Computing Statist.*, T. A. Atchison and H. F. Martz, Jr., eds. Texas Tech Univ. Math. Ser., Texas Tech. University, Lubbock, Tex.

[2] Cogburn, R. (1967). *Ann. Math. Statist.*, **38**, 447–464.

[3] Cox, D. R. (1974). In *Perspectives in Probability and Statistics*, J. Gani, ed. pp. 47–55. Applied Probability Trust, London

[4] Efron, B. and Morris, C. (1972). *J. Amer. Statist. Ass.*, **67**, 130–139.

[5] Efron, B. and Morris, C. (1977). *Sci. Amer.*, **236**(5), 119–127.

[6] Ferguson, T. S. (1973). *Ann. Statist.*, **1**, 209–230.

[7] Gilliland, D. C. and Hannan, J. (1977). *Ann. Statist.*, **5**, 516–521.

[8] Gilliland, D., Hannan, J., and Vardeman, S. (1979). Rm-401, Department of Statistics and Probability, Michigan State University, East Lansing, MI.

[9] Hannan, J. (1957). In *Contributions to the Theory of Games*, Vol. 3. Princeton University Press, Princeton, N.J., pp. 97–139.

[10] Johns, M. V., Jr. and Van Ryzin, J. (1971). *Ann. Math. Statist.*, **42**, 1521–1539.

[11] Johns, M. V. Jr. and Van Ryzin, J. (1972). *Ann. Math. Statist.*, **43**, 934–947.

[12] Korwar, R. and Hollander, M. (1976). *Ann. Statist.*, **4**, 581–588.

[13] Lord, F. M. and Cressie, N. (1975). *Sankhyā B*, **37**, 1–9.

[14] Lwin, T. and Maritz, J. S. (1976). *Scand. Acturial J.*, 185–196.

[15] Maritz, J. S. (1970). *Empirical Bayes Methods.* Methuen's Monogr. Appl. Prob. Statist. Methuen, London.

[16] Neyman, J. (1962). *Rev. Int. Statist. Inst.*, **30**, 112–127.

[17] Phadia, E. G. (1980). *Ann. Statist.*, **8**, 226–229.

[18] Robbins, H. (1951). *Proc. 2nd Berkeley Symp. Math. Statist. Prob.* University of California Press, Berkeley, Calif., pp. 131–148.

[19] Robbins, H. (1956). *Proc 3rd Berkeley Symp. Math. Statist. Prob.* University of California Press, Berkeley, Calif., pp. 157–163.

[20] Robbins, H. (1963). *Rev. Int. Statist. Inst.*, **31**, 195–208.

[21] Robbins, H. (1964). *Ann. Math. Statist.*, **35**, 1–20.

[21a] Stein, C. (1956). *Proc. Third Berkeley Symp. Math. Statist. Prob.*, **1**, 197–206.

[22] Susarla, V. and Van Ryzin, J. (1978). *Ann. Statist.*, **6**, 740–754.

[23] Van Houwelingen, J. C. (1974). *Statist. Neerlandica*, **28**, 209–221.

[24] Van Houwelingen, J. C. (1976). *Ann. Statist.*, **4**, 981–989.

[25] Van Ryzin, J. (1970). *Proc. First. Int. Symp. Nonparametric Tech. Stat. Inference*, M. L. Puri, ed. Cambridge University Press, Cambridge, pp. 585–603.

[26] Van Ryzin, J. (1971). In *Statistical Decision Theory and Related Topics*, J. Yaekel and S. S. Gupta, eds. Academic Press, New York, pp. 181–205.

[27] Van Ryzin, J. and Susarla, V. (1977). *Ann. Statist.*, **5**, 172–181.

Acknowledgment

Thanks are due to Professors J. Blum, D. Gilliland, and J. Hannan and J. Van Ryzin for their comments and suggestions which were incorporated into the article. The author is also grateful to Professor H. Robbins for suggesting the last problem mentioned in the concluding remarks section, and other useful suggestions.

(BAYESIAN INFERENCE DECISION THEORY

ESTIMATION, POINT JAMES–STEIN ESTIMATORS PRIOR DISTRIBUTION)

V. SUSARLA

EMPIRICAL DENSITY FUNCTION (EDF)

The empirical density function $g_n(x)$ with parameter $\lambda(>0)$ is defined as

$$g_n(x) = \frac{F_n(x + \lambda) - F_n(x - \lambda)}{2\lambda},$$

where $F_n(x)$ is the empirical distribution function. It is an approximate derivative of $F_n(x)$, and is also called the naive estimator* of the density function.

It is easy to verify that if λ is kept constant

$$\lim_{n \to \infty} g_n(x) = g(x) = \frac{F(x + \lambda) - F(x - \lambda)}{2\lambda}$$

with probability 1, where $F(\cdot)$ is the underlying CDF.

By making λ depend on n, an estimate of the PDF $F'(\cdot) = f(\cdot)$ is obtained. A recommended choice for $\lambda = \lambda_n$, using the expected mean square criterion to make it proportional, is $n^{-1/5}$ (under the assumption that the third derivative of the density exists) [1].

Empirical density functions are only the simplest of a large class of (nonparametric) density estimators.

For more information on this topic, see Waterman and Whiteman [2] and Wegman [3].

References

[1] Rosenblatt, M. (1956). *Ann. Math. Statist.*, **27**, 832–837.

[2] Waterman, M. S. and Whiteman, D. E. (1978). *Int. J. Math. Educ. Sci. Tech.*, **9**, 127–137.

[3] Wegman, E. J. (1972) *Technometrics*, **14**, 533–546.

(DENSITY FUNCTION ESTIMATION)

EMPIRICAL DISTRIBUTION FUNCTION *See* EDF STATISTICS

Table 1 Illustrative Index Computation

Step1. Establishment-occupation sample weights are applied to the occupational earnings to obtain weighted average earnings for each occupation in each of the 62 SIC industry cells for the survey periods:

	Straight-Time Hourly Earnings			Sample Weight	Computation of Weighted Average Earnings			
Sample Occupation	September (1)	December (2)	March (3)	(4)	September (1) × (4)	December (2) × (4) (a)	December (2) × (4) (b)	March (3) × (4)
Electricians:								
Establishment 1	$5.30	$5.31	$5.40	1.0	$ 5.30	$ 5.31	$ 5.31	$ 5.40
Establishment 2	5.20	5.22	5.38	2.0	10.40	10.44	10.44	10.76
Establishment 3	5.16	5.18[a]	[b]	3.0	15.48	15.54	[b]	[b]
Total				6.0	31.18	31.29	15.75	16.16
Weighted average					5.196	5.215	5.250	5.387
Carpenters:								
Establishment 1	4.90	4.90	4.94	1.5	7.35	7.35	7.35	7.41
Establishment 2	4.80	4.86	4.94	2.5	12.00	12.15	12.15	12.35
Establishment 3	4.86	4.89[a]	[b]	3.5	17.01	17.12	[b]	[b]
Total				7.5	36.36	36.62	18.65	19.76
Weighted average					4.848	4.883	4.875	4.940

[a]Indicates imputed data.
[b]Indicates that Establishment 3, classified as a temporary nonrespondent in December is a dropout in March. In (a) imputed data from Establishment 3 are used in calculation; in (b) only data from Establishments 1 and 2 are used.

Step 2. These weighted average earnings are multiplied by base weight period employment from the decennial census to obtain wage bills for each occupation-industry cell for the survey periods:

Sample Occupation	Weighted Average Earnings				Occupational Weight—1970 Census Employment (4)	Wage Bills			
	September (1)	December (2a)[a]	December (2b)	March (3)		September (1) × (4)	December (2a)[a] × (4)	December (2b) × (4)	March (3) × (4)
Electricians	$5.196	$5.215	$5.250	$5.387	3,000	$15,588	$15,645	$15,750	$16,131
Carpenters	4.848	4.883	4.875	4.940	2,000	9,696	9,776	9,750	9,880
Total wage bill, craft and kindred workers, Industry I						25,284	25,421	25,500	26,041

[a]This column includes imputed figures from Establishment 3 (see step 1); Col. (2b) excludes Establishment 3.

Step 3. The wage bills are summed across all cells to obtain total wage bills for the survey periods:

Industry	Wage Bills			
	September (1)	December (2a)	December (2b)	March (3)
1.	$ 25,284	$ 25,421	$ 25,500	$ 26,041
62.				
Total	285,125	287,345	287,650	291,600

Step 4. The aggregate wage bill for the current survey period is divided by the wage bill for the prior period to obtain ratios:
December = 287,345 ÷ 285,125 = 1.0078   March = 291,600 ÷ 287,650 = 1.0137

Step 5. The ratios are converted to quarterly percentage changes:
December = 0.78 %   March = 1.37%
Eventually, the ratios will serve as link relatives to move the index from quarter to quarter:
September index = 100.00   December index = 100.00 × 1.0078 = 100.78   March index = 100.78 × 1.0137 = 102.16

Reproduced from U.S. Department of Labor, Bureau of Labor Statistics, BLS Handbook of Methods for Surveys and Studies, Bulletin 1910 (1976), p. 190.

505

Table 2 Rate of Wage and Salary Changes in Employment Cost Index

(Percent changes)

Series	3 Months Ended					12 Months Ended			
	December 1978	March 1979	June 1979	September 1979	December 1979	March 1979	June 1979	September 1979	December 1979
All private nonfarm workers	1.5	2.0	1.9	2.1	2.4	7.8	7.6	7.7	8.7
Workers, by occupational group									
White-collar workers	1.2	1.9	1.7	2.3	2.4	7.3	7.0	7.4	8.6
Professional and technical workers	1.5	1.9	1.1	2.7	2.8	7.1	6.8	7.5	8.8
Managers and administrators	1.5	2.4	1.5	2.0	1.4	7.2	6.9	7.6	7.4
Sales workers	0.1	− 0.2	4.2	0.7	3.9	7.9	6.7	4.8	8.8
Clerical workers	1.3	2.7	1.4	2.9	2.1	7.4	7.4	8.5	9.4
Blue-collar workers	1.9	1.9	2.3	2.0	2.5	8.3	8.4	8.4	9.0
Craft and kindred workers	1.6	2.1	2.1	2.2	1.9	8.5	8.2	8.3	8.6
Operatives, except transport	2.3	1.9	2.2	1.7	3.1	7.8	8.6	8.3	9.2
Transport equipment operatives	1.7	1.5	3.5	2.4	2.4	9.3	8.3	9.5	10.2
Nonfarm laborers	2.4	1.5	2.7	1.7	2.9	7.9	8.6	8.5	9.1
Service workers	0.6	3.2	0.9	1.1	1.8	8.3	7.2	5.9	7.2
Workers, by industry division									
Manufacturing	2.7	1.7	1.8	1.8	3.1	8.0	8.2	8.1	8.6
Durables	2.7	1.8	1.5	2.1	3.3	8.4	8.2	8.4	9.0
Nondurables	2.5	1.5	2.3	1.2	2.7	7.4	8.2	7.6	7.8
Nonmanufacturing	0.8	2.2	2.0	2.3	2.0	7.6	7.2	7.5	8.8
Construction	1.1	1.3	2.6	2.0	1.1	7.9	7.6	7.2	7.2

Transportation and public utilities	1.9	2.6	1.6	2.9	2.0	8.9	8.4	9.2	9.4
Wholesale and retail trade	1.1	2.1	2.4	1.9	1.3	7.3	7.0	7.7	7.9
Wholesale trade	2.0	1.6	2.9	1.2	2.1	6.6	7.7	7.8	7.9
Retail trade	0.8	2.3	2.3	2.1	1.0	7.6	6.7	7.7	7.9
Finance, insurance, and real estate	a	3.1	3.2	1.9	4.3	a	a	a	13.2
Services	0.7	2.1	1.1	2.6	2.5	7.0	6.4	6.6	8.5
Workers, by region									
Northeast	1.8	1.5	1.7	1.7	2.1	6.8	6.9	6.9	7.3
South	1.6	2.6	1.7	1.7	2.4	8.8	8.3	7.7	8.5
North Central	1.5	1.9	2.5	2.0	2.6	7.5	7.8	8.2	9.4
West	1.5	2.0	2.0	2.5	1.8	7.9	7.7	8.2	8.5
Workers, by bargaining status									
Union	2.0	1.8	2.1	2.2	2.6	8.2	8.3	8.4	9.0
Manufacturing	2.7	1.7	2.0	1.9	3.4	8.7	9.1	8.6	9.4
Nonmanufacturing	1.4	1.9	2.2	2.5	1.7	7.7	7.6	8.2	8.5
Nonunion	1.1	2.1	1.9	1.9	2.3	7.5	7.2	7.3	8.5
Manufacturing	2.6	1.7	1.7	1.6	2.7	7.3	7.4	7.7	7.9
Nonmanufacturing	0.5	2.3	2.0	2.1	2.1	7.6	7.1	7.1	8.8
Workers, by area									
Metropolitan areas	1.6	2.0	2.0	2.2	2.5	7.8	7.7	7.9	8.9
Other areas	1.3	2.1	2.1	1.6	1.9	7.6	7.4	7.3	7.9

[a] Not available.

EMPLOYMENT COST INDEX

The Employment Cost Index (ECI), now being developed by the Bureau of Labor Statistics*, U.S. Department of Labor, is designed to provide an accurate, timely, and comprehensive measure of change in the "price" of labor. Although data releases as of February 1980 showed percent changes in pay rates over quarterly and annual periods, the long-run objective is to present findings in index-number form. The ECI unit of observation is an occupation within an establishment.[1] Occupations are studied rather than individual workers, because pay rates are generally set for jobs performed rather than individuals filling them. The "price" of labor is defined as employer expenditures for worker compensation—wage rates plus supplementary benefits—and is expressed as payments per hour worked. The Bureau measures employer payments rather than worker receipts, because this approach is consistent with basic uses anticipated for the series, that is, analyses of relations between pay changes and shifts in other economic variables, such as the price level, productivity, employment, and unemployment.

In computing the ECI, the items studied and their weights are held constant over time, so that measured changes essentially reflect price changes, and not shifts in either units of observation or their relative importance. The ECI thus provides a measure of pay change largely unaffected by the employment shifts that influence most existing statistical series. It presents changes in the price of a standardized mix of purchased labor services, much as the Bureau's Consumer Price Index (CPI)* presents changes in the price of a standardized market basket of consumer goods and services. For analyses of the ECI rationale and conceptual framework, see Samuels [3] and Sheifer [4].

The basic sample design of the Employment Cost Index involves a two-stage sampling process. The first stage was a mail survey of approximately 10,000 establishments to determine the number of persons employed in each of 23 occupations per establishment. Controlled selection was used to subsample this first stage to obtain a probability sample of 2000 establishments reporting on 9000 occupations.[2] It represents the private nonfarm economy, excluding households—the industrial coverage of the series as of early 1980. Both blue- and white-collar and nonsupervisory and supervisory occupations are included.

In developing quarterly ECI statistics, sample weights are applied to collected occupational compensation data to obtain a weighted average for each studied occupation, by industry. These averages are multiplied by base-period employment weights from the 1970 Census of Population to obtain wage bills for occupation–industry cells. Wage bills are summed across cells and percent changes are determined by dividing current quarter by prior-quarter wage bills. Percent changes over more than two survey periods are obtained by compounding successive changes for individual quarters. Index computation procedures are illustrated in Table 1 and are described more fully in ref. 8. For detailed discussions of ECI methodology, see Sheifer et al. [7].

ECI data were first issued in June 1976, covering the fourth quarter of 1975 and the first quarter of 1976. Statistics have since been issued on a quarterly basis, approximately 2 months after the survey months of March, June, September, and December. The release for the fourth quarter of 1979 provided, in addition to an overall measure of pay change, detail for 11 occupational groups, 11 industrial categories, 4 geographic regions, workers covered by union contracts and those not covered, and workers in or outside metropolitan areas (see Table 2).

Data are initially issued in press releases and are also published in the monthly BLS periodical *Current Wage Developments*. Articles either describing the series or analyzing findings appear occasionally in the *Monthly Labor Review*. For examples, see Kohler [2] and Sheifer [6].

As of February 1980, ECI reports were limited to changes in average straight-time hourly earnings in the sampled occupational units, such earnings serving as a proxy for wage rates. Beginning later in 1980, the ECI will measure changes in total compensation through inclusion of employer expenditures for fringe benefits. Benefit data will be collected and processed to reflect changes in underlying expenditure rates and not such usage fluctuations as variations in extent of overtime work. It is anticipated that government employees will be added to ECI coverage in 1981 and the remaining currently excluded industrial sectors—agriculture and households—in subsequent years.

NOTES

1. An establishment is an economic unit where business is conducted, e.g., factory, mine, warehouse, or group of retail stores. It is not necessarily the same as an enterprise or company, which may have one or more establishments.

2. The ECI uses the census occupational classification system's three-digit code level of detail (accountants, carpenters, etc.); this classification system encompasses all specific jobs in the economy within 441 occupations. (Not all, of course, will be found in a given establishment.) The survey design calls for obtaining earnings data for from 1 to 10 occupations per establishment. The current universe consists of approximately 4 million establishments.

References

[1] Kohler, D. M. (1978). *Monthly Labor Rev.*, Jan., 22–23. (Summarizes ECI findings through 1977.)

[2] Kohler, D. M. (1979). *Monthly Labor Rev.*, July, 28–31. (A more detailed review of ECI findings, covering the period through 1978.)

[3] Samuels, N. J. (1971). *Monthly Labor Rev.*, Mar., 3–8. (Discusses the rationale underlying the development of the series.)

[4] Sheifer, V. J. (1975). *Monthly Labor Rev.*, July, 3–12. (A statement of the ECI conceptual framework.)

[5] Sheifer, V. J. (1977). *Statist. Rep.*, Jan. 101–114. (A generalized description of the Employment Cost Index.)

[6] Sheifer, V. J. (1978). *Monthly Labor Rev.*, Jan., 18–21, 24–26. (Describes a method for incorporating supplementary benefits, consistent with the ECI conceptual framework.)

[7] Sheifer, V. J. et al. (1978). *1978 Proc. Sect. Surv. Res. Methods, Amer. Statist. Ass.*, pp. 683–711. (A series of papers covering the ECI conceptual framework, procedures for calculating change, survey design, estimation procedures, and variance estimation.)

[8] U.S. Department of Labor, Bureau of Labor Statistics (1976). *BLS Handbook of Methods for Surveys and Studies.* Bull. 1910, pp. 184–191. (A generalized description of the ECI, emphasizing sampling aspects and index computation.)

(INDEX NUMBERS
LABOR STATISTICS)

Victor J. Sheifer

ENGINEERING STATISTICS

NEED FOR STATISTICAL METHODS IN ENGINEERING

Results of experiments and industrial production are not constant and fully predictable, even when all input conditions are controlled as well as possible. Instead, they exhibit variation from trial to trial, time to time, or piece to piece. Wherever there is variation in results, the problems of estimation, prediction, decision making, and control involve varying data, that is, statistics. Thus one should be using appropriate statistical methods, which have been developed for all sorts of problems involving variation. These methods provide for efficient experimental design and data collection, analysis and interpretation, and decision making.

PROBLEMS AND METHODS

The following lists some kinds of engineering problems for which there are available statistical methods.

1. Pattern of variation or distribution model descriptive of the way results vary, whether measurements or counts of events, such as successes and failures. There are a

great many models available, such as the normal* and gamma* distributions for measurements, and the binomial* and Poisson* for counts. Analysis depends upon the model appropriate.

2. Is a proposed population model feasible, or must it be discarded? Tests for goodness of fit to the model are available.

3. Estimation of the parameter(s) for a feasible population model. It can be a point estimate or an interval estimate. Typical population parameters are the average and variability for measurement models, or the proportion effective in a count or attribute model.

4. Could the observed sample of results have reasonably arisen from a more or less completely specified population model assumed for some specified conditions? Or is the observed sample incompatible with the assumed population?

5. In a sample of observations, is an apparently outlying observation compatible with the rest of the sample? Or should it be regarded as having arisen from a different population than the others? Example: Can a set of measurements be regarded as homogeneous?

6. We have two samples of data taken under possibly differing sets of conditions.

a. Could the population *types* be regarded as compatible?

b. Are the two samples compatible as to parameters, such as averages? Or is the difference between the sample averages, say, too great for chance to explain, so that we must conclude that the population averages differ?

c. If there is a difference in parameters, we can estimate the difference or ratio.

d. Examples: control versus test condition; uniformity of measurement conditions.

e. Estimation of limits to the difference in parameters.

7. A series of samples taken under possibly differing conditions, such as times of production or sets of experimental conditions.

a. Could all the samples readily have arisen from the same population? If so, in industrial production of the process would be called "in control," that is, stable. If not, the process would be "out of control." Similarly for sets of experimental conditions: conclude "no evidence of differences," or "there are real differences in conditions."

b. Test for adequacy of control of a measuring technique.

c. Choice of the "best" set of experimental conditions, or at least to group into classes of indistinguishable sets of conditions.

d. Test for homogeneity of samples before lumping all results into one big sample.

e. A control condition versus several sets of experimental conditions.

8. Systematic varying of input conditions, with one or several observations per combination of conditions, to study the separate "effects" and "joint effects" of the input conditions. (*See* ANALYSIS OF VARIANCE.)

9. Any of a large number of experimental designs for efficiency and full use of every observation. (*See* DESIGN OF EXPERIMENTS.)

10. Observations taken in sequence, after each one of which we can take action A, say, acceptance, or action B, say, rejection, or request another observation. Leads to reliable decisions on a small average number of observations. (*See* SEQUENTIAL ANALYSIS.)

11. Decision making on a lot or process, so as to control the risk or probability of a wrong decision. Can be for counts of events (attributes) or for measurements (continuous data).

12. Measurement error* estimation, accuracy, and precision. Calibration curve for a measurement technique. Homogeneity of results.

13. Control of industrial production process. (*See* QUALITY CONTROL.)

14. Determination of process capability in industry. Comparison to requirements or specifications.

15. Determination of needed sample size for desired assurance, that is, for control of risks of wrong decisions.

16. Study of relationship between variables, for determining which variables to control and at what levels, or for estimation and prediction. Determining type of relation: straight line or plane, or nonlinear. Estimation of necessary constants or parameters. (*See* CORRELATION; CURVE FITTING; REGRESSION.)

a. Just two variables "dependent" and "independent."

b. k variables.

c. Seeking most influential or most strongly related "independent" variables.

d. Seeking the simplest model, eliminating variables.

e. Estimation or prediction with appropriate error. Point or interval estimate.

f. Making allowance or correction for uncontrolled variables in the experiment.

17. Determination of tolerance limits to a distribution, so that at least a proportion P lies below the limit, or between two limits.

18. Estimation of median lethal dose through sequential experimentation.

19. Probability analysis of a system. (*See* SYSTEM ANALYSIS.)

20. Reliability of components or a system. (*See* RELIABILITY.)

21. Optimization of a production process by sequential experimentation on process conditions.

Bibliography

Barlow, R. E. and Proschan, F. (1975). *Statistical Theory of Reliability and Life Testing Probability Models.* Holt, Rinehart and Winston, New York. (A somewhat more mathematical book than Lloyd and Lipow, particularly useful for model construction.)

Burr, I. W. (1974). *Applied Statistical Methods.* Academic Press, New York.

Burr, I. W. (1976). *Statistical Quality Control Methods.* Marcel Dekker, New York.

Duncan, A. J. (1974). *Quality Control and Industrial Statistics.* Richard D. Irwin, Homewood, Ill.

Lloyd, D. K. and Lipow, M. (1962). *Reliability.* Prentice-Hall, Englewood Cliffs, N.J.

Owen, D. B. (1962). *Handbook of Statistical Tables.* Addison-Wesley, Reading, Mass.

(CONTROL CHARTS
CUMULATIVE SUM CONTROL
 CHARTS
DESIGN OF EXPERIMENTS
QUALITY CONTROL
RELIABILITY)

IRVING W. BURR

ENGLISH BIOMETRIC SCHOOL

The term relates to the focus of interest (biometry)* of the early workers in what was to evolve into the mainstream of classical statistical theory: namely, the situation where precise prior distributional assumptions, often of normality, apart from parameters, are made in respect of data, and their consequences developed ("parametric" statistics). The methods of exact measurement and statistical analysis in biological settings as practiced by F. Galton*, through Galton's book *Natural Inheritance* (1889) inspired the biologist W. F. R. Weldon* and applied mathematician and philosopher of science K. Pearson* (together at University College, London, from 1891) to turn their attention to this area. (F. Y. Edgeworth may also be regarded as being on the periphery of this group.) This led to the founding of the still-prominent journal *Biometrika*, which first appeared in 1901, founded by Galton, Pearson, and Weldon, and in its early years was co-edited by the two younger men (Pearson and Weldon). Although it was initially very much concerned with biometry, after K. Pearson's editorship it came progressively to publish almost exclusively papers on mathematical statistics, the biometric function passing to other organs.

After those already mentioned, eminent authors of the formative period of the tradition of the "English biometric school" include W. S. Gosset* ("Student"), G. U. Yule*, R. A. Fisher*, E. S. Pearson and J. Neyman, and M. G. Kendall. In their work a strong biometric flavor persisted. It is notable that Fisher was as eminent in population genetics as in statistical theory; and it is necessary also to mention J. B. S. Haldane, who, although primarily a geneticist, made a number of statistical contributions.

Distinct from this stream of development is the "Continental direction" of statistics, which had its origins in dispersion theory. It is said that this direction sought to give test criteria of more universal applicability by avoiding precise distributional assumptions. This philosophical standpoint is embodied in the field now known as "nonparametric statistics." The outstanding figure in the synthesis of the two philosophical streams was A. A. Chuprov*.

Bibliography

[1] Pearson, E. S. (1965). *Biometrika*, **52**, 3–18 (Also as pp. 323–338 of ref. 4.)

[2] Pearson, E. S. (1967). *Biometrika*, **54**, 341–355. (Also as pp. 339–354 of ref. 4.)

[3] Pearson, E. S. (1968). *Biometrika*, **55**, 445–457. (Also as pp. 405–418 of ref. 4.)

[4] Pearson, E. S. and Kendall, M. G., eds. (1970). *Studies in the History of Statistics and Probability*. Charles Griffin, London. (Contains papers that give a quite comprehensive historical picture of the English biometric school, apart from refs. 1, 2, 3, and 5 cited here, including work on Gosset, Yule, K. Pearson, Fisher, and the Neyman–Pearson cooperation.)

[5] Pearson, K. (1906). *Biometrika*, **5**, 1–52. (Also as pp. 265–322 of ref. 4. Contains a description of the interaction between K. Pearson and Weldon.)

(BIOMETRIKA
BIOMETRY
GENETICS, STATISTICS IN
NONPARAMETRIC STATISTICS)

E. SENETA

ENTROPY

The entropy of the random variable **X**, denoted by $H(\mathbf{X})$, was introduced by Claude E. Shannon [13] in a historically significant article as a measure of choice, uncertainty, and information*. Let

$$H(\mathbf{X}) = -E\big[\log p_{\mathbf{X}}(x)\big],$$

where **X** is a discrete random variable and $p_{\mathbf{X}}(x) = \Pr[\mathbf{X} = x]$.

If $p_i = \Pr[\mathbf{X} = x_i] \geqslant 0$, $i = 1, 2, \ldots, n$, and $\sum_{i=1}^n p_i = 1$, then

$$H(\mathbf{X}) = H(p_1, p_2, \ldots, p_n)$$

$$= -\sum_{i=1}^n p_i \log p_i,$$

($0 \log 0$ is defined as $\lim_{p \to 0} p \log p = 0$) is the customary representation for the entropy of **X**, when **X** can assume only a finite number of values with positive probability. This extends readily to any finite number of discrete random variables whose joint distribution has positive probability on a finite set of points. Specifically, let $(\mathbf{X}, \mathbf{Y})$ be a pair of random variables with $P\{\mathbf{X} = x_i, \mathbf{Y} = y_j\} = p_{ij} > 0$, $i = 1, 2, \ldots, m$, $j = 1, 2, \ldots, n$; $\sum_{i=1}^m \sum_{j=1}^n p_{ij} = 1$. Then,

$$H(\mathbf{X}, \mathbf{Y}) = H(p_{11}, p_{12}, \ldots, p_{mn})$$

$$= -\sum_{i=1}^m \sum_{j=1}^n p_{ij} \log p_{ij}.$$

Similarly, the conditional entropy, $H(\mathbf{X} \mid \mathbf{Y})$, is defined as $\sum_{j=1}^n p_{\cdot j} H(\mathbf{X} \mid \mathbf{Y} = y_j)$, where

$$H(\mathbf{X} \mid \mathbf{Y} = y_j) = -\sum_{i=1}^m P[\mathbf{X} = x_i \mid \mathbf{Y} = y_j]$$

$$\times \log P[\mathbf{X} = x_i \mid \mathbf{Y} = y_j]$$

and $p_{\cdot j} = \sum_{i=1}^m p_{ij}$, $p_{i \cdot} = \sum_{j=1}^n p_{ij}$.

PROPERTIES OF ENTROPY AND UNIQUENESS

To justify the use of $H(p_1, p_2, \ldots, p_n)$ as a measure of uncertainty or information, it is desirable to introduce the "uniqueness theorem" of Shannon [13].

Uniqueness Theorem. Let $\mathbf{X}$ and $\mathbf{Y}$ be arbitrary discrete random variables, whose joint distribution is $P\{\mathbf{X} = x_i, \mathbf{Y} = y_j\} = p_{ij}$, $i = i$, $2, \ldots, m$, $j = 1, 2, \ldots, n$, and marginal distributions specified by $p_{i\cdot}$, $p_{\cdot j}$, respectively. For every positive integer n and every vector $(p_1, p_2, \ldots, p_n)$ with $p_i \geqslant 0$, $\sum_{i=1}^{n} p_i = 1$, let $f(p_1, p_2, \ldots, p_n)$ be a real-valued function. If

a. $f(p_1, p_2, \ldots, p_n) \leqslant f(1/n, 1/n, \ldots, 1/n)$, $n = 1, 2, \ldots$,

b. $f(p_{11}, \ldots, p_{mn}) = f(p_{\cdot 1}, p_{\cdot 2}, \ldots, p_{\cdot n})$
$+ \sum_{j=1}^{n} p_{\cdot j} \sum_{j=1}^{m} f((p_{ij}/p_{\cdot j}), \ldots,$
$(p_{mj}/p_{\cdot j}))$,

c. $f(p_1, p_2, \ldots, p_n, 0) = f(p_1, p_2, \ldots, p_n)$,

d. For every n, $f(p_1, p_2, \ldots, p_n)$ is a continuous function of $p_1, p_2, \ldots, p_n$,

then $f(p_1, p_2, \ldots, p_n) = cH(p_1, \ldots, p_n)$ and

$$\sum_{j=1}^{m} p_{\cdot j} \sum_{i=1}^{m} f\left(\frac{p_{ij}}{p_{\cdot j}}, \ldots, \frac{p_{mj}}{p_{\cdot j}}\right) = cH(\mathbf{X} \mid \mathbf{Y}),$$

for some positive constant c, or $f(p_1, p_2, \ldots, p_n)$ is identically zero.

Thus any such function agrees with $H(p_1, p_2, \ldots, p_n)$ up to a constant multiple. Note, however, that we have not specified the base of the indicated logarithms. Conventionally, one defines $H(\frac{1}{2}, \frac{1}{2}) = -(\frac{1}{2}\log\frac{1}{2} + \frac{1}{2}\log\frac{1}{2}) = 1$, which is equivalent to using base 2 logarithms. This convention amounts to a choice of the scale of measurement for entropy and the units are referred to as bits (of information); "bit" is a contraction of binary digit.

Now $H(p_1, p_2, \ldots, p_n) = 0$ if and only if $p_i = 1$ for some i and $H(1/n, 1/n, \ldots, 1/n) = \log n$. If $p_i = 1$ for some i, there is no "uncertainty" and if each $p_i = 1/n$, this is the most uncertain situation (for fixed n) in the sense that the outcome of such an experiment is the hardest to predict. Properties (a) and (c) imply the assertion that the uncertainty is a monotonically increasing function of n, the number of alternatives, when the

alternatives are equally likely. From property (b), if $\mathbf{X}_1, \mathbf{X}_2, \ldots, \mathbf{X}_t$ are independent random variables, $H(\mathbf{X}_1, \mathbf{X}_2, \ldots, \mathbf{X}_t) = \sum_{i=1}^{t} H(\mathbf{X}_i)$. If, in addition, the $\mathbf{X}_i$'s are identically distributed, $H(\mathbf{X}_1, \mathbf{X}_2, \ldots, \mathbf{X}_t) = tH(\mathbf{X}_1)$.

To identify entropy as a measure of information, envision an experiment in which the random variable $\mathbf{X}$ is observed by an observer, who subsequently transmits the observed value to you. Before you are provided with this information, the uncertainty is $H(\mathbf{X})$. Upon receipt of the information, the uncertainty is zero. Interpreting "information" as the removal of uncertainty, $H(\mathbf{X})$ is a natural measure of information.

Using bits as the unit for entropy, $H(\mathbf{X})$ is a lower bound to the average number of binary digits (or responses to yes–no inquiries) needed to determine the value of $\mathbf{X}$. Some examples will serve to illustrate this.

Example 1. Let $\mathbf{X}_1, \mathbf{X}_2, \ldots, \mathbf{X}_t$ be independent Bernoulli random variables with $P\{\mathbf{X}_i = 1\} = p_i$ and $P\{\mathbf{X}_i = 0\} = q_i$, $i = 1, 2, \ldots, t$. Then

$$H(\mathbf{X}_1, \mathbf{X}_2, \ldots, \mathbf{X}_t)$$
$$= -\sum_{i=1}^{t} (p_i \log p_i + q_i \log q_i)$$

and if each $p_i = \frac{1}{2}$, $H(\mathbf{X}_1, \mathbf{X}_2, \ldots, \mathbf{X}_t) = t$. The yes-or-no questions in this case are: Is the ith observation a "1"?

Example 2. Twelve coins and a balance scale are provided. One coin is known to be counterfeit. The probability that a given coin is counterfeit is $\frac{1}{12}$; hence the uncertainty is $H(\frac{1}{12}, \frac{1}{12}, \ldots, \frac{1}{12}) = -12(\frac{1}{12}\log\frac{1}{12})$ ~ 3.585. Each weighing using the balance has three possible outcomes: left side lighter, right side lighter, both equal. Thus it is possible to observe $\log 3$ units of information per weighing or 1.585 bits of information, if we can select the coins to be weighed so that these three cases are equally likely. This suggests that $3.585/1.585 \sim 2.2$ trials are needed. Since 2.2 is not an integer, 3 trials are required. We proceed as follows. Place 4 coins selected at random in left pan, 4 in right pan. The first weighing acquires 1.585 bits of information and identifies the counterfeit coin as a member of a set of 4 coins. Divide those 4 coins as follows: 1 to left balance, 1 to right balance, and 2 unweighed. This then provides $-2(\frac{1}{4}\log\frac{1}{4}) - \frac{1}{2}\log\frac{1}{2} = 1.5$ bits of information, or a total of 3.085 after two weighings. With probability $\frac{1}{2}$ no further weighings are needed; otherwise 0.5 bit of information is obtainable through a last weighing of the two remaining coins. Note that the average number of weighings using this scheme is $2(0.5) + 3(0.5) = 2.5 < 3$. Note also that the total information acquired is $1.585 + 1.500 + 0.500 = 3.585$, the initial amount of uncertainty. See WEIGHING DESIGNS.

An important property of entropy is given by the (noiseless) coding theorem. Let **X** be a random variable with $P\{\mathbf{X} = x_i\} = p_i \geqslant 0$, $i = 1, 2, \ldots, n$, $\sum_{i=1}^{n} p_i = 1$. A collection of sequences of zeros and ones is said to be a binary code for **X** if each x_i has a corresponding sequence that is not the initial part of a sequence for some x_j with $j \neq i$, $1 \leqslant i$, $j \leqslant n$. This last condition is essential to permit decoding. For example, for $n = 3$, the set of sequences $(0, 10, 11)$ permits decoding but $(0, 1, 00)$ does not, since if 00 is observed, it is not evident whether this denotes x_1 followed by x_1 or denotes x_3. In particular, the binary code for **X** of minimal expected length $E[L]$ satisfies $E[L] \geqslant H(\mathbf{X})$, or the entropy is a lower bound for the expected length of such codes.

Example 3. Let $\mathbf{X} = (\mathbf{X}_1, \mathbf{X}_2, \mathbf{X}_3)$ where the $\mathbf{X}_i$'s are independent, identically distributed Bernoulli* random variables with $P[X_i = 1] = p = 1 - P[X_i = 0]$. If $p = \frac{1}{2}$, then $H(X) = 3$ and the sequence of outcomes is the optimal code, attaining the lower bound. If $p = \frac{3}{4}$, $H(\mathbf{X}) = 2.434$. A suggested code is given in Table 1.

The indicated code has $E[L] = 2.649$, which is less than the three digits used to describe the possible outcomes.

Table 1

Outcomes	Probability	Code
000	0.016	1111111
001	0.047	1111110
010	0.047	111110
100	0.047	11110
011	0.141	1110
101	0.141	110
110	0.141	10
111	0.422	0

The definition of entropy extends immediately to arbitrary discrete random variables. However, it is easy to construct discrete variables **X** with $H(\mathbf{X}) = \infty$.

ENTROPY AS A DESCRIPTIVE STATISTICAL PARAMETER

The entropy of **X** or equivalently of the probability mass function $p_X(x)$ may be regarded as a descriptive quantity, just as the median*, mode*, variance*, and coefficient of skewness* may be regarded as descriptive parameters.

The entropy is a measure of the extent to which the probability is concentrated on a few points or dispersed over many points. Thus the entropy is a measure of dispersion, somewhat like the standard deviation. However, there are significant differences, in that the entropy does not depend on the outcomes having ordered values. Thus, the three random variables given in Table 2 will have identical entropies but substantially different variances, as indicated.

More specifically, the entropy is invariant under any one-to-one transformation of the values x_i, $i = 1, 2, \ldots, n$. This makes the entropy a particularly useful descriptive measure for random variables that assume nonnumerical values. As such, it is often used for linguistic data, such as the frequencies of words in a language. In terms of probabilistic models, the entropy is best interpreted as a measure of heterogeneity for multinomial distributions* (with unordered cells).

Table 2

x	$P_{X_1}(x)$	$P_{X_2}(x)$	$P_{X_3}(x)$
-2	0	$\frac{1}{4}$	0
-1	$\frac{1}{4}$	0	0
0	$\frac{1}{2}$	$\frac{1}{2}$	$\frac{1}{4}$
1	$\frac{1}{4}$	0	$\frac{1}{4}$
2	0	$\frac{1}{4}$	$\frac{1}{2}$

$H(\mathbf{X}_1) = H(\mathbf{X}_2) = H(\mathbf{X}_3) = 1.5,\ E(\mathbf{X}_1) = E(\mathbf{X}_2)$
$= 0,\ E(\mathbf{X}_3) = 1.25,$
$\sigma_{X_1}^2 = 0.5,\ \sigma_{X_2}^2 = 2,\ \sigma_{X_3}^2 = 0.6875.$

The utility of entropy as a descriptive measure justifies an interest in the statistical estimation of the entropy from observed data. The maximum likelihood estimator* has been studied by Bašarin [3]. A more detailed treatment of the estimation problem is given in Harris [5].

ENTROPY OF CONTINUOUS RANDOM VARIABLES

For continuous random variables, the entropy of the random variable **x** is defined by analogy as $H(\mathbf{X}) = - E[\log f_{\mathbf{X}}(\mathbf{X})]$, where $f_{\mathbf{X}}(x)$ denotes the probability density function of the random variable **X**. Here it is convenient to use a different scale of measurement and use natural logarithms instead of logarithms to base 2.

Thus, for the exponential distribution* with

$$f_{\mathbf{X}}(x) = \lambda e^{-\lambda x}, \qquad x > 0,\ \lambda > 0,$$

$H(\mathbf{X}) = \log(e/\lambda) = 1 - \log \lambda.$

Similarly, for the normal distribution with mean zero, that is,

$$f_{\mathbf{X}}(x) = (2\pi)^{-1/2}\sigma^{-1}e^{-x^2/2\sigma^2}, \qquad \sigma > 0,$$

we have

$$H(\mathbf{X}) = \log\left[(2\pi e)^{1/2}\sigma\right]$$
$$= \log\left[(2\pi)^{1/2}\sigma\right] + \tfrac{1}{2}.$$

The examples above make it evident that the invariance property of the entropy for discrete random variables no longer applies. However, many of the other properties still apply or have natural analogs. Some of these are enumerated below.

1. Let $f_{\mathbf{X}}(x)$ be any probability density function* with $P[a \leqslant X \leqslant b] = 1$. Then $H(\mathbf{X})$ is a maximum if $f_{\mathbf{X}}(x)$ is the uniform distribution* on $[a, b]$.

2. If **X**, **Y** are any random variables with a continuous joint probability density function $f_{\mathbf{X},\mathbf{Y}}(x, y)$, then

   $$H(\mathbf{X},\mathbf{Y}) \leqslant H(\mathbf{X}) + H(\mathbf{Y})$$

   with equality if and only if **X** and **Y** are independent.

3. If X is a random variable with $P\{\mathbf{X} > 0\} = 1$ and $E[X] = \alpha$ is given, then the entropy is a maximum if

   $$f_{\mathbf{X}}(x) = \alpha^{-1}e^{-x/\alpha}, \qquad \alpha > 0,\ x > 0.$$

4. If **X** is a random variable with $E[\mathbf{X}] = 0$ and given variance σ^2, then the entropy is a maximum if and only if **X** is a normally distributed random variable. This last observation has been used to construct a test of normality* by Vasicek [15].

MISCELLANEOUS ASPECTS

Some measures closely related to entropy have been found useful in statistical inference. Among the more extensively employed of these measures are the Kullback–Leibler information numbers $I(1, 2)$ and $J(1, 2)$, defined by

$$I(1,2) = \int \log\left[f_{\mathbf{X}_1}(x)/f_{\mathbf{X}_2}(x) \right] f_{\mathbf{X}_1}(x)\,dx$$

and

$$J(1,2) = I(1,2) + I(2,1).$$

Defining the quantity $- E[\log g(\mathbf{X})]$, where $g(x)$ is a probability density function*, and **X** is distributed by $f_{\mathbf{X}}(x)$ as the "inaccuracy," then it is well known that

$$- E[\log g(X)] \geqslant - E[\log f(X)],$$

with the equality if and only if $g(x) = f(x)$. Thus $I(1, 2)$ is the difference of the inaccuracy and the entropy. If this difference is large, there are substantial statistical differences between the two populations whose distributions are denoted by $f_{X_1}(x), f_{X_2}(x)$. If this difference is small, it will be difficult to tell whether data are from one population or the other. Thus the power* of statistical tests is related to the magnitude of $I(1, 2)$. It should be noted that $I(1, 2)$ is the expected value of the logarithm of the likelihood ratio*; $J(1, 2)$ is a symmetrized measure of this notion of a statistical distance* between the two populations.

Rényi [11] introduced the notion of entropies of order α defined by

$$H_\alpha(p_1, p_2, \ldots, p_n) = (1 - \alpha)^{-1} \log\left(\sum p_i^\alpha\right),$$

$$\alpha \neq 1.$$

As $\alpha \to 1$, $H_\alpha(p_1, p_2, \ldots, p_n)$ reduces to $H(p_1, p_2, \ldots, p_n)$. $H_2(p_1, p_2, \ldots, p_n)$ is frequently used as a measure of diversity. Entropies of order α are compared with various other descriptive measures as measures of economic diversity in Hart [6]. The reader should also see Kemp [7] concerning the utility of second-order entropy as a descriptive measure.

The notion of entropy is also widely used in communication theory*, where it is employed as a means of measuring the amount of information that can be transmitted over noisy communication channels.

Further source material is given in refs. 1, 2, 4, 8–10, 12, and 14, as well as in the references previously cited, and in the various related entries in this encyclopedia.

References

[1] Abramson, N. (1963). *Information Theory and Coding*. McGraw-Hill, New York. (An elementary introduction.)

[2] Aczél, J. and Daróczy, Z. (1975). *On Measures of Information and Their Characterizations*. Academic Press, New York. (A highly technical treatment of the measurement of information and the properties of the various information measures.)

[3] Bašarin, G. P. (1959). *Teor. verojatn. ee primen.*, **4**, 361–364.

[4] Guiaşu, S. (1977). *Information Theory with Applications*. McGraw-Hill, New York. (A general theoretical treatment of information theory and its applications, including communication theory, statistical inference, coding theory, and pattern recognition.)

[5] Harris, B. (1975). *Colloq. Math. Societatis János Bolyai*, **16**, 323–355.

[6] Hart, P. E. (1975). *J. R. Statist. Soc. A*, **138**, 423–434. (Contains an extensive list of references to papers that discuss various descriptive measures of diversity.)

[7] Kemp, A. W. (1973). *Bull. Inst. Int. Statist.*, **45**, 45–51.

[8] Kemp, A. W. (1975). *Bull. Inst. Int. Statist.*, **46**, 446–452.

[9] Khinchin, A. I. (1957). *Mathematical Foundations of Information Theory*. Dover, New York. (A rigorous development of Shannon's original ideas.)

[10] Kullback, S. (1959). *Information Theory and Statistics*. Wiley, New York. (The uses of information measures in statistical inference, with particular emphasis on contingency tables and multivariate analysis.)

[11] Rényi, A. (1966). *Wahrscheinlichkeitsrechnung mit einem Anhang über Informationstheorie*. VEB Deutscher Verlag der Wissenschaften, Berlin. (A probabilistic treatment of information measures and their properties in an appendix to a standard textbook on probability theory.)

[12] Reza, F. M. (1961). *An Introduction to Information Theory*. McGraw-Hill, New York. (A highly readable text-book on information theory, written at the intermediate level of difficulty.)

[13] Shannon, C. E. (1948). *Bell Syst. Tech. J.*, **27**, 379–423, 623–656. (The historical paper in which the notions of the entropy, channel capacity, and coding were developed by Shannon.)

[14] Shannon, C. E. and Weaver, W. (1959). *The Mathematical Theory of Communication*. University of Illinois Press, Champaign, Ill. (Reprint of article by Shannon [13] with an added commentary by W. Weaver.)

[15] Vasicek, O. (1976). *J. R. Statist. Soc. B*, **38**, 54–59.

(COMMUNICATION THEORY, STATISTICAL DIVERSITY INDICES INFORMATION THEORY AND CODING THEORY)

B. HARRIS

ENVIRONMENTAL STATISTICS

The central problem of environmental statistics is to determine how our quality of life is affected by our surroundings, in particular by such factors as air and water pollution, solid wastes, hazardous substances, foods, and drugs. Statisticians, in collaboration with scientists and other research workers in this field, try to elucidate the complex interrelationships that exist in our global ecosystem. Basic to this task is the collection, cleaning, organization, and analysis of vast amounts of environmental information, including pollution, health, biological, meteorological, and other data. Naturally, the full range of statistical methods find application, from simple informal graphical procedures to complicated formal quantitative techniques. The purpose of the following discussion is to highlight some of the important special characteristics of statistical problems that arise in the area of environmental statistics.

ENVIRONMENTAL DATA

Some characteristics of typical sets of environmental data are given in Table 1. Especially since the advent of equipment for the automatic and rapid recording of measurements, data sets exceeding several million observations are now commonplace. Obviously, an important task is to keep such data in an organized fashion, check them carefully for outliers*, and remove bad values. With large data sets, the time and effort spent on this preliminary step of data cleaning can be considerable. Whether the data set is large or small, constant efforts must be made to ensure that the quality of the data is as high as possible. A recurrent, frustrating difficulty is that for environmental data, unlike simple weights and measures for which standards exist, problems of standardization and calibration* often make it difficult to compare confidently the recorded levels of a particular quantity in two different loca-

Table 1   Some Characteristics of Typical Sets of Environmental Data

1. Large quantities of data.
2. Aberrant values—either bad data or unusual occurrences.
3. Lack of standardized methods of measurement.
4. Measurement errors, both random and systematic.
5. Missing data.
6. Serial correlation* (autocorrelations) among observations.
7. Seasonal fluctuations.
8. Complex cause-and-effect relationships.
9. Lurking variables—those that exert an influence but are not measured.
10. Nonnormal distribution of observations.
11. Observational* data rather than data from designed experiments.

tions, or at two different time periods at the same location. Sometimes, for instance, the method for measuring a particular pollutant is changed and the measured values before and after the change are not comparable. Systematic errors, in this situation and others, can be more important than random errors. Data resulting from locating, counting, and identifying tumors or determining cause of death in environmental health studies, for example, can be subject to gross error. Because of equipment failures and other reasons, gaps may occur in the records (the problem of missing data*).

Environmental data are often collected sequentially (e.g., air and water pollution measurements) and hence are frequently serially correlated, i.e., autocorrelated (*see* TIME SERIES). Most statistical theory, however, is based on the assumption that data are not serially correlated; using such methods can be seriously misleading. Ordinary regression analysis*, which rests on this assumption, for instance, is sometimes misused in analyzing environmental data. One reason for high serial correlation* is the rapid rate at which data are sometimes collected. Usually, data taken closely together in time or space will be positively serially correlated.

A related point is that environmental data are often periodic. Ambient temperature at a given location, for example, typically goes through yearly and daily cycles. Certain types of air pollution will have more severe health effects in the summer than in the winter. Biological phenomena, such as migration patterns, exhibit seasonal rhythms. Models used in the analysis of such data usually incorporate appropriate known periodic variations. Especially with regard to many types of biological data (e.g., aquatic life), erratic and quite pronounced random fluctuations will be superimposed on more or less regular seasonal fluctuations. *See also* PERIODICITY.

Given the existence of myriad cause-and-effect relationships, many of which are unknown (the fabric of the ecosystem), it is not surprising that variables that have an important influence are sometimes overlooked and unrecorded or, even if recorded, are not used. The presence of such lurking variables can confuse investigators who are trying to unravel the intricate linkages and delicate balances in the environment.

Although the statisticians' favorite distribution for representing continuous data is the normal*, frequently environmental data are positively skewed and such that the lognormal distribution* gives a better approximation. Because regulations are often written in terms of the highest recorded value for some pollutant measured during some specified time period, statisticians working in this field frequently become involved in extreme value theory*. Regrettably, some regulations that concern the level of pollutants that vary stochastically in time make no allowance whatsoever for the statistical nature of the data and state flatly (and unrealistically in some cases) that the level of that pollutant shall *never* exceed a certain fixed value.

Most environmental data are observational rather than the results of randomized designed experiments (*see* RANDOMIZATION). This fact creates problems in making causal inferences, which are the most useful inferences to be able to make. This point is discussed below in the section entitled "Observational Data." *See also* CAUSATION.

Depending on the situation, some or all of the factors above will have to be considered by a statistician concerned with how best to collect and analyze data to answer the questions at issue. In the data collection phase, statistical problems may involve design of experiments* and clinical trials*—involving human beings, animals, plants, or elements of the physical environment, sample surveys*, epidemiological* studies, or even a census*. In the analysis phase, statistical techniques will be most effective when used innovatively as an adjunct to rather than a replacement for information and methods available from relevant subject-matter fields. In particular, since most important environmental problems are multidisciplinary, effective analysis often requires that statisticians work cooperatively with specialists in health, engineering, biology, chemistry, meteorology, economics, and law. (*See* entries on STATISTICS IN these subjects.)

GRAPHICAL METHODS

The most effective statistical techniques for analyzing environmental data are graphical methods. They are useful *in the initial stage* for checking on the quality of the data (e.g., spotting outliers), highlighting interesting features of the data, and generally suggesting what statistical analyses should be done next. Interestingly enough, after intermediate quantitative analyses have been completed, they are again useful *in the final stage* for providing compact and readily understood summaries of the main findings of investigations (e.g., see Fig. 1).

An innovative plot in Tiao et al. [10] shows the average daily variation of the concentration of oxidants, primarily ozone, in downtown Los Angeles from 1955 to 1973. This plot clearly shows that the situation improved over that period. Within a particular year, the concentration of oxidants was generally higher from June through October than during the remainder of the year. The

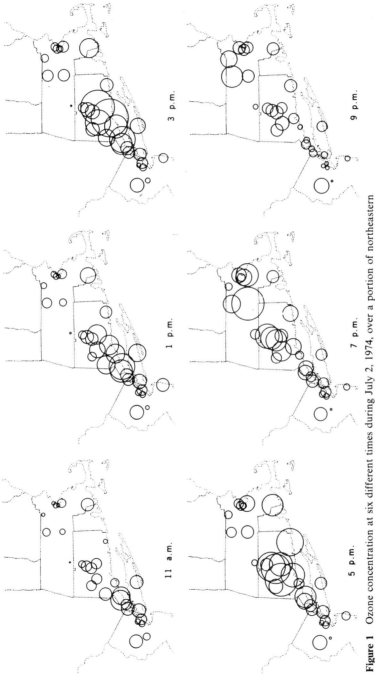

Figure 1 Ozone concentration at six different times during July 2, 1974, over a portion of northeastern United States. Concentration is proportional to diameter of each circle.

highest readings occurred during the middle of the day, approximately from 9:00 A.M. to 3:00 P.M.

A model that is consistent with these data has the following ingredients:

1. Pollutants are produced by heavy morning traffic and other sources.
2. Sunlight is of great enough strength and duration to trigger a photochemical reaction, involving these pollutants and oxygen in the air, that yields ozone.
3. This reaction is slow (it typically requires 3 to 5 hours after initiation to yield significant amounts of ozone).
4. There is an inversion height during the day (but not at night) to keep a "lid" on this mixture so that it does not disperse before it has the time to react.

Figure 1, from Cleveland et al. [5] shows data collected on ozone on July 2, 1974, a day when there was a prevailing wind from the southwest. The diameter of each circle is proprotional to the concentration. This graphical display shows how photochemical air pollution resulting from primary emissions in the New York City metropolitan area was transported into Connecticut and Massachusetts. Federal, state, and local standards for air pollutants typically include limits that should not be exceeded. The existence or the suspected existence of such transport phenomena, however, obviously complicates the process of setting and enforcing reasonable regulations. Furthermore, it points to wider international problems; pollutants from one country can clearly show up in others.

MODEL-BUILDING*

Empirical and mechanistic models have been developed and used extensively, e.g., in preparing environmental impact statements and in deciding on matters of public policy. Some countries, for instance, have enacted legislation that restricts the manufacture of fluorocarbons, on the basis of mechanistic

models that predict a higher incidence of skin cancer because of the depletion of stratospheric ozone by these chemicals.

Figure 2 illustrates the iterative process of building a model, which is a set of one or more equations that adequately describes some phenomena of interest. Upon confronting the available data (box a) with the tentative model (box b), which is based on subject-matter knowledge—biology, chemistry, physics, engineering—one logically asks first whether there is any evidence that the model is inadequate (box 1). If so, one must repair it (feedback loop I) and perhaps collect more data (feedback loop II). If the model is adequate, one then estimates the parameters (constants) (box 2) and assesses their precision (box 3), usually using regression techniques. If the precision is not high enough, one can return to the field or laboratory to obtain additional data.

Empirical models are those that have no particular basis in scientific theory, but are typically used for reasons of convenience (e.g., linear regression* equations). Theoretical or mechanistic models, on the other

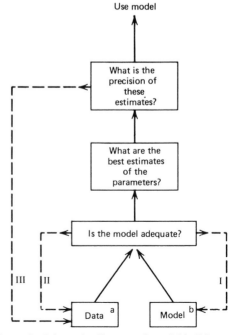

Figure 2 Schematic diagram of model-building process.

hand, have some basis in theory (e.g., meteorology or chemical kinetics). To maintain the quality of our environment, it is crucial to assess accurately the impact of our actions on it and, where necessary, to control them. Researchers can use models in helping to make such assessments. Decision makers can be aided in the selection of appropriate control strategies because models, if they are adequate, can be used to predict what will probably happen in the event various different policies are implemented. The advantages of mechanistic models are that they usually contribute better to scientific understanding, provide a better basis for extrapolation, and provide a more parsimonious representation of the response function. But for reasons of simplicity an empirical graduating function is often used (see GRADUATION). For further references on mathematical model-building as it relates to environmental systems, see Cleveland and Graedel [4], Phadke et al. [7], Rickert and Hines [8], Tiao and Hillman [9], Tiao et al. [11], and van Belle and Schneiderman [12].

OBSERVATIONAL DATA

If investigators must rely on observational data rather than designed experiments in which randomization* is employed, they will have difficulty in making inferences in terms of cause and effect. The crux of the matter is that correlation observed in observational data (no matter how striking or "significant" it may be) does not necessarily imply causation*. Most environmental data are such that it would have been impossible to collect them using randomized experiments (e.g., see Tiao et al. [10] and refer to Fig. 1; *see also* DESIGN OF EXPERIMENTS). In interpreting environmental data, then, one must always bear in mind the possibility that an observed correlation between two or more quantities may not be the result of a cause-and-effect linkage between them but may be merely a correlation caused by the action of one or more lurking variables.

The process of scientific inference can be facilitated by an understanding of underlying biological or physical mechanisms. Suppose that a certain correlation is detected in a collection of historical data and it suggests that a certain factor causes increased incidence of some disease. If some other sound research has discovered a mechanism by which this factor might induce the disease, one would be more confident in asserting a causal link than if no such knowledge were available.

One problem mentioned previously is that, since environmental data are often collected sequentially in time, the observations are frequently serially correlated. In such circumstances, intervention analysis* can be useful. See Tiao et al. [10] for an illustration that shows how a quantitative measure can be obtained of the decrease in the concentration of ozone in downtown Los Angeles in 1960. It is about 1.10 pphm, with a standard deviation of 0.10 pphm. The reason this calculation was done is that around 1960, Rule 63 was implemented, which mandated a reduction of reactive hydrocarbons in gasoline sold in the area; the Golden State Freeway was also opened at that time. Thus a question of considerable interest was whether it was possible to detect a decrease in this pollutant at that time, which could plausibly be attributed to these factors.

Intervention analysis takes into account that observations may not be independent. If, however, when analyzing historical environmental data, one uses ordinary standard statistical methods that rest on the assumption of independence (e.g., standard *t*-tests or regression analysis), serious mistakes can be made, such as declaring results "significant" when they are not, and vice versa (see Box et al. [2, pp. 21–86, 487–498]).

SUMMARY

Table 2 lists some important statistical questions associated with the analysis of environmental data. For purposes of illustration in this article, examples concerned with air pollution have been emphasized. The same

Table 2   Some Important Statistical Questions Associated with the Analysis of Environmental Data

1. National and global monitoring of environmental quality involves many laboratories. What is the best way to ensure standardization so that the data are consistent and comparable? For concerns related to public health, what is the best way to ensure consistency and comparability of data collected in different parts of the world? What data should be collected? What is the best way to cope with the problem of long latency periods?

2. Environmental data often occur as multiple time series. What is the best way to analyze such data? Mathematical models for environmental data should often be stochastic and three-dimensional and take time into account. How can such models best be created, fitted, and verified?

3. Designed experiments are often impossible. All the important variables are rarely known anyway (or if they are known, they are too difficult or impossible to measure). Surrogate and lumped measurements are often made. What is the best way to analyze such data, realizing that lurking and interacting variables may confuse the answers?

4. To evaluate the effect on human health of the presence of one or more constituents in the environment, it is common to take data on animals at high doses and extrapolate these results first to low doses, then to extrapolate from animals to human beings. Attempts are sometimes made to predict human mutagenicity on the basis of short-term tests (e.g., the Ames test). What is the best way to analyze such data, together with those from epidemiological studies and other sources, to estimate risks to human beings?

principles apply to other areas of application. For example, for a statistical overview of hazardous substances, the environment, and public health, see Hunter and Crowley [6]; for a discussion of statistical aspects associated with water pollution, see Berthouex et al. [1]. The essential nature of the problem in environmental statistics was aptly summarized by Rachel Carson [3]:

> When one is concerned with the mysterious and wonderful functioning of the human body, cause and effect are seldom simple and easily demonstrated relationships. They may be widely separated both in space and time. To discover the agent of disease and death depends on a patient piecing together of many seemingly distinct and unrelated facts developed through a vast amount of research in widely separated fields.

Bibliography

See the following works, as well as the references given below, for more information on the topic of environmental statistics.

Analytical Studies for the U.S. Environmental Protection Agency, Vol. 4: *Environmental Monitoring* (1977). Report of the Study Group on Environmental Monitoring, Committee on National Statistics, Environmental Studies Board, Numerical Data Advisory Board, Assembly of Mathematical and Physical Sciences, National Research Council, National Academy of Sciences, Washington, D.C. (Contains valuable information on the existence and management of data banks, as well as detailed discussion of many practical aspects of environmental monitoring.)

Box, G. E. P. and Tiao, G. C. (1975). *J. Amer. Statist. Ass.,* **70**, 70–79.

Breslow, N. and Whittemore, A., eds. (1979). *Energy and Health.* SIAM, Philadelphia.

Federal Register. From time to time contains definitive information on legal and procedural matters concerning environmental matters, e.g., see Vol. 44, No. 111, Thursday, June 7, 1979, Rules and Regulations, pp. 32854–32956 (on the National Pollutant Discharge Elimination System), and Vol. 44, No. 116, Thursday, June 14, 1979, Proposed Rules pp. 34244–34416 (on disposal of hazardous wastes and other matters).

Love, L. B. and Seskin, E. P. (1976). In *Statistics and the Environment, Proc. 4th Symp. Mar. 3–5 1976.* American Statistical Association, 1977. (See also the remarks by J. L. Whittenberger and J. W. Tukey that follow this paper.)

Pratt, J. W., ed. (1974). *Statistical and Mathematical Aspects of Pollution Problems*. Marcel Dekker, New York.

Proceedings of the Symposia on Statistics and the Environment, cosponsored by the American Statistical Association and other organizations. Third Symposium published by the Journal of the Washington Academy of Sciences, June, 1974. Fourth Symposium published by the American Statistical Association, 1977. Fifth Symposium published 1979 in *Environ. Health Perspect.*, **32**, 241. (These proceedings contain discussion of environmental problems from many different viewpoints, e.g., from those of the statistician, toxicologist, epidemiologist, and lawyer.)

Rhodes, R. C. and Hochheiser, S., eds. (1979). *Data Validation Conference Proceedings*. U.S. Environmental Protection Agency, Research Triangle Park, N.C. (Eighteen papers concerned with such matters as data checking, cleaning, editing, screening, auditing, and verification.)

Whittemore, A., ed. (1977). *Environmental Health: Quantitative Methods*. SIAM SIMS, Philadelphia, 1977.

Willgoose, C. E. (1979). *Environmental Health: Commitment for Survival*. W. B. Saunders, Philadelphia. (A basic, nonstatistical introduction to environmental health, emphasizing the interrelationship of all living things to one another and to their environment.)

References

[1] Berthouex, P. M., Hunter, W. G., and Pallesen, L. (1978). *J. Quality Tech.*, **10**, 139–149. (Contains 38 references.)

[2] Box, G. E. P., Hunter, W. G., and Hunter, J. S. (1978). *Statistics for Experimenters: An Introduction to Design, Data Analysis and Model-Building*. Wiley, New York.

[3] Carson, R. (1962). *Silent Spring*. Houghton-Mifflin, Boston.

[4] Cleveland, W. S. and Graedel, T. E. (1979). *Science*, **204**, 1273–1278. (An excellent comprehensive summary, which contains many references of work done by research workers related to a large-scale problem of air pollution in the United States.)

[5] Cleveland, W. S., Kleiner, B., McRae, J. E., and Warner, J. L. (1976). *Science*, **191**, 179–181.

[6] Hunter, W. G. and Crowley, J. J. (1979). *Environ. Health Perspect.* (in press). (Contains 66 references.)

[7] Phadke, M. S., Box, G. E. P., and Tiao, G. C. (1976). In *Statistics and the Environment, Proc. of the 4th Symp. Mar. 3–5, 1976*. American Statistical Association, 1977.

[8] Rickert, D. A. and Hines, W. G. (1978). *Science*, **200**, 1113–1118.

[9] Tiao, G. C. and Hillmer, S. C. (1978). *Environ. Sci. Tech.*, **12**, 430–435.

[10] Tiao, G. C., Box, G. E. P., and Hamming, W. J. (1975). *J. Air Pollut. Control Ass.*, **25**, 260–268.

[11] Tiao, G. C., Phadke, M. S., and Box, G. E. P. (1976). *J. Air Pollut. Control Ass.*, **26**, 485–490.

[12] van Belle, G. and Schneiderman, M. (1973). *Int. Statist. Rev.*, **41**, 315–331.

(BIOSTATISTICS
CAUSATION
DATA COLLECTION
EDITING STATISTICAL DATA
GRAPHICAL REPRESENTATION OF
  DATA
OBSERVATIONAL STUDIES)

WILLIAM G. HUNTER

EPIDEMIOLOGICAL STATISTICS

Epidemiology means, of course, the science of epidemics. But in practice it is no longer about infectious diseases only. The short *Oxford English Dictionary* (1971 edition, corrected 1975) quotes the Sidney Society Lexicon:

Epidemic, adjective. (1) Of a disease, prevalent among a people or community at a special time and produced by some special causes not generally present in the affected locality. (2) Widely prevalent; universal. The literal meaning is "upon the population." This includes much of "demography," which is given as "statistics of births, death, diseases, etc." "Epidemiology" as it has grown and matured has reverted to this literal meaning. It is about population studies in medicine in the broadest sense.

In 1849, Snow collected statistics of cholera cases in central London, and related the incidence to where their drinking water came from [28]. Little was then known about infected water. We would now say that he was doing epidemiology. The whole process of how the disease is transmitted in polluted water would now be called the etiology of the disease. Once this became known, the pioneer studies had achieved their ends and were forgotten over the years.

Similarly, it was an empirical finding that dairymaids working with cows did not get smallpox. They contracted a mild disease, cowpox, which made them immune. All those vaccinated because of this discovery were inoculated with cowpox.

The empirical stage of our knowledge is certainly not always short lived. The evidence that "(cigarette) smoking may damage your health" is only epidemiological 30 years after the first results were widely publicized. It is strong evidence because it is made up of many different strands, which separately are breakable. The etiology of bronchitis, emphysema, and lung cancer, as related to cigarette smoking, is still complex and largely unknown.

In other situations, the combined evidence is less strong. "On the basis of epidemiological findings we are now being advised, as ordinary citizens, to avoid tobacco and alcohol; to limit our consumptions of sugar, milk, dairy products, eggs, and fatty meats; to rid ourselves of obesity, and to engage in violent—although some suggest moderate—physical exercise . . . ." [3]

In the discussion to this paper the seconder of the vote of thanks [26] commented: "There is nothing wrong with this method of formulating hypotheses [i.e., theories of causation of diseases based on epidemiological associations] but too often the authors of such studies succeed in implying that a tenable hypothesis, developed from a careful epidemiological survey, is a proven one. Attribution of this step, by the public, to a statistical argument, is a certain means of discrediting our science."

ROLE OF THE STATISTICIAN

Statistics and statisticians have always played a major role, but this has changed. It used to be almost entirely in descriptive as opposed to theoretical statistics, and observational rather than inferential. Now the processes can best be described as descriptive statistics plus modeling. However, "It is *descriptive* statistics and scientific method which have to become fully one" [14].

Typically, the classic book by Bradford Hill [20] consists largely of elementary statistics and, for its time, advanced scientific method. It/he provided very many examples of misuses of statistics comparable with those in a well-known popular exposition [25].

Small-sample statistics and "exact" inference have helped in epidemiology, for example when the data being interpreted consist of very many small groups. Usually, however, there are disturbing factors that make rigorous inference impossible. Changes in the underlying assumptions, and in the models, make far more difference than that between a moderately good and an optimal method of data analysis or inference.

Experimental design* of the kind that was so brilliantly successful in agriculture is generally impracticable for human subjects; even methods used in clinical trials* have little place. But there have been exceptional situations where experiments are practically and ethically allowable. For example there were the pioneer studies on coal miners in the Rhondda Fach begun before 1950. This is an isolated valley in the south Wales mining area in Great Britain. It was thought that a disabling disease, pulmonary massive fibrosis, was brought about by a tubercular infection. The coal workers in this population were therefore provided with intensive medical examinations to detect and treat tuberculosis at a very early stage. In a control valley (Aberdare) the corresponding routine medical checks were given. In both populations chest x-rays and various lung physiological variables were measured for many years in the same individuals. This was therefore a *longitudinal* study. Follow-up* studies are still being published. Much more information has been obtained than the original plan provided for [4].

A completely contrasting example is interesting. Immediately after World War II, large industrial areas in Europe became very short of food. In an area under British occupation, a population of schoolchildren were allocated different kinds of extra rations at random, especially of bread, and the effect on their growth was studied. It appeared that all kinds of bread were equally effective. The question remained, of course, whether results under such abnormal conditions throw any light on normal food values.

Epidemiological methods are used to determine what is generally "normal." In the Rhondda Fach population mentioned above, blood pressures, systolic and diastolic, were measured regularly. These were also compared with those in a representative sample of adults in a rural district not far away, in the Vale of Glamorgan.

This is a good instance where the statistician has more to contribute. The blood pressure of an individual increases with age, and also with his or her weight and arm girth. For much of adult life, evidence from *cross-sectional* studies is that the increase with age is accounted for by the increase in arm girth. These are studies in which the population is measured at many or all ages, but at the same time. In a longitudinal study the same individuals are measured over the years. If they are born within a few years of one another, this population is called a *cohort*. The blood pressure of an individual fluctuates even over a few hours or less. The distribution of blood pressures in people of the same age varies itself with age; the variance increases, and the distribution is of log-normal type. But among old people the distribution is truncated: very high blood pressure reduces the chance of survival.

The example of blood pressures is one where epidemiologists need to know what normal values are. This can be a major problem when lung variables are being investigated, such as forced expiratory volume (FEV) and forced vital capacity (FVC). These are measured and assessed by performance tests; for FEV the subject blows as hard as he can into a bag, after one or two practice blows. Sometimes he improves with practice; sometimes he tires. There have been many arguments and analyses on the question: Is his best performance the best measure? (Or should it be the mean of the best three, or the mean of all except the first two practice blows?) Fortunately it seems to make little difference which is chosen. However, the actual *distribution* of the perfor-

mances ought to yield extra information. The object is to assess all kinds of external or environmental* factors on the lungs, e.g., industrial dusts, atmospheric pollution, cigarette smoking, the weather, the climate, and the season. We also want to detect early signs of chronic bronchitis and asthma. Again we need to know what happens in normal individuals. Age, weight, and height are important, and interesting empirical laws have been found (see, e.g., Fletcher et al. [16]) relating these to the lung variables.

Asthma is also a childhood disease, and polluted air can affect people of all ages. Measurements on growing children lead to additional problems (see Fig. 6). Their FEVs are well correlated with their height but possibly during the pubic spurt the FEV lags behind. Without going into details, all this should give an idea of what epidemiological statisticians have to be doing now.

COMPUTING

In 1955 Hollerith punched cards were more indispensable than the hand and the electric calculating machines of this period. Apparatus consisting essentially of many long rods selected cards that had their holes in the required places. The first electronic computers did not bring about much change. Their memories were not large enough to enable moderately complex calculations to be programmed easily, and large amounts of data could not possibly be stored. It was years before programs were available by which existing punched cards could be read into the computer. In the present generation of computers*, input, output, memory capacity, and storage capacity are incomparably greater. At the same time the first (or is it the second?) generation of table computers takes care of small-scale work. But it is more necessary than ever to understand what is in a statistical package program. It is so tempting to use an existing program when what is really needed is statistical common sense and instinct. But on balance so much more has become practicable in epidemiological statistics through these new facilities that they are worth all the extra problems they bring with them; *see* STATISTICAL SOFTWARE.

New problems cannot be ignored. Information on individuals that is stored in computers can be misused. The epidemiologist was never in an ivory tower, but he is further away from it now than ever.

LONGITUDINAL AND CROSS-SECTIONAL DATA AND GROWTH

Data on growth are needed for many problems. They provide a clear example of the difference between cross-sectional and longitudinal data. At the beginning of puberty the growth rate of a child slows down, and after this there is a spurt. This then slowly flattens out. But the time of onset of puberty varies. Its mean is earlier in girls, but for both girls and boys in defined populations it is a distribution. The consequence is that a cross-sectional curve of mean height against age, for populations of boys and girls separately, shows almost no sign of puberty: the structure shown in individual curves is lost [37]. However, the curve for medians* may show some structure.

In a related example, it is generally believed that the onset of puberty became earlier in many countries during periods when the general standard of living rose. This would show up in cross-sectional mean heights for different years, for, say, 14-year-olds, although from such data alone, they could simply be becoming taller at all ages. In fact, evidence from countries that have or had compulsory military service indicates that both may be true. The poorer children are shorter as teenagers, they go on growing for longer, but they do not catch up completely.

SMALL RISKS UNDERGONE BY MILLIONS OF PEOPLE: HUMAN RADIATION HAZARDS

The fact that large doses of ionizing radiation could induce serious illnesses was found

in those working with radium, and in radiologists using the x-ray apparatus of the time. Epidemiology was scarcely needed for implicating such diseases, in particular skin tumors, cancers of the bone, digestive organs, lungs, and pharynx, and leukemia.

Leukemia is still a rare disease; 30 per million per year was a typical rate round about 1950, but it appeared then to be increasing rapidly, especially in children and in old people.

When the risk to an individual is small, but millions of people are undergoing this risk, this provides good grounds for "retrospective" or backward-looking research. The population under study may be defined to be those with the disease in question who live in a particular region, perhaps a whole country in a particular year. Then the medical histories of everyone in this study population would be obtained, as well as histories of exposure to suspected causes. But this can never prove anything until we know what the unaffected people were typically exposed to. Hence the same questions must be asked of *matched controls*. These should be people comparable in as many ways as possible with the patient, e.g., having about the same age and sex, living in the same district under similar conditions.

Even so, rigorous scientific inference is difficult to achieve with such data. In what is now a classical example, the population under study consisted of young children attacked by a malignant disease in England and Wales in the years 1953, 1954, and 1955. One of such diseases was acute leukemia, which lasted a short time and was usually fatal. Among their mothers, about 10% had undergone diagnostic x-rays while pregnant and about 5% of control mothers (of healthy child controls) had been so x-rayed [34].

Did the x-rays, then, double the risk to the child? This interpretation was generally rejected; it was argued that the mothers were x-rayed because something was possibly going wrong during their pregnancies and the ones with leukemic children were more likely to have some symptoms, and so to be x-rayed, than the mothers of controls.

Broad subdivisions in terms of reasons for x-raying the mothers were therefore made, and cases and controls were compared within each subgroup. In general, the same pattern emerged. The main early papers reporting these findings are inevitably tedious to read, because so much space had to be devoted to all these subcomparisons.

Later, more evidence was obtained by bringing in many more years; in the meantime a number of statisticians have been working on different quantitative interpretations of case-control data. One refinement, which is peculiar to the leukemia data, is worth mentioning, because it is a very instructive example of back-and-forth play among observation, theory, modeling, and medical hypotheses which are then tested on more data. In the firstborn children, the distribution of ages of onset of (acute) leukemia has a definite maximum at $3\frac{1}{2}$ to 4 years. (In adult leukemia, also, there is a clear maximum in the fourth year after ionizing radiation that could have caused it. In the cases where the leukemia arises from damage to the fetus that is *not* caused by x-rays, it is well known that the fetus is most vulnerable soon after conception. However, the pregnant mother is most often x-rayed shortly before the child is born—about 8 months later. If, then, the leukemia caused by radiation is initiated 8 months later, is the modal time of its onset about 8 months later than that of nonradiogenic leukemia? Refined analyses that were practicable only when data for many more years were available showed that this was the case. This is indirect evidence, but it is independent of control data and of any possible bias due to the reasons for the x-rays [21, 22, 32, 33].

In later analyses, attempts were made to estimate the doses, or at least the number of times the pregnant mother was x-rayed, to try to quantify the risk factor. It appeared that such doses varied widely between hospitals, and when the first finds were published, these doses tended to be reduced. The original estimate of risk factor of 2 : 1 was possibly too high, and it was at best an average from very heterogeneous data.

An important negative finding was that in any case the increase in the use of x-rays was not nearly enough to explain the secular increase in childhood leukemia. In particular, the peak in 3- and 4-year-olds was not present at all 20 years earlier. The same group of researchers provided an explanation that depends mainly and convincingly on other epidemiological evidence. The hypothesis was that the new victims of leukemia—at all ages—would have died of pneumonia in the conditions of 20 years earlier. This was supported by evidence from the incidence of deaths from pneumonia at all ages, the secular change in this incidence, and its geographical variation. There is no space to give details of these fascinating studies in which the main contribution of the epidemiological statistics is to biological understanding [21, 22, 31].

HUMAN RADIATION HAZARDS AND PROSPECTIVE OR FORWARD-LOOKING STUDIES

It can always be rightly said that epidemiology and controlled experiments can never go together. But these objections can be partly met in *prospective* studies. For irradiated pregnant mothers this was in fact tried. The population under study consisted of all such pregnant mothers irradiated in a few large hospitals. Then the medical histories of their children were followed for several years afterward. Their incidence of leukemias was slightly *below* the national average, which was in contradiction to all the findings from the retrospective surveys [7]. Several conflicting explanations were proposed:

1. There were inevitable biases in the retrospective surveys.
2. The numbers in the prospective study were too small.
3. Combined with 2, those particular hospitals had "safer" x-ray apparatus than most others.

Probably the third explanation was right. It was supported by long-term evidence from the retrospective studies mentioned above, in which it was concluded that the risk to the child was decreasing.

A different prospective study yielded very clear results. The population consisted of 13,000 patients who had "poker back" (ankylosing spondylitis; articulation between the joints of the spinal column is gradually lost). Radiotherapy alleviated the symptoms, and there were complete records of doses and the dates they received them. About 40 patients developed leukemia; the expected number from their age and sex distribution was about 4. The calculations were complex because most patients had several treatments, but information was obtained relating doses to increased risks, and on the distribution of time intervals between dose and onset of the disease [6,41].

If data are available over a very long period, many of the problems simplify. For the workers exposed to ionizing radiation at Hanford, Washington, there have been health records available since 1945. Recently, these have been analyzed statistically. The findings have caused a stir, because they suggest that most existing models relating dose to effect underestimate the extra risk from various diseases, which, it is suspected, can be caused by small chronic doses (i.e., over long periods) of ionizing radiations, although the risk remains small [24]. This interpretation has, however, led to much argument (see, e.g., Darby [8], Brodsky [1], and Lyon et al. [23]).

For further discussion relevant to some prospective studies, *see* COHORT ANALYSIS.

EPIDEMIOLOGY AND CIGARETTE SMOKING

Before 1950 epidemiologists were asking why the incidence of lung cancer in men in so many countries had been increasing so much. Increases in most other cancers seemed to be attributable to populations aging; age-standardized rates showed much less change. So they looked for some causal

factor that had increased during the twentieth century, and cigarette smoking seemed to be the only plausible one. There appeared to be a time lag of decades between the supposed cause and its effect, but this was reasonable. Also, most smokers who began young continued to smoke over many years. Women, too, began to smoke more cigarettes and their lung cancer incidence increased; both occurred much later than with men [36]. Figure 1 shows such data for Great Britain for most of this century.

In a pioneer prospective study all British medical doctors were asked to answer a questionnaire; this was kept short and simple and about 70% responded. The study was continued for many years. A clear association was found between incidence of lung cancer and rate of smoking cigarettes. Another feature was that many doctors in this population reduced their smoking, whereas comparable groups in other professions did not. Many years after the study started, incidence among doctors seemed to be becoming increasingly *less* than that of these other groups [12,13].

Another factor that had been increasing over the years was atmospheric pollution. Evidence comparing clean and polluted districts suggested that this does increase the risk of lung cancer. However, miners working underground mostly breathed far more polluted air than city dwellers do. They had a lower risk. As a population they smoked less, being forbidden to smoke underground.

Retrospective studies based on mortality statistics have been done in many countries. All show a high ratio, smokers to nonsmokers, for men and women separately. The ratio increases with the rate they smoke cigarettes. There are large differences between countries. It is also possible that the risk among individuals, even in the same environment within one country, and with the same smoking habits varies greatly.

A majority of medical doctors and medical statisticians are now convinced that cigarette smokers have increased risks of getting many diseases. The majority of the rest of the population, 25 years after the first major findings appeared, have, however, gone on

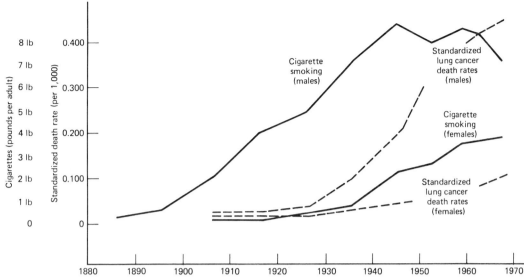

Figure 1 Cigarette smoking and age-standard death rates plotted at the middle of 10- and 5-year periods from 1880 to 1970. Reproduced by permission from Townsend [40], with thanks to the author and the Royal Statistical Society.

smoking. Many of the rates of increase seem to be leveling off and the effect per cigarette has possibly fallen appreciably.

UNSOLVED PROBLEMS ON THE EFFECTS OF CIGARETTE SMOKING

Considering the amount and variety of all this evidence, and much more, it is remarkable that so much is still unknown. Many arguments of the critics, the defenders of cigarettes, are instructive to epidemiologists. This writer's view is that they are often right, but that there are still more weaknesses in every alternative to the hypothesis that cigarettes are a causal factor.

Various factors obviously confuse the picture. Any separate groups of people who smoke less are almost certainly different in other ways. One such group studied was a population of Mormons, especially in the state of Utah. They have a low incidence of many diseases that have been related to cigarette smoking. This religious group also drinks no alcohol, tea, or coffee (see discussion in Burch [3]).

Groups responding to questionnaires probably differ from nonrespondents, even if they answer correctly. Estimates of amounts smoked were thought unreliable. There was complementary evidence from different countries relating standardized incidence of lung cancer to consumption of cigarettes per head of population, say 0, 10, 20, or 30 years earlier. There was a positive correlation, which was much higher with the 20-year time lag than with no time lag; but the individual variations are extremely large.

In the Doll–Hill population of doctors, those saying they inhaled when smoking cigarettes were found to be less at risk than noninhalers smoking at the same rate. This suggests that any effect caused by cigarettes could not be due to direct delivery of an insult to the lung. However, a probable answer to the inhalation paradox, which has not attracted much attention, came from a modest epidemiological study in which it was found that inhalers threw away longer cigarette ends—which were known to contain more carcinogens—than noninhalers. The subjects were asked to supply butt ends but were not told why [17, 19].

Other factors are involved, for nonsmokers do get lung cancer. According to Burch [3], one of these factors is inherited, so that those with it are more likely both to smoke

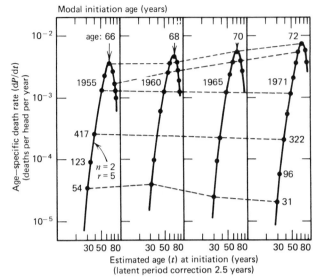

Figure 2 Age-specific recorded death rates (dP/dt) from lung cancer [International classification of diseases (ICD) 162 and 163] in men in England and Wales against estimated age at initiation for the years 1955, 1960, 1965, and 1971 $(t - \lambda$ in discussion[a], age $t)$. Log log plots: some actual numbers of cases are shown for 1955 and 1971. Fits to $y(t)$ as given[a]. Reproduced by permission from Burch [3] with thanks to the author and the Royal Statistical Society.

[a] See the section "Unsolved Problems on the Effects of Cigarette Smoking."

heavily and to get various diseases. Most epidemiologists disagree, but the arguments are worth studying.

All the major difficulties relate to quantitative forecasting and modeling. Compared with the hours spent collecting data, much less has been done on playing with the data to locate empirical patterns. There is, in any case, a rapid increase of incidence with age. However, the age of a regular smoker of a constant number of cigarettes per day is perfectly correlated with the total amount he or she has smoked! Unfortunately, there is not much evidence from *ex*-smokers. The overall age-specific death rates of lung cancer, in several countries, and in different years fits:

$$y(t) = nrk\theta\tau^{r-1}\exp(-k\tau^r)$$

$$\times \{1 - \exp(-k\tau^r)\}^{n-1}$$

where $\tau = t - \lambda$, $n = 5$, $r = 2$, $\lambda = 2.5$ but different θ and k in the various populations and years; $t =$ age ([2]; see the next section for interpretations of these parameters).

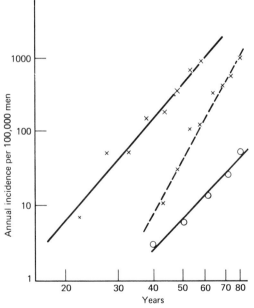

Figure 3 Lung cancer incidence by age and duration of smoking (From Doll, [11]). ×--× = cigarette smokers by age; ×—× = cigarette smokers by duration of smoking; O—O = nonsmokers by age. Reproduced by permission from Townsend [40], with thanks to the author and the Royal Statistical Society.

These curves have modes that vary around $t = 75$ (Fig. 2). These are not cohort data. They correspond in any case to a positive power law relating incidence to age, namely,

$$y = \theta nrk^n(t - \lambda)^{rn-1}.$$

Such positive power laws are also found when cigarette smokers and nonsmokers are considered separately, and duration of smoking appears to act differently from age, as Fig. 3 [39] shows. Some models that have been considered are based on

$$I_x = bNK(x - w)^{K-1},$$

where I_x is the age-specific death rate at age x, N cigarettes are smoked per week, over a period $x - w$. In Fig. 3, b is the same for smokers and nonsmokers, and K is about 5.

This equation has been extended by making N a time variable, which is estimated

from data on total cigarette consumption in 5-year periods. Townsend [39] made some other refinements, and compared observed and predicted mortality rates such as those in Fig. 4. It is interesting that the maxima are predicted, although the agreement is not exact. The interpretation of such maxima is different from that in Fig. 2 (see the following section).

FORM OF AGE INCIDENCE CURVES IN GENERAL

We have been moving from small to large populations. Data on age-incidence curves for many diseases does not need to be obtained from an epidemiological study, but from annual mortality statistics. These are, of course, cross-sectional; we would expect

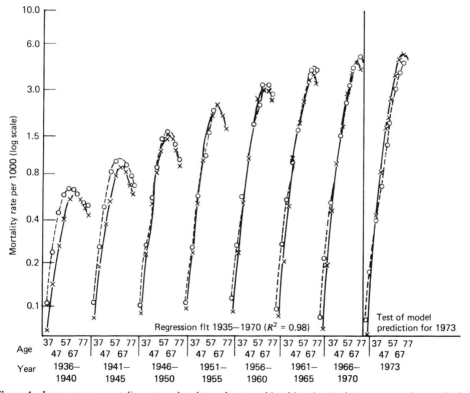

Figure 4 Lung cancer mortality rates related to cohort smoking histories (males). $\times$ = actual mortality[a]. Reproduced by permission from Townsend [40] with thanks to the author and the Royal Statistical Society.

[a] See the end of the section "Unsolved Problems on the Effects of Cigarette Smoking" for details.

cohort data to be more realistically related to a model. Despite this, these positive power laws, which can be valid over several decades of age, are found everywhere. Burch's collection contains over 200 examples. Mostly these are given in the cumulative form

$$G(t) = \theta \left[1 - \exp\left\{ -k(t - \lambda)^r \right\} \right]^n.$$

Here θ is the proportion of the population that is susceptible, $G(t)$ the proportion that gets the disease at age t or earlier, and λ is assumed to be a constant latent period between the end of an initiation or promotion period and the event actually observed. In this theory n and r are integers, and are related to numbers of clones of cells and numbers of mutations, which finally lead to this event. Often either n or $r = 1$.

Whether Burch's underlying theory is right or not, a great variety of data are described economically in this way; they can be used as input in testing almost anything relevant. In epidemiology it is well-nigh impossible to ignore age distributions in the populations under study.

There are several related models in which observed powers are related to numbers of mutations. It seems certain that a summation of random time intervals is involved. An alternative interpretation for those positive power laws is that they are simply parts of the increasing part of probability curves. Then there is no restriction to integral powers. However, it seems possible to fit the same data to a function like

$$b(t - v)^k,$$

with different values of v, so that k depends on v. On both interpretations the curves have maxima, and the best evidence comes when a decrease at high ages is actually seen, but of course at high ages, say over 60, such a survival population can be very different from that for the age range of, say 50 to 60 [5, 10, 11].

AN EXAMPLE WHERE AGE MUST BE ALLOWED FOR: SIDE EFFECTS FROM ADDING FLUORIDE TO DRINKING WATER

Obviously, whether fluoridation is effective in preventing dental caries depends on longitudinal population studies of the teeth of children drinking the altered water over several years. There has also been a heated controversy over possible side effects. In the United States, data were available from 1950 to 1970 for 20 large cities, of which 10 had fluoridated water (F^+) and 10 did not (F^-). The crude death rates from all malignant neoplasms showed a greater increase, starting from near equality, in the F^+ cities. The two groups of cities, however, had different distributions of sex, age, and race, which *changed* differently during these 20 years. Allowing for these, the excess cancer rate increased during these 20 years by 8.8 per 100,000 in F^+ and by 7.7 in F^-; but in proportional terms the excess cancer rate increased by 1% in F^+ and 4% in F^-. Then it was found that 215 too many cancer deaths had been entered for Boston, F^+, because those for Suffolk county, containing Boston, had been transcribed instead [27].

After this had been put right, the relative increase was 1% less for F^+ than for F^-. On the whole this is in line with other evidence. This example shows, however, how obvious and trivial pitfalls are interspersed with ones that are far from obvious.

This controversy is still going on [28, 30]. It was brought out that data for intervening years should be used if possible, but the way the distributions of sex, age, and race, which were not fully available, changed during these 20 years could be critical. In particular, linear and nonlinear interpolations yield different estimates of expected numbers of deaths.

EPIDEMIOLOGY AND HEART DISEASES

Heart diseases are the most common cause of death in all "developed" countries. There is probably more data than for all other topics mentioned so far put together. The data can be both reliable and very dirty. Factors that have been correlated, for example, with death rates in particular (middle or old) age groups include what is eaten, drunk, and smoked and how much, way of life in general, including exercise taken, climate, and population density (see, e.g., Tarpeinin [38] and Townsend and Meade [40]). There are many comparative studies within countries and between countries, which are too numerous to mention. One very experienced group of epidemiologists has based a major study entirely upon comparative statistics for 18 developed countries [35]. Figure 5 shows the kind of data in broad outline. In this study various combinations of variables were tried, principally in multiple regression* analyses. One factor that was strongly negatively correlated with the mortality rates was wine consumption per head. No common third variable was found as an alternative explanation for this association. It is particularly strong within Europe, except for Scotland and Finland, but other factors are needed to explain the far higher rates for the

United States, Canada, Australia, and New Zealand.

In complete contrast to this approach are prospective longitudinal studies from small areas. Unlike those for leukemia, there are now enough cases. A well-known name is Framingham*, a small town in Massachusetts. A population of about 6500 people aged 30 to 62 in 1948 were studied every 2 years for 20 years; a small number of factors were investigated in relation to the incidence of coronary heart disease: namely, age, weight in relation to height, systolic and diastolic blood pressure, smoking habits, and serum cholesterol level [9, 18].

In a more recent study in The Hague, Holland [15], the population was about 1750 of all ages 5 to 75; an additional factor was whether the women took oral contraceptives. (So far, for those who did, the risk of coronary heart disease seemed about the same, but their blood pressure was higher.)

Population studies have been going on for several years in Vlaardingen (a town in a large industrial/port area near Rotterdam, Holland) and in Vlagdwedde (a wholly rural area in northeast Netherlands). The selected populations are of all ages; in particular, many different measurements on lungs, involving blowing and breathing, are made besides the measurements given above. A control group had 117 male conscripts aged 19 to 20 years. This is part of an extensive field survey on chronic nonspecific lung disease, but much statistical and biological information should be valuable apart from this purpose (see, e.g., Sterk et al. [29]).

This kind of data can also tell us more about the range of "normal" physiological variables as functions of age; then criteria for early signs of heart diseases can be detected. Typically, the data are both cross-sectional and longitudinal, for the same individuals are measured at, say, 6-month intervals.

Our final example gives data from growing children at one school, but it illustrates all these points. Figure 6 shows their forced vital capacities, which are related to the

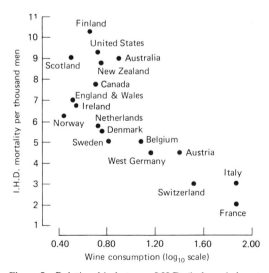

Figure 5 Relationship between I.H.D. (ischaemic heart disease) mortality-rate in men aged 55–64 and wine consumption. Reproduced from St. Leger, Cochrane, and Moore [35], with thanks to the authors and *The Lancet*.

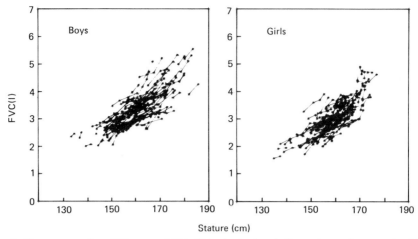

Figure 6 Data on growing children aged $11\frac{1}{2}$–$13\frac{1}{2}$ obtained at half-year intervals at a school in The Hague (St. Aloysius College, Netherlands); FVC = forced vital capacity (lungs). With acknowledgements to Ch. Schrader (Physiology Laboratory, Leiden).

growth of their lungs, as a function of height. Each separate line corresponds to one child, measured at 6-month intervals. We want to know what is the normal course of the growth of the lung in an individual. It is not necessarily given by a curve giving the best fit $y(x)$ for all the points.

The problem of describing and modeling such data is typical for epidemiologists today. If the population was of middle-aged people, the x-axis could be, say, blood pressure.

It is also only recently that the computer plotters have become more generally available on smaller machines. It makes it possible to look at far more data quickly before doing any detailed analyses.

SCOPE OF EPIDEMIOLOGY IN THE 1980s

On any live problem in this field, active researchers disagree. Such controversies are usually well aired. It is less obvious when a major question is whether the field belongs to epidemiology at all. In relation to typical risk problems, for example on the long-term effects of ionizing radiation or of taking "soft" drugs, one may have to decide whether to study human populations only or

also to experiment on animals on a large scale. It is probably the case that in all countries with good welfare services and medical records, studies on human beings are less costly than those on animals. "Record linkage," e.g., for records for an individual obtained at different times and places, is in its early stages, and involves ethical problems of secrecy.

The tendency is now more toward social and "preventive medicine," that is, observing human subjects before they become ill.

Table 1 reminds us that in the age group 15 to 25, road accidents now provide the largest single cause of death in most of the countries listed. Interpreting accident data seems to belong far more to epidemiology than to, say, traffic engineering.

Epidemiological statisticians studying such data seem less likely to be misled by inappropriate indices. An example is the number of fatal accidents to passengers traveling by air per million miles traveled. But most such accidents take place at or near takeoff or landing; hence the increase in risk for a long flight compared with a short one is very little. On the other hand, such an index is probably reasonable for railway passengers, and possibly so for car drivers. For them we have to distinguish between acci-

Table 1   Correlation Coefficients Between Death-Rates and Certain Variables[a]

| | Ages 55–64 | | | | | | | | Age 25–34 | |
| | Hypertensive Disease | | Ischaemic Heart Disease | | Cerebro-vascular Disease | | Bronchitis | | Road Accidents | |
	M	F	M	F	M	F	M	F	M	F
Doctors	− 0.38	0.03	− 0.51	− 0.46	− 0.69	− 0.36	− 0.42	− 0.42	0.42	0.15
Nurses	0.10	− 0.17	0.65	0.64	− 0.20	0.11	0.14	0.25	− 0.15	0.06
G.N.P.	− 0.62	− 0.22	− 0.17	− 0.26	− 0.69	− 0.77	− 0.67	− 0.65	0.10	0.19
Population density	− 0.05	0.04	− 0.44	− 0.45	− 0.02	− 0.16	0.13	− 0.08	0.00	− 0.08
Cigarettes	0.26	0.23	0.28	0.44	0.08	0.22	0.44	0.35	0.47	0.34
Alcohol										
Total	− 0.09	0.18	− 0.70	− 0.58	0.15	− 0.16	− 0.29	− 0.28	0.38	0.21
Wine	− 0.01	0.23	− 0.70	− 0.61	0.13	− 0.14	− 0.32	− 0.27	0.13	0.02
Beer	− 0.03	− 0.09	0.23	0.31	0.14	0.17	0.35	0.22	0.37	0.25
Spirits	− 0.38	− 0.15	− 0.26	− 0.32	− 0.35	− 0.51	− 0.57	− 0.47	0.30	0.37
Calories	0.43	0.06	0.51	0.61	− 0.02	0.36	0.57	0.67	− 0.20	− 0.03
Total fat	− 0.16	− 0.40	0.45	0.46	− 0.45	− 0.16	0.11	0.17	− 0.18	0.01
Saturated fat	0.10	− 0.17	0.64	0.62	− 0.16	0.15	0.30	0.35	− 0.08	0.05
Mono-unsaturated fat	− 0.11	− 0.42	0.60	0.60	− 0.35	− 0.05	0.11	0.15	− 0.28	− 0.17
Poly-unsaturated fat	− 0.45	− 0.28	− 0.48	− 0.47	− 0.51	− 0.60	− 0.33	− 0.30	− 0.05	− 0.11
Keys' prediction[b]	0.04	− 0.19	0.70	0.69	− 0.10	0.19	0.24	0.27	0.06	0.14

Source. St. Leger, Cochrane, and Moore [35], with thanks to the authors and *The Lancet.*

[a]Obtained from data from the 17 countries shown in Fig. 5.

[b]Keys' prediction (Keys et al. [20a]) is a predictive equation for serum-cholesterol.

dents where another vehicle is or is not involved. Where it is involved, there are more accidents in urban areas where traffic is dense than in rural areas or on motorways. This leads in turn to another misuse of such statistics, in that in towns, motorists have to drive more slowly—namely, to a doubtful conclusion that it is not more risky to drive faster. Naturally, effects of speed have to be compared when other things are comparable; data for motorways and streets should not be pooled.

It is even more difficult to set up reliable comparative statistics for cyclists and pedestrians.

This nonmedical example shows up pitfalls that epidemiological statisticians are always being confronted with; consider the population at risk, consider third variables correlated with both of two other variables, ask whether the underlying model is reasonable, with any set of comparative statistics.

In conclusion: the main object in this account has been to give instructive examples of what epidemiological statisticians

have been and are doing. As such it is far from complete; the geographical and reference coverage is even less so. However, there are very many references in the books and articles in the further reading list, which follows.

FURTHER READING

Books

A good introduction with a wealth of data clearly presented is Abraham M. Lilienfeld's *Foundations of Epidemiology* (Oxford University Press, New York, 1976). This can be complemented by Mervyn Susser's *Causal Thinking in the Health Sciences: Concepts and Strategies of Epidemiology* (Oxford University Press, New York, 1973).

Despite its title, at least half the essays in Alvan R. Feinstein's *Clinical Biostatistics* (C. V. Mosby, St. Louis, Mo., 1977) are relevant to epidemiology. They are selected from feature articles that appear about every 3 months in *Clinical Pharmacology and Therapeutics*. The author's approach is controversial; the articles are very long but his style is fresh and stimulating and they contain many thorough discussions; those on terminology are particularly recommended.

A more specialized work, but one that contains a great deal on the practical aspects of organizing good population surveys and questionnaires, and on the statistical methods and problems, is *The Natural History of Chronic Bronchitis and Emphysema: An Eight-Year Study of Early Chronic Obstructive Lung Disease in Working Men in London* by Charles Fletcher, Richard Peto, Cecily Tinker, and Frank E. Speizer (Oxford University Press, New York, 1976).

Articles

Two review articles, mainly on statistical methods, are to be recommended also for the very full references (225 and 165 + 26, respectively). They are Charles J. Kowalsky and Kenneth E. Guire's 1974 article "Longitudinal data analysis" (*Growth* **38**, 131–169) and Sonja M. McKinlay's 1975 contribution "The design and analysis of observational studies—a review" (*J. Amer. Statist. Ass.* **70**, 503–520, with comment by Stephen E. Fienberg, pp. 521–523).

For methods currently used, some useful recent references are:

Breslow, N. and W. Powers (1978). *Biometrics*, **34**, 100–105.

Holland, W. W. (coordinator), with Armitage, P., Kassel, N., and Premberton, J., eds. (1970). *Data Handling in Epidemiology*. Oxford University Press, New York.

Kemper, H. C. G. and van 't Hof, M. A. (1978). *Pediatrics*, **129**, 147–155.

McKinlay, S. M. (1975). *J. Amer. Statist. Ass.*, **70**, 859–864.

McKinlay, S. M. (1977). *Biometrics*, **33**, 725–735.

Prentice, R. (1976). *Biometrics*, **32**, 599–606.

van 't Hof, M. A., Roede, M. J., and Kowalski, C. J. (1977). *Hum. Biol.*, **49**, 165–179.

Some of these references were provided by P. I. M. Schmitz (Department of Biostatistics, Erasmus University, Rotterdam)—to whom many thanks.

References

[1] Brodsky, A. (1979). *Health Phys.*, **36**, 611–628.

[2] Burch, P. R. J. (1976). *The Biology of Cancer: A New Approach*. MIP Press, Lancaster, England.

[3] Burch, P. R. J. (1978). *J. R. Statist. Soc. A*, **141**, 437–458.

[4] Cochrane, A. L., Haley, T. J. L., Moore, F., and Hole, D. (1979). *Brit. J. Ind. Med.*, **36**, 15–22.

[5] Cook, P. J., Doll, R., and Fellingham, S. A. (1969). *Brit. J. Cancer*, **4**, 93–112.

[6] Court-Brown, W. M. and Doll, R. (1965). *Brit. Med. J.*, **ii**, 1327–1332.

[7] Court-Brown, W. M., Doll, R., and Hill, A. B. (1960). *Brit. Med. J.*, **ii**, 1539–1545.

[8] Darby, S. C. (1979). *Radiat. Prot. Bull.*, **28**, 7–10.

[9] Dawber, T. R., Kannel, W. B., and Lyell, L. P. (1963). *Ann. N.Y. Acad. Sci.*, **107**, 539–556.

[10] Defares, J. G., Sneddon, I. N., and Wise, M. E. (1973). *An Introduction to the Mathematics of Medicine and Biology* 2nd ed. North-Holland, Amsterdam/Year Book Medical Publishers, Chicago, pp. 589–601.

[11] Doll, R. (1971). *J. R. Statist. Soc. A*, **134**, 133–166.

[12] Doll, R. and Peto, R. (1976). *Brit. Med. J.* **i**, 1525–1536.

[13] Doll, R. and Pike, M. C. (1972). *J. R. Coll. Phys.* **6**, 216–222.

[14] Ehrenberg, A. S. C. (1968). *J. R. Statist. Soc. A*, **131**, 201.

[15] Erasmus University (1980). Annual Report of the Institute of Epidemiology.

[16] Fletcher, C., Peto, R., Tinker, C., and Speizer, F. E. (1976). *The Natural History of Chronic Bronchitis and Emphysema: An Eight-Year Study of Early Chronic Obstructive Lung Disease in Working Men in London*. Oxford University Press, New York.

[17] Good, I. J. (1962). In *The Scientist Speculates*, I. J. Good, A. J. Mayne, and J. Maynard Smith, eds. Heinemann, London.

[18] Gordon, T. and Kannel, W. B. (1970). In *The Community as an Epidemiologic Laboratory: A Case Book of Community Studies*, I. I. Kessler and M. L. Levia, eds. Johns Hopkins Press, Baltimore, Md., pp. 123–146.

[19] Higgins, I. T. T. (1964). *Brit. J. Ind. Med.*, **21**, 321–323.

[20] Hill, A. B. (1971). *Principles of Medical Statistics*, 9th ed. Oxford University Press, London.

[20a] Keys, A., Anderson, J. T., and Grande, F. (1957). *Lancet*, **ii**, 959.

[21] Kneale, G. W. (1971). *Biometrics*, **27**, 563–590.

[22] Kneale, G. W. (1971). *Brit. J. Prev. Soc. Med.*, **25**, 152–159.

[23] Lyon, J. L., Klauber, M. R., Gardner, J. W., and Udall, K. S. (1979). *N. Eng. J. Med.*, **300**, 397–402.

[24] Mancuso, T. F., Stewart, A. H. and Kneale, G. W. (1977). *Health Phys.* **33**, 369–385.

[25] Moroney, M. J. (1965). *Facts from Figures*, 3rd ed. Penguin, Baltimore, Md.

[26] Oldham, P. D. (1978). *J. R. Statist. Soc. A*, **141**, 460–462.

[27] Oldham, P. D. and Newell, D. J. (1977). *Appl. Statist.*, **26**, 125–135.

[28] Snow, J. (1936). In *Snow on Cholera*. Commonwealth Fund, New York, pp. 1–175.

[29] Sterk, P. J., Quanjer, Ph. H., van der Maas, L. L. J., Wise, M. E., and van der Lende, R. (1980). *Bill. Europ. Physiopath. Resp.*, **16**, 195–213.

[30] Stern, G. J. A. (1980). *Appl. Statist.* **29**, 93.

[31] Stewart, A. M. (1972). In *Clinics in Haematology*, S. Roath, ed. W. B. Saunders, London, pp. 3–22.

[32] Stewart, A. M. and Kneale, G. W. (1970). *Lancet*, **i**, 1185–1188.

[33] Stewart, A. M. and Kneale, G. W. (1970). *Lancet*, **ii**, 4–18.

[34] Stewart, A. M., Webb, J. W., and Hewitt, D. (1958). *Brit. Med. J.*, **i**, 1495–1508.

[35] St. Leger, A. S., Cochrane, A. L., and Moore, F. (1979). *Lancet*, **i**, 1017–1020.

[36] Stocks, P. (1970). *Brit. J. Cancer*, **24**, 215–225.

[37] Tanner, J. M., Whitehouse, R. H. and Takaishi, M. (1965). *Arch. Dis. Child.*, **41**, 454–571.

[38] Tarpeinen, O. (1979). *Cancer*, **59**, 1–7.

[39] Townsend, J. L. (1978). *J. R. Statist. Soc. A*, **141**, 95–107.

[40] Townsend, J. L. and Meade, T. W. (1979). *J. Epidemiol. Commun. Health*, **33**, 243–247.

[41] Wise, M. E. (1962). In *Physicomathematical Aspects of Biology*. (Italian Physical Society), N. Rashevsky, ed. Academic, New York.

(BIOSTATISTICS
COHORT ANALYSIS
DEMOGRAPHY
FRAMINGHAM: AN EVOLVING
    LONGITUDINAL STUDY
VITAL STATISTICS)

M. E. WISE

EQUALIZATION OF VARIANCE

Transformation of data to make the variance less dependent upon some other parameter—usually the expected value. As an example for a variable X with the Poisson distribution*, the variance is equal to the expected value. For the transformed variable $\sqrt{X}$, however, dependence of variance on the expected value is much less marked, although still existent. A formula suggesting an appropriate approximately variance-equalizing transformation, when the variance of X is $V(\xi)$, where ξ is the expected value of X, is

$$g(X) \propto \int^X \{V(\xi)\}^{-1/2} d\xi.$$

This formula can be obtained by the delta method*. By further analysis, improvements can be effected. For example, in the case of the Poisson distribution, $\sqrt{X + \frac{3}{8}}$ is rather better than $\sqrt{X}$ [1].

For more information on this topic, see Bartlett [2].

References

[1] Anscombe, F. J. (1948). *Biometrika*, **35**, 246–254.

[2] Bartlett, M. S. (1947). *Biometrics*, **3**, 39–52.

(COEFFICIENT OF VARIATION
FREEMAN–TUKEY
    TRANSFORMATIONS)

EQUINORMAL DISTRIBUTION

A distribution with CDF

$$F_X(x) = \frac{1}{\sqrt{2\pi}\,\sigma} \int_{-\infty}^{x} \int_0^1 t^{-1/2} \exp\left[-\frac{1}{2} y^2 (t\sigma^2)^{-1} \right] dt\, dy.$$

The distribution is symmetrical about zero; its variance is $\frac{1}{2}\sigma^2$. It belongs to a class of modified normal distributions constructed by Romanowski [1, 2].

References

[1] Romanowski, M. (1964). *Bull. Géod.*, **73**, 195–216.

[2] Romanowski, M. (1965). *Metrologie*, **4**(2), 84–86.

(LINEO-NORMAL DISTRIBUTION

QUADRI-NORMAL DISTRIBUTION
RADICO-NORMAL DISTRIBUTION)

EQUIVALENT PLANS   *See* DOUBLE SAM-
PLING

EQUIVARIANT ESTIMATORS

An equivariant estimator of a parameter or
parametric function is an estimator that is
transformed in the same way as the parameter or the parametric function when the observations are transformed in a certain way.
Examples:

Example 1: Location Parameter* Models.
Let $x_1, \ldots, x_n$ be independent observations
of an rv $X = \mu + Z$, where the real parameter μ is unknown but Z has a fixed distribution (completely or at least partially known).
For any real a the transformation g_a:
$(x_1, \ldots, x_n) \to (x_1 + a, \ldots, x_n + a)$ transforms the original observations into new observations $x_j' = x_j + a$ of the rv $X' = \mu +
a + Z$. The form of the original model is not
affected, but the parameter μ is transformed
into $\mu + a$. The model is invariant under the
transformations considered. An equivariant
estimator μ^* of μ thus satisfies, for all a and
all $x_1, \ldots, x_n$, $\mu^*(x_1 + a, \ldots, x_n + a)
= \mu^*(x_1, \ldots, x_n) + a$. Every such estimator
is of the form $\bar{x} + h(x_1 - \bar{x}, \ldots, x_n - \bar{x})$ for
some function h. It is easy to see that any
linear combination $\sum c_j x_{j:n}$ of the ordered
observations with $\sum c_j = 1$ is an equivariant
estimator of μ.
   More generally, let $\mathbf{x}$ be an observation of
a random n-dimensional (column) vector
$\mathbf{X} = \mathbf{A}\boldsymbol{\mu} + \mathbf{Z}$, where $\mathbf{Z}$ has a fixed distribution (with not necessarily independent com-
ponents) and the $n|m$ matrix $\mathbf{A}$ is known
and has full rank but the m-dimensional
(column) vector $\boldsymbol{\mu}$ is unknown. The transfor-
mation $g_{\mathbf{a}} : \mathbf{x} \to \mathbf{x} + \mathbf{A}\mathbf{a}$ leaves the model invariant and corresponds to the transforma-
tion $\bar{g}_{\mathbf{a}} : \boldsymbol{\mu} \to \boldsymbol{\mu} + \mathbf{a}$ on the parameter space
and induces thus the transformation $\phi
\to \phi + \mathbf{C}^T \mathbf{a}$ of a linear (real) parametric func-
tion $\phi = \mathbf{C}^T \boldsymbol{\mu}$. An equivariant estimator of ϕ

satisfies $\phi^*(\mathbf{x} + \mathbf{A}\mathbf{a}) = \phi^*(\mathbf{x}) + \mathbf{C}^T \mathbf{a}$ for all $\mathbf{a}$
and all $\mathbf{x}$.   ■

**Example 2: Location and Scale Parameter
Models*.**   Here $x_1, \ldots, x_n$ are observations
of an rv $X = \mu + \sigma Z$, where both μ and σ
$(\sigma > 0)$ are unknown. A transformation of
the form $g_{a,b} : (x_1, \ldots, x_n) \to (bx_1 +
a, \ldots, bx_n + a)$ (with $b > 0$) induces on the
parameter space the transformation $\bar{g}_{a,b} : (\mu,
\sigma) \to (b\mu + a, b\sigma)$. Thus equivariant estima-
tors of μ and σ satisfy, for all real a and
all $b > 0$,

$$\mu^*(bx_1 + a, \ldots, bx_n + a) = b\mu^*(x_1, \ldots, x_n) + a,$$

$$\sigma^*(bx_1 + a, \ldots, bx_n + a) = b\sigma^*(x_1, \ldots, x_n).$$

Any equivariant estimator of μ has the form
$\bar{x} + s \cdot h((x_1 - \bar{x})/s, \ldots, (x_n - \bar{x})/s)$,
where s^2 is the sample variance.
   More generally, let $\mathbf{x}$ be an observation of
a random vector $\mathbf{X} = \mathbf{A}\boldsymbol{\mu} + \sigma \mathbf{Z}$. The transformation $\mathbf{x} \to b\mathbf{x} + \mathbf{A}\mathbf{a}$ induces the transformation $(\boldsymbol{\mu}, \sigma) \to (b\boldsymbol{\mu} + \mathbf{a}, b\sigma)$ on the parameter space. An estimator ϕ^* of a linear para-
metric function $\phi = \mathbf{C}^T \boldsymbol{\mu} + c_{m+1}\sigma$ ($\boldsymbol{\mu}$ is m-
dimensional) is equivariant if, for all $\mathbf{a}$ and
all $b > 0$,

$$\phi^*(bx_1 + \mathbf{A}\mathbf{a}, \ldots, bx_n + \mathbf{A}\mathbf{a})$$

$$= b\phi^*(x_1, \ldots, x_n) + \mathbf{C}^T \mathbf{a}.   ■$$

**Example 3: Multivariate normal distribu-
tion*.**   Let $\mathbf{x}_1, \ldots, \mathbf{x}_n$ be a sample from a
multivariate normal distribution $\mathfrak{N}(\boldsymbol{\mu}, \boldsymbol{\Sigma})$ in
R^m, where $\boldsymbol{\mu}$ and the possibly singular co-
variance matrix $\boldsymbol{\Sigma}$ are both unknown. The
transformation $g_{\mathbf{a}, B} : (\mathbf{x}_1, \ldots, \mathbf{x}_n) \to (\mathbf{B}\mathbf{x}_1 +
\mathbf{a}, \ldots, \mathbf{B}\mathbf{x}_n + \mathbf{a})$, where the $m|m$ matrix $\mathbf{B}$ is
nonsingular, leaves the model invariant and
induces on the parameter space the transfor-
mation $\bar{g}_{\mathbf{a},\mathbf{B}} : (\boldsymbol{\mu}, \boldsymbol{\Sigma}) \to (\mathbf{B}\boldsymbol{\mu} + \mathbf{a}, \mathbf{B}\boldsymbol{\Sigma}\mathbf{B}^T)$. The
estimators $\boldsymbol{\mu}^* = \sum \mathbf{x}_i / n = \bar{\mathbf{x}}$ and $\boldsymbol{\Sigma}^* =
\text{constant} \cdot \sum (\mathbf{x}_i - \bar{\mathbf{x}})(\mathbf{x}_i - \bar{\mathbf{x}})^T$ are equivari-
ant since

$$\boldsymbol{\mu}^*(\mathbf{B}\mathbf{x}_1 + \mathbf{a}, \ldots, \mathbf{B}\mathbf{x}_n + \mathbf{a}) = \mathbf{B}\boldsymbol{\mu}^*(\mathbf{x}_1, \ldots, \mathbf{x}_n) + \mathbf{a},$$

$$\boldsymbol{\Sigma}^*(\mathbf{B}\mathbf{x}_1 + \mathbf{a}, \ldots, \mathbf{B}\mathbf{x}_n + \mathbf{a}) = \mathbf{B}\boldsymbol{\Sigma}^*(\mathbf{x}_1, \ldots, \mathbf{x}_n)\mathbf{B}^T.   ■$$

Equivariant estimation can only be used for
statistical models with invariant structure,

i.e., when there is given on the sample space a class of transformations g, which form a group $\mathcal{G}$ under composition, such that if the sample variable $\mathbf{X}$ (usually a vector of high dimension) is distributed as P_θ, $\theta \in \Theta$, then $g(\mathbf{X})$ is distributed as $P_{\bar{g}(\theta)}$, where $\bar{g}(\theta) \in \Theta$. The transformations $\bar{g}$ also form a group $\bar{\mathcal{G}}$. In what follows $g(\mathbf{X})$ and $\bar{g}(\theta)$ are written $g\mathbf{X}$ and $\bar{g}\theta$, respectively. An important special case occurs when, as in Examples 1 and 2 (and also in Example 3 if only nonsingular Σ's are considered), $\mathbf{X} = g\mathbf{Z}$, where $g \in \mathcal{G}$ is unknown and $\mathbf{Z}$ has a fixed distribution. In this case $\mathcal{G}$, $\bar{\mathcal{G}}$, and the parameter space Θ can be identified with each other. With the additional assumption that $\mathcal{G}$ is unitary, i.e., that different transformations g_1 and g_2 map a sample point on different points, this is Fraser's [3] structural model.

The parameter θ is estimated equivariantly by θ^* if $\theta^*(\mathbf{x}) \in \Theta$ and $\theta^*(g\mathbf{x}) = \bar{g}\theta^*(\mathbf{x})$. A parametric function $\phi = \phi(\theta)$ can be estimated equivariantly only if, for all $\bar{g}$, $\phi(\bar{g}\theta_1) = \phi(\bar{g}\theta_2)$ whenever $\phi(\theta_1) = \phi(\theta_2)$. Then to every $\bar{g}$ there exists a transformation $\bar{\bar{g}}$ on Φ, the range space of ϕ, such that $\phi(\bar{g}\theta) = \bar{\bar{g}}\phi(\theta)$. An estimator ϕ^* of ϕ is equivariant if $\phi^*(g\mathbf{x}) = \bar{\bar{g}}\phi^*(\mathbf{x})$.

The equivariant estimators were introduced by Pitman [8] for location and scale parameters and he called them "proper" estimators. The term "equivariant" was used by Berk [1] and Wijsman [9] and replaces the older somewhat misleading term "invariant." Other names are "cogredient estimator" (indicated in Lehmann [7, p. 1:17]) and "transformation variable" [3, p. 51], while the name "covariant estimator" does not seem to have been used. A statistic $T(\mathbf{x})$ can be called equivariant, with respect to $\mathcal{G}$, if $T(g\mathbf{x}_1) = T(g\mathbf{x}_2)$ for all g whenever $T(\mathbf{x}_1) = T(\mathbf{x}_2)$.

A standard motivation for using only equivariant estimators is that the result of the statistical analysis should not depend on the choice of units, e.g., on whether the Celsius or Fahrenheit system is used for temperature measurements. Concerning Examples 1 and 2 this is a strong argument, but it is less relevant in the case of Example 3.

However, to restrict attention to equivariant estimators is to lean upon the principle of invariance*.

Most common estimators of parametric functions that can be estimated equivariantly for models with invariant structure are equivariant. For example, maximum likelihood estimators* are always equivariant, and so are linear unbiased estimators of linear parametric functions for linear models.

To single out a best equivariant estimator of an equivariantly estimable parametric function $\phi = \phi(\theta)$ (maybe $\phi = \theta$), a loss function $L(\theta, d)$ is needed. Here d takes values in Φ and the loss function should satisfy $L(\bar{g}\theta, \bar{\bar{g}}d) = c(\bar{g})L(\theta, d)$, where $c(\bar{g})$ is a constant depending only on $\bar{g}$. Under the assumption that $\bar{\bar{g}}$ is transitive, i.e., that, for any θ, there exists $\bar{g}_\theta \in \bar{\mathcal{G}}$ such that $\theta = \bar{g}_\theta\theta_0$, where θ_0 is fixed, it follows that the risk function $R(\theta, \phi^*) = E_\theta[L(\theta, \phi^*(\mathbf{X}))]$ associated with an equivariant estimator ϕ^* of ϕ is independent of θ. This important result shows that under transitivity of $\bar{\mathcal{G}}$, which holds in particular for Fraser's structural model, a locally best (i.e., minimum risk) equivariant estimator is also uniformly best, and hence a best equivariant estimator usually exists. It is appropriate to point out that when $\bar{\mathcal{G}}$ is transitive and unitary, the original loss function $L(\theta, d)$ can be replaced by the new equivalent loss function $W(\theta, d) = L(\theta_0, (\bar{\bar{g}}_\theta)^{-1}d)$, which satisfies $W(\bar{g}\theta, \bar{\bar{g}}d) = W(\theta, d)$. For example, in Example 2 the loss functions $L((\mu, \sigma), d) = (d - \mu)^2$ and $W((\mu, \sigma), d) = (d - \mu)^2/\sigma^2$ for estimation of μ are equivalent.

In Example 1 the best equivariant estimator of μ with respect to quadratic loss—the Pitman estimator* of μ—is given by $\mu_P^* = \bar{x} - E_0[\bar{x} \mid x_1 - \bar{x}, \ldots, x_n - \bar{x}]$. It is unbiased. Here $\bar{x}$ can be replaced by any other equivariant estimator of μ. This expression for μ_P^* was given by Girshick and Savage [4], whereas Pitman [8] gave μ_P^* as $\mu_P^* = \int \mu f(x_1 - \mu) \cdot \ldots \cdot f(x_n - \mu) \, d\mu / \int f(x_1 - \mu) \cdot \ldots \cdot f(x_n - \mu) \, d\mu$, where f is the density of Z. The latter expression is of considerable interest, as it shows that μ_P^* is formally the mean of the posterior distribution of μ given $x_1, \ldots, x_n$ provided that the (improper) prior density for μ is uniform on $(-\infty, \infty)$. Use of μ_P^* presupposes complete knowledge of f. If f is not completely known, one must usually restrict attention to estimators in some subclass of the class of all equivariant estimators. For example, if only the moments of Z up to order $2N$ are known and the loss is quadratic, it is natural to seek a best estimator in the class of equivariant polynomial estimators of degree N.

In Example 2, the best equivariant estimators of μ and σ (quadratic loss) are given by $\mu_P^* = \bar{x} - s \cdot E_{(0,1)}[\bar{x} \cdot s \mid U]/E_{(0,1)}[s^2 \mid U]$ and $\sigma_P^* = s \cdot E_{(0,1)}[s \mid U]/E_{(0,1)}[s^2 \mid U]$, where $U = ((x_1 - \bar{x})/s, \ldots, (x_n - \bar{x})/s)$. These estimators can also be written as ratios of integrals (see below). They are not unbiased but there exist constants c_1, c_2 so that $\mu_P^* + c_1 \sigma_P^*$ and $c_2 \sigma_P^*$ are the best unbiased estimators of μ and σ, respectively [2].

For Fraser's structural model with certain additional assumptions of regularity type, the following results hold [5]: If the loss function is taken as $W(\theta, d)$ above and if $\phi^*(\mathbf{x})$ minimizes $E^{\theta|\mathbf{x}}[W(\theta, \phi^*(\mathbf{x}))]$ and is unique, then ϕ^* is an equivariant estimator of ϕ and is the best one. Here $E^{\theta|\mathbf{x}}$ denotes expectation with respect to the posterior distribution of θ given $\mathbf{x}$ with the prior distribution for θ equal to the right-invariant Haar measure* on Θ ($= \mathcal{G}$, which is assumed to be locally compact). Moreover, if ϕ is real and $W(\theta, d) = w(\theta)(d - \phi)^2$ for some function $w(\theta)$, then the best equivariant estimator of ϕ is given by

$$\phi_P^* = E^{\theta|\mathbf{x}}\left[w(\theta)\phi(\theta)\right]/E^{\theta|\mathbf{x}}\left[w(\theta)\right].$$

Under rather general conditions a best equivariant estimator is minimax* but more rarely admissible*; see, e.g., James and Stein [6] and Zacks [10, Chaps. 7 and 8].

The best equivariant estimators are mathematically the natural estimators for Fraser's structural model in the same way as the best unbiased estimators are natural for (complete) exponential families*. Unfortunately,

they are usually difficult to calculate numerically and have been little used in practice.

Finally, to those interested in estimation when the group $\bar{\mathcal{G}}$ is not transitive, the possibility of Bayes equivariant estimation should be mentioned; for details, see Zacks [10, Chap. 7] and also Zidek [11].

References

[1] Berk, R. H. (1967). *Ann. Math. Statist.*, **38**, 1436–1445.

[2] Bondesson, L. (1979). *Scand. J. Statist.*, **6**, 22–28.

[3] Fraser, D. A. S. (1968). *The Structure of Inference*. Wiley, New York.

[4] Girshick, M. A. and Savage, L. J. (1951). *Proc. 2nd Berkeley Symp. Math. Statist. Prob.*, Vol 1. University of California Press, Berkeley, Calif., pp. 53–74.

[5] Hora, R. B. and Buehler, R. J. (1966). *Ann. Math. Statist.*, **37**, 643–656. (This is a well-written mathematical paper with a good list of references.)

[6] James, W. and Stein, C. (1961). *Proc. 4th Berkeley Symp. Math. Statist. Prob.*, Vol 1. University of California Press, Berkeley, Calif., pp. 361–379.

[7] Lehmann, E. L. (1950). Notes on the Theory of Estimation. Mimeographed lecture notes, University of California, Berkeley.

[8] Pitman, E. J. G. (1939). *Biometrika*, **30**, 391–421. (This is historically an important paper in which expressions for the best equivariant estimators of location and scale parameters are derived. For a modern reader, however, these derivations are somewhat tedious.)

[9] Wijsman, R. A. (1967). *Proc. 5th Berkeley Symp. Math. Statist. Prob.*, Vol 1. University of California Press, Berkeley, Calif., pp. 389–400.

[10] Zacks, S. (1971). *The Theory of Statistical Inference*. Wiley, New York. (This is the only textbook that so far pays much attention to equivariant estimation. A reader of the book must be prepared to correct errors, however.)

[11] Zidek, J. V. (1969). *Ann. Inst. Statist. Math. Tokyo.* **21**, 291–308.

(INVARIANCE
INVARIANCE PRINCIPLES IN
  STATISTICS
STRUCTURAL ESTIMATION)

LENNART BONDESSON

ERDÖS–RÉNYI LAW OF LARGE NUMBERS

Given a sequence $\{X_n : 1 \leqslant n < \infty\}$ of i.i.d. random variables $[E(X_i) = 0$ and $\text{var}(X_i) = 1]$ with a moment generating function* $\phi(t)$, Erdös and Rényi [4] investigate the behavior of the maximum $\sum(N, K)$ of $N - K + 1$ successive K-point averages of the form $K^{-1}(S_{n+K} - S_n)$ for $0 \leqslant n \leqslant N - K$, when $S_n = \sum_1^n X_i, S_0 \equiv 0$. This is a problem of deviations of the sample mean from the population mean, which is of importance in statistical inference in connection with asymptotic distributions of various test statistics*. More specifically, if $\psi(x)$ is a function of real variables such that $(\log \log x)^{-1}\psi(x) \to \infty$ as $x \to \infty$, then from the law of the iterated logarithm* it follows that $\lim(n\psi(n))^{-1/2}S_n = 0$ a.s. Consequently, for the $N - K + 1$ successive partial sums $S_{n+K} - S_n$ with $0 \leqslant n \leqslant N - K$, we have for a fixed n,

$$\lim(K\psi(K))^{-1/2}(S_{N+K} - S_n) = 0 \quad \text{a.s.}$$

Define

$$\sum_\psi(N, K) = \max_{0 \leqslant n \leqslant N-K}\left[(K\psi(K))^{-1/2}\right.$$

$$\left. \times (S_{N+K} - S_n)\right],$$

where K depends on N. The Erdös–Rényi law of large numbers studies the asymptotic behavior of $\sum_\psi(N, K)$ for various choices of K between 1 and N, as $N \to \infty$. By choosing K properly, one can make $\sum_\psi(N, K)$ converge almost surely to *any* given positive λ on the whole R_1^+ or some bounded subinterval of it. Generalizations of the Erdös–Rényi law of large numbers were investigated by Book [2, 3].

For more information on this topic, see Bahadur and Ranga Rao [1].

References

[1] Bahadur, R. R. and Ranga Rao, R. (1960) *Ann. Math. Statist.*, **31**, 1015–1027.

[2] Book, S. A. (1975). *Proc. Amer. Math. Soc.*, **48**, 438–446.

[3] Book, S. A. (1976). *Canad. J. Statist.*, **4**, 185–211.

[4] Erdös, P. and Rényi, A. (1970). *J. Anal. Math.*, **23**, 103–111.

(ASYMPTOTIC NORMALITY
LAWS OF THE ITERATED
  LOGARITHM
LIMIT THEOREMS
LIMIT THEOREMS, CENTRAL)

ERGODIC THEOREMS

Let $X_1, X_2, \ldots$ be a sequence of independent, identically distributed random variables. If we want to estimate the associated distribution function $F(x)$ for all x, we can do so observing infinitely many X's. Indeed, if $F_n(x)$ is the proportion of X's among $X_1, \ldots, X_n$ which do not exceed x, then it follows from the strong law of large numbers* that $F_n(x)$ converges to $F(x)$ for every x, with probability 1. More generally, we can estimate any appropriate characteristic of $F(x)$ in the same manner.

However, if we assume some form of dependence* among the random variables, the situation becomes considerably more complicated. If we assume only that the random variables have a common distribution function $F(x)$, then it may not be possible to estimate $F(x)$ by observing the process X_1, $X_2, \ldots$ without further restrictions on the joint distributions of the process.

One such assumption which is frequently made is that of *stationarity* or time invariance. We shall say the process is stationary* if for every n, the joint distribution of $X_1, \ldots, X_n$ is the same as the joint distribution of $X_{1+k}, \ldots, X_{n+k}$ for all positive integers k. What this means is that the probability structure of the process remains invariant under time shifts. Suppose, then, that $\{X_n\}_{n=1}^\infty$ is a stationary process and let f be a function which is integrable with respect to the common distribution function F. Then, as in the case of independent, identically distributed random variables, we can say a good deal about the limiting behavior of averages of the $(1/n)\sum_{i=1}^{n} f(X_i)$. In fact, as we shall see below, the limit of such a sequence of averages will always exist with probability 1.

To discuss such problems, it will be useful to cast them in the language of ergodic theory. Let Ω be a space, $\mathfrak{F}$ a σ-algebra of subsets of Ω, and μ a probability measure* defined on the sets in $\mathfrak{F}$. Let T be a transformation* mapping Ω onto Ω which is measurable; i.e., $A \in \mathfrak{F}$ if and only if both $TA \in \mathfrak{F}$ and $T^{-1}A \in \mathfrak{F}$. We shall say that T is measure preserving if for every $A \in \mathfrak{F}$, we have $\mu(T^{-1}A) = \mu(A)$. If, in addition, T is one-to-one, then we will also have $\mu(TA) = \mu(A)$. We shall refer to such a structure as the measure-preserving system $(\Omega, \mathfrak{F}, \mu, T)$ (m.-p. system).

To see how this serves as a model for stationary processes, consider such a process $\{X_n\}_{n=1}^{\infty}$. Let μ be the probability measure induced on the space Ω of infinite sequences of real numbers by the multivariate distributions of the process, and define T as follows. If $\omega = (\omega_1, \omega_2, \dots)$ is an infinite sequence of real numbers, then $\omega' = T(\omega)$ is defined by $\omega' = (\omega_2, \omega_3, \dots)$. T defined in this way is called the shift transformation*. It can be easily verified that T is measure-preserving (m.-p.). Conversely if $(\Omega, \mathfrak{F}, \mu, T)$ is an m.-p. system and f is a measurable function defined on Ω, then the process $\{X_n\}_{n=0}^{\infty}$ is a stationary process, where $X_0 = f(\omega)$, $X_1 = f(T\omega), \dots, X_n = f(T^n\omega)$, etc. Thus we see that stationary processes can be studied within the framework of m.-p. systems.

The study of m.-p. systems is the object of a flourishing area of modern mathematics called ergodic theory. It is interesting to note that while the results of ergodic theory have been widely used to solve many problems in probability theory, the original motivation came from statistical and celestial mechanics. In 1899, Poincaré published a theorem that seems to have initiated the modern study of ergodic theory. Let $(\Omega, \mathfrak{F}, \mu, T)$ be an m.-p. system, and let $A \in \mathfrak{F}$. If $x \in A$, we can look at the orbit of x under T, i.e., the set of points $x, Tx, T^2x, \dots$. Let us say that x is *recurrent* if infinitely many of the ele-

ments of the orbit of x return to A, in other words, if for infinitely many positive integers n, we have $T^nx \in A$. We then have

Theorem 1 (Poincaré, 1899]. Almost every point of A is recurrent.

By "almost every point" we mean that the set of points of A which are not recurrent has μ-measure zero. One immediate consequence of the Poincaré recurrence theorem is the fact that if $A \in \mathfrak{F}$ and $\mu(A) > 0$, then there exists a positive integer n such that $\mu(A \cap T^nA) > 0$. Recently, Furstenberg [9] proved a remarkable generalization of this fact. Namely

Theorem (Furstenberg, 1977). Let $A \in \mathfrak{F}$ with $\mu(A) > 0$. Then for for every integer $k \geqslant 2$, there is an integer n such that

$$\mu(A \cap T^nA \cap T^{2n}A \cap \dots \cap T^{(k-1)n}A) > 0.$$

Furstenberg's theorem is important not only for itself but because it has been used to prove several deep and difficult results in number theory.

One of the important questions in ergodic theory is the degree to which the transformation T "mixes" the space. From now on, let us assume that every set that we discuss is an element of $\mathfrak{F}$. Clearly, if there exists a set A that is invariant under T, i.e., a set for which $A = TA$, then if $x \in A$, we have $T^nx \in A$ for all n, and if $x \in A^c$, we have $T^nx \in A^c$ for all n. So in this case, T does not mix the space too well. Let us call T *ergodic* (from the Greek: wandering) if whenever $A = TA$, we have $\mu(A) = 0$ or $\mu(A) = 1$. If T is ergodic, then T does mix the space in the sense that if $\mu(A) > 0$ and $\mu(B) > 0$, then there exists a positive integer such that $\mu(T^nA \cap B) > 0$; in other words, the orbit of the set A will eventually "hit" every set B of positive measure. In fact, more is true. We can also determine the limiting relative frequency with which the orbit of A hits any set B. One of the consequences of the ergodic theorems that we discuss below is the following:

Theorem 3. Let $(\Omega, \mathcal{F}, \mu, T)$ be a m.-p. system. Then T is ergodic if and only if for every pair of sets A and B we have

$$\lim_{n \to \infty} (1/n) \sum_{j=1}^{n} (T^j A \cap B) = \mu(A) \mu(B).$$

Note that this theorem tells us that if $\mu(A) > 0$, $\mu(B) > 0$, then $\mu(T^n A \cap B) > 0$ for infinitely many n, and in fact the set of integers n for which $\mu(T^n A \cap B) > 0$ cannot be too "thin".

Actually, we can talk about stronger "mixing" properties of T than those provided by ergodicity. Let us say that T is weakly mixing if for every pair of sets A and B we have

$$\lim_{n \to \infty} (1/n) \sum_{j=1}^{n} |\mu(T^j A \cap B)$$
$$- \mu(A) \mu(B)| = 0.$$

It follows at once that a weakly mixing transformation is ergodic, and it is relatively easy to give examples of ergodic transformations that are not weakly mixing. Finally, we shall say that T is strongly mixing if for every pair of sets A and B, we have $\lim_{n \to \infty} \mu(T^n A \cap B) = \mu(A)\mu(B)$. Again, we see that strong mixing implies weak mixing and again we can give examples where the converse is false. Notice that strong mixing looks like asymptotic independence between the set B and the sequence of sets $T^n A$. In fact, all the mixing conditions are in some sense asymptotic independence conditions.

We now turn to the ergodic theorems proper. In addition to the system $(\Omega, \mathcal{F}, \mu, T)$, consider also $L_1(\mu)$, the class of measurable functions defined on Ω such that $\int_\Omega |f| \, d\mu < \infty$. In many applications, including probability theory, we may be interested in knowing $\int |f| \, d\mu$ without knowing μ.

It was long conjectured that, in some sense, one should be able to estimate the "space average" of a function by its

"time average," that is, one should be able to estimate $\int_\Omega f \, d\mu$ by

$$\lim_{n \to \infty} (1/n) \sum_{j=0}^{n-1} f(T^j \omega)$$

for some suitably chosen ω. That this cannot be true in general can be seen from the following example. Suppose that T is not ergodic. Then there is a set A with $0 < \mu(A) < 1$ such that $TA = A$. Now if $\omega \in A$, we have $T^n \omega \in A$ for all n, and if $\omega \in A^c$, we have $T^n \omega \in A^c$ for all n. Let $f(\omega) = X_A(\omega)$, the indicator function of A; then

$$\frac{1}{n} \sum_{j=0}^{n-1} f(T^j \omega) = 1 \quad \text{for all } n \quad \text{for all } \omega \in A,$$

and if $\omega \in A^c$, we have $(1/n) \sum_{j=0}^{n-1} (T^j \omega) = 0$ for all n. So at the very least, we must require ergodicity in order to be able to estimate $\int_\Omega f \, d\mu$.

It turns out that this is sufficient. One of the most fundamental theorems in ergodic theory is the so-called "individual ergodic theorem" due to G. D. Birkhoff [2].

Theorem 4 (Birkhoff, 1931). Let $(\Omega, \mathcal{F}, \mu, T)$ be añ m.-p. system, and let $f \in L_1(\mu)$. Then

(a) $\lim_{n \to \infty} (1/n) \sum_{j=0}^{n-1} f(T^j \omega) = f^*(\omega)$ exists for almost all ω.

(b) $f^* \in L_1(\mu)$, and $\int f^* \, d\mu = \int f \, d\mu$.

(c) f^* is invariant, i.e., $f^*(T\omega) = f^*(\omega)$ for almost all ω, and if T is ergodic, $f^*(\omega)$ is constant for almost all ω, and hence $f^* = \int_\Omega f \, d\mu$.

Here again, the statement "for almost all ω" means that the assertion is true except possibly on a set of μ-measure zero. So Birkhoff did show that we can estimate time averages by space averages if T is ergodic. Note that the Birkhoff theorem contains the familiar strong law of large numbers*. For if $\{X_n\}_{n=1}^{\infty}$ is a sequence of independent, identically distributed random variables with $E|X_i| < \infty$, we can set up the m.-p. system $(\Omega, \mathcal{F}, \mu, T)$, where Ω consists of all infinite sequences of real numbers, $\mathcal{F}$ is the σ-algebra determined by the process $\{X_n\}$, and μ is the appropriate probability measure on the process. If we let T be the shift transformation, it is easily verified that T is ergodic. Finally, for $\omega = (\omega_1, \omega_2, \ldots)$, define $f(\omega) = \omega_1$. Then

$$(1/n) \sum_{j=0}^{n-1} f(T^j) = (1/n) \sum_{i=1}^{n} X_i$$

and the strong law follows. Of course, the ergodic theorem applies to considerably more general situations. In particular, a number of problems in analysis and continued fractions may be solved by applying the ergodic theorem.

At about the same time that Birkhoff was working on the individual ergodic theorem, John von Neumann [13] proved the "mean ergodic theorem." Let $L_2(\mu)$ be the space of measurable functions defined on Ω for which $\int_\Omega |f|^2 d\mu < \infty$. If $\{f_n\}_{n=1}^\infty$ and f are functions in $L_2(\mu)$, we shall say that f_n converges to f "in the mean" if $\int |f_n - f|^2 d\mu \to 0$. Now for $f \in L_2(\mu)$, define $(Uf)(\omega) = f(T\omega)$. One can easily show that $f \in L_2(\mu)$ implies that $Uf \in L_2(\mu)$, and in fact $\int_\Omega |Uf|^2 d\mu = \int |f|^2 d\mu$; i.e., U is what is called a unitary operator. Whereas every measure-preserving transformation gives rise to a unitary operator, the converse is not in general true; that is, there exist unitary operators on $L_2(\mu)$ which are not generated by measure-preserving transformations. Nevertheless, von Neumann's ergodic theorem applies to every unitary operator.

Theorem 5 (von Neumann, 1932). Let U be a unitary operator on $L_2(\mu)$. Then for each $f \in L_2(\mu)$, there exists $f^* \in L_2(\mu)$ such that $(1/n) \sum_{j=1}^{n} U^j f$ converges in the mean to f^*. Moreover, $Uf^* = f^*$.

In particular, if U is induced by a measure-preserving transformation, then f^* is the same limiting function as in the individual ergodic theorem.

Many generalizations of the individual and the mean ergodic theorem have been proved. Among the most important of these are the Chacon–Ornstein theorem and the

Yosida theorem. A good discussion of these may be found in the book by Brown [5].

There are still many interesting unsolved problems concerning ergodic theorems. One obvious question concerns various methods of averaging. Classical ergodic theorems are concerned with averaging methods of the form $(1/n) \sum_{j=1}^{n} f(T^j \omega)$, so called Cesaro averaging. Many other averaging methods can be used. For example, if $\{k_1, k_2, \ldots\}$ is an infinite sequence of positive integers, we can ask whether $(1/n) \sum_{j=1}^{n} f(T^{k_j} \omega)$ converges, either in the mean or almost everywhere when $f \in L_1(\mu)$. For mean convergence, this question has been answered. For example, it was shown [4] that mean convergence holds for *every* infinite subsequence if and only if T is strongly mixing. Also, necessary and sufficient conditions are known for a given infinite sequence of positive integers so that the mean ergodic theorem will hold for every m.-p. system [3]. For the individual ergodic theorem, the results are still quite sketchy. It is known that for every m.-p. system, there is an infinite sequence of integers $\{k_1, k_2, \ldots\}$, and an $f \in L_1(\mu)$ such that $(1/n) \sum_{j=1}^{n} f(T^{k_j} \omega)$ fluctuates infinitely often between 0 and 1 for almost every $\omega \in \Omega$ [11]. On the other hand, examples of infinite sequence of integers have been given so that the individual ergodic theorem does hold for every m.-p. system [6]. Research is currently continuing concerning these and other questions.

Literature

In addition to the references cited in the text, several basic introductory books are also included. The earliest and still very useful and interesting is Halmos [10]. An excellent introduction is Friedman [8]. More recent and both quite up to date are Brown [5] and Walters [14]. A follow-up to these is Denker et al. [7]. A survey article on recent progress in ergodic theorems is Krengel [12].

References

[1]   Billingsley, P. (1965). *Ergodic Theory and Information*. Wiley, New York.

[2] Birkhoff, G. D. (1931). *Proc. Natl. Acad. Sci. USA*, **17**, 656–660.

[3] Blum, J. R. and Eisenberg, B. (1974). *Proc. Amer. Math. Soc.*, **42**, 423–429.

[4] Blum, J. R. and Hanson, D. L. (1960). *Bull. Amer. Math. Soc.*, **66**, 308–311.

[5] Brown, J. R. (1976). *Ergodic Theory and Topological Dynamics*, Academic Press, New York.

[6] Brunel, A. and Keane, M. (1969). *Z. Wahrscheinlichkeitsth. verwend. Geb.*, **12**, 231–240.

[7] Denker, M., Grillenberger, C., and Sigmund, K. (1976). Ergodic Theory on Compact Spaces, *Lect. Notes Math.*, **527**.

[8] Friedman, A. (1970). *Introduction to Ergodic Theory*. Van Nostrand Reinhold, New York.

[9] Furstenberg, H. (1977). *J. Anal. Math.*, **31**, 204–256.

[10] Halmos, P. R. (1956). *Lectures on Ergodic Theory*, Chelsea, New York.

[11] Krengel, U. (1971). *Ann. Math. Statist.*, **42**, 1091–1095.

[12] Krengel, U. (1977). *Soc. Math. Fr.*, 151–192.

[13] von Neumann, J. (1932). *Proc. Natl. Acad. Sci. USA*, **18**, 263–266.

[14] Walters, P. (1975). Ergodic Theory–Introductory Lectures, *Lect. Notes Math.* 458.

(LAWS OF LARGE NUMBERS LIMIT THEOREMS, CENTRAL STOCHASTIC PROCESSES)

J. R. BLUM

ERLANG DISTRIBUTION

A standard gamma distribution* with (positive) integer-valued shape parameters.

ERROR ANALYSIS

Our understandings of the world in which we live and the decisions that are made in conducting the affairs of society depend largely on numerical values. Much depends on the size of the error associated with those values. A single measurement can be the basis for actions taken to maintain our health, safety, or the quality of our environment. It is important, therefore, that the errors of such measurements be small enough so that the actions taken are only negligibly affected by those errors. In any governmental regulatory action or measurements involved in legal actions, it is also obvious that the shadow of doubt surrounding the measurements should be suitably small. But this is no less true for measurements in all branches of science and industry, because even though legal action may not be involved, the validity of scientific inference, the effectiveness of process control, or the quality of production may depend on adequate measurement.

This article deals with the perspective by which one can give operational meaning to the "shadow of doubt" or "uncertainty" generated by errors of measurement. Some commonly used procedures are given for the error analysis by which one determines the possible extent of the errors affecting his or her system of measurement and hence of the output of that system—the measured values themselves.

THE REFERENCE BASE TO WHICH MEASUREMENTS MUST BE RELATED

Whatever quantity is being measured, there exists in our mind the idea that it has a correct value. We recognize that we will not achieve this sought-after value due to imperfections in our understanding of the quantity being measured (model error) and our inability to make measurements without error. However, we cannot usefully define error as the difference from an unknowable "true value."

As a guide to finding a suitable practical alternative, it is instructive to contemplate what might happen if a measurement were to become an inportant element in a legal controversy. Two essential features should arise. First, that the contending parties would have to agree on what (actually realizable) measurement would be mutually acceptable. The logic of this seems unassailable—if one cannot state what measurement system would be accepted as "correct," then one would have no defensible way of developing specifications or regulations involving such measurements. Second, the uncertainty to be attached to the measurement would be established by a form of "cross-examination" by which one would determine the shadow of doubt relative to this acceptable value.

Such a consensus or generally accepted value can be given a particularly simple meaning in dealing with measurements of such quantities as length, mass, voltage, resistance, or temperature. One may require that uncertainties be expressed relative to the standards as maintained by a local laboratory or, when appropriate, by the National Bureau of Standards*. In other cases, nationally accepted artifacts, standard reference materials, or in some cases a particular measurement process may constitute a reference base (e.g., see ref. 5). One basic quality of all these examples should not be overlooked—all are operationally realizable. The confusion engendered by introducing the term "true value" as the correct but unknowable value is thus avoided.

PROPERTIES OF MEASUREMENT

In discussing uncertainty, we must account for two characteristics of measurement processes. First, repeated measurements of the same quantity by the same measurement process will disagree; and second, the limiting means of measurements by two different processes will disagree. These facts lead to a perspective from which to view measurement, namely that the measurement be regarded as the "output" of a process analogous to an industrial production process. In defining the process, one must state the conditions under which a "repetition" of the measurement would be made, analogous to defining the conditions of manufacture in an industrial process.

The need for this specification of the process becomes clear if one envisions the cross-examination process. One would begin with such questions as:

1. Within what limits would an additional measurement by the same instrument agree?

2. Would the agreement be poorer if the time interval between repetitions were increased?

3. What if different instruments were used?

To these can be added questions about the conduct of the measurement.

1. What effect do environmental conditions have on the item being measured or on the instruments making the measurement?

2. Is the result dependent on the procedure used (e.g., do operator differences exist)?

3. Are there instrumental biases or differences due to reference standards or calibrations*?

There are also the obvious questions relating to the representativeness of the item being measured.

1. If the item being measured is a sample from a larger aggregate, what sampling* errors could occur?

2. Are biases introduced because of the (small) size of the sample, or from the handling, storage, or other factors?

Note that one can replace the term "measurement" in the paragraphs above by the term "experiment" as one encounters it in agricultural, biological, and social sciences and the analogous questions still apply.

THE CONCEPT OF A REPETITION

Every measurement has a set of conditions in which it is presumed to be valid. At a very minimum, it is the set of repeated measurements with the same instrument–operator–procedure configuration. (This is the type of repetition expected in some process control operations.) If the measurement is to be interchangeable with one made at another location, the repetition would involve different instrument–operator–procedure–environment configurations. (This type of repetition occurs in producing items to satisfy a specification and in manufacturing generally.)

To evaluate a measurement process, some redundancy needs to be built into the system to determine the process parameters. This redundancy should be representative of the set of repetitions to which the uncertainty statement is to apply. In traditional statistical literature on the design of experiments*, the importance of a "control" has long been recognized.

The essential characteristic needed to establish the validity of measurement is predictability of the process, i.e., that the variability remains at the same level and that the process is not drifting or shifting abruptly from its established values. The evidence of predictability must come from redundant measurement of a control or reference item of known value which has properties similar to those of the regular work load in order to verify this condition. It is not enough that a control be present in agricultural experiments, clinical trials*, etc., but the response of the control group itself should behave as an outcome from an ongoing stable process.

In measuring an "unknown," one gets a single value, but one is still faced with the need to make a statement that allows for the anticipated scatter of the results. If we had a sufficiently long record of measurements, we could predict the limits within which we were fairly certain that the next measurement would lie. Such a statement should be based on a collection of independent determinations, each one similar in character to the new observation, that is, so that each observation of the collection and also the new observation can be considered as random drawings from the same probability distribution. These conditions will be satisfied if the collection of points is from a sufficiently broad set of environmental and operating conditions to allow all the random effects to which the process is subject to have a chance to exert their influence on the variability. Suitable data collections can be obtained by incorporating an appropriate reference measurement into routine measurement procedures, provided that such measurements are representative of the same

variability to which the unknown is subject. The statistical procedures for expressing the results will depend on the structure of the data, but they cannot overcome deficiencies in the representativeness of the values used.

Results from the control item provide the basis for determining the parameters of the measurement process and verifying that the properties are transferable to measurements on test items. One is saying, in effect, "If we could have measured the unknown again and again, a sequence of values such as those for the control item would have been obtained." Whether our single value is above or below the mean we cannot say, but we are fairly certain it would not differ by more than the bounds to the scatter of the values on the control item.

ERROR ANALYSIS

Error analysis can be looked upon as the methodology by which one arrives at a numerical value for the "shadow of doubt" or uncertainty associated with his measurement. The goal is to be able to determine the uncertainty to be attached to an isolated measurement on a possibly transient phenomenon. The focus is therefore directed to the process that generated the measurement, and one has the problem of "sampling" the output of the measuring process so as to determine its operating characteristics—its variability, dependence on enrivonmental and other factors, and its possible offset from the reference base for such measurements.

The determination of nonsampling errors in survey and census data (see, e. g., ref. 3), or the selection of mathematical models in all branches of science are examples of special problems related to error analysis which cannot reasonably be treated apart from the subject matter itself. However, by contemplating the cross-examination that might ensue if one's data were the subject of controversy, one should be able to set forth the type of information needed for an adequate "defense"—i.e., what type of redundancy is needed in the system, and what tests are needed to demonstrate the adequacy of the assumed model relative to the intended use of the measurements.

PARAMETERS OF A MEASUREMENT PROCESS: RANDOM VARIATION

The parameters of a measurement process fall into two classes: those related to random variation and those related to the offset of the process from the reference base or inadequacies in the assumed model.

Repeated careful measurements of the same quantity taken close together in time generally agree with each other much better than those separated by a long time interval. If the ith measurement on the jth occasion can be represented by the equation

$$y_{ij} = \mu + \eta_i + \epsilon_{ij},$$

where η_i is independently distributed with mean 0 and variance σ_B^2 and ϵ_{ij} are independently distributed with mean 0 and variance σ_w^2 and are independent of the η's, then a sequence of k_i measurements per occasion will lead to values $\hat{\sigma}_w$ and $\hat{\sigma}_B$ for these parameters. (*See* VARIANCE COMPONENTS.) The variance of the average of k values on the ith occasion is var$(\bar{y}_i) = \sigma_B^2 + \sigma_w^2/k_i$.

When the measurements are part of a regular work load, a work item can often be measured a second time. If this can be done on different occasions giving the values x and y, then the variability of the differences, d_i, where

$$d_i = x - y,$$

should be $2\sigma_B^2 + 2\sigma_w^2$. The consistency of the variance value from a sequence of values of d is a check on the use of the formula for var$(\bar{y}_i)$ on test items. Remember that the values of $\hat{\sigma}_B$ and $\hat{\sigma}_w$ arose from measurements on a control which may behave differently than the test items.

In many chemical processes one finds a nested series of variance components arising from partitioning the original sample for dilution, or other such aliquoting. For such cases there are a number of experimental arrangements for minimizing cost for an al-

lowable level of variability in the determination of the variance component.

When the quantities of interest arise from a least-squares* analysis of a set of data, the residuals provide a value for the standard deviation, which can be used to determine the standard deviation of each coefficient. If the data were collected over a short span of time so that any between-occasion component is suppressed, the observed variability will not be predictive of that actually to be found when the between-occasion variability is indeed present.

When the result can be shown to vary with operator, instrument, procedural, or environmental factors, or some other background variable, one has the problem of whether to regard these as random variation or to treat them as systematic effects to be handled as part of the process offset. Despite the fact that an analysis of variance* of a designed experiment may produce a "between-instrument variance" when one has only a few instruments, one would probably be better off to determine the offset for each instrument and apply a correction to eliminate the instruments bias relative to the reference base for the measurements. There will be other cases where it is expedient to randomize the selection of the instrument and include an extra component of variance in the error model. In either case one has to have an appropriate answer to the question about the effect of possible differences between instruments on the result.

Once agreement is reached as to what constitutes a repetition, then the error model can be constructed and the proper redundancy incorporated to produce values for the components of variance and covariance in the model. One is then faced with the problem of assigning a limit for the effects of random error, by which we mean a limit that would only rarely be exceeded as long as the process is in a state of statistical control. Thus if y is a function of the random variable, $x_1, x_2, \ldots, x_n$ such that

$$y = y(x_1, x_2, \ldots, x_n),$$
$$\text{var}(y) = h[\text{var}(x_1), \text{var}(x_2), \text{cov}(x_1, x_2), \ldots]$$
$$= \sigma^2,$$

then one will use

$$\text{limit of random error} = R(\sigma).$$

The selection of the rule for setting the limits for random error is somewhat arbitrary. Even if the form of the probability distribution were known, one has the choice as to what percentage points should define "rare event." Perhaps the most widely used rule is to use $R = \pm 3\sigma$ as the limit. This rule has been used successfully in industrial quality control work when the distributions are reasonably symmetric and not too different from the normal distribution, for which it gives a coverage of 99.7%. Asymmetrical limits would be appropriate for highly skewed distributions. An evaluation of the appropriateness of the rule can be made by studying the operating characteristics of the system, e.g., correctness of decisions based on the values and their reported uncertainties.

PROPAGATION OF ERROR

When the quantity of interest is a function of several measured quantities, i.e., $y = F(x_1, x_2, x_3, \ldots)$, one has the problem of determining the variability in y from knowledge of the variability in the x's. This procedure, which usually is called *propagation of error*, is based on the formula

$$\text{var}(y) = \sum \left(\frac{\partial F}{\partial x_i} \right)^2 \text{var}(x_i)$$

$$+ 2 \sum_{i,j} \left(\frac{\partial F}{\partial x_j} \right) \left(\frac{\partial F}{\partial x_j} \right) \text{cov}(x_i x_j),$$

the partial derivatives and variances being the long run or "true" values for the quantities, but in practice one uses the values available from existing data. (For a discussion of the limitations of the formula, see ref. 1, pp. 352ff., and ref. 4.) The formula is based on the Taylor series expansion of the function $F(x_1, x_2, \ldots)$ about its mean $F(\mu_1, \mu_2, \ldots)$ and is usually satisfactory if the partial derivatives are small and var(x_i) is small relative to x_i.

The following examples are among the most commonly occurring:

Quantity	Approximate Variance
$ax + by$	$a^2 \text{var}(x) + b^2 \text{var}(y) + 2ab\, \text{cov}(x, y)$
x/y	$\left(\frac{x}{y} \right)^2 \left(\frac{\text{var}(x)}{x^2} + \frac{\text{var}(y)}{y^2} - 2 \frac{\text{cov}(x, y)}{xy} \right)$
xy	$y^2 \text{var}(x) + x^2 \text{var}(y) + 2xy\, \text{cov}(x, y)$
x^2	$4x^2 \text{var}(x)$
$\ln(x)$	$\dfrac{\text{var}(x)}{x^2}$
$\sqrt{x}$	$\dfrac{\text{var}(x)}{4x}$

One is faced with the problem of demonstrating by some form of redundancy that one is entitled to regard the random variation attributed to the function y by these propagation-of-error procedures as adequately representing the actual variation.

SYSTEMATIC ERROR: OFFSET FROM REFERENCE BASE

By defining the offset of a measurement process (or equivalently its systematic error) as the limiting mean of the difference between the results of the process and that given by the reference base to which measurements are referred, one has a means of operationally determining such an offset or at least determining a bound for its possible extent. As with variability, one must build some redundancy into the system for the purpose of measuring the offset.

In work involving physical standards (mass, volt, etc.), reference standards or carefully measured artifacts constitute an ideal way of comparing measurement processes. The possession of a standard calibrated by a central laboratory is not sufficient evidence that one's measurement process is correct—the measurement procedures must also be checked. A transfer standard having measurement characteristics similar to that of the items of interest, measured in both the reference laboratory and by the process, can be used to determine the offset and as an independent check on the variability in both processes. (A program of this type is provided by the National Bureau of Standards for length standards [2].) In such cases the measured offset can be used as a bias correction, c, to the regular measurement. If the offset is determined and the correction made with each measurement, the random variation in the correction becomes part of the random error of the process and the offset may be regarded as zero.

When the offset is determined only infrequently (say, once a year), then whatever error there is in the correction will persist until the next redetermination, and all measurements will be offset by the error in the correction. One can set a bound to the possible error in the correction and even though this bound is due almost entirely to random error, it has to be treated as a systematic error as long as the correction is used. Our ignorance of the quantity does not make it a random variable. There is no way one can reduce the size of the error in his or her own organization, and in a cross-examination it would have to be admitted that the error could be as large as the stated limit and that there were no repetitions available by which the effect could be reduced.

Despite the exchange of artifacts with a reference laboratory, there may be systematic errors in one's measurements. Even if one regularly calibrates (*see* CALIBRATION) one's instrument, there is the question of linearity of response—possible difference in instrument response between the items used for calibration and that of the test items (e.g., in chemical analysis the reference items are usually free of contaminants, whereas the test samples frequently are not).

The propagation-of-error formula for expressing a bound to the systematic error, $|\Delta y|$, in a function $y = F(x_1, x_2, \ldots)$ due to errors $\Delta x_1, \Delta x_2, \ldots$ in the variables is given by

$$|\Delta y| = \left| \frac{\partial F}{\partial x_1} \Delta x_1 \right| + \left| \frac{\partial F}{\partial x_2} \Delta x_2 \right| + \cdots$$

with the partial derivatives being evaluated at the observed value of the quantities in-

volved. It remains to be verified experimentally that the formula expresses all the effects of the variables.

UNCERTAINTY OF A REPORTED VALUE

The uncertainty to be associated with a measurement will include an allowance for the possible offset of the measurement process from the reference base plus an amount to account for the random error. Thus uncertainty, U, can be written for symmetrical error limits

$$U = \pm (R + E),$$

where R is the limit (e.g., 3σ) for the random error and E is the possible offset of the process. Because E remains fixed (although unknown), it cannot be combined in quadrature with E. (The modification for asymmetrical limits is obvious.)

TEST OF THE MODEL

It is the predictability of the process—its being in a state of statistical control*—upon which the validity of the error analysis depends. A system of monitoring the output of the measurement process is, therefore, necessary to update the parameters of the process and to test the adequacy of the assumed model. The values for standards, reference artifacts, or other controls should be studied to see if they exhibit any time-dependent trends, a dependence on environment, or any other factors. The controls also provide a check on the adequacy of the random error limits. The percentage of values for the controls that fall outside the error limits should be compared with that nominally assumed for the system (e.g., 1.0 or 0.3%). If the controls exceed their limits too frequently, it is not safe to assume that the regular work load has a better rate, and further, the parameter values and even the validity of the model is in question.

THE ISOLATED EXPERIMENT

A single isolated experiment has to be looked upon as a truncated measurement process. One needs evidence that the process measures unambiguously the sought-after quantity—i.e., that the errors associated with the final value are small enough relative to the intended use. Once the universe of discourse for which the result is expected to be valid has been defined, the type of repetitions that must be accounted for in the random error model can be specified. In determining the bounds for possible systematic error, one does not have a reference base available and must devise experimental procedures to overcome this lack (e.g., extrapolation to a value at zero impurity from results at a sequence of impurity levels). Sometimes this cannot be done and no similar experiment has been done elsewhere. One might express an opinion on the magnitude of the error from this factor, but if one contemplates a cross-examination, he or she will not combine this with those limits determined from experimentation.

OPERATING CHARACTERISTICS OF THE METHOD OF ERROR ANALYSIS

Often one will have only a few degrees of freedom* available in the determination of a variance component, or the measurement of the offset of one's measurements relative to the national reference standards may be based on only a few measurements. The limits of error are thus themselves subject to error. If the poorly determined variance component involves a component of minor importance, the functioning of the overall system as measured by the percentage of correct decisions in the regulation of a process, protection of the environment, or in health or safety may be little affected.

The adequacy of the uncertainty statements should be evaluated by their effect on the success of the endeavor in which they are used. The operating characteristics* of the system under various alternatives, such as departure from the assumed distribution or the presence of correlation* when independence was assumed, needs to be investigated. One needs to develop an operationally verifiable model in which the risks and costs of wrong decisions are taken into account.

Error analysis is the methodology by which quality control* is brought to measurement. And like quality control, it is not just the conformity and exactness of the numbers that is at issue—their appropriateness relative to their end use must also be examined. Ideally, such a study should enable one to state what additional (or different) measurement efforts would improve the process of which the measurements are adjunct.

References

[1] Cramér, H. (1946). *Mathematical Methods of Statistics*. Princeton University Press, Princeton, N. J.

[2] Croarkin, C., Beers, J., and Tucker, C. (1979). Measurement Assurance for Gage Blocks. *Natl. Bur. Stand. Monogr.* 163, U.S. Government Printing Office, Washington, D. C.

[3] Gonzalez, M. E., Ogus, J. L., Shapiro, G., and Tepping, B. J. (1975). *J. Amer. Statist. Ass.*, **70**, 5–23.

[4] Ku, H. H. (1966). *J. Res. Natl. Bur. Stand.*, **70C**, 283–273.

[5] Nuclear Regulatory Commission (1975). Measurement Control Program for Special Nuclear Materials Control and Management (10 CFR 70.57), *Fed. Reg.*, **40**(155), 33651–33653.

(ANALYSIS OF VARIANCE
COLLECTION OF DATA
CONTROL CHARTS
EDITING STATISTICAL DATA
GENERAL LINEAR MODEL
QUALITY CONTROL
STATISTICAL DIFFERENTIALS)

J. M. CAMERON

ERROR OF THE THIRD KIND   *See* CONSULTING, STATISTICAL

ESTIMABILITY

Any observable random variable Y (or any function of it) can be regarded as an estimate of a parameter θ that occurs in its probability distribution, and in this sense θ can be considered "estimable." However, in statistics "estimable" is used in a special sense. A parameter θ is called (linearly) estimable if there exists a linear function of observations, whose expectation is equal to θ for all values of θ in a given set. Some conditions of estimability in the context of the general linear model are described here.

The notion of estimability in the general linear model* theory was introduced by R. C. Bose [1,2] (see also Chakrabarti [3], Scheffé [5], and Rao [4]).

Let $Y_1, Y_2, \ldots, Y_n$ be a set of n random variables with

$$\mathbf{Y} = \mathbf{A}'\boldsymbol{\theta} + \boldsymbol{\epsilon}, \tag{1}$$

where $\mathbf{Y}$ is the $n \times 1$ column vector of $(Y_1, \ldots, Y_n)$, $\boldsymbol{\theta}$ is a $m \times 1$ vector of unknown parameters, $\mathbf{A}'$ is a known $n \times m$ matrix of rank r and $\boldsymbol{\epsilon}$ is a column vector of n random variables $\epsilon_1, \ldots, \epsilon_n$, with

$$E(\boldsymbol{\epsilon}) = \mathbf{0}, \tag{2}$$

$$\operatorname{var} \boldsymbol{\epsilon} = \sigma^2 \mathbf{I}. \tag{3}$$

[Equations (1), (2), and (3) define what is known as a "fixed-effects" model.]

A linear parametric function $\mathbf{l}'\boldsymbol{\theta}$ is called *estimable* if there exists a linear function of the observations $\mathbf{c}'\mathbf{Y}$ such that

$$E(\mathbf{c}'\mathbf{Y}) = \mathbf{l}'\boldsymbol{\theta} \tag{4}$$

for $\boldsymbol{\theta}$ in a given Θ which is in general m-dimensional Euclidean space.

A necessary and sufficient condition for (4) to hold is

$$\mathbf{Ac} = \mathbf{l}, \tag{5}$$

that is, $\mathbf{l}$ belongs to the vector space generated by the column vectors of $\mathbf{A}$. Alternatively,

$$\operatorname{rank} \mathbf{A} = \operatorname{rank}(\mathbf{A} : \mathbf{l}). \tag{6}$$

Since rank $\mathbf{A}$ is assumed to be r, the vector space of estimable linear functions is of dimension r.

Suppose that $\mathbf{l}'\boldsymbol{\theta}$ is estimable and $\mathbf{c}'\mathbf{Y}$ is an unbiased estimate of $\mathbf{l}'\boldsymbol{\theta}$. Now $\mathbf{c}$ can be expressed as a unique sum

$$\mathbf{c} = \mathbf{d} + \mathbf{e} = \mathbf{A}'\mathbf{q} + \mathbf{e}, \tag{7}$$

where $\mathbf{d} = \mathbf{A}'\mathbf{q}$ belongs to the vector space $V_c(\mathbf{A}')$ generated by the column vectors of $\mathbf{A}'$ and $\mathbf{e}$ to its orthogonal complement. Then $\mathbf{d} = \mathbf{A}'\mathbf{q}$ is called the orthogonal projection of $\mathbf{c}$ on $V_c(\mathbf{A}')$ and one can write $\mathbf{d}$ as

$$\mathbf{d} = \mathbf{A}'(\mathbf{A}\mathbf{A}')^- \mathbf{A}_\mathbf{c}, \tag{8}$$

where $(\mathbf{A}\mathbf{A}')^-$ is a generalized inverse* of $\mathbf{A}\mathbf{A}'$ and $\mathbf{P} = \mathbf{A}'(\mathbf{A}\mathbf{A}')^- \mathbf{A}$ is a symmetric and idempotent* matrix of rank r and projects orthogonally on the $V_c(\mathbf{A}')$.

Then $\mathbf{d}'\mathbf{Y} = \mathbf{q}'\mathbf{A}\mathbf{Y} = \mathbf{c}'\mathbf{P}\mathbf{Y}$ is an unbiased* estimate of $\mathbf{l}'\boldsymbol{\theta}$ and among all linear unbiased estimates of $\mathbf{l}'\boldsymbol{\theta}$ it has the minimum variance. It is called the *best linear unbiased estimate** (b.l.u.e.) of $\mathbf{l}'\boldsymbol{\theta}$. From this result it follows that every $\mathbf{d}'\mathbf{Y}$, where $\mathbf{d}'$ is of the form $\mathbf{A}'\mathbf{q}$, is the best linear unbiased estimate of its expectation*. The vector space generated by the column vectors of $\mathbf{A}'$, $V_c(\mathbf{A}')$, is called the *estimation space* and its orthogonal complement $V_c^\perp(\mathbf{A}')$ is called the *error space*. Its dimension is $n - r$. If $\hat{\boldsymbol{\theta}}$ is a solution of the least-squares normal equations

$$\mathbf{A}\mathbf{A}'\boldsymbol{\theta} = \mathbf{A}\mathbf{Y}, \tag{9}$$

then

$$\mathbf{l}'\hat{\boldsymbol{\theta}} = \mathbf{d}'\mathbf{Y}. \tag{10}$$

Thus the least-squares estimate of the estimable linear function is its best linear unbiased estimate for the linear model defined by (1), (2), and (3).

$$\operatorname{var}(\mathbf{d}'\mathbf{Y}) = \operatorname{var}(\mathbf{l}'\hat{\boldsymbol{\theta}}) = \mathbf{d}'\mathbf{d}\sigma^2$$
$$= \mathbf{l}'(\mathbf{A}\mathbf{A}')^- \mathbf{l}\sigma^2 = \mathbf{l}'\mathbf{q}\sigma^2, \tag{11}$$

where $\mathbf{d} = \mathbf{A}'\mathbf{q}$.

Suppose that $\mathbf{q}_1'\mathbf{A}\mathbf{Y}$ and $\mathbf{q}_2'\mathbf{A}\mathbf{Y}$ are, respectively, the b.l.u.e.'s of $\mathbf{l}_1'\boldsymbol{\theta}$ and $\mathbf{l}_2'\boldsymbol{\theta}$; then

$$\operatorname{cov}(\mathbf{q}_1'\mathbf{A}\mathbf{Y}, \mathbf{q}_2'\mathbf{A}\mathbf{Y}) = \mathbf{q}_1'\mathbf{A}\mathbf{A}'\mathbf{q}_2\sigma^2$$
$$= \mathbf{l}_1'(\mathbf{A}\mathbf{A}')^- \mathbf{l}_2\sigma^2$$
$$= \mathbf{l}_1'\mathbf{q}_2\sigma^2$$
$$= \mathbf{l}_2'\mathbf{q}_1\sigma^2. \tag{12}$$

Suppose that e' belongs to the error space; then

$$E(e'Y) = e'A'\theta = 0 \qquad \text{for every } \theta$$

and

$$E(e'Y)^2 = e'e\sigma^2.$$

Suppose that $e_1, \ldots, e_{n-r}$ are $n - r$ orthonormal vectors belonging to the error space. Then

$$E\left[\sum_{j=1}^{n-r} (e_j'Y)^2\right] = (n - r)\sigma^2 \qquad (13)$$

and if the ϵ's are normally distributed, $\sum_{j=1}^{n-r}(e_j'Y)^2/\sigma^2$ is distributed as χ^2 with $(n - r)$ degrees of freedom. $S_e^2 = \sum_{j=1}^{n-r}(e_j'Y)^2$ is called the *error sum of squares* and can be calculated as

$$S_e^2 = Y'Y - \hat{\theta}'AY$$
$$= \min_\theta (Y - A'\theta)'(Y - A'\theta), \qquad (14)$$

where $\hat{\theta}$ is any solution of the normal equations (9).

The sum of squares due to a set of linear functions CY is defined to be the square of the length of the orthogonal projection of Y on the vector space generated by the column vectors of C', that is,

$$Y'C(CC')^- CY. \qquad (15)$$

Then

$$EY'C'(CC')^- CY$$
$$= s\sigma^2 + \theta'AC'(CC')^- CA'\theta, \qquad (16)$$

where $s = \text{rank } C$.

For testing a linear hypothesis

$$H : G\theta = g,$$

provided that all the linear functions $G\theta$ are estimable, one can first calculate the sum of squares due to the b.l.u.e.'s $CY = G\hat{\theta}$ [$\hat{\theta}$ a solution of the normal equations (9)] as

$$S_H^2 = (CY - g)'(CC')^- (CY - g). \qquad (17)$$

Then a test of the hypothesis H is based on

$$T = \frac{S_H^2/t}{S_e^2/n - r}, \qquad (18)$$

where t is the rank of G, and T is distributed

as F on t and $n - r$ degrees of freedom when H is true.

The case when all the linear functions in $G\theta$ are not estimable is discussed under TESTABILITY.

When (3) is modified to

$$\text{var } \epsilon = \sigma^2 \Sigma, \qquad (19)$$

Σ a known positive definite matrix, corresponding results for b.l.u.e.'s can be derived by first making a linear transformation (see, e.g., Rao [4])

$$Z = TY, \qquad (20)$$

where $\Sigma = T^{-1}(T^{-1})'$. Then (1) becomes

$$E(Z) = TA'\theta \qquad (21)$$

and

$$\text{var } Z = \sigma^2 I. \qquad (22)$$

In a design with n experimental units grouped into b disjoint blocks, and v treatments, let k_j denote the number of experimental units in the jth block, $j = 1, \ldots, b$, let r_i denote the number of experimental units that have received the ith treatment, $i = 1, 2, \ldots, v$, and let n_{ij} denote the number of experimental units that are in the jth block and have received the ith treatment. A fixed-effects additive model for the observations $Y_u, u = 1, \ldots, n$, for such a design can be written as

$$E(Y) = \mu j + A'\alpha + L'\beta \qquad (23)$$

$$\text{var } Y = \sigma^2 I. \qquad (24)$$

Y is the $n \times 1$ vector of observations from the n experimental units, μ a constant, α is the $v \times 1$ vector of treatment effects and β is the $b \times 1$ vector of block effects, j is the $n \times 1$ vector of 1's, and $A' = (a_{iu})$, where $a_{iu} = 1$ if the uth experimental unit is in the jth block, $l_{ju} = 0$ otherwise. Then

$$N = (n_{ij}) = AL'. \qquad (25)$$

In order to find out which linear functions of treatment effects are estimable, Bose [2] introduced the notion of *connectedness* of a design. A treatment and a block are said to be *associated* if the treatment has been assigned to an experimental unit that is in the given block. Two treatments, two blocks, or a treatment and a block are defined to be *connected* if these two entities are linked by a chain in the design, consisting alternately of blocks and treatments such that any two adjacent members of the chain are associated. A design (or a component of a design) is said to be a *connected design* (or a connected component of a design) if every pair of the design (or portion of the design) is connected. In general, a design must always decompose into a number of connected portions such that a block or a treatment belonging to any one portion will be unconnected with a block or a treatment belonging to another component. Then one can prove the following:

Theorem. The necessary and sufficient condition that the linear function

$$l_1\theta_1 + \cdots + l_v\theta_v \qquad (26)$$

is estimable under the model (23), (24) is that the sum of the coefficients of the treatment effects in each connected component should be equal to zero. In particular, $t_1 + t_2 + \cdots + t_v$ is nonestimable.

The linear function (25) is called a *contrast** if $\sum_{i=1}^v l_i = 0$.

For a connected design, every treatment contrast is estimable. In this case one can show that the rank of the matrix

$$\mathbf{C} = \mathbf{D(r)} - \mathbf{N D^{-1}(k) N'} \qquad (27)$$

is $v - 1$; $\mathbf{D(r)}$ is the diagonal matrix with $(r_1, \ldots, r_v)$ in the principal diagonal, and $\mathbf{D(k)}$ is also a diagonal matrix with $(k_1, \ldots, k_l)$ in the diagonal. Further, one can show that if rank $\mathbf{C} = v - 1$, then all treatment contrasts are estimable and thus all block contrasts are also estimable. Thus one can alternatively define a design to be connected [for the model (23), (24)] if rank $\mathbf{C} = v - 1$.

Further, if rank $\mathbf{C} = s$, then only s linearly independent treatment contrasts are estimable.

References

[1]   Bose, R. C. (1944). *Proc. 31st Indian Sci. Congr.*, **3**, 5–6.

[2]   Bose, R. C. (1947). *Proc. 34th Indian Sci. Congr.*, Delhi, Part 11, 1–25.

[3]   Chakrabarti, M. C. (1962). *Mathematics of Design and Analysis of Experiments*. Asia Publishing House, Bombay.

[4]   Rao, C. R. (1973). *Linear Statistical Inference and Its Applications*. Wiley, New York, Chap. 4.

[5]   Scheffé, H. (1959). *The Analysis of Variance*. Wiley, New York.

(BLOCKS, BALANCED INCOMPLETE BLOCKS, RANDOMIZED DESIGN OF EXPERIMENTS ESTIMATION, POINT GENERAL LINEAR MODEL LEAST SQUARES UNBIASEDNESS)

I. M. CHAKRAVARTI

ESTIMATION, POINT

THE MEASUREMENT PROBLEM

Point estimation is the statistical term for an everyday activity: making an educated guess about a quantity that is unknown but concerning which some information is available —a distance, weight, temperature, or intelligence quotient; the size of a population or the age of an artifact; the time it will take to complete a job.

The prototype of such a problem is that of n measurements $x_1, \ldots, x_n$ of an unknown quantity θ and the question of how to combine them to obtain a good value for θ. The most common answer is now (but was not always; see Plackett [26]): the *average* (or arithmetic mean*) $\bar{x}$ of the n measurements. The *median** is sometimes proposed as an alternative. The theory of point estimation is concerned with questions such as: Which is better, mean or median? Or rather: under what circumstances is the mean better than the median, and vice versa? And beyond that: In a given situation, what is the best way of combining the x's?

An early proposal was to estimate θ by means of the value a that minimizes the sum of squared differences $\sum(x_i - a)^2$. The resulting value is the *least-squares* estimate of θ. This minimizing value turns out to be $a = \bar{x}$, i.e., $\bar{x}$ is the least-squares estimate of θ. An alternative to least squares is to minimize instead the sum of absolute values $\sum |x_i - a|$. The solution of this problem is the median of the x's. (The general method of least squares* is due to Gauss* and Legendre; for some history, see Plackett [27].)

The mean and median cannot be justified by these derivations as reasonable estimates of the true value θ since no explicit assumptions have been made connecting the x's with θ. Such a connection can be established by thinking of the measurement process as a repeatable operation, capable of producing different values. This leads to considering the outcome $(X_1, \ldots, X_n)$ of the process as random variables taking on values $(x_1, \ldots, x_n)$ with frequencies specified by their joint probability distribution, which depends on θ. This step, which is the cornerstone of the theory of estimation (and more generally of statistical inference*), is attributed by Eisenhart [11] to Simpson in 1755.

It frequently will be reasonable to assume that each of the X_i has the same distribution, and that the variables $X_1, \ldots, X_n$ are independent, so that the X's are independent, identically distributed (i.i.d). If the common distribution of the errors, $X_i - \theta$, does not depend on θ and has probability density f, the joint density of the X's is then

$$f(x_1 - \theta) \cdot \cdots \cdot f(x_n - \theta). \qquad (1)$$

Here one often assumes that f is known, except possibly for an unknown scale factor, that the X's have, for example, a normal*, double exponential*, or Cauchy* distribution. Such assumptions will have their origin in previous experience with similar data. A family (1) with given f is called a *location family** and θ a *location parameter**.

The densities (1) constitute a model for the measurement problem. Other statistical problems require different models which, however, share the following features with (1). Observations are taken (the *data*), which are the values of random variables $X = (X_1, \ldots, X_n)$, and the X's have a joint probability density (or probability) $p_\theta(x)$ $= p_\theta(x_1, \ldots, x_n)$, depending on a parameter $\theta = (\theta_1, \ldots, \theta_s)$.

To estimate θ or more generally a real- or vector-valued function $g(\theta)$, one calculates a corresponding function of the observations, a *statistic*, say $\delta = \delta(X_1, \ldots, X_n)$. To distinguish the statistic, which is a random variable or vector, from the value $\delta(x_1, \ldots, x_n)$ it takes on in a particular case, some authors refer to the first as an *estimator* and the second as an *estimate* of θ. (This is the terminology adopted here). Others apply the term "estimate" to both concepts. The problem of point estimation is to find an estimator δ that tends to be close to θ, and to say something about how close.

In this formulation, the subject can be viewed as a special case of *decision theory**. Other aspects are treated in *Bayesian inference** and *descriptive statistics*, respectively. The theory should, of course, be considered not in isolation but together with its applications to such diverse areas of statistical methodology as *variance components**, *contingency tables**, *density estimation**, *linear models**, *Markov chains* (see MARKOV PROCESSES), *multivariate analysis**, *survey sampling**, and *time series**.

Many of the key ideas of the theory of estimation go back to Laplace*, particularly his *Théorie analytique des probabilitiés* 3rd ed. (1820); and to Gauss's work on the method of least squares, particularly the *Theoria combinationis observationum erroribus minimis obnoxiae* (1821, 1823), English translations of which are published in his collected works. The vigorous development of the theory during the last decades owes its principal impetus to Fisher* ([13] and later papers) and Wald* [32].

MAXIMUM LIKELIHOOD*

The most widely used method for generating estimators is the method of maximum likelihood. Considered for fixed $x = (x_1, \ldots, x_n)$ as a function of θ, $p_\theta(x)$ is called the *likelihood** of θ, and the value $\hat{\theta} = \hat{\theta}(X_1, \ldots, X_n)$ of θ which maximize $p_\theta(X)$ constitutes the *maximum likelihood estimator* (MLE) of θ. The MLE of a function $g(\theta)$ of θ is $g(\hat{\theta})$. (For a survey of the literature of MLE's see Norden [23]).

Example 1

(a) If X has the binomial distribution

$$P(X = x) = \binom{n}{x} p^x (1 - p)^{n-x},$$

$$x = 0, 1, \ldots, n,$$

the MLE of p is $\hat{p} = X/n$.

(b) If $X_1, \ldots, X_n$ are i.i.d. according to the normal distribution with mean ξ and variance σ^2, the MLE of (ξ, σ^2) is $(\sum X_i / n, \sum (X_i - \bar{X})^2 / n)$.

(c) If $X_1, \ldots, X_n$ are i.i.d. according to the uniform distribution* on $(0, \theta)$, the MLE of θ is $\max(X_1, \ldots, X_n)$.

The theory of MLEs was initiated by Edgeworth and Fisher. (Priority issues are discussed by Pratt [28].) It centers on three properties that MLEs possess in the i.i.d. case, as $n \to \infty$. For $s = 1$, these are as follows.

1 **Consistency.** For most families of distributions encountered in practice, including those of Example 1, $\hat{\theta}_n$ (the MLE of θ based on n observations) is *consistent*, i.e., $P(|\hat{\theta}_n - \theta| < a) \to 1$ for every $a > 0$.

2 **Asymptotic normality*.** If $\hat{\theta}_n$ is consistent and $p_\theta(x)$ satisfies mild regularity conditions due to Cramér [9], which hold in parts (a) and (b) of Example 1 but not in (c), the $\hat{\theta}_n$ are *asymptotically normal*. More precisely, the distribution of $\sqrt{n}(\hat{\theta}_n - \theta)$ tends to the normal distribution with mean zero and variance $1/I(\theta)$, where

$$I(\theta) = E\left[(\partial/\partial\theta) \log p_\theta(X_i) \right]^2$$

is the *information* (sometimes called Fisher information*) that a single observation X_i contains about θ.

3 **Asymptotic efficiency*.** If θ_n^* is any other sequence of estimators for which $\sqrt{n}(\theta_n^* - \theta)$ tends to a normal distribution with mean zero and variance, say,

$\tau^2(\theta)$ and which satisfies some weak additional assumption [e.g., that $\tau^2(\theta)$ is continuous], then $\tau^2(\theta) \geqslant 1/I(\theta)$. Thus no other asymptotically normal estimator has smaller asymptotic variance, and in this sense is $\hat{\theta}_n$ *asymptotically efficient*. Without some additional restriction on the competing estimator, the result is not true, but in any case the set of points θ for which $\tau^2(\theta) < 1/I(\theta)$, called points of *superefficiency* of the competing estimator, must have Lebesgue measure zero (see LeCam [21]) and Bahadur [2]).

The results described above require some amplification.

1. Cramér's conditions for property 2 do not ensure consistency of $\hat{\theta}_n$. Although consistency holds very commonly, a general theorem to that effect requires much stronger assumptions (see Wald [31] and LeCam [21]).

2. On the other hand, there does exist under Cramér's assumptions, with probability tending to 1, a solution $\hat{\theta}_n$ of the likelihood equation $\partial \log p_\theta(x)/\partial \theta = 0$, which is consistent and hence asymptotically normal and efficient. In the case of multiple roots, the correct choice may, however, be difficult.

3. Result 2 implies that the errors $\hat{\theta}_n - \theta$ tend to zero at the rate $1/\sqrt{n}$. This is not the case in situations like that of part (c) of Example 1, where the set of x's with $p_\theta(x) > 0$ changes with θ. It is then much easier to estimate θ, and in part (c), for instance, it is $n(\hat{\theta}_n - \theta)$ which tends to a (nonnormal) limit distribution, so that the error of the MLE tends to 0 at the rate of $1/n$.

4. Results 1 to 3 can be extended to the case that $\theta = (\theta_1, \ldots, \theta_s)$ with $s > 1$. The only major change is that the earlier definition of $I(\theta)$ must be replaced by that of the *information matrix* $I(\theta)$ whose i, jth element is

$$E\left[\frac{\partial}{\partial\theta_i} \log p_\theta(X_k) \frac{\partial}{\partial\theta_j} \log p_\theta(X_k) \right].$$

The vector $\sqrt{n}(\hat{\theta}_n - \theta)$ is then asymptotically normally distributed with mean zero and covariance matrix $I^{-1}(\theta)$.

5. The situation is quite different when the number of parameters tends to infinity with n. If, for example, the X_{ij}'s $(j = 1, \ldots, n; \; i = 1, \ldots, r)$ are independently normally distributed with mean ξ_i and common variance σ^2, then as $r \to \infty$, the MLE of σ^2, although corresponding to the unique root of the likelihood equations, is not consistent ([22]; see also Kiefer and Wolfowitz [20]).

Solutions of the likelihood equation are not the only asymptotically efficient estimators. General classes of asymptotically efficient estimators are the Bayes estimators, discussed in a later section, and the BAN (best asymptotically normal) estimators of Neyman (see Wijsman [35]). Related efficiency results not requiring the restriction to asymptotically normal estimators hold for the maximum probability estimators of Weiss and Wolfowitz [34] and the asymptotically median unbiased estimators considered by Pfanzagl [24].

SMALL-SAMPLE THEORY; SUFFICIENT STATISTICS

The method of the preceding section is asymptotically efficient, i.e., for large samples not much improvement is possible. However, nothing is said about how large the sample size n has to be for this to be the case, and the result provides no information about the performance of the estimators for fixed n. How to measure this performance is less clearcut for fixed n than it is asymptotically. In the limit (as $n \to \infty$), the estimators of the preceding section have a distribution with common shape (normal), so that their accuracy is described completely by the variance of the limiting distribution. For fixed n, on the other hand, different estimators have distributions of widely varying shapes, which

makes it more difficult to describe their accuracy and to make comparisons.

The accuracy of an estimator δ of a scalar quantity $g(\theta)$ can be measured, for example, by its expected squared or absolute error, or by the probability $P(|\delta(X) - g(\theta)| \leq a)$ for some a. As was pointed out by Gauss, the choice is fairly arbitrary. For reasons of mathematical convenience the most common choice is expected squared error. A somewhat different attitude is taken in *decision theory**, where it is assumed that the consequences of an estimated value d when the true value is θ can be measured by an ascertainable *loss function* $L(\theta, d)$. The expected loss $EL(\theta, \delta(X))$, for varying θ, is then called the *risk function* of δ and is used as a yardstick of its performance.

In these terms, the problem is that of the choice of a "best" estimator (see the following two sections) as one that minimizes the risk. This problem can often be greatly simplified by first removing from the data those parts that carry no information about θ. A statistic T is said to be *sufficient** if the conditional distribution of X given T is independent of θ. For any estimator $\delta(X)$ it is then possible to construct an estimator depending only on T which has the same distribution, and hence for any L the same risk function, as $\delta(X)$; *see* SUFFICIENCY.

This assertion applies quite generally to any decision problem. However, for estimation problems a stronger statement is possible when the loss function is a strictly convex function of d, e.g., when it is squared error. Then for any estimator $\delta(X)$ that is not already a function of T, there exists another estimator, namely the conditional expectation of $\delta(X)$ given T, which depends only on T and which for all θ has a *smaller* risk than $\delta(X)$. This result is known as the Rao–Blackwell theorem*.

In part (b) of Example 1, the statistics $[\bar{X} = n^{-1}\sum X_i, \; \sum(X_i - \bar{X})^2]$ together are sufficient for ξ and σ^2; in (c), $\max(X_1, \ldots, X_n)$ is sufficient for θ. Such far-reaching reductions are not always possible. For example, if $X_1, \ldots, X_n$ are i.i.d. according to a Cauchy distribution* centered at θ, sufficiency only reduces the sample $X, \ldots, X_n$ to the ordered sample $X_{(1)} < \cdots < X_{(n)}$.

UNBIASEDNESS AND EQUIVARIANCE

Suppose now that the accuracy of an estimator $\delta = \delta(X_1, \ldots, X_n)$ of the scalar quantity $g(\theta)$ is measured by the expected squared error

$$R(\theta, \delta) = E[\delta - g(\theta)]^2, \qquad (2)$$

the risk function of δ. Much of the theory of estimation can be extended to other risk functions but for the sake of simplicity attention will be restricted here to (2).

A best estimator would be one that minimizes (2) for all θ. However, for any given value θ_0 of θ, it is always possible to reduce (2) to zero by putting $\delta \equiv g(\theta_0)$, and therefore no uniformly best estimator exists. This difficulty can be avoided by imposing on the estimators some condition that forces them to a certain degree of impartiality with respect to the different possible values of θ. Two such conditions are unbiasedness and equivariance.

Unbiasedness*

An estimator is *unbiased* for estimating $g(\theta)$ if $E(\delta) = g(\theta)$ for all θ, so that on the average δ will estimate the right value. For unbiased estimators (2) reduces to the variance of δ. For a large class of problems it turns out that among all unbiased estimators there exists one that uniformly minimizes the variance and which is therefore UMVU* (uniformly minimum variance unbiased). The most important situation for which this is the case is that of an *exponential family** of distributions in which $p_\theta(x)$ is of the form

$$p_\theta(x) = e^{\sum_{i=1}^{k} \eta_i(\theta) T_i(x) - B(\theta)} h(x). \qquad (3)$$

For such a family any $g(\theta)$ for which an unbiased estimator exists has a UMVU estimator provided that, as θ varies, the set of points $(\eta_1(\theta), \ldots, \eta_k(\theta))$ contains a k-dimensional rectangle. The UMVU estima-

tor is in fact the unique unbiased estimator which depends on x only through $(T_1(x), \ldots, T_k(x))$.

The families of parts (a) and (b) in Example 1 are both exponential families. In the binomial case, $k = 1$, $T_1(x) = x$, and hence X/n is the UMVU estimator of p. More generally,

$$\frac{X(X - 1) \cdots (X - r + 1)}{n(n - 1) \cdots (n - r + 1)}$$

is UMVU for p^r $(r \leqslant n)$. In the normal case, $k = 2$ and one can put $T_1(x) = \bar{x}$ and $T_2(x) = \sum(x_i - \bar{x})^2$. It follows that $\bar{X}$ and $\sum(X_i - \bar{X})^2/(n - 1)$ are UMVU for ξ and σ^2, respectively.

In general, every $g(\theta)$ for which an unbiased estimator exists has a UMVU estimator if and only if the given family of distributions has a *sufficient statistic* that is *complete**. This condition is satisfied by the family (3) under the additional condition stated there. It also holds for the family of part (c) in Example 1. On the other hand, if the X's are i.i.d. according to the uniform distributions on $(\theta - \frac{1}{2}, \theta + \frac{1}{2})$, no complete sufficient statistic exists, and no UMVU estimator exists for θ, although θ has many unbiased estimators.

Unbiasedness is an attractive property since it ensures a correct average value in the long run. However, it is no cure-all. UMVU estimators or even any unbiased estimator may not exist for a given $g(\theta)$. When δ is UMVU, it need not be a desirable estimator. It may, for example, take on values outside the range of $g(\theta)$, or it may depend too strongly on the assumptions of the model. Even when the model is trusted, there may exist an alternative estimator with a small bias but with much smaller risk. It is therefore important to consider other approaches also.

Equivariance

To illustrate this concept, suppose that in the measurement problem of the first section, θ is a temperature that is measured by taking

n readings $X_1, \ldots, X_n$ on a thermometer calibrated on the Celsius scale (C). It has been decided to estimate θ by means of the estimator $\delta(X_1, \ldots, X_n)$. Another worker is used to the Kelvin scale (K = C + 273), and therefore converts the readings into $X_i' = X_i + 273$ to estimate $\theta' = \theta + 273$ by means of $\delta(X_1', \ldots, X_n')$. It seems desirable that the resulting estimate of θ' should be just 273 more than the original estimate of θ, i.e., that for $a = 273$,

$$\delta(x_1 + a, \ldots, x_n + a) = \delta(x_1, \ldots, x_n) + a. \tag{4}$$

An analogous argument suggests that in fact (4) should hold for all a. An estimator δ satisfying (4) for all $x_1, \ldots, x_n$ and all a is called *equivariant* or by some authors *invariant* under translations. (*See* EQUIVARIANT ESTIMATORS).

Condition (4) is satisfied by the mean and median, and by many other estimators to be considered later in the section "Admissibility and Stein Estimation." The class of all equivariant estimators can be characterized as follows. Let δ_0 be any equivariant estimator, for example the mean. Then δ is equivariant if and only if

$$\delta = \delta_0 + h(y_1, \ldots, y_{n-1}), \tag{5}$$

where $y_i = x_i - x_n$, and h is arbitrary.

That (4) is successful as an impartiality requirement is seen by the fact that for any equivariant estimator, the risk (2) is equal to a constant independent of θ, i.e., an equivariant estimator estimates all value of θ with equal accuracy. The best such estimator is the one for which this constant risk is smallest. It is known as the *Pitman estimator** of θ and is given by

$$\delta(x_1, \ldots, x_n)$$
$$= \frac{\int \theta f(x_1 - \theta) \cdots f(x_n - \theta) \, d\theta}{\int f(x_1 - \theta) \cdots f(x_n - \theta) \, d\theta}. \tag{6}$$

An analogous approach applies in the case that θ is a scale parameter, or that both location and scale are unknown [25] and in a number of other cases.

BAYES AND MINIMAX ESTIMATORS

Instead of restricting the class of estimators in the hope of finding a uniformly minimum rish estimator in the restricted class, one can undertake, without imposing restrictions, the less ambitious task of minimizing some overall aspect of the risk function. Two natural such criteria are the average risk and the maximum risk.

Bayes' Estimation

Consider the weighted average risk

$$\int R(\theta, \delta) w(\theta) \, d\theta, \qquad \int w(\theta) \, d\theta = 1 \tag{7}$$

for a given weight function w, and with R given by (2). The estimator minimizing (7) is

$$\delta_w(x) = \int \theta w(\theta) p_\theta(x) \, d\theta \bigg/ \int w(\theta) p_\theta(x) \, d\theta. \tag{8}$$

As mentioned at the end of the section "Maximum Likelihood," the estimators (8) have a remarkable large-sample property. If $w(\theta) > 0$ for all θ and under suitable regularity conditions, the estimator δ_w is asymptotically efficient. In fact, if $\hat{\theta}_n$ denotes the consistent root of the likelihood equation, then $\sqrt{n}(\hat{\theta}_n - \delta_w)$ tends to zero in probability as $n \to \infty$. The first who appears to have realized this was Laplace. For a rigorous treatment and some history, see LeCam [21].

The estimators δ_w can be given an interesting alternative interpretation. Suppose that θ itself is a random variable, unobservable but distributed with known probability density $w(\theta)$. Then (7) represents the overall average value of the squared error when the variation of both X and θ is taken into account, and δ_w is the *Bayes estimator*, which minimizes this average. In this context, δ_w is the expected value of θ given x. Interpreting w as the *prior distribution** of θ (before the data are in), the conditional distribution of θ given x is the *posterior distribution** of θ (the prior distribution of θ modified in the light of the data) and its expectation is the best estimate of the unobservable θ. As $n \to \infty$, the influence of the prior distribution becomes weaker and in the limit it disappears. This accords with the asymtotic efficiency result mentioned above, which shows that asymptotically δ_w is independent of w.

The principal problem in the Bayes approach is the assessment and interpretation of the prior distribution. For a discussion of these issues, *see* BAYESIAN INFERENCE. Bayes estimators play a role not only by being asymptotically efficient and as solutions of the estimation problem under the assumption of a known prior distribution for θ, but also as a fundamental tool in decision theory*.

Minimax Estimation*

Instead of minimizing the average risk (7), one may wish to minimize the maximum risk. The resulting *minimax* estimator provides the best protection even under the worst circumstances.

The principal method for deriving minimax estimators is based on a result in Wald's decision theory, which in turn is adapted from the Fundamental Theorem of game theory*. The theorem says roughly that the minimax estimator is a Bayes estimator with respect to that prior distribution (called *least favorable**) for which the Bayes risk $\int R(\theta, \delta_w)w(\theta)\,d\theta$ is largest. (This is the prior distribution which makes the Bayesian statistician's task as hard as possible.) A simple sufficient condition for a Bayes estimator to be minimax and w least favorable is that the risk of δ_w is constant, independent of θ.

As stated, the method is not quite general enough to cover many of the most interesting cases. Consider, for instance, the location problem of the section on equivariance. The Pitman estimator is a constant risk estimator which has the form (8) of a Bayes estimator, but with a weight function $w(\theta) \equiv 1$, which is not a probability density since its integral is infinite. The theory can, however, be extended to cover the case of such "improper" priors. In this way one can show, for example, that the Pitman estima-

tor (6) is minimax [6]. For generalizations, see Kiefer [19].

Minimax estimators, by concentrating on the maximum possible risk, sometimes view the situation in too gloomy a light. In part (a) of Example 1, for instance, it turns out that the minimax estimator for p is

$$\delta(X) = \frac{X}{n} \cdot \frac{\sqrt{n}}{1 + \sqrt{n}} + \frac{1}{2(1 + \sqrt{n})},$$

which has constant risk $\frac{1}{4}(1 + \sqrt{n})^{-2}$. A comparison with the standard estimator X/n, whose risk is $p(1 - p)/n$, shows that for large n, X/n is better than δ except in an interval about $\frac{1}{2}$ whose length tends to zero as $n \to \infty$, and that the ratio of the maxima of the two risk functions $\to 1$ as $n \to \infty$. In the limit, which is never reached, X/n would be uniformly better. This can be summarized by saying that X/n is asymptotically subminimax.

ADMISSIBILITY* AND STEIN ESTIMATION

An estimator δ of $g(\theta)$ is *inadmissible* if there exists a better estimator δ', i.e., one that dominates δ in the sense that $R(\theta, \delta') \leqslant R(\theta, \delta)$ for all θ with strict inequality holding for some θ. An estimator is *admissible* if it is not inadmissible. Admissibility is a very weak condition and constitutes a minimal requirement that one might impose on an estimator.

If $a \leqslant g(\theta) \leqslant b$ for all θ, a simple necessary condition for admissibility of δ is that $a \leqslant \delta(X) \leqslant b$ with probability 1. Otherwise, δ can be improved by replacing it by a or b when $\delta(X)$ is $< a$ of $> b$, respectively. A simple sufficient condition for admissibility is that δ is the unique Bayes estimator for some prior distribution.

Example 2. Suppose that $X_1, \ldots, X_n$ are i.i.d. according to the normal distribution with mean θ and unit variance and with parameter space (a) $-\infty < \theta < \infty$; (b) $0 \leqslant \theta$; (c) $-a \leqslant \theta \leqslant a$ $(a < \infty)$. Then it turns out that $\overline{X}$ is minimax and admissible for (a); minimax but inadmissible for (b); and neither minimax nor admissible for (c).

A body of new statistical methods is growing up around a surprising discovery by Stein [30]. Let $X = (X_1, \ldots, X_s)$ be independently normally distributed with mean $\theta = (\theta_1, \ldots, \theta_s)$ and unit variance. For the problem of estimating θ with risk $\sum R(\theta_i, \delta_i)/s$, where δ_i is the estimator of θ_i, Stein found that X is inadmissible when $s \geq 3$. [It is admissible for s = 1 by part (a) of Example 2, and is also admissible for s = 2.] A typical example of an estimator dominating X (when $s \geq 4$) is given by

$$\delta_i = \overline{X} + \left(1 - \frac{s-3}{S}\right)(X_i - \overline{X}), \quad (9)$$

where $S = \sum(X_i - \overline{X})^2$. This estimator is obtained from X by shrinking it toward the point $(\overline{X}, \ldots, \overline{X})$. Its risk is substantially smaller than that of X when $\theta_1 = \cdots = \theta_s$. The improvement decreases as the parameter point moves away from the line $\theta_1 = \cdots = \theta_s$, and tends to zero as $\sum(\theta_i - \overline{\theta})^2 \to \infty$. See STEIN ESTIMATORS.

Such "shrinkage" estimators dominate the standard estimators of vector-valued parameters in many situations. There is, however, a price for the improvement in the average risk of the s component estimation problems: namely, a substantial increase in the largest of the maximum risks of the component problems. Whether a shrinkage estimator is appropriate depends, therefore, at least in part, on how important it is to protect these individual risk components.

For a discussion of various aspects of Stein estimation and a guide to the literature, see Efron and Morris [10] and Brown [7]. Other shrinkage methods, motivated by somewhat different considerations, are discussed in RIDGE REGRESSION.

ROBUST ESTIMATION

Consider once more the measurement problem of the first section. If the X's are i.i.d. according to a normal distribution with mean θ and variance σ^2, the sample mean $\overline{X}$ is asymptotically efficient. For any fixed n, it is UMVU, best equivariant, minimax, and admissible, Suppose, however, that the assumption of normality is erroneous and that the X's come, in fact, from a Cauchy distribution. Then $\overline{X}$ has the same distribution as a single observation X_1. Thus its accuracy does not improve with n and it is not even consistent. Typically, neither the normal nor the Cauchy distribution is realistic. Outlying observations tend to occur not as frequently as predicted by Cauchy; however, *gross errors* resulting from misreading, misrecording, and other exigencies of experimental reality tend to vitiate the assumption of normality. The mean and other least-squares estimators are sensitive to such outliers, so that it is desirable to look for more robust alternatives.

An extremely robust estimator of θ is the median, which discards all of the ordered observations except the central one(s). To compare it with the mean, or to compare the performance of any two asymptotically normal estimators, consider their *asymptotic relative efficiency** (ARE). This is the reciprocal of the ratio of the numbers of observations required by the two estimators to obtain the same asymptotic variance, which turns out to be the reciprocal of the ratio of their asymptotic variances. The ARE of median to mean is ∞ when the X's are Cauchy and is $2/\pi \sim 0.64$ in the normal case. The latter is rather low and suggests that the median perhaps goes too far in discarding observations. A compromise between mean and median is the α-trimmed mean* $\overline{X}_\alpha$ $(0 < \alpha < \frac{1}{2})$, which is the average of the observations remaining after the upper and lower $100\alpha\%$ of the observations have been discarded. (The mean and median correspond to the limiting cases $\alpha = 0$ and $\alpha = 0.5$, respectively.)

To compare these estimators, consider distributions intermediate to normal* and Cauchy*, for example the t-distributions* with ν degrees of freedom (t_1 = Cauchy, $t_\nu \to$ normal as $\nu \to \infty$). The ARE of $\overline{X}_\alpha$ to $\overline{X}$ for several values of α and ν is shown in Table 1.

Table 1   ARE of $\overline{X}_\alpha$ to $\overline{X}$ for t-Distribution with ν Degrees of Freedom

ν	α 0.05	0.125	0.25	0.375	0.5
3	1.70	1.91	1.97	1.85	1.62
5	1.20	1.24	1.21	1.10	0.96
∞	0.99	0.94	0.84	0.74	0.64

This table supports the idea that a moderate amount of trimming can provide much better protection than $\overline{X}$ against fairly heavy tails, as represented for example by t_3, while at the same time giving up little in the normal case. The table suggests that $\overline{X}_{0.125}$, for example, does quite well for both t_3 and t_5 with little efficiency loss at the normal.

A large number of other robust estimators have been proposed. Particularly suited to generalizations beyond the location problem are the *M*-estimators* of Huber. An important tool in the comparison of different robust estimators is the *influence curve** of Hampel. A comparison of many alternative estimators under a variety of different distributions is reported in Andrews et al. [1]. Recent surveys of the field stressing different aspects of the problem are provided by Bickel [4], Huber [16], and Hogg [15].

The efficiencies considered above have been relative to a fixed estimator such as the mean. One can also consider the *absolute asymptotic efficiency* (AAE) of an estimator δ, i.e., for each density f the ARE of δ relative to an estimator which is asymptotically efficient for that f. By definition, the AAE is always $\leqslant 1$. Surprisingly, it is possible to construct estimators for which it equals 1 for symmetric f satisfying certain regularity conditions. (Theoretically, this can be done, for example, by using part of the sample, say $\sqrt{n}$ observations, to estimate f, and then using an estimator of θ which is best for the estimated f.) The idea of such estimators was first proposed by Stein [30] and has since led to a large literature on *adaptive** procedures, which use the observations to adapt themselves to the unknown distribution or other unknown aspects of the model. A survey of such procedures is given by Hogg [14]; see also Beran [3].

Literature

Many general books on statistical inference include a discussion of estimation theory, among them those of Bickel and Doksum [5], Cox and Hinkley [8], Ferguson [12], Kendall and Stuart [18], Rao [29], and Zacks [36]. A book-length treatment is Wasan [33]. Specific estimators of the parameters of the most important parametric models are discussed in Johnson and Kotz [17].

References

[1] Andrews, D. F., Bickel, P. J., Hampel, F. R., Huber, P. J., Rogers, W. H., and Tukey, J. W. (1972). *Robust Estimates of Location: Survey and Advances*. Princeton University Press, Princeton, N.J.

[2] Bahadur, R. R. (1964). *Ann. Math. Statist.*, **35**, 1545–1552.

[3] Beran, R. (1978). *Ann. Statist.*, **6**, 292–313.

[4] Bickel, P. J. (1976). *Scand. J. Statist.*, **3**, 145–168.

[5] Bickel, P. J. and Doksum, K. A. (1977). *Mathematical Statistics: Basic Ideas and Selected Topics*. Holden-Day, San Francisco.

[6] Blackwell, D. and Girshick, M. A. (1954). *Theory of Games and Statistical Decisions*. Wiley, New York.

[7] Brown, L. D. (1975). *J. Amer. Statist. Ass.*, **70**, 417–427.

[8] Cox, D. R. and Hinkley, D. V. (1974). *Theoretical Statistics*. Chapman & Hall, London.

[9] Cramér, H. (1946). *Mathematical Methods of Statistics*. Princeton University Press, Princeton, N.J.

[10] Efron, B. and Morris, C. (1975). *J. Amer. Statist. Ass.*, **70**, 311–319.

[11] Eisenhart, C. (1964). *J. Wash. Acad. Sci.*, **54**, 24–33.

[12] Ferguson, T. S. (1967). *Mathematical Statistics*. Academic Press, New York.

[13] Fisher, R. A. (1922). *Philos. Trans. R. Soc. Lond. A*, **222**, 309–368.

[14] Hogg, R. V. (1974). *J. Amer. Statist. Ass.*, **69**, 909–923.

[15] Hogg, R. V. (1979). *Amer. Statist.*, **33**, 108–115.

[16] Huber, P. J. (1977). *Robust Statistical Procedures*, SIAM, Philadelphia.

[17] Johnson, N. L. and Kotz, S. (1969–1972). *Distributions in Statistics*, 4 vols. Wiley, New York.

[18] Kendall, M. G. and Stuart, A. (1979). *The Advanced Theory of Statistics*, Vol. 2, 3rd ed. Charles Griffin, New York.

[19] Kiefer, J. (1957). *Ann. Math. Statist.*, **28**, 573–601.

[20] Kiefer, J. and Wolfowitz, J. (1956). *Ann. Math. Statist.*, **27**, 887–906.

[21] LeCam, L. (1953). *Univ. Calif. Publ. Statist.*, **1**, 277–329.

[22] Neyman, J. and Scott, E. L. (1948). *Econometrica*, **16**, 1–32.

[23] Norden, R. H. (1972). *Rev. Int. Inst. Statist.*, **40**, 329–354; *ibid.* **41**, 39–58 (1973).

[24] Pfanzagl, J. (1974). *Proc. Prague Symp. Asymptotic Statist.*, **1**, 201–272.

[25] Pitman, E. J. G. (1939). *Biometrika*, **30**, 391–421.

[26] Plackett, R. L. (1958). *Biometrika*, **45**, 130–135.

[27] Plackett, R. L. (1972). *Biometrika*, **59**, 239–251.

[28] Pratt, J. W. (1976). *Ann. Statist.*, **4**, 501–514.

[29] Rao, C. R. (1973). *Linear Statistical Inference and Its Applications*, 2nd ed. Wiley, New York.

[30] Stein, C. (1956). *Proc. 3rd Berkeley Symp. Math. Statist. Prob.*, Vol. 1. University of California Press, Berkeley, Calif., pp. 187–196.

[31] Wald, A. (1949). *Ann. Math. Statist.*, **20**, 595–601.

[32] Wald, A. (1950). *Statistical Decision Functions*. Wiley, New York.

[33] Wasan, M. T. (1970). *Parametric Estimation*. McGraw-Hill, New York.

[34] Weiss, L. and Wolfowitz, J. (1974). Maximum Probability Estimators and Related Topics. *Lect. Notes Math.*, **424**.

[35] Wijsman, R. A. (1959). *Ann. Math. Statist.*, **30**, 185–191, 1268–1270.

[36] Zacks, S. (1971). *The Theory of Statistical Inference*. Wiley, New York.

(ADAPTIVE METHODS
ADMISSIBILITY
ARITHMETIC MEAN
BAYESIAN INFERENCE
COMPLETENESS
CONTINGENCY TABLES
DECISION THEORY
DENSITY ESTIMATION
EFFICIENCY
EQUIVARIANT ESTIMATORS
EXPONENTIAL FAMILY
FISHER, R. A.
GAUSS'S CONTRIBUTIONS
   TO STATISTICS
GENERAL LINEAR MODEL
INFLUENCE CURVE
LEAST SQUARES

MARKOV PROCESSES
MAXIMUM LIKELIHOOD
MAXIMUM PROBABILITY
   ESTIMATION
MEDIAN
MINIMAX METHOD
MULTIVARIATE ANALYSIS
NORMAL DISTRIBUTION
OUTLIERS
PITMAN ESTIMATORS
RAO–BLACKWELL THEOREM
ROBUSTNESS
SUFFICIENCY
SUPEREFFICIENCY
SURVEY SAMPLING
TIME SERIES
UNBIASEDNESS)

E. L. LEHMANN

ESTIMATION EFFICIENCY FACTOR
See BLOCKS, BALANCED INCOMPLETE

ETHICAL PROBLEMS   *See* CLINICAL TRIALS; CONSULTING, STATISTICAL; LAW, STATISTICS IN; PRINCIPLES OF PROFESSIONAL STATISTICAL PRACTICE

EULER–MACLAURIN EXPANSION

An approximate expression for the integral of a function $f(x)$ over the range a to $b(= a + nh)$ in terms of the values of $f(x)$ at $x = a, a + h, \ldots, a + nh$ and the derivatives of $f(x)$ at $x = a$ and $x = b$:

$$\int_a^{a+nh} f(x)\,dx = \tfrac{1}{2} f(a) + f(a+h) + \cdots$$
$$+ f(a + (n-1)h) + \tfrac{1}{2} f(b)$$
$$- \tfrac{1}{12} h\big(f'(b) - f'(a)\big)$$
$$+ \tfrac{1}{720} h^3\big(f'''(b) - f'''(a)\big)$$
$$- \cdots .$$

The general term in the expansion is

$$- B_{2r}\big(f^{(2r-1)}(b) - f^{(2r-1)}(a)\big),$$

where B_{2r} is the $(2r)$th Bernoulli number*.

As written, the formula is a quadrature* formula. It can also be used as an approximate summation formula if the integral can not be evaluated explicitly.

Bibliography

Kellison, S. G. (1975). *Fundamentals of Numerical Analysis*, Richard D. Irwin, Homewood, Ill.

Milne-Thomson, L. M. (1933). *Calculus of Finite Differences*. Macmillan, London.

(FINITE DIFFERENCES
QUADRATURE)

EULER POLYNOMIAL

The rth Euler polynomial $E_r(y)$ is defined by

$$2e^{yx}(e^x + 1)^{-1} = \sum_{r=0}^{\infty} E_r(y) \frac{x^r}{r!} \qquad (|x| < \pi).$$

$E_r(0)$ is called the rth Euler number, and written E_r. For r odd, $E_r = 0$.

EULER'S CONJECTURE

That there does not exist a Graeco-Latin square* with dimension which is neither a prime nor a power of a prime—i.e., composite.

The conjecture proved correct for 6×6 squares, but not for squares of any other dimension. It was shown by Bose et al. [1] that there are Graeco-Latin squares of any dimension except 2×2 and 6×6.

Reference

[1] Bose, R. C., Parker, E. T., and Shrikhande, S. S. (1960). *Canad. J. Math.*, **12**, 189–203.

EULER'S CONSTANT

The limit, as n approaches infinity, of $1 + \frac{1}{2} + \cdots + 1/n - \log n$. It is conventionally denoted by γ. The numerical value is $0.577216 \ldots$.

In distribution theory, Euler's constant arises in connection with Type I extreme-value* distributions.

EVENNESS   *See* DIVERSITY INDICES

EVERETT'S CENTRAL DIFFERENCE INTERPOLATION FORMULA

The formula is (with $\phi = 1 - \theta$)

$$u_{x+\theta} = \phi u_x + \theta u_{x+1} - \tfrac{1}{6}\phi(1 - \phi^2)\delta^2 u_x$$
$$- \tfrac{1}{6}\theta(1 - \theta^2)\delta^2 u_{x+1} + \tfrac{1}{120}(1 - \phi^2)$$
$$\times (4 - \phi^2)\delta^4 u_x$$
$$+ \tfrac{1}{120}\theta(1 - \theta^2)(4 - \theta^2)\delta^4 u_{x+1} \cdots.$$

The general formula for terms in central differences* of order $2r$ is

$$\epsilon_{2r}(\phi)\delta^{2r}u_x + \epsilon_{2r}(\theta)\delta^{2r}u_{x-1}$$

with

$$\epsilon_{2r}(\phi) = \frac{1}{(2r + 1)!}\phi\prod_{j=1}^{r}(j^2 - \phi^2).$$

If u_x is a polynomial in x of degree $(2r + 1)$, the value of $u_{x+\theta}$ obtained by stopping at differences of order $2r$ is exact. Thus the formula is correct to order one *greater* than the highest order of central differences used. Greater accuracy is often attainable by using modified differences, as described in THROW-BACK.

Bibliography

Fox, L. (1958). Tables of Everett Interpolation Coefficients. *Natl. Phys. Lab.* (*U.K.*) *Math Tables*, **2**.

(CENTRAL DIFFERENCES
FINITE DIFFERENCES
INTERPOLATION
THROWBACK)

EVOLUTIONARY OPERATION (EVOP)

An efficient industrial process should be run to produce not only a product, but also information on how to improve that product.

This is the basic philosophy of evolutionary operation (EVOP), an operating technique that has been widely and successfully applied to the process industries since the 1950s and has been of value in many types of manufacturing. It is applied to the daily operation of the plant. It needs no special staff and can be handled by the regular plant personnel after a brief training period.

Evolutionary operation is a management tool in which a continuous investigating routine becomes the basic mode of operation and replaces normal static operation. It is not a substitute for fundamental investigation, but does often indicate an area that needs more fundamental study.

The fact that EVOP is conducted in the full-scale plant rather than in the laboratory or on a pilot plant has two advantages. First, we do not need to allow for scale-up effects, so the results have immediate relevance. Often, the modifications that provide these improvements on the plant scale cannot even be simulated on the small scale. Second, little additional cost is involved in making manufacturing runs, which must be made in any case.

When a process is first started up, empirical adjustments are usually needed and, in later routine operation, further modification may occur through chance discovery and sometimes from brief special experimental investigations. Advance in this way is typically very slow, however. The effect of EVOP is greatly to increase the speed at which progress takes place, and frequently to lead to improvements that otherwise would not have occurred.

HOW EVOP WORKS

EVOP institutes a set of rules for normal plant operation so that, without serious danger of loss through manufacture of unsatisfactory material, an evolutionary force is at work which steadily moves the process toward its optimum conditions.

Normally, routine production is conducted by running a plant at a set of rigidly defined operating conditions called the *works process*. The works process embodies the known best conditions of operation. A manufacturing procedure in which the plant operator aims always to reproduce the works process exactly is called static operation. This method of operation, if strictly adhered to, precludes the possibility of evolutionary development. The objectives that static operation sets out to achieve are, nevertheless, essential to successful manufacture, for in practice we are interested not only in the productivity of the process but also in the physical properties of the product that is manufactured. If *arbitrary* deviations from the works process were allowed, these physical properties might fall outside specification limits. In the EVOP method, deviations are made, but not arbitrarily. A carefully planned cycle of minor variants on the works process is agreed upon. The routine of plant operation then consists of running each of the variants in turn, typically in a factorial* pattern like that in Fig. 1. In this particular arrangement, the process operating variables temperature (*A*), concentration (*B*), and pressure (*C*) have been varied, each at two levels, and the measured response is by-product yield, which it is desired to *reduce*. The changes in levels of the

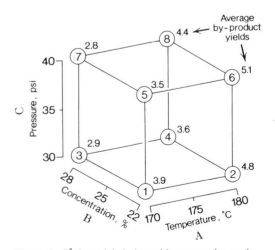

Figure 1 2^3 factorial design with average by-product yields after three cycles.

variables are chosen to be small enough so that production specifications are not likely to be violated. (In this example, the standard operating conditions had been $A = 175°C$, $B = 25\%$, $C = 35$ psi, and the process engineers considered that saleable product would still be produced with operating conditions within the ranges A: 170 to 180°C; B: 22 to 28%; C: 30 to 40 psi.)

Because the changes are deliberately chosen so that they will not affect the response variables very much, the effects produced by the changes are usually not detectable in just one cycle around the design. However, the persistent repetition of the cycle of variants, and the examination of the (less variable) *average* responses thus generated, can show whether, by how much, and in what direction properties of the product change as conditions are changed in the immediate vicinity of the best conditions known so far. In this way, routine manufacture produces not only a product, but also the information needed to improve it. Once favorable differences in response reveal themselves, process conditions may be changed appropriately. In Fig. 1, for example, if we could rely on the reality of the apparent differences shown in the by-product yield averages, we should evidently lower temperature and increase concentration to achieve lower by-product levels.

THE INFORMATION BOARD

The EVOP results must be continuously presented to the process superintendent in a way that he or she can easily understand. This is best done by the use of a prominently displayed information board, kept up to date by a person to whom the duty is specifically assigned. The information board shows the current estimated effects of changing the input variables, and the error limits associated with these estimated effects. Repetition of the cycle reduces the error limits so that, at any given stage, the process superintendent can see what weight of evidence exists for moving the scheme of variants to some new point, or for abandoning the

study of any unproductive process variables in favor of new ones. The superintendent can also see what types of changes are undesirable from the standpoint of producing material of inferior quality. It is also possible to calculate how much the scheme costs to run.

HOW DOES EVOP DIFFER FROM STANDARD EXPERIMENTATION?

A permanent change in the routine of plant operation is very different from the performance of specialized experiments. Experiments last a limited time, during which special facilities can be made available; and some manufacture of substandard material is to be expected and will be allowed for in the budget. Evolutionary operation, however, is virtually a permanent method of running the plant and can, therefore, demand few special facilities and concessions. In particular, this implies that changes in the levels of the variables must cause effects that are expected to be virtually undetectable *in individual runs* and that the techniques used must be simple enough to be run continuously by works personnel under actual operating conditions.

THE EVOP COMMITTEE

Obvious variants for study in a chemical process are the levels of temperature, concentration, pressure, etc., but there would be many less obvious ways in which a manufacturing procedure could be tentatively modified. Frequently, instances of marked improvement are discovered in a process that has been running for many years due to some innovation that had not been considered previously. It will be seen that EVOP differs from the natural evolutionary process in one vital respect. In nature, variants occur spontaneously, but in this artificial evolutionary process, they must actually be introduced. Therefore, to make the artificial evolutionary process effective, a situation must be set up in which useful ideas are continually forthcoming. An atmosphere for the generation of such ideas is perhaps best induced by bringing together, at suitable intervals, a group of people with special, but different, technical backgrounds. In addition to plant personnel, obvious candidates for such a group are, for example, a research person with an intimate knowledge of the chemistry of similar reactions to that being considered, and a chemical engineer with special knowledge of the type of plant in question. The intention should be to have complementary rather than common disciplines represented.

These people should form the nucleus of a small EVOP committee. Meeting perhaps once a month, they help and advise the process superintendent in the performance of EVOP. The major task of such a group is to discuss the implications of current results and to make suggestions for future phases of operation. Their deliberations will frequently lead to the formulation of theories and ideas which, in turn, suggest new leads that can be pursued profitably.

This scientific feedback occurring in EVOP is perhaps its most important aspect. The installation of an EVOP committee ensures that, at regular intervals, the process data are examined and the process is discussed by intelligent technical people having a wide range of knowledge and experience in several different areas. This provides constant scientific analysis of results leading to a flow of new ideas to be incorporated into the investigation. Without this stimulus, EVOP may cease to be effective and fail.

Note that EVOP is run by plant personnel. The fact that specialists join the EVOP committee does not alter this principle. In practice, the time spent by the specialists is perhaps one afternoon a month. The ultimate responsibility for running the EVOP scheme still rests with the process superintendent and not with the specialists, who serve only in an advisory capacity.

The establishment of the EVOP committee completes a practical method of process improvement that requires no special equipment and can be applied to almost any manufacturing process, whether the plant is simple or elaborate. Its mode of operation is shown diagramatically in Fig. 2.

EVOP EDUCATION PROGRAMS

For EVOP to be properly effective, it must be understood, appreciated, and supported by people at various levels in a company. Thus some sort of educational and orientational program is needed so that everyone

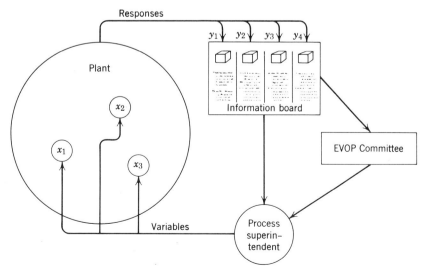

Figure 2 Complete EVOP process represented as a closed feedback loop involving plant, information board, and plant superintendent, guided by the EVOP committee.

can appreciate what EVOP can do, and what is involved. Because different people need different levels of understanding according to their position in the company, three types of education must be available: for (1) higher management; (2) supervisory personnel, e.g., plant engineers, chemists, and superintendents; and (3) plant operators. Each of these three groups has a different role in the EVOP program and, consequently, a different approach is needed for each.

Considerable experience is now available in the successful organization of this educational effort. It has been found that a well-organized presentation lasting, perhaps, 45 minutes is usually sufficient to acquaint higher management with the basic ideas of EVOP. A two- or three-day course with suitable teaching and simulation aids can provide adequate training for supervisory personnel, and about a half-day of instruction is sufficient for the initial training of plant operators, whose goodwill and continuing efforts play an essential role in the successful running of EVOP schemes.

One, perhaps unexpected, bonus of EVOP is the part it has been found to play in acquainting users with some of the more important basic ideas in statistics, such as variation in response variables, limits of error for the effects of the variables, basic rules of good experimental design*, interaction* between variables, blocking, and geometrical representation of response* functions.

Many of the people who run an EVOP program become enthusiastic. They were taught the minimum amount of statistics and design, but they have applied it and have seen it come to life before their eyes. Such people are usually anxious to hear more, and they will have the confidence and motivation to apply what they hear. Many companies have special process development groups, experimental groups, troubleshooters, and operations research* groups. To be useful, these groups must be staffed with people who are imaginative and enterprising. How are people of this kind found? One way

is to seek out those who have conducted successful EVOP programs.

TWO CASE HISTORIES

The case histories below are abstracted from ref. 3.

Decreasing Product Cost per Ton

In a study described by Jenkins [5], the objective was to decrease the cost per ton of a certain product of a petrochemical plant. At one stage of the investigation two variables believed to be important were (1) the reflux ratio of the distillation column, and (2) the ratio of recycle flow to purge flow. It was expected that changes of the magnitude planned would introduce transient effects that would die out in about 6 hours. A further 18 hours of steady operation was then allowed to make the necessary measurements for each run.

The appearance of the information board at the end of each of three phases is shown in Fig. 3. The response was the average cost per ton, recorded only to the nearest unit. The design employed was a 2^2 factorial with a center point. The results at the end of phase I, shown on the left of Fig. 3, are averages obtained after five repetitions (cycles). At this point it was decided that there was sufficient evidence to justify a move to the lower reflux ratio and higher recycle/purge ratio used in phase II. The results from phase II, which was terminated after five further cycles, confirmed that lower costs indeed result from this move and suggested that still higher values of the recycle/purge ratio should be tried, leading to phase III. This phase, brought to an end after four cycles, led to the conclusion that the lowest cost was obtained with the reflux ratio close to 6.3 and the recycle/purge ratio about 8.5, where a cost per ton of about £80 was achieved.

The program described took $4\frac{1}{2}$ months to complete at a cost of about £6000. The savings resulting from reducing the cost per ton

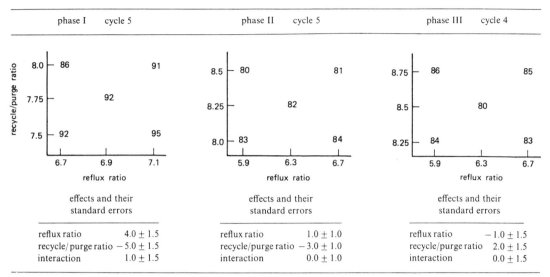

Figure 3 Appearance of the information board at the end of phases I, II, and III, petrochemical plant data. The response is manufacturing cost per ton of product; the objective is to reduce cost.

from £92 to £80 were about £100,000 per year.

Increasing Plant Throughput

The chemical product in this case was a polymer latex. The initial objective was to determine the conditions that, subject to a number of other constraints, gave the *lowest optical densities* for a given batch of raw material. Later, when it was realized that shorter addition times were possible, resulting in great savings because of increased throughput, the objective shifted to *increasing plant throughput* subject to obtaining satisfactory optical densities (less than 40).

Three process variables were studied in this EVOP program: addition time (t) of one of the reactants, temperature (T), and stirring rate (S). Although it would have been possible to use all three variables simultaneously, it was decided, for the sake of simplicity of operation, to study these factors in different combinations two at a time. The data below have been coded for proprietary reasons.

Initially, the standard operating conditions were $t = 3$ hours, $T = 120°$F, $S = 88$ rpm. In phase I, temperature was held at its standard condition, and time and stirring

rate were varied as shown in Fig. 4. The ranges adopted for these two factors were, however, extremely narrow because production personnel feared that substandard product might be produced. Twelve cycles were performed in this first phase, during which a total of $12 \times 5 = 60$ batches were manufactured. It will be seen that even after this long period of production, no effects of any kind were detected. This result, at first disappointing, was actually quite helpful. It convinced those in charge of production that changing the two selected variables over wider ranges would be unlikely to produce substandard material. Wider ranges were therefore allowed in phase II, and after 18 cycles a stirring effect was apparent (1.6 ± 0.7), indicating that lower optical densities might be obtained at lower stirring rates. The engineers in charge, however, did not choose to lower the stirring rate at this stage. For them the most important finding was the unexpected one that no deleterious effect resulted from reducing the addition time. The unit had been believed to be operating at full capacity. A saving of 15 minutes in addition time would result in an 8% increase in throughput and a considerable rise in profitability. To investigate further the possibilities for increased throughput, addition time

optical density reading

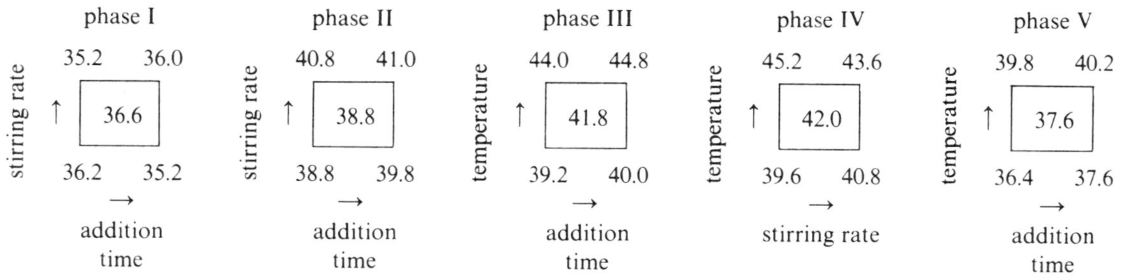

Phase I (stirring rate × addition time)

	addition time →	
35.2		36.0
	36.6	
36.2		35.2

Phase II (stirring rate × addition time)

	addition time →	
40.8		41.0
	38.8	
38.8		39.8

Phase III (temperature × addition time)

	addition time →	
44.0		44.8
	41.8	
39.2		40.0

Phase IV (temperature × stirring rate)

	stirring rate →	
45.2		43.6
	42.0	
39.6		40.8

Phase V (temperature × addition time)

	addition time →	
39.8		40.2
	37.6	
36.4		37.6

settings

	phase I	phase II	phase III	phase IV	phase V
temperature	120	120	118 120 122	116 118 120	114 116 118
stirring rate	83 88 93	78 88 98	88	68 78 88	78
addition time	2:50 3:00 3:10	2:35 3:00 3:25	2:30 2:45 3:00	2:45	2:15 2:30 2:45

effects with standard errors

	phase I	phase II	phase III	phase IV	phase V
temperature			4.8 ± 1.2	4.2 ± 0.7	3.0 ± 0.4
stirring rate	−0.4 ± 0.5	1.6 ± 0.7		−0.2 ± 0.7	0.8 ± 0.4
addition time	0.2 ± 0.5	0.6 ± 0.7	0.8 ± 1.2		−0.4 ± 0.4
interaction	0.6 ± 0.5	−0.4 ± 0.7	0.0 ± 1.2	−1.4 ± 0.7	

Figure 4 Progress of an EVOP study through five phases. The response is the optical density. The initial objective was to reduce optical density, but this was changed after phase II to achieving the lowest cost while maintaining the optical density below 40.

and temperature were varied simultaneously in phase III. Continuing to contradict established belief, the results confirmed that increased throughput did not result in an inferior product and suggested that optical density might be reduced by a reduction in temperature. Phase IV used slightly lower temperatures, while stirring rate was varied about a lower value and addition time was held cautiously at the reduced value of 2 hours 45 minutes (2:45).

In phase V both temperature and addition time were varied about lower values (116°F, 2:30), without deleterious effect, and the stirring rate was held at 78. The results indicated that temperature could be reduced still further.

The net result of the EVOP program up to this point was that throughput was increased by a factor of 25%, stirring rate was reduced from 88 to 78, and temperature was lowered from 120 to 114°F. All of these changes, but especially the first, resulted in economies.

One iteration that occurred in this example involved a change in the objective desired. Originally, improvement in the optical density was sought. What was actually found was the profitable but unexpected discovery that the unit, while yielding a product of about the *same* optical density, could give dramatically increased production.

As exercises, the reader may wish to (1) plot all the data from the first case history on a single grid with reflux ratio and recycle/purge ratio as coordinates, and (2) sketch a three-dimensional grid with coordinates covering appropriate ranges of temperature, stirring rate, and addition time and plot all the Fig. 4 data onto this grid. These figures will give additional insight about the shape of the basic underlying response surface*.

WHY EVOP WORKS

At any fixed set of input conditions, response data come from a statistical distribution with a specific "location" (mean μ) and a specific "spread" (standard deviation σ). The parameters μ and σ are, typically, unknown. A basic result in statistics is that, if taking n observations can be stimulated by n independent drawings from a distribution with mean μ and standard deviation σ, the mean of the n observations can be regarded as from a distribution with mean μ and standard deviation $\sigma/n^{1/2}$. This basic fact is used to make EVOP work. Figure 5a shows distributions of single observations on a response y when a predictor (or input) variable X is changed. As we see from the projection, these distributions overlap to such an extent that it would not be clear from single observations (one from each distribution) which true response level (value of μ) was higher, and which lower. When the distributions of averages are considered, however, as shown in Fig. 5b, the spreads are much narrower and differences can be distinguished. This simple basic principle, combined with the theory of factorial designs*, provides the statistical mechanism for the success of EVOP.

Further Reading

An excellent review of industrial applications of EVOP has been presented by Hunter and Kittrell [4]. This review, which lists 68 references, discusses applications of EVOP in a wide variety of environments. Among these, applications in the chemical industry are most numerous, with special references to uses by the Dow Chemical Company, American Cyanamid Company, Imperial Chemical Industries Limited, and Tennessee Eastman Company. Several applications in the food industry are also mentioned, in particular use by Swift and Company, Canadian Packers Limited, and the A. E. Staley Manufacturing Company. Also discussed are uses of EVOP by the canning industry, and the Maumee Chemical Company in such diverse projects as the production of saccharin, a biocide for sea lampreys, isatoic anhydride, anthranilic acid, and benzotriazole. Other applications are mentioned

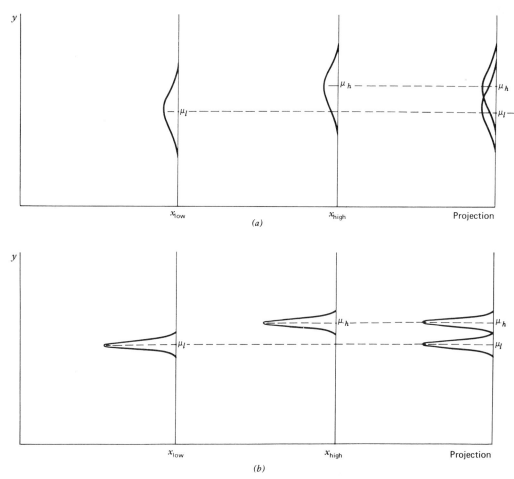

Figure 5 (*a*) Distributions of response observations at two levels of X; with single observations, the different mean values μ_l and μ_h will not be detected. (*b*) Distributions of the means of n observations; with mean observations the different mean values μ_l and μ_h will be detectable if n is large enough.

in the automotive industry. In particular EVOP was applied to resistance welding of automotive sheet metal.

Complete details of EVOP techniques are given in ref. 2. The original paper is ref. 1.

References

[1]   Box, G. E. P. (1957). *Appl. Statist.*, **6**, 81–101.

[2]   Box, G. E. P. and Draper, N. R. (1969). *Evolutionary Operation*. Wiley, New York.

[3]   Box, G. E. P., Hunter, W. G., and Hunter, J. S. (1978). *Statistics for Experimenters*. Wiley, New York.

[4]   Hunter, W. G. and Kittrell, J. R. (1966). *Technometrics*, **8**, 389–397.

[5]   Jenkins, G. M. (1969). *J. Syst. Eng.*, **1**, 90–101.

Acknowledgment

This article is based on a paper previously written by the authors for the journal *Industrial Engineering*. We are grateful to *Industrial Engineering* for permission to reuse this material.

(CHEMISTRY, STATISTICS IN DESIGN OF EXPERIMENTS RESPONSE SURFACES)

GEORGE E. P. BOX
NORMAN R. DRAPER

EVOP   *See* EVOLUTIONARY OPERATION (EVOP)

EXCESSIVE FUNCTION   *See* DYNAMIC PROGRAMMING

EXCHANGEABILITY

HISTORICAL REMARKS

The concept of exchangeability was introduced by de Finetti in his classical work in 1930 [7]. After a long pause in its development, exchangeability has received much attention ever since the 1950s. The most basic work in this new era of development is the one by Hewitt and Savage [12]. See also the monograph by Phelps [19] concerning Choquet's work of 1956 for its relevance to exchangeability.

The first limit theorem appears in the original work of de Finetti (the law of large numbers* contained in Theorem 1). The central limit problem for infinite sequences of exchangeable variables was first treated by Chernoff and Teicher [4], for which a more systematic development is given in the recent book by Chow and Teicher [5]. (Although the title of this book refers to exchangeability, it gives a very limited coverage of the theory. It hardly goes beyond Theorem 2 and the central limit problem.) An extension of the theory of Chernoff and Teicher to a functional limit theorem is given by Billingsley [3, p. 212] and the asymptotic normality* of the sum of a special finite sequence of exchangeable variables is proved by Moran [17].

General results on the asymptotic distribution of extremes for exchangeable variables were first obtained by Berman [2], which results have been extended both to the finite case and to other dependence structures through Theorem 4 by the present author. See Chapter 3 in the book by Galambos [11]. Interesting Poisson limit theorems can be found in Kendall [15] and Riddler-Rowe [21].

Each of Rényi and Révész [20] and Kendall [15] gives a different proof for Theorem

2. The theory of stochastic processes* with exchangeable increments had a significant development; see Kallenberg [14] and Takács [23] and their references. For the zero–one law, see Aldous and Pitman [1] and the references therein.

de Finetti [8] gives an interesting exposition on the role of exchangeability in the foundations of probability theory. This book also provides an extensive list of references.

Current research is very active in the field of extending all aspects of the theory of exchangeable variables to random variables whose values fall into an abstract space. However, we shall not discuss such extensions here.

DEFINITIONS

The concept of exchangeability is best exemplified by two sequences, one of which is associated with selection without replacement from a finite population and the other with the Bayesian interpretation of statistical sampling.

Suppose that an urn contains a fixed number N of balls, out of which M are red and $N-M$ are white. We select n balls ($n \leqslant N$) from the urn at random, one by one and without replacement. Let A_j be the event that the jth ball drawn is red. Then an easy combinatorial argument yields that, for $k \geqslant 1$ and for $1 \leqslant i_1 < i_2 < \cdots < i_k \leqslant n$,

$$P\left(\bigcap_{j=1}^{k} A_{i_j} \right) = \frac{M(M-1)\cdots(M-k+1)}{N(N-1)\cdots(N-k+1)} .$$

(1)

The noteworthy fact in this formula is that the right-hand side does not depend on the actual subscripts i_j but only on the number k of subscripts involved. This property is the basic condition in the following definition of exchangeability of events.

Definition 1. A finite or denumerably infinite sequence $A_1, A_2, \ldots$ of events is called exchangeable if, for any $k \geqslant 1$ and for all relevant indices $1 \leqslant i_1 < i_2 < \cdots < i_k$,

$$p_k = P\left(\bigcap_{j=1}^k A_{i_j} \right) \qquad (2)$$

does not depend on the actual subscripts i_j but only on the number k of terms in the intersection.

Notice that independent events with a common probability of occurrence are also exchangeable. However, as formula (2) shows, exchangeable events are not necessarily independent.

There is a significant difference between the theory of finite and infinite sequences of exchangeable events, which will be seen in the next section.

For introducing the concept of exchangeable random variables, let us recall the "Bayesian" viewpoint of sampling. Let $X_1, X_2, \ldots, X_n$ be observations on a random variable X with distribution function $F(x, \theta)$, where θ is a parameter. The Bayesian statistician assumes that θ is a random variable whose distribution is independent of the X_j. For a given value of θ, the observations $X_1, X_2, \ldots, X_n$ are also assumed to be independent and identically distributed. Thus in such an interpretation, the actual distribution of the sample is

$$P(X_1 \leqslant x_1, X_2 \leqslant x_2, \ldots, X_n \leqslant x_n)$$
$$= E\left[\prod_{j=1}^n F(x_j, \theta) \right] \qquad (3)$$

where the expectation E is with respect to the distribution of θ. Now, for any permutation of $x_1, x_2, \ldots, x_n$, the right-hand side of (3) is unchanged and thus, by letting some x_j tend to infinity, we get that, for $k \geqslant 1$ and for any integers $1 \leqslant i_1 < i_2 < \cdots < i_k \leqslant n$, the joint distribution of the vector $(X_{i_1}, X_{i_2}, \ldots, X_{i_k})$ does not depend on the actual subscripts i_j but only on the dimension k. Random variables with this distributional property are called "exchangeable." This is stated in a more formal way as:

Definition 2. A finite or denumerably infinite sequence $X_1, X_2, \ldots$ of random vari-

ables is called exchangeable if, for any $k \geqslant 1$ and for all relevant indices $1 \leqslant i_1 < i_2 < \cdots < i_k$, the distribution of the vector $(X_{i_1}, X_{i_2}, \ldots, X_{i_k})$ does not depend on the actual subscripts $i_1, i_2, \ldots, i_k$, but only on the dimension k.

Notice that for exchangeable random variables $X_1, X_2, \ldots$ the events $A_j = A_j(x) = \{X_j \leqslant x\}$ are exchangeable for each real x. Hence remarks on exchangeable events apply to exchangeable random variables as well.

EXISTENCE THEOREMS AND REPRESENTATIONS

Let us first discuss the case of infinite sequences of events A_j. Let I_j be the indicator variable of A_j; i.e., $I_j = 1$ or 0 according as A_j occurs or fails to occur. Now, if

$$Y_n = (1/n) \sum_{j=1}^n I_j,$$

then, for $m > n$,

$$E\left[(Y_n - Y_m)^2 \right] = \frac{m-n}{mn}(p_1 - p_2)$$
$$< (1/n)(p_1 - p_2),$$

where p_1 and p_2 are defined at (2) (for detailed calculations, see Galambos [10, p. 103]. This last inequality implies that the sequence Y_n is a Cauchy sequence in the set of random variables with finite variance. A basic theorem of functional analysis therefore yields (see Loève, [16, p. 161] that there is a random variable U with finite variance for which the expectation of $(Y_n - U)^2$ converges to zero. From this fact, on the other hand, it follows that, with probability 1, $0 \leqslant U \leqslant 1$, and thus all moments of Y_n converge to the corresponding moments of U. But by the definition of Y_n and p_k [see (2)], it easily follows that, for each fixed $k \geq 1$, as $n \to +\infty$, $E(Y_n^k)$ converges to p_k. Thus we got the following representation, due to de Finetti [7].

Theorem 1. For an infinite sequence A_1, $A_2, \ldots$ of exchangeable events, there is a random variable U such that $P(0 \leqslant U \leqslant 1)$ $= 1$, $E[(Y_n - U)^2] \to 0$ as $n \to +\infty$, and, for each $k > 1$,

$$p_k = E(U^k), \qquad (4)$$

where p_k is defined at (2).

The converse of Theorem 1 is also true. If U is a random variable with $P(0 \leqslant U \leqslant 1) = 1$, then there is an infinite sequence A_1, A_2, ... of exchangeable events such that (4) is valid for the numbers defined at (2). One can simply construct such a sequence A_1, A_2, ... by constructing conditionally independent events $A_1, A_2, \ldots$, given $U = x$, which satisfy $P(A_j \mid U = x) = x$ (almost surely).

The construction described in the preceding paragraph is not just a special example. It can be deduced from Theorem 1 that the events A_1, A_2, ... are always conditionally independent given U. Let us give an outline of proof of this fact in the more general case of exchangeable random variables X_1, $X_2, \ldots$. We apply Theorem 1 with the events $A_j = A_j(x) = \{X_j \leqslant x\}$, where x is an arbitrary real number. The corresponding random variables $Y_n = Y_n(x)$ and $U = U(x)$ also depend on x. Now, Theorem 1, through the Chebyshev inequality*, yields that $Y_n(x)$ converges in probability to $U(x)$. Therefore, for any fixed t and for $x_1, x_2, \ldots, x_t$, on the one hand, the boundedness of $Y_n(x)$ and $U(x)$ imply that

$$E\big[Y_n(x_1) Y_n(x_2) \cdots Y_n(x_t) I_B \big]$$
$$\to E\big[U(x_1) U(x_2) \cdots U(x_t) I_B \big],$$

while, on the other hand, the invariance under the permutation of the x_j leads to the limit

$$E\big[Y_n(x_1) Y_n(x_2) \cdots Y_n(x_t) I_B \big]$$
$$\to P\big(X_1 \leqslant x_1, X_2 \leqslant x_2, \ldots, X_t \leqslant x_t, I_B \big)$$

where I_B is the indicator variable of an event B that is measurable with respect to the smallest sigma-(σ-)field $\mathscr{F}$ on which each $U(x)$, x real, is measurable. These last two limits, and the definition of conditional dis-

tributions*, give that

$$P(X_1 \leqslant x_1, X_2 \leqslant x_2, \ldots, X_t \leqslant x_t \mid \mathscr{F})$$
$$= U(x_1) U(x_2) \cdots U(x_t)$$

almost surely. By a standard measure-theoretic argument one can guarantee that $U(x)$ is almost surely a distribution function in x, and thus, by taking expectation, we arrive at a formula similar to (3). We thus get

Theorem 2. An infinite sequence X_1, $X_2, \ldots$ of random variables is exchangeable if, and only if, there is a sigma field $\mathscr{F}$ such that, given $\mathscr{F}$, $X_1, X_2, \ldots$ are almost surely independent and identically distributed. Consequently, the n-dimensional distribution of the X_j can always be expressed as at (3).

Since Theorem 2 is a direct consequence of Theorem 1, it is frequently quoted in the literature as the de Finetti theorem. However, its actual proof appeared in the literature only much later than de Finetti's original work. The most convenient reference to a proof is Loève [16, p. 365]. See also Olshen [18] concerning the possible forms of the σ-field $\mathscr{F}$.

Before turning to the discussion of finite sequences, let us establish that, for infinite sequences of exchangeable events A_j, $p_2 \geq p_1^2$ [see (2)]. Namely, if $f_n(A)$ denotes the number of those A_j, $1 \leqslant j \leqslant n$, which occur, then

$$0 \leqslant V[f_n(A)] = 2S_2 - S_1^2 + S_1$$
$$= n(n-1)p_2 - n^2 p_1^2 + np_1,$$

where

$$S_1 = \sum_{j=1}^{n} P(A_j) = np_1 \qquad \text{and}$$

$$S_2 = \sum_{j=2}^{n} \sum_{i=1}^{j-1} P(A_i \cap A_j) = \frac{1}{2} n(n-1)p_2.$$

Hence, $n^2 p_1^2 - np_1 \leqslant n(n-1)p_2$, from which the claimed inequality follows upon dividing by n^2 and letting $n \to +\infty$.

Now, if $A_1, A_2, \ldots, A_n$ are exchangeable and $p_2 < p_1^2$, then these events cannot constitute a finite segment of an infinite sequence of exchangeable events. Consequently, the representation (4) is not applicable in such cases. Since the probabilities at (1) indeed satisfy $p_2 < p_1^2$, we got that Theorems 1 and 2 cannot apply to finite sequences of exchangeable variables.

For the finite case, Kendall [15] obtained the following representation.

Theorem 3. Let $A_1, A_2, \ldots, A_n$ be exchangeable events. Then there is an integer-valued random variable V such that $P(1 \leqslant V \leqslant N) = 1$ for some N and (1) applies to $P(\cap_{j=1}^{k} A_{i_j} \mid V = M)$ for $k \leqslant M$, $1 \leqslant M \leqslant N$. Hence

$$ p_k = E\left\{ \frac{V(V-1) \cdots (V-k+1)}{N(N-1) \cdots (N-k+1)} \right\}. $$

(5)

By the classical method of inclusion-exclusion*, Theorems 1 and 3 lead to the following

Corollary. Let $A_1, A_2, \ldots, A_n$ be exchangeable events. Let $f_n(A)$ denote the number of those A_j which occur. Then

$$ P(f_n(A) = t) = E\left\{ \binom{V}{t}\binom{N-V}{n-t} \bigg/ \binom{N}{n} \right\}, $$

(6)

where V and N are as in Theorem 3. Furthermore, if the A_j can be extended into an infinite sequence of exchangeable events, then the representation

$$ P(f_n(A) = t) = E\left\{ \binom{n}{t} U^t (1-U)^{n-t} \right\} $$

(7)

also applies, where U is as in Theorem 1.

Assume now that $A_1, A_2, \ldots, A_n, A_{n+1}, \ldots, A_m$ are exchangeable. Even though it is not known that the A_j can be extended into an infinite sequence of exchangeable events, the representations at (6) and (7) are close uniformly in t if n/m is small. An actual estimate for the "distance" between (6) and

(7) is given by Diaconis and Freedman [6] in their Theorem (3), which makes the claim of the preceding sentence explicit; see also Riddler-Rowe [21] and Stam [22] in this connection.

As it turns out, (6) is not a characteristic property of exchangeability. Galambos [9] proved that (6) applies whatever be the dependence structure of the A_j. As a matter of fact, the following result was obtained by Galambos.

Theorem 4. For an arbitrary sequence $A_1, A_2, \ldots, A_n$ of events, there are exchangeable events $C_1, C_2, \ldots, C_n$ such that

$$ P(f_n(A) = t) = P(f_n(C) = t), \qquad 0 \leqslant t \leqslant n. $$

Consequently, (6) applies.

This result has the interesting philosophical implication that a probability theory based on exchangeability alone leads to the same limiting properties for frequencies as a theory permitting arbitrary dependence. Theorem 4 also shows that exchangeability has a prominent role in the theory of "order statistics"* for dependent random variables $X_1, X_2, \ldots, X_n$, since the distribution of "order statistics" reduces to the distribution of $f_n(A)$ through the special sequence $A_j = \{ X_j \leqslant x \}$. For the case of extremes, see Chapter 3 in the book by Galambos [11].

The proof of Theorem 4 is based on the following nonprobabilistic characterization of the numbers p_k for exchangeable events, which result is also due to Kendall [15].

Theorem 5. A sequence $1 = p_0 \geqslant p_1 \geqslant \cdots \geqslant p_n \geqslant 0$ can be associated as at (2) with a set of n exchangeable events if and only if $\delta^r p_{n-r} \geqslant 0$, $0 \leqslant r \leqslant n$, and

$$ \sum_{r=0}^{n} \binom{n}{r} \delta^r p_{n-r} = 1, $$

where $\delta = 1 - D$ and $Dp_j = p_{j+1}$.

APPLICATIONS

We have mentioned in the second section that in "Bayesian statistical inference"* the observations on a random variable are usually considered exchangeable. Hence the theory of exchangeable variables is directly applicable in "Bayesian statistics." In addition to the other two areas mentioned earlier (foundations of probability theory and the asymptotic theory of "order statistics" for dependent random variables), we mention two more areas of actual or potential applications.

Several probabilistic and statistical problems can be formulated in terms of urn models, which in turn are closely related to exchangeable sequences of events. The most notable is the "Pólya urn model"* and several of its extensions which have a large variety of applications. In this model, the events of drawing a ball of a given color at any stage are exchangeable; hence limit theorems* for exchangeable events would immediately imply several classical limit theorems for this model. However, most known proofs of results for urn models* avoid the theory of exchangeable events mainly due to the classical nature of these problems. For a large variety of results for urn models, see Johnson and Kotz [13], who also discuss the relation of some models to exchangeable events.

Another area where exchangeability can fruitfully be applied is "nonparametric inference"*. Namely, the "ranks"* of observations from a population with a continuous distribution form an exchangeable sequence. For a more detailed discussion, see the entry NONPARAMETRIC INFERENCE.

Bibliography

Koch, G. (ed.) (1982). *Exchangeability in Probability and Statistics*. North Holland, Amsterdam.

References

[1] Aldous, D. and Pitman, J. (1979). *Ann. Prob.*, **7**, 704–723.

[2] Berman, S. M. (1962). *Ann. Math. Statist.*, **33**, 894–908.

[3] Billingsley, P. (1968). *Convergence of Probability Measures*. Wiley, New York.

[4] Chernoff, H. and Teicher, H. (1958). *Ann. Math. Statist.*, **29**, 118–130.

[5] Chow, Y. S. and Teicher, H. (1978). *Probability Theory: Independence, Interchangeability, Martingales*. Springer-Verlag, New York.

[6] Diaconis, P. and Freedman, D. (1980). *Ann. Prob.*, **8**, 745–764.

[7] Finetti, B. de (1930). *Atti Accad. Naz. Lincei Cl. Sci. Fiz. Mat. Nat. Rend.*, **4**, 86–133.

[8] Finetti, B. de (1972). *Probability, Induction and Statistics*. Wiley, New York.

[9] Galambos, J. (1973). *Duke Math. J.*, **40**, 581–586.

[10] Galambos, J. (1977). In ref. 13, appendix to Chap. 2.

[11] Galambos, J. (1978). *The Asymptotic Theory of Extreme Order Statistics*. Wiley, New York.

[12] Hewitt, E. and Savage, L. J. (1955). *Trans. Amer. Math. Soc.*, **80**, 470–501.

[13] Johnson, N. L. and Kotz, S. eds. (1977). *Urn Models and their Application*. Wiley, New York.

[14] Kallenberg, O. (1975). *Trans. Amer. Math. Soc.*, **202**, 105–121.

[15] Kendall, D. G. (1967). *Stud. Sci. Math. Hung.*, **2**, 319–327.

[16] Loève, M. (1963). *Probability Theory*, 3rd ed. D. Van Nostrand, New York.

[17] Moran, P. A. P. (1973). *J. Appl. Prob.*, **10**, 837–846.

[18] Olshen, R. (1974). *Z. Wahrscheinlichkeitsth. verwand. Geb.*, **28**, 317–321.

[19] Phelps, R. R. (1966). *Lectures on Choquet's Theorem*. Princeton University, Princeton, N. J.

[20] Rényi, A. and Révész, P. (1963). *Publ. Math. Debrecen*, **10**, 319–325.

[21] Riddler-Rowe, C. J. (1967). *Stud. Sci. Math. Hung.*, **2**, 415–418.

[22] Stam, A. J. (1978). *Statist. Neerlandica*, **32**, 81–91.

[23] Takács, L. (1975). *Adv. Appl. Prob.*, **7**, 607–635.

Acknowledgment

This work was supported by Grant AFOSR 78-3504.

(BAYESIAN INFERENCE
FOUNDATIONS OF PROBABILITY)

JANOS GALAMBOS

EXPECTATION *See* EXPECTED VALUE

EXPECTED PROBABILITY

If there are k mutually exclusive possible outcomes of an "experiment" with probabilities $p_1, p_2, \ldots, p_k$, the expected value of the probability associated with an observed outcome is $\sum_{j=1}^{k} p_j^2$. Fry [1] calls this the *expected probability* for the experiment. This concept is used as a basis for indices representing the unexpectedness of a particular outcome. (*See* PLAUSIBILITY.) The reciprocal of expected probability is the *total plausibility*. Since $\sum_{j=1}^{k} p_j^2 \leqslant 1$, the total plausibility cannot be less than 1.

For any discrete distribution, the expected probability equals the probability that two independent random variables, each having this distribution, are equal to each other.

Reference

[1]   Fry, T. C. (1964). *Probability and Its Engineering Uses*, 2nd ed. D. Van Nostrand, Princeton, N. J.

EXPECTED VALUE

The *expected value* of a random variable is a theoretically defined quantity intended to be an analog of the (observable) arithmetic mean*. From the point of view of frequency theory of probability, it can be regarded as corresponding to a limit of the arithmetic mean as sample size increases indefinitely. If the random variable X has cumulative distribution function* $F_x(x)$, the expected value can be defined as the Lebesgue–Stieltjes integral

$$E[X] = \int_{-\infty}^{\infty} x \, dF_X(x) \, dx. \qquad (1)$$

The symbol "$E[\cdot]$" means "expected value of"; note that $E[X]$ is not a mathematical function of X. It is sometimes referred to, loosely, as the expected value "of the distribution $F_X(x)$." Note that $E[X]$ need not be a possible value of X.

Calculation of expected values commonly uses the more specific formulas

$$E[X] = \int_{-\infty}^{\infty} xf_X(x) \, du \qquad (2)$$

for an (absolutely) continuous variable with density function $f_X(x) = dF_X(x)/dx$; and

$$E[X] = \sum_j x_j p_j \qquad (3)$$

for a discrete variable taking values $\{x_j\}$ with probabilities $\{p_j\}$.

The expected value of a function $g(X)$ is defined as

$$\int_{-\infty}^{\infty} g(x) \, dF_X(x)$$

with similar modifications of (2) and (3) for absolutely continuous and discrete random variables, respectively.

Among many other formulas for expected values we mention here the formula

$$E[X] = \int_0^{\infty} \{1 - F_X(x)\} \, dx$$

valid for positive random variables [i.e., when $F_X(0) = 0$].

(CHEBYSHEV'S INEQUALITY
CUMULANTS
MOMENTS)

EXPERIENCE

A term employed to describe the results of actual observation as opposed to calculation from a theoretical model. One of the most common uses is in actuarial* work, where "mortality experience" refers to the data provided by records of policyholders in life insurance companies.

In work on clinical trials, the experience is often described as a "life table," although this term is more usually regarded as referring to a model distribution.

(CLINICAL TRIALS
LIFE TABLES
MORTALITY DATA)

EXPERIMENTAL ERROR   *See* ANALYSIS OF VARIANCE; COEFFICIENT OF DETERMINATION; ERROR ANALYSIS; MEASUREMENT ERROR

EXPERIMENTWISE ERROR RATE

When performing several posterior tests* on differences in the case of an ANOVA with significant F-ratio, the experimentwise error rate is defined as

$$\alpha' = \frac{\left\{\begin{array}{l}\text{number of experiments having one} \\ \text{or more Type I errors with respect} \\ \text{to tests on sample differences}\end{array}\right\}}{\text{number of experiments}}.$$

(POSTERIOR TESTS
SIMULTANEOUS INFERENCE
TUKEY'S SIMULTANEOUS
    COMPARISON PROCEDURE)

EXPLORATORY DATA ANALYSIS

Statisticians, as well as others who apply statistical methods to data, have often made preliminary examinations of data in order to explore their behavior. In this sense, exploratory data analysis has long been a part of statistical practice. Since about 1970, "exploratory data analysis" has most often meant the attitude, approach, and techniques developed, primarily by John W. Tukey, for flexible probing of data before any probabilistic model is available.

BROAD PHASES OF DATA ANALYSIS

One description of the general steps and operations that make up practical data analysis identifies two broad phases: an exploratory phase and a confirmatory phase. Exploratory data analysis is concerned with isolating patterns and features of the data and with revealing these forcefully to the analyst. It often provides the first contact with the data, preceding any firm choice of models for either structural or stochastic components; but it also serves to uncover unexpected departures from familiar models. An important element of the exploratory approach is flexibility, both in tailoring the analysis to the structure of the data and in responding to patterns that successive steps of analysis uncover.

Confirmatory data analysis concentrates on assessing the reproducibility of the observed patterns or effects. Its role is closer to that of traditional statistical inference in providing statements of significance and confidence; but the confirmatory phase also encompasses (among others) the step of incorporating information from an analysis of another, closely related body of data and the step of validating a result by collecting and analyzing new data.

In brief, exploratory data analysis emphasizes flexible searching for clues and evidence, while confirmatory data analysis stresses evaluating the available evidence. The rest of this article describes the basic concepts of exploratory data analysis and illustrates two simple techniques.

FOUR THEMES

Throughout exploratory data analysis, four main themes appear and often combine. These are resistance, residuals, re-expression, and display.

Resistance

Resistance is a matter of insensitivity to misbehavior in data. More formally, an analysis or summary is *resistant* if an arbitrary change in any small part of the data produces only a small change in the analysis or summary. This attention to resistance reflects the understanding that "good" data seldom contain less than about 5% gross errors, and protection against the adverse effects of these should always be available.

It is worthwhile to distinguish between resistance and the related notion of robustness (*see* ROBUSTNESS). Robustness generally implies insensitivity to departures from assumptions surrounding an underlying probabilistic model. (Some discussions regard resistance as one aspect of "qualitative robustness.")

In summarizing the location of a sample, the median* is highly resistant. (In terms of efficiency, it is not so highly robust because other estimators achieve greater efficiency* across a broader range of distributions.) By contrast, the mean is highly nonresistant. A number of exploratory techniques for more structured forms of data provide resistance because they are based on the median.

Residuals

*Residuals** are what remain of the data after a summary or fitted model has been subtracted out, according to the schematic equation

$$residual = data - fit.$$

For example, if the data are the pairs (x_i, y_i) and the fit is the line $\hat{y}_i = a + bx_i$, then the residuals are $r_i = y_i - \hat{y}_i$.

The attitude of exploratory data analysis is that an analysis of a set of data is not complete without a careful examination of the residuals. This emphasis reflects the tendency of resistant analyses to provide a clear separation between dominant behavior and unusual behavior in the data. When the bulk of the data follows a consistent pattern, that pattern determines a resistant fit. The residuals then contain any drastic departures from the pattern, as well as the customary chance fluctuations. Unusual residuals suggest a need to check on the circumstances surrounding those observations. As in more traditional practice, the residuals can warn of systematic difficulties with the data— curvature, nonadditivity, and nonconstancy of variability.

Re-expression

Re-expression involves the question of what scale would help to simplify the analysis of the data. Exploratory data analysis emphasizes the benefits of considering, at an early stage, whether the scale in which the data are originally expressed is satisfactory. If not, a re-expression into another scale may help to promote symmetry, constancy of variability, straightness of relationship, or additivity of effect, depending on the structure of the data.

The re-expressions most often used in exploratory data analysis come from the fam-

ily of functions known as power transformations, which take y into y^p (almost always with a simple value of p such as $\frac{1}{2}$, -1, or 2), together with the logarithm (which, for data analysis purposes, fits into the power family at $p = 0$). For example, one investigation of the relationship between gasoline mileage and the characteristics of automobiles gained substantially from re-expressing mileage in a reciprocal scale and working with gallons per 100 miles instead of miles per gallon.

Displays

Displays meet the analyst's need to see behavior—of data, of fits, of diagnostic measures, and of residuals—and thus to grasp the unexpected features as well as the familiar regularities.

A major contribution of the developments associated with exploratory data analysis has been the emphasis on visual displays and the variety of new graphical techniques. Brief examples can serve to illustrate two of these, the *stem-and-leaf display* and the *schematic plot*. See also GRAPHICAL REPRESENTATION OF DATA.

Example 1. The extremely high interest rates early in 1980 attracted many investors to the so-called money market funds. One listing of 50 such funds showed annual yields (based on the 30-day period ended April 9, 1980) ranging from 13.7 to 17.1%. The stem-and-leaf display in Fig. 1 shows these 50 rates of return. In overall outline,

13 ·	798887
14 ∗	332
14 ·	8776666699668778
15 ∗	2430403222
15 ·	9979
16 ∗	203030
16 ·	755
17 ∗	11

stem | leaves

Figure 1 Stem-and-leaf display for yields of money market funds (unit = 0.1%; i.e., all entries are multiples of 0.1%). (The leaves have been written down in the order in which the data values were encountered.)

this display resembles a histogram with interval width 0.5%, but it is able to show the full detail of these data values by using the last digit of each observation instead of simply enclosing a standard amount of area. This feature makes it easy to go back from a part of the display to the individual data values (and their identities) in the listing of the data. Noticeable features of Fig. 1 are that just over half the funds were yielding between 14.6 and 15.4%, that another group had yields around 16%, and that six funds had yields from 13.7 to 13.9%. The names of these last six indicate that at least three invest only in securities of the U.S. government. More detailed probing might uncover some common feature in the portfolios of the funds whose yields were around (or above) 16%.

Several alternatives and refinements permit the stem-and-leaf display to accommodate a wide variety of data patterns. The interval width (i.e., the range of data values represented by one line in the display) may be 2, 5, or 10 times a power of 10; and stray observations may be listed separately at each end so as not to distort the scale. The use of positive and negative stems readily handles sets of residuals.

Example 2. Seventy-nine men participated in a 5-mile road race. Table 1 shows the time that each runner in each of the four age divisions took to complete the race, and Fig. 2 uses four parallel schematic plots to compare the times in the four divisions.

In a schematic plot (also called a "boxplot") the box extends from the lower hinge (an approximate quartile) to the upper hinge and has a line across it at the median. The dashed lines show the extent of the data, except for observations that are apparent strays (defined according to a rule of thumb based on the hinges). Such strays appear individually—in order to focus attention on them—as the time of 55.53 minutes in the "25 and Under" division does in Fig. 2. The general intent is to indicate the median, outline the middle half of the data, and show the range, with more detail at the ends if needed.

Table 1   Times (minutes) for Runners to Complete a Men's 5-Mile Road Race

25 and Under	26–35	36–45		46 and Over
27.03	29.95	30.12	39.97	33.08
27.68	31.02	30.18	40.48	34.52
27.83	31.40	30.72	40.88	36.97
30.30	32.10	31.77	41.13	37.05
32.63	32.57	34.62	42.00	38.07
33.22	32.82	34.83	42.93	41.63
33.33	35.53	35.20	43.67	42.08
33.87	35.55	35.73	44.07	48.47
34.37	38.85	35.78	44.38	49.55
34.62	39.92	36.10	44.47	
35.17	40.30	36.70	45.65	
35.55	41.27	36.72	47.63	
41.33	41.45	38.53	48.08	
41.82	42.60	39.03	49.50	
41.93	44.35	39.22		
43.08	49.07	39.30		
43.10	49.78	39.43		
43.37	49.92	39.85		
55.53		39.87		

The impressions that one gets from Fig. 2 are mixed. The fastest time, lower hinge, and median tend (as expected) to increase with increasing age, but the upper hinge and slowest time are more nearly constant. In the "25 and Under" division the dashed line at the upper end is rather short (relative to the length of the box), suggesting (correctly) that the box may conceal grouping instead of smoothly decreasing frequency of deviation from a central value.

Literature

The first published presentation of exploratory data analysis (1970–1971) was the preliminary edition [6] of the book by John W. Tukey, whose 1977 book [9] represents the definitive account of the subject. Other expository accounts containing substantial discussions of such attitudes and techniques include Erickson and Nosanchuk [1], Koopmans [2], Leinhardt and Wasserman [3], McNeil [4], Mosteller and Tukey [5], and Velleman and Hoaglin [11].

Broader discussions of the roles of exploratory and confirmatory data analysis in scientific inquiry appear in Tukey [7, 10].

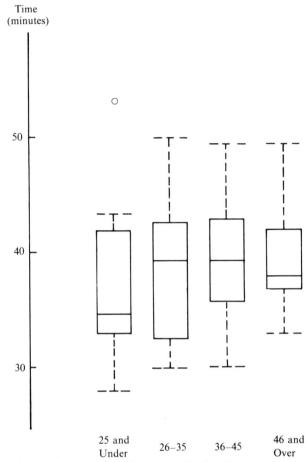

Figure 2   Parallel schematic plots for times of runners in the four divisions of a men's 5-mile road race.

Tukey [8] describes and illustrates a number of the most important displays.

Published computer programs for selected techniques of exploratory data analysis are available in Velleman and Hoaglin [11] and, in a rudimentary form, in McNeil [4].

References

[1]   Erickson, B. H. and Nosanchuk, T. A. (1977). *Understanding Data*. McGraw-Hill Ryerson, Toronto.

[2]   Koopmans, L. H. (1981). *Introduction to Contemporary Statistics*. Duxbury Press, Boston.

[3]   Leinhardt, S. and Wasserman, S. S. (1978). In *Sociological Methodology, 1979*, Karl Schuessler, ed. Jossey-Bass, San Francisco, pp. 311–372.

[4]   McNeil, D. R. (1977). *Interactive Data Analysis*. Wiley, New York.

[5]   Mosteller, F. and Tukey, J. W. (1977). *Data Analysis and Regression*. Addison-Wesley, Reading, Mass.

[6]   Tukey, J. W. (1970–1971). *Exploratory Data Analysis*, limited prelim. ed. Addison-Wesley, Reading, Mass. (Available from University Microfilms.)

[7]   Tukey, J. W. (1972). *Quart. Appl. Math.*, **30**, 51–65.

[8]   Tukey, J. W. (1972). In *Statistical Papers in Honor of George W. Snedecor*, T. A. Bancroft, ed. Iowa State University Press, Ames, Iowa, pp. 293–316.

[9]   Tukey, J. W. (1977). *Exploratory Data Analysis*. Addison-Wesley, Reading, Mass.

[10]   Tukey, J. W. (1980). *Amer. Statist.*, **34**, 23–25.

[11]   Velleman, P. F. and Hoaglin, D. C. (1981). *Applications, Basics, and Computing of Exploratory Data Analysis*. Duxbury Press, Boston.

(GRAPHICAL REPRESENTATION
  OF DATA
RESIDUALS)

DAVID C. HOAGLIN

EXPONENTIAL DISTRIBUTION

DEFINITIONS AND HISTORY

This article deals with a continuous univariate distribution whose density is defined below in (1). A separate entry is devoted to the multivariate case (*see* MULTIVARIATE EXPONENTIAL).

The distribution of a continuous random variable is called exponential (sometimes negative exponential) if its density function is of the form

$$p(x) = \sigma^{-1}\exp\left[-(x-a)/\sigma\right],$$
$$x \geqslant a, \sigma > 0. \quad (1)$$

This density also belongs to the "exponential family*" as well as to the families of "gamma*", "chi square*", and "Weibull*" distributions. However, because of the importance and appeal of several of its basic properties, which are not shared by other members of these families, the exponential distribution is accepted and treated in the statistical literature as one of the basic distributions rather than just a member of a larger family.

The special case of $a = 0$ and $\sigma = 1$ in (1) is called the unit exponential distribution (also known as the standard exponential distribution). The importance of the unit exponential distribution lies in the fact that if the random variable X is exponentially distributed with density as in (1) then

$$Y = (X - a)/\sigma \quad (2a)$$

has the unit exponential distribution. This fact is a special case of the following more general monotonic transformation. Let X be a random variable with a continuous distribution function $F(x)$. Then

$$Y = -\log\{1 - F(X)\} \quad (2b)$$

has the unit exponential distribution (here

we take $\log 0 = -\infty$). Notice that if the density of $F(x)$ is as it is in (1) then (2b) becomes (2a). Because (2a, b) is a monotonic transformation, it is of particular importance in the theory of "order statistics." Namely, it permits one to study the order statistics* of samples from a population with an arbitrary continuous distribution through the order statistics of exponential variables that exhibit very appealing properties (see Property 5 in the next section).

An exponential distribution truncated from below is still an exponential distribution with the same scale parameter. This fact is due to the "lack of memory" property, stated in the next section as Property 1. On the other hand, easy calculations indicate that an exponential distribution truncated by the exclusion of values exceeding the value T has the density function

$$p^*(x) = A\exp\left[-(x-a)/\sigma\right],$$
$$a \leqslant x \leqslant T,$$

where $A = A(a, \sigma, T)$ is a well defined constant.

We make only a few remarks on the history of the exponential distribution because much of it is extensively covered in the books by Galambos and Kotz [8, pp. 1–5] and Johnson and Kotz [11, pp. 209–210]. What we wish to emphasize here is that the actual application of stochastic models with an exponential population distribution achieved acceptance in the statistical literature only after World War II. Scattered applications did occur in applied papers but very little, if any, theoretical considerations accompanied their discussions. A much forgotten but important paper was published in 1937 by P. V. Sukhatme, who discovered the following important property of the exponential distribution: the consecutive differences of order statistics of a sample from an exponential population are independent exponential variables. This property, which was rediscovered by Rényi (1953) and by Epstein and Sobel [6], became a basic property from which the asymptotic theory of order statistics (mainly quantiles*) can be developed. It also served as a fundamental theorem in life tests (which approach was initiated by the paper by Epstein and Sobel). The fact that a property of the exponential distribution can serve as a basic theorem in a general theory, in which the underlying population distribution is an arbitrary continuous distribution, is due to the exponentiality of Y in the transformation (2b). Although Sukhatme could not break the dominance of the assumption of normality in statistical analysis, his vision of the importance of this property proved right. Indeed, the quoted papers of Rényi and Epstein and Sobel, together with a paper by Davis [3], initiated a very active research field.

The reader can find very good bibliographies on exponentiality in the books by Galambos and Kotz [8] and Johnson and Kotz [11, pp. 228–232]. Some additional references, with emphasis on the developments in the past decade, are discussed in the third section of this article.

BASIC PROPERTIES

The location and scale parameters a and σ, respectively, have the following practical meaning. If the density of the random variable X is as it is in (1), then the location parameter a is a lower bound for X; that is, $P(X \geqslant a) = 1$. In addition, $E(X) = a + \sigma$ and $V(X) = \sigma^2$. For presenting the basic properties, one can therefore assume $a = 0$, which we shall do in the following section.

A property of a distribution can serve to determine whether a distributional assumption is in agreement with reality when a model is adopted for a specific practical problem. Furthermore, if a property is characteristic to a distribution then, through such a property, the subjectivity of model building can be avoided and a scientific reasoning will assure the practitioner about the proper and unique choice of the model. With this in mind, we list those five basic properties of the exponential distribution (assuming $a = 0$) that are most frequently applied in justifying the choice of an expo-

nential model (i.e., that the underlying population distribution is assumed exponential).

PROPERTY 1: THE LACK OF MEMORY. An easy calculation shows that, if the random variable X has density (1) with $a = 0$, then the conditional distribution of $X - u$, given $X \geqslant u$, does not depend on u. Indeed,

$$
\begin{aligned}
P(X - u &\geqslant z \mid X \geqslant u) \\
&= P(X \geqslant u + z)/P(X \geqslant u) \\
&= e^{-(u+z)/\sigma} e^{u/\sigma} = e^{-z/\sigma} \\
&= P(X \geqslant z), \quad \text{all } u > 0. \quad (3)
\end{aligned}
$$

This property is known in the literature under several names, one of which is the "lack of memory" property. Others are the "old is as good as new" property and the "no-aging" property. It is true (see CHARACTERIZATIONS OF DISTRIBUTIONS) that only the density at (1) can satisfy the extreme sides of equation (3).

PROPERTY 2: CONSTANT HAZARD (MORTALITY) RATE. While the lack of memory expresses the fact that the future is not influenced by the past, the constant hazard rate property relates the present to the past. Namely, given that $X \geqslant u$, the present can be viewed to be represented by $u \leqslant X < u + \Delta u$, where Δu is arbitrarily small. Now, if the density of X exists and if it is given by (1) with $a = 0$, then

$$
\begin{aligned}
P(u \leqslant X &< u + \Delta u \mid X \geqslant u) \\
&= (e^{-u/\sigma} - e^{-(u+\Delta u)/\sigma})/e^{-u/\sigma} \\
&= 1 - e^{-\Delta u/\sigma} \\
&= \sigma^{-1}\Delta u + o(\Delta u),
\end{aligned}
$$

where the last term means that the error is of smaller order of magnitude than Δu as Δu becomes arbitrarily small (tends to zero). Hence, as $\Delta u \to 0$,

$$
\lim \frac{1}{\Delta u} P(u \leqslant X < u + \Delta u \mid X \geqslant u) = \frac{1}{\sigma}
$$

the same value for all $u > 0$. This last property is called the "constant hazard rate," or "mortality rate," property. This is again characteristic to the density at (1) ($a = 0$).

See also HAZARD RATE CLASSIFICATION OF DISTRIBUTIONS.

PROPERTY 3: CONSTANT EXPECTED RESIDUAL LIFE. If, instead of distributional properties, one is interested in expectations, the validity of the equation

$$E(X - u \mid X \geq u) = E(X), \qquad \text{all } u > 0,$$
(4)

can easily be checked for the density (1) with $a = 0$. It evidently expresses the same physical property in terms of expectations that was expressed by (3) in terms of distributions. Again, no other distribution can satisfy (4). When X represents a random life length then the residual life is a conditional concept, namely, $X - u$, given $X \geq u$. Hence the name of constant expected residual life for the property at (4).

PROPERTY 4: SAMPLE SIZE TIMES THE MINIMUM IS DISTRIBUTED AS THE POPULATION. Let $X_1, X_2, \ldots, X_n$ be independent observations on the random variable X. Let $X_{1:n}$ be the smallest of the X_j. Then, if the distribution of X is exponential, so is $nX_{1:n}$. More precisely, if the density of X is as it is at (1) with $a = 0$, then $nX_{1:n}$ has the same distribution as X. Just as the previous ones, this property is also characteristic to the exponential distribution.

We finally record the property mentioned in connection with the history of the exponential distribution.

PROPERTY 5: Let $X_1, X_2, \ldots, X_n$ be independent observations on an exponential random variable X. Let $X_{1:n} \leq X_{2:n} \leq \cdots \leq X_{n:n}$ be the order statistics of the X_j. Then the (normalized) differences $(n - j + 1)(X_{j:n} - X_{j-1:n})$, $1 \leq j \leq n$, where $X_{0:n} = 0$, are independent and identically distributed exponential variates.

As mentioned earlier, this property forms the basis of several life tests* and also provides a convenient mathematical tool for proving the asymptotic normality* of sample quantiles.

One more property will be formulated in the next section which is widely applied as a test statistic in goodness of fit tests.

For the interrelation of Properties 1–4 and for several extensions of them, together with an analysis of the relevant literature, see the book by Galambos and Kotz [8], particularly Chapters 1–3. Chapter 4 of this same book discusses the role of the exponential distribution in the theory of renewal processes*.

STATISTICAL INFERENCE

It has been mentioned that the parameters a and σ are related to the first two moments through the relations $E(X) = a + \sigma$ and $V(X) = \sigma^2$, where X is a random variable with density (1). Additional moment relations are easily obtained from the generating function* $E(e^{tX}) = (1 - \sigma t)^{-1}e^{at}$, or from the central moment generating function $E(e^{t(X-a-\sigma)}) = (1 - \sigma t)^{-1}e^{-t\sigma}$. It immediately follows from this last form that all central moments of X are proportional to integer powers of σ. Therefore, general methods for estimating moments of a distribution provide estimators for $a + \sigma$ and for powers of σ. Through the median*, another linear form of a and σ, namely $a + \sigma \log 2$, can be estimated.

Based on a sample $X_1, X_2, \ldots, X_n$, the maximum likelihood* estimators of a and σ are $a^* = X_{1:n}$ and $\sigma^* = n^{-1}\sum_{i=1}^{n} X_i - X_{1:n}$, respectively. The simple distributional property of the estimator a^* is expressed in Property 4 of the previous section. $X_{1:n}$ remains the maximum likelihood estimator of a even if σ is known, while the maximum likelihood estimator of σ is the sample mean minus a if a is known. It may be mentioned that these estimators, with the exception of the last one, are biased.

Notice that another expression for σ^* is

$$\sigma^* = n^{-1}\sum_{i=2}^{n}(n - i + 1)(X_{i:n} - X_{i-1:n}).$$

Hence, in view of Property 5 of the previous

section, $2n\sigma^*/\sigma$ is a chi square variable and is independent of a^*. This shows a remarkable similarity to the case of the "normal distribution" so far as the estimators of their parameters are concerned. Thus, with the same technique as for the normal case, one can construct confidence intervals and tests of significance for a and σ, starting with the statistics $n(X_{1:n} - a)/\sigma^*$ and $2n\sigma^*/\sigma$, respectively.

Testing for exponentiality has drawn the attention of several statisticians. Because most goodness of fit* tests are insensitive for deciding between two distributions that are uniformly close to each other, further attempts to improve goodness of fit tests can be expected.

Let us first mention those goodness of fit tests that are based on complete samples (as opposed to censored ones). One general approach is to transform the data X_1, $X_2, \ldots, X_n$ to a set of uniformly distributed variables. This can be done in several ways. If the parameters of the population are 'known,' then the usual transformation is $Y_j = F(X_j)$, where, in our case, the density of $F(x)$ is given at (1). On the other hand, if only the location parameter a is known (and thus $a = 0$ can be achieved by simple subtraction), then the transformation $Z_j = S_j/S_n$, where $S_j = X_1 + X_2 + \cdots + X_j$, is applicable. Namely, for a sample from the density at (1) with $a = 0$, the set Z_j, $1 \leqslant j \leqslant n - 1$, has the same multivariate distribution as the order statistics of a sample of size $n - 1$ from a population uniformly distributed over the interval $(0, 1)$. When the location parameter is also unknown, then a transformation similar to Z_j is possible in which the role of X_j is taken over by the normalized difference

$$(n - j + 1)(X_{j:n} - X_{j-1:n}).$$

To the transformed variables one can then apply the "Kolmogorov–Smirnov test*." The original data can, of course, be used directly and tests can be developed based on the empirical distribution function. For this latter case, see Lilliefors [15] and Durbin [4], while for the mentioned transformations, see

Sherman [20], Moran [16], Epstein [5], Pyke [18] and Csörgő, Seshadri, and Yalovsky [1], who give further references. Gail and Gastwirth [7] propose another test for exponentiality, which is scale free and is based on the Lorenz curve*. See also Chapter 6 in David [2], which is also an appropriate reference to the next paragraph on censoring.

A very effective method of dealing with a sample from an exponential population is to use censored data. We mention here the following censoring* only. Assume that n items (or individuals) are under test and experimentation stops as soon as r items fail ($r < n$). If $X_1, X_2, \ldots, X_n$ are the actual times for the n items until failure, then, at termination of the experiment, we actually observe the order statistics $X_{1:n} \leqslant X_{2:n} \leqslant \cdots \leqslant X_{r:n}$. In view of Property 5, if the population distribution is exponential, the normalized differences

$$(n - j + 1)(X_{j:n} - X_{j-1:n}), \qquad 1 \leqslant j \leqslant r,$$

can be used as effectively as in the case of a complete sample. The obvious advantage is the potential saving of time in experimentation. We again refer to Epstein [5] for a good exposition of life tests using censored data. In regard to estimation and confidence intervals*, based on censored data, we mention (in addition to the already quoted Chapter 6 of David [2]). Perng [17] (parameter estimation and tests of significance), Rasch [19] (sample size considerations) and the following group of papers dealing with prediction* intervals for $X_{s:n}$, based on the censored sample $X_{r:n}$, $1 \leqslant r < s$: Lawless [13], Lingappaiah (1973), Hahn and Nelson [10], Likes [14], and Kaminsky [12].

Guenther, Patil, and Uppuluri [9] discuss tolerance* bounds and intervals when the population is exponential.

The exponential distribution is very popular in the field of reliability theory*, where all the five properties formulated in the previous section have useful and practical interpretation. The survey by Shimi and Tsokos [21] gives a very thorough discussion of this subject. Although their survey is not limited to exponentiality, the interested reader will find it very useful to see the variety of methods available for reliability estimates. It could be stimulating to make comparisons of conclusions under different model assumptions.

References

[1] Csörgö, M., Seshadri, V., and Yalovsky, M. (1975). In *Statistical Distributions in Scientific Work*, Vol. 2, G. P. Patil, et al., eds. Reidel, Dordrecht, pp. 79–90.

[2] David, H. A. (1981). *Order Statistics*, 2nd ed. Wiley, New York.

[3] Davis, D. J. (1952). *J. Amer. Statist. Assoc.*, **47**, 113–150.

[4] Durbin, J. (1975). *Biometrika*, **62**, 5–22.

[5] Epstein, B. (1960, a, b). *Technometrics*, **2**, 83–101; *ibid.* **2**, 167–183.

[6] Epstein, B. and Sobel, M., (1953). *J. Amer. Statist. Assoc.*, **48**, 486–502.

[7] Gail, M. H. and Gastwirth, J. L. (1978). *J. Amer. Statist. Assoc.*, **73**, 787–793.

[8] Galambos, J. and Kotz, S. (1978). *Characterizations of Probability Distributions*. Lecture Notes in Mathematics, Vol. 675, Springer Verlag, Heidelberg. (This book covers a large variety of problems on the theory of the exponential distribution. On pp. 1–5, there is a detailed historical account; a large variety of characterizations leading to the exponential distribution are discussed; the relation of point processes to exponentiality is given in Chapter 4; and it has an up-to-date bibliography on characterizations.)

[9] Guenther, W. C., Patil, S. A. and Uppuluri, V. R. R. (1976). *Technometrics*, **18**, 333–340.

[10] Hahn, G. J. and Nelson, W. B. (1973). *J. Qual. Tech.*, **5**, 178–188.

[11] Johnson, N. L. and Kotz, S. (1970). *Continuous Univariate Distributions I*. Wiley, New York. (Chapter 18 is devoted to the exponential distribution. Main emphasis is on the statistical problems related to the exponential distribution—such as estimation of parameters, tests of significance, generation of random numbers, and goodness of fit tests. It contains a good historical account and a detailed bibliography on the statistical aspects of exponentiality.)

[12] Kaminsky, K. S. (1977). *Technometrics*, **19**, 83–86.

[13] Lawless, J. F. (1971). *Technometrics*, **13**, 725–730.

[14] Likes, J. (1974). *Technometrics*, **16**, 241–244.

[15] Lilliefors, H. W. (1969). *J. Amer. Statist. Assoc.*, **64**, 387–389.

[16] Moran, P. A. P. (1951). *J. R. Statist. Soc. Ser. B*, **13**, 147–150.

[17] Perng, S. K. (1977). *Comm. Statist. A*, **6**, 1399–1407.

[18] Pyke, R. (1965). *J. R. Statist. Soc. Ser. B*, **27**, 395–436; Discussion 437–449.

[19] Rasch, D. (1977). *Biom. Z.*, **19**, 521–528.

[20] Sherman, B. (1950). *Ann. Math. Statist.*, **21**, 339–361.

[21] Shimi, I. N. and C. P. Tsokos (1977). In *The Theory and Applications of Reliability*, Vol. I, C. P. Tsokos and I. N. Shimi, eds., Academic Press, New York, pp. 5–47.

Acknowledgment

This work was supported by the Air Force Office of Scientific Research under a grant (number 78-3504) to Temple University.

(CHARACTERIZATIONS OF DISTRIBUTIONS
CHI-SQUARE DISTRIBUTION
EXPONENTIAL FAMILIES
GAMMA DISTRIBUTION
HAZARD RATE CLASSIFICATION OF DISTRIBUTIONS
KOLMOGOROV-SMIRNOV TESTS
MULTIVARIATE EXPONENTIAL DISTRIBUTION
NORMAL DISTRIBUTION
ORDER STATISTICS)

JANOS GALAMBOS

EXPONENTIAL FAMILIES

There are two main types of parametric families of distributions in statistics, the *transformation* (or *group*) *families* and the *exponential families*. The transformation families are those generated from a single probability measure by a group of transformations on the sample space, the key example being any location-scale family* $\sigma^{-1}f((x - \mu)/\sigma)$, where f is a known probability density function*. The exponential families are characterized by having probability (point or density) function of the form

$$p(x; \omega) = a(\omega)b(x)\exp\{\theta(\omega) \cdot t(x)\}, \quad (1)$$

where ω is a parameter (both x and ω may, of course, be multidimensional), $\theta(\omega)$ and $t(x)$ are vectors of common dimension, k

say, and $\cdot$ denotes inner product, i.e., $\theta(\omega) \cdot t(x) = \sum_{i=1}^{k} \theta_i(\omega) t_i(x)$. Remarkably, the basics of the present statistical theories of both location-scale families and exponential families were given in a single paper by Fisher [19]. Distributional families which are both exponential and of the transformation type possess particularly nice properties. Their general structure has been studied by Roy [26], Rukhin [27, 28], and Barndorff–Nielsen et al. [6].

A great many of the commonly occurring families of distributions are exponential. Examples of such families are the binomial*, multinomial*, Poisson*, geometric*, logarithmic*, (multivariate) normal*, Wishart*, gamma*, beta*, Dirichlet*, and von Mises–Fisher*. Moreover, if $x_1, \ldots, x_m$ are independent observations, each following an exponential family, then the model for $x = (x_1, \ldots, x_m)$ is also exponential, so that, for instance, factorial and regression experiments often have an exponential model.

Sometimes a family is not exponential in its entirety, but the subfamilies obtained by fixing a (one- or multi-dimensional) component of the parameter are exponential. For instance, the negative binomial* family is not exponential, but for any fixed value of the shape parameter one has an exponential family. In a considerable number of other cases a distributional family for observed data y, although not necessarily exponential itself, may be thought of as derivable from an exponential family as follows. There exists a real or fictitious data set x and an exponential family of distributions for x such that y can be viewed as a function of x with the actual distributional model for y being equal to that derived from the exponential model for x. A grouped empirical distribution y obtained by grouping a sample x from the normal distribution provides an example of this. In such incompletely observed exponential situations the underlying exponential structure can often be taken to advantage in a statistical analysis; see further at the end of this article.

The exponential families share a large number of important and useful properties

which often make an incisive statistical analysis feasible. The theory of exponential families, outlined in the following, is concerned with such general properties. The theory is, in fact, quite rich and only its more salient and practically useful features can be indicated here. Where no references are given below the reader may consult the monograph by Barndorff-Nielsen [5] for details and additional information. Further guidance to the literature on exponential families is given in the bibliography to the present article.

FIRST PROPERTIES

Let $\mathcal{P}$ denote the exponential family of distributions with probability functions $p(x; \omega)$. The right-hand side of (1) is said to be an exponential representation of $\mathcal{P}$ or $p(x; \omega)$, and the vectors $\theta = \theta(\omega)$ and $t = t(x)$ are called the *canonical parameter* and the *canonical statistic*. (Sometimes the term "natural" is used instead of canonical.) The smallest k for which an exponential representation of $\mathcal{P}$ with θ and t of dimension k is possible is the *order* of $\mathcal{P}$, and such a representation is said to be *minimal*. The canonical statistic t is a sufficient statistic for $\mathcal{P}$ and it is minimal sufficient if the exponential representation is minimal. (Essentially, the exponential families are the only models that allow a sufficient reduction of the data.) For theoretical purposes it is mostly convenient to work with minimal exponential representations and it will be assumed from now on that (1) is minimal. Also, θ can be assumed to be in one-to-one correspondence with ω, and when a is considered as a function of θ, we simply write $a(\theta)$ for $a(\omega(\theta))$. Similarly, we may write $p(x; \theta)$ instead of $p(x; \omega)$, etc. Set $c(\theta) = a(\theta)^{-1}$.

Example 1. The multinomial distribution* on r cells has probability function

$$p(x; \pi) = \frac{n!}{x_1! \cdots x_r!} \pi_1^{x_1} \cdots \pi_r^{x_r}.$$

Here $x_1 + \cdots + x_r = n$ and $\pi_1 + \cdots + \pi_r$

= 1. The family of distributions obtained by letting $\pi_1, \ldots, \pi_r$ vary freely except for the restrictions $\pi_1 > 0, \ldots, \pi_r > 0$ is the multinomial family. This family is obviously exponential with $t = x = (x_1, \ldots, x_r)$ as canonical statistic and $\theta = \ln \pi = (\ln \pi_1, \ldots, \ln \pi_r)$ as canonical parameter. However, due to the affine constraint $x_1 + \cdots + x_r = n$, the exponential representation with this t and θ as the canonical variates is not minimal. A minimal representation is obtained by taking $t = (x_1, \ldots, x_{r-1})$ and $\theta = (\ln(\pi_1/\pi_r), \ldots, \ln(\pi_{r-1}/\pi_r))$ and rewriting (1) as

$$p(x; \theta) = (1 + e^{\theta_1} + \cdots + e^{\theta_{r-1}})^{-n}$$
$$\times \frac{n!}{x_1! \cdots x_r!} e^{\theta_1 x_1 + \cdots + \theta_{r-1} x_{r-1}}. \quad (2)$$

The probability functions $p(x; \omega)$ are all densities with respect to one and the same measure μ, which is, typically, either a counting measure or Lebesgue measure. Let Ω be the domain of variation for ω and let $\Theta = \theta(\Omega)$ denote the canonical parameter domain for $\mathcal{P}$. Furthermore, let $\tilde{\Theta} = \{\theta : \int b(x) e^{\theta \cdot t(x)} d\mu < \infty\}$, which is a convex subset of R^k. Then $\mathcal{P}$ is said to be *full* if $\Theta = \tilde{\Theta}$, and $\mathcal{P}$ is *regular* if it is full and if Θ is an open subset of R^k. All of the examples of exponential families mentioned above are regular, and regular families are particularly well behaved (see below). If Θ is a smooth manifold in $\tilde{\Theta}$, then $\mathcal{P}$ is said to be a *curved exponential family*.

Example 2. Let $x_1, \ldots, x_m$ and $y_1, \ldots, y_n$ be two independent samples, the first from the normal distribution $N(\xi, \sigma^2)$, the second from the $N(\eta, \tau^2)$ distribution. If all four parameters are unknown, then the model for the full set of observations is a regular exponential family of order 4. The submodel determined by equating the means ξ and η (which is considered in the Behrens–Fisher* situation) is still of order 4 although it involves only three independent parameters. It is nonregular, but constitutes a curved exponential family.

Example 3. A random sample of n individuals from a human population has been classified according to the ABO blood group system, the observed numbers of individuals of the various phenotypes being x_A, x_B, x_{AB}, and x_{OO}. Under the standard genetical assumptions and with p, q, and r denoting the theoretical gene frequencies, the probability of the observation is

$$\frac{n!}{x_A! x_B! x_{AB}! x_{OO}!} (p^2 + 2pr)^{x_A}$$
$$\times (q^2 + 2qr)^{x_B} (2pq)^{x_{AB}} (r^2)^{x_{OO}}. \quad (3)$$

The model (3) is a curved exponential family of order 3. (It is assumed here that p, q, and r are all positive and that they vary freely except for the constraint $p + q + r = 1$.) A minimal canonical parameter is given by $\theta = (\ln\{(p/r)^2 + 2p/r\}, \ln\{(q/r)^2 + 2q/r\}, \ln\{2(p/r)(q/r)\})$, and this parameter varies over a two-dimensional manifold of $\tilde{\Theta} = R^3$.

Suppose that $\mathcal{P}$ is full and let $\theta \in \text{int } \Theta$, the interior of Θ. The function $c(\theta + \zeta)/c(\theta)$, considered as a function of ζ, is then the Laplace transform for the statistic t under the distribution determined by $p(x; \theta)$. It follows that the cumulants* of t can be obtained by differentiation of $\kappa(\theta) = -\ln a(\theta)$, and in particular one has

$$E_\theta t = \frac{\partial \kappa}{\partial \theta} \quad \text{and} \quad V_\theta t = \frac{\partial^2 \kappa}{\partial \theta \, \partial \theta'} \quad (4)$$

(vectors are taken to be row vectors and transposition is indicated by $'$). The mean value mapping defined on $\text{int } \Theta$ by $\theta \to \tau$, where $\tau(\theta) = E_\theta t$, is one-to-one and both ways continuously differentiable. The range $\mathcal{T}$ of τ is a subset of $\text{int } C$, where C denotes the closed convex hull of the support S of the distribution of t. The sets Θ and C (and also S and $\mathcal{T}$) are important for visualizing the theory of exponential families. It is best to think of Θ and C as being subsets of two different k-dimensional Euclidean spaces, with the exponential character of the statistical model establishing a duality relation between Θ and C. A particular aspect of this is that the powerful theory of convex analysis can be fruitfully applied to the study of exponential families, the most central reason being that $\kappa(\theta)$ is a strictly convex function on Θ. *See* GEOMETRY IN STATISTICS: CONVEXITY.

Example 1 (continued). The multinomial family is a regular exponential family of order $r - 1$, and with $\Theta = R^{r-1}$ and C equal to the simplex $\{(w_1, \ldots, w_{r-1}) : w_1 \geq 0, \ldots, w_{r-1} \geq 0, w_1 + \cdots + w_{r-1} \leq n\}$. Moreover, by (2),

$$\kappa(\theta) = n\ln(1 + e^{\theta_1} + \cdots + e^{\theta_{r-1}}).$$

Example 4. The inverse Gaussian distribution* is a distribution on the positive part of the real axis with probability function

$$P(x; \chi, \psi) = \frac{\sqrt{x}}{\sqrt{2\pi}} e^{\sqrt{\chi\psi}} x^{-3/2} e^{-(1/2)(\chi x^{-1} + \psi x)}.$$

Taking $t = (-\frac{1}{2}x^{-1}, -\frac{1}{2}x)$, $\theta = (\chi, \psi)$, and $\Theta = \{(\chi, \psi) : \chi > 0, \psi \geq 0\}$, one has an exponential family of order 2 which is full, but not regular, since Θ includes some of its own boundary points. Here $C = \{(w_1, w_2) : w_1 < 0, w_2 < 0, w_1 w_2 \leq \frac{1}{4}\}$ and

$$\kappa(\chi, \psi) = -\tfrac{1}{2}\ln\chi - \sqrt{\chi\psi}. \tag{5}$$

Hence, for $\psi > 0$ one finds by differentiation of (5) formulae such as

$$E_{(\chi,\psi)}x = \sqrt{\chi/\psi}, \quad E_{(\chi,\psi)}x^{-1} = \chi^{-1} + \sqrt{\psi/\chi}$$

and

$$V_{(\chi,\psi)}(x, x^{-1})$$
$$= \begin{bmatrix} \chi^{1/2}\psi^{-3/2} & -(\chi\psi)^{-1/2} \\ -(\chi\psi)^{-1/2} & 2\chi^{-2} + \chi^{-3/2}\psi^{1/2} \end{bmatrix};$$

see (4).

Various important operations on exponential families lead again to exponential families. In particular, the distribution of the canonical statistic t is exponential, and if $x_1, \ldots, x_n$ is a random sample from a population governed by $\mathscr{P}$, then the joint distribution of $x_1, \ldots, x_n$ is exponential with expo-

nential representation

$$a(\theta)^n b(x_1) \cdots b(x_n)$$
$$\times \exp\{\theta \cdot (t(x_1) + \cdots + t(x_n))\}. \tag{6}$$

Moreover, if $\theta = (\theta^{(1)}, \theta^{(2)})$ and $t = (t^{(1)}, t^{(2)})$ are similar partitions of θ and t, then the conditional distribution of $t^{(2)}$ given $t^{(1)}$ is exponential with $\theta^{(2)}$ and $t^{(2)}$ as canonical parameter and statistic, respectively. Thus conditioning on $t^{(1)}$ eliminates $\theta^{(1)}$, a fact that is extremely important from the viewpoint of conditional inference*.

Example 5. In the item analysis model the observations x_{ij} ($i = 1, \ldots, r; j = 1, \ldots, s$) are assumed to be independent 0–1 variates (*see* PSYCHOLOGICAL TESTING THEORY) with

$$\Pr[x_{ij} = 1] = e^{\alpha_i + \beta_j}/(1 + e^{\alpha_i + \beta_j}).$$

For instance, i may indicate a person and j may indicate a question, while x_{ij} is 1 or 0 according as person i answers question j correctly or not. The parameters α_i and β_j describe, respectively, the ability of person i and the difficulty of question j. This is an exponential model with $(x_1, \ldots, x_r, x_{.1}, \ldots, x_{.s})$, the set of row and column sums, as a canonical statistic and with $(\alpha_1, \ldots, \alpha_r, \beta_1, \ldots, \beta_s)$ as the corresponding canonical parameter. Inference on the row parameters, say, may be performed in the conditional distribution of $(x_1, \ldots, x_r)$ given $(x_{.1}, \ldots, x_{.s})$, which depends on $(\alpha_1, \ldots, \alpha_r)$ only. For most purposes, conditional inference on $(\alpha_1, \ldots, \alpha_r)$ will be superior to unconditional inference, especially if the number of columns s in the table of x_{ij}'s is large compared to the number of rows r. In particular, if r is fixed while s tends to infinity, then the unconditional maximum likelihood estimate* of $(\alpha_1, \ldots, \alpha_r)$ is not even consistent, whereas the conditional estimate is consistent and asymptotically normal (under mild regularity assumptions).

The conditionality phenomenon discussed in Example 5 is not particular to that example but similar conclusions hold in wide generality; see Andersen [2–4].

The *conjugate family**** of a full exponential family $a(\theta)b(x)\exp\{\theta \cdot t\}$ is the family of distributions on Θ having probability functions

$$p(\theta; \gamma, \chi) = d(\gamma, \chi)e^{\chi \cdot \theta - \gamma\kappa(\theta)};$$

i.e., θ is here considered as a random variable, γ and χ are parameters, and $d(\gamma, \chi)$ is a norming constant which makes the integral of $p(\theta; \gamma, \chi)$, relative to Lebesgue measure, equal to 1. Clearly, the conjugate family is also exponential. When it is appropriate to view θ as a random variable and $a(\theta)b(x)$ $\exp\{\theta \cdot t\}$ as the conditional distribution of x given θ, it will in many cases be convenient to choose the conjugate family as the model for θ. The classical construction of the negative binomial distribution as a mixture of the Poisson family with respect to a gamma distribution can be taken as an exemplification of this. Note also that if θ follows a distribution from the conjugate family, then so does the conditional distribution of θ given x (the posterior distribution of θ).

If Θ belongs to the interior of $\tilde{\Theta}$, in particular if $\mathscr{P}$ is regular, then the mean value parameter τ can be used instead of ω or θ as the parameter of $\mathscr{P}$. Also, in this case, the mixed parameter $(\tau^{(1)}, \theta^{(2)})$, where $\tau^{(1)} = E_\theta t^{(1)}$, affords a parametrization of $\mathscr{P}$. The latter parametrization has the property that if $\mathscr{P}$ is regular, then $\tau^{(1)}$ and $\theta^{(2)}$ are variation independent.

Any subfamily of an exponential family is, of course, exponential. For a full exponential family $\mathscr{P} = \{P_\theta : \theta \in \Theta\}$ those subfamilies $\mathscr{P}_0 = \{P_\theta : \theta \in \Theta_0\}$ that are *affine* (*linear*), i.e., for which Θ_0 is of the form $\Theta_0 = \Theta \cap L$ with L an affine (linear) subspace of R^k, are of special importance. In fact, the log-linear models**** are precisely of this kind. By a suitable choice of the exponential representation of $\mathscr{P}$ it can always be arranged that L is a linear subspace and moreover, if convenient, that Θ_0 is determined by $\Theta_0 = \{\theta : \theta^{(2)} = 0\}$. If $\mathscr{P}_0$ is linear and if $\mathbb{P}$ denotes the projection onto L, then $\mathscr{P}_0$ has a minimal exponential representation such that $\mathbb{P}\theta$ and $\mathbb{P}t$, interpreted as vectors of the dimension of L, are the canonical parameter and the canonical statistic in this representation.

Example 6. Most of the manageable models and submodels for one- or multidimensional normal random variables are affine in the sense described above.

Example 7. Let x_{ij} $(i = 1, \ldots, r; j = 1, \ldots, s)$ be independent 0–1 random variables and denote the probability that x_{ij} equals 1 by $\pi_{ij}(0 < \pi_{ij} < 1)$. The matrix x of these observations follows an exponential model of order $r \cdot s$ and with $\Theta = R^{r \cdot s}$. The item analysis model discussed in Example 5 is the linear subfamily determined by $\Theta_0 = \Theta \cap L = L$, where L is the linear subspace of $R^{r \cdot s}$ such that a point θ belongs to L if and only if $\theta_{ij} = \alpha_i + \beta_j$ for some $\alpha_i \in R$ and $\beta_j \in R(i = 1, \ldots, r; j = 1, \ldots, s)$. The projection $\mathbb{P}x$ of x on L is given by $(\mathbb{P}x)_{ij} = x_{i\cdot} + x_{\cdot j} - \bar{x}\ldots$ Thus $(\bar{x}.., x_1. - \bar{x}..,$ $\ldots, x_{r-1}. - \bar{x}.., x._1 - \bar{x}., \ldots, x._{s-1} - \bar{x}..)$ is a minimal canonical statistic for the item analysis model. This statistic is in one-to-one affine correspondence with the canonical (but not minimal canonical) statistic $(x_1., \ldots, x_r., x._1, \ldots, x._s)$.

Example 8. A full m-dimensional Poisson contingency table x is a set of independent Poisson random variables $x_{i_1 i_2 \ldots i_m}$ $(i_\nu = 1, \ldots, d_\nu, \nu = 1, \ldots, m)$ with freely varying mean value parameters $\lambda_{i_1 i_2 \ldots i_m}$. In this exponential model x is a minimal canonical statistic with θ given by $\theta_{i_1 i_2 \ldots i_m} = \ln \lambda_{i_1 i_2 \ldots i_m}$ as the corresponding canonical parameter. The hierarchical submodels are linear exponential subfamilies such that $\mathbb{P}t$ is in one-to-one affine correspondence with the so-called minimal set of fitted marginals of the table x.

If Θ contains an open subset of R^k, then the family of distributions of t is complete****, a fact that is useful, in particular, in relation to Basu's theorem*.

Let u be an arbitrary statistic such that u possesses a probability function $p(u; \theta)$ with respect to some measure ν. Then, assuming for simplicity that $0 \in \Theta$ and $a(0) = 1$ (as can always be arranged), one has

$$p(u; \theta) = a(\theta)E_0(e^{\theta \cdot t} | u)p(u; 0). \quad (7)$$

The problem of determining an explicit expression for the probability function for u is thus in effect solved if one can find such an expression for just one element of $\mathcal{P}$.

Example 9.   The von Mises–Fisher distribution* is the distribution on the unit sphere in d-dimensional Euclidean space whose density with respect to the uniform distribution on the sphere is of the form

$$a(\kappa)e^{\kappa \mu \cdot x} \quad (8)$$

where $\kappa \geqslant 0$ while x and μ are unit vectors in R^d. The norming constant depends on κ only and the class of von Mises–Fisher distributions on the unit sphere in R^d is an exponential family of order d and with x and $\theta = \kappa\mu$ as canonical variates.

Let $x_1, \ldots, x_n$ be a sample of n observations from the distribution (8). The model for $x_1, \ldots, x_n$ is again exponential with the resultant vector $x_. = x_1 + \cdots + x_n$ as canonical statistic and with $\theta = \kappa\mu$ as canonical parameter. By (7) the length r of the resultant $x_.$ has a density with respect to Lebesgue measure of the form

$$a(\kappa)^n E_0(e^{\kappa \mu \cdot x_.} | r)p(r; 0). \quad (9)$$

The advantage here is that the second and third factors in this expression are to be calculated under the uniform distribution for $x_1, \ldots, x_n$. In particular, observing that the unit vector in the direction of the resultant $x_.$ must also follow the uniform distribution when $\theta = 0$, one sees that $E_0(e^{\kappa \mu \cdot x_.} | r) = a(\kappa r)^{-1}$.

The log-likelihood function for ω based on a single observation x with distribution (1) is

$$l(\omega) = l(\omega; x) = \theta(\omega) \cdot t(x) - \kappa(\theta(\omega)), \quad (10)$$

or, in shorthand notation,

$$l(\theta) = l(\theta; t) = \theta \cdot t - \kappa(\theta).$$

This is a strictly concave function of θ provided that the canonical parameter domain Θ is convex.

Note that for a random sample of n observations $x_1, \ldots, x_n$ the log-likelihood function is of the form

$$l(\theta) = n\big(\theta \cdot \bar{t} - \kappa(\theta)\big), \quad (11)$$

where $\bar{t} = (t(x_1) + \cdots + t(x_n))/n$. By (4) the first- and second-order derivatives of (10) may be written

$$\frac{\partial l}{\partial \omega} = (t - \tau)\frac{\partial \theta'}{\partial \omega} \quad (12)$$

and

$$\frac{\partial^2 l}{\partial \omega \, \partial \omega'} = -\frac{\partial \theta}{\partial \omega'}V_\theta t \frac{\partial \theta'}{\partial \omega} + (t - \tau) \cdot \frac{\partial^2 \theta}{\partial \omega \, \partial \omega'}. \quad (13)$$

[The second term on the right-hand side of (13) is to be interpreted as the sum over i of $(t_i - \tau_i)\partial^2\theta_i/(\partial \omega \, \partial \omega')$.] From the latter formula it follows that the Fisher (or expected) information* function $i(\omega)$ is given by

$$i(\omega) = \frac{\partial \theta}{\partial \omega'}V_\theta t \frac{\partial \theta'}{\partial \omega}$$

and that this is related to the observed information function $j(\omega)$, i.e., minus the left-hand side of (13), by

$$i(\omega) = j(\omega) + (t - \tau) \cdot \frac{\partial^2 \theta}{\partial \omega \, \partial \omega'}. \quad (14)$$

In general, the maximum likelihood estimate $\hat{\theta}$ (or $\hat{\omega}$) has to be found by numerical iteration, although in a number of important special cases $\hat{\theta}$ can be expressed explicitly in terms of t. The Newton–Raphson algorithm* and the Davidon–Fletcher–Powell algorithm are both, ordinarily, very efficient for determining $\hat{\theta}$. For certain special types of models, notably log-linear models for contingency tables*, the so-called method of iterative scaling (or Deming–Stephan algorithm) provides a convenient procedure for the computation of $\hat{\theta}$ (see Darroch and Ratcliff [13]). *See* ITERATED MAXIMUM LIKELIHOOD ESTIMATOR.

The exponential families met in practice are mainly either regular or a curved subfamily of a regular family; and for the further discussion $\mathscr{P}$ will be assumed to be of one of these types and the two types will be considered in turn.

REGULAR EXPONENTIAL FAMILIES

For a regular exponential family $\mathscr{P}$ the maximum likelihood estimate exists, and is then —by the strict concavity of $l(\theta)$—unique if and only if the likelihood equation, which may be written [see (12)]

$$E_\theta t = t,$$

has a solution $\hat{\theta}$. This happens precisely when $t \in \operatorname{int} C$.[1] (In case the distributions of $\mathscr{P}$ are discrete, there is therefore a positive probability that $\hat{\theta}$ does not exist, as t will fall on the boundary of C with positive probability.)

It follows by (14) that $i(\hat{\omega}) = j(\hat{\omega})$; i.e., at the maximum likelihood point the observed information equals the expected information.

Suppose that $\mathscr{P}_0$ is a linear (or affine) subfamily of $\mathscr{P}$. If the maximum likelihood estimate exists under $\mathscr{P}$, then it also exists under $\mathscr{P}_0$. The converse does not hold true in general. The precise condition for the maximum likelihood estimate to exist under $\mathscr{P}_0$ is that, in the previously established notation, the likelihood equation $\mathbb{P}\tau(\theta) = \mathbb{P}t$ has a solution in Θ_0.

In repeated sampling the maximum likelihood estimate $\hat{\theta}$ is determined from $E_\theta t = \bar{t}$ [see (11)]. Since $\tau(\theta) = E_\theta t$ is a one-to-one smooth function it follows trivially from the asymptotic normality of $\bar{t}$ and from (4) that $\hat{\theta}$ is asymptotically normal with mean θ and variance (matrix) $(nV_\theta t)^{-1}$. More refined approximations to the distributions of $\bar{t}$ and $\hat{\theta}$ may be obtained from Edgeworth or saddlepoint expansions (see Barndorff-Nielsen and Cox [7]). When the order of $\mathscr{P}$ is 1, simple transformations are available for improving the approximate normality or for variance or spread stabilization. Specifically, for a fixed

$\lambda \in [0, 1]$, let

$$\phi(\theta) = \int \{ \kappa''(\theta) \}^\lambda d\theta$$

(indefinite integration), and note that $\bar{t} = \hat{\tau}$. Then

a. For $\lambda = \frac{1}{3}$, the transformation $\phi(\theta)$ improves normality relative to that of $\hat{\theta}$, and it symmetrizes the log-likelihood function by making the third derivative of this function equal to 0 at the maximum likelihood point $\hat{\phi}$.

b. For $\lambda = \frac{1}{2}$, $\phi(\theta)$ stabilizes the variance relative to that of $\hat{\theta}$, and it makes the second derivative of the log-likelihood function at $\hat{\phi}$—and hence the information function $j(\phi) = i(\phi)$—constant.

c. For $\lambda = \frac{1}{2}$, $\phi(\tau)$ stabilizes the variance relative to that of $\hat{\tau}$.

d. For $\lambda = \frac{2}{3}$, $\phi(\tau)$ improves normality relative to that of $\hat{\tau}$.

[Recall that $\phi(\tau)$ is an abbreviated notation for $\phi(\theta(\tau))$.]

Example 10. In the case of the gamma distribution

$$p(x; \theta) = \frac{\theta^\kappa}{\Gamma(\kappa)} x^{\kappa-1} e^{-\theta x}$$

with known shape parameter κ, these transformations are (a) $\phi = \theta^{1/3}$, (b) $\phi = \ln\theta$, (c) $\phi = \ln\tau$, and (d) $\phi = \tau^{1/3}$. Here $\tau = E_\theta x$ and multiplicative constants have been dropped from the expressions for ϕ.

CURVED EXPONENTIAL FAMILIES

Curved exponential models occur not seldomly in practice, and they are of great importance in discussions of various key concepts and methods of statistical inference (see, e.g., Efron [16] and Efron and Hinkley [18]). Instances of such models are given in Examples 2, 3, and 11.

In exponential models of this kind the canonical parameter domain Θ is a smooth manifold in R^k. Assume more specifically that Ω is an open subset of R^m where $m < k$, and that the $k \times m$ matrix $\partial\theta/\partial\omega'$ of partial derivatives of $\theta(\omega)$ is a continuous function of ω and has rank m for all $\omega \in \Omega$. The form of the score function (12) shows that, in general, the maximum likelihood estimate is determined as that value $\hat{\theta} \in \Theta$ for which $t - \tau(\hat{\theta})$ is orthogonal to the tangent plane of Θ at $\hat{\theta}$. The set of values of t that give rise to one and the same estimate $\hat{\theta}$ thus belongs to a hyperplane of dimension $k - m$. Without further assumptions it may be shown that asymptotically $\hat{\omega}$ exists, and is consistent and normally distributed with asymptotic variance $(ni(\omega))^{-1}$. Moreover, if $\mathcal{P}_0 = \{P_\theta : \theta \in \Theta_0\}$ is a curved subfamily of $\mathcal{P}$ with Θ_0 a manifold as above, the dimension of the manifold being $q < m$, then the likelihood ratio test statistic for $\mathcal{P}_0$ under $\mathcal{P}$ is asymptotically χ^2-distributed with $m - q$ degrees of freedom.

Example 11. A pure birth process* x_t, with birth intensity $\lambda \in (0, \infty)$ and initial population size 1, has been observed continuously over a fixed time interval $[0, t]$. The family of distributions for $\{x_s : 0 \leq s \leq t\}$ is curved exponential, of order 2, with $(\ln \lambda, -\lambda)$ as canonical parameter and (b_t, s_t) as the corresponding canonical statistic. Here b_t is the number of births in the interval $[0, t]$ and s_t is the total lifetime,

$$s_t = \int_0^t x_s \, ds.$$

The log-likelihood function is

$$l(\lambda) = b_t \ln \lambda - \lambda s_t$$

and thus

$$\partial l/\partial \lambda = b_t/\lambda - s_t.$$

On comparing with formula (12) one sees that the mean vector $\tau = E_\lambda(b_t, s_t)$ must be proportional to $(\lambda, 1)$; indeed,

$$\tau = \{(e^{\lambda t} - 1)/\lambda\}(\lambda, 1).$$

Moreover, the pairs (b_t, s_t) which give rise to one and the same maximum likelihood estimate $\hat{\lambda}$ are those on the half-line $\{w(\hat{\lambda}, 1) : w > 0\}$. (For details on this model, see the contribution by Keiding to the discussion of Efron [16].)

INCOMPLETELY OBSERVED EXPONENTIAL SITUATIONS

Suppose that the exponential model (1) holds but that only the value u of some function of x—and not x itself—has been observed. A variety of often-occurring statistical situations can usefully be viewed in this way, (see Sundberg [31], Haberman [20], and Pedersen [24]). In particular, many frequency tables for which some of the individuals or items studied are only partly classified fall within the present framework. Phenotype classifications in genetics are commonly of this kind.

By (7), the log-likelihood function based on u is

$$l(\theta) = \kappa(\theta \mid u) - \kappa(\theta)$$

where $\kappa(\theta \mid u) = \log E_0(\exp\{\theta \cdot t\} \mid u)$. If the exponential model is regular, then the first- and second-order derivatives of l with respect to θ take the form

$$\frac{\partial l}{\partial \theta} = E_\theta(t \mid u) - E_\theta t$$

and

$$\frac{\partial^2 l}{\partial \theta \, \partial \theta'} = V_\theta(t \mid u) - V_\theta t.$$

Thus the likelihood equation is

$$E_\theta t = E_\theta(t \mid u). \qquad (15)$$

(It may furthermore be noted that the conditional distribution of t given u is again exponential.) Cyclic iteration in this likelihood equation will usually converge to the maximum likelihood estimate $\hat{\theta}$, but convergence is generally slow (see Dempster, et al. [14]). However, in genetical applications the algorithm is often convenient. (In genetics*, this whole procedure for determining maximum likelihood estimates is known as the gene counting method and was developed by Cepellini, et al. [9] and Smith [29, 30]).

Example 3 (continued). Let x_{AA}, x_{AB}, x_{BB}, x_{BO}, x_{AO}, and x_{OO} denote the numbers of individuals in the sample of the various possible genotypes. These numbers follow a regular exponential family of order 2 and one may view the actually observed phenotype numbers $x_A = x_{AA} + x_{AO}$, $x_B = x_{BB} + x_{BO}$, x_{AB}, x_{OO} as deriving from incomplete observation of the genotypes. It now follows from (15), essentially without calculation, that the likelihood equations are

$$2np = x_A + x_{AB} + x_A \frac{p}{p + 2r},$$

$$2nq = x_B + x_{AB} + x_B \frac{q}{q + 2r}.$$

CONCLUDING REMARKS

The larger part of parametric statistical models occurring in the theory and applications of statistics are exponential or have a partially exponential structure, and the many useful properties shared by exponential families enhance their methodological importance. For obtaining a further appreciation of the role of exponential models in developing and illustrating principles and methods of statistics, the reader is referred to the books by Barndorff-Nielsen [5], Cox and Hinkley [11], and Lehmann [23]; see also Efron and Hinkley [18]. The shared properties mentioned above also imply that wide classes of exponential or partially exponential models can be handled by compact, integrated computer programs, such as in GLIM* or GENSTAT (*see* STATISTICAL SOFTWARE).

Further Reading

In addition to the references given in the article itself, some further bibliographical notes will be presented here. The many important properties of exponential families, beyond that of yielding sufficient reduction, have been discovered rather gradually and through the contributions of many research workers. Only a short bibliography will be presented here, but guidance to the large majority of works in the field is available via the references actually given here; see in particular Barndorff-Nielsen [5], Barndorff-Nielsen and Cox [7], Efron and Hinkley [18], and Lehmann [23].

Fisher's [19] indication that exponential families are the only families of distributions that yield nontrivial sufficiency reductions was quickly taken up by Darmois [12], Koopman [21], and Pitman [25], who sought rigorous mathematical formulations of this indication. Sometimes, therefore, the terms Darmois–Koopman family or Darmois–Koopman–Pitman family* are used instead of exponential family. The first fully satisfactory result in this direction is due to Dynkin [15].

In statistical mechanics* certain exponential families appear, by derivation from various assumptions, as models of the probability distributions of local properties in large physical systems; see, for instance, Kubo [22]. Models of this kind originated with Maxwell*, Boltzmann*, and Gibbs at the end of the last century.

A rather comprehensive account of the exact, as opposed to asymptotic, theory of exponential families is given in Barndorff-Nielsen [5]; see also Chentsov [10] and Efron [17]. Asymptotic properties, including the limiting behavior of maximum likelihood estimators, are discussed in Andersen [1], Berk [8], Sundberg [31], and Barndorff-Nielsen and Cox [7].

The relationships between ancillarity and sufficiency and exponential families are treated in Barndorff-Nielsen [5] and Efron and Hinkley [18].

An interesting class of models, the *factorial series* families, which possess a number of properties similar to those of exponential families, have been introduced and studied by Berg; see Barndorff-Nielsen [5]. The observation x and parameter ω corresponding to a factorial series family are both k-dimensional vectors whose coordinates are nonnegative integers, and the probability function is of the form

$$p(x; \omega) = a(\omega)b(x)\omega^{(x)},$$

where $\omega^{(x)} = \omega_1^{(x_1)} \omega_2^{(x_2)} \ldots \omega_k^{(x_k)}$ [the notation $n^{(m)}$ indicating the descending factorial, i.e., $n^{(m)} = n(n - 1) \cdots (n - m + 1)$]. Such distributions arise as the result of certain sampling procedures employed to obtain data for inference on the sizes of various classes of elements or individuals, the parameters $\omega_1, \ldots, \omega_k$ denoting these sizes.

NOTE

1. More generally, if $\mathcal{P}$ is full but not necessarily regular, then the maximum likelihood estimate* $\hat{\theta}$ exists if and only if $t \in \operatorname{int} C$; and $\hat{\theta}$ is unique. If, in addition, $E_\theta t = t$ has a solution in the interior of Θ, then this solution equals $\hat{\theta}$. For the maximum likelihood estimate to be always a solution of $E_\theta t = t$ it is therefore necessary and sufficient that $\mathcal{T} = \operatorname{int} C$. The latter condition is equivalent to $\kappa(\theta)$ being *steep*, which means that $|\partial \kappa / \partial \theta|$, the length of the gradient of $\kappa(\theta)$, tends to ∞ for θ tending to the boundary of Θ. This occurs, in particular, for regular families. The inverse Gaussian family of distributions provides an example of a steep, nonregular exponential family.

References

[1] Andersen, A. H. (1969). *Bull. Int. Statist. Inst.*, **43**, Bk. 2, 241–242.

[2] Andersen, E. B. (1970). *J. R. Statist. Soc. B*, **32**, 283–301.

[3] Andersen, E. B. (1971). *J. Amer. Statist. Ass.*, **66**, 630–633.

[4] Andersen, E. B. (1973). *Conditional Inference and Models for Measuring*. Mentalhygiejnisk Forlag, Copenhagen.

[5] Barndorff-Nielsen, O. (1978). *Information and Exponential Families*. Wiley, Chichester, England.

[6] Barndorff-Nielsen, O., Blaesild, P., Jensen, J. L., and Jørgensen, B. (1981–1982). *Proc. R. Soc. London* A. To appear.

[7] Barndorff-Nielsen, O. and Cox, D. R. (1979). *J. R. Statist. Soc. B*, **41** 279–312.

[8] Berk, R. H. (1972). *Ann. Math. Statist.*, **43**, 193–204.

[9] Ceppellini, R., Siniscalco, M., and Smith, C. A. B. (1955). *Ann. Hum. Genet. Lond.*, **20**, 97–115.

[10] Chentsov, N. N. (1972). *Statistical Decision Rules and Optimal Conclusions*. Nauka, Moscow (in Russian).

[11] Cox, D. R. and Hinkley, D. V. (1974). *Theoretical Statistics*. Chapman & Hall, London.

[12] Darmois, G. (1935). *C. R. Acad. Sci. Paris*, **260**, 1265–1266.

[13] Darroch, J. N. and Ratcliff, D. (1972). *Ann. Math. Statist.*, **43**, 1470–1480.

[14] Dempster, A. P., Laird, N. M., and Rubin, D. B. (1977). *J. R. Statist. Soc. B*, **39**, 1–38.

[15] Dynkin, E. B. (1951). *Select. Trans. Math. Statist. Prob.*, **1**, 23–41. (1961: English transl.; original in Russian).

[16] Efron, B. (1975). *Ann. Statist.*, **3**, 1189–1242.

[17] Efron, B. (1978). *Ann. Statist.*, **6**, 362–376.

[18] Efron, B. and Hinkley, D. V. (1978). *Biometrika*, **65**, 457–487.

[19] Fisher, R. A. (1934). *Proc. R. Soc. Lond. A*, **144**, 285–307.

[20] Haberman, S. J. (1974). *Ann. Statist.*, **2**, 911–924.

[21] Koopman, L. H. (1936). *Trans. Amer. Math. Soc.*, **39**, 399–409.

[22] Kubo, R. (1965). *Statistical Mechanics*. North-Holland, Amsterdam.

[23] Lehmann, E. L. (1959). *Testing Statistical Hypotheses*. Wiley, New York.

[24] Pedersen, J. G. (1978). *Ann. Hum. Genet. London*, **42**, 231–237.

[25] Pitman, E. J. G. (1936). *Proc. Camb. Philos. Soc.*, **32**, 567–579.

[26] Roy, K. K. (1975). *Sankhyā* A, **37**, 82–92.

[27] Rukhin, A. L. (1974). *Zap. Naucvn. Sem. Leningrad. Otdel. Mat. Inst. Steklov*, **43**, 59–87. *J. Sov. Math.*, **9**, 886–910. (1978: English transl.).

[28] Rukhin, A. L. (1975). In *Statistical Distributions in Scientific Work*, Vol. 3, G. P. Patil, S. Kotz, and J. K. Ord, eds. D. Reidel, Dordrecht, Holland.

[29] Smith, C. A. B. (1957). *Ann. Hum. Genet. Lond.*, **21**, 254–276.

[30] Smith, C. A. B. (1967). *Ann. Hum. Genet. Lond.*, **31**, 99–107.

[31] Sundberg, R. (1974). *Scand. J. Statist.*, **1**, 49–58.

O. BARNDORFF-NIELSEN

EXPONENTIAL SCORES *See* LOGRANK SCORES

EXPONENTIAL SMOOTHING

Exponential smoothing methods are widely used in operations research*, business, economic, and engineering contexts, particularly in the areas of inventory and production control. See, for example, Smith [12], McKenzie [9], Bradshaw [2], Dyer [5], Little [8], Brennan [3], Wu [13], and Pandit, Burney and Wu [10]. Perhaps the primary and most definitive reference to exponential smoothing per se is Brown [4, Chaps. 7 and 12]. In general, the concept of exponential smoothing is seen by contemporary statistical authors as a special case of a more general modeling procedure. We open our discussion with the classical motivation for exponential smoothing.

Consider a time series* $X_1, \ldots, X_n$ which may not be sufficiently smooth so that we may make appropriate inferences. The classic moving average smoother can be represented as

$$Y_t = \frac{1}{m} \sum_{j=0}^{m-1} X_{t-j},$$

where Y_t represents the average of the m most recent observations. In general, this moving average can be represented recursively as

$$Y_t = Y_{t-1} + \frac{X_t - X_{t-m}}{m}, \qquad (1)$$

which for large complex systems of data is desirable as it minimizes recomputations. One further motivation for the development of an exponential smoother is to recognize that, although the recursive formulation of the moving average minimizes recomputation, it still requires the storage of past values of X_t. What is desired is a smoother that requires only the most recent value of the smoothed data, namely Y_{t-1}, and the current value of the unsmoothed data, X_t. Clearly, in formula (1) the only other quantity involved in the computation of Y_t is X_{t-m}. If X_{t-m} is unknown, a best estimate of it would be Y_{t-1} so that (1) becomes

$$Y_t = Y_{t-1} + \frac{X_t - Y_{t-1}}{m}$$

or, rewriting,

$$Y_t = \left(1 - \frac{1}{m}\right) Y_{t-1} + \frac{1}{m} X_t.$$

In general, then, the exponentially smoothed version of X_t is

$$Y_t = \alpha X_t + (1 - \alpha) Y_{t-1}, \qquad 0 < \alpha < 1. \qquad (2)$$

Clearly, by recursively substituting in (2),

$$Y_t = \alpha X_t + (1 - \alpha)\alpha X_{t-1} + (1 - \alpha)^2 Y_{t-2}$$

$$= \alpha \sum_{k=0}^{\infty} (1 - \alpha)^k X_{t-k}. \qquad (3)$$

Since the weights on the observations X_{t-k} are exponentially decreasing, the smoother is called an exponential smoother. (Strictly speaking, the series is a geometric series so that geometric smoothing would be a more appropriate term.) In essence, an exponential smoother is an infinite moving average filter with weights declining exponentially or, more properly, geometrically.

Clearly, for a finite time series there is a problem in computing an infinite moving average. In practice the required initial value of Y_t is obtained from an initial average of the first observations X_t. Exponential smoothing is quite common in some literatures. For example, in the field of systems engineering, it is the simplest case of proportional control where an estimate is corrected with each new observation in proportion to the difference between the previous estimate and the new observation. In terms of inventory and production control, it is, in general, not desirable to change production schedules dramatically each month since this would be disruptive to the function of the labor force. Thus production Y_t, scheduled in month t is a weighted average of the production Y_{t-1} scheduled in the previous month and the demand X_t for the month t. Thus a production control system would follow equation (2) so that the production is the exponentially smoothed demand. Similarly, psychologists model the learning process by an exponential smoother. For example, one's comprehension level Y_t at time t is a weighted average of previous comprehension level Y_{t-1} and the new experience X_t; see Brown [4].

Readers familiar with Box-Jenkins models* (see Box and Jenkins [1]) will recognize a relationship of the exponential smoother to the ARIMA models. We let a_t represent an i.i.d. sequence of random variables (the shocks), B, a backward shift operator, and ∇, a difference operator. (See Box and Jenkins [1] for complete notation, and FINITE DIFFERENCES, CALCULUS OF.) Then the so-called ARIMA (0, 1, 1) is represented by

$$\nabla X_t = (1 - \theta B)a_t, \qquad -1 < \theta < 1. \quad (4)$$

Let us suppose we can write X_t in the inverted form

$$\Pi(B)X_t = a_t, \qquad (5)$$

where

$$\Pi(B) = I - \sum_{j=1}^{\infty} \Pi_j B^j.$$

Substituting (5) in (4),

$$\Pi(B)(I - \theta B) = (I - B) = \nabla.$$

Equating coefficients of B, we have

$$\Pi_j = (1 - \theta)\theta^{j-1}$$

or

$$\Pi_j = \alpha(1 - \alpha)^{j-1}$$

with

$$\alpha = 1 - \theta.$$

Thus

$$X_t = \sum_{j=1}^{\infty} \Pi_j X_{t-j} + a_t$$

$$= \alpha \sum_{j=1}^{\infty} (1 - \alpha)^{j-1} X_{t-j} + a_t \qquad (6)$$

The process Y_t defined by

$$Y_t = \alpha \sum_{j=1}^{\infty} (1 - \alpha)^{j-1} X_{t-j}$$

is an exponentially weighted moving average (exponential smoother) and it is not hard to show that

$$Y_t = \alpha X_t + (1 - \alpha)Y_{t-1}, \qquad 0 < \alpha < 2,$$
$$(7)$$

which is essentially equation (2). Since $-1 < \theta < 1$, it follows that $0 < \alpha < 2$ rather than $\alpha < 1$ as in (2). The process Y_t is called, in Box-Jenkins, the "level" of the process X_t. Noting from (6) that

$$X_t = Y_t + a_t,$$

and substituting this in equation (7), we have that

$$Y_t = Y_{t-1} + \alpha a_t. \qquad (8)$$

In general, the process X_t is nonstationary. Y_t, being an exponentially smoothed version of X_t, should be smoother than X_t and, indeed, equation (8) tells us that the change in level is not the full value of a_t but the value of a_t tempered or not by α. For $1 < \alpha < 2$, the "level" changes faster than the process and, of course, for $0 < \alpha < 1$, it changes

more slowly. *See* also AUTOREGRESSIVE-INTEGRATED MOVING AVERAGE MODELS.

There is also an intimate relation of exponential smoothers to digital filters. A digital filter is simply a linear time, invariant filter in discrete time. A digital filter is said to be recursive if the output at time t depends linearly on a fixed number of input and output values, that is, if

$$Y_t = - \sum_{j=1}^{p} \phi_j Y_{t-j} + \sum_{k=0}^{q} \sigma_k X_{t-k}.$$

(Again readers familiar with Box-Jenkins models will recognize this as an ARIMA (p, q) model except that the X_t process is not necessarily white noise.) Thus if $p = 1$, $q = 0$, $-\phi_1 = 1 - \alpha$ and $\theta_0 = \alpha$, it is clear that the exponential smoothing operation is simply a very special case of a recursive filter. Theory for recursive filters is fairly well developed especially in the electrical engineering literature. See Koopmans [7] and Rabiner and Rader [11]. Perhaps the most well known special case of the recursive filter is the so-called Kalman filter*; see ref. 6.

References

[1]   Box, G. E. P. and Jenkins, G. M. (1976). *Time Series Analysis Forecasting and Control*, Holden-Day, San Francisco. The classic text on ARIMA models. Easy mathematics generally with somewhat heuristic, nonrigorous arguments.

[2]   Bradshaw, A. (1977). *Int. J. Syst. Sci.*, **8**, 1123–1134. Illustrative of production control modeling using exponential smoothing. Shows that exponential smoothing leads to poor transient behavior of the production system.

[3]   Brennan, J. M. (1977). *J. Syst. Manage.*, **28**, 39–45. Popularized article on statistical inventory management based on exponential smoothing.

[4]   Brown, R. G. (1962). *Smoothing, Forecasting and Prediction of Discrete Time Series*, Prentice Hall, Englewood Cliffs, NJ. One of the earliest treatments of the forecasting of time series and probably the definitive statement of exponential smoothing. Contains two relevant chapters, one entitled, "Exponential Smoothing," the second, "General Exponential Smoothing." The mathematical level is generally easy.

[5]   Dyer, T. G. J. (1977). *Q.J.R. Meteorol. Soc.*, **103**, 177–189. An illustration of the use of exponential

smoothing as well as other techniques for short- and long-term rain forecasting.

[6] Kalman, R. E. (1960). *J. Basic Eng.*, **82**, 34–45. The paper which introduced the Kalman filter. Followed a year later by a joint paper with Bucy in the same journal. These two established many of the fundamental properties.

[7] Koopmans, L. H. (1974). *The Spectral Analysis of Time Series*, Academic Press, New York. An excellent treatment of filters in general and recursive filters in particular. A moderate level of mathematical difficulty with good rigor.

[8] Little, J. D. C. (1977). *IEEE Trans. Autom. Control*, **AC-22**, 187–195. An illustration of an adaptive control approach to marketing featuring the use of exponential smoothers. The optimal control turns out to involve rather simple exponential smoothing rules.

[9] McKenzie, E. (1978). *J. Oper. Res. Soc.*, **29**, 449–458. A more technical paper for tracking the error in one-step-ahead forecasts. The error process is obtained for the most general form of exponential smoothed systems used in optimal conditions.

[10] Pandit, S. M., Burney, F. A., and Wu, S. M. (1976). *J. Eng. Indust. Trans. ASME (B)*, **98**, 614–619. An illustration of the use of exponential smoothing in a rather specific engineering context.

[11] Rabiner, L. R. and Rader, C. M., eds. (1972). *Digital Signal Processing*, IEEE Press, New York. An assortment of papers republished by the IEEE on various aspects of filtering and spectral analysis. Contains several papers on recursive filters.

[12] Smith, L. D. (1978). *Omega*, **6**, 83–88. Illustrates the use of exponential smoothers on enrollment projections for a public school system.

[13] Wu, S. M. (1977). *J. Eng. Indust. Trans. ASME (B)*, **99**, 708–714. Related to Pandit, Burney, and Wu above, but illustrates the use of exponential smoothing in a more general dynamically changing system.

(AUTOREGRESSIVE-INTEGRATED MOVING AVERAGE MODELS GEOMETRIC MOVING AVERAGE GRADUATION)

E. J. Wegman

EXPOSED TO RISK

A concept that is useful in the analysis of incomplete mortality data*. Such data arise when individuals enter and/or leave an experience during the period of observations, from causes other than death. They are "exposed to risk" (of death) just for the period they are in the experience.

"Exposed to risk" is measured in units of individual $\times$ time (e.g., person-years). It is a numerical quantity obtained by summing the periods of exposure in a certain category (e.g., age last birthday) of all individuals in experience.

Bibliography

Batten, R. W. (1978). *Mortality Table Construction.* Prentice-Hall, Englewood Cliffs, N. J.

Elandt-Johnson, R. C. and Johnson, N. L. (1980). *Survival Models and Data Analysis.* Wiley, New York, Chap. 3.

Gershenson, H. (1961). *Measurement of Mortality.* Society of Actuaries, Chicago.

(ACTUARIAL STATISTICS, LIFE LIFE TABLES MORTALITY DATA)

EXTRAPOLATION

Estimating the expected value of a dependent variable* outside the range of values of predicting variables on which the estimation is based.

A common form of extrapolation arises with time series* when a formula is fitted to values $u_1, u_2, \ldots, u_n$ observed at times $t_1 < t_2 < \cdots < t_n$ and the fitted formula is used to predict the value at t, with t greater than t_n.

More generally, a regression function* of a variable Y on a predicting variable X, based on observations (X_i, Y_i) $(i = 1, \ldots, n)$ may be used to predict values of Y corresponding to values of X outside the range $X_1, X_2, \ldots, X_n$ [i.e., less than $\min(X_1, \ldots, X_n)$ or greater than $\max(X_1, \ldots, X_n)$].

Usually, considerable caution is needed in using extrapolation, because it is based on the assumption that the model used in fitting remains valid outside the range of observation. The greater the reliance that can be placed on this assumption, the greater is the

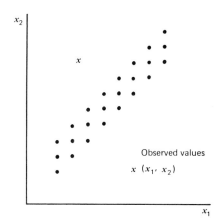

Figure 1   Hidden extrapolation.

confidence in the accuracy of the results of extrapolation.

It is sometimes overlooked that in multiple regression* situations where Y is predicted from k different variables X_1, $\ldots, X_k$, using observations on n individuals $(Y_i, X_{1i}, \ldots, X_{ki})$ $(i = 1, \ldots, n)$, a particular set of values $(X_1, X_2, \ldots, X_k)$ may be such that each X_h is well within the range of observed values $\min(X_{hi}) - \max(X_{hi})$, and yet the point $(X_1, X_2, \ldots, X_k)$ is well outside the region of observed sets of values $(X_{1i}, X_{2i}, \ldots, X_{ki})$. When a fitted multiple regression* formula is used to estimate the expected value of Y corresponding to $(X_1, X_2, \ldots, X_k)$, this is an application of extrapolation, even though each X_h is within the range of its observed values. Figure 1 exemplifies such a situation for $k = 2$.

(**INTERPOLATION**
PREDICTION)

EXTREMAL PROCESSES

Extreme values and records have traditionally been of great importance in connection with structural strengths and sufficiency of design in life testing* and reliability* and also in climatological studies connected with floods, droughts, pollution, and earthquakes. Classical contexts are discussed in Gumbel's 1958 book [15]. Glick [13] gives a very enter-

taining review of the relation of records to traffic modeling; testing for randomness*, constructing tolerance limits* for failure distributions, and more. (See also Galambos [12] and Barr [2].)

For the study of the stochastic behavior of maxima and records, it is useful to have at one's disposal a tool called extremal processes. In this survey we review the properties and uses of extremal processes.

Suppose that $\{X_n, n \geqslant 1\}$ are independent, identically distributed (i.i.d.) real random variables with common CDF $F(x)$. Set $Y_n = \bigvee_{i=1}^{n} X_i$. (Here and in what follows $\bigvee_{i=1}^{n} X_i$ means $\max\{X_1, \ldots, X_n\}$ and $\bigwedge_{i=1}^{n} X_i$ means $\min\{X_1, \ldots, X_n\}$.) If we are interested in the properties of $\{Y_n, n \geqslant 1\}$ as a process, we need the finite-dimensional (fi di) distributions and a little reflection shows that for integers $1 \leqslant t_1 < t_2 < \ldots < t_k$ and reals $x_1, \ldots, x_k$, we have for $P[Y_{t_i} \leqslant x_i, i \leqslant k]$ the expression

$$F_{t_1, \ldots, t_k}(x_1, \ldots, x_k)$$
$$= F^{t_1}\left(\bigwedge_{i=1}^{k} x_i\right) F^{t_2 - t_1}\left(\bigwedge_{i=2}^{k} x_i\right) \times$$
$$\cdots \times F^{t_k - t_{k-1}}(x_k). \qquad (1)$$

If we remove the restriction that the t's be integers and require only that $0 < t_1 < \cdots < t_k$, then (1) defines a consistent family of fi di distributions. By the Kolmogorov extension theorem there exists a continuous-time process $\{Y(t), t > 0\}$ with these fi di's and such a process is called an *extremal process* generated by F.

An immediate result of the construction is that

$$\{Y_n, n \geqslant 1\} \stackrel{d}{=} \{Y(n), (n \geqslant 1)\}. \qquad (2)$$

(Here $\stackrel{d}{=}$ means that the two sequences have the same distribution.) So the original discrete-time process $\{Y_n\}$ may be considered to be embedded (*see* EMBEDDED PROCESSES) in the continuous time $Y(\cdot)$. This embedding allows properties discovered in continuous time to be transferred to the discrete-time process $\{Y_n\}$.

Extremal processes arise in the following ways:

1. Let N be the counting measure of a Poisson process* on $R_+ \times R$ with points $\{t_k, j_k\}$ and mean measure $EN(dt \times dx) = dt \times m(dx)$, where there must exist some x_0 such that if $x > x_0$, then $m(x, \infty) < \infty$. Define $Y(t) = \sup\{j_k : t_k \leqslant t\}$. Then

$$P\big[\, Y(t) \leqslant x \,\big] = P\big[\, N((0, t) \times (x, \infty)) = 0 \,\big]$$
$$= \exp\{-EN((0, t] \times (x, \infty))\}$$
$$= \exp\{-tm(x, \infty)\}.$$

From the fact that $\{N(A_i), i \leqslant k\}$ are independent Poisson variables when $\{A_i, i \leqslant k\}$ are disjoint subsets of $R_+ \times R$, we get (1) with

$$F(x) = \exp\{-m(x, \infty)\}.$$

Conversely, any extremal-F process may be realized by constructing a Poisson process on $R_+ \times R$ with mean measure $dt \times m(dx)$, where $m(x, \infty) = -\log F(x)$ and applying the functional indicated above.

2. Let $\{X(t), t \geqslant 0\}$ be a stochastically continuous process with stationary*, independent increments with paths that are right continuous with left limits and suppose that the Lévy measure* is m. Define $Y(t) = \sup\{X(s) - X(s-): s \leqslant t\}$. Then Y is extremal generated by $F(x) = \exp\{-m(x, \infty)\}$. Since the jump times and jump sizes of X form a Poisson process on $R_+ \times R$ with mean measure $dt \times m(dx)$ (see Breiman [4, Chap. 14]), the reasoning of I above applies [10].

3. *Weak convergence*: Suppose that $\{X_n, n \geqslant 1\}$ are i.i.d. and there exist $a_n > 0$, b_n such if $Y_n = \bigvee_{i=1}^{n} X_i$,

$$\lim_{n \to \infty} P\big[\, Y_n \leqslant a_n x + b_n \,\big] = G(x), \quad (3)$$

nondegenerate for all $x \in R$. Then G is a classical extreme value distribution* [5, 14] and a functional limit theorem follows: For $t > 0$ define

$$Y_n(t) = a_n^{-1}(Y_{[nt]} - b_n).$$

Then

$$Y_n(\cdot) \xrightarrow{\mathcal{D}} Y(\cdot) \quad (4)$$

where $Y(\cdot)$ is extremal generated by G [8, 16, 22, 32–34]. In (4), $\xrightarrow{\mathcal{D}}$ means convergence in distribution of the stochastic processes in the function space $D(0, \infty)$, the space of real-valued functions which are right continuous with finite left limits existing on $(0, \infty)$, and $D(0, \infty)$ is metrized by the Skorokhod metric*. (See Billingsley [3] for a discussion of this and other matters related to invariance principles; see also Lindvall [17].) This convergence implies but is stronger than convergence of the fi di distributions of $Y_n(\cdot)$ to those of $Y(\cdot)$. Through the magic of the invariance principle* [3], if maxima of non-i.i.d. sequences obey a functional limit theorem such as (4), then asymptotic behavior of such maxima will be the same as that of $\{Y_n\}$ satisfying (3) and (4). *See also* CONVERGENCE OF SEQUENCES OF RANDOM VARIABLES.

From the representation in I it follows that Y is stochastically continuous with nondecreasing paths. In fact, $\lim_{t \to \infty} \uparrow Y(t) = \sup\{x : F(x) < 1\}$ and $\lim_{t \to 0} \downarrow Y(t) = \inf\{x : F(x) > 0\}$. From the form of the fi di's in (1) we see that Y is a Markov process* (see Breiman, [4, Chap. 15]) and in fact Y is easily seen to be a Markov jump process*. Let $\{\tau_n, -\infty < n < \infty\}$ be the times when Y jumps. Then the range of Y, $\{Y(\tau_n), -\infty < n < \infty\}$, is the embedded Markov process* of states visited and one quickly checks

$$P\big[\, Y(\tau_{n+1}) \geqslant y \mid Y(\tau_n) = x \,\big]$$
$$= \begin{cases} \dfrac{m(y, \infty)}{m(x, \infty)} & \text{if } y \geqslant x \\ 0 & \text{if } y < x \end{cases} \quad (5)$$

(see Breiman [4, Secs. 15.5, 15.6]); remember that $m(x, \infty) = -\log F(x)$. In the special case that $F(x) = \Lambda(x) = \exp\{-e^{-x}\}$, we get

$$P\big[\, Y(\tau_{n+1}) \leqslant y \mid Y(\tau_n) = x \,\big]$$
$$= \begin{cases} 1 - e^{-(y-x)}, & y \geqslant x \\ 0, & y < x, \end{cases}$$

which we recognize as the transition function of the Markov process consisting of sums of i.i.d. exponential random variables. Thus $\{Y(\tau_n)\}$ has the same distribution as the sequence of points of a homogeneous Poisson process. For continuous but otherwise general F, we find by a probability integral transform method that $\{Y(\tau_n)\}$ is the point sequence of a nonhomogeneous Poisson process and the expected number of points in $(a, b]$ is

$$-\log(-\log F(b)) - (-\log(-\log F(a))).$$

In fact, the two-dimensional point process with points $\{(Y(\tau_n), \tau_{n+1} - \tau_n)\}$ is also Poisson.

There is a discrete-time analog of these results which can be obtained by the embedding (2). If $\{X_n, n \geq 1\}$ is an i.i.d. sequence with common continuous CDF $F(x)$ and $Y_n = \bigvee_{i=1}^{n} X_i$, then $\{Y_n, n \geq 1\}$ is Markov and the succession of states visited $\{Y_{L(n)}\} = \{X_{L(n)}\}$ is the *record value sequence* and $\{L(n)\}$ is the sequence of indices where records* occur. Put another way, set $L(0) = 1$, $L(1) = \inf\{j : Y_j > Y_1\}$ and $L(n + 1) = \inf\{j > L(n): Y_j > Y_{L(n)}\}$. We find that $\{X_{L(n)}\}$ is the point sequence of a Poisson process and the mean number of points in $(a, b]$ is

$$-\log(1 - F(b)) - (-\log(1 - F(a))). \quad (6)$$

Furthermore, $\{(X_{L(n)}, L(n + 1) - L(n))\}$ are the points of a two-dimensional Poisson process.

Back to continuous time: We seek information about the structure of the jump times $\{\tau_n\}$ of Y. When F is continuous, $\{\tau_n\}$ is the point sequence of a nonhomogeneous Poisson process with intensity t^{-1}. This can be obtained from already discussed results about the range of Y as follows. Set $Q(x) = -\log F(x)$ and suppose that $\{E_j\}$, $\{U_j\}$ are two i.i.d. sequences with $P[E_j > x] = e^{-x}$, $x > 0$ and $P[U_j \leq x] = x$, $0 \leq x \leq 1$. Finally, define $T(t) = \inf\{n \mid U_n \leq t\}$. Then

$$\{Y(t), 0 < t \leq 1\}$$

$$\overset{d}{=} \left\{ Q^{-1}\left(\sum_{i=1}^{T(t)} E_i \right), 0 < t \leq 1 \right\},$$

$$(7)$$

as can be seen by verifying (1) for the process on the right. (Alternatively, one may return to the Poisson process of I, and order the points in $[0, 1) \times R$ according to decreasing second components. These points can then be represented as $\{(U_n, Q^{-1}(\sum_{i=1}^{n} E_i))\}$.) The number of times Y jumps for $t \in (a, b] \subset (0, 1]$ is the number of jumps of $T(t)$, $t \in (a, b]$ by (7). From the definition of $T(t)$, this is the number of records from the i.i.d. sequence $\{U_n^{-1}\}$ falling in $[b^{-1}, a^{-1})$. We know records form a Poisson process, so the number of jumps of Y in $(a, b]$ is Poisson and from (6) the mean is

$$-\log P[U_1^{-1} > a^{-1}]$$

$$-(-\log P[U_1^{-1} > b^{-1}]) = \log\frac{b}{a}.$$

The assertion about $\{\tau_n\}$ being a Poisson process now follows readily.

Previously, we saw that $\{Y(\tau_n)\}$ and $\{Y_{L(n)}\}$ were both point sequences of Poisson processes and we just saw that $\{\tau_n\}$ is also Poisson. The sequence $\{L(n)\}$ is not Poisson, but it does behave asymptotically like a Poisson process in a manner we now describe. Define ν and μ as the counting functions of the point processes with points $\{\tau_n\}$ and $\{L(n)\}$, respectively:

$$\nu(a, b] = \sum_{n=-\infty}^{\infty} 1_{(a,b]}(\tau_n)$$

$$\mu(a, b] = \sum_{n=0}^{\infty} 1_{(a,b]}(L(n)).$$

If we consider $\{Y_n\}$ as embedded in $\{Y(t), t > 0\}$, which we may do because of (2), then ν, μ are both defined on the same space and so may be directly compared. Observe that

$$[\mu(n - 1, n] = 1] = [\nu(n - 1, n] > 0]$$

and thus ν counts jumps that μ misses since μ only checks to see whether or not $Y(n) > Y(n - 1)$ but is not sensitive to all jumps of Y in $(n - 1, n]$. If $\nu(n - 1, n] > 1$ for infinitely many n, then μ and ν will not be related in a useful way. Fortunately,

$$P[\nu(n - 1, n] > 1 \quad \text{i.o.}] = 0$$

since $\sum P[\nu(n - 1, n] > 1] < \infty$ [recall that

$E\nu(n - 1, n] = \log(n/(n - 1))]$. Therefore, there exists a random variable j such that for all n sufficiently large,

$$\nu[1, n] - \mu[1, n] = j, \qquad (8)$$

since eventually $\nu(n - 1, n] = 1$ or 0 only and $\mu(n - 1, n] = 1$ iff $\nu(n - 1, n] = 1$. The variable j represents the number of early multiple jumps in intervals of length 1 which are seen by ν but missed by μ.

The usual transformation technique shows that $\nu(1, e^t]$, $t > 0$, is the counting function of a homogeneous Poisson process and hence

$$\lim_{t \to \infty} \frac{\nu(1, t]}{\log t} = 1 \quad \text{a.s.}$$

$$\lim_{t \to \infty} P\left[\frac{\nu(1, t] - \log t}{\sqrt{\log t}} \leqslant x \right] = N(0, 1, x)$$

$$\lim_{t \to \infty} \sup \frac{\nu(1, t] - \log t}{\sqrt{\log t \log_3 t}} = 1 \quad \text{a.s.}$$

and because of (8) the same results hold with $\nu(1, t]$ replaced by $\mu(1, n]$. In a similar way strong laws, central limit theorems and iterated logarithm results may be proven for $\{L_n\}$ and $\{L_n - L_{n-1}, n \geqslant 1\}$. (see Resnick [20], Shorrock [24–26], Williams [36], Westcott [35], and Renyi [19].)

The invariance principle described in (4) says that whenever h is a continuous mapping from an appropriate subset of $D(0, \infty)$ into a range that is a complete, separable metric space, we have $hY_n \overset{\mathcal{D}}{\to} hY$. This invariance principle now has some teeth, since by capitalizing on our detailed knowledge of the structure of Y we can evaluate the distribution of hY for a variety of useful and interesting mappings h. Here is one application of how (4) can be used:

For a jump function $x(\cdot)$ let $\delta^+ x$ be the length of time between $t = 1$ and the first jump past $t = 1$. Similarly, $\delta^- x$ is the length of time between the last jump preceding $t = 1$ and $t = 1$. If (3) and (4) hold, then by the continuous mapping theorem [3, p. 30],

$$\delta^{\pm} Y_n(\cdot) \overset{\mathcal{D}}{\to} \delta^{\pm} Y(\cdot). \qquad (9)$$

Note that $\delta^{\pm} Y = \delta^{\pm} \nu$ and since we know that ν is nonhomogeneous Poisson, it will be easy to compute the distribution of either $\delta^+ \nu$ (forward recurrence time at $t = 1$) or $\delta^- \nu$ (backward recurrence time at $t = 1$). To evaluate $\delta^+ Y_n(\cdot)$ we have

$$\delta^+ Y_n(\cdot) = \inf\{t > 1 : Y_{[nt]} > Y_1\} - 1$$
$$= \inf\left\{\frac{k}{n} > 1 : Y_k > Y_1\right\} - 1$$
$$= (\inf\{k > n : Y_k > Y_1\} - n)/n$$
$$= (L(\mu[1, n] + 1) - n)/n$$

and the numerator of the ratio represents the number of observations past n necessary to obtain a new record. For $\delta^- Y_n(\cdot)$ we have

$$\delta^- Y_n(\cdot) = (n - L(\mu[1, n]))/n$$

and the numerator is the elapsed time between the last record time before n and n. From (9) for $x > 0$:

$$\lim_{n \to \infty} P\left[(L(\mu[1, n] + 1) - n)/n \leqslant x \right]$$
$$= x(1 + x)^{-1}$$

$$\lim_{n \to \infty} P\left[(n - L(\mu[1, n]))/n \leqslant x \right]$$
$$= x \wedge 1.$$

These results hold even when (3) fails as long as F is continuous, since the behavior of $\{L(n)\}$ is distribution-free* by the probability integral transform method. Other applications of the invariance principle are given in Resnick [22].

We close with a discussion of multivariate problems. If $\{X_n\}$ are real random variables, then X_j is a record if $X_j > \bigvee_{i=1}^{j-1} X_i$. This is a natural definition and few would argue with it. If $\{\mathbf{X}_n\} = \{X_n', X_n''\}$ are i.i.d. R^2-valued random vectors, then it is less obvious how to define what a record is. Here are some candidates:

1. $\mathbf{X}_n$ is a record if either $X_n' > \bigvee_{i=1}^{n-1} X_i'$ or $X_n'' > \bigvee_{i=1}^{n-1} X_i''$.

2. $\mathbf{X}_n$ is a record if $X_n' > \bigvee_{i=1}^{n-1} X_i'$ and $X_n'' > \bigvee_{i=1}^{n-1} X_i''$.

3. Given $h: R^2 \to R$, say $\mathbf{X}_n$ is a record if $h(\mathbf{X}_n) > \bigvee_{i=1}^{n-1} h(\mathbf{X}_i)$.

An obvious choice for h is $h(\mathbf{x}) = \|\mathbf{x}\|$, the Euclidean distance from the origin.

Definition 3 is appealing because it is easy* but it is also cheap; it is a one-dimensional definition and one-dimensional results will apply. For instance, if $h(\mathbf{X}_1)$ has a continuous distribution, then $\{\mathbf{X}_n : \mathbf{X}_n$ is a record according to definition 3$\}$ are the points of a Poisson process on R^2. This was observed by Deken [7], who also computes the mean measure.

In R^1, extremal processes illuminated the structure of records and we hope the same thing will be true in R^2. Definition (1) seems well suited to study the extremal process methods and we can try to define multivariate extremal processes by mimicking (1). So given F, a CDF on R^2, and $t_1 < \cdots < t_k$ and $\mathbf{x}_1 \leqslant \mathbf{x}_2 \leqslant \cdots \leqslant \mathbf{x}_k$ ($\mathbf{x} \leqslant \mathbf{y}$ if and only if $x' \leqslant y'$ and $x'' \leqslant y''$), we can try to define a multivariate process $\mathbf{Y}(\cdot)$ by

$$P\big[\mathbf{Y}(t_i) \leqslant \mathbf{x}_i, i \leqslant k\big]$$
$$= F^{t_1}(\mathbf{x}_i)F^{t_2 - t_1}(\mathbf{x}_2) \cdots F^{t_i - t_{k-1}}(\mathbf{x}_k).$$

One encounters an immediate problem: If F is a CDF on R^2, F^t may not be. Call F max-infinitely divisible (max id) if F^t is a CDF for all $t > 0$ (see Balkema and Resnick [1]). Then F is max id iff there exists a measure m on $[-\infty, \infty)^2$ satisfying

$$m(R \times [-\infty, \infty)) = m([-\infty, \infty) \times R) = \infty,$$

$$m\big(([-\infty, x'] \times [-\infty, x''])^c\big) < \infty$$

for some $\mathbf{x}$ and

$$F(\mathbf{x}) = \exp\big\{-m\big(([-\infty, x']$$
$$\times [-\infty, x''])^c\big)\big\}.$$

This is the case if there exists a Poisson process on $R_+ \times [-\infty, \infty)^2$ with counting function N, and points $\{(t_k, \mathbf{j}_k)\}$ such that

$$F(\mathbf{x}) = P\bigg[\bigg(\bigvee_{t_k \leqslant 1} j_k', \bigvee_{t_k \leqslant 1} j_k''\bigg) \leqslant \mathbf{x}\bigg]$$

and then setting

$$\mathbf{Y}(t) = \bigg(\bigvee_{t_k \leqslant t} j_k', \bigvee_{t_k \leqslant t} j_k''\bigg)$$

gives the multivariate extremal process. The scheme of I much earlier reappears here.

If $\{\mathbf{X}_n\}$ are i.i.d. R^2-valued vectors set

$$\mathbf{Y}_n = \bigg(\bigvee_{j=1}^n X_j', \bigvee_{j=1}^n X_j''\bigg).$$

Questions about possible limit laws of $\{Y_n\}$ and their domains of attraction* are satisfactorily resolved in de Haan and Resnick [6]. (See also Galambos [12, Chap. 5].) There is no trouble extending the multivariate analog of (3) to an invariance principle. However, because of the absence of detailed information about the structure of multivariate extremal processes, such an invariance principle is not presently very useful. Some preliminary properties of R^k-valued extremal processes are given in de Haan and Resnick [6].

Literature

The interested reader is referred to the list of references that follows. The papers of Shorrock [24–28] are especially instructive and offer alternative approaches to the ones offered here. In particular, they contain very clear discussions of the distribution theory of processes related to records and extremes. The survey by Glick [13] is light and very entertaining. Also useful are the works by Vervaat [31] and Galambos [12]. Papers dealing with the structure of extremal processes are Dwass [9–11], Pickands [18], Resnick [21], Resnick and Rubinovitch [23], Tiago de Oliveira [29], and Weissman [34]. For information about records, see Dwass [9], Rényi [19], Resnick [20, 21], Shorrock [24–26], Vervaat [30], Westcott [35], and Williams [36]. The basic reference to the theory of weak convergence is the fine book by Billingsley [3]; see also Lindvall [17]. Weak convergence for extremes is discussed in Durrett and Resnick [8], Gnedenko [14], de Haan [5], Lamperti [16], Resnick [22], and Weissman [32–34]. De Haan [5] and Vervaat [31] relate the theory of regular variation to limit law problems for extremes. Multivariate problems are treated in Balkema and Resnick [1], Galambos [12], de Haan and Resnick [6], and Weissman [32–34]. The relationship of extreme value problems and Poisson processes is considered in Balkema and Resnick [1], Deken [7], Durrett and Resnick [8], Dwass [10, 11], Pickands [18], Resnick [21, 22], Shorrock [27, 28], and Weissman [32–34].

References

[1] Balkema, A. and Resnick, S. (1977). *J. Appl. Prob.*, **14**, 309–319.

[2] Barr, D. (1972). *Math. Mag.*, **45**, 15–19.

[3] Billingsley, P. (1968). *Convergence of Probability Measures*. Wiley, New York.

[4] Breiman, L. (1968). *Probability*. Addison-Wesley, Menlo Park, Calif.

[5] de Haan, L. (1970). *On Regular Variation and Its Application to the Weak Convergence of Sample Extremes*. Math. Centre Tract. No. 32, Amsterdam.

[6] de Haan, L. and Resnick, S. (1977). *Z. Wahrscheinlichkeitsth. verwend. Geb.* **40**, 317–337.

[7] Deken, J. (1976). *On Records: Scheduled Maxima Sequences and Longest Common Subsequences*. Ph.D. thesis, Dept. of Statistics, Stanford University.

[8] Durrett, R. and Resnick, S. (1978). *Ann. Prob.*, **6**, 829–846.

[9] Dwass, M. (1964). *Ann. Math. Statist.*, **35**, 1718–1725.

[10] Dwass, M. (1966). *Ill. J. Math.*, **10**, 381–391.

[11] Dwass, M. (1974). *Bull. Inst. Math. Acad. Sin.*, **2**, 255–265.

[12] Galambos, J. (1978). *The Asymptotic Theory of Extreme Order Statistics*. Wiley, New York.

[13] Glick, N. (1978). *Amer. Math. Monthly*, **85**, 2–26.

[14] Gnedenko, B. V. (1943). *Ann. Math.*, **44**, 423–453.

[15] Gumbel, E. (1958). *Statistics of Extremes*. Columbia University Press, New York.

[16] Lamperti, J. (1964). *Ann. Math. Statist.*, **35**, 1726–1737.

[17] Lindvall, T. (1973). *J. Appl. Prob.*, **10**, 109–121.

[18] Pickands, J. (1971). *J. Appl. Prob.*, **8**, 745–756.

[19] Rényi, A. (1962). *Colloq. Comb. Meth. Prob. Theory*. Mathematisk Institut, Aarhus Universitet, Aarhus, Denmark, pp. 104–115.

[20] Resnick, S. (1973). *J. Appl. Prob.*, **10**, 863–868.

[21] Resnick, S. (1974). *Adv. Appl. Prob.*, **6**, 392–406.

[22] Resnick, S. (1975). *Ann. Prob.*, **3**, 951–960.

[23] Resnick, S. and Rubinovitch, M. (1973). *Adv. Appl. Prob.*, **5**, 287–307.

[24] Shorrock, R. (1972). *J. Appl. Prob.*, **9**, 219–223.

[25] Shorrock, R. (1972). *J. Appl. Prob.*, **9**, 316–326.

[26] Shorrock, R. (1973). *J. Appl. Prob.*, **10**, 543–555.

[27] Shorrock, R. (1974). *Adv. Appl. Prob.*, **6**, 392–406.

[28] Shorrock, R. (1975). *J. Appl. Prob.*, **12**, 316–323.

[29] Tiago de Oliveira, J. (1968). *Publ Inst. Statist. Univ. Paris*, **17**, 25–36.

[30] Vervaat, W. (1972). *Success Epochs in Bernoulli Trials with Applications in Number Theory*. Math. Centre Tract No. 42, Amsterdam.

[31] Vervaat, W. (1973). *Limit Theorems for Partial Maxima and Records*. Math. Dept., Catholic University of Nijmegen, Nijmegen, Netherlands.

[32] Weissman, I. (1975). *Ann. Prob.*, **3**, 172–177.

[33] Weissman, I. (1975). *J. Appl. Prob.*, **12**, 477–487.

[34] Weissman, I. (1975). *Ann. Prob.*, **3**, 470–473.

[35] Westcott, M. (1977). *Proc. R. Soc. Lond. A*, **356**, 529–547.

[36] Williams, D. (1973). *Bull. Lond. Math. Soc.*, **5**, 235–237.

(JUMP PROCESSES
RECORDS
STOCHASTIC PROCESSES)

SIDNEY I. RESNICK

EXTREME STUDENTIZED DEVIATE

If X has a normal $N(\xi, \sigma^2)$ distribution*, then $Z = (X - \xi)/\sigma$ has a unit normal distribution*. Replacing σ in this formula by a statistic S, independent of Y and distributed as σ (chi with ν degrees of freedom*)$/\sqrt{\nu}$, we obtain

$$T = (Y - \eta)/S,$$

which is distributed as Student's t* with ν degrees of freedom*. The procedure of replacing σ by S is called studentization*.

By analogy, the extreme deviate in a sample of n values $X_1 X_2, \ldots, X_n$ which is

$$D = \max_{j=1, \ldots, n} |X_j - \overline{X}|,$$

where $\overline{X} = n^{-1} \sum_1^n X_i$, can be studentized by dividing by S, giving $D^* = D/S$.

A distinction has to be made whether S is calculated from some other samples—in which case it is independent of D—or from the same sample—in which case it is equal to $\{(n-1)^{-1} \sum_{i=1}^n (X_i - \overline{X})^2\}^{1/2}$ and it is not independent of D.

In either case it is possible to use D^* in (1) tests for outliers*, (2) procedures for grouping means (e.g., following analysis of variance*). To do this, tables of percentage points of the distribution of D^* are needed. These depend on both n, the sample size, and ν, the number of degrees of freedom of S. Such tables are included in Pearson and Hartley [1].

Reference

[1]   Pearson, E. S. and Hartley, H. O. (1954). *Biometrika Tables for Statisticians*, Vol. 1 Cambridge University Press, Cambridge. Table 26.

(CHAUVENET CRITERION
MULTIPLE COMPARISONS
OUTLIERS
STUDENTIZATION)

EXTREME-VALUE DISTRIBUTIONS

The theory of extreme values, and the extreme-value distributions, play an important role in theoretical and applied statistics. For example, extreme-value distributions arise quite naturally in the study of size effect on material strengths, the occurrence of floods and droughts, the reliability of systems made up of a large number of components, and in assessing the levels of air pollution. Other applications of extreme-value distributions arise in the study of what are known as "record values" and "breaking records" (*see* RECORDS). For an up-to-date and a fairly complete reference on the the theory of extreme values, we refer the reader to the recent book by Galambos [9]. For a more classical, and still useful treatise on the subject, we refer to Gumbel [11].

PRELIMINARIES

Suppose that $X_1, X_2, \ldots, X_n$ are independent and identically distributed random variables from a distribution $F(x)$ which is assumed to be continuous. The theory of extreme values primarily concerns itself with the distribution of the smallest and largest values of $X_1, X_2, \ldots, X_n$. That is, if

$$X_{1:n} = \min(X_1, X_2, \ldots, X_n) = X_{(1)} \quad (1)$$

and

$$X_{n:n} = \max(X_1, X_2, \ldots, X_n) = X_{(n)}, \quad (2)$$

then knowing $F(x)$, we would like to say something about $L_n(x) = \Pr[X_{(1)} \leqslant x]$ and $H_n(x) = \Pr[X_{(n)} \leqslant x]$. The random variables $X_{(1)}$ and $X_{(n)}$ are also known as the extreme values.

In order to give some motivation as to why the random variables, $X_{(1)}$ and $X_{(n)}$, and their distribution functions are of interest to us, we shall consider the following situations:

1. Consider a chain that is made up of n links; the chain breaks when any one of its links break. The first link to break is the weakest link, i.e., the one that has the smallest strength. It is meaningful to assume that the strength of the ith link, say X_i, $i = 1, 2, \ldots, n$, is a random variable with distribution function $F(x)$. Since the chain breaks when its weakest link fails, the strength of the chain is therefore described by the random variable $X_{(1)} = \min(X_1, X_2, \ldots, X_n)$.

2. Consider an engineering or a biological system that is made up of n identical components, all of which may function simultaneously. For example, a large airplane may contain four identical engines which could be functioning simultaneously, or the human respiratory system, which consists of two identical lungs. The system functions as long as any one of the n components is functioning. Such systems are known as *parallel-redundant* systems and occur quite often in practice. Suppose that the time to failure (the life length) of the ith component, say X_i, $i = 1, 2, \ldots, n$, is a random variable with distribution function $F(x)$. Since the system fails at the time of failure of the last component, the life length of the system is described by the random variable $X_{(n)} = \max(X_1, X_2, \ldots, X_n)$.

It easy to envision several other physical situations in which the random variables $X_{(1)}$ and $X_{(n)}$ arise quite naturally. For instance, the use of $X_{(n)}$ for setting air pollution standards is discussed by Singpurwalla [27] and by Mittal [25]; and the use of $X_{(1)}$ in studying the time for a liquid to corrode through a surface having a large number of small pits is discussed in Mann et al. [24, p. 130].

DISTRIBUTION OF THE EXTREME VALUES

Even though our assumption that X_1, $X_2, \ldots, X_n$ are independent is hard to justify in practice, we shall, in the interest of simplicity and an easier exposition, continue to retain it. Note that

$$L_n(x) = \Pr[X_{(1)} \leqslant x] = 1 - \Pr[X_{(1)} > x]$$
$$= 1 - \Pr[X_1 > x, X_2 > x,$$
$$\ldots, X_n > x],$$

since the probability that the smallest value is larger than x is the same as the probability that all the n observations exceed x. Because of independence

$$L_n(x) = 1 - \prod_{i=1}^{n} \Pr[X_i > x]$$
$$= 1 - (1 - F(x))^n, \qquad (3)$$

since all the n observations have a common distribution $F(x)$. Using analogous arguments we can show that

$$H_n(x) = \Pr[X_{(n)} \leqslant x] = (F(x))^n. \quad (4)$$

Thus under independence, when $F(x)$ is completely specified, we can, in principle, find the distribution of $X_{(1)}$ and $X_{(n)}$. Often the distribution functions, $L_n(x)$ and $H_n(x)$, take simple forms. For example, if $F(x)$ is an exponential distribution with a scale parameter $\lambda > 0$, that is, if $F(x) = 1 - e^{-\lambda x}$, $x \geqslant 0$, then $L_n(x) = 1 - e^{-n\lambda x}$—again an exponential distribution with a scale parameter $n\lambda$.

Despite the simplicity of the foregoing results, there are two considerations that motivate us to going beyond (3) and (4). The first

consideration pertains to the fact that in many cases $L_n(x)$ or $H_n(x)$ do not take simple and manageable forms, and the second consideration is motivated by the fact that in many practical applications of the extreme-value theory, n is very large. For example, if $F(x) = 1 - e^{-\lambda x}$, then $H_n(x) = (1 - e^{-\lambda x})^n$, and when $F(x)$ is the distribution function of a standard normal variate, then $H_n(x) = (\int_{-\infty}^{x} (1/\sqrt{2\pi}) e^{-s^2/2} ds)^n$. It so happens that under some very general conditions on $F(x)$, the distributions of $X_{(1)}$ and $X_{(n)}$ when n becomes large take simple forms. The distributions $L_n(x)$ and $H_n(x)$, when $n \to \infty$, are known as the *asymptotic* (or the *limiting*) *distribution of extreme values*, and the associated theory that enables us to study these is known as the asymptotic theory of extremes; the word "asymptotic" describes the fact that n is getting large.

Asymptotic Distribution of Extremes

The key notion that makes the asymptotic distributions of $X_{(1)}$ and $X_{(n)}$ of interest is that for some constants α_n, $\beta_n > 0$, γ_n, and $\delta_n > 0$, the quantities $(X_{(1)} - \alpha_n)/\beta_n$ and $(X_{(n)} - \gamma_n)/\delta_n$ become more and more independent of n. The α_n, β_n, γ_n, and δ_n are referred to as the *normalizing constants*. A goal of the asymptotic theory of extreme values is to specify the conditions under which the normalizing constants exist, and to determine the limiting distribution functions $L(x)$ and $H(x)$, where

$$\lim_{n \to \infty} \Pr\left[\frac{X_{(1)} - \alpha_n}{\beta_n} < x \right]$$
$$= \lim_{n \to \infty} L_n(\alpha_n + \beta_n x) = L(x) \quad (5)$$

and

$$\lim_{n \to \infty} \Pr\left[\frac{X_{(n)} - \gamma_n}{\delta_n} < x \right]$$
$$= \lim_{n \to \infty} H_n(\gamma_n + \delta_n x) = H(x). \quad (6)$$

Since
$$\max[\min](X_1, X_2, \ldots, X_n)$$
$$= -\min[\max](-X_1, -X_2, \ldots, -X_n),$$
$$(7)$$

the theory for the largest extreme is identical to the theory for the smallest extreme, and vice versa. However, we shall, for the sake of completeness, give the pertinent results for both the maxima and the minima.

The fundamental result in the theory of extreme values was discovered in 1927 by Fréchet [8] and in 1928 by Fisher and Tippett [7] and was formalized in 1943 by Gnedenko [10]. It states that if

$$(X_{(n)} - \gamma_n)/\delta_n$$

has a limiting distribution $H(x)$, then $H(x)$ *must have one of the three possible forms.* An analogous result holds for $(X_{(1)} - \alpha_n)/\beta_n$. The immediate implication of this result is that irrespective of what the original distribution F is, the asymptotic distribution of $X_{(n)}$ (if it exists) is any one of three possible forms. Thus the asymptotic distribution of the extreme values is in some sense akin to the normal distribution for the sample mean. This property of the asymptotic distribution of the extremes is another motivation for our study of the limiting distributions.

We summarize the foregoing results via the following theorem of Gnedenko.

Theorem 1 (Gnedenko). Let $X_1, X, \ldots, X_n$ be independent and identically distributed with distribution function F, and let $X_{(n)} = \max(X_1, X_2, \ldots, X_n)$. Suppose that for some sequences of normalizing constants $\{\gamma_n\}$, and $\{\delta_n > 0\}$, and some other constants $a \geqslant 0, b > 0$

$$\lim_{n \to \infty} \Pr\left\{ \frac{X_{(n)} - \gamma_n}{\delta_n} \leqslant \frac{x - a}{b} \right\}$$

$$= H\left(\frac{x - a}{b} \right) \qquad (8)$$

for all continuity points of x, where $H(\cdot)$ is a nondegenerate distribution function. Then $H(\cdot)$ must belong to one of the following three "extreme-value types":

I(largest) (*see* GUMBEL DISTRIBUTION):

$$H^{(1)}\left(\frac{x - a}{b} \right) = \exp\left[-\exp\left(-\frac{x - a}{b} \right) \right],$$

$$-\infty < x < \infty, \qquad (9)$$

II(largest):

$$H^{(2)}\left(\frac{x - a}{b} \right) = \begin{cases} 0, & x < a \\ \exp\left[-\left(\frac{x - a}{b} \right)^{-\alpha} \right], \\ x \geqslant a, \quad \alpha > 0, \end{cases}$$

$$(10)$$

III(largest):

$$H^{(3)}\left(\frac{x - a}{b} \right) = \begin{cases} \exp\left[-\left(-\frac{x - a}{b} \right)^{\alpha} \right], \\ x < a, \quad \alpha > 0 \\ 1 \quad x \geqslant a. \end{cases}$$

$$(11)$$

Whenever (9), or (10), or (11) holds for some sequences $\{\gamma_n\}$ and $\{\delta_n > 0\}$, we shall say that F belongs to the *domain of attraction** of $H^{(i)}$, $i = 1, 2, 3$, and write $F \in \mathcal{D}(H^{(i)})$. Furthermore, it is not necessary for us to know the exact form of F in order to determine to which domain of attraction it belongs. A useful feature of the extreme-value theory·is that it is just the behavior of the tail of $F(x)$ that determines its domain of attraction. Thus a good deal can be said about the asymptotic behavior of $X_{(n)}$ based on a limited knowledge about F. We shall formalize the foregoing facts by giving below the necessary and sufficient conditions for $F \in \mathcal{D}(H^{(i)})$, $i = 1, 2, 3$.

Theorem 2 (Gnedenko). Let $x_0 \leqslant \infty$ be such that $F(x_0) = 1$, and $F(x) < 1$ for all $x < x_0$. Then:

1. $F \in \mathcal{D}(H^{(1)})$ if and only if there exists a continuous function $A(x)$ such that $\lim_{x \to x_0} A(x) = 0$, and such that for all h,

$$\lim_{x \to x_0} \frac{1 - F\{x(1 + hA(x))\}}{1 - F(x)} = e^{-h}.$$

2. $F \in \mathcal{D}(H^{(2)})$ if and only if

$$\lim_{x \to \infty} \frac{1 - F(x)}{1 - F(kx)} = k^{\alpha}$$

for each $k > 0$ and $\alpha > 0$.

3. $F \in \mathcal{D}(H^{(3)})$ if and only if

$$\lim_{h \to 0} \frac{1 - F(x_0 - kh)}{1 - F(x_0 - h)} = k^\alpha$$

for each $k > 0$, and $\alpha > 0$.

We note from the theorem the role played by x_0, the tail point of F, i.e., the point where $F = 1$.

Using the criteria given in Theorem 2, we can verify that if F is either an exponential*, or a normal*, or a Weibull* distribution $[F(x) = 1 - \exp(-x^\alpha), \ x \geqslant 0, \ \alpha > 0]$, then F belongs to the domain of attraction of $H^{(1)}$, whereas if F is a uniform distribution, then $F \in \mathcal{D}(H^{(3)})$; this conclusion, of course, is for the largest values. Another property exhibited by the extreme-value-type distributions $H^{(i)}(\cdot)$, $i = 1, 2, 3$, is that they belong to their own domain of attraction. That is, $H^{(i)} \in \mathcal{D}(H^{(i)})$, for $i = 1, 2, 3$; this is also referred to as the self-locking property.

Methods for determining the constants γ_n and δ_n involve some additional notation and detail, and these can be found in Gnedenko [10] or Galambos [9].

Analogous to the three "extreme-value types" for the largest values given in Theorem 1, we have three extreme-value types for the smallest values $X_{(1)}$. That is, if

$$\lim_{n \to \infty} \Pr \left\{ \frac{X_{(1)} - \alpha_n}{\beta_n} \leqslant \frac{x - a}{b} \right\} = L \left(\frac{x - a}{b} \right),$$

then $L(\cdot)$ must belong to one of the following:

I(smallest):

$$L^{(1)} \left(\frac{x - a}{b} \right) = 1 - \exp \left[-\exp \left(\frac{x - a}{b} \right) \right],$$
$$-\infty < x < \infty, \quad (12)$$

II(smallest):

$$L^{(2)} \left(\frac{x - a}{b} \right) = \begin{cases} 0, & x < a \\ 1 - \exp \left[-\left(\frac{x - a}{b} \right)^\alpha \right], \\ & x \geqslant a, \quad \alpha > 0. \end{cases}$$
$$(13)$$

III(smallest):

$$L^{(3)} \left(\frac{x - a}{b} \right) = \begin{cases} 1 - \exp \left[-\left(-\frac{x - a}{b} \right)^{-\alpha} \right], \\ & x < a, \quad \alpha > 0, \\ 1, & x \geqslant a. \end{cases}$$
$$(14)$$

Using criteria that are analogous to Theorem 2, we can verify that if F is a normal distribution, then $F \in \mathcal{D}(L^{(1)})$, whereas if F is an exponential, a uniform, or a Weibull, then $F \in \mathcal{D}(L^{(2)})$. Here again, the distributions $L^{(i)}$ are self-locking. By way of a comment, we note that $L^{(2)}(\cdot)$ is, in fact, the Weibull distribution which was mentioned before and which is quite popular in reliability theory*.

Current research in extreme value theory is being vigorously pursued from the point of view of dropping the assumption of independence and considering dependent sequences $X_1, X_2, \ldots, X_n$. One widely used class of dependent random variables is the *exchangeable* one (*see* EXCHANGEABILITY).

Definition.   (Galambos [9, p. 127] The random variables $X_1, X_2, \ldots, X_n$ are said to be exchangeable if the distribution of the vector $(X_{i_1}, X_{i_2}, \ldots, X_{i_n})$ is identical to that of $(X_1, \ldots, X_n)$ for all permutations $(i_1, i_2, \ldots, i_n)$ of the subscripts $(1, 2, \ldots, n)$.

Generalizations of Gnedenko's results when the sequence $X_1, \ldots, X_n$ is exchangeable are given in Chapter 3 of Galambos [9]. For an excellent and a very readable, albeit mathematical, survey of results when the sequence $X_1, \ldots, X_n$ is dependent, we refer the reader to Leadbetter [16].

Another aspect of the current research in extreme value theory pertains to multivariate extreme-value distributions*.

ESTIMATION OF THE PARAMETERS OF THE ASYMPTOTIC DISTRIBUTIONS

In order for us to discuss methods for estimating the parameters a, b and α of the distributions $H^{(i)}$ and the $L^{(i)}$, $i = 1, 2, 3$, it will be helpful if we recognize several relationships that exist between them.

For example, if we denote the asymptotic distribution of $X_{(n)}$, $H^{(1)}((x-a)/b)$ by $H^{(1)}(a,b)$, and the asymptotic distribution of $X_{(1)}$, $L^{(1)}((x-a)/b)$ by $L^{(1)}(a,b)$, then it can be verified that $Y_{(n)} \overset{\text{def}}{=} -X_{(n)}$ has the distribution $L^{(1)}(-a,b)$. We shall denote this relationship by writing

$$H^{(1)}(a,b) \overset{-X_{(n)}}{\longrightarrow} L^{(1)}(-a,b).$$

In a similar manner, if we denote $H^{(i)}((x-a)/b)$ and $L^{(i)}((x-a)/b)$ by $H^{(i)}(a,b,\alpha)$ and $L^{(i)}(a,b,\alpha)$ respectively, for $i=2,3$, then

$$H^{(2)}(a,b,\alpha) \overset{-X_{(n)}}{\longrightarrow} L^{(3)}(-a,b,\alpha),$$

$$H^{(3)}(a,b,\alpha) \overset{-X_{(n)}}{\longrightarrow} L^{(2)}(-a,b,\alpha).$$

If, however, $Y_{(n)} \overset{\text{def}}{=} X_{(n)}^{-1}$, and the location parameter $a=0$, then

$$H^{(2)}(0,b,\alpha) \overset{X_{(n)}^{-1}}{\longrightarrow} L^{(2)}(0,b^{-1},\alpha),$$

$$H^{(3)}(0,b,\alpha) \overset{X_{(n)}^{-1}}{\longrightarrow} L^{(3)}(0,b^{-1},\alpha).$$

Other transformations that are of interest are $Y_{(n)} = e^{-X_{(n)}}$ and $Y_{(1)} = \ln X_{(1)}$; these give us

$$H^{(1)}(a,b) \overset{e^{-X_{(n)}}}{\longrightarrow} L^{(2)}(0,e^{-a},b^{-1}),$$

$$L^{(2)}(0,b,\alpha) \overset{\ln X_{(1)}}{\longrightarrow} L^{(1)}(\ln b,\alpha^{-1}).$$

If we suppress the arguments of the $H^{(i)}$ and the $L^{(i)}$, $i=1,2,3$, then the following illustration, suggested to us by M. Y. Wong, is a convenient summary of the relationships described above.

It is easy to verify that in Fig. 1, the reverse relationships also hold. For example, if $Y_{(n)} \overset{\text{def}}{=} -X_{(1)}$, then

$$L^{(3)}(a,b,\alpha) \overset{-X_{(1)}}{\longrightarrow} H^{(2)}(-a,b,\alpha),$$

and so on. In view of this last relationship, and the relationships implied by Fig. 1, it follows that we need only consider the distribution $L^{(1)}(a,b)$. All the other distributions considered here can be transformed to the

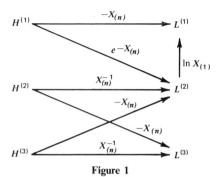

Figure 1

distribution $L^{(1)}(a,b)$, either by a change of variable or by a change of variable with a setting of the location parameter equal to zero. It is because of this fact that some of the literature on the Weibull distribution* with a location parameter of 0 ($L^{(2)}(0,b,\alpha)$) appears under the heading of "an extreme-value distribution," which is a common way of referring to the distribution $L^{(1)}(\cdot,\cdot)$.

When the location parameter a associated with the distributions $H^{(i)}$ and $L^{(i)}$, $i=2,3$, cannot be set equal to zero, most of the relationships mentioned before do not hold, and thus we cannot be content by just considering the distribution $L^{(1)}(a,b)$, We will have to consider both $H^{(2)}(a,b,\alpha)$ and $H^{(3)}(a,b,\alpha)$ or their duals $L^{(3)}(-a,b,\alpha)$ and $L^{(2)}(-a,b,\alpha)$, respectively. Estimation of the parameters a (or $-a$), b, and α is discussed in the next section.

Estimation for the Three-Parameter Distributions

The standard approach for estimating the three parameters associated with $H^{(i)}$ and $L^{(i)}$, $i=2,3$, is the one based on the method of maximum likelihood*. Because of the popularity of the Weibull distribution, the case $L^{(2)}(a,b,\alpha)$ has been investigated very extensively. We give below an outline of the results for this case, and guide the reader to the relevant references.

Let $X_{(1)} \leqslant X_{(2)} \leqslant \cdots \leqslant X_{(n)}$ be the smallest ordered observations in a sample of size n from the distribution $L^{(2)}(a,b,\alpha)$. Harter and Moore [12] and also Mann et al.

[24, p. 186] give the three likelihood equations* and suggest procedures for an iterative solution of these. They also give suggestions for dealing with problems that arise when the likelihood function* increases monotonically in $(0, X_{(1)})$.

Lemon [17] modified the likelihood equations so that one need iteratively solve only two equations for estimates of the location parameter* a and the shape parameter* α, which then specify an estimate of the scale parameter b.

Mann et al. [24] discuss, as well, the graphical method of estimation, quick initial estimates proposed by Dubey [3], and iterative procedures involving linear estimates as leading to a median unbiased estimate of a. (A recent result of Somerville [28] suggests that in iteratively obtaining a median unbiased estimate of a Weibull location parameter, the quantity k, defined in Mann et al. [24, p. 341], should be approximately $r/5$.)

Rockette et al. [26] have conjectured that there are never more than two solutions to the likelihood equations. They show that if there exists a solution that is a local maximum, there is a second solution that is a saddle point. They also show that, even if a solution $(\hat{a}, \hat{b}, \hat{\alpha})$ is a local maximum, the value of the likelihood function $L(\hat{a}, \hat{b}, \hat{\alpha})$ may be less than $L(a_0, b_0, \alpha_0)$, where $a_0 = x_{(1)}$, $\alpha_0 = 1$, and $b_0 = $ maximum likelihood estimate of the mean of a two-parameter exponential distribution.

Estimation for the Two-Parameter Distributions

When the location parameter a associated with the distributions $H^{(i)}$ and $L^{(i)}$, $i = 2, 3$, is known, or can be set equal to zero, there are several approaches that can be used to obtain good point estimators of the parameters b and α. The same is also true when we are interested in the parameters a and b of $H^{(1)}$ and $L^{(1)}$. These approches involve an iterative solution of the maximum likelihood equations, and the use of linear estimation techniques.

MAXIMUM LIKELIHOOD ESTIMATION. The maximum likelihood method has the advantage that it can be applied efficiently to any sort of censoring of the data.

For all the extreme-value distributions, the order statistics are the sufficient statistics. Thus, unless there are only two observations, the sufficient statistics are not complete and no small-sample optimality properties hold for the maximum likelihood estimators. The maximum likelihood estimators of the two parameters are, however, asymptotically unbiased as well as asymptotically normal and asymptotically efficient. One can use the maximum likelihood estimates with tables of Thoman et al. [29] and of Billman et al. [1] to obtain confidence bounds on the parameters.

LINEAR ESTIMATION* TECHNIQUES. Linear techniques allow for the estimation of the two parameters of interest without the necessity of iteration (see Mann et al. [24], and WEIBULL DISTRIBUTION), by best linear invariant (BLI) estimators and best linear unbiased (BLU) estimators. Tables of Mann et al. [23] and Mann and Fertig [20] can be used with either the BLI or BLU estimates to obtain confidence and tolerance bounds for censored samples of size n, $n = 3(1)25$. See also Mann et al. [24, p. 222] for tables with $n = 3(1)13$. Thomas and Wilson [30] compare the BLU and BLI estimators with other approximately optimal linear estimators based on all the order statistics.

If samples are complete and sample sizes are rather large, one can use tables of Chan and Kabir [2] or of Hassanein [13] to obtain linear estimates of a and b based on from 2 through 10 order statistics. These tables apply to weights and spacings for the order statistics that define estimators that are asymptotically unbiased with asymptotically smallest variance. Hassanein's results have the restriction that the spacings are the same for both estimators, but he also considers samples with 10% censoring. Tables of Mann and Fertig [22] allow for removal of small-sample bias from Hassanein's estimators and give exact variances and covariances. This enables one to calculate approximate confidence bounds from these estimators.

For samples having only the first r of n possible observations, the unbiased linear estimator of Engelhardt and Bain [5] for parameter b, $b_{r,n}^{**} = \sum_{i=1}^{r} |X_{(s)} - X_{(i)}| (nk_{r,n})^{-1}$ is very efficient, especially for heavy censoring*. To obtain $b_{r,n}^{**}$, one need only know a tabulated value of $k_{r,n}$ and an appropriate value for s; s is a function of r and n.

A corresponding estimator for a is then given by $a_{r,n}^{**} = X_{(s)} - E(Z_s) b_{r,n}^{**}$, where $Z_s = (X_{(s)} - a)/b$.

Mann et al. [24, pp. 208–214, 241–252] give tables and references to additional tables for using these estimators. More recent references that aid in the use of these estimators are given in Engelhardt [4].

The estimators $b_{r,n}^{**}$ and $a_{r,n}^{**}$ approximate the BLU estimators and can be converted easily to approximations to the BLI estimators, which in turn approximate results obtained by maximum likelihood procedures.

The estimator $b_{r,n}^{**}$ has the property that $2b_{r,n}^{**}/[\text{var}(b_{r,n}^{**})/b]$ is very nearly a chi-squared variate with $2/\text{var}(b_{r,n}^{**}/b)$ degrees of freedom. This property holds for any efficient unbiased estimator of b, including a maximum likelihood estimator corrected for bias. Because the BLI estimator so closely approximates the maximum likelihood estimators of b, tables yielding biases for the BLI estimators can be used to correct the maximum likelihood estimators for bias.

The fact that unbiased estimators of b are approximately proportional to chi-square* variates has been used to find approximations to the sampling distributions of functions of estimators of a and other distribution percentiles. Mann et al. [24] describe an F-approximation that can be used with complete samples to obtain confidence bounds on very high (above or below 90%), or very low distribution percentiles, or with highly censored data to obtain a confidence bound for a. The precision of this approximation is discussed by Lawless [14] and Mann [18, 19]. Engelhardt and Bain [6] have suggested the use of a $\ln \chi^2$ approximation, the regions of utility of which tend to complement those of the F-approximation. Lawless [15] reviews methods for constructing confidence intervals or other characteristics of the Weibull or extreme-value distribution.

References

[1] Billman, B. R., Antle, C. L., and Bain, L. J. (1971). *Technometrics*, **14**, 831–840.

[2] Chan, L. K. and Lutful Kabir, A. B. M. (1969). *Naval Res. Logist. Quart.*, **16**, 381–404.

[3] Dubey, S. D. (1966). *Naval Res. Logist. Quart.*, **13**, 253–263.

[4] Engelhardt, M. (1975). *Technometrics*, **17**, 369–374.

[5] Engelhardt, M. and Bain, L. J. (1973). *Technometrics*, **15**, 541–549.

[6] Engelhardt, M. and Bain, L. J. (1977). *Technometrics*, **19**, 323–331.

[7] Fisher, R. A. and Tippett, L. H. C. (1928). *Proc. Camb. Philos. Soc.*, **24**, 180–190.

[8] Frechet, M. (1927). *Ann. Soc. Pol. Math. Cracovie*, **6**, 93–116.

[9] Galambos, J. (1978). *The Asymptotic Theory of Extreme Order Statistics*. Wiley, New York.

[10] Gnedenko, B. V. (1943). *Ann Math.*, **44**, 423–453.

[11] Gumbel, E. J. (1958). *Statistics of Extremes*. Columbia University Press, New York.

[12] Harter, H. L. and Moore, A. H. (1965). *Technometrics*, **7**, 639–643.

[13] Hassanein, K. M. (1972). *Technometrics*, **14**, 63–70.

[14] Lawless, J. F. (1975). *Technometrics*, **17**, 255–261.

[15] Lawless, J. F. (1978). *Technometrics*, **20**, 355–364.

[16] Leadbetter, M. R. (1975). In *Stochastic Processes and Related Topics*, Vol. 1, M. L. Puri, ed. Academic Press, New York.

[17] Lemon, G. H. (1974). *Technometrics*, **17**, 247–254.

[18] Mann, N. R. (1977). *Naval Res. Logist. Quart.*, **24**, 187–196.

[19] Mann, N. R. (1978). *Naval Res. Logist. Quart.*, **25**, 121–128.

[20] Mann, N. R. and Fertig, K. W. (1973). *Technometrics*, **15**, 87–101.

[21] Mann, N. R. and Fertig, K. W. (1975). *Technometrics*, **17**, 361–378.

[22] Mann, N. R. and Fertig, K. W. (1977). *Technometrics*, **19**, 87–93.

[23] Mann, N. R., Fertig, K. W., and Scheuer, E. M. (1971). Confidence and Tolerance Bounds and a New Goodness-of-Fit Test for Two-Parameter Weibull or Extreme Value Distributions. *ARF*

71-0077, Aerospace Research Laboratories, Wright Patterson Air Force Base, Oh.

[24] Mann, N. R., Schafer, R. E., and Singpurwalla, N. D. (1974). *Methods for Statistical Analysis of Reliability and Life Data*. Wiley, New York.

[25] Mittal, Y. (1978). *Ann. Statist.*, **6**, 421–432.

[26] Rockette, H., Antle, C., and Klimko, L. A. (1974). *J. Amer. Statist. Ass.*, **69**, 246–249.

[27] Singpurwalla, N. D. (1972). *Technometrics*, **14**, 703–711.

[28] Somerville, P. N. (1977). In *The Theory and Applications of Reliability*, Vol. I, P. Tsokos and I. N. Shimi, eds. Academic Press, New York, pp. 423–432.

[29] Thoman, D. R., Bain, L. J., and Antle, C. E. (1970). *Technometrics*, **12**, 363–371.

[30] Thomas, D. R. and Wilson, W. M. (1972). *Technometrics*, **14**, 679–691.

Acknowledgments

Nancy R. Mann's research was supported by the Office of Naval Research, Contract N00014-76-C-0723.

Nozer D. Singpurwalla's research was supported by the Nuclear Regulatory Commission under Contract NRC-04-78-239, with George Washington University.

(GUMBEL DISTRIBUTION
LIMIT THEOREMS
WEIBULL DISTRIBUTION)

NANCY R. MANN
NOZER D. SINGPURWALLA

EYE ESTIMATE

Strictly speaking, this term means an estimate in which the only measuring instrument is the human eye. By customary usage, it includes all estimates in which there is an element of judgment, so that the value of the estimate is not predetermined exactly by the available data. As an example, consider the use of graphical methods of fitting a distribution with probability plotting* paper. A straightedge will be used to actually draw a straight line among the plotted points, although the eye provides judgment as to where the straightedge should be placed. Further, once the line has been drawn and its slope and intercept measured, these latter values may be used in further numerical calculation (without relying any more on the eye alone). Nevertheless, the resulting values are termed "eye estimates."

The term *may* be used even when the eye does not intervene at all: for example, in measurements by the senses of touch, smell, or taste. However, this is not commonly done, and it is a practice not to be recommended, as it can be confusing and there are available other, and more apt, terms.

(RANKED SET SAMPLING)